Male Infertility

Contemporary Clinical Approaches, Andrology, ART & Antioxidants

Sijo J. Parekattil • Ashok Agarwal
Editors

Male Infertility

Contemporary Clinical Approaches, Andrology, ART & Antioxidants

Springer

Editors
Sijo J. Parekattil
Director of Urology, Winter Haven Hospital
University of Florida
Winter Haven, Florida, USA

Ashok Agarwal
Director, Center for Reproductive Medicine
Cleveland Clinic
Cleveland, Ohio, USA

ISBN 978-1-4614-3334-7 ISBN 978-1-4614-3335-4 (eBook)
DOI 10.1007/978-1-4614-3335-4
Springer New York Heidelberg Dordrecht London

Library of Congress Control Number: 2012938258

Printed on acid-free paper

Springer is part of Springer Science+Business Media (www.springer.com)

Foreword

Drs. Parekattil (University of Florida) and Agarwal (Cleveland Clinic) have combined their expertise in the clinical and laboratory science, respectively, of male reproduction to create a novel and valuable textbook. Dr. Parekattil's urologic fellowships at the Cleveland Clinic in both male infertility/microsurgery and laparoscopic/robotic surgery added considerably to his vision of the field of male fertility. He has pioneered the area of robotic microsurgical procedures for correction of obstructions in the male reproductive tract, varicocelectomy, neurolysis of the spermatic cord for relief from scrotal and groin pain, and for other procedures. Dr. Agarwal is a recognized authority in the study of the effect of reactive oxygen species (ROS) in the semen and their impact on sperm function. The compilation of a textbook that contains chapters that are written by so many international experts in both clinical and laboratory areas of male reproduction is a prodigious undertaking and the editors are to be congratulated for their efforts to create this book.

Drs. Wiser, Sandlow, and Köhler (University of Wisconsin at Milwaukee) offer an encyclopedic discussion on a wide variety of causes of male infertility. They consider genetic, anatomic, endocrine, environmental, coital, immunologic, and surgically correctable factors that may affect a man's fertility potential, and options for treatment other than in vitro fertilization (IVF)/ICSI of many of these conditions. Drs. Haywood, Laborde, and Brannigan (Northwestern University) offer a detailed discussion of various endocrinopathies that may cause male infertility. Drs. Agarwal and du Plessis (Tygerberg, South Africa) offer a chapter concerning the impact of obesity on male fertility. Drs. Mori and Sabanegh (Cleveland Clinic) present a detailed consideration of laboratory tests that are used for the diagnosis of male infertility. Dr. Claudio Teloken and associates (Brazil) discuss various environmental factors that may cause male infertility. The effects of various malignancies and of irradiation therapy and chemotherapy used to treat these malignancies on spermatogenesis are reviewed by Daniel Williams, IV (University of Wisconsin), who also discusses the use of sperm that were cryopreserved before treatment for assisted reproductive technologies when spermatogenesis does not resume after chemotherapy.

Drs. Faasse and Niederberger (University of Illinois at Chicago) present a detailed discussion of the epidemiological considerations of male infertility, including considerations about the incidence and prevalence of infertility in developed countries and in the developing world. They consider the relationship between semen parameters and male infertility and whether or not semen parameters predict fertility and assisted reproductive technology (ART) outcomes. Their chapter also concerns novel assays for the evaluation of male infertility and whether or not male semen quality is declining, as well as health care resource utilization for male infertility in the USA. Dr. Henkel (Bellville, South Africa) gives a thorough discussion of infections of the male genitourinary tract according to the etiologic agent and the location of the infection within the genitourinary tract.

Drs. Inci and Gunay (Turkey) discuss medications and other exogenous agents that may decrease the male fertility potential. Drs. Cohen and Moseley (University of Florida) present four fascinating cases of ethical issues that may arise and create problems to both physicians and patients when making decisions about the management of fertility problems. Dr. Vaithinathan et al. (Puducherry, India) consider the various causes of apoptosis that may affect a man's fertility status, while Drs. Misro and Ramya (New Delhi, India) offer an elegant view of fuel energy sources for spermatozoa. Drs. Hsieh and Shin (Georgetown University) discuss nutritional pathways to protect male reproductive health. Their chapter primarily concerns the use of various antioxidants for infertile men and the impact of obesity on men's fertility potential. Dr. Pasqualotto et al. (Brazil) discuss the effect of the aging process on male fertility status and the effect of increased DNA abnormalities that occur with increasing age on the genomic pattern of the offspring of aged men.

Dr. Kumar (New Delhi, India) discusses various causes of azoospermia in India, some of which are rarely, if ever, seen in the USA. He also presents his techniques for microsurgical vasoepididymostomy and for microsurgical varicocelectomy. Drs. Cassidy, Jarvi, Grober, and Lo (Toronto, Canada) discuss their use of a "mini-incision" to perform vasectomy reversal using no-scalpel techniques.

Drs. Hsiao and Goldstein (Cornell Weill School of Medicine) discuss the historical development of microsurgical vasoepididymostomy. The latter author's extensive experience with that procedure and the evolution of his anastomotic technique are described in detail. Drs. Stember and Schlegel (also at Cornell Weill) review Schlegel's extensive experience with microsurgical dissection for testicular sperm extraction (microTESE). All aspects of dealing with patients who have non-obstructive azoospermia, including genetic findings, testicular histology and hormonal values in these patients are considered. The details of the microTESE procedure, including sperm processing methods and results, are clearly described.

Various discussions of the use of robotic microsurgery are contained in this book. Drs. Brahmbhatt and Parekattil discuss robotic microsurgery to perform vasovasostomy, varicocelectomy, as well as to perform spermatic cord denervation and robotic single port neurolysis to relieve scrotal and groin pain caused by various etiologies. They describe newly available and prototype instrumentation for robotic microsurgery. Drs. Shu and Wang (Houston, Texas) discuss their method of robotic microsurgical varicocelectomy. They deliver the spermatic cord externally in the normal operative manner, but then use the DaVinci robot to perform dissection within the cord. They found a short learning curve to perform this procedure and now prefer this method for performing varicocelectomy due to the complete lack of tremor and the ease of identification of the various structures within the spermatic cord. Dr. De Wil et al. (Belgium and Switzerland) present interesting arguments that robotic vasovasostomy is useful to reduce tremor and to enhance dexterity when this procedure is performed. They also discuss the background of robotic surgery and describe the details of their method of robotic vasovasostomy. While many may question the value of robots for urologic microsurgical procedures, the total elimination of tremor when the robot is used certainly should improve the results of microsurgical procedures. The question that will need to be answered in the future is whether or not cost/benefit considerations support the use of robots for urologic microsurgery except for the unusual situation in which repair of vasal obstruction deep within the pelvis may be required.

Drs. Rosevear and Wald (University of Iowa) review experimental procedures to develop vas deferens allografts to bridge the gap when the distance between the two ends of the vas is too long for reconstruction to be performed with microsurgical vasovasostomy. A chapter by Drs. Ignatov and Turek (San Francisco) concerns the diagnosis and management of various types of ejaculatory dysfunction, as well as the surgical treatment of ejaculatory duct obstruction. Drs. Cocuzza (Sao Paulo, Brazil) and Parekattil discuss the various uses of radiology in the diagnosis and treatment of male infertility.

There are detailed discussions about IVF and intracytoplasmic sperm injection (ICSI) in the treatment of male infertility. Drs. Rhoton-Vlasak and Drury (University of Florida) discuss their institution's multidisciplinary clinic for the evaluation of couples as potential candidates for ART and describe in great detail the laboratory aspects of ARTs, including the indications for, and use of, antioxidants in vitro during ART procedures.

Dr. Antinori (Italy) discusses motile sperm organellar morphology examination (MSOME) to reach a final video magnification of 6,600×. This allows the detection of subcellular abnormalities of sperm. The author assumes that certain subcellular microscopic changes in sperm are a result of exposure to reactive oxygen species. Subsequently, the intracytoplasmic morphologically selected sperm injection (IMSI) procedure is described as a new method of sperm selection to treat couples with previously failed ICSI cycles. This technology currently is so expensive that it falls into the category of experimental therapy, but it offers promise for the future. Drs. Grunewald and Paasch (Leipzig, Germany) also elaborate on subcellular and other methods of sperm selection for ICSI.

The discussions about the effect of ROS on sperm quality and male fertility and on the use of antioxidants to treat male infertility are virtually encyclopedic in nature. Dr. Lampiao et al. review oxidative stress, the physiologic role of ROS in normal sperm physiology, the pathologic effects of ROS on sperm function, and the prevention of, and therapy for, impaired fertility due to seminal fluid ROS. Dr. Balercia et al. from Italy discuss the role of coenzyme Q10 and L-carnitine in the treatment of ROS damage to sperm, and Drs. Iremashvili, Brackett, and Lynne (University of Miami) discuss the sources, consequences, and possibility for therapy of ROS in the semen of spinal cord injured men. Drs. Said (Toronto, Canada) and Agarwal contribute information about the use of antioxidants in sperm cryopreservation. Dr. Lanzafame et al. (Syracuse and Catania, Italy) present detailed information about the use and value of a host of antioxidants to treat ROS-induced sperm damage. As a balance to these considerations of antioxidant therapy, the harmful effects and risks of antioxidant therapy are discussed by Drs. Stewart and Kim (University of Tennessee). Dr. Garrido et al. (Valencia, Spain) discuss the role of antioxidants in ICSI.

Drs. Celik-Ozenci (Antalya, Turkey) and Huszar (Yale University) add other aspects of oxidative stress on sperm and mention the fact that medical therapy for this problem has not proved to be significantly successful. They show the benefits of the sperm–hyaluronic acid binding assay and hyaluronic acid-mediated selection of single sperm without DNA chain fragmentation in order to select sperm for ICSI. Drs. Hwang and Lamb (Baylor at Houston) offer an explanation for the molecular mechanisms of antioxidants in male infertility. Dr. Tremellen (Australia) contributes a critical review of the literature concerning the effectiveness of antioxidant therapy for the enhancement of male reproductive health. The author should be congratulated on presenting an overview of the common methodological weaknesses in antioxidant trials. Dr. Ko and others (Cleveland Clinic) conclude that there is a need for randomized controlled trials regarding the value of medical treatment of infertility due to ROS.

Dr. Alvarez (Barcelona) discusses the effects on sperm function after thawing that are caused by the loss of intracellular antioxidant enzyme activity during cryopreservation. Dr. Varghese et al. (Montreal, Canada and Cleveland Clinic) discuss the use of antioxidants during embryo culture and during embryo cryopreservation to prevent lipid peroxidation of the cell membrane of the developing embryo.

A unique feature of the book is the inclusion of a section of "Expert Commentary" at the end of many of the chapters. This treatise contains discussions of virtually all areas of the diagnosis and treatment of male infertility and also contains information about newer evolving therapies to treat male infertility. The chapters on ethical considerations in male infertility and on the management of chronic inguinal and scrotal pain are welcome additions that are not usually found in textbooks concerning male infertility. The addition of the various laboratory considerations of male infertility makes the book valuable for the libraries of not only clinicians, but also of laboratory scientists.

Louisville, KY　　　　Arnold Belker, MD

Preface

The field of male infertility truly illustrates the need for a multispecialty approach to the effective diagnosis and management of such conditions. From the initial referral possibly from a reproductive endocrinology gynecologist and embryologist to the male infertility urologist, andrologist, nurses, researchers, and alternative medicine specialists, this multidisciplinary team needs to work as a cohesive unit to provide our patients with the most effective and highest quality care.

This book was an attempt to gather experts from each of these fields and present an integrated management approach with detailed descriptions of topics ranging from the initial clinical diagnosis, management, new treatment options, and scientific rational for the various approaches. The book initially focuses on the clinical aspects of male infertility and then dives into the use of antioxidants as adjunctive therapy and the current state of affairs in antioxidant research. The authors come from leading institutions from around the globe in an attempt to capture a wide range of techniques and approaches. We are hoping that this text may serve as a reference guide for specialists across this team to further enhance dialogue, discussion, and refinement in our multidisciplinary approach.

We would like to thank the authors for their contributions and our families for their patience in allowing us to put together this project. We would like to acknowledge the Glickman Urological Institute at the Cleveland Clinic Foundation and the Department of Urology at University of Florida for institutional support for this endeavor as well. We are grateful to Richard Lansing, executive editor, for his support and advice, and to Margaret Burns, developmental editor, for her tireless efforts in reviewing and editing each of the manuscripts. We hope that this book will provide a concise, consolidated reference for antioxidants and male infertility.

Winter Haven, FL, USA Sijo J. Parekattil
Cleveland, OH, USA Ashok Agarwal

Contents

Contributors

Ashok Agarwal, Ph.D. Center for Reproductive Medicine, Glickman Urological and Kidney Institute, Cleveland Clinic Foundation, Cleveland, OH, USA

Ignacio S. Alvarez, Ph.D. Department of Cell Biology, School of Life Sciences, University of Extremadura, Badajoz, Spain

Juan G. Alvarez, M.D., Ph.D. Centro de Infertilidad Masculina Androgen, La Coruña, Harvard Medical School, Boston, Spain

Monica Antinori, M.D. Department of Infertility, RAPRUI Day Surgery, Rome, Italy

Giancarlo Balercia, M.D. Andrology Unit, Endocrinology, Department of Internal Medicine and Applied Biotechnologies, Umberto I Hospital, School of Medicine, Italy

Department of Clinical and Molecular Sciences, Marche Polytechnic University, Italy

Edson Borges Jr., M.D. Department of Fertility, Center for Assisted Fertilization, Sao Paulo, Brazil

Kasonde Bowa, M.Sc., M.Med., F.R.C.S., F.A.C.S., F.C.S. Department of Surgery, University Teaching Hospital, Lusaka, Zambia and Ndola Central Hospital, Zambia

Nancy L. Brackett, Ph.D., H.C.L.D. The Miami Project to Cure Paralysis, University of Miami Miller School of Medicine, Lois Pope Life Center, Miami, FL, USA

Jamin V. Brahmbhatt, M.D. Department of Urology, University of Tennessee Health Science Center, Memphis, TN, USA

Robert E. Brannigan, M.D. Department of Urology, Northwestern University Feinberg School of Medicine, Chicago, IL, USA

Orhan Bukulmez, M.D. Division of Reproductive Endocrinology and Infertility, Department of Obstetrics and Gynecology, University of Texas Southwestern Medical Center, Dallas, TX, USA

Stephanie Cabler, B.Sc. Center for Reproductive Medicine, Glickman Urological and Kidney Institute, Cleveland Clinic Foundation, Cleveland, OH, USA

Aldo E. Calogero, M.D. Department of Medical and Pediatrics Sciences, University of Catania, Catania, Italy

Darby J. Cassidy, M.D. Department of Urology, University Hospital of Northern British Columbia, Prince George, BC, Canada

Ciler Celik-Ozenci, D.D.S., Ph.D. Department of Histology and Embryology School of Medicine, Akdeniz University, Campus, Antalya, Turkey

Marcello Cocuzza, M.D. Department of Urology, University of Sao Paulo, Sao Paulo, Brazil

Marc S. Cohen, M.D. Department of Urology, Shands at the University of Florida, Gainesville, FL, USA

George A. de Boccard, M.D. Clinique Generale Beaulieu, Robot Assisted Laparoscopy Surgery Center, Geneva, Switzerland

Shereen Cynthia D'Cruz, M.Phil. Department of Biochemistry and Molecular Biology, Pondicherry University, Kalapet, Pondicherry, India

Peter Frank De Wil, M.D. Urological Department, Kliniek Sint Jan, Brussels, Belgium

Fnu Deepinder, M.D. Department of Endocrinology, Diabetes, and Metabolism, Cedars Sinai Medical Center, Los Angeles, CA, USA

Kenneth C. Drury, Ph.D. In Vitro Fertilization and Andrology Laboratory, Department of Obstetrics and Gynecology, University of Florida College of Medicine, Gainesville, FL, USA

Stefan S. du Plessis, Ph.D., M.B.A. Department of Medical Physiology, Faculty of Health Sciences, Stellenbosch University, Tygerberg, Western Cape, South Africa

Sandro C. Esteves, M.D., Ph.D. ANDROFERT, Andrology and Human Reproduction Clinic, Center for Male Reproduction, Campinas, SP, Brazil

Mark A. Faasse, M.D. Department of Urology, University of Illinois at Chicago, Chicago, IL, USA

Vincenzo Ficarra, M.D., Ph.D. Department of Oncological and Surgical Sciences, Urology Clinic, University of Padua, Padua, Italy

Sandra García-Herrero, Ph.D. IVIOMICS, Valencia, Spain

Nicolas Garrido, Ph.D. Andrology Laboratory and Sperm Bank, Instituto Universitario IVI Valencia, Valencia, Spain

Marc Goldstein, M.D. Cornell Institute for Reproductive Medicine, New York Presbyterian Hospital/Weill Cornell Medical Center, Weill Cornell Medical College of Cornell University, New York, NY, USA

Túlio M. Graziottin, M.D., Ph.D. Andrology Section of Fertilitat – Human Reproduction Center, Department of Urology, Santa Casa Hospital and Federal University of Health Sciences, Porto Alegre, RS, Brazil

Ethan D. Grober, M.D., M.Ed. Department of Surgery, Division of Urology, University of Toronto, Mount Sinai Hospital, Toronto, ON, Canada

Sonja Grunewald, M.D. Department of Dermatology, Venerology and Allergology, European Training Center of Andrology, University of Leipzig, Leipzig, Germany

Levent Mert Gunay, M.D. Department of Urology, Hacettepe University School of Medicine, Ankara, Turkey

Sam Haywood, B.A. Department of Urology, Northwestern University Feinberg School of Medicine, Chicago, IL, USA

Ralf Henkel, B.Ed., Ph.D. Department of Medical Biosciences, University of the Western Cape, Bellville, South Africa

Wayland Hsiao, M.D. Assistant Professor in Urology, Emory University, Department of Urology, Atlanta, GA, USA

Tung-Chin Hsieh, M.D. Department of Urology, Baylor College of Medicine, Houston, TX, USA

Gabor Huszar, M.D. Sperm Physiology Laboratory, Yale School of Medicine, New Haven, CT, USA

Kathleen Hwang, M.D. Department of surgery (Urology), Brown University, Providence, RI, TX, USA

Aleksey P. Ignatov, B.S. The Turek Clinic, San Francisco, CA, USA

Kubilay Inci, M.D. Department of Urology, Hacettepe University School of Medicine, Ankara, Turkey

Viacheslav Iremashvili, M.D., Ph.D. Department of Urology, University of Miami Miller School of Medicine, Miami, FL, USA

Keith Jarvi, M.D. Department of Surgery, Division of Urology, University of Toronto, Mount Sinai Hospital, Toronto, ON, Canada

Samuel Juncal, M.S., M.D. Andrology Section of Fertilitat – Human Reproduction Center, Federal University of Health Sciences, Porto Alegre, RS, Brazil

John S. Kachimba, M.Med, FCS, FCS(Urol) Department of Surgery, University Teaching Hospital, Levy Mwanawasa General Hospital, Lusaka, Zambia

John C. Kefer, M.D., Ph.D. Alpine Urology, Boulder, CO, USA

Edward D. Kim, M.D. Division of Urology, Department of Surgery, Center for Reproductive Medicine, University of Tennessee Medical Center, Knoxville, TN, USA

Edmund Y. Ko, M.D. Section of Male Infertility, Department of Urology, Glickman Urological and Kidney Institute, Cleveland Clinic, Cleveland, OH, USA

Tobias S. Köhler, M.D., M.Ph. Division of Urology, Southern Illinois University, Springfield, IL, USA

Rajeev Kumar, M.Ch. Department of Urology, All India Institute of Medical Sciences, New Delhi, India

Sandro La Vignera, M.D. Department of Medical and Pediatrics Sciences, University of Catania, Catania, Italy

Eric L. Laborde, M.D. Department of Urology, Northwestern University Feinberg School of Medicine, Chicago, IL, USA

Dolores J. Lamb, Ph.D. Scott Department of Urology, Baylor College of Medicine, Houston, TX, USA

Fanuel Lampiao, Ph.D. Department of Medical Physiology, Faculty of Health Sciences, Stellenbosch University, Tygerberg, Western Cape, South Africa

Francesco Lanzafame, M.D. Centro Territoriale de Andrologia, Siracusa, Italy

Gian Paolo Littarru, M.D. Dipartimento di Scienze Cliniche Specialistiche ed Odontostomatologiche, Universita Politecnica delle Marche, Ancona, Italy

Kirk C. Lo, M.D., C.M. Division of Urology, Department of Surgery, University of Toronto, Mount Sinai Hospital, Toronto, ON, Canada

Charles M. Lynne, M.D. Department of Urology, University of Miami Miller School of Medicine, Miami, FL, USA

Antonio Mancini, M.D. Department of Internal Medicine, Division of Endocrinology, The Catholic University of the Sacred Heart, Rome, Italy

Francisco Javier Martin-Romero, Ph.D. Department of Biochemistry and Molecular Biology, School of Life Sciences, University of Extremadura, Badajoz, Spain

P.P. Mathur, Ph.D. Department of Biochemistry & Molecular Biology and Center for Bioinformatics, School of Life Sciences, Pondicherry University, Kalapet, Pondicherry, India

Marcos Meseguer, Ph.D. Clinical Embryology Laboratory, Instituto Universitario IVI Valencia, Valencia, Spain

M.M. Misro, Ph.D. Department of Reproductive Biomedicine, National Institute of Health and Family Welfare, New Delhi, India

Ricardo Miyaoka, M.D. ANDROFERT, Andrology & Human Reproduction Clinic, Center for Male Reproduction, Campinas, SP, Brazil

Ryan Mori, M.D., M.S. Cleveland Clinic Lerner College of Medicine and Glickman Urological and Kidney Institute, Cleveland clinic Foundation, Cleveland, OH, USA

Ray E. Moseley, Ph.D. Department of Community Health and Family Medicine, University of Florida Health Science Center, Gainesville, FL, USA

Alexandre Mottrie, M.D., Ph.D. Department of Urology, O.L.V. Clinic Aalst, Aalst, Belgium

Craig S. Niederberger, M.D. F.A.C.S. Department of Urology, University of Illinois at Chicago College of Medicine and Department of Bioengineering, University of Illinois at Chicago College of Engineering, Chicago, IL, USA

C. J. Opperman, B.Sc. Department of Medical Physiology, Faculty of Health Sciences, Stellenbosch University, Tygerberg, South Africa

Uwe Paasch, M.D., Ph.D. Division of Dermatopathology, Division of Aesthetics and Laserdermatology, Department of Dermatology, Venerology and Allergology, European Training Center of Andrology, Univeristy of Leipzig, Leipzig, Germany

Sijo J. Parekattil, M.D. Division of Urology, Winter Haven Hospital, University of Florida, Winter Haven, FL, USA

Eleonora Pasqualotto, M.D., Ph.D. CONCEPTION—Center for Human Reproduction, Caxias do Sul, RS, Brazil

Department of Clinical Medicine, Center of Health Science, University of Caxias do Sul, Caxias do Sul, RS, Brazil

Fabio Pasqualotto, M.D., Ph.D. Department of Anatomy and Urology, Institute of Biotechnology, University of Caxias do Sul, Caxias do Sul, RS, Brazil
CONCEPTION—Center for Human Reproduction, Caxias do Sul, RS, Brazil

Antonio Pellicer, M.D. Department of Gynecology and Obstetrics, School of Medicine, Universidad de Valencia, Assisted Reproduction Unit, Instituto Universitario IVI Valencia, Valencia, Spain

Eulalia Pozo-Guisado, Ph.D. Department of Biochemistry and Molecular Biology, School of Life Sciences, University of Extremadura, Badajoz, Spain

T. Ramya, M.Sc. Department of Reproductive Biomedicine, National Institute of Health and Family Welfare, New Delhi, India

José Remohí, M.D. Department of Gynecology and Obstetrics, School of Medicine, Universidad de Valencia, Assisted Reproduction Unit, Instituto Universitario IVI Valencia, Valencia, Spain

Alice Rhoton-Vlasak, M.D. Department of Obstetrics and Gynecology, Shands Hospital, Gainesville, FL, USA

Laura Romany, M.D. Clinical Embryology Laboratory, Instituto Universitario IVI Valencia, Valencia, Spain

Henry M. Rosevear, M.D. Department of Urology, University of Iowa, Iowa City, IA, USA

Edmund Sabanegh, M.D. Department of Urology, Glickman Urological and Kidney Institute, Center for Reproductive Medicine, Cleveland Clinic, Cleveland, OH, USA

Tamer Said, M.D., Ph.D. Andrology Laboratory and Reproductive Tissue Bank, The Toronoto Institute for Reproductive Medicine, Toronto, ON, Canada

Jay Sandlow, M.D. Department of Urology, Medical College of Wisconsin, Milwaukee, WI, USA

Maria San Gabriel, Ph.D. Division of Urology, Department of Surgery, Royal Victoria Hospital, McGill University, Montreal, QC, Canada

Peter Schlegel, M.D. Department of Urology, Weill Cornell Medical College, New York, NY, USA

Paul Shin, M.D. Urologic Surgeons of Washington, Washington, DC, USA

Tung Shu, M.D. Clinical Assistant Professor of Urology, Baylor College of Medicine, Center for Kidney Health at the Vanguard Urologic Institute, TX, USA

Aspinder Singh, B.S. Center for Reproductive Medicine, Glickman Urological and Kidney Institute, Cleveland Clinic Foundation, Cleveland, OH, USA

Doron Sol Stember, M.D. Division of Urology, Department of Surgery, Memorial Sloan-Kettering Cancer Center, New York, NY, USA

Adam F. Stewart, M.D. Division of Urology, Department of Surgery, University of Tennessee Medical Center, Knoxville, TN, USA

Claudio Telöken, M.D., Ph.D. Andrology Section of Fertilitat – Human Reproduction Center, Department of Urology, Santa Casa Hospital and Federal University of Health Sciences, Porto Alegre, RS, Brazil

Kelton Tremellen, M.B.B.S. (Hons.), Ph.D., F.R.A.N.Z.C.O.G., C.R.E.I. School of Pharmacy and Medical Sciences, University of South Australia, Dulwich, SA, Australia

Paul J. Turek, M.D., F.A.C.S., F.R.S.M. The Turek Clinic, San Francisco, CA, USA

S. Vaithinathan, Ph.D. Department of Biochemistry and Molecular Biology, Pondicherry University, Kalapet, Pondicherry, India

Alex C. Varghese, Ph.D. Montreal Reproductive Centre, Montreal, QC, Canada

Moshe Wald, M.D. Department of Urology, University of Iowa, Iowa City, IA, USA

Run Wang, M.D., F.A.C.S. Department of Urology, University of Texas Medical School at Houston and MD Anderson Cancer Center, Houston, TX, USA

MD Anderson Cancer Center, Houston, TX, USA

Daniel H. Williams IV, M.D. Department of Urology, University of Wisconsin Hospital and Clinics, Madison, WI, USA

Herbert J. Wiser, M.D. Division of Urology, Southern Illinois University, Springfield, IL, USA

Armand Zini, M.D. Department of Surgery, St. Mary's Hospital, Montreal, QC, Canada

Part I

Male Infertility Diagnosis and Management

Causes of Male Infertility

1

Herbert J. Wiser, Jay Sandlow, and Tobias S. Köhler

Abstract

Of all sexually active couples, 12–15% are infertile. When broken down by gender, a male component can be identified 50% of the time either in isolation or in combination with a female factor. The majority of the causes of male infertility are treatable or preventable, so a keen understanding of these conditions is paramount. Despite advancements in assisted reproductive technologies, the goal of a male infertility specialist is not simply to retrieve sperm. Instead, the male infertility specialist attempts to optimize a male's reproductive potential and thereby allow a couple to conceive successfully through utilization of less invasive reproductive techniques. Often, this involves the use of sperm or testicular tissue cryopreservation prior to fertility insult. At the same time, the male fertility specialist is wary of underlying or causal, potentially serious medical or genetic conditions that prompted reproductive evaluation. Previous research in a US male fertility clinic analyzing 1,430 patients identified causes of infertility from most to least common: varicocele, idiopathic, obstruction, female factor, cryptorchidism, immunologic, ejaculatory dysfunction, testicular failure, drug effects/radiation, endocrinology, and all others. The focus of this book on the role of reactive oxygen species (ROS) is easily applied to the majority of the listed conditions (described in detail in later chapters) which comprise this chapter's overview of pre-testicular, testicular, and post-testicular causes of male infertility.

Keywords

Causes of male infertility • Pre-testicular causes • Testicular causes • Pharmacologic causes • Varicocele • Cryptorchidism • Testicular cancer • Medications and infertility

Of all sexually active couples, 12–15% are infertile [1]. When broken down by gender, a male component can be identified 50% of the time either in isolation or in combination with a female factor [2]. The majority of the causes of male infertility are treatable or preventable, so a keen understanding of these conditions is paramount. Despite advancements in assisted reproductive technologies, the goal of a male infertility specialist is not simply to retrieve sperm. Instead, the male infertility specialist attempts to optimize a male's reproductive potential and thereby allow a couple to conceive successfully through utilization of less invasive reproductive techniques. Often, this involves the use of sperm or testicular tissue cryopreservation prior to fertility insult. At the same time, the male fertility specialist is wary of underlying or causal, potentially serious medical or genetic conditions that prompted reproductive evaluation. Previous research in a US male fertility clinic analyzing

H.J. Wiser, MD (✉) • T.S. Köhler, MD, MPh
Division of Urology, Southern Illinois University, 301 N 8th Street, Floor 4, Springfield, IL, USA
e-mail: hwiser@siumed.edu

J. Sandlow, MD
Department of Urology, Medical College of Wisconsin, 9200 West Wisconsin Avenue, Milwaukee, WI, USA

S.J. Parekattil and A. Agarwal (eds.), *Male Infertility: Contemporary Clinical Approaches, Andrology, ART & Antioxidants*,
DOI 10.1007/978-1-4614-3335-4_1,

1,430 patients identified causes of infertility from most to least common: varicocele, idiopathic, obstruction, female factor, cryptorchidism, immunologic, ejaculatory dysfunction, testicular failure, drug effects/radiation, endocrinology, and all others [3]. The focus of this book on the role of reactive oxygen species (ROS) is easily applied to the majority of the listed conditions (described in detail in later chapters) which comprise this chapter's overview of pre-testicular, testicular, and post-testicular causes of male infertility.

Causes of Male Infertility

Pre-testicular

Hypogonadotropic Hypogonadism

Hypogonadotropic hypogonadism affects fertility at multiple levels. Sperm production is deleteriously affected by a lack of testosterone and a lack of a stimulatory effect on the Sertoli/germ cell complex. Sexual function is also negatively impacted with effects seen at the level of erectile function, ejaculatory function, and sexual desire. There are many etiologies of hypogonadotropic hypogonadism. The most common are elevated prolactin, medications, illicit drugs, and pituitary damage. Kallmann syndrome is another, albeit rare, cause of hypogonadotropic hypogonadism.

Elevated Prolactin

Elevated prolactin may cause hypogonadism by suppressing the release of GnRH. Symptoms of hypogonadism, especially erectile dysfunction and loss of libido, are the most common presenting symptoms in males with hyperprolactinemia, though galactorrhea and gynecomastia may also be evident [4].

Elevated prolactin may be secondary to various etiologies. The most common of these is a prolactinoma, which typically arises from the pituitary. Because prolactinomas in men are more likely to manifest through mass effect, visual disturbances and headaches are more likely to be present when compared to women with prolactinomas [5].

There are other significant causes for hyperprolactinemia as well. Prolactin is elevated in renal failure, as well as in patients with hypothyroidism and cirrhosis. Prolactin levels may also be elevated in certain systemic diseases such as systemic lupus erythematosus, rheumatoid arthritis, celiac disease, and systemic sclerosis. Many drugs elevate prolactin levels, especially those which block the effects of dopamine, such as antipsychotics [6].

Pharmacologic

Various medications may cause hypogonadotropic hypogonadism. Estrogens and progestins may cause a decrease in testosterone levels via negative feedback to the hypothalamic-pituitary-gonadal axis. Marijuana is known to decrease testosterone levels by working on the endocannabinoid receptors present at multiple levels of the hypothalamic-pituitary axis [7]. Both ethanol and cannabinoids suppress GnRH secretion at the level of the hypothalamus. Endocannabinoid receptors have also been found in the pituitary and so may also affect the hypothalamic-pituitary axis at that level as well [8]. LHRH agonists and antagonists are used for the treatment of prostate cancer, precocious puberty, and gender reassignment surgeries. In the male, both induce profound hypogonadism. LHRH antagonists directly and intuitively decrease LH and FSH levels. LHRH agonists produce a tonically stimulated state which, unlike the physiologic circadian rhythmicity of normal LHRH stimulation, acts to decrease LH and FSH secretion. Narcotics may also produce profound hypogonadism. Nearly 40% of men using methadone were found to have total testosterone levels less than 230 ng/dL [9].

Kallmann Syndrome

Kallmann syndrome affects between one in 8,000–10,000 males [10, 11]. It is a spectrum of disease in which the primary manifestations are anosmia and hypogonadotropic hypogonadism which leads to an absence of puberty. Multiple genetic defects can lead to Kallmann syndrome [12]. These most commonly manifest through the same mechanism whereby GnRH secreting neurons fail to migrate to the hypothalamus. Lack of these neurons in the hypothalamus results in a lack of GnRH secretion and thus hypogonadism.

Hypergonadotropic Hypogonadism

One of the most common causes of hypergonadotropic hypogonadism is Klinefelter syndrome (Klinefelter's). Klinefelter's affects male fertility by altering spermatogenesis both directly and indirectly by altering the hormonal milieu [13–15]. Interestingly, sex hormone levels are normal until puberty. During puberty, they do rise to low-normal levels, but plateau. By adulthood, serum testosterone levels are typically below normal. Histologic studies demonstrate gradual degeneration of the testes with development, with hyperplasia of poorly functioning Leydig cells [16]. Klinefelter's also directly affects spermatogenesis, as discussed later in this chapter.

Testicular

Varicocele

A varicocele is a dilation of the pampiniform plexus likely caused by the absence or incompetence of the venous valves of the internal spermatic vein. Varicoceles have long been associated with infertility. The first written description is attributed to Celsius who noticed the association between the varicocele and testicular atrophy [17]. In the 1800s, surgical correction was seen to improve semen quality. It is currently seen to be the most common surgically correctable cause of male infertility. Roughly 12% of all men have a varicocele, but this number jumps to 25% in men with abnormal semen parameters [18].

Varicoceles affect multiple semen parameters; total sperm count, sperm motility, and sperm morphology are all negatively affected [19, 20]. There are many theories about the underlying pathophysiology of a varicocele, with heat, renal metabolites, and hormonal abnormalities all playing a role. However, most agree that disruption of the countercurrent heat exchange mechanism in the testis, causing hyperthermia, is the most likely mechanism.

Scrotal temperature in humans is variable during the day, but remains 1–2°C lower than core body temperature at 33–36°C [21]. Thermoregulation of the gonads at a temperature lower than that of body temperature is a trait that is well preserved in homeotherms and especially in mammals [22]. Nearly all mammals have a scrotum. Other mechanisms, such as the efficient heat exchange system in whales, have developed in animals in environments where the scrotum would not be efficient at keeping gonadal heat at a few degrees below body temperature. Besides this teleologic evidence that lower body temperature is necessary for testicular function, numerous studies point to impaired sperm production and a decrease in semen quality when scrotal temperatures are elevated [23–28]. One study showed that men with scrotal skin temperatures above 35°C for >75% of the day had sperm concentrations of 33 million/mL as compared with men with scrotal skin temperatures greater than 35°C for <50% of the day who had sperm concentrations of 92 million/mL [27]. The mechanism by which heat causes decreased sperm counts is poorly understood, but one hypothesis is that increased temperature could increase the metabolic rate of testicular and epididymal sperm, secondarily increasing the amount of oxidative damage to both the structure and the DNA of the spermatocytes and spermatids [22].

Varicoceles are noted to be associated with higher scrotal temperatures [29], and cooling of the scrotum has been shown to improve semen parameters [30]. Interestingly, the temperature of the contralateral testis is also elevated in men with unilateral varicoceles. So the cause of elevated testicular temperatures, which intuitively would seem to be an impaired countercurrent flow mechanism, is less clear [31]. Hormonal abnormalities are a similarly controversial area, with no consistent hormonal changes associated with the presence of a varicocele. Testosterone, SHBG, FSH, and LH have all been examined, and different studies have produced opposing results [17].

Cryptorchidism

Cryptorchidism is well known to affect fertility. The severity of its effect on fertility is directly proportional to the severity of the cryptorchidism, with bilateral cryptorchidism having more severe effects than unilateral, and with higher testes having worse function than lower testes [32–35].

Similarly, orchidopexy has been shown to improve fertility, with the best results obtained with fixation at a young age, especially prior to 1 year of age [36]. Fixation after age 10 may not improve fertility, or may improve it only modestly, suggesting that permanent and progressive damage is done to the testis while in an abnormal position, and this is supported by histologic studies [37, 38]. Actual paternity rates in men who underwent orchidopexy for unilateral cryptorchidism are 89%, slightly less than the non-cryptorchid group, which had a 94% paternity rate. Bilaterally cryptorchid men post-orchidopexy had markedly lower paternity rates, at 62% [34, 35].

The pathophysiology of the effects of cryptorchidism is complex, with heat likely playing a partial but significant role [39, 40]. A number of other factors are also likely to come in to play, including the underlying genetics, hormonal milieu, and environmental exposures which originally led to the cryptorchidism [41–43].

Testicular Cancer

Testicular cancer is strongly associated with infertility. There are multiple ways in which testicular cancer is related to and can contribute to reduced fertility. Both testicular cancer and impaired spermatogenesis may be related in their etiology of embryologic testicular dysgenesis. The testicular dysgenesis syndrome is a spectrum of disease that may involve cryptorchidism, hypospadias, decreased spermatogenesis, and testis cancer. In this syndrome, it is thought that all of these share an origin of abnormal fetal testis development. As a result of this developmental anomaly, any number of these manifestations may be present in a boy [44]. Testicular tumors may also directly contribute to infertility by secreting hormones, which can downregulate sperm production in the contralateral testis [45–48]. This is uncommon but has been seen with Leydig and Sertoli cell tumors as well as seminomas. Tumors may also directly disrupt spermatogenesis by mass effect or by the effects of the inflammatory reaction to the tumor [49]. Cancer treatments may also decrease fertility.

At presentation, roughly 10% of men will be azoospermic, and roughly 50% will be oligospermic. While orchiectomy will result in a rebound in semen parameters in roughly 90% of these men [50], further treatment with surgery, chemotherapy, or radiation can further decrease fertility.

Ionizing Radiation

Excellent data on the effects of ionizing radiation is available from two similar studies, which are unlikely to be repeated. Researchers in these studies prospectively irradiated the testes of prisoners with single or multiple doses of radiation up to 600 cGy [51, 52]. Sperm counts were followed, and serial testicular biopsies were done. These studies showed that sperm counts declined when testes were irradiated and that decline was dose dependant. At low doses of ~7.5 cGy, a mild decline of sperm counts was seen, and this decline increased to severe oligospermia by 30–40 cGy and azoospermia by 78 cGy. The time to recovery was also seen to be dose dependant, with those receiving 20 cGy beginning to have a recovery of sperm counts by 6 months, those with 100 cGy at 7 months, 200 cGy at 11 months, and 600 cGy at 24 months. The percent of men achieving a complete recovery and time to achieve a complete recovery also declined with increasing radiation doses.

Decline to the nadir of sperm counts was seen at roughly 64 days, corresponding roughly to the time required for sperm cell production from spermatogonia. More rapid declines were seen with higher radiation doses, indicating increased damage to the more highly differentiated cells undergoing spermatogenesis. Biopsy results from these studies showed that spermatogonia numbers nadired at much lower levels with higher doses of radiation and that these nadirs took longer to achieve than those which had received lower doses of radiation.

These studies provide excellent information into the biology of the effects of radiation on spermatogenesis on the healthy young testis. Clinically, however, the effects we see are often more pronounced given the setting of the radiation, namely, cancer patients undergoing radiotherapy. Fractionated radiation has been shown to be more damaging than single dose radiation [53]. One report showed that fractionated radiation with a total dose of 200 cGy may cause permanent azoospermia [54].

As would be expected, Leydig cells are more resistant to radiation than the germinal epithelium [51]. Doses of 20 Gy are known to cause declines in testosterone [55]. Doses on the order of 2 Gy do not cause appreciable drops in testosterone [56].

Chemotherapy

Chemotherapy typically targets rapidly dividing cells and thus has profound effects on the germinal epithelium. As such, the expected outcome of acute chemotherapy is a decline in spermatogenesis, and this has been well documented since the late 1940s [57]. The mechanism by which chemotherapeutics decrease fertility and the rates of recovery is both drug and dose dependent [58–61].

Bleomycin, etoposide, and cisplatin or carboplatin (BEP) is the most commonly used chemotherapy regimen for testicular germ cell tumors. The decrease in fertility seen post-BEP chemotherapy is likely the result of a direct reduction of spermatogenesis and not as a result of any change in the hormonal milieu. Indeed, testosterone levels are not seen to be significantly reduced at 12 months post-chemotherapy, and FSH levels are appropriately elevated. FSH levels decline as spermatogenesis returns over the following 2–4 years [62]. It should be noted, however, that return of spermatogenesis is not guaranteed. In patients who were normospermic prior to chemotherapy, fewer than 4 cycles of BEP have not typically been associated with high rates of permanent infertility [63]. However, high-dose BEP is associated with approximately a 50% of permanent infertility in one study [64]. Notably, even in azoospermic men post-high dose BEP chemotherapy, nests of spermatogenesis have been found on TESE [65].

Genetic Azoospermia/Oligospermia

It is estimated that 2–8% of infertile men have an underlying genetic abnormality, with this number rising to 15% in azoospermic men [66]. Although the majority of male infertility does not have an identifiable genetic cause, two potential etiologies are Y chromosome microdeletions and karyotypic abnormalities. The two most common karyotypic abnormalities are Klinefelter's (47,XXY) and chromosomal translocations.

Y chromosome microdeletions are a common cause of these, occurring in 11–18% of azoospermic men and 4–14% of oligospermic men [67]. Currently, research is focused on the azoospermia factor (AZF) region on the long arm of the Y chromosome at Yq11. This area itself contains three separate regions, AZFa, AZFb, and AZFc, and microdeletions of these areas lead to slightly different phenotypes [68]. Deletions in the AZFa and AZFb regions both cause azoospermia, but histologically, they are different with AZFa deletions resulting in Sertoli cell-only syndrome and AZFb deletions causing an arrest of spermatogenesis at the primary spermatocyte stage [66]. AZFc deletions are the most common of the Y chromosome microdeletions and are found in 5–7% of oligospermic men [68]. Unlike the AZFa/AZFb deletions, they do not uniformly result in azoospermia; rather, a spectrum of phenotypes are seen with partial deletions being found in normospermic men, from oligospermia to azoospermia in some full deletions [66]. In men undergoing micro-TESE sperm extraction with azfC deletions, about 35% have sperm found successfully [69].

Classic and mosaic Klinefelter's are common karyotypic abnormalities found in infertile men. Klinefelter's has a prevalence of one in 660 males; thus, it is the most common genetic cause of male infertility as 75–90% of men with Klinefelter's will be azoospermic, with some with mosaic Klinefelter's being mainly oligospermic [13, 70, 71]. Studies which show higher prevalences of azoospermic men with AZF deletions than Klinefelter's are likely flawed by a selection bias as men with the obvious stigmata of Klinefelter's are not tested and included in these studies [72, 73].

As would be expected, Klinefelter's has much broader effects than Y chromosome microdeletions and affects fertility through two routes, direct effects on spermatogenesis and indirect hormonal effects on spermatogenesis [13–15].

As far as altered spermatogenesis is concerned, the majority of Klinefelter's patients actually do produce sperm, as is witnessed by the 69% TESE sperm retrieval rates [74]; however, the quantity of sperm produced is typically very low. Biopsy studies of Klinefelter's testes have demonstrated that spermatogenesis is halted pre-pachytene in the vast majority of aneuploid cells and that meiosis was seen mainly in cells with normal karyotypes [15].

Robertsonian translocations are a third significant genetic cause of infertility. They occur in 0.8% of infertile men, and this number rises to 1.6% in oligospermic men [75]. Phenotypes are highly variable given the possibilities of recombination [66].

Environmental Factors

Hyperthermia is considered to be a major contributor in the pathogenesis of infertility in men with varicocele and cryptorchidism. Many lifestyle factors also have the potential to increase scrotal temperatures, including underwear type, heated car seats, and occupational heat exposure. The role of underwear style in male infertility has been investigated. One small study of 14 normospermic men, a tight polyester scrotal support, when worn day and night, was shown to make all azoospermic at a mean time of 140 days. After removal of the scrotal support, all men regained function at a mean time of 157 days [76]. However, normal underwear, i.e., boxer or brief style, has not been shown to exert a significant influence on semen parameters [77]. Other types of heat exposure, such as occupational heat exposure in a group of welders, have been shown to decrease semen quality [78]. Sedentary posture, heated car seats, and sauna and hot tub use are all lifestyle factors that increase scrotal temperature as well and may contribute to a decline in fertility [79].

Recently, cell phones have been implicated as possibly playing a role in decreasing male fertility, and several studies show that there may be some basis for this. One observational study assessed semen parameters and cell phone usage in 361 men who presented to an infertility clinic. Sixty percent of the men in this study had greater than 2 h of cell phone use per day, with 30% using their cell phones for more than 4 h per day. They found that sperm counts, motility, viability, and morphology all worsened with increasing cell phone use [80]. The mechanism by which cell phones affect semen parameters has not yet been elucidated, but one hypothesis is that cell phone-generated electromagnetic radiation (CPEMR) alters mitochondrial function and acts to increase reactive oxygen species. This is somewhat corroborated by one study which looked at the effects of CPEMR on semen parameters and found increased levels of reactive oxygen species with decreased viability and motility in the sperm exposed to CPEMR [81].

Tobacco use has been implicated in the pathogenesis of numerous cancers and medical diseases. While the use of tobacco significantly impacts female fertility, its impact on male fertility is less clear. Semen parameters, including sperm density, motility, and morphology, have all been shown to be worsened with tobacco use [82–85]. However, a significant reduction in fertility has not yet been proven.

Testicular Injury

Injury to the testicle can be sustained either directly or indirectly. Direct trauma to the testis is typically managed by debridement of devitalized seminiferous tubules and closure of the tunica albuginea [86]. The resultant loss of volume of seminiferous tubules and possible obstruction from scarring is one possible cause of decreased fertility. Reports on testicular salvage after bilateral trauma indicated that preserved volume of testis is the key to preserving fertility [86–91].

Indirect damage to the testis may be sustained by exposure to infection or inflammation of the testis. The classic infectious agent causing infertility is mumps. Orchitis occurs in roughly 20% of postpubertal males with mumps [92] and is bilateral in 30% of these. Of those postpubertal males with bilateral mumps orchitis, 25% will have resultant infertility. In other words, 1.5% of postpubertal males with mumps may become infertile as a result of the disease. In nations where immunization against mumps is common, this is a rare phenomenon. The mechanism by which mumps causes orchitis is via pressure atrophy. Infection of the testis with the mumps virus causes inflammation and swelling, which is limited by the tunica albuginea; this in turn leads to atrophy [93].

Other bacterial and viral pathogens may also cause infertility at the testicular level, most commonly; this is the result of spread of infection from the epididymis [94, 95]. The mechanism for infertility in these cases may be persistent inflammation which suppresses testicular function or obstruction secondary to resultant sclerosis.

Primary Ciliary Dyskinesia

Ultrastructural defects that affect sperm motility are described under the grouping of primary ciliary dyskinesia (PCD). PCD is a rare and heterogeneous genetic disease which affects one in 20,000–60,000 [96]. Many components of cilia and flagella are affected, though the defect is found in the dynein in over 80% of cases [97]. The key clinical finding is chronic respiratory infections leading to bronchiectasis. When situs inversus is present in addition to the other components, it is termed Kartagener's syndrome. Male infertility secondary to sperm dysmotility is related to the dysfunction of the flagellate tail of the sperm. It is a common finding, though not universal, and this is likely related to the heterogeneity of the genetics.

Sertoli Cell-Only Syndrome

Sertoli cell-only syndrome may be either primary or secondary, and attempts have been made to distinguish these histologically [98]. The primary form is hypothesized to result from a lack of migration of the germ cells to the seminiferous tubules during embryologic development. The secondary form is due to a gonadotoxic insult to the testis after birth. While the different etiologies of these would intuitively suggest a higher likelihood of finding sperm in biopsies of testes with secondary Sertoli cell-only syndrome, this is not borne out in the literature [99].

Antisperm Antibodies

In the normal male, sperm reside in an immunoprivileged site. The blood-testis barrier prevents proteins from the sperm from interacting with the immune system and setting up an immune reaction against them. Trauma, infection, and inflammation all may disrupt this barrier and result in immunity against the germinal epithelium and spermatozoa.

Antisperm antibodies (ASA) are very common, with 8–17% of men and 1–22% women in infertile couples testing positive for serum ASA [100, 101]. As expected, ASA are heterogeneous in their binding sites and, as such, have wide ranging effects on sperm function. Some ASA will not significantly affect fertility, and 0.9–2.5% of fertile men will test positive for serum ASA [102, 103]. ASA targeted against proteins on the head region are more likely to affect zona binding and sperm penetration, whereas ASA targeted against the tails of spermatozoa are more likely to decrease motility and cervical mucus penetration and cause sperm agglutination [104]. Antibody type also plays a significant role in the degree of reduction of fertility. In a study of ASA in men who had undergone vasectomy reversal, IgA ASA were associated with a much more significant reduction in fertility than IgG ASA [105]. While ASA clearly may affect fertility in some cases, serum ASA positivity is not a strong predictor of infertility.

DNA Damage

There are many etiologies of sperm DNA damage. Radiation, toxins, genital tract inflammation, varicocele, advanced paternal age, and testicular hyperthermia all induce significant DNA damage [106, 107] and will be discussed at length elsewhere in this book.

Post-testicular

Absence of the Vas Deferens

Congenital bilateral absence of the vas deferens (CBAVD) is a condition strongly related to cystic fibrosis (CF) and has even been considered as a diagnostic criterion for CF. However, while current dogma states that nearly all patients with CF have bilaterally absent vasa, there is little data to support this. Indeed, two recent articles suggest that CBAVD is present in half or less of CF patients. One series looking at children with CF who were undergoing inguinal hernia repair reported only a 24% (6/25) rate of CBAVD [108]. A series of 20 adults with CF and a mean age around 30 years old had a CBAVD rate of 55% [109]. In this latter series, only one man had a semen analysis consistent with possible fertility, and the more constant finding was atrophy of the seminal vesicles, which was seen in 18/20.

Nevertheless, CBAVD is strongly associated to CF, and the same genetics, mutation of the CFTR, are typically responsible for both phenomenons [110]. So, while men with CF do not necessarily have CBAVD, most men with CBAVD do have a CFTR mutation [111–113]. The pathophysiology of CBAVD thus clearly involves altered chloride transport in the majority of cases, and, like the respiratory and pancreatic sequelae seen with CF, there is evidence that the genital abnormalities and pathology seen are a progressive disease. Namely, intentionally aborted CF fetuses demonstrate normal vas deferens, albeit with secretions filling their lumens. This suggests that the mechanism for CBAVD is atresia, and not aplasia, when a CFTR mutation is present [114]. One interesting sequela of this is that renal agenesis is not associated with CBAVD [115].

Congenital unilateral absence of the vas deferens (CUAVD) is a different entity altogether [116]. While there is still a significant rate of CFTR mutations in men with CUAVD, especially when the obstructive azoospermia is present [110], the majority of CUAVD is the result of an embryologic Wolffian duct aberrancy [117]. As such, renal agenesis is often seen with CUAVD, though CUAVD is not always seen in men with unilateral renal agenesis, as there are many other embryologic missteps that may occur to result in renal agenesis. While there is only a 20% rate of CUAVD seen in those with a unilateral renal agenesis, there is a 79% rate of unilateral renal agenesis seen in men with CUAVD. Since CUAVD not associated with a CFTR mutation is usually a unilateral and isolated phenomenon, fertility is often preserved.

Young's Syndrome

Young's syndrome is a rare disorder which presents clinically as obstructive azoospermia and chronic sinopulmonary infections [118]. Thus, it can be difficult to differentiate clinically from cystic fibrosis variants and primary ciliary dyskinesia. Indeed, definitive diagnosis of Young's syndrome requires negative CFTR genetic testing as well as investigation of ciliary ultrastructure to rule out primary ciliary dyskinesia [119]. Normal spermatogenesis is seen, and the obstructive azoospermia is due to inspissated secretions in the vas deferens.

The etiology of Young's syndrome is unclear with childhood mercury exposure having been postulated to play a role in the past [120]. Interestingly, the incidence of Young's syndrome has plummeted from estimates of one in 500 in the

1980s down to case reports and articles which question the existence of Young's syndrome today [121]. The observation that the reduced incidence over the last 50 years coincides with a decrease in mercury use and poisoning is tempered by the fact that our knowledge of genetics has rapidly advanced. Thus, the decreased incidence of Young's syndrome is more likely due to the increased correct genetic diagnosis of CF spectrum disease.

EjDO/Seminal Vesicle Dysfunction

Ejaculatory duct obstruction is a common etiology of male infertility, occurring in 1–5% of men presenting with infertility [122]. There are many causes of ejaculatory duct obstruction, including cystic fibrosis spectrum disease, Wolffian or Muellerian origin cysts, calcifications, tuberculosis and other GU infections, calculi, and urinary tract instrumentation [123, 124]. Additionally, chronic ejaculatory duct obstruction may affect the seminal vesicle in a manner analogous to the effect of bladder outlet obstruction on the bladder. Namely, with longstanding obstruction, the seminal vesicles may lose contractility, and resolution of the anatomical obstruction may not improve seminal vesicle emptying during ejaculation. Seminal vesicle dysfunction may also be seen in the absence of previous obstruction. This can be secondary to multiple sclerosis, diabetes, spinal cord injury or other neurologic insult, and medications. One interesting physical finding seen in 25–50% of men with spinal cord injuries likely related to seminal vesicle dysfunction is brown semen [125]. This brown coloration is not derived from heme and is not related to semen stasis per se.

Vasectomy and Vasectomy Reversal

Vasectomy is a procedure that is intended to produce infertility, and it is successful in over 90% of cases [126]. Some of the key determinants of success are related to aspects of surgical technique. The main reason for a correctly performed vasectomy to fail is recanalization of the vas deferens, a finding that has been histologically verified [127, 128]. Some debate remains as to which techniques provide the lowest recanalization rates. The manner of ligating the ends, non-ligation versus clipping versus suture ligation, length of vas removed, as well as whether to fold vas ends are all controversial [129, 130]. Two maneuvers which do seem to provide significant benefits are luminal cauterization and fascial interposition [131, 132].

Vasectomy reversal may be performed in an attempt to return fertility to the sterilized man. The outcomes of vasectomy reversal are dependent on a number of factors. Surgical technique is one factor, with use of a microscope significantly improving pregnancy rates over loupe-assisted vasovasostomy [133]. Time elapsed since fertility also plays a significant role with a 97% patency rate and 76% pregnancy rate being achieved if surgery is performed at less than 3 years since vasectomy. Patency and pregnancy rates decline as time elapses, with patency and pregnancy rates of 79% and 44%, respectively, if the vasectomy was between 9 and 14 years prior and 71% and 30% if greater than 15 years elapsed [134]. Type of vasectomy reversal also plays a role with vasoepididymostomy (VE) having lower patency and pregnancy rates than vasovasostomy (VV) [135]. Presence and type of antisperm antibodies also may reduce fertility rates in men after vasectomy reversal [105]. Sperm granulomas were previously thought to decrease testicular pressure and so portend better vasectomy reversal outcomes. Though better sperm quality has been found in the vasa of men with sperm granulomas at the time of surgery, patency and pregnancy rates are not significantly different [134, 136]. Similarly, a testicular vasal remnant of 2.7 cm or longer predicts finding whole sperm in the vasal fluid [137], though research is not available as to its effect on patency and pregnancy rates. Though repeat attempts at vasectomy reversal would intuitively seem less likely to succeed, high success rates have been reported with combined VV/VE patency rates of 89% and pregnancy rates of 58% if the interval of obstruction was less than 10 years [138].

Nerve Injury

Nervous injury affecting ejaculation may occur at many levels and have a diverse etiology ranging from spinal cord injury to neural damage during retroperitoneal or pelvic surgery to neuropathy from systemic diseases. Ejaculatory dysfunction is present in 90% of spinal cord injury patients [139]. The type and severity of ejaculatory dysfunction are dependent on the level and extent of the injury. Higher cord lesions often result in an intact reflex arc which allows for penile vibratory stimulation to induce ejaculation. Men with sacral lesions or lesions of the efferent parasympathetic nerves are often not responsive to penile vibratory stimulation and may require endorectal electrical stimulation to induce ejaculation [140].

Retroperitoneal lymph node dissection (RPLND) for testicular cancer resulted in a high rate of ejaculatory dysfunction until the development of methods to spare the sympathetic nerve fibers. Both emission and bladder neck contraction are mediated by the sympathetic nervous system, and damage to the sympathetic chain and the hypogastric plexus overlying the great vessels results in a high degree of ejaculatory dysfunction. In the past, RPLND was associated with a 55–60% chance of ejaculatory dysfunction [141, 142]. Modified templates have helped to reduce the rates of retrograde ejaculation, with one study demonstrating an 82% rate of antegrade ejaculation with a modified unilateral template [143]. Another study using a modified bilateral template demonstrated an 88% rate of preservation of antegrade ejaculation [144].

Nerve sparing RPLND, developed in the late 1980s, has reduced the incidence of retrograde ejaculation even further to 0–7% [145, 146]. Nerve sparing RPLND may also be done after chemotherapy, though only 136 of 341 men qualified for this as compared to standard RPLND in one series [147]. Rates of ejaculatory dysfunction were also higher at 21%.

Medications

Medications affecting ejaculation do so by altering adrenergic signaling. This is most clearly seen with alpha-1 antagonists. Tamsulosin and silodosin, especially, are known to cause ejaculatory dysfunction [148, 149]. Previously, this was thought to be retrograde ejaculation. Recent studies have shown that the ejaculatory dysfunction induced by alpha-1 antagonists is actually a failure of emission [150, 151].

Antipsychotics have long been associated with sexual dysfunction, including ejaculatory dysfunction. Antipsychotics have effects on many different neurotransmitters including dopamine, norepinephrine, acetylcholine, and serotonin. Predictably, altered ejaculatory function with antipsychotics use correlates with anti-adrenergic actions of the antipsychotics [152]. Even atypical antipsychotics like risperidone may affect ejaculation [153, 154].

Resection of the Prostate

Surgery of the prostate is well known to cause retrograde ejaculation. Transurethral resection of the prostate as well as the laser photovaporization and enucleation all have a high likelihood of inducing retrograde ejaculation since removal of the proximal prostatic urethra severely diminishes the resistance to backflow of semen.

Coital

Abnormal coital practices may play a role in infertility when they interfere with semen deposition in the vagina or affect their timing with the female reproductive cycle. Similarly, erectile dysfunction and penile abnormalities such as hypospadias and chordee may interfere with semen deposition and thus may play a role in infertility.

Lubricants are commonly used by infertile couples, and many vaginal lubricants have been shown to negatively affect fertility. Many synthetic lubricants not only affect sperm motility but have also been shown to increase the DNA fragmentation index. In one study, FemGlide, Replens, and Astroglide all affected sperm motility, and FemGlide and K-Y jelly increased DNA fragmentation. One lubricant that has not been shown to have a significant impact on sperm motility or DNA fragmentation is Pre-Seed [155]. Another study showed similar findings with decreased motility in sperm exposed to K-Y jelly and Touch. Non-viability was seen in sperm exposed to Replens and Astroglide which was comparable to the non-viability seen when sperm were exposed to the spermicide nonoxylnol-9 [156]. In this study, canola oil was not found to affect sperm motility or viability. Yet another study showed that K-Y jelly, saliva, and olive oil all reduced sperm motility, while baby oil did not significantly affect motility [157].

Expert Commentary

This chapter has described the pre-testicular, testicular, and post-testicular spectrum of conditions known to affect male fertility. Many of the listed causes stem from or are subject to further degradation from reactive oxygen species. Pre-testicular causes often alter the normal hormonal milieu for sperm development, providing a suboptimal environment for sperm and perhaps a greater exposure or sensitivity to free radical damage. Testicular causes of infertility such as radiation, toxins, genital tract inflammation, varicocele, and testicular hyperthermia all induce significant DNA damage and thus increase reactive oxygen species. Finally, post-testicular causes of male infertility often affect sperm transit time, increasing likelihood of free radical damage of sperm. Despite our understanding of many of the conditions leading to male infertility, idiopathic infertility still comprises a large portion of the men evaluated for problems with reproduction. The proportion of idiopathic infertility will likely decrease with further understanding of the role of reactive oxygen species and clarification of the role of DNA integrity assays.

The male partners of all couples presenting with infertility must be examined and evaluated. Infertility itself is an independent risk factor for testicular cancer and genetic disease. Indeed, men of reproductive years often forego visiting physicians, and the infertility visit offers a viable platform for general health screening and recommendations. It must be remembered that the majority of the causes of male infertility are either preventable or treatable. Treatment success goes beyond simply harvesting sperm for assisted reproductive techniques. Facilitating pregnancy through intrauterine insemination with varicocele repair should be viewed with the same regard as facilitating natural pregnancy with vasectomy reversal. Finally, the importance of sperm banking cannot be understated, as cryopreservation of reproductive tissue prior to reproductive insult from chemotherapy or surgery is simple and often is the only chance of preserving future fertility.

Five-Year View

Although details and further understanding of some of the causes of male fertility conditions of male described have been elucidated in recent years, previous and future chapters in reproductive textbooks are and will be very similar to this one. However, the development and refinement of DNA integrity tests and determination of the relative importance of reactive oxygen species will likely obviate the need for male infertility evaluation. For example, if varicocele repair is definitively proven to reduce DNA damage to sperm, and decreased DNA damage to sperm is definitively proven to improve success rates with assisted reproductive techniques, referral from female fertility specialists will likely increase. Additional public and provider education on the rationale for male infertility referral and messages on the need for sperm banking will also increase the need for specialists knowledgeable in the causes and treatment of male infertility. This in itself may bring challenges because in relation to female assisted reproductive technology centers, large areas are greatly underserved by male fertility specialists [158].

Key Issues

- The majority of the causes of male infertility are either preventable or treatable.
- Male infertility is an independent risk factor for testicular cancer and genetic diseases.
- Sperm banking should be utilized liberally prior to potential gonadotoxic exposure.
- Pre-testicular causes of male infertility exert their negative effect via imbalances in the hormonal milieu of sperm production. Sexual function is also negatively impacted with effects seen at the level of erectile function, ejaculatory function, and sexual desire.
- Medications can negatively impact pre-testicular, testicular, and post-testicular function.
- Varicocele is the most common cause of male infertility.
- For fertility potential, cryptorchidism is best treated early, especially if bilateral.
- After testicular trauma, fertility is most dependent on operative testicular volume preservation.
- Severe oligospermia or azoospermia requires genetic screening, given their high associated prevalence of Klinefelter's syndrome, karyotypic abnormalities, and microdeletion of the Y chromosome.
- CBAVD is not always seen with cystic fibrosis; evaluation for renal agenesis in CUAVD is essential.
- Many testicular causes of male infertility (radiation, toxins, environmental factors, genital tract inflammation, varicocele, testicular hyperthermia) lead directly to sperm DNA damage.
- Several post-testicular causes of male infertility stem from surgery.

References

1. Mosher WE. Reproductive impairments in the United States, 1965–1982. Demography. 1985;22:415–30.
2. Tielemans E, Burdorf A, te Velde E, Weber R, van Kooij R, Heederik D. Sources of bias in studies among infertility clients. Am J Epidemiol. 2002;156:86–92.
3. Sigman M. Male Infertility. Med Health R I. 1997;80(12):406–9.
4. Buvat J. Hyperprolactinemia and sexual function in men: a short review. Int J Impot Res. 2003;15(5):373–7.
5. Carter JN, Tyson JE, Tolis G, et al. Prolactin-screening tumors and hypogonadism in 22 men. N Engl J Med. 1978;299(16):847–52.
6. Patel SS, Bamigboye V. Hyperprolactinaemia. J Obstet Gynaecol. 2007;27(5):455–9.
7. Fasano S, Meccariello R, Cobellis G, et al. The endocannabinoid system: an ancient signaling involved in the control of male fertility. Ann N Y Acad Sci. 2009;1163:112–24.
8. Rettori V, De Laurentiis A, Fernandez-Solari J. Alcohol and endocannabinoids: neuroendocrine interactions in the reproductive axis. Exp Neurol. 2010;224(1):15–22.
9. Hallinan R, Byrne A, Agho K, et al. Hypogonadism in men receiving methadone and buprenorphine maintenance treatment. Int J Androl. 2009;32(2):131–9.
10. Dodé C, Hardelin JP. Kallmann syndrome. Eur J Hum Genet. 2009;17:139–46.
11. Fechner A, Fong S, McGovern P. A review of Kallmann syndrome: genetics, pathophysiology, and clinical management. Obstet Gynecol Surv. 2008;63(3):189–94.
12. Hardelin JP, Dode C. The complex genetics of Kallmann syndrome: KAL1, FGFR1, FGF8, PROKR2, PROK2, et al. Sex Dev. 2008;2:181–93.
13. Kamischke A, Baumgardt A, Horst J, et al. Clinical and diagnostic features of patients with suspected Klinefelter Syndrome. J Androl. 2003;24:41–8.
14. Blanco J, Egozcue J, Vidal F. Meiotic behavior of the sex chromosomes in three patients with sex chromosome abnormalities (47, XXY, mosaic 46, XY/47, XXY, and 47, XYY) assessed by flourescence in-situ hybridization. Hum Reprod. 2001;16(5):887–92.
15. Bergere M, Wainer R, Nataf V, et al. Biopsied testis cells of four 47, XXY patients: fluorescence in-situ hybridization and ICSI results. Hum Reprod. 2002;17:32–7.
16. Wikström AM, Dunkel L. Testicular function in Klinefelter syndrome. Horm Res. 2008;69(6):317–26.
17. Nagler HM, Grotas AB. Varicocele. In: Lipshultz LI, Howards SS, Niederberger CS, editors. Infertility in the male. 4th ed. New York City, NY: Cambridge University; 2009.
18. World Health Organization. The influence of varicocele on parameters of fertility in a large group of men presenting to infertility clinics. Fertil Steril. 1992;57:1289–93.
19. MacLeod J. Seminal cytology in the presence of varicocele. Fertil Steril. 1965;16(6):735–57.
20. Paduch DA, Niedzielski J. Semen analysis in young men with varicocele: preliminary study. J Urol. 1996;156:778–90.
21. Hjollund NH, Storgaard L, Ernst E, et al. The relation between daily activities and scrotal temperature. Reprod Toxicol. 2002; 16(3):209–14.
22. Ivell R. Lifestyle impact and the biology of the human scrotum. Reprod Biol Endocrinol. 2007;5:15.
23. Paul C, Murray AA, Spears N, et al. A single, mild, transient scrotal heat stress causes DNA damage, subfertility and impairs formation of blastocysts in mice. Reproduction. 2008;136(1):73–84.
24. Dada R, Gupta NP, Kucheria K. Spermatogenic arrest in men with testicular hyperthermia. Teratog Carcinog Mutagen. 2003;S1: 235–43.

25. Esfandiari N, Saleh RA, Blaut AP, et al. Effects of temperature on sperm motion characteristics and reactive oxygen species. Int J Fertil Womens Med. 2002;47(5):227–33.
26. Bedford JM. Effects of elevated temperature on the epididymis and testis: experimental studies. Adv Exp Med Biol. 1991;286:19–32.
27. Hjollund NH, Bonde JP, Jensen TK, et al. Diurnal scrotal skin temperature and semen quality. The Danish first pregnancy planner study team. Int J Androl. 2000;23(5):309–18.
28. Wang C, McDonald V, Leung A, et al. Effect of increased scrotal temperature on sperm production in normal men. Fertil Steril. 1997;68(2):334–9.
29. Zorgniotti AW, MacLeod J. Studies in temperature, human semen quality, and varicocele. Fertil Steril. 1973;24(11):854–63.
30. Jung A, Eberl M, Schill WB. Improvement of semen quality by nocturnal scrotal cooling and moderate behavioral change to reduce genital heat stress in men with oligoasthenoteratozoospermia. Reproduction. 2001;121(4):595–603.
31. Goldstein M, Eid JF. Elevation of intratesticular and scrotal skin surface temperature in men with varicocele. J Urol. 1989;142(3):743–5.
32. Trsinar B, Muravec UR. Fertility potential after unilateral and bilateral orchidopexy for cryptorchidism. World J Urol. 2009;27(4):513–9.
33. Gracia J, Sánchez Zalabardo J, Sánchez García J, et al. Clinical, physical, sperm and hormonal data in 251 adults operated on for cryptorchidism in childhood. BJU Int. 2000;85(9):1100–3.
34. Lee PA, O'Leary LA, Songer NJ, et al. Paternity after unilateral cryptorchidism: a controlled study. Pediatrics. 1996;98:676–9.
35. Lee PA, O'Leary LA, Songer NJ, et al. Paternity after bilateral cryptorchidism. A controlled study. Arch Pediatr Adolesc Med. 1997;151(3):260–3.
36. Canavese F, Mussa A, Manenti M, et al. Sperm count of young men surgically treated for cryptorchidism in the first and second year of life: fertility is better in children treated at a younger age. Eur J Pediatr Surg. 2009;19(6):388–91.
37. Wiser A, Raviv G, Weissenberg R, et al. Does age at orchidopexy impact on the results of testicular sperm extraction? Reprod Biomed Online. 2009;19(6):778–83.
38. Cooper ER. The histology of the retained testis in the human subject at different ages and its comparison to the testis. J Anat. 1929;64:5–10.
39. Murphy F, Paran TS, Puri P. Orchidopexy and its impact on fertility. Pediatr Surg Int. 2007;23(7):625032. Epub 13 Mar 2007.
40. Setchell BP. The Parkes Lecture: heat and the testis. J Reprod Fertil. 1998;114(2):179–94.
41. Leissner J, Filipas D, Wolf HK, et al. The undescended testis: considerations and impact on fertility. BJU Int. 1999;83(8):885–91.
42. Hadziselimovic F, Zivkovic D, Bica DTG, et al. The importance of mini-puberty for fertility in cryptorchidism. J Urol. 2005;174:1536–9.
43. Kurpisz M, Havryluk A, Nakonechnyj A, et al. Cryptorchidism and long-term consequences. Reprod Biol. 2010;10(1):19–35.
44. Jørgensen N, Meyts ER, Main KM, Skakkebaek NE. Testicular dysgenesis syndrome comprises some but not all cases of hypospadias and impaired spermatogenesis. Int J Androl. 2010;33(2):298–303. Epub 4 Feb 2010.
45. Abe T, Takaha N, Tsujimura A, et al. Leydig cell tumor of the testis presenting male infertility: a case report. Hinyokika Kiyo. 2003;49(1):39–42.
46. Shiraishi Y, Nishiyama H, Okubo K, et al. Testicular Leydig cell tumor presenting as male infertility: a case report. Hinyokika Kiyo. 2009;55(12):777–81.
47. Chovelidze S, Kochiashvili D, Gogeschvili G, et al. Cases of Leydig cell tumor in male infertility. Georgian Med News. 2007;143:76–9.
48. Hayashi T, Arai G, Hyochi N, et al. Suppression of spermatogenesis in ipsilateral and contralateral testicular tissues in patients with seminoma by human chorionic gonadotropin beta subunit. Urology. 2001;58(2):251–7.
49. Ho GT, Gardner H, DeWolf WC, et al. Influence of testicular carcinoma on ipsilateral spermatogenesis. J Urol. 1992;148(3):821–5.
50. Carmignani L, Gadda F, Paffoni A, et al. Azoospermia and severe oligospermia in testicular cancer. Arch Ital Urol Androl. 2009;81(1):21–3.
51. Rowley MJ, Leach DR, Warner GA, et al. Effect of graded doses of ionizing radiation on the human testis. Radiat Res. 1974;59(3):665–78.
52. Paulsen CA. The study of radiation effects on the human testis: including histologic, chromosomal and hormonal aspects. Final progress report of AEC contract AT(45-1)-2225, Task Agreement 6. RLO-2225-2. 1973.
53. Speiser B, Rubin P, Casarett G. Aspermia following lower truncal irradiation in Hodgkin's disease. Cancer. 1973;32(3):692–8.
54. Ash P. The influence of radiation on fertility in man. Br J Radiol. 1980;53:271–8.
55. Giwercman A, von der Maase H, Berthelsen JG, et al. Localized irradiation of testes with carcinoma in situ: effects of Leydig cell function and eradication of malignant germ cells in 20 patients. J Clin Endocrinol Metab. 1991;73(3):596–603.
56. Shapiro E, Kinsella TJ, Makuch RW, et al. Effects of fractionated irradiation on endocrine aspects of testicular function. J Clin Oncol. 1985;3(9):1232–9.
57. Spitz S. The histological effects of nitrogen mustard on human tumours and tissues. Cancer. 1948;1(3):383–98.
58. Watson AR, Rance CP, Bain J. Long term effects of cyclophosphamide on testicular function. BMJ. 1985;291:1457–60.
59. Pryzant RM, Meistrich ML, Wilson G, et al. Long-term reduction in sperm count after chemotherapy with and without radiation therapy for non-Hodgkin's lymphomas. J Clin Oncol. 1993;11(2):239–47.
60. da Cunha MF, Meistrich ML, Fuller LM, et al. Recovery of spermatogenesis after treatment for Hodgkin's disease: limiting dose of MOPP chemotherapy. J Clin Oncol. 1984;2(6):571–7.
61. Meistrich ML, Chawla SP, Da Cunha MF, et al. Recovery of sperm production after chemotherapy for osteosarcoma. Cancer. 1989;63(11):2115–23.
62. Pectasides D, Pectasides M, Farmakis D. Testicular function in patients with testicular cancer treated with Bleomycin-Etoposide-Carboplatin (BEC90) combination chemotherapy. Eur Urol. 2004;45(2):187–93.
63. Pont J, Albrect W. Fertility after chemotherapy for testicular germ cell cancer. Fertil Steril. 1997;68:1–5.
64. Ishikawa T, Kamidono S, Fujisawa M. Fertility after high-dose chemotherapy for testicular cancer. Urology. 2004;63:137–40.
65. Sakamoto H, Oohta M, Inoue K, et al. Testicular sperm extraction in patients with persistent azoospermia after chemotherapy for testicular germ cell tumor. Int J Urol. 2007;14(2):167–70.
66. Ferlin A, Raicu F, Gatta V, Zuccarello D, Palka G, Foresta C. Male infertility: role of genetic background. Reprod Biomed Online. 2007;14(6):734–45.
67. Foresta C, Moro E, Ferlin A. Y chromosome microdeletions and alterations of spermatogenesis. Endocr Rev. 2001;22(2):226–39.
68. Vogt PH. Azoospermia factor (AZF) in Yq11: towards a molecular understanding of its function for human male fertility and spermatogenesis. Reprod Biomed Online. 2005;10(1):81–93.
69. Stahl PJ, Masson P, Mielnik A, et al. A decade of experience emphasizes that testing for Y microdeletions is essential in American men with azoospermia and severe oligozoospermia. Fertil Steril. 2010;94(5):1753–6.
70. Bojesen A, Gravholt CH. Klinefelter syndrome in clinical practice. Nat Clin Pract Urol. 2007;4(4):192–204.

71. Ferlin A, Garolla A, Foresta C. Chromosome abnormalities in sperm of individuals with constitutional sex chromosomal abnormalities. Cytogenet Genome Res. 2005;111:310–6.
72. Zhou-Cun A, Yang Y, Zhang SZ, et al. Chromosomal abnormality and Y chromosome microdeletion in Chinese patients with azoospermia or severe oligozoospermia. Yi Chuan Xue Bao. 2006;33(2):111–6.
73. Foresta C, Garolla A, Bartoloni L, Bettella A, Ferlin A. Genetic abnormalities among severely oligospermic men who are candidates for intracytoplasmic sperm injection. J Clin Endocrinol Metab. 2005;90(1):152–6.
74. Schiff JD, Palermo GD, Veeck LL, et al. Intracytoplasmic sperm injection in men with Klinefelter syndrome. J Clin Endocrinol Metab. 2005;90(11):6263–7.
75. O'FlynnO'Brien KL, Varghese AC, Agarwal A. The genetic causes of male factor infertility: a review. Fertil Steril. 2010;93(1):1–12.
76. Shafik A. Contraceptive efficacy of polyester-induced azoospermia in normal men. Contraception. 1992;45(5):439–51.
77. Munkelwitz R, Gilbert BR. Are boxer shorts really better? A critical analysis of the role of underwear type in male subfertility. J Urol. 1998;160(4):1329–33.
78. Bonde JP. Semen quality in welders exposed to radiant heat. Br J Ind Med. 1992;49(1):5–10.
79. Jung A, Schuppe HC. Influence of genital heat stress on semen quality in humans. Andrologia. 2007;39:203–15.
80. Agarwal A, Deepinder F, Sharma RK, et al. Effect of cell phone usage on semen analysis in men attending infertility clinic: an observational study. Fertil Steril. 2008;89:124–8.
81. Agarwal A, Desai NR, Makker K, et al. Effects of radiofrequency electromagnetic waves (RF-EMW) from cellular phones on human ejaculated semen: an in vitro pilot study. Fertil Steril. 2009;92:1318–25.
82. Collodel G, Capitani S, Pammolli A, et al. Semen quality of male idiopathic infertile smokers and nonsmokers: an ultrastructural study. J Androl. 2010;31:108–13.
83. Calogero A, Polosa R, Perdichizzi A, et al. Cigarette smoke extract immobilizes human spermatozoa and induces sperm apoptosis. Reprod Biomed Online. 2009;19:564–71.
84. Gaur DS, Talekar MS, Pathak VP. Alcohol intake and cigarette smoking: impact of two major lifestyle factors on male fertility. Indian J Pathol Microbiol. 2010;53:35–40.
85. Künzle R, Mueller MD, Hänggi W, et al. Semen quality of male smokers and nonsmokers in infertile couples. Fertil Steril. 2003;79:287–91.
86. Brandes SB, Buckman RF, Chelsky MJ, et al. External genitalia gunshot wounds: a ten-year experience with fifty-six cases. J Trauma. 1995;39:266–71.
87. Cass AS, Ferrara L, Wolpert J, et al. Bilateral testicular injury from external trauma. J Urol. 1988;140:1435–6.
88. Kuhlmann J, Bohme H, Tauber R. Bilateral testicular gunshot injuries. Urologe A. 2005;44:918–20.
89. Tomomasa H, Oshio S, Amemiya H, et al. Testicular injury: late results of semen analyses after uniorchiectomy. Arch Androl. 1992;29:59–63.
90. Lin WW, Kim ED, Quesada ET, et al. Unilateral testicular injury from external trauma: evaluation of semen quality and endocrine parameters. J Urol. 1998;159:841–3.
91. Kukadia AN, Ercole CJ, Gleich P, et al. Testicular trauma: potential impact on reproductive function. J Urol. 1996;156:1643–6.
92. Philip J, Selvan D, Desmond A. Mumps orchitis in the non-immune postpubertal male: a resurgent threat to male fertility? BJU Int. 2006;97:138–41.
93. Masarani M, Wazait H, Dinneen M. Mumps orchitis. J R Soc Med. 2006;99:573–5.
94. Osegbe DN. Testicular function after unilateral bacterial epididymo-orchitis. Eur Urol. 1991;19:204–8.
95. Schuppe HC, Meinhardt A, Allam JP, et al. Chronic orchitis: a neglected cause of male infertility? Andrologia. 2008;40:84–91.
96. Zariwala MA, Knowles MR, Omran H. Genetic defects in ciliary structure and function. Annu Rev Physiol. 2007;69:423–50.
97. Leigh MW, Pittman JE, Carson JL, et al. Clinical and genetic aspects of primary ciliary dyskinesia/Kartagener syndrome. Genet Med. 2009;11:473–87.
98. Terada T, Hatakeyama S. Morphological evidence for two types of idiopathic "Sertoli-cell-only" syndrome. Int J Androl. 1991;14(2):117–26.
99. Weller O, Yogev L, Yavetz H, et al. Differentiating between primary and secondary Sertoli-cell-only syndrome by histologic and hormonal parameters. Fertil Steril. 2005;83(6):1856–8.
100. Collins JA, Burrows EA, Yeo J, et al. Frequency and predictive value of antisperm antibodies among infertile couples. Hum Reprod. 1993;8(4):592–8.
101. Menge AC, Medley NE, Mangione CM, et al. The incidence and influence of antisperm antibodies in infertile human couples on sperm-cervical mucus interactions and subsequent fertility. Fertil Steril. 1982;38:439–46.
102. Sinisi AA, Di Finizio B, Pasquali D, et al. Prevalence of antisperm antibodies by SpermMARtest in subjects undergoing a routine sperm analysis for infertility. Int J Androl. 1993;16:311–4.
103. Heidenreich A, Bonfig R, Wilbert DM, et al. Risk factors for antisperm antibodies in infertile men. Am J Reprod Immunol. 1994;31:69–76.
104. Walsh T, Turek P. Immunologic infertility. In: Lipshultz LI, Howards SS, Niederberger CS, editors. Infertility in the male. 4th ed. New York City, NY: Cambridge University; 2009.
105. Meinertz H, Linnet L, Fogh-Andersen P, et al. Antisperm antibodies and fertility after vasovasostomy: a follow-up study of 216 men. Fertil Steril. 1990;54:315–21.
106. Hammiche F, Laven J, Boxmeer J, et al. Semen quality decline among men below 60 years of age undergoing IVF or ICSI treatment. J Androl. 2010;32:70–6. Epub ahead of print.
107. Belloc S, Benkhalifa M, Junca AM, et al. Paternal age and sperm DNA decay: discrepancy between chromomycin and aniline blue staining. Reprod Biomed Online. 2009;19:264–9.
108. Escobar MA, Grosfeld JL, Burdick JJ, et al. Surgical considerations in cystic fibrosis: a 32-year evaluation of outcomes. Surgery. 2005;138:560–71.
109. Wilschanski M, Corey M, Durie P, et al. Diversity of reproductive tract abnormalities in men with cystic fibrosis. JAMA. 1996;276:607–8.
110. Lissens W, Mercier B, Tournaye H, et al. Cystic fibrosis and infertility caused by congenital bilateral absence of the vas deferens and related clinical entities. Hum Reprod. 1996;S4:55–78.
111. Donat R, McNeill AS, Fitzpatrick DR, et al. The incidence of cystic fibrosis gene mutations in patients with congenital bilateral absence of the vas deferens in Scotland. Br J Urol. 1997;79:74–7.
112. Sokol RZ. Infertility in men with cystic fibrosis. Curr Opin Pulm Med. 2001;7:421–6.
113. Dörk T, Dworniczak B, Aulehla-Scholz C, et al. Distinct spectrum of CFTR gene mutations in congenital absence of vas deferens. Hum Genet. 1997;100:365–77.
114. Gaillard DA, Carré-Pigeon F, Lallemand A. Normal vas deferens in fetuses with cystic fibrosis. J Urol. 1997;158:1549–52.
115. Radpour R, Gourabi H, Gilani M, et al. Correlation between CFTR gene mutations in Iranian men with congenital absence of the vas deferens and anatomical genital phenotype. J Androl. 2008;29:35–40.
116. Donohue RE, Fauver HE. Unilateral absence of the vas deferens. A useful clinical sign. JAMA. 1989;261:1180–2.
117. Shapiro E, Goldfarb DA, Ritchey ML. The congenital and acquired solitary kidney. Rev Urol. 2003;5:2–8.
118. Handelsman DJ, Conway AJ, Boylan LM, et al. Young's syndrome. Obstructive azoospermia and chronic sinopulmonary infections. N Engl J Med. 1984;310:3–9.

119. Domingo C, Mirapeix RM, Encabo B, et al. Clinical features and ultrastructure of primary ciliary dyskinesia and Young syndrome. Rev Clin Esp. 1997;197:100–3.
120. Goeminne PC, Dupont LJ. The sinusitis-infertility syndrome: Young's saint, old devil. Eur Respir J. 2010;35:698.
121. Arya AK, Beer HL, Benton J, et al. Does Young's syndrome exist? J Laryngol Otol. 2009;123:477–81.
122. Smith JF, Walsh TJ, Turek PJ. Ejaculatory duct obstruction. Urol Clin North Am. 2008;35:221–7.
123. Paick JS, Kim SH, Kim SW. Ejaculatory duct obstruction in infertile men. BJU Int. 2000;85:720–4.
124. Carson CC. Transurethral resection for ejaculatory duct stenosis and oligospermia. Fertil Steril. 1984;41:482–4.
125. Wieder JA, Lynne CM, Ferrell SM, et al. Brown-colored semen in men with spinal cord injury. J Androl. 1999;20:594–600.
126. Labrecque M, Nazerali H, Mondor M, et al. Effectiveness and complications associated with 2 vasectomy occlusion techniques. J Urol. 2002;168:2495–8.
127. Freund MJ, Weidmann JE, Goldstein M, et al. Microrecanalization after vasectomy in man. J Androl. 1989;10:120–32.
128. Cruickshank B, Eidus L, Barkin M. Regeneration of vas deferens after vasectomy. Urology. 1987;30:137–42.
129. Hallan RI, May AR. Vasectomy: how much is enough? Br J Urol. 1988;62:377–9.
130. Adams CE, Wald M. Risks and complications of vasectomy. Urol Clin North Am. 2009;36:331–6.
131. Sokal DC, Labrecque M. Effectiveness of vasectomy techniques. Urol Clin North Am. 2009;36:317–29.
132. Cook LA, Van Vliet H, Lopez LM, et al. Vasectomy occlusion techniques for male sterilization. Cochrane Database Syst Rev. 2007;2:CD003991.
133. Jee SH, Hong YK. One-layer vasovasostomy: microsurgical versus loupe-assisted. Fertil Steril. 2010;94(6):2308–11. Epub 13 Jan 2010.
134. Belker AM, Thomas Jr AJ, Fuchs EF, et al. Results of 1,469 microsurgical vasectomy reversals by the Vasovasostomy Study Group. J Urol. 1991;145:505–11.
135. Nagler HM, Jung H. Factors predicting successful microsurgical vasectomy reversal. Urol Clin North Am. 2009;36:383–90.
136. Magheli A, Rais-Bahrami S, Kempkensteffen C, et al. Impact of obstructive interval and sperm granuloma on patency and pregnancy after vasectomy reversal. Int J Androl. 2010;41(1):52–7.
137. Witt MA, Heron S, Lipshultz LI. The post-vasectomy length of the testicular vasal remnant: a predictor of surgical outcome in microscopic vasectomy reversal. J Urol. 1994;151:892–4.
138. Hollingsworth MR, Sandlow JI, Schrepferman CG, et al. Repeat vasectomy reversal yields high success rates. Fertil Steril. 2007;88:217–9.
139. Talbot HS. The sexual function in paraplegia. J Urol. 1955;73: 91–100.
140. Utida C, Truzzi JC, Bruschini H, et al. Male infertility in spinal cord trauma. Int Braz J Urol. 2005;31:375–83.
141. Narayan P, Lange PH, Fraley EE. Ejaculation and fertility after extended retroperitoneal lymph node dissection for testicular cancer. J Urol. 1982;127:685–8.
142. Lange PH, Narayan P, Vogelzang NJ, et al. Return of fertility after treatment for nonseminomatous testicular cancer: changing concepts. J Urol. 1983;129:1131–5.
143. Pizzocaro G, Salvioni R, Zanoni F. Unilateral lymphadenectomy in intraoperative stage I nonseminomatous germinal testis cancer. J Urol. 1985;134:485–9.
144. Richie JP. Clinical stage 1 testicular cancer: the role of modified retroperitoneal lymphadenectomy. J Urol. 1990;144:1160–3.
145. Donohue JP, Foster RS, Rowland RG, et al. Nerve-sparing retroperitoneal lymphadenectomy with preservation of ejaculation. J Urol. 1990;144:287–91.
146. Heidenreich A, Albers P, Hartmann M, et al. Complications of primary nerve sparing retroperitoneal lymph node dissection for clinical stage I nonseminomatous germ cell tumors of the testis: experience of the German Testicular Cancer Study Group. J Urol. 2003;169:1710–4.
147. Pettus JA, Carver BS, Masterson T, et al. Preservation of ejaculation in patients undergoing nerve-sparing postchemotherapy retroperitoneal lymph node dissection for metastatic testicular cancer. Urology. 2009;73:328–31.
148. Hellstrom WJ, Sikka SC. Effects of acute treatment with tamsulosin versus alfuzosin on ejaculatory function in normal volunteers. J Urol. 2006;176:1529–33.
149. Marks LS, Gittelman MC, Hill LA, et al. Rapid efficacy of the highly selective alpha1A-adrenoceptor antagonist silodosin in men with signs and symptoms of benign prostatic hyperplasia: pooled results of 2 phase 3 studies. J Urol. 2009;181:2634–40.
150. Hisasue S, Furuya R, Itoh N, et al. Ejaculatory disorder caused by α-1 adrenoceptor antagonists is not retrograde ejaculation but a loss of seminal emission. Int J Urol. 2006;13:1311–6.
151. Kobayashi K, Masumori N, Hisasue S, et al. Inhibition of seminal emission is the main cause of an ejaculation induced by a new highly selective α1A-blocker in normal volunteers. J Sex Med. 2008;5:2185–90.
152. Smith SM, O'Keane V, Murray R. Sexual dysfunction in patients taking conventional antipsychotic medication. Br J Psychiatry. 2002;181:49–55.
153. Loh C, Leckband SG, Meyer JM, et al. Risperidone-induced retrograde ejaculation: case report and review of the literature. Int Clin Psychopharmacol. 2004;19:111–2.
154. Haefliger T, Bonsack C. Atypical antipsychotics and sexual dysfunction: five case-reports associated with risperidone. Encéphale. 2006;32:97–105.
155. Agarwal A, Deepinder F, Cocuzza M, et al. Effect of vaginal lubricants on sperm motility and chromatin integrity: a prospective comparative study. Fertil Steril. 2008;89:375–9.
156. Kutteh WH, Chao CH, Ritter JO, et al. Vaginal lubricants for the infertile couple: effect on sperm activity. Int J Fertil Menopausal Stud. 1996;41:400–4.
157. Anderson L, Lewis SE, McClure N. The effects of coital lubricants on sperm motility in vitro. Hum Reprod. 1998;13: 3351–6.
158. Nangia AK, Likosky DS, Wang D. Distribution of male infertility specialists in relation to the male population and assisted reproductive technology centers in the United States. Fertil Steril. 2010;94(2):599–609.

Further Reading

Blau H, Freud E, Mussaffi H, et al. Urogenital abnormalities in male children with cystic fibrosis. Arch Dis Child. 2002;87: 135–8.

Chen-Mok M, Bangdiwala SI, Dominik R, et al. Termination of a randomized controlled trial of two vasectomy techniques. Control Clin Trials. 2003;24:78–84.

Fejes I, Závaczki Z, Koloszár S, et al. Hypothesis: safety of using mobile phones on male fertility. Arch Androl. 2007;53:105–6.

Laboratory Evaluation for Male Infertility

2

Ryan Mori and Edmund Sabanegh Jr.

Abstract

The diagnosis and treatment options for male infertility have recently undergone a sea of change as advancements in technology and understanding in the fields of molecular biology, genetics, and laboratory medicine have grown. Further, advancements in assisted reproductive technologies (ART) have rendered previously subfertile and infertile couples with various options for pregnancy. Such changes in the understanding and treatment of fertility and infertility have necessitated a much more detailed assessment of the couple presenting with infertility. At the basis of this assessment is a sophisticated and methodological evaluation of male factor infertility including laboratory assessment of urine, serum, and semen, as well as radiological and genetic studies.

Keywords

Male infertility evaluation • Assisted reproductive technologies • Laboratory evaluation • Semen analysis • Cryptorchidism • Varicocele • Macroscopic assessment • Computer-assisted semen analysis

The diagnosis and treatment options for male infertility have recently undergone a sea of change as advancements in technology and understanding in the fields of molecular biology, genetics, and laboratory medicine have grown. Further, advancements in assisted reproductive technologies (ART) have rendered previously subfertile and infertile couples with various options for pregnancy. Such changes in the understanding and treatment of fertility and infertility have necessitated a much more detailed assessment of the couple presenting with infertility. At the basis of this assessment is a sophisticated and methodological evaluation of male factor infertility including laboratory assessment of urine, serum, and semen, as well as radiological and genetic studies.

Conception requires a balanced coordination between the endocrinologic and reproductive systems of both the male and the female partners. Studies in normal individuals demonstrate that within 1 year of unprotected intercourse, 60–75% of couples will achieve conception, whereas 90% will achieve conception after 1 year [1]. Based on such studies, the currently accepted definition of infertility by the American Society for Reproductive Medicine (ASRM) is the absence of conception after 12 months of regular, unprotected intercourse [2].

The workup and diagnosis of infertility is unique in medicine in that it involves multiple organ systems of two individuals. Pathology is often difficult to isolate given this complexity. Isolated male factor has been shown to be causative in 20% of infertility cases and is a contributing factor in

R. Mori, MD, MS (✉)
Cleveland Clinic Lerner College of Medicine and Glickman Urological and Kidney Institute, Cleveland Clinic Foundation, Cleveland, OH, USA

E. Sabanegh Jr., MD
Department of Urology, Glickman Urological and Kidney Institute, Center for Reproductive Medicine, Cleveland Clinic, 9500 Euclid Avenue/Q10-1, Cleveland, OH 44195, USA
e-mail: sabanee@ccf.org

S.J. Parekattil and A. Agarwal (eds.), *Male Infertility: Contemporary Clinical Approaches, Andrology, ART & Antioxidants*,
DOI 10.1007/978-1-4614-3335-4_2,

conjunction with female factor pathology in an additional 30% of cases [3]. These estimates have changed little over time despite great diagnostic advancements [4, 5].

A workup of couples presenting for evaluation of failure to conceive after 12 months of unprotected intercourse should consist of concurrent male and female partner evaluation. Further, as recommended by the practice committees of both the American Urological Association (AUA) and the ASRM, workup for infertility should be started earlier than 12 months in the setting of (1) male risk factors for infertility, (2) advanced maternal age (>35y), or (3) there is concern about male factor infertility [2]. The initial workup of the male partner should be basic, methodological, and cost-effective. Isolated pathology should be isolated and treated when possible given the high cost of ART. The initial evaluation of the male may suggest the need to proceed with more costly advanced testing. Many treatments of male factor infertility and subfertility allow pregnancy using the patient's own sperm with or without ART. Options for uncorrectable causes of male factor infertility include donor sperm insemination in a healthy female partner as well as adoption.

Clinical and Laboratory Evaluation of Male Factor Infertility

Initial Evaluation of Male Factor Infertility

The initial evaluation of a male presenting with infertility mandates a thorough general history, physical exam, and review of systems as well as a focused and targeted reproductive history and physical exam. A plethora of general medical conditions may contribute to infertility or altered sexual function and may be undiagnosed prior to urological evaluation (Table 2.1). Up to 1.3% of men undergoing evaluation for infertility are diagnosed with a significant and potentially life-threatening general medical condition [6]. Basic laboratory testing is a critical component of the initial evaluation of male factor and includes urinalysis, basic semen analysis, and a routine serum hormone analysis. Data from the initial evaluation will guide more advanced testing in infertility and should proceed in a methodological and cost-effective fashion.

History and Review of Systems

A complete medical history is important to obtain as a variety of medical conditions can contribute to abnormal fertility and sexual function in the male, as detailed in Table 2.1. Recent acute systemic illness such as viremia or fever should be noted. The human spermatogenesis cycle has a length of 64 days with an additional 5–10 days needed for epididymal sperm transit [7–9]. Thus, any insult to spermatogenesis such as a febrile illness may not be manifested for 2–3 months in semen analysis. In addition, medical conditions such as diabetes mellitus, hypertension, diseases of the thyroid, certain neoplasms, and diseases of the central and peripheral nervous systems may have substantial impact on fertility, erectile or ejaculatory function.

A reproductive history should focus on identification of primary versus secondary infertility, details of prior conceptions of both partners, previous fertility treatments, and evaluation of libido, erectile function, and ejaculatory function.

Table 2.1 Pertinent components of the history for male infertility evaluation

Past medical history
• Infertility Previous conceptions Duration Previous evaluations/treatments Female partner fertility status: previous conceptions/outcomes, evaluation, previous treatments
• Sexual Erectile/ejaculatory function Lubrications Intercourse timing/knowledge
• Childhood Infectious: mumps orchitis, sexually transmitted infections/urethritis Trauma: groin/testicular trauma, torsion, prior inguinal surgery Onset of puberty
• Adult General/systemic: obesity, hypertension Metabolic/endocrinologic: DM, metabolic syndrome, thyroid function Infectious: sexually transmitted infections/urethritis, urinary tract infections, epididymo-orchitis/prostatitis Neoplasms: treatments (radiation, chemotherapy) Neurological: spinal cord, MS Trauma: testicular, CNS/PNS
Past surgical history
• Inguinal: orchidopexy, herniorrhaphy
• Pelvic/retroperitoneal: prostate, bladder/bladder neck, RPLND
• Scrotal: vasectomy, hydrocele
Social history
• Environmental/occupational exposures
• Tobacco use
• Alcohol use
• Recreational drugs: marijuana, cocaine, anabolic steroids
Family history
• Chromosomal abnormalities: Klinefelter's syndrome
• Infertility
• Cystic fibrosis
Medications
• Ejaculatory dysfunction: antihypertensives, alpha-blockers
• Erectile dysfunction: antidepressants, psychotropics
• Hypogonadism: anabolic steroids
• Spermatogenesis: antibiotics

Primary infertility is by definition the absence of previous conception, whereas secondary infertility represents conception in the past with the current or previous partner. A sexual history should be addressed and include timing and frequency of coitus. As sperm survive within cervical mucus for 2–5 days [10], optimal timing of intercourse is at least every 48 h during the periovulatory period [11]. Commercially available ovulation prediction kits aid the couple in determining accurate time of ovulation. Attention should also be paid to types of lubricants used as many commercially available sexual lubricants have been shown to adversely affect sperm quality [12, 13].

Attention should be paid to the past medical history of the patient as well as a number of childhood diseases and conditions which may adversely affect future fertility including but not limited to mumps orchitis, cryptorchidism, testicular torsion or trauma, and previous inguinal surgery. Studies suggest that paternity rates for unilateral cryptorchidism are only slightly decreased, but significantly reduced in cases of bilateral cryptorchidism [14]. Controversy exists regarding future sperm quality and paternity after orchidopexy. The timing of the onset of puberty should be assessed as either delayed or precocious puberty may be indicative of underlying endocrinologic abnormalities.

A thorough review of systems and family medical history can identify genetic diseases which may affect fertility including Klinefelter's syndrome, Kallmann syndrome, and cystic fibrosis. The existence of male siblings with infertility may suggest Y chromosome microdeletions or other chromosomal abnormalities, although most genetic diseases present as de novo rather than inherited mutations. Cystic fibrosis (CF) is associated with congenital absence of the vas deferens bilaterally (CBAVD). Genetic causes of infertility can be transmissible, especially with the advent of ART. With the growing use of ART, we can expect the incidence of genetically derived infertility to increase in the future.

Past surgical history should focus on surgery involving the male genitourinary tract, the retroperitoneum, or the inguinal region as such surgeries can be associated with ejaculatory dysfunction or obstruction or erectile dysfunction. Previous exposure to ionizing radiation should be noted as this has been shown to affect sperm quality. Social history should identify any hazardous occupational exposures as well as ingestion of potentially gonadotoxic substances including ETOH, tobacco, marijuana, and other recreational drugs. Use of anabolic steroids or other performance-enhancing drugs should be assessed in appropriate patient populations. A review of medications is valuable and should include both prescribed substances as well as over-the-counter and herbal substances. Specific attention should be given to medications that can lead to impaired ejaculation (alpha-blockers, antihypertensives) or sexual dysfunction (antidepressants, antipsychotic agents).

Physical Examination

The physical examination consists of a general examination as well as a detailed genital examination. The general overall appearance and degree of virilization can offer clues about possibly androgen deficiency. Assessment of body habitus, hair growth patterns, and gynecomastia should be noted and may suggest underlying endocrinological or hormonal abnormalities. A man with a history of primary infertility presenting with disproportionately long extremities and low volume testes is highly suggestive of Klinefelter's syndrome.

The genital exam includes a thorough evaluation of the phallus and testes as well as the paratesticular structures. The phallus should be evaluated for potential causes of altered deposition of ejaculate including penile curvature, hypospadias, or meatal stenosis. The testes should be examined in both the supine and standing positions. The exam is often facilitated by warming of the scrotum via either room temperature or a scrotal warming pack to prevent retraction of the testes from the cremasteric reflex. The testes should be palpated for masses, and attention should be paid to the volume of the testis and any discrepancies in symmetry. Normal adult testis volume should be at least 20 cm^3, or 4×3 cm [15], and an orchidometer or calipers can assist in measurement. Enlargement or tenderness of the epididymides may suggest obstruction or inflammation. The spermatic cord should be evaluated with the patient in the upright position to aid in identification of abnormally dilated spermatic veins, which by definition is a varicocele. Classically, a prominent varicocele is described as having a "bag of worms" feel on physical exam. Examination during a Valsalva maneuver is required to correctly grade the varicocele if present. A grade I varicocele is only detectable during the Valsalva maneuver. Grade II varicoceles are palpable without Valsalva maneuver, and grade III varicoceles are visible through the scrotal skin. Varicoceles are common findings, present in up to 15% of normal males. In men presenting for evaluation of infertility, 19–41% have been shown to have varicoceles. Ninety percent of unilateral varicoceles are left sided, thought to be secondary to the more acute insertion of the left gonadal vein into the renal vein. An isolated moderate to severe unilateral right varicocele raises suspicion of a retroperitoneal process obstructing the insertion of the gonadal vein more proximally such as a retroperitoneal mass or large renal mass with a vein thrombus.

During examination of the spermatic cord, the vas deferens should be palpated. Absence of the vas deferens raises suspicion of genetic causes of infertility, such as mutation in cystic fibrosis transmembrane regulator gene (CFTR). The prostate should be palpated for midline cysts, or Mullerian duct cysts, which can be associated with ejaculatory duct obstruction. The seminal vesicles are not normally palpable, but may be in the setting of obstruction.

Table 2.2 Comparison of 1999 and 2010 WHO normal reference values for semen parameters

Semen parameter	Normal values: 1999 WHO	2010 WHO lower reference limit (5th centile + 95% CI)
Semen volume (mL)	2–6	1.5 (1.4–1.7)
Total motility PR + NP (%)	50+	40 (38–42)
Progressive motility PR (%)	25+	32 (31–34)
Vitality (% live spermatozoa)	50+	58 (55–63)
Sperm number (10^6 sperm/ejaculate)	>40	39 (33–46)
Sperm concentration (10^6 sperm/mL)	20	15 (12–16)
Morphology (% normal)	>30	4 (3.0–4.0)

Data from World Health Organization. WHO Laboratory Manual for the Examination of Human Semen and Sperm-Cervical Mucus Interaction, 1999; World Health Organization, Department of Reproductive Health and Research. WHO Laboratory Manual for the Examination and Processing of Human Semen, 5th edn. 2010; Cooper TG, Noonan E, von Eckardstein S, et al. World Health Organization reference values for human semen characteristics. Hum Reprod Update. 2010;16(3):231–45

Table 2.3 WHO nomenclature related to semen quality

Aspermia	No semen
Asthenospermia	<32% Progressively motile spermatozoa
Asthenoteratospermia	Percentages of both progressively motile and morphologically normal spermatozoa below the reference limits
Azoospermia	No spermatozoa present in the ejaculate
Cryptozoospermia	Spermatozoa not found in fresh preparations but found in centrifuged pellet
Hematospermia	Erythrocytes in ejaculate
Leukospermia	Presence of leukocytes in ejaculate > threshold (1 mil/cm^3)
Necrospermia	Low percentage of live and high percentage of immotile spermatozoa in ejaculate
Normozoospermia	Within normal limits for spermatozoa number and motility
Oligoasthenospermia	Total number of spermatozoa and percentage of progressively motile spermatozoa below lower reference limits
Oligoasthenoteratospermia	Total number of, percentage of progressively motile, and percentage of morphologically normal spermatozoa below the lower reference limits
Oligoteratospermia	Total number of and percentage of progressively motile spermatozoa below the lower reference limits
Oligospermia	Total number of spermatozoa below the lower reference limit
Teratospermia	Percentage of morphologically normal spermatozoa below the reference limit

Adapted from Appendix 1 Reference values and semen nomenclature. WHO Laboratory manual for the Examination and Processing of Human Semen, 5th edn. 2010, with permission

Laboratory Evaluation of Male Factor Infertility

Basic Testing

The initial assessment of a man presenting for an evaluation of male factor infertility should include a basic laboratory assessment of the semen. Further testing such as serum endocrine assays or genetic tests is suggested and guided by the results of the physical and laboratory evaluations.

Semen Analysis. A semen analysis is a critical component of the initial workup of male infertility. It can offer great insight into the etiology of infertility or subfertility; however, it serves as a surrogate rather than a true measure of fertility. An abnormal semen analysis can yield a viable pregnancy, and normal semen parameters can be associated with failure to conceive. As few as 15% of men presenting with infertility have currently recognizable abnormalities on semen analysis [16]. Table 2.2 reports a distribution of abnormalities in semen analyses in men presenting for infertility evaluation, and Table 2.3 lists the currently accepted nomenclature of diagnoses set forth by the World Health Organization (WHO). Beyond just the appearance of semen and spermatozoa microscopically, it is often necessary to obtain functional studies of sperm to assess the true fertilizing potential. In addition, guided by the basic semen analysis, more advanced testing may be needed and will be explored in greater depth below.

Evaluation of Basic Semen Analyses

Collection. Prior to analysis, semen must be properly collected into a sterile container. Sperm count and semen volume are variable from day to day; thus, it is essential to evaluate at least 2 samples to characterize baseline data for a patient [17]. Semen parameters and ejaculate volume can also vary widely based on the frequency of ejaculation, and it is currently recommended that a period of 2–5 days of ejaculatory abstinence precede sample collection. This period of abstinence should remain a constant with further semen samples to maintain comparability.

Optimally, the sample should be collected via self-stimulation in a private room near the laboratory to reduce the time between collection and analysis and to ensure constant temperature of the specimen. Lubricants should be avoided if possible as they may lead to altered sperm motility. The semen sample should be complete, and the man should be cautioned to collect all fractions of the ejaculate, as the first fraction contains sperm-rich prostatic fluids [18]. Coitus interruptus should be avoided as this first fraction is often loss. Semen collection devices that are free of spermicidal

Table 2.4 Components of the macroscopic semen assessment

Macroscopic variable	Normal quality
Liquefaction	Homogenous, <60 min
Appearance	Homogenous, gray opalescent
Viscosity	Thread < 2 cm
Volume	>1.5 mL
pH	>7.2

lubricants are less than ideal, but may be necessary if the patient has barriers to conventional collection methods. Semen should be analyzed within 1 h of collection to prevent alteration in semen parameters secondary to delayed analysis. Patients unable to provide a sample on location should be instructed to bring the sample to the lab at a constant temperature near body temperature for analysis within 1 h.

Some patients will be unable to provide a semen sample due to erectile or ejaculatory dysfunction and may require oral or intracavernosal injection therapy to produce a specimen. Discussion of such techniques is beyond the scope of this chapter.

Macroscopic Assessment. Semen analysis begins with a macroscopic assessment after liquefaction occurs. The time to liquefaction should be noted and considered abnormal if greater than 60 min. The five macroscopic variables are semen volume, viscosity, color, coagulation, and pH (Table 2.4).

The semen volume is best measured by weight, but can also be measured directly. The volume of the ejaculate is supplied mainly by the seminal vesicles and prostate gland, with some contribution from the bulbourethral glands and the epididymides. The viscosity can be measured by drawing the sample into a wide-bore 1.5-mm pipette and letting the semen drop. A thread length of greater than 2 cm is abnormally viscous. While viscosity is a consistently measured parameter, the significance of an abnormal assessment is controversial with many experts discounting the importance of this finding. A normal liquefied semen sample is described as homogenous and gray opalescent in color. Seminal pH results from the balance of acidic prostatic secretion and alkaline seminal vesicular secretions. This principle can be clinically valuable in the setting of low semen volume and abnormal pH, as the location of obstruction can be logically deduced to be at the level of the ejaculatory duct.

Microscopic Assessment. A wet prep of the semen next examined under light microscopic magnification. A basic microscopic semen analysis assesses agglutination/aggregation, sperm count, motility, morphology, and presence of non-sperm cells. The World Health Organization has set forth "normal" reference ranges for each of the above semen parameters. There has been some lack of consensus as to the utility and applicability of previously accepted lower limits of normal. The lower reference values have been recently updated based on analysis of semen samples of men with proven fertility and time to pregnancy less than 12 months [19]. Distributions of semen parameters across these men were obtained using standardized WHO laboratory criteria, and the lower reference limit, taken as the 5th centile, is presented in comparison with previous lower limits from the 1999 WHO criteria [20] in Table 2.5.

Table 2.5 Distribution of semen analysis parameters in men presenting for initial evaluation of infertility

Semen parameter	Incidence (%)
Any abnormality	37
Motility	26
Asthenospermia	24
Oligospermia	8
Agglutination	2
Volume	2
Morphology	1
Azoospermia	8
Normal semen analysis	55

From Lipshultz L. Subfertility. In: Kaufman JJ, editor. Current urologic therapy. Philadelphia: WB Saunders. 1980, with permission of Elsevier

- Sperm aggregation and agglutination: Semen is first analyzed as a wet prep specimen. Agglutination or clumping of spermatozoa with either sperm or non-sperm semen elements is noted as well as site of binding (head to head, tail to tail, or mixed fashion). Though some degree of sperm agglutination is considered normal, more considerable amounts could represent the presence of antisperm antibodies (ASA) [21]. Aggregation refers to the clumping together of nonmotile spermatozoa. The presence of agglutination may also be suggestive of ASA. Sperm agglutination with non-sperm semen elements may occur in the presence of infection. Thus, the presence of agglutination on microscopic assessment should trigger further testing with seminal white blood cell assessment and antisperm antibody measurement.
- Motility: The degree of progressive sperm motility is related to pregnancy rates [22]. Sperm are classified as either being motile or immotile. Further, motile sperm are assessed for their degree of progressive motility. Movement is considered progressively motile (PR) if the spermatozoa is moving actively either linearly or in a large circle. Nonprogressive motility (NP) refers to motion with absence of progression such as movement in small circles. Total motility (PR + NP) and progressive motility (PR) are reported in the basic semen analysis as percentages of total sperm. Complete absence of motility may be suggestive of ultrastructural cilia abnormalities such as Kallmann syndrome as well as necrospermia. Necrospermia is assessed via sperm vitality testing as detailed below.

- Sperm vitality: Sperm vitality is measured as a function of the membrane integrity of sperm cells and is expressed as a percentage of total sperm. This measurement is especially important when a large number of immotile sperm are present in order to rule out necrospermia [16]. The 2 methods commonly employed to assess the integrity of the cell membrane are the dye exclusion test and hypotonic swelling. The dye exclusion test is based on the principle that an intact membrane will not take in dye. In the latter test, a hypotonic solution is utilized, and spermatozoa with intact membrane swell within 5 min. This test is particularly useful if viable sperm are being identified for use in ICSI. Recently published lower reference limit for sperm vitality is 58% which is in agreement with previous assessments [19].
- Sperm count and concentration: Of paramount importance in the evaluation of male factor infertility is the presence of sperm in the ejaculate. Azoospermia, or the absence of sperm in the ejaculate, may occur from ejaculatory dysfunction, obstruction of the reproductive tract, or as a result of abnormal sperm production. Both the number of sperm per ejaculate and the sperm concentration have been correlated with both time to pregnancy and pregnancy rates [23]. Sperm concentration is directly measured and expressed in terms of millions per milliliter, whereas total sperm number is a calculated value based on semen volume and concentration. The total number of spermatozoa in the ejaculate has been correlated with testis volume; however, the concentration is heavily influenced by the volume of the glandular secretions. The normal sperm concentration is commonly accepted as >20 million sperm per mL semen based on the 1999 WHO cutoff [20]. However, according to this threshold, 20% of 18-year-old males would be classified as oligospermic [24]. As demonstrated in Table 2.5, more recent data by the WHO sets the lower reference limit of sperm concentration to 15 million/mL in a population of fertile men [19]. Oligospermia is still commonly accepted by definition as a sperm concentration of less than 20 million sperm per mL. For an exhaustive list of current WHO infertility nomenclature and definition, refer to Table 2.3.
- Non-sperm cells: The number of non-sperm cells present in a semen sample should be estimated and may have implications regarding underlying pathology. The most commonly encountered non-sperm cells in semen are epithelial cells, immature germ cells, and leukocytes [25]. The latter two cell types are referred to collectively as round cells due to their appearance microscopically and are not easily differentiated by microscopy alone. If the estimated round cell concentration exceeds 1×10^6/mL, then further testing should be done to assess the nature of the cell types. This is most reliably done by using immunohistochemistry to stain for specific leukocyte markers. However, the Endtz test allows a cost-effective method of identifying leukocytes by measuring peroxidase enzyme activity visualized with orthotoluidine dye [26]. The presence of leukocytes in semen may have detrimental effects on semen quality including reductions in sperm motility and DNA integrity [27, 28] as well as elevations in seminal reactive oxygen species [29].
- Sperm morphology: Sperm cells by nature have varying morphologies during maturation. Assessment is operator dependent and thus more subjective than other components of the basic semen analysis. Studies of sperm taken from postcoital cervical mucus or zona pellucida offer insight about the morphology of sperm with fertilizing potential [30–32]. Multiple classification schemata exist, but the currently most widely used are the WHO criteria and Kruger's strict criteria. Kruger and colleagues demonstrated that in men with sperm concentrations greater than 20 million sperm per mL and greater than 30% motility, fertilization rates were significantly higher for those men with greater than 14% normal sperm by the rigid criteria [13]. By WHO criteria, teratospermia represents less than 15% of sperm with normal morphology [20]. There remains substantial controversy regarding the predictive implications of abnormal morphology on assisted reproductive outcomes with both intrauterine insemination and intracytoplasmic sperm injection.

Normal spermatozoa consist of a head, midpiece, and a tailpiece (Fig. 2.1). A spermatozoa must have both a normal head and tail to be considered morphologically normal. The head should be a smooth ovoid shape with a well-defined acrosomal region (more lightly stained region) comprising about 40–70% of the head volume and containing no large vacuoles. The head should taper into a midpiece aligned along the same axis of the head. The tail should be of uniform caliber and approximately 10 times the length of the head. Looping of the tail is acceptable, but any sharp abnormalities are abnormal. Excess residual cytoplasm is also a morphologic abnormality. Common abnormalities of spermatozoa head, midpiece, and tail are demonstrated in Fig. 2.1.

Abnormal sperm morphology is commonly seen with defective spermatogenesis or with certain epididymal pathologies. They generally have less fertilizing potential and have been linked with increased DNA fragmentation [33] and chromosomal abnormalities [34].

All intact spermatozoa per area surveyed should be assessed, and the percentage of normal sperm should be recorded. Recently updated lower reference limits for normal forms in fertile men is 4% by strict criteria [19, 21], which is considerably lower than the previously accepted reference range. Some spermatozoa may have multiple defects; thus, percentages of specific defects should be based on total sperm count.

Computer-Assisted Semen Analysis (CASA). With advances in technology, it is now feasible to measure sperm motility,

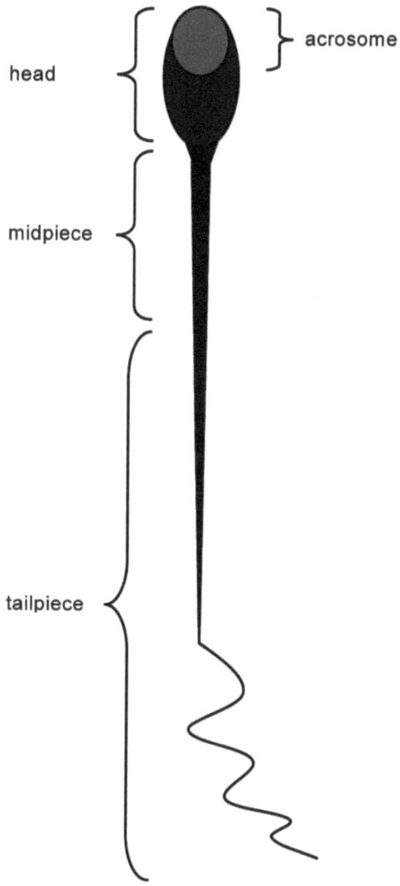

Normal Characteristics	Common Abnormalities
Head	Microcephalic head, megalocephalic head, tapered head, round sperm
Smoothe oval shape	
Acrosome 40-70% of surface	
3-5mm in length	
2-3mm in width	
Midpiece	Elongated, thin, distended, bent,
1.5X length of head	
<1mm in width	
Tailpiece	Short, multiple, bent, broken, coiled
Uniform, uncoiled	
10X length of head	

Fig. 2.1 Sperm morphology (Kruger Strict Criteria)

kinematics, and concentration using computer-aided sperm analysis (CASA). The potential advantages over more traditional methods are its precision and ability to quantify the kinematic properties of the spermatozoa. At present, this technology requires costly equipment and training and is more commonly a tool used in research settings rather than basic pathology laboratories.

Functional Sperm Testing

The fertilizing potential of a sperm cell cannot always be inferred from a basic semen analysis. Functional sperm tests assess various processes in the normal fertilization cycle to include sperm transit, penetration of the zona pellucida of the oocyte, and, ultimately, fertilization. Sperm-mucous interaction can be assessed by a cervical mucus migration assay where the rate of sperm transit through mucous is measured. The sperm penetration assay (SPA) measures the ability of a human sperm to penetrate a specially prepared hamster egg that has been stripped of the outer zona pellucida, allowing cross-species fertilization. This yields useful information about a sperm's ability to successfully undergo capacitation, acrosome reaction, membrane fusion with oocytes, and chromatin decondensation. A good result on SPA suggests proceeding with a trial of spontaneous conception or intrauterine insemination (IUI), whereas a poor SPA result might suggest the need for IVF with ICSI.

The acrosome reaction test measures the ability of the sperm cell to mount an effective acrosome reaction and may be useful in the setting of profound teratospermia with head predominant abnormalities where IVF is not successful. Currently, this test may be recommended in the setting of abnormal sperm head morphology or failure to fertilize oocytes in conventional IVF cycles. However, this test has been difficult to standardize, limiting its role at the present time.

Additional/Advanced Semen Testing

Antisperm Antibody Testing. Under normal anatomical circumstances, the seminiferous tubules are immunologically protected from the humoral environment as the tight junctions between the Sertoli cells form the blood-testis barrier. However, following a breech of this barrier such as after orchitis, scrotal trauma, or surgery, the immune system may be exposed to these "foreign" sperm antigens, and antisperm antibodies (ASA) may develop. The presence of sperm agglutination, especially head to head, should suggest the possibility of ASA. In addition, low sperm motility in the setting of previous injury or surgery (i.e., vasectomy), leukocytospermia, or otherwise unexplained infertility should also lower the clinician's threshold for evaluating for ASA.

Antisperm antibodies may be found in the serum and in the seminal plasma and bound to the sperm themselves. They can cause agglutination and immotility or can be spermatotoxic depending on the type of antibody. In previous reports, up to 10% of infertile men present with ASA versus only 2% of fertile men [35]. However, men can have normal semen parameters with presence of ASA as well [36]. The direct ASA test detects sperm-bound immunoglobulins, while indirect testing detects the biological activity of circulating ASA. Direct assays of sperm-bound immunoglobulins are preferred since most agree that sperm-bound antibodies are the most clinically relevant. Serum ASA testing, once widely used, has been largely discarded due to the much greater sensitivity of semen testing.

Sperm DNA Damage. Sperm DNA damage has been shown to have a positive correlation with abnormal semen parameters. [37]. Various factors and agents have been associated with sperm DNA damage including testicular cancers, tobacco use, and certain chemotherapeutic agents [27]. The etiology of sperm DNA damage at this time is thought to involve abnormal chromatin packing, elevated reactive oxygen species [27], and apoptosis [38]. There are several tests currently available to assess sperm DNA integrity by assessing strand breaks in situ [39]. The single cell gel electrophoresis assay, or comet assay, uses gel electrophoresis to assess DNA fragmentation. With the terminal deoxynucleotidyl transferase-mediated deoxyuridine triphosphate nick-end labeling (TUNEL) assay, a fluorescent-labeled nucleotide is transferred to the hydroxyl end of a broken DNA strand, and flow cytometry is used to assess DNA nicks. The sperm chromatin structure assay (SCSA) uses low pH to denature sperm DNA at the site of DNA breaks and is followed by acridine orange staining and flow cytometry to measure percentage of denatured DNA. Meta-analyses have shown that couples with sperm DNA fragmentation indices (DNI) less than 30% [40, 41] are twice as likely to achieve pregnancy using IVF. Although tests of DNA integrity are not part of the basic semen analysis, they can be useful especially in the setting of unexplained infertility but normal bulk semen parameters.

ROS Testing

Reactive oxygen species (ROS), also known as free radicals, are by-products of normal intercellular and intracellular metabolism and have been implicated in the etiology of multiple diseases across multiple organ systems. They are a necessary result of oxygen metabolism, which takes place in all healthy tissues, and are formed by the addition of unpaired electrons to the oxygen molecule through the process of reduction. The addition of an electron to molecular O_2 produces a superoxide anion radical (O_2–) that is reactive. Secondary ROS include hydroxyl radical (OH), peroxyl radical (ROO), and hydrogen peroxide (H_2O_2). ROS are critical to normal cell physiology, but excessive levels are detrimental to cell survival and function. In excess, they induce cellular damage by the oxidation of cellular and cell membrane components and cause DNA damage by modification of bases, deletions, frameshifts, and chromosomal relocations.

Oxidative stress by definition refers to the imbalance between reactive oxygen species and their scavengers, known generally as antioxidants. Antioxidants are the crux of the host defense against ROS and are formed via both enzymatic and nonenzymatic pathways. The primary enzymatic antioxidants include superoxide dismutases, catalase, and glutathione peroxidase. The superoxide dismutases convert superoxide into O_2 and H_2O_2. Catalase and glutathione peroxidase further degrade H_2O_2 into water and O_2. Glutathione is the main nonenzymatic antioxidant, and its cysteine subunit contains a sulfhydryl group which directly scavenges free radicals. Vitamins E and C are also important nonenzymatic scavengers of free radicals.

Reactive oxygen species have been shown to be critical throughout normal spermatogenesis and conception, specifically during sperm capacitation [42] and sperm-oocyte fusion [43]. Seminal fluid contains superoxide dismutase, catalase, glutathione peroxidase, and glutathione reductase. High levels of ROS are correlated with poor sperm quality and function. Spermatozoa which have been incubated with ROS overnight have increased lipid peroxidation. Further, addition of free radical scavengers like alpha tocopherol has been shown to revive sperm motility both in vitro [11] and in vivo.

Sperm are exposed to free radicals both by intrinsic production, extrinsic production in the semen, and via external and environmental sources including cigarette smoking, exposure to certain industrial compounds, and increased intrascrotal temperatures.

In normal semen, there is a low level of oxidative stress, as the free radicals necessary for certain cell signaling processes are kept in check by the antioxidants in order to avoid cellular damage. However, this balance is lost in conditions where there is either increased production of free radicals or decreased buffering capacity by available antioxidants. Leukocytes and spermatozoa are both significant sources of free radicals in the semen. Normal semen contains some white blood cells, predominantly neutrophils. Neutrophils exert their normal cytotoxic function partly by releasing high concentrations of ROS. Although the relationship is incompletely defined, leukospermia, defined as peroxidase positive leukocytes at a concentration greater than 1×10×6 per mL by the WHO [21], has been associated with altered semen parameters including decreased sperm concentration, motility, and morphology [4]. Although there is a correlation between seminal leukocytes and infertility [10], some studies have failed to show an alteration of sperm parameters in the presence of leukospermia [5]. However, several studies have linked proinflammatory cytokines including IL-6, IL-8, and TNF-alpha to altered sperm function [38, 44, 45].

In addition to seminal leukocytes, spermatozoa themselves are source of ROS. As sperm mature, they extrude cytoplasm rich in reducing molecules. Abnormalities in sperm maturation lead to retention of cytoplasm and increased levels of ROS in semen. The effects of the ROS on over all sperm function may correlate with the site of ROS production, intrinsic or extrinsic to the sperm. Some authors suggest that higher levels of extrinsic ROS, such as that produced by leukospermia, have greater effects on sperm count, motility, and morphology, whereas increased intrinsic ROS production is associated with higher levels of DNA fragmentation [37].

Testing for reactive oxygen species is not a part of the standard initial evaluation of male factor infertility. However, there is increasing research being conducted in this area, and despite the lack of randomized control trials, it is becoming evident that men presenting with infertility with increased oxidative stress in the semen may benefit from antioxidant therapy [46, 47]. Additionally, infertile men with varicoceles have been shown to have elevated seminal ROS [48] with levels that correlate with varicocele grade [49], and several studies have shown that surgical varicocelectomy is correlated with decreased seminal oxidative stress, increased seminal antioxidants, and improved sperm quality [20, 50, 51].

Various direct and indirect testing modalities exist to determine the level of oxidative stress in the semen. The most commonly employed method of ROS testing is a chemiluminescence probe assay, which is used to quantify redox activities of spermatozoa [52]. Using this technique, a luminol probe (5-amino-2,3-dihydro-1,4-phthalazinedione), used to measure both intracellular and extracellular ROS, or a lucigen probe measuring superoxide radicals released extracellularly is employed. Direct assays of oxidative stress are available, but cost and practicality issues have rendered these assays tools of research with limited clinical application at this time.

Hormonal Assessment

In addition to a basic semen analysis, many patients may warrant a serum hormonal analysis at time of presentation. The goal of such an analysis is to evaluate the hypothalamic-pituitary-gonadal axis and to rule out an underlying endocrinopathy or primary testicular failure. Controversy exists as to the necessity of a hormonal evaluation in all patients presenting with infertility. However, there is a consensus that basic hormonal testing including serum follicle-stimulating hormone (FSH) and early morning serum total testosterone should be done in the setting of (1) abnormally low sperm count (<10 million/cm^3), (2) impaired sexual function, and (3) clinical findings suggestive of an endocrine abnormality such as reduced testicular volume or gynecomastia [1, 53].

The pituitary and gonadal hormones are released in a pulsatile fashion, and this rhythm is orchestrated by the hypothalamus, which receives diffuse input from multiple cortical and subcortical brain regions. The hypothalamus communicates to the anterior pituitary both via neuronal input as well as via the portal vascular system which allows direct delivery of hypothalamic hormones in high concentration to the anterior pituitary. The most important of these hypothalamic hormones is luteinizing hormone-releasing hormone (LHRH) or gonadotropin-releasing hormone (GNRH) which stimulates the secretion of luteinizing hormone (LH) and FSH. GNRH secretion is modulated by many factors and is under direct negative feedback control by circulating gonadal hormones including testosterone and inhibin.

LH, FSH, and prolactin are the primary hormones released from the anterior pituitary into the systemic circulation. LH stimulates the Leydig cells in the testes to produce testosterone, while FSH acts on the Sertoli cells of the testes and promotes development and growth of the seminiferous.

The testes are composed primarily of Leydig cells, Sertoli cells, and seminiferous tubules. The bulk of the testicular volume is comprised of seminiferous tubules and germinal elements; thus, reduced testicular size suggests impaired spermatogenesis [54]. Within the testes, Leydig cells are responsible for androgen synthesis. Testosterone, the primary circulating male androgen, is secreted in a pulsatile fashion with a regular circadian cycle peaking in the early morning. Only 2% of the serum testosterone is free in the systemic circulation with the remainder bound in roughly equal proportions to sex hormone-binding globulin (SHBG) and albumin. Alterations in serum SHBG will increase free testosterone in the serum. SHBG levels are influenced by a number of conditions including liver and thyroid disease, medications, advanced age, and obesity. Peripherally, testosterone is reduced to dihydrotestosterone (DHT) by 5 alpha-reductase and is also converted to estradiol by aromatases.

Sertoli cells line the seminiferous tubules and are linked by tight junctions, forming the blood–testis barrier which provides an immunologically naïve environment for spermatogenesis. The Sertoli cells are under the control of FSH and produce multiple paracrine factors important in stimulating and supporting spermatogenesis. Inhibin B is released from Sertoli cells in response to FSH stimulation and is important as regulator of the HGP axis via negative feedback at the pituitary and hypothalamus. The Sertoli cells also express androgen-binding protein in response to FSH stimulation and allow for very high intraluminal testosterone levels to support spermatogenesis via paracrine mechanisms

The most common abnormality encountered on hormonal analysis in the infertile man is an elevated serum FSH. This finding is suggestive of impaired spermatogenesis; however, the finding of elevated FSH is not always present in cases of testicular failure [55]. Abnormal preliminary screening tests should prompt a more involved endocrinologic workup consisting of total and free testosterone, prolactin, TSH, LH, and FSH. Clinical findings of headaches or visual field changes or findings of an elevated prolactin level necessitate an MRI of the sella turcica to evaluate for a potential macroadenoma of the pituitary.

Genetic Evaluation

In patients who present with azoospermia or severe oligospermia, where obstructive etiologies have been ruled out, evaluation of chromosome number and structure becomes important. A karyotype will rule out genetic conditions most commonly associated with infertility, including Klinefelter's syndrome (47, XXY), 46XX, 47XXY, and Noonan's syndrome. The chromosome structure of the Y chromosome should be assessed as well. This is done with the Y chromosome-linked microdeletion assay. Disruptions or deletions in various loci of the Y chromosome have been associated with severe defects in spermatogenesis. Early studies identified an area on the short arm of the Y chromosome critical for spermatogenesis referred to as the azoospermic factor (AZF) [56]. Subsequently, this has been divided into 3 regions: AZFa, AZFb, and AZFc [57]. Deletions in the regions of AZFa and AZFb are less common and usually associated with poor sperm retrieval rates for ART. AZFc microdeletions are the most commonly found microdeletion in azoospermic men and are associated with the most promising sperm retrieval rates [36]. Although Y chromosome microdeletions have no apparent impact on the health of the patient, the possibility of an inheritable form of infertility in male offspring via ART should prompt appropriate genetic counseling.

Expert Commentary

The purpose of this chapter was to delineate the evaluation of a man presenting with infertility with a focus on the laboratory evaluation. In the new era of the human genome project and with a better understanding of molecular biology and genetics, advances in both the diagnostic capabilities and the management options for male factor infertility are quickly advancing the field. In addition, enhancements in assisted reproductive technology are allowing many couples the opportunity to conceive when it would have been otherwise impossible. As conception is more reliant on laboratory technology, the laboratory evaluation of a man or couple presenting with infertility will remain of paramount importance.

Five-Year View

Conventional semen parameters continue to provide poor prognostic information regarding both frequency and quality of conception. Over the near term, marked research efforts will be dedicated to further elucidating sperm function. Emphasis will continue on the role of oxidative stress and sperm DNA fragmentation on subfertility. Novel research efforts in the field of metabolomics, metabolic profiling of semen, hold promise in elucidating the cause of heretofore idiopathic infertility.

Key Issues

- Infertility is the absence of conception after 12 months or regular unprotected intercourse, and male factor plays a role in up to 50% of cases.
- The initial clinical evaluation of male factor infertility should include a detailed and complete history and physical examination as well as a focused sexual history and genitourinary examination.
- Initial assessment should include a basic microscopic and macroscopic assessment of 2 separate semen analyses collected properly. Further testing should be based on the findings of the initial evaluation.
- New data on lower reference limits of semen parameters in fertile males suggests that there is a broader range of normal parameters than previously thought.
- Advanced semen testing should not be routine, but may be warranted based on the evaluation and includes antisperm antibody detection, assays of sperm DNA damage, and analysis of seminal oxidative stress and seminal antioxidant levels.
- Further testing with basic and extensive hormonal analysis as well as genetic testing may be warranted based on the initial evaluation.
- Reactive oxygen species derived from both intrinsic and extrinsic sources are being increasingly implicated in many cases of subfertility and infertility and have been shown to affect semen quality through several mechanisms including DNA fragmentation and lipid peroxidation. Antioxidant therapy as well as modification of exposures to extrinsic sources of ROS may have a role in the management of infertility.

References

1. The Optimal Evaluation of the Infertile Male: American Urological Society Best Practice Statement. American Urological Society, Education and Research Inc. 2010.
2. Said TM, Agarwal A, Sharma RK, Thomas AJ, Sikka SC. Impact of sperm morphology on DNA damage caused by oxidative stress induced by beta-nicotinamide adenine dinucleotide phosphate. Fertil Steril. 2005;83:95–103.
3. Slama R, et al. Time to pregnancy and semen parameters: a cross-sectional study among fertile couples from four European Cities. Hum Reprod. 2002;17(2):503–15.
4. Moskovtsev SI, Willis J, White J, Mullen JB. Leukocytospermia: relationship to sperm deoxyribonucleic acid integrity in patients evaluated for male factor infertility. Fertil Steril. 2007;88(3):737–40. Epub 6 Mar 2007.
5. Tomlinson MJ, Barratt CL, Cooke ID. Prospective study of leukocytes and leukocyte subpopulations in semen suggests they are not a cause of male infertility. Fertil Steril. 1993;60:1069–75.
6. Kolettis PN, Sabanegh ES. Significant medical pathology discovered during a male infertility evaluation. J Urol. 2001;166(1):178–80.
7. Franca LR, Avelar GF, Almeida FFL. Spermatogenesis and sperm transit through the epididymis with emphasis on pigs. Theriogenology. 2005;63:300–18.
8. Mortimer D, Leslie EE, Kelly RW, Templeton AA. Morphological selection of human spermatozoa in vivo and in vitro. J Reprod Fertil. 1982;64:391–9.
9. Clermont Y, Heller C. Spermatogenesis in man: an estimate of its duration. Science. 1963;140:184–5.
10. Wolff H. The biologic significance of white blood cells in semen. Fertil Steril. 1995 Jun;63(6):1143–57. Review.
11. Verma A, Kanwar KC. Effect of vitamin E on human sperm motility and lipid peroxidation in vitro. Asian J Androl. 1999;1(3):151–4.
12. Agarwal A, Deepinder F, Cocuzza M, Short RA, Evenson DP. Effect of vaginal lubricants on sperm motility and chromatin integrity: a prospective comparative study. Fertil Steril. 2008;89(2):375–9.
13. Kruger TF, Acosta AA, Simmons KF, et al. Predictive value of abnormal sperm morphology in in vitro fertilization. Fertil Steril. 1988;49:112–7.
14. Cendron M, Keating MA, Huff DS, et al. Cryptorchidism, orchiopexy and infertility: a critical long-term retrospective analysis. J Urol. 1989;142:559–62.
15. Charny CW. The spermatogenic potential of the undescended testis before and after treatment. J Urol. 1960;83:697.
16. Misell LM, Holochwost D, Boban D, et al. A stable isotope/mass spectrometric method for measuring the kinetics of human spermatogenesis in vivo. J Urol. 2006;175:242–6.
17. Carlsen E, et al. Effects of ejaculatory frequency and season on variations in semen quality. Fertil Steril. 2004;82(2):358–66.
18. Bjorndahl L, Kvist U. Sequence of ejaculation affects the spermatozoon as a carrier and its message. Reprod Biomed Online. 2003;7(4):440–8.
19. Cooper TG, Noonan E, von Eckardstein S, Auger J, et al. World Health Organization reference values for human semen characteristics. Hum Reprod Update. 2010;16(3):231–45.
20. Zini A, Blumenfield A, Libman J, Willis J. Beneficial effect of microsurgical varicocelectomy on human sperm DNA integrity. Hum Reprod. 2005;20:1018–21.
21. World Health Organization. World Health Organization: WHO Laboratory Manual for the Examination of Human Semen and Sperm-Cervical Mucus Interaction. 1999.
22. Jouannet P, Ducot B, Feneux D, Spira A. Male factors and the likelihood of pregnancy in infertile couples. I. Study of sperm characteristics. Int J Androl. 1988 Oct;11(5):379–94.
23. Spira A. Epidemiology of human reproduction. Hum Reprod. 1986;1:111–5.
24. Andersen AG, Jensen TK, Carlsen E, Jørgensen N, Andersson AM, Krarup T, Keiding N, Skakkebaek NE. High frequency of sub-optimal semen quality in an unselected population of young men. Hum Reprod. 2000 Feb;15(2):366–72.
25. Fedder J. Nonsperm cells in human semen: with special reference to seminal leukocytes and their possible influence on fertility. Arch Androl. 1996;36(1):41–65.
26. Sigma M, Jarow J. Male infertility. In: Wein AJ, Kavoussi LR, Novick AC, Partin AW, Peters CA, editors. Campbell-Walsh urology. 9th ed. Philadelphia: Saunders Elsevier; 2007.
27. Agarwal A, Saleh RA, Bedaiwy MA. Role of reactive oxygen species in the pathophysiology of human reproduction. Fertil Steril. 2003;79:829–43.
28. Zini A, Libman J. Sperm DNA damage: importance in the era of assisted reproduction. Curr Opin Urol. 2006;16(6):428–34.
29. Athayde KS, et al. Development of normal reference values for seminal reactive oxygen species and their correlation with leukocytes and semen parameters in a fertile population. J Androl. 2007;28(4):613–20.
30. Fredricsson B, Bjork G. Morphology of postcoital spermatozoa in the cervical secretion and its clinical significance. Fertil Steril. 1977;28:841–5.
31. Mosher WD, Pratt WF. Fecundity and infertility in the United States: incidence and trends. Fertil Steril. 1991;56:192.
32. Marks JL, McMahon R, Lipshultz LI. Predictive parameters of successful varicocele repair. J Urol. 1986;136:609–12.
33. Gandini L, Lombardo F, Paoli D, Caponecchia L, Familiari G, Verlengia C, Dondero F, Lenzi A. Study of apoptotic DNA fragmentation in human spermatozoa. Hum Reprod. 2000;15(4):830–9.
34. Lee JD, Kamiguchi Y, Yanagimachi R. Analysis of chromosome constitution of human spermatozoa with normal and aberrant head morphologies after injection into mouse oocytes. Hum Reprod. 1996 Sep;11(9):1942–6.
35. Guzick DS, Overstreet JW, Factor-Litvak P, Brazil CK, Nakajima ST, Coutifaris C, Carson SA, Cisneros P, Steinkampf MP, Hill JA, Xu D, Vogel DL, National Cooperative Reproductive Medicine Network. Sperm morphology, motility, and concentration in fertile and infertile men. N Engl J Med. 2001;345(19):1388–93.
36. Oates RD, Silber S, Brown LG, Page DC. Clinical characterization of 42 oligospermic or azoospermic men with microdeletion of the AZFc region of the Y chromosome, and of 18 children conceived via ICSI. Hum Reprod. 2002;17(11):2813–24.
37. Agarwal A, Said TM. Role of sperm chromatin abnormalities and DNA damage in male infertility. Hum Reprod Update. 2003;9(4):331–45.
38. Sanocka D, et al. Male genital tract inflammation: The role of selected interleukins in regulation of pro-oxidant and antioxidant enzymatic substances in seminal plasma. J Androl. 2003;24:448–55.
39. Evenson DP, Wixon R. Clinical aspects of sperm DNA fragmentation detection and male infertility. Theriogenology. 2006;65:979–91.
40. Evenson DP, Larson KL, Jost LK. Sperm chromatin structure assay: Its clinical use for detecting sperm DNA fragmentation in male infertility and comparisons with other techniques. J Androl. 2002;23:25–43.
41. Li Z, Wang L, Cai J, Huang H. Correlation of sperm DNA damage with IVF and ICSI outcomes: a systematic review and meta-analysis. J Assist Reprod Genet. 2006;23(9–10):367–76. Epub 4 Oct 2006.
42. Griveau JF, Grizard G, Boucher D, Le Lannou D. Influence of oxygen tension on function of isolated spermatozoa from ejaculates of oligozoospermic patients and normozoospermic fertile donors. Hum Reprod. 1998;13(11):3108–13.
43. O'Flaherty C, de Lamirande E, Gagnon C. Positive role of reactive oxygen species in mammalian sperm capacitation: triggering and

modulation of phosphorylation events. Free Radic Biol Med. 2006;41(4):528–40.
44. Camejo MI, Segnini A, Proverbio F. Intgerleukin-6 in seminal plasma of infertile men, and lipid peroxidation in their sperm. Arch Androl. 2001;47:97–101.
45. Martinez P, Proverbio F, Camejo MI. Sperm lipid peroxidation and pro-inflammatory cytokines. Asian J Androl. 2007;9:102–7.
46. Silver EW, et al. Effect of antioxidant intake on sperm chromatin stability in healthy non-smoking men. J Androl. 2005;26:550–6.
47. Keskes-Ammar L, Feki-Chakroun N, Rebai T, et al. Sperm oxidative stress and the effect of an oral vitamin E and selenium supplement on semen quality in infertile men. Arch Androl. 2003;49: 83–94.
48. Köksal IT, Tefekli A, Usta M, Erol H, Abbasoglu S, Kadioglu A. The role of reactive oxygen species in testicular dysfunction associated with varicocele. BJU Int. 2000 Sep;86(4):549–52.
49. Allamaneni SS, Naughton CK, Sharma RK, Thomas Jr AJ, Agarwal A. Increased seminal reactive oxygen species levels in patients with varicoceles correlate with varicocele grade but not with testis size. Fertil Steril. 2004;82(6):1684–6.
50. Mostafa T, et al. Varicocelectomy reduces reactive oxygen species levels and increases antioxidant activity of seminal plasma from infertile men with varicocele. Int J Androl. 2001;24:261–5.
51. Evers JLH, Collins JA. Assessment of efficacy of varicocele repair for male subfertility: A systematic review. Lancet. 2003;361: 1849–52.
52. Baker MA, Aitken RJ. Reactive oxygen species in spermatozoa: methods for monitoring and significance for the origins of genetic disease and infertility. Reprod Biol Endocrinol. 2005;3:67.
53. Report on optimal evaluation of the infertile male. AUA Best Practice Policy and ASRM Practice Committee Report, Volume 1, 2001. http://www.auanet.org. Accessed 20 Dec 2009.
54. Lipshultz LI, Corriere Jr JN. Progressive testicular atrophy in the varicocele patient. J Urol. 1977 Feb;117(2):175–6.
55. Turek PJ, Kim M, Gilbaugh 3rd JH, Lipshultz LI. The clinical characteristics of 82 patients with Sertoli cell-only testis histology. Fertil Steril. 1995;64(6):1197–200.
56. Tiepolo L, Zuffardi O. Localization of factors controlling spermatogenesis in the nonfluorescent portion of the human Y chromosome long arm. Hum Genet. 1976;34:119–24.
57. Vogt PH. Human chromosome deletions in Yq11, AZF candidate genes and male infertility: history and update. Mol Hum Reprod. 1998;4(8):739–44.

Further Reading

Griveau JF, Renard P, Le Lannou D. Superoxide anion production by human spermatozoa as a part of the ionophore-induced acrosome reaction process. Int J Androl. 199;18(2):67–74.

Kefer JC, Agarwal A, Sabanegh EC. Role of antioxidants in the treatment of male infertility. Int J Urol. 2009;16:449–57.

Kutteh WH, Chao CH, Ritter JO, et al. Vaginal lubricants of the infertile couple: effect on sperm activity. Int J Fertil Menopausal Stud. 1996;41:400–4.

Lampiao F, du Plessis SS. TNF-alpha and IL-6 affect human sperm function by elevating nitric oxide production. Reprod Biomed Online. 2008;17(5):628–31.

Liu DY, Baker HW. Morphology of spermatozoa bound to the zona pellucida of human oocytes that failed to fertilize in vitro. J Reprod Fertil. 1992;94:71–84.

McLachlan RI, et al. Semen analysis: its place in modern reproductive medical practice. Pathology. 2003;35(1):25–33.

Munuce MJ, Berta CL, Pauluzzi F, Caille AM. Relationship between antisperm antibodies, sperm movement, and semen quality. Urol Int. 2000;65(4):200–3.

Sakkas D, Mariethoz E, Manicardi G, Bizzaro D, Bianchi PG, Bianchi U. Origin of DNA damage in ejaculated human spermatozoa. Rev Reprod. 1999;4(1):31–7.

Shekarriz M, Sharma RK, Thomas Jr AJ, Agarwal A. Positive myeloperoxidase staining (Endtz test) as an indicator of excessive reactive oxygen species formation in semen. J Assist Reprod Genet. 1995;12(2):70–4.

Simmons FA. Human infertility. New Engl J Med. 1956;255:1140.

Thonneau P, Marchand S, Tallec A, et al. Incidence and main causes of infertility in a resident population (1,850,000) of three French regions (1988–1989). Hum Reprod. 1991;6:811–6.

Tur-Kaspa I, Maor Y, Levran D, et al. How often should infertile men have intercourse to achieve conception? Fert Steril. 1994;62(2): 370–5.

Wilcox AJ, Weinberg CR, Baird DD. Timing of sexual intercourse in relation to ovulation. Effects on the probability of conception, survival of the pregnancy, and sex of the baby. New Engl J Med. 1995;333(23):1517–21.

World Health Organization, Department of Reproductive Health and Research. WHO Laboratory Manual for the Examination and Processing of Human Semen, 5th edn. 2010.

Imaging Modalities in the Management of Male Infertility

3

Marcello Cocuzza and Sijo J. Parekattil

Abstract

Infertility affects approximately 15% of couples desiring conception, and male infertility underlies almost half of the cases. Assisted reproductive techniques (ART) are increasingly being used to overcome multiple sperm deficiencies and because of their effectiveness have been suggested by some to represent the treatment for all cases of male factor infertility regardless of etiology. Although the use of these technologies may allow infertile couples to achieve pregnancy rapidly, associated higher cost, potential safety issues, and considering the fear of transferring the unnecessary burden of invasive treatment on healthy female partners weigh down this treatment option heavily. Diagnostic imaging techniques may be indicated as part of the complete male fertility evaluation. Productive therapy can be instituted only after completion of a thorough evaluation that begins with a detailed, direct history and physical examination. Due to the introduction and enhancement of newer imaging modalities, reliable adjuncts to clinical examination can be obtained to diagnose a variety of causes of male infertility including varicocele, epididymal blockage, testicular microlithiasis, seminal vesicle agenesis, and ejaculatory obstruction. Imaging plays a key role in the evaluation of the hypospermia or azoospermic man. It can detect correctable abnormalities, which can lead to a successful conception. It can also reveal potentially life-threatening disorders in the course of an infertility evaluation as testicular tumors. The goal of this chapter is to provide the reader with a foundation for a comprehensive evaluation of the male partner as well as emerging technologies that can improve the treatment of correctable causes of male infertility.

Keywords

Imaging in male fertility • Assisted reproductive techniques • Testicular tissue imaging • Doppler duplex flow imaging • MRI spectral imagine • Testicular artery mapping • Varicocelectomy • Ejaculatory duct imaging • Testicular microlithiasis

Infertility affects approximately 15% of couples desiring conception, and male infertility underlies almost half of the cases. Assisted reproductive techniques (ART) are increasingly being used to overcome multiple sperm deficiencies and because of their effectiveness have been suggested by some to represent the treatment for all cases of male factor infertility regardless of etiology. Although the use of these technologies may allow infertile couples to achieve pregnancy rapidly, associated higher cost, potential safety issues, and the fear of transferring the unnecessary burden

M. Cocuzza, MD (✉)
Department of Urology, University of Sao Paulo (USP), Ave. Dr. Eneas de Carvalho Aguiar, 255, 7 andar, Sala 710F, Sao Paulo, Brazil
e-mail: mcocuzza@uol.com.br

S.J. Parekattil, MD
Division of Urology, Winter Haven Hospital University of Florida, Winter Haven, 200 Avenue F, N. E., Winter Haven, FL 33881, USA
e-mail: sijo.parekattil@winterhavenhospital.org

S.J. Parekattil and A. Agarwal (eds.), *Male Infertility: Contemporary Clinical Approaches, Andrology, ART & Antioxidants*,
DOI 10.1007/978-1-4614-3335-4_3, © Springer Science+Business Media, LLC 2012

of invasive treatment on healthy female partners weigh down this treatment option heavily.

Diagnostic imaging techniques may be indicated as part of the complete male fertility evaluation. Productive therapy can be instituted only after completion of a thorough evaluation that begins with a detailed, direct history and physical examination. Due to the introduction and enhancement of newer imaging modalities, reliable adjuncts to clinical examination can be obtained to diagnose a variety of causes of male infertility including varicocele, epididymal blockage, testicular microlithiasis, seminal vesicle agenesis, and ejaculatory obstruction. Imaging plays a key role in the evaluation of the hypospermia or azoospermic man. It can detect correctable abnormalities, which can lead to a successful conception. It can also reveal potentially life-threatening disorders in the course of an infertility evaluation such as testicular tumors. The goal of this article is to provide the reader with a foundation for a comprehensive evaluation of the male partner as well as emerging technologies that can improve the treatment of correctable causes of male infertility.

Testicular Tissue Imaging for Guided Sperm Retrieval

Doppler Duplex Flow Imaging

Testis biopsy with cryopreservation of sperm is a procedure performed in men with possible nonobstructive azoospermia (NOA), and at times, in men with a previous vasectomy (who do not want a reversal) and men with spinal cord injuries (who fail electro-ejaculation or vibratory ejaculation). Recent studies have illustrated that it is likely to find active spermatogenesis in areas with good blood supply within the testicle. These studies utilized detailed color Doppler ultrasonography and needle guidance techniques to localize possible areas of spermatogenesis within the testicle. We have been exploring the efficacy of using a percutaneous handheld Doppler phase shift measurements of the testicle at the time of biopsy to localize areas of spermatogenesis.

A prospective blinded controlled trial of six patients who underwent testis biopsy from September 2008 to August 2009 for NOA (two men), previous vasectomy (two men), and spinal cord injury (two men) was performed. Percutaneous handheld Doppler (Vascular Technology™, Nashua, NH) blood flow shift measurements were taken from 12 different marked regions of one testicle (one testicle scanned on each patient the larger testicle was chosen). The surgeon then obtained 12 biopsies from the same testicle in these marked regions (blinded to the Doppler analysis). The findings from the biopsies were then compared to the pre-biopsy Doppler phase shift mapping to assess if the Doppler readings had any predictive value in detecting spermatogenesis.

The Doppler phase shift readings were analyzed and an algorithm developed to identify areas of the testicle with specific flow patterns. These flow patterns were then analyzed to assess for any correlation with spermatogenesis. A predictive model optimized to identify areas of sperm production was then created. The model was 85% accurate (ROC 0.8, 95% CI 0.6–0.9) in identifying areas within the testicle that had sperm based on Doppler phase shift readings.

Our preliminary evaluation of handheld Doppler phase shift flow mapping of the testicle at the time of biopsy appears to have promise in detecting areas of spermatogenesis. However, further data analysis on more patients has now shown that the predictive value was not as high as we initially had expected, and so there is still more work that has to be done in this area before conclusions can be made about the use of this technology for this application.

MRI Spectral Imaging

Patients with idiopathic oligospermia or azoospermia, especially those with normal serum gonadotrophins and physical examination, always present a diagnostic dilemma. Both situations can represent a ductal obstruction or a testicular failure, but they have completely different prognoses. Testicular functions are currently evaluated in rather indirect ways, by seminal parameters and hormonal assays. Histological analysis, which involves obtaining specimens by biopsies or surgical explorations, can directly evaluate testicular tissue. However, it cannot be widely used in clinical situations due to possible damage to testicular functions and its invasive character. Therefore, noninvasive techniques for evaluating testicular functions in vivo are needed [1].

Ultrasound is the initial radiological method that is used to evaluate the testis. However, the increased availability of magnetic resonance imaging (MRI) has allowed this noninvasive diagnostic tool to further evaluate testicular function. This technique was already used in a number of experimental studies on testis [1, 2]. On humans, there are few magnetic resonance spectroscopy (MRS) reports; one describes its application on a patient with testicular non-Hodgkin's lymphoma to monitor response to irradiation [3], and the other reveals the in vivo tissue characterization of the testis in patients with carcinoma in situ [4]. Also, differentiation between normal healthy testes and those with markedly decreased spermatogenesis presenting with oligospermia or azoospermia in whom spermatogenesis is completely absent was achieved [5, 6].

MRI spectroscopy is a noninvasive technique for obtaining metabolic information from living tissue based upon differences in the ratio of peaks of lipid and choline levels [7]. These metabolites may be used to evaluate the state of fertility and to investigate ischemia–reperfusion disorders.

More recently, in vivo hydrogen MR spectroscopy using stimulated echo acquisition mode measurements was performed with a short echo time, improving the detection of signals from low-molecular-weight metabolites including glutamate, choline, creatinine, and glycine not only in the normal state but also in diseased conditions such as ischemia [8]. In addition to these metabolites, a lactate signal could be observed in the ischemic testis. The presence of a lactate signal in the H spectra could be utilized to distinguish between normal and ischemic testes [8].

MR spectroscopy is a sensitive tool for assessment of testicular metabolic integrity and differentiation of normal testicles from those with markedly decreased spermatogenesis. MRS may improve sperm retrieval rates by better identifying isolated foci of spermatogenesis during testicular sperm retrieval in men presenting with nonobstructive azoospermia. MR spectroscopy of the testis might be a promising new modality that warrants further clinical studies to assess its diagnostic and therapeutic capability.

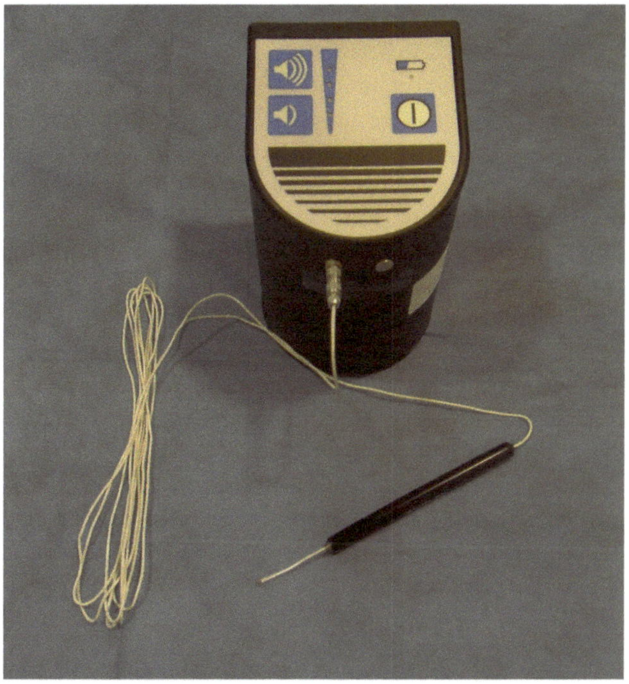

Fig. 3.1 VTI surgical Doppler 9.3 MHz and a disposable probe flow detector (Vascular Technology Int., USA)

Testicular Artery Mapping During Varicocelectomy

Current data supports the statement that varicocele repair does indeed have a beneficial effect in reversing the harmful effects of varicocele upon testicular function in selected patients by improving seminal parameters in the majority of controlled studies [9]. A diversity of open surgical techniques has been used to repair this condition, including retroperitoneal, inguinal, and subinguinal. Recently, open microsurgical inguinal or subinguinal varicocelectomy techniques have been shown to result in higher spontaneous pregnancy rates and fewer recurrences and postoperative complications than conventional varicocelectomy techniques in infertile men [10]. The subinguinal approach has the same principles as the inguinal approach but is performed through an incision below the external inguinal obviating to the need to open the aponeurosis of the external oblique causing less postoperative pain.

The microsurgical subinguinal varicocelectomy is the preferred approach for most experts. The use of an operating microscope allows the preservation of the testicular artery and lymphatics vessels, resulting in lower recurrence rates as well as hydrocele after the procedure [11]. On the other hand, the spermatic cord at the subinguinal level has a greater number of internal spermatic veins and an increased likelihood of encountering multiple spermatic arteries [12]. Previous studies reported that multiple spermatic arteries are identified in approximately 40% of the spermatic cords during microsurgical varicocelectomy at the subinguinal level [12, 13]. The recognition of the main spermatic artery can be confirmed by visualization of clear pulsatile movement and/or evidence of antegrade, pulsatile blood flow with gentle lifting and partial occlusion of the vessel. However, the identification of tiny secondary arteries is not at all times apparent, and a sterile intraoperative probe attached to a 9.3-MHz VTI surgical Doppler and a disposable probe flow detector (Vascular Technology Inc., USA) (Fig. 3.1) have been used at this point [14, 15]. Consequently, it is possible that an inadvertent unrecognized ligation of a small internal spermatic artery occurs more frequently than reported [16]. Following are some of the reasons that could explain how the injury occurs: First, the size of the arteries may be so small that the pulsation is difficult to identify. Second, aggressive manipulation of the vessels during dissection can lead to spasm, making it difficult to identify arterial pulsation. Third, the arteries tend to be in close proximity to or buried under complex branches of veins [16]. In all these situations, the use of vascular Doppler may help to preserve the arterial branches (Fig. 3.2). Even though there is no agreement about the necessity to preserve all testicular arterial branches during varicocele surgery [17, 18], not doing so might be responsible for suboptimal improvement in seminal parameters in some cases [19].

A recent study showed that concomitant use of intraoperative vascular Doppler during subinguinal varicocelectomy allows a higher number of arterial branches to be identified and therefore preserved [20]. Data concerning surgery using the Doppler vascular device and without it show that a solitary artery is identified in 45.5% and 69.5% of cords, respectively, 2 arteries are identified in 43.5% and 28.5%, respectively, and 3 or more arteries are identified in 11% and 2%, respectively. Also, the authors reported that a higher number of internal spermatic veins were ligated when Doppler was used (Table 3.1). The use of intraoperative Doppler gives the surgeon more confidence during dissection of a dense complex of adherent veins surrounding the artery present in 95% of cases when the subinguinal approach is used [12]. Accidental artery ligation documented by a pulsatile twitching of the ligated vessel stump under magnification is less common when Doppler is applied [20].

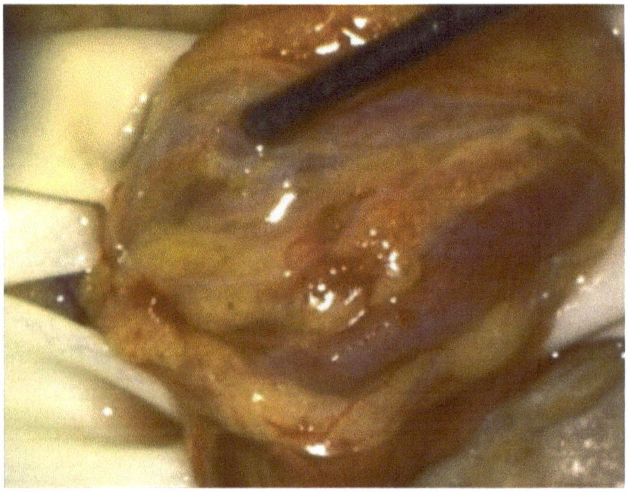

Fig. 3.2 Using a vascular Doppler probe to preserve all testicular arterial branches during varicocele repair

The clinical implication of these findings can be supported by recent studies showing that the total number of veins ligated was significantly positive correlated with improvements in total sperm motility and sperm concentration [21, 22]. These results suggest that ligating a larger number of veins should decrease reflux, which in turn would lead to diminished insult to spermatogenesis.

Management of Testicular Lesions in Infertile Patients

Organ-Sparing Microsurgical Resection of Testicular Tumors in Infertile Patients

Epidemiologic studies have given attention to a worldwide possible increase in testicular cancer incidence in the last two decades, particularly in industrialized developed countries [23]. One of the possible explanations for this augmented detection of testicular lesions is the widespread use of ultrasound as a screening method in all fields of medical practice, including scrotal ultrasonography in urology [24, 25].

In patients presenting with bilateral tumors (Fig. 3.3) or tumors in a solitary testis, the gold-standard procedure is to perform a radical orchiectomy which leads to permanent sterility, lifelong dependence on androgen replacement therapy, and psychological problems of castration at a young age [26]. As a result, organ-sparing surgery has been reported as a safe procedure in selected patients, especially for infertile men desiring to preserve their fertility [27–29]. The German Testicular Cancer Study Group established a guideline for organ-sparing surgery for testis tumors that includes cold ischemia during spermatic cord clamping, restriction of the procedure for organ-confined tumors of less than 20 mm that do not infiltrate the rete testis, performance of multiple biopsies of the tumor bed, and application of adjuvant local radiotherapy to eradicate carcinoma in situ and avoid local recurrence [29].

Table 3.1 Intraoperative evaluation of internal spermatic veins ligated, number of lymphatic spared, and arteries preserved and injured in 377 spermatic cord dissections during microsurgical subinguinal varicocele repair with and without vascular Doppler

Variable	With Doppler (no. spermatic cords = 225)	Without Doppler (no. spermatic cords = 152)	*P* value
Number of veins ligated[a]	8.0 (3.1)	7.3 (2.8)	0.02
Number of arteries preserved[a]	1.6 (0.6)	1.3 (0.5)	<0.01
Number of arteries injured[b]	0	2 (1.1%)	0.06
Number of lymphatics spared[a]	2.2 (1.2)	2.0 (1.5)	0.21
Operative time unilateral repair (min)[a]	52.8 ± 17.8	53.0 ± 36.7	0.98
Operative time bilateral repair (min)[a]	101.0 ± 16.2	101.9 ± 16.3	0.37

[a]Values are mean and SD. Compared using student's unpaired *t* test

[b]Data presented as number (percentage) of patients. Compared using chi-square test. $P < 0.05$ was considered statistically significant

From [20], with permission of Elsevier, Inc.

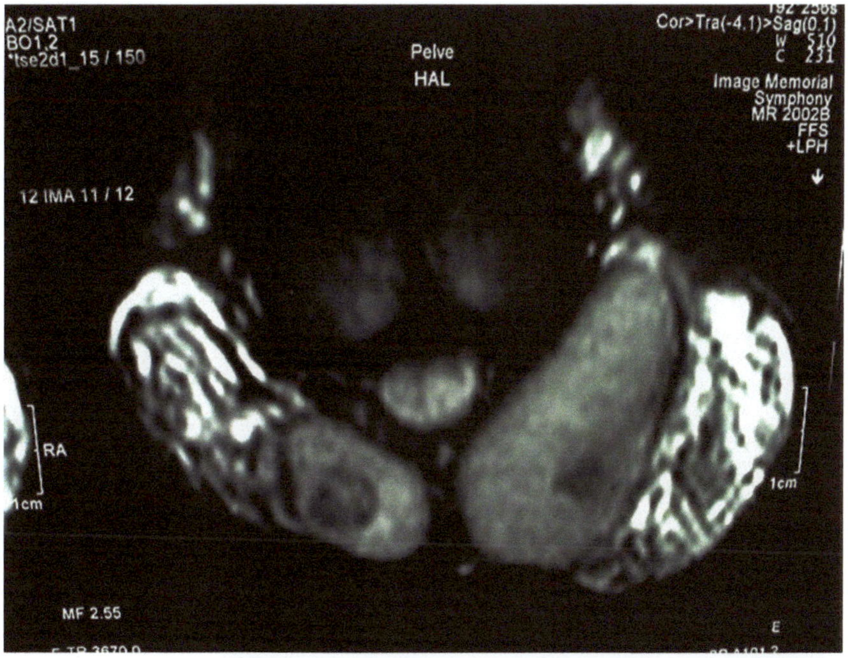

Fig. 3.3 Magnetic resonance showing a solid bilateral testicular lesion

Testicular germ cell tumors are the most common type of malignancy during the reproductive age [30, 31]. However, as the majority of incidental testicular nonpalpable lesions with negative markers diagnosed in scrotal ultrasonography performed during andrology investigation show a benign histology, surgical approach must be as conservative as possible for the testicular parenchyma [32]. In such cases, the most important step is the confirmation that frozen section analysis is an oncologically useful method for assessing small incidental testicular tumors when performed by an experienced pathologist [33]. The high degree of oncological efficiency achieved by the frozen section analysis during resection of testicular masses supports organ-sparing approaches that reduce the chances of facing difficult decisions intraoperatively [34].

Due to the advances of microsurgical techniques, partial orchiectomy was appointed as first-line therapy in infertile men even for nonpalpable small testicular lesions [27, 35, 36]. Also, infertile patient presenting with azoospermia and incidental testicular lesions now can experience the chance to father their own genetic offspring [37]. The combination of organ-sparing surgery, microdissection for TESE, cryopreservation, and assisted reproductive technologies represents a powerful tool to preserve fertility even for azoospermic men [38]. The complete procedure was meticulously described by Hallak et al. as follows [35]: During the procedure, the testis can be delivered through the inguinal incision, respecting principles for oncological procedures to avoid any potential spillage of tumor cells. Vas deferens must be carefully isolated from the spermatic cord and blood circulation interrupted by a delicate vascular clamp placed across the spermatic cord (Fig. 3.4). Slugged ice may be used to prevent warm ischemia and a temperature probe inserted far from the tumor location controls temperature at 12–15°C (Figs. 3.5 and 3.6). A linear ultrasound transducer at 15 MHz guides real-time intraoperative placement of a 30-gauge 10-cm-long stereotaxic hook-shaped needle (Guiding-Marker System; Hakko, Tokyo, Japan) adjacent to the tumor to guide microsurgical resection (Figs. 3.7 and 3.8). Using a surgical operating microscope, the tumor can be gently dissected and removed along with the adjoining parenchymal

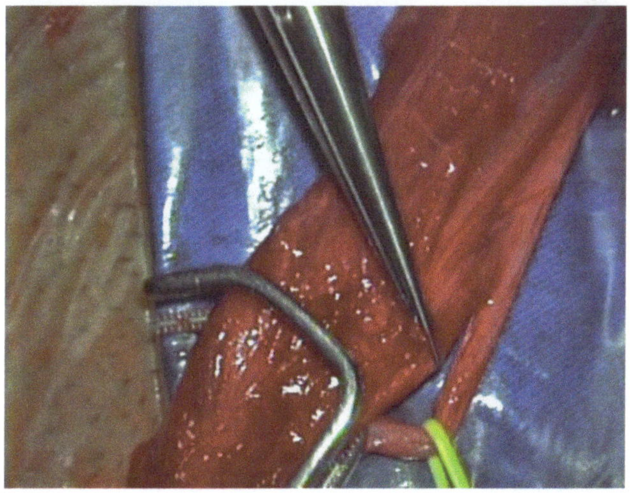

Fig. 3.4 Blood circulation was interrupted by a delicate vascular clamp placed across the spermatic cord after the vas deferens was carefully isolated

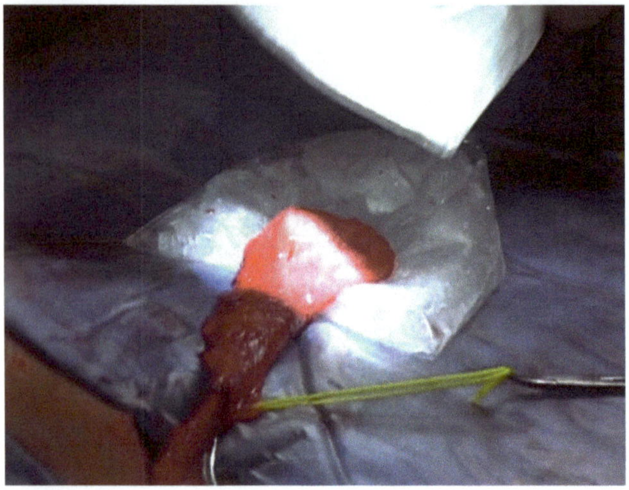

Fig. 3.5 Slugged ice wrapped the testicle, thus preventing from warm ischemia

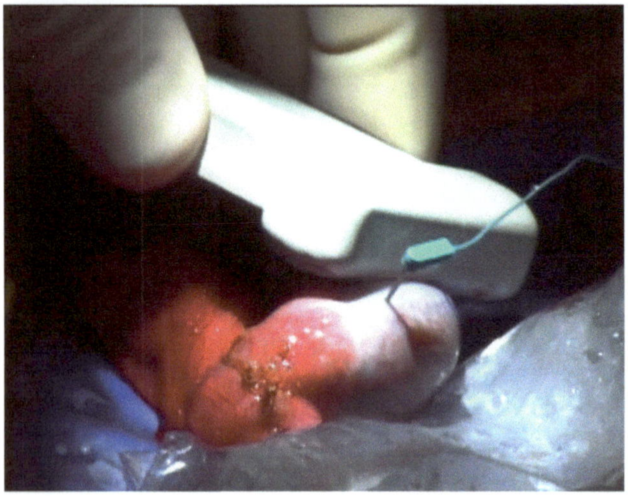

Fig. 3.6 A temperature probe was inserted far from the tumor location in the upper pole of the testicle

tissue (Fig. 3.9). Frozen section studies must be performed, and if malignancy is confirmed, biopsies of the tumor cavity margins and remaining parenchyma must be obtained to ensure absence of residual tumors. After biopsies are sent for frozen section, the testicular parenchyma must be meticulously microdissected for identification of functioning seminiferous tubules, as reported by Schlegel [38]. After excision of selected enlarged and opaque tubules, viable spermatozoa can be retrieved for cryopreservation in 80% of cases [35]. This procedure integrates modern skills accumulated in the field of male infertility, combining knowledge in testicular vascular anatomy, oncology, microsurgery, organ preservation, tissue preparation, and sperm cryopreservation.

The possibility of intracytoplasmic sperm injection with cryopreserved testicular sperm has given infertile azoospermic

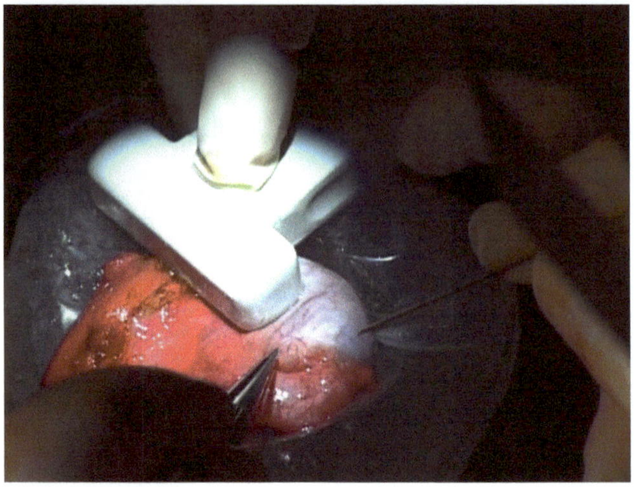

Fig. 3.7 A linear ultrasound transducer at 15-MHz guides real-time intraoperative placement of the stereotaxic hook-shaped needle adjacent to the nodule. The 30-gauge stereotaxic hook-shaped needle that permits the hook to be completely contained in and pass through the needle lumen. The introducing needle can thus be optimally positioned prior to engaging the hook. The hook is ejected and released from the needle tip reforming to hook and anchoring into the tissue adjacent to the tumor

men the chance to have their own genetic offspring [37]. It is very important to keep in mind that this approach is only appropriate in centers experienced in managing testicular cancer for patients who want to preserve fertility.

Testicular Microlithiasis

Testicular microlithiasis (TM) is an entity of unknown etiology that results in the formation of intratubular calcifications. TM is detected in 0.6% of adult males with clinical indications for scrotal ultrasonography [39]. Although TM is uncommon, it has been associated with several conditions such as Klinefelter's syndrome, cryptorchidism, varicocele, testicular atrophy, torsion, tumors, and infertility [40–43]. The mechanism by which microlithiasis affects spermatogenesis is unknown [43]. Scrotal ultrasound is diagnostic, and typically, there are small echogenic foci (1–3 mm).

The clinical significance of TM is debatable since inconsistencies exist in the literature. Therefore, there are conflicting recommendations regarding the appropriate interval and duration of subsequent surveillance of patients with testicular microlithiasis, as well as initial management strategies. It is of concern to the urologist due to its possible association with testis cancer [44]. Although commonly present in patients with germ cell tumors, there appears to be no definitive association with TM and cancer [45]. Therefore, follow-up at this time should be dictated based on risk factors for developing testis cancer more than on the presence of

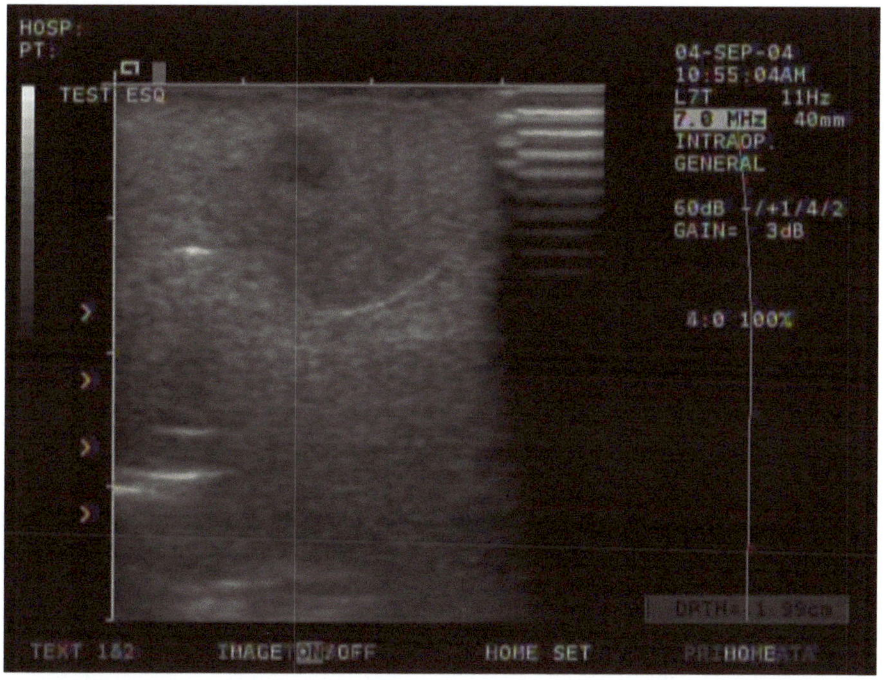

Fig. 3.8 Intraoperative ultrasound image showing a nonpalpable hypoechoic intratesticular lesion to guide real-time needle placement

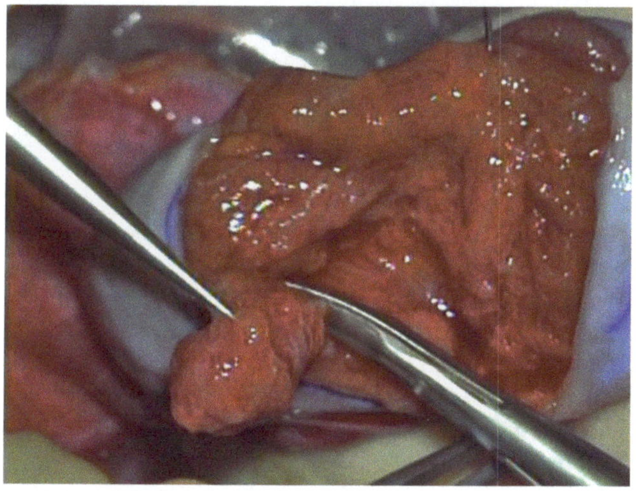

Fig. 3.9 After the tunica albuginea is incised in an avascular region using an operating microscope, dissection respecting testicular lobules and arteries is conducted. Seminiferous tubules were separated carefully by blunt dissection following the needle until the lesion that is excised using micro-instruments is found, leaving 2–3-mm borders as safe margins around the nodule

TM [46]. Casteren et al. suggested taking a testicular biopsy in a selected patient population with at least one additional risk factor for testicular germ cell tumors [47]. Similarly, there is no definitive proof that TM by itself can cause infertility. Probably, decrease seminal function is not directly due to the TM but rather to an underlying testicular abnormality or associated condition such as cryptorchidism [48].

Seminal Vesicle and Ejaculatory Duct Imaging

Turek et al. [49] recently presented a technique of transrectal vasodynamics (pressure-flow study of the seminal vesicle) to assess for partial ejaculatory duct obstruction (EDO). We have been exploring the use of a 3D transrectal ultrasound imaging and needle targeting system to perform vasodynamics for the treatment of necrospermia with partial EDO (Target Scan Touch, Envisioneering, St. Louis, MO).

A patient with necrospermia (viability staining of sperm) and low ejaculatory volume (<0.5 cc) was evaluated for EDO by vasodynamics performed utilizing a 3D transrectal ultrasound needle guidance system (TargetScan™, Envisioneering) in January 2008. This system was utilized to access and maintain a flexible needle within the seminal vesicle (SV) during vasodynamics. The decision to perform transurethral unroofing of the ejaculatory duct would be based on the pressure readings obtained during vasodynamics.

The patient had no visible left SV and a dilated right SV (>1.5 cm). The system provided an easy stable guided platform to place and maintain a flexible needle within the right SV. A pressure reading of 41 cm H_2O was achieved with delayed emptying of the right SV (on simultaneous cystoscopic examination of the ejaculatory ducts within the urethra). Transurethral (TUR) unroofing of the right ejaculatory duct was performed, and the post-TUR SV pressure reading dropped to 31 cm H_2O. The right SV drained promptly post-TUR on cystoscopic examination. Postoperatively, the patient initially had retrograde ejaculation. At 1 year post-op, the patient now has an ejaculate volume of 1 cc with 70 million sperm/hpf with 64% motility.

The use of a 3D transrectal ultrasound needle guidance system to perform vasodynamics is feasible for the treatment of patients with necrospermia and partial EDO, and it enhances a surgeon's ease in performing vasodynamics. Further testing and evaluation is needed.

Expert Commentary

In approximately half of infertility cases, a male factor is involved. Thus, identifying the pathology and treating the male may allow couples to regain fertility and conceive through natural intercourse. The goal of infertility management is to diagnose reversible causes of infertility and treat them to achieve seminal improvement and pregnancy. Despite advancements in the diagnostic workup of infertile men, up to 25% of patients exhibit abnormal semen analyses for which no etiology can be identified [50]. This condition is referred to as idiopathic male infertility, and nonspecific treatments are usually applied that are based on theoretical concepts. A variety of empiric medical therapies have been recommended to treat these patients. However, the majority of these therapies have been shown to be effective in repeated controlled randomized studies. Even though the available assisted reproduction techniques can help overcome severe male factor infertility, the application of these methods in infertile couples classified as *idiopathic* infertility would definitely represent overtreatment.

We look toward the future with excitement and hope that the advanced technologies discussed will not only provide new treatment options but will reduce the number of couples without a definitive diagnosis of the cause of failure to conceive. There is no hesitation that all technical advances such as those explained in this chapter will drive the development of pioneering approaches to the management of the infertile male by andrologists. The use of imaging in the field of infertility is no longer just for diagnosis but also as part of the arsenal that provides more precise surgery procedures as described above.

Five-Year View

Numerous advances have been made in reproductive medicine in the last few years. Infertile couples who previously were considered untreatable now have a chance at genetic paternity. ART provide a great opportunity to families with infertility, and their use has become routine in the treatment of infertile couples. The increasing use of ICSI as an efficient therapy for cases of male infertility has become an applicable means to overcome multiple sperm deficiencies. Even men with potentially treatable causes of infertility can be treated with ART instead of a specific therapy.

The objective of this chapter was to discuss the potential role of imaging modalities in the management of male infertility. Collaborations with radiologists have provided a rich opportunity to explore and expand imaging techniques for intraoperative use during urologic surgery. Due to the continual improvement of a variety of imaging and tissue characterization modalities, the surgeons of tomorrow will have a number of tools at their disposal to improve intraoperative surgical decision-making. Only through the application of evidence-based assessment and evaluation, however, will there be a firm understanding of the true impact of these new technologies on the field of urology. However, further studies will be needed to confirm whether or not these techniques can evolve into widespread clinical practice.

As a result, the application of technological advances during varicocelectomy including optical magnification, microsurgery skills and vascular Doppler may offer patients maximal preservation of the arterial blood supply to the testes. However, additional research is needed to better clarify whether the use of Doppler during varicocelectomy is likely to improve testicular function and seminal parameters.

New methods for testicular screening, such as MRS, appear to be promising for testicular sperm extraction procedures. By identifying testicular locations that are likely to contain viable spermatozoa in the testes of nonobstructive azoospermic men, the potential for testicular damage may be reduced.

Key Issues

MR spectroscopy is a sensitive tool for assessment of testicular metabolic integrity and differentiation of normal testicles from those with markedly decreased spermatogenesis. MRS may improve sperm retrieval rates by better identifying isolated foci of spermatogenesis during testicular sperm extraction in men presenting with nonobstructive azoospermia. MR spectroscopy of the testis might be a promising new contribution that warrants further clinical studies to assess its full diagnostic and therapeutic capability.

Microsurgical technique remains the gold-standard procedure for the varicocele repair, but the concomitant use of intraoperative Doppler should be seriously considered as a tool to improve surgical outcome and safety.

Use of intraoperative vascular Doppler during microsurgical varicocelectomy allows a higher number of arterial branches to be preserved, and more internal spermatic veins are likely to be ligated.

Microsurgical organ-sparing testicular tumor resection associated with microdissection for TESE and tissue cryopreservation techniques may be considered an attractive option for infertile patients presenting with azoospermia and incidental testicular lesions, especially for those with solitary testicles and bilateral tumors. The technique described in the chapter in a precise manner can be easily reproduced by others.

Despite greater awareness of TM, a clear definition is currently missing, and the etiology is still obscure. This causes confusion in management and follow-up. There is no convincing evidence that TM alone is premalignant. However, when it accompanies other potentially premalignant features, we recommend annual ultrasound follow-up. With longer follow-up of patients with TM, its true likelihood of leading to cancer will be elucidated, and more evidence-based guidelines can be established.

Acknowledgments The authors are grateful to Ashok Agarwal, Cleveland Clinic, for his support and encouragement.

References

1. Sasagawa I, Nakada T, Kubota Y, Ishigooka M, Uchida K, Doi K. In vivo 31P magnetic resonance spectroscopy for evaluation of testicular function in cryptorchid rats. J Urol. 1995;154(4):1557–9.
2. Srinivas M, Degaonkar M, Chandrasekharam VV, et al. Potential of MRI and 31P MRS in the evaluation of experimental testicular trauma. Urology. 2002;59(6):969–72.
3. Kiricuta IC, Bluemm RG, Ruhl J, Beyer HK. 31-P MR spectroscopy and MRI of a testicular non-Hodgkin lymphoma recurrence to monitor response to irradiation. A case report. Strahlenther Onkol. 1994;170(6):359–64.
4. Thomsen C, Jensen KE, Giwercman A, Kjaer L, Henriksen O, Skakkebaek NE. Magnetic resonance: in vivo tissue characterization of the testes in patients with carcinoma-in-situ of the testis and healthy subjects. Int J Androl. 1987;10(1):191–8.
5. Chew WM, Hricak H, McClure RD, Wendland MF. In vivo human testicular function assessed with P-31 MR spectroscopy. Radiology. 1990;177(3):743–7.
6. van der Grond J, Laven JS, te Velde ER, Mali WP. Abnormal testicular function: potential of P-31 MR spectroscopy in diagnosis. Radiology. 1991;179(2):433–6.
7. Navon G, Gogol E, Weissenberg R. Phosphorus-31 and proton NMR analysis of reproductive organs of male rats. Arch Androl. 1985;15(2–3):153–7.
8. Yamaguchi M, Mitsumori F, Watanabe H, Takaya N, Minami M. In vivo localized 1 H MR spectroscopy of rat testes: stimulated echo acquisition mode (STEAM) combined with short TI inversion recovery (STIR) improves the detection of metabolite signals. Magn Reson Med. 2006;55(4):749–54.
9. Agarwal A, Deepinder F, Cocuzza M, et al. Efficacy of varicocelectomy in improving semen parameters: new meta-analytical approach. Urology. 2007;70(3):532–8.
10. Cayan S, Shavakhabov S, Kadioglu A. Treatment of palpable varicocele in infertile men: a meta-analysis to define the best technique. J Androl. 2009;30(1):33–40.
11. Goldstein M, Gilbert BR, Dicker AP, Dwosh J, Gnecco C. Microsurgical inguinal varicocelectomy with delivery of the testis: an artery and lymphatic sparing technique. J Urol. 1992;148(6):1808–11.
12. Hopps CV, Lemer ML, Schlegel PN, Goldstein M. Intraoperative varicocele anatomy: a microscopic study of the inguinal versus subinguinal approach. J Urol. 2003;170(6 Pt 1):2366–70.
13. Grober ED, O'Brien J, Jarvi KA, Zini A. Preservation of testicular arteries during subinguinal microsurgical varicocelectomy: clinical considerations. J Androl. 2004;25(5):740–3.
14. Minevich E, Wacksman J, Lewis AG, Sheldon CA. Inguinal microsurgical varicocelectomy in the adolescent: technique and preliminary results. J Urol. 1998;159(3):1022–4.
15. Wosnitzer M, Roth JA. Optical magnification and Doppler ultrasound probe for varicocelectomy. Urology. 1983;22(1):24–6.
16. Chan PT, Wright EJ, Goldstein M. Incidence and postoperative outcomes of accidental ligation of the testicular artery during microsurgical varicocelectomy. J Urol. 2005;173(2):482–4.
17. Matsuda T, Horii Y, Yoshida O. Should the testicular artery be preserved at varicocelectomy? J Urol. 1993;149(5 Pt 2):1357–60.
18. Student V, Zatura F, Scheinar J, Vrtal R, Vrana J. Testicle hemodynamics in patients after laparoscopic varicocelectomy evaluated using color Doppler sonography. Eur Urol. 1998;33(1):91–3.
19. Penn I, Mackie G, Halgrimson CG, Starzl TE. Testicular complications following renal transplantation. Ann Surg. 1972;176(6):697–9.
20. Cocuzza M, Pagani R, Coelho R, Srougi M, Hallak J. The systematic use of intraoperative vascular Doppler ultrasound during microsurgical subinguinal varicocelectomy improves precise identification and preservation of testicular blood supply. Fertil Steril. 2010;93(7):2396–9.
21. Belani JS, Yan Y, Naughton CK. Does varicocele grade predict vein number and size at microsurgical subinguinal repair? Urology. 2004;64(1):137–9.
22. Pasqualotto FF, Lucon AM, de Goes PM, et al. Relationship between the number of veins ligated in a varicocelectomy with testicular volume, hormonal levels and semen parameters outcome. J Assist Reprod Genet. 2005;22(6):245–9.
23. Huyghe E, Matsuda T, Thonneau P. Increasing incidence of testicular cancer worldwide: a review. J Urol. 2003;170(1):5–11.
24. Horstman WG, Haluszka MM, Burkhard TK. Management of testicular masses incidentally discovered by ultrasound. J Urol. 1994;151(5):1263–5.
25. Carmignani L, Gadda F, Mancini M, et al. Detection of testicular ultrasonographic lesions in severe male infertility. J Urol. 2004;172(3):1045–7.
26. Fossa SD, Opjordsmoen S, Haug E. Androgen replacement and quality of life in patients treated for bilateral testicular cancer. Eur J Cancer. 1999;35(8):1220–5.
27. Colpi GM, Carmignani L, Nerva F, Guido P, Gadda F, Castiglioni F. Testicular-sparing microsurgery for suspected testicular masses. BJU Int. 2005;96(1):67–9.
28. Yossepowitch O, Baniel J. Role of organ-sparing surgery in germ cell tumors of the testis. Urology. 2004;63(3):421–7.
29. Heidenreich A, Weissbach L, Holtl W, et al. Organ sparing surgery for malignant germ cell tumor of the testis. J Urol. 2001;166(6):2161–5.
30. Bokemeyer C, Schmoll HJ, Schoffski P, Harstrick A, Bading M, Poliwoda H. Bilateral testicular tumours: prevalence and clinical implications. Eur J Cancer. 1993;29A(6):874–6.
31. Coleman MP, Esteve J, Damiecki P, Arslan A, Renard H. Trends in cancer incidence and mortality. IARC Sci Publ. 1993;121:1–806.
32. Carmignani L, Gadda F, Gazzano G, et al. High incidence of benign testicular neoplasms diagnosed by ultrasound. J Urol. 2003;170(5):1783–6.
33. Leroy X, Rigot JM, Aubert S, Ballereau C, Gosselin B. Value of frozen section examination for the management of nonpalpable incidental testicular tumors. Eur Urol. 2003;44(4):458–60.
34. Steiner H, Holtl L, Maneschg C, et al. Frozen section analysis-guided organ-sparing approach in testicular tumors: technique, feasibility, and long-term results. Urology. 2003;62(3):508–13.
35. Hallak J, Cocuzza M, Sarkis AS, Athayde KS, Cerri GG, Srougi M. Organ-sparing microsurgical resection of incidental testicular tumors plus microdissection for sperm extraction and cryopreservation

in azoospermic patients: surgical aspects and technical refinements. Urology. 2009;73(4):887–91. discussion 891–882.
36. Binsaleh S, Sircar K, Chan PT. Feasibility of simultaneous testicular microdissection for sperm retrieval and ipsilateral testicular tumor resection in azoospermic men. J Androl. 2004;25(6): 867–71.
37. Oates RD, Mulhall J, Burgess C, Cunningham D, Carson R. Fertilization and pregnancy using intentionally cryopreserved testicular tissue as the sperm source for intracytoplasmic sperm injection in 10 men with non-obstructive azoospermia. Hum Reprod. 1997;12(4):734–9.
38. Schlegel PN. Testicular sperm extraction: microdissection improves sperm yield with minimal tissue excision. Hum Reprod. 1999;14(1):131–5.
39. Hobarth K, Susani M, Szabo N, Kratzik C. Incidence of testicular microlithiasis. Urology. 1992;40(5):464–7.
40. Ganem JP, Workman KR, Shaban SF. Testicular microlithiasis is associated with testicular pathology. Urology. 1999;53(1):209–13.
41. Aizenstein RI, DiDomenico D, Wilbur AC, O'Neil HK. Testicular microlithiasis: association with male infertility. J Clin Ultrasound. 1998;26(4):195–8.
42. Vrachliotis TG, Neal DE. Unilateral testicular microlithiasis associated with a seminoma. J Clin Ultrasound. 1997;25(9):505–7.
43. Qublan HS, Al-Okoor K, Al-Ghoweri AS, Abu-Qamar A. Sonographic spectrum of scrotal abnormalities in infertile men. J Clin Ultrasound. 2007;35(8):437–41.
44. Ringdahl E, Claybrook K, Teague JL, Northrup M. Testicular microlithiasis and its relation to testicular cancer on ultrasound findings of symptomatic men. J Urol. 2004;172(5 Pt 1):1904–6.
45. Rashid HH, Cos LR, Weinberg E, Messing EM. Testicular microlithiasis: a review and its association with testicular cancer. Urol Oncol. 2004;22(4):285–9.
46. Dagash H, Mackinnon EA. Testicular microlithiasis: what does it mean clinically? BJU Int. 2007;99(1):157–60.
47. van Casteren NJ, Looijenga LH, Dohle GR. Testicular microlithiasis and carcinoma in situ overview and proposed clinical guideline. Int J Androl. 2009;32(4):279–87.
48. Miller RL, Wissman R, White S, Ragosin R. Testicular microlithiasis: a benign condition with a malignant association. J Clin Ultrasound. 1996;24(4):197–202.
49. Eisenberg ML, Walsh TJ, Garcia MM, Shinohara K, Turek PJ. Ejaculatory duct manometry in normal men and in patients with ejaculatory duct obstruction. J Urol. 2008;180(1):255–60. discussion 260.
50. Greenberg SH, Lipshultz LI, Wein AJ. Experience with 425 subfertile male patients. J Urol. 1978;119(4):507–10.

Endocrinopathies

4

Sam Haywood, Eric L. Laborde,
and Robert E. Brannigan

Abstract

Spermatogenesis depends on an intricate interplay of hormonal factors both centrally and in the testis. Centrally, the hypothalamus releases gonadotropin-releasing hormone (GnRH), which acts on the anterior pituitary to cause secretion of luteinizing hormone (LH) and follicle-stimulating hormone (FSH). At the level of the testis, FSH acts on Sertoli cells to induce the maturation process in spermatogonia. LH exerts its effect on Leydig cells, stimulating production of testosterone. Effective spermatogenesis requires local testosterone concentrations to be much higher than serum concentrations. This intratesticular testosterone then acts indirectly to stimulate germ cell maturation through actions on Sertoli cells. Although endocrinopathies only account for a small minority of cases of male infertility, about 1–2%, the treatment of these conditions offers patients a strategy of directed therapy. Broad classification of endocrinopathies involves two main categories: hormonal deficiency and hormonal excess, with specific hormonal abnormalities falling under each of the above categorizations.

Keywords

Male infertility • Spermatogenesis • Endocrinopathy • Hypogonadotropic hypogonadism • Hypergonadotropic hypogonadism • Hyperprolactinemia • Hyperthyroidism • Androgen excess • Estrogen excess

Spermatogenesis depends on an intricate interplay of hormonal factors both centrally and in the testis. Centrally, the hypothalamus releases gonadotropin-releasing hormone (GnRH), which acts on the anterior pituitary to cause secretion of luteinizing hormone (LH) and follicle-stimulating hormone (FSH). At the level of the testis, FSH acts on Sertoli cells to induce the maturation process in spermatogonia. LH exerts its effect on Leydig cells, stimulating production of testosterone [1]. Effective spermatogenesis requires local testosterone concentrations to be much higher than serum concentrations [2]. This intratesticular testosterone then acts indirectly to stimulate germ cell maturation through actions on Sertoli cells [1, 2].

Although endocrinopathies only account for a small minority of cases of male infertility, about 1–2% [3], the treatment of these conditions offers patients a strategy of directed therapy. Broad classification of endocrinopathies involves two main categories: hormonal deficiency and hormonal excess, with specific hormonal abnormalities falling under each of the previously mentioned categorizations.

S. Haywood, BA (✉) • E.L. Laborde, MD • R.E. Brannigan, MD
Department of Urology, Northwestern University Feinberg School of Medicine, Galter Pavilion, Suite 20-150 675 N. Saint Clair Street, Chicago, IL 60611, USA
e-mail: r-brannigan@northwestern.edu

S.J. Parekattil and A. Agarwal (eds.), *Male Infertility: Contemporary Clinical Approaches, Andrology, ART & Antioxidants*,
DOI 10.1007/978-1-4614-3335-4_4,

Hormonal Deficiency

Hypogonadotropic Hypogonadism

As the name suggests, hypogonadotropic hypogonadism is a state of testosterone deficiency associated with subnormal levels of gonadotropins (FSH and LH). Etiologies of hypogonadotropic hypogonadism can be numerous and are divided into congenital and acquired causes.

Kallmann syndrome is one identified congenital etiology of hypogonadotropic hypogonadism. Inherited in an X-linked recessive fashion, Kallmann syndrome can arise due to a variety of mutations, the most prevalent of which involves the KAL1 gene. Features include hypogonadism as well as anosmia, facial defects, renal agenesis, and neurologic abnormalities [4]. The hypogonadism and associated clinical sequelae (delayed puberty, infertility) result from a failure of migration of GnRH-secreting neurons. This failure of migration leads to absence of GnRH secretion which in turn leads to absent LH and FSH secretion [3].

Hypogonadotropic hypogonadism can also be acquired, as in the case of pituitary insufficiency resulting from pituitary tumors, surgery, infarct, or infiltrative disease. Regardless of the myriad etiologies of hypogonadotropic hypogonadism, the underlying disturbance is low gonadotropin levels, and treatment can be affected through pharmacologic replacement.

Treating hypogonadotropic hypogonadism involves replacement of the deficient hormones through gonadotropin therapy. Agents used in this therapy include human chorionic gonadotropin (hCG), human menopausal gonadotropin (hMG), and recombinant follicle-stimulating hormone (rFSH). Human chorionic gonadotropin use stems from its properties as an LH analogue, acting at the Leydig cell to stimulate androgen secretion. Human menopausal gonadotropin is a product purified from the urine of postmenopausal women that contains both LH and FSH. Regimens of gonadotropin therapy for men with hypogonadotropic hypogonadism typically begin with hCG administration alone for 3–6 months. Dosages range from 1,000 to 1,500 USP units either IM or SC three times per week. Adequacy of therapy can be assessed by measuring serum testosterone levels, with the goal of achieving sustained normal levels. Although the pertinent goal for spermatogenesis is adequate intratesticular testosterone concentrations, this value is not normally assessed in gonadotropin replacement therapy. However, intratesticular testosterone levels show linear correlation with administered hCG dosage [5]. After titration to sustained normal testosterone levels, usually after 3–6 months of hCG monotherapy, therapy is initiated to replace FSH levels. One method of FSH replacement involves hMG given at doses of 75–150 IU IM/SC three times a week at a separate injection site. Alternatively, rFSH can be used at dosages of 150 IU SC three times a week [6]. Relative efficacy of hMG versus rFSH has been studied to some extent in women undergoing IVF, but comparisons in male patients are lacking. Replacing gonadotropins in this manner has shown promising results as more than 90% of treated males experience spermatogenesis [3]. The time to spermatogenesis can be quite variable, with the average response occurring in about 6–9 months. However, therapy may be required for up to 1–2 years before a response may occur, and some individuals unfortunately never respond to this modality [7]. An Australian study of 38 men with hypogonadotropic hypogonadism found that median time to first sperm in the ejaculate was 7.1 months, while median time to conception was 28.2 months [8].

While spermatogenesis occurs in a strong majority of patients, sperm concentrations achieved through gonadotropin therapy still sometimes fall below goal ranges (<20 million sperm/mL). Despite this, fertility outcomes with gonadotropin therapy are very good. In a study of 24 men with hypogonadotropic hypogonadism treated with gonadotropin therapy, 22 men achieved pregnancy, even though mean sperm concentration was 16.7 million sperm/mL [9]. A retrospective study of Japanese men found sperm production in 71% of men treated with hCG (3,000 IU) and hMG (75 IU), provided testicular size was greater than prepubertal sizes (>4 mL) [10]. A recently published Saudi Arabian paper studied 87 infertile men with hypogonadotropic hypogonadism treated with IM gonadotropins for a median of 26 months, with the primary outcome of fertility. Overall, 35 of the 87 patients (40%) were able to achieve pregnancy [11].

An important area of newer research focuses on determining predictors of response to gonadotropin therapy. The aforementioned long-term study in Japanese men found a correlation between testicular size pretreatment and response to gonadotropin therapy. Men with testicular size >4 mL had a 71% chance of responding to treatment, whereas men with testicular size <4 mL had only a 36% chance of responding to treatment [10]. In addition, the above Saudi Arabian study found that only pretreatment testicular size was predictive of conception. In particular, responders to treatment had a mean testicular pretreatment volume of 9.0 ± 3.6 mL, while the pretreatment testicular volume of nonresponders was only 5.7 ± 2.0 mL. Interestingly, there was no significant difference in conception rates between men with hypogonadotropic hypogonadism due to congenital or acquired etiologies [11]. Larger baseline testicular size has also been shown as an independent predictor of response time to gonadotropin therapy, and achieving summed testis volume >20 mL after treatment increased the odds of achieving both goal sperm parameters and pregnancy by at least twofold [8]. It is worth noting that the lower sperm concentrations found in these studies, while below traditional goals of infertility management, may allow for pregnancy with adjunctive use of assisted reproductive therapies such as intrauterine insemination or in vitro fertilization. Additionally, such medical treatment may allow increased efficacy of surgical sperm extraction.

Another method of treatment for men with hypogonadotropic hypogonadism involves the use of antiestrogen agents. These agents competitively bind to estrogen receptor sites in the hypothalamus. Normally, estradiol acts via negative feedback at this endocrine center to inhibit gonadotropin secretion. By binding at these sites, antiestrogen agents block estradiol's feedback inhibition of the hypothalamus and thus increase the hypothalamic secretion of GnRH. The increased secretion of GnRH leads to increased pituitary secretion of gonadotropins, which thereby stimulates an increase in intratesticular testosterone production. The most commonly used agent in this class is clomiphene citrate, but similar agents include tamoxifen, raloxifene, and toremifene. These drugs have been previously studied in the setting of empiric therapy for idiopathic infertility with mixed results [7]. However, the directed use of clomiphene in patients with proven hypogonadotropic hypogonadism has shown to be useful in limited settings. An American study treated four men with hypogonadotropic hypogonadism with clomiphene citrate 50 mg three times a week and found improved testosterone levels and semen parameters in three of these patients. Subsequently, two of these three men achieved documented pregnancy [12]. Similar success at the biochemical level has also been described in case reports, although fertility was not a goal of these treatments [13, 14]. Clomiphene treatment of male infertility can be associated with such side effects as visual disturbances, GI upset, weight gain, hypertension, and insomnia [7].

It is worth noting that exogenous GnRH treatment represents another avenue of medical therapy for hypogonadotropic hypogonadism. Synthetic analogues of GnRH can be administered to stimulate secretion of gonadotropins. However, the short half-life of these agents combined with the necessary pulsatile release to recreate normal physiology requires a method of frequent administration, such as frequent injections, nasal sprays, or an implantable pump. These methods are obviously less convenient, and further, studies have not shown this treatment has a strong benefit for hypogonadotropic hypogonadism [15].

Hypergonadotropic Hypogonadism

In hypergonadotropic hypogonadism, the main perturbation is an inadequate or absent function of the testes. Gonadotropins are appropriately elevated secondary to lack of negative feedback from estradiol, testosterone, and inhibin B from the testis. Without appropriate androgen secretion, spermatogenesis is impaired. These men also typically have significant testicular atrophy with fibrosis and markedly reduced germ cell number, also leading to abnormally low levels of spermatogenesis. Hypergonadotropic hypogonadism can occur as a result of genetic etiologies (e.g., Klinefelter syndrome) or from acquired conditions. Acquired etiologies of hypergonadotropic hypogonadism include destruction of normal gonadal tissue from chemotherapy or radiation, trauma, mumps orchitis, or androgen decline in the aging male. Men with hypergonadotropic hypogonadism not desiring fertility can be treated with exogenous testosterone therapy, but men trying to conceive should generally not be given exogenous testosterone. The treatment for men trying to conceive is less well characterized. Aromatase inhibitors have been suggested as treatment for men with Klinefelter syndrome [4]. A small cohort of patients with Klinefelter syndrome treated with aromatase inhibitors showed hormonal improvements with treatment, although the study did not comment on semen parameters in the Klinefelter subset. In particular, for this subset of patients, testolactone therapy was more efficacious with respect to hormonal levels than anastrazole [16].

It is important to mention the additional potential advantage in the setting of surgical sperm extraction after adjuvant medical therapy in men with Klinefelter syndrome. Surgical sperm extraction alone has resulted in successful retrieval in up to 50% of attempts [17]. Ramasamy et al. retrospectively studied 68 azoospermic men with Klinefelter syndrome. Of these 68, 56 men were treated for low testosterone levels (<300 ng/dL) with a combination of medical therapies (aromatase inhibitors, hCG, clomiphene) before microdissection TESE. Of the 56 men receiving medical therapy before TESE, 28 received testolactone alone, 12 received testolactone and weekly hCG, 9 received anastrozole alone, 1 received anastrozole and hCG, and 4 received hCG alone. Three patients total received clomiphene citrate. While there was no difference among specific agents in terms of successful sperm extraction, these medical regimens collectively resulted in improved sperm retrieval when patients responded to medical therapies with posttreatment testosterone >250 ng/dL. More specifically, successful sperm extraction was seen in 77% of men with posttreatment testosterone >250 ng/dL versus 55% of men with posttreatment testosterone <250 ng/dL [17].

Hypothyroidism

Thyroid hormones are essential in organ development and routine metabolism. However, there have been few studies evaluating hypothyroidism and male reproduction. Hypothyroidism has long been associated with diminished libido and erectile dysfunction [18]. Additionally, a recent study by Meeker et al. revealed a correlation between thyroxine (T4) level and sperm concentration, with higher T4 being correlated with better sperm concentrations [19]. Sperm concentration may not be the only parameter affected, as a study by Krassas et al. showed that men with hypothyroidism have a lower than normal percentage of sperm with normal morphology. Correcting the hypothyroidism resulted in 76% of the patients having a normal morphology [20]. Overall, there is a relative scarcity of data regarding hypothyroidism and semen parameters. Nonetheless, these studies do suggest a link between thyroid function and spermatogenesis.

Hormonal Excess

Androgen Excess

Within the hypothalamic–pituitary–testis axis, testosterone exerts negative feedback inhibition on the hypothalamic secretion of GnRH. This effect is indirect and thought to occur via aromatization of testosterone to estradiol. Acting in this manner, excess circulating testosterone can suppress this axis and cause inhibition of spermatogenesis. Testosterone excess can result from exogenous testosterone administration or from endogenous production. Therapeutic administration can inadvertently result in testosterone excess, but testosterone excess can also result from the illicit use of anabolic steroids. Regardless of the cause, exogenous androgens typically suppress gonadotropin secretion with resultant decreased levels of intratesticular testosterone and decreased spermatogenesis. Diagnosis is suggested by normal to high serum testosterone levels with suppressed gonadotropins. The first step in treating a male with suspected androgen excess is to remove the exogenous source. Return of spermatogenesis usually occurs within 4 months but in some instances can take up to 3 years [21, 22]. If sperm parameters do not improve adequately or are slow to improve, some evidence suggests beneficial effects of gonadotropin therapy in improving intratesticular testosterone levels [22, 23]. If response to treatment remains suboptimal after a trial of gonadotropin therapy, limited evidence suggests a possible use of clomiphene in reestablishing the hypothalamic–pituitary–testis axis [24].

Androgen excess can also result from endogenous androgen production. The most common endogenous source is congenital adrenal hyperplasia, although functional tumors (adrenal or testicular) and androgen insensitivity syndromes could also be responsible [3]. These etiologies have their own treatment strategies that will not be discussed here, but consideration should be given to their presence for complete patient care.

Estrogen Excess

As mentioned earlier, testosterone's ability to inhibit GnRH secretion at the hypothalamus is mediated through conversion to estrogens. A primary excess of estrogens can act similarly to inhibit the hypothalamic–pituitary–testis axis and thus contribute to decreased fertility. While estrogens are produced in the testis along with testosterone, the main source of estrogens in males is peripheral aromatization of testosterone by the enzyme aromatase, found in adipose tissue. The rising prevalence of obesity in our society puts more men at risk for estrogen excess. In particular, the ratio of testosterone to estradiol (T:E2) appears to be an important measure of estrogen excess, with a goal ratio >10:1 sought by many clinicians. Pavlovich et al. examined a cohort of infertile men and found significantly reduced T:E2 ratios in the infertile men compared to a fertile control group (6.9 vs. 14.5) [25].

Treatment of relative estrogen excess involves inhibitors of the aromatase enzyme. There are two main classes of aromatase inhibitors: steroidal agents (e.g., testolactone) and nonsteroidal agents (e.g., anastrozole). Both have shown utility in treatment of infertile men with low T:E2 ratios. The above Pavlovich study treated 63 men with male factor infertility and low T:E2 ratios with testolactone, 50–100 mg twice daily. Treatment was effective in improving both T:E2 ratio and sperm quality, as defined by concentrations and motility [25]. A more recent study by Raman and Schlegel treated 140 infertile men with abnormal T:E2 ratios with either testolactone (100–200 mg daily) or anastrozole (1 mg daily). Both treatment arms showed improvement in T:E2 ratio as well as improved sperm concentration and motility. Further, the study did not show any significant difference between the two classes of aromatase inhibitors in terms of hormonal profile or semen analysis, except in the setting of Klinefelter syndrome, where testolactone was superior in treating the abnormal T:E2 ratios [16]. These studies combined show a clear role for aromatase inhibitors in infertile men with abnormal T:E2 ratios. This treatment strategy may be of particular importance in obese patients [26].

Thyroid Excess

As was touched upon earlier, the role of thyroid hormones in spermatogenesis is not entirely clear. However, hyperthyroidism appears to adversely affect semen parameters. Abalovich et al. found that patients with hyperthyroidism have lower bioavailable testosterone, higher sex-hormone-binding globulin, and higher LH levels compared to controls [27]. Hyperthyroid patients were reported to have markedly impaired semen parameters, including low motility, low ejaculate volume, low sperm concentration, and abnormal morphology. The authors noted that 85% of the seminal abnormalities normalized on semen testing conducted 7–19 months after achievement of euthyroid status. A more recent study also found that hyperthyroidism can impair semen parameters [28]. The authors of this study reported that hyperthyroid patients had significantly lower sperm motility than controls. Motility was improved after euthyroid status was achieved with medical thyroid ablation. Just as with hypothyroidism, there is a scarcity of data regarding hyperthyroidism and spermatogenesis. However, the available studies seem to suggest that hyperthyroidism can adversely affect semen parameters.

Prolactin Excess

Hyperprolactinemia, an excess of the hormone prolactin, is another hormonal etiology of male infertility. The diagnosis is relatively straightforward, as hyperprolactinemia can be detected on routine serum testing, but determination of a particular etiology can be more challenging. Hyperprolactinemia can occur in the case of hypothyroidism, liver disease, stress, use of certain medications (i.e., phenothiazines, tricyclic antidepressants), and with functional pituitary adenomas (prolactinomas). Clinical suspicion must be high for excess prolactin, as the manifestations can range from asymptomatic in many patients to galactorrhea or hypoandrogenic states (i.e., low libido, erectile dysfunction) in affected patients. Patients with pituitary adenomas may also present with bilateral temporal visual field defects. This state, known as bitemporal hemianopsia, is the result of the close anatomic proximity of the pituitary gland to the optic chiasm. Growth of the pituitary tumor compresses the optic nerve, leading to visual field deficits.

Hyperprolactinemia can cause male infertility through its inhibitory effects on the hypothalamus. The high levels of prolactin suppress secretion of GnRH from the hypothalamus, which subsequently impairs the release of gonadotropins, the production of testosterone, and spermatogenesis. The multiple effects on the hypothalamic–pituitary–testicular axis can result in a patient presenting with multiple problems such as decreased libido, inability to achieve erection, and abnormal semen parameters.

Once the diagnosis of hyperprolactinemia is made, the practitioner should obtain an MRI study focusing on the pituitary gland. If a prolactinoma is found, it can be characterized based on its size and appearance. The main differentiation is between microadenomas, lesions <10 mm, and macroadenomas, which are lesions >10 mm. If a prolactinoma is discovered, medical therapy focuses on blocking the secretion of prolactin through the use of dopamine agonists. Examples of these agents include bromocriptine, cabergoline, pergolide, and quinagolide, with the most well-characterized agents being bromocriptine and cabergoline. These agonists make use of the natural inhibition of prolactin secretion by dopamine. This can actually cause regression of the tumor, although the process generally occurs over months. Possible side effects of dopamine agonists include nausea, vomiting, and postural hypotension. While inhibition of excess prolactin secretion prevents the disruption of the hypothalamic–pituitary axis, there have been few studies specifically elucidating the effects of these dopamine agonists on spermatogenesis and fertility. A 1974 study treated men with functional prolactinomas and hypogonadism with bromocriptine and found no increase in sperm motility [29]. However, more recently, DeRosa and colleagues compared bromocriptine and cabergoline in such patients. Both treatments showed overall improvements in sperm number, motility, rapid progression, and morphology within 6 months of therapy [30]. A subsequent study from the same institution compared seminal fluid parameters between men with prolactinomas and control men. After 24 months of treatment with cabergoline (initial dose 0.5 mg weekly, subsequently titrated to PRL levels), two-thirds of men showed restored gonadal function as compared against healthy control men [31].

Comparing cabergoline and bromocriptine, it appears that cabergoline is more efficacious at normalization of prolactin levels and regressing tumor burden [32]. Further, a higher percentage of patients show a clinical response to cabergoline when compared with bromocriptine. Finally, there is a higher overall rate of permanent remission and fewer side effects with cabergoline compared to bromocriptine [32]. Considering all of these findings, cabergoline is often the first therapy utilized in treating men with prolactinomas.

While treatment of prolactinomas with dopamine agonists can be effective in many cases, a significant percentage of men may still remain persistently hypogonadotropic. Recent research suggests that clomiphene citrate may be an effective treatment for these men. Ribiero and Abucham treated 14 persistently hypogonadal men with clomiphene (50 mg/day for 12 weeks) and noted both improved testosterone levels and sperm motility [33].

Ablative therapy for prolactinomas—in the form of radiation therapy or transsphenoidal resection—is also available. Ablative therapy is typically reserved for those who fail medical management. Ablative treatments remove the source of prolactin and thus the inhibition of GnRH secretion. Measurement of the patient's gonadotropin levels post-treatment remains important, as further intervention with exogenous gonadotropins may be necessary to optimize therapeutic benefit.

Expert Commentary

Medical treatments for male infertility have traditionally centered on empiric approaches to enhance spermatogenesis. Over the past two decades, improved insight has been gained into the pathophysiology of male infertility and the outcomes associated with empiric therapy. With this insight, more targeted and directed use of medical agents has been described. As such, "empiric therapy" is used with less frequency now than was the case 20 years ago. Many available medical therapies for infertile men are used to optimize the hormonal milieu and thus optimize spermatogenesis. This has indeed been the focus of this chapter. However, numerous other medical agents are used routinely to address other specific pathophysiologic conditions leading to male factor infertility. These agents include antimicrobial drugs, anti-inflammatory medications, and sympathetic agonists. Each of these classes of drugs has clear indications for use in specifically targeted subgroups of infertile men, and they are each described in other specific chapters of this text.

One important point clearly delineated in the literature over recent years is this: empiric medical therapy generally has limited utility and benefit in the treatment of infertile men. While randomized, double-blinded, placebo-controlled studies are costly in terms of time and money, they remain the proving ground for effective medical therapies. More than one agent has failed to pass this test in recent years, but this is a good thing. While the armamentarium of available medical agents for the treatment of male infertility is somewhat limited, this fact should push us to strive harder to gain enhanced insight into the pathophysiological mechanisms leading to decreased male reproductive potential. It is with this enhanced insight into the fundamental problems leading to male infertility that we will develop additional, effective medical therapies.

Five-Year View

The future for medical therapies for male infertility truly hinges on enhanced understanding of the root causes of impaired male reproduction. Much effort is now being expended by leading investigators throughout the world to gain insight into the following:

- The genetic basis of male infertility
- Environmental factors that interfere with male reproductive potential
- Reactive oxygen species and their detrimental effects on sperm DNA/sperm function
- Unexplained infertility (clarifying the pathophysiology)

Each of these broad categories certainly has potential to provide additional targets for directed medical therapies, but this will not come until we gain enhanced understanding of each of these enumerated categories.

An emerging health issue here in the United States and throughout much of the world is metabolic syndrome. As has been clearly shown in numerous studies, metabolic syndrome can negatively affect male reproductive health in a myriad of ways [34]. In addition to promoting lifestyle changes in affected patients, physicians will likely be increasingly called upon to medically address certain problems associated with this syndrome that can hinder male reproduction, such as hypogonadism, diminished libido, erectile dysfunction, and a "proinflammatory" state that may set the stage for abnormally high sperm DNA damage.

Key Issues

Targeted medical therapies for male infertility generally address endocrinopathies, infection, inflammation, and disorders of erection and ejaculation. Empiric medical therapy has gradually been replaced by medical treatments aimed at addressing specific underlying medical problems causing decreased male reproductive potential.

The armamentarium of available medical therapies for male infertility is somewhat limited in terms of numbers of patients suitable for treatment. However, enhanced understanding of the pathophysiological mechanisms causing male factor infertility should broaden opportunities for the development of specifically targeted, effective agents.

References

1. Alukal J, Lamb D, Niederberger C, Makhlouf A. Spermatogenesis in the adult. In: Lipshultz L, Howards S, Niederberger C, editors. Infertility in the male. New York: Cambridge University; 2009. p. 74–89.
2. Caroppo E. Male hypothalamic-pituitary-gonadal axis. In: Lipshultz L, Howards S, Niederberger C, editors. Infertility in the male. New York: Cambridge University; 2009. p. 14–28.
3. Kim HH, Schlegel PN. Endocrine manipulation in male infertility. Urol Clin North Am. 2008;35:303–18.
4. Sussman EM, Chudnovsky A, Niederberger CS. Hormonal evaluation of the infertile male: has it evolved? Urol Clin North Am. 2008;35:147–55.
5. Coviello AD, Matsumoto AM, Bremner WJ, et al. Low-dose human chorionic gonadotropin maintains intratesticular testosterone in normal men with testosterone-induced gonadotropin suppression. J Clin Endocrinol Metab. 2005;90:2595–602.
6. Bouloux PM, Nieschlag E, Burger HG, et al. Induction of spermatogenesis by recombinant follicle-stimulating hormone (puregon) in hypogonadotropic azoospermic men who failed to respond to human chorionic gonadotropin alone. J Androl. 2003;24: 604–11.
7. Schiff JD, Ramirez ML, Bar-Chama N. Medical and surgical management male infertility. Endocrinol Metab Clin North Am. 2007;36:313–31.

8. Liu PY, Baker HW, Jayadev V, Zacharin M, Conway AJ, Handelsman DJ. Induction of spermatogenesis and fertility during gonadotropin treatment of gonadotropin-deficient infertile men: predictors of fertility outcome. J Clin Endocrinol Metab. 2009;94:801–8.
9. Clark ABR, Vantman D, Sherins R. A low sperm concentration does not preclude fertility in men with isolated hypogonadotropic hypogonadism after gonadotropin therapy. Fertil Steril. 1988;50: 343–7.
10. Miyagawa Y, Tsujimura A, Matsumiya K, et al. Outcome of gonadotropin therapy for male hypogonadotropic hypogonadism at university affiliated male infertility centers: a 30-year retrospective study. J Urol. 2005;173:2072–5.
11. Farhat R, Al-Zidjali F, Alzahrani AS. Outcome of gonadotropin therapy for male infertility due to hypogonadotrophic hypogonadism. Pituitary. 2010;13(2):05–10.
12. Whitten SJ, Nangia AK, Kolettis PN. Select patients with hypogonadotropic hypogonadism may respond to treatment with clomiphene citrate. Fertil Steril. 2006;86:1664–8.
13. Ioannidou-Kadis S, Wright PJ, Neely RD, Quinton R. Complete reversal of adult-onset isolated hypogonadotropic hypogonadism with clomiphene citrate. Fertil Steril. 2006;86:1513e5–9.
14. Burge MR, Lanzi RA, Skarda ST, Eaton RP. Idiopathic hypogonadotropic hypogonadism in a male runner is reversed by clomiphene citrate. Fertil Steril. 1997;67:783–5.
15. Boyle K. Nonsurgical treatment: empiric therapy. In: Lipshultz L, Howards S, Niederberger C, editors. Infertility in the male. New York: Cambridge University; 2009. p. 438–53.
16. Raman JD, Schlegel PN. Aromatase inhibitors for male infertility. J Urol. 2002;167:624–9.
17. Ramasamy R, Ricci JA, Palermo GD, Gosden LV, Rosenwaks Z, Schlegel PN. Successful fertility treatment for Klinefelter's syndrome. J Urol. 2009;182:1108–13.
18. Griboff SI. Semen analysis in myxedema. Fertil Steril. 1962;13: 436–43.
19. Meeker JD, Godfrey-Bailey L, Hauser R. Relationships between serum hormone levels and semen quality among men from an infertility clinic. J Androl. 2007;28:397–406.
20. Krassas GE, Papadopoulou F, Tziomalos K, Zeginiadou T, Pontikides N. Hypothyroidism has an adverse effect on human spermatogenesis: a prospective, controlled study. Thyroid. 2008;18:1255–9.
21. Dohle GR, Smit M, Weber RF. Androgens and male fertility. World J Urol. 2003;21:341–5.
22. Turek PJ, Williams RH, Gilbaugh 3rd JH, Lipshultz LI. The reversibility of anabolic steroid-induced azoospermia. J Urol. 1995;153:1628–30.
23. Menon DK. Successful treatment of anabolic steroid-induced azoospermia with human chorionic gonadotropin and human menopausal gonadotropin. Fertil Steril. 2003;79 Suppl 3: 1659–61.
24. Tan RS, Vasudevan D. Use of clomiphene citrate to reverse premature andropause secondary to steroid abuse. Fertil Steril. 2003;79:203–5.
25. Pavlovich CP, King P, Goldstein M, Schlegel PN. Evidence of a treatable endocrinopathy in infertile men. J Urol. 2001;165: 837–41.
26. Roth MY, Amory JK, Page ST. Treatment of male infertility secondary to morbid obesity. Nat Clin Pract Endocrinol Metab. 2008;4:415–9.
27. Abalovich M, Levalle O, Hermes R, et al. Hypothalamic-pituitary-testicular axis and seminal parameters in hyperthyroid males. Thyroid. 1999;9:857–63.
28. Krassas GE, Pontikides N, Deligianni V, Miras K. A prospective controlled study of the impact of hyperthyroidism on reproductive function in males. J Clin Endocrinol Metab. 2002;87:3667–71.
29. Thorner MO, McNeilly AS, Hagan C, Besser GM. Long-term treatment of galactorrhoea and hypogonadism with bromocriptine. Br Med J. 1974;2:419–22.
30. De Rosa M, Colao A, Di Sarno A, et al. Cabergoline treatment rapidly improves gonadal function in hyperprolactinemic males: a comparison with bromocriptine. Eur J Endocrinol. 1998;138: 286–93.
31. De Rosa M, Ciccarelli A, Zarrilli S, et al. The treatment with cabergoline for 24 month normalizes the quality of seminal fluid in hyperprolactinaemic males. Clin Endocrinol (Oxf). 2006;64: 307–13.
32. Gillam MP, Molitch ME, Lombardi G, Colao A. Advances in the treatment of prolactinomas. Endocr Rev. 2006;27:485–534.
33. Ribeiro RS, Abucham J. Recovery of persistent hypogonadism by clomiphene in males with prolactinomas under dopamine agonist treatment. Eur J Endocrinol. 2009;161:163–9.
34. Kasturi SS, Tannir J, Brannigan RE. The metabolic syndrome and male infertility. J Androl. 2008;29:251–9.

Molecular Mechanisms of Antioxidants in Male Infertility

5

Kathleen Hwang and Dolores J. Lamb

Abstract

The failure to conceive within 1 year occurs in approximately 15% of couples, and approximately 50% of problems related to conception is either caused entirely by the male or is a combined problem with the male and his partner. Male infertility continues to be a clinical challenge of increasing significance. While the etiology of suboptimal semen quality is currently not completely understood, oxidative stress has demonstrated ability to affect fertility. Oxidative stress is induced by reactive oxygen species (ROS). ROS in the form of superoxide anion, hydrogen peroxide, and hydroxyl radical are formed as a by-product of oxygen metabolism. The presence of excess ROS can cause oxidative damage to lipids, proteins, and DNA. Spermatozoa, like any other aerobic cell, are constantly trying to maintain that fine balance. Abnormal ROS formation is found in up to 40% of infertile patients, with some reports suggesting an inverse relationship between seminal ROS levels and spontaneous pregnancy outcomes of infertile couples. Many studies have attempted to define the relationship between seminal ROS and IVF but have met with conflicting results. Nevertheless, a growing body of knowledge on ROS and fertility makes testing for oxidants in the semen an important part of the infertile male evaluation.

Keywords

Male infertility • Reactive oxygen species • Oxidative stress • Oxidative metabolism • Varicocele • Antioxidant supplementation • Molecular defects • Chemiluminescence assay • Leukocytospermia

The failure to conceive within 1 year occurs in approximately 15% of couples [1], and approximately 50% of problems related to conception is either caused entirely by the male or is a combined problem with the male and his partner. Male infertility continues to be a clinical challenge of increasing significance. While the etiology of suboptimal semen quality is currently not completely understood, oxidative stress has demonstrated ability to affect fertility [2, 3]. Oxidative stress is induced by reactive oxygen species (ROS). ROS in the form of superoxide anion, hydrogen peroxide, and hydroxyl radical are formed as a by-product of oxygen metabolism. The presence of excess ROS can cause oxidative damage to lipids, proteins, and DNA [4–6]. Spermatozoa, like any other aerobic cell, are constantly trying to maintain that fine balance [7]. Abnormal ROS formation is found in up to 40% of infertile patients [8], with some reports suggesting an inverse relationship between seminal ROS levels and spontaneous pregnancy outcomes of infertile couples [9]. Many studies have attempted to define the relationship between seminal

K. Hwang, MD (✉)
Department of Surgery (Urology), Brown University, 2 Dudley Street, Ste 175, Providence, RI 02905, e-mail: Kathleen_hwang@brown.edu

D.J. Lamb, PhD
Scott Department of Urology, Baylor College of Medicine, One Baylor Plaza, N730, Houston, TX 77030, USA
e-mail: dlamb@bcm.edu

S.J. Parekattil and A. Agarwal (eds.), *Male Infertility: Contemporary Clinical Approaches, Andrology, ART & Antioxidants*,
DOI 10.1007/978-1-4614-3335-4_5,

ROS and IVF [10, 11] but have met with conflicting results. Nevertheless, a growing body of knowledge on ROS and fertility makes testing for oxidants in the semen an important part of the infertile male evaluation.

Reactive Oxygen Species and Antioxidants

ROS are formed via several mechanisms but essentially are products of normal cellular metabolism. Most of the energy generated in the body is produced by the enzymatically controlled reaction of oxygen with hydrogen in oxidative phosphorylation occurring within the mitochondria [12]. Free radicals are formed during the enzymatic reduction of oxygen to produce energy [13]. A free radical is an oxygen molecule containing one or more unpaired electrons, which makes this structure highly susceptible to radical formation. The primary form of ROS, superoxide anion radical, is formed with the addition of one electron to dioxygen. Secondary ROS include the hydroxyl radical, peroxyl radical, and hydrogen peroxide. Furthermore, there is another class of free radicals derived from nitrogen, which includes nitrous oxide, peroxynitrile, nitroxyl anion, and peroxy-nitrous acid.

Oxidative stress occurs when the production of ROS overwhelms the natural antioxidant defense mechanisms leading to cellular damage. Cellular damage occurs when the free radicals donate their unpaired electron onto nearby cellular structures. To prevent cellular damage by excessive production of ROS, enzymatic and nonenzymatic antioxidant pathways scavenge surplus ROS and allow a balance to be achieved. Oxidative stress leads to the activation of transcription factors and signaling pathways, partly through the activation of the innate immune response. Such activation leads to the release of cytokines and chemokines.

These cytokines are synthesized and secreted on demand, for example, during an infection. Cytokines are also involved in gonadal and sperm function. A multitude of cytokines have been shown to be present in human semen, such as TGF-α [14] and TNF-α [15]. Various cytokines have demonstrated the ability to decrease spermatozoan motility as well as increase the generation of ROS in semen [16, 17].

Antioxidant Pathways

Fortunately, the human body has developed several antioxidant strategies to protect itself from ROS damage. These mechanisms allow for normal oxidative metabolism to occur without damaging cells while still allowing for normal cellular responses such as destruction of infectious pathogens [13]. Superoxide dismutase (SOD) and catalase are enzymatic metal-containing antioxidants, which inactivate the superoxide anion and peroxide radicals, converting them into oxygen and water. Glutathione peroxidase also participates in the reduction of hydrogen peroxide using glutathione as an electron donor.

Nonenzymatic antioxidants present within semen include vitamins E and C, glutathione, flavenoids, and albumin [18]. The majority of these participants primarily act by directly neutralizing free radical activity chemically. Several researchers have reported a significant reduction in nonenzymatic activity in seminal plasma of infertile men [19–21]. Under normal conditions, these antioxidants act to maintain this delicate balance to achieve an overall low level of oxidative stress in the semen.

Vitamin E is a major chain-breaking antioxidant and acts by trapping organic free radicals and/or deactivating excited oxygen molecules to prevent tissue damage. It specifically acts as a peroxyl and alkoxyl radical scavenger in lipid environments and thus prevents lipid peroxidation in lipoproteins and particularly membranes [22]. Similarly, vitamin C has powerful antioxidant capabilities as it neutralizes hydroxyl, superoxide, and hydrogen peroxide radicals and prevents sperm agglutination [22]. As a water-soluble antioxidant, it has also been shown to recycle oxidized vitamin E.

Sources of Reactive Oxygen Species

Oxidative stress and its role in the origins of male infertility was first established in 1943, when MacLeod demonstrated that catalase could support the motility of human spermatozoa incubated under aerobic conditions [23]. This observation then prompted him to suggest that sperm must produce hydrogen peroxide during normal oxidative metabolism. Human spermatozoa are capable of generating low levels ROS associated with the positive physiological event of sperm capacitation necessary for fertility.

During spermatogenesis, a low-level production of free radicals by sperm plays a necessary role in preparation for capacitation. Hydrogen peroxide stimulates the acrosome reaction, sperm hyperactivation [24], and tyrosine phosphorylation [25], leading to binding to the zona pellucida. Cytoplasm is normally extruded from the spermatozoa prior to release of the germinal epithelium. This cytoplasmic residue is a source of high levels of the enzyme glucose-6-phosphate dehydrogenase, which generates nicotinamide adenine dinucleotide phosphate (NADPH). NADPH then generates ROS via NADPH oxidase within the sperm membrane [26, 27].

The two main sources of free radicals within semen are leukocytes and sperm. The rate of production of ROS by leukocytes is reported to be 1,000 times higher than that of spermatozoa at capacitation [28]. This finding seems to implicate leukocytes as the likely dominant producer of seminal ROS; however, further evidence suggests that oxidative stress and consequent sperm injury is as much related to location as well as concentration. To better characterize seminal ROS production, the sources are separated into intrinsic (sperm) and extrinsic (leukocytes). Henkel et al. suggested that there is a stronger relationship for intrinsic ROS production as the more important variable in terms of fertility potential with increased levels of DNA fragmentation [29].

Although several studies have attempted to characterize the relationship between the presence of leukocytes in semen and male infertility, it still remains incompletely defined. Leukocytospermia has long been associated with decreased sperm concentration, motility, and morphology and defective fertilization; however, these clinical, epidemiological, and experimental studies have reported inconsistent results [30–32], leaving this relationship still controversial.

Lifestyle

Despite the fact that smoking is a lifestyle hazard for both the active and passive smoker, almost one-third of Americans between the ages of 18 and 24 smoke, and the number continues to rise [33]. While much is known of the carcinogenic properties of tobacco and its resultant effects on organs such as lung and bladder, the impact on fertility still remains less defined. Numerous studies, however, have established the fact that toxins in cigarette smoke reach the male reproductive system and their effects are mainly due to their direct interaction with the components of seminal fluid [34]. This interaction has led to a greater presence of ROS and increased leukocytes and rounds cells, as well as a higher frequency of DNA fragmentation in comparison to nonsmokers [35–37]. In smokers, either male or female, there was noted to be a significant delay of over 6 months in natural conception in comparison to nonsmokers [38].

Nicotine and its metabolites are present in the spermatozoa of smokers, and furthermore, these toxicants are found in embryos resulting from in vitro fertilization cycles with male smokers [39]. Cigarette smoking results in a 48% increase in seminal leukocyte concentration [37], and smokers have decreased levels of seminal plasma antioxidants such as vitamin E and vitamin C, placing their sperm at additional risk for oxidative damage [40, 41].

Ethanol is one of the most abused substances worldwide. It has been shown to have deleterious effects at all levels of the male reproductive system. Excessive alcohol consumption causes an increase in systemic oxidative stress as ethanol stimulates the production of ROS, confounded by the antioxidant-deficient diets that most alcohol abusers maintain [42, 43]. However, despite the demonstrated link, there has been no study to date that has examined the direct link between alcohol intake and sperm oxidative damage.

Environmental

Phthalates are among the most widely used man-made chemicals released into the environment over the last several decades. They are primarily used as plasticizers in the manufacture of flexible vinyl, which is found in medical devices, toys, floor and wall coverings, personal care products, and food packaging. Ubiquitous use of phthalates results in exposure mainly through dietary consumption, dermal absorption, or inhalation and has been linked with impaired spermatogenesis and increased sperm DNA damage [44, 45]. Lee et al. reported an increase in the generation of ROS within the testis after oral administration of phthalate esters to rats, leading to a concomitant decrease in antioxidant levels and subsequent impaired spermatogenesis [46].

Furthermore, several environmental pesticides [47, 48] as well as heavy metal exposure [49] have been linked with testicular oxidative stress and subsequent sperm oxidative damage. Patients employed in industries with high levels of exposure risks should be counseled to take aggressive precautionary measures to avoid future contact as well as be tested for heavy metals.

Infection/Inflammation

The invasion of microorganisms into tissue results in a natural defense mechanism which includes the oxidative burst of leukocytes and macrophages [50]. Mazzilli et al. demonstrated significantly elevated superoxide anion generation in patients with sperm cultures positive for aerobic bacteria compared with fertile controls [51]. Men prone to genitourinary infection, such as paraplegics, have been found to have high degrees of sperm oxidative pathology [52] and were negatively correlated with sperm motility and independent of the method of ejaculation [53]. *Chlamydial* infections, current or past, have also been linked with an increase in oxidative damage to sperm [54].

Up to 50% of men at some point in their lives will experience prostatitis, and chronic nonbacterial prostatitis will account for in excess of 90% of all cases [55]. It is reported that an adverse autoimmune response to seminal or prostate antigens is responsible for the pathology leading to an increase in proinflammatory cytokines and activated ROS [56, 57]. These antigens will stimulate a release of T lymphocytes followed by a subsequent liberation of cytokines such as TNF-α, IFN-γ, and IL-1β that stimulates chemotaxis and activation of further leukocytes leading to the resultant oxidative stress [56]. Hence, it is not so surprising that many studies link chronic nonbacterial prostatitis with a reduction in sperm density, motility, and morphology [58, 59].

Testicular

Oxidative stress is now believed to be the principal underlying pathology linking varicoceles with male infertility [60–62].

Clinical varicoceles are one of the most common causes of male infertility. There is a strong correlation between the increase in varicocele-related ROS production and subsequent reduction in sperm DNA integrity when assessed by terminal deoxynucleotidyl transferase dUTP nick end labeling (TUNEL) [62]. Several researchers have also proposed that oxidative stress plays a significant role for infertility following vasectomy reversal. Many believe that vasectomy disrupts the normal blood–testis barrier, leading to a loss of immune integrity and activation of immune responses [63]. Multiple studies have demonstrated an increase in seminal leukocytes, proinflammatory cytokines, and free radical production following vasectomy reversal [64, 65].

Cryptorchidism has been established as a common cause for male factor infertility wherein the primary pathology is hypospermatogenesis due to deficient maturation of gonocytes [66]. It has also recently been reported that men with cryptorchidism surgically treated with orchidopexy early in life still have markedly elevated sperm ROS production and DNA fragmentation compared with fertile controls [67].

Prolonged periods of ischemia followed by spontaneous or surgical restoration of blood flow results in an influx of leukocytes into both testicles [68] and a subsequent increase in generation of free radicals [69]. The annual incidence of torsion of the spermatic cord is 1 in 4,000 males and has been long recognized as a cause of male infertility. This ischemic-reperfusion injury model leads to necrosis of the germinal cells with resulting subfertility or infertility.

Molecular Defects

Experimental evidence is emerging that ROS are involved in several fundamental mechanisms of sperm physiology. However, at present, the precise mechanisms are still not completely defined and still under investigation. Nonetheless, there has been convincing data reported in animal models that may highlight specific potential molecular relationships.

Mitochondria

Mitochondria are responsible for the generation of the majority of adenosine triphosphate (ATP) and are crucial as human cells rely on ATP for growth, differentiation, cellular homeostasis, and several physiological functions. They also serve as an important source of ROS within most mammalian cells [70]. Spermatozoa have mitochondria uniquely located around the midpiece to be precisely at the site of maximal energy requirement. Mitochondrial DNA (mtDNA) accumulates polymorphisms and mutations about 10–17 times faster than nuclear DNA [71]. Several studies have reported that human cells harboring mutated mtDNA have lower respiratory function and show increased production of superoxide anions, hydroxyl radicals, and hydrogen peroxide [72, 73].

Another investigator reported a correlation between ROS and mitochondria in apoptosis, where high levels of ROS were found to disrupt the inner and outer mitochondrial membrane resulting in a release of cytochrome C [74]. Cytochrome C protein activates the caspases and induces apoptosis, which was also noted to be higher in infertile men with elevated levels of ROS. A recent study in mice carrying different proportions of pathogenic mtDNA showed respiratory chain defects that lead to meiotic arrest during spermatogenesis and subsequent oligospermia, asthenospermia, and teratospermia [75]. An increased number of mtDNA mutations may also lead to abnormal sperm morphology and ultrastructural defects [76].

DNA Damage

Cells are constantly being exposed to environmental and endogenous stressors, such as alkylating agents, ROS, and other active metabolites that are capable of causing DNA damage. During replication and repair, lesions induced by these toxic compounds can produce alternate nucleotides, potentially leading to permanent alterations in genetic material. Fortunately, there are intrinsic repair mechanisms in place such as base excision repair (BER). Although DNA polymerase β is known to be the main polymerase in the BER pathway, there are multiple other DNA polymerases that participate in this process as well.

Braithwaite et al. [77] recently examined the interrelationship between these enzymes in mammalian cells and their effect on oxidative stress-induced mutagenesis. They focused in on DNA polymerase β and DNA polymerase λ and were able to generate double knockout mouse embryonic fibroblast cell lines. These cell lines were examined for sensitivity to 3 DNA-damaging agents: an alkylating agent, hydrogen peroxide, and a thymidine analog. The double knockout cell lines demonstrated a hypersensitivity to hydrogen peroxide in comparison to wild-type mice indicating significant roles of DNA polymerases β and λ in protection of these cells against toxic agents, particularly hydrogen peroxide.

OGG1 and MYH are two major enzymes involved in mammalian oxidative DNA damage repair, most commonly preventing G to T mutations [78, 79]. Xie et al. [80] examined the roles of MYH and OGG1 in the protection against oxidative stress. They generated double knockout mice and developed embryonic fibroblast cell lines and examined their phenotypes associated with oxidative stress. The cells were exposed to hydrogen peroxide, *cis*-platinum, and γ-irradiation. The cells had significantly increased sensitivity to hydrogen peroxide exposure. Their findings that having both these deficiencies contribute to centrosome amplification and multinuclear formation, suggesting that *myh* and *ogg1* are likely required for normal cell cycle progression and cell division under oxidative stress.

Impact on Semen Quality

The structure of the sperm membrane is unique as it contains large amounts of unsaturated fatty acids rendering them particularly vulnerable to oxidative stress. These unsaturated fatty acids provide fluidity necessary for membrane fusion events; however, these molecules also make them more susceptible to free radical attack. Seminal fluid is an important source of antioxidants and is key in protecting spermatozoa from oxidative injury [81, 82]. SOD, catalase, glutathione peroxidase, vitamin E, and vitamin C are all contained in seminal fluid. This is particularly important as spermatozoa, unlike most cells, have little cytoplasmic fluid and subsequently contain only minimal amounts of the critical ROS scavenging enzymes and pathways, leaving them virtually no capacity for protein synthesis and little antioxidant capacity [81].

Elevated levels of ROS have been detected in the semen of 25% of infertile men [8, 81, 83] (Table 5.1). These elevated levels of ROS have been correlated with sperm DNA damage, although no ROS threshold above which sperm DNA damage is detected has been established [61, 84, 85]. DNA damage in male germline cells is associated with poor fertilization rates following IVF, defective preimplantation embryonic development, and high rates of miscarriage. Several studies have demonstrated that ROS can induce sperm DNA damage in vitro, supporting that ROS may play a role in the etiology of sperm DNA damage in infertile men [18, 86]. There has also been evidence to suggest that ROS generation has a central role in the pathophysiology of age-related decrease in male fertility (reviewed in Desai et al. [87]). Several studies have also evaluated the relationship between semen antioxidant levels and sperm DNA damage and have reported conflicting results.

Increased ROS levels have also been associated with decreased sperm motility [88–90]. Proposed mechanisms by which ROS reduces sperm motility is by decreasing axonemal protein phosphorylation as well as lipid peroxidation [91]. The ability to revive sperm motility both in vivo and in vitro with antioxidants, such as vitamin E, confirms the evidence that lipid peroxidation is a major cause of motility loss in spermatozoa [92].

Detection of Oxidative Stress in Male Infertility

There are various methods to identify seminal ROS; however, the most common method is via the indirect chemiluminescence assay. These tests have been designed to quantify the level of oxidative stress in men undergoing evaluation for infertility with the goal of deriving a therapeutic plan to decrease oxidative stress levels and improve overall sperm quality. The chemiluminescence assay utilizes a luminometer to measure chemical reactions between ROS found in human semen and a chemiluminescent probe, such as luminol or lucigenin. Luminol is an uncharged particle that is cell membrane permeable and therefore can react extracellularly and intracellularly with hydrogen peroxide, hydroxyl anions, and superoxide anions. In contrast, lucigenin is a positively charged particle that is membrane impermeable and reacts with superoxide anions in the extracellular space [93]; therefore, luminol is used more commonly because of its ability to measure extracellular as well as intracellular levels of ROS.

Leukocyte contamination in the semen has been shown to negatively impact fertility [31]. Leukocytes have been shown to be responsible for a significant proportion of ROS activity in the semen [32, 94]; therefore, these assays should be coupled with selective leukocyte removal strategies if pyospermia is present [95]. Not doing so would lead to a falsely elevated ROS value. Other factors that may spuriously increase ROS results include repeated centrifugation of the sample and the use of certain oxidase-containing buffers for sample preparation [96, 97]. In contrast, prolonged time from preparation to

Table 5.1 Linking oxidative stress and male infertility

Author	Year	*n*	Findings
Athayde et al. [83]	2007	47	Infertile men with significantly ↑ levels of ROS compared to fertile controls
Mostafa et al. [41]	2006	68	Infertile men with significantly ↓ levels of protective antioxidants in semen compared to fertile controls
Twigg et al. [18]	1998	0	In vitro generation of ROS associated with evidence of sperm peroxidation, ↓ motility
Saleh et al. [61]	2003	92	Seminal ROS in infertile men correlated with ↓ motility and ↑ sperm membrane oxidation
Keskes-Ammar et al. [107]	2003	54	Antioxidant treatment may improve sperm motility
Twigg et al. [18]	1998	0	In vitro generation of ROS associated with ↑ in sperm DNA damage
Henkel et al. [29]	2005	63	Infertile men with seminal ROS is correlated with ↑ in sperm DNA damage
Greco et al. [108]	2005	64	Antioxidant treatment in infertile men may improve sperm DNA quality
Tremellen et al. [110]	2007	60	The use of antioxidant supplements by infertile men may ↑ partner's chances of spontaneous or IVF assisted pregnancy

analysis of the sample can artificially decrease the ROS identifiable in the semen [98]. For this reason, it is recommended that testing be performed within an hour of sperm preparation [99]. Finally, poor liquefaction of the sample can interfere with the normal oxidative process resulting in a falsely lower ROS value [97].

Management of Oxidative Stress in Male Infertility

As our clinical understanding of the impact of oxidative stress on male fertility expands, the natural desire to define and provide more effective therapeutic options for these patients also increases. Primarily, treatment should be aimed at identification and amelioration of any underlying cause before implementing any antioxidant treatment. While many trials have been performed to address this exact situation, many of these studies are difficult to interpret. However, the low cost and toxicity range of the majority of these antioxidant therapies offers a huge appeal for both patients and clinicians.

Lifestyle Modifications

Lifestyle choices such as smoking, poor diet, alcohol abuse, obesity, and even exposure to environmental toxins have all been linked with oxidative stress. First and foremost, patients should be counseled on making positive lifestyle changes such as a diet high in fruits and vegetables, and a reduction or cessation in smoking and alcohol intake. However, there have been conflicting results in the literature regarding the benefits of dietary supplements or individual vitamins [100, 101]. Patients exposed to occupational or environmental toxins should be counseled and reeducated on proper ventilation and use of personal protective equipment at work.

Treatment of Infection/Inflammation

Chlamydia and *Ureaplasma* infections within the semen and male accessory glands have been definitively linked to an increase in oxidative stress [51, 54]. Antibiotics can effectively treat both of these pathogens, and recent studies have confirmed the ability of appropriate antibiotic treatment to improve sperm quality by reducing sperm oxidative stress [102, 103]. Men with either *Chlamydia* or *Ureaplasma* were randomized to receive no treatment vs. antibiotic treatment for 3 months. The antibiotic-treated group demonstrated a significant reduction in seminal leukocyte and ROS production compared with the controls. They further exhibited an improvement in sperm motility and natural conception [103].

Nonsteroidal anti-inflammatory (NSAID) drugs may also reduce free radical production by seminal leukocytes. Gambera et al. reported that a 1-month course of a COX-2 inhibitor was able to significantly improve sperm motility, morphology, and viability while reducing the sperm leukocyte count [104].

Antioxidant Supplementation

Oral antioxidant supplements have been reported to augment the scavenging capacity of seminal plasma thereby reducing levels of ROS within the semen. The oral antioxidants vitamin E, vitamin C, β-carotene, and acetyl-cysteine are all potent scavengers of ROS and have all been shown to reduce seminal ROS levels [105]. A randomized control study evaluated 3 months of vitamin E treatment with placebo, confirming this reduction in seminal ROS levels [106]. Keskes-Ammar et al. randomized 54 men to either vitamin E and selenium or vitamin B for 3 months, examining the semen for quantitative levels of MDA, a lipid peroxidation marker. While less than half of the patients completed the study, results revealed that vitamin E and selenium supplementation produced a significant decrease in MDA concentrations with improved sperm quality in comparison to vitamin B [107]. Most recently, a well-designed prospective, randomized controlled trial of 2 months treatment with vitamin C and vitamin E demonstrated a very significant reduction in percent of DNA-fragmented spermatozoa [108]; however, no differences in basic sperm parameters were noted.

While many studies have shown significant improvements in sperm motility with antioxidant supplementation [107, 109, 110], other authors have not found any changes in semen parameters nor any pregnancies initiated during this period in prospective randomized trials [111]. Although these data seemingly conflict, it is difficult to directly compare the results due to varying study design. While the data supporting improvements in sperm quality with antioxidant treatment is more established, the ability of these changes to translate into improved chances of pregnancy is less clear. Nevertheless, the majority of the studies provide persuasive evidence to support the efficacy of antioxidants, particularly the vitamins, on improving overall sperm quality.

Varicocele

Several studies have concluded that surgical treatment of a varicocele is highly effective and can reduce seminal ROS levels and improve sperm DNA integrity [112, 113]. Using the Cochrane Menstrual Disorders and Subfertility Group register, Evers et al. performed a meta-analysis which did not show any benefit [114]. However, the most recent meta-analysis examining the effect of varicocele repair on spontaneous conception by Marmar et al. noted that men undergoing varicocelectomy demonstrated lower oxidative stress and a significant benefit in spontaneous conception in comparison to the control group [115].

Key Issues

Over the past decade, there has been an expanding body of evidence that now supports a role for oxidative stress as a significant cause of male infertility, as well as refining the relationship between the two. Despite our increased understanding of the role of ROS on male fertility, many questions still remain unanswered. Well-designed randomized controlled trials will be required to truly assess the potential of these antioxidants alone or in combination to be able to derive consistent clinical guidelines of therapy. Nonetheless, the evaluation of oxidative stress in male infertility patients should be routine in clinical practice.

Acknowledgments This study was supported in part by the 1K12DK083014 Multidisciplinary K12 Urology Research Career Development Program at Baylor to D.J.L. and K.H. from the National Institute of Kidney and Digestive Diseases.

References

1. Thonneau P, Marchand S, Tallec A, et al. Incidence and main causes of infertility in a resident population (1,850,000) of three French regions (1988–1989). Hum Reprod. Jul 1991;6(6):811–6.
2. Aitken RJ. A free radical theory of male infertility. Reprod Fertil Dev. 1994;6(1):19–23. discussion 23–14.
3. Vine MF. Smoking and male reproduction: a review. Int J Androl. Dec 1996;19(6):323–37.
4. Aitken RJ, Baker MA, Sawyer D. Oxidative stress in the male germ line and its role in the aetiology of male infertility and genetic disease. Reprod Biomed Online. 2003;7(1):65–70.
5. Griveau JF, Le Lannou D. Reactive oxygen species and human spermatozoa: physiology and pathology. Int J Androl. 1997;20(2): 61–9.
6. Henkel R, Hajimohammad M, Stalf T, et al. Influence of deoxyribonucleic acid damage on fertilization and pregnancy. Fertil Steril. 2004;81(4):965–72.
7. Sies H. Strategies of antioxidant defense. Eur J Biochem. 1993; 215(2):213–9.
8. Iwasaki A, Gagnon C. Formation of reactive oxygen species in spermatozoa of infertile patients. Fertil Steril. 1992;57(2):409–16.
9. Aitken RJ, Irvine DS, Wu FC. Prospective analysis of sperm-oocyte fusion and reactive oxygen species generation as criteria for the diagnosis of infertility. Am J Obstet Gynecol. 1991;164(2): 542–51.
10. Sukcharoen N, Keith J, Irvine DS, Aitken RJ. Prediction of the in-vitro fertilization (IVF) potential of human spermatozoa using sperm function tests: the effect of the delay between testing and IVF. Hum Reprod. 1996;11(5):1030–4.
11. Hammadeh ME, Radwan M, Al-Hasani S, et al. Comparison of reactive oxygen species concentration in seminal plasma and semen parameters in partners of pregnant and non-pregnant patients after IVF/ICSI. Reprod Biomed Online. 2006;13(5): 696–706.
12. Tremellen K. Oxidative stress and male infertility–a clinical perspective. Hum Reprod Update. 2008;14(3):243–58.
13. Valko M, Leibfritz D, Moncol J, Cronin MT, Mazur M, Telser J. Free radicals and antioxidants in normal physiological functions and human disease. Int J Biochem Cell Biol. 2007;39(1):44–84.
14. Yie SM, Lobb DK, Clark DA, Younglai EV. Identification of a transforming growth factor alpha-like molecule in human seminal plasma. Fertil Steril. 1994;61(1):129–35.
15. Hussenet F, Dousset B, Cordonnier JL, et al. Tumour necrosis factor alpha and interleukin 2 in normal and infected human seminal fluid. Hum Reprod. 1993;8(3):409–11.
16. Fedder J, Ellerman-Eriksen S. Effect of cytokines on sperm motility and ionophore-stimulated acrosome reaction. Arch Androl. 1995;35(3):173–85.
17. Rajasekaran M, Hellstrom WJ, Naz RK, Sikka SC. Oxidative stress and interleukins in seminal plasma during leukocytospermia. Fertil Steril. 1995;64(1):166–71.
18. Twigg J, Fulton N, Gomez E, Irvine DS, Aitken RJ. Analysis of the impact of intracellular reactive oxygen species generation on the structural and functional integrity of human spermatozoa: lipid peroxidation, DNA fragmentation and effectiveness of antioxidants. Hum Reprod. 1998;13(6):1429–36.
19. Song GJ, Norkus EP, Lewis V. Relationship between seminal ascorbic acid and sperm DNA integrity in infertile men. Int J Androl. 2006;29(6):569–75.
20. Lewis SE, Sterling ES, Young IS, Thompson W. Comparison of individual antioxidants of sperm and seminal plasma in fertile and infertile men. Fertil Steril. 1997;67(1):142–7.
21. Fraga CG, Motchnik PA, Shigenaga MK, Helbock HJ, Jacob RA, Ames BN. Ascorbic acid protects against endogenous oxidative DNA damage in human sperm. Proc Natl Acad Sci USA. 1991;88(24):11003–6.
22. Agarwal A, Nallella KP, Allamaneni SS, Said TM. Role of antioxidants in treatment of male infertility: an overview of the literature. Reprod Biomed Online. 2004;8(6):616–27.
23. MacLeod J. The role of oxygen metabolism and motility of human spermatozoa. Am J Physiol. 1943;138:512–8.
24. de Lamirande E, Gagnon C. Human sperm hyperactivation and capacitation as parts of an oxidative process. Free Radic Biol Med. 1993;14(2):157–66.
25. Aitken RJ, Buckingham DW, Brindle J, Gomez E, Baker HW, Irvine DS. Analysis of sperm movement in relation to the oxidative stress created by leukocytes in washed sperm preparations and seminal plasma. Hum Reprod. 1995;10(8):2061–71.
26. Fisher HM, Aitken RJ. Comparative analysis of the ability of precursor germ cells and epididymal spermatozoa to generate reactive oxygen metabolites. J Exp Zool. 1997;277(5):390–400.
27. Said TM, Agarwal A, Sharma RK, Thomas Jr AJ, Sikka SC. Impact of sperm morphology on DNA damage caused by oxidative stress induced by beta-nicotinamide adenine dinucleotide phosphate. Fertil Steril. 2005;83(1):95–103.
28. Plante M, de Lamirande E, Gagnon C. Reactive oxygen species released by activated neutrophils, but not by deficient spermatozoa, are sufficient to affect normal sperm motility. Fertil Steril. 1994;62(2):387–93.
29. Henkel R, Kierspel E, Stalf T, et al. Effect of reactive oxygen species produced by spermatozoa and leukocytes on sperm functions in non-leukocytospermic patients. Fertil Steril. 2005;83(3):635–42.
30. Moskovtsev SI, Willis J, White J, Mullen JB. Leukocytospermia: relationship to sperm deoxyribonucleic acid integrity in patients evaluated for male factor infertility. Fertil Steril. 2007;88(3): 737–40.
31. Wolff H. The biologic significance of white blood cells in semen. Fertil Steril. 1995;63(6):1143–57.

32. Saleh RA, Agarwal A, Kandirali E, et al. Leukocytospermia is associated with increased reactive oxygen species production by human spermatozoa. Fertil Steril. 2002;78(6):1215–24.
33. Orleans CT. Increasing the demand for and use of effective smoking-cessation treatments reaping the full health benefits of tobacco-control science and policy gains–in our lifetime. Am J Prev Med. 2007;33(6 Suppl):S340–348.
34. Kulikauskas V, Blaustein D, Ablin RJ. Cigarette smoking and its possible effects on sperm. Fertil Steril. 1985;44(4):526–8.
35. Sepaniak S, Forges T, Fontaine B, et al. Negative impact of cigarette smoking on male fertility: from spermatozoa to the offspring. J Gynecol Obstet Biol Reprod (Paris). 2004;33(5):384–90.
36. Trummer H, Habermann H, Haas J, Pummer K. The impact of cigarette smoking on human semen parameters and hormones. Hum Reprod. 2002;17(6):1554–9.
37. Saleh RA, Agarwal A, Sharma RK, Nelson DR, Thomas Jr AJ. Effect of cigarette smoking on levels of seminal oxidative stress in infertile men: a prospective study. Fertil Steril. 2002;78(3):491–9.
38. Hull MG, North K, Taylor H, Farrow A, Ford WC. Delayed conception and active and passive smoking. The Avon Longitudinal Study of Pregnancy and Childhood Study Team. Fertil Steril. 2000;74(4):725–33.
39. Zenzes MT, Puy LA, Bielecki R, Reed TE. Detection of benzo[a] pyrene diol epoxide-DNA adducts in embryos from smoking couples: evidence for transmission by spermatozoa. Mol Hum Reprod. 1999;5(2):125–31.
40. Fraga CG, Motchnik PA, Wyrobek AJ, Rempel DM, Ames BN. Smoking and low antioxidant levels increase oxidative damage to sperm DNA. Mutat Res. 1996;351(2):199–203.
41. Mostafa T, Tawadrous G, Roaia MM, Amer MK, Kader RA, Aziz A. Effect of smoking on seminal plasma ascorbic acid in infertile and fertile males. Andrologia. 2006;38(6):221–4.
42. Villalta J, Ballesca JL, Nicolas JM, Martinez de Osaba MJ, Antunez E, Pimentel C. Testicular function in asymptomatic chronic alcoholics: relation to ethanol intake. Alcohol Clin Exp Res. 1997;21(1):128–33.
43. Wu D, Cederbaum AI. Alcohol, oxidative stress, and free radical damage. Alcohol Res Health. 2003;27(4):277–84.
44. Hauser R, Meeker JD, Singh NP, et al. DNA damage in human sperm is related to urinary levels of phthalate monoester and oxidative metabolites. Hum Reprod. 2007;22(3):688–95.
45. Duty SM, Silva MJ, Barr DB, et al. Phthalate exposure and human semen parameters. Epidemiology. 2003;14(3):269–77.
46. Lee E, Ahn MY, Kim HJ, et al. Effect of di(*n*-butyl) phthalate on testicular oxidative damage and antioxidant enzymes in hyperthyroid rats. Environ Toxicol. 2007;22(3):245–55.
47. Chitra KC, Sujatha R, Latchoumycandane C, Mathur PP. Effect of lindane on antioxidant enzymes in epididymis and epididymal sperm of adult rats. Asian J Androl. 2001;3(3):205–8.
48. Latchoumycandane C, Chitra KC, Mathur PP. 2,3,7,8-Tetrachlorodibenzo- p-dioxin (TCDD) induces oxidative stress in the epididymis and epididymal sperm of adult rats. Arch Toxicol. 2003;77(5):280–4.
49. Hsu PC, Guo YL. Antioxidant nutrients and lead toxicity. Toxicology. 2002;180(1):33–44.
50. Roos D. The involvement of oxygen radicals in microbicidal mechanisms of leukocytes and macrophages. Klin Wochenschr. 1991;69(21–23):975–80.
51. Mazzilli F, Rossi T, Marchesini M, Ronconi C, Dondero F. Superoxide anion in human semen related to seminal parameters and clinical aspects. Fertil Steril. 1994;62(4).862–8.
52. Brackett NL, Ibrahim E, Grotas JA, Aballa TC, Lynne CM. Higher sperm DNA damage in semen from men with spinal cord injuries compared with controls. J Androl. 2008;29(1):93–9. discussion 100–101.
53. Padron OF, Brackett NL, Sharma RK, Lynne CM, Thomas Jr AJ, Agarwal A. Seminal reactive oxygen species and sperm motility and morphology in men with spinal cord injury. Fertil Steril. 1997;67(6):1115–20.
54. Segnini A, Camejo MI, Proverbio F. Chlamydia trachomatis and sperm lipid peroxidation in infertile men. Asian J Androl. 2003;5(1):47–9.
55. Schaeffer AJ. Epidemiology and demographics of prostatitis. Andrologia. 2003;35(5):252–7.
56. Motrich RD, Maccioni M, Molina R, et al. Reduced semen quality in chronic prostatitis patients that have cellular autoimmune response to prostate antigens. Hum Reprod. 2005;20(9):2567–72.
57. Batstone GR, Doble A, Gaston JS. Autoimmune T cell responses to seminal plasma in chronic pelvic pain syndrome (CPPS). Clin Exp Immunol. 2002;128(2):302–7.
58. Henkel R, Ludwig M, Schuppe HC, Diemer T, Schill WB, Weidner W. Chronic pelvic pain syndrome/chronic prostatitis affect the acrosome reaction in human spermatozoa. World J Urol. 2006;24(1): 39–44.
59. Krieger JN, Berger RE, Ross SO, Rothman I, Muller CH. Seminal fluid findings in men with nonbacterial prostatitis and prostatodynia. J Androl. 1996;17(3):310–8.
60. Barbieri ER, Hidalgo ME, Venegas A, Smith R, Lissi EA. Varicocele-associated decrease in antioxidant defenses. J Androl. 1999;20(6):713–7.
61. Saleh RA, Agarwal A, Nada EA, et al. Negative effects of increased sperm DNA damage in relation to seminal oxidative stress in men with idiopathic and male factor infertility. Fertil Steril. 2003;79 Suppl 3:1597–605.
62. Smith R, Kaune H, Parodi D, et al. Increased sperm DNA damage in patients with varicocele: relationship with seminal oxidative stress. Hum Reprod. 2006;21(4):986–93.
63. Filippini A, Riccioli A, Padula F, et al. Control and impairment of immune privilege in the testis and in semen. Hum Reprod Update. 2001;7(5):444–9.
64. Kolettis PN, Sharma RK, Pasqualotto FF, Nelson D, Thomas Jr AJ, Agarwal A. Effect of seminal oxidative stress on fertility after vasectomy reversal. Fertil Steril. 1999;71(2):249–55.
65. Shapiro RH, Muller CH, Chen G, Berger RE. Vasectomy reversal associated with increased reactive oxygen species production by seminal fluid leukocytes and sperm. J Urol. 1998;160(4):1341–6.
66. Huff DS, Hadziselimovic F, Snyder III HM, Blyth B, Duckett JW. Early postnatal testicular maldevelopment in cryptorchidism. J Urol Aug. 1991;146(2 (Pt 2)):624–6.
67. Smith R, Kaune H, Parodi D, et al. Extent of sperm DNA damage in spermatozoa from men examined for infertility. Relationship with oxidative stress. Rev Med Chil. 2007;135(3):279–86.
68. Turner TT, Bang HJ, Lysiak JL. The molecular pathology of experimental testicular torsion suggests adjunct therapy to surgical repair. J Urol. 2004;172(6 Pt 2):2574–8.
69. Filho DW, Torres MA, Bordin AL, Crezcynski-Pasa TB, Boveris A. Spermatic cord torsion, reactive oxygen and nitrogen species and ischemia-reperfusion injury. Mol Aspects Med. 2004;25(1–2): 199–210.
70. Balaban RS, Nemoto S, Finkel T. Mitochondria, oxidants, and aging. Cell. 2005;120(4):483–95.
71. Bogenhagen DF. Repair of mtDNA in vertebrates. Am J Hum Genet. 1999;64(5):1276–81.
72. Liu CY, Lee CF, Hong CH, Wei YH. Mitochondrial DNA mutation and depletion increase the susceptibility of human cells to apoptosis. Ann N Y Acad Sci. 2004;1011:133–45.
73. Taylor RW, Turnbull DM. Mitochondrial DNA mutations in human disease. Nat Rev Genet. 2005;6(5):389–402.
74. Wang X, Sharma RK, Gupta A, et al. Alterations in mitochondria membrane potential and oxidative stress in infertile men: a prospective observational study. Fertil Steril. 2003;80 Suppl 2:844–50.
75. Nakada K, Sato A, Yoshida K, et al. Mitochondria-related male infertility. Proc Natl Acad Sci USA. 2006;103(41):15148–53.

76. Shamsi MB, Kumar R, Bhatt A, et al. Mitochondrial DNA Mutations in etiopathogenesis of male infertility. Indian J Urol. 2008;24(2):150–4.
77. Braithwaite EK, Kedar PS, Stumpo DJ, et al. DNA polymerases beta and lambda mediate overlapping and independent roles in base excision repair in mouse embryonic fibroblasts. PLoS One. 2010;5(8):e12229.
78. Yang H, Clendenin WM, Wong D, et al. Enhanced activity of adenine-DNA glycosylase (Myh) by apurinic/apyrimidinic endonuclease (Ape1) in mammalian base excision repair of an A/GO mismatch. Nucleic Acids Res. 2001;29(3):743–52.
79. Slupska MM, Luther WM, Chiang JH, Yang H, Miller JH. Functional expression of hMYH, a human homolog of the Escherichia coli MutY protein. J Bacteriol. 1999;181(19):6210–3.
80. Xie Y, Yang H, Miller JH, et al. Cells deficient in oxidative DNA damage repair genes Myh and Ogg1 are sensitive to oxidants with increased G2/M arrest and multinucleation. Carcinogenesis. 2008;29(4):722–8.
81. Zini A, de Lamirande E, Gagnon C. Reactive oxygen species in semen of infertile patients: levels of superoxide dismutase- and catalase-like activities in seminal plasma and spermatozoa. Int J Androl. 1993;16(3):183–8.
82. Jeulin C, Soufir JC, Weber P, Laval-Martin D, Calvayrac R. Catalase activity in human spermatozoa and seminal plasma. Gamete Res. 1989;24(2):185–96.
83. Athayde KS, Cocuzza M, Agarwal A, et al. Development of normal reference values for seminal reactive oxygen species and their correlation with leukocytes and semen parameters in a fertile population. J Androl. 2007;28(4):613–20.
84. Barroso G, Morshedi M, Oehninger S. Analysis of DNA fragmentation, plasma membrane translocation of phosphatidylserine and oxidative stress in human spermatozoa. Hum Reprod. 2000;15(6): 1338–44.
85. Irvine DS, Twigg JP, Gordon EL, Fulton N, Milne PA, Aitken RJ. DNA integrity in human spermatozoa: relationships with semen quality. J Androl. 2000;21(1):33–44.
86. Sawyer DE, Mercer BG, Wiklendt AM, Aitken RJ. Quantitative analysis of gene-specific DNA damage in human spermatozoa. Mutat Res. 2003;529(1–2):21–34.
87. Desai N, Sabanegh Jr E, Kim T, Agarwal A. Free radical theory of aging: implications in male infertility. Urology. 2010;75(1): 14–9.
88. Lenzi A, Lombardo F, Gandini L, Alfano P, Dondero F. Computer assisted sperm motility analysis at the moment of induced pregnancy during gonadotropin treatment for hypogonadotropic hypogonadism. J Endocrinol Invest. 1993;16(9):683–6.
89. Agarwal A, Ikemoto I, Loughlin KR. Relationship of sperm parameters with levels of reactive oxygen species in semen specimens. J Urol. 1994;152(1):107–10.
90. Armstrong JS, Rajasekaran M, Chamulitrat W, Gatti P, Hellstrom WJ, Sikka SC. Characterization of reactive oxygen species induced effects on human spermatozoa movement and energy metabolism. Free Radic Biol Med. 1999;26(7–8):869–80.
91. Agarwal A, Makker K, Sharma R. Clinical relevance of oxidative stress in male factor infertility: an update. Am J Reprod Immunol. 2008;59(1):2–11.
92. Suleiman SA, Ali ME, Zaki ZM, el-Malik EM, Nasr MA. Lipid peroxidation and human sperm motility: protective role of vitaminE. J Androl. 1996;17(5):530–7.
93. McKinney KA, Lewis SE, Thompson W. Reactive oxygen species generation in human sperm: luminol and lucigenin chemiluminescence probes. Arch Androl. 1996;36(2):119–25.
94. Aitken RJ, West KM. Analysis of the relationship between reactive oxygen species production and leucocyte infiltration in fractions of human semen separated on Percoll gradients. Int J Androl. 1990;13(6):433–51.
95. Aitken RJ, Buckingham DW, West K, Brindle J. On the use of paramagnetic beads and ferrofluids to assess and eliminate the leukocytic contribution to oxygen radical generation by human sperm suspensions. Am J Reprod Immunol. 1996;35(6):541–51.
96. Shekarriz M, DeWire DM, Thomas Jr AJ, Agarwal A. A method of human semen centrifugation to minimize the iatrogenic sperm injuries caused by reactive oxygen species. Eur Urol. 1995;28(1):31–5.
97. Aitken RJ, Baker MA, O'Bryan M. Shedding light on chemiluminescence: the application of chemiluminescence in diagnostic andrology. J Androl. 2004;25(4):455–65.
98. Aitken RJ, Clarkson JS. Cellular basis of defective sperm function and its association with the genesis of reactive oxygen species by human spermatozoa. J Reprod Fertil. 1987;81(2):459–69.
99. Kobayashi H, Gil-Guzman E, Mahran AM, et al. Quality control of reactive oxygen species measurement by luminol-dependent chemiluminescence assay. J Androl. 2001;22(4):568–74.
100. Piomboni P, Gambera L, Serafini F, Campanella G, Morgante G, De Leo V. Sperm quality improvement after natural anti-oxidant treatment of asthenoteratospermic men with leukocytospermia. Asian J Androl. 2008;10(2):201–6.
101. Silver EW, Eskenazi B, Evenson DP, Block G, Young S, Wyrobek AJ. Effect of antioxidant intake on sperm chromatin stability in healthy nonsmoking men. J Androl. 2005;26(4):550–6.
102. Omu AE, Al-Othman S, Mohamad AS, Al-Kaluwby NM, Fernandes S. Antibiotic therapy for seminal infection. Effect on antioxidant activity and T-helper cytokines. J Reprod Med. 1998;43(10):857–64.
103. Vicari E. Effectiveness and limits of antimicrobial treatment on seminal leukocyte concentration and related reactive oxygen species production in patients with male accessory gland infection. Hum Reprod. 2000;15(12):2536–44.
104. Gambera L, Serafini F, Morgante G, Focarelli R, De Leo V, Piomboni P. Sperm quality and pregnancy rate after COX-2 inhibitor therapy of infertile males with abacterial leukocytospermia. Hum Reprod. 2007;22(4):1047–51.
105. Comhaire FH, Christophe AB, Zalata AA, Dhooge WS, Mahmoud AM, Depuydt CE. The effects of combined conventional treatment, oral antioxidants and essential fatty acids on sperm biology in subfertile men. Prostaglandins Leukot Essent Fatty Acids. 2000;63(3):159–65.
106. Kessopoulou E, Powers HJ, Sharma KK, et al. A double-blind randomized placebo cross-over controlled trial using the antioxidant vitamin E to treat reactive oxygen species associated male infertility. Fertil Steril. 1995;64(4):825–31.
107. Keskes-Ammar L, Feki-Chakroun N, Rebai T, et al. Sperm oxidative stress and the effect of an oral vitamin E and selenium supplement on semen quality in infertile men. Arch Androl. 2003;49(2): 83–94.
108. Greco E, Iacobelli M, Rienzi L, Ubaldi F, Ferrero S, Tesarik J. Reduction of the incidence of sperm DNA fragmentation by oral antioxidant treatment. J Androl. 2005;26(3):349–53.
109. Lenzi A, Sgro P, Salacone P, et al. A placebo-controlled double-blind randomized trial of the use of combined l-carnitine and l-acetyl-carnitine treatment in men with asthenozoospermia. Fertil Steril. 2004;81(6):1578–84.
110. Tremellen K, Miari G, Froiland D, Thompson J. A randomised control trial examining the effect of an antioxidant (Menevit) on pregnancy outcome during IVF-ICSI treatment. Aust N Z J Obstet Gynaecol. 2007;47(3):216–21.
111. Rolf C, Cooper TG, Yeung CH, Nieschlag E. Antioxidant treatment of patients with asthenozoospermia or moderate oligoasthenozoospermia with high-dose vitamin C and vitamin E: a randomized, placebo-controlled, double-blind study. Hum Reprod. 1999;14(4): 1028–33.

112. Zini A, Blumenfeld A, Libman J, Willis J. Beneficial effect of microsurgical varicocelectomy on human sperm DNA integrity. Hum Reprod. 2005;20(4):1018–21.
113. Werthman P, Wixon R, Kasperson K, Evenson DP. Significant decrease in sperm deoxyribonucleic acid fragmentation after varicocelectomy. Fertil Steril. 2008;90(5):1800–4.
114. Evers JL, Collins JA. Assessment of efficacy of varicocele repair for male subfertility: a systematic review. Lancet. 2003;361(9372):1849–52.
115. Marmar JL, Agarwal A, Prabakaran S, et al. Reassessing the value of varicocelectomy as a treatment for male subfertility with a new meta-analysis. Fertil Steril. 2007;88(3):639–48.

Surgical Treatment for Male Infertility

6

Sandro C. Esteves and Ricardo Miyaoka

Abstract

Infertility complaint is common in the urologic office. The role of the urologist in this context cannot be underestimated, since he/she is trained to diagnose, to counsel, to provide medical or surgical treatment whenever possible, or to correctly refer the male patient for assisted conception. The urologist can also be part of the multi-professional reproductive team in the assisted reproduction unit, being responsible for the above-cited tasks as well as for the sperm surgical retrieval from the epididymis or testicle. Two major breakthroughs occurred in the area of male infertility with regard to treatment. The first was the development of microsurgery which increased success rates for reconstruction of the reproductive tract. The second was the development of intracytoplasmic sperm injection (ICSI) and the demonstration that spermatozoa retrieved from either the epididymis or the testis were capable of fertilization and pregnancy. Thereafter, several sperm retrieval methods have been developed to collect epididymal and testicular sperm for ICSI in azoospermic men. Microsurgery was incorporated to this armamentarium, either for collection of sperm from the epididymis in men with obstructive azoospermia or from the testicle in those with nonobstructive azoospermia. This chapter describes the most common surgical treatments for male infertility. It includes not only the reconstructive interventions for the male reproductive system but also the sperm retrieval techniques to be used in cases of obstructive (OA) and nonobstructive azoospermia (NOA). A critical commentary, based on the authors' experience in the surgical management of infertile males, and a review of important publications from the last 5 years are included. Finally, a list of key issues is provided to summarize the current knowledge in this area.

Keywords

Male infertility • Varicocele repair • Vasovasostomy • Vasoepididymostomy • Sperm retrieval techniques • Microsurgery • Surgical treatments for infertility • Intracytoplasmic sperm injection

S.C. Esteves, MD, PhD (✉)
ANDROFERT, Andrology and Human Reproduction Clinic, Center for Male Reproduction, Av. Dr. Heitor Penteado, 1464, Campinas, SP 13075-460, Brazil
e-mail: s.esteves@androfert.com.br

R. Miyaoka, MD
ANDROFERT, Andrology & Human Reproduction Clinic, Center for Male Reproduction, Av. Dr. Heitor Penteado, 1464, Campinas, SP 13075-460, Brazil
e-mail: rmiyaoka@androfert.com.br

S.J. Parekattil and A. Agarwal (eds.), *Male Infertility: Contemporary Clinical Approaches, Andrology, ART & Antioxidants*, DOI 10.1007/978-1-4614-3335-4_6, © Springer Science+Business Media, LLC 2012

Infertility complaint is common in the urologic office. Approximately 8% of men in reproductive age seek for medical assistance for fertility-related problems. Of these, 1–10% carries conditions that compromise the reproductive potential [1]. The role of the urologist in this context cannot be underestimated, since he/she is trained to diagnose, to counsel, to provide medical or surgical treatment whenever possible, or to correctly refer the male patient for assisted conception. The urologist can also be part of the multi-professional reproductive team in the assisted reproduction unit, being responsible for the previously cited tasks as well as for the sperm surgical retrieval from the epididymis or testicle.

In a group of 2,875 infertile couples attending our tertiary center for male reproduction, potentially surgical correctable conditions were identified in 34.4% of the male partners. About 1/3 of these individuals were azoospermic. Although reconstructive surgery would be possible in only 30% of this subgroup, most of the remaining would be candidates for sperm retrieval techniques if enrolled in assisted reproduction programs. Therefore, surgical management can be offered to more than 50% of our patient population in daily practice (Table 6.1).

Table 6.1 Distribution of diagnostic categories of couples seeking infertility evaluation in a male infertility clinic

Category	*N*	%
Varicocele	629	21.9
Infectious	72	2.5
Hormonal	54	1.9
Ejaculatory dysfunction	28	1.0
Systemic diseases	11	0.4
Idiopathic	289	10.0
Normal/female factor	492	17.1
Immunologic	54	1.9
Obstruction	359	12.5
Cancer	11	0.4
Cryptorchidism	342	11.9
Genetic	189	6.6
Testicular failure	345	11.9
Total	2,875	

Two major breakthroughs occurred in the area of male infertility with regard to treatment. The first was the development of microsurgery, which increased success rates for reconstruction of the reproductive tract. The second was the development of intracytoplasmic sperm injection (ICSI) and the demonstration that spermatozoa retrieved from either the epididymis or the testis were capable of fertilization and pregnancy [2, 3]. Thereafter, several sperm retrieval methods have been developed to collect epididymal and testicular sperm for ICSI in azoospermic men. Microsurgery was incorporated to this armamentarium, either for collection of sperm from the epididymis in men with obstructive azoospermia or from the testicle in those with nonobstructive azoospermia (NOA) [2, 4].

This chapter describes the most common surgical treatments for male infertility. It includes not only the reconstructive interventions for the male reproductive system but also the sperm retrieval techniques to be used in cases of obstructive (OA) and NOA. A critical commentary, based on the authors' experience in the surgical management for the infertile male, and a review of important publications from the last 5 years are included. Finally, a list of key issues is provided to summarize the current knowledge in this area.

Surgical Treatments

Varicocele Repair

Varicoceles can be identified in up to 35% of the male population with infertility complaints [1]. The etiology of varicocele formation is likely to be multifactorial, and several theories aim to explain the impact of varicoceles on testicular function, none of them fully elucidating the variable effect of varicocele on human spermatogenesis and male fertility [5–8]. The association between varicocele and infertility is still a matter of debate. However, there is an increased incidence of this condition among infertile men [9]. Moreover, varicocele is associated with reduced semen parameters and testicular size [10]. Lastly, it has been shown that surgical treatment of clinical varicoceles improves semen quality and increases the likelihood of pregnancy [11, 12]. Despite these facts, it is still unclear why most men with varicocele retain fertility and why fertility status is not always improved after treatment [13].

Preoperative Planning and Patient Evaluation

Treatment of varicocele in infertile men aims to restore or improve testicular function. Current recommendations suggest that treatment should be offered for couples with documented infertility whose male partner has a clinically palpable varicocele and abnormal semen analysis. A detailed medical history must be taken and prognostic factors identified. Physical examination with the patient standing in a warm room is the preferred diagnostic method. Varicoceles diagnosed by this method are termed "clinical" and may be graded according to the size. Large varicoceles (grade III) are varicose veins seen through the scrotal skin. Moderate (grade II) and small-sized varicoceles (grade I) are dilated veins palpable without and with the aid of the Valsalva maneuver, respectively [14]. In the presence of bilateral palpable varicocele, it is recommended to perform surgery on both sides at the same operative time [15].

Physical examination may be inconclusive or equivocal in cases of low-grade varicocele and in men with a history of previous scrotal surgery, concomitant hydroceles, or obesity.

Imaging studies may be recommended during the evaluation of infertile men for varicocele if physical examination is inconclusive. When a varicocele is not palpable but a retrograde blood flow is detected by other diagnostic methods such as venography, Doppler examination, ultrasonography, scintigraphy, or thermography, the varicocele is termed subclinical [16, 17]. However, the role of subclinical varicocele as a cause of male infertility remains debatable, and current evidence does not support the recommendation for treating infertile men with subclinical varicocele [18, 19].

Preoperative hormone profile including follicle-stimulating hormone (FSH) and testosterone is recommended. Testicular volume should be assessed using a measurement instrument such as the Prader orchidometer or a pachimeter. At least two semen analyses must be obtained and evaluated according to the World Health Organization guidelines [20]. It seems that infertile men either with higher preoperative semen parameters or undergoing varicocele repair for large varicoceles are more likely to show postoperative semen parameters improvement [21]. On the other hand, reduced preoperative testicular volume, elevated serum FSH levels, diminished testosterone concentrations, and subclinical varicocele are negative predictors for fertility improvement after surgery [18, 22–27].

Men with clinical varicoceles presenting with azoospermia may be candidates for surgical repair. In such cases, genetic evaluation including Giemsa karyotyping and polymerase chain Yq microdeletion screening for AZFa, AZFb, and AZFc regions is recommended. A testis biopsy (open or percutaneous) may be obtained to assess testicular histology, which has been shown to be the only valid prognostic factor for restoration of spermatogenesis [20, 28]. The benefit of varicocelectomy in azoospermic men with genetic abnormalities is doubtful and should be carefully balanced. The same caution is valid for patients with atrophic testes and/or history of cryptorchidism, testicular trauma, orchitis, and systemic or hormonal dysfunction due to the fact that varicocele in such cases may not be the cause of infertility but merely coincidental [29].

As for all surgical reconstructive procedures, the evaluation of the female partner's reproductive potential is recommended before an intervention is indicated, and the alternatives to varicocele repair discussed.

Anesthesia and Techniques

Varicocele repair may be carried out using local, regional, or general anesthesia, according solely with the surgeon and patient's preferences. In the authors' practice, we routinely perform microsurgical subinguinal varicocele repair using short-acting propofol intravenous anesthesia associated with the blockage of the spermatic cord using 10 mL of a 2% lidocaine hydrochloride on an outpatient basis [20].

Varicoceles are surgically treated either by open (with or without magnification) or laparoscopic approaches. The main concept is the occlusion of the dilated veins of the pampiniform plexus. The high retroperitoneal and laparoscopic approaches are performed for internal spermatic vein ligation, while the inguinal and subinguinal approaches allow the ligation of the internal and external spermatic and cremasteric veins that may contribute to the varicocele.

Retroperitoneal Techniques

High retroperitoneal open varicocele repair involves incision medial to the anterior superior iliac spine at the level of the internal inguinal ring (Fig. 6.1a). The external oblique muscle is split, the internal oblique muscle is retracted and the peritoneum is teased away. Exposure of the internal spermatic artery and vein is carried out retroperitoneally near the ureter. At this level, only one or two internal spermatic veins are present, but the internal spermatic artery may not be easy to identify. The veins are ligated near to the point of drainage into the left renal vein. Neither the parallel inguinal and retroperitoneal collateral veins that exit the testis and bypass the retroperitoneal area of ligation nor the cremasteric veins can be identified in the retroperitoneal approach. It is believed that these collaterals are the primary cause of recurrence seen in retroperitoneal varicocelectomy. The surgical approach on the right side may be more difficult because the right gonadic vein drains in the inferior vena cava. Laparoscopic varicocelectomy is a retroperitoneal approach using high magnification. The spermatic artery and the lymphatics are easily identified and spared; collateral veins can also be clipped or coagulated. However, external spermatic veins cannot be treated, thus leading to a recurrence rate of approximately 5% [30]. Laparoscopy varicocele repair is more invasive and costly, and it is associated with higher complication rates than open procedures [30–32].

Inguinal and Subinguinal Techniques

The classic approach to the inguinal varicocelectomy involves a 5–10-cm incision over the inguinal canal, opening of the external oblique aponeurosis and isolation of the spermatic cord (Fig. 6.1a). The internal spermatic veins are dissected and ligated. An attempt is made to positively identify and spare the testicular artery and the lymphatics. External spermatic veins running parallel to the spermatic cord or perforating the floor of the inguinal canal can be identified and ligated. Although internal and external spermatic veins can be identified macroscopically, the use of magnification facilitates identification and preservation of internal spermatic artery and lymphatics, which may prevent testicular atrophy and hydrocele formation, respectively [33].

Microsurgical varicocelectomy can be performed via an inguinal or subinguinal approach. The main advantage of the subinguinal over the inguinal approach is that the former obviates the need to open the aponeurosis of the external

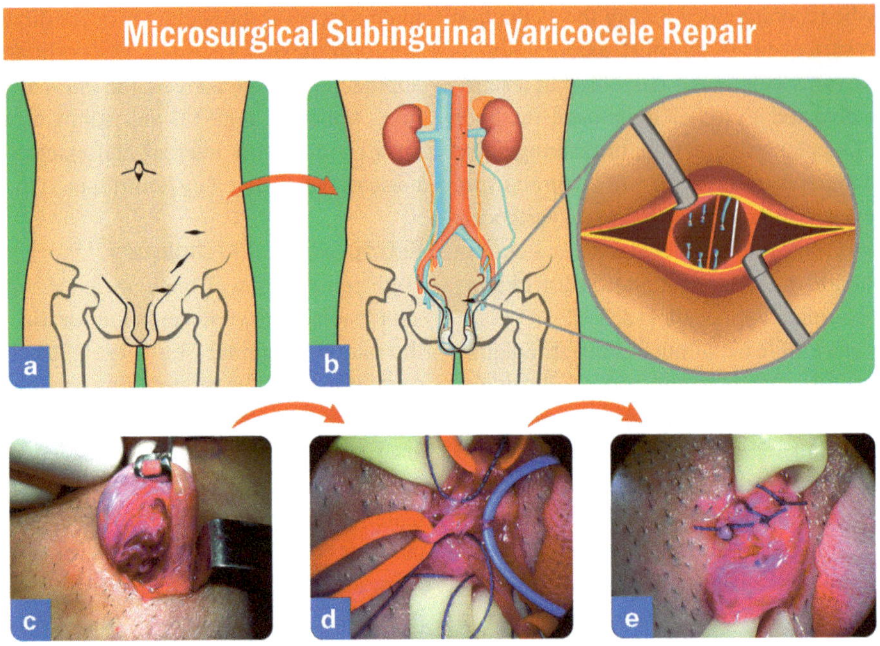

Fig. 6.1 Microsurgical subinguinal varicocele repair. (**a**) Incision sites commonly used for subinguinal, inguinal, and retroperitoneal varicocele repair. (**b**) In the subinguinal approach, a transverse incision is made just below the level of the external inguinal ring. (**c–e**) Intraoperative photographs of the spermatic cord. (**c**) Dilated cremasteric veins are identified by elevating the spermatic cord with a Babcock clamp. (**d**) Testicular artery (*blue vessel loop*), lymphatics (*blue cotton suture*), and dilated varicose veins (*red vessel loops*) are demonstrated. (**e**) Final surgical aspect of varicose veins transected and ligated with nonabsorbable sutures

oblique, which usually results in more postoperative pain and a longer time before the patient return to work. In our practice, varicoceles are treated using a testicular artery and lymphatic-sparing subinguinal microsurgical repair [12, 20]. Briefly, a 2.5-cm skin incision is made over the external inguinal ring (Fig. 6.1a, b). The subcutaneous tissue is separated until the spermatic cord is exposed. The cord is elevated with a Babcock clamp, and the posterior cremasteric veins are ligated and transected (Fig. 6.1c). A Penrose drain is placed behind the cord without tension. The cremasteric fascia is then opened to expose the cord structures, and the dissection proceeds using the operating microscope with magnification ranging from 6 to 16×. Dilated cremasteric veins are ligated and transected. Lymphatics and arteries are identified and preserved (Fig. 6.1d). Whenever needed, the cord structures are sprayed with papaverine hydrochloride to increase the arterial beat. All dilated veins of the spermatic cord are identified, tagged with vessel loops, then ligated using nonabsorbable sutures and transected (Fig. 6.1e). Vasal veins are ligated only if they exceed 2 mm in diameter. Sclerosis of small veins is not used.

Postoperative Care

Local dressing and scrotal supporter are kept for 48–72 h and 1 week, respectively. Scrotal ice packing is always recommended to control local edema for the first 48 postoperative hours. Patients are counseled to restrain from physical activity and sexual intercourse for 2–3 weeks. Oral analgesics usually suffice to control postoperative pain. Postoperative follow-up aims to evaluate improvement in semen parameters, complications, and spontaneous or assisted conception. Semen analysis should be performed every 3 months until the semen parameters stabilize or pregnancy occurs.

Reconstructive Surgery of the Vas Deferens and Epididymis

Vasovasostomy and vasoepididymostomy are surgical procedures designed to bypass an obstruction in the male genital tract. Approximately 13% of married men aged 15–44 years reported having had a vasectomy in the United States [34]. As expected, vasectomy frequency increases with older age and greater number of biological children. The number of men seeking for vasectomy reversal due to changes in marital status or reproductive goals has increased and vary from 2% to 6% [35]. In Brazil, it is estimated that 200,000 and 7,000 vasectomies and reversals are annually performed, respectively [36]. While the vast majority of vasovasostomy and vasoepididymostomy procedures are to reverse intentional obstructions, other indications include correction of epididymal or vasal obstructions due to genital infections, iatrogenic injuries related to inguinal or scrotal surgery, especially during the early childhood and postvasectomy pain syndrome [37].

Preoperative Planning and Patient Evaluation

A detailed medical history must be taken and prognostic factors identified. Obstruction intervals from vasectomy to reversal are believed to play a major role in determining surgery outcomes. Patency and pregnancy rates for obstruction intervals up to 15 years are approximately 74% and 40%, respectively [38]. Obstruction intervals longer than 15 years are associated with lower patency and pregnancy rates. Long obstruction intervals are associated with higher incidence of epididymal obstruction, and as a result, vasoepididymostomy (VE) is likely to be required. A computer model based on obstructive interval and patient age was created to determine the need for VE. The model was designed to be 100% sensitive in detecting patients requiring VE. In the test group, the model was 100% sensitive in predicting VE with a specificity of 58.8% [39].

A history of a previous vasectomy reversal attempt does not preclude a new one. Patency and pregnancy rates of 79% and 31%, respectively, are reported for repeated reversals [40]. These authors found that the history of conception with the current partner had been the only significant predictor for a successful pregnancy. History of genital/inguinal surgery should raise concern about the possibility of iatrogenic surgical obstruction. Repair of obstruction in the inguinal canal or retroperitoneum can be technically challenging.

A detailed physical examination should be carried out. Small and soft testes may indicate impaired spermatogenesis. Indurate, irregular epididymis and the presence of hydrocele are often associated with epididymal obstruction and may suggest the need for vasoepididymostomy. Palpation of a granuloma in the vas deferens should be interpreted as a favorable prognostic sign. Its presence means that sperm has leaked at the vasectomy site preventing overpressure within the epididymis tubules and rupture [35, 38, 41]. If a vasal gap is detected, the patient should be advised that a larger incision into the inguinal region may be needed in order to allow a tension-free anastomosis to be performed. Specific laboratory tests are not necessary before reconstructive surgeries. Serum FSH testing is recommended as a marker of testicular reserve only if testicular damage is suspected on physical examination. The usefulness of antibody testing remains controversial, and evidence suggests that late failures following reversals are likely to be technical rather than immunological [42, 43]. Besides, overall conception rates are acceptably high, and the presence of antisperm antibodies does not correlate closely with postsurgical fecundability [44].

The female partner fertility has to be carefully assessed before indicating reconstruction procedures, and alternatives to vasectomy reversal should be discussed. It has been shown that reversal outcomes in men with the same partners are significantly better than those with new partners. The proven fecundity as a couple, shorter obstructive interval, and stronger emotional dedication to achieving conception may act as possible factors for the higher success rate [35, 45]. Female age greater than 40 years is a negative predictor for success [46, 47].

Anesthesia and Techniques

Vasovasostomy and vasoepididymostomy may be safely performed using local, regional, or general anesthesia. In the authors' practice, procedures are carried out on an outpatient basis. Continuous propofol intravenous anesthesia coupled with the blockage of the spermatic cord using 10–20 mL of a 1% lidocaine hydrochloride solution is our preferred anesthetic method.

Incision

2-centimeter longitudinal scrotal incisions are placed in the anterior aspect of the scrotum on each side. The incision is made onto the palpable granuloma or onto the identified vasal gap. Only the vas ends are delivered through the skin incision. The incision may be extended to the inguinal region in case the vasectomy has been performed high in the scrotum, or a large segment has been removed, and also in repeat reconstructions with difficult vasal mobilization. The testis is delivered only if a vasoepididymostomy or a robotic-assisted anastomosis is to be performed.

Approaching the Vas

Microsurgical dissection is carried out onto the region of the prior vasectomy site to free the vas and its vascular pedicle from surrounding scar tissue. Hemostasis is obtained with great care using either bipolar or handheld thermal cautery units. After the vas has been mobilized and its scarred ends excised, patency of the abdominal vas end is confirmed by the introduction of a 24-gauge blunt-tipped angiocatheter into its lumen and injection of 20-mL sterile saline through the catheter. The vas ends must be adequately mobilized in order to allow a complete tension-free anastomosis. Either a microsurgical clamp or holding sutures can be used to accomplish this step according to the surgeon's preference.

Vasal Fluid Examination

Fluid from the testicular vas end is examined both macroscopically and under the optical microscope for the presence of sperm. The presence of copious, clear, watery, or cloudy fluid and motile sperm is associated with excellent patency rates of 94% as opposed to only 60% when no sperm is found in the vasal fluid [35]. Thick toothpaste-like vasal fluid is suggestive of epididymal obstruction [35, 48]. The quality of sperm found in the intravasal fluid and the surgeon's

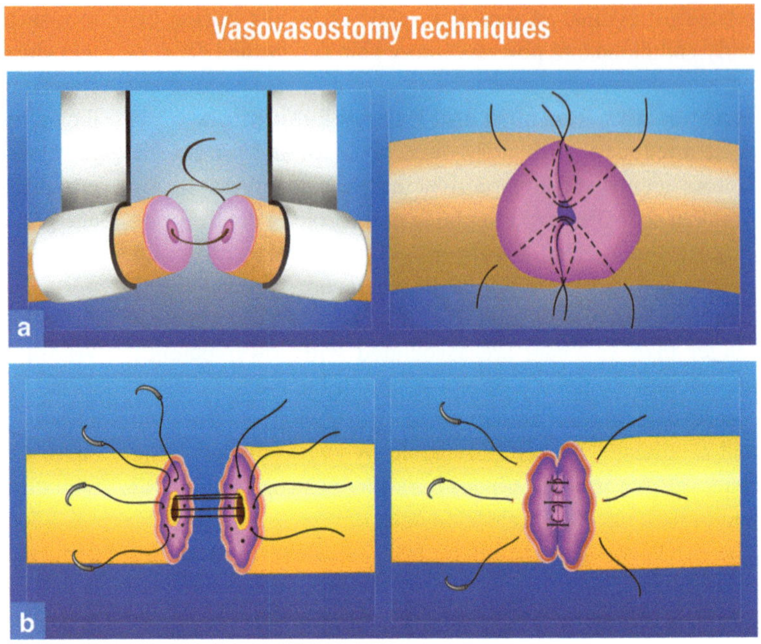

Fig. 6.2 Microsurgical vasovasostomy techniques. Illustration of the modified one-layer (**a**) and the multilayer microdot technique (**b**) (see text for detailed description)

microsurgical skills are the most important factors to determine the type of reconstructive technique. Typically, the presence of sperm or sperm parts, and even a "dry" vas, is associated with adequate patency rates of about 70–80% following vasovasostomies [49, 50]. The vasoepididymostomy is a challenging surgical procedure that should only be attempted by experienced microsurgeons. Meticulous microsurgical technique and high magnification are required for a precise anastomosis of the vas (luminal diameter of 300–400 μm) to the epididymal tubule (luminal diameter of 150–250 μm). Intraoperative sperm harvesting and cryopreservation can be offered during vasoepididymostomy [51].

Vasovasostomy Techniques

Attention to surgical details directly affects the success of reconstructive microsurgeries. These include the accurate mucosa-to-mucosa approximation, a watertight tension-free anastomosis, preservation of the vasal blood supply and healthy tissue (mucosa and muscularis), and an adequate microscopic atraumatic technique.

Modified One-Layer Technique

The modified one-layer technique described by Sharlip is the authors' choice for vasectomy reversal [52]. Our preference is to perform the anastomosis using a 9-0 nylon suture mounted on a taper-pointed needle and with the aid of a vas clamp (ASSI, catalog# MSPK-3678). The operation is performed entirely with the surgeon located on the patient's right side. The first suture is placed in the medial surface of the right vas (zero degree position) (Fig. 6.2a). This suture is placed through the full thickness of the vas wall on the testicular side first taking a generous bite of adventitia and muscularis and a tiny portion of the mucosa. The suture is then passed into the corresponding zero-degree position of the abdominal side taking a bite at the edge of the mucosa and a large portion of the muscularis/adventitia layer. This suture is tied and cut long, so it is easily identified as the procedure continues. The second suture is placed 180° opposite to the first, and again it takes the full aspect of the vas wall, firstly on the testicular side and then on the abdominal one. This suture is also tied and cut long. A third full thickness suture is placed at the 60° position, one third of the distance from the first to the second sutures. Before it is tied, a fourth suture is placed at the 120° position, two thirds of the distance from the first and second sutures. The fourth suture is then tied after careful inspection of their proper placement (Fig. 6.2a). A fifth suture is placed between these two at the 90° position, but only superficially through the muscularis. This completes the anastomosis of the anterior portion of the vas. At this point, four full thickness stitches and one muscular suture have been placed, and half of the total circumference of the vas wall is closed. The vas clamp is then rotated 180°, and verification of accidental back-walling and proper position of full thickness sutures is carried out. After rotation of the vas, two full thickness sutures are then placed at the 240 and 300° positions. These sutures are inserted and inspected before being tied. A final suture is placed in the muscularis at the 270° position. This complete the anastomosis, summing up 8 sutures in total instead of 12 as first described by Sharlip. Upon anastomosis completion, the surrounding loose fibrous tissue is sutured over the anastomotic site alleviating tension. Scrotal incision is closed in the routine usual manner.

Two-Layer Technique

This technique, described by Belker, involves the placement of five to eight interrupted 10-0 nylon sutures in the inner mucosal layer and eight to ten 9-0 nylon sutures in the outer muscular and adventitial layer [53]. The use of an approximating clamp and a holding suture is recommended to stabilize the vas ends for the anastomosis. Before suturing begins, the vas ends are placed parallel to each other in such a way that allows the surgeon to look straight down into their lumens. As suturing proceeds, the transected ends of the vas bend toward each other, bringing the suture together without tension. First, three posterior muscular layer sutures are placed in a row so that the knots are outside. At this point, 90° of the circumference is approximated, leaving full access to the mucosa. Then, three posterior mucosal sutures are placed and tied. Subsequently, the far-corner and near-corner sutures are placed and tied alternately until space remains for only two or three sutures in the anterior aspect of the anastomosis. These remaining stitches are then placed and left long and untied until back-walling can be safely ruled out. The sutures are then tied, and the muscular layer is sutured with caution to avoid penetration of the lumen by the outer-layer sutures. Placement of these sutures is easier to perform from the assistant side toward the surgeon's side. Closure of scrotal incision is performed in the usual manner.

Multilayer Microdot Technique

This method, originally described by Goldstein, is adequate to treat markedly discrepant vas diameters in straight or convoluted vas [54]. Vasal ends are prepared with a 90° right angle cut, and methylene blue stain can be used to better visualize the mucosal rings. Planned needle exit points can be marked with microtip marking (Fig. 6.2b). Polypropylene monofilament 10-0 double-armed sutures (70-μm-diameter) with taper-pointed needles are used for the anastomosis. Sutures are placed in an inside-out fashion eliminating the possibility for accidental back-walling. The mucosa and about one-third thickness of the muscularis should be included in each bite, symmetrically on each side of vas ends. Four initial sutures are placed in the anterior aspect of the vas and tied up (Fig. 6.2b). Three 9-0 sutures are then placed exactly in between the previously placed mucosal sutures, just above but not through the mucosa, sealing the gap between the mucosal sutures. The vas is then rotated 180°, and four additional 10-0 sutures are placed completing the mucosal part of the anastomosis. Just prior to tying the last mucosal knot, vas lumen is irrigated with heparinized saline solution to prevent formation of clots. After completion of mucosal layer, 9-0 sutures are placed in between each mucosal suture, again avoiding penetrating the mucosa itself (Fig. 6.2b). Superficial additional adventitia 9-0 sutures should be placed only if necessary. Procedure is completed by approximating the vas sheath with four to six 6-0 sutures.

Robotic-Assisted Technique

Recent reports have shown the feasibility of performing the classic above described techniques using robotic assistance. The robot can offer additional benefits of enhanced imaging (up to ×100 magnification) and control of physiologic tremor [55, 56].

Vasoepididymostomy Techniques

The procedure starts with the placement of a longitudinal incision in the upper scrotum. The testis is delivered through the incision, and the testis and epididymis are thoroughly inspected. The site of obstruction can be often seen as an area where the epididymis changes from a firm, wide caliber to a smaller, softer structure. The distal end of the vas deferens is mobilized in a similar fashion as that described for the vasovasostomy, but usually a longer length is required to perform an epididymal anastomosis. At this point, the microscope is brought into the operating field to aid the surgeon perform the anastomosis. Currently, three variations of the technique have been used for precise approximation of the vas deferens lumen to a single epididymal tubule: end-to-end, end-to-side, and end-to-side intussusception techniques. Prior to the anastomosis, a dilated epididymal tubule must be identified immediately above the level of obstruction. The tubule must be opened and the fluid inspected for the presence of motile sperm. If no sperm are identified, a more proximal site of the epididymis will be required for the anastomosis.

End-to-End Technique

First described by Silber, the end-to-end VE is the most difficult anastomosis to perform [57]. It involves the dissection of a single epididymal tubule, its complete transection, and then its anastomosis to the vas lumen. The epididymis is dissected off the testis for 3–5 cm to provide an adequate length to achieve a tension-free anastomosis. Initially, two 9-0 nylon sutures are placed at the 5 and 7 o'clock positions of the seromuscular surface of the vas, to secure the cut end of the distal vas to the epididymal tunica. Next, four 10-0 nylon sutures mounted in double-armed 70-μm fishhook-shaped taper-pointed needles are placed in a quadrant fashion between the vas mucosa and the epididymal tubule (Fig. 6.3a). These sutures are not tied until all of them have been positioned. The anastomosis is completed by placing several interrupted 9-0 nylon sutures to approximate the seromuscular layer of the vas to the epididymal tunic layer.

End-to-Side Technique

The end-to-side VE, popularized by Thomas, is performed by creating a small window in a loop of the epididymal tubule

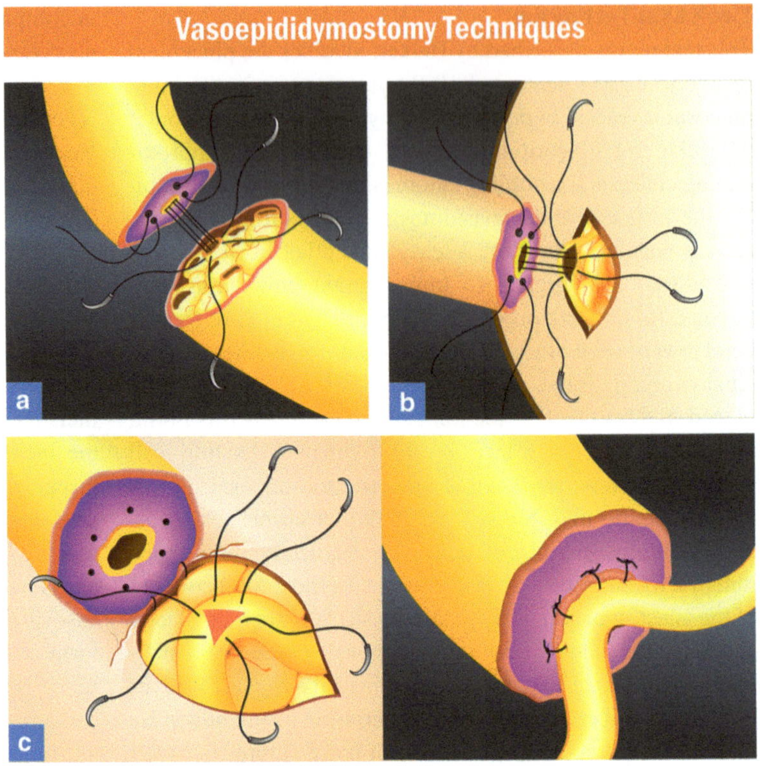

Fig. 6.3 Microsurgical vasoepididymostomy techniques. Illustration of the end-to-end (**a**), end-to-side (**b**), and triangulation end-to-side (**c**) anastomoses (see text for detailed description)

proximal to the obstruction and by suturing the end of the vas lumen to the open window [58]. The advantages over the end-to-end anastomosis include less dissection and bleeding during the anastomosis because hemostasis can be secured before opening the tubule. Moreover, only one tubule is opened making the identification of the patent tubule more precise and easy. With the tubule opened and sperm presence confirmed, three or four double-armed 10-0 nylon sutures are placed in a quadrant fashion through the edge of the epididymal tubule (Fig. 6.3b). The sutures are placed in the corresponding quadrant of the vasal mucosa and tied. The anastomosis is completed with additional 9-0 nylon sutures between the epididymal tunic and the seromuscular layer of the vas deferens. Lastly, several 9-0 nylon sutures are used to anchor the vas deferens to the parietal layer of the tunica vaginalis. These final sutures are designed to prevent tension on the anastomosis and are placed well away from the vasoepididymostomy site.

Triangulation End-to-Side Vasoepididymostomy

This technique was introduced by Berger [59] with subsequent modifications by others [60, 61]. It is the simplest and fastest among the three techniques described in this chapter and the authors' choice for vasoepididymostomy. The intention is to combine the precision of the conventional end-to-side anastomosis with a simplified microsuture placement. Rather than a direct approximation of the epididymal tubule to the vas, this method involves pulling the epididymal tubule into the vas lumen. An opening window is made in the epididymal tunic corresponding to the vas diameter. Two 9-0 sutures are used to secure the muscular layer of the vas to the epididymal tunic to avoid tension on the anastomosis site. Three double-armed 10-0 nylon sutures are placed equidistantly in a triangular configuration in the desired epididymal tubule (Fig. 6.3c). Then, the epididymal tubule is carefully opened with microscissors or microknife between the positioned sutures. Once sperm is confirmed in the epididymal fluid, the needles are passed through the lumen of the vas in an inside-out fashion. The sutures are then tied, creating an invagination of the epididymal tubule into the vasal lumen (Fig. 6.3c). Finally, additional 9-0 nylon sutures are placed to approximate the seromuscular layer of the vas to the epididymal tunic.

Recently, a modification of the triangulation end-to-side VE was described by Marmar [60]. In this technique, a single epididymal tubule is exposed, and two 10-0 nylon sutures mounted on double-armed 70-μm bicurve needles are placed on the field. A needle from each suture is mounted on a styrofoam block and positioned parallel to the other with sufficient room for passage of the tip of a microblade between them. A microneedle holder is used to grasp both needles simultaneously and move them from the block to the field while maintaining the parallel arrangement. The tips of both needles are passed through a selected tubule at once. The two sutures are retracted laterally, and a tubulotomy is cut between them with a microknife. Then, all four needles from the epididymal sutures are individually placed into the mucosal lumen of the vas and out through the muscularis on the cut end. Needles are placed at the 8 and 10 o'clock positions on the left side and at the 2 and 4 o'clock positions on the right side. Sutures are tied allowing the epididymal tubule to invaginate into the vas lumen. The anastomosis is completed with 3–4 additional 9-0 nylon sutures through the muscularis of the vas and epididymal tunic.

Postoperative Care

Local dressing and scrotal supporter are kept for 48–72 h and 2 weeks, respectively. Scrotal ice packing is always recommended to control local edema for the first 72 postoperative hours. Patients are counseled to restrain from physical activity and sexual intercourse for 1 or 2 months in cases of vasovasostomy and vasoepididymostomy, respectively. Oral analgesics usually suffice to control postoperative pain. Postoperative follow-up is aimed at evaluating improvement in semen parameters, complications, and spontaneous or assisted conception. Semen analysis should be performed every 2 months after surgery until the semen parameters stabilize or pregnancy occurs.

Transurethral Resection of the Ejaculatory Duct

Ejaculatory duct obstruction (EDO) is a potential surgically correctable cause of male infertility. Congenital obstructions are caused by atresia or stenosis of the ejaculatory ducts, or utricular, mullerian, and wolffian duct cysts. Acquired obstructions may be secondary to trauma or infectious/inflammatory etiologies. Traumatic damage to the ejaculatory ducts may occur after removal of seminal vesicle cysts, pull-through operations for imperforate anus, and even prolonged catheterization or instrumentation. Genital or urinary infection and prostatic abscess may cause stenosis or complete obstruction of the ducts [62]. Prostatic infection may also result in calculus formation and secondary obstruction, while tuberculosis produces genital devastation.

Preoperative Planning and Patient Evaluation

Diagnostic criteria typically include history, physical examination, semen analyses, and transrectal ultrasound evaluation (TRUS). The clinical presentation may be highly variable, and in addition to a history of infertility, complaints may include painful ejaculation, hemospermia, and perineal and/or testicular pain, although patients can be completely asymptomatic. Physical examination is usually normal. Occasionally, the seminal vesicles or a mass is palpable on rectal examination. Prostatic tenderness and/or epididymal enlargement may exist. Hormone profiles are usually normal.

Semen analyses may reveal oligozoospermia or azoospermia, decreased motility, and decreased ejaculate volume. The presence of a low-volume (<1.5 mL) acidic (pH<7.0) azoospermic ejaculate with absent fructose, palpable vasa, and epididymal thickening is virtually pathognomonic. However, the typical clinical picture may be complicated in cases obstruction is unilateral, partial, or functional [62]. Postejaculate urinalyses are often performed to exclude retrograde ejaculation in patients with low-volume ejaculates.

High-resolution TRUS using with a 5–7 MHz biplanar transducer is recommended in all cases of suspected EDO. The exact definition of obstruction on TRUS, however, is still a matter of debate due to marked variability in the size and shape of the vas deferens, seminal vesicles, and ejaculatory ducts in fertile and infertile men. Common ultrasound findings include dilation of the seminal vesicles (defined as a cross-sectional width of greater than 1.5 cm) or ejaculatory ducts (defined as an internal duct diameter of greater than 2.0 mm), calcifications or calculi in the region of the ejaculatory duct or verumontanum, and midline or eccentrically located prostatic cysts [63–65]. Ultrasound-guided transrectal seminal vesiculography has been shown to provide excellent radiographic visualization of the ejaculatory ducts [66]. Also, TRUS-guided seminal vesicle aspiration and the presence of motile sperm in the aspirates seem to be a useful diagnostic tool since the seminal vesicles are not sperm reservoirs [67]. A testicular biopsy can be done to document the presence of normal spermatogenesis. The authors' preference is to perform a "wet prep" using the percutaneous testicular sperm aspiration (TESA) technique either before or at the time of surgery. In such cases, the presence of motile sperm is highly indicative of obstruction.

Anesthesia and Technique

Transurethral resection of the ejaculatory duct (TURED) is performed using regional or general anesthesia. Our choice is to perform the procedure, as originally described by Farley and Barnes [68], with minor modifications [62]. First, the obstruction is documented using intraoperative vasotomy and vasography. The vas is delivered using a small scrotal incision and dissected free of the associated perivasal vessels. A mixture of injectable saline and radiographic contrast material in a 1:1 ratio is injected into the abdominal end of the vas, together with methylene blue dye, by direct vas puncture with a 30-gauge lymphangiogram needle [62]. Vasography confirms obstruction, whereas dye injection confirms patency by visualization of the effluxing dye mixture during TURED. A 9-0 nylon suture is placed at the muscular layer of the vas to close the vasotomy site. TURED is performed with the patient in the dorsal lithotomy position. A resectoscope with 24-French loop is used to remove a strip of tissue on the floor of the prostate just proximal to and including a portion of the verumontanum (Fig. 6.4a, b). The ducts are considered adequately opened by visualizing its dilated portion and the dye efflux. An 18-French Foley catheter is left in place for 24 h, and the patient is discharged the next day.

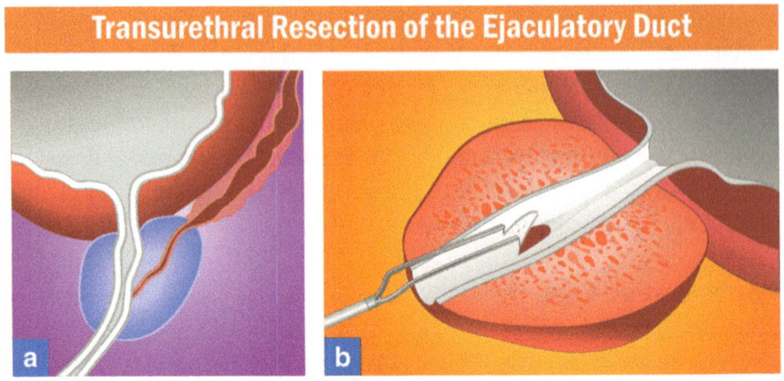

Fig. 6.4 Transurethral resection of the ejaculatory duct. Schematic representation of the ejaculatory duct entering into the prostatic urethra (**a**). A resectoscope loop is used to remove a strip of tissue on the floor of the prostate just proximal to and including a portion of the verumontanum (**b**)

Postoperative Care

An indwelling catheter is kept in place for 24–48 h, and patients are discharged the following day. Oral quinolone antibiotics and anti-inflammatory medication are prescribed for 5 days. Scrotal supporter is recommended for 1 week to avoid scrotal edema due to vasotomy. Frequent ejaculation is recommended 3–4 weeks postoperatively, and patients are monitored with monthly semen analyses.

Sperm Retrieval Techniques

Azoospermia, defined as the absence of spermatozoa in the ejaculate after centrifugation, is found in 1–3% of the male population and approximately 10% of infertile males [67]. In this scenario, two different clinical situations exist, i.e., obstructive and nonobstructive azoospermia (NOA). In obstructive azoospermia (OA), spermatogenesis is normal, but a mechanical blockage exists in the genital tract, somewhere between the epididymis and the ejaculatory duct, or the epididymis and vas deferens are totally or partially absent. Acquired OA may be due to vasectomy, failure of vasectomy reversal, postinfectious diseases, surgical procedures in the scrotal, inguinal, pelvic, or abdominal regions, and trauma. Congenital causes of OA include cystic fibrosis, congenital absence of the vas deferens (CAVD), ejaculatory duct or prostatic cysts, and Young's syndrome [67]. NOA comprises a spectrum of testicular histopathology resulting from various causes that include environmental toxins, medications, cryptorchidism, genetic and congenital abnormalities, varicocele, trauma, viral orchitis, endocrine disorders, and idiopathic. In both OA and NOA, pregnancy may be achieved through in vitro fertilization associated to intracytoplasmic sperm injection (ICSI) [2, 3].

The goals of surgical sperm retrieval are threefold: (1) to retrieve an adequate number of sperm both for immediate use and cryopreservation, (2) to obtain the best quality sperm possible, and (3) to minimize damage to the reproductive tract as to not jeopardize future attempts of sperm retrieval and testicular function. Several surgical methods have been developed to retrieve epididymal and testicular sperm from azoospermic men. Either percutaneous (PESA) [69] or microsurgical epididymal sperm aspiration (MESA) [2] can be successfully used to retrieve sperm from the epididymis in men with obstructive azoospermia. TESA can be used to retrieve sperm from the testes both in men with OA who fail PESA as well as in those with NOA [70]. Testicular sperm extraction (TESE) using single or multiple open biopsies [71, 72], and more recently microsurgery (micro-TESE), is indicated for men with NOA [4, 71–74]. Sperm can be easily obtained from infertile men with OA, whereas individuals exhibiting NOA have historically been the most difficult to treat.

Preoperative Planning and Patient Evaluation

It is important to distinguish whether the lack of sperm in the ejaculate is from an obstructive or nonobstructive process since the choice of the retrieval method is based on the type of azoospermia. History, physical examination, and hormonal analysis (FSH, testosterone) are undertaken to define the type of azoospermia. Together, these factors provide a ~90% prediction of its type (obstructive or nonobstructive) [75]. Men with obstructive azoospermia usually have normal testes and endocrine profile. Occasionally, the epididymis or the seminal vesicles are enlarged, or a cyst may be palpable on rectal examination. The presence of a low-volume (<1.5 mL) acidic (pH<7.0) azoospermic ejaculate with absent or low fructose and epididymal thickening is pathognomonic of obstructive azoospermia due to either congenital bilateral absence of the vas deferens or EDO; the differential diagnosis would be the presence of the vas in the latter. Approximately two thirds of men with CAVD have mutations of the cystic fibrosis transmembrane conductance regulator (CFTR) gene. Failure to identify a CFTR abnormality in a man with CBAVD does not rule out the presence of a mutation, since some are undetectable by routine testing methods. In such cases, the female partner should be offered CF testing before proceeding with treatments that utilize the sperm because of the high risk of the male being a cystic fibrosis (CF) carrier. If a CFTR gene mutation is identified (approximately 4% of female partners are carriers), testing should be offered to the male as well and counseling is recommended before proceeding with sperm retrieval and ICSI due to the risk of the transmission of cystic fibrosis to the offspring [75, 76]. Azoospermic men with idiopathic obstruction and men with a clinical triad of chronic sinusitis, bronchiectasis, and obstructive azoospermia (Young syndrome) may be at higher risk for CF gene mutations as well.

The serum FSH is a critical factor in determining whether a diagnostic testicular biopsy is needed to differentiate the type of azoospermia in men with normal semen volume. Elevated FSH and small testis are indicative of testicular failure (NOA); therefore, a testicular biopsy is not necessary for diagnostic purposes [76]. However, if sperm retrieval is being considered, a biopsy may be performed to determine whether spermatozoa are likely to be retrievable by future TESA or extraction. The absence of sperm in a biopsy specimen taken from a man with NOA, however, does not absolutely predict whether sperm are present elsewhere within the testicle [4, 77]. Conversely, men with FSH levels and semen volume within normal ranges may have either nonobstructive or obstructive azoospermia [78]. In such cases, there is no noninvasive method to differentiate obstructive from NOA, and a testicular biopsy is usually required to provide a definitive diagnosis. Testicular biopsy can be performed by a standard open incision technique or by percutaneous methods. Histology evaluation of testicular specimens may indicate the presence of normal spermatogenesis in cases of OA, while hypospermatogenesis, maturation arrest or Sertoli cell-only syndrome is seen in NOA ones.

All men with primary testicular failure of unknown origin should be offered karyotyping and Y microdeletion testing. The frequency of karyotypic abnormalities is reported to be 10–15% in men with NOA, and Klinefelter syndrome accounts for approximately two-thirds of the cases [79]. Genetic testing may provide prognostic information for sperm retrieval [75]. In contrast to patients with either partial or complete AZFc deletion, in whom sperm can be found within the testis, the chance of finding sperm in men with complete AZFa or AZFb deletions is unlikely [80, 81]. Genetic counseling should be offered to the male whenever a genetic abnormality is detected prior to performing ICSI with his sperm.

Sperm retrieval from the epididymis is indicated in obstructive cases only. Testicular sperm retrieval can be performed either in OA and NOA cases. In OA, testicular retrievals are carried out after failed epididymal retrieval or as a primary retrieval procedure in cases of absent epididymis or intense epididymal fibrosis. In NOA, testicular sperm retrievals are the only option to collect sperm.

Anesthesia

Sperm retrieval techniques can be safely performed using local, regional, or general anesthesia. In the authors' practice, percutaneous sperm retrievals are carried out under local anesthesia alone or in association with intravenous bolus infusion of a short-acting hypnotic agent (propofol). In both cases, a 10–15 mL solution of 2% lidocaine hydrochloride is injected around the spermatic cord near the external inguinal ring. In cases of using intravenous anesthesia, local injection of the anesthetic is performed after patient unconsciousness is achieved. Microsurgical sperm retrievals are performed under either local anesthesia, as described above, in association with continuous infusion of propofol using a syringe-drive automated-pump device, or epidural anesthesia.

Techniques

Percutaneous sperm retrievals can be carried out with short duration anaesthesia, and are associated with less postoperative discomfort. They are easily repeatable and less expensive than open macro- or microsurgical techniques. Moreover, the procedures do not require microsurgery training. Microsurgical techniques, on the other hand, are associated with better quality and higher numbers of sperm retrieved per attempt, which optimizes the opportunity to cryopreserve sperm for future ICSI procedures.

Percutaneous Sperm Retrieval

Typically, percutaneous sperm retrieval is performed on an outpatient basis using a needle attached to a syringe. The standard procedures are described below. Minor technical modifications have been added, but irrespective of the technique the main goal is to aspirate either epididymal fluid or testicular parenchyma for diagnostic or therapeutic purposes. Loupe magnification may be used to avoid injuring small vessels seen through the scrotal skin.

- *Epididymal Sperm Aspiration* (*PESA*). The epididymis is stabilized between the index finger, thumb, and forefinger while the testis is held with the palm of the hand. A 26-gauge needle attached to a 1-mL tuberculin syringe is inserted into the epididymis through the scrotal skin (Fig. 6.5a). Negative pressure is created, and the tip of the needle is gently moved in and out within the epididymis until fluid enters the syringe. The amount of epididymal fluid obtained during aspiration is often minimal (~0.1 mL), except in cases of CAVD in which 0.3–1.0 mL may be aspirated. The needle is withdrawn from the epididymis, and the aspirate is flushed into a 0.5–1.0 mL 37°C sperm medium. The tube containing the epididymal aspirate is transferred to the laboratory for microscopic examination. PESA is repeated at a different site of the same epididymis (from cauda up to the caput) and/or at the contralateral side until adequate number of motile sperm is retrieved. If PESA fails in retrieving motile sperm for ICSI, TESA is performed at the same operative time.
- *Testicular Sperm Aspiration* (*TESA*). The epididymis is stabilized between the index finger, thumb, and forefinger

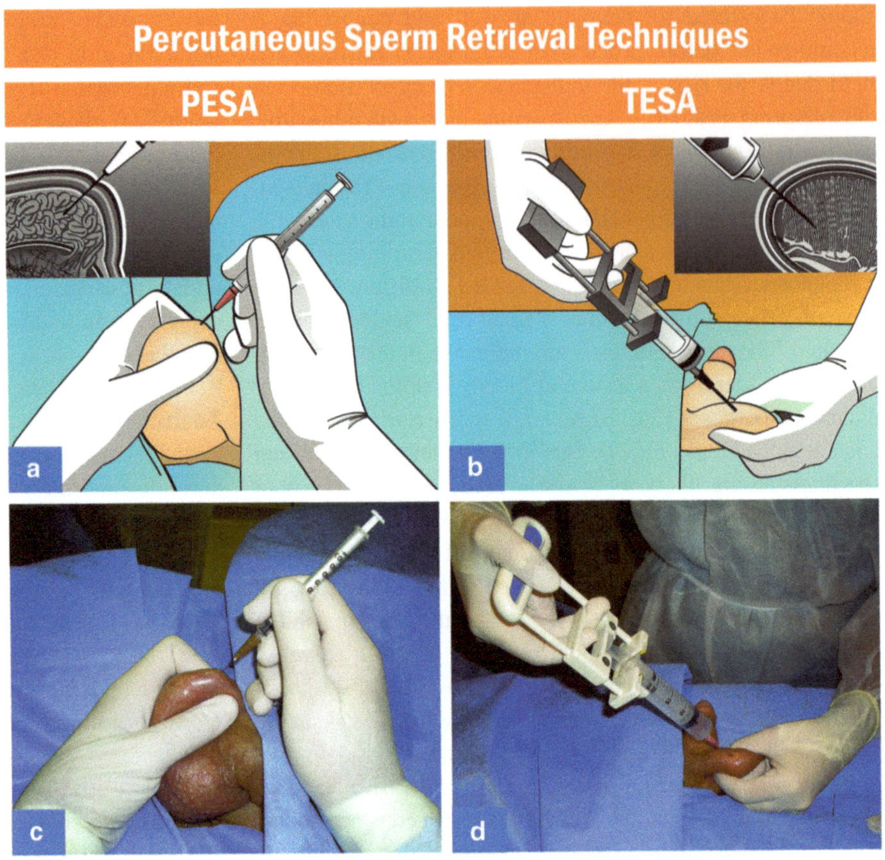

Fig. 6.5 Percutaneous sperm retrieval techniques [schematic representation (**a**) and (**b**) and intraoperative photographs (**c**) and (**d**)]. (**a**) Percutaneous epididymal sperm aspiration (PESA). Epididymis is stabilized between the index finger, thumb, and forefinger. A needle attached to a tuberculin syringe is inserted into the epididymis through the scrotal skin, and fluid is aspirated. (**b**) Testicular sperm aspiration (TESA). A 20-mL needled syringe connected to a holder is percutaneous inserted into the testis. Negative pressure is created and the tip of the needle is moved within the testis to disrupt the seminiferous tubules and sample different areas

while the anterior scrotal skin is stretched. A 18-gauge needle attached to a 20-mL syringe is connected to a syringe holder and is inserted through the stretched scrotal skin into the anteromedial or anterolateral portion of the superior testicular pole, in an oblique angle toward the medium and lower poles (Fig. 6.5b). Negative pressure is created by pulling the syringe holder while the tip of the needle is moved in and out within the testis in an oblique plane to disrupt the seminiferous tubules and sample different areas. When a small piece of testicular tissue is aspirated, the needle is gently withdrawn from the testis while the negative pressure is maintained. A pair of microsurgery forceps is used to grab the seminiferous tubules that exteriorize from the scrotal skin, thus aiding in the removal of the specimen. The specimen is flushed into a tube containing 0.5–1.0 mL warm sperm medium and is transferred to the laboratory for microscopic examination. TESA or TESE may be performed at the contralateral testis if insufficient or no sperm are obtained.

Microsurgical Sperm Retrieval

The microsurgical approach allows direct visualization of epididymal and seminiferous tubules with high magnification. These techniques have been associated with retrieval of higher sperm numbers and of better quality in MESA and higher retrieval success rates in micro-TESE. MESA does not compromise the success of future reconstructive procedure, if desired, since the damage to the epididymal tubule is minimal. The amount of testicular parenchyma removed in micro-TESE is low compared to open biopsy, which is particularly important to preserve testicular androgen production in men with NOA who already have compromised testicular function.

- *Epididymal Sperm Aspiration* (*MESA*). A transverse incision (2-3 cm) is made through the anesthetized scrotal layers, and the testis is exteriorized. The epididymis is examined and its tunica incised. An enlarged tubule is dissected and opened with sharp microsurgical scissors. Fluid exuding from the tubule is aspirated with a silicone tube or blunted needle attached to a 1-mL tuberculin syringe (Fig. 6.6). The aspirate is flushed into a 0.5–1.0 mL 37°C sperm medium. The tube containing the epididymal aspirate is transferred to the laboratory for microscopic examination. MESA is repeated at a different site of the same epididymis (from cauda up to the caput) and/or at the contralateral side until adequate number of motile sperm is retrieved. If MESA fails to retrieve motile sperm, TESA or TESE may be performed at the same operative time.
- *Testicular Sperm Extraction* (*micro-TESE*). The testis is delivered in the same fashion as described for MESA. A single, large, midportion incision is made in an avascular area of the tunica albuginea under 6–8× magnification, and the testicular parenchyma is widely exposed

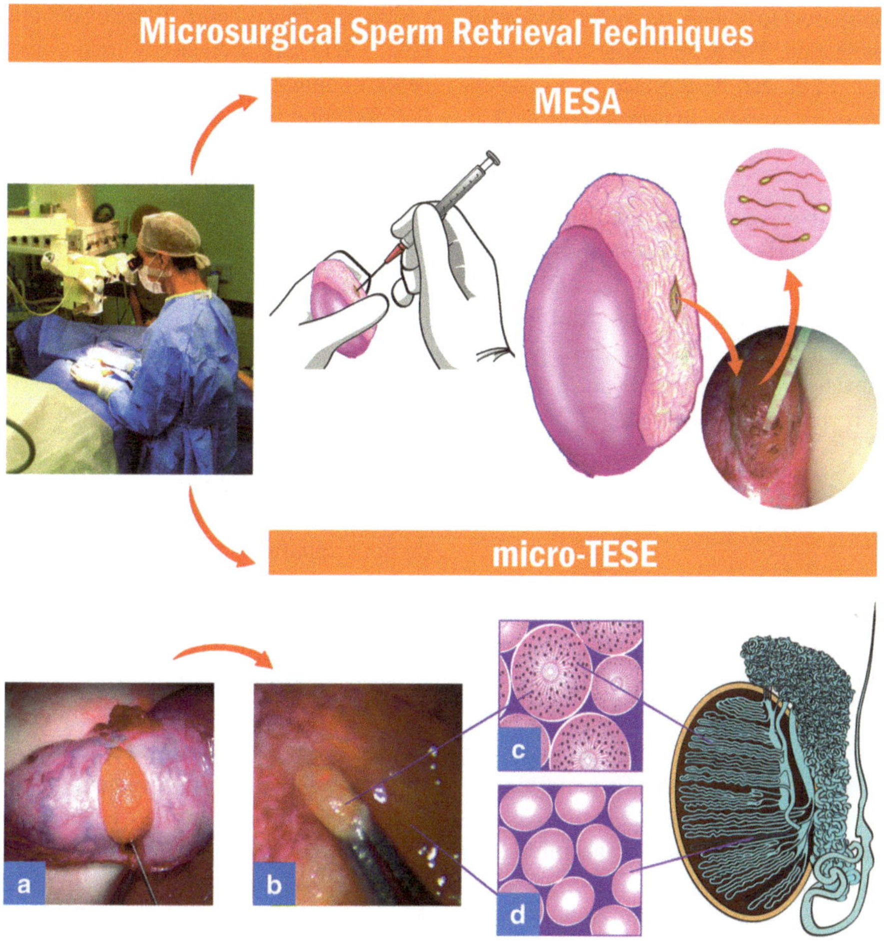

Fig. 6.6 Microsurgical sperm retrieval techniques. Operating microscope and microsurgical technique are used throughout the procedures. On the top: microsurgical epididymal sperm aspiration (MESA). After exposure of testis and epididymis, a dilated epididymal tubule is dissected and opened. Fluid is aspirated, diluted with sperm medium, and sent to the laboratory for examination. On the bottom: microsurgical testicular sperm extraction (micro-TESE). (**a**) After the testicle is exteriorized, a single and large incision is made in an avascular area of the albuginea to expose the seminiferous tubules. (**b**) Dilated tubules are identified and removed with microforceps (intraoperative photograph at ×25 magnification). (**c**) Illustration of the histopathology cross section of a dilated seminiferous tubule with active spermatogenesis. (**d**) Illustration of the histopathology cross section of a thin tubule with Sertoli cell-only syndrome

(Fig. 6.6a). Dissection of the testicular parenchyma is carried out at ×16–25 magnification searching for enlarged seminiferous tubules. The superficial and deep testicular regions may be examined, if needed, and microsurgical-guided testicular biopsies are performed by removing enlarged tubules which are more likely to harbor active spermatogenesis (Fig. 6.6b–d). If enlarged tubules are not seen, then any tubule different than the remaining ones in size is excised. If all tubules are identical in appearance, random micro-biopsies (at least three at each testicular pole) are performed. Testicular tissue specimens are placed at an outer-well dish containing sperm media. Specimens are washed grossly to remove blood clots and are sent to the laboratory for processing and search for sperm. The albuginea and scrotal layers are closed using nonabsorbable and absorbable sutures, respectively.

Conventional Testicular Open-Biopsy Sperm Extraction (TESE)

Single or multiple open testicular biopsies may be taken to obtain sperm in both OA and NOA, but mainly in cases of NOA. TESE can be also used as a diagnostic tool to obtain testicular parenchyma for histopathology analysis and search of sperm prior to the ICSI cycle. A transverse 2-cm incision is made through the anesthetized scrotal skin, cremaster, and parietal tunica vaginalis. Conventional TESE is carried out without magnification. A small self-retaining eyelid retractor is placed to improve exposure of the tunica albuginea since the testis is not exteriorized. The tunica albuginea is incised for approximately 0.5–1 cm. Gentle pressure is made onto the testis to extrude testicular parenchyma out of the small incision. A fragment of approximately 5×5×5 mm is excised with sharp scissors and placed promptly in sperm culture medium. The specimen is sent to the laboratory for processing and microscopic examination. The albuginea is closed using nonabsorbable sutures. TESE can be repeated in a different testicular pole if the multiple biopsies approach is selected.

Postoperative Care

Patients are discharged at the same day and can return to normal activities 1 and 3 days after percutaneous and open techniques, respectively. Scrotal ice packing and supporter is recommended to control edema and alleviate pain. Patients should restrain from ejaculation and strenuous physical activity for approximately 7–10 days. Oral analgesics are prescribed, but pain complaint is often minimal.

Expert Commentary

Varicocele Repair

The surgical treatment of varicocele should aim to achieve the highest improvement in the male fertility status with lower complication rates. Increase in the likelihood of spontaneous pregnancy after treatment is difficult to ascertain due to a variety of factors that include the lack of a uniform post-treatment follow-up interval and the female factor parameters, such as age and reproductive health. Therefore, we believe that the ultimate treatment goal is to improve the male fertility status regardless of the method to be used for conception (unassisted or assisted). The ideal surgical technique should aim for ligation of all internal and external spermatic and cremasteric veins, with preservation of spermatic arteries and lymphatics. This only can be achieved by using the inguinal or subinguinal microsurgical techniques. The urologist who opts to treat varicocele using microsurgery should obtain appropriate training. It is also important to have adequate microsurgical instruments and a binocular operating microscope with foot-control zoom magnification. Microsurgical varicocelectomy, either using inguinal or subinguinal approaches, requires more skill as compared to other surgical modalities because a higher number of internal spermatic vein channels and smaller-diameter artery are seen at the level of the inguinal canal. However, the routine use of microsurgery during varicocele repair may aid the urologist to master his/her microsurgical skills, which will be of great benefit when performing more demanding and less frequent microsurgical procedures, such as vasovasostomies and vasoepididymostomies. In our practice, varicocelectomy is performed on both sides at the same operative time in the presence of bilateral palpable varicocele. When a clinically palpable varicocele is identified in one side, the contralateral cord is examined using a pencil-probe Doppler (9 MHz) stethoscope to determine if a subclinical varicocele exists. If so, it is treated at the same time as the coexistent clinical varicocele. This is based on the observation that altered blood flow after varicocelectomy may unmask an underlying venous anomaly and result in clinical varicocele formation [82, 83]. Loupe magnification may be used to facilitate the ligation of the dilated varicose veins; however, it is often insufficient for proper identification of both testicular arteries and lymphatics. In our early experience with loupe-assisted varicocele repair, papaverine instillation was needed in most cases to aid in the identification of arterial pulsations.

In our hands, the subinguinal microsurgical varicocelectomy using the operating microscope is the method of choice to treat varicocele-associated infertility. In a group of 384 treated individuals, the mean operative time was 113 and 90 minutes, and the number of veins ligated was 11 and 6, in unilateral and bilateral varicocelectomies, respectively. We do not use intraoperative Doppler to identify testicular arteries. Artery identification is based on microscopic examination using 10–16× magnification. Only occasionally papaverine instillation is used to facilitate positive identification of arterial pulsation. Using this approach, we were able to identify at least one artery in 97.6% of the cases. Recurrence was reported in 4 cases (1.0%), and no hydrocele formation occurred. Improvement (≥15% change from baseline preoperative values) in at least one semen parameter (count, motility, and morphology) was observed in 68.5% of the individuals. Spontaneous pregnancies were achieved in 58 couples (33.7%) whose male partners had been treated in a follow-up period of 18 months.

Reconstructive Surgery of the Vas Deferens and Epididymis

Both microsurgical reconstruction and sperm retrieval combined with IVF/ICSI are effective treatments for infertility due to obstructive azoospermia. A choice between the two must be based not only on the needs and preferences of the individual couple but also on the couple's clinical profile taking into account the cause of azoospermia and any coexisting factors in the female partner. Consequently, both partners should be evaluated thoroughly before making a specific treatment recommendation. Cost issues also play a role in the decision-making process since ART is seldom reimbursed by health insurance companies in most countries. Most importantly, infertility clinics and doctors should not limit couple's options for treatment based on their own technical limitations but always provide all treatment options available for that particular case scenario.

In experienced hands, reconstructive surgery of the male reproductive tract can be highly successful. Microsurgical techniques are clearly superior compared to macrosurgical or loupe-assisted anastomoses [35, 84]. Sperm return to the ejaculate after microsurgery in 50–95% of patients, and 30–75% of couples can be expected to achieve pregnancy without assisted reproductive techniques (ART). The likelihoods of sperm appearance in the semen and of pregnancy after microsurgical vasectomy reversal are inversely related to the duration of time since vasectomy [35]. Other factors that affect success rates include the gross appearance of vas fluid at the time of surgery, the presence or absence of sperm in the vas fluid and their quality, the length of the remaining segment adjacent to the epididymis, the age of the female partner, and the experience of the microsurgeon.

Currently, several programs offer microsurgical training for urology residents. Short-term microsurgery courses are of limited value; however, they can help urologists acquire the initial skills needed to use microsurgery in a routine basis. It is important to emphasize that microsurgical procedures for male infertility may be very demanding; therefore, one should only embark on performing either vasovasostomies or vasoepididymostomies after mastering microsurgical skills in the microsurgery laboratory using animals or synthetic models. Among several predictors for a successful microsurgical reconstruction of the male reproductive system, surgeon's skills are the most relevant for treatment outcomes. Surgeon's skills are crucial when vasoepididymostomies are needed, which frequently cannot be anticipated. Therefore, mastering both vasovasostomy and vasoepididymostomy techniques allows for real-time decision-making without compromising clinical results. Even with the advent of robotic surgery and its application in the infertility field with an expectation to "level the playing field" for surgeons with modest experience, training is still necessary.

Microsurgical vasovasostomy (VV) is the method of choice for vasectomy reversal. We prefer the end-to-end modified one-layer anastomosis using a vas clamp in such cases. Our approach is to use the intussusception technique whenever vasoepididymostomy (VE) is required. The reasons are that these techniques are less time-consuming and yield similar patency rates compared to other approaches. If no fluid effluxes from the testicular vasal end, VV is still our preferred approach in cases the if epididymis is soft on palpation and vasectomy interval is less than 15 years. VE is left for cases in which thick toothpaste-like and no-sperm vasal fluid is noted, particularly in vasectomy with vasectomy intervals greater than 15 years. The surgical principles for a successful anastomosis are (1) adequate mucosa-to-mucosa approximation, (2) watertight and tension-free anastomosis, (3) adequate blood supply and healthy mucosa and muscularis, and (4) atraumatic anastomotic technique. It is not advisable to perform varicocele repair at the same time of vasectomy reversal. In vasectomized men, vasal veins are often compromised which would jeopardize venous return after ligation of internal and external spermatic veins. If necessary, varicocelectomy may be performed 6 months later when new venous and arterial channels are formed around the anastomosis. Although vasectomy reversal can be performed after percutaneous epididymal aspiration (PESA), we advise our patients that reconstructive surgery is likely to be unsuccessful in such cases. Therefore, if an undecided patient ultimately opts for sperm retrieval coupled with ICSI, we prefer to retrieve sperm from the testis without jeopardizing the chances of a future reconstruction.

In our case series involving 180 vasectomy reversals, overall patency following the modified one-layer vasovasostomy ($N=126$) and the intussusception vasoepididimostomy ($N=54$) techniques were 79% and 44%, respectively. Spontaneous pregnancy rates were respectively 42% and 33% in a mean follow-up period of 15 months. Three out of five men with patent anastomosis achieved normal semen analysis postoperatively. We consider that the reconstructive procedure failed when no sperm is found in the semen analysis by 6 months after vasovasostomy or by 18 months after vasoepididymostomy. In our case series, late obstructions after patency occurred in 2% and 25% after vasovasostomies and vasoepididymostomies, respectively. Interestingly, all patients who had late obstructions were cigarette smokers and/or had obstruction intervals greater than 8 years. In our case series, most pregnancies occurred between 8 and 13 months postoperatively (range 3–24 months).

Transurethral Resection of Ejaculatory Ducts

Resection of the ejaculatory ducts is a hazardous procedure. The typical patient with EDO is young and has a small prostate. Therefore, TURED is carried out very close to the bladder neck, rectum, and sphincter. If a midline cyst is present, resection is performed to completely unroof the cyst. If not, the prostatic side of the verumontanum is resected until a dilated portion of the ejaculatory duct is seen. Resection is performed with pure cutting to avoid thermal injury to the proximal ejaculatory duct. We feel more comfortable placing a finger in the patient rectum to prevent rectal injury during resection and having methylene blue injected through the vasotomy site. Resection is completed by positive identification of free dye efflux into the urethra.

In our series involving 25 patients with complete or partial obstructions, TURED results varied according to the etiology of obstruction, congenital, or acquired. Semen quality improvement (ejaculate volume, sperm count, and motility) was observed in 85% of the patients in the congenital group. Spontaneous pregnancy was obtained in 60% of these patients an average of 7 months postoperatively. In the group of patients with secondary EDO, seminal improvement was detected in only 30% of the individuals, and only one spontaneous pregnancy was achieved. Two individuals (one in each group conceived via assisted reproduction techniques). Complication rates were similar in both groups (35%) and included reflux of urine to the unroofed cyst cavity with consequent impairment of the semen parameters, retrograde ejaculation, and one case of epididymitis with obstruction. Rectal injury or incontinence was not reported in our case series.

Sperm Retrieval Techniques

The adoption of strict criteria to diagnose OA is crucial for obtaining high success retrieval rate in the range of 90–100% using percutaneous techniques. Using PESA, our approach is to perform the first aspiration at the corpus epididymis, then proceed up to the caput if necessary since aspirates from the cauda are usually rich in poor-quality senescent spermatozoa, debris, and macrophages. Most cases of PESA failures are not necessarily technical failures because immotile spermatozoa are found. However, in certain cases of epididymal fibrosis due to multiple PESA attempts or postinfection, PESA may be ineffective to retrieve sperm. In these cases, PESA can be attempted at the contralateral epididymis, or TESA can be applied successfully if there is normal spermatogenesis. Some authors advocate that MESA allows the collection of larger and cleaner quantities of sperm than PESA, but this debate seems trivial. In our series of 142 men with OA, cumulative successful retrieval rates of PESA and/or TESA was 97.9%, and an adequate number of motile sperm for cryopreservation was obtained in approximately one-third of the cases [85]. In rare circumstances, we perform MESA for sperm retrieval in OA men with coagulation disorders. Successful sperm retrieval (SRR) is obtained in over 85% of the cases using PESA alone, but more than one aspiration is required in several cases. In cases of failed PESA, TESA is adequate to obtain sperm in practically all cases. Motile spermatozoa is obtained in approximately 73% of the cases after the first or second PESA aspirations, and TESA is carried out as a rescue procedure after failed PESA in about 14% of the individuals. In our series, success in sperm retrieval using percutaneous techniques and pregnancy outcomes by ICSI were similar in the vasectomy, CBAVD, and postinfectious etiology categories [85]. Either epididymal or testicular spermatozoa retrieved from these men exhibits similar reproductive potential. In our series of OA, live birth rates by ICSI were 40.2%. Also, our data indicated that ICSI outcomes using fresh epididymal and testicular spermatozoa retrieved from men with OA are comparable to those obtained with ejaculated sperm [86]. We routinely freeze epididymal and testicular spermatozoa which are left over from the ICSI cycle.

In cases of NOA, our approach is to retrieve sperm by TESA only in the favorable prognosis cases such as the ones with a previously successful TESA attempt or those with testicular biopsy result showing hypospermatogenesis. If TESA fails, however, we neither perform a second aspiration in the same testis, at the same operative time, nor convert it to an open procedure to avoid the risk of hematoma and testicular injury. Extensive bleeding is often seen during a rescue TESE after a failed TESA, and enlarged seminiferous tubules are difficult to identify even using the operating microscope. In these occasions, we opt to perform TESA or TESE at the contralateral testis. For NOA patients without previous diagnostic testicular biopsy or TESA attempt, our choice is to perform sperm extraction using micro-TESE. In the IVF laboratory, it is far less technically demanding and labor intensive to extract spermatozoa from small volume specimens than large pieces of testicular tissue that must be dissected, red blood cells lysed, and the rare spermatozoa searched for in a tedious fashion under an inverted microscope. TESE/micro-TESE may be scheduled either for the day of oocyte collection or the day before. In a previous study, we observed that optimal fertilization with ICSI using surgically retrieved sperm had been obtained when the time interval from hCG administration to microinjection did not exceed 44 h [87]. Testicular tissue sperm processing, searching, and selection of viable spermatozoa for ICSI may take several hours in NOA cases. Our laboratory takes approximately 11.6 min to handle a single testicular spermatozoon from processing to microinjection in NOA, and only 5.5 min in OA. In other words, the average time required to perform ICSI in a typical NOA treatment cycle involving 8–12 metaphase-II oocytes is approximately 2 h. For these reasons, we elect to perform micro-TESE on the day before oocyte collection when a busy next day IVF laboratory workload is anticipated. According to our experience involving approximately 200 individuals with NOA, SRR rate was 55.7% and sperm could be obtained in similar rates in all etiology categories of cryptorchidism, orchitis, genetic, radio-/chemotherapy, and idiopathic. Testicular histopathology results were predictive of sperm collection using both TESA and micro-TESE [74, 77]. SRR by TESA was 100% and 82.3% in our group of NOA presenting with either hypospermatogenesis or a history of previous successful TESA attempt. Using micro-TESE, SRR rates were significantly higher than TESA in cases of SCO and maturation arrest (39.2% vs. 22.8%) [77]. Both methods yielded similar SRR of 100% in cases of hypospermatogenesis. According to our data, the chances of retrieving spermatozoa (odds ratio [OR] = 43.0; 95% confidence interval [CI]: 10.3–179.5) and of achieving a live birth by ICSI (OR = 1.86; 95% CI: 1.03–2.89) were significantly increased in couples whose male partner had obstructive rather than NOA [88]. These findings indicate that the reproductive potential of infertile men undergoing ART is related to the type of azoospermia.

Five-Year View

Varicocele Repair

In a recent systematic review comparing different surgical modalities to treat varicocele for male infertility [30], it was concluded that open microsurgical inguinal or subinguinal varicocelectomy techniques resulted in higher spontaneous pregnancy rates and fewer recurrences and postoperative complications than laparoscopic, radiologic embolization and macroscopic inguinal or retroperitoneal varicocelectomy techniques. Postoperative complications vary with surgical techniques. Hydrocele formation is the most common complication of varicocelectomy, with the incidence ranging from 0% to 10%. The lowest and highest reported hydrocele formation rates are seen in the microsurgical and in the high retroperitoneal series, respectively. Recurrences range from 0% to 35% varying with varicocelectomy techniques. Overall recurrence rates are low for microsurgical varicocelectomy and high for retroperitoneal and macrosurgical inguinal approaches [30]. Accidental testicular artery ligation during microsurgical varicocelectomy has been reported in about 1% of the cases, and it may cause testicular atrophy [89]. It has been recently demonstrated that the concomitant use of intraoperative vascular Doppler during microsurgical varicocelectomy allows more arterial branches to be preserved, and more internal spermatic veins are likely to be ligated [90]. Table 6.2 summarizes treatment results, postoperative recurrence, hydrocele formation, and spontaneous pregnancy rates among different techniques.

Varicocelectomy studies report significant improvements in one or more semen parameters in approximately 65% of men [91]. The mean time for semen improvement and spontaneous pregnancy after surgery is approximately 5 and 7 months, respectively [92]. However, it is still unknown why fertility potential is not always improved after varicocelectomy. Studies evaluating predictors for successful varicocele repair indicate that infertile men either with higher preoperative semen parameters or undergoing varicocele repair for large varicoceles are more likely to show postoperative semen parameters improvement [91, 93]. It was also shown that men who achieved a postoperative total motile sperm count greater than 20 million and decreased sperm DNA fragmentation after surgical varicocelectomy were more likely to initiate a pregnancy either spontaneously or via assisted conception [94, 95]. The distribution of antioxidant enzyme genes genotype of infertile men with varicocele has been recently determined. It has been suggested that genetic polymorphisms in the glutathione S-transferase T1 gene may affect individual response to varicocelectomy [96]. Conversely, reduced preoperative testicular volume, elevated serum FSH levels, diminished testosterone concentrations, subclinical varicocele, as well as the presence of Y chromosome microdeletions seem to be negative predictors for fertility improvement after surgery [27, 97].

It is recommended to treat bilateral clinical varicoceles at the same operative time [15]. The management of infertile men with a unilateral clinical varicocele and a subclinical one at the contralateral side, on the other hand, is a matter of debate. Zheng et al. found that bilateral varicocelectomy had no benefit over the left clinical varicocelectomy [98]. However, the authors used a retroperitoneal approach for vein ligation which has been shown to be associated with high recurrence rate. Elbendary et al., in a prospective trial, observed that the magnitude of change in sperm count and motility, and the spontaneous pregnancy rates were significantly higher in the group of men who had bilateral varicocele repair [99].

It is still debatable whether varicoceles can cause or contribute to azoospermia. A recent meta-analysis reported appearance of sperm in ejaculates of 39% of azoospermic individuals whose varicoceles had been treated [28]. In the aforesaid study, testicular histopathology results were predictive of success. Postoperatively, appearance of sperm in the ejaculates was increased 9.4-fold in patients with biopsy-proven hypospermatogenesis (HS) or maturation arrest (MA) than in Sertoli cell-only syndrome. Although the use of motile ejaculated sperm is preferred for ICSI [86], persistent azoospermia after varicocele repair is still a potential problem and sperm extraction before ICSI will be inevitable for many individuals. In these circumstances, successful sperm retrieval rates of 60% have been reported using testicular microdissection (micro-TESE) sperm extraction [100]. It has been suggested that varicocele repair may maximize the chances of retrieving sperm for ICSI in azoospermic men with clinical varicoceles [101].

Table 6.2 Treatment results for varicocele repair in infertile men. Postoperative recurrence, hydrocele formation, and spontaneous pregnancy rates among different techniques[a]

Technique	Recurrence rate (%)	Hydrocele formation rate (%)	Spontaneous pregnancy rate (%)
Retroperitoneal high ligation [30, 32]	7–35	6–10	25–55
Laparoscopic [30, 32]	2–7	0–9	14–42
Macroscopic inguinal [30, 32]	0–37	7	34–39
Microscopic inguinal or subinguinal [15, 30, 32, 93]	0–0.3	0–1.6	33–56

[a]Values are expressed as range

In cases that a clinical palpable varicocele coexists with impaired semen quality, its surgical repair has been shown to be the best treatment option [11]. Recent meta-analyses demonstrated a beneficial effect of varicocelectomy on the fertility status of infertile men with clinical varicocele [11, 102, 103]. Overall, sperm concentration, motility, and morphology are increased by 9.7 million/mL, 10% and 3%, respectively, after varicocelectomy [104]. Sperm DNA integrity is also increased after varicocele repair [95, 105, 106]. Spontaneous pregnancy rates are higher in men with treated (33–36%) as compared to untreated varicoceles (15–20%) [11, 103]. Our group has recently demonstrated that treatment of clinical varicoceles may improve the outcomes of ICSI in couples with varicocele-related infertility [12]. In our study, the chances of live birth were significantly increased by 1.9-fold, while the chances of miscarriage were reduced by 2.3-fold if the varicocele had been treated before assisted conception.

Reconstructive Surgery of the Vas Deferens and Epididymis

Over the past two decades, treatment options for couples with reconstructible obstructive azoospermia had a marked improvement. Refinements in microsurgical reconstruction as well as advances in ART, specifically sperm retrieval techniques for ICSI, have led to improved outcomes and cost-effectiveness. According to the most recent data, microsurgical reconstruction of the vas remains a cost-effective, reliable, and effective means of restoring fertility in the majority of men who have previously undergone vasectomy when the reconstruction is performed by an experienced microsurgeon [107–111]. However, data comparing surgical reconstruction versus sperm retrieval/ICSI are neither randomized nor homogenous. Therefore, a comprehensive understanding of the factors that can affect outcomes, overall cost, and the morbidity associated with each treatment modality, taking into account the institution providing the treatment, is recommended.

Overall, patency/pregnancy rates following microsurgical vasovasostomy and vasoepididimostomy are 92/55% and 78/40%, respectively [35, 38, 45, 48, 50, 57–60, 112] (Table 6.3). Most pregnancies occur within 24 months after surgery. Pregnancy rates are related to the time elapsed from vasectomy and reversal and female age. Although female partner's age obviously does not affect patency rates after vasectomy reversal, it does affect pregnancy rates (14% in women aged >40 years vs. 56% in those aged <39 years) [47]. Pregnancy rates are also lower after longer duration of vasal obstruction. Approximately 30–40% of couples achieve pregnancy following surgical reconstructions performed in obstruction intervals greater than 15 years as compared to >50% in shorter intervals [41, 48]. Vasectomy reversal has been shown to be feasible in patients who failed percutaneous epididymal aspiration (PESA). Marmar et al. showed that PESA procedures cause limited trauma to the epididymis, and up to 50% pregnancy rates may be obtained in vasectomy reversal after PESA; however, success is higher for couples whose female partners have 37 years old or less [113].

Table 6.3 Treatment results for vasovasostomy and vasoepididymostomy

Author	Patients (*N*)	Technique	Patency rate (%)	Pregnancy rate (%)
Vasovasostomy				
Belker et al. [35]	1,247	Modified one-layer	89	57
		Two-layer	86	51
Boorjian and Lipkin [38]	159	Two-layer	95	83
Chan and Goldstein [45]	1,048	Two-layer	99	54
Kolettis et al. [50]	34	Both	76	35
Vasoepididimostomy				
Silber [57]	139	End-to-end	78	56
Thomas [58]	137	End-to-side	79	50
Berger [59]	12	Triangulation intussusception	92	NR
Marmar [60]	9	Modified intussusception	78	22
Chan et al. [112]	68	Triangulation intussusception	84	40
Schiff et al. [48]	153	End-to-end	73	NR
		End-to-side	74	
		3-Suture intussusception	84	
		2-Suture intussusception	80	

Type of anastomosis, patency, and spontaneous pregnancy rates using different techniques. *NR* not reported

As patency and pregnancy rates yielded by the existing technically demanding surgical procedures do not reach 100%, efforts continue to be made in order to widen the options for reconstructive repair. Several modifications have been suggested and include intussusception vasoepididymostomy anastomotic techniques, the use of novel biomaterials/sealants and absorbable and nonabsorbable stents, and the use of robotics [48, 56, 112, 114–117]. Recent modifications to the conventional vasoepididymostomy techniques simplified the anastomoses. In a prospective study, Chan et al. reported overall patency and pregnancy rates of 84% and 40% using the intussusception technique [114]. These findings were confirmed by Schiff et al. who reported patency and pregnancy rates of approximately 82% and 45%, respectively, using simplified intussusception techniques [48]. It is suggested that anastomoses are more watertight by using intussusception techniques; therefore, granuloma formation is decreased. Since pregnancy rates following vasoepididymostomy are below 50% and late failures occur in approximately 20% of the cases, it is tempting to retrieve sperm intraoperatively for cryopreservation, particularly in cases of difficult reconstruction. However, a recent cost-analysis study demonstrated that sperm harvesting and cryopreservation during vasectomy reversal is not cost-effective [51]. The rationale of using sealants around the anastomotic site is to decrease operative time and to simplify the procedure without compromising success rates. Fibrin sealant can stimulate the coagulation cascade producing a fibrin seal around the anastomosis. When mixed with thrombin and calcium, fibrinogen is converted to fibrin monomer which in turn is converted to a stable cross-linked fibrin polymer [115]. Ho et al. achieved 85% patency rates and 23% pregnancy rates using three transmural 9-0 sutures and fibrin glue in a mean follow-up of 6.2 months [115]. There are concerns, however, about the potential contact of the glue with the vas lumen, which may result in possible obstruction, and also about transmission of viral disease because fibrin glue is derived from pooled plasma [112]. The use of nonabsorbable polymeric stent has been reported only in animal model. Preliminary results showed 100% patency rates in a follow-up of 39–47 weeks, and the total sperm count was significantly higher in the stented group [116]. The use of robotics for microsurgical procedures is also a novel concept. The rationale to add this technology to the already existing armamentarium relies on the possibility of controlling the physiologic static tremor, enhancing visual magnification (up to 100× when using a digital microscopic camera), and improving the ergonomics [117]. Animal studies suggest that robotic-assisted vasectomy reversal is easier to perform and yields better pregnancy rates than microsurgical reversal [118]. In a preliminary experience in humans, Parekattil et al. reported shorter operative time and higher postoperative sperm count for robot-assisted vasectomy reversal as compared to the microsurgical technique [115]. However, the advantages of the robot over an experienced microsurgeon are yet to be proven in larger series. A robotic system costs more than 1 million dollar, and its annual maintenance surpasses U$100,000. These cost issues will certainly represent a barrier to the widely adoption of robotics into microsurgical urologic practice.

Transurethral Resection of Ejaculatory Ducts

EDO is a treatable cause of male infertility, but the diagnosis is difficult to make, particularly in cases of partial obstruction. Transrectal ultrasound is valuable but not specific. It is suggested that adjunctive procedures, such as magnetic resonance imaging, chromotubation, seminal vesicle aspiration, seminal vesicle scintigraphy, and ejaculatory duct manometry, are more sensitive for diagnosis [119–121]. Transurethral resection of ejaculatory ducts (TURED) remains the treatment of choice. However, less invasive approaches using balloon dilation coupled or not with transurethral incision of the ejaculatory ducts have been yielded with similar results with fewer complications than TURED [122, 123].

Sperm Retrieval Techniques

The best technique for sperm retrieval in men with obstructive and NOA is yet to be determined. To date, no randomized controlled trial has compared the efficiency of these strategies, and thus, current recommendations are based on cumulative evidence provided by descriptive, observational, and controlled studies [124]. Percutaneous epididymal aspiration can be performed without surgical scrotal exploration, repeatedly, easily, at low cost, without an operating microscope or expertise in microsurgery, under local anesthesia, and it is associated with minimal postoperative discomfort. Microsurgical aspiration has the advantage of retrieving larger number of sperm which facilitates cryopreservation, and it is associated with a reduced risk of hematoma [109]. Meta-analysis results demonstrated no significant difference in any outcome measure between the use of epididymal or testicular sperm in men with OA [125]. The etiology of the obstruction and the use of fresh or frozen-thawed epididymal/testicular sperm do not seem to affect ICSI outcomes in terms of fertilization, pregnancy, or miscarriage rates [85, 109, 126]. In cases of NOA, the efficiency of TESA for retrieving spermatozoa is lower than TESE [127–129], except in the favorable cases of men with previous successful TESA or testicular histopathology showing hypospermatogenesis. In these circumstances, SRR rates may be as high as 100% [77]. In a recent systematic review, the mean reported SRR for TESE was 49.5% [128]. TESE with multiple biopsies resulted in higher SRR than fine-needle aspiration, a variation of TESA, especially in cases of Sertoli cell-only (SCO) syndrome and maturation arrest [128]. In NOA, current evidence suggests that micro-TESE performs better than conventional TESE or TESA in cases of SCO, where tubules containing active focus of spermatogenesis can be positively identified using microsurgery. Sperm retrieval rates ranging from 35% to 77% have been reported with micro-TESE [74, 128–132]. To allow for adequate healing and the resumption of spermatogenesis, the minimum recommended interval between sperm retrieval procedures in NOA is 3–6 months [129–131].

Postoperative complications of sperm retrieval techniques include persistent pain, swelling, infection, hydrocele, and hematoma [127–134]. The development of intratesticular hematoma has been observed in most patients undergoing TESE with single or multiple biopsies based on postoperative ultrasound results but they often resolve spontaneously without compromising testicular function [132]. In the larger-volume standard testicular biopsy, the risk of transient or even permanent testicular damage (such as complete devascularization) can result in decreased serum testosterone levels [129, 131]. Less invasive techniques, such as TESA and micro-TESE, aim to reduce the incidence of short and long-term complications. Several studies have documented a lower frequency of complications following micro-TESE compared with the conventional technique [128–131, 133]. Using micro-TESE, proper identification of testicular vessels under the tunica albuginea is made prior to the placement of an incision in the albuginea. The use of optical magnification and microsurgery technique allows the preservation of intratesticular blood supply, as well as the identification of tubules more likely to harbor sperm production [131]. Therefore, efficacy of sperm retrieval is improved while the risks of large tissue removal are minimized. The small amount of tissue extracted also facilitates sperm processing [130]. In certain groups of patients such as those with Klinefelter syndrome, who already have diminished androgen production, a significant decrease on serum testosterone has been documented following micro-TESE [130]. However, testosterone levels return to the presurgical values in most Klinefelter men in a 12-month follow-up period. It is recommended that sperm retrieval should be performed by surgeons who have training in the procedures because of the potential serious postoperative complications [129].

It has been suggested that the clinical outcomes of ICSI using testicular sperm extracted by TESA or micro-TESE in NOA are significantly lower than those obtained with either ejaculated or epididymal/testicular sperm from men with OA [86, 88, 125]. Testicular spermatozoa of men with severely impaired spermatogenesis seems to have decreased fertility potential and have a higher tendency to carry deficiencies such as the ones related to the centrioles and genetic material, which ultimately affect the capability of the male gamete to activate the egg and trigger the formation and development of a normal zygote and a viable embryo [134]. From the limited data available, it is suggested that the sperm retrieval technique itself has no impact on ICSI success rates [128]. However, frozen-thawed surgically retrieved sperm from NOA men have significantly impaired reproductive potential than fresh ones [125, 134]. Meta-analysis results showed that fertilization rates by ICSI remained similar, but implantation was significantly higher (by 73%) with the use of fresh compared to frozen-thawed testicular sperm [135].

The question of whether or not ICSI using sperm retrieved from men with either OA or NOA might be associated with increased risk for birth defects is still unresolved. In vitro fertilization, in general, is associated with multiple gestation and an increased risk of congenital abnormalities (including hypospadias) [136]. ICSI, in particular, carries an increased risk of endocrine abnormalities, as well as epigenetic imprinting effects [136]. Although the absolute risk of any of these conditions remains low [136–139], current data is limited and study populations are heterogenic. It is therefore recommended that well-defined groups of ICSI with ejaculated sperm, ICSI with epididymal sperm and ICSI with testicular sperm, and a control group of naturally conceived children are closely followed up.

Key Issues

- Varicocele treatment is indicated for men with clinically palpable varicocele and abnormal semen parameters. Open microsurgical inguinal or subinguinal techniques are considered the best treatment modalities because they result in higher spontaneous pregnancy rates and fewer recurrences and postoperative complications than laparoscopic, radiologic embolization and macroscopic inguinal or retroperitoneal varicocelectomy techniques. There are no absolute predictive factors for a successful varicocele repair, and existing evidence does not support the recommendation for treating infertile men with subclinical varicocele.
- Surgical repair of varicocele improves semen parameters and functional markers of oxidative stress and DNA integrity. The chance for either spontaneous or assisted conception is increased after repair of clinical varicoceles. Recovery of spermatogenesis can be achieved after repair of clinical varicocele in infertile men with NOA. Testicular histopathology is predictive of success, and men with maturation arrest and hypospermatogenesis are more likely to ejaculate motile spermatozoa after surgery. Also, the chance of retrieving testicular sperm for ICSI is optimized in non-obstructed azoospermic men with treated clinical varicocele.
- Men with obstructive azoospermia may father children either by surgical correction of the obstruction, which may allow the couple to conceive naturally, or retrieval of sperm directly from the epididymis or testis, followed by ICSI.
- Best results with reconstructive surgery of the male reproductive tract are achieved by surgeons who have the necessary microsurgical training and ongoing experience with the procedures. Ideally, procedures should be performed by surgeons who have the capability to perform both vasovasostomy and vasoepididymostomy because, in many cases, the need of the latter cannot be anticipated.
- In experienced hands, microsurgical vasectomy reversal is highly successful. Sperm return to the ejaculate in 70–95% of the patients after surgery, and 30–75% of

couples can be expected to achieve unassisted pregnancy. Patency and pregnancy after microsurgical vasectomy reversal are inversely related to the interval of obstruction since vasectomy. Other factors that affect success rates include the intraoperative appearance of vasal fluid, the presence or absence of sperm in the vasal fluid and their quality, the length of the remaining segment adjacent to the epididymis, the age of the female partner, and the experience of the microsurgeon.

- Vasoepididymostomy should be performed only by those having the requisite training and experience in reproductive microsurgery. When treatment requires a complex reconstruction of the male reproductive tract, cryopreservation of retrieved sperm should be considered because surgery may not be successful.
- EDO is a potentially treatable cause of male infertility, and TURED is the treatment of choice. After TURED, sperm return to the ejaculate in approximately 50–75% of the men, and approximately 20% of couples achieve spontaneous pregnancy. However, results are highly variable and depend on the etiology (acquired or congenital) and type (partial or complete) of obstruction. Complications of TURED occur in approximately 20% of men, including hematuria, hematospermia, urinary tract infection, epididymitis, and a watery ejaculate due to reflux of urine.
- In obstructive azoospermia, sperm production is normal and gametes can be easily retrieved from epididymis or testes in most cases, irrespective of the technique. PESA or TESA are simple and efficient methods for retrieving epididymal or testicular spermatozoa in men with OA. In NOA, TESE with or without magnification (micro-TESE) is the preferred approach, and sperm can be retrieved in approximately 50% of the cases. The use of microsurgery during TESE may improve the efficacy of sperm extraction with significantly less tissue removed, which ultimately facilitates sperm processing. Testicular histology results, if available, may be useful to predict the chances to retrieve sperm in men with NOA. However, sperm can be obtained even in the worst case scenario except in the cases of Y chromosome infertility with complete AZFa and/or AZFb microdeletions.
- The goals of sperm retrieval are to obtain the best quality sperm possible, to retrieve adequate numbers of sperm for immediate use and cryopreservation, and to minimize damage to the reproductive tract. In both OA and NOA, sperm retrieval technique itself seems to have no impact on IVF/ICSI success rates. The reproductive potential of infertile men undergoing ART is related to the type of azoospermia. The chances of retrieving spermatozoa and of achieving a live birth by ICSI are increased in couples whose male partner had obstructive rather than NOA. Children conceived using sperm retrieved from men with OA and NOA should be followed up because it is still unclear if there is an increased risk of birth defects when ICSI is carried out with non-ejaculated sperm.

Acknowledgments The authors are indebted to Mrs. Fabiola Bento for her editorial assistance and to Dr. Marcelo Coccuza for providing his personal observations regarding reconstructive surgery.

References

1. Vital and Health Statistics, series 23, no.26, CDC. http://www.cdc.gov. Accessed 10 Dec 2009.
2. Silber S, Nagy ZP, Liu J, et al. Conventional in-vitro fertilization versus intracytoplasmic sperm injection for patients requiring microsurgical sperm aspiration. Hum Reprod. 1994;9:1705–9.
3. Devroey P, Liu J, Nagy ZP, et al. Pregnancies after testicular extraction (TESE) and intracytoplasmic sperm injection (ICSI) in non-obstructive azoospermia. Hum Reprod. 1995;10:1457–60.
4. Schlegel PN. Testicular sperm extraction: microdissection improves sperm yield with minimal tissue excision. Hum Reprod. 1999;14:131–5.
5. Goldstein M, Eid JF. Elevation of intratesticular and scrotal skin surface temperature in men with varicocele. J Urol. 1989; 142:743–5.
6. Chehval MJ, Purcell MH. Varicocelectomy: incidence of external vein involvement in the clinical varicocele. Urology. 1992;39: 573–5.
7. Nistal M, Gonzalez-Peramato P, Serrano A, et al. Physiopathology of the infertile testicle. Etiopathogenesis of varicocele. Arch Esp Urol. 2004;57:883–904.
8. Agarwal A, Prabakaran S, Allamaneni SS. Relationship between oxidative stress, varicocele and infertility: a meta-analysis. Reprod Biomed Online. 2006;12:630–3.
9. World Health Organization. The influence of varicocele on parameters of fertility in a large group of men presenting to infertility clinics. Fertil Steril. 1992;57:1289–93.
10. Jarow JP. Effects of varicocele on male fertility. Hum Reprod Update. 2001;7:59–64.
11. Marmar JL, Agarwal A, Prabaskan S, et al. Reassessing the value of varicocelectomy as a treatment for male subfertility with a new meta-analysis. Fertil Steril. 2007;88:639–48.
12. Esteves SC, Oliveira FV, Bertolla RP. Clinical outcomes of intracytoplasmic sperm injection in infertile men with treated and untreated clinical varicocele. J Urol. 2010;184:1241–586.
13. Redmon JB, Carey P, Pryor JL. Varicocele-the most common cause of male factor infertility? Hum Reprod Update. 2002 ;8:53–8.
14. Esteves S. Infertilidade masculina. In: Rhoden EL, editor. Urologia no consultório. 1ªth ed. Porto Alegre, RS: Artmed Editora; 2009. p. 470–500.
15. Libman J, Jarvi K, Lo K, Zini A. Beneficial effect of microsurgical varicocelectomy is superior for men with bilateral versus unilateral repair. J Urol. 2006;176:2602–5.
16. Gat Y, Bachar GN, Zukerman Z, et al. Physical examination may miss the diagnosis of bilateral varicocele: a comparative study of 4 diagnostic modalities. J Urol. 2004;172:1414–7.
17. Geatti O, Gasparini D, Shapiro B. A comparison of scintigraphy, thermography, ultrasound and phlebography in grading of clinical varicocele. J Nucl Med. 1991;32:2092–7.
18. Yamamoto M, Hibi H, Hirata Y, et al. Effect of varicocelectomy on sperm parameters and pregnancy rate in patients with subclinical varicocele: a randomized prospective controlled study. J Urol. 1996;155:1636–8.

19. Kantartzi PD, Goulis ChD, Goulis GD, et al. Male infertility and varicocele: myths and reality. Hippokratia. 2007;11:99–104.
20. Esteves SC, Glina S. Recovery of spermatogenesis after microsurgical subinguinal varicocele repair in azoospermic men based on testicular histology. Int Braz J Urol. 2005;31:541–8.
21. Steckel J, Dicker AP, Goldstein M. Relationship between varicocele size and response to varicocelectomy. J Urol. 1993;149:769–71.
22. Marmar JL. The pathophysiology of varicoceles in the light of current molecular and genetic information. Hum Reprod Update. 2001;7:461–72.
23. Marks JL, McMahon R, Lipshultz LI. Predictive parameters of successful varicocele repair. J Urol. 1986;136:609–12.
24. Yoshida K, Kitahara S, Chiba K, et al. Predictive indicators of successful varicocele repair in men with infertility. Int J Fertil. 2000;45:279–84.
25. Cayan S, Lee D, Black LD, et al. Response to varicocelectomy in oligospermic men with and without defined genetic infertility. Urology. 2001;57:530–5.
26. Pryor JL, Kent-First M, Muallem A, et al. Microdeletions in the Y chromosome of infertile men. N Engl J Med. 1997;336:534–9.
27. Kondo Y, Ishikawa T, Yamaguchi K, et al. Predictors of improved seminal characteristics by varicocele repair. Andrologia. 2009; 41:20–3.
28. Weedin JW, Khera M, Lipshultz LI. Varicocele Repair in Patients with Nonobstructive Azoospermia—A Meta-Analysis. J Urol. 2010;183:2309–15.
29. Esteves SC. Editorial comment. J Urol. 2010;183:2315.
30. Cayan S, Shavakhabov S, Kadioglu A. Treatment of palpable varicocele review in infertile men: a meta-analysis to define the best technique. J Androl. 2009;30:33–40.
31. Sautter T, Sulser T, Suter S, et al. Treatment of varicocele: a prospective randomized comparison of laparoscopy versus antegrade sclerotherapy. Eur Urol. 2002;41:398–400.
32. Al-Kandari AM, Shabaan H, Ibrahim HM, et al. Comparison of outcomes of different varicocelectomy techniques: open inguinal, laparoscopic, and subinguinal microscopic varicocelectomy: a randomized clinical trial. Urology. 2007;69:417–20.
33. Hopps CV, Lemer ML, Schlegel PN, et al. Intraoperative varicocele anatomy: a microscopic study of the inguinal versus subinguinal approach. J Urol. 2003;170:2366–70.
34. Anderson JE, Warner L, Jamieson DJ, et al. Contraception. 2010;82:230–5.
35. Belker AM, Thomas AJ, Fuchs EF, et al. Results of 1469 microsurgical vasectomy reversals by the Vasovasostomy Study Group. J Urol. 1991;145:505–11.
36. Vasectomia no Brasil. Veja online; http://veja.abril.com.br/041000/p_084.html. Accessed 3 Oct 2010.
37. Lipshultz LI, Rumohr JA, Bennet RC. Techniques for vasectomy reversal. Urol Clin N Am. 2009;36:375–832.
38. Boorjian S, Lipkin M, Goldstein M. The impact of obstructive interval and sperm granuloma on outcome of vasectomy reversal. J Urol. 2004;171:304–6.
39. Parekattil SJ, Kuang W, Agarwal A, et al. Model to predict if a vasoepididymostomy will be required for vasectomy reversal. J Urol. 2005;173:1681–4.
40. Hernandez J, Sabanegh ES. Repeat vasectomy reversal after initial failure. J Urol. 1999;161:1153–6.
41. Bolduc S, Fischer MA, Deceunik G, et al. Factors predicting overall success: a review of 74 microsurgical vasovasostomies. Can Urol Assoc J. 2007;1:388–91.
42. Carbone Jr DJ, Shah A, Thomas Jr AJ, Agarwal A. Partial obstruction, not antisperm antibodies, causing infertility after vasovasostomy. J Urol. 1998;159:827–30.
43. Chawla A, O'Brien J, Lisi M, et al. Should all urologists performing vasectomy reversal be able to perform vasoepididymostomy if required? J Urol. 2004;172:829–30.
44. Eggert-Kruse W, Christmann M, Gerhard I, et al. Circulating antisperm antibodies and fertility prognosis: a prospective study. Hum Reprod. 1989;4:513–20.
45. Chan PT, Goldstein M. Superior outcomes of microsurgical vasectomy reversal in men with the same female partners. Fertil Steril. 2004;81:1371–4.
46. Hinz S, Rais-Bahrami S, Kempkensteffen C, et al. Fertility rates following vasectomy reversal: importance of age of the female partner. Urol Int. 2008;81:416–20.
47. Gerrard Jr ER, Sandlow JI, Oster RA, et al. Effect of female partner age on pregnancy rates after vasectomy reversal. Fertil Steril. 2007;87:1340–4.
48. Schiff J, Chan P, Li PS, et al. Outcome and late failures compared in 4 techniques of microsurgical vasoepididymostomy in 153 consecutive men. J Urol. 2005;174:651–5.
49. Sharlip I. Absence of fluid during vasectomy reversal has no prognostic significance. J Urol. 1996;155:365–9.
50. Kolettis PN, Burns JR, Nangia AK, et al. Outcomes for vasovasostomy performed when only sperm parts are present in the vasal fluid. J Androl. 2006;27:565–7.
51. Boyle KE, Thomas Jr AJ, Marmar JL, et al. Sperm harvesting and cryopreservation during vasectomy reversal is not cost effective. Fertil Steril. 2006;85:961–4.
52. Sharlip ID. Microsurgical vasovasostomy: modified one-layer technique. In: Goldstein M, editor. Surgery of male infertility. 1st ed. New York, NY: WB Saunders; 1995. p. 67–76.
53. Belker AM. Microsurgical vasovasostomy: two-layer technique. In: Goldstein M, editor. Surgery of male infertility. 1st ed. New York, NY: WB Saunders; 1995. p. 61–76.
54. Goldstein M. Vasovasostomy: surgical approach, decision making, and multilayer microdot technique. In: Goldestein M, editor. Surgery of male infertility. 1st ed. New York, NY: WB Saunders; 1995. p. 46–60.
55. Fleming C. Robot-assisted vasovasostomy. Urol Clin N Am. 2004;31:769–72.
56. Parekattil SJ, Cohen MS. Robotic surgery in male infertility and chronic orchialgia. Curr Opin Urol. 2010;20:75–9.
57. Silber S. Microscopic vasoepididymostomy: specific microanastomosis to the epididymal tubule. Fertil Steril. 1978;30:565–71.
58. Thomas Jr AJ. Vasoepididymostomy. Urol Clin North Am. 1987;14:527–38.
59. Berger RE. Triangulation end-to-side vasoepididymostomy. J Urol. 1998;159:1951–3.
60. Marmar JL. Modified vasoepididymostomy with simultaneous double needle placement, tubulotomy and tubular invagination. J Urol. 2000;163:483–6.
61. Chan PT, Li PS, Goldstein M. Microsurgical vasoepididymostomy: a prospective randomized study of 3 intussusception techniques in rats. J Urol. 2003;169:1924–9.
62. Netto Jr NR, Esteves SC, Neves PA. Transurethral resection of partially obstructed ejaculatory ducts: seminal parameters and pregnancy outcomes according to the etiology of obstruction. J Urol. 1998;159:2048–53.
63. Meacham RB, Hellerstein DK, Lipshultz LI. Evaluation and treatment of ejaculatory duct obstruction in the infertile male. Fertil Steril. 1993;59:393–7.
64. Carter SS, Shinohara K, Lipshultz LI. Transrectal ultrasonography in disorders of the seminal vesicles and ejaculatory ducts. Urol Clin N Am. 1989;16:773–90.
65. Hellerstein DK, Meacham RB, Lipshultz LI. Transrectal ultrasound and partial ejaculatory duct obstruction in male infertility. Urology. 1992;39:449–52.
66. Jones TR, Zagoria RJ, Jarow JP. Transrectal US-guided seminal vesiculography. Radiology. 1997;205:276–8.
67. Jarow JP, Espeland MA, Lipshultz LI. Evaluation of the azoospermic patient. J Urol. 1989;142:62–5.

68. Farley S, Barnes R. Stenosis of ejaculatory ducts treated by endoscopic resection. J Urol. 1973;109:664–6.
69. Craft I, Tsirigotis M, Bennett V, et al. Percutaneous epididymal sperm aspiration and intracytoplasmic sperm injection in the management of infertility due to obstructive azoospermia. Fertil Steril. 1995;63:1038–42.
70. Craft I, Tsirigotis M. Simplified recovery, preparation and cryopreservation of testicular spermatozoa. Hum Reprod. 1995;10:1623–7.
71. Okada H, Dobashi M, Yamazaki T, et al. Conventional versus microdissection testicular sperm extraction for nonobstructive azoospermia. J Urol. 2002;168:1063–7.
72. Tsujimura A, Matsumiya K, Miyagawa Y, et al. Conventional multiple or microdissection testicular sperm extraction: a comparative study. Hum Reprod. 2002;17:2924–9.
73. Ramasamy R, Lin K, Gosden LV, et al. High serum FSH levels in men with nonobstructive azoospermia does not affect success of microdissection testicular sperm extraction. Fertil Steril. 2009;92:590–3.
74. Esteves SC, Verza Jr S, Gomes AP. Successful retrieval of testicular spermatozoa by microdissection (micro-TESE) in nonobstructive azoospermia is related to testicular histology. Fertil Steril. 2006;86:S354.
75. Schlegel PN. Causes of azoospermia and their management. Reprod Fertil Dev. 2004;16:561–72.
76. Sharlip ID, Jarow J, Belker AM, et al. Report on Evaluation of the Azoospermic Male. AUA Best Practice Policy and ASRM Practice Committee Report. American Urological Association, April 2001.
77. Esteves SC, Verza S, Prudencio C, Seol B. Sperm retrieval rates (SRR) in nonobstructive azoospermia (NOA) are related to testicular histopathology results but not to the etiology of azoospermia. Fertil Steril. 2010;94(Suppl):S132.
78. Male Infertility Best Practice Policy Committee of the American Urological Association, Practice Committee of the American Society for Reproductive Medicine. Report on evaluation of the azoospermic male. Fertil Steril. 2006;86 Suppl 1:S210–215.
79. De Braekeleer M, Dao TN. Cytogenetic studies in male infertility: a review. Hum Reprod. 1991;6:245–50.
80. Brandell RA, Mielnik A, Liotta D, et al. AZFb deletions predict the absence of spermatozoa with testicular sperm extraction: preliminary report of a prognostic genetic test. Hum Reprod. 1998; 13:2812–5.
81. Hopps CV, Mielnik A, Goldstein M, et al. Detection of sperm in men with Y chromosome microdeletions of the AZFa, AZFb and AZFc regions. Hum Reprod. 2003;18:1660–5.
82. Nagler HM, Luntz RK, Martinis FG. Varicocele. In: Lipshultz LI, Howards SS, editors. Infertility in the Male. 3rd ed. St Louis, Missouri: Mosby; 1997. p. 336–59.
83. Dhabuwala CB, Hamid S, Moghisi KS. Clinical versus subclinical varicocele: improvement in fertility after varicocelectomy. Fertil Steril. 1992;57:854–7.
84. Jee SH, Hong YK. One-layer vasovasostomy: microsurgical versus loupe-assisted. Fertil Steril. 2010;94(6):2308–11.
85. Esteves SC, Verza S, Prudencio C, Seol B. Success of percutaneous sperm retrieval and intracytoplasmic sperm injection (ICSI) in obstructive azoospermic (OA) men according to the cause of obstruction. Fertil Steril. 2010;94(Suppl):S233.
86. Verza Jr S, Esteves SC. Sperm defect severity rather than sperm source is associated with lower fertilization rates after intracytoplasmic sperm injection. Int Braz J Urol. 2008;34:49–56.
87. Schneider DT, Gomes AP, Verza Jr S, et al. Optimal time interval for intracytoplasmic sperm injection after administration of human chorionic gonadotrophin in severe male factor infertility. Fertil Steril. 2006;86:S155.
88. Prudencio C, Seol B, Esteves SC. Reproductive potential of azoospermic men undergoing intracytoplasmic sperm injection is dependent on the type of azoospermia. Fertil Steril. 2010; 94(Suppl):S232–3.
89. Chan PT, Wright EJ, Goldstein M. Incidence and postoperative outcomes of accidental ligation of the testicular artery during microsurgical varicocelectomy. J Urol. 2005;173:482–4.
90. Cocuzza M, Pagani R, Coelho R, et al. The systematic use of intraoperative vascular Doppler ultrasound during microsurgical subinguinal varicocelectomy improves precise identification and preservation of testicular blood supply. Fertil Steril. 2010;93:2396–9.
91. Schlesinger MH, Wilets IF, Nagler HM. Treatment outcome after varicocelectomy. A critical analysis. Urol Clin North Am. 1994;21:517–29.
92. Colpi GM, Carmignani L, Nerva F, et al. Surgical treatment of varicocele by a subinguinal approach combined with antegrade intraoperative sclerotherapy of venous vessels. BJU Int. 2006;97:142–5.
93. Shindel AW, Yan Y, Naughton CK. Does the number and size of veins ligated at left-sided microsurgical subinguinal varicocelectomy affect semen analysis outcomes? Urology. 2007;69:1176–80.
94. Matkov TG, Zenni M, Sandlow J, et al. Preoperative semen analysis as a predictor of seminal improvement following varicocelectomy. Fertil Steril. 2001;75:63–8.
95. Smit M, Romijn JC, Wildhagen MF, et al. Decreased sperm DNA fragmentation after surgical varicocelectomy is associated with increased pregnancy rate. J Urol. 2010;183:270–4.
96. Jeng SY, Wu SM, Lee JD. Cadmium accumulation and metallothionein overexpression in internal spermatic vein of patients with varicocele. Urology. 2009;73:1231–5.
97. Cocuzza M, Cocuzza MA, Bragais FM, Agarwal A. The role of varicocele repair in the new era of assisted reproductive technology. Clinics (Sao Paulo). 2008;63:395–404.
98. Zheng YQ, Gao X, Li ZJ, et al. Efficacy of bilateral and left varicocelectomy in infertile men with left clinical and right subclinical varicoceles: a comparative study. Urology. 2009;73: 1236–40.
99. Elbendary MA, Elbadry AM. Right subclinical varicocele: how to manage in infertile patients with clinical left varicocele? Fertil Steril. 2009;92:2050–3.
100. Schlegel PN, Kaufmann J. Role of varicocelectomy in men with nonobstructive azoospermia. Fertil Steril. 2004;81:1585–8.
101. Inci K, Hascicek M, Kara O, et al. Sperm retrieval and intracytoplasmic sperm injection in men with nonobstructive azoospermia, and treated and untreated varicocele. J Urol. 2009;182:1500–5.
102. Meng MV, Greene KL, Turek PJ. Surgery or assisted reproduction? A decision analysis of treatment costs in male infertility. J Urol. 2005;174:1926–31.
103. Ficarra V, Cerruto MA, Liguori G, et al. Treatment of varicocele in subfertile men: the cochrane review—a contrary opinion. Eur Urol. 2006;49:258–63.
104. Agarwal A, Deepinder F, Cocuzza M, et al. Efficacy of Varicocelectomy in Improving Semen Parameters: new meta-analytical approach. Urology. 2007;70:532–8.
105. Zini A, Blumenfeld A, Libman J, et al. Beneficial effect of microsurgical varicocelectomy on human sperm DNA integrity. Hum Reprod. 2005;20:1018–21.
106. Moskovtsev SI, Lecker I, Mullen JB, et al. Cause-specific treatment in patients with high sperm DNA damage resulted in significant DNA improvement. Syst Biol Reprod Med. 2009;55:109–15.
107. Lee R, Li PS, Goldstein M, Tanrikut C, et al. A decision analysis of treatments for obstructive azoospermia. Hum Reprod. 2008;23:2043–9.
108. Robb P, Sandlow JI. Cost-effectiveness of vasectomy reversal. Urol Clin North Am. 2009;36:391–6.

109. Male Infertility Best Practice Policy Committee of the American Urological Association, Practice Committee of the American Society for Reproductive Medicine. Report on the management of infertility due to obstructive azoospermia. Fertil Steril. 2008;90 Suppl 3:S121–4.
110. Malizia BA, Hacker MR, Penzias AS. Cumulative live-birth rates after in vitro fertilization. N Engl J Med. 2009;360:236–43.
111. Hsieh MH, Meng MV, Turek PJ. Markov modeling of vasectomy reversal and ART for infertility: how do obstructive interval and female partner age influence cost effectiveness? Fertil Steril. 2007;88:840–6.
112. Chan PT, Brandell RA, Goldstein M. Prospective analysis of outcomes after microsurgical intussusceptions vasoepididymostomy. BJU Int. 2005;96:598–601.
113. Marmar JL, Sharlip I, Goldstein M. Results of vasovasostomy or vasoepididymostomy after failed percutaneous epididymal sperm aspirations. J Urol. 2008;179:1506–9.
114. Kolettis PN. Restructuring Reconstructive Techniques—advances in reconstructive techniques. Urol Clin N Am. 2008;35:229–34.
115. Ho KLV, Witte MN, Bird ET, et al. Fibrin glue assisted 3-suture vasovasostomy. J Urol. 2005;174:1360–363.
116. Vrijhof EJ, De Bruine A, Zwinderman A, et al. The use of newly designed nonabsorbable polymeric stent in reconstructing the vas deferens: a feasibility study in New Zealand white rabbits. BJU Int. 2005;95:1081–5.
117. Parekattil SJ, Atalah HN, Cohen MS. Video technique for human robot-assisted microsurgical vasovasostomy. J Endourol. 2010;24:511–4.
118. Schiff J, Li PS, Goldstein M. Robotic microsurgical vasovasostomy and vasoepididymostomy in rats. Int J Med Robot. 2005;1:122–6.
119. Eisenberg ML, Walsh TJ, Garcia MM, et al. Ejaculatory duct manometry in normal men and in patients with ejaculatory duct obstruction. J Urol. 2008;180:255–60.
120. Orhan I, Duksal I, Onur R, et al. Technetium Tc 99 m sulphur colloid seminal vesicle scintigraphy: a novel approach for the diagnosis of the ejaculatory duct obstruction. Urology. 2008;71:672–6.
121. Onur MR, Orhan I, Firdolas F, et al. Clinical and radiological evaluation of ejaculatory duct obstruction. Arch Androl. 2007;53:179–86.
122. Lawler LP, Cosin O, Jarow JP, et al. Transrectal US-guided seminal vesiculography and ejaculatory duct recanalization and balloon dilation for treatment of chronic pelvic pain. J Vasc Interv Radiol. 2006;17:169–73.
123. Manohar T, Ganpule A, Desai M. Transrectal ultrasound- and fluoroscopic-assisted transurethral incision of ejaculatory ducts: a problem-solving approach to nonmalignant hematospermia due to ejaculatory duct obstruction. Endourol. 2008;22:1531–5.
124. Van Peperstraten A, Proctor ML, Johnson NP, et al. Techniques for surgical retrieval of sperm prior to ICSI for azoospermia. Cochrane Database Syst Rev. 2006;3:CD002807.
125. Nicopoullos JD, Gilling-Smith C, Almeida PA, et al. Use of surgical sperm retrieval in azoospermic men: a meta-analysis. Fertil Steril. 2004;82:691–701.
126. Kamal A, Fahmy I, Mansour R, et al. Does the outcome of ICSI in cases of obstructive azoospermia depend on the origin of the retrieved spermatozoa or the cause of obstruction? A comparative analysis. Fertil Steril. 2010 Nov;94(6):2135–40.
127. Hauser R, Yogev L, Paz G, et al. Comparison of efficacy of two techniques for testicular sperm retrieval in nonobstructive azoospermia: multifocal testicular sperm extraction versus multifocal testicular sperm aspiration. J Androl. 2006;27:28–33.
128. Donoso P, Tournaye H, Devroey P. Which is the best sperm retrieval technique for non-obstructive azoospermia? A systematic review. Hum Reprod Update. 2007;13:539–49.
129. Carpi A, Sabanegh E, Mechanick J. Controversies in the management of nonobstructive Azoospermia. Fertil Steril. 2009; 91:963–70.
130. Schiff JD, Palermo GD, Veeck LL, et al. Success of testicular sperm injection and intracytoplasmic sperm injection in men with Klinefelter syndrome. J Clin Endocrinol Metab. 2005; 90:6263–7.
131. Ramasamy R, Yagan N, Schlegel PN. Structural and functional changes to the testis after conventional versus microdissection testicular sperm extraction. Urology. 2005;65:1190–4.
132. Carpi A, Menchini Fabris F, Palego F, et al. Fine-needle and large needle percutaneous aspiration biopsy of the testicle in men with nonobstructive azoospermia: safety and diagnostic performance. Fertil Steril. 2005;83:1029–33.
133. Turunc T, Gul U, Haydardedeoglu B, et al. Conventional testicular sperm extraction combined with the microdissection technique in nonobstructive azoospermic patients: a prospective comparative study. Fertil Steril. 2010;94(6):2157–60. Epub 20 Feb 2010.
134. Tesarik J. Paternal effects on cell division in the human preimplantation embryo. Reprod Biomed Online. 2005;10:370–5.
135. Schlegel PN, Liotta D, Hariprashad J, et al. Fresh testicular sperm from men with nonobstructive azoospermia works best for ICSI. Urology. 2004;64:1069–71.
136. Alukal JP, Lamb DJ. Intracytoplasmic sperm injection (ICSI)—what are the risks? Urol Clin North Am. 2008;35:277–88.
137. Knoester M, Helmerhorst FM, Vandenbroucke JPM, et al. Artificial Reproductive Techniques Follow-up Project. Cognitive development of singletons born after intracytoplasmic sperm injection compared with in vitro fertilization and natural conception. Fertil Steril. 2008;90:289–96.
138. Belva F, Henriet S, Liebaers I, et al. Medical outcome of 8-year-old singleton ICSI children and a spontaneously conceived comparison group. Hum Reprod. 2007;22:506–15.
139. Woldringh GH, Besselink DE, Tillema AH, et al. Karyotyping, congenital anomalies and follow-up of children after intracytoplasmic sperm injection with non-ejaculated sperm: a systematic review. Hum Reprod Update. 2010;16:12–9.

Microsurgery for Male Infertility: The AIIMS Experience

7

Rajeev Kumar

Abstract

The diseases responsible for male infertility and for cases undergoing exploration for microsurgical reconstruction at the All India Institute of Medical Sciences possibly differ from those reported in most Western literature. The largest number of cases surgically explored is for primary infertility with obstruction of unknown etiology. This has resulted in lower percentage of patients with a successful outcome. However, the socioeconomic factors around infertility management in our country dictate an attempt at reconstruction even when the expected outcomes are poor. Infertility is a major social issue in the Indian culture with immense pressure on married couples to have children. The stigma surrounding infertility is such that couples prefer seeking treatment through discreet unqualified practitioners who offer quick cures rather than visit a public hospital, thus delaying meaningful intervention. Coupled with the low per capita income and lack of adequate health insurance, social issues have a major impact on the management of male infertility at AIIMS. Fortunately, in patients with favorable prognostic factors, the outcomes are generally good.

Keywords

Microsurgery for male infertility • Congenital bilateral absence of the vas deferens • Genital tuberculosis • Filariasis • Vaso-epididymal anastomosis • Microsurgical varicocelectomy • Vasovasal anastomosis • Azoospermia

The All India Institute of Medical Sciences (AIIMS), New Delhi, is a government funded tertiary care medical institution that provides medical education, undertakes research and provides medical care at nominal costs. All citizens, and even non-citizens, can avail themselves of the services. Academically, it is consistently ranked the best among over 300 medical colleges in the country. The institution is equipped with state-of-the-art technologies including two da Vinci®(Intuitive Surgicals, CA) surgical robots. The low cost along with exceptional facilities results in a high demand for services and leads to significant waiting periods for most elective surgical procedures. Understanding this basic nature of the institution is essential to understanding the development and experience of microsurgery for male infertility at AIIMS.

Infertility is a major social issue in the Indian culture. Parenthood is considered an essential function in life with immense social pressure on married couples to have children. The stigma surrounding infertility is such that couples prefer seeking treatment through discreet unqualified practitioners who offer quick cures rather than visit a public hospital. Coupled with the low per capita income and lack of adequate health insurance, social issues have a major impact on the management of male infertility at AIIMS.

First, there is significant delay in seeking treatment for the male partner. In a survey, we found that 8–10% of all patients

R. Kumar, MCh (✉)
Department of Urology, All India Institute of Medical Sciences, Ansari Nagar, New Delhi 110029, India
e-mail: rajeev02@gmail.com

S.J. Parekattil and A. Agarwal (eds.), *Male Infertility: Contemporary Clinical Approaches, Andrology, ART & Antioxidants*, DOI 10.1007/978-1-4614-3335-4_7,

presenting to the urology clinic had infertility, the average age of the male partner at presentation to us was 27 years, the mean duration of infertility was 6.2 years, 88% had previously consulted medical practitioners, and 84% had received medical therapy, most often of unknown nature (unpublished data). Interestingly, in no case did we find the male presenting to us on his own, without a referral from the physician evaluating the wife. This delayed presentation increases the need for early intervention and quick results.

The second major impact of the social conditions is the need for cost-effective interventions. There is no financial support for undergoing in vitro fertilization (IVF), which continues to be expensive. Patients are willing to accept even minimal success rates if surgery can help prevent IVF. Less than 2% of our patients choose IVF over microsurgical reconstruction.

Evaluation and Etiologies

Evaluation of the infertile male aims at identifying the etiology, reversibility, and potential underlying medical conditions that may have manifest as infertility. Microsurgery is currently offered for bypassing an obstruction in the epididymis or the vas deferens and for correction of a varicocele. The evaluation and investigations are tailored to diagnose suitability for microsurgery and maximize successful outcomes.

All men provide a detailed history of their condition. A history of previous vasectomy, hernia surgery, retroperitoneal surgery, hydroceles/scrotal surgery, and tuberculosis suggests obstructive azoospermia. Low volume ejaculate is determined from direct questioning of "few drops" versus a "spoonful," while a history of orchitis, radiation or chemotherapy would indicate testicular failure. Previous conceptions, work environment, and smoking histories are recorded. The physical examination is used to confirm the presence of vas deferens, testicular size, epididymis, secondary sexual characteristics and record any abnormalities in the genitalia. Examination for varicocele is performed in the erect and supine positions. Since all men come to us with a referral from another physician, most carry a semen analysis report that is used for focusing further evaluation. At least two semen samples are used to categorize the abnormality.

Congenital bilateral absence of the vas deferens (CBAVD) is diagnosed clinically in the majority of patients. A combination of low volume ejaculate, absent vas on palpation, hemi-epididymis, and acidic azoospermia are considered sufficient evidence [1]. Patients with a low volume ejaculate and a palpable vas undergo transrectal ultrasound (TRUS) for determining dilatation of the seminal vesicles and ejaculatory ducts secondary to an ejaculatory duct obstruction. A post-ejaculate urine examination for sperms is performed to rule out retrograde ejaculation if the TRUS findings are inconclusive. These men are generally not considered candidates for microsurgical reconstruction.

Men with normal volume azoospermia, normal secondary sexual characteristics, normal size testis, and palpable vas deferens are evaluated to determine suitability for microsurgery. The primary aim in these men is to differentiate obstructive from nonobstructive azoospermia. Unless a history suggestive of testicular insult is present, these men undergo estimation of serum follicle stimulating hormone levels (FSH) and fine needle aspiration cytology of both the testis (FNAC). Normal testicular spermatogenesis in such men is considered diagnostic of obstruction. FNAC is performed using a 23 G butterfly needle attached to a 10 ml syringe, with multiple passes into the testis through a single skin puncture. A minimum of 2,000 cells are evaluated on an air-dried slide stained with the May–Grunwald–Giemsa stain. Similar to biopsy histology, these are reported as normal spermatogenesis, hypospermatogenesis, maturation arrest, or only Sertoli cells seen [2]. We rely on the FNAC reports to determine normal sperm production and do not perform diagnostic biopsies [3]. The FSH levels are used to prognosticate the outcome and an evaluation of outcomes when the FSH and FNAC are discordant is currently being performed.

We are able to determine the etiology of obstruction in less than one in five of men diagnosed with obstructive azoospermia [4]. This is in stark contrast to the under 20% incidence of idiopathic cases in most reported series of microsurgical reconstruction [5, 6]. Vasectomy reversals form a very small proportion of men undergoing microsurgery for infertility at our center primarily because vasectomy is an uncommon method of contraception among Indian couples. Vasectomy reversals, on the other hand, form the commonest indications for microsurgery in western populations [5, 6]. We believe that the most common cause of obstruction in our population is undiagnosed genital infection [4]. Three conditions specific to this population are tuberculosis, filariasis, and small pox.

Tuberculosis

Tuberculosis (TB) continues to be a significant public health problem in developing countries. Genital tuberculosis is an extrapulmonary form of tuberculosis seen most often in men in the reproductive age group [7]. The epididymis is one of the favored sites for genital TB and may be the primary site of involvement in over three-fourths of cases [8]. The propensity for involvement of the epididymis probably stems from the high vascularity of the globus minor [9]. Other sites that may be involved include the prostate, seminal vesicles, and the penile shaft [10]. The pathophysiology of TB induced infertility is most often obstructive. Obstruction may arise either directly by the granulomatous lesions or due to distortion of normal architecture and scarring [11].

Evaluation for TB requires a high degree of suspicion. Most men do not provide a definitive history of past TB and physical signs may be minimal. Involvement of scrotal structures most often presents with normal volume azoospermia. This may be accompanied by palpable nodules within the epididymis or a beaded feel to the vas deferens. Rarely, scrotal abscesses or sinuses may be present. Involvement of the prostate/seminal vesicles and the ejaculatory ducts is usually contiguous. It may manifest as low volume azoospermia due to an obstruction of the ejaculatory ducts. A tissue diagnosis may be attempted but does not always yield positive results and such men may be classified as idiopathic obstruction. Discrete obstruction of the ejaculatory ducts may be treated with a transurethral resection of the ejaculatory ducts [10]. Involvement of the vas and the epididymis with nodules is rarely amenable to microsurgical reconstruction. If discrete nodules in the vas are felt, an attempt may be made at vasovasal anastomosis, bypassing the obstructing nodule. Unfortunately, the outcomes of surgery in such patients are usually poor and they are rarely correctable.

Filariasis

Filarial involvement of the lymphatic system can involve the scrotum and the epididymis. Filariasis is endemic in a large part of north-central India. Infertility due to filariasis is probably a result of inflammatory scarring and, at times, iatrogenic following interventions to treat filarial hydroceles. A number of studies have described the association between filariasis and male infertility [12, 13]. The most common clinical manifestation of scrotal filariasis is hydrocele. While the hydrocele itself may not cause infertility, this is often associated with dense adhesions around the epididymis and tunical calcifications. The tunica vaginalis is extremely thickened and the underlying tubules of the epididymis are often thin and flimsy [4]. The adhesions and scarring may be so dense that identification of the epididymis itself is not feasible, making it prone to injury during corrective hydrocele surgery. We generally discourage men who have had previous hydrocele surgeries from attempting microsurgical reconstruction because of the poor outcomes in such men.

Small Pox

Despite the eradication of small pox from India in the 1970s, we still occasionally come across men in their early 40s with clinical stigmata of the disease. Small pox resulted in obstructive azoospermia with a much higher incidence than would be expected from population studies [14]. Microsurgical reconstruction in these men is often gratifying because the disease tended to affect the terminal part of the epididymis, allowing a vaso-epididymostomy in the more proximal dilated tubules.

Microsurgical Procedures

Microsurgical reconstruction is offered at AIIMS for vasectomy reversal, vaso-epididymal obstruction, and varicoceles (Table 7.1). All procedures are performed as day-care surgery under regional or light general anesthesia. An operating microscope with face-to-face attachments for surgeon and assistant along with an offset arm with facility for camera attachment and transmission is used. There has been a steady evolution in case selection and techniques over the years and this has been accompanied by an improvement in outcomes.

Vaso-Epididymal Anastomosis

Scrotal explorations for vaso-epididymal anastomosis form the most common microsurgical procedure performed. Nearly all such procedures are performed for men with idiopathic obstruction. An obstruction is suspected in men with normal volume azoospermia with palpable vas deferens and normal spermatogenesis on testicular FNAC. These men are counseled about the options of scrotal exploration and IVF. A presumptive prognosis for success of reconstruction is provided based on age, testicular size, serum FSH, and presence of scrotal pathology.

In the 1990s, we used to perform a non-mucosal anastomosis between the vas deferens and the epididymal tubules, often with large 6-0 or 7-0 sutures. This was primarily due

Table 7.1 Common surgeries and their indications

Vasoepididymostomy
Idiopathic
Inflammatory
Tuberculosis
Filarial
Unknown
Post-vasectomy
Vasovasostomy
Post-vasectomy
Inflammatory nodules
Trauma
Hernia surgery
Hydrocele surgery
Varicocelectomy
Clinically palpable varicocele

Table 7.2 Evolution in vasoepididymostomy technique

Early 1990s: 6-0/7-0 sutures, non-mucosal anastomosis
Poor expertise in microsurgery
Limited equipment
Late 1990s: Berger's technique [15] of three suture intussusception
Thicker needles, inadequate space for three needles
Early 2000s: Marmar's technique [16] of two suture intussusception
Thicker needles
Thin tubules
Current: Modified, longitudinal technique of two suture intussusception [18]

to a lack of microsurgical expertise. The results of this "fistula" technique were universally poor with a rare patient reporting patency. Two publications toward the end of the last century resulted in major changes in our approach (Table 7.2). Berger's paper on the triangulation intussusception technique of end-to-side vasoepididymostomy was the first of these publications [15]. The use of three double-armed sutures instead of multiple independent sutures for the inner layer, coupled with our increasing familiarity with an operating microscope allowed us to attempt this technique in a number of patients with some success. A major problem that we faced with this approach was with the size of needles for the 10-0 suture. The 10-0 double-armed polyamide suture available to us was swaged on 200-μm needles. Placing three such needles in the small epididymal tubule proved difficult and we would often manage to get only two sutures in place.

Marmar's article in 2000 on the two-suture technique of vasoepididymostomy proved to be the turning point in our approach to this procedure [16]. It was simpler to perform and afforded excellent results. We began using this technique in 2002 and have been using our modifications of it since then [17, 18].

In 2003, Chan et al. published a comparative report on three different intussusception techniques in rats and concluded longitudinal placement of sutures in the epididymal tubule resulted in higher patency rates [19]. Difficulty in placing our thicker needles transversely in the tubules had led us to this modification in our patients at about the same time and we commented on our findings in a reply to their publication [20].

Our Surgical Technique

During the initial part of our experience, we performed microsurgical reconstruction unilaterally. Prior to using the two-suture technique, our results had been poor and we wished to keep one side untouched in case the patient wished to seek treatment for our failure elsewhere. Our initial microscope had a relatively high minimal height and the height of the operating table and the eyepiece of the microscope made it impossible to sit and operate. Thus, all our microsurgeries were performed with the surgeon standing up. This initial practice has been continued and we still perform all such surgeries standing up even after changing our microscope in 2009 to a Zeiss Opmi Vario® (*Carl Zeiss Micro Imaging GmbH, Germany*) with an S-88 stand. Our basic microsurgical instruments set consists of curved and straight microneedle holders without ratchet, curved microscissors, straight iris scissors, Jeweler's toothed, and non-toothed microforceps with a platform to assist suture tying and Adson's toothed and non-toothed forceps. We do not use clamps for holding the epididymis or the vas deferens.

With the patient supine, a longitudinal incision is made in the scrotum at its anteriolateral edge toward the upper end of the testis. The incision is deepened to expose the tunica vaginalis, which is incised to deliver the testis. Any adhesions within the tunical layer are divided to expose the anterior surface of the epididymis. The surface is inspected for calcifications/nodules and visibly dilated tubules. In patients with a previous hydrocelectomy, the tunical layer is obliterated and an attempt is made, by feel, to identify the epididymis and incise layers of tissue above it to expose the epididymal tunic. The spermatic cord is palpated posterior-lateral to the testis to confirm the presence of the vas deferens and get a visual impression of its diameter. Occasionally, in men with a more distal obstruction, the vas may feel dilated and thick fluid in the lumen may be visible through its wall.

Before dividing the vas deferens, we inspect the epididymis through its tunica to identify dilated tubules. In men with clearly dilated tubules, we proceed to preparing the vas. However, if the epididymis does not show any dilated tubules and feels flabby in its entire length, we make an incision in the epididymal tunic and observe the individual tubules under the microscope. If there still appears to be a doubt about the presence of obstruction, one of the distal loops of the tubule is incised and the fluid is examined for sperms. If sperms are confirmed, a more proximal site is selected for the anastomosis.

The vas is isolated from the remaining cord structures with blunt dissection, maintaining a mesentery of blood vessels to the vas in its entire length. The junction of the convoluted and straight vas is identified and a small segment of the vas at this site is elevated from its mesentery on a small hemostat. Using a sharp straight knife, the vas is hemisectioned at this level and inspected for any fluid that may suggest a more distal block. A 24 G angiocatheter is carefully inserted into the lumen of the distal vas and flushed slowly with up to 20 cm^3 of saline. Free flow of saline with no regurgitation is used as an indicator of distal patency and the section of the vas is completed. If there is resistance to flow of saline, a 3-0 polyamide suture is threaded into the lumen to determine the level of block. If the block is at a short distance, the vas is exposed to the expected site of

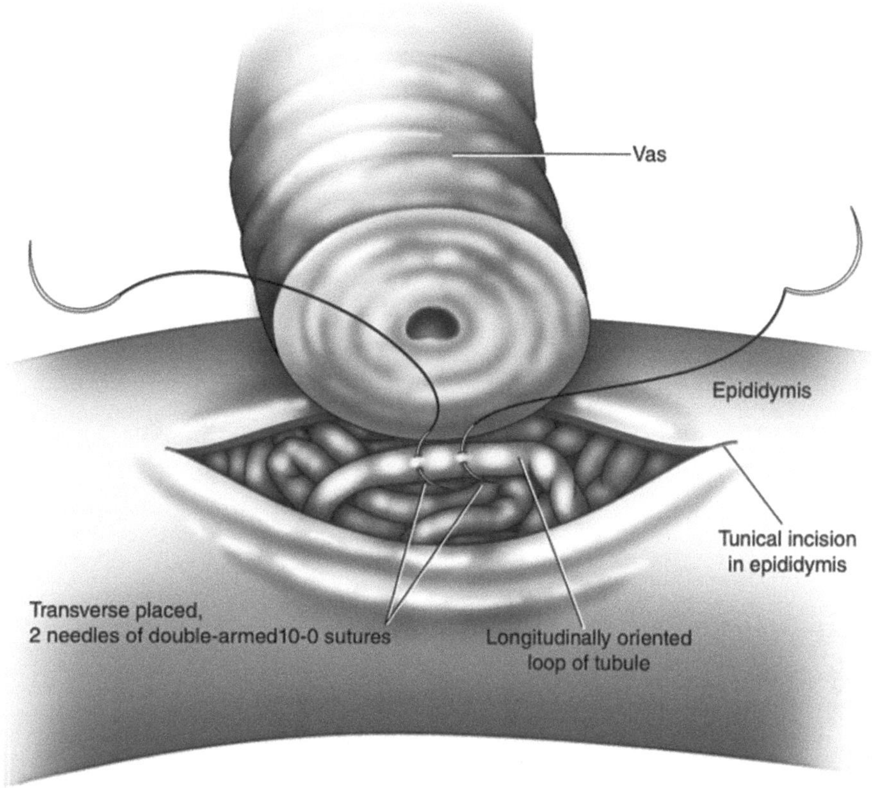

Fig. 7.1 Transverse suture placement in the epididymal tubule for a vasoepididymostomy

block and a fresh vasotomy is made at that site and the procedure repeated. For more distant blocks, a formal vasography is performed.

Once distal patency of the vas is confirmed, further mobilization of the mesentery is performed, keeping a good amount of tissue and blood vessels around the vas. The periadventitial tissue is held gently with a hemostat until the first seromuscular sutures are placed.

The epididymis is reinspected and its tunica incised over the most visibly dilated tubules. Individual loops of the tubule are gently separated from their surrounding alveolar tissue until the selected tubule bulges above the rest. In our earlier cases, the sutures were placed transversally in the loop in a manner that they could be tied without crossing over of the sutures [19, 20] (Fig. 7.1). In our more recent cases, we place the sutures longitudinally, again in a manner such that they can be tied to each other without crossing over [17] (Fig. 7.2).

The end of the vas is maneuvered to the site of the tunical incision. An 8-0 polyamide suture is placed from outside-in at the 5 o'clock position in the seromuscular layer of the vas. The needle is then passed through the epididymal tunic from inside out and tied. The procedure is repeated with another suture at the 7 o'clock position. The direction of placement of these sutures is important in that placing the suture along the direction of the vas helps provide traction to the needle as it enters the thick vas. Placement inside-out on the epididymal tunic allows the needle to lift the tunic from the underlying tubules, preventing inadvertent injury to the tubules.

Once the vas is secured to the epididymis, the needle of a 10-0 polyamide double-armed suture is placed in the loop of the epididymal tubule. The needle is inserted at the end closer to the vas and exits at the opposite end. This needle may be placed transversally or longitudinally in the loop depending on the selected procedure. A second similar suture is placed, parallel to the first. The tubule between the two needles is incised using a microknife. It is important to ensure that the length of this incision does not extend beyond the entry and exit points of the two sutures [18]. Fluid from the tubules is directly collected on a sterile microslide for examination under a light microscope. The procedure is continued if sperms are seen in the fluid. The needles are pulled through and kept separate from each other. The other ends of the sutures with their needles are now closer to the vas and are placed first into the vas lumen. These needles are placed at the 5 o'clock and 7 o'clock positions in the vas from inside the lumen, out through the muscular layer but not the full thickness of the vas. The two needles away from the vas are now placed similarly into the vas lumen at the 1 o'clock and 11 o'clock positions ensuring that the sutures are not

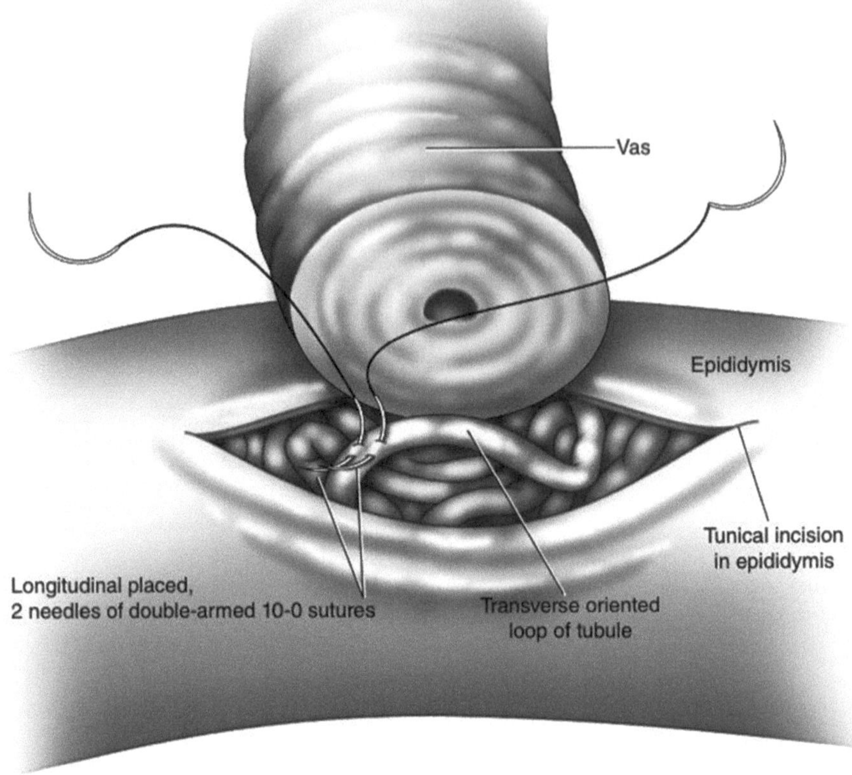

Fig. 7.2 Longitudinal suture placement in the epididymal tubule for a vasoepididymostomy

entangled. The needles exiting at 7 o'clock and 11 o'clock in the vas belong to the same suture as do the 5 o'clock and 1 o'clock needles. The two needles of each suture are held together and pulled to "hitch" up the tubule into the vas lumen. Both ends of the same suture are then tied to each other, intussuscepting the loop into the vas lumen. Additional 8-0 polyamide sutures are placed in the anterior layer of the vas and the epididymal tunic to secure the anastomosis. Two to three 8-0 polyamide sutures are also placed proximally in the serosal layer of the vas and the epididymal tissue to protect the anastomosis from cremasteric contractions. In case there are no sperms in the epididymal fluid, the sutures are removed and the procedure repeated at a more proximal site. All patients receive 3–5 days of antibiotics and are reviewed for suture removal after a week. A semen analysis is performed at 6 weeks and repeated every 3 months until patency is demonstrated. Patients who do not have a patent anastomosis by 1 year are advised to begin evaluation for IVF though they are counseled that delayed patency may occur even at 18 months. The modifications in our technique are described in Table 7.3.

Table 7.3 Modifications to the vasoepididymostomy technique

1.	Vas fixed to epididymal tunic prior to tubular sutures
	Advantage: Limited manipulations required once the mucosal sutures are in place
	Disadvantage: Vas sutures need to be taken down if there are no sperms in the epididymal fluid
2.	Mucosal sutures placed sequentially, not simultaneously
	Difficult to hold both 200-μm needles together in the micro-needle holder
3.	Use of 200-μm needles on 10-0 suture instead of 70-μm
	Cost and availability
4.	Incision of tubule with needles in situ
	Avoids inadvertent division of the suture material

Outcomes

Despite our stringent inclusion criteria for a VEA, we are able to perform an anastomosis in only about 60% of all explorations [4]. Even among men where we are able to find sperms within the epididymis, in some the tubules are extremely thin with minimal dilatation. In such cases, particularly where the sperms are found only in tubules in the head, a single tubule mucosal anastomosis is not feasible and a non-mucosal anastomosis between the vas lumen and epididymal tunic with an incision in the tubules needs to be performed.

Among cases where a single tubule anastomosis is feasible, our success rates, on average, have been 50% in the form of a patent anastomosis, documented by return of sperms in the ejaculate [17, 18]. These rates have been higher at around 80% when the surgery has been performed bilaterally using the longitudinal suture placement technique and in men with motile sperms in the epididymal fluid. Further, technical satisfaction with the procedure was associated with higher patency rates [21].

There are a number of reasons why our outcomes differ from those reported by other centers. The most important among these is an unclear etiology of obstruction. Most of our patients have primary infertility and the diagnosis of obstruction at the vasoepididymal junction is one of exclusion. Another potential reason is the use of thicker needles on the sutures. The 200-μm needles that we use are five times less expensive than the standard 70-μm needles and this difference is often a major concern for our patients. Our attempts at preserving one side and operating unilaterally may also be causative since our outcomes among bilateral procedures have been much better. Finally, short follow-up of patients is a problem that plagues all our procedures. It is well known that patency may become apparent many months after a procedure but most of our patients provide only one or at most two semen samples after the surgery. This may also be related to the previously discussed issues of social pressure and need for early outcomes. A number of these men possibly opt for assisted reproduction soon after the surgery, not willing to wait for a successful surgery.

Vasovasal Anastomosis

Vasectomy is an uncommon form of contraception in India, accounting for fewer than 5% of all contraceptive methods [22]. A decision to undergo a vasectomy is usually taken after careful thought and requests for reversal are rare. The most frequent reason for seeking a reversal is the loss of a child [23]. This statistic has two important implications. First, the stress on success is likely to be greater in this group of patients than in those who already have living children and second, most patients are interested in early patency during which they may father a child without significant concern about delayed closures. The low volumes of procedures also mean that there is inadequate training of surgeons. These factors have been instrumental in our attempts at simplifying the surgical technique of vasectomy reversal.

Our Surgical Technique

A semen analysis is obtained in all men seeking vasectomy reversal to confirm azoospermia. No additional investigations are requested if the testicular size is normal and both vas are palpable. A longitudinal incision is placed in the scrotal skin, as for the VEA procedure. The testis and the spermatic cord are delivered out of the scrotum and the site of vasectomy is identified. A Babcock forceps is used to hold the vas deferens above and below the site of vasectomy and a short segment of the vas is stripped of its adventitial tissue. The distal vas is sharply divided and flushed with saline to test for patency, similar to the procedure previously described for a vasoepididymostomy. The proximal end of the vas is then divided and the fluid examined for sperms. In the absence of any fluid, a gentle barbotage of the proximal vas is performed with saline and this fluid is then examined. If no fluid or sperms are demonstrable, the epididymis is examined to look for a secondary block and a vasoepididymostomy is considered.

The adventitial tissue of the two ends of the vas is held with the tip of small hemostats to appose them to each other. Two 8-0 polyamide sutures are placed at the 5 and 7 o'clock positions in the seromuscular layers of the two ends and tied. The suture ends are left long and held in rubber shod hemostat clamps. The hemostats applied to the adventitia are removed. A double-armed 10-0 polyamide suture is placed, inside out at 6 o'clock in the mucosa of the distal vas and the second needle is placed at the corresponding position of the proximal vas. The suture is tied. Three additional 10-0 polyamide sutures are placed at the 3, 9, and 12 o'clock positions in the mucosa of both ends of the vas. These three sutures are tied sequentially once all have been placed. Two additional 8-0 polyamide sutures are placed in the seromuscular layer (Fig. 7.3). Additional sutures may be placed in the seromuscular layer or the adventitia to stabilize the anastomosis [24]. The postoperative advice and follow-up are as described for the vasoepididymostomy procedure.

Outcomes

Vasectomy reversals have traditionally been a very satisfying procedure with excellent patency rates, usually above 90% [25]. Our own practice of vasectomy reversals has been sporadic with fewer than 8–10 cases per year. The results were felt to be good but a detailed record is not available. Since 2008, we began using the 4×4 technique described above exclusively following a study protocol. We have found this technique simple to perform with excellent outcomes. All eight patients described in our recent report have a patent anastomosis [24]. Cases performed after the last accrual in this report are also patent. One of the potential problems with this technique is the possibility of sperm leakage and delayed closure. While this is certainly a theoretical possibility, as we stated earlier, most of our patients are keen on an immediate patency with little interest in delayed results. This procedure is simple and may be performed by a greater number of surgeons than the more difficult microdot technique. We feel the simplicity of the procedure with excellent early results, but unknown long-term outcomes may be an acceptable trade off with the more difficult techniques that have proven long-term outcomes.

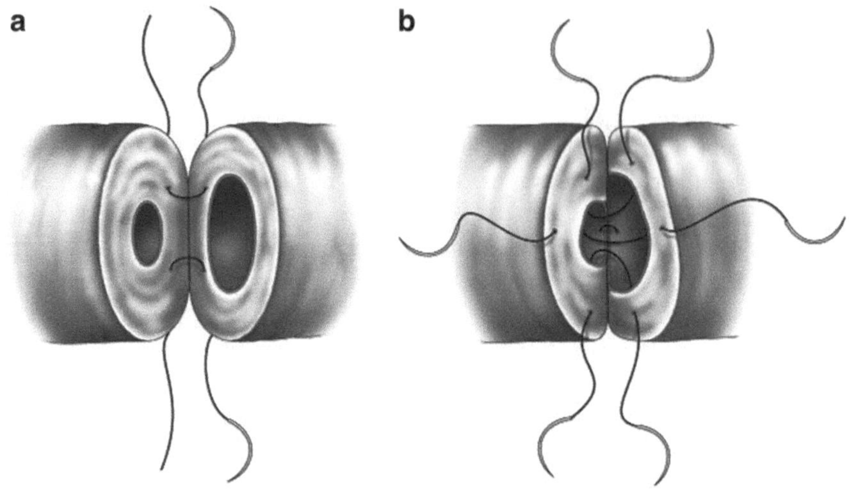

Fig. 7.3 Suture placement technique for a 4×4 vasovasostomy

Microsurgical Varicocelectomy

We maintain a high threshold for case selection for varicocelectomy. Only men with clinically palpable varicoceles are accepted. This policy began even before the AUA and ASRM guidelines were published but were reinforced by these guidelines [26, 27]. Surgery is performed for the side where the varicocele is palpable and a Doppler of the scrotum is requested only where a clinical examination is doubtful.

Our Surgical Technique

The external inguinal ring is identified by insinuating a finger through the scrotal skin into the inguinal region. A 2 cm transverse incision is made in the skin overlying the external ring. Under magnification, this incision is deepened to expose the spermatic cord. The tissue around the cord and over the external ring is separated and a small incision is made in the external ring to make it wider. This incision is directed along the inguinal canal. The cord is held with a Babcock's forceps and, under the microscope, is freed from the bed and surrounding tissues. The cord is then brought out onto the surface. The fossa is examined for any visibly dilated veins that are ligated and divided. The cord is held on the surface over a hemostat. The superficial layers of the spermatic fascia are divided longitudinally, closer to the cranial end of the cord. The artery is identified and isolated. All visible veins are individually identified and ligated and divided. The cord is then taken onto the surgeon's nondominant hand and scanned inside and outside the spermatic fascia to confirm that all major veins have been divided. We do not deliver the testis into the wound or ligate the gubernacular vessels.

Outcomes

We have previously reported our outcomes for subinguinal varicocelectomy [26]. Briefly, one-third of patients are able to father a child through natural conception after the surgery and a majority will show improvements in seminal parameters. The procedure has resulted in downgrading the ART techniques required in a number of patients who have undergone this procedure [28].

Training and Credentials

Andrology and microsurgery procedures are performed as a part of the standard urological services available to all patients. Urology residents, during their 3 years of training, are posted to assist in these surgical procedures. There are no fellowship programs in male infertility or andrology and few would return for dedicated specialized training in microsurgery. This results in very little postresidency acquisition of skills or even skills maintenance because the numbers of procedures performed in urologic practice in the community is very limited.

Five-Year View and Key Issues

The major issues concerning microsurgery for male infertility in our practice are lack of awareness and inadequate training facilities. The general perception among treating gynecologists and IVF specialists is that the outcomes of surgical intervention are poor. We are focusing our energies toward improving this outlook by delivering lectures and publishing our results so that a greater number of men with obstructive azoospermia may be counseled about surgical options. With regards surgical training, the opportunities still remain limited, primarily due to inadequate infrastructure and individuals performing these surgeries. Over the next few years, we hope to train a significant number of urologists in microsurgical techniques so that these may become available at more centers.

Another important aspect of our practice is the inability to diagnose the etiology of obstruction in the majority of cases. While infection remains the most likely suspect, we have been unable to document this. We have initiated studies to evaluate etiology using molecular methods and hope to be able to reach some conclusion over the next 5 years.

Conclusions

The diseases responsible for male infertility and for cases undergoing exploration for microsurgical reconstruction at AIIMS possibly differ from those reported in most Western literature. The largest majority of cases explored are for primary infertility with obstruction of unknown etiology. This has resulted in lower percentage of patients with a successful outcome. However, the socioeconomic factors around infertility management in our country dictate an attempt at reconstruction even when the expected outcomes are poor. In patients with favorable prognostic factors, the outcomes are generally good.

References

1. Kumar R, Thulkar S, Kumar V, Jagannathan NR, Gupta NP. Contribution of investigations to the diagnosis of congenital vas aplasia. ANZ J Surg. 2005;5:807–9.
2. Meng MV, Cha I, Ljung BM, Turek PJ. Testicular fine-needle aspiration in infertile men: correlation of cytologic pattern with biopsy histology. Am J Surg Pathol. 2001;25:71–9.
3. Kumar R, Gautam G, Gupta NP, Aron M, Dada R, Kucheria K, et al. Role of testicular fine-needle aspiration cytology in infertile men with clinically obstructive azoospermia. Nat Med J Ind. 2006;19:18–9.
4. Kumar R. Surgery for azoospermia in the Indian patient: why is it different? Indian J Urol. 2011;27(1):98–101.
5. Schiff J, Chan P, Li PS, Finkelberg S, Goldstein M. Outcome and late failures compared in 4 techniques of microsurgical vasoepididymostomy in 153 consecutive men. J Urol. 2005;174:651–5.
6. Chan PT, Brandell RA, Goldstein M. Prospective analysis of outcomes after microsurgical intussusception vasoepididymostomy. BJU Int. 2005;96:598–601.
7. Viswaroop B, Johnson P, Kurian S, Chacko N, Kekre N, Gopalakrishnan G. Fine-needle aspiration cytology versus open biopsy for evaluation of chronic epididymal lesions: a prospective study. Scand J Urol Nephrol. 2005;39:219–21.
8. Orakwe JC, Okafor PI. Genitourinary tuberculosis in Nigeria; a review of thirty-one cases. Niger J Clin Pract. 2005;8:69–73.
9. Lenk S, Schroeder J. Genitourinary tuberculosis. Curr Opin Urol. 2001;11:93–8.
10. Kumar R. Reproductive tract tuberculosis and male infertility. Indian J Urol. 2008;24:92–5.
11. Kumar R, Hemal AK. Bilateral epididymal masses with infertility. ANZ J Surg. 2004;74:391.
12. Ekwere PD. Filarial orchitis: a cause of male infertility in the tropics–case report from Nigeria. Cent Afr J Med. 1989;35: 456–60.
13. Eyquem A, Heuze D, Schwartz J, Languillat G. Implications of heterophile antigens in immunological infertility in males. Arch Androl. 1978;1:241–8.
14. Phadke AM, Samant NR, Dewal SD. Smallpox as an etiologic factor in male infertility. Fertil Steril. 1973;24:802–4.
15. Berger RE. Triangulation end-to-side vasoepididymostomy. J Urol. 1998;159:1951–3.
16. Marmar JL. Modified vasoepididymostomy with simultaneous double needle placement, tubulotomy and tubular invagination. J Urol. 2000;163:483–6.
17. Kumar R, Gautam G, Gupta NP. Early patency rates following the two-stitch invagination technique of vasoepidiymal anastomosis for idiopathic obstruction. BJU Int. 2006;97:575–7.
18. Kumar R, Mukherjee S, Gupta NP. Intussusception vasoepididymostomy with longitudinal suture placement for idiopathic obstructive azoospermia. J Urol. 2010;183:1489–92.
19. Chan PT, Li PS, Goldstein M. Microsurgical vasoepididymostomy: a prospective randomized study of 3 intussusception techniques in rats. J Urol. 2003;169:1924–9.
20. Kumar R. Re: microsurgical vasoepididymostomy: a prospective randomized study of 3 intussusception techniques in rats. J Urol. 2004;171:810–1.
21. Gautam G, Kumar R, Gupta NP. Factors predicting the patency of two-stitch invagination vasoepididymal anastomosis for idiopathic obstruction. Indian J Urol. 2005;21:112–5.
22. Dhillon BS, Chandhiok N, Kambo I, Saxena NC. Induced abortion and concurrent adoption of contraception in the rural areas of India (an ICMR task force study). Indian J Med Sci. 2004;58: 478–84.
23. Jina RP, Kumar V. Recanalisation of vas. J Indian Med Assoc. 1979;72:30–2.
24. Kumar R, Mukherjee S. "4×4 vasovasostomy": a simplified technique for vasectomy reversal. Indian J Urol. 2010;26(3): 350–2.
25. Belker AM, Thomas Jr AJ, Fuchs EF, Konnak JW, Sharlip ID. Results of 1,469 microsurgical vasectomy reversals by the Vasovasostomy Study Group. J Urol. 1991;145:505–11.
26. Kumar R, Gupta NP. Subinguinal microsurgical varicocelectomy: evaluation of results. Urol Int. 2003;71:361–7.
27. Sharlip ID, Jarow JP, Belker AM, Lipshultz LI, Sigman M, Thomas AJ, Schlegel PN, Howards SS, Nehra A, Damewood MD, Overstreet JW, Sadovsky R. Best practice policies for male infertility. Fertil Steril. 2002;77:873–82.
28. Kumar R, Gupta NP. Varicocele and the urologist. Indian J Urol. 2006;22:98–104.

Advanced Techniques of Vasoepididymostomy

8

Wayland Hsiao and Marc Goldstein

Abstract

With advances in surgical technique and the development of microsurgical techniques, modern vasoepididymostomy allows us to accurately approximate the mucosa of a single epididymal tubule to the mucosa of the vasal lumen. With this increased precision, we have been able to achieve even higher patency and pregnancy rates. Microsurgical vasoepididymostomy, however, is the most technically demanding procedure in all of microsurgery. In virtually no other operation are results so dependent upon technical perfection. Thus, microsurgical vasoepididymostomy should only be attempted by an experienced microsurgeon who performs a sufficient volume of microsurgery. Modern IVF–ICSI has opened up reproductive options for those couples desiring fertility. This has caused some to question the need for advanced reconstructive reproductive tract surgery. However, in the hands of experienced microsurgeons, vasoepididymostomy is a safe, effective method of reconstruction for patients who do not want to undergo IVF or desire multiple children.

Keywords

Vasoepididymostomy techniques • Epididymal obstruction • Obstructive azoospermia • Anastomosis • Anastomotic technique • End-to-side intussusception • Anastomotic site • Transseptal vasoepididymostomy

The first vasoepididymostomy (VE) was reported in 1902 by Dr. Edward Martin at the University of Pennsylvania. His technique involved slashing across multiple epididymal tubules and anastomosis of the vas to the epididymal tunic in a side-to-side manner with four fine silver wires [1, 2]. Patency depended on the formation of a fistula. In 1909, Martin reported in a series of 11 patients with epididymal obstruction a patency rate of 64% and a pregnancy rate of 27% [3]. He proved that vasoepididymostomy was technically feasible and his approach is the foundation on which subsequent work was based.

With advances in surgical technique and the development of microsurgical techniques, modern vasoepididymostomy allows us to accurately approximate the mucosa of a single epididymal tubule to the mucosa of the vasal lumen [4]. With this increased precision, we have been able to achieve even higher patency and pregnancy rates [5, 6]. Microsurgical vasoepididymostomy, however, is the most technically demanding procedure in all of microsurgery. In virtually no other operation are results so dependent upon technical perfection. Thus, microsurgical vasoepididymostomy should only be attempted by an experienced microsurgeon who performs a sufficient volume of microsurgery.

W. Hsiao, MD (✉)
Assistant Professor in Urology, Emory University, Department of Urology, 1365 Clifton Road, NE Building B, Suite 1400 Atlanta, GA 30322, USA
e-mail: whsiao@emory.edu

M. Goldstein, MD
Cornell Institute for Reproductive Medicine, New York Presbyterian Hospital/Weill Cornell Medical Center, Weill Cornell Medical College of Cornell University, 525 East 68th Street, Box 580, New York, NY 10021, USA
e-mail: mgoldst@med.cornell.edu

S.J. Parekattil and A. Agarwal (eds.), *Male Infertility: Contemporary Clinical Approaches, Andrology, ART & Antioxidants*, DOI 10.1007/978-1-4614-3335-4_8, © Springer Science+Business Media, LLC 2012

Vasoepididymostomy

Vasoepididymostomy is indicated in patients with obstructive azoospermia, and the decision to perform a vasoepididymostomy rather than a vasovasostomy is an intraoperative decision. During vasectomy reversal, the testicular end of the vas is cut and the intravasal fluid is evaluated both grossly as well as with the aid of a 400× bench microscope. The presence of thick toothpaste like fluid devoid of sperm, scant fluid in a patient without a sperm granuloma, or scant fluid with no spermatozoa seen on barbitage, constitute indications for vasoepididymostomy. For non-vasectomy-related obstruction, vasoepididymostomy is indicated when a testis biopsy reveals complete spermatogenesis and transaction of the proximal vas reveals no sperm even with barbitage.

The evolution of modern single tubule vasoepididymostomy techniques has progressed from the original end-to-end anastomoses described by Silber, end-to-side anastomoses described by Wagenknecht and Fogdestam, and end-to-side intussusception anastomoses first described by Berger. In all of these, the initial exposure and setup are similar. A high vertical scrotal incision is made about 3–4 cm in length aimed toward the external ring of the inguinal canal. In cases with inadequate length of the vas, the incision can be extended over the external ring and inguinal dissection of the vas performed. After incision through the skin and dartos fascia, the testicle is delivered with the tunica vaginalis intact. Using a Babcock clamp, the vas is isolated and surrounded with a Penrose drain. The operating microscope is brought into the field. The junction of the straight and convoluted vas is identified and isolated. The vas is then dissected free of its investing sheath and blood vessels under the operating microscope to expose a clean segment of bare vas. The bare segment of vas is hemitransected with a 15° ultrasharp knife until the lumen is visualized. The vasal fluid is then sampled using a bench microscope at 400× magnification. If no spermatozoa are seen then an additional 0.1–0.2 ml of fluid is injected into the testicular end and that fluid is expressed back out by squeezing the testis and epididymis and the fluid examined under the bench microscope. Absence of vasal sperm on microscopic exam in a man with either a normal testis biopsy or a positive antisperm antibody assay [7] confirms the diagnosis of epididymal obstruction.

At this point, the abdominal end of the vas is checked for patency by cannulating the abdominal end of the vas with a 24-gage angiocather and injecting 1 ml of lactated ringers. Smooth injection without resistance or backflow confirms patency of the abdominal end of the vas. If further confirmation is desired, then a Foley catheter can be inserted after injecting indigo carmine and the color of the urine inspected. Green urine or blue urine confirms the patency of the abdominal vas as well as the ejaculatory ducts.

Once epididymal obstruction is confirmed and the need for a vasoepididymostomy verified, the vas is prepared for anastomosis by complete transection using an ultrasharp knife drawn through a slotted 2, 2.5, or 3 mm nerve holding clamp. This gives the surgeon a perfect 90° cut of healthy vasal tissue. The cut surface of the testicular end of the vas deferens is inspected using 15–25 power magnification and should look like a bullseye with the three vasal layers distinctly visible. A healthy white mucosal ring should be seen which springs back immediately after gentle dilation. This layer is surrounded by muscularis which should appear smooth and homogeneous. A gritty-looking muscularis layer may indicate the presence of scar/fibrosis. Healthy bleeding should be noted from both the cut edge of the mucosa and the surface of the muscularis. If the blood supply is poor or the muscularis is gritty, the vas is recut until healthy tissue is found. The vasal artery and vein are ligated with 6-0 vicryl. Small bleeders are controlled with a microbipolar forceps set at low power. At this point, the tunica vaginalis is opened and the epididymis is inspected.

In patients with previous vasectomy, there are some minor variations, but the overall approach is similar. In these patients, the testicular and abdominal ends of the vas are identified and dissected free. The abdominal end is transected and checked for patency. After confirmation of abdominal end patency, the testicular end is then inspected, sectioned, and intravasal fluid microscopically inspected. Serial sectioning and microscopic evaluation are performed until either sperm are seen or one has reached the convoluted vas. Once, sectioning has progressed well into the convoluted vas and barbitage reveals no spermatozoa, the need for vasoepididymostomy is confirmed. The tunica vaginalis is opened and the epididymis is inspected under the operating microscope. At this point, it is time to determine the site of anastomosis.

End-to-End Anastomosis

This is the original microsurgical technique introduced by Silber and it is the first technique to allow the anastomosis of a specific epididymal tubule to the vas. At its introduction, it was far superior to any method previously described. In this technique, the epididymis is dissected down to it junction with the convoluted vas. The epididymis is then serially sectioned until a large rush of fluid is noted (Fig. 8.1), indicating that the area of obstruction has been bypassed. The single tubule with gushing fluid is identified and anastomosed to the vas with 3–5 interrupted 10-0 nylon sutures. The outer layer of the vas is anastomosed to the tunica of the epididymis with 9-0 nylon sutures (Fig. 8.2).

The advantages of this technique include the ability to dissect off the epididymis and rotate it to gain additional

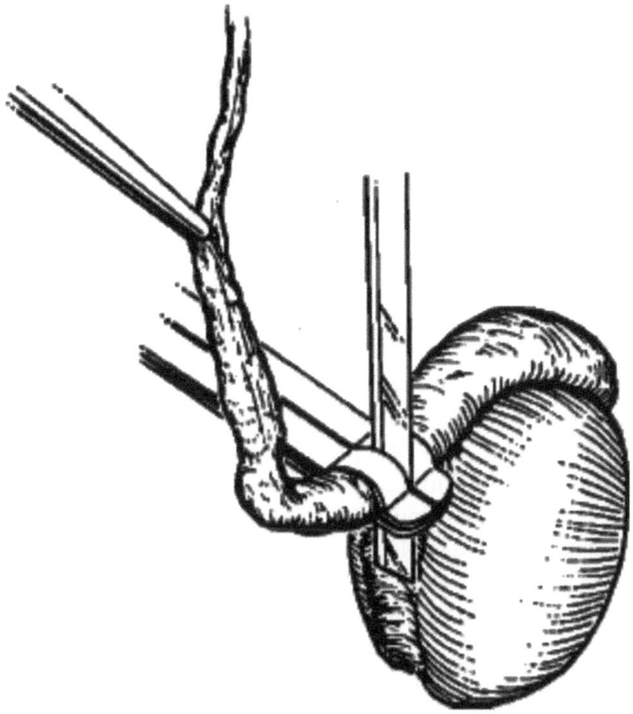

Fig. 8.1 Sectioning method employed in the end-to-end technique (from Goldstein M. Surgical management of male infertility. In: Wein AJ, editor, Campbell-Walsh urology, 9th ed. St. Louis: WB Saunders; 2006, with permission of Elsevier)

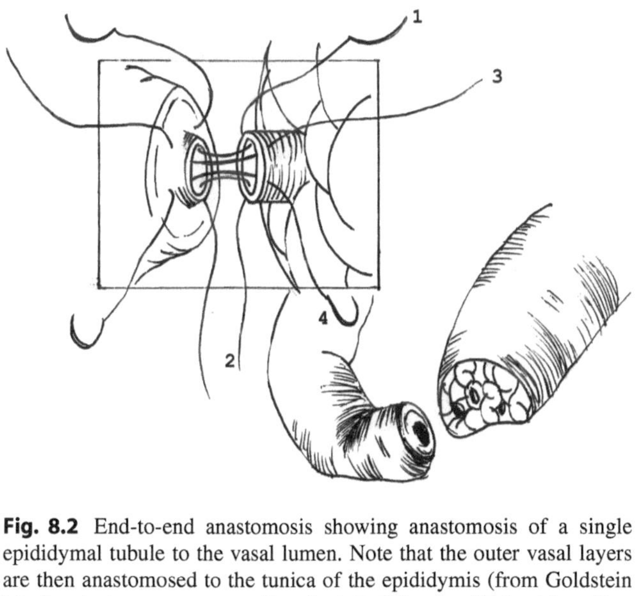

Fig. 8.2 End-to-end anastomosis showing anastomosis of a single epididymal tubule to the vasal lumen. Note that the outer vasal layers are then anastomosed to the tunica of the epididymis (from Goldstein M. Surgical management of male infertility. In: Wein AJ, editor, Campbell-Walsh urology. 9th ed., St. Louis: WB Saunders; 2006, with permission of Elsevier)

length if there are issues with short vasal length. A major disadvantage of this technique is that the outer diameter of the epididymal tunica is far larger than the outer diameter of the vas deferens, making a water-tight closure exceedingly difficult. Also, given that the blood supply is invariably affected during transaction, it becomes more difficult to clean, blood-free sperm for cryopreservation than it is with the end-to-side technique.

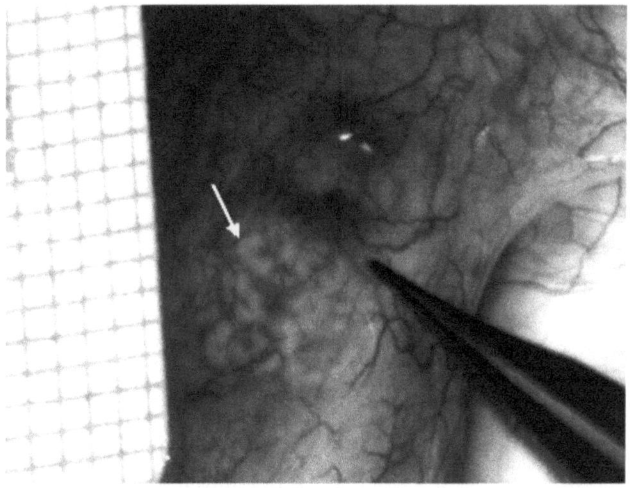

Fig. 8.3 Example of dilated epididymal tubules seen under the operating microscope

End-to-Side Techniques

End-to-side techniques of vasoepididymostomy improved on the end-to-end technique and have the advantage of being relatively bloodless and less traumatic to the delicate epididymis [8–11]. It requires minimal dissection of the tubule and allows the surgeon to easily tailor the size of the opening in the epididymal tubule. Also, this method allows for the preservation of the epididymal branches of the testicular artery. Thereby, if another vasovasostomy is required, the blood supply to the intervening segment of vas can be preserved. In cases where the integrity of the testicular artery is in doubt (previous orchiopexy, previous varicocelectomy, or hernia repair), preservation of the deferential artery may be required for the maintenance of testicular blood supply.

The selection of an anastomotic site is a bit more involved with the end-to-side technique when compared with the end-to-end technique. After the vas has been prepared, the tunica vaginalis is opened and the testis delivered. Inspection of the epididymis under the operating microscope may reveal a clearly delineated demarcation above which epididymal tubules are markedly dilated and below which the tubules are collapsed. Often, a discrete yellow sperm granuloma is noted, above which the epididymis is indurated and the tubules dilated and below which the epididymis is soft and the tubules collapsed (Fig. 8.3). If the level of obstruction is not clearly delineated a 70-μm tapered needle from the 10-0 nylon microsuture is used to puncture the epididymal tubule beginning as distal as possible and fluid sampled from the puncture site until sperm are found. At that level, the puncture is sealed with microbipolar forceps and the anastomosis is performed proximal to the puncture site.

An anastomotic site is selected where the epididymal tubules are clearly dilated. An avascular area is grasped with jeweler's forceps and the epididymal tunica tented upwards. A 3–4 mm buttonhole is made in the tunica with microscissors to match the outer diameter of the vas. The epididymal tubules are then gently dissected until dilated loops of tubules are clearly exposed.

At this point an opening is made in the tunica vaginalis and the vas deferens end is brought through and secured to the tunic with 2–3 interrupted 6-0 prolene sutures to ensure that the vasal lumen reaches the opening in the epididymal tunica without tension and with some length to spare. The posterior edge of the epididymal tunica is then approximated to the posterior edge of the vas muscularis and adventitia with 2–3 interrupted suture of double-armed 9-0 nylon. At the end of this step, the vasal lumen should be in close approximation to the epididymal tubule selected as the site for anastomosis. The proper positioning of the vasal segment as well as proper setup are critical to the creation of a long-lasting tension-free anastomosis.

Anastomotic Technique

Once setup for the anastomosis is complete, the surgeon has a choice of anastomotic techniques which vary by the number of sutures placed, the order of suture placement, and intussusception of the tubule. We will discuss the classic end-to-side anastomosis as well as the various intussusception techniques.

Original End-to-Side

The classic end-to-side approach involves creation of a longitudinal incision along the selected epididymal tubule. This is done under 25–32× magnification. The intratubular fluid is microscopically inspected with the bench microscope. If no sperm are seen on microscopic exam, then the tubule is closed with a 10-0 suture and the overlying tunica closed with 9-0 nylon. A more proximal location is then identified and the setup for anastomosis is repeated. If sperm are found on microscopic inspection, it is safe to continue with the procedure. The extruded epididymal fluid is aspirated into glass capillary tubes and flushed into media for cryopreservation [12]. Diluted indigo carmine is applied to the field to highlight the edges of the epididymal tubule as well as the mucosal edges of the vas segment. Of note, we have previously shown that methylene blue and radiographic contrast are toxic to spermatozoa, while diluted indigo carmine is not [13]. Thus, it is our preference to use indigo carmine diluted 50% with lactate ringers for all vasograms and for emphasis of the mucosal edges.

Constant irrigation with saline or lactated ringers is required to keep the delicate epididymal tubule open and to visualize the edges. The posterior mucosal edge of the cut epididymal tubule is approximated to the posterior edge of the vasal mucosa with 2 interrupted sutures of 10-0 nylon double-arm sutures with 70-μm diameter tapered needles. After these mucosal sutures are tied, the anterior mucosal anastomosis is completed with 2–4 additional 10-0 interrupted sutures. The outer muscularis and adventitia of the vas is then approximated to the cut edge of the epididymal tunica with 6–10 additional interrupted sutures of 9-0 nylon double armed with 100-μm diameter needles. The vasal sheath is secured to the epididymal tunica with 3–5 sutures of 9-0 nylon allowing for a straight course without kinks. The tunica vaginalis is then closed with 5-0 Vicryl and the dartos reapproximated with absorbable suture. The skin is closed in a subcuticular fashion.

End-to-Side Intussusception Technique

The next advance in vasoepididymostomy techniques came with the development of intussusception techniques. This method was first introduced by Berger in 1998 [14]. The setup is identical to that for the classic procedure. After the vas is fixed to the opening in the epididymal tunica, six microdots are placed on the cut surface of the vas to mark the sites of needle exit. The microdot technique ensures precise suture placement by exact mapping of each planned suture. The microdot method separates the planning from the placement of sutures [15]. Much as a civil engineer is consulted before workmen commence construction on a bridge, the microdot method allows the surgeon to completely focus on each individual task at hand. This results in substantially improved accuracy in suture placement as well as better suture spacing. Next, the epididymal tubule selected for anastomosis is dissected until it is free of surrounding tissue and displays prominently. Indigo carmine is applied to highlight the tubule. Using double-arm 10-0 nylon sutures with 70μm tapered needles, three sutures are placed in the epididymal tubule in a triangular configuration. The needles are left in situ, creating a triangle of needles (Fig. 8.4). It must be remembered that the needle of the 10-0 suture is 70 μm in diameter while the suture material itself is only 17 μm. Thus, if the needles are pulled through prematurely, epididymal fluid and sperm would immediately leak though the suture hole causing the tubules to collapse, making placement of subsequent sutures and opening the tubules more difficult. Leaving the needles in situ also prevents accidentally cutting sutures when making the opening in the epididymal tubule.

After all three needles are properly placed, Berger originally described using a 9-0 cutting needle to lift the tubule

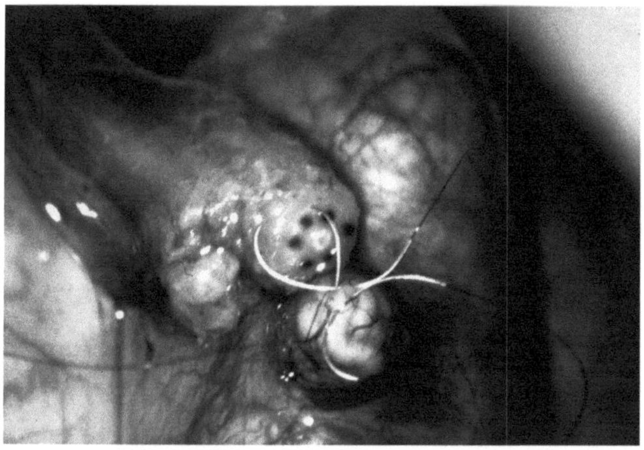

Fig. 8.4 Triangle of needles formed during the triangulation end-to-side intussusception technique introduced by Berger

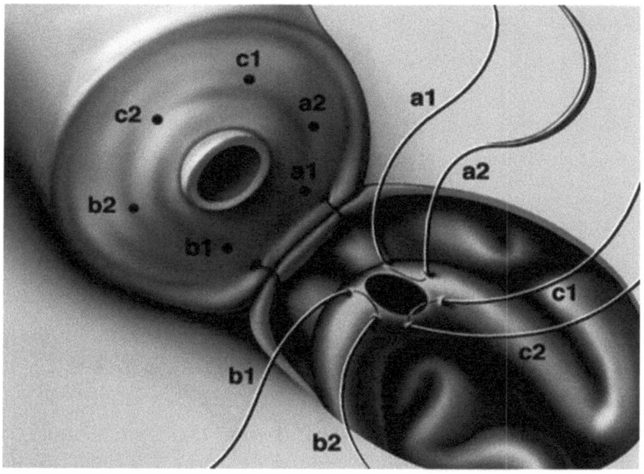

Fig. 8.5 End-to-side triangulation intussusception technique introduced by Berger (from Goldstein M. Surgical management of male infertility. In: Wein AJ, editor, Campbell-Walsh urology. 9th ed., St. Louis: WB Saunders; 2006, with permission of Elsevier)

and tear an opening in it. We prefer a 15° microknife to make an opening is made in the epididymal tubule in the center of the triangle. The three needles are then pulled though. The six needles are now laid out to avoid tangling. The extruded fluid is inspected under the microscope for sperm. If sperm are seen, then the six needles are passed inside out the vas deferens exiting through the six previously placed microdots (Fig. 8.5). The sutures are then tied and intussuscepting the epididymal tubule into the vas lumen and thereby creating a watertight closure. Intussusception also allows the flow of fluid from the epididymal tubule into the vas to push the edges epididymal tubule against the vasal mucosa, further reinforcing the watertight nature of this anastomosis. The edges of the vas are then closed with interrupted 9-0 nylon sutures (Fig. 8.6). Limitations of the triangulation technique include the need for a relatively large

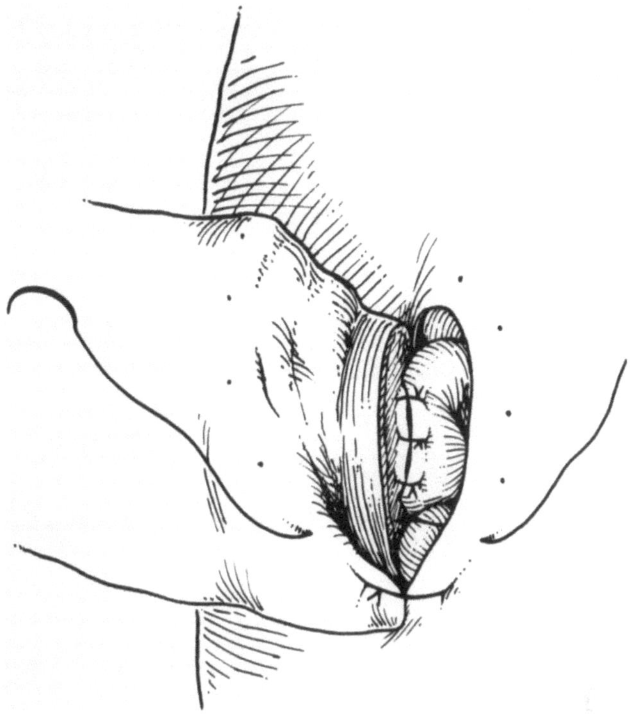

Fig. 8.6 Closure of the epididymal tunica should be done with 9-0 nylon sutures with particular attention paid to avoid incorporating any underlying tubules into the closure (from Goldstein M. Surgical management of male infertility. In: Wein AJ, editor, Campbell-Walsh urology. 9th ed., St. Louis: WB Saunders; 2006, with permission of Elsevier)

tubule for the three needles to fit. Thus, this technique is not suitable for anastomosis to the efferent ductules or the proximal caput epididymis where the tubule is smaller.

Two-Stitch Variation of Vasoepididymostomy

This is our currently preferred method of VE which allows for a two–stitch intussuscepted anastomosis. In this technique, four microdots are made on the vasal end. Two needles from two separate 10-0 double-arm sutures are then placed longitudinally in the tubule with care not to pull the needles completely through. The opening in the epididymal tubule is then made with a 15° microknife between the needles. After microscopic confirmation of the presence of spermatozoa, the needles are passed. The four needles are then passed through the vasal lumen and exiting the microdots (inside to outside). A 9-0 suture is placed to pull the anterior vas and adventitia toward the opening in the epididymal tubule bringing the vas mucosa into close approximation to the opening in the epididymal tubule. The lumen is irrigated with heparinized saline just prior to tying the mucosal sutures. Finally, the mucosal sutures are tied down (Fig. 8.7), allowing for the intussusception of the epididymal tubule. The outer layer is closed with interrupted 9-0 nylon sutures careful to not inadvertently incorporate any epididymal tubules when placing these sutures

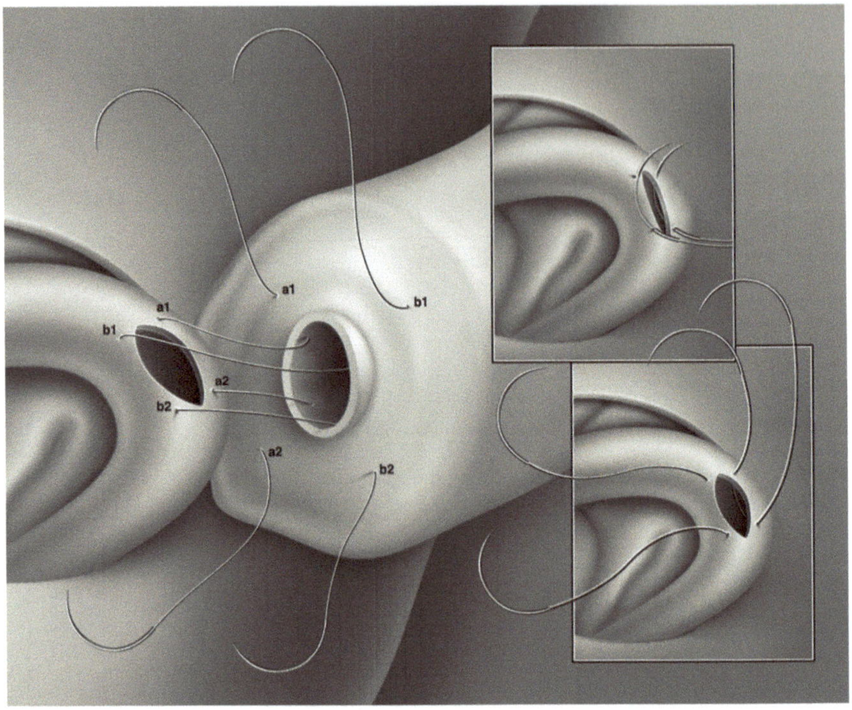

Fig. 8.7 Longitudinal intussuscepted vasoepididymostomy technique. Mucosal suture placement (from Goldstein M. Surgical management of male infertility. In: Wein AJ, editor, Campbell-Walsh urology. 9th ed., St. Louis: WB Saunders; 2006, with permission of Elsevier)

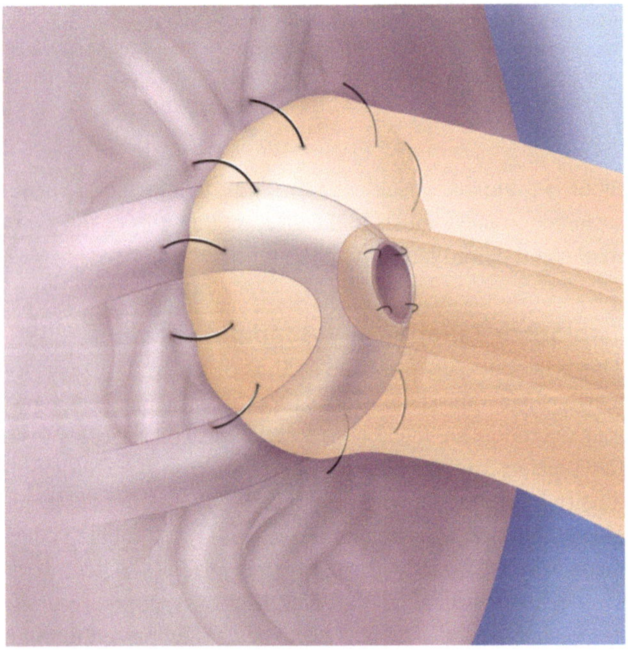

Fig. 8.8 Completed anastomosis for longitudinal intussuscepted technique (from Goldstein M. Surgical management of male infertility. In: Wein AJ, editor, Campbell-Walsh urology. 9th ed., St. Louis: WB Saunders; 2006, with permission of Elsevier)

(Fig. 8.8). Again, by not passing the needles, one keeps a distended tubule which makes suture placement more accurate and reliable. Variations in this technique include mounting two needles in a single needle holder and placing them simultaneously transversely in the tubule as suggested by Marmar.

Of note, the cost of double-arm sutures can be high. In response to this, we have developed a single-arm technique of VE which we have found to be almost as effective as its double-arm counterpart [16]. It begins with the standard setup for VE. We then place four microdots in the vasal end. Two 10-0 single-arm nylon sutures are then passed through the microdots and exiting the vasal lumen (outside to inside). After this, the same two sutures are placed longitudinally in the selected tubule and the needles are not completely passed. After opening the tubule and confirming the presence of spermatozoa, the needles are pulled through and the needles passed through the vasal lumen and exiting the microdot (inside to outside) (Fig. 8.9). The sutures are then tied allowing the intussusceptions of the epididymal tubule. The outer sheath of the vas deferens is then approximated to the tunic of the epididymis with 2–4 interrupted 9-0 nylon sutures, removing all tension from the anastomosis.

Techniques When Vasal Length Is Severely Compromised

One of the most common problems that arise during vasoepididymostomy is inadequate vasal length, often due to a very destructive vasectomy. When there is inadequate length of the vas deferens to reach the dilated epididymal tubule without tension, a number of surgical techniques may be employed involving any of the following: increasing length of the epididymis, increasing the vasal length, fixation of the testis, or use of the contralateral vas deferens.

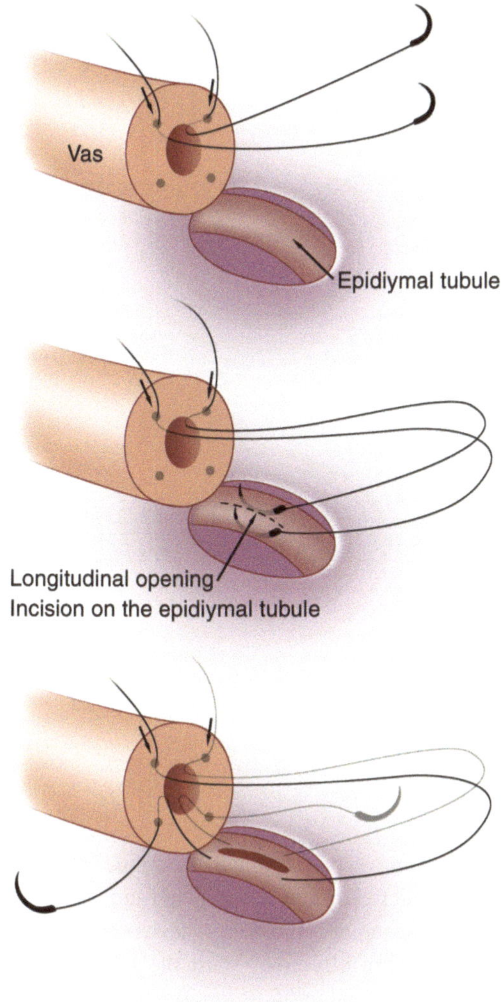

Fig. 8.9 Technique of single-arm vasoepididymostomy technique. Needles are passed outside in on the vas deferens. The needles are then passed longitudinally in the selected epididymal tubule and the cut made in the epididymal tubule. The needles are then passed and then placed inside out in the vas deferens (from Goldstein M. Surgical management of male infertility. In: Wein AJ, editor, Campbell-Walsh urology. 9th ed., St. Louis: WB Saunders; 2006, with permission of Elsevier)

To gain length on the epididymis, the cauda and corpus epididymis can be dissected down to the vasoepididymal junction and then dissected off the testes as in the end-to-end operation. The epididymis is encircled with a small Penrose drain at the level of obstruction and dissected off of the testis up to the level of obstruction, yielding sufficient length to perform the anastomosis. Usually an avascular plane can be found right on the tunica albuginea of the testis between the epididymis and testis and injury to the epididymal blood supply can be avoided. The inferior and, if necessary, middle epididymal branches of the testicular artery are ligated and divided to free up an adequate length of epididymis. The superior-epididymal branches entering the epididymis at the caput are always preserved and since the epididymis has a dual blood supply, this is adequate blood supply for the entire epididymis.

If the epididymis is indurated and dilated throughout its length, the epididymis is dissected all the way past the vasoepididymal junction. This dissection is often facilitated by first dissecting the convoluted vas to the vasoepididymal junction from below, and then, after encircling the epididymis with a Penrose drain, dissecting the epididymis to the vasoepididymal junction from above. In this way, the entire vasoepididymal junction can be freed up. This will allow preservation of maximal epididymal length in cases of distal obstruction near the vasoepididymal junction. After the epididymis is dissected off of the testis and flipped-up, a two-stitch longitudinal end-to-side intussusception anastomosis can be performed as described previously (Fig. 8.10).

Increasing vasal length can be done with extensive blunt dissection of the vas deferens off the spermatic cord toward the inguinal ring. Dissection can also be carried into the inguinal canal to the internal ring with a finger sweeping motion. In extreme situations, the vas deferens can be rerouted medial to the vessels similar to the Prentiss maneuver employed during difficult orchiopexies [17]. An opening in the floor of the inguinal canal is made and the vas rerouted medially under the floor of the canal and right over the pubis.

It is also possible to perform an orchiopexy positioning the testicle in a horizontal or even upside down configuration to decrease the length needed. One must be careful to make sure the cord has no kinks in it and that the stitches do not damage the blood supply to the testis.

In cases where there is a unilateral atrophic testis or the contralateral testis is missing, it is possible to perform a crossed transseptal vasoepididymostomy. This is even more attractive if there is an ipsilateral hernia repair or where there is a second obstruction in the inguinal or abdominal vas. In this procedure, the contralateral vas is harvested as close to the vasoepididymal junction as possible. If vasal length is still inadequate, then the testicle can be pexed in the contralateral scrotal compartment to facilitate a tension-free anastomosis.

Long-Term Follow-up Evaluation and Results

Microsurgical vasoepididymostomy in the hands of experienced skilled microsurgeon will result in the appearance of sperm in the ejaculate in 50–85% of men. Classic end-to-side or older end-to-end methods result inpatency rates about 70% with a 43% pregnancy rate with a follow-up of 2 years [5, 18]. With intussusception techniques, patency rates are 70–90% with pregnancy rates of 40–45% [6, 14, 19–22]. Regardless of technique, pregnancy rates are higher the more distal the anastomosis is performed [23]. Therefore, one

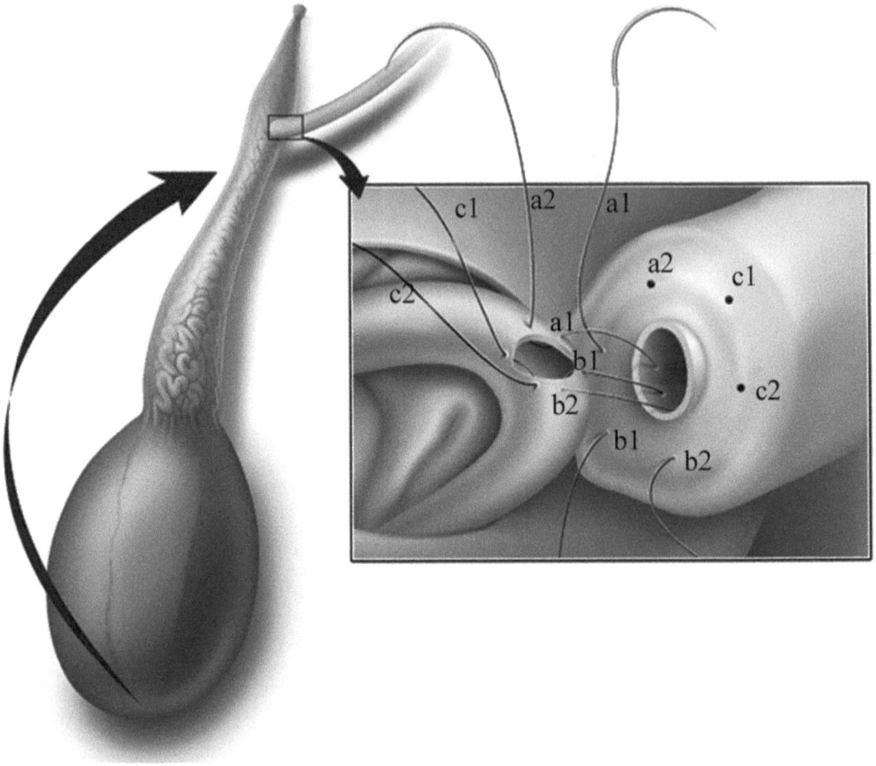

Fig. 8.10 Technique of dissecting the corpus and cauda epididymis to gain further length in cases of short vasal length. This is most helpful when the entire epididymis is dilated (from Goldstein M. Surgical management of male infertility. In: Wein AJ, editor, Campbell-Walsh urology. 9th ed., St. Louis: WB Saunders; 2006, with permission of Elsevier)

should always strive to make the anastomosis as distal as possible on the epididymis.

Another vexing problem is that of late anastomotic failure. With the older end-to-end or end-to-side methods, at 14 months after surgery 25% of initially patent anastomoses have shut down [12]. With intussusception techniques, the late shut down rates appear to be less than 10%, but long-term follow-up with these techniques have not yet been reported. Nevertheless, we recommend banking sperm both intraoperatively [24] and as soon as motile sperm appear in the ejaculate postoperatively after vasoepididymostomy, regardless of technique employed. In men with very low counts or poor sperm quality postoperatively and men who remain azoospermic, the sperm intraoperatively cryopreserved can be used for IVF with intracytoplasmic sperm injection. Persistently azoospermic men without cryopreserved sperm can opt for either a redo-vasoepididymostomy and/or microscopic epididymal sperm aspiration combined with IVF and intracytoplasmic sperm injection.

Expert Commentary

The modern evolution of vasoepididymostomy has been a remarkable journey. Since Martin's first attempts over 100 years ago, we have continued to make significant strides in the refinement of this surgical technique. Most recently, adoption of microsurgical techniques and intussusception methods of vasoepididymostomy have made this surgery progressively more effective. With the introduction of the two-stitch longitudinal intussusception method, anastomoses have become simpler and easier to teach with a decreasing risk of technical error.

Modern IVF–ICSI has opened up reproductive options for those couples desiring fertility. This has caused some to question the need for advanced reconstructive reproductive tract surgery. However, in the hands of experienced microsurgeons, vasoepididymostomy is a safe, effective method of reconstruction for patients who do not want to undergo IVF or desire multiple children. In addition, vasoepididymostomy skills are crucial to have because of the possibility of finding secondary epididymal obstruction at the time of vasectomy reversal. It is of our opinion that any reproductive surgeon what performs vasal reconstruction must be capable of performing a vasoepididymostomy.

Five-Year View

While vasoepididymostomy is already associated with good outcomes we look forward to the future. Further technical refinements will most likely focus on the simplification of the vasoepididymostomy procedure, decreasing operative times and making the procedure more accessible to more surgeons. These developments will come from microsurgical models and animal models.

On the other hand, refinements in molecular genetics will continue to elucidate the pathophysiology of idiopathic infertility and allow us to identify those patients that will most benefit from advanced surgical reconstruction. The focus of this research must be on translation of these genetic preoperative predictors into treatments that increase postoperative success.

Key Issues

- Good anastomoses relies on healthy tissue and an accurate watertight mucosa to mucosa opposition in a tension-free anastomosis; setup is key.
- Anastomotic techniques include: end-to-end, end-to-side, and end-to-side intussuscepted.
- Vasal length can be increased on the epididymal end or the vasal end or both. If these two maneuvers prove insufficient, orchiopexy should be considered as well as a crossed septal vasoepididymostomy in cases of unilateral testicular atrophy or absence.
- Our preferred anastomosis is the two-suture longitudinal, end-to-side intussuscepted technique.
- We have developed a single-arm version of vasoepididymostomy, which is useful when double-arm sutures are difficult to obtain.
- Vasoepididymostomy is the most challenging of all microsurgery and should only be performed by surgeons with sufficient training and adequate volume of microsurgery.

Acknowledgments Dr. Hsiao is supported by a grant from the Frederick J. and Theresa Dow Wallace Fund of the New York Community Trust.

References

1. Jequier AM. Edward Martin (1859–1938). The founding father of modern clinical andrology. Int J Androl. 1991;14(1):1–10.
2. Martin E, Carnett JB, Levi JV, et al. The surgical treatment of sterility due to obstruction at the epididymis; together with a study of the morphology of human spermatozoa. Univ Pa Med Bull. 1902;15:2.
3. Martin E. The operation of epididymo-vasostomy for the relief of sterility. Therapeutic Gazette. 1909;1–19.
4. Silber SJ. Microscopic vasoepididymostomy: specific microanastomosis to the epididymal tubule. Fertil Steril. 1978;30(5):565–71.
5. Schlegel PN, Goldstein M. Microsurgical vasoepididymostomy: refinements and results. J Urol. 1993;150(4):1165–8.
6. Chan PT, Brandell RA, Goldstein M. Prospective analysis of outcomes after microsurgical intussusception vasoepididymostomy. BJU Int. 2005;96(4):598–601.
7. Lee R, Goldstein M, Ullery BW, Ehrlich J, Soares M, Razzano RA, et al. Value of serum antisperm antibodies in diagnosing obstructive azoospermia. J Urol. 2009;181(1):264–9.
8. Wagenknecht LV, Klosterhalfen H, Schirren C. Microsurgery in andrologic urology. I. Refertilization. J Microsurg. 1980;1(5):370–6.
9. Krylov VS, Borovikov AM. Microsurgical method of reuniting ductus epididymis. Fertil Steril. 1984;41(3):418–23.
10. Fogdestam I, Fall M, Nilsson S. Microsurgical epididymovasostomy in the treatment of occlusive azoospermia. Fertil Steril. 1986;46(5):925–9.
11. Thomas Jr AJ. Vasoepididymostomy. Urol Clin North Am. 1987;14(3):527–38.
12. Matthews GJ, Schlegel PN, Goldstein M. Patency following microsurgical vasoepididymostomy and vasovasostomy: temporal considerations. J Urol. 1995;154(6):2070–3.
13. Sheynkin YR, Starr C, Li PS, Goldstein M. Effect of methylene blue, indigo carmine, and Renografin on human sperm motility. Urology. 1999;53(1):214–7.
14. Berger RE. Triangulation end-to-side vasoepididymostomy. J Urol. 1998;159(6):1951–3.
15. Goldstein M, Li PS, Matthews GJ. Microsurgical vasovasostomy: the microdot technique of precision suture placement. J Urol. 1998;159(1):188–90.
16. Monoski MA, Schiff J, Li PS, Chan PT, Goldstein M. Innovative single-armed suture technique for microsurgical vasoepididymostomy. Urology. 2007;69(4):800–4.
17. Prentiss RJ, Weickgenant CJ, Moses JJ, Frazier DB. Surgical repair of undescended testicle. Calif Med. 1962;96:401–5.
18. Pasqualotto FF, Agarwal A, Srivastava M, Nelson DR, Thomas Jr AJ. Fertility outcome after repeat vasoepididymostomy. J Urol. 1999;162(5):1626–8.
19. Kolettis PN, Thomas Jr AJ. Vasoepididymostomy for vasectomy reversal: a critical assessment in the era of intracytoplasmic sperm injection. J Urol. 1997;158(2):467–70.
20. Schiff J, Chan P, Li PS, Finkelberg S, Goldstein M. Outcome and late failures compared in 4 techniques of microsurgical vasoepididymostomy in 153 consecutive men. J Urol. 2005;174(2):651–5. quiz 801.
21. Marmar JL. Modified vasoepididymostomy with simultaneous double needle placement, tubulotomy and tubular invagination. J Urol. 2000;163(2):483–6.
22. Brandell AR, Goldstein M. Reconstruction of the male reproductive tract using the microsurgical triangulation technique for vasoepididymostomy. J Urol. 1999;161(Suppl):350.
23. Silber SJ. Role of epididymis in sperm maturation. Urology. 1989;33(1):47–51.
24. Matthews GJ, Goldstein M. A simplified method of epididymal sperm aspiration. Urology. 1996;47(1):123–5.

Grafting Techniques for Vasectomy Reversal

9

Henry M. Rosevear and Moshe Wald

Abstract

Grafting of the male reproductive tract is an exciting new area of tissue engineering which may allow natural conception for patients with significant lengths of obstructed vas deferens. While stents had a significant and important role in increasing patency and pregnancy rates in the pre-microsurgical era, their role in the modern era of microsurgical two-layered anastomosis remains to be defined. To date, if the vasal obstruction is amenable to a primary watertight, tension-free anastomosis, microsurgical non-stented techniques remain the gold standard. Cases where a tension-free anastomosis is not possible because of the physical length of the obstruction remain problematic, but further research into tissue engineering in the form of implantable conduits holds much promise.

Keywords

Partial vasal agenesis • Congenital prostatic cysts • Obstruction of the vas deferens • Two-layered vasovasostomy • Microsurgical two-layered anastomosis • Spontaneous recanalization • Tuberculous epididymitis • Gonorrheal urethritis • Obstructive epididymitis • Chlamydial epididymitis

Surgical reconstruction of the vas deferens is performed to remove an obstructive lesion that is present along its course. Obstruction can exist at various parts of the vas deferens and can be the result of a prior vasectomy, congenital anomaly, inflammation secondary to a urogenital tract infection, trauma, or a surgical misadventure during prior inguinal, pelvic, or scrotal surgery. While no official reporting system exists in the United States to monitor the number of vasectomies performed each year, a survey in 2002 estimated this to be 526,501, which is approximately consistent with data reported in 1991 and 1995 [1, 2]. An estimated 2–6% of all men, and up to 11% of men aged 20–24 at the time of vasectomy, request a vasectomy reversal [3]. It has been estimated that between 30,000 and 80,000 vasectomy reversals are performed annually in the USA, though as with vasectomies, reporting requirements are not standardized so the exact number is unknown [4].

Congenital anomalies which can lead to obstruction of the vas deferens include congenital absence of the vas deferens, which is commonly associated with cystic fibrosis [5]. Partial vasal agenesis and congenital prostatic cysts can also lead to obstruction of the vas deferens [6, 7]. One example of a genetic disorder which can lead to obstructive azoospermia is Young's syndrome, which is characterized by chronic sinusitis and bronchiectasis as well as obstructive azoospermia [8]. In Young's syndrome, the obstruction usually occurs at the junction of the caput to corpus epididymis due to inspissated secretions. Inflammatory causes of obstruction are rare in the antibiotic era but include tuberculous epididymitis, gonorrheal urethritis progressing to obstructive epididymitis, and chlamydial epididymitis [9].

The success of a vasectomy reversal depends on several factors, only some of which can be controlled at the time of surgery. Factors which are independent of the method of

H.M. Rosevear, MD (✉) • M. Wald, MD
Department of Urology, University of Iowa,
200 Hawkins Drive, 3RCP, Iowa City, IA 52242-1089, USA
e-mail: henry-rosevear@uiowa.edu; moshe-wald@uiowa.edu

S.J. Parekattil and A. Agarwal (eds.), *Male Infertility: Contemporary Clinical Approaches, Andrology, ART & Antioxidants*,
DOI 10.1007/978-1-4614-3335-4_9,

reversal but which may influence subsequent conception include age and fertility potential of the patient's partner, length of obstructive interval, presence of antisperm antibodies, and high intravasal and epididymal pressure after the original obstruction [10–13]. Some factors which influence the success of the reversal are directly related to the technique chosen and include rate of stricture or scar development and granuloma formation [14]. The most commonly cited reason for these specific complications is an anastomosis made under tension, devascularization of the wall of the vas deferens, or a technical problem with the anastomosis leading to sperm leakage [15].

The current gold standard to surgically correct an obstructed vas deferens is a microscope-assisted two-layered vasovasostomy, but this has not always been so [16, 17]. Given the complex and time-consuming nature of this operation, new surgical techniques including robotics, modifications, and tools are continuously being explored. Some of these techniques include the use of fibrin glue, laser soldering, absorbable and nonabsorbable stents, and artificial conduits with or without specific growth factors added [18]. This chapter highlights the development of surgical grafting techniques for vasectomy reversal, including the use of stents and grafts, as well as the current clinical application of these devices and areas where further research is required.

Grafting Techniques in Reconstruction of the Male Reproductive Tract

Stents

As previously stated, several factors related to the success of a vasectomy reversal can be controlled at the time of surgery. From the 1950s to the mid-1970s, macrosurgical techniques for vasovasostomy were common and accepted as the gold standard. This technique allowed a primary anastomosis of the vas deferens to be created but was plagued with, by today's standards, low patency and pregnancy rates. According to a survey of American Urological Association (AUA) members published in 1973, members at that time practicing vasovasostomy reported a 38% patency rate and a 19.5% pregnancy rate [19]. It should be remembered that the microsurgical techniques which are common today were not developed until the mid-1970s, and, as such, stricture resulting in either partial or complete vasal obstruction was the most pressing technical complication of that period. To address this common complication, approximately 90% of urologists performing vasovasostomies at that time employed stents, with either silver wire or nylon suture being the most commonly reported [19]. The reason for the widespread use of stents can also be found in the 1973 AUA survey. Members reported that the pregnancy rate for non-stented reversals was significantly lower, at 10.9% compared to 19.9–26% for stented procedures, depending on the stent used. Numerous techniques had been developed by 1973 which maximized both patency and pregnancy rates. The main variation between these techniques was the use of loupes for magnification and/or the use of stent [20].

In the lexicon of the modern urologist, a stent most commonly refers to the hollow silicon tube that is used in the ureter for treatment of either intrinsic or extrinsic ureteral obstruction. A vasovasostomy stent, as it was originally used, was quite different. A stent in that sense was any foreign body, usually a piece of suture, the purpose of which was to maintain patency of the lumen of the vas deferens during and immediately after a macrosurgical (either with or without the supplemental use of loupes) primary anastomosis of the vas deferens. The simple goal of the stent was to prevent obstruction at the anastomosis site either because of a poorly placed suture at the time of surgery or as prevention of stricture or scar formation in the immediate postoperative period. In one example of this technique, a short section of 2-0 nylon suture is used as a stent to bridge the anastomosis while 6-0 Prolene is used to actually complete the anastomosis [21]. In this technique, the nylon suture is removed before the operation is completed, and its purpose is to ensure that the vas deferens lumen remains patent during the procedure. In another variation on this theme, described by Dorsey, a zero monofilament suture is fed through a hollow needle introduced approximately 1 cm proximal to the site of the intended anastomosis [22]. This suture is then fed into the distal vas deferens. The anastomosis is then completed using 6-0 Ethiflex, and the proximal end of the stenting suture is brought through the scrotal skin and removed in 12–14 days. The goal of this stent in this technique is to ensure patency of the anastomotic site both during the procedure and in the immediate postoperative healing period. The success rates of these procedures were reported to be over 80% patency, which contrasts with the success rates reported in the 1973 AUA survey.

Even with the improvement in both patency and pregnancy rates reported by clinicians using stents during vasovasostomy, there were numerous known disadvantages of stents, especially with the use of exteriorized stents such as described by Dorsey [20]. The point of exit for the stent is a theoretical source of infection as well as a location where sperm can leave the lumen of the vas [20, 23]. Additionally, the location where the exteriorized stent left the lumen of the vas was identified in a publication by Fernandes as a common site of subsequent luminal obstruction (often instead of the primary anastomosis itself) [24]. To avoid the problem of an exteriorized stent, some groups have experimented with absorbable intravasal suture as a stent to bridge the anastomosis—the theoretical advantage being that these stents would slowly dissolve, maintaining the patency of the anastomosis both during the procedure itself and the postoperative healing period without need to be removed. In one experiment in a canine model, Montie et al. compared three groups: no stent, a Dexon intravasal stent, and a chromic intravasal stent [23]. Three to six months after the vasectomy reversal procedure, retrograde vasography was used to identify patency rates. Both absorbable stent groups had higher patency rates than the control (no stent) group, with the chromic group having the highest overall patency at 70% vs. 60% for the Dexon group and 50% for the no stent group. This concept was tested in a human clinical model by Rowland and colleagues a few years later, who found that intravasal absorbable stents (using 3-0 chromic) had higher patency rates than a group using exteriorized silkworm gut stents (86% vs. 67%, respectively) [25].

In 1975, Silber reported the first use of microsurgical vasovasostomy in humans [26]. His work, along with independent work by Owen, led the way to the modern microsurgical two-layered anastomosis [26, 27]. From a historical standpoint, it should be noted that it was Silber and his group who found through histologic and electron microscopic work that stricture was more common than originally thought with macrosurgical anastomosis techniques [28]. Silber also popularized the two-layered closure based on his observation that this technique provided a better watertight mucosal approximation given the common discrepancies between proximal and distal luminal diameters of the vas deferens [29]. The techniques developed by these investigators have allowed the microsurgical two-layered vasovasostomy anastomosis to become the gold standard for vasectomy reversal, with success rates dependent on time since obstruction. These success rates may be as high as 97% patency and 76% pregnancy when the obstructive interval is less than 3 years, and 71% patency and 30% pregnancy after 15 years of obstruction [13].

The reproducible success of this new technique resulted in a dearth of research into alternative techniques for many years. Even with its success, Silber's microsurgical technique was not perfect. The downside was that microsurgical anastomosis was a time-consuming operation best done by surgeons with specialized training using expensive operating microscopes. This prompted research into new techniques that would simplify the technique while maintaining the high patency and pregnancy rates. In 1989, Flam et al. reported work in a rat model on a hollow, absorbable polyglycolic acid tube [30]. In their experiment, they inserted a 10-mm long by 0.5-mm outer diameter hollow stent into the lumen of the vas at the site of anastomosis on one side and completed the anastomosis with a single layer of suture (Fig. 9.1). On the contralateral side, they performed standard microsurgical anastomosis. They showed a trend toward improved patency in the stented vas deferens. Flam emphasized in his paper that sperm leakage at the site of the anastomosis could lead to

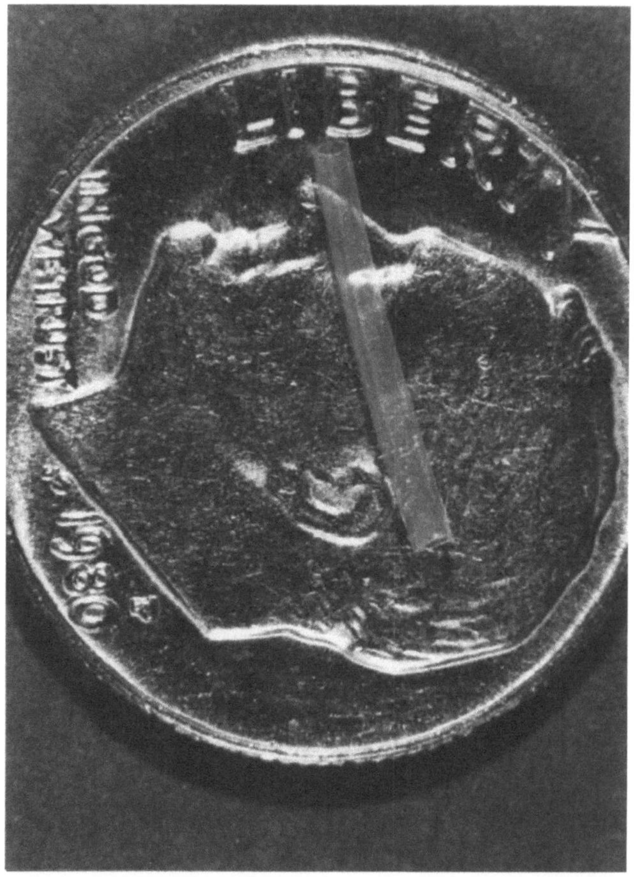

Fig. 9.1 Hollow polyglycolic acid stent (0.5-mm outer diameter) shown on a dime. Used with permission from Flam et al. Experimental study of hollow, absorbable polyglycolic acid tube as stent for vasovasostomy (From [30], with permission)

secondary stricture and should be avoided. This work led to a clinical trial using absorbable intravasal stents conducted by Rothman and colleagues in 1996 [31]. This randomized study compared conventional two-layered microsurgical anastomosis to a modified approach using an absorbable polyglycolic acid stent without intraluminal sutures (Fig. 9.2). While the operative time was significantly reduced in the stented group (118 min vs. 137 min), both the patency and the pregnancy rates were lower in the stented groups (81% vs. 89% and 22% vs. 51%, respectively), and the authors concluded that intravasal stents should not be used.

More recently, Vrijhof et al. reported on a nonabsorbable stent in a rabbit model [14]. They theorized that the absorbable nature of the previously reported stents allowed strictures to develop at the site of the anastomosis once the stent had dissolved and that a nonreactive nonabsorbable stent would bypass this problem while simplifying the operation. Their stent was made of a biocompatible material designed to have both hydrophilic and hydrophobic characteristics. The stent also had a transverse ridge which was designed to

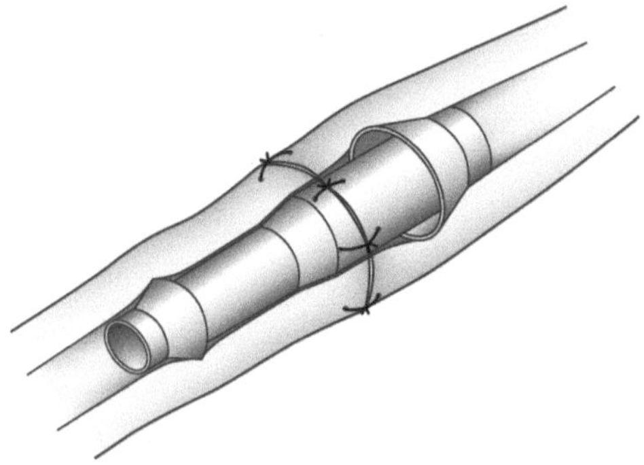

Fig. 9.2 Absorbable self-retaining polyglycolic acid stent

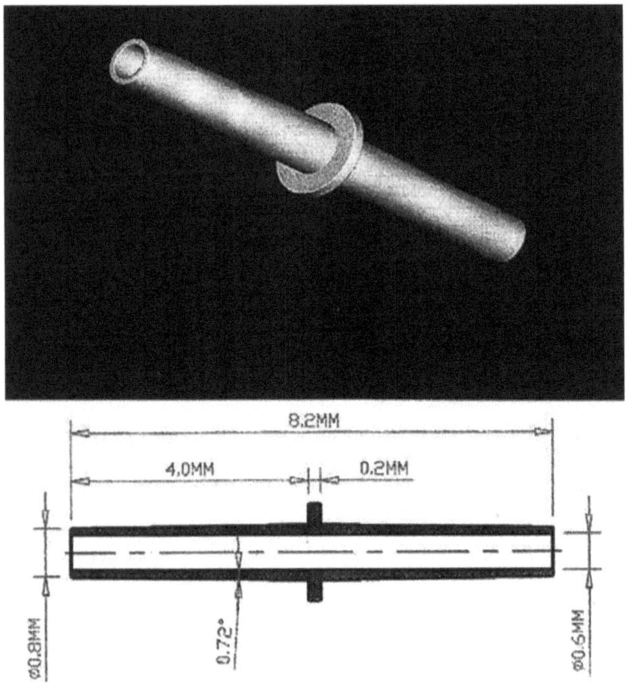

Fig. 9.3 Nonabsorbable polymeric stent with transverse ridge to minimize migration (From [14], with permission)

minimize migration from the anastomotic site (Fig. 9.3). This group reported that all vasa were patent at the end of their study (39–47 weeks) and that total sperm count was higher in the stented group. No human data is available on this type of stent.

In the era of cost-conscious medicine, especially when many patients must pay out of pocket for vasectomy reversals, further research into efforts that simplify the present gold standard is appropriate with the caveat that patency and pregnancy rates should not be compromised. It is important to note that all of the absorbable and nonabsorbable stents which have been used to date in human studies have been well tolerated with no side effects and little to no inflammatory response.

In summary, stents were investigated as a method to improve patency rates in the era of macrosurgical vasovasostomy but were eclipsed by the application of the operating microscope to the field and the introduction of microsurgical two-layered vasovasostomy. Efforts to improve and simplify the microsurgical operation using absorbable stents have not improved overall patency or pregnancy rates. Recent efforts using nonabsorbable stents show promise in animal models but have not been tested in humans so their utility remains unproven. The ideal stent would, at a minimum, maintain the patency and pregnancy rates achieved through a conventional two-layer microsurgical anastomosis while decreasing the operative time, training, and cost required to achieve these results.

Conduits

The preferred method to bypass an obstructed portion of the vas deferens, regardless of the etiology of the obstruction, is surgical excision or exclusion of the obstructed segment and reanastomosis of the vas deferens using a microsurgical two-layered anastomosis. The goal of the operation is a watertight, tension-free, widely patent anastomosis. As described previously, numerous techniques have been suggested in an attempt to simplify this procedure while preserving its patency and pregnancy rates. The assumption in all of the previously described techniques is that the vas deferens could be sufficiently mobilized to allow a tension-free anastomosis. Unfortunately, cases exist where, due to the physical length of the obstruction, the vas deferens cannot be reconstructed in a watertight, tension-free manner. These cases present a clinical challenge because the resulting obstructive azoospermia is theoretically amenable to surgical correction. Presently, the only reproductive option available for these patients is surgical sperm retrieval. The technique of retrieving sperm from either the testicle or epididymis has been successfully reported in cases of obstructive azoospermia that is not surgically correctable but must be coupled with in vitro fertilization [32]. The hormonal manipulation, surgical interventions, risk of multiple gestations, and increased financial cost of in vitro fertilization make this solution less than ideal and creates an intriguing field of research into reconstruction of the male reproductive tract.

Grafting of the male reproductive tract theoretically can take one of three forms. The first option is to use transplanted vas deferens with all of the complications, both technical and immunological, associated with such a procedure. The second option is to replace the obstructed segment of vas deferens with a tubular structure, the sole purpose of which is to simply allow passage of sperm in a distal direction. An analogous clinical problem can be found in vascular surgery where surgeons often replace diseased segments of vessels with either endogenous grafts such as the long saphenous vein or exogenous grafts such as a Teflon-coated endovascular stent. The third option involves tissue engineering. Tissue engineering as it applies to reconstruction of the male reproductive system involves the concept of creating an artificial conduit which serves as a scaffolding for the regrowth of the vas deferens itself. In a different biological system, polymer scaffoldings have been shown to facilitate peripheral nerve regeneration in segments as long as 1 cm [33]. Regardless of the method chosen to graft over the obstructed segment of vas deferens, the goal is to reestablish continuity of the male reproductive tract, allowing sperm to be present in the ejaculate and eliminating the need for assisted reproductive techniques (ART). It should be noted that even small amounts of ejaculated sperm could be a significant improvement, as this may allow for less invasive forms of ART [34].

The first reported experiment on the use of grafts in reconstruction of the male reproductive tract was by Romero-Maroto and colleagues in 1989 [35]. This group reported successfully autotransplanting a pediculated segment of vas deferens from one side to the contralateral in rabbits. They reported good patency rates, but no data on pregnancies was noted. The clinical use of this technique is likely limited, as these subjects would likely be candidates for crossover vasovasostomies, a rare procedure with a high reported success rate [36], and given the questionable feasibility of harvesting a long vasal segment for reconstruction of the contralateral side.

Regarding the second option for grafting the male reproductive tract, Carringer et al. in 1995 reported patency rates in rats after either a vasal or vascular graft obtained from either the contralateral side of the same animal or from female rats, respectively [37]. In this study, three different lengths of grafts were used (0.5, 1.0, and 1.5 cm), corresponding to approximately 10%, 20%, and 30% of the entire vas deferens length. Patency was confirmed by direct examination of the graft 4 weeks postoperatively. The authors found an overall patency rate of approximately 40% in both surgical groups (vasal and vascular graft) with higher rates in the shorter segments. Pregnancy rates were not evaluated. No human clinical trials have been reported using either of these techniques. Questions on long-term patency of extensive artificial grafts remain unanswered, even in animal models, and should be further investigated.

The lack of a suitable allograft in humans for vasal reconstruction has led to research on the potential for biocompatible degradable polymer scaffolding for tissue engineering. As mentioned earlier, this model has been successfully applied to the clinical problem of peripheral nerve regeneration [33]. Additions to this technology, including micro-patterned (grooved) inner lumens as well as target-specific growth factors, can increase the efficacy of this technology [38, 39]. The vas deferens is a good target for investigation because it has been shown to undergo spontaneous recanalization at the site of vasectomy [40].

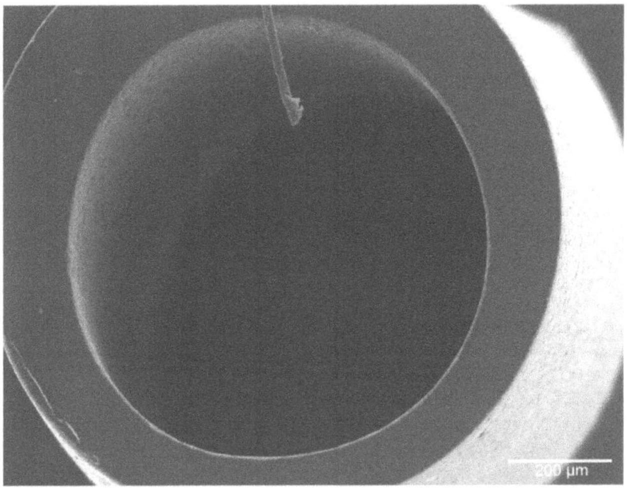

Fig. 9.4 Scanning electron microscope image of PDLA conduit. Bar = 200 μm (From [42], with permission)

Further evidence that may support tissue engineering of the vas deferens is the demonstration of elevated levels of selected growth factors at the vasectomy site in an animal model. Previous examination of vasectomy sites in rats using real-time polymerase chain reaction, enzyme-linked immunosorbent assay, and histopathological analysis demonstrated a 12-fold increase in platelet-derived growth factor beta and a ninefold increase in transforming growth factor beta [41].

Using the peripheral nerve regeneration model as a guide, biodegradable conduits made of D,L-lactide were studied for reconstruction of the reproductive tract in a rat model [42]. Biodegradable conduits with micro-patterned grooves on the inner surface were implanted in 47 rats following vasectomy (Fig. 9.4, scanning electron microscopy image). At 8 weeks postimplantation of the conduits, no evidence of recanalization was found. However, at 12 weeks, evidence of recanalization was noted in three of the remaining rats, with one showing a microcanal spanning the entire 0.5-cm conduit and the other two showing distinct epithelialized vas deferens microcanals at the conduit edges (Fig. 9.5) [42].

Following the demonstration of microrecanalization of the vas deferens in this biodegradable graft model, attempts were made to identify ways to maximize this response (unpublished data). Based on the identification of elevated growth factor levels at the site of vasectomy, the effect of local microparticle-delivered growth factors on the rate of vasal recanalization in a biodegradable conduit model was examined. Delivering growth factors to a specific location in the body over a sustained period of time is not a simple task. Effective supplementation of growth factors selectively at the site of the grafted vas deferens may be compromised by the fact that the ability of growth factors to perform their

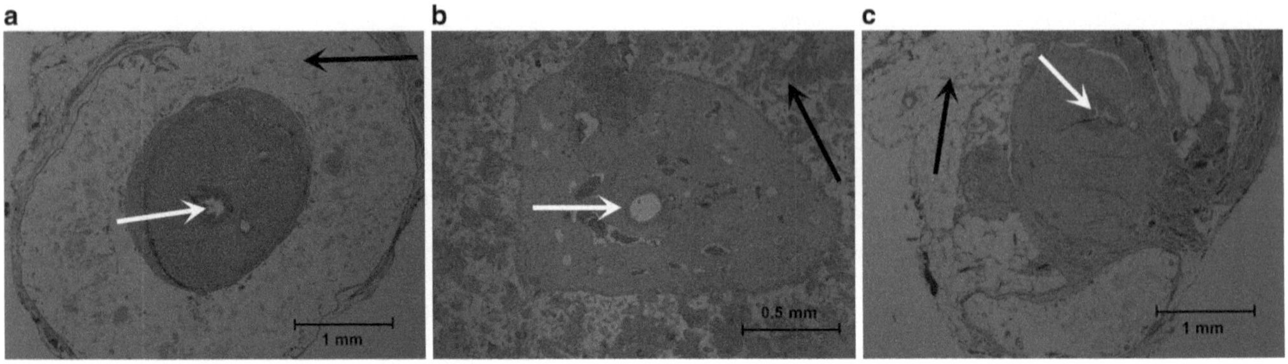

Fig. 9.5 (**a**) Evidence of microrecanalization at the midpoint of a 0.5-cm poly-(D,L-lactide) (PDLA) graft (magnification ×40). Bar = 1 mm. (**b**) Microcanal at the midpoint of a 0.5-cm PDLA graft (magnification ×200) Bar = 0.5 mm. (**c**) Microcanal at the interface zone of a 0.5-cm PDLA graft (magnification ×40). Bar = 1 mm. *All panels*: *white arrows* microcanals, *black arrows* graft (From [42], with permission)

function depends on their tertiary structure, which is susceptible to degradation if it is not protected from the local environment. Thus, delivery of a locally sustained concentration of growth factors requires the use of microspheres. The goal of a microsphere is to sequester the biologically active molecule and allow a controlled, sustained release of the molecule. The exact timing of the sustained release is a function of the characteristics of the microsphere into which it is placed. With these considerations in mind, a poly-(D,L-lactide) material was chosen for construction of the microspheres. As the biodegradable conduits used in this study were constructed of the identical material, noncovalent binding was assumed to keep the microspheres near the conduit. Reconstruction of surgically induced vasal gaps using biodegradable conduits soaked in microspheres containing TGF-beta and PDGF revealed an increase in the number of new microcanals in the graft but not in their length at 12 weeks postoperatively.

In an effort to further optimize the conditions for vasal recanalization, methods to increase the vascularity of the reconstructed vas deferens were investigated, based on an observation suggesting that neovascularization increased with time at the conduit to vasal border (unpublished data). To bolster this neovascularization and potentially increase the rate of recanalization, the effect of oral sildenafil citrate on recanalization in the biodegradable graft model was examined. Sildenafil citrate is a type-5 phosphodiesterase inhibitor that has been shown to promote neovascularization in other systems [43]. Rats received a daily dose of 5 mg/kg of oral sildenafil citrate following reconstruction of the vas deferens with a biodegradable graft. At 16 weeks, the rats on sildenafil citrate had a significantly increased number of microcanals (29 vs. 4) though the average length of the canals was constant at 2 mm. This observation was confirmed by an increase in staining for CD31, an endothelial marker. An ongoing study involves combining both oral sildenafil citrate with increasing the local concentration of TGF-beta and platelet-derived growth factors via microspheres. Areas of future research into this field include examining different substrates of which the conduit itself is composed and embedding the growth factors directly into the conduit to maximize local concentration.

Expert Commentary

Grafting of the male reproductive tract is an exciting new area of tissue engineering which may allow natural conception for patients with significant lengths of obstructed vas deferens. While stents had a significant and important role in increasing patency and pregnancy rates in the pre-microsurgical era, their role in the modern era of microsurgical two-layered anastomosis remains to be defined. To date, if the vasal obstruction is amenable to a primary watertight, tension-free anastomosis, microsurgical non-stented techniques remain the gold standard. Cases where a tension-free anastomosis is not possible because of the physical length of the obstruction remain problematic, but further research into tissue engineering in the form of implantable conduits holds much promise.

Five-Year View

The potential role of implantable conduits in reconstruction of the male reproductive tract remains to be fully investigated. Currently, biodegradable conduits have been shown limited success in guiding regrowth of vasal tissue elements along the entire length of a 5-mm long segment though further research into supplementing naturally occurring growth factors in the hopes of maximizing regrowth is ongoing. As the obstructed segment of vas deferens that is bridged by bioengineered conduits lengthens, the barriers to success mount. One such barrier is that oxygen and other nutrients necessary for cell survival can diffuse approximately 2–3 mm with the need for angiogenesis. The 5-mm segment under current investigation approaches the length that could be bridged without angiogenesis. The use of sildenafil citrate to increase angiogenesis is one possible approach to minimize this problem. Other specific research problems that need to be addressed include maximizing local delivery of growth factors. While current research employs nanoparticles, another possibility is the impregnation of the conduit itself with growth factors. The specific design and engineering issues that this approach raises are under investigation.

Key Issues

- Microsurgical two-layered anastomosis to correct vasal obstruction remains the gold standard.
- The role of vasal stenting in the era of microsurgical two-layered anastomosis is limited.
- Biodegradable conduits to bridge segments of obstructed vas deferens where a primary microsurgical two-layered anastomosis is impossible remain investigational but hold significant long-term promise.

Acknowledgment We would like to thank Kris Greiner for her editorial assistance in preparing this chapter.

References

1. Barone MA, Hutchinson PL, Johnson CH, et al. Vasectomy in the United States, 2002. J Urol. 2006;176:232–6.
2. Magnani RJ, Haws JM, Morgan GT, et al. Vasectomy in the United States, 1991 and 1995. Am J Public Health. 1999;89:92–4.
3. Holman CD, Wisniewski ZS, Semmens JB, et al. Population-based outcomes after 28,246 in-hospital vasectomies and 1,902 vasovasostomies in Western Australia. BJU Int. 2000;86:1043–9.
4. Schiff J, Li PS, Goldstein M. Toward a sutureless vasovasostomy: use of biomaterials and surgical sealants in a rodent vasovasostomy model. J Urol. 2004;172:1192–5.
5. Castaldo G, Tomaiuolo R, Vanacore B, et al. Phenotypic discordance in three siblings affected by atypical cystic fibrosis with the F508del/D614G genotype. J Cyst Fibros. 2006;5:193–5.
6. Stricker H, Kunin J, Faerber G. Congenital prostatic cyst causing ejaculatory duct obstruction: management by transrectal cyst aspiration. J Urol. 1993;149:1141–3.
7. Engin G, Kadioglu A, Orhan I, et al. Transrectal US and endorectal MR imaging in partial and complete obstruction of the seminal duct system. A comparative study. Acta Radiol. 2000;41:288–95.
8. Handelsman DJ, Conway AJ, Boylan LM, et al. Young's syndrome. Obstructive azoospermia and chronic sinopulmonary infections. N Engl J Med. 1984;310(1):3–9.
9. Thomas AH, Sabanegh Jr ES. Microsurgical treatment of male infertility. In: Lipshultz LI, Howards SS, Niederberger CS, editors. Infertility in the Male. 4th ed. New York: Cambridge University; 2009. p. 392–406. A very well written summary of current microsurgical surgical techniques.
10. Kolettis PN, Woo L, Sandlow JI. Outcomes of vasectomy reversal performed for men with the same female partners. Urology. 2003; 61:1221–3.
11. Silber SJ. Microscopic vasectomy reversal. Fertil Steril. 1977;28:1191–202.
12. Vrijhof HJ, Delaere KP. Vasovasostomy results in 66 patients related to obstructive intervals and serum agglutinin titres. Urol Int. 1994;53:143–6.
13. Belker AM, Thomas Jr AJ, Fuchs EF, et al. "Results of 1,469 microsurgical vasectomy reversals by the Vasovasostomy Study Group. J Urol. 1991;145:505–11.
14. Vrijhof EJ, de Bruine A, Zwinderman A, et al. New nonabsorbable stent versus a microsurgical procedure for vasectomy reversal: evaluating tissue reactions at the anastomosis in rabbits. Fertil Steril. 2005;84:743–8.
15. Carbone Jr DJ, Shah A, Thomas Jr AJ, et al. Partial obstruction, not antisperm antibodies, causing infertility after vasovasostomy. J Urol. 1998;159:827–30.
16. Practice Committee of the American Society for Reproductive Medicine. Vasectomy reversal. Fertil Steril. 2008;90:S78–82.
17. Lipshultz LI, Rumohr JA, Bennett RC. Techniques for vasectomy reversal. Urol Clin North Am. 2009;36:375–82. The most recent definitive treatise on surgical techniques.
18. Kolettis PN. Restructuring reconstructive techniques—advances in reconstructive techniques. Urol Clin North Am. 2008;35: 229–34.
19. Derrick Jr FC, Yarbrough W, D'Agostino J. Vasovasostomy: results of questionnaire of members of the American Urological Association. J Urol. 1973;110:556–7.
20. Kim HH, Goldstein M. History of vasectomy reversal. Urol Clin North Am. 2009;36:359–73. A concise and fascinating summary of the history of the vasectomy reversal.
21. Amelar RD, Dubin L. Vasectomy reversal. J Urol. 1979;121:547–50.
22. Dorsey JW. Surgical correction of post-vasectomy sterility. J Urol. 1973;110:554–5.
23. Montie JE, Stewart BH, Levin HS. Intravasal stents for vasovasostomy in canine subjects. Fertil Steril. 1973;24:877–83.
24. Fernandes M, Shah KN, Draper JW. Vasovasostomy: improved microsurgical technique. J Urol. 1968;100:763–6.
25. Rowland R, Nanninga JB, O'Connor VJ. Improved results in vasovasostomies using internal plain catgut stents. Urology. 1977;10: 260–2.
26. Silber SJ. Microsurgery in clinical urology. Urology. 1975;6:150–3.
27. Owen ER. Microsurgical vasovasostomy: a reliable vasectomy reversal. Aust N Z J Surg. 1977;47:305–9.
28. Silber SJ, Galle J, Friend D. Microscopic vasovasostomy and spermatogenesis. J Urol. 1977;117:299–302.
29. Silber SJ. Vasectomy and vasectomy reversal. Fertil Steril. 1978;29:125–40.
30. Flam TA, Roth RA, Silverman ML, et al. Experimental study of hollow, absorbable polyglycolic acid tube as stent for vasovasostomy. Urology. 1989;33:490–4.
31. Rothman I, Berger RE, Cummings P, et al. Randomized clinical trial of an absorbable stent for vasectomy reversal. J Urol. 1997;157:1697–700.
32. Nudell DM, Conaghan J, Pedersen RA, et al. The mini-micro-epididymal sperm aspiration for sperm retrieval: a study of urological outcomes. Hum Reprod. 1998;13:1260–5.
33. Miller C, Shanks H, Witt A, et al. Oriented Schwann cell growth on micropatterned biodegradable polymer substrates. Biomaterials. 2001;22:1263–9.
34. Kamischke A, Nieschlag E. Analysis of medical treatment of male infertility. Hum Reprod. 1999;14 Suppl 1:1–23.

35. Romero-Maroto J, Escribano G, Egea L, et al. Transplant of a pediculate segment of vas deferens. Experimental study. Eur Urol. 1989;16:133–7.
36. Gilis J, Borovikov AM. Treatment of vas deferens large defects. Int Urol Nephrol. 1989;21:627–34.
37. Carringer M, Pedersen J, Schnurer LB. Experimental vas replacement by either vas or a vascular graft. Scan J Urol Neprhol. 1995; 29:97–102.
38. Rutkowski GE, Miller CA, Jeftinija S, et al. Synergistic effects of micropatterned biodegradable conduits and Schwann cells on sciatic nerve regeneration. J Neural Eng. 2004;1:151–7.
39. Miller C, Jeftinija S, Mallapragada S. Synergistic effects of physical and chemical guidance cues on neurite alignment and outgrowth on biodegradable polymer substrates. Tissue Eng. 2002;8:367–78.
40. Labrecque M, Hays M, Chen-Mok M, et al. Frequency and patterns of early recanalization after vasectomy. BMC Urol. 2006; 6:25.
41. Stahl BC, Ratliff TL, De Young BR, et al. Involvement of growth factors in the process of post-vasectomy micro-recanalization. J Urol. 2008;179:376–80.
42. Simons CM, De Young BR, Griffith TS, et al. Early microrecanalization of vas deferens following biodegradable graft implantation in bilaterally vasectomized rats. Asian J Urol. 2009;11: 373–8.
43. Koneru S, Varma Penumathsa S, Thirunavukkarasu M, et al. Sildenafil-mediated neovascularization and protection against myocardial ischaemia reperfusion injury in rats: role of VEGF/angiopoietin-1. J Cell Mol Med. 2008;12:2651–64.

10 Mini-Incision Vasectomy Reversal Using the No-Scalpel Vasectomy Instruments and Principles

Darby J. Cassidy, Keith Jarvi, Ethan D. Grober, and Kirk C. Lo

Abstract

A wide variety of techniques have been described in the literature for vasectomy reversal, and as with most techniques, they continue to evolve in current surgical practice. We have developed a mini-incision technique at the University of Toronto for vasectomy reversal using the principles and instruments used for the no-scalpel vasectomy. By taking advantage of the compliance of the scrotal wall, a substantial length of vas deferens can safely be delivered through a mini-incision for an efficient approach to vasectomy reversal. By avoiding a longer incision, delivery of the testicle and more extensive tissue dissection, one can significantly reduce postoperative pain and hasten return to daily activities without compromising patient outcomes as evidenced by our experience.

Keywords

Vasoepididymostomy • Vasovasostomy • No-scalpel vasectomy • Microsurgical vasovasostomy • Mini-incision vasectomy reversal • Anastomosis • Microscopic bipolar electrocautery

A wide variety of techniques have been described in the literature for vasectomy reversal, and as with most techniques, they continue to evolve in current surgical practice. We have developed a mini-incision technique at the University of Toronto for vasectomy reversal using the principles and instruments used for the no-scalpel vasectomy.

History

Like most surgical procedures, vasectomy reversal techniques are in constant evolution. The genesis of the techniques used for vasectomy reversal dates back to the work of Dr. Edward Martin at the University of Pennsylvania in the early 1900s. In 1902 Martin performed the first documented vasoepididymostomy for a man with obstructive azoospermia secondary to gonorrhea [1]. In 1909 he published a series of 11 azoospermic men who underwent vasoepididymostomies with a patency rate of 64% and pregnancy rate of 27%. Martin's publication and the demonstrated effectiveness of his vasoepididymostomy technique served to dispel the earlier and widely held belief that such anastomoses were hardly worth pursuing given their technical challenges and expected low success rates. Francis Hagner subsequently reproduced Martin's outcomes in his series of 33 patients with reported patency and pregnancy rates of 64% and 48% respectively, solidifying its role as an effective technique in the management of obstructive azoospermia [1].

Quinby reported the first successful vasovasostomy in 1919 on a man who underwent vasectomy 8 years earlier [1]. O'Connor, Quinby's former assistant, subsequently published a series of 14 vasectomy reversals using Quinby's technique in 14 men with an overall patency rate of 64%,

D.J. Cassidy, MD (✉)
Department of Urology, University Hospital of Northern British Columbia, Prince George, BC, Canada

K. Jarvi, MD • E.D. Grober MD, MEd • K.C. Lo, MD, CM
Department of Surgery, Division of Urology, University of Toronto, Mount Sinai Hospital, Toronto, ON, Canada
e-mail: klo@mtsinai.on.ca

S.J. Parekattil and A. Agarwal (eds.), *Male Infertility: Contemporary Clinical Approaches, Andrology, ART & Antioxidants*, DOI 10.1007/978-1-4614-3335-4_10, © Springer Science+Business Media, LLC 2012

once again demonstrating vasovasostomy as an effective technique for vasectomy reversal [1].

As the interest in family planning and the role of women in the workplace evolved over the ensuing decades, the rates of vasectomy increased substantially as did the inevitable demand for vasectomy reversal that followed.

Vasectomy remains the most commonly performed urologic procedure in North America, with over 500,000 vasectomies performed annually in the United States alone [1–3]. In 1974, the no-scalpel vasectomy was introduced and provided surgeons with a technique that minimized discomfort and postprocedural morbidity without compromising patient outcomes [4–6]. Between 2 and 11% of vasectomized men will ultimately request a reversal of their vasectomy for a variety of reasons such as a new partner or death of a child [3]. As a result, vasectomy reversal is a very commonly requested procedure for the urologist, and the demand continues to grow.

Vasectomy Reversal Techniques

The original descriptions for vasectomy reversal were macrosurgical techniques with or without the use of loupe magnification. The early techniques used in the 1900s used thin silver wire for the vasal anastomosis, which ultimately evolved into the use of nonabsorbable 4-0 to 6-0 sutures during the 1970s [2]. The reported patency rates for these techniques ranged from 79 to 88% with pregnancy rates of 34–50%. These techniques were largely abandoned when the operating microscope became widely available in most centers.

Silber and Owen are both credited with the first description of a microsurgical vasovasostomy in humans in 1977, although several authors had previously described this technique using animal models [2]. These anastomoses were performed with an operating microscope using 16–25 times magnification and 9-0 nylon sutures in a one- or two-layer closure. In the two-layer technique, the initial three sutures are placed through the mucosa and the immediately adjacent muscular layer along the anterior wall. Once these were tied, the vas was then rotated to expose the posterior wall, allowing the placement and tying of the remaining three mucosal sutures. The 9-0 nylon is then used to separately place the second layer of six seromuscular sutures. The one-layer closure is done with 6–7 full thickness 9-0 nylon sutures.

Further evolution and refinements of the original techniques have been taken place over time. In the 1980s, the two-layer closure technique evolved to use 10-0 nylon for the mucosal anastomosis and 8 or 9-0 nylon for the seromuscular layer. Marc Goldstein invented and introduced the microspike vas approximator clamp and microdot suture placement technique which allowed for greater stabilization of the vasal ends and more precise placement of the 10-0 anastomotic sutures by following the preplaced microdots, especially when the vasal lumens are of disparate caliber [1, 7].

The current technique for vasovasostomy has been very well described by Larry Lipshultz and his group and is commonly used by today's microsurgeons [2]. With this technique, the testicles and spermatic cords are delivered via a single 4–6 cm midline or bilateral 4–6 cm paramedian scrotal incision(s). The vasectomy site is identified, and the healthy testicular and abdominal ends of the vas are mobilized using iris scissors and Jacobsen mosquitoes, with care taken to preserve as much perivasal adventitia and vasal blood supply as possible so as not to compromise the vasal ends; 5-0 absorbable stay sutures are placed superficially on both the testicular and abdominal vas 1–2 cm from the intended transection sites. The testicular vas is then transected sharply in a guillotine fashion with a fresh scalpel blade, and the expressed fluid is examined immediately with light microscopy at 100–400 times magnification to confirm patency by identifying the presence of sperm or sperm parts in the fluid. The abdominal vas is transected in an identical fashion and its patency confirmed with saline vasogram or methylene blue vasography with temporary insertion of a Foley catheter. Hemostasis is managed with bipolar electrocautery to minimize vasal injury. If the intraoperative findings are suitable for vasovasostomy (copious thin fluid and/or the presence of sperm or sperm parts and normal vasography), the two vasal ends are approximated and stabilized either by placement within a vas approximator clamp or microvascular clamp or by placing 1–2 adventitial holding stitches at the 6 o'clock position. The operating microscope is now brought into the field, and a two-layer anastomosis is begun. A double-armed 10-0 nylon mucosal suture is placed and tied at the 6 o'clock position. Three to five additional 10-0 mucosal sutures are placed around the circumference of the vasal lumen and tied. Interrupted single-armed 9-0 nylon sutures are placed in the seromuscular layer circumferentially to complete the second layer. A common modification to this technique eliminates the delivery of the testicle in an effort to minimize postoperative morbidity [7]. With this technique, a 4–6-cm incision is made in the upper scrotum angled toward the external inguinal ring along the path of the vas deferens on each side. This allows for the easy identification of the vasectomy site and mobilization of the testicular and abdominal vasal ends. The anastomosis is then performed in an identical fashion to the previous description.

The patency and pregnancy rates reported in the literature are widely variable and dependent on a number of preoperative, operative, and postoperative factors that may or may not have been controlled for. It is universally accepted that the microsurgical approach yields superior patency and pregnancy rates compared to macrosurgical anastomoses given the ability to more precisely place more delicate sutures [3]. Unpublished data also demonstrates that surgeons with microsurgery training experience superior outcomes with an average patency rate of 89% compared to 53% in inexperienced hands [3]. The Vasovasostomy Study Group reviewed the outcomes of 1,469 contemporary microsurgical vasovasostomies [3, 8]. They demonstrated a 97% patency rate and 76% pregnancy rate in men less than 3 years from their vasectomy. As the interval from vasectomy increases, the rates decline with a 71% patency rate and 30% pregnancy rate when 15 years or more from the vasectomy. The Vasovasostomy Study Group also determined that there was no statistically significant difference in patency or pregnancy rates between one-layer and two-layer anastomosis, with the decision based on surgeon preference and experience [1, 8].

The morbidity of vasovasostomy has been poorly examined with no published studies discussing the morbidity associated with the various techniques for vasectomy reversal. Postoperative pain, swelling, bruising, and activity limitation are all commonly seen after vasectomy reversal especially if the testicle and its investing tunica vaginalis are delivered. Most men are counseled to wear supportive briefs or scrotal supports for 2 weeks, to take 1–2 weeks off work, and to limit themselves to light physical exertion for 3–4 weeks.

We have already discussed many technical modifications that have been developed in an attempt to improve surgical outcomes, but none were specifically developed to help reduce the morbidity of the procedure.

Mini-Incision Vasectomy Reversal

In an effort to maintain the established effectiveness of microsurgical vasovasostomy and to reduce postoperative morbidity, Keith Jarvi and his group at the University of Toronto applied the established gold standard technique used for vasectomy (no-scalpel vasectomy) to vasovasostomy [4, 9]. The mini-incision and no-scalpel techniques for performing vasectomy are familiar to most practicing urologists and have been shown to reduce complication rates and decrease recovery times without compromising vasectomy outcomes, making these techniques attractive to urologists performing vasectomy reversals.

Technique of Mini-Incision Vasectomy Reversal

The instruments required for this technique are those used for no-scalpel vasectomy and the traditional vasectomy reversal. These include two no-scalpel ring vasectomy clamps, one no-scalpel sharpened snap clamp, multiple fresh 15-blade scalpels, two nontoothed Adson forceps, two to three jeweler's forceps, two microscopic tying forceps, one nonlocking microscopic needle driver, microscopic scissors, Wexel sponges, and the appropriate sutures.

In an identical fashion to the no-scalpel vasectomy, the vas deferens is palpated, manipulated, and stabilized through the scrotal skin in the mid to upper scrotum lateral to the median scrotal raphe using the three-finger technique previously described (Fig. 10.1). It is important to bring the vas deferens at least 1 cm lateral to the midline to be in the more pliable portion of the scrotal skin. The no-scalpel vasectomy ring clamp is then used to grasp the vas deferens approximately 5 mm from the previous vasectomy site or directly onto the site of vasectomy occlusion if possible in an effort to minimize vasal injury (Fig. 10.2). Using the ring forceps, the abdominal vas is gently elevated to just below the scrotal skin and a 15-blade scalpel or no-scalpel sharpened snap is used to make a small subcentimeter skin incision directly over the vas (Fig. 10.3). This incision is then carried down through the skin and dartos muscle layer using the sharpened snap, being careful not to injure the underlying vas. Once the vas is exposed, the second ring clamp is used to regrasp the exposed vas within the incision and elevate it gently out of wound (Figs. 10.4 and 10.5). The vas is then carefully mobilized and a perivasal window is created with a combination of blunt and sharp dissection for a length of approximately

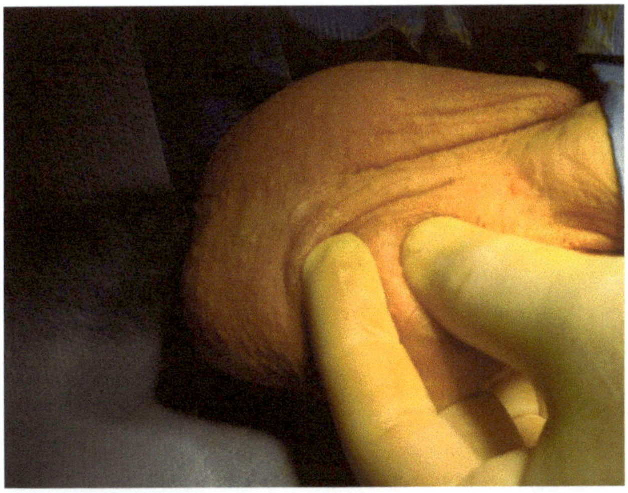

Fig. 10.1 The vas deferens is identified and secured using the three-finger technique

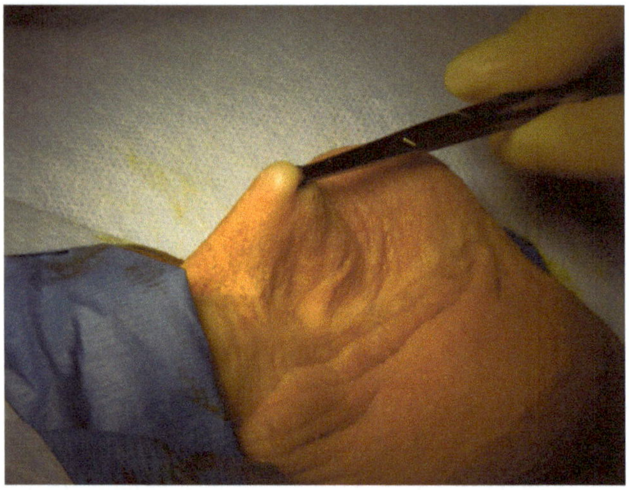

Fig. 10.2 Abdominal end of the vas deferens is grasped with a ring vasectomy clamp approximately 5 mm away from vasectomy defect and elevated

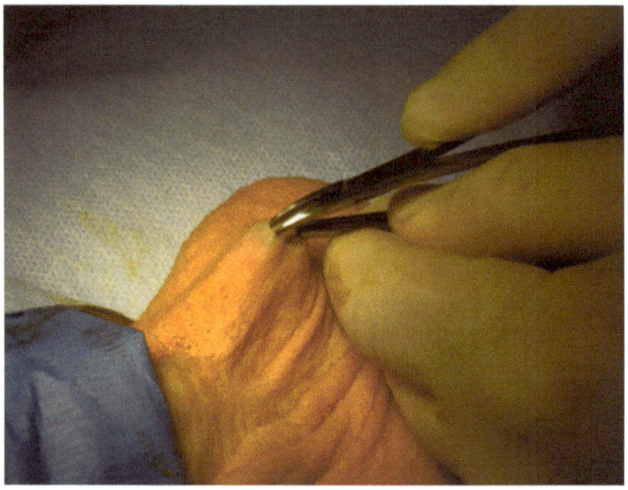

Fig. 10.3 Using no-scalpel vasectomy techniques, the skin and dartos muscle layers are opened directly over the vas deferens for a length of 8–10 mm

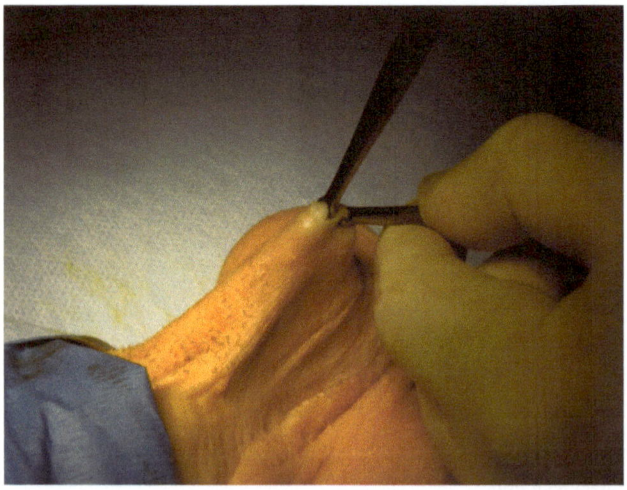

Fig. 10.4 A second no-scalpel vasectomy ring clamp is used to grasp the vas within the incision

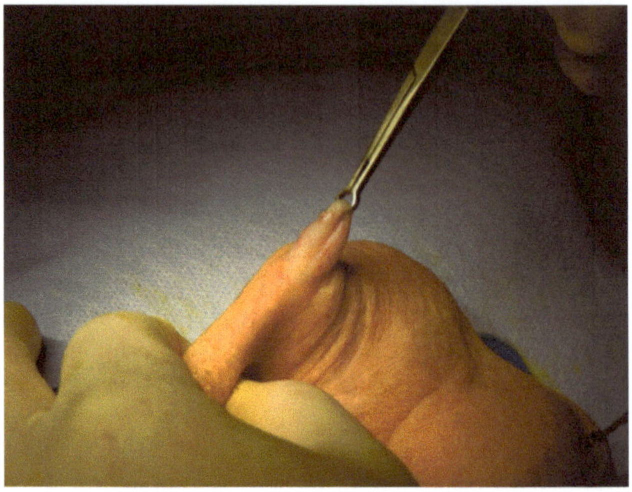

Fig. 10.5 The abdominal end of the vas is gently delivered out the incision

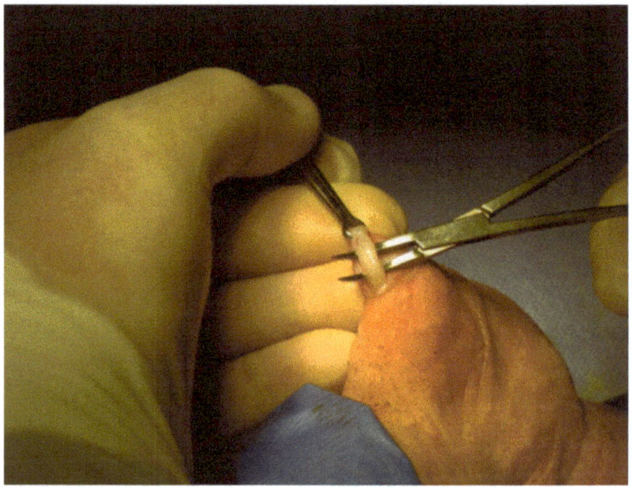

Fig. 10.6 A sharpened snap is used to create a small 1–1.5 cm perivasal window

1 cm with care taken to preserve the vasculature within the perivasal adventitia and finally secured with a vessel loop (Figs. 10.6 and 10.7). At every step, meticulous hemostasis is essential, particularly since the blood vessels in the dartos and subcutaneous layer may be difficult to control through the mini-incision after they retract into the scrotum. This is best achieved with the judicious use of microscopic bipolar electrocautery. With the abdominal end of the vas mobilized and secured, the testicular end of the vas is palpated through the incision beyond the identified vasectomy site and is regrasped with the ring forceps through the mini-incision in the scrotal skin and gently pulled out of the incision, mobilized, and secured in a manner identical to the abdominal vas (Fig. 10.8). Using this technique, a substantial portion of the vas can be delivered through the mini-incision given the inherent compliance of the scrotal skin (Fig. 10.9). Care must

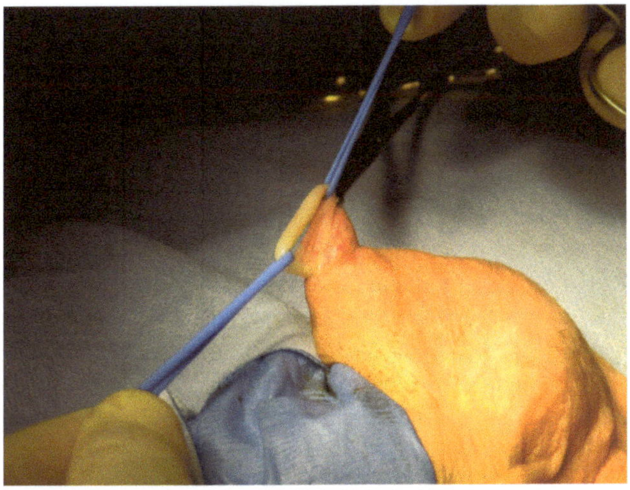

Fig. 10.7 The abdominal end of the vas is secured with vessel loops

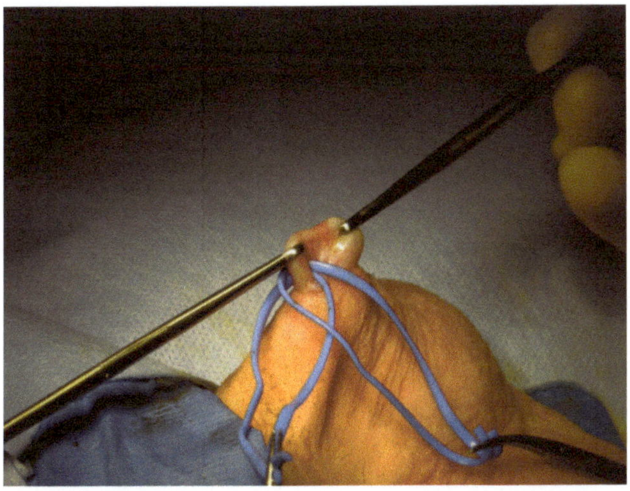

Fig. 10.8 With the abdominal end of the vas secured, the testicular end of the vas is delivered through the same incision and mobilized in an identical fashion to the abdominal end (note the vasectomy defect between the clamps)

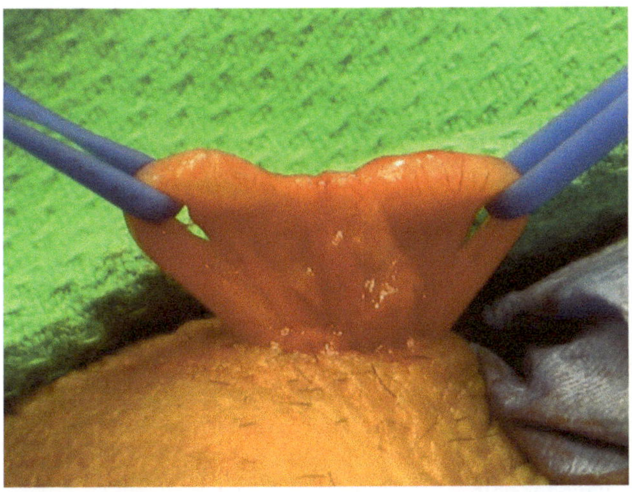

Fig. 10.9 Both vasal ends easily delivered through the mini-incision

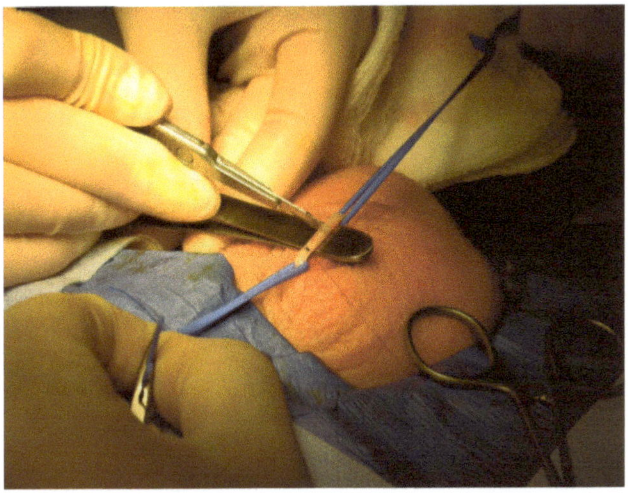

Fig. 10.10 Stay sutures of 5-0 Biosyn or PDS are placed superficially into the seromuscular layer of the vasal ends approximately 1 cm from the intended transection site, and the ends are then transected with a fresh 15 scalpel blade in a guillotine fashion

be taken during the mobilization of the testicular vas as the convoluted portion is often encountered and is very thin and easily injured. Stay sutures of 5-0 Biosyn or PDS are carefully placed through the superficial seromuscular layer of the vas approximately 5–10 mm away from the anticipated transection site on both the abdominal and testicular vas. These stay sutures, once tied, function to allow both control of the vasal ends to prevent retraction into the incision but also can be tied after the microsurgical anastomosis is complete to relieve tension on the repair. With both vasal ends secured, each vas is then transected sharply using a fresh 15-blade scalpel in a guillotine fashion stabilizing the vas over a nontoothed Adson forceps (Fig. 10.10). A commercially available vas approximator clamp or small vascular clamp is then used to control the vasal ends and bring them into close proximity to each other just outside the incision. Finally, a solid, high-contrast backing is placed beneath both the vasal ends and the vas approximator clamp to improve visualization of the sutures and provide support during the microscopic anastomosis (Fig. 10.11). The anastomosis is then performed under operating microscope magnification in a standard fashion. We begin by placing 4 10-0 double-armed nylon sutures through the mucosal and smooth muscle layer anteriorly in an inside out fashion on both vasa. Once all four anterior sutures are placed, they are tied down using microscopic tying forceps after cutting the needles off. The second anterior layer is then completed by placing 3 9-0 single-armed nylon sutures between the tied 10-0 sutures incorporating the seromuscular layers only. Once the 9-0 sutures are tied down, the vas approximator is rotated to expose the posterior wall of the vasa. The patency of the two vasal lumens is easy to assess visually under magnification and can also be

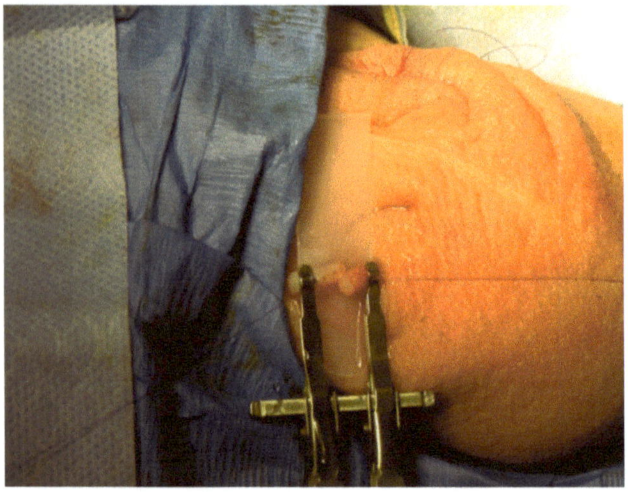

Fig. 10.11 The vasal ends are then secured in place in a vas approximating clamp and the plastic backboard in placed under the vasa and clamp

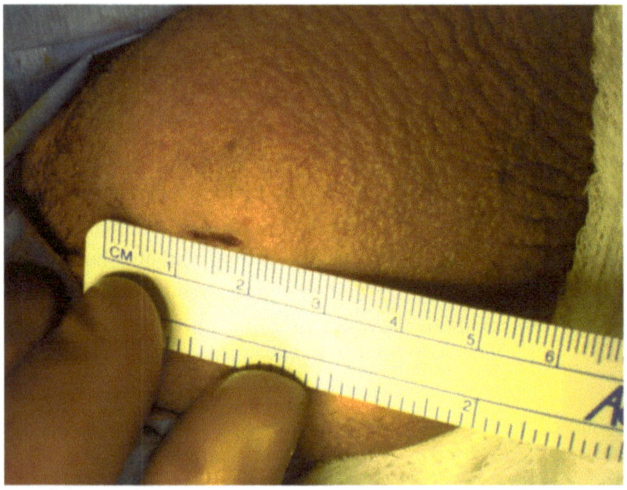

Fig. 10.12 Final incision length of <1 cm

confirmed by gentle probing with a jeweler forceps. Two to three additional 10-0 nylon sutures are then placed through the posterior mucosal layer depending on luminal size disparity between the two ends. Once they are tied down, three additional 9-0 nylon seromuscular sutures are placed between the 10-0 sutures to complete the two-layer anastomosis. As previously mentioned, the 5-0 stay sutures can be tied loosely together to prevent tension on the anastomosis, or they can be removed at this point. The vas is then returned to the scrotum and the operating microscope removed from the operative field. Hemostasis of the skin edges and dartos muscle is managed with electrocautery. Typically, only one stitch is needed for closure, and in many cases, no stitch is needed at all. The opening in the skin is typically 8–10 mm in length (Fig. 10.12). Intraoperatively, a local incision block is performed using 5 cm^3 of 0.25% bupivacaine on each side at the conclusion of the case. All patients are discharged home the same day with a prescription for 30 tablets of narcotic analgesia and are counseled to use a scrotal support for 7 days, refrain from sexual intercourse for 2 weeks, and avoid strenuous exercise and heavy lifting for 3 weeks. Office follow-up is arranged for 4–6 weeks time, and semen analysis is arranged for 2 and 4 months post-op and then every 3 months until pregnancy is achieved.

For primary vasectomy reversal or redo vasectomy reversals, the mini-incision approach is technically feasible in the majority of men. Occasionally, scarring around the vas deferens or indistinct anatomy may preclude the use of the mini-incision technique and the larger, traditional surgical incision is then required.

Outcome of Mini-Incision Vasectomy Reversal

To date over 200 mini-incision vasectomy reversals have been performed by three different surgeons at the University of Toronto. Using a single surgeon's data, 164 consecutive vasectomy reversals from 2004 to 2010 were reviewed [9]. All patients were followed-up in the office or by telephone call 4 weeks after surgery. All postoperative complications were recorded, and pain scores were documented using a validated postvasectomy pain scale subsequently adapted to vasectomy reversals [10]. Patients were also asked to quantify the number of days required for return of work and resumption of daily activities after surgery. Semen analysis was also carried out at 2 and 4 months postoperatively and evaluated according to WHO 1992 criteria [11].

Of the 164 men, 139 underwent bilateral vasectomy reversal with 55% having a mini-incision technique. The patency rate for the mini-incision technique was 96% and was not statistically different from the patency rate of men who had undergone the traditional incision vasectomy reversal. Mean semen parameters also did not differ between the mini-incision technique and the traditional incision.

Fifty-three men completed the pain and recovery assessment including 20 men who underwent mini-incision vasectomy reversal. Reported pain severity in the mini-incision group was significantly less during the first 48 h postprocedure compared to men who underwent vasectomy reversal using the traditional incision, although at 1 week, there was no statistically significant difference in pain scores (Fig. 10.13).

Following the mini-incision vasectomy reversal patients returned to self-reported "normal everyday activities" 2 days earlier compared to men following traditional incision vasectomy reversal. Time to return to work however was not different between the two groups and averaged 5 days for both.

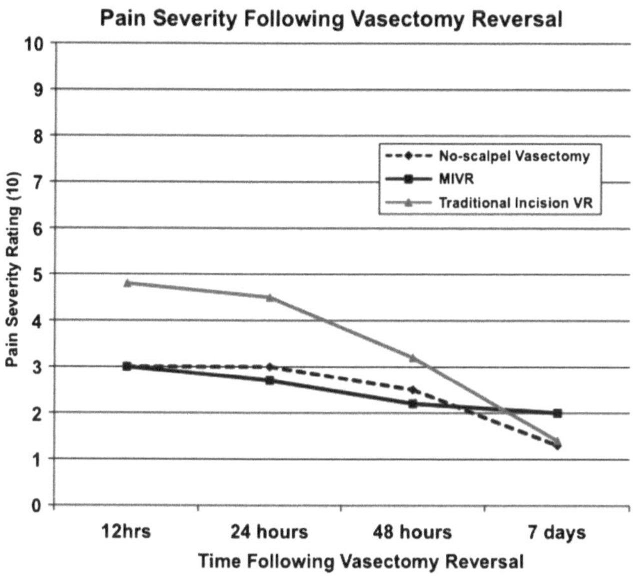

Fig. 10.13 Pain severity during the first 48 h following surgery was less among patients who received a bilateral MIVR compared to patients who received a traditional incision VR

Key Issues

The mini-incision vasectomy reversal technique developed at the University of Toronto uses techniques and instruments familiar to most urologists. By taking advantage of the compliance of the scrotal wall, a substantial length of vas deferens can safely be delivered through a mini-incision for an efficient approach to vasectomy reversal. By avoiding a longer incision, delivery of the testicle and more extensive tissue dissection, one can significantly reduce postoperative pain and hasten return to daily activities without compromising patient outcomes as evidenced by our experience.

References

1. Kim H, Goldstein M. History of vasectomy reversal. Urol Clin North Am. 2009;36(3):359–73.
2. Lipshultz LI, Rumohr JA, Bennett RC. Techniques for vasectomy reversal. Urol Clin North Am. 2009;36(3):375–82.
3. Nagler HM, Jung H. Factors predicting successful microsurgical vasectomy reversal. Urol Clin North Am. 2009;36(3): 383–90.
4. Jarvi K, Grober ED, Lo KC, et al. Mini-incision microsurgical vasectomy reversal using no-scalpel vasectomy techniques and instruments. Urology. 2008;72(4):913–5.
5. Li SQ, Goldstein M, Zhu J, Huber D. The no-scalpel vasectomy. J Urol. 1991;145(2):341–4.
6. Weiss RS, Li PS. No-needle jet anesthetic technique for no-scalpel vasectomy. J Urol. 2005;173(5):1677–80.
7. Goldstein M, Li PS, Matthews GJ. Microsurgical vasovasostomy: the microdot technique of precision suture placement. J Urol. 1998;159(1):188–90.
8. Belker AM, Thomas Jr AJ, Fuchs EF, et al. Results of 1,469 microsurgical vasectomy reversals by the Vasovasostomy Study Group. J Urol. 1991;145(3):505–11.
9. Grober E, Jarvi K, Lo K, Shin EJ. Mini-incision vasectomy reversal using no-scalpel vasectomy principles: efficacy and postoperative pain compared with traditional approaches to vasectomy reversal. Urology. 2011;77(3):602–6.
10. Aggarwal H, Chiou RK, Siref LE, Sloan SE. Comparative analysis of pain during anesthesia and no-scalpel vasectomy procedure among three different local anesthetic techniques. Urology. 2009;74(1):77–81.
11. WHO Laboratory Manual for the examination of human semen and sperm-cervical mucus interaction, 4th ed., World Health Organization, 1999.

Robotic Microsurgery for Male Infertility and Chronic Orchialgia

11

Jamin V. Brahmbhatt and Sijo J. Parekattil

Abstract

Since the use of the operating microscope for microsurgery in 1975, there has been a steady increase in the use of such technology in the operative management of male infertility and chronic testicular or groin pain. Added to the reports relating to greater patency rates and fertility rates of vasovasostomy performed with the operating microscope, the concepts of magnification have been successfully applied to vasoepididymostomy and varicocele ligation. More recently, microscopic spermatic cord neurolysis has demonstrated applicability to the treatment of groin and testicular discomfort. These techniques require varying degrees of microsurgical skills and an array of supporting technology, neither of which may be part of many urologist's personal or technical armamentarium. The melding of improved visualization with magnification to an ergonomic platform that can be operated remotely has a significant application to testicular and reproductive surgery. Robotic assistance during surgical procedures has been utilized in a wide array of surgical fields with the above mentioned benefits. This chapter covers the latest developments in the robotic microsurgical platform, robotic microsurgical tools, and current evaluations of various robotic microsurgical applications for male infertility and patients with chronic testicular or groin pain.

Keywords

Robotic microsurgery • Vasovasostomy • Chronic orchialgia • Male infertility • Intraoperative vascular Doppler ultrasound • Spermatic cord denervation • Spermatic cord neurolysis • Varicocele ligation

Since the use of the operating microscope for microsurgery in 1975 [1], there has been a steady increase in the use of such technology in the operative management of male infertility and chronic testicular or groin pain [1–11]. Added to the reports relating to greater patency rates and fertility rates of vasovasostomy performed with the operating microscope [12], the concepts of magnification have been successfully applied to vasoepididymostomy and varicocele ligation. More recently, microscopic spermatic cord neurolysis has demonstrated applicability to the treatment of groin and testicular discomfort [13, 14]. These techniques require varying degrees of microsurgical skills and an array of supporting technology, neither of which may be part of many urologists' personal or technical armamentarium. The melding of improved visualization with magnification to an ergonomic platform that can be operated remotely has a significant application to testicular and reproductive surgery. Robotic assistance during surgical procedures has been utilized in a wide array of surgical fields with the above mentioned benefits [15–19]. This chapter

J.V. Brahmbhatt, MD
Department of Urology, University of Tennessee Health Science Center, Memphis, TN, USA
e-mail: jaminbrahmbhatt@gmail.com

S.J. Parekattil, MD (✉)
Division of Urology, Winter Haven Hospital, University of Florida, 200 Avenue F. N.E., Winter Haven, FL 33881, USA
e-mail: sijo.parekattil@winterhavenhospital.org

S.J. Parekattil and A. Agarwal (eds.), *Male Infertility: Contemporary Clinical Approaches, Andrology, ART & Antioxidants*,
DOI 10.1007/978-1-4614-3335-4_11, © Springer Science+Business Media, LLC 2012

covers the latest developments in the robotic microsurgical platform, robotic microsurgical tools, and current evaluations of various robotic microsurgical applications for male infertility and patients with chronic testicular or groin pain.

Novel Equipment

With any new field, the development of novel tools or instruments that can enable surgeons to create new solutions for existing clinical needs is of paramount importance. Below are some new products that enhance the ability to perform robotic-assisted microsurgery.

New Robotic Surgical Platform

Intuitive Surgical (Sunnyvale, CA) now offers an enhanced four-arm da Vinci-type Si robotic system with high-definition digital visual magnification that allows for greater magnification than the standard robotic system (up to 10–15×). The enhanced magnification capability allows the surgeon to position the camera 6–7 cm away from the operative field to avoid any local tissue effects from the heat emitted from the camera lighting (this was a problem with the older system, where the camera had to be placed within 2–3 cm of the operative field for microsurgery). This new system (Fig. 11.1) allows greater range of motion and better microsurgical instrument handling. The additional fourth arm has improved range of motion and positioning capabilities to provide the microsurgeon with one additional tool during procedures. The robot is positioned from the right side of the patient for microsurgical cases as illustrated in Fig. 11.2.

Refined Robotic Doppler Flow Probe

Cocuzza et al. [20] have shown that the systematic use of intraoperative vascular Doppler ultrasound during microsurgical subinguinal varicocelectomy improves precise identification and preservation of testicular blood supply. During robotic microsurgical cases, the standard Doppler probe has to be held by a surgical assistant and cannot be manipulated readily with the robotic graspers. A new revised micro-Doppler flow probe (MDP) has been developed by Vascular Technology Inc. (Nashua, NH) that is designed specifically for use with the robotic platform (Fig. 11.3). This new probe allows for easy manipulation of the probe with the fourth arm and allows the surgeon to perform real-time Doppler monitoring of the testicular artery during cases such as robotic-assisted microscopic varicocelectomy

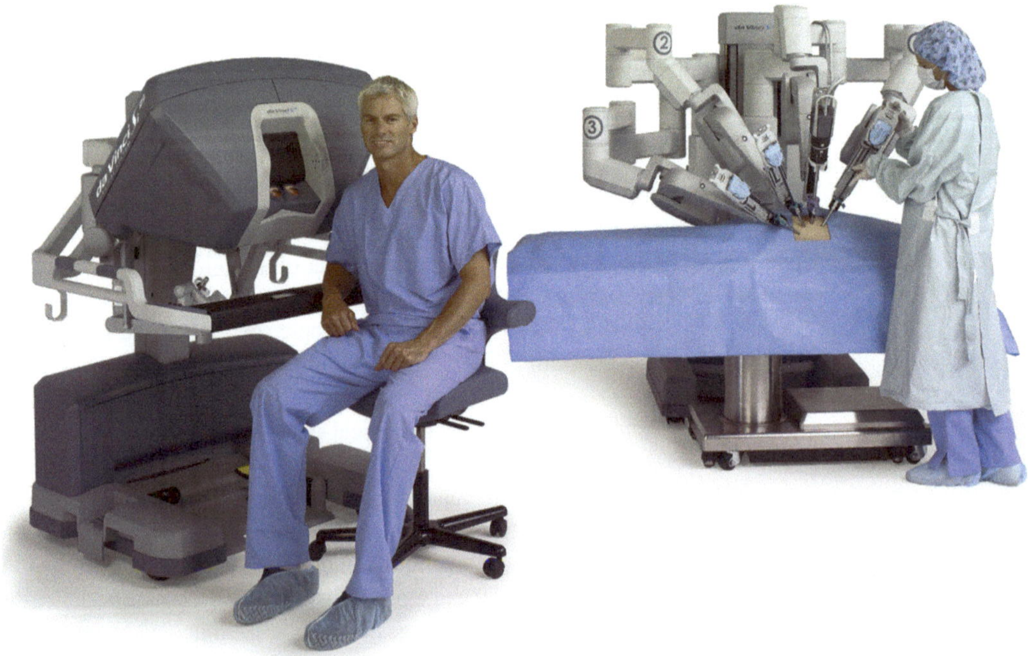

Fig. 11.1 daVinci Si Robotic platform (courtesy of Intuitive Surgical © 2010)

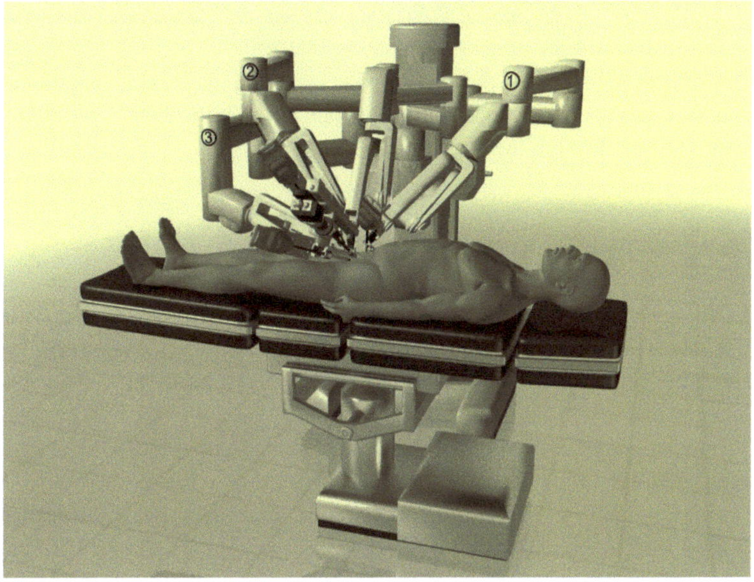

Fig. 11.2 Robotic platform positioning for microsurgical cases

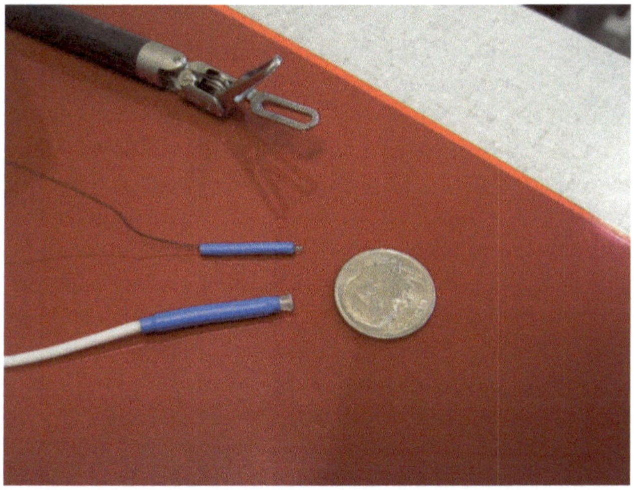

Fig. 11.3 Robotic micro-Doppler probe

(RAVx) and robotic-assisted microscopic denervation of the spermatic cord (RMDSC). This allows the surgeon to hear the testicular artery flow while dissecting out the veins and nerves with the other two robotic arms.

A recent prospective randomized control trial of the MDP in 273 robotic microsurgical cases from Jul 2009 to Sept 2010 was performed: 67 robotic subinguinal varicocelectomies (RVx) and 206 robotic spermatic cord denervation procedures (RMDSC). The use of the MDP was randomized to 5 RVx and 20 RMDSC procedures. The primary end point was operative time, and secondary end point was surgeon ease in testicular artery localization and robotic grasper maneuverability. Operative duration was not affected by utilization of the MPD ($p=0.5$). The MDP was effective in identifying all testicular arteries within the spermatic cord in all cases. Due to the compact size of the MDP, maneuverability using the robotic grasper was significantly improved over the standard handheld Doppler probe. MDP allowed for full range of motion of the robotic arms allowing the surgeon to easily scan vessels from a wide range of angles. No complications from use of the MDP occurred. The new micro-Doppler probe for robotic microsurgical procedures appears to have performed effectively in this study.

Vascular Technology Inc. (Nashua, NH) has recently developed an even smaller microprobe that can detect flow through vessels at about 0.5-mm diameter (Fig. 11.3). This just expands further potential applications for this technology.

Enhanced Digital Visual Magnification

The miniaturization and development of advanced digital microscopic cameras (100–250×) allow even greater magnification than the standard robotic (10–15×) and microscopic (10–20×) magnification in use at this time. Our group is currently involved in clinical trials of a 100× digital camera (Digital Inc., China) that can be utilized via the TilePro™ da Vinci Si robotic system (Intuitive Surgical, Sunnyvale, CA) to allow the surgeon to toggle or use simultaneous 100× and 10–15× visualization. This provides the surgeon with unparalleled visual acuity for complex microsurgical procedures.

Karl Storz (El Segundo, CA) also offers a robotic arm platform to hold an optical mini-scope that offers 16–20× magnification that can then be used during the da Vinci robotic cases to provide an additional enhanced magnification view (routed through the da Vinci console).

New Saline-Enhanced Electrosurgical Micro Blade

Saline-enhanced electrosurgical resection (SEER) has been utilized for cauterization of vessels in a number of liver and renal applications. It is a form of electrosurgical cautery and resection that leads to minimal smoke and scar formation. A new robotic SEER micro blade is in development (Bovie Inc., Clearwater, FL) that may provide an option for ablation of veins during robotic-assisted subinguinal varicocelectomy.

Robotic Microsurgical Procedures

Robotic-Assisted Microscopic Vasectomy Reversal

A number of groups have developed robotic-assisted techniques to perform robotic-assisted microscopic vasectomy reversal (RAVV) in animal and ex vivo human models [21–25]. Some studies suggest that robotic-assisted reversal may have advantages over microsurgical reversal in terms of ease of performing the procedure and improved patency rates [23, 24]. A few groups have actually performed human robotic-assisted vasovasostomies using the initial da Vinci robotic system [26] (Intuitive Surgical, Sunnyvale, CA).

These efforts have been recently confirmed in human RAVV cases performed using the new da Vinci Si system [27, 28]. A recent prospective control study of robotic versus pure microsurgical vasovasostomy in 90 vasectomy reversal cases performed from Aug 2007 to Sept 2010 by a single fellowship trained microsurgeon was performed (this was an extension to a previously published study) [29]. The primary end point was operative duration. The secondary end point was total motile sperm count at 2, 5, 9, and 12 months postoperatively. Case breakdown was as such: 45 robotic-assisted cases and 45 pure microsurgical cases. Selection of approach (robotic vs. pure microscopic) was based on patient choice (robotic was more expensive than microscopic). Preoperative patient characteristics were similar in both groups. The same suture material and suturing technique (2 layer 10-0 and 9-0 nylon anastomosis) was used in both approaches.

Median clinical follow-up was 14 months (range 1–37 months). Median duration from vasectomy in the robotic group was 8 years (1–19) and in the microscopic group 6.5 years (1–19). Ninety-four percentage overall patency was achieved in the RAVV cases and 79% in MVV (>1 million sperm/high power field). Median operative duration was significantly decreased in RAVV at 90 min (60–180) compared to MVV at 120 min (60–180), $p=0.004$. Mean postoperative total motile sperm counts were not significantly higher in RAVV versus MVV, but the rate of postoperative sperm count recovery was significantly greater in RAVV.

The use of robotic assistance in microsurgical vasovasostomy may have potential benefit over MVV with regard to decreasing operative duration and improving the rate of recovery of postoperative total motile sperm counts. Further evaluation and longer follow-up are needed to assess its clinical potential and the true cost–benefit ratio.

Robotic-Assisted Microscopic Varicocelectomy

Although reports of robotic-assisted laparoscopic intra-abdominal varicocelectomy have been published [30], there are a number of publications that suggest that microscopic subinguinal varicocelectomy(MVx) may provide superior outcomes compared to intra-abdominal varicocelectomy [31–34]. Shu, Wang et al. were the first to publish on robotic-assisted microsurgical subinguinal varicocelectomy (RAVx) [35]. They compared standard microsurgical to robotic-assisted varicocelectomy and found that the robotic approach provided advantages in terms of slightly decreasing operative duration and complete elimination of surgeon tremor.

To further explore these findings, we performed a prospective randomized control trial of MVx to RAVx in a canine varicocele model by a fellowship-trained microsurgeon. The surgeon performed cord dissection and ligation of three veins with 3-0 silk ties. Twelve canine varicocelectomies were randomized into two arms of six: MVV versus RAVx. Procedure duration, vessel injury, and knot failures were recorded. The RAVx mean duration (9.5 min) was significantly faster than MVV (12 min), $p=0.04$. The duration for robot setup and microscope setup was not significantly different. There were no vessel injuries or knot failures in either group.

A review of our prospective clinical database of 97 RAVx cases from Jun 2008 to Sept 2010 (median follow-up 11 months: range 1–27) is as follows. The median duration per side was 30 min (10–80). Indications for the procedure were the presence of a grade two or three varicocele and the following conditions: 10 with azoospermia, 42 with oligoospermia, and 49 with testicular pain (with or without oligoospermia, and failed all other conservative treatment options). Three-month follow-up was available for 81 patients: 75% with oligoospermia had a significant improvement in sperm count or motility; one with azoospermia was converted to oligoospermia. For testicular pain, 92% had complete resolution of pain (targeted neurolysis of the spermatic cord had

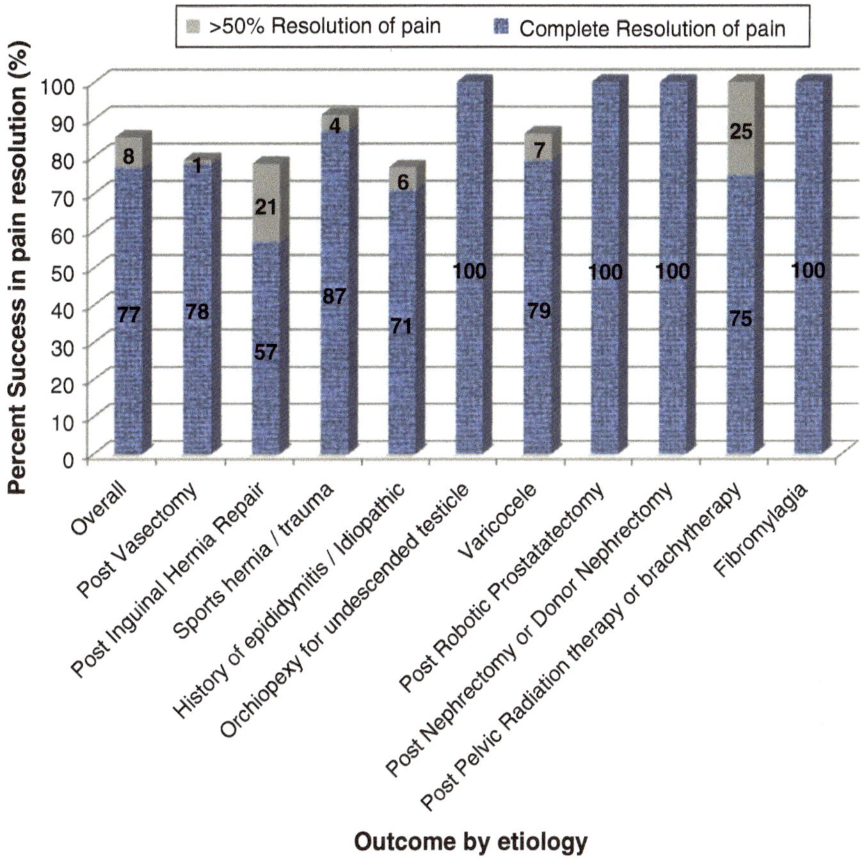

Fig. 11.4 Pain elimination outcomes for robotic-assisted microsurgical denervation of the spermatic cord for various groin or testicular pain etiologies

been performed in addition to varicocelectomy). One recurrence or persistence of a varicocele occurred (by physical and ultrasound exam), one patient developed a small postoperative hydrocele, and two patients had small postoperative scrotal hematomas (treated conservatively). The fourth robotic arm allowed the surgeon to control one additional instrument during the cases decreasing reliance on the microsurgical assistant. The fourth arm also enabled the surgeon to perform real-time intraoperative Doppler mapping of the testicular arteries while dissecting the veins with the other arms if needed.

Robotic-assisted microsurgical subinguinal varicocelectomy appears to be safe, feasible, and efficient. The preliminary human results appear promising. Further evaluation and comparative effectiveness studies are warranted.

Robotic-Assisted Microscopic Denervation of the Spermatic Cord

Recent studies by Levine et al. [13] and Oliveira et al. [14] have shown that microscopic denervation of the spermatic cord is an effective treatment option for men with chronic testicular pain. Our group has been developing a robotic-assisted microsurgical approach for the denervation of the spermatic cord (RMDSC) to assess if there may be any potential benefit over the standard microscopic technique.

A review of our initial 230 RMDSC cases from Oct 2008 to Sept 2010 was performed (median follow-up of 8 months). Selection criteria for patients were as such: chronic testicular pain (>6 months), failed standard pain management treatments, and negative urologic workup. A robotic-assisted subinguinal, inguinal, or intra-abdominal approach was utilized based on the location of pain. Pain was assessed utilizing a standardized validated tool (PIQ-6). The median operative duration was 20 min (7–150). The fourth robotic arm allowed the surgeon to control one additional instrument leading to less reliance on the microsurgical assistant.

Postoperatively, 77% (176) patients had complete resolution of pain and 8% (19) had a 50% decrease in pain. RMDSC was successful in eliminating testicular and or groin pain for a number of possible etiologies: postvasectomy pain syndrome (PVPS), postinguinal hernia pain, sports hernia or groin trauma pain, chronic epididymitis or idiopathic pain, varicocele pain, postrobotic prostatectomy groin or testicular pain, postnephrectomy or donor nephrectomy groin or testicular pain, postpelvic radiation or brachytherapy groin or testicular pain, and fibromyalgia groin pain. Figure 11.4 illustrates the outcomes in these various pain categories.

Single Port and Abdominal Robotic Microsurgical Neurolysis

Chronic groin pain can be debilitating for patients. Microsurgical subinguinal denervation of the spermatic cord (MDSC) is a treatment option for this pain. However, there are limited further options for patients who fail this treatment or who have phantom pain after orchiectomy. Our goal was to develop a single port and abdominal robotic microsurgical neurolysis technique to ligate the genitofemoral and inferior hypogastric nerve fibers within the abdomen above the internal inguinal ring.

We performed a prospective study of patients with chronic groin pain who had either failed previous MDSC or had phantom pain after orchiectomy. Primary end point was impact of pain on quality of life (PIQ-6 pain impact questionnaire from RAND) and secondary end point was operative robotic duration. PIQ-6 scores were collected pre-op and at 1, 3, 6, and 12 months post-op.

We completed 30 cases (five single port) from Jun 2009 to Sept 2010. Elimination of pain occurred in 60% (18 cases), and a greater than 50% reduction in pain occurred in an additional 13% (4 cases) within 1 month post-op. Two of the failures were patients that had pain elimination for 6 months, but then pain returned thereafter. Median OR duration was 10 min (5–30). There were three complications: one post-op scrotal hematoma that resolved with conservative measures, one patient had pain at one of the port sites, and one patient had pain that shifted from the groin to the leg. Single port and abdominal robotic microsurgical neurolysis appears to be an option for treatment in this difficult patient population. Further follow-up and evaluation is warranted.

Expert Commentary

The use of robotic assistance during microsurgical procedures is expanding. The application of this technology in other microsurgery fields apart from urology is also expanding, such as ophthalmology, hand surgery, and plastic and reconstructive microsurgery. The advantages of a stable microsurgical platform, ergonomic surgeon instrument controls, elimination of tremor, and magnified immersive 3D vision are all intuitively apparent. Further comparative effectiveness studies are ongoing and will be forthcoming on the true applicability of this new surgical platform. However, the preliminary results so far are quite impressive.

Five-Year View

The use of robotic assistance during microsurgical procedures is likely to expand in all areas of microsurgery. In order to provide a structured evidence-based platform to develop these procedures and protect patient safety, a dedicated society has been formed (Robotic-Assisted Microsurgical and Endoscopic Society: RAMSES—www.roboticmicrosurgeons.org). This society has developed a core curriculum for robotic microsurgical training that could be utilized in the fields of urology, hand surgery, plastic/reconstructive, ophthalmology, and vascular surgery. The goal of this group is to further the prudent and scientific use of such techniques. Such organizations may be able to influence the direction of instrument/equipment development by industry.

Key Issues

- Robotic assistance during microsurgery provides microsurgeons with many advantages: improved operative efficiency, elimination of tremor, scaling of motion, and enhanced imaging.
- Improved clinical outcomes appear likely with robotic assistance, and preliminary studies appear to support this concept.
- Novel treatment options for men with chronic testicular or groin pain, postvasectomy pain, sports hernia pain, postnephrectomy, donor nephrectomy, and phantom groin pain are now available with this technology.
- Structured evidence-based platforms for the scientific development of this technology are necessary to protect patient safety. Groups such as RAMSES may provide guidance.

Acknowledgments We would like to thank Dr. Johannes Vieweg, Dr. Li-Ming Su, Dr. Philip Li, Dr. Hany Atalah, Katy Lyall, David Regan, Dr. Rachana Suchdev, Intuitive Surgical and Vascular Technology Inc. for their continued support in the pursuit and refinement of robotic microsurgical techniques and tools.

References

1. Silber SJ. Microsurgery in clinical urology. Urology. 1975;6:150–3.
2. Marmar JL. Modified vasoepididymostomy with simultaneous double needle placement, tubulotomy and tubular invagination. J Urol. 2000;163:483–6.
3. Berger RE. Triangulation end-to-side vasoepididymostomy. J Urol. 1998;159:1951–3.
4. Chan PT, Li PS, Goldstein M. Microsurgical vasoepididymostomy: a prospective randomized study of 3 intussusception techniques in rats. J Urol. 2003;169:1924–9.
5. Fogdestam I, Fall M. Microsurgical end-to-end and end-to-side epididymovasostomy to correct occlusive azoospermia. Scand J Plast Reconstr Surg. 1983;17:137–40.
6. Marmar JL, Kim Y. Subinguinal microsurgical varicocelectomy: a technical critique and statistical analysis of semen and pregnancy data. J Urol. 1994;152:1127–32.
7. Owen ER. Microsurgical vasovasostomy: a reliable vasectomy reversal. Aust N Z J Surg. 1977;47:305–9.
8. Schlegel PN. Testicular sperm extraction: microdissection improves sperm yield with minimal tissue excision. Hum Reprod. 1999;14: 131–5.

9. Schultheiss D, Denil J. History of the microscope and development of microsurgery: a revolution for reproductive tract surgery. Andrologia. 2002;34:234–41.
10. Silber SJ. Microscopic vasoepididymostomy: specific microanastomosis to the epididymal tubule. Fertil Steril. 1978;30:565–71.
11. Thomas Jr AJ. Vasoepididymostomy. Urol Clin North Am. 1987; 14:527–38.
12. Owen ER. Microsurgical vasovasostomy: a reliable vasectomy reversal. J Urol. 1977;47:305–9.
13. Levine LA. Microsurgical denervation of the spermatic cord. J Sex Med. 2008;5:526–9.
14. Oliveira RG, Camara C, Alves Jde M, et al. Microsurgical testicular denervation for the treatment of chronic testicular pain initial results. Clinics (Sao Paulo). 2009;64:393–6.
15. Bourla DH, Hubschman JP, Culjat M, et al. Feasibility study of intraocular robotic surgery with the da vinci surgical system. Retina. 2008;28:154–8.
16. Casale P. Robotic pediatric urology. Expert Rev Med Devices. 2008;5:59–64.
17. Colombo Jr JR, Santos B, Hafron J, et al. Robotic assisted radical prostatectomy: surgical techniques and outcomes. Int Braz J Urol. 2007;33:803–9.
18. Guru KA, Wilding GE, Piacente P, et al. Robot-assisted radical cystectomy versus open radical cystectomy: assessment of postoperative pain. Can J Urol. 2007;14:3753–6.
19. Rodriguez E, Chitwood Jr WR. Minimally invasive, robotic cardiac surgery. Ann Thorac Surg. 2008;85:357–8.
20. Cocuzza M, Pagani R, Coelho R, et al. The systematic use of intraoperative vascular doppler ultrasound during microsurgical subinguinal varicocelectomy improves precise identification and preservation of testicular blood supply. Fertil Steril. 2010;93(7):2396–9.
21. Kuang W, Shin PR, Matin S, Thomas Jr AJ. Initial evaluation of robotic technology for microsurgical vasovasostomy. J Urol. 2004; 171:300–3.
22. Kuang W, Shin PR, Oder M, Thomas Jr AJ. Robotic-assisted vasovasostomy: a two-layer technique in an animal model. Urology. 2005;65:811–4.
23. Schiff J, Li PS, Goldstein M. Robotic microsurgical vasovasostomy and vasoepididymostomy: a prospective randomized study in a rat model. J Urol. 2004;171:1720–5.
24. Schiff J, Li PS, Goldstein M. Robotic microsurgical vasovasostomy and vasoepididymostomy in rats. Int J Med Robot. 2005; 1:122–6.
25. Schoor RA, Ross L, Niederberger C. Robotic assisted microsurgical vasal reconstruction in a model system. World J Urol. 2003;21:48–9.
26. Fleming C. Robot-assisted vasovasostomy. Urol Clin North Am. 2004;31:769–72.
27. Parekattil SJ, Moran ME. Robotic instrumentation: evolution and microsurgical applications. Indian J Urol. 2010;26:395–403.
28. Parekattil S, Cohen M, Vieweg J. Human robotic assisted bilateral vasoepididymostomy and vasovasostomy procedures: Initial safety and efficacy trial. Proc SPIE. 2009;7161:71611L.
29. Parekattil S, Atalah H, Cohen M. Video technique for human robotic assisted microsurgical vasovasostomy. J Endourol. 2010;24: 511–4.
30. Corcione F, Esposito C, Cuccurullo D, et al. Advantages and limits of robot-assisted laparoscopic surgery: preliminary experience. Surg Endosc. 2005;19:117–9.
31. Chen XF, Zhou LX, Liu YD, et al. comparative analysis of three different surgical approaches to varicocelectomy. Zhonghua Nan Ke Xue. 2009;15:413–6.
32. Cayan S, Shavakhabov S, Kadioglu A. Treatment of palpable varicocele in infertile men: a meta-analysis to define the best technique. J Androl. 2009;30:33–40.
33. Al-Said S, Al-Naimi A, Al-Ansari A, et al. Varicocelectomy for male infertility: a comparative study of open, laparoscopic and microsurgical approaches. J Urol. 2008;180:266–70.
34. Al-Kandari AM, Shabaan H, Ibrahim HM, et al. Comparison of outcomes of different varicocelectomy techniques: open inguinal, laparoscopic, and subinguinal microscopic varicocelectomy: a randomized clinical trial. Urology. 2007;69:417–20.
35. Shu T, Taghechian S, Wang R. Initial experience with robot-assisted varicocelectomy. Asian J Androl. 2008;10:146–8.

Robot-Assisted Vasectomy Reversal

12

Peter Frank De Wil, Vincenzo Ficarra, George A. de Boccard, and Alexandre Mottrie

Abstract

In the early part of the twentieth century, the vasectomy gained momentum with the application for eugenic, punitive, and therapeutic purposes. Today, with the exception of its use as prophylaxis procedure after transurethral resection of prostate or adenomectomy for benign prostatic hyperplasia (BPH) to prevent urogenital infections, vasectomy is widely used as a contraceptive tool, and it is calculated that about 50 million men have relied on vasectomy for family planning. Above all, in the United States, vasectomy is employed as contraceptive method by nearly 11% of married couples. With the upcoming success of vasectomy, a new surgical procedure, the vasectomy reversal was born.

Keywords

Robot-assisted vasectomy reversal • Robot-assisted surgery • Vasoepididymostomy • Epididymitis • Benign prostatic hyperplasia • Vasectomy reversal

Since the early part of the twentieth century, vasectomy was used for eugenic, punitive, and therapeutic purposes. Nowadays, this procedure is widely used as a contraceptive tool, and it is calculated that about 50 million men have relied on vasectomy for family planning. More than in Europe, in the United States, vasectomy is employed as contraceptive method by nearly 11% of married couples [1]. Second indication for vasectomy is after transurethral resection of prostate or adenomectomy for benign prostatic hyperplasia (BPH) to prevent urogenital infections.

In the early years of past century, Dr. Edward Martin, surgeon at the University of Pennsylvania was the first to perform reversal procedures. He performed vasoepididymostomies in men who had obstruction secondary to epididymitis and not vasectomy.

This work is what led Jeequier to label Martin as the "father of andrology" [2].

The number of vasectomy reversals is rising because of the rising popularity of vasectomy combined with the continuing upward trend of divorce and remarriage, especially in industrialized countries. In order to plan the more appropriate intervention to reach pregnancy, the chances of success in terms of patency of vas and/or pregnancy should be counseled based on the personal experience of the surgeon, patient's health history, age, results of physical examination, and reproductive potential of his partner. A potential alternative is sperm extraction for IVF or ICSI. In most men, vasectomy reversal is technically feasible, but success rates depend on a variety of factors determined by both male and female fertility factors [3].

P.F. De Wil, MD (✉)
Urological Department, Kliniek Sint Jan, Kruidtuinlaan 32, 1000, Brussels, Belgium
e-mail: peter_dewil@hotmail.com

V. Ficarra, MD, PhD
Department of Oncological and Surgical Sciences, Urology Clinic, University of Padua, Monoblocco Ospedaliero, IV Floor, Via Giustiniani 2, 35100 Padua, Italy
e-mail: vincenzo.ficarra@unipd.it

G.A. de Boccard, MD
Clinique Generale Beaulieu, Robot-Assisted Laparoscopic Surgery Center, 20 Chemin Beau Soleil, 1206 Geneve, Switzerland
e-mail: boccard@iprolink.ch

A. Mottrie, MD, PhD
Department of Urology, O.L.V. Clinic Aalst, Aalst, Belgium

S.J. Parekattil and A. Agarwal (eds.), *Male Infertility: Contemporary Clinical Approaches, Andrology, ART & Antioxidants*, DOI 10.1007/978-1-4614-3335-4_12,

Vasectomy reversal is a technically challenging procedure, and a number of microsurgical advances have enhanced outcomes (e.g., use of the operating microscope with optical magnification). A number of techniques where developed using microsurgical technique [4, 5], one-layer [6–8], microdot [9], stents [10], and oblique technique [11]. This chapter reviews the application of robotic-assisted microsurgery for vasectomy reversal.

Robotic Vasovasostomy

So why use robotic assistance for microsurgical vasectomy reversal? Robotic assistance provides a number of benefits for the microsurgeon: elimination of tremor, scaling of motion (1:5), and enhanced magnification—all in a fixed stable microsurgical platform. It obviates the need for specialized tissue support platforms and reduces the reliance of the microsurgeon on a skilled microsurgical assistant. These are all important advantages for technically challenging procedures.

From an economical point of view, only centers already equipped with a da Vinci robot should perform robotic vasovasostomy because the additional cost is only a few hundred euro/dollars [12]. Five studies have reported the results of robot-assisted vasectomy reversal in experimental animal models and in humans: Kuang et al. [13, 14], Schiff et al. [15], Parekattil et al. [16], and De Boccard et al. [17]. This chapter focuses on our technique.

Personal Surgical Technique

The personal surgical technique has been previously described in detail [18]. The surgical technique in preparing the vas deferens for vasovasostomy is identical to the microsurgical technique. A scrotal median raphe approach is utilized (incision about 5 cm in length). The previous vasectomy sites are identified, and the proximal and distal ends of the vas deferens are dissected free. The distal end of the vas (testicular side) is transected, and the fluid from this end is evaluated. The proximal vas end is flushed with 2 cm^3 of saline to assess for patency. The two ends of the vas deferens are now loosely approximated using either a 4-0 or 7-0 absorbable suture to create a tension-free anastomosis.

Once the vas ends have been prepared, the robotic platform (da Vinci Si system, Intuitive Surgical, Sunnyvale, CA) is brought into the field. The robotic system is placed in between the legs of the patient as in a robotic prostatectomy placement. The first assistant remains on the left of the patient, and the primary surgeon moves from the right of the patient to the surgical console. A 0° lens is utilized and placed at an 80° angle to the patient. The right-side anastomosis is performed first, and the camera is positioned in such a manner that the camera can be moved to the left anastomosis at the same angle (by simply sliding it over to the left by a few centimeters). Black Diamond microforceps are utilized in arms 1 and 2. They are positioned at 45° to the camera lens. Arm 3 carries the Potts scissors and is positioned in front of the camera at 30° to the patient. The standard robotic trocars are placed for all arms even though the procedure in not being performed intracorporeally in order to stabilize the instruments.

The anastomosis is now completed utilizing 8-0 or 9-0 nylon interrupted sutures. Three posterior full-thickness (muscularis and mucosa) sutures are placed and then tied down to secure the posterior plate of the anastomosis. Five full-thickness sutures are placed anteriorly and then tied down to complete the anastomosis. Two or three additional sutures may be placed to secure the supporting muscularis. The use of the third arm with the Potts scissors improves the efficiency of the procedure and obviates the need for a skilled microsurgical assistant (a scrub nurse is sufficient). After completion of the anastomosis on both sides, the robot is now undocked and the scrotal incision is closed with a 4-0 absorbable suture in two layers.

Our experience in robotic-assisted microsurgical vasovasostomy includes 11 cases performed from January 2006 to December 2009 in Aalst, Belgium, and 14 cases performed in Geneva. We have only had one scrotal hematoma that was treated conservatively, and all cases were patent 6 months postsurgically. Our mean total sperm count was 48.6 million/ml after RAVV (robotic-assisted vasovasostomy) and 22.8 million/ml after MAVV (microsurgical-assisted vasovasostomy). The RAVV cases performed in Belgium where compared to 17 cases who underwent MAVV at that same institution, and so far, the pregnancy rates for RAVV vs. MAVV where 36–41%, respectively.

Expert Commentary

The advantages of the da Vinci system include improved visibility with a three-dimensional view, a comfortable and ergonomically superior position during surgery, and increased degrees of freedom of motion of instruments. Another important advantage is improved stability during suturing as a result of the motion reduction feature in the robot.

For all these reasons, the da Vinci surgical robotic system was applied in the fields of general, vascular, cardiac, pediatric, gynecologic, and urologic surgery in only a few years time. Recently, a number of reports have shown that the robot allows surgeons to significantly simplify the more complex reconstructive steps of the laparoscopic procedures.

There has been an explosive increase in the number of urologic procedures being attempted using da Vinci assistance, and we believe that robotic technology represents the future of minimally invasive surgery, and applications for the robot will expand as more centers report their results. As reported by Fleming et al. in 2004 [12], robot-assisted vasectomy reversal is an attractive alternative to both traditional microsurgical vasovasostomy and vasoepididymostomy for several reasons:

- Greater ease and precision of suture placement are possible with elimination of the physiological tremor. While the magnification of the robotic camera (10–15×) is not as high as that of the operating microscope (up to 25×), enhanced control with motion reduction compensated for this difference. Data coming from the literature showed that the precision of suture placement resulted in a more rapid suture placement and watertight anastomosis. Moreover, the robotic technology minimizes the differences between placing sutures with the left or right hand, which further facilitates suturing. These concepts find clinical confirmation in the percentages of patent anastomosis and in the percentages of the presence of sperm granuloma.
- The training period is probably shorter than traditional microscopic techniques and benefits achieved with the surgical robot were acquired with a short learning curve. The learning curve period was considered nonexistent for experienced microsurgeons and probably fewer cases are necessary to become competent also for robotic surgeons without specific microsurgical training. As reported by Fleming et al., a surgeon without expertise in microsurgical technique should participate in a rat microsurgery course and should have extensive lab animal microsurgery experience, performing at least five to eight cases [12]. Vice versa, experienced microsurgeons need to learn to use the robot approximatively 30 min in a "dry lab," practicing suturing with a piece of Gore-tex vascular graft and 9-0 nylon suture [12]. However, to be critical, no studies yet demonstrated the exact impact of learning curve for vasectomy reversal procedures.
- More surgeons with expertise in robotic surgery will be able to provide high-quality surgical care for their patients. Although data coming from literature are promising, the robot-assisted reversal vasectomy is still in its feasibility phase and most of the available studies were performed in experimental models. Our personal experience confirms the good results reported in Literature in terms of patent anastomoses rate and mean sperm count and pregnancy rate in comparison with the current gold standard treatment represented by microsurgical-assisted techniques.

Potential drawbacks to use a robot for reversal of vasectomy are the suboptimal instrumentation available, not designed for microsurgery, and the lack of tactile feedback. Another critical issue could be represented by the costs of the procedure. However, it is clear that RAVV must be considered an optional procedure to perform only in centers performing routinely robotic surgery. We believe that the contribution of the robotic surgery to the microsurgical technique has the potential for a more profound impact.

Five-Year View

Robot-assisted vasovasostomy is a promising and attractive alternative to the microsurgical reversal vasectomy procedures. In the next years, this technique must reach a more acceptable level of evidence to justify its use. Only seven records are retrieved when we perform a Medline database search using as search terms *vasovasostomy* and *robot*. This shows this procedure is in its infancy or feasibility step. The wide diffusion in the United States and Europe of the da Vinci device allows us to believe that in the next years also the RAVV will be performed in an increasing number of cases and urologic centers. Further studies with adequate sample size and longer follow-up are needed to assess the clinical impact of the robotic surgery in the reversal vasectomy. Parallel, it would be useful that Intuitive Surgical Systems provides more dedicated instruments to further improve the clinical results.

Key Issues

- Conventional microscope-assisted reversal vasectomy (MAVV) (vasovasostomy and vasoepididymostomy) is a technically difficult procedure that is most successful in the hands of well-trained microsurgeons.
- Robot-assisted vasovasostomy (RAVV) is an attractive and promising procedure alternative to traditional microscopic techniques.
- Data of literature coming from few experimental studies performed in animal models and preliminary human clinical studies.
- In the last decade, the feasibility of robot-assisted vasovasostomy becomes more realistic.
- Preliminary results showed that RAVV is associated with low complications, high patent anastomosis rate, and good pregnancy rate in comparison with MAVV.
- The training period is probably shorter than traditional microscopic techniques, and benefits achieved with the surgical robot were acquired with a short learning curve.
- Further studies with adequate sample size and longer follow-up are needed to assess the clinical impact of the robotic surgery in the reversal vasectomy.

References

1. Sandlow JI, Nagler HM. Vasectomy and vasectomy reversal: important issues. Urol Clin North Am. 2009;36(3):xiii–xiv.
2. Jequier AM. Edward Martin (1859–1938). The founding father of modern clinical andrology. Int J Androl. 1991;14(1):1–10.
3. Belker AM, Thomas Jr AJ, Fuchs EF, et al. Results of 1469 microsurgical vasectomy reversals by the Vasovasostomy Study Group. J Urol. 1991;145:505.
4. Yarbro ES, Howards SS. Vasovasostomy. Urol Clin North Am. 1987;14(3):515–26.
5. Meacham RB, Niederberger CS. Use of a moderated international Internet information exchange in the study of male reproduction. Urology. 1996;48(1):3–6.
6. Thomas Jr AJ, Pontes JE, Buddhdev H, Pierce Jr JM. Vasovasostomy: evaluation of four surgical techniques. Fertil Steril. 1979;32(3): 324–8.
7. Fuse H, Kimura H, Katayama T. Modified one-layer microsurgical vasovasostomy in vasectomized patiens. Int Urol Nephrol. 1995; 27(4):451–6.
8. Fischer MA, Grantmyre JE. Comparison of modified one- and two-layer microsurgical vasovasostomy. BJU Int. 2000;85(9):1085–8.
9. Goldstein M, Li PS, Matthews GJ. Microsurgical vasovasostomy: the microdot technique of precision suture placement. J Urol. 1998;159(1):188–90.
10. Rothman I, Berger RE, Cummings P, et al. Randomized clinical trial of an absorbable stent for vasectomy reversal. J Urol. 1997; 157(5):1697–700.
11. Hendry WF. Vasectomy and vasectomy reversal. Br J Urol. 1994; 73(4):337–44.
12. Fleming C. Robot-assisted vasovasostomy. Urol Clin North Am. 2004;31:769–72.
13. Kuang W, Shin PR, Matin S, et al. Initial evaluation of robotic technology for microsurgical vasovasostomy. J Urol. 2004;171(1):300–3.
14. Kuang W, Shin PR, Oder M, et al. Robotic assisted vasovasostomy: a two-layer technique in an animal model. J Urol. 2005;65(4):811–4.
15. Schiff J, Li PS, Goldstein M. Robotic microsurgical vasovasostomy and vasoepididymostomy: a prospective randomized study in a rat model. J Urol. 2004;171:1720.
16. Parekattil SJ, Atalah HN, Cohen MS. Video technique for human robot-assisted microsurgical vasovasostomy. J Endourol. 2010; 24(4):511–4.
17. De Boccard G-A. Robotic vasectomy reversal, video, posted YouTube. 25 Oct 2009. http://www.youtube.com/watch?v=4zHRyFHC7nE.
18. De Boccard G-A, Mottrie A. Robotic surgery in male infertility, in Robotics in Genitourinary Surgery, vol. 7. New York: Springer; 2011. p. 617–23.

Further Reading

Boorjian S, Lipkin M, Goldstein M. The impact of obstructive interval and sperm granuloma on outcome of vasectomy reversal. J Urol. 2004;171:304–6.

Carbone Jr DJ, Shah A, Thomas Jr AJ, et al. Partial obstruction, not antisperm antibodies, causing infertility after vasovasostomy. J Urol. 1998;159:827–30.

Centers for Disease Control and Prevention. National summary and Fertility clinic reports. www.cdc.gov/ART. Accessed 1 June 2011.

Chan PT, Goldstein M. Superior outcomes of microsurgical vasectomy reversal in men with the same female partners. Fertil Steril. 2004;81:1371–4.

Hinz S, Rais-Bahrami S, Kempkensteffen C, et al. Fertility rates following vasectomy reversal: importance of age of the female partner. Urol Int. 2008;81:416–20.

Kim HH, Goldstein M. History of vasectomy reversal. Urol Clin North Am. 2009;36:359–73.

Kolettis PN. Is physical examination useful in predicting epididymal obstruction? Urology. 2001;57:1138–40.

Meinertz H, Linnet L, Fogh-Andersen P, et al. Antisperm antibodies and fertility after vasovasostomy: a follow-up study of 216 men. Fertil Steril. 1990;64:315–8.

Parekattil SJ, Kuang W, Kolettis PN, et al. Multi-institutional validation of vasectomy reversal predictor. J Urol. 2006;175:247–9.

Potts JM, Pasqualotto FF, Nelson D, Thomas AJ, Agarwal A. Patient characteristics associated with vasectomy reversal. J Urol. 1999; 161:1835–9.

13 Robotic-Assisted Varicocelectomy

Tung Shu and Run Wang

Abstract

Several theories have been proposed to explain the observed pathophysiology of varicoceles. Semen quality uniformly declines in animals with induced varicoceles, even when there is only a left varicocele. The reduction in scrotal temperature after varicocele ligation supports a causative role of increased temperature on the infertility produced by varicocele. It has been hypothesized that varicoceles cause hypoxia, which might play a role in altering spermatogenesis in the varicocele patients. A higher frequency of sperm cells with fragmented DNA has been reported in the ejaculate of subjects with varicocele, in comparison with fertile donors, a phenomenon that might be correlated with an increase in reactive oxygen species. Numerous studies have reported the significant benefits on semen parameters with surgical treatment of varicocele. Currently, there are several surgical approaches available for the treatment of varicocele, including the retroperitoneal approach, (high ligation via open, laparoscopic, retroperitoneoscopic, single-incision laparoscopic, or robotic-assisted), the inguinal approach (open), and the subinguinal approach (open microscopic). Of these approaches, the subinguinal microscopic approach offers the best outcomes, including shorter hospital stays, preservation of the testicular arteries and lymphatics, least number of postoperative complications, recurrence, and a higher number of pregnancies. The microscopic assistance, however, takes longer time to perform due to surgeons who are unaccustomed to use microinstruments, two-dimensional vision, and inability to see their own hands.

Keywords

Varicocele • Robot-assisted varicocelectomy • Pampiniform plexus • Retroperitoneal approach • Retroperitoneoscopic • Subinguinal microscopic approach • Subinguinal varicocelectomy

A varicocele is defined as a meshwork of distended blood vessels in the scrotum. It is usually left-sided resulting from the dilatation of the spermatic veins or pampiniform plexus. It is currently the most common surgically correctable finding identified in men being evaluated for infertility and is observed in 8–16.2% of the normal male population and in 21–39% of infertile men [1, 2].

Several theories have been proposed to explain the observed pathophysiology of varicoceles. Semen quality uniformly declines in animals with induced varicoceles, even

T. Shu, MD (✉)
Clinical Assistant Professor of Urology, Baylor College of Medicine, Center for Kidney Health at the Vanguard Urologic Institute, 6400 Fannin, Suite 2300, Houston, TX 77030, USA
e-mail: tung.shu@vanguardurology.com

R. Wang, MD, FACS
Department of Urology, University of Texas Medical School at Houston and MD Anderson Cancer Center, 6431 Fannin Street, MSB 6.018, Houston, TX 77030, USA

MD Anderson Cancer Center, Houston, TX, USA
e-mail: run.wang@uth.tmc.edu

S.J. Parekattil and A. Agarwal (eds.), *Male Infertility: Contemporary Clinical Approaches, Andrology, ART & Antioxidants*,
DOI 10.1007/978-1-4614-3335-4_13, © Springer Science+Business Media, LLC 2012

when there is only a left varicocele. The reduction in scrotal temperature after varicocele ligation supports a causative role of increased temperature on the infertility produced by varicocele. It has been hypothesized that varicoceles cause hypoxia, which might play a role in altering spermatogenesis in the varicocele patients [3]. A higher frequency of sperm cells with fragmented DNA has been reported in the ejaculate of subjects with varicocele, in comparison with fertile donors, a phenomenon that might be correlated with an increase in reactive oxygen species [4].

Numerous studies have reported the significant benefits on semen parameters with surgical treatment of varicocele [4–8]. Currently, there are several surgical approaches available for the treatment of varicocele [9–12], including the retroperitoneal approach, (high ligation via open, laparoscopic, retroperitoneoscopic, single-incision laparoscopic, or robotic-assisted), the inguinal approach (open), and the subinguinal approach (open microscopic). Of these approaches, the subinguinal microscopic approach offers the best outcomes, including shorter hospital stays, preservation of the testicular arteries and lymphatics, least number of postoperative complications, recurrence, and a higher number of pregnancies [2, 12]. The microscopic assistance, however, takes longer time to perform due to surgeons who are unaccustomed to use microinstruments, two-dimensional vision, and inability to see their own hands.

The da Vinci® Surgical System has helped surgeons overcome the limitations for both traditional open and conventional minimally invasive surgeries. With miniaturized wristed instruments, 3D camera, and computer technologies, the da Vinci® Surgical System filters and translates surgeon's hand movements seamlessly into precise micromovements of the da Vinci instruments. With added experience from using the da Vinci® Surgical System, dramatic improvements in tissue handling, time, and skill can be achieved [13, 14]. With preliminary experience, we have used the da Vinci® Surgical System to perform robotic-assisted subinguinal varicocelectomy in comparison to the standard microscopic approach for the treatment of varicocele [15].

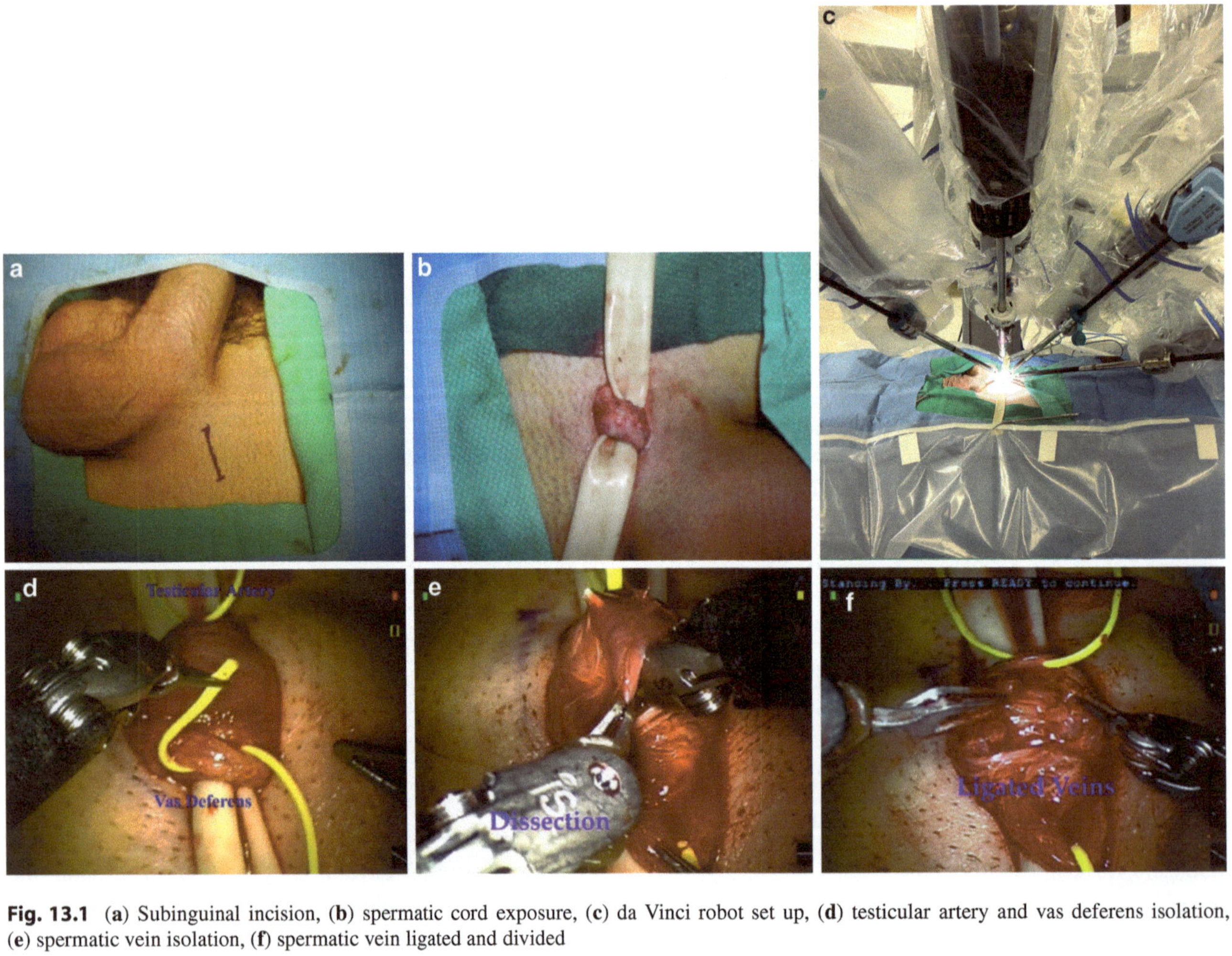

Fig. 13.1 (**a**) Subinguinal incision, (**b**) spermatic cord exposure, (**c**) da Vinci robot set up, (**d**) testicular artery and vas deferens isolation, (**e**) spermatic vein isolation, (**f**) spermatic vein ligated and divided

Materials and Methods

Eight patients aged 29.1 ± 12.5 years underwent microscopic subinguinal varicocelectomies: seven patients with left-sided repair and one patient with bilateral repair. Eight patients aged 22.0 ± 8.0 years underwent robot-assisted varicocelectomies: seven patients with left-sided repair and one patient with bilateral repair.

All varicocelectomies were performed through inguinal incisions (Fig. 13.1a). The spermatic cord was exposed and delivered out of the wound with a Penrose drain placed underneath the cord structures (Fig. 13.1b). At this time, the da Vinci® Surgical System or operating microscope was brought in and placed above the surgical field (Fig. 13.1c). The testicular artery and vas deferens with vassal artery and small vassal veins were identified and isolated (Fig. 13.1d). Other veins within the cord were isolated (Fig. 13.1e) and ligated with 5-0 Vicryl sutures and divided (Fig. 13.1f). At the completion of the varicocelectomy, only the testicular artery, lymphatics, and vas deferens with its vessels remained.

Results

The average operative time for the microscopic subinguinal varicocelectomy was 73.9 ± 12.2 min, whereas the robotic-assisted approach took 71.1 ± 21.1 min. Average follow-up time for the patients in the microscopic group was 34.3 ± 6.4 months, whereas the robotic-assisted group was 10.9 ± 7.1 months (Table 13.1).

In our experience, there was no difficulty in identifying and isolating vessels and the vas deferens with the robotic-assisted approach. A short learning curve for tying with 5-0 sutures was required because of the lack of tactile sensation when using the da Vinci® Surgical System. Patients in both groups were able to resume daily activities on the day of surgery and full activities within 2 weeks. There was no intraoperative or postoperative complication. No recurrence of varicocele was observed in either group of patients.

Table 13.1 Data of microscopic and robot-assisted varicocelectomies

	Age (years)	Average operative time (min)	Follow-up time (months)
Microscopic technique 8 patients, 9 varicocelectomies	29.1 ± 12.5	73.9 ± 12.2	34.3 ± 6.4
Robot-assisted technique 8 patients, 9 varicocelectomies	22.0 ± 8.0	71.1 ± 21.1	10.9 ± 7.1

Expert Commentary

In 2006, we perform the first robotic-assisted subinguinal varicocelectomy using the da Vinci® Surgical System [15]. From our continuing experience, we believe that the robotic-assisted varicocelectomy can be safely and effectively performed when compared to traditional microscopic approach. There was not a significant difference with regards to operative time; however, with increased experience with the da Vinci® Surgical System, the time should decrease. When compared to the microscopic approach, the tremor is completely eliminated with the robot. More importantly, the advantage of decreased intraoperative and postoperative complications experienced with microsurgical approach is maintained with the robotic approach.

We are currently studying the cost-effectiveness and efficacy with regard to the improvement of semen quality and pregnancy for patients with infertility with our described robotic-assisted subinguinal varicocelectomy.

Five-Year View

Given the advantages of decreased hand tremor, scaling of motion, and enhanced magnification, robotic assistance for microsurgical procedures appears to be a natural advantage for microsurgeons. This is yet to be proven, but as the technology improves and more surgeons adapt these techniques, this evidence is likely to be forthcoming.

Key Issues

- The subinguinal microscopic approach offers the best outcomes, including shorter hospital stays, preservation of the testicular arteries and lymphatics, least number of postoperative complications, recurrence, and a higher number of pregnancies [2, 12].
- Robotic assistance provides the microsurgeon with advantages in terms of decreased hand tremor, scaling of motion, and enhanced magnification.
- The average operative time for the microscopic subinguinal varicocelectomy was 73.9 ± 12.2 min, whereas the robotic-assisted approach took 71.1 ± 21.1 min.
- When compared to the microscopic approach, the tremor is completely eliminated with the robot. More importantly, the advantage of decreased intraoperative and postoperative complications experienced with microsurgical approach is maintained with the robotic approach.

References

1. Jarow JP. Effects of varicocele on male fertility. Hum Reprod Update. 2001;7:59–64.
2. Watanabe M, Nagai A, Kusumi N, Tusboi H, Nasu Y, Kumon H. Minimal invasiveness and effectivity of subinguinal microscopic varicocelectomy: a comparative study with retroperitoneal high and laparoscopic approaches. Int J Urol. 2005;12:892–8.
3. Pasqualotto FF, Sobreiro BP, Hallak J, Pasqualotto EB, Lucon AM. Induction of spermatogenesis in azoospermic men after varicocelectomy repair: an update. Fertil Steril. 2006;85:635–9.
4. Enciso M, Muriel L, Fernandez J, Goyanes V, Segrelles E, Marcos M, et al. Infertile men with varicocele show a high relative proportion of sperm cells with intense nuclear damage level, evidenced by the sperm chromatin dispersion test. J Androl. 2006;27:106–11.
5. Daitch J, Bedaiwy M, Pasqualatto E, Hendin B, Hallak J, Falcone T, et al. Varicocelectomy improves intrauterine insemination success rates in men with varicocele. J Urol. 2001;165:1510–3.
6. Lee JS, Park HJ, Seo JT. What is the indication of varicocelectomy in men with nonobstructive azoospermia? Urology. 2007;69:352–5.
7. Zucchi A, Mearini L, Mearini E, Fioretti F, Bini V, Poreno M. Varicocele and fertility: relationship between testicular volume and seminal parameters before and after treatment. J Androl. 2006;27:548–51.
8. Gat Y, Zukerman Z, Chakraborty J, Gornish M. Varicocele, hypoxia and male infertility. Fluid mechanics analysis of the impaired testicular venous drainage system. Hum Reprod. 2005;20:2614–9.
9. Palomo A. Radical cure of varicocele by a new technique: preliminary report. J Urol. 1949;152:1127–32.
10. Belloli G, D'Agostino S, Zen F, Loverno E. Fertility rates after successful correction of varicocele in adolescence and adulthood. Eur J Pediatr Surg. 1995;5:216–8.
11. Kaouk JH, Palmer JS. Single-port laparoscopic surgery: initial experience in children for varicocelectomy. BJU Int. 2008;102:97–9.
12. Al-Kandari AM, Shabaan H, Ibrahim HM, Elshebiny YH, Shokeir AA. Comparison of outcomes of different varicocelectomy techniques: open inguinal, laparoscopic, and subinguinal micro-scopic varicocelectomy: a randomized clinical trial. Urology. 2007;69:417–20.
13. Nguyen MM, Das S. The evolution or robotic urologic surgery. Urol Clin North Am. 2004;31:653–8.
14. Corcione F, Esposito C, Cuccurullo D, Settembre A, Miranda N, Amato F, et al. Advantages and limits of robot-assisted laparoscopic surgery: preliminary experience. Surg Endosc. 2005;19:117–9.
15. Shu T, Taghechian S, Wang R. Initial experience with robot-assisted varicocelectomy. Asian J Androl [serial online]. 2008;10(1):146–8.

Epidemiological Considerations in Male Infertility

14

Mark A. Faasse and Craig S. Niederberger

Abstract

The diagnosis of male infertility indicates impairment in male reproductive potential. Unfortunately, there is no gold standard test for assessing male reproductive potential. The diagnostic accuracy of semen analysis is compromised by substantial overlap between the distributions of semen characteristics in empirically fertile and infertile men. Novel assays, including seminal reactive oxygen species levels, may prove to be of greater clinical utility for identifying men with below-average reproductive potential. Efforts to identify and treat modifiable risk factors should ideally be targeted toward these individuals. Although economic analyses have demonstrated that pathology-directed treatment for male infertility is often more cost-effective than immediate use of assisted reproductive technology, there is a declining trend in surgical treatment of male infertility.

Keywords

Male infertility • Incidence and prevalence of infertility • Infertility in developing world • Epidemiology of infertility • Reactive oxygen species • Novel diagnostic assays • Semen studies

This chapter primarily aims to discuss the epidemiologic relationship between infertility and male reproductive potential. It will focus in detail on the accuracy and diagnostic value of semen studies. Controversial reports of declining sperm counts during the twentieth century will also be addressed, followed by a review of trends in health-care resource utilization and cost analysis models pertinent to the management of male infertility.

M.A. Faasse, MD (✉)
Department of Urology, University of Illinois at Chicago, 840 South Wood Street, M/C 955, Chicago, IL 60616, USA
e-mail: mfaasse@sbcglobal.net

C.S. Niederberger, MD, FACS
Department of Urology, University of Illinois at Chicago College of Medicine and Department of Bioengineering, University of Illinois at Chicago College of Engineering, Chicago, IL 60616, USA
e-mail: craignied@gmail.com

The subject of this textbook, antioxidant therapy for male infertility, has an epidemiologic basis in the understanding that reactive oxygen species (ROS) contribute to sperm damage and are present in higher levels in the semen of infertile men [1–4]. It may be surprising for some to encounter an epidemiologic approach to male infertility in the modern era, given the increasing availability and use of assisted reproductive technology (ART) [5]. Indeed, the relative ease of surgical sperm retrieval in cases of azoospermia and severe oligozoospermia has seemingly rendered comprehensive evaluation and treatment of infertile men less relevant. Even though such an evaluation may identify one or more modifiable risk factors for infertility (as well as potentially serious underlying or coexisting illnesses and genetic abnormalities), the outcome of pathology-directed treatment is always uncertain and may not be realized until several additional months or years have elapsed.

However, risk and uncertainty are also attendant to the use of ART. Recent cost-effectiveness studies have demonstrated

S.J. Parekattil and A. Agarwal (eds.), *Male Infertility: Contemporary Clinical Approaches, Andrology, ART & Antioxidants*, DOI 10.1007/978-1-4614-3335-4_14, © Springer Science+Business Media, LLC 2012

that a straight-to-ART approach is less efficient than pathology-directed treatment in many situations involving male infertility. Therefore, renewed emphasis is being placed on the paradigm of intervention aimed at correction of modifiable male risk factors [6].

Epidemiology of Infertility

It is important to distinguish reproductive potential, or capacity, from actual reproductive performance, or outcome [7, 8]. Infertility is ultimately defined by a reproductive *outcome* (i.e., childlessness). By contrast, "male infertility" is a diagnosis of relative impairment in male reproductive *potential.* Before discussing male infertility in more detail, we will review the epidemiology of infertility in general.

Incidence and Prevalence of Infertility in Developed Countries

Childlessness within long-standing marriage has been observed to occur in roughly 10% of couples in developed countries [9]. This figure represents a measure of reproductive outcomes, but it does not exclude couples who remain voluntarily childless or do not have regular intercourse during the fertile phase of the female's menstrual cycle. In such couples, reproductive potential may never be fully tested [5].

In order to capture the concept of impairment in reproductive potential within the definition of infertility, the World Health Organization (WHO) has specified infertility as "no conception after at least 12 months of unprotected intercourse" [10]. By this criterion, the lifetime *incidence* of unwanted infertility is approximately 15% among couples in Western countries [11]. Two thirds of these cases are primary, i.e., in couples who have never previously conceived, while one third are secondary [9].

It is not uncommon to see infertility defined by other durations of time, such as 2 or 5 years. Also, while the outcome of interest in the WHO's definition is conception, others prefer to regard infertility as the absence of live birth [8]. This distinction is important, since 10–25% of recognized pregnancies end in miscarriage [12].

The annual *prevalence* of infertility among married US women aged 15–44 years was 7.4% in 2002, down from 11.2% in 1965, and 8.5% in 1982 [13]. This trend may reflect improvements in reproductive potential, perhaps through better awareness of ideal timing for intercourse, which may be attributable to at-home ovulation tests. However, other factors could also be involved, such as more couples utilizing infertility treatment prior to reaching the 12-month threshold necessary for inclusion in the rate's numerator. Declining marriage rates among lower socioeconomic classes may also be disproportionately removing women at greater risk of infertility from the denominator [13]. A popular misconception is that infertility is synonymous with, or virtually approximates, sterility [7]. In fact, only 3–5% of couples are sterile [9, 14].

Infertility in the Developing World

According to the WHO, more than 25% of married women in developing countries were experiencing primary or secondary infertility of at least 5 years duration in 2002 [15]. In contrast to Western societies, the vast majority of infertility in developing countries is secondary (Table 14.1). Sexually transmitted infections and postpartum complications are the key factors contributing to this disparity [16]. The WHO found no evidence of an association between the prevalence of HIV infection and infertility.

The rate of childlessness in developing countries among sexually experienced women who are beyond childbearing age is approximately 3% [15]. This figure is comparable to the estimated prevalence of sterility in Western societies.

High cost represents an important barrier to the accessibility of most infertility treatment, and specifically ART, in the developing world [16]. Therefore, further identification of avoidable gonadotoxins, as well as clarification of the role of less-expensive therapies, would be especially welcome [17].

Table 14.1 Prevalence of infertility of 5 years duration in various regions of the developing world and the prevalence of infertility of 12 months duration in the USA

	Primary or secondary infertility (%)	Primary infertility (%)	Secondary infertility (%)
Sub-Saharan Africa	30.0	2.6	27.5
Near East/North Africa	20.4	2.8	17.6
South Central Asia	28.3	2.6	25.8
Southeast Asia	23.5	1.9	21.6
Latin America/Caribbean	16.0	2.8	13.2
USA	7.4	3.0	4.4

Adapted from [15], with permission. US data were reported in [13]

Table 14.2 Cumulative spontaneous pregnancy rates of couples in five hypothetical categories of reproductive potential

	MFR (%)	Cumulative pregnancy rate after: 6 months (%)	12 months (%)	24 months (%)	60 months (%)
Highest fecundability	60	100	–	–	
Average fecundability	20	74	93	100	
Below-average fecundability	5	26	46	71	95
Severely impaired fecundability	1	6	11	21	45
Sterile	0	0	0	0	0

MFR monthly fecundity rate; cumulative pregnancy rate $= 1-(1-\text{MFR})^{\#\text{ of months}}$
From [20], with permission of Elsevier

Table 14.3 Hypothetical model of the proportion of couples with varying degrees of reproductive potential in the residual population, dependent on the duration of infertility

	MFR (%)	Composition of residual nonpregnant couples after: 0 months (%)	6 months (%)	12 months (%)	24 months (%)	60 months (%)
Highest fecundability	60	3	–	–	–	–
Average fecundability	20	79	58	30	–	–
Below-average fecundability	5	10	21	30	30	8
Severely impaired fecundability	1	5	13	24	40	44
Sterile	0	3	8	16	30	48

MFR monthly fecundity rate
From [20], with permission of Elsevier

Reproduction—a Matter of Chance: The Natural History of Infertility

Reproduction has been described as "a matter of chance depending on the subtle balance between success or failure of complex, mostly poorly understood, sequential processes that may lead to a pregnancy and eventually to the birth of a healthy child" [14]. Failure of a couple to reproduce is a unique medical problem in that it occurs *between* rather than *within* individuals [8].

Individuals' reproductive potential is a continuous, as opposed to dichotomous, variable. It reflects the influence of many factors, including age. Reproductive potential declines in members of both sexes over 30 years old, but female age has the most profound effect on the likelihood of conception [18, 19].

Since a couple's reproductive capacity is the composite of its individual members' reproductive potentials, it, too, is a continuous variable. Impairment of a male's reproductive potential may be compensated for—or compounded—by that of his female partner, and vice versa. This concept is illustrated by the fact that artificial insemination with donor semen is more often successful in partners of azoospermic men than in partners of men with oligozoospermia [20].

Couples' monthly, or cycle-wise, likelihood of conception falls along a spectrum of probability. This is referred to as fecundability, or monthly fecundity rate (MFR). The average MFR for human couples having regular, unprotected intercourse is approximately 20%, and the overall distribution of human MFRs is believed to range from 0 to 60% [21, 22]. The variable likelihood of pregnancy at 6, 12, 24, and 60 months has been calculated for couples with different MFRs (Table 14.2). Based on these values, a hypothetical model has been constructed of the proportion of couples with varying degrees of reproductive potential (MFR) among residual nonpregnant couples after specified durations of infertility (Table 14.3) [23].

Table 14.2 demonstrates that couples with average fecundability have a better than 90% chance of conceiving within 12 months. However, 30% of couples who do not conceive within 12 months are still of average reproductive potential (see Table 14.3). Population-based studies have found that couples who are infertile after 12 months retain a roughly 50% likelihood of achieving unassisted pregnancy by 24 months [24]. Thereafter, the odds of conception decline precipitously, as illustrated in Fig. 14.1. The proportion of couples that are sterile increases with the duration of infertility.

An obvious but nevertheless critically important factor in determining reproductive potential is the timing of sexual intercourse relative to ovulation (Fig. 14.2) [25]. In a cohort of 340 German couples who received natural family planning education intended to improve their timing of intercourse, the monthly probability of achieving pregnancy averaged 38%, markedly higher than the average human MFR of 20% [26].

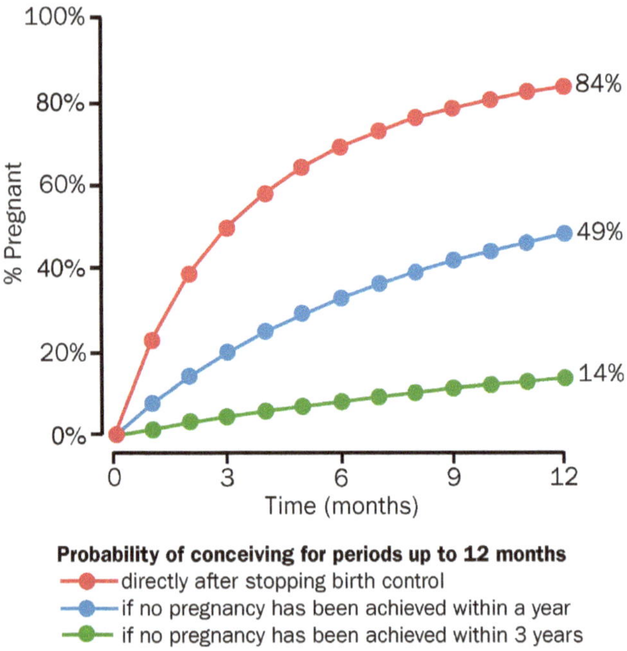

Fig. 14.1 Cumulative probability of conception in couples having unprotected intercourse (from [14], with permission)

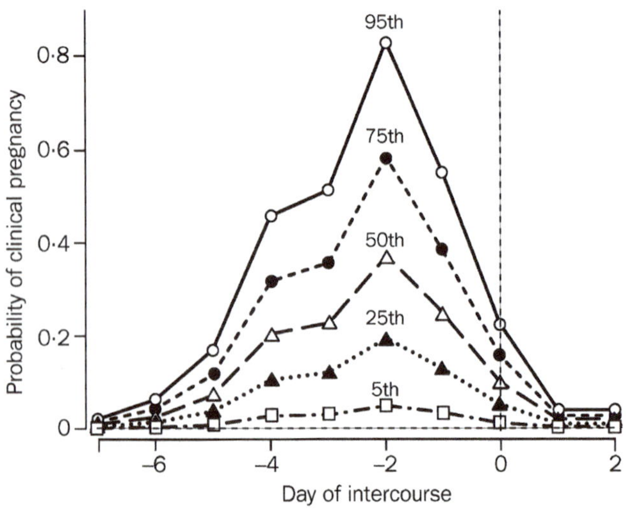

Fig. 14.2 Probability of clinical pregnancy after intercourse on a particular day relative to ovulation (day 0) for couples at specified percentiles of the population distribution of reproductive potential (from [20], with permission of Elsevier; adapted from [25], with permission of Oxford University Press)

Diagnostic Accuracy and Utility of Semen Studies

Male infertility is diagnosed in approximately 50% of couples presenting for evaluation of infertility [11]. Basic semen analysis remains the most widely utilized laboratory study for this purpose. However, the limitations of semen analysis should be clearly understood.

The parameters usually assessed by semen analysis include ejaculate volume, sperm concentration (density), sperm motility, and sperm morphology. From the ejaculate volume, concentration, and percentage of motile sperm, the total sperm count and total motile count are calculated. Semen pH, viscosity, white blood cell concentration, and the degree of sperm agglutination may also be reported [27].

More specialized studies that are undertaken on a case-by-case basis include evaluation of sperm viability, antisperm antibodies, and functional assays, such as assessment of sperm–cervical mucus interaction, capacitation, and sperm penetration of a zona-free hamster oocyte [28]. Relatively recent developments have included the introduction of tests for seminal ROS levels and sperm DNA fragmentation [29, 30].

Relationship Between Semen Parameters and Male Infertility

Semen analysis definitively confirms male factor infertility in men who are found to have semen characteristics at the negative extremes, i.e., azoospermia, nonmotile sperm, or the severest cases of teratozoospermia. However, only a small fraction of men who present for evaluation of infertility have such findings [11]. The present section explores the relationship between male infertility and the entire spectrum of semen quality.

Since 1980, the WHO has published reference values for human semen parameters. These values have been changed periodically (Table 14.4) [31]. As of 2010, they represent the fifth percentile in the distribution of semen parameters from a population of men with proven fertility [32]. They provide no information regarding the distribution of semen parameters in men who are infertile.

Although the WHO criteria are commonly used as thresholds for designation of male infertility, the diagnostic picture in a clinical setting is considerably more complex. This is because of substantial overlap between the distributions of semen characteristics in empirically fertile men and those with infertility whose female partners have had a normal fertility evaluation (Fig. 14.3).

Basic performance measures of a test such as semen analysis include calculation of its sensitivity and specificity. These concepts are illustrated in Tables 14.5 and 14.6. If a sperm concentration below 15 million/mL is the criterion for a "positive" result (i.e., diagnosis of male infertility), then men who are actually fertile will be correctly classified 95% of the time. In other words, the threshold of 15 million/mL has a specificity of 95%. However, 85% of infertile men also have sperm concentrations above 15 million/mL, so the sensitivity of the test at this threshold is only 15% [33]. Many men with impaired reproductive potential will not be recognized as such.

Table 14.4 WHO reference values for analysis of semen parameters

	1992	1999	2010
Ejaculate volume (mL)	≥2.0	≥2.0	≥1.5
Sperm concentration (10^6/mL)	≥20	≥20	≥15
Total sperm count (10^6/ejaculate)	≥40	≥40	≥39
Sperm motility (% motile)	≥50 (*a*+b)[a]	≥50 (*a*+*b*)	≥40 (*a*+*b*+*c*)
Sperm morphology (% normal)	≥30	≥14[b]	≥4
Sperm viability/vitality (% live)	≥75	≥75	≥58
White blood cells (10^6/mL)	<1.0	<1.0	<1.0

From [31], with permission
[a]Motility is graded as follows: *a*=rapid progressive motility (>25 μm/s); *b*=slow/sluggish progressive motility (5–25 μm/s); *c*=nonprogressive motility; *d*=immotility
[b]Kruger (Tygerberg) strict criteria were adopted by the WHO in 1999

If we assume that 50% of couples presenting for an evaluation of infertility have a contributing male factor, then the predictive value of a positive result (i.e., its likelihood of being correct) would be 75%. The predictive value of a negative, or normal, result would be only 53% (see Table 14.6).

Of course, diagnostic thresholds can be changed. If they are increased, sensitivity improves, but there is a reciprocal decline in specificity, and vice versa. One way to assess the diagnostic accuracy of a test across all thresholds is by a receiver operating characteristic (ROC) curve [28, 34]. ROC curves are constructed by plotting the probability of detecting true positives (sensitivity) against that of detecting false positives (1 − specificity) at each threshold.

The total area under the ROC curve (AUC) for a particular test represents its overall discriminatory capability. A perfect test has an AUC of 1.0, while a test is useless if the likelihood of a true positive matches that of a false positive at every threshold, resulting in the no-discrimination line from (0, 0) to (1, 1) and an AUC of 0.5. The amount by which a test's ROC curve diverges from this line—and by extension, to which its AUC exceeds 0.5—is the degree to which it is diagnostically helpful. An AUC that exceeds 0.9 is considered excellent, while an AUC of less than 0.7 is poor.

Figure 14.4 represents an example of an ROC curve generated from MacLeod's data on sperm concentration in fertile and infertile men [35]. Its AUC is only 0.59, indicating that the overall accuracy of sperm concentration for diagnosis of male infertility just narrowly exceeds that of random chance. In Guzick and colleagues' series, which was depicted in Fig. 14.3, the AUC for sperm density, motility, and morphology were 0.60, 0.59, and 0.66, respectively [36].

The ROC curve alone does not provide information regarding the likelihood that a specific patient's positive or negative test result is correct. This probability is dependent both on test performance and the prevalence of disease in the

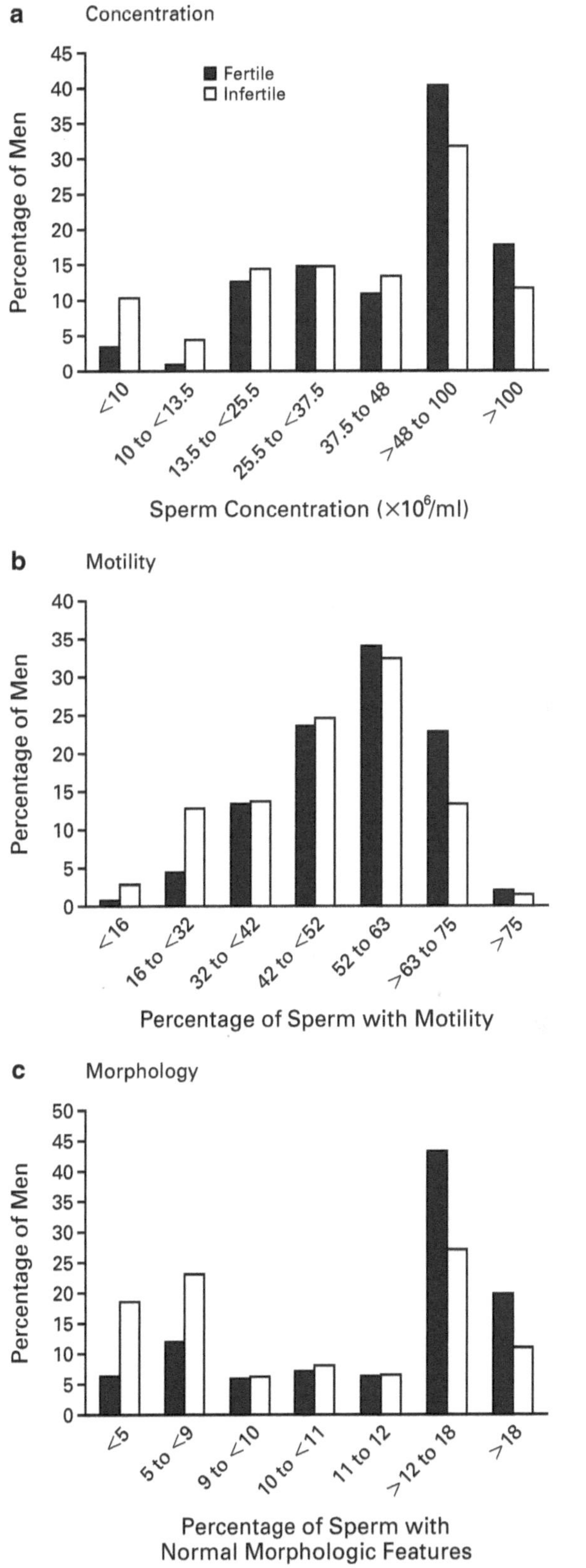

Fig. 14.3 Frequency histograms depicting the percentage of men from fertile (*shaded bars*) and infertile (*unshaded bars*) couples with sperm concentration (**a**), sperm motility (**b**), and sperm morphology (**c**) within specified ranges. Data were obtained from 696 fertile men and 765 men from infertile couples, whose female partners had an unrevealing fertility evaluation (from [33], Copyright 2001, Massachusetts Medical Society. All rights reserved)

Table 14.5 2×2 Table depicting actual fertility status vs. test results using the 2010 WHO reference value for sperm concentration as a threshold for diagnosis of male infertility, in a hypothetical population of 200 men presenting for evaluation of infertility

		Actual fertility status	
		Infertile	Fertile
Test results	Infertile	15 (TP)	5 (FP)
	Fertile	85 (FN)	95 (TN)

TP true positives, *FP* false positives, *FN* false negatives, *TN* true negatives

Table 14.6 Accuracy metrics based on Table 14.5, using the 2010 WHO reference value for sperm concentration (15 million/mL) as a threshold for diagnosis of male infertility

Accuracy metric	Formula for calculation	Result (%)
Sensitivity	TP/(TP+FN)	15
Specificity	TN/(TN+FP)	95
Positive predictive value	TP/(TP+FP)	75
Negative predictive value	TN/(TN+FN)	53
Classification accuracy	(TP+TN)/N	55

TP true positives, *FP* false positives, *FN* false negatives, *TN* true negatives; *N*=total number

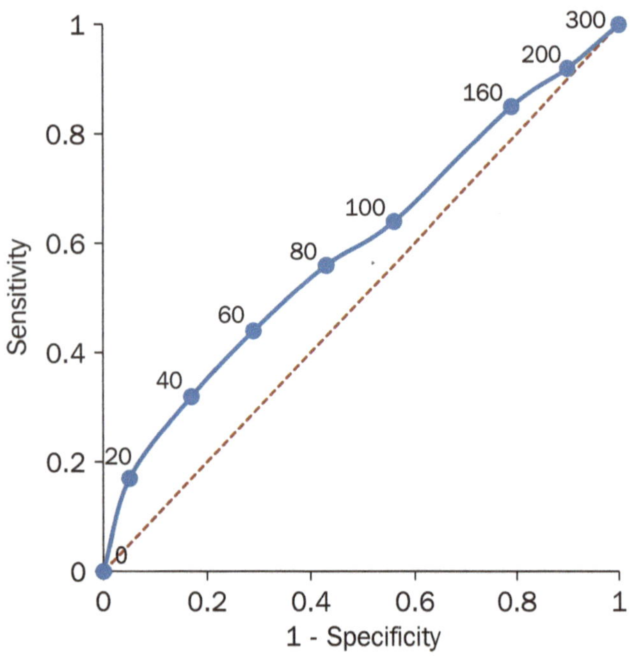

Fig. 14.4 Receiver operating characteristic (ROC) curve analysis of sperm concentration for diagnosis of male infertility at specified thresholds (data from [35]). The no-discrimination line is shown in *red* (from [36], with permission of Elsevier)

relevant population. As previously illustrated, a diagnosis of male infertility based on sperm concentration below 15 million/mL may be correct 75% of the time in the population of males from infertile couples. False-positive results for infertility are relatively rare in this population. By comparison, if semen analysis were performed on men in the general population (e.g., to assess sperm donors with no prior reproductive history), its positive predictive value would be considerably less, on account of a much lower prevalence of male infertility.

Four studies smaller than that of Guzick and colleagues have reported higher AUCs for sperm density, motility, and morphology, as described in Table 14.7 [37–40]. In general, motility and morphology demonstrated greater discriminatory capability than sperm concentration. Although the authors of these studies reported "optimal thresholds" for discrimination of fertile and infertile men, some of these thresholds have been criticized for having an unacceptably low positive predictive value in the setting of an infertile population [41].

Guzick and colleagues took a different approach to selecting diagnostic thresholds, using classification and regression tree (CART) analysis to determine *two* thresholds for each parameter that define the upper and lower boundaries of an indeterminate range lying between the fertile and infertile ranges (see Table 14.7). Unfortunately, a large number of men presenting for infertility evaluation fall into the indeterminate range, leaving unanswered the question of whether they warrant intervention for modifiable risk factors. The odds of male infertility multiply if more than one semen parameter is within the infertile range [33].

Do Semen Parameters Prospectively Predict Fertility and Assisted Reproductive Technique Outcomes?

Only a handful of studies have attempted to prospectively identify variables associated with male reproductive potential. One such project included 200 couples, some of whom had had prior pregnancies [42]. After discontinuing contraception, the couples were followed for up to 12 months. Seventy-eight percent conceived during the study period, and both sperm motility and morphology were significantly associated with fertility. The difference in sperm concentration between fertile and infertile couples remained statistically insignificant.

Bonde and colleagues investigated 430 Danish couples between 20 and 35 years old who had never previously been or tried to become pregnant, following them for up to six menstrual cycles after discontinuation of contraception [43]. Sixty percent of couples became pregnant. The probability of conception increased up to a threshold sperm concentration of 40 million/mL, but there was no additional likelihood of pregnancy at higher sperm densities (Fig. 14.5). This finding has led some reproductive specialists to argue that the appropriate threshold of sperm concentration for diagnosis of male

Table 14.7 Reported areas under the ROC curve for various seminal characteristics

	Population	AUC density	AUC motility	AUC morphology (strict)
MacLeod [35][a]	1,000 fertile; 800 infertile[b] (ages NR)	0.59	NR	NR
Guzick et al. [33][a]	696 fertile (avg age, 33.5±5.0); 765 infertile (avg age, 34.7±4.9)	0.60 (13.5, 48)	0.59 (32, 63)	0.66 (9, 11)
Ombelet et al. [37]	144 fertile; 143 infertile (ages NR)	0.69 (34)	0.61 (45)	0.78 (10)
Gunalp et al. [38]	61 fertile (avg age, 29.9); 62 infertile (avg age, 31.3)	0.56 (34)	0.71 (42)	0.70 (12)
Menkveld et al. [39]	107 fertile (avg age, 33.8±4.3); 103 infertile[b] (avg age, 33.7±3.9)	NR	0.79 (45)	0.78 (4)
Jedrzejczak et al. [40]	113 fertile (avg age, 31±4.7); 109 infertile (avg age, 32.2±4.1)	0.80 (45–50)	0.91 (24)	0.82 (11)

AUC area under the ROC curve, *NR* not reported; ages of study populations are reported as mean±SD; values in parentheses are the "optimal thresholds" identified by the respective studies for discrimination of fertile from infertile men—see text for additional details

[a]For data from these studies, the ROC curve was plotted and AUC was reported by Niederberger [36]

[b]These studies did not specifically report female partners as having had a negative fertility evaluation

infertility should be 40 instead of 15 million/mL [44]. A change of this sort, however, would also increase the number of false-positive diagnoses, prompting unnecessary evaluation and treatment.

Leushuis and colleagues have published an incisive review of prediction models in reproductive medicine, including several that use one or more semen characteristics to predict conception by infertile couples [45]. One such model, which has been externally validated in a population excluding men with total motile sperm counts of less than three million, takes account of sperm motility as well as characteristics of the female partner and the duration of the couple's infertility; it is available online at www.freya.nl/probability.php [46]. Another model utilizes inputs of sperm concentration, motility, morphology, and hypoosmotic swelling to assess the likelihood of pregnancy, with a reported accuracy greater than 85% [40]. Of note, however, the AUCs for each variable in this study substantially exceeded those published in the other reports described in Table 14.7, and the predictive model has yet to be validated.

Sperm quality also affects ART outcomes, at least to some degree. Several investigations have demonstrated a positive correlation between semen characteristics, including ROS levels, and in vitro fertilization/intracytoplasmic sperm injection (IVF/ICSI) success rates, yet there was no apparent association with clinical pregnancy rates [47, 48]. Studies of intrauterine insemination (IUI), by contrast, have shown a correlation between successful outcomes and sperm concentration, motility, and morphology [41, 49–51].

Novel Assays for Diagnosis of Male Infertility

Andrologists often lament the lack of more accurate studies for the diagnosis of male infertility. Several tests are currently in various stages of development, including genomic,

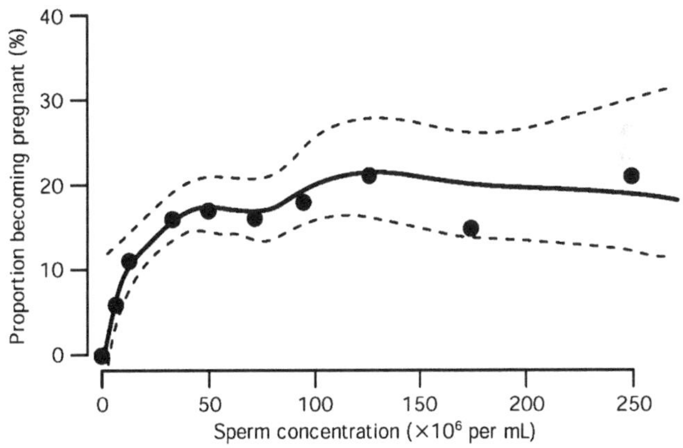

Fig. 14.5 Probability of pregnancy per menstrual cycle relative to sperm concentration (from [43], with permission of Elsevier)

proteomic, glycomic, lipidomic, and metabolomic analyses [52, 53]. Testing for seminal ROS levels was found to have an AUC of 0.82 in a study of 105 patients [54]. Further evaluation in a larger cohort of patients is necessary.

Are Sperm Counts Declining?

One of the most controversial issues in reproductive medicine during the past 20 years has been a purported decline in semen quality during the twentieth century. In 1992, a widely publicized meta-analysis of 61 studies appeared to demonstrate a worldwide decrease in average sperm concentration from 113 million/mL in 1940 to 66 million/mL in 1990 [55]. This report was echoed by additional publications, raising the question of whether exposure to environmental toxins, such as estrogenic compounds like diethylstilboestrol (DES), was adversely affecting testicular function [56, 57].

Important methodological shortcomings have since been identified in Carlsen and colleagues' analysis, as well as in the quality of many studies that were included in their review [58, 59]. Geographical differences were found to be the source of much of the variance in sperm density [60]. All studies included in the review from before 1970 were performed in the USA; however, since US studies generally reported higher sperm concentrations than those conducted elsewhere, the review was biased toward an apparent decline in sperm concentration by inclusion of international studies post-1970.

In subsequent investigations, Fisch and colleagues found no evidence of a decline in sperm density in the USA [61, 62]. They did identify substantial differences between the average sperm counts of men from different states. A review of worldwide studies also did not reveal a global decline in semen quality, although more limited, locoregional trends could not be excluded [58].

Health-Care Resource Utilization for Male Infertility

Medical intervention for male infertility may take the form of outpatient care, surgical procedures, and ART. The Urologic Diseases in America (UDA) Project, which published its first report in 2007, has facilitated a better understanding of the scope of health-care resource utilization for male infertility in the USA.

Office Visits and Ambulatory Surgery Cases

According to data from the National Ambulatory Medical Care Survey (NAMCS), which are summarized in the UDA Project's report, the average number of physician office visits for male infertility in the USA exceeded 150,000 annually between 1992 and 2000, with little variability [63]. However, there was a 24% decline in ambulatory surgery cases for male infertility from 1994 to 2002. Reasons for this trend are unclear but may include preferential use of ART.

Age-wise, the greatest utilization of health-care services for male infertility was among men 25–34 years old. Varicocele was the most commonly identified diagnostic code, accounting for 53% of office visits and 67% of ambulatory surgeries.

Other data from the NAMCS demonstrate substantial regional variation in resource utilization for male infertility. Men living in the Northeast USA had a rate of ambulatory surgery visits associated with a diagnosis of infertility of 104 per 100,000, while those in the Midwest, South, and West had rates of just 72, 50, and 29 per 100,000, respectively. The explanation for this variability is probably multifactorial, reflecting a combination of patient demand and availability of services.

Assisted Reproductive Technology

In 2007, 142,435 ART cycles were performed, an increase of 60% since 1998 [64]; and 43,412 live births resulted from these cycles, representing a cumulative success rate of 30.5%. A total of 18.5% of couples who had one or more ART cycles in 2007 carried a diagnosis of male factor infertility. Their overall likelihood of live birth per ART cycle was 35.8%, with a strongly inverse correlation of success to maternal age.

Sixty-three percent of ART procedures in 2007 involved ICSI, compared to 54% in 2000 [65]. Surprisingly, although ICSI was originally developed specifically to overcome severe oligozoospermia or azoospermia, a diagnosis of male infertility was recorded for only 48% of couples undergoing ICSI.

Cost of Treatment for Male Infertility

The overall economic burden of health care for male infertility is difficult to estimate with precision. While expenditures for office and ambulatory surgery visits were reported by the NAMCS to be $17 million in 2000, this figure does not account for IVF/ICSI or out-of-pocket expenditures [63]. If the assumed cost per ART cycle is $15,000, then expenditures on ART alone for male infertility exceeded $650 million in 2007, given the percentage of ART procedures involving ICSI that were associated with a diagnosis of male factor infertility.

Cost Analysis Models for Management of Male Infertility

When a male risk factor for infertility is identified, the couple is often faced with the choice of using ART or having pathology-directed treatment. Evaluating the economic efficiency of these alternatives is the domain of cost analysis.

Cost analysis is only meaningful with respect to treatments that have previously been demonstrated to be effective. Our intention in this chapter is not to review the studies that have established the effectiveness of the interventions discussed here—or, in some cases, the controversies surrounding them. Rather, our focus is limited to a brief introduction to cost analysis studies as they pertain to male infertility.

Two types of cost analysis are utilized in reproductive medicine: cost-minimization analysis and cost-effectiveness analysis. The first, cost-minimization analysis, is also known as cost identification. It involves assessment (and comparison) of the costs associated with particular treatments. Direct and/or indirect costs may be taken into account—direct costs being health-care expenditures and indirect costs being "downstream" burdens such as transportation expenses, lost wages, etc. In well-conducted economic analyses, future costs should be appropriately discounted to present values by a factor of 3–5% per year.

When the outcomes of alternative interventions are not equivalent, cost-effectiveness analysis is useful to compare them, as it involves not only identifying the costs that accrue but also expressing them relative to the probability of a particular result (e.g., dollars per pregnancy or live birth). Decision analysis, such as Markov modeling, is the most common technique employed for this purpose in the field of male infertility. Details of relevant methodology are covered in *Methods for the Economic Evaluation of Health Care Programmes* [66].

Comparison of cost-effectiveness has been applied to several management options that may be encountered in male reproductive medicine, including varicocele treatment vs. immediate ART (with or without surgical sperm retrieval), vasectomy reversal vs. ART, and hormonal therapy vs. ART for hypogonadotropic hypogonadism. Pathology-directed intervention has almost invariably been found to be more cost-effective than a straight-to-ART approach [67, 68]. One exception is when a varicocele is present in the setting of nonobstructive azoospermia; in this situation, microsurgical testicular sperm extraction (TESE) for ICSI is more cost-effective than varicocelectomy [69].

Every cost-effectiveness analysis in the arena of male infertility is sensitive to assumptions about treatment costs, success and complication rates, and the subsequent management of couples for whom first-line intervention is unsuccessful [70, 71]. Precise characterization of the clinical scenario(s) is very important and should be considered when determining the generalizability of results. For instance, the age of the female partner has a significant influence on the relative cost-effectiveness of vasectomy reversal and ART [72].

Expert Commentary

This chapter examined a number of issues that are commonly misunderstood and/or misrepresented with respect to the epidemiology of infertility, in general, and male infertility, in particular. First, infertility should not be confused with sterility—or even below-average fecundability. Approximately 10% of couples with average reproductive potential will not conceive within 12 months of unprotected intercourse and are therefore designated as infertile. Such couples may comprise a substantial proportion of those who present for evaluation of infertility (see Table 14.3). Their odds of spontaneous, unassisted conception remain high.

From a clinical standpoint, it would be ideal to accurately identify men with below-average reproductive potential. However, while semen analysis is the most common test employed for this purpose, its utility is limited by substantial overlap between the distributions of semen characteristics in empirically fertile and infertile men. Using relatively low diagnostic thresholds, such as the 2010 WHO reference values, carries the advantage of high specificity and perhaps a relatively decent positive predictive value in the setting of an infertility clinic; unfortunately, the negative predictive value only narrowly exceeds that of a coin flip.

The diagnostic inaccuracy of semen analysis is a fundamental problem for several additional reasons. From an epidemiologic perspective, the absence of a gold standard test for male infertility means that we have, at best, an uncertain grasp of its actual prevalence, let alone its association with putative risk factors, such as varicocele, cryptorchidism, sexually transmitted infections, etc. Moreover, if the true prevalence of male infertility is unclear, so, too, is the predictive value of tests employed for its diagnosis.

Finally, there is the question of how to counsel infertile men whose semen parameters exceed diagnostic thresholds. Given the poor predictive value of these "negative" results, should we advise consideration of further evaluation and treatment for modifiable risk factors identified in these individuals? An affirmative response to this query carries the risk of unnecessary treatment and costs, while the alternative may deprive some couples of an improvement in reproductive potential that facilitates natural conception, allows the use of IUI instead of IVF, or increases the odds of successful IVF/ICSI [73]. Well-designed studies to address this question are necessary.

Five-Year View

An indisputable need exists for more accurate diagnostic tests in the field of male infertility. Promising data have been published for analyses of seminal ROS levels and sperm DNA integrity, and the development of these assays is contributing to the advancement of an epidemiologic approach to the management of male factor infertility. Further improvements in understanding of the pathophysiology of male infertility will hopefully give rise to additional and more cost-effective diagnostic and treatment options as well.

Of course, rigorous validation of diagnostic tests in this arena is fraught with challenges, none greater than trying to establish males' actual reproductive potential. Time-to-pregnancy outcomes would seem to be best suited for this purpose, but numerous factors conspire against their reliability, including the play of chance, temporal fluctuations in semen parameters, confounding influences of the female partner's fecundability, and misattributed paternity.

Multivariate models, including those discussed in this chapter, may eventually prove to be valuable for diagnosis of male infertility and construction of nomograms to predict natural conception and live birth [74]. In theory, this would permit a more precise utilization of semen analysis and other test results as “continuous function parameters” that reflect a particular likelihood of male infertility, rather than as dichotomous variables with arbitrary thresholds [75]. This strategy would be analogous to the contemporary use of serum prostate-specific antigen (PSA) testing for prediction of an individual’s risk of biopsy-detectable prostate cancer and would similarly permit “individualized decision making” regarding management options [76].

Key Issues

- There is a 15% lifetime incidence of infertility among couples in Western countries; only 3–5% of couples are sterile.
- If the timing of sexual intercourse relative to ovulation is deliberately controlled, the monthly probability of pregnancy may approach 40%; however, the human average for monthly fecundity is 20%.
- Couples who are infertile after 12 months retain a roughly 50% chance of natural conception within the following year. These odds are lower if either partner is over 30 years old; female age, in particular, has a profound effect on fecundability.
- The absence of a gold standard test for male infertility is a fundamental problem, preventing an accurate understanding of its epidemiology and of the predictive value of tests employed for its diagnosis.
- WHO reference values for semen parameters represent the fifth percentile in men with proven fertility; however, there are many infertile men whose semen quality exceeds these thresholds. Basic semen analysis does not capture all of the variables relevant to male reproductive potential.
- Sperm concentration correlates with the probability of conception up to a density of 40 million/mL, but there is no additional likelihood of pregnancy at higher sperm densities.
- Purported declines in semen quality during the twentieth century appear to be an artifact of bias from geographical differences in sperm counts.
- Economic analyses suggest that pathology-directed treatment of male infertility is generally more cost-effective than proceeding directly to ART. Further studies are necessary to clarify which patients will benefit from evaluation and treatment of modifiable risk factors.

Acknowledgments The author would like to thank Kelli M. Mulder-Westrate and Ranya N. Sweis for critically reviewing several drafts of this chapter and offering helpful advice. Jonathan L. Faasse and Kristin M. Faasse provided valuable assistance with the figures.

References

1. Tremellen K. Oxidative stress and male infertility—a clinical perspective. Hum Reprod Update. 2008;14:243–58.
2. Sharma RK, Said T, Agarwal A. Sperm DNA damage and its clinical relevance in assessing reproductive outcome. Asian J Androl. 2004;6:139–48.
3. Athayde KS, Cocuzza M, Agarwal A, et al. Development of normal reference values for seminal reactive oxygen species and their correlation with leukocytes and semen parameters in a fertile population. J Androl. 2007;28:613–20.
4. Agarwal A, Sharma RK, Nallella KP, Thomas AJ, Alvarez JG, Sikka SC. Reactive oxygen species as an independent marker of male factor infertility. Fertil Steril. 2006;86:878–85.
5. Lumley J. Epidemiological approaches to infertility. Reprod Fertil Dev. 1998;10:17–21.
6. Alukal JP, Lipshultz LI. Why treat the male in the era of assisted reproduction? Semin Reprod Med. 2009;27:109–14.
7. Habbema JDF, Collins J, Leridon H, Evers JLH, Lunenfeld B, te Velde ER. Towards less confusing terminology in reproductive medicine: a proposal. Hum Reprod. 2004;19:1497–501.
8. Davies MJ, De Lacey SL, Norman RJ. Towards less confusing terminology in reproductive medicine: clarifying medical ambiguities to the benefit of all. Hum Reprod. 2005;20:2669–71.
9. Greenhall E, Vessey M. The prevalence of subfertility: a review of the current confusion and a report of two new studies. Fertil Steril. 1990;54:978–83.
10. Rowe PJ, Comhaire FH, Hargreave TB, Mahmoud AMA. WHO manual for the standardized investigation, diagnosis and management of the infertile male. New York, NY: Cambridge University Press; 2000.
11. Sigman M, Jarow JP. Male infertility. In: Wein AJ, Kavoussi LR, et al., editors. Campbell-Walsh urology. 9th ed. Philadelphia, PA: Saunders Elsevier; 2007. p. 609–53.
12. Simpson JL, Jauniaux ERM. Pregnancy loss. In: Gabbe SG, Niebyl JR, Simpson JL, editors. Obstetrics: normal and problem pregnancies. 5th ed. Philadelphia, PA: Churchill Livingstone Elsevier; 2007. p. 628–49.
13. Stephen EH, Chandra A. Declining estimates of infertility in the United States: 1982–2002. Fertil Steril. 2006;86:516–23.
14. te Velde ER, Eijkemans R, Habbema HDF. Variation in couple fecundity and time to pregnancy, an essential concept in human reproduction. Lancet. 2000;355:1928–9.
15. Rutstein SO, Shah IH. Infecundity, infertility, and childlessness in developing countries. DHS Comparative Reports No. 9. Calverton, MD: ORC Macro and the World Health Organization; 2004.
16. Lunenfeld B, van Steirteghem A. Infertility in the third millennium: implications for the individual, family and society: condensed meeting report from the Bertarelli Foundation’s Second Global Conference. Hum Reprod Update. 2004;10:317–26.
17. Kefer JC, Agarwal A, Sabanegh E. Role of antioxidants in the treatment of male infertility. Int J Urol. 2009;16:449–57.
18. van Noord-Zaadstra BM, Looman CWN, Alsbach H, Habbema JDF, te Velde ER, Karbaat J. Delaying childbearing: effect of age on fecundity and outcome of pregnancy. BMJ. 1991;302:1361–5.
19. Ford WCL, North K, Taylor H, Farrow A, Hull MGR, Golding J. Increasing paternal age is associated with delayed conception in a large population of fertile couples: evidence for declining fecundity in older men. The ALSPAC Study Team. Hum Reprod. 2000;15:1703.
20. Evers JLH. Female subfertility. Lancet. 2002;350:151–9.
21. Spira A. Epidemiology of human reproduction. Hum Reprod. 1986;1:111–5.

22. Leridon H, Spira A. Problems in measuring the effectiveness of infertility therapy. Fertil Steril. 1984;41:580–6.
23. Evers JL, te Velde ER. Vruchtbaarheidsstoornissen. In: Heineman MJ, Bleker OP, Evers JL, Heintz AP, editors. Obstetrie en Gynaecologie, de voortplanting van de mens. Maarssen: Elsevier Science; 2001. p. 435–71.
24. Bongaarts J. A method for estimation of fecundability. Demography. 1975;12:645–60.
25. Dunson DB, Colombo B, Baird DD. Changes with age in the level and duration of fertility in the menstrual cycle. Hum Reprod. 2002;17:1399–403.
26. Gnoth C, Godehardt D, Godehardt E, Frank-Herrmann PF, Freundl G. Time to pregnancy: results of the German prospective study and impact on the management of infertility. Hum Reprod. 2003;188: 1959–66.
27. Sharlip ID, Jarow JP, Belker AM, et al. Best practice policies for male infertility. Fertil Steril. 2002;77:873–82.
28. Muller CH. Rationale, interpretation, validation, and uses of sperm function tests. J Androl. 2000;21:10–30.
29. Kobayashi H, Gil-Guzman E, Mahran AM, et al. Quality control of reactive oxygen species measurement by luminol-dependent chemiluminescence assay. J Androl. 2001;22:568–74.
30. Barratt CLR, Aitken RJ, Björndahl L, et al. Sperm DNA: organization, protection and vulnerability: from basic science to clinical applications—a position report. Hum Reprod. 2010;25: 824–38.
31. Menkveld R. Clinical significance of the low normal sperm morphology value as proposed in the fifth edition of the WHO Laboratory Manual for the Examination and Processing of Human Semen. Asian J Androl. 2010;12:47–58.
32. Cooper TG, Noonan E, von Eckardstein S, et al. World Health Organization reference values for human semen characteristics. Hum Reprod Update. 2010;16:231–45.
33. Guzick DS, Overstreet JW, Factor-Litvak P, et al.; for the National Cooperative Reproductive Medicine Network. Sperm morphology, motility, and concentration in fertile and infertile men. NEJM. 2001; 345:1388–93.
34. Zweig MH, Campbell G. Receiver-operating characteristic (ROC) plots: a fundamental evaluation tool in clinical medicine. Clin Chem. 1993;39:561–77.
35. MacLeod J. Semen quality in one thousand men of known fertility and in eight hundred cases of infertile marriage. Fertil Steril. 1951;2:115–39.
36. Niederberger CS. Understanding the epidemiology of fertility treatments. Urol Clin North Am. 2002;29:829–40.
37. Ombelet W, Bosmans E, Janssen M, et al. Semen parameters in a fertile versus subfertile population: a need for change in the interpretation of semen testing. Hum Reprod. 1997;12:987–93.
38. Gunalp S, Onculoglu C, Gurgan T, Kruger TF, Lombard CJ. A study of semen parameters with emphasis on sperm morphology in a fertile population: an attempt to develop clinical thresholds. Hum Reprod. 2001;16:110–4.
39. Menkveld R, Wong WY, Lombard CJ, et al. Semen parameters, including WHO and strict criteria morphology, in a fertile and subfertile population: an effort towards standardization of in-vivo thresholds. Hum Reprod. 2001;16:1165–71.
40. Jedrzejczak P, Taszarek-Hauke G, Hauke J, Pawelczyk L, Duleba AJ. Prediction of spontaneous conception based on semen parameters. Int J Androl. 2008;31:499–507.
41. van der Merwe FH, Kruger TF, Oehninger SC, Lombard CJ. The use of semen parameters to identify the subfertile male in the general population. Gynecol Obstet Invest. 2005;59:86–91.
42. Zinaman MJ, Brown CC, Selevan SG, Clegg ED. Semen quality and human fertility: a prospective study with healthy couples. J Androl. 2000;21:145–53.
43. Bonde JPE, Ernst E, Jensen TK, et al. Relation between semen quality and fertility: a population-based study of 430 first-pregnancy planners. Lancet. 1998;352:1172–7.
44. Skakkebaek N. Normal reference ranges for semen quality and their relations to fecundity. Asian J Androl. 2010;12:95–8.
45. Leushuis E, van der Steeg JW, Steures P, et al. Prediction models in reproductive medicine: a critical appraisal. Hum Reprod Update. 2009;15:537–52.
46. van der Steeg JW, Steures P, Eijkemans MJC, et al. Pregnancy is predictable: a large-scale prospective external validation of the prediction of spontaneous pregnancy in subfertile couples. Hum Reprod. 2007;22:536–42.
47. Zorn B, Vidmar G, Meden-Vrtovec H. Seminal reactive oxygen species as predictors of fertilization, embryo quality and pregnancy rates after conventional in vitro fertilization and intracytoplasmic sperm injection. Int J Androl. 2003;26:279–85.
48. Mercan R, Lanzendorf SE, Mayer J, Nassar A, Muasher SJ, Oehninger S. The outcome of clinical pregnancies following intracytoplasmic sperm injection is not affected by semen quality. Andrologia. 1998;30:91–5.
49. Ombelet W, Deblaere K, Bosmans E, et al. Semen quality and intrauterine insemination. Reprod Biomed Online. 2003;7:485–92.
50. Allamaneni SS, Bandaranayake I, Agarwal A. Use of semen quality scores to predict pregnancy rates in couples undergoing intrauterine insemination with donor sperm. Fertil Steril. 2004;82: 606–11.
51. Shibahara H, Obara H, Ayustawati, et al. Prediction of pregnancy by intrauterine insemination using CASA estimates and strict criteria in patients with male factor infertility. Int J Androl. 2004; 27:63–8.
52. Aitken RJ. Whither must spermatozoa wander? The future of laboratory seminology. Asian J Androl. 2010;12:99–103.
53. Deepinder F, Chowdary HT, Agarwal A. Role of metabolomic analysis of biomarkers in the management of male infertility. Expert Rev Mol Diagn. 2007;7:351–8.
54. Desai N, Sharma R, Makker K, Sabanegh E, Agarwal A. Physiologic and pathologic levels of reactive oxygen species in neat semen of infertile men. Fertil Steril. 2009;92:1626–31.
55. Carlsen E, Giwercman A, Keiding N, Skakkebaek NE. Evidence for decreasing quality of semen during past 50 years. BMJ. 1992;305:609–13.
56. de Kretser DM. Declining sperm counts: environmental chemicals may be to blame. BMJ. 1996;312:457–8.
57. Sharpe RM, Skakkebaek NE. Are oestrogens involved in falling sperm counts and disorders of the male reproductive tract? Lancet. 1993;341:1392–5.
58. Fisch H. Declining worldwide sperm counts: disproving a myth. Urol Clin N Am. 2008;35:137–46.
59. Sherins RJ. Are semen quality and male fertility changing? NEJM. 1995;332:327–8.
60. Fisch H, Goluboff ET. Geographic variations in sperm counts: a potential cause of bias in studies of semen quality. Fertil Steril. 1996;65:1044–6.
61. Fisch H, Goluboff ET, Olson JH, Feldshuh J, Broder SJ, Barad DH. Semen analyses in 1,283 men from the United States over a 25-year period: no decline in quality. Fertil Steril. 1996;65:1009–14.
62. Saidi JA, Chang DT, Goluboff ET, Bagiella E, Olsen G, Fisch H. Declining sperm counts in the United States? A critical review. J Urol. 1999;161:460–2.
63. Niederberger CS, Joyce GF, Wise M, Meacham RB. Male infertility. In: Litwin MS, Saigal CS, editors. Urologic diseases in America. US Department of Health and Human Services, Public Health Service, National Institutes of Health, National Institute of Diabetes and Digestive and Kidney Diseases. Washington, DC: U.S. Government Printing Office; 2007, NIH Publication No. 07-5512. p. 461–81.

64. Centers for Disease Control and Prevention (CDC), American Society for Reproductive Medicine, Society for Assisted Reproductive Technology. 2007 assisted reproductive technology success rates: national summary and fertility clinic reports. Atlanta: U.S. Department of Health and Human Services, Centers for Disease Control and Prevention; 2009.
65. Wright VC, Schieve LA, Reynolds MA, Jeng G. Assisted reproductive technology surveillance—United States, 2000. MMWR Surveill Summ. 2003;52(No. SS-9):1–16.
66. Drummond MF, Schulpher MJ, Torrance GW, O'Brien BJ, Stoddart GL. Methods for the economic evaluation of health care programmes. 3rd ed. New York, NY: Oxford University Press; 2005. p. 277–322.
67. Shin D, Honig SC. Economics of treatments for male infertility. Urol Clin N Am. 2002;29:841–53.
68. Robb P, Sandlow JI. Cost-effectiveness of vasectomy reversal. Urol Clin N Am. 2009;36:391–6.
69. Lee R, Li PS, Goldstein M, Schattman G, Schlegel PN. A decision analysis of treatments for nonobstructive azoospermia associated with varicocele. Fertil Steril. 2009;92:188–96.
70. Penson DF, Paltiel AD, Krumholz HM, Palter S. The cost-effectiveness of treatment for varicocele related infertility. J Urol. 2002;168: 2490–4.
71. Meng MV, Greene KL, Turek PJ. Surgery or assisted reproduction? A decision analysis of treatment costs in male infertility. J Urol. 2005;174:1926–31.
72. Hsieh MH, Meng MV, Turek PJ. Markov modeling of vasectomy reversal and ART for infertility: how do obstructive interval and female partner age influence cost effectiveness? Fertil Steril. 2007;88:840–6.
73. Esteves SC, Oliveira FV, Bertolla RP. Clinical outcome of intracytoplasmic sperm injection in infertile men with treated and untreated clinical varicocele. J Urol. 2010;184:1442–6.
74. Boyd JC. Defining laboratory reference values and decision limits: populations, intervals, and interpretations. Asian J Androl. 2010; 12:83–90.
75. Niederberger C. Responses to semen analysis CART report. J Androl. 2003;24:329–31.
76. Thompson IM, Ankerst DP, Etzioni R, Wang T. It's time to abandon an upper limit of normal for prostate specific antigen: assessing the risk of prostate cancer. J Urol. 2008;180:1218–22.

Male Infertility from the Developing Nation Perspective

15

Kasonde Bowa and John S. Kachimba

Abstract

Infertility is an important problem worldwide that is of particular importance in societies of the developing world. In many of these societies, marriage is largely a vehicle for procreation and children a valuable economic resource and insurance for immortality. The problem would appear to be most pronounced in parts of sub-Saharan Africa, in countries that lie in the "infertility belt." The region includes Cameroon, Central African Republic, Gabon, Democratic Republic of the Congo, Togo, Sudan, Kenya, and Tanzania. In these countries, as many as one-third of couples are unable to conceive. A need exists for more studies to evaluate the incidence, risk factors, and prevention of this disease process in many parts of Africa, and the WHO has been at the forefront of this need for more research into infertility in Africa and developing countries of the Asian subcontinent. Though many infertile unions worldwide suffer from primary infertility, the predominance of secondary infertility in sub-Saharan Africa has been widely accepted. Furthermore, with the high prevalence of the human immunodeficiency virus infection in this part of Africa, the prevalence rates of tuberculosis, sexually transmitted infections, and septic complications associated with deliveries will rise and fuel this epidemic of secondary infertility. There remains a paucity of information on male infertility in sub-Saharan Africa. Though women bear the brunt of infertility socially, research on the continent has shown that a significant proportion of infertility is a result of male factor infertility.

Keywords

Male factor infertility • Sub-Saharan Africa • Secondary infertility • Reproductive tract infections • Hemoglobinopathies • Genitourinary tuberculosis • Varicocelectomy • Prevention of male infertility • Iatrogenic causes • Bilharziasis

Infertility is defined by the World Health Organization as the inability of a sexually active couple to conceive after regular intercourse at least twice a week for a period of 12 months [1]. Infertility is an important problem worldwide that is of particular importance in societies of the developing world [2]. In many of these societies, marriage is largely a vehicle for procreation and children a valuable economic resource and insurance for immortality. The problem would appear to be most pronounced in parts of sub-Saharan Africa, in countries that lie in the "infertility belt" [3]. The is a region spanning from Cameroon in the west to Tanzania in the east and includes the following countries: Cameroon, Central African

K. Bowa, MSc, M.Med, FRCS, FACS, FCS (✉)
Department of Surgery, University Teaching Hospital, Lusaka, Zambia and Ndola Central Hospital, 6th Floor Eastwing, Corner of Broadway and Nkana ST., Ndola, Zambia
e-mail: kbowa@yahoo.com

J.S. Kachimba, M.Med, FCS, FCS(Urol)
Department of Surgery, University Teaching Hospital, Levy Mwanawasa General Hospital, P.O.Box 50263, Ridgeway, Lusaka, Zambia
e-mail: jskachimba@yahoo.com

S.J. Parekattil and A. Agarwal (eds.), *Male Infertility: Contemporary Clinical Approaches, Andrology, ART & Antioxidants*,
DOI 10.1007/978-1-4614-3335-4_15,

Republic, Gabon, Democratic Republic of the Congo, Togo, Sudan, Kenya, and Tanzania. In these countries, as many as one-third of couples are unable to conceive [3].

A need exists for more studies to evaluate the incidence, risk factors, and prevention of this disease process in many parts of Africa, and the WHO has been at the forefront of this need for more research into infertility in Africa and developing countries of the Asian subcontinent [4, 5]. Though many infertile unions worldwide suffer from primary infertility, the predominance of secondary infertility in sub-Saharan Africa has been widely accepted [4]. This is also noted in developing countries of the Asian continent and South America. Furthermore, with the high prevalence of the human immunodeficiency virus infection in this part of Africa, the prevalence rates of tuberculosis, sexually transmitted infections, and septic complications associated with deliveries will rise and fuel this epidemic of secondary infertility. There remains a paucity of information on male infertility in sub-Saharan Africa. Though women bear the brunt of infertility socially, research on the continent has shown that a significant proportion of infertility is a result of male factor infertility. Male factor causes are generally present in 20% of cases of infertility [6]. Ikechebelu showed that a large proportion of couples in southeastern Nigeria presenting with infertility had male factor causation alone (42.4%), as compared to 25.8% resulting from female factors alone [2]. It is accepted that male factor causes are present in 20% of cases of infertile couples, while in approximately 30–40% of couples, both male and female factors are responsible for infertility [7]. Hence, in half of the couples presenting to a health facility, there is a male factor at play as a cause of infertility.

Prevalence of Male Infertility in Developing Countries

The prevalence of male infertility has not been comprehensively studied in most of Africa. Most views on male infertility are extrapolated from studies on infertility in women [4]. The first comprehensive epidemiologic study of infertility in the sub-Saharan Africa was conducted by Larsen. This study showed the prevalence of infertility in countries of this region to be 12.5–16% [2] and countries of the so-called infertility belt having significantly higher prevalences [2, 4].

The prevalence of infertility in men increases with age due to decreased sperm function, an increase in genetic abnormalities, and various secondary causes such as sexually transmitted infections [8, 9]. Most infertile men resort to polygamy as a solution to their infertility with all the attendant risks of multiple partner relationships [6]. The key determinants of fertility are economic, social, cultural, and medical. The medical determinants of fertility in Africa are HIV infection, sexually transmitted infections, tuberculosis, and, in areas of Africa with high prevalence, hemoglobinopathies (such as sickle cell disease and thalassemia). Unlike developed countries where there is a high number of urology specialists as recommended by WHO (1:100,000 population), who are able to manage and prevent male infertility, this is not the case in Africa. Aside from South Africa and Egypt, there are low numbers of urologists across Africa. The early detection of undescended testis, testicular torsion, and varicocoele means that these treatable conditions will not result in male infertility. The management of established infertility through medical or surgical means is also limited by this manpower constraint. Furthermore, there are only a few centers in Egypt and South Africa that provide various methods of assisted reproductive treatments. The total fertility rates are only a proxy of male fertility rates. A more direct determinant of male fertility rates are sperm counts. Over the last 40 years, there has been a decline in sperm counts worldwide [10].

Etiology of Male Infertility in Africa

In sub-Saharan Africa, the principal causes of male infertility are acquired and result in secondary infertility. This results from sexually transmitted diseases and iatrogenic causes due to poor surgical care in many areas [3]. There is a small percentage of men who experience infertility as a result of anatomic, genetic, hormonal, and immunologic problems. These form a small percentage of infertile men that is reflected in virtually all populations and estimated to form approximately 5% of the infertile male population [11]. In sub-Saharan Africa, sexually transmitted diseases are the main cause of male infertility [12].

Reproductive Tract Infections

Neisseria Gonorrhoeae

Gonorrhea is a worldwide infection that has seen resurgence in prevalence in the last few decades. Approximately 22% of men develop gonorrhea after a single episode of vaginal intercourse with an infected female [13, 14]. The infection starts as a cervicitis in females that is largely asymptomatic and facilitates its spread. In the male, gonorrhea begins as a urethritis that is symptomatic. Dunn et al. in a study of rural men in India found a high level of reproductive health problems [5]. More than 10% of these men had urethral discharge that they did not seek medical treatment and had continued to have sexual relationships without the use of condoms. This is a scenario that is common in many settings in sub-Saharan Africa. Untreated urethritis ascends and involves the prostate, seminal vesicles, and epididymis. Chronic seminal vesiculitis and chronic epididymitis are associated with abnormal sperm parameters and later complicated into an obstructive azoospermia.

Chlamydia Trachomatis

Chlamydia trachomatis causes nongonococcal urethritis in sexually active men and frequently complicates into epididymitis. This is an indolent infection that goes unrecognized for long periods and is a common finding in infertile men [15].

Human Immunodeficiency Virus Infection

HIV infection is a global epidemic infecting millions of people worldwide. The majority of infections occur in the third world. Sub-Saharan Africa's deaths due to HIV/AIDS were estimated to account for 76% of the 2.1 million deaths occurring worldwide as a result of this pandemic [16]. HIV/AIDS is the leading cause of death in sub-Saharan Africa. The Joint United Nations Programme on HIV/AIDS (UNAIDS) estimates that in 2007, there were 2.5 million new infections with HIV/AIDS, 68% of these occurred in sub-Saharan Africa [16]. Men infected with HIV are at risk of infertility secondary to altered spermatogenesis. Though the exact mechanism remains unclear, it is postulated to occur by a direct effect of the infection and an increased susceptibility to sexually transmitted infections and their occlusive sequelae.

Genital Tuberculosis

Tuberculosis continues to be a worldwide problem, and in Africa, HIV infection has led to a dramatic increase in the numbers of patients. Alongside the high HIV prevalence, there is a high incidence of TB. The WHO reports that over 1.9 billion people are currently infected with TB [17]. Most of these are in the low- to middle-income countries of Asia and Africa. It has been reported that there has been an annual increase in TB of 10% in developing countries as a result of HIV [17]. There is a high coinfection with TB in patients with HIV infection, and this occurs mainly when the CD4 count falls below 200. Extrapulmonary tuberculosis accounts for approximately 15% of cases, of which 14% are patients presenting with genitourinary tuberculosis [17]. In approximately a third of patients with genitourinary TB, the infection affects the genital tract only. Genitourinary TB affects more men than women (ratio 2:1). The highest incidence is in the age group 30–50 years. The prostate, epididymis, and seminal vesicles are the commonly affected sites. The most common site is the epididymis. Sexual transmission from female to male has been described in 3.9% of cases [17]. The diagnosis is difficult, and a high index of suspicion is required to diagnose this condition early and commence therapy before obstructive complications occur. Uncharacteristic urologic complaints, sterile pyuria, persistent cystitis, infertility without overt genital lesions, or sperm anomalies on evaluation should be investigated for the exclusion of genitourinary TB. TB of the genital tract can occur as an isolated disease leading to obstruction of the seminal ducts and vasa leading to infertility.

Genitourinary Bilharziasis

Bilharziasis is a communicable disease of major public health importance in the developing world that unfortunately remains neglected. It is estimated that over 200 million people are infected with this parasite and >80% live in Africa [18]. Bilharziasis of the genitourinary tract is due to infection with *Schistosoma haematobium*. Bilharzial infection leading to infertility affects the prostate, seminal vesicles, spermatic cord, and epididymis. The testis has remarkable immunity against bilharzial infection. Disease in these structures frequently affects the reproductive age group from 20 to 40 years with chronic vague symptoms that may include perineal discomfort, low back pain, dysuria, and hemospermia. As the infestation enters its chronic stage, the process of fibrosis predominates. The process of fibrosis in the seminal vesicles is marked to a degree that is unsurpassed by any other organ, including the bladder [18]. The extensive fibrosis in these structures leads to obstructive azoospermia if not diagnosed and treated early.

Lepromatous Leprosy

Leprosy occurs predominantly in Asia, Africa, and South America. Leprosy is a chronic mycobacterial disease of man caused by the *Mycobacterium leprae* that affects the peripheral nerves mainly and secondarily involves the skin and other organs including the testes. The testes are the only internal organ outside the body, and *M. leprae* has a preference for the lower temperatures in these organs [19]. Atrophy of the testes is common in patients with lepromatous leprosy with resultant infertility and impotence [19, 20].

Sickle Cell Disease

In areas of Africa where malaria is endemic, there is a high prevalence of sickle cell disease (SCD). Malaria is a parasitic infection which is the leading cause of mortality and morbidity in children under 5 years of age. The sickle cell trait reduces the severity of malaria and incidence of malaria in carriers. Therefore, the trait has persisted at a high prevalence of sickle cell disease and other hemoglobinopathies because of this protective effect. The prevalence of SCD/thalassemias is 15–20% in the populations of Central, Western, and North Africa [21]. In SCD patients, 10–15% of patients will have recurrent priapism that may result in erectile dysfunction and affect fertility. Men with SCD and those with thalassemias have been shown to have reduced sperm counts compared to matched controls. This infertility is the result of an acquired hypogonadotropic hypogonadism at the level of the pituitary. In addition, it is worsened by recurrent infections and infarctions in the testes resulting in further reduction in testicular function [21].

Iatrogenic Causes of Male Infertility in Africa

While reviews of iatrogenic causes of infertility in Western Europe reveal that this contributes approximately 5% of cases of infertility in both men and women, in Africa, this rate is higher [22]. While most evidence highlights this cause in women, it is also a cause of infertility in men. Kuku and Osegbe showed a pattern of male infertility due to vascular injuries sustained during hernia repairs in Nigeria [22].

Management of Male Infertility in Africa

Frequently in the African culture, infertility is blamed on the woman and results in divorce. The male seeks fertility services reluctantly and often after a number of failed marriages. The male client will typically have attempted several local remedies before coming to seek medical advice. The longer the period of infertility that lapses, the lower the chances of conceiving. If the duration of infertility is over 4 years of unprotected sexual intercourse, the conception rate per month is only 1.5% [23]. Thus, it is very important to start investigating and treating infertility at the earliest opportunity.

Evaluating and treating infertile couples is a costly procedure, and in Africa, a cost-effective approach is adopted. The couple is approached and treated as a whole, and investigation is undertaken in both to provide the best chance for a successful outcome. A successful outcome is the delivery of a healthy child. Secondary infertility is much more common in this region, and certain basics are followed to minimize expense on time and investigations. A detailed history from both partners along with a simple semen analysis is usually sufficient to make a clinical diagnosis. A history that looks at previous STIs, hernia surgery, infection with any endemic infections like TB, bilharziasis, and HIV which lead to infertility should reduce the need for more complex and invariably expensive investigations. This can be followed by hormonal profile where indicated. Postejaculatory urine and ultrasonography are additional methods of evaluation depending on the etiology. In many centers in Africa, the clinician will rely on the history and a semen analysis. The parameters for semen analysis are assessed using WHO guidelines. Two samples given at separate occasions, approximately 6–8 weeks apart, are required to make a diagnosis.

The treatment is divided into medical, surgical, and ARTs. In many cases where there is a demonstrable obstruction to a reproductive structure or a varicocele, surgery is the only treatment option available. Varicocelectomy has shown success in treating infertility in the absence of other causes. Up to two thirds of patients experience some improvement in sperm quality after varicocelectomy [23, 24]. Surgical repair of an obstructed vas has provided a much greater challenge in sub-Saharan Africa where the expertise and technology is lacking. Medical therapy consists of several empirical treatments with varying success rates. Common drugs in use include clomiphene which has antiestrogen properties and works by lowering estrogen while raising the levels of follicular stimulating hormone (FSH) which in turn increases spermatogenesis. Tamoxifen is used on similar grounds. Another drug available for the medical treatment of the infertile male is pentoxifylline that works by improving circulation and has been shown to improve sperm motility. Several antioxidant vitamins such as vitamins A, E, and C are used as they have been shown to improve sperm motility and quality.

The next logical measure in the management of infertility in Africa will be the introduction of assisted reproductive techniques (ARTs) [25]. Though the indications will be no different from developing countries, the demand will be higher due to the prevalence of tubal obstructive causes and male factor infertility. Needless to say, the establishment of ART services in Africa is faced with enormous obstacles because these are expensive services to establish and maintain. It was estimated that to set up an ART unit in Nigeria, running it would raise that health facility's budget by 50% making it a difficult service to support in limited resource countries [12]. Despite all these obstacles, ART services have been set up in South Africa and Egypt and are being accessed by the affluent portions of the population. Their impact on the treatment of infertility in this region is minimal as the majority of patients come from economically disadvantaged backgrounds. As countries in the region look to ARTs in the management of infertility, intrauterine insemination may be a logical first step as it is a procedure that is not so expensive.

Prevention of Male Infertility in Africa

There are numerous reasons that justify mounting a preventive campaign as a major part of managing male infertility in developing countries. Other than the fact that reproductive tract infections are a major cause of male infertility and are preventable, other reasons include:

- The difficulties in diagnosis, the complex medical and surgical treatments, and their poor long-term outcomes make it better to prevent these conditions from complicating into infertility. Further, these costly diagnostic modalities and treatments are concentrated in the few tertiary centers available and not accessible to many couples.
- The lack of specialist expertise to manage infertile couples and the virtual absence of research data on infertility in developing countries.
- The failure of most health authorities to prioritize infertility care and integrate it into the reproductive health care services.

Prevention strategies should involve a number of components that include:

- Ensuring all health-care personnel are continually updated on STIs management to ensure prompt diagnosis and effective treatment is instituted.
- A countrywide education campaign targeting young people on the nature and consequences of STIs, their control, and treatment. This campaign should stress the importance of abstinence and condom use to prevent infection with STIs.
- Extensive education campaigns to ensure all persons have knowledge of and have access to treatment for STIs.

Expert Commentary

Male infertility in sub-Saharan Africa, as illustrated by southern African state of Zambia, presents a very difficult clinical condition to manage. The expectations of the numerous patients that attend health facilities for treatment are unfortunately at odds with the goals of many health authorities that place emphasis on prevention of infertility rather than treatment. Further, the traditional view of reproductive health as predominantly a "women's health issue" permeates the decision makers in government that results in minimal funding for male reproductive health. Zambia has a population of 12 million with an HIV prevalence of 14.3% and a per capita income of 803 USD. The average fertility rate is 6.2. There has been a general decline in the fertility rate in Zambia as in other African countries for the last 20 years. In the Zambia Health and Demographics Survey of 2007, there was a decline in fertility rates from 7.2 in 1980 to 6.2 [26]. This decline is in all the nine provinces but most marked in the two most urbanized provinces (Copperbelt and Lusaka). A general look at the provinces shows the higher the HIV prevalence, the lower the total fertility rate (Fig. 15.1).

It is a known fact that HIV infection has some effect on reproduction. In the female, this includes abortions and pathological disease in the reproductive system [27]. In the male, HIV causes erectile dysfunction, testicular atrophy, and various sperm dysfunctions [28]. HIV infection is also associated with testicular tumors and tuberculosis of the testis. All of these adversely affect male fertility.

Despite numerous limitations in ancillary investigations and laboratory facilities, many hospitals in the region, through collaborative research agreements, are now able to evaluate the hormonal profile for patients and obtain histological diagnosis in azoospermatic men undergoing testicular biopsy. At the Lusaka University Teaching Hospital in Zambia, a fivefold increase in testicular biopsies for infertility was recorded over a 10-year period [26]. Azoospermia was the only indication for testicular biopsy. In a review of the relationship between HIV and male infertility done during the same period, it was noted that among men presenting to the urology clinic, there was a high HIV prevalence of 26% compared to the general population of 13% [29]. There was a high prevalence both of TB and STIs in this group of patients as well. TB is the most common opportunistic infection seen at the UTH.

Medical treatment is largely empirical and with the challenges in follow-up and poor long-term patient compliance, treatment efficacy has not been determined. The main treatment modality is the use of antiestrogens. It takes advantage of their effect of increasing pituitary output of luteinizing hormone (LH) and follicle-stimulating hormone (FSH) and hence stimulating testosterone production and spermatogenesis. The antiestrogens used are clomiphene citrate and tamoxifen citrate. The other nonsurgical treatment commonly employed is the use of antioxidants. These supplements, especially vitamins E and C, are utilized for their known effects of improving sperm function and protecting sperm DNA from oxidative stress, respectively.

Surgical treatments offered are very limited in scope. Varicocelectomy is performed in patients presenting with infertility, an abnormal semen analysis, and a clinically demonstrable varicocele. Surgical treatment options are not available to the majority of infertile men in the region as these require specialist expertise and equipment that is lacking in sub-Saharan Africa. The microscopic techniques required in unblocking obstructed seminal tubes or vasa require a huge investment in manpower training and equipment. For most countries in the region, it is an investment deemed not prudent to make in the present economic climate. The other option for treating infertile couples in this region, especially with improvement in HIV outcomes that are now evident with improved HAART, is the use of assisted reproductive technologies (ARTs) [12, 30]. These ARTs are presently beyond the economies of countries in this region.

Five-Year View

Over the next 5 years, the management of male infertility in the poorer countries of the world will be influenced by a number of factors. The increasing spread of HIV, TB, and STIs will continue to negatively impact on male fertility. The increasingly successful public health campaigns will see a huge demand for infertility care from the millions affected. In response, many hospitals will strengthen research collaborations both regionally and internationally that will improve patient assessment and diagnosis.

Though the poor economic climate will make it difficult to broaden the scope of surgical treatment options, there will be a move to a more accurate delineation of obstructive lesions. There will be greater cooperation in utilizing available surgical expertise in providing definitive surgical

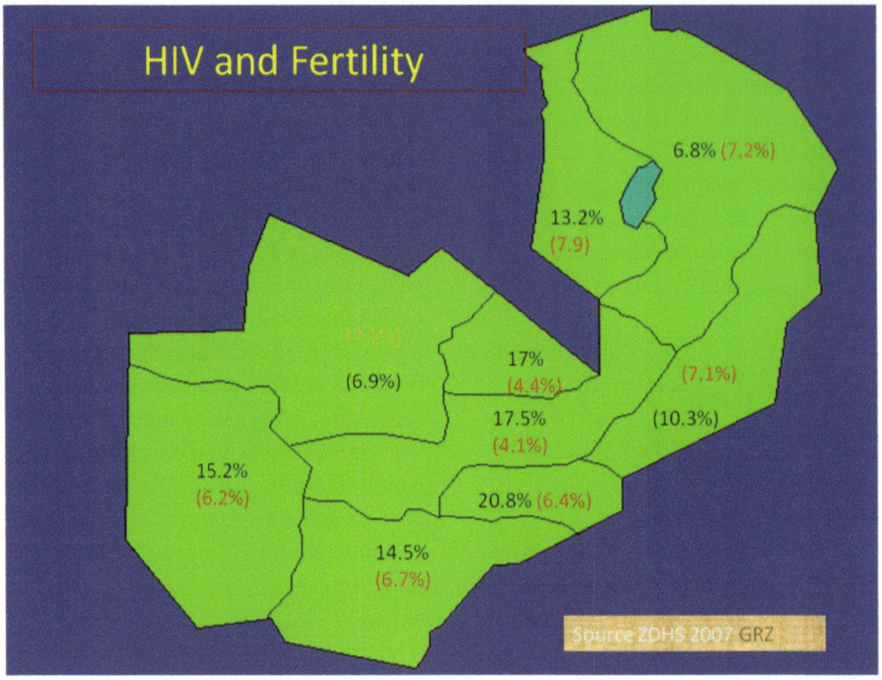

Fig. 15.1 The nine provinces of Zambia and the HIV prevalence rates by province (*black*) and fertility rates by province (*red*) (data from [26])

treatments for patients. The medical treatments available will see a surge in use with antiestrogens and antioxidants becoming more widespread.

With high HIV infection in Africa, fertility becomes a high risk activity in spreading HIV infection. There is also a high level of discordant couples in Africa for HIV up to 15–20% in many parts of Africa. The dual epidemic of HIV and TB poses a threat to fertility in Africa. An innovative solution to this is assisted fertility and sperm processing in HIV-positive men with HIV-negative women. These services would be needed in Africa in the coming years to maintain fertility without increasing the risk of HIV transmission. Finally, adoption, which is rarely mentioned as a treatment option in many developing countries due to strong extended family ties, will have to be addressed. Adoption policies in many of these countries will have to be amended to provide an option for family life for the many infertile couples that are unable to have children.

Key Issues

- Infertility is a huge public health problem in many developing countries. In a large part of sub-Saharan Africa, this problem is more pronounced as in these societies marriage is largely a vehicle for procreation with the resultant children being a valuable economic resource and socioanthropological insurance for immortality.
- In many of countries of the developing world, there is a high prevalence of HIV infection, TB, and STIs that are the principal cause of the more common secondary infertility.
- The evaluation and management of infertile couples is a costly procedure, and in most developing countries, a cost-effective approach is adopted following the WHO guidelines. Adopting this approach, many of the hurdles of limited human and laboratory resources are overcome.
- Though surgical treatments in patients with correctable causes are limited to varicocelectomy, empirical medical treatments and antioxidant therapies are readily available and widely in use.
- Most developing nations are mounting massive public awareness campaigns aimed at preventing reproductive tract infections that are a major cause of secondary infertility.

Acknowledgments We gratefully acknowledge the assistance rendered by Dr. Linda Kalilani-Phiri, Dr. Nenad Spasojevic, and Dr Joseph Yikona.

References

1. Sigman M, Jarow JP. Male infertility. In: Wein AJ, Kavoussi LR, Novick AC, Partin AW, Peters CA, editors. Campbell-Walsh Urology. 9th ed. Philadelphia: Saunders; 2007. p. p609–50.
2. Ikechebelu JI, Adinma JIB, Orie EF, Ikegwuonu SO. High prevalence of male infertility in southeastern Nigeria. J Obstet Gynaecol. 2003;23(6):657–9.
3. Aboulghar MA. The importance of fertility treatment in the developing world. BJOG. 2005;112:1174–6.
4. Larsen U. Primary and secondary infertility in sub-Saharan Africa. Int J Epidemiol. 2000;29:285–91.
5. Dunn KM, Das S, Das R. Male reproductive health: a village based study of camp attenders in rural India. Reprod Health. 2004;1:7. doi:10.1186/1742-4755-1-7.

6. Heyns CF, Bornman MS. Men's health in Africa (part 1): reproductive urogenital diseases and human immunodeficiency virus infection. J Mens Health. 2008;5(2):127–32.
7. Drury PL, Howlett TA. Endocrine disease. In: Kumar P, Clark M, editors. Kumar and clark clinical medicine. 5th ed. Edinburgh: WB Saunders; 2002. p. 999–1068.
8. Ng KK, Donat R, Chan L, Lalak A, Di Pierro I, Handelsman DJ. Sperm output of older men. Hum Reprod. 2004;19:1811–5.
9. Sloter E, Nath J, Eskenasi B, Wyrobek AJ. Effects of male age on the frequencies of germinal and heritable chromosomal abnormalities in humans and rodents. Fertil Steril. 2004;81:925–43.
10. Chakroun Feki N, Abid N, Rebai A, et al. Semen quality decline among men in infertile relationships: experience over 12 years in the south of Tunisia. J Androl. 2009;30:541–7.
11. World Health Organization. Infertility: a tabulation of available data on prevalence of primary and secondary infertility. Geneva: Programme on maternal and child health and family planning, Division of Family Health, WHO; 1991.
12. Cates W, Farley TMM, Rowe PJ. Worldwide patterns of infertility: is Africa different? Lancet. 1985;2(8455):596–8.
13. Belsey MA. The epidemiology of infertility: a review with particular reference to sub-Saharan Africa. Bull World Health Organ. 1976;54:319–41.
14. Kamel R, Lumley J. Textbook of tropical surgery. London: Westminster; 2004. p. 885–9.
15. Sheir MI, Abdalla S. Chlamydiae. In: Kamel R, Lumley J, editors. Textbook of tropical surgery. London: Westminster; 2004. p. 878–80.
16. UNAIDS: AIDS epidemic update. 2007. http://data.unaids.org. Accessed 19 Apr 2011.
17. Khalaf IM, Ashamallah AK. Genitourinary tuberculosis. In: Kamel R, Lumley J, editors. Textbook of tropical surgery. London: Westminster; 2004. p. 663–70.
18. Khalaf IM, Koraitim M. Genitourinary schistosomiasis and urinary bladder carcinoma. In: Kamel R, Lumley J, editors. Textbook of tropical surgery. London: Westminster; 2004. p. 673–80.
19. Hussein S. Mycobacteriae leprae. In: Kamel R, Lumley J, editors. Textbook of tropical surgery. London: Westminster; 2004. p. 926–49.
20. Leekassa R, Lockwood D. Leprosy. In: Parry E, Godfrey R, Mabey D, Gill G, editors. Principles of medicine in Africa. 3rd ed. Cambridge: Cambridge University Press; 2004. p. 575–90.
21. Adeyoju AB, Olujohungbe AB, Morris J, et al. Priapism in sickle-cell disease; incidence, risk factors and complications—an international multicentre study. BJU Int. 2002;90:898–902.
22. Kuku SF, Osegbe DN. Oligo/azoospermia in Nigeria. Arch Androl. 1989;22:233–8.
23. Dohle GR, Jungwirth A, Kopa Z, Giwercman A, Diemer T, Hargreave TB. Guidelines on male infertility. EAU guidelines. The Netherlands: European Association of Urology; 2009.
24. American Urological Association. American Society for Reproductive Medicine. Report on varicocele and infertility. 2001. http://www.auanet.org. Accessed 24 Jun 2011.
25. Sharma S, Mittal S, Aggarwal P. Management of infertility in low resource countries. BJOG. 2009;116 Suppl 1:77–83.
26. Central Statistical Office (CSO), Ministry of Health (MOH), Tropical Diseases Research Centre (TDRC), University of Zambia, and Macro International Inc. Zambia Demographic and Health Survey 2007. Calverton: CSO and Macro International Inc.; 2009.
27. World Health Organization. WHO laboratory manual for the examination of human semen and sperm-cervical mucus interaction. Cambridge: Cambridge University Press; 1999.
28. Bowa K, Labib MA. A review of HIV in Urology. Med J Zambia. 2004;5(2):77–9.
29. Bowa K, Labib M, Munalula KB, Mudenda V, Chikwenya M. HIV and male infertility at the University Teaching Hospital Lusaka. East Cent Afr J Surg. 2008;13(2):65–8.
30. Inhorn MC. Global infertility and the globalization of new reproductive technologies: illustrations from Egypt. Soc Sci Med. 2003;56:1937–51.

Ethical Dimensions of Urology: Four Case Studies

16

Marc S. Cohen and Ray E. Moseley

Abstract

In recent years, the progression of medical science has led to what used to be extremely rare ethical problems on the cutting edge of urological management of male infertility to become a more routine part of urological practice. These ethical problems, even though they have become more common in urological practice, still cause significant issues for the physician. In this chapter we discuss four cases that illustrate the ethical problems associated with identification and framing of the ethical problem, analysis of the problem, and the resolution of the problem. The basic purpose of these case discussions is that it is much easier to avoid and/or satisfactorily resolve an ethical problem when it is recognized early before a crisis mode arises or before poor ethical decisions are made and have to be unraveled after the fact. These cases represent four possible scenarios but are by no means exhaustive of the possible ethical issues that may arise from the current urological practice.

Keywords

Urology case studies • Ethics in urology • Male infertility • Patient assent • Ethical dilemmas in urology • Cryopreservation • Legal concerns

In recent years, the progression of medical science has led to what used to be extremely rare ethical problems on the cutting edge of urological management of male infertility to become a more routine part of urological practice. These ethical problems, even though they have become more common in urological practice, still cause significant issues for the physician. In this chapter we discuss four cases that illustrate the ethical problems associated with identification and framing of the ethical problem, analysis of the problem, and the resolution of the problem. The basic purpose of these case discussions is that it is much easier to avoid and/or satisfactorily resolve an ethical problem when it is recognized early before a crisis mode arises or before poor ethical decisions are made and have to be unraveled after the fact. These cases represent four possible scenarios but are by no means exhaustive of the possible ethical issues that may arise from the current urological practice.

Physicians are traditionally trained to find and fix problems. Ethical dilemmas are not so directly rectified. It is hoped that by posing the following case scenarios and suggesting the various ethical subtleties and concerns, physicians will begin to appreciate the spectrum of responses that each scenario contains. By appreciating these subtleties, it is hoped that physicians and nonphysicians will appreciate the complex tapestry that each poses and the value of considering these dilemmas and solutions before such events occur. What follows, therefore, is not a set of fixed answers for those particular cases as no single answer or combinations of responses exists. Rather, we hope this chapter will provide a framework for consideration of these and similar scenarios. As opposed to dealing with clinical pathology, there is no true "standard

M.S. Cohen, MD (✉)
Department of Urology, Shands at the University of Florida, 1600 SW Archer Boulevard, Box 100247 HSC, Gainesville, FL 32610, USA
e-mail: cohenms@urology.ufl.edu

R.E. Moseley, PhD
Department of Community Health and Family Medicine, University of Florida Health Science Center, P.O. Box 100237, Gainesville, FL 100234, USA

S.J. Parekattil and A. Agarwal (eds.), *Male Infertility: Contemporary Clinical Approaches, Andrology, ART & Antioxidants*,
DOI 10.1007/978-1-4614-3335-4_16,

of care" other than to do the right thing, which may, depending on the situation, be a matter of conjecture and prospective. The physician must now consider not just a disease but the unique environment and circumstances superimposed on this disease. To the end, institutional policies may exist (or be created) to address issues that are controversial.

One of the particular and frequent recurring themes in these scenarios is the need to recognize and reflect not only on the disease and the patient but also the need to address the spectrum and mores of the family and society. In doing so, the focus of the physician must frequently shift from the effect of the disease on the patient to the effect on the family.

Case Studies

Case 1

A 11-year-old male patient recently diagnosed with leukemia is to be started on chemotherapy shortly. The parents are interested in fertility preservation for their child. At this point, there is no data on human spermatogonia being matured to produce mature sperm. There have been some animal studies that have done this in a lab setting (in vitro), which indicate that this might be possible. There are no established guidelines on the cryopreservation of testicular tissue and spermatogonia in prepubertal males for future fertility preservation. However, the technology is advancing at a pace such that this may be possible in the near future.

Analysis

The major ethical issue facing the urologist is "Should the urologist honor the wishes of the parents and cryopreserve this tissue in preparation for the possible future ability of the 11-year-old patient to use it to produce offspring?" If the answer is "yes" then there are several procedural/ethical issues that will arise. These include: What is the appropriate role of the 11-year-old patient in this decision? and Should "assent" by the patient be required before this procedure is undertaken.

In analyzing the first ethical issues, the urologist must first determine if this procedure is in the appropriate scope of medical practice. In other words, would the results of this procedure, i.e., cryopreservation, storage conditions and options, and the future medical use of this tissue have a reasonable chance of actually being able to be used to produce future offspring? This entails both a practice issue, since physicians should not ethically or professionally engage in medical procedures that offer no reasonable chance of success and an informed consent issue, since any appropriate informed consent should detail the chance of success for a procedure and help the decision makers evaluate the options. Many physicians and medical ethicists would argue that this type of procedure at this early stage of development should not be offered outside of an approved research trial or without the prerequisite scientific studies validating the safety and efficacy of this type of procedure in producing future offspring [1]. Another powerful argument to consider is that this will be a very expensive procedure with long-term storage costs and will be made by parents who are under significant stress of caring for their very ill child, who may be inappropriately swayed by guilt or manipulated by for profit cryopreservation businesses.

A more straightforward ethical issue is whether or not the patient should be asked, whether he agrees or "assents" to having the procedure, and whether this should be encouraged or required. Assent is the concept that persons with diminished capacity who may not fully understand all of the relevant information about a medical procedure should still be asked if they wish to participate in the medical procedure and should agree to participation. The determining factor is the level of capacity and understanding of the child. The literature on child assent (mostly involving the issue of assent to participate in medical research) suggests that a relatively mature 11-year-old should be asked for agreement [2]. This, however, would involve a relatively in-depth discussion to assess the child's capacity and agreement by the parents. What is clear is that as the child matures, assuming that the treatment for the leukemia is successful, he should at the minimum be informed of this cryopreserved tissue, and he should eventually become the decision maker over the use of this tissue, and when he reaches 18 will have the legal authority to decide what to do with the tissue.

There are additional ethical issues related to this case. These include: Whether this procedure is an appropriate allocation of medical resources?, Who owns this tissue?, and Whether it is ethically permissible to use the tissue in the event the patient dies from the leukemia? This second set of ethical issues are certainly not separate from health policy issues and legal issues, and they clearly are not just decisions that should be made by the individual physician on a case-by-case basis. The danger for the individual urologist facing this case would be, if he or she were to unduly influence the decision based on their own personal views, of whether this is an appropriate allocation of general medical resources. This would run the risk of violating the formal principle of justice of treating similar cases in a similar manner, since different urologists would have widely divergent views on this aspect and there is no clear professional or ethical policy that was enlightened by transparent and open debate.

Case 2

A 26-year-old man, recently engaged, has met brain death criteria and has now been pronounced dead after a severe MVA.

His parents and fiancée want to cryopreserve his testicular tissue and sperm so that the fiancée can eventually have his children.

Analysis

The major ethical issue is "Whether the parents and/or the fiancée have the right to either make this decision to cryopreserve his testicular tissue and sperm, and/or to speak for their dead loved one, if they believe that this is what he would want?" In analyzing this issue, the urologist should consider a number of subquestions that should have bearing on the ethical resolution of this issue. For example, it would be nice to know what the recently deceased person thought about this issue. Admittedly, it is unlikely that a 26-year old would have given much thought to this issue; however, he may have expressed values and views which may be related and would give some insight on what he would want. For example, the range of things this deceased person may have believed or done could include previously expressed his wishes on this subject. Who knows for sure if he had seen a television show that discussed this very issue. He may have clearly expressed his desire to have children some day, or he may have expressed his wishes in writing given that more and more persons are writing Living Wills with specific wishes detailed. He may also have expressed the views that having a family and/or passing on his family name was extremely important to him.

The answers to these questions could certainly make us more confident (or less confident) that the deceased would actually want to have his tissue harvested and cryopreserved; however, this would not necessarily help in the resolution of this issue unless we hold the position that the dead have any rights at all, and if next of kin and loved ones have the moral authority to make these decisions. Traditionally, next of kin make burial decisions and help make sure that a dead person's estate is fairly distributed to survivors according to the deceased wishes. However, harvesting this type of tissue for this purpose seems pretty far afield from these traditional and established roles of family and loved ones.

There are also a number of possible scenarios that could follow from such a decision that involve both legal and ethical issues. Who does the sperm belong to—parents, fiancée, or both? Does the one who pays for the cryopreservation have any bearing on this? If the fiancée marries someone does she lose authority over the cryopreserved sperm? Do any children who are produced from this sperm have a claim on the estate of the "father"?

From the perspective of the urologist, this decision should require some careful thought and a lot of information for the decision makers. The general ethical principle involved in this type of case is that a person, and in this case a dead person, should generally not be used as a means to some end especially if the person does not or has not agreed to this use. In this case, the deceased person's sperm should not be used to produce a child that he may or may not have wanted to produce. It would seem at minimum that there should be some pretty good indications that the deceased person would have wanted this cryopreservation in this type of situation, before cryopreservation is undertaken.

This case illustrates a highly complex ethical decision superimposed on a highly time-sensitive issue. There are truly no right answers to the acute emotional issues it poses. What it does demonstrate is the need to seriously consider the "what if…" issues, inevitably, of these unnerving acute episodes. Because of the inevitable nature of such events an institutional policy is a recommended suggestion. That being said, these policies tend to be a legal defense mechanism, founded in the environment of litigious society that without emotion gives the physician the protection from having to make a decision usually based on a family member's or romantic partner's opinion of the patient's unexpressed and undocumented thoughts of their own personal legacy. These policies are frequently a way for obvious (and perhaps legally appropriate) reasons "wash its hands" of responsibility.

Case 3

A 19-year-old White man with attention deficit disorder schedules an appointment to discuss elective vasectomy. The patient is a junior in college. He visits you on one occasion accompanied by his mother. At the time of the original visit where he was accompanied by his mother, the patient notes that he has no physical problems. The patient states that he is in love with his fiancée and they both have agreed that they do not want children as a couple. That patient's mother is upset by the idea of a vasectomy and hoping that you will convince the patient otherwise. The patient claims that he is "sure" that he does not want to have children under alternative circumstances. The visit by the patient and his fiancée reveals a very upset 18-year-old woman who has been intolerant to birth control pills and hormone implants. She is not willing to use condoms, foam, or diaphragm, because of concerns related to efficiency. The fiancée feels that vasectomy would offer better efficiency and would represent less of a surgical risk as compared to tubal ligation. The patient wishes to support his fiancée in this decision.

Analysis

This case illustrates the importance of communication that inevitably exists in fertility issues. As has been previously suggested, physician training is founded in identifying and fixing problems. When superimposed onto a hostile environment a patient or partner in distress may be seen as a stereotype of that environment; in other words, the patient may be seen as the problem, rather than an unfortunate victim.

When so perceived and inappropriately managed, the outcomes can be disastrous.

Approaching such scenarios requires the physician to reorder his or her priorities. Rather than focusing on "finding the problem" a more useful approach in these types of cases may be as Stephen Covey ("Seven Habits of Highly Effective People") suggests, "Seek first to understand, then to be understood" [3]. As suggested by a number of patient-centered interview processes, it is important to acknowledge the emotional aspects of this volatile situation of the client and the physician and seek first to find meaning (if possible) for both. Fortunately, as compared to Case 2, the only time restraints superimposed here are self-imposed. There is time to address these issues if the physician chooses to do so. Recognizing and acknowledging the emotional aspects of these encounters can be equally important being able to recognize and call upon external support can be of the utmost importance (e.g., in this instance Planned Parenthood Organizations) can provide additional support.

Should the physician honor the patient's request for a vasectomy? Is this request "outside" of the range of reasonable medical alternatives for the patient's situation? A cornerstone of modern medical ethics is that adult competent patients generally, and except in very rare circumstances, have an unrestricted right to choose between medical procedures that are viable and appropriate medical alternatives for a legitimate medical need. This presumes that the patient receives information about the acceptable medical alternatives, understands those alternatives, and can make a free and uncoerced decision.

The first thing the physician must decide is whether the patient is requesting a medical procedure that is not just technically possible but that is also one of the appropriate medical alternatives for the patient's medical situation. Although most physicians would probably not recommend the requested intervention as the first choice, it does appear to be in the range of possible alternatives. The contraindications for the procedure all appear to be social concerns and not strictly medical concerns. Even though the patient is an adult he is only 19 years old, and persons of that age commonly change their minds about issues such as marriage and family. The significance of this concern is magnified since the results of the procedure are difficult if not impossible to reverse. For this reason alone, extra care and time should be taken with the patient to insure that this is the best alternative and that he carefully considers the alternatives and long-term consequences.

An additional feature of this case that adds to the ethical difficulty is that the patient seems to be making a choice of medical interventions based, not on a direct medical need, but mainly on the medical issues and personal preferences of his partner. For this reason, in some real sense both involved individuals are patients. Although the fiancée was present during one visit, it is not clear that this would constitute a through workup or exploration of her contraception concerns. This issue should be thoroughly explored by the physician (namely that his fiancée has received a through medical workup of her medical issues). For example, it is far from clear that this couple understands the data on the effectiveness of birth control pills when used correctly.

Additionally, in this case the physician should be concerned about whether the patient really understands the ramifications of choosing this medical procedure as a solution to his problem, and whether the patient is making a free and uncoerced decision. This patient is clearly under significant and conflicting pressure from both his mother and his fiancée, and this conflict alone should lead to the physician offering advice to not rush into this procedure. A valid informed consent requires not just that the appropriate information is given to the patient but also that the patient understands the information and that the patient's decision is not coerced [4].

So what can the physician do to be reassured that the patient really understands the requested procedure and that the patient is not being somehow forced into this decision? The first thing to note is that this is not an emergency situation, and that the patient should be given adequate time to thoroughly consider this decision. Secondly, the patient should be informed about how many people change their minds about desiring family and the chances of a successful vasectomy reversal in that event [5]. Also, the physician should probably discuss and consider recommending the possible cryopreservation of the patient's sperm.

Case 4

A 30-year-old "mentally challenged" man is brought to your office by his mother. The patient has been known to be sexually active and his mother, who is his legal guardian, has been concerned that this may result in an unwanted child. The patient has limited understanding secondary to his mental state; however, the patient does seem to understand that this would mean he would not be able to have children, but he voices no objections. You want to help the family, but have concerns regarding your ethical/legal positions regarding a request by the mother to perform a vasectomy.

Analysis

This case raises a number of significant ethical and legal concerns. Before any of the ethical and legal issues can even be framed and addressed additional medical information is required. Exactly what is the mental status of the 30-year old patient? What is the nature of the "known sexual activity?" What are the risks of pregnancy and are there less restrictive and permanent ways to achieve birth control? The answer to these questions will frame the appropriate ethical questions and possible resolutions.

For this case, we will presume that this patient lives in a group facility for persons with a similar mental disability. We will also presume that this is a well-run and appropriate structured environment. However, even in the safest facility it is possible and even likely that sexual activity will occur between residents of this facility. It should be noted that, if the sexual activity is between staff or other nonresidents and the patient, it raises additional serious legal and ethical concerns that should be reported to the proper authorities. In a similar manner, if the patient lives alone or at home with his mother, this possible sexual activity raises the possibility of important safety and supervision concerns, which would need to be in turn explored by the physician or possibly referred to appropriate state agencies.

Before considering whether to proceed, with even considering a permanent surgical procedure, the relative risk of the patient having a child should be evaluated. If our presumed scenario is accurate, it would be relevant to know if the sexual activity and a possible resulting child was just a potential fear of the mother or if there was ongoing sexual activity. It should be noted that it is not uncommon for people with mental disability to be in long-term relationships with fellow facility residents. If, for example, this was the case, then it would be important to know if the patient's significant other could even become pregnant.

Should a parent be able to consent to sterilization of a mentally incompetent adult child? In most states, legal guardians for those who have mental disabilities have the legal authority to make these types of medical decisions [6]. However, it should be noted that these decisions are often controversial and some types of decisions have been challenged by disability rights advocates [7]. The important question for the physician involved in this type of case is, what is the potential for abuse? For example, is this procedure being for the convenience of the mother/guardian, or so that the facility does not need to so closely monitor the activities of residents? As opposed to what is in the best interest of the resident, or simply because of some misguided social Darwinian bias? We would suggest that any physician who is asked to participate in this type of permanent medical procedure should seek guidance and support by consulting with a second physician, and considering requesting judicial or ethics committee review. As an additional safeguard, we would suggest that the physician make every effort if possible to seek assent from the patient prior to any medical procedure.

Expert Commentary

We have tried to illustrate through the use of four clinical scenarios some of the ethical dilemmas that may arise as part of the day-to-day interactions involving male patients with infertility-/fertility-related issues. These cases, by necessity are meant to be thought productive but not all inclusive. The basic, key steps to approaching these problems are as follows in Sect. 4.

Five-Year View

As stated in the beginning of this chapter, what used to be rare has become commonplace and routine. This text itself is a testament to the rapid advancement in our understanding of male infertility and the exponential technological changes that have taken place to meet the challenges posed; it is intuitive that these advances are only the beginning. As these technologies continue to "push the envelope" over the ensuing 5 years, it is also understood that the ethical dilemmas will continue to increase.

Key Issues

General Steps to Resolving Ethical Problems

1. Identify the ethical problem both in terms of the decisions to be made and the conflicting values present in the case.
2. *Critically* analyze the reasons for and against the various possible courses of action, paying special attention not to overstate or understate the reasons.
3. Clearly state the resolution or the range of acceptable options, both in terms of the decision to be made and of the prevailing supporting values.

Potential Mistakes to Avoid

1. Using value-laden language in both the description of the ethical problem and the analysis, thus introducing unacknowledged bias.
2. Not adequately taking into account long-term consequences when analyzing the various courses of actions.
3. Failure to verify facts or to recognize crucial lack of factual information, thus making unsupported assumptions.

References

1. Neal MS, Nagel K, Duckworth J, Bissessar H, Fischer MA, Portwine C, Tozer R, Barr RD. Effectiveness of sperm banking in adolescents and young adults with cancer: a regional experience. Cancer. 2007;110:1125–30.
2. Miller VA, Drotar D, Kodish E. Children's competence for assent and consent: a review of empirical findings. Ethics Behav. 2004; 14:255–96.

3. Covey Stephen R. The seven habits of highly effective people: powerful lessons in personal change. New York: Simon and Schuster Inc; 1990. p. 237.
4. Goldstein MM. Health information technology and the idea of informed consent. J Law Med Ethics. 2010;38:27–36.
5. Howard G. Who asks for vasectomy reversal and why? Br Med J (Clin Res Ed). 1982;285:490–3.
6. Lachance D. In re Grady: the mentally retarded individual's right to choose sterilization. Am J Law Med. 1981;6:559–91.
7. Zuba-Ruggieri R. Making links, making connections: internet resources for self-advocates and people with developmental disabilities. Intellect Dev Disabil. 2007;45:209–16.

Management of Fertility in Male Cancer Patients

17

Daniel H. Williams IV

Abstract

Cancer survival rates have improved dramatically over the last couple of decades due to advances in diagnostic techniques and therapies. Roughly 15% of cases of newly diagnosed cancer in men are in those younger than 55 years of age, and about one quarter of them are younger than age 20. Consequently, the population of young cancer survivors has grown, and the focus of cancer treatments has shifted from one of survival alone to that of survival *and* quality of life after treatment. For many men and their families, the maintenance and preservation of fertility during and after treatment is important. However, antineoplastic agents, radiation, and surgical therapies can all pose significant threats to a man's fertility potential, as can the presence of cancer itself. Male infertility due to cancer treatments may be temporary or permanent and can range from mild to severe. Because it is difficult—if not impossible—to predict the exact impact of cancer therapy on an individual man's ability to father a biological child, sperm cryopreservation *prior* to therapy remains the cornerstone of fertility preservation in this patient population. Unfortunately, in many cases, sperm cryopreservation remains underutilized. When sperm has not been banked prior to treatment and when men are azoospermic by their cancer treatment, surgical sperm retrieval in conjunction with advanced reproductive technologies (ART) is offered.

Keywords

Antineoplastic agents • Fertility in male cancer patients • Fertility preservation • Cryopreservation • Chemotherapy and male fertility • Surgical sperm retrieval • Recovery of spermatogenesis • Gonadotoxicity and testicular dysfunction • Sperm banking

Cancer survival rates have improved dramatically over the last couple of decades due to advances in diagnostic techniques and therapies [1–3]. Roughly 15% of cases of newly diagnosed cancer in men are in those younger than 55 years of age, and about one quarter of them are younger than age 20 [4]. Consequently, the population of young cancer survivors has grown, and the focus of cancer treatments has shifted from one of survival alone to that of survival *and* quality of life after treatment.

For many men and their families, the maintenance and preservation of fertility during and after treatment is important [5–11]. However, antineoplastic agents, radiation, and surgical therapies can all pose significant threats to a man's fertility potential, as can the presence of cancer itself. Male infertility due to cancer treatments may be temporary or permanent and can range from mild to severe. Because it is difficult—if not impossible—to predict the exact impact of cancer therapy on an individual man's ability to father a biological child, sperm cryopreservation *prior* to therapy remains the cornerstone of fertility preservation in this

D.H. Williams IV, MD (✉)
Department of Urology, University of Wisconsin Hospital and Clinics, 1685 Highland Avenue, Madison, WI 53719, USA
e-mail: williams@urology.wisc.edu

S.J. Parekattil and A. Agarwal (eds.), *Male Infertility: Contemporary Clinical Approaches, Andrology, ART & Antioxidants*,
DOI 10.1007/978-1-4614-3335-4_17, © Springer Science+Business Media, LLC 2012

patient population [12–20]. Unfortunately, in many cases, sperm cryopreservation remains underutilized [6–9]. When sperm has not been banked prior to treatment and when men are azoospermic by their cancer treatment, surgical sperm retrieval in conjunction with advanced reproductive technologies (ART) is offered.

Effects of Cancer on Male Fertility

The causes of poor semen quality in patients with cancer are not well understood, and multiple factors are likely involved. Some of these factors include preexisting defects in germ cells, local tumor effects, endocrine disturbances, and autoimmune and systemic effects of cancer [21–24].

A number of studies report that cancer adversely affects semen quality. However, published results of large studies are conflicting. Some suggest that cancer adversely affects semen quality [14, 25, 26], while others have found no differences between semen analyses of men with and without cancer [27]. Additionally, some studies suggest that the type of malignancy impacts semen quality [14, 28–30] whereas others do not [31, 32].

Ragni et al. reported that 11.6% of men who wished to cryopreserve sperm at their institution were azoospermic [28]. This ranged from 3.9% of men with non-Hodgkin's lymphoma to 15.3% of men with testicular tumors. Lass et al. reported that 10.5% of untreated men were azoospermic including 9.6% with testicular tumors, 13.3% with leukemia or lymphoma, and 3.7% of men with other malignancies [14]. Colpi et al. reported normal semen parameters according to WHO criteria in only 40% of men with lymphoma, 37% with testicular cancer, and 37% with other tumors [26]. Men with Hodgkin's disease usually present with poor semen parameters [33, 34]. Likewise, Lass et al. reported that 50% of men with cancer who cryopreserved at their institution had fewer than ten million motile sperm per ejaculate [14]. Finally, men with testicular cancer had semen parameters that were inferior to those of normal controls [25, 30]. In contrast, Rofeim and Gilbert compared semen parameters of 214 men with a variety of cancers to 22 men without cancer and found no significant differences between the groups [27].

Some studies suggest that the type of malignancy impacts semen quality. A large study of 776 men with cancer demonstrated that sperm density was significantly reduced in men with testicular cancer but that sperm quality did not vary significantly among men with other malignancies [28]. Similarly, a study of 314 patients with cancer found that men with testicular cancer had the lowest pretreatment sperm concentrations compared to those with other malignant neoplasms [29]. A number of studies have found that men with testicular tumors had significantly lower sperm quality compared to those with hematological or other malignancies [14, 25, 30, 33, 35]. Sperm DNA integrity has also been shown to be worse in men with cancer prior to treatment compared to fertile controls [36].

However, there is also evidence to suggest that the type of malignancy does not impact semen quality. Meseguer et al. reviewed semen parameters of 184 men who banked sperm before cancer treatment and found no significant differences in total sperm counts among men with different malignancies [32]. Likewise, Chung et al. found that sperm counts and motility did not differ by type of cancer in 97 patients who froze sperm before the initiation of cancer therapy [31].

Effects of Chemotherapy on Male Fertility

Chemotherapy negatively affects spermatogenesis, either transiently or permanently [37–39]. These drugs directly damage proliferating cells, so early differentiating sperm cells are exquisitely sensitive to these agents. However, even the relatively quiescent sperm precursors can be damaged due to cumulative effects of multiple doses of chemotherapy [40]. Later-stage germ cells, namely spermatocytes and spermatids, are less sensitive to chemotherapy since they are not dividing, and this accounts for the finding of some sperm immediately following chemotherapy with a slow decline in counts over the ensuing months. Leydig cell function appears to be less effected by chemotherapy.

Improved chemotherapy regimens have resulted in lower rates of infertility; however, azoospermia after treatment continues to be a concern [41]. When men are rendered completely azoospermic after treatment, some report that only 20–50% of these men will have some recovery of spermatogenesis [42], while others report up to 80% recovery depending on the type of cancer and chemotherapeutic regimen [39].

Alkylating agents including cisplatin are widely used for testicular cancer and have a high risk of azoospermia, particularly when coupled with ifosfamide, and the risk of permanent azoospermia seems to be dose and agent dependent [26, 39, 43, 44]. Likewise, most regimens for Hodgkin's disease also put men at high risk for azoospermia [45]. The impact of newer chemotherapeutic agents like taxanes and monoclonal antibodies remains unknown [46]. Age at treatment may play a role in recovery of spermatogenesis; however, this remains unclear [47].

Efforts have been made to explore strategies that may offer protection to the germinal epithelium during cancer therapy. One such approach has been the use of luteinizing hormone-releasing hormone analogs during gonadotoxic therapies have been examined in men. While these medications held promise in some animal studies, they did not significantly protect against spermatogenic failure in humans [48, 49].

Effects of Radiation on Male Fertility

Radiation therapy negatively affects spermatogenesis, either transiently or permanently by directly inducing DNA damage [38, 50]. A number of variables can affect the deleterious effect of radiation therapy on gonadal function, including total dose, source of radiation, gonadal protection, scatter radiation, and individual susceptibility [26, 51]. Gonadal shielding should be routinely employed; however, small amounts of scatter radiation are inevitable. As little as 0.15 Gy can result in impaired sperm production [52, 53]. Doses over 0.5 Gy typically result in reversible azoospermia [38]. Semen parameters often reach their nadir 4–6 months after treatment. Doses over 2.5 Gy place men at risk for prolonged or permanent azoospermia [38, 51]. Leydig cell function is affected when doses reach >15 Gy [26, 39]. Regimens for malignancies such as testicular leukemia and for total body irradiation prior to bone marrow transplants usually result in irreversible damage to the spermatogonia and permanent sterility [44]. Newer radiotherapy strategies may result in less gonadal toxicity but results are pending.

Effects of Surgery on Male Fertility

Surgical procedures such as retroperitoneal lymph node dissection in men with testicular cancer can cause infertility as a result of ejaculatory dysfunction due to damage of the pelvic plexus [54]. Both anejaculation and retrograde may result. Modified RPLND templates have been shown to reduce the risk of ejaculatory dysfunction in these men [8, 55, 56].

While more prevalent in older men, prostate cancer may affect younger men of reproductive age. Additionally, many men are waiting until later in life to father children or begin second families. In removing an important male reproductive organ, radical prostatectomy renders these men sterile. Likewise, bilateral orchiectomy and cystectomy can put men at risk for reproductive failure. Low-anterior or abdomino-perineal approaches to gastrointestinal malignancies may also put men at risk for ejaculatory failure [57].

Recovery of Spermatogenesis Following Cancer Treatment

Gonadotoxicity and testicular dysfunction are well-known side effects of cancer therapies since chemotherapy, radiotherapy, and surgery can all affect fertility potential [15, 41, 52, 58, 59]. Many men are rendered azoospermic following treatment. Spermatogenesis often returns in these men; however, the timing of the return (ranging from months to years) and the sperm quality when it returns are variable [29, 60–62]. Approximately 15–30% of childhood cancer survivors are permanently sterile following therapy [63].

A number of factors may influence the recovery of spermatogenesis following cancer therapy. In addition to the treatment regimen, the individual's pretreatment fertility potential and the influence of the cancer itself on the man's overall health can both impact posttreatment fertility [51, 64]. While the data are somewhat conflicting, certain malignancies, including testicular cancer and Hodgkin's disease, seem to influence pretreatment fecundity [14, 30].

Among testis cancer survivors, most were successful in achieving pregnancies, ranging from 71 to 82%, although successful paternity took many years in some cases and depended on the intensity of the chemo- or radiation therapy [65–67]. Men with low stage seminoma rarely become azoospermic after orchiectomy and radiation [68, 69]. High-dose testicular radiation for testicular intraepithelial neoplasia usually results in infertility [70].

Survivors of Hodgkin's lymphoma typically experience azoospermia after treatment. Depending on the chemotherapy and radiotherapy regimens, many patients recover some degree of spermatogenesis, but this may take up to 5–10 years [34, 71–73]. Non-Hodgkin's lymphoma regimens seem to be less gonadotoxic than those used to treat Hodgkin's disease [39]. Life-saving bone marrow transplantation strategies can also impair fertility, with azoospermia rates ranging from 10 to 70% depending on the agents, doses, and body irradiation templates employed [74, 75].

Men with genitourinary malignancies make up a unique subset of patients with cancer since their treatment directly and structurally impacts the male reproductive tract. Although few men undergoing prostate cancer treatment cite fertility as a concern, the prevalence of prostate cancer in young men is growing [76]. In a large review of over 14,000 men undergoing radical prostatectomy, 476 were 45 years of age or younger at the time of surgery [77]. For these men and their families, fertility needs may not yet have been met. If sperm are not cryopreserved preoperatively, testicular and/or epididymal sperm extraction, in conjunction with in vitro fertilization (IVF) and intracytoplasmic sperm injection (ICSI), is the only chance to father offspring. Regarding radiotherapy for prostate cancer, external beam therapy seems to have a greater negative impact on spermatogenesis than does brachytherapy [78, 79]. Like prostate cancer, fertility potential for most men with bladder cancer is not a significant concern due to age at diagnosis; however, young men may also develop urothelial carcinoma and elect to undergo various therapies. Smaller studies of men undergoing intravesical chemotherapy showed greater changes in semen parameters in men treated with BCG as opposed to mitomycin C [80]. A study of prostate-sparing cystectomy reported sperm in post-ejaculate urine samples [81].

Fertility following treatment of thyroid cancer with radioactive iodine is excellent; however, the treatment of pediatric sarcomas and rectal cancers results in a high rate of testicular dysfunction [82–84].

Risk of Malignancy in the Offspring of Men with Cancer

Two main concerns have been raised regarding risk of cancer in the children of cancer patients: does having cancer increase the risk of passing on this risk to one's offspring, and do cancer therapies portend any mutagenic risk? In particular, in the era of assisted reproductive technologies (ART) such as IVF/ICSI, is there any greater risk than in the past?

Prior to IVF/ICSI, studies did not demonstrate any increased risk of malignancy in the offspring of cancer survivors except in cases of known heritable diseases [85–87]. However, immediately following chemotherapy, there is a risk of sperm chromosomal abnormalities and aneuploidy which seems to lessen over time [88–92]. Chromosome analyses of testicular cancer patients after chemotherapy demonstrated no significant difference in the frequency of chromosomal abnormalities before and after therapy [93].

Semen Collection for Male Cancer Patients

Semen for sperm cryopreservation is generally obtained by masturbation. For many men, this may be an embarrassing or uncomfortable process. It is critical that men understand how to collect semen and that they be offered a private and relaxing environment to do so. Alternatively, men may collect a semen sample at home or another location than the clinic, providing that they keep the specimen at body temperature and return it to the lab within approximately 45 min to an hour after collection. Lubricants should be avoided as they can contaminate the specimens. The entire specimen should be collected, particularly in light of the fact that more sperm are present at the beginning of the ejaculate than at the end [94]. Wide-mouth specimen containers should be tested by each laboratory to ensure that they are compatible with semen collection and are not harmful to sperm.

Some men have difficulty providing semen specimens with masturbation. An alternative is to collect the sample with a condom. However, the condom must be approved by the laboratory as commercially available condoms generally contain spermicides that will kill sperm. This method typically results in fewer sperm being collected but may be necessary for some men. Additionally, anxiety, religious beliefs, pain, medications, and other factors may make semen collection challenging.

The adolescent male population is one in which extremely careful counseling and tactful, age-appropriate instructions are necessary, as these patients are at risk for emotional distress from this process. Parents should be included in discussions, although separate sessions with the adolescent are oftentimes useful. Unfortunately, no guidelines exist for the best approach to semen cryopreservation in the adolescent male, but individual institutional strategies are available [44].

At the time of diagnosis, many cancer patients are inpatients, and it may be at this time that many are first offered sperm cryopreservation. The logistics of collecting a specimen in this setting can be challenging, given the disruptions and interruptions that can occur in an inpatient room or bathroom. Additionally, some men are quite ill and debilitated by their cancer at the time of presentation and are unable to produce a sample. In these cases, surgical sperm retrieval can be offered.

Semen analyses are performed on all samples prior to cryopreservation. Semen parameters should be documented in accordance with the WHO guidelines [94]. Sometimes multiple collections are recommended depending on the number of motile sperm seen, the time since the last ejaculation, and individual variability.

Sperm Cryopreservation

The freezing of spermatozoa was first described as early as the eighteenth century, but modern techniques made cryopreservation practical and feasible in the mid-1900s with the development of sperm cryoprotectants. Today, common uses of sperm cryopreservation include banking sperm prior to vasectomy, at the time of vasectomy reversal for backup, prior to men engaging in potentially life-threatening activities (e.g., military deployment), and—pertinent to this review—prior to gonadotoxic, life-saving cancer treatments.

When semen is cryopreserved, a small aliquot of it is frozen separately, thawed, and reanalyzed after the initial freeze. This "test thaw" allows the postthaw survival to be determined as it can vary among individuals and even among different ejaculates from the same person [44]. Postthaw sperm motility is a good representation of the entire ejaculate and gives a reliable estimate of the total motile sperm count for that sample in the future [95].

While paternity with cryopreserved sperm clearly is possible, the freeze–thaw process may either negatively affect sperm quality and/or enhance any underlying sperm defects [96, 97]. Currently, sperm are kept in vials with cryoprotective agents like glycerol in combination with test yolk buffer, and these vials can be stored indefinitely in liquid nitrogen. Future techniques of dry storage may afford less damage to sperm [98].

The number of specimens that should be cryopreserved will differ for each patient. Determining factors include age, number of previous children, and semen quality. Abstinence of at least 48 h will typically maximize the yield of sperm per sample [94]. Even with time constraints and pressing health issues, men should be encouraged to consider sperm banking. Particularly in the era of ART, it is now possible to cryopreserve samples with low sperm counts that in the past were considered inadequate for freezing [13]. Poor semen quality has not been shown to affect fertilization or pregnancy rates after cryopreservation and IVF–ICSI, as long as live sperm can be recovered [99].

It has been my practice to encourage patients to initiate and complete sperm cryopreservation *before* starting any cancer therapy that affects the reproductive system. For example, in the setting of radical orchiectomy for testicular cancer, it is easier for a patient to ejaculate without a fresh inguinal incision; also, if that patient is found to have azoospermia, arrangements can be made to perform surgical sperm retrieval under the same anesthetic. If chemotherapy or radiation treatment has already been initiated, cryopreservation of semen is still possible during treatment, at least until the patient becomes azoospermic [42]. It should be noted that the effects of these gonadotoxic agents on sperm are largely unknown. Animal studies demonstrate a high incidence of mutagenic effects in offspring from matings that take place during immediately following treatment of the male with chemotherapy or radiation [100]. Increased frequency of sperm aneuploidy has also been reported after the initiation of chemotherapy and may persist up to 18 months or longer [89]. While the clinical impact of such effects in humans is unknown, sperm cryopreservation should ideally be performed before initiation of chemotherapy or radiotherapy. Otherwise, men are advised to wait 12–18 months after the completion of therapy before pursuing fertility treatments [101].

Attitudes About Sperm Banking

Over half of cancer patients desire future fertility, including over three quarters of those without children at the time of their cancer diagnosis [11, 102]. Currently, sperm banking is the only pretreatment strategy for male cancer patients to preserve their future fertility [103]. However, less than a quarter of cancer patients bank sperm, and the most common reason for not doing so is lack of information [102]. Only two-thirds of men awaiting cancer therapy are aware of sperm banking [104]. Schover et al. also showed that over 90% of responding oncologists felt that sperm banking should be offered to all men before treatment, but almost half failed to do so [105]. Reasons for this included time, high costs, and lack of convenient facilities. Only 10% claimed that they offered sperm banking to all eligible men despite evidence suggesting at least 50% of young men with cancer are interested in doing so [106].

Reebals et al. addressed oncology nurse practice issues in determining whether newly diagnosed adolescent male patients are offered the option of sperm banking before undergoing chemotherapy treatment. They distributed questionnaires to nurses and nurse practitioners who care for adolescent male cancer patients at the time of diagnosis, during chemotherapy, and during follow-up care. Over 95% of respondents agreed that all male patients undergoing cancer treatment should be offered sperm banking. Oncologists and nurse practitioners were seen as appropriate professionals to discuss this option. The authors concluded that lack of knowledge regarding sperm banking limited nurses' willingness to discuss this topic, and education regarding cryopreservation could improve knowledge and practice patterns [107].

Saito et al. reported a positive psychological effect in 80% of interviewed cancer patients who banked sperm. They found that in particular, if sperm was banked on the patient's own initiative, that doing so offered encouragement during therapy [11].

Obstacles to Sperm Banking

There are a variety of reasons why a patient may choose not to cryopreserve semen prior to starting cancer treatments, including modesty of both the patient and health-care provider, privacy, discomfort, cost, urgency to begin treatment, and access to sperm banking facilities. Schover et al. found that the most common reason patients did not bank sperm was because of the lack of information [102].

In 1995, Koeppel reported over 50,000 new cases of cancer in men under the age of 35 and realized that with rising survival rates and the harmful effects of treatment on fertility potential that semen cryopreservation should be offered to these patients [108]. The author acknowledged the controversies regarding the practicality and usage of sperm banking including the challenges faced by health-care professionals in discussing such sensitive issues with patients. Oncology nurses were identified as key members of the treatment teams who could discuss infertility and sperm banking with patients at the most opportune time, before initiation of chemotherapy. It was recognized that improved knowledge would reinforce the importance of offering sperm banking to circumvent treatment-induced infertility.

Finding a sperm bank for a patient should not be a barrier in discussing the option. Information about sperm banks is readily available online. Most banks will have mail kits available that allow patients to collect sample at home and ship them to the sperm bank. This approach allows for privacy and convenience for the patient.

Cost has been identified as an obstacle for patients. Schover et al. demonstrated that both physicians and patients are under the impression that sperm cryopreservation is too costly [105]. Although cost varies by facility, it is estimated that initial processing fees are approximately $350 with monthly storage fees ranging from $10 to $50 per month. Insurance coverage is variable, but some will cover a portion of the cost, particularly in the setting of cancer treatment. National agencies such as the American Cancer Society may also have financial aid programs. Many sperm banks also offer payment plans based on need and income.

Canada and Schover acknowledged the limited time oncologists have with each patient and suggested that training oncology nurses, social workers, and nurse practitioners to discuss infertility with new cancer patients is a reasonable approach to this barrier [109]. Educational materials including patient education sheets and interactive computer programs for patients and their families are useful. Educating health-care providers via lectures, grand rounds, and in-service presentations is encouraged.

Developing an efficient, seamless system to provide this service to cancer patients during such an emotional time is also critical. Phone numbers and protocols should be readily available on inpatient wards and in outpatient clinics. Semen collection rooms should be readily accessible to patients.

Although semen collection is recommended prior to starting treatment, the urgency to start therapy sometimes trumps the ability to provide a sample for cryopreservation. In these cases, collection is possible after starting therapy; however, the impact of chemotherapeutic and radiotherapy regimens on the risk of genetic defects in the offspring remains unknown. Patients and their families must be counseled as such. Some authors report that samples collected within 10–14 days of starting treatment may still be safe to use for future ART, based on sperm transit times through the reproductive tract [110].

Lastly, there may be legal considerations surrounding sperm banking that need to be addressed. As summarized by Leonard et al., the law surrounding cryopreservation of semen is still uncertain [44]. It remains unclear if semen is categorized as property, person, or a unique material that is neither person nor property. Additionally, the disposition of cryopreserved specimens in the event of a dispute remains unclear [111]. Consent forms and contracts are important supporting documents for sperm banking, and they should address to whom the sperm belongs, what will happen in the event of death, and how payments for these services will be handled. Confounding factors may include cases of minors or in instances where there is potential for secondary gains (e.g., inheritance).

Semen Cryopreservation in Adolescent Male Cancer Patients

While adult male cancer patients may be more willing to accept the notion of sperm banking to preserve future fertility, adolescents may be intimidated and embarrassed by the concept. Their fertility wishes may not be realized for many years, and the long-term psychosocial impact of infertility on survivors of childhood cancer remains largely unknown [112]. In addition, opinions vary regarding the most appropriate age for discussing sperm banking and who should be responsible for addressing this issue.

The exact age at which sperm production first begins is unknown and probably varies based on individual factors. Enlargement of the testes represents a transition from Tanner stage I to II, and it is around and after this time that spermatogenesis likely begins, even prior to the adolescent growth spurt [113, 114]. Nevertheless, adolescent males with cancer, ranging from age 14 to 17 years, have been found to be good candidates for sperm banking [115, 116].

Ginsberg et al. examined the feasibility of offering sperm banking to young male cancer patients and determined the beliefs and decision-making processes of these patients and their parents. Of the 68 patients in their study who collected semen samples, 50 of them completed the study. They found that 80% of the patients made the decision to bank sperm with their parents and that all of the patients who banked sperm felt that they were making the right decision to do so. Patients and parents alike wanted information about semen cryopreservation. The authors concluded that because semen quality was dramatically reduced, even by one course of gonadotoxic therapy, sperm banking should be offered to all eligible patients prior to therapy. Parents played an important role in the decision to bank sperm [117].

Klosky et al. assessed sperm cryopreservation among males newly diagnosed with cancer aged 13 years and older. Oncologists assigned infertility risk to patients and reported whether their patients engaged in sperm cryopreservation. Less than 30% of their patients banked sperm. The authors found that the decision to cryopreserve semen was associated with a number of factors including a diagnosis of central nervous system malignancy or non-central nervous system solid tumor diagnosis, higher socioeconomic status, and not being a member of an Evangelical religious group. They concluded that sperm banking was underutilized by adolescent males and that newer strategies were needed to increase the number of these patients who participated in this fertility-preserving activity [118].

Emotional maturity is another important concept when discussing sperm banking in adolescent males. Boys who are not physically mature may still be able to collect sperm.

Conversely, a physically mature adolescent may not be emotionally or sexually mature to perform a semen collection by masturbation.

Surgical Sperm Retrieval

A low—but still clinically significant—percentage of men with cancer who present for sperm banking will be azoospermic or will be unable to collect a semen sample. In such cases, surgical sperm retrieval techniques may be offered. Oftentimes, these procedures require a concerted and coordinated effort between the urologist and the fertility laboratory. Time pressures are typically present as these patients generally need to begin urgent therapy. The various approaches are discussed below, and they may be able to be scheduled concomitantly with any oncologically related procedures such as vascular access, lymph node sampling, or bone marrow biopsies.

Testicular sperm extraction (TESE) refers to an incisional testicular biopsy performed to obtain sperm for cryopreservation. Sperm obtained with this approach may only be used for ART [119]. For these men, it is difficult to predict success rates for sperm retrieval, although roughly half of azoospermic men with testicular cancer or malignant lymphomas will have sperm found on TESE [120, 121]. Additionally, men with testicular cancer undergoing radical orchiectomy may have microdissection TESE performed on the removed testicle [122, 123]. These men may also be scheduled for a simultaneous sperm retrieval procedure under anesthesia on the contralateral testicle.

For male cancer survivors who are azoospermic and who did not cryopreserve sperm prior to their cancer therapy, testicular sperm retrieval techniques in conjunction with ART can be offered [63, 124]. Microdissection TESE affords retrieval rates of approximately 50% in men with postchemotherapy azoospermia [125–127].

Microsurgical epididymal sperm aspiration is a procedure used to obtain sperm from the epididymides in the setting of obstructive azoospermia. An example where this approach would be indicated in a cancer patient with azoospermia is following a radical prostatectomy which from a reproductive perspective is similar to postvasectomy patient. In this patient population, testicular function is usually preserved, and cryopreserved sperm have been shown to be suitable for ART [128, 129].

Some patients with cancer have undergone surgical procedures that affect their ejaculatory function. Retroperitoneal lymph node dissections for testicular cancer and low-anterior and abdominoperineal resections for gastrointestinal malignancies can put men at risk for ejaculatory failure, despite improvements in surgical techniques. When medical treatment fails to improve emission and ejaculation, then electroejaculation (EEJ) may be offered. EEJ has been shown to be an effective way to retrieve sperm for ART [130, 131]. Sperm quality tends to be impaired in these patients, and pregnancy rates are better when these sperm are used for IVF/ICSI rather than intrauterine insemination [131–133]. EEJ should be used with caution in the setting of thrombocytopenia or leukopenia given the potential risks of bleeding or infection.

Outcomes of Using Cryopreserved Sperm for ART

While it is generally accepted that cancer and cancer therapies adversely affect a man's reproductive potential, the outcomes of ART up until recently have only been addressed in case reports and small studies. This is due, in part, to the advances in ICSI which has revolutionized the treatment of male infertility due to the need for only a few sperm either in the ejaculate or testicular tissue. Cryopreserved sperm may be used for intrauterine insemination (IUI) and/or IVF with ICSI. How to best use frozen sperm for ART depends on the quantity and quality of the sperm, how well the sperm survive the freeze–thaw process, the presence of any female factors, and patient/couple preference.

Sanger et al. reviewed the literature from an era prior to widespread use of ICSI. Fifty-four deliveries resulting from cryopreserved semen of male cancer survivors from fertility clinics and another 61 deliveries resulting from the use of cryopreserved semen from male cancer survivors were reported from sperm banks [13].

Naysmith et al. assessed the effect of cancer treatments on the natural and assisted reproductive potentials of men. Semen samples were analyzed before and after cancer therapy. Twenty-seven percent of the men had abnormal semen parameters before treatment. Following treatment, 68% of the samples were abnormal. Twenty-three percent of men developed azoospermia after treatment. Pretreatment sperm cryopreservation improved the fertility potential of 55% of their patients. The authors commented that improving awareness and education of patients and providers on the impact of cancer and cancer treatments on fertility is essential. They also stressed that with the advent of ICSI, all men with cancer should be offered pretreatment sperm cryopreservation as even men with very low sperm concentrations the chance of conception is very reasonable [15].

Tryde-Schmidt et al. reported their experience with couples referred for ART because of male-factor infertility due to cancer and cancer treatment. Most of their patients had testicular cancer and lymphomas. Ninety percent of the men had adjuvant treatment with chemo- and/or radiation therapy. Perhaps, most impressively, semen was cryopreserved in

82% of their men prior to treatment. Following cancer therapy, 43% of the men had motile spermatozoa in the ejaculate, while 57% were azoospermic. Both fresh and cryopreserved sperm were used, and the clinical pregnancy rates per cycle were 14.8% after IUI, 38.6% after ICSI, and 25% after ICSI–frozen embryo transfer, with corresponding delivery rates of 11.1, 30.5, and 21%. Cryopreserved semen was used in 58% of the pregnancies. Of note, the delivery rate per cycle was similar after use of fresh or cryopreserved sperm. The authors concluded that male cancer survivors have a good chance of fathering a child by using either fresh ejaculated sperm or cryopreserved sperm and that ICSI be used as a first choice, given the better success rates with ICSI as well as the need for overall higher total motile sperm counts for IUI which are not always available postthaw [134].

These reports of successful pregnancies with cryopreserved sperm in male cancer survivors are supported by numerous other studies [3, 14, 32, 135–139].

Van Casteren et al. reported their experience with ART using cryopreserved semen of cancer patients. Five hundred and fifty-seven male cancer patients banked 749 semen samples. Out of the total group of 557 men who cryopreserved semen, 218 (39%) returned for semen analysis after cancer treatment. Motile sperm were found in 155 (71.1%) of these 218 men. Twenty of these 218 men reported a spontaneous pregnancy. While only 42 of the cancer survivors (9.6%) ultimately requested the use of their banked semen, these men would have been unable to father their own child if their sperm had not been banked prior to therapy. Half of these men were successful in having live births using IVF/ICSI [140].

Conceptually, there could be differences in ICSI success rates when using fresh versus cryopreserved sperm; however, current studies indicate no difference in pregnancy outcomes between the two [141–143].

A few studies have looked at the utilization of cryopreserved sperm by male cancer survivors. In one study of 258 men, only 18 returned for treatment [16]. Ginsburg et al. found that at their fertility center, 19 male cancer survivors underwent a total of 35 IVF cycles, and 11 of these cycles used cryopreserved semen [138]. In a larger study, Magelssen et al. looked at posttreatment paternity in 1,388 testicular cancer survivors. Four hundred and twenty-two of these men had cryopreserved semen *after* orchiectomy. Overall, only 29 men (7%) used their cryopreserved semen for ART, while 67 men (17%) fathered at least one child with fresh semen [64, 106].

Lastly, according to a study by Saito et al., if male cancer survivors had return of spermatogenesis following treatment, none would choose to use their cryopreserved sperm. Even if the cryopreserved sperm was not used, as in most cases, a positive psychological effect of having banked sperm was achieved [11].

Five-Year View and Key Issues

An exciting new direction for fertility preservation in men with cancer is implementing stem cell technologies for germ cell transplantation and testicular grafting. Spermatogonial stem cells may be used in the future for preservation of testicular tissue and fertility preservation in men and boys prior to treatment, as these cells are capable of self-renewal, proliferation, and repopulation of the seminiferous tubules [101].

Schlatt et al. [144] recently reviewed the physiology of spermatogonial stem cells in rodent and primate testes and concluded that while germ cell transplantation has become an important research tool in rodents and other animal models [145–152], the clinical application in humans remains experimental. Regarding testicular grafting as another exciting strategy for fertility preservation in males prior to gonadotoxic therapy, both autologous and xenologous transfer of immature tissue revealed a high regenerative potential of immature testicular tissue and generation of sperm in rodents and primates. Like germ cell transplantation, however, further research is needed before an application in humans can be considered safe and efficient. Despite current limitations in regard to generation of sperm from cryopreserved male germ line cells and tissues, and since future improvements of germ cell transplantation and grafting approaches are likely, retrieval and cryopreservation of testicular tissue prior to therapy should be offered to young men with cancer who are at high risk of fertility loss, as this could be their only option to maintain their fertility potential after treatment [153, 154]. Additionally, prepubertal testicular tissue from boys facing gonadotoxic treatment may be cryopreserved under special conditions. Doing so may offer fertility preservation for young patients in the future [155].

A potential concern about using spermatogonial stem cells and testicular grafts is the theoretical risk of restoring cancer cells back into the recipient. This effect has been demonstrated in leukemic rat models [156]. But efforts have been made to reduce this risk using telomerase in culture [157]. The use of embryonic stem cell technology to treat infertile men is also under investigation; however, significantly more translational research is needed, before these technologies are applied to the treatment of human male infertility [158].

Summary

Improvements in cancer treatments have resulted in more men living into their reproductive years, and fertility is an important measure of quality of life in this patient population. However, *all* cancer therapies—chemotherapy, radiation, and surgery—are potential threats to a man's reproductive potential. The type of treatment(s) and individual susceptibilities to the deleterious effects of these treatments make it next to impossible to predict whether or not a man will recover spermatogenesis after therapy and what his sperms' potential is to safely fertilize an egg. Stem cell transplantation technologies may hold promise in the future but are unavailable for use in humans at this time. Advances in ART now provide more men with opportunities to become biological fathers, even in the setting of poor semen parameters. Thus, sperm cryopreservation *prior* to initiating life-saving cancer treatment offers men and their families hope and the best chances to father biologically related children in the future. It is a safe and effective means of preserving a man's fertility and should be offered to all men with cancer before treatment. Posttreatment male infertility also may be treated with ART and advances in surgical sperm retrieval. Barriers to sperm banking still exist, but the sensitive nature of many of these can be overcome by patient and provider education, as well as deliberate, coordinated strategies at comprehensive cancer care centers to make fertility preservation for male cancer patients a priority during pretreatment planning.

References

1. Mcvie JG. Cancer treatment: the last 25 years. Cancer Treat Rev. 1999;25:323–31.
2. Lass A, Akagbosu F, Brinsden P. Sperm banking and assisted reproduction treatment for couples following cancer treatment of the male partner. Hum Reprod Update. 2001;7:370–7.
3. Agarwal A, Ranganathan P, Kattal N, Pasqualotto F, Hallak J, Khayal S, et al. Fertility after cancer: a prospective review of assisted reproductive outcome with banked semen specimens. Fertil Steril. 2004;81:342–8.
4. Steliarova-Foucher E, Stiller C, Kaatsch P, Berrino F, Coebergh JW, Lacour B, et al. Geographical patterns and time trends of cancer incidence and survival among children and adolescents in Europe since the 1970s (the ACCISproject): an epidemiological study. Lancet. 2004;364:2097–105.
5. Gritz ER, Wellisch DK, Wang HJ, Siau J, Landsverk JA, Cosgrove MD. Long-term effects of testicular cancer on sexual functioning in married couples. Cancer. 1989;64:1560–7.
6. Rieker PP, Fitzgerald EM, Kalish LA, Richie JP, Lederman GS, Edbril SD, et al. Psychosocial factors, curative therapies, and behavioral outcomes. A comparison of testis cancer survivors and a control group of healthy men. Cancer. 1989;64:2399–407.
7. Rieker PP, Fitzgerald EM, Kalish LA. Adaptive behavioral responses to potential infertility among survivors of testis cancer. J Clin Oncol. 1990;8:347–55.
8. Hartmann JT, Albrecht C, Schmoll HJ, Kuczyk MA, Kollmannsberger C, Bokemeyer C. Long-term effects on sexual function and fertility after treatment of testicular cancer. Br J Cancer. 1999;80:801–7.
9. Schover LR, Rybicki LA, Martin BA, Bringelsen KA. Having children after cancer. A pilot survey of survivors' attitudes and experiences. Cancer. 1999;86:697–709.
10. Mackie E, Hill J, Kondryn H, Mcnally R. Adult psychosocial outcomes in long-term survivors of acute lymphoblastic leukaemia and Wilms' tumour: a controlled study. Lancet. 2000;355:1310–4.
11. Saito K, Suzuki K, Iwasaki A, Yumura Y, Kubota Y. Sperm cryopreservation before cancer chemotherapy helps in the emotional battle against cancer. Cancer. 2005;104:521–4.
12. Fossa SD, Aass N, Molne K. Is routine pre-treatment cryopreservation of semen worthwhile in the management of patients with testicular cancer? Br J Urol. 1989;64:524–9.
13. Sanger WG, Olson JH, Sherman JK. Semen cryobanking for men with cancer—criteria change. Fertil Steril. 1992;58:1024–7.
14. Lass A, Akagbosu F, Abusheikha N, Hassouneh M, Blayney M, Avery S, et al. A programme of semen cryopreservation for patients with malignant disease in a tertiary infertility centre: lessons from 8 years' experience. Hum Reprod. 1998;13:3256–61.
15. Naysmith TE, Blake DA, Harvey VJ, Johnson NP. Do men undergoing sterilizing cancer treatments have a fertile future? Hum Reprod. 1998;13:3250–5.
16. Audrins P, Holden CA, Mclachlan RI, Kovacs GT. Semen storage for special purposes at Monash IVF from 1977 to 1997. Fertil Steril. 1999;72:179–81.
17. Agarwal A. Semen banking in patients with cancer: 20-year experience. Int J Androl. 2000;23 Suppl 2:16–9.
18. Kelleher S, Wishart SM, Liu PY, Turner L, Di Pierro I, Conway AJ, et al. Long-term outcomes of elective human sperm cryostorage. Hum Reprod. 2001;16:2632–9.
19. Bahadur G, Ling KL, Hart R, Ralph D, Riley V, Wafa R, et al. Semen production in adolescent cancer patients. Hum Reprod. 2002;17:2654–6.
20. Saito K, Suzuki K, Noguchi K, Ogawa T, Takeda M, Hosaka M, et al. Semen cryopreservation for patients with malignant or nonmalignant disease: our experience for 10 years. Nippon Hinyokika Gakkai Zasshi. 2003;94:513–20.
21. Petersen PM, Skakkebaek NE, Rorth M, Giwercman A. Semen quality and reproductive hormones before and after orchiectomy in men with testicular cancer. J Urol. 1999;161:822–6.
22. Petersen PM, Skakkebaek NE, Vistisen K, Rorth M, Giwercman A. Semen quality and reproductive hormones before orchiectomy in men with testicular cancer. J Clin Oncol. 1999;17:941–7.
23. Rueffer U, Breuer K, Josting A, Lathan B, Sieber M, Manzke O, et al. Male gonadal dysfunction in patients with Hodgkin's disease prior to treatment. Ann Oncol. 2001;12:1307–11.
24. Agarwal A, Allamaneni SS. Disruption of spermatogenesis by the cancer disease process. J Natl Cancer Inst Monogr. 2005;34:9–12.
25. Hallak J, Kolettis PN, Sekhon VS, Thomas Jr AJ, Agarwal A. Sperm cryopreservation in patients with testicular cancer. Urology. 1999;54:894–9.
26. Colpi GM, Contalbi GF, Nerva F, Sagone P, Piediferro G. Testicular function following chemo-radiotherapy. Eur J Obstet Gynecol Reprod Biol. 2004;113 Suppl 1:S2–6.
27. Rofeim O, Gilbert BR. Normal semen parameters in cancer patients presenting for cryopreservation before gonadotoxic therapy. Fertil Steril. 2004;82:505–6.
28. Ragni G, Somigliana E, Restelli L, Salvi R, Arnoldi M, Paffoni A. Sperm banking and rate of assisted reproduction treatment: insights from a 15-year cryopreservation program for male cancer patients. Cancer. 2003;97:1624–9.
29. Bahadur G, Ozturk O, Muneer A, Wafa R, Ashraf A, Jaman N, et al. Semen quality before and after gonadotoxic treatment. Hum Reprod. 2005;20:774–81.
30. Williams DHT, Karpman E, Sander JC, Spiess PE, Pisters LL, Lipshultz LI. Pretreatment semen parameters in men with cancer. J Urol. 2009;181:736–40.
31. Chung K, Irani J, Knee G, Efymow B, Blasco L, Patrizio P. Sperm cryopreservation for male patients with cancer: an epidemiological analysis at the University of Pennsylvania. Eur J Obstet Gynecol Reprod Biol. 2004;113 Suppl 1:S7–11.

32. Meseguer M, Molina N, Garcia-Velasco JA, Remohi J, Pellicer A, Garrido N. Sperm cryopreservation in oncological patients: a 14-year follow-up study. Fertil Steril. 2006;85:640–5.
33. Hendry WF, Stedronska J, Jones CR, Blackmore CA, Barrett A, Peckham MJ. Semen analysis in testicular cancer and Hodgkin's disease: pre- and post-treatment findings and implications for cryopreservation. Br J Urol. 1983;55:769–73.
34. Viviani S, Bonfante V, Santoro A, Zanini M, Devizzi L, Di Russo AD, et al. Long-term results of an intensive regimen: VEBEP plus involved-field radiotherapy in advanced Hodgkin's disease. Cancer J Sci Am. 1999;5:275–82.
35. Berthelsen JG, Skakkebaek NE. Gonadal function in men with testis cancer. Fertil Steril. 1983;39:68–75.
36. Stahl O, Eberhard J, Cavallin-Stahl E, Jepson K, Friberg B, Tingsmark C, et al. Sperm DNA integrity in cancer patients: the effect of disease and treatment. Int J Androl. 2009;32:695–703.
37. Spitz S. The histological effects of nitrogen mustards on human tumors and tissues. Cancer. 1948;1:383–98.
38. Apperley JF, Reddy N. Mechanism and management of treatment-related gonadal failure in recipients of high dose chemoradiotherapy. Blood Rev. 1995;9:93–116.
39. Howell SJ, Shalet SM. Spermatogenesis after cancer treatment: damage and recovery. J Natl Cancer Inst Monogr. 2005;34:12–7.
40. Schrader M, Muller M, Straub B, Miller K. The impact of chemotherapy on male fertility: a survey of the biologic basis and clinical aspects. Reprod Toxicol. 2001;15:611–7.
41. Meirow D, Schenker JG. Cancer and male infertility. Hum Reprod. 1995;10:2017–22.
42. Carson SA, Gentry WL, Smith AL, Buster JE. Feasibility of semen collection and cryopreservation during chemotherapy. Hum Reprod. 1991;6:992–4.
43. Pont J, Albrecht W. Fertility after chemotherapy for testicular germ cell cancer. Fertil Steril. 1997;68:1–5.
44. Leonard M, Hammelef K, Smith GD. Fertility considerations, counseling, and semen cryopreservation for males prior to the initiation of cancer therapy. Clin J Oncol Nurs. 2004;8:127–31, 145.
45. Viviani S, Santoro A, Ragni G, Bonfante V, Bestetti O, Bonadonna G. Gonadal toxicity after combination chemotherapy for Hodgkin's disease. Comparative results of MOPP vs ABVD. Eur J Cancer Clin Oncol. 1985;21:601–5.
46. Lee SJ, Schover LR, Partridge AH, Patrizio P, Wallace WH, Hagerty K, et al. American Society of Clinical Oncology recommendations on fertility preservation in cancer patients. J Clin Oncol. 2006;24:2917–31.
47. Kenney LB, Laufer MR, Grant FD, Grier H, Diller L. High risk of infertility and long term gonadal damage in males treated with high dose cyclophosphamide for sarcoma during childhood. Cancer. 2001;91:613–21.
48. Kreuser ED, Klingmuller D, Thiel E. The role of LHRH-analogues in protecting gonadal functions during chemotherapy and irradiation. Eur Urol. 1993;23:157–63. discussion 163-154.
49. Cespedes RD, Peretsman SJ, Thompson Jr IM, Jackson C. Protection of the germinal epithelium in the rat from the cytotoxic effects of chemotherapy by a luteinizing hormone-releasing hormone agonist and antiandrogen therapy. Urology. 1995;46: 688–91.
50. Lushbaugh CC, Casarett GW. The effects of gonadal irradiation in clinical radiation therapy: a review. Cancer. 1976;37:1111–25.
51. Trottmann M, Becker AJ, Stadler T, Straub J, Soljanik I, Schlenker B, et al. Semen quality in men with malignant diseases before and after therapy and the role of cryopreservation. Eur Urol. 2007;52: 355–67.
52. Speiser B, Rubin P, Casarett G. Aspermia following lower truncal irradiation in Hodgkin's disease. Cancer. 1973;32:692–8.
53. Leiper AD, Grant DB, Chessells JM. Gonadal function after testicular radiation for acute lymphoblastic leukaemia. Arch Dis Child. 1986;61:53–6.
54. Fossa SD, Ous S, Abyholm T, Norman N, Loeb M. Post-treatment fertility in patients with testicular cancer. II. Influence of cis-platin-based combination chemotherapy and of retroperitoneal surgery on hormone and sperm cell production. Br J Urol. 1985;57:210–4.
55. Donohue JP. Evolution of retroperitoneal lymphadenectomy (RPLND) in the management of non-seminomatous testicular cancer (NSGCT). Urol Oncol. 2003;21:129–32.
56. Large MC, Sheinfeld J, Eggener SE. Retroperitoneal lymph node dissection: reassessment of modified templates. BJU Int. 2009;104: 1369–75.
57. Jones OM, Stevenson AR, Stitz RW, Lumley JW. Preservation of sexual and bladder function after laparoscopic rectal surgery. Colorectal Dis. 2009;11:489–95.
58. Jacobsen KD, Olsen DR, Fossa K, Fossa SD. External beam abdominal radiotherapy in patients with seminoma stage I: field type, testicular dose, and spermatogenesis. Int J Radiat Oncol Biol Phys. 1997;38:95–102.
59. Giwercman A, Petersen PM. Cancer and male infertility. Baillieres Best Pract Res Clin Endocrinol Metab. 2000;14:453–71.
60. Huyghe E, Matsuda T, Daudin M, Chevreau C, Bachaud JM, Plante P, et al. Fertility after testicular cancer treatments: results of a large multicenter study. Cancer. 2004;100:732–7.
61. Gandini L, Sgro P, Lombardo F, Paoli D, Culasso F, Toselli L, et al. Effect of chemo- or radiotherapy on sperm parameters of testicular cancer patients. Hum Reprod. 2006;21:2882–9.
62. Spermon JR, Ramos L, Wetzels AM, Sweep CG, Braat DD, Kiemeney LA, et al. Sperm integrity pre- and post-chemotherapy in men with testicular germ cell cancer. Hum Reprod. 2006;21: 1781–6.
63. Tournaye H, Liu J, Nagy PZ, Camus M, Goossens A, Silber S, et al. Correlation between testicular histology and outcome after intracytoplasmic sperm injection using testicular spermatozoa. Hum Reprod. 1996;11:127–32.
64. Magelssen H, Brydoy M, Fossa SD. The effects of cancer and cancer treatments on male reproductive function. Nat Clin Pract Urol. 2006;3:312–22.
65. Lampe H, Horwich A, Norman A, Nicholls J, Dearnaley DP. Fertility after chemotherapy for testicular germ cell cancers. J Clin Oncol. 1997;15:239–45.
66. Brydoy M, Fossa SD, Klepp O, Bremnes RM, Wist EA, Wentzel-Larsen T, et al. Paternity following treatment for testicular cancer. J Natl Cancer Inst. 2005;97:1580–8.
67. Huddart RA, Norman A, Moynihan C, Horwich A, Parker C, Nicholls E, et al. Fertility, gonadal and sexual function in survivors of testicular cancer. Br J Cancer. 2005;93:200–7.
68. Joos H, Sedlmayer F, Gomahr A, Rahim HB, Frick J, Kogelnik HD, et al. Endocrine profiles after radiotherapy in stage I seminoma: impact of two different radiation treatment modalities. Radiother Oncol. 1997;43:159–62.
69. Nalesnik JG, Sabanegh Jr ES, Eng TY, Buchholz TA. Fertility in men after treatment for stage 1 and 2A seminoma. Am J Clin Oncol. 2004;27:584–8.
70. Classen J, Dieckmann K, Bamberg M, Souchon R, Kliesch S, Kuehn M, et al. Radiotherapy with 16 Gy may fail to eradicate testicular intraepithelial neoplasia: preliminary communication of a dose-reduction trial of the German Testicular Cancer Study Group. Br J Cancer. 2003;88:828–31.
71. Da Cunha MF, Meistrich ML, Fuller LM, Cundiff JH, Hagemeister FB, Velasquez WS, et al. Recovery of spermatogenesis after treatment for Hodgkin's disease: limiting dose of MOPP chemotherapy. J Clin Oncol. 1984;2:571–7.
72. Marmor D, Duyck F. Male reproductive potential after MOPP therapy for Hodgkin's disease: a long-term survey. Andrologia. 1995;27:99–106.
73. Tal R, Botchan A, Hauser R, Yogev L, Paz G, Yavetz H. Follow-up of sperm concentration and motility in patients with lymphoma. Hum Reprod. 2000;15:1985–8.

74. Jacob A, Barker H, Goodman A, Holmes J. Recovery of spermatogenesis following bone marrow transplantation. Bone Marrow Transplant. 1998;22:277–9.
75. Anserini P, Chiodi S, Spinelli S, Costa M, Conte N, Copello F, et al. Semen analysis following allogeneic bone marrow transplantation. Additional data for evidence-based counselling. Bone Marrow Transplant. 2002;30:447–51.
76. Boyd BG, Mccallum SW, Lewis RW, Terris MK. Assessment of patient concern and adequacy of informed consent regarding infertility resulting from prostate cancer treatment. Urology. 2006;68:840–4.
77. Magheli A, Rais-Bahrami S, Humphreys EB, Peck HJ, Trock BJ, Gonzalgo ML. Impact of patient age on biochemical recurrence rates following radical prostatectomy. J Urol. 2007;178:1933–7. discussion 1937-1938.
78. Daniell HW, Tam EW. Testicular atrophy in therapeutic orchiectomy specimens from men with prostate carcinoma: association with prior prostate bed radiation and older age. Cancer. 1998;83:1174–9.
79. Mydlo JH, Lebed B. Does brachytherapy of the prostate affect sperm quality and/or fertility in younger men? Scand J Urol Nephrol. 2004;38:221–4.
80. Raviv G, Pinthus JH, Shefi S, Mor Y, Kaufman-Francis K, Levron J, et al. Effects of intravesical chemotherapy and immunotherapy on semen analysis. Urology. 2005;65:765–7.
81. Colombo R, Bertini R, Salonia A, Da Pozzo LF, Montorsi F, Brausi M, et al. Nerve and seminal sparing radical cystectomy with orthotopic urinary diversion for select patients with superficial bladder cancer: an innovative surgical approach. J Urol. 2001;165:51–5. discussion 55.
82. Hyer S, Vini L, O'connell M, Pratt B, Harmer C. Testicular dose and fertility in men following I(131) therapy for thyroid cancer. Clin Endocrinol (Oxf). 2002;56:755–8.
83. Longhi A, Macchiagodena M, Vitali G, Bacci G. Fertility in male patients treated with neoadjuvant chemotherapy for osteosarcoma. J Pediatr Hematol Oncol. 2003;25:292–6.
84. Mansky P, Arai A, Stratton P, Bernstein D, Long L, Reynolds J, et al. Treatment late effects in long-term survivors of pediatric sarcoma. Pediatr Blood Cancer. 2007;48:192–9.
85. Hawkins MM, Draper GJ, Smith RA. Cancer among 1,348 offspring of survivors of childhood cancer. Int J Cancer. 1989;43:975–8.
86. Winther JF, Boice Jr JD, Mulvihill JJ, Stovall M, Frederiksen K, Tawn EJ, et al. Chromosomal abnormalities among offspring of childhood-cancer survivors in Denmark: a population-based study. Am J Hum Genet. 2004;74:1282–5.
87. Sankila R, Olsen JH, Anderson H, Garwicz S, Glattre E, Hertz H, et al. Risk of cancer among offspring of childhood-cancer survivors. Association of the Nordic Cancer Registries and the Nordic Society of Paediatric Haematology and Oncology. N Engl J Med. 1998;338:1339–44.
88. Robbins WA, Meistrich ML, Moore D, Hagemeister FB, Weier HU, Cassel MJ, et al. Chemotherapy induces transient sex chromosomal and autosomal aneuploidy in human sperm. Nat Genet. 1997;16:74–8.
89. De Mas P, Daudin M, Vincent MC, Bourrouillou G, Calvas P, Mieusset R, et al. Increased aneuploidy in spermatozoa from testicular tumour patients after chemotherapy with cisplatin, etoposide and bleomycin. Hum Reprod. 2001;16:1204–8.
90. Frias S, Van Hummelen P, Meistrich ML, Lowe XR, Hagemeister FB, Shelby MD, et al. NOVP chemotherapy for Hodgkin's disease transiently induces sperm aneuploidies associated with the major clinical aneuploidy syndromes involving chromosomes X, Y, 18, and 21. Cancer Res. 2003;63:44–51.
91. Thomas C, Cans C, Pelletier R, De Robertis C, Hazzouri M, Sele B, et al. No long-term increase in sperm aneuploidy rates after anticancer therapy: sperm fluorescence in situ hybridization analysis in 26 patients treated for testicular cancer or lymphoma. Clin Cancer Res. 2004;10:6535–43.
92. Wyrobek AJ, Schmid TE, Marchetti F. Relative susceptibilities of male germ cells to genetic defects induced by cancer chemotherapies. J Natl Cancer Inst Monogr. 2005;34:31–5.
93. Martin R. Human sperm chromosome complements in chemotherapy patients and infertile men. Chromosoma. 1998;107: 523–7.
94. World Health Organisation. WHO laboratory manual for the examination of human semen and sperm-cervical mucus interaction. 4th ed. Cambridge University Press: Cambridge; 1999.
95. Padron OF, Sharma RK, Thomas Jr AJ, Agarwal A. Effects of cancer on spermatozoa quality after cryopreservation: a 12-year experience. Fertil Steril. 1997;67:326–31.
96. Bolten M, Weissbach L, Kaden R. Cryopreserved human sperm deposits: usability after decades of storage. Urologe A. 2005;44: 904–8.
97. Gandini L, Lombardo F, Lenzi A, Spano M, Dondero F. Cryopreservation and sperm DNA integrity. Cell Tissue Bank. 2006; 7:91–8.
98. Meyers SA. Dry storage of sperm: applications in primates and domestic animals. Reprod Fertil Dev. 2006;18:1–5.
99. Kuczynski W, Dhont M, Grygoruk C, Grochowski D, Wolczynski S, Szamatowicz M. The outcome of intracytoplasmic injection of fresh and cryopreserved ejaculated spermatozoa—a prospective randomized study. Hum Reprod. 2001;16:2109–13.
100. Meistrich ML. Potential genetic risks of using semen collected during chemotherapy. Hum Reprod. 1993;8:8–10.
101. Shin D, Lo KC, Lipshultz LI. Treatment options for the infertile male with cancer. J Natl Cancer Inst Monogr. 2005;34:48–50.
102. Schover LR, Brey K, Lichtin A, Lipshultz LI, Jeha S. Knowledge and experience regarding cancer, infertility, and sperm banking in younger male survivors. J Clin Oncol. 2002;20:1880–9.
103. Dohle GR, Colpi GM, Hargreave TB, Papp GK, Jungwirth A, Weidner W. Eau guidelines on male infertility. Eur Urol. 2005;48:703–11.
104. Edge B, Holmes D, Makin G. Sperm banking in adolescent cancer patients. Arch Dis Child. 2006;91:149–52.
105. Schover LR, Brey K, Lichtin A, Lipshultz LI, Jeha S. Oncologists' attitudes and practices regarding banking sperm before cancer treatment. J Clin Oncol. 2002;20:1890–7.
106. Magelssen H, Haugen TB, Von During V, Melve KK, Sandstad B, Fossa SD. Twenty years experience with semen cryopreservation in testicular cancer patients: who needs it? Eur Urol. 2005;48:779–85.
107. Reebals JF, Brown R, Buckner EB. Nurse practice issues regarding sperm banking in adolescent male cancer patients. J Pediatr Oncol Nurs. 2006;23:182–8.
108. Koeppel KM. Sperm banking and patients with cancer. Issues concerning patients and healthcare professionals. Cancer Nurs. 1995;18:306–12.
109. Canada AL, Schover LR. Research promoting better patient education on reproductive health after cancer. J Natl Cancer Inst Monogr. 2005;34:98–100.
110. Chatterjee R, Haines GA, Perera DM, Goldstone A, Morris ID. Testicular and sperm DNA damage after treatment with fludarabine for chronic lymphocytic leukaemia. Hum Reprod. 2000;15:762–6.
111. Schuster TG, Hickner-Cruz K, Ohl DA, Goldman E, Smith GD. Legal considerations for cryopreservation of sperm and embryos. Fertil Steril. 2003;80:61–6.
112. Zebrack BJ, Zeltzer LK. Quality of life issues and cancer survivorship. Curr Probl Cancer. 2003;27:198–211.
113. Hirsch M, Lunenfeld B, Modan M, Ovadia J, Shemesh J. Spermarche—the age of onset of sperm emission. J Adolesc Health Care. 1985;6:35–9.

114. Nielsen CT, Skakkebaek NE, Richardson DW, Darling JA, Hunter WM, Jorgensen M, et al. Onset of the release of spermatozoa (spermarche) in boys in relation to age, testicular growth, pubic hair, and height. J Clin Endocrinol Metab. 1986;62:532–5.
115. Kliesch S, Behre HM, Jurgens H, Nieschlag E. Cryopreservation of semen from adolescent patients with malignancies. Med Pediatr Oncol. 1996;26:20–7.
116. Bahadur G, Ling KL, Hart R, Ralph D, Wafa R, Ashraf A, et al. Semen quality and cryopreservation in adolescent cancer patients. Hum Reprod. 2002;17:3157–61.
117. Ginsberg JP, Ogle SK, Tuchman LK, Carlson CA, Reilly MM, Hobbie WL, et al. Sperm banking for adolescent and young adult cancer patients: sperm quality, patient, and parent perspectives. Pediatr Blood Cancer. 2008;50:594–8.
118. Klosky JL, Randolph ME, Navid F, Gamble HL, Spunt SL, Metzger ML, et al. Sperm cryopreservation practices among adolescent cancer patients at risk for infertility. Pediatr Hematol Oncol. 2009;26:252–60.
119. Vanderzwalmen P, Zech H, Birkenfeld A, Yemini M, Bertin G, Lejeune B, et al. Intracytoplasmic injection of spermatids retrieved from testicular tissue: influence of testicular pathology, type of selected spermatids and oocyte activation. Hum Reprod. 1997;12:1203–13.
120. Kim ED, Gilbaugh 3rd JH, Patel VR, Turek PJ, Lipshultz LI. Testis biopsies frequently demonstrate sperm in men with azoospermia and significantly elevated follicle-stimulating hormone levels. J Urol. 1997;157:144–6.
121. Schrader M, Muller M, Sofikitis N, Straub B, Krause H, Miller K. "Onco-tese": testicular sperm extraction in azoospermic cancer patients before chemotherapy-new guidelines? Urology. 2003;61: 421–5.
122. Baniel J, Sella A. Sperm extraction at orchiectomy for testis cancer. Fertil Steril. 2001;75:260–2.
123. Binsaleh S, Sircar K, Chan PT. Feasibility of simultaneous testicular microdissection for sperm retrieval and ipsilateral testicular tumor resection in azoospermic men. J Androl. 2004;25:867–71.
124. Devroey P, Liu J, Nagy Z, Goossens A, Tournaye H, Camus M, et al. Pregnancies after testicular sperm extraction and intracytoplasmic sperm injection in non-obstructive azoospermia. Hum Reprod. 1995;10:1457–60.
125. Chan PT, Palermo GD, Veeck LL, Rosenwaks Z, Schlegel PN. Testicular sperm extraction combined with intracytoplasmic sperm injection in the treatment of men with persistent azoospermia postchemotherapy. Cancer. 2001;92:1632–7.
126. Damani MN, Master V, Meng MV, Burgess C, Turek P, Oates RD. Postchemotherapy ejaculatory azoospermia: fatherhood with sperm from testis tissue with intracytoplasmic sperm injection. J Clin Oncol. 2002;20:930–6.
127. Meseguer M, Garrido N, Remohi J, Pellicer A, Simon C, Martinez-Jabaloyas JM, et al. Testicular sperm extraction (TESE) and ICSI in patients with permanent azoospermia after chemotherapy. Hum Reprod. 2003;18:1281–5.
128. Oates RD, Lobel SM, Harris DH, Pang S, Burgess CM, Carson RS. Efficacy of intracytoplasmic sperm injection using intentionally cryopreserved epididymal spermatozoa. Hum Reprod. 1996;11:133–8.
129. Janzen N, Goldstein M, Schlegel PN, Palermo GD, Rosenwaks Z, Hariprashad J. Use of electively cryopreserved microsurgically aspirated epididymal sperm with IVF and intracytoplasmic sperm injection for obstructive azoospermia. Fertil Steril. 2000;74:696–701.
130. Ohl DA, Denil J, Bennett CJ, Randolph JF, Menge AC, Mccabe M. Electroejaculation following retroperitoneal lymphadenectomy. J Urol. 1991;145:980–3.
131. Ohl DA, Wolf LJ, Menge AC, Christman GM, Hurd WW, Ansbacher R, et al. Electroejaculation and assisted reproductive technologies in the treatment of anejaculatory infertility. Fertil Steril. 2001;76:1249–55.
132. Chung PH, Verkauf BS, Mola R, Skinner L, Eichberg RD, Maroulis GB. Correlation between semen parameters of electroejaculates and achieving pregnancy by intrauterine insemination. Fertil Steril. 1997;67:129–32.
133. Schatte EC, Orejuela FJ, Lipshultz LI, Kim ED, Lamb DJ. Treatment of infertility due to anejaculation in the male with electroejaculation and intracytoplasmic sperm injection. J Urol. 2000;163:1717–20.
134. Schmidt KL, Larsen E, Bangsboll S, Meinertz H, Carlsen E, Andersen AN. Assisted reproduction in male cancer survivors: fertility treatment and outcome in 67 couples. Hum Reprod. 2004;19:2806–10.
135. Khalifa E, Oehninger S, Acosta AA, Morshedi M, Veeck L, Bryzyski RG, et al. Successful fertilization and pregnancy outcome in in-vitro fertilization using cryopreserved/thawed spermatozoa from patients with malignant diseases. Hum Reprod. 1992;7:105–8.
136. Palermo G, Joris H, Devroey P, Van Steirteghem AC. Pregnancies after intracytoplasmic injection of single spermatozoon into an oocyte. Lancet. 1992;340:17–8.
137. Rosenlund B, Sjoblom P, Tornblom M, Hultling C, Hillensjo T. In-vitro fertilization and intracytoplasmic sperm injection in the treatment of infertility after testicular cancer. Hum Reprod. 1998;13:414–8.
138. Ginsburg ES, Yanushpolsky EH, Jackson KV. In vitro fertilization for cancer patients and survivors. Fertil Steril. 2001;75:705–10.
139. Zorn B, Virant-Klun I, Stanovnik M, Drobnic S, Meden-Vrtovec H. Intracytoplasmic sperm injection by testicular sperm in patients with aspermia or azoospermia after cancer treatment. Int J Androl. 2006;29:521–7.
140. Van Casteren NJ, Van Santbrink EJ, Van Inzen W, Romijn JC, Dohle GR. Use rate and assisted reproduction technologies outcome of cryopreserved semen from 629 cancer patients. Fertil Steril. 2008;90:2245–50.
141. Ulug U, Bener F, Karagenc L, Ciray N, Bahceci M. Outcomes in couples undergoing ICSI: comparison between fresh and frozen-thawed surgically retrieved spermatozoa. Int J Androl. 2005;28:343–9.
142. Wald M, Ross LS, Prins GS, Cieslak-Janzen J, Wolf G, Niederberger CS. Analysis of outcomes of cryopreserved surgically retrieved sperm for IVF/ICSI. J Androl. 2006;27:60–5.
143. Borges Jr E, Rossi LM, Locambo De Freitas CV, Guilherme P, Bonetti TC, Iaconelli A, et al. Fertilization and pregnancy outcome after intracytoplasmic injection with fresh or cryopreserved ejaculated spermatozoa. Fertil Steril. 2007;87:316–20.
144. Schlatt S, Ehmcke J, Jahnukainen K. Testicular stem cells for fertility preservation: preclinical studies on male germ cell transplantation and testicular grafting. Pediatr Blood Cancer. 2009;53: 274–80.
145. Brinster RL, Avarbock MR. Germline transmission of donor haplotype following spermatogonial transplantation. Proc Natl Acad Sci USA. 1994;91:11303–7.
146. Dobrinski I. Germ cell transplantation. Semin Reprod Med. 2005;23:257–65.
147. Dobrinski I. Germ cell transplantation and testis tissue xenografting in domestic animals. Anim Reprod Sci. 2005;89:137–45.
148. Avarbock MR, Brinster CJ, Brinster RL. Reconstitution of spermatogenesis from frozen spermatogonial stem cells. Nat Med. 1996;2:693–6.
149. Izadyar F, Matthijs-Rijsenbilt JJ, Den Ouden K, Creemers LB, Woelders H, De Rooij DG. Development of a cryopreservation protocol for type a spermatogonia. J Androl. 2002;23:537–45.
150. Nagano M, Patrizio P, Brinster RL. Long-term survival of human spermatogonial stem cells in mouse testes. Fertil Steril. 2002;78:1225–33.

151. Honaramooz A, Behboodi E, Megee SO, Overton SA, Galantino-Homer H, Echelard Y, et al. Fertility and germline transmission of donor haplotype following germ cell transplantation in immunocompetent goats. Biol Reprod. 2003;69: 1260–4.
152. Zhang X, Ebata KT, Nagano MC. Genetic analysis of the clonal origin of regenerating mouse spermatogenesis following transplantation. Biol Reprod. 2003;69:1872–8.
153. Goossens E, Tournaye H. Testicular stem cells. Semin Reprod Med. 2006;24:370–8.
154. Jahnukainen K, Ehmcke J, Schlatt S. Testicular xenografts: a novel approach to study cytotoxic damage in juvenile primate testis. Cancer Res. 2006;66:3813–8.
155. Keros V, Hultenby K, Borgstrom B, Fridstrom M, Jahnukainen K, Hovatta O. Methods of cryopreservation of testicular tissue with viable spermatogonia in pre-pubertal boys undergoing gonadotoxic cancer treatment. Hum Reprod. 2007;22:1384–95.
156. Jahnukainen K, Hou M, Petersen C, Setchell B, Soder O. Intratesticular transplantation of testicular cells from leukemic rats causes transmission of leukemia. Cancer Res. 2001;61: 706–10.
157. Feng LX, Chen Y, Dettin L, Pera RA, Herr JC, Goldberg E, et al. Generation and in vitro differentiation of a spermatogonial cell line. Science. 2002;297:392–5.
158. Toyooka Y, Tsunekawa N, Akasu R, Noce T. Embryonic stem cells can form germ cells in vitro. Proc Natl Acad Sci USA. 2003;100:11457–62.

Genetic Aspects of Male Infertility

18

Orhan Bukulmez

Abstract

Genetic abnormalities may account for 15–30% of male factor infertility. Genes and genomic regulation involved in male genital tract development, gonadal development, and function including those related to spermatogenesis may be involved with male infertility. Although many of the genetic factors are still to be elucidated and many of the predicted genetic perturbations are yet to find their place for clinical applications, it is essential to appraise the current information on genetic basis of male reproductive system disorders.

Keywords

Genetics in male infertility • Genomic regulation of male sexual development • Sertoli cells • Testis formation • Hypospadias • Peritubular myoid cells • Spermatogenic genes • Male genital tract development • Spermatogenesis

Genetic abnormalities may account for 15–30% of male factor infertility [1]. Genes and genomic regulation involved in male genital tract development, gonadal development, and function including those related to spermatogenesis may be involved with male infertility. Although many of the genetic factors are still to be elucidated and many of the predicted genetic perturbations are yet to find their place for clinical applications, it is essential to appraise the current information on genetic basis of male reproductive system disorders.

Genomic Regulation of Male Sexual Development

The gender-specific development simply relies on whether testes or ovaries form in the embryo from paired omnipotent structures known as genital ridges. Male developmental pathway depends on the presence and correct function of a male-determining gene from Y chromosome called *Sry* (sex-determining region Y) which functions in a specific set of genital ridge cells to stimulate them to differentiate as Sertoli cells, which are the cells that interact with and nurture the germ cells. Sertoli cells play a role in orchestrating the differentiation of other cell types required for testis formation such as germ cells and steroid hormone-producing cells [2]. In the absence of *Sry* function, the developmental pathway proceeds to female differentiation, although in reality the sexual development is more complex involving multiple networks of molecular signals and fragility of these pathways are reflected to the fact that disorders of sexual development are among the most common birth defects. These disorders range from hypospadias to complete sexual ambiguity and sex reversal which are often associated with infertility.

O. Bukulmez, MD (✉)
Division of Reproductive Endocrinology and Infertility, Department of Obstetrics and Gynecology, University of Texas Southwestern Medical Center, 5323 Harry Hines Boulevard, Dallas, TX 75390-9032, USA
e-mail: orhan.bukulmez@utsouthwestern.edu

Testicular Development

Early formation of indifferent genital ridges is a requirement before testicular development. Studies in mice have shown that several transcription factor genes are required

S.J. Parekattil and A. Agarwal (eds.), *Male Infertility: Contemporary Clinical Approaches, Andrology, ART & Antioxidants*, DOI 10.1007/978-1-4614-3335-4_18, © Springer Science+Business Media, LLC 2012

for this process. These genes include empty spiracles homologue 2 (Emx2), GATA-binding protein-4 (GATA4), Lim homeobox protein 9 (Lhx9), steroidogenic factor 1 (SF-1/NR5A1), dosage-sensitive sex reversal, adrenal hypoplasia critical region, X chromosome, gene 1 (DAX-1/Nr0b1), and Wilms' tumor 1 (WT-1) [3–8]. WT-1 and SF-1 are crucial for formation of genital ridges in humans. Furthermore, they are both important in sex-specific gonadal development [9, 10]. DAX-1 levels and expression thresholds on the other hand are important for male versus female functional development [8].

Timely expression of *Sry* is important for testicular development [11]. Spatially dynamic *Sry* expression begins in waves within the genital ridge, peaks for a short time, and then declines as demonstrated in mice [12, 13]. The mechanism of this specific and tight regulation of *Sry* regulation remains elusive. Decreased expression of *Sry* was noted in the presence of splice variant mutants of WT-1, GATA4, friend of GATA (FOG2), and insulin receptor family [14, 15]. Delayed expression of *Sry* has been implicated in XY sex reversals, unilateral or bilateral ovotestes, to delayed testis formation [16, 17]. It is thought that *Sry* expression must reach a certain threshold within a specific temporal window of competence in the precursors of supporting cells for proper testicular development. *Sry* encodes for a nuclear high-mobility group (HMG) domain protein that binds and bends DNA.

Downstream of *Sry*, other factors such as SRY-box containing gene 9 (SOX9), SOX8, DAX-1, and fibroblast growth factor 9 (FGF9) have important roles in Sertoli cell differentiation and function [2]. SOX9 is considered as an early acting and essential component of male development pathway.

Peritubular myoid cells of testis are necessary for testicular cord development and integrity. Differentiation of these cells correlates with Sertoli cell-specific secretion of desert hedgehog (DHH). DHH receptor patched (PTC) is expressed on peritubular myoid cells and Leydig cells. It has been demonstrated that null mutation of DHH in mice leads to impaired differentiation of peritubular myoid cells and Leydig cells leading to feminized males [18, 19]. DHH mutations in humans lead to partial or pure XY gonadal dysgenesis accompanied by impaired cord formation and decreased testosterone levels [20, 21].

Fetal Leydig cell development is essential for male sexual differentiation. The candidate genes important in Leydig cell differentiation include aristaless-related homeobox gene (ARX), α-thalassemia/mental retardation syndrome, X-linked (ATRX) and platelet-derived growth factors (PDGFs), and their receptor PDGFRA [22–24].

Primordial germ cells migrate from their origin at the posterior part of the embryo through hindgut to populate genital ridges where they interact with somatic cells to form primitive sex cords. Germ-cell migration is facilitated by interferon-induced transmembrane proteins 1 and 3 (IFITM1 and IFITM3) [25]. Stromal cell-derived factor 1 (SDF1) and its receptor CXCR4 function in the colonization of genital ridges [26].

Testicular Descent

Testicular descent occurs in two phases. Transabdominal phase of testicular descent happens at 8–15 weeks of gestation in humans and controlled by insulin-like 3 (INSL3) hormone produced by Leydig cells, acting via its receptor LGR8 (also known as GREAT) [27]. Inguinoscrotal phase is usually completed by 35th week of gestation and is facilitated by the neurotransmitter calcitonin gene-related peptide (CGRP or CALCA) released by genitofemoral nerve under the influence of androgens. Mutations in genes involved with androgen signaling and those encode transcription factors homeobox A10 (HOXA10), HOXA11, and developmentally and sexually retarded with transient immune abnormalities (DESRT) lead to second-stage arrest in testicular descent [28].

Spermatogenesis

There are many genes and molecules involved in spermatogenesis. For instance, the number of sperm-specific membrane proteins alone has been estimated to be greater than 200 [29]. Through the use of cDNA microarrays in humans, over 100 genes seem to be involved in the regulation of spermatogenesis [30]. One study showed that of 1,652 genes whose expression increased with onset of meiosis, 351 of them were expressed only in the male germ line [31]. There are many genes involved in DNA condensation, sperm maturation, adhesion, and motility as well. The genes and proteins involved with male germ cells and spermatogenesis have been reviewed extensively [32]. This review presents 178 genes associated with spermatogonia, spermatocytes, and Sertoli cells [32]. These genes are located on both autosomal and sex chromosomes.

In patients with azoospermia, deleted regions on Yq has attracted clinical attention. Deleted in azoospermia (DAZ) gene on Y chromosome belongs to a family of three members: DAZ, BOULE, and DAZ-like. Their proteins contain a highly conserved RNA-binding motif [33]. DAZ proteins bind to RNAs and may be involved in posttranscriptional regulation of mRNA expression [34]. The DAZ gene family is expressed exclusively in germ cells.

Many forms of partial deletions occur on Y chromosome in some male infertility cases [35]. These infertility-related deletions are not considered to be inherited [36], and most infertile males do not show any mutations or deletions on the

Yq. Majority of genes involved with spermatogenesis are actually located on autosomal chromosomes [32, 37]. For instance, mutations in protamine (PRM) and transition protein (TPN) genes involved with histone to protamine replacement located on autosomes have been found in 1/200–1/300 of male infertility cases in Japan [37].

As shown in mice, X chromosome is also enriched for spermatogenic genes functioning both in premeiotic and in postmeiotic germ cells [38]. In human, the examples of genes in X chromosome important in spermatogenesis include structural maintenance of chromosomes 1A (component of meiotic cohesion complex, SMC1A) on Xp11.22–p11.21 and testis-expressed 11 (binding protein expressed only in male germ cells, TEX11) on Xq13.1. The X chromosome seems to play more important role in the premeiotic stages of mammalian spermatogenesis.

Male Genital Tract Development

The Wolffian (mesonephric) ducts (WDs) lead to mature male genital tract. Urogenital sinus contributes to the genital tract by developing prostate. In XY embryo, Mullerian ducts degenerate in an active process facilitated by anti-Mullerian hormone (AMH) which is secreted by Sertoli cells. AMH binds to its receptor AMHR2 on the surface of Mullerian duct mesenchymal cells inducing secretion of matrix metalloproteinase 2 (MMP2) leading to apoptosis of Mullerian duct epithelial cells [39]. Failure of this process in humans results in persistent Mullerian duct syndrome (PMDS), an autosomal recessive condition which can lead to male infertility [40, 41].

Wolffian ducts differentiate under the influence of testosterone into epididymis, vas deferens, and seminal vesicle [42]. Mice lacking androgen receptor (AR) show agenesis of epididymis, vas deferens, and seminal vesicles. Bone morphogenic protein 4 (BMP4), BMP7, BMP8, HOXA10, and HOXA11 genes play important roles in epididymis development. Fibroblast growth factor 10 (FGF10) and growth and differentiation factor 7 (GDF7) are essential in proper development of seminal vesicles [2].

Male External Genitalia Development

Male external genitalia largely depends on the expression of 5α-reductase in the genital tubercle mesenchyme-converting testosterone to 5α-dihydrotestosterone (DHT) which is the most potent ligand for AR. Mutations of 5α-reductase result in abnormalities in male external genitalia and prostate development.

Complete androgen insensitivity syndrome due to X-linked AR gene mutation leads to XY sex reversal, but partial forms may present with various phenotypes ranging from ambiguous genitalia to male infertility. Other mediators important in male external genitalia include cell surface molecules like ephrins and their receptors (Ephs), Wnts, FGFs, BMPs, noggin, and Hox genes [2, 43]. Urethral fusion defects lead to hypospadias. HOXA13 and HOXD13 gene mutations reported in hand-foot-genital syndrome suggest that these genes are important in pathogenesis of hypospadias [44].

Genetic Defects Associated with Male Infertility

Numerical and Structural Chromosomal Abnormalities

Infertile men have an eight- to tenfold higher prevalence of chromosomal abnormalities than fertile men [45]. Chromosomal abnormalities can be detected in about 5% of infertile men, and the same frequency in azoospermic men was estimated to at 15% [1]. In a review of studies involving 9,766 azoospermic and severely oligospermic men, sex and autosomal chromosomal anomalies were found in 4.2% and 1.5% of infertile men, as compared with 0.14% and 0.25%, respectively, in control population [46]. It should also be noted that 0.37% of sperm donors with normal sperm parameters have been reported to have chromosomal translocations [47].

Aneuploidy is the most common error resulting from chromosomal anomalies in infertile men [48]. Although many autosomal and sex chromosomes can be involved, most common of those are Klinefelter syndrome, XYY syndrome, XX male syndrome, mixed gonadal dysgenesis, autosomal translocations, and Y-chromosome microdeletions. Especially, men with nonobstructive azoospermia present with high incidence of aneuploidy of up to 13.7% being predominantly numerical or structural defects [49]. In men with oligospermia, a 4.6% prevalence of autosomal translocations and inversions was reported [50]. Sex chromosomal aneuploidy may account for approximately two-thirds of chromosomal abnormalities observed in infertile men [51].

Klinefelter Syndrome

Klinefelter syndrome is seen in about 1 in 500 male live deliveries and is the most common known genetic cause of azoospermia accounting for up to 14% of all cases [52]. It results from X-chromosomal aneuploidy in which 90% of cases carry an extra X chromosome (47, XXY) and 10% are mosaics as 47XXY/46XY. In about half of the Klinefelter syndrome cases, extra X chromosome is paternally derived. The classic triad associated with the syndrome includes small and firm testes, azoospermia, and gynecomastia. It is also associated with eunuchoid body habitus with increased height, low intelligence quotation scores, varicosities, obesity, diabetes, increased incidence of extragonadal germ-cell tumors, leukemia, and breast cancer. There is high phenotypic variation, and many patients may not demonstrate these classic findings. The only invariant finding of non-mosaic form is that of small testes volume of 2–4 ml. The laboratory findings include severe oligospermia or azoospermia and low testosterone levels with increased LH and FSH. Testicular histopathology is consistent with seminiferous tubular sclerosis and hyalinization and sometimes Sertoli cell only. In some cases, remarkable small islands of spermatogenesis can be observed creating opportunity to obtain testicular sperm [53, 54].

Mosaic forms of the syndrome are associated with spontaneous fertility. Testicular sperm extraction (TESE) with intracytoplasmic sperm injection (ICSI) has been successful in achieving successful pregnancies in non-mosaic forms. The success with TESE in those cases ranges from 27% to 69% [54, 55]. Interestingly, 80–100% of mature sperm obtained from 47XXY patients show normal haploid sex chromosome with either X or Y [56, 57]. This may be either due to somatic germ line mosaicism or abnormal germ cells that simply do not develop due to meiotic arrest. Nevertheless, the rates of aneuploid sperm, although low in absolute terms, are increased in men with Klinefelter syndrome as is the prevalence of aneuploid embryos (both sex and autosomal) making genetic and preimplantation genetic diagnosis (PGD) counseling an important component of management [58–60].

XYY Syndrome

This syndrome is seen in 1/1,000 live male births. Phenotypic characteristics include increased height, decreased intelligence, higher risk of some malignancies like leukemia, and aggressive or antisocial behavior [52]. XYY syndrome is associated with severe oligospermia or azoospermia with elevated FSH but normal testosterone and LH levels. Testis biopsies were found consistent with maturation arrest or Sertoli cell only. Although, as similar to Klinefelter syndrome, majority of the sperm obtained from these patients show normal haploid sex chromosomes, higher rates of both sex and autosomal chromosomal imbalances have been reported in 47XYY men [61, 62].

XX Male Syndrome

Its main characteristics are gynecomastia at puberty and azoospermia. It is less frequent than Klinefelter or XYY syndromes with a frequency of 1 in 20,000 live male births. Patients usually show elevated FSH and LH levels with low testosterone. Testicular histology shows absent spermatogenesis with hyalinization of seminiferous tubules, fibrosis, and Leydig cell clumping [52]. It is thought that the translocation of *Sry* to X chromosome results in testes development; however, there is no spermatogenesis since Yq is totally lacking [63]. Since there is no spermatogenesis, any surgical or medical treatment will not be successful for fertility purposes. These patients may require testosterone treatment for hypogonadism.

Mixed Gonadal Dysgenesis

This is a rare condition with male or female phenotype usually with a unilateral testis and a contralateral streak gonad. Patients may have ambiguous genitalia and abdominal testes showing Sertoli cell only. The gonads are predisposed to malignant germ-cell tumors and need to be removed prior to puberty. The karyotype may be 45X/46XY or 46XY. The mutations of Sry have not been detected in majority of the cases with some suspect genes downstream to *Sry* [52].

Translocations and Inversions

Translocations or inversions of autosomal chromosomes can be detected in 1 in 600 to 1 in 1,000 live deliveries. Exchanges between chromosomes may interrupt important genes at the break point or may interfere with normal chromosomal pairing during meiosis. Robertsonian translocations involving chromosomes 13, 14, 15, 21, and 22 and reciprocal translocations are at least eightfold more common in infertile men [52].

Robertsonian translocations occur when two acrocentric chromosomes fuse with loss of the short arm material, hence the chromosome number will be 45. They are the most common chromosomal abnormalities in humans seen in 0.1% of newborns. Most commonly, they involve chromosomes 13;14 and 14;21 (Fig. 18.1). Robertsonian translocations may be seen in 1.5% of oligospermic and 0.2% of azoospermic men [1, 64, 65]. Furthermore, carriers of Robertsonian translocations are at risk of pregnancies with miscarriages or birth defects. Interestingly, within some families, fertility is unaffected despite the same apparent translocation of t(13;14)(q10;q10) [47]. Of those sperm produced, most have a normal balanced chromosomal complement, but an unbalanced karyotype can be present in 4–40% of sperm [66, 67]. Therefore, there is increased risk of trisomies or uniparental disomy. Since the carriers of Robertsonian translocations may pass translocation of imbalanced chromosomal abnormalities to the offspring, genetic counseling and PGD is recommended if sperm from ejaculate or testis is used for ICSI.

Reciprocal translocations are due to the exchange of material between autosomes or between the X or Y chromosome and an autosome and occur in 0.7% of severely oligospermic or azoospermic men [68]. The chromosomal number is normal, and the chromosomes and the break points may be unique to that particular family involved. Depending on the chromosomal material lost, the phenotype may vary. When the sperm is produced, more than 50% are

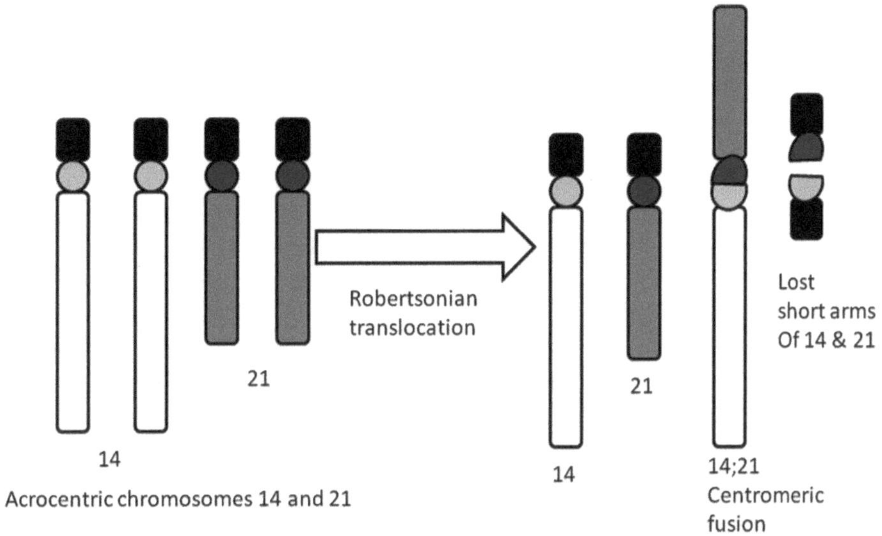

Fig. 18.1 A Robertsonian translocation involves centromeric fusion of the long arms of the acrocentric chromosomes, while the short arms are lost

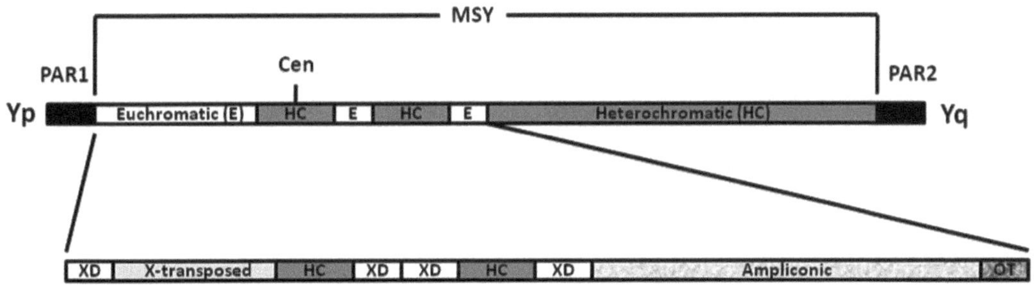

Fig. 18.2 Y chromosome: MSY, Yq, Yp, centromere (Cen), pseudoautosomal (PAR1 and PAR2), and heterochromatic (HC) regions followed by the enlarged view of euchromatic (E) region extending from boundary between PAR1 and Yp to the heterochromatic region of Yq. *XD* X-degenerate, *OT* other (from [72], with permission)

chromosomally unbalanced [66, 69], making genetic counseling even more important.

Chromosomal inversions may involve the centromere (pericentric) or a peripheral segment of the chromosome (paracentric). Although many inversions are harmless, they may have pathological implications according to the chromosome and the site involved. For example, inversion of chromosome 9 is more frequently observed in infertile men. Due to formation of abnormal loops during chromosomal pairing, chromosomal unbalance can occur effecting spermatogenesis or the resulting embryo [66, 70].

Y Chromosome

The human Y chromosome is 60 megabases (Mb) in length with the least number of genes but the highest copy number of the repetitive sequences as compared to autosomal chromosomes [71]. About 104 coding genes encode about 48 proteins. Among these proteins, 16 proteins have been discovered in the azoospermia factor (AZF) region [72]. The much smaller pseudoautosomal regions (PAR1 2.6 Mb and PAR2 320 bp) (Fig. 18.2) which pair with X chromosome during meiosis are located at both ends of the Y chromosome. The region outside of PARs previously known as the non-recombining region of Y chromosome (NRY) is now called male-specific Y (MSY) which compromises of 95% of the chromosome's length. It has also been shown that MSY may also be somewhat involved in X–Y crossing-over during male meiosis while MSY is flanked on both sides by PARs [73]. But again, the vast majority of the Y chromosome, including MSY, is less recombining and transmitted as a single block from generation to generation with functional variants and neutral polymorphisms being linked [72].

MSY is made up of a combination of three classes of gene-rich euchromatic (X-transposed, X-degenerate, and ampliconic) and heterochromatic sequences (Fig. 18.2). MSY encodes about 27 proteins, and within MSY, the X-transposed sequences which only encode for two genes (3.4 Mb) are 99% identical to the DNA sequences in Xq21 [72]. X-degenerate sequences are surviving relics of ancient autosomes and encode 16 proteins of MSY. Amplicons which encode nine proteins are sequence of nucleotides that are nearly identical large repeats reading in the same (direct) or opposite (inverted) directions. Genes in ampliconic segments may be replicated by recombination between the repetitive sequences. Most Y-chromosome genes expressed in the testes are located in the ampliconic regions. The Yp (Yp11) and the proximal part of the Yq (Yq11 subdivided into Yq11.1, 11.21, 11.22, 11.23) consist of euchromatin, while the distal part of Yq is made up of heterochromatin which is one-half to two-thirds of the Yq (Yq12) [73, 74] (see Fig. 18.2). Loci identified in Y chromosome are thought to be involved in the production and differentiation of the sperm since microdeletions of these loci are associated with severe oligospermia or nonobstructive azoospermia. Accordingly, seven deletion intervals were described both on Yp and Yq [75].

Male infertility affects 1 in 20 men, and primary spermatogenic failure accounts for about 50% of the cases [72]. Yq microdeletions were detected in 5–15% of males with spermatogenic failure. More specifically, these deletions occur in 6–8% of severely azoospermic men and in 3–15% of azoospermic men [76]. These deletions include the total Yq12 heterochromatin block and the part of Yq at Yq11.23. Consequently, it was suggested that at least one genetic Y factor essential for spermatogenesis is located in the distal Yq11 called as azoospermia factor (AZF), although about 6% of severe oligospermia cases occur with deletions outside the AZF region [74].

AZF Region

AZF region on Yq is most thoroughly studied male fertility locus in humans [77]. AZF region is further divided into three regions defined as AZFa, AZFb, and AZFc [78] (Fig. 18.3). While AZFa region is truly separate and distinct, the AZFb and AZFc regions actually overlap one another and are simply different stretches of Yq within one much longer, encompassing expanse [79]. It is thought that AZF microdeletions result from intrachromosomal recombination events between homologous repetitive sequence blocks in Yq11 [79]. As there is no counterpart in the genome for mitotic pairing and meiotic recombination of the MSY, this repetitive palindromic sequence structure might have evolved to protect the long-term genetic integrity of Y chromosome by allowing MSY to pair with and to repair itself. However, on rare occasions, this nonallelic homologous recombination may go wrong when two spatially separate ampliconic regions permanently stick together during Y-chromosome replication, resulting in loss of all chromosomal material in the intervening portion. In some cases, this may occur due to the deficiency of enzymes necessary for DNA repair. P8–P1 represents eight palindromes ordered nearest to farthest from the centromere within the euchromatic region of Yq. P5–P1 region is statistically more prone to nonallelic homologous recombination due to its unique molecular structure [80].

- AZFa: AZFa microdeletion is responsible for azoospermia in 1% of males with nonobstructive azoospermia. AZFa region is not palindromic, 792 kb in length, and is located in proximal Yq. AZFa region's candidate genes include USP9Y (ubiquitin-specific protease 9, Y chromosome), or DFFRY (Drosophila fat facets-related Y), DBY (DEAD box on the Y), and UTY (ubiquitous TPR motif on the Y) [78, 81–84]. USPY9 occupies less than half of the AZFa interval and is mostly involved with microdeletions. However, minority of patients may have a de novo point mutation in this region [83, 85]. DBY consists of 17 exons and encodes for a putative ATP-dependent RNA helicase that shuttles between nucleus and cytoplasm whose specific function in spermatogenesis remains elusive. In AZFa deletions, most frequent presentation is Sertoli cell-only syndrome (SCO). In SCO I cases, there are no germ cells in the seminiferous tubules. In SCO II associated with partial AZFa deletion, some germ cells with incomplete differentiation and maturation and degeneration can be seen.
- AZFb: An AZFb microdeletion is 6.2 Mb long and begins in P5 palindrome and ends in the proximal portion of P1, hence the name P5/proximal P1 microdeletion. The so-called AZFb/AZFc microdeletion is also named as P5/distal P1 microdeletion since it also starts in P5 but spans a larger area of 7.7 Mb and ends in distal P1. AZFb or AZFb/AZFc microdeletions are observed in 1–2% of males with nonobstructive azoospermia. AZFb candidate genes include EIF1AY (translation-initiation factor 1A, Y isoforms) and RBMY (RNA-binding motif on the Y). For the former, no deletion specifically removing EIF1AY has been reported. Complete AZFb deletions are associated with maturation arrest at the primary spermatocyte or spermatid stages.
- AZFc: AZFc region stretches from the distal portion of the P3 palindrome to the distal portion of P1 and is 3.5 Mb in length (Fig. 18.3). Initially, it was believed that AZFb and AZFc were non-overlapping areas, but subsequent studies demonstrated that both the AZFb and AZFb/AZFc regions overlap AZFc region. AZFc microdeletion is also called as b2/b4 (Fig. 18.3) since it occurs when nonallelic homologous recombination happens between the b2 and b4 amplicons in P3–P1 with loss of all intervening material [79]. AZFc microdeletions are the most common microdeletions found in men with nonobstructive azoospermia, occurring in up to 13% of cases with azoospermia and 6% of men with severe oligospermia. Deleted in azoospermia (DAZ) cluster is the primary candidate gene in the AZFc region. Several other genes in addition to DAZ were mapped in this region including CDY1 (chromodomain Y1), BPY2

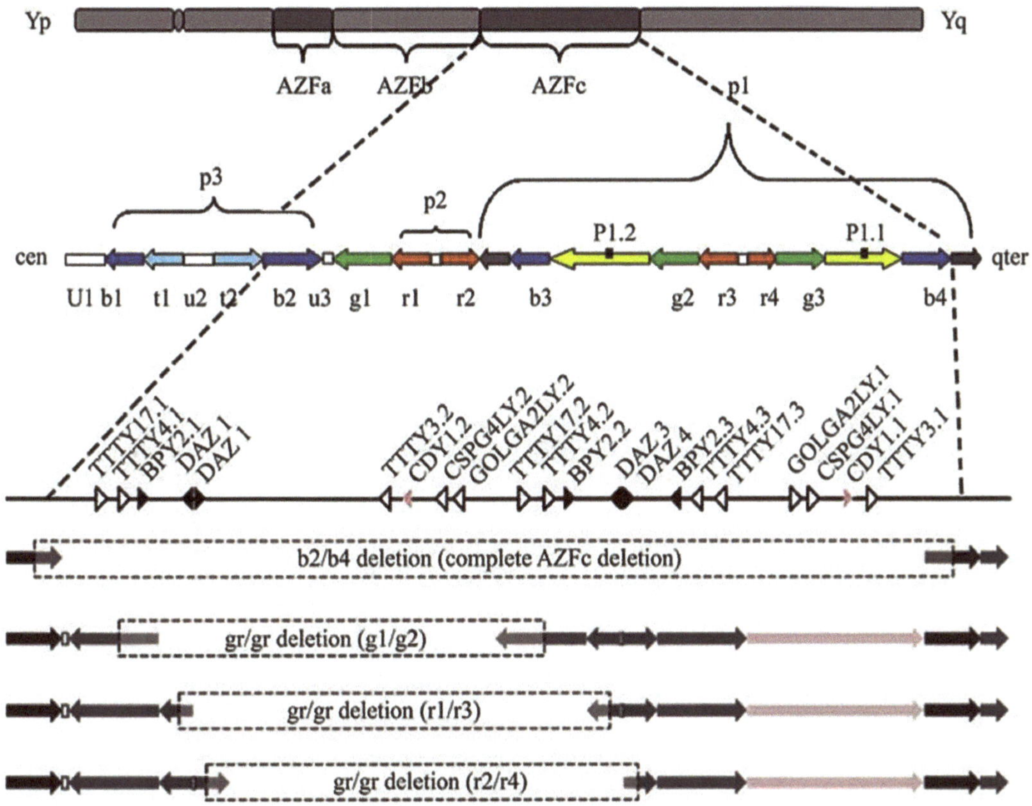

Fig. 18.3 AZFc is compromised of distinct families of nearly identical amplicons. Amplicon b, g, u, r, and t have 4, 3, 3, 4, and 2 repeats, respectively, in the haplotype. Palindromes are indicated as P1, P2, and P3. Repetitive sequences are direct repeats and inverted repeats as the direction of the *arrows* show. AZFc contains 12 families of transcription units which are all expressed in testis such as BPY2, DAZ, CDY1, and others. AZFc microdeletion subtypes include complete deletion b2/b4 and partial deletion g1/g2, r1/r3, and r2/r4 (from [72], with permission)

(basic protein Y2), PRY (PTA-BL related Y), and TTY2 (testis transcript Y2). AZFc deletions are associated with a wide range of conditions from azoospermia to mild to severe oligospermia. This is reflected in testicular histology being consistent with hypospermatogenesis or SCO II in which there is a better chance of finding focal areas of spermatogenesis.

Among all cases with Yq microdeletions, deletions involving DAZ seem to be the most frequent one. Some reports suggest that DAZ deletions may be encountered up to 13% of cases of male infertility [77]. DAZ was thought to be acquired by Y chromosome from an autosomal homologue DAZL (DAZ-like) on chromosome 3p24 which shows a single DAZ repeat. DAZ gene cluster on Y chromosome consists of seven copies of DAZ of which four copies are located relatively close together within a deletion interval on Yq11. DAZ encodes an RNA-binding protein exclusively expressed in early germ cells and is thought to be responsible for activation of silent mRNAs during pre-meiosis stages. It was reported that AZFc may not be critical for meiotic recombination, whereas absence of AZFc regions results in extension of the zygotene stage and reduction of chromosomal condensation [86]. Although most deletions involve all four DAZ genes, an absence of only two is also associated with defective spermatogenesis [33].

Some other partial microdeletions detected in AZFc region such as b2/b3, b1/b3, and gr/gr do not seem to have any clinical significance, although race-specific various phenotypic outcomes were reported [87, 88]. Actually, microdeletions of the AZFc may present with spermatogenic failure in Dutch, Spanish, Chinese, and Italians. However, the presence of AZFc deletion in healthy French, Germans, and Han Chinese questions its importance in male infertility [89, 90]. A partial AZFc microdeletion such as gr/gr can be passed from father to son, but again, its clinical significance is still debated [91, 92]. Whereas frequent deletion of DAZ gene cluster in male infertility cases suggests its importance in spermatogenesis, the variable penetration of AZFc deletions in general suggests some functional redundancies in its function. Perhaps microdeletion is not an independent event but is compensated by the activation of other genes by gene replication or dosage compensation [72].

Yq Microdeletions in Clinical Practice

In infertile men with nonobstructive azoospermia or severe oligospermia, it is critical that the patients know the results of Y-chromosomal microdeletion analysis before TESE and ICSI. Briefly, complete AZFa, AZFb, and AZFb/AZFc microdeletions predict that TESE will be unsuccessful since sperm will not be found and presently there is no available treatment [93]. Men with Y-chromosomal microdeletions rarely have a sperm density above five million per milliliter. Although vast majority of AZFc microdeletions are de novo which means that the father of the patient is not affected, rare cases of natural transmission were also reported [79].

As noted above, it is thought that finding of ejaculated sperm in complete AZFa and AZFb deletions is highly unusual [35]. Combined deletions of two or more regions that include AZFb are associated with SCO or maturation arrest histology. Patients with AZFc deletions have the best prognosis for finding testicular sperm during TESE. Many reports suggest that 50–60% of azoospermic AZFc-deleted men will have testicular sperm enough for ICSI [35]. In patients with complete AZFa or AZFb microdeletions, the probability of finding sperm during TESE attempts is extremely low if not impossible. Successful TESE is often not possible in cases with deletions involving one or more regions that include AZFa or AZFb as well.

Small studies are not clear for the outcome of ICSI when testicular sperm is used in men with AZF deletions. Some reported normal fertilization rates but poorer embryo quality as compared to those without AZF deletion and other demonstrated comparable fertilization and pregnancy rates [52]. There is also evidence that men harboring AZFc microdeletions may show time-dependent decline in sperm production. Therefore, counseling patients for sperm cryopreservation for future use is essential.

Men with AZF deletions who conceive via assisted reproduction are likely to pass on the Yq deletion to male offspring [94, 95]. However, the children conceived by testicular sperm seem to be somatically healthy, and their AZFc deletion has not been shown to be altered, although it is expected that male offspring will suffer from similar deficiencies of spermatogenesis.

Since microdeletions cannot be detected by conventional cytogenetic methods, Yq analysis is performed on peripheral blood lymphocytes via polymerase chain reaction (PCR) where various center-specific primers were used to amplify sequence-tagged sites (STSs) of DNA which makes evaluation of data challenging. By PCR amplification of STSs which are specific to each of the AZF regions under review, deletions are identified by the absence of one or more ampliconic products. This multiplex PCR is restricted to the analysis of a set number of loci and cannot detect novel Yq microdeletions or mutations. To minimize such limitations of PCR, other genomic techniques such as array comparative genomic hybridization has been proposed although the cost of this technology limits its introduction to clinical medicine [96]. Difficulties in reporting may occur due to the fact that euchromatic Yq consists of large, nearly identical amplicon repeats and palindromes, and furthermore, the description of microdeleted regions has not always correlated with known gene deletions. It is also possible that some cases may show mosaicism when comparing to Yq deletions in leukocytes to sperm [97].

Other Y-Chromosome Conditions

Massive palindromes in the human Y chromosome harbor mirror-image gene pairs essential for spermatogenesis. These gene pairs have been maintained by intrapalindrome, arm-to-arm recombination. Isodicentric Y (idicY) chromosomes may be formed by homologous crossing-over between the opposing arms of palindromes in sister chromatids. This event is usually associated in mosaic event with 45X cell line giving a karyotype of 45X/46XidicY. These patients retain two Yps and two SRYs although the latter may not be functional due to mitotic instability leading to female phenotype. In cases with male phenotype, the end result would be azoospermia possibly due to Yq segment loss at break points [98]. It remains elusive if successful TESE and ICSI with PGD can be performed in these cases.

Short arm of Y chromosome also harbors genes related to spermatogenesis. TSPY gene is one of these genes with copies on Yq as well [99]. A study of copy number variation of TSPY demonstrated that more copies were found in infertile men [100].

X Chromosome

Many X-chromosome genes influence male infertility. From rodent studies, it was suggested that X chromosome may play an important role in premeiotic stages of mammalian spermatogenesis [101]. Deletions, translocations, and inversions of X chromosome may result in severe infertility and azoospermia [102–104]. For example, paracentric inversion involving Xq12–25 or deletion of a portion of Xp may result in a phenotype consistent with Klinefelter syndrome. Several X-linked gene mutations were reported in infertile men with oligospermia or azoospermia. These genes include SOX3 (sex-determining region Y box 3), FATE, and ZFX [105–108].

Androgen receptor (AR) gene is located on Xq11–12. Knockout mice studies have suggested that AR signaling in Sertoli cells plays an important role in meiosis I during spermatogenesis [109]. Lack of AR in Leydig cells may lead to spermatogenic arrest at the round spermatid stage [110]. The functional AR in germ cells however was not found to be essential in spermatogenesis [111]. Therefore, androgens control spermatogenesis, but germ cells themselves do not express a functional AR. Androgen regulation is thought to be mediated by Sertoli and peritubular myoid cells. Studies in mice with selective AR knockout in Sertoli cells have suggested that AR plays an important role in meiosis and progression of spermatocytes to round spermatids [112]. While complete AR gene mutations are associated with androgen insensitivity syndrome with a female phenotype, incomplete forms of AR gene mutations were detected in more frequently in infertile men [113].

X-linked spinal and bulbar muscular atrophy or Kennedy's disease is caused by expansion of a CAG repeat in the first exon of AR gene. The CAG repeat encodes a polyglutamine tract in AR protein. The greater the expansion of the CAG repeat, the greater the polyglutamine repeat expansion and the earlier the disease onset and the more severe the disease manifestations [114]. Glutamine repeat motif in the first exon of AR gene (polyQ region) is polymorphic in general population numbering between 10 and 36 repeats. In Kennedy's disease, polyQ region is expanded between 40 and 62 repeats [114].

The CAG repeat expansion mutation in AR gene does not affect sexual differentiation. The repeat expansion likely causes a toxic accumulation of mutated AR in nuclei and cytoplasm of motor neurons, resulting in their degeneration and loss [115]. Patients present with amyotrophic, proximal, or distal weakness and wasting of the facial, bulbar (dysphagia and dysarthria), and limb muscles; occasionally sensory disturbances; and endocrinologic disturbances, such as androgen resistance, gynecomastia, elevated testosterone, and reduced fertility, due to defects in spermatogenesis and testicular atrophy. The onset of neurological symptoms is between the ages of 30 and 50 years. In parallel to those observations, CAG repeat polymorphism has been investigated as possible as a possible cause for male infertility [116]. However, it is still controversial if longer or shorter CAG repeats are associated with higher or lower sperm quality [117–119].

USP26 (Xq26.2) and TAF7L (Xq22.1) genes are expressed in testis, and single-nucleotide polymorphisms (SNPs) for both genes were investigated for potential association with male factor infertility [120, 121]. Although some studies have suggested potential associations between these SNPs and abnormal spermatogenesis, since the effects of SNPs are largely influenced by ethnicity, the precise roles of these findings in relation to male infertility remain elusive [122–124].

X-linked forms of Kallmann syndrome are related to the deletions in KAL-1 gene located in the short arm of X chromosome (Xp22.32). This gene codes for a cell adhesion protein, anosmin-1, which is involved in the migration of gonadotropin-releasing hormone (GnRH) neurons during embryonic development. The condition is associated with hypogonadotropic hypogonadism with sexual infantilism due to the deficiency of GnRH, anosmia, or hyposmia due to the absence or hypoplasia of olfactory bulbs and tracts, cognitive and ocular abnormalities, and even with mid-facial clefts and renal agenesis [125]. While X-linked KAL-1 mutations are responsible for 30–70% of the condition, the remaining cases are related to the deletions in fibroblast growth factor receptor 1 (FGFR1) gene on chromosome 8 which shows autosomal-dominant inheritance [126]. With treatment, favorable reproductive outcomes can be attained in addition to maturation of secondary sex characteristics.

Congenital Bilateral Absence of the Vas Deferens and Cystic Fibrosis

Congenital bilateral absence of the vas deferens (CBAVD) is estimated to occur in 1/1,000 to 1:10,000 and may be encountered in 1–2% of cases with male infertility. It is detected in 9.6% of cases with obstructive azoospermia [127]. It is thought to result from abnormal development of WD although it is not clear if absence of vas deferens is always congenital. Absence of distal WD derivates has been related to the early obstruction of these ducts by viscous secretions rather than an embryonic developmental defect. Approximately 80% of CBAVD cases are caused by mutations on both alleles of the cystic fibrosis transmembrane conductance regulator (CFTR) gene. CBAVD is usually associated with absence of body and tail of epididymis, vas deferens, and seminal vesicles, but the head of epididymis is intact (Table 18.1).

Given the fact that almost all CF male patients are infertile due to CBAVD, it was investigated if CFTR was also involved in infertility because of CBAVD alone. In an earlier small study, it was reported that 41% of azoospermic men with CBAVD was found to be heterozygous for F508del CFTR mutation as compared to the population risk of 2.8% [128]. Later, R117H mutation was also found at a higher frequency in CBAVD cases [129]. In more recent large study, analyzing 7,420 alleles of CFTR gene showed that a CFTR mutation can be identified in 78.9% of patients with CBAVD. In French men with CBAVD, about 71% of the CBAVD patients had a mutation on both CFTR genes and about 16%

Table 18.1 Clinical detection of CFTR mutation-related CBAVD

Azoospermia
Low seminal fluid volume (<2.0 ml)
Biochemical features of the semen: pH < 7.2, absent or decreased fructose, and α1–4 glucosidase (markers of properly functioning seminal vesicles and epididymis, respectively)
Absence of palpable vas deferens
On transrectal ultrasound: absence of the intra-abdominal tract of the vas deferens, globus major, and different degrees of hypoplasia of the seminal vesicles
Normal plasma follicle-stimulating hormone (FSH), luteinizing hormone (LH), and testosterone levels
Normal karyotype
Normal testicular size

From [134], with permission

Table 18.2 Cystic fibrosis detection and carrier rates before and after testing

Ethnic group	Detection rate (%)	Carrier rate before testing	Carrier risk after negative test result
Ashkenazi Jewish	94	1/24	1/400
Non-Hispanic Caucasian	88	1/25	1/208
Hispanic American	72	1/46	1/164
African American	65	1/65	1/186
Asian American	49	1/94	1/184

Modified from [138], with permission of Lippincott Williams & Wilkins

had a mutation on one CFTR genes and the remaining 13% of CBAVD patients had no mutation [130].

In 20% of CBAVD patients, the absence of vas deferens is associated with renal malformations [131]. Although a minority of CBAVD patients may have a mild lung disease or a positive sweat test, most CBAVD patients do not have lung disease. It is possible that a mild mutant of CFTR protein with partial chloride channel activity can sustain a normal non-diseased phenotype except for proper functioning and maintenance of vas deferens after its development [132].

The CFTR gene spans about 190 kb and contains 27 exons (chromosome 7q31). The CFTR protein is a glycosylated transmembrane protein which functions as a chloride channel. In the CFTR gene, currently 1,719 sequence variations have been reported in populations with various geographic location and ethnicity (http://www.genet.sickkids.on.ca/cftr/StatisticsPage.html) [133]. These are found in both CF and related phenotypes called as CFTR-related disorders (CFTR-RD). These are clinical disorders with CFTR dysfunction where the diagnosis of CF cannot be established. These entities include CBAVD, disseminated bronchiectasis, chronic pancreatitis, and chronic rhinosinusitis [134]. As CF is inherited in a recessive way, CF will develop when deleterious mutations are found on both CFTR alleles. If the mutation is only on one allele, the individual is a CF carrier. 1 in 2,500 newborns have CF, and 1 in 25 Caucasians is a CF carrier [132] (Table 18.2).

More than 1,200 CF-causing CFTR mutations have been identified. A CF patient may carry two identical or different mutations; the latter condition is called compound heterozygosity for two CFTR mutations. Most mutations are point mutations. The distribution of these mutations varies according to ethnicity. The most common mutation F508del is seen in 70% of Northern European populations, but it is seen in lower frequencies in Southern Europeans [132]. About 1–5% of mutations remain undetermined in CF patients and even more in patients with atypical presentations. Frequency of undetected mutations increases from Northern to Southern European populations [134]. Undetected mutations may lie within the introns or regulatory regions which are not routinely analyzed. Besides F508del, other mutations exist in most populations each reaching frequencies of about 1–2% such as G5542X, G551D, R553X, W1282X, and N1303K. Therefore, in most populations, these mutations and some ethnic-specific mutations compromise 85–95% of all CFTR mutations. The remaining mutations are rare and sometimes can be found in a single family or particular population. Depending on the effect at protein level which predicts the severity of the clinical condition, CF mutations are divided in several arbitrary classes [132]. One large study reported that CF patients had two severe mutations (88%) or one severe and one mild/variable mutation (12%), whereas CBAVD men had either a severe and a mild/variable (88%) or two mild/variable (12%) mutations [130].

Most commercial genetic tests for CFTR mutations screen only for the most frequent CF-causing mutations and not the milder mutations, therefore resulting in a CFTR mutation detection rate of only about 60% in CBAVD patients [130]. The most common CFTR mutation presenting with a mild phenotype found in CBAVD patients is the 5T polymorphism (variant) [135].

The 5T splicing variant of the intron 8 acceptor splice site is not considered a CF-causing mutation, but it may be associated with CFTR-RD [134]. At the polypyrimidine tract of intron 8 acceptor splice site, the variants are named according to the number of thymidines as 5T, 7T, and 9T. The lower the number, the lower the efficiency of exon 9 splicing. The extent of splicing is further related to the number of adjacent TG repeats; the higher the number of TGs, the lower the efficiency of splicing. Patients with (TG)13/5T in trans with a CF-causing mutation may have mild CF. R117H CFTR-RD mutation can be found in cis with 5T or 7T. R117H;5T is considered a mild CF mutation, but R117H;7T is considered as a CFTR-RD mutation. If R117H;5T is trans with a F508del severe CF mutation, the signs of CF may be present. The presence of R117H;7T in the same configuration have been shown to be symptom-free [136].

When 5T is found in compound heterozygosity with a severe CFTR mutation, or even with another 5T, CBAVD can be observed. However, not all men compound heterozygous for a severe CFTR mutation and 5T develop CBAVD, such as fathers of some CF children [132]. Hence, 5T polymorphism is a mutation with partial penetrance. Again, the R117H mutation can either result in CF or CBAVD by being associated by either 5T or 7T allele [137]. Its association with 7T allele may result in CBAVD, and R117H;5T may result in CF. F508del mutation is found at a higher frequency in CF patients as compared with those with CBAVD, while R117H is more frequently observed in patients with CBAVD.

Initial CF screening guidelines included 25 pan-ethnic mutations that were present in at least 0.1% of patients with CF. The commercially automated methodology uses PCR with allele-specific oligonucleotide primers. After evaluation of the data, two mutations 1078delT and I148T were removed from the panel since they were not increasing the sensitivity for CF carrier state or diagnosis [138]. Some screening panels may identify 5T, 7T, and 9T variants, although they are not offered in routine CF carrier screening. As detailed above, the presence of 5T allele may decrease mRNA stability affecting exon 9 [70]. Since CF may occur when 5T is on the same chromosome (cis) with R117H missense mutation along with a CFTR mutation on the other chromosome, reflex 5T testing is done if R117H is detected in the screening panel. Since males with 5T allele on both chromosomes are at increased risk for CBAVD, 5T testing should be ordered in CBAVD cases. CBAVD patients with two 5T variants and female with a R117H mutation with 5T variant in cis position need genetic counseling to discuss the risk of having an offspring with CF. It should always be remembered that the primary goal of universal CF screening test is to detect CF not CBAVD, with a reasonable sensitivity. Therefore, at times, complete analysis of CFTR gene by DNA sequencing may be necessary in patients with CF and in patients with CBAVD, tested negative with commercial CFTR screening test results.

Spermatozoa of CBAVD patients used in an ICSI program may transmit CFTR mutation to the offspring. Most CBAVD patients may carry a severe CF-causing CFTR mutation with 50% chance of transmitting that mutation. If the carrier risk of a Caucasian female is 1/25 (0.04), with herself having a 50% chance of transmitting the gene to the offspring, the risk of having a child with CF would be 1 in 100 ($0.5 \times 0.04 \times 0.5 = 0.01$) as compared with a risk of 1/2,500 in general population. Therefore, partner testing with genetic counseling would be of utmost importance. Since the commercial genetic tests have about 90% of sensitivity (see Table 18.1) when no mutation is found in the test, the partner still has a risk of 1/250 being a carrier of an untested mutation; therefore, the CBAVD couple may still run a risk of 1/1,000 of having a child with CF [132].

Genes Involved in Meiotic Recombination

Although the incidence of chromosome abnormalities is about ten times higher in infertile males than in the general population, most infertile men have normal karyotype. However, these patients may show an increased incidence of aneuploid sperm and diploid sperm in their ejaculate or in sperm obtained from testes [139, 140]. It has also been shown that the risk of aneuploidy or diploidy in sperm correlates with decreasing numbers of sperm and total progressive motility [141]. Many times, meiotic disturbances are the culprits in those cases.

Meiotic recombination in germ cell occurs in prophase of meiosis and involves the induction of double-strand DNA breaks, the pairing of parental homologous chromosomes, followed by the repair of double-strand breaks using the intact homologous chromosome as a template. Several studies have suggested significantly lower rates of meiotic recombination and impaired synapsis in infertile men [142–145]. Faulty meiotic recombination can also cause fertility problems especially if the meiotic errors cannot be corrected; meiotic checkpoint molecules activate apoptotic pathways leading to testicular failure. Furthermore, it was estimated that 5–10% of cases of nonobstructive azoospermia may be due to meiotic arrest [145]. There are many genes involved in meiotic recombination, investigation of which has been mostly relevant to etiology of many cancers [146]. Among these gene products, the absence of type II topoisomerase, Spo11 (chromosome 20q13.2-13.3), and synaptonemal complex protein, SYCP3 (chromosome 12q23.2) is shown to be associated with nonobstructive azoospermia in humans [146–148].

Gene Mutations Associated with Sperm Functional Defects

The primary ciliary dyskinesia presents with immotile but viable sperm along with varying degrees of respiratory tract dysfunction, situs inversus totalis (Kartagener's syndrome), and hydrocephalus. Its frequency is 1 in 20,000–60,000 live births [149]. Most of the genetically characterized primary ciliary dyskinesia variants exhibit mutations in the genes dynein, axonemal, heavy chain 5 [DNAH5, 5p15.2]; dynein, axonemal, heavy chain 11 [DNAH11, 7p21]; dynein, axonemal, intermediate chain 1 [DNAI1, 9p13.3]; and dynein, axonemal, intermediate chain 2 [DNAI2, 17q25] that encode axonemal dynein-arm components responsible for ciliary-beat generation and sperm-specific thioredoxin domain containing 3 [TXNDC3, 7p14.1] encoding a thioredoxin [150]. Some cases may have mutations in retinitis pigmentosa GTPase regulator (RPGR) gene on Xp11.4 which is also associated with retinitis pigmentosa [151]. The patients initially may present with severe asthenospermia. In these cases, other clinical signs of primary ciliary dyskinesia should be sought. The majority of patients have sperm in the ejaculate, and ICSI has been successfully used in those cases [152].

Mutations of aurora kinase C (AURKC, serine/threonine kinase 13, 19q13) have been shown to cause male infertility in the region of Northwest Africa (Morocco, Tunisia, Algeria). In those countries, the carrier rate of the mutation can be as high as 1 in 50. Men homozygous for the mutation demonstrate sperm with a 4N chromosomal complement, large heads, and multiple tails [153]. In mice knockout studies, absence of AURKC was only associated with a high rate of teratospermia with retained fertility [154]. It remains to be elucidated if screening for AURKC mutations is needed in other populations.

Copy Number Variations

These are pieces of 1 kb or longer DNA segments that vary in number between individuals which are considered as submicroscopic duplications and/or deletions of the genome [155]. Copy number variations (CNVs) can be detectable by higher-resolution genome-wide microarray comparative genomic hybridization assays and can be further confirmed by PCR-based methods. The complexity about CNVs lies on the facts that their presence can cause overt disease, a predisposition to a disease, or may have no effects at all. In general, CNVs may affect up to 20% of human genome [70]. CNVs have been investigated in many medical disorders; however, the data on male infertility are scarce, and more investigation is needed.

Mitochondrial Genetics

Mitochondrial DNA is a double-stranded circular DNA molecule coding for 2 rRNAs, 22 tRNAs, and 13 polypeptides essential for respiratory enzyme complexes involved in oxidative phosphorylation [156]. Mitochondrial DNA has no introns, and it mutates at 10–20 times higher rates than nuclear DNA due to its unique structure and replication system [157]. Mid-piece of mammalian sperm contains about 80 mitochondria with single copy of DNA in each organelle. Spermatozoa are dependent on mitochondria for energy needed for rapid progressive motility. Mitochondrial DNA mutations caused by the oxidative damage induced by reactive oxygen species or free radicals may lead to male infertility [158]. In general, about 85% of sperm samples may contain various mitochondrial DNA deletions which may partly explain the age-related decline in fertility in males.

The presence of multiple mitochondrial mutations have been reported to be associated with oligoasthenoteratospermia [159, 160].The key nuclear enzyme involved in the elongation and repair of mitochondrial DNA strands is DNA polymerase gamma (POLG). The catalytic subunit of POLG is encoded by POLG gene on chromosome 15q24 which includes a CAG repeat region [161]. POLG mutations are associated with mutations in mitochondrial genome which subsequently affects ATP production and sperm function. Expanded CAG repeats in the region of POLG gene are also associated with several neuromuscular disorders which may be associated with male factor infertility as well. Many of these disorders like Huntington's disease are transmitted in an autosomal-dominant fashion.

Epigenetic Alterations

Epigenetics refers to the alterations of the gene expression without any change in DNA nucleotide sequence. Epigenetic mechanisms are associated with the way in which the genome is packed affecting the ability of genes to be activated. It is involved mostly with regulation of transcription or translation.

The most established epigenetic mechanism heritable through germ line is DNA methylation. This is a post-replicative modification in which a methyl group is covalently added to CpG (cytosine–guanine) dinucleotide residues of DNA by DNA methyltransferases [162]. Other well-known epigenetic mechanisms include chromatin condensation and histone modifications. The regions of chromatin can be transiently condensed or uncondensed leading to variation in gene expression through transcriptional suppressors, functional RNAs, or interaction with various proteins [163]. Histones are subjected to modifications like phosphorylation, acetylation, methylation, ubiquitination, carbonylation, and such affecting gene expression [164]. Small noncoding RNAs like micro (mi)-RNAs or Piwi-interacting (pi) RNAs are the newest epigenetic mechanisms acting through transcriptional or translational regulation [165, 166]. In mice, the global loss of miRNAs in Dicer (RNase III endonuclease playing central role in miRNA biogenesis) knockout mice has detrimental effects on spermatogenesis [166].

Reprogramming of methylation patterns in mammals occurs usually after fertilization (preimplantation stage) and during fetal development of the germ line (gametogenesis) especially during germ line differentiation [167]. Allelic differences in methylation which is characteristics of imprinted genes are also delineated during the germ-cell line establishment [168]. Imprinted genes conserve their methylation patterns through generations. Therefore, if methylation changes are induced in imprinted genes or new methylation sites are established during germ-cell differentiation or after fertilization, heritable factors can either diminish or persist affecting the ultimate phenotype in the offspring [169]. Most endocrine disruptors or environmental factors do not promote DNA sequence mutation but induce modifications of DNA without altering nucleotide composition, i.e., epigenetic changes [170, 171].

Imprinting abnormalities associated with Angelman, Prader–Willi, Beckwith–Wiedemann, and Silver–Russell syndromes have been associated with assisted reproductive technologies (ART), whereas correlations have been found to be weak. It seems that these imprinting syndromes may be associated with infertility factors associated with preexisting methylation aberrations rather than ART itself [172]. Favoring this assumption, it has been reported that epigenetic abnormalities are common in sperm of men with severe oligospermia [173, 174]. Rodent studies also revealed that perturbation of DNA methyltransferases or DNA methylation in male germ cells can affect fertility and sperm function [175]. Abnormal sperm chromatin packaging may have a role in proper establishment of methylation patterns and abnormal protamine 1 and protamine 2 ratios (P1:P2 equals to unity in fertile men), which were detected in infertile men [176–178], and may lead to changes in imprinted genes [179].

One recent study has reported aberrant imprinting patterns of paternally demethylated genes in frozen sperm specimens of men with oligospermia and in those specimens showing normal sperm concentration but abnormal P1:P2 ratios [180]. It is proposed that the erasure and resetting of DNA methylation that takes place in primordial germ cells is an important stage to prevent DNA methylation defects; however, the effectors and modulators of these steps need further investigation before the applicability of this information to clinical practice [181].

Malignancy Risks Associated with Genetic Perturbations in Infertile Men

Multiple studies have suggested an association between male infertility and testicular germ-cell tumor which is the most common malignancy in men aged 15–35 years [182, 183]. Infertility probably precedes the development of occult testicular cancer. This association suggests common genetic and environmental factors in both infertility and testicular germ cell tumors. Increased risk of these cancers have been associated with factors related with genetics and epigenetics which include cryptorchidism, chromosome 12 aneuploidy, DNA mismatch repair gene defects, Y-chromosome instability, and stem cell dysregulation via abnormal RNA interference [182, 184, 185]. Cryptorchidism itself may be associated with mutations in HOXA10, INSL3 and INSL3 receptor LGR8/GREAT, AR, estrogen receptor (ER)α, and SF-1 gene mutations [186]. While elucidation of all these factors and pathways are required, all men evaluated for infertility needs to be adequately assessed and screened for testicular tumors.

Male Genetic Testing in Clinical Practice

Many genetic potential causes of both spermatogenic failure and obstructive azoospermia have been demonstrated, but despite much progress in animal data [187], the validation of the proposed genetic tests have been slow [188]. Therefore, only a limited number of tests are currently recommended in the evaluation of infertile men. Arbitrarily defining severe oligospermia as sperm concentration less than five million per milliliter and azoospermia as the sperm density below the detection limit, infertile men with those two conditions are recommended to undergo genetic testing.

Men with congenital unilateral or bilateral absence of vas deferens should be tested for a CFTR mutation which also includes 5T variant. Almost all men with clinical CF have CBAVD. At least two-thirds of men with CBAVD have mutations of the CFTR gene. However, failure to identify a CFTR mutation by commercial tests in a man with CBAVD does not rule out a mutation, since they may still harbor a mutation undetectable with the currently recommended screening panel. Even, it has been recommended that patients with CBAVD should be assumed to have a CFTR mutation [51]. Although most men with CBAVD have normal spermatogenesis, coexisting spermatogenesis defects should always be ruled out before harvesting sperm for ICSI [189]. Furthermore, since about 25% of men with unilateral absence of vas deferens and 10% of men with CBAVD may have unilateral renal agenesis, an abdominal ultrasound is also required [190].

There is a need to routinely karyotype infertile men with persistent or severe oligospermia (<10 million depending on the male phenotype or <5 million per ml for all) or nonobstructive azoospermia [66] (Fig. 18.3). Yq deletions are more frequent in azoospermia cases than men with severe oligospermia. Nevertheless, routine Yq microdeletion STS-PCR testing is required before testicular sperm harvesting or before ART in men with nonobstructive azoospermia or severe oligospermia. Chromosomal abnormalities may result in impaired testicular function, while Y-chromosome microdeletions result in isolated spermatogenic failure.

At present, testing sperm for aneuploidies or inversions in men with abnormal sperm analysis or men with abnormal karyotype is not recommended since the exact data set establishing threshold levels for the percentage of sperm with abnormal karyotype to assist in clinical and PGD decision making is lacking [70]. Currently, routine assessment of CNVs, epigenetic assessment of infertile men, or their sperm do not find any support in the clinical practice.

Gene Therapy for Male Infertility

Around 15% of the male infertility patients are azoospermic, and normal sperm production with obstructive cause accounts for 40% of these cases. In the remaining 60% of the cases with spermatogenesis defects, approximately half of them may have low levels of sperm which may be obtained through TESE techniques to be used for ICSI [191, 192]. Only for cases without any viable sperm in the testes gene therapy may be considered.

There are many challenges to gene therapy for male infertility. Karyotypic abnormalities like Klinefelter syndrome and Y-chromosome deletions involve additions or deletions of large amounts of DNA. Currently, there is no technology to manipulate large amounts of DNA for gene therapy. Furthermore, although we mentioned some genetic causes of male infertility in this article, many men with severe infertility do not have any identifiable genetic defects even if the genetic cause is likely. Knowledge of the exact gene(s) involved is essential to proceed with any gene therapy. In humans, at least 178 genes may be involved with spermatogenesis only [32], and mice models for male infertility suggests over 150 single gene defects leading to male infertility [193, 194].

Another obstacle is that gene therapy for male infertility may involve both somatic (Sertoli, Leydig cells) and germ cells needing more complex approaches. Serious ethical and safety concerns exist regarding inducing genetic alterations in the germ line, and currently, germ line genetic therapy is prohibited. There are suggested approaches like using episomes which are not integrated into the genome to be used in that respect [195]. For somatic cells, using viral vectors to integrate wild type of gene carries the risk of insertional mutagenesis or carcinogenesis since technology does not allow selected site insertion [196]. There are trials on nonviral approaches to gene therapy like cationic lipids, gold, nanoparticles, and electroporation of naked DNA, although like viral vectors, these methods may still induce immune response in the host and their efficiency is currently lower than the viral vectors [197].

New suggested approaches to deal with the obstacles include the use of embryonic stem cells, transplantation, and xenotransplantation [198–201]. Enucleated oocyte can be combined with patient's somatic cell nucleus, nucleus can be reprogrammed, and oocyte is stimulated to become blastocyst since it contains diploid chromosomes. Then stem cells can be differentiated into germ cell lineage, and these cells can be transplanted back to patient's seminiferous tubules or further development of germ cells can be achieved in vitro to be used for ICSI. In xenotransplantation, spermatogonia can be harvested from patient's testis and genetically corrected in vitro and then transferred into the testes of nonhuman primate whose seminiferous tubules are devoid of primate germ cells, to achieve spermatogenesis. Sperm obtained from xenograft can be used for ICSI. There are many safety and ethical concerns about these procedures as well.

Five-Year View and Key Issues

Primary spermatogenic failure accounts for about 50% of the cases of male infertility. Utilizing the clinically applicable tests, in only a minority of these men, the condition may be explained by genetic perturbations and the rest is called as "idiopathic" many of which may have yet unknown genetic mechanisms playing a role. Progress in the knowledge of genetics, epigenetics, and gene regulation should result in a greater understanding of male infertility. Considering the fact that the majority of infertile men may have an underlying genetic perturbation of some sort, general health surveillance of the male should become a priority before any specific treatment or assisted reproduction is attempted. The health of the offspring resulting from assisted reproductive technologies should also be followed to detect any clinical picture associated with genetic or epigenetic dysregulation. Finding a cure for spermatogenic failure will continue to be a challenge. It is fortunate that there are many new and evolving technologies to help identify the etiology of male infertility which may lead to targeted therapeutics. Although emerging animal studies offer some hope, the investigation should be directed toward how this data can find application in humans while avoiding any adverse effects. For the reasons above, male factor infertility research and treatment should incorporate a multidisciplinary approach. In this regard, collaboration between the reproductive endocrinologists, assisted reproduction laboratory staff, urologists, clinical geneticists, pediatricians, research scientists, and other members of the infertility team is essential.

Future Directions

Approaches such as microarray profiling, comparative genomic hybridization, and mutagenesis screening will have important roles in efforts to understand genetic origins of male infertility. Incorporating genomics, proteomics, and metabolomics into male infertility research may help to identify the population-specific complete role of genes involved with fertility [202–205]. These technologies provide vast amount of data with plenty of background "noise" at times. Therefore, the results from advanced genomic, proteomic, and metabolomic techniques should be confirmed by PCR, Western blot, flow cytometry, mass spectroscopy and chromatography, and protein function assays. These approaches will lead to better preconceptional counseling and more directed approaches for PGD. Future directions should also involve identifications of single gene defects associated with male infertility and advancing the field of gene therapy further to address safety and ethical concerns before these technologies could be applied in humans.

References

1. Ferlin A, Raicu F, Gatta V, Zuccarello D, Palka G, Foresta C. Male infertility: role of genetic background. Reprod Biomed Online. 2007;14:734–45.
2. Wilhelm D, Koopman P. The makings of maleness: towards an integrated view of male sexual development. Nat Rev Genet. 2006;7:620–31.
3. Miyamoto N, Yoshida M, Kuratani S, Matsuo I, Aizawa S. Defects of urogenital development in mice lacking Emx2. Development. 1997;124:1653–64.
4. Tevosian SG, Albrecht KH, Crispino JD, Fujiwara Y, Eicher EM, Orkin SH. Gonadal differentiation, sex determination and normal Sry expression in mice require direct interaction between transcription partners GATA4 and FOG2. Development. 2002;129: 4627–34.
5. Birk OS, Casiano DE, Wassif CA, Cogliati T, Zhao L, Zhao Y, et al. The LIM homeobox gene Lhx9 is essential for mouse gonad formation. Nature. 2000;403:909–13.
6. Achermann JC, Ito M, Hindmarsh PC, Jameson JL. A mutation in the gene encoding steroidogenic factor-1 causes XY sex reversal and adrenal failure in humans. Nat Genet. 1999;22:125–6.

7. Achermann JC, Ozisik G, Ito M, Orun UA, Harmanci K, Gurakan B, et al. Gonadal determination and adrenal development are regulated by the orphan nuclear receptor steroidogenic factor-1, in a dose-dependent manner. J Clin Endocrinol Metab. 2002;87: 1829–33.
8. Ludbrook LM, Harley VR. Sex determination: a 'window' of DAX1 activity. Trends Endocrinol Metab. 2004;15:116–21.
9. Ozisik G, Achermann JC, Jameson JL. The role of SF1 in adrenal and reproductive function: insight from naturally occurring mutations in humans. Mol Genet Metab. 2002;76:85–91.
10. Englert C. WT1—more than a transcription factor? Trends Biochem Sci. 1998;23:389–93.
11. Sinclair AH, Berta P, Palmer MS, Hawkins JR, Griffiths BL, Smith MJ, et al. A gene from the human sex-determining region encodes a protein with homology to a conserved DNA-binding motif. Nature. 1990;346:240–4.
12. Wilhelm D, Martinson F, Bradford S, Wilson MJ, Combes AN, Beverdam A, et al. Sertoli cell differentiation is induced both cell-autonomously and through prostaglandin signaling during mammalian sex determination. Dev Biol. 2005;287:111–24.
13. Bullejos M, Koopman P. Spatially dynamic expression of Sry in mouse genital ridges. Dev Dyn. 2001;221:201–5.
14. Hammes A, Guo JK, Lutsch G, Leheste JR, Landrock D, Ziegler U, et al. Two splice variants of the Wilms' tumor 1 gene have distinct functions during sex determination and nephron formation. Cell. 2001;106:319–29.
15. Nef S, Verma-Kurvari S, Merenmies J, Vassalli JD, Efstratiadis A, Accili D, et al. Testis determination requires insulin receptor family function in mice. Nature. 2003;426:291–5.
16. Bullejos M, Koopman P. Delayed Sry and Sox9 expression in developing mouse gonads underlies B6-Y(DOM) sex reversal. Dev Biol. 2005;278:473–81.
17. Taketo T, Lee CH, Zhang J, Li Y, Lee CY, Lau YF. Expression of SRY proteins in both normal and sex-reversed XY fetal mouse gonads. Dev Dyn. 2005;233:612–22.
18. Clark AM, Garland KK, Russell LD. Desert hedgehog (Dhh) gene is required in the mouse testis for formation of adult-type Leydig cells and normal development of peritubular cells and seminiferous tubules. Biol Reprod. 2000;63:1825–38.
19. Bitgood MJ, Shen L, McMahon AP. Sertoli cell signaling by desert hedgehog regulates the male germ line. Curr Biol. 1996;6: 298–304.
20. Canto P, Vilchis F, Soderlund D, Reyes E, Mendez JP. A heterozygous mutation in the desert hedgehog gene in patients with mixed gonadal dysgenesis. Mol Hum Reprod. 2005;11:833–6.
21. Canto P, Soderlund D, Reyes E, Mendez JP. Mutations in the desert hedgehog (DHH) gene in patients with 46, XY complete pure gonadal dysgenesis. J Clin Endocrinol Metab. 2004;89: 4480–3.
22. Kitamura K, Yanazawa M, Sugiyama N, Miura H, Iizuka-Kogo A, Kusaka M, et al. Mutation of ARX causes abnormal development of forebrain and testes in mice and X-linked lissencephaly with abnormal genitalia in humans. Nat Genet. 2002;32:359–69.
23. Tang P, Park DJ, Marshall Graves JA, Harley VR. ATRX and sex differentiation. Trends Endocrinol Metab. 2004;15: 339–44.
24. Brennan J, Tilmann C, Capel B. Pdgfr-alpha mediates testis cord organization and fetal Leydig cell development in the XY gonad. Genes Dev. 2003;17:800–10.
25. Tanaka SS, Yamaguchi YL, Tsoi B, Lickert H, Tam PP. IFITM/Mil/fragilis family proteins IFITM1 and IFITM3 play distinct roles in mouse primordial germ cell homing and repulsion. Dev Cell. 2005;9:745–56.
26. Molyneaux KA, Zinszner H, Kunwar PS, Schaible K, Stebler J, Sunshine MJ, et al. The chemokine SDF1/CXCL12 and its receptor CXCR4 regulate mouse germ cell migration and survival. Development. 2003;130:4279–86.
27. Adham IM, Agoulnik AI. Insulin-like 3 signalling in testicular descent. Int J Androl. 2004;27:257–65.
28. Ivell R, Hartung S. The molecular basis of cryptorchidism. Mol Hum Reprod. 2003;9:175–81.
29. Quill TA, Ren D, Clapham DE, Garbers DL. A voltage-gated ion channel expressed specifically in spermatozoa. Proc Natl Acad Sci USA. 2001;98:12527–31.
30. Sha J, Zhou Z, Li J, Yin L, Yang H, Hu G, et al. Identification of testis development and spermatogenesis-related genes in human and mouse testes using cDNA arrays. Mol Hum Reprod. 2002;8: 511–7.
31. Schultz N, Hamra FK, Garbers DL. A multitude of genes expressed solely in meiotic or postmeiotic spermatogenic cells offers a myriad of contraceptive targets. Proc Natl Acad Sci USA. 2003;100: 12201–6.
32. Hermo L, Pelletier RM, Cyr DG, Smith CE. Surfing the wave, cycle, life history, and genes/proteins expressed by testicular germ cells. Part 1: background to spermatogenesis, spermatogonia, and spermatocytes. Microsc Res Tech. 2010;73(4):241–78.
33. Saxena R, de Vries JW, Repping S, Alagappan RK, Skaletsky H, Brown LG, et al. Four DAZ genes in two clusters found in the AZFc region of the human Y chromosome. Genomics. 2000;67: 256–67.
34. Yen PH. Putative biological functions of the DAZ family. Int J Androl. 2004;27:125–9.
35. Hopps CV, Mielnik A, Goldstein M, Palermo GD, Rosenwaks Z, Schlegel PN. Detection of sperm in men with Y chromosome microdeletions of the AZFa, AZFb and AZFc regions. Hum Reprod. 2003;18:1660–5.
36. Silber SJ, Repping S. Transmission of male infertility to future generations: lessons from the Y chromosome. Hum Reprod Update. 2002;8:217–29.
37. Nishimune Y, Tanaka H. Infertility caused by polymorphisms or mutations in spermatogenesis-specific genes. J Androl. 2006;27:326–34.
38. Mueller JL, Mahadevaiah SK, Park PJ, Warburton PE, Page DC, Turner JM. The mouse X chromosome is enriched for multicopy testis genes showing postmeiotic expression. Nat Genet. 2008;40:794–9.
39. Roberts LM, Visser JA, Ingraham HA. Involvement of a matrix metalloproteinase in MIS-induced cell death during urogenital development. Development. 2002;129:1487–96.
40. Belville C, Marechal JD, Pennetier S, Carmillo P, Masgrau L, Messika-Zeitoun L, et al. Natural mutations of the anti-Mullerian hormone type II receptor found in persistent Mullerian duct syndrome affect ligand binding, signal transduction and cellular transport. Hum Mol Genet. 2009;18:3002–13.
41. Belville C, Van Vlijmen H, Ehrenfels C, Pepinsky B, Rezaie AR, Picard JY, et al. Mutations of the anti-Mullerian hormone gene in patients with persistent Mullerian duct syndrome: biosynthesis, secretion, and processing of the abnormal proteins and analysis using a three-dimensional model. Mol Endocrinol. 2004;18: 708–21.
42. Hannema SE, Hughes IA. Regulation of Wolffian duct development. Horm Res. 2007;67:142–51.
43. Dravis C, Yokoyama N, Chumley MJ, Cowan CA, Silvany RE, Shay J, et al. Bidirectional signaling mediated by ephrin-B2 and EphB2 controls urorectal development. Dev Biol. 2004;271: 272–90.
44. Goodman FR, Bacchelli C, Brady AF, Brueton LA, Fryns JP, Mortlock DP, et al. Novel HOXA13 mutations and the phenotypic spectrum of hand-foot-genital syndrome. Am J Hum Genet. 2000;67:197–202.
45. Chandley AC. Genetic contribution to male infertility. Hum Reprod. 1998;13 Suppl 3:76–83. discussion 4-8.
46. Johnson MD. Genetic risks of intracytoplasmic sperm injection in the treatment of male infertility: recommendations for genetic counseling and screening. Fertil Steril. 1998;70:397–411.

47. Ravel C, Berthaut I, Bresson JL, Siffroi JP. Prevalence of chromosomal abnormalities in phenotypically normal and fertile adult males: large-scale survey of over 10,000 sperm donor karyotypes. Hum Reprod. 2006;21:1484–9.
48. O'Flynn O'Brien KL, Varghese AC, Agarwal A. The genetic causes of male factor infertility: a review. Fertil Steril. 2010;93:1–12.
49. Palermo GD, Colombero LT, Hariprashad JJ, Schlegel PN, Rosenwaks Z. Chromosome analysis of epididymal and testicular sperm in azoospermic patients undergoing ICSI. Hum Reprod. 2002;17:570–5.
50. Van Assche E, Bonduelle M, Tournaye H, Joris H, Verheyen G, Devroey P, et al. Cytogenetics of infertile men. Hum Reprod. 1996;11 Suppl 4:1–24. discussion 5-6.
51. American Society for Reproductive Medicine. Evaluation of the azoospermic male. Fertil Steril. 2008;90:S74–7.
52. Walsh TJ, Pera RR, Turek PJ. The genetics of male infertility. Semin Reprod Med. 2009;27:124–36.
53. Friedler S, Raziel A, Strassburger D, Schachter M, Bern O, Ron-El R. Outcome of ICSI using fresh and cryopreserved-thawed testicular spermatozoa in patients with non-mosaic Klinefelter's syndrome. Hum Reprod. 2001;16:2616–20.
54. Schiff JD, Palermo GD, Veeck LL, Goldstein M, Rosenwaks Z, Schlegel PN. Success of testicular sperm extraction [corrected] and intracytoplasmic sperm injection in men with Klinefelter syndrome. J Clin Endocrinol Metab. 2005;90:6263–7.
55. Denschlag D, Tempfer C, Kunze M, Wolff G, Keck C. Assisted reproductive techniques in patients with Klinefelter syndrome: a critical review. Fertil Steril. 2004;82:775–9.
56. Blanco J, Egozcue J, Vidal F. Meiotic behaviour of the sex chromosomes in three patients with sex chromosome anomalies (47, XXY, mosaic 46, XY/47, XXY and 47, XYY) assessed by fluorescence in-situ hybridization. Hum Reprod. 2001;16:887–92.
57. Bergere M, Wainer R, Nataf V, Bailly M, Gombault M, Ville Y, et al. Biopsied testis cells of four 47, XXY patients: fluorescence in-situ hybridization and ICSI results. Hum Reprod. 2002;17:32–7.
58. Levron J, Aviram-Goldring A, Madgar I, Raviv G, Barkai G, Dor J. Sperm chromosome analysis and outcome of IVF in patients with non-mosaic Klinefelter's syndrome. Fertil Steril. 2000;74:925–9.
59. Staessen C, Tournaye H, Van Assche E, Michiels A, Van Landuyt L, Devroey P, et al. PGD in 47, XXY Klinefelter's syndrome patients. Hum Reprod Update. 2003;9:319–30.
60. Tournaye H, Staessen C, Liebaers I, Van Assche E, Devroey P, Bonduelle M, et al. Testicular sperm recovery in nine 47, XXY Klinefelter patients. Hum Reprod. 1996;11:1644–9.
61. Shi Q, Martin RH. Multicolor fluorescence in situ hybridization analysis of meiotic chromosome segregation in a 47, XYY male and a review of the literature. Am J Med Genet. 2000;93:40–6.
62. Gonzalez-Merino E, Hans C, Abramowicz M, Englert Y, Emiliani S. Aneuploidy study in sperm and preimplantation embryos from nonmosaic 47, XYY men. Fertil Steril. 2007;88:600–6.
63. Van der Auwera B, Van Roy N, De Paepe A, Hawkins JR, Liebaers I, Castedo S, et al. Molecular cytogenetic analysis of XX males using Y-specific DNA sequences, including SRY. Hum Genet. 1992;89:23–8.
64. De Braekeleer M, Dao TN. Cytogenetic studies in male infertility: a review. Hum Reprod. 1991;6:245–50.
65. Chantot-Bastaraud S, Ravel C, Siffroi JP. Underlying karyotype abnormalities in IVF/ICSI patients. Reprod Biomed Online. 2008;16:514–22.
66. Martin RH. Cytogenetic determinants of male fertility. Hum Reprod Update. 2008;14:379–90.
67. Ogur G, Van Assche E, Vegetti W, Verheyen G, Tournaye H, Bonduelle M, et al. Chromosomal segregation in spermatozoa of 14 Robertsonian translocation carriers. Mol Hum Reprod. 2006;12:209–15.
68. Mau-Holzmann UA. Somatic chromosomal abnormalities in infertile men and women. Cytogenet Genome Res. 2005;111:317–36.
69. Estop AM, Van Kirk V, Cieply K. Segregation analysis of four translocations, t(2;18), t(3;15), t(5;7), and t(10;12), by sperm chromosome studies and a review of the literature. Cytogenet Cell Genet. 1995;70:80–7.
70. McLachlan RI, O'Bryan MK. Clinical review#: state of the art for genetic testing of infertile men. J Clin Endocrinol Metab. 2010;95:1013–24.
71. Rozen S, Skaletsky H, Marszalek JD, Minx PJ, Cordum HS, Waterston RH, et al. Abundant gene conversion between arms of palindromes in human and ape Y chromosomes. Nature. 2003;423:873–6.
72. Li Z, Haines CJ, Han Y. "Micro-deletions" of the human Y chromosome and their relationship with male infertility. J Genet Genomics. 2008;35:193–9.
73. Skaletsky H, Kuroda-Kawaguchi T, Minx PJ, Cordum HS, Hillier L, Brown LG, et al. The male-specific region of the human Y chromosome is a mosaic of discrete sequence classes. Nature. 2003;423:825–37.
74. Foresta C, Moro E, Ferlin A. Y chromosome microdeletions and alterations of spermatogenesis. Endocr Rev. 2001;22:226–39.
75. Vollrath D, Foote S, Hilton A, Brown LG, Beer-Romero P, Bogan JS, et al. The human Y chromosome: a 43-interval map based on naturally occurring deletions. Science. 1992;258:52–9.
76. Kleiman SE, Bar-Shira Maymon B, Yogev L, Paz G, Yavetz H. The prognostic role of the extent of Y microdeletion on spermatogenesis and maturity of Sertoli cells. Hum Reprod. 2001;16:399–402.
77. Poongothai J, Gopenath TS, Manonayaki S. Genetics of human male infertility. Singapore Med J. 2009;50:336–47.
78. Vogt PH. Genetics of idiopathic male infertility: Y chromosomal azoospermia factors (AZFa, AZFb, AZFc). Baillieres Clin Obstet Gynaecol. 1997;11:773–95.
79. Sadeghi-Nejad H, Oates RD. The Y chromosome and male infertility. Curr Opin Urol. 2008;18:628–32.
80. Repping S, Skaletsky H, Lange J, Silber S, Van Der Veen F, Oates RD, et al. Recombination between palindromes P5 and P1 on the human Y chromosome causes massive deletions and spermatogenic failure. Am J Hum Genet. 2002;71:906–22.
81. Brown GM, Furlong RA, Sargent CA, Erickson RP, Longepied G, Mitchell M, et al. Characterisation of the coding sequence and fine mapping of the human DFFRY gene and comparative expression analysis and mapping to the Sxrb interval of the mouse Y chromosome of the Dffry gene. Hum Mol Genet. 1998;7:97–107.
82. Chuang RY, Weaver PL, Liu Z, Chang TH. Requirement of the DEAD-Box protein ded1p for messenger RNA translation. Science. 1997;275:1468–71.
83. Foresta C, Ferlin A, Moro E. Deletion and expression analysis of AZFa genes on the human Y chromosome revealed a major role for DBY in male infertility. Hum Mol Genet. 2000;9:1161–9.
84. Moro E, Ferlin A, Yen PH, Franchi PG, Palka G, Foresta C. Male infertility caused by a de novo partial deletion of the DAZ cluster on the Y chromosome. J Clin Endocrinol Metab. 2000;85:4069–73.
85. Sun C, Skaletsky H, Birren B, Devon K, Tang Z, Silber S, et al. An azoospermic man with a de novo point mutation in the Y-chromosomal gene USP9Y. Nat Genet. 1999;23:429–32.
86. Geoffroy-Siraudin C, Aknin-Seiffer I, Metzler-Guillemain C, Ghalamoun-Slaimi R, Bonzi MF, Levy R, et al. Meiotic abnormalities in patients bearing complete AZFc deletion of Y chromosome. Hum Reprod. 2007;22:1567–72.
87. Navarro-Costa P, Pereira L, Alves C, Gusmao L, Proenca C, Marques-Vidal P, et al. Characterizing partial AZFc deletions of the Y chromosome with amplicon-specific sequence markers. BMC Genomics. 2007;8:342.

88. Wu B, Lu NX, Xia YK, Gu AH, Lu CC, Wang W, et al. A frequent Y chromosome b2/b3 subdeletion shows strong association with male infertility in Han-Chinese population. Hum Reprod. 2007;22:1107–13.
89. Zhang F, Li Z, Wen B, Jiang J, Shao M, Zhao Y, et al. A frequent partial AZFc deletion does not render an increased risk of spermatogenic impairment in East Asians. Ann Hum Genet. 2006;70:304–13.
90. Zhu XB, Liu YL, Zhang W, Ping P, Cao XR, Liu Y, et al. Vertical transmission of the Yq AZFc microdeletion from father to son over two or three generations in infertile Han Chinese families. Asian J Androl. 2010;12:240–6.
91. Giachini C, Nuti F, Marinari E, Forti G, Krausz C. Partial AZFc deletions in infertile men with cryptorchidism. Hum Reprod. 2007;22:2398–403.
92. Zhang F, Lu C, Li Z, Xie P, Xia Y, Zhu X, et al. Partial deletions are associated with an increased risk of complete deletion in AZFc: a new insight into the role of partial AZFc deletions in male infertility. J Med Genet. 2007;44:437–44.
93. Ferlin A, Arredi B, Speltra E, Cazzadore C, Selice R, Garolla A, et al. Molecular and clinical characterization of Y chromosome microdeletions in infertile men: a 10-year experience in Italy. J Clin Endocrinol Metab. 2007;92:762–70.
94. Mulhall JP, Reijo R, Alagappan R, Brown L, Page D, Carson R, et al. Azoospermic men with deletion of the DAZ gene cluster are capable of completing spermatogenesis: fertilization, normal embryonic development and pregnancy occur when retrieved testicular spermatozoa are used for intracytoplasmic sperm injection. Hum Reprod. 1997;12:503–8.
95. Oates RD, Silber S, Brown LG, Page DC. Clinical characterization of 42 oligospermic or azoospermic men with microdeletion of the AZFc region of the Y chromosome, and of 18 children conceived via ICSI. Hum Reprod. 2002;17:2813–24.
96. Osborne EC, Lynch M, McLachlan R, Trounson AO, Cram DS. Microarray detection of Y chromosome deletions associated with male infertility. Reprod Biomed Online. 2007;15:673–80.
97. Ferlin A, Moro E, Onisto M, Toscano E, Bettella A, Foresta C. Absence of testicular DAZ gene expression in idiopathic severe testiculopathies. Hum Reprod. 1999;14:2286–92.
98. Lange J, Skaletsky H, van Daalen SK, Embry SL, Korver CM, Brown LG, et al. Isodicentric Y chromosomes and sex disorders as byproducts of homologous recombination that maintains palindromes. Cell. 2009;138:855–69.
99. Lardone MC, Parodi DA, Valdevenito R, Ebensperger M, Piottante A, Madariaga M, et al. Quantification of DDX3Y, RBMY1, DAZ and TSPY mRNAs in testes of patients with severe impairment of spermatogenesis. Mol Hum Reprod. 2007;13:705–12.
100. Vodicka R, Vrtel R, Dusek L, Singh AR, Krizova K, Svacinova V, et al. TSPY gene copy number as a potential new risk factor for male infertility. Reprod Biomed Online. 2007;14:579–87.
101. Wang PJ, McCarrey JR, Yang F, Page DC. An abundance of X-linked genes expressed in spermatogonia. Nat Genet. 2001;27: 422–6.
102. Cantu JM, Diaz M, Moller M, Jimenez-Sainz M, Sandoval L, Vaca G, et al. Azoospermia and duplication 3qter as distinct consequences of a familial t(X;3) (q26;q13.2). Am J Med Genet. 1985;20:677–84.
103. Lee S, Lee SH, Chung TG, Kim HJ, Yoon TK, Kwak IP, et al. Molecular and cytogenetic characterization of two azoospermic patients with X-autosome translocation. J Assist Reprod Genet. 2003;20:385–9.
104. Nemeth AH, Gallen IW, Crocker M, Levy E, Maher E. Klinefelter-like phenotype and primary infertility in a male with a paracentric Xq inversion. J Med Genet. 2002;39:E28.
105. Olesen C, Silber J, Eiberg H, Ernst E, Petersen K, Lindenberg S, et al. Mutational analysis of the human FATE gene in 144 infertile men. Hum Genet. 2003;113:195–201.
106. Raverot G, Lejeune H, Kotlar T, Pugeat M, Jameson JL. X-linked sex-determining region Y box 3 (SOX3) gene mutations are uncommon in men with idiopathic oligoazoospermic infertility. J Clin Endocrinol Metab. 2004;89:4146–8.
107. Schneider-Gadicke A, Beer-Romero P, Brown LG, Mardon G, Luoh SW, Page DC. Putative transcription activator with alternative isoforms encoded by human ZFX gene. Nature. 1989;342: 708–11.
108. Schneider-Gadicke A, Beer-Romero P, Brown LG, Nussbaum R, Page DC. ZFX has a gene structure similar to ZFY, the putative human sex determinant, and escapes X inactivation. Cell. 1989;57:1247–58.
109. Wang RS, Yeh S, Tzeng CR, Chang C. Androgen receptor roles in spermatogenesis and fertility: lessons from testicular cell-specific androgen receptor knockout mice. Endocr Rev. 2009;30:119–32.
110. Xu Q, Lin HY, Yeh SD, Yu IC, Wang RS, Chen YT, et al. Infertility with defective spermatogenesis and steroidogenesis in male mice lacking androgen receptor in Leydig cells. Endocrine. 2007;32: 96–106.
111. Tsai MY, Yeh SD, Wang RS, Yeh S, Zhang C, Lin HY, et al. Differential effects of spermatogenesis and fertility in mice lacking androgen receptor in individual testis cells. Proc Natl Acad Sci USA. 2006;103:18975–80.
112. De Gendt K, Swinnen JV, Saunders PT, Schoonjans L, Dewerchin M, Devos A, et al. A Sertoli cell-selective knockout of the androgen receptor causes spermatogenic arrest in meiosis. Proc Natl Acad Sci USA. 2004;101:1327–32.
113. Ferlin A, Vinanzi C, Garolla A, Selice R, Zuccarello D, Cazzadore C, et al. Male infertility and androgen receptor gene mutations: clinical features and identification of seven novel mutations. Clin Endocrinol (Oxf). 2006;65:606–10.
114. Greenland KJ, Zajac JD. Kennedy's disease: pathogenesis and clinical approaches. Intern Med J. 2004;34:279–86.
115. Finsterer J. Bulbar and spinal muscular atrophy (Kennedy's disease): a review. Eur J Neurol. 2009;16:556–61.
116. Lazaros L, Xita N, Kaponis A, Zikopoulos K, Sofikitis N, Georgiou I. Evidence for association of sex hormone-binding globulin and androgen receptor genes with semen quality. Andrologia. 2008;40:186–91.
117. Rajpert-De Meyts E, Leffers H, Petersen JH, Andersen AG, Carlsen E, Jorgensen N, et al. CAG repeat length in androgen-receptor gene and reproductive variables in fertile and infertile men. Lancet. 2002;359:44–6.
118. Yong EL, Loy CJ, Sim KS. Androgen receptor gene and male infertility. Hum Reprod Update. 2003;9:1–7.
119. Dowsing AT, Yong EL, Clark M, McLachlan RI, de Kretser DM, Trounson AO. Linkage between male infertility and trinucleotide repeat expansion in the androgen-receptor gene. Lancet. 1999;354:640–3.
120. Stouffs K, Lissens W, Tournaye H, Van Steirteghem A, Liebaers I. Possible role of USP26 in patients with severely impaired spermatogenesis. Eur J Hum Genet. 2005;13:336–40.
121. Nuti F, Krausz C. Gene polymorphisms/mutations relevant to abnormal spermatogenesis. Reprod Biomed Online. 2008;16: 504–13.
122. Akinloye O, Gromoll J, Callies C, Nieschlag E, Simoni M. Mutation analysis of the X-chromosome linked, testis-specific TAF7L gene in spermatogenic failure. Andrologia. 2007;39: 190–5.
123. Ravel C, El Houate B, Chantot S, Lourenco D, Dumaine A, Rouba H, et al. Haplotypes, mutations and male fertility: the story of the testis-specific ubiquitin protease USP26. Mol Hum Reprod. 2006;12:643–6.
124. Stouffs K, Willems A, Lissens W, Tournaye H, Van Steirteghem A, Liebaers I. The role of the testis-specific gene hTAF7L in the aetiology of male infertility. Mol Hum Reprod. 2006;12:263–7.

125. Fechner A, Fong S, McGovern P. A review of Kallmann syndrome: genetics, pathophysiology, and clinical management. Obstet Gynecol Surv. 2008;63:189–94.
126. Kim SH, Hu Y, Cadman S, Bouloux P. Diversity in fibroblast growth factor receptor 1 regulation: learning from the investigation of Kallmann syndrome. J Neuroendocrinol. 2008;20: 141–63.
127. Stuhrmann M, Dork T. CFTR gene mutations and male infertility. Andrologia. 2000;32:71–83.
128. Dumur V, Gervais R, Rigot JM, Lafitte JJ, Manouvrier S, Biserte J, et al. Abnormal distribution of CF delta F508 allele in azoospermic men with congenital aplasia of epididymis and vas deferens. Lancet. 1990;336:512.
129. Gervais R, Dumur V, Rigot JM, Lafitte JJ, Roussel P, Claustres M, et al. High frequency of the R117H cystic fibrosis mutation in patients with congenital absence of the vas deferens. N Engl J Med. 1993;328:446–7.
130. Claustres M, Guittard C, Bozon D, Chevalier F, Verlingue C, Ferec C, et al. Spectrum of CFTR mutations in cystic fibrosis and in congenital absence of the vas deferens in France. Hum Mutat. 2000;16:143–56.
131. Daudin M, Bieth E, Bujan L, Massat G, Pontonnier F, Mieusset R. Congenital bilateral absence of the vas deferens: clinical characteristics, biological parameters, cystic fibrosis transmembrane conductance regulator gene mutations, and implications for genetic counseling. Fertil Steril. 2000;74:1164–74.
132. Cuppens H, Cassiman JJ. CFTR mutations and polymorphisms in male infertility. Int J Androl. 2004;27:251–6.
133. Cystic fibrosis mutation database. http://www.genet.sickkids.on.ca/cftr/StatisticsPage.html. Last updated 1 Apr 2010.
134. Dequeker E, Stuhrmann M, Morris MA, Casals T, Castellani C, Claustres M, et al. Best practice guidelines for molecular genetic diagnosis of cystic fibrosis and CFTR-related disorders—updated European recommendations. Eur J Hum Genet. 2009;17:51–65.
135. Chillon M, Casals T, Mercier B, Bassas L, Lissens W, Silber S, et al. Mutations in the cystic fibrosis gene in patients with congenital absence of the vas deferens. N Engl J Med. 1995;332: 1475–80.
136. Scotet V, Audrezet MP, Roussey M, Rault G, Dirou-Prigent A, Journel H, et al. Immunoreactive trypsin/DNA newborn screening for cystic fibrosis: should the R117H variant be included in CFTR mutation panels? Pediatrics. 2006;118:e1523–9.
137. Kiesewetter S, Macek Jr M, Davis C, Curristin SM, Chu CS, Graham C, et al. A mutation in CFTR produces different phenotypes depending on chromosomal background. Nat Genet. 1993;5: 274–8.
138. American College of Obstetricians and Gynecologists. Update on carrier screening for cystic fibrosis. ACOG Committee Opinion Number 325, December 2005. Obstet Gynecol. 2005;106: 1465–8.
139. Egozcue J, Blanco J, Anton E, Egozcue S, Sarrate Z, Vidal F. Genetic analysis of sperm and implications of severe male infertility—a review. Placenta. 2003;24(Suppl B):S62–5.
140. Bernardini L, Gianaroli L, Fortini D, Conte N, Magli C, Cavani S, et al. Frequency of hyper-, hypohaploidy and diploidy in ejaculate, epididymal and testicular germ cells of infertile patients. Hum Reprod. 2000;15:2165–72.
141. Vegetti W, Van Assche E, Frias A, Verheyen G, Bianchi MM, Bonduelle M, et al. Correlation between semen parameters and sperm aneuploidy rates investigated by fluorescence in-situ hybridization in infertile men. Hum Reprod. 2000;15:351–65.
142. Gonsalves J, Sun F, Schlegel PN, Turek PJ, Hopps CV, Greene C, et al. Defective recombination in infertile men. Hum Mol Genet. 2004;13:2875–83.
143. Sun F, Greene C, Turek PJ, Ko E, Rademaker A, Martin RH. Immunofluorescent synaptonemal complex analysis in azoospermic men. Cytogenet Genome Res. 2005;111:366–70.
144. Sun F, Turek P, Greene C, Ko E, Rademaker A, Martin RH. Abnormal progression through meiosis in men with nonobstructive azoospermia. Fertil Steril. 2007;87:565–71.
145. Topping D, Brown P, Judis L, Schwartz S, Seftel A, Thomas A, et al. Synaptic defects at meiosis I and non-obstructive azoospermia. Hum Reprod. 2006;21:3171–7.
146. Sanderson ML, Hassold TJ, Carrell DT. Proteins involved in meiotic recombination: a role in male infertility? Syst Biol Reprod Med. 2008;54:57–74.
147. Miyamoto T, Hasuike S, Yogev L, Maduro MR, Ishikawa M, Westphal H, et al. Azoospermia in patients heterozygous for a mutation in SYCP3. Lancet. 2003;362:1714–9.
148. Aarabi M, Modarressi MH, Soltanghoraee H, Behjati R, Amirjannati N, Akhondi MM. Testicular expression of synaptonemal complex protein 3 (SYCP3) messenger ribonucleic acid in 110 patients with nonobstructive azoospermia. Fertil Steril. 2006;86:325–31.
149. Meeks M, Bush A. Primary ciliary dyskinesia (PCD). Pediatr Pulmonol. 2000;29:307–16.
150. Loges NT, Olbrich H, Becker-Heck A, Haffner K, Heer A, Reinhard C, et al. Deletions and point mutations of LRRC50 cause primary ciliary dyskinesia due to dynein arm defects. Am J Hum Genet. 2009;85:883–9.
151. Moore A, Escudier E, Roger G, Tamalet A, Pelosse B, Marlin S, et al. RPGR is mutated in patients with a complex X linked phenotype combining primary ciliary dyskinesia and retinitis pigmentosa. J Med Genet. 2006;43:326–33.
152. Gerber PA, Kruse R, Hirchenhain J, Krussel JS, Neumann NJ. Pregnancy after laser-assisted selection of viable spermatozoa before intracytoplasmatic sperm injection in a couple with male primary cilia dyskinesia. Fertil Steril. 1826;2008(89):e9–12.
153. Dieterich K, Zouari R, Harbuz R, Vialard F, Martinez D, Bellayou H, et al. The Aurora Kinase C c.144delC mutation causes meiosis I arrest in men and is frequent in the North African population. Hum Mol Genet. 2009;18:1301–9.
154. Kimmins S, Crosio C, Kotaja N, Hirayama J, Monaco L, Hoog C, et al. Differential functions of the Aurora-B and Aurora-C kinases in mammalian spermatogenesis. Mol Endocrinol. 2007;21: 726–39.
155. Lee C, Iafrate AJ, Brothman AR. Copy number variations and clinical cytogenetic diagnosis of constitutional disorders. Nat Genet. 2007;39:S48–54.
156. Anderson S, Bankier AT, Barrell BG, de Bruijn MH, Coulson AR, Drouin J, et al. Sequence and organization of the human mitochondrial genome. Nature. 1981;290:457–65.
157. Yakes FM, Van Houten B. Mitochondrial DNA damage is more extensive and persists longer than nuclear DNA damage in human cells following oxidative stress. Proc Natl Acad Sci USA. 1997;94:514–9.
158. Wei YH, Kao SH, Lee HC. Simultaneous increase of mitochondrial DNA deletions and lipid peroxidation in human aging. Ann N Y Acad Sci. 1996;786:24–43.
159. Rovio AT, Marchington DR, Donat S, Schuppe HC, Abel J, Fritsche E, et al. Mutations at the mitochondrial DNA polymerase (POLG) locus associated with male infertility. Nat Genet. 2001;29:261–2.
160. St John JC, Jokhi RP, Barratt CL. The impact of mitochondrial genetics on male infertility. Int J Androl. 2005;28:65–73.
161. Ropp PA, Copeland WC. Cloning and characterization of the human mitochondrial DNA polymerase, DNA polymerase gamma. Genomics. 1996;36:449–58.
162. Surani MA. Reprogramming of genome function through epigenetic inheritance. Nature. 2001;414:122–8.
163. Wallace JA, Orr-Weaver TL. Replication of heterochromatin: insights into mechanisms of epigenetic inheritance. Chromosoma. 2005;114:389–402.

164. Margueron R, Trojer P, Reinberg D. The key to development: interpreting the histone code? Curr Opin Genet Dev. 2005;15: 163–76.
165. Kim VN. Small RNAs just got bigger: Piwi-interacting RNAs (piRNAs) in mammalian testes. Genes Dev. 2006;20:1993–7.
166. Papaioannou MD, Nef S. microRNAs in the testis: building up male fertility. J Androl. 2010;31:26–33.
167. Reik W, Dean W, Walter J. Epigenetic reprogramming in mammalian development. Science. 2001;293:1089–93.
168. Ferguson-Smith AC, Surani MA. Imprinting and the epigenetic asymmetry between parental genomes. Science. 2001;293: 1086–9.
169. Guerrero-Bosagna C, Sabat P, Valladares L. Environmental signaling and evolutionary change: can exposure of pregnant mammals to environmental estrogens lead to epigenetically induced evolutionary changes in embryos? Evol Dev. 2005;7:341–50.
170. MacPhee DG. Epigenetics and epimutagens: some new perspectives on cancer, germ line effects and endocrine disrupters. Mutat Res. 1998;400:369–79.
171. Guerrero-Bosagna CM, Skinner MK. Epigenetic transgenerational effects of endocrine disruptors on male reproduction. Semin Reprod Med. 2009;27:403–8.
172. Bukulmez O. Does assisted reproductive technology cause birth defects? Curr Opin Obstet Gynecol. 2009;21:260–4.
173. Marques CJ, Carvalho F, Sousa M, Barros A. Genomic imprinting in disruptive spermatogenesis. Lancet. 2004;363:1700–2.
174. Marques CJ, Costa P, Vaz B, Carvalho F, Fernandes S, Barros A, et al. Abnormal methylation of imprinted genes in human sperm is associated with oligozoospermia. Mol Hum Reprod. 2008;14: 67–74.
175. Trasler JM. Epigenetics in spermatogenesis. Mol Cell Endocrinol. 2009;306:33–6.
176. Aoki VW, Emery BR, Carrell DT. Global sperm deoxyribonucleic acid methylation is unaffected in protamine-deficient infertile males. Fertil Steril. 2006;86:1541–3.
177. Aoki VW, Emery BR, Liu L, Carrell DT. Protamine levels vary between individual sperm cells of infertile human males and correlate with viability and DNA integrity. J Androl. 2006;27: 890–8.
178. Aoki VW, Liu L, Carrell DT. A novel mechanism of protamine expression deregulation highlighted by abnormal protamine transcript retention in infertile human males with sperm protamine deficiency. Mol Hum Reprod. 2006;12:41–50.
179. Paldi A. Genomic imprinting: could the chromatin structure be the driving force? Curr Top Dev Biol. 2003;53:115–38.
180. Hammoud SS, Purwar J, Pflueger C, Cairns BR, Carrell DT. Alterations in sperm DNA methylation patterns at imprinted loci in two classes of infertility. Fertil Steril. 2010;94(5):1728–33.
181. Barratt CL, Aitken RJ, Bjorndahl L, Carrell DT, de Boer P, Kvist U, et al. Sperm DNA: organization, protection and vulnerability: from basic science to clinical applications—a position report. Hum Reprod. 2010;25:824–38.
182. Hotaling JM, Walsh TJ. Male infertility: a risk factor for testicular cancer. Nat Rev Urol. 2009;6:550–6.
183. Walsh TJ, Croughan MS, Schembri M, Chan JM, Turek PJ. Increased risk of testicular germ cell cancer among infertile men. Arch Intern Med. 2009;169:351–6.
184. Dieckmann KP, Pichlmeier U. Clinical epidemiology of testicular germ cell tumors. World J Urol. 2004;22:2–14.
185. Nathanson KL, Kanetsky PA, Hawes R, Vaughn DJ, Letrero R, Tucker K, et al. The Y deletion gr/gr and susceptibility to testicular germ cell tumor. Am J Hum Genet. 2005;77:1034–43.
186. Kojima Y, Mizuno K, Kohri K, Hayashi Y. Advances in molecular genetics of cryptorchidism. Urology. 2009;74:571–8.
187. Yan W. Male infertility caused by spermiogenic defects: lessons from gene knockouts. Mol Cell Endocrinol. 2009;306:24–32.
188. Matzuk MM, Lamb DJ. The biology of infertility: research advances and clinical challenges. Nat Med. 2008;14:1197–213.
189. Meng MV, Black LD, Cha I, Ljung BM, Pera RA, Turek PJ. Impaired spermatogenesis in men with congenital absence of the vas deferens. Hum Reprod. 2001;16:529–33.
190. Schlegel PN, Shin D, Goldstein M. Urogenital anomalies in men with congenital absence of the vas deferens. J Urol. 1996;155: 1644–8.
191. Kim ED, Gilbaugh 3rd JH, Patel VR, Turek PJ, Lipshultz LI. Testis biopsies frequently demonstrate sperm in men with azoospermia and significantly elevated follicle-stimulating hormone levels. J Urol. 1997;157:144–6.
192. Schlegel PN, Palermo GD, Goldstein M, Menendez S, Zaninovic N, Veeck LL, et al. Testicular sperm extraction with intracytoplasmic sperm injection for nonobstructive azoospermia. Urology. 1997;49:435–40.
193. Brugh 3rd VM, Maduro MR, Lamb DJ. Genetic disorders and infertility. Urol Clin North Am. 2003;30:143–52.
194. O'Bryan MK, de Kretser D. Mouse models for genes involved in impaired spermatogenesis. Int J Androl. 2006;29:76–89. discussion 105-8.
195. Manzini S, Vargiolu A, Stehle IM, Bacci ML, Cerrito MG, Giovannoni R, et al. Genetically modified pigs produced with a nonviral episomal vector. Proc Natl Acad Sci USA. 2006;103: 17672–7.
196. Boekelheide K, Sigman M. Is gene therapy for the treatment of male infertility feasible? Nat Clin Pract Urol. 2008;5:590–3.
197. Lamb DJ. Would gene therapy for the treatment of male infertility be safe? Nat Clin Pract Urol. 2008;5:594–5.
198. Brinster RL. Germ line stem cell transplantation and transgenesis. Science. 2002;296:2174–6.
199. Brinster RL. Male germ line stem cells: from mice to men. Science. 2007;316:404–5.
200. Nayernia K, Nolte J, Michelmann HW, Lee JH, Rathsack K, Drusenheimer N, et al. In vitro-differentiated embryonic stem cells give rise to male gametes that can generate offspring mice. Dev Cell. 2006;11:125–32.
201. Ryu BY, Orwig KE, Oatley JM, Lin CC, Chang LJ, Avarbock MR, et al. Efficient generation of transgenic rats through the male germ line using lentiviral transduction and transplantation of spermatogonial stem cells. J Androl. 2007;28:353–60.
202. Deepinder F, Chowdary HT, Agarwal A. Role of metabolomic analysis of biomarkers in the management of male infertility. Expert Rev Mol Diagn. 2007;7:351–8.
203. He Z, Chan WY, Dym M. Microarray technology offers a novel tool for the diagnosis and identification of therapeutic targets for male infertility. Reproduction. 2006;132:11–9.
204. Lin YH, Lin YM, Teng YN, Hsieh TY, Lin YS, Kuo PL. Identification of ten novel genes involved in human spermatogenesis by microarray analysis of testicular tissue. Fertil Steril. 2006;86:1650–8.
205. Martinez-Heredia J, de Mateo S, Vidal-Taboada JM, Ballesca JL, Oliva R. Identification of proteomic differences in asthenozoospermic sperm samples. Hum Reprod. 2008;23:783–91.

Novel Approaches in the Management of Klinefelter Syndrome

19

Fnu Deepinder

Abstract

Klinefelter syndrome is the most common form of hypogonadism in men and is the leading genetic cause of male infertility. It also represents the most prevalent chromosomal aneuploidy in human beings. It is characterized by the presence of an extra X chromosome in a phenotypic male. The most abundant karyotype is 47, XXY, although other patterns including mosaicism (47, XY/47, XXY) and higher grade chromosomal aneuploidies containing supranumerous X chromosomes (48, XXXY, 49, XXXXY) are not uncommon. The latter phenotypes are more severely affected in terms of physical and mental development than men with classic 47, XXY karyotypes. The genetic cause is either from meiotic nondisjunction leading to failure of separation of the chromosome pair during the first or second division of gametogenesis or from mitotic nondisjunction in the developing zygote. Increasing maternal age has been reported to raise the risk of Klinefelter syndrome. The estimated prevalence of Klinefelter syndrome in men is 1:500 or 0.1–0.2% of the general population. However, extremely large discrepancies have been reported between prenatal and postnatal prevalence suggesting high rates of under diagnosis. A large Danish national registry study observed only 25% of the expected patients diagnosed after birth, and less than 10% of the expected diagnoses were made before puberty. Some of the major reasons for this underdiagnosis are thought to be the variable phenotype of Klinefelter syndrome and low awareness of the disease among medical professionals. Recognition of clinical features is hence important for early detection of this syndrome.

Keywords

Klinefelter syndrome • Chromosomal aneuploidy • Hypogonadism • 47, XXY karyotypes • Osteoporosis • Incomplete pubertal development • Mosaicism • Barr-body analysis

Klinefelter syndrome [1] is the most common form of hypogonadism in men and is the leading genetic cause of male infertility. It also represents the most prevalent chromosomal aneuploidy in human beings [2, 3]. It is characterized by the presence of an extra X chromosome in a phenotypic male. The most abundant karyotype is 47, XXY, although other patterns including mosaicism (47, XY/47, XXY) and higher grade chromosomal aneuploidies containing supranumerous X chromosomes (48, XXXY, 49, XXXXY) are not uncommon. The latter phenotypes are more severely affected in terms of physical and mental development than men with classic 47, XXY karyotypes [4]. The genetic cause is either meiotic nondisjunction leading to failure of separation of the chromosome pair during first or second division of gametogenesis or from mitotic nondisjunction in the developing zygote. Increasing maternal age has been reported to raise the risk of Klinefelter syndrome [5].

F. Deepinder, MD (✉)
Department of Endocrinology, Diabetes and Metabolism, Cedars Sinai Medical Center, 8700 Beverly Boulevard, B-131, Los Angeles, CA 90048, USA
e-mail: fnu.deepinder@cshs.org

S.J. Parekattil and A. Agarwal (eds.), *Male Infertility: Contemporary Clinical Approaches, Andrology, ART & Antioxidants*,
DOI 10.1007/978-1-4614-3335-4_19, © Springer Science+Business Media, LLC 2012

The estimated prevalence of Klinefelter syndrome in men is 1:500 or 0.1–0.2% of the general population [2, 3]. However, extremely large discrepancies have been reported between prenatal and postnatal prevalence suggesting high rates of under diagnosis. A large Danish national registry study observed only 25% of the expected patients diagnosed after birth, and less than 10% of the expected diagnoses were made before puberty [6]. Some of the major reasons for this underdiagnosis are thought to be the variable phenotype of Klinefelter syndrome and low awareness of the disease among medical professionals. Recognition of clinical features is hence important for early detection of this syndrome.

Diagnosis

A suspected diagnosis is based on a combination of typical clinical findings and laboratory investigations. Clinical signs and symptoms vary by age.

Prenatal

Although uncommon, fetus can be prenatally diagnosed to have 47, XXY karyotype by routine amniocentesis in high risk pregnancies especially in the presence of advanced maternal age. If confirmed, professional genetic counseling should be offered to the parents regarding the prognosis of the baby.

At Birth

Klinefelter syndrome is associated with several major and minor congenital abnormalities at birth including cleft palate, inguinal hernia, testis retention, clinodactyly, hypospadias, and microphallus [7]. Although these are not specific, it should prompt chromosomal evaluations to detect sex-chromosome aneuploidy, and if such screening is suggestive of Klinefelter syndrome, confirmatory karyotyping tests should follow [8].

School Age

Male children in this age group present with learning disability, language delay, and behavioral problems that often lead to chromosomal evaluation thus leading to the diagnosis of Klinefelter syndrome.

Adolescence

Adolescent boys present with delayed or incomplete pubertal development. These children may have varying signs and symptoms of androgen deficiency including eunuchoid body habitus with long legs, sparse body hair, gynecomastia, small phallus, and small firm testes [9].

Adults

Adults are often recognized through evaluation of decreasing libido, potency, and infertility. The development of secondary sexual characteristics such as beard growth, muscle bulk, and secondary body hair is either reduced or delayed. In addition, these men have several long-term consequences of hypogonadism including osteoporosis, glucose intolerance, and metabolic syndrome [8]. The incidence of breast cancer, mediastinal germ cell tumors, and non-Hodgkin lymphomas has also been found to be grossly elevated in these men [10].

A quick and reliable screening test for Klinefelter syndrome is Barr-body analysis. If suggestive, Klinefelter syndrome can be confirmed by chromosomal analysis in the lymphocytes. In certain cases such as chromosomal mosaicism, testicular biopsy is needed which generally reveals hyalinization and fibrosis of seminiferous tubules, absence of spermatogenesis, and relative hyperplasia of the Leydig cells [11].

All men with Klinefelter syndrome should have a full hormonal workup including testosterone, luteinizing hormone (LH), follicle-stimulating hormone (FSH), estradiol, prolactin, cortisol, sex hormone-binding globulins (SHBG), inhibin-B, thyroid function tests, and insulin-like growth factor-1 (IGF-1). These patients have normal levels of FSH, LH, and testosterone during prepubertal period; however, the serum testosterone declines after puberty and the levels of LH and FSH rise in most cases. The serum concentration of SHBG is also high causing further decline in free testosterone levels. The estrogen levels are usually higher than in normal men. The inhibin-B which is a marker of Sertoli cell function has demonstrated to be a better marker of spermatogenesis than FSH. The levels of inhibin-B decrease significantly after puberty in boys with Klinefelter syndrome reflecting loss of Sertoli cells [12]. At least three semen samples should be collected and analyzed, which usually reveal azoospermia. Cortisol levels should also be routinely measured to rule out any coexisting adrenal insufficiency [13].

In addition to the reproductive workup, it is important to monitor fasting blood glucose and lipids as these men have increased risk of diabetes and metabolic syndrome [8]. They also have increased risk of deep vein thrombosis and pulmonary embolism thus necessitating routine hematocrit checks to detect increased viscosity [14]. In addition, bone densitometry by dual-energy X-ray absorptiometry (DEXA) scan should be performed at 2- to 3-year intervals due to increased risk of osteoporosis, and their vitamin D status should be monitored. It is also suggested to do echocardiogram as a large proportion of these patients have been found to have mitral valve prolapse and their cardiovascular mortality is high [15].

Treatment

Treatment and care of Klinefelter males is a multidisciplinary approach and depends upon the patient's age.

Children

The most important problems associated with this syndrome in early childhood are delayed speech and learning disabilities. Parents should be counseled to anticipate communication problems in their child so as to avoid negative interactions. In addition, referral to speech therapist should be considered as soon the child shows signs of delayed speech. Language therapy is often required for these children in order to develop skills to better understand and deliver complex language [9].

Adolescents

At puberty, most patients with Klinefelter syndrome experience decline in testosterone and rise in LH and FSH. Testosterone treatment should be initiated in these boys around the age 12 years if their gonadotropin levels are elevated, even if their testosterone levels are in the lower limit of normal range [8]. Androgen replacement promotes development of secondary sexual characteristics, normalization of body proportions, prevention of gynecomastia, and improvement in energy, mood, and concentration among these adolescents [16, 17]. This aids in the development of normal male self-image and helps them build relationships with other people of same or opposite sex. The goal of testosterone therapy should be normalization of LH and testosterone levels in the age-appropriate mid-normal range [8].

Adult Males

Treatment in adult men with Klinefelter syndrome can be broadly classified into two categories.

Androgen Replacement

The patients who are not interested in fertility should be on lifelong testosterone-replacement therapy in order to prevent long-term manifestation of androgen deficiency such as osteoporosis, obesity, diabetes, and metabolic syndrome. In addition, androgen therapy improves mood, behavior, and self esteem and reduces fatigue and irritability [17]. Transdermal testosterone in either 5–10 mg/day patch or 5–10 g/day 1% gel is the preferred form of testosterone replacement as they lead to better steady state levels of serum testosterone than the injectable forms like testosterone cypionate or testosterone enanthate at doses of 50–400 mg intramuscular every 2–4 weeks. Other available testosterone preparations include buccal testosterone and subdermal implants. All patients on androgen replacement should have routine prostate examinations and measurement of PSA and hematocrit levels every 6 months.

Fertility

Until last decade, men with Klinefelter syndrome were considered sterile. However, recent literature suggests that Klinefelter syndrome males are born with spermatogonia which undergo massive apoptosis during early puberty [18, 19]. Spermatozoa have been found in the testes of these men, and in a minority of patients, viable sperm can also be seen in the ejaculate. With the advent of testicular sperm extraction (TESE) and intracytoplasmic sperm injection (ICSI), now it is possible to reproduce even when the spermatozoa are not present in the ejaculate but only in the testes.

Testicular sperm retrieval rates by microsurgical techniques in Klinefelter syndrome patients have been reported to be as high as 40–70% [20, 21]. Various studies have evaluated the parameters predicting sperm recovery in these men with varying results. While a small study by Madgar et al. reported that testicular volume, testosterone concentration, and the hCG test predicted sperm recovery in Klinefelter men [22], a recent study by Ramasamy et al. found no predictive value of serum FSH, LH, and testicular volume for sperm recovery [23].

Although optimal hormonal therapy prior to sperm retrieval has not been established so far, it is a common practice to stop testosterone replacement prior to any intervention due to possible deleterious effects of exogenous testosterone in suppressing spermatogenesis. Aromatase inhibitor such as anastrozole is used for 6 months prior to sperm extraction in order to decrease intratesticular estradiol and increase testosterone [13, 24]. In addition, few centers use human chorionic gonadotropin (hCG) and/or clomiphene citrate to stimulate endogenous testosterone production and spermatogenesis. However, hCG should be used along with aromatase inhibitors to prevent concomitant rise in estrogen levels [13]. Men with hypogonadism who respond to medical therapy with a resultant testosterone levels of greater than 250 ng/dL have been found to have a better chance of sperm retrieval than men who did not [23]. It is unclear if the rise in endogenous testosterone improves spermatogenesis or normalization of serum testosterone levels with medical therapy just occurs in men with greater potential for spermatogenesis without a cause and effect relationship [23].

Staessen et al. reported 20% live birth rate among 20 couples who underwent ICSI due to underlying Klinefelter syndrome in male partners [25]. Concerns are raised about any increased risk of chromosomal aberrations among offspring of Klinefelter syndrome patients born with assisted reproduction. A number of studies have reported an increased overall risk of both sex-chromosome and autosomal aneuploidies [26, 27]. However, most infants born have normal karyotype which is likely due to a high proportion of chromosomally normal spermatozoa in these men [28]. Nevertheless, couples should be counseled before undergoing any assisted reproductive procedure for all possible genetic risks. Furthermore, preimplantation genetic diagnostic techniques like embryo biopsy can be used as a tool for embryo selection by identifying good quality embryos prior to implantation [12].

Future Prospects

Klinefelter syndrome leads to infertility in eventually most of the men. Hence, children and adolescent boys with Klinefelter syndrome may be offered fertility preservation before they present with infertility. As of today, sperm cryopreservation is the only effective method of fertility preservation in men whereas fertility preservation options in prepubertal males are still experimental. Semen banking can be offered to those postpubertal boys with Klinefelter syndrome in whom spermatozoa can be retrieved from the ejaculate. Otherwise, microsurgical testicular biopsy can be utilized for sperm recovery. The optimal timing for testicular biopsy is the time of spermarche, i.e., production of spermatozoa so that motile sperm can be retrieved. Various techniques including scrotal ultrasound and magnetic resonance spectroscopy have been used in some centers to determine optimal timing for testicular biopsy in these adolescents [13].

The absence of spermatozoa and spermatids in testes of prepubertal boys prevents them from benefitting from the technique of sperm freezing. However, the spermatogonial stem cells are often present in prepubertal testicular tissue of patients with Klinefelter syndrome [18, 19] and can be isolated and successfully cryopreserved with almost 70% cells surviving freezing and thawing as demonstrated in animal experiments [29]. Although at present, it is impossible to generate haploid male gametes from diploid germ cells with the existing in vitro approaches, these stem cells can be either retransplanted autologously into the testes at a later time when fertility is desired or transplanted into the animals. The techniques are known as spermatogonial stem cell auto- and xenotransplantation, respectively [30–32]. There are concerns though with the later procedure that animal infectious agents like retroviruses may be introduced in human germ line when these cells are used to procure conception [33]. Some of the other challenges encountered with this technology include the ischemic damage to the transplanted testicular tissue, in vitro enrichment of stem cell spermatogonia, and noninvasive transfer of germ cell suspensions into the rete testis. Hence, in spite of latest developments suggesting bright prospects for fertility preservation, the technique needs to be developed for isolation, storage, and reinfusion of spermatogonial stem cells in humans and creating an in vitro culture system that supports full spermatogenesis.

Another investigational technique involves removal of testicular tissue from boys with Klinefelter syndrome and cryopreserving the tissue if spermatogonia are found. The cryopreserved testicular tissue can be transplanted to an ectopic site such as under the skin at a time when fertility is desired, known as ectopic autografting of testicular tissue, or into animals, called ectopic xenografting of testicular tissue. The grafted testicular tissue revascularizes in the ectopic site producing complete spermatogenesis [34, 35]. Sperm retrieval from grafted tissue can then be used to generate healthy offspring with assisted fertilization techniques. Grafting has been successfully demonstrated in animals including mice, hamsters, goats, calves, and monkeys [36–38].

Although recent advances in medicine have brightened the prospects of preserving paternity in males suffering from Klinefelter syndrome, at present, only sperm cryopreservation is considered accepted standard clinical practice. Cryopreservation of testicular tissue and spermatogonial stem cell transplantation should only be offered as a part of an approved research protocol after thorough counseling of patients and/or their family members as there are still many unresolved issues related to these technologies.

Five-Year View and Key Issues

Klinefelter syndrome is the most common sex-chromosome aberration in men but remains underdiagnosed. There is no published randomized, placebo-controlled trial on the effects of testosterone-replacement therapy in patients with Klinefelter syndrome. Such studies should be performed to evaluate the efficacy of testosterone replacement as compared to placebo and determine the appropriate doses and formulations that restore normal testosterone levels in men with Klinefelter syndrome.

Over the last decade, with advancements in assisted reproductive techniques and successful delivery of healthy children from men with Klinefelter syndrome, intense research has started to investigate optimal methods of hormonal manipulations, preservation of fertility in adolescents, and development of universal early screening programs. In some states, screening programs for Klinefelter syndrome are already in place, which might increase the number of such patients seen by endocrinologists and urologists in the near future. Development of randomized clinical trials comparing different forms of interventions in men and children with Klinefelter syndrome will hopefully provide the evidence that is essential to allow optimization of treatment in these patients.

References

1. Klinefelter HF, Reifenstein EC, Albright F. Syndrome characterized by gynecomastia, aspermatogenesis without Leydigism, increased, excretion of follicle stimulating hormone. J Clin Endocrinol Metab. 1942;2:615–27.
2. Philip J, Lundsteen C, Owen D, Hirschhom K. The frequency of chromosome aberrations in tall men with special reference to 47, XYY and 47, XXY. Am J Hum Genet. 1976;28:404–11.
3. Perwein E. Incidence of Klinefelter's syndrome. In: Bandmann HJ, Breit R, editors. Klinefeleter's syndrome. Berlin: Springer; 1984. p. 8–11.
4. Samango-Sprouse C. Mental development in polysomy X Klinefelter syndrome (47, XXY; 48, XXXY): effects of incomplete X inactivation. Semin Reprod Med. 2001;19:193–202.
5. Hook EB. Rates of chromosome abnormalities at different maternal ages. Obstet Gynecol. 1981;58:282–5.
6. Bojesen A, Juul S, Hojbjerg Gravholt C. Prenatal and postnatal prevalence of Klinefelter syndrome: a national registry study. J Clin Endocrinol Metab. 2003;88:622–6.
7. Robinson A, Lubs HA, Nielsen J, Sørensen K. Summary of clinical findings: profiles of children with 47, XXY, 47, XXX and 47, XYY karyotypes. Birth Defects Orig Artic Ser. 1979;15(1):261–6.
8. Bojesen A, Gravholt CH. Klinefelter syndrome in clinical practice. Nat Clin Pract Urol. 2007;4(4):192–204.
9. Visootsak J, Aylstock M, Graham Jr JM. Klinefelter syndrome and its variants: an update and review for the primary pediatrician. Clin Pediatr (Phila). 2001;40(12):639–51.
10. Aguirre D, Nieto K, Lazos M, Peña YR, Palma I, Kofman-Alfaro S, Queipo G. Extragonadal germ cell tumors are often associated with Klinefelter syndrome. Hum Pathol. 2006;37(4):477–80.
11. Kamischke A, Baumgardt A, Horst J, Nieschlag E. Clinical and diagnostic features of patients with suspected Klinefelter syndrome. J Androl. 2003;24(1):41–8.
12. Lanfranco F, Kamischke A, Zitzmann M, Nieschlag E. Klinefelter's syndrome. Lancet. 2004;364(9430):273–83.
13. Paduch DA, Fine RG, Bolyakov A, Kiper J. New concepts in Klinefelter syndrome. Curr Opin Urol. 2008;18(6):621–7.
14. Campbell WA, Price WH. Venous thromboembolic disease in Klinefelter's syndrome. Clin Genet. 1981;19(4):275–80.
15. Fricke GR, Mattern HJ, Schweikert HU, Schwanitz G. Klinefelter's syndrome and mitral valve prolapse. An echocardiographic study in twenty-two patients. Biomed Pharmacother. 1984;38(2):88–97.
16. Myhre SA, Ruvalcaba RH, Johnson HR, Thuline HC, Kelley VC. The effects of testosterone treatment in Klinefelter's syndrome. J Pediatr. 1970;76(2):267–76.
17. Nielsen J, Pelsen B, Sørensen K. Follow-up of 30 Klinefelter males treated with testosterone. Clin Genet. 1988;33(4):262–9.
18. Lin YM, Huang WJ, Lin JS, Kuo PL. Progressive depletion of germ cells in a man with nonmosaic Klinefelter's syndrome: optimal time for sperm recovery. Urology. 2004;63(2):380–1.
19. Wikström AM, Raivio T, Hadziselimovic F, Wikström S, Tuuri T, Dunkel L. Klinefelter syndrome in adolescence: onset of puberty is associated with accelerated germ cell depletion. J Clin Endocrinol Metab. 2004;89(5):2263–70.
20. Schiff JD, Palermo GD, Veeck LL, Goldstein M, Rosenwaks Z, Schlegel PN. Success of testicular sperm extraction [corrected] and intracytoplasmic sperm injection in men with Klinefelter syndrome. J Clin Endocrinol Metab. 2005;90(11):6263–7.
21. Friedler S, Raziel A, Strassburger D, Schachter M, Bern O, Ron-El R. Outcome of ICSI using fresh and cryopreserved-thawed testicular spermatozoa in patients with non-mosaic Klinefelter's syndrome. Hum Reprod. 2001;16(12):2616–20.
22. Madgar I, Dor J, Weissenberg R, Raviv G, Menashe Y, Levron J. Prognostic value of the clinical and laboratory evaluation in patients with nonmosaic Klinefelter syndrome who are receiving assisted reproductive therapy. Fertil Steril. 2002;77(6):1167–9.
23. Ramasamy R, Ricci JA, Palermo GD, Gosden LV, Rosenwaks Z, Schlegel PN. Successful fertility treatment for Klinefelter's syndrome. J Urol. 2009;182(3):1108–13.
24. Raman JD, Schlegel PN. Aromatase inhibitors for male infertility. J Urol. 2002;167(2 Pt 1):624–9.
25. Staessen C, Tournaye H, Van Assche E, Michiels A, Van Landuyt L, Devroey P, Liebaers I, Van Steirteghem A. PGD in 47, XXY Klinefelter's syndrome patients. Hum Reprod Update. 2003;9(4):319–30.
26. Hennebicq S, Pelletier R, Bergues U, Rousseaux S. Risk of trisomy 21 in offspring of patients with Klinefelter's syndrome. Lancet. 2001;357(9274):2104–5.
27. Morel F, Bernicot I, Herry A, Le Bris MJ, Amice V, De Braekeleer M. An increased incidence of autosomal aneuploidies in spermatozoa from a patient with Klinefelter's syndrome. Fertil Steril. 2003;79 Suppl 3:1644–6.
28. Levron J, Aviram-Goldring A, Madgar I, Raviv G, Barkai G, Dor J. Sperm chromosome analysis and outcome of IVF in patients with non-mosaic Klinefelter's syndrome. Fertil Steril. 2000;74(5):925–9.
29. Izadyar F, Matthijs-Rijsenbilt JJ, den Ouden K, Creemers LB, Woelders H, de Rooij DG. Development of a cryopreservation protocol for type A spermatogonia. J Androl. 2002;23(4):537–45.
30. Orwig KE, Schlatt S. Cryopreservation and transplantation of spermatogonia and testicular tissue for preservation of male fertility. J Natl Cancer Inst Monogr. 2005;34:51–6.
31. Nagano M, Patrizio P, Brinster RL. Long-term survival of human spermatogonial stem cells in mouse testes. Fertil Steril. 2002;78(6):1225–33.
32. Sofikitis N. Transplantation of human spermatogonia into the seminiferous tubules (STs) of animal testicles results in the completion of the human meiosis and the generation of human motile spermatozoa. Fertil Steril. 1999;72 suppl 1:S83–4.
33. Patience C, Takeuchi Y, Weiss RA. Infection of human cells by an endogenous retrovirus of pigs. Nat Med. 1997;3(3):282–6.
34. Brinster RL, Zimmermann JW. Spermatogenesis following male germ-cell transplantation. Proc Natl Acad Sci USA. 1994;91(24):11298–302.
35. Ogawa T. Spermatogonial transplantation technique in spermatogenesis research. Int J Androl. 2000;23 Suppl 2:57–9.
36. Honaramooz A, Snedaker A, Boiani M, Scholer H, Dobrinski I, Schlatt S. Sperm from neonatal mammalian testes grafted in mice. Nature. 2002;418(6899):778–81.
37. Schlatt S, Kim SS, Gosden R. Spermatogenesis and steroidogenesis in mouse, hamster and monkey testicular tissue after cryopreservation and heterotopic grafting to castrated hosts. Reproduction. 2002;124(3):339–46.
38. Oatley JM, de Avila DM, Reeves JJ, McLean DJ. Spermatogenesis and germ cell transgene expression in xenografted bovine testicular tissue. Biol Reprod. 2004;71(2):494–501.

ART and Its Impact on Male Infertility Management

20

Alice Rhoton-Vlasak and Kenneth C. Drury

Abstract

Approximately 12–15% of sexually active couples are infertile. The etiology of infertility is likely multifactorial. Previous work has estimated that 50% of infertility is attributed to the female, 30% to the male, and 20% to both the male and female. Recent advances, mainly in the assisted reproductive technologies, have allowed some couples with severe male factor infertility to establish a pregnancy. The most significant advance is in vitro fertilization (IVF) with intracytoplasmic sperm injection (ICSI). Previously, these couples had at best only a remote chance of establishing a pregnancy due to severely reduced sperm concentration, compromised sperm function, or uncorrectable obstruction. As a result of the high success of IVF/ICSI, and as a by-product of the fact that fertility visits are often initiated by the female partner, the practice of modern assisted reproductive technology (ART) can often times proceed without a complete evaluation of the male partner. Instead, almost any couple even those with a severe male factor, after being evaluated by only a reproductive endocrinologist, can theoretically become pregnant using IVF/ICSI.

Keywords

Male factor infertility • Urological evaluation • Male infertility workup • Klinefelter's syndrome • Obstructive azoospermia • Cystic fibrosis membrane regulator gene • Assisted reproductive technology • Urology • Reproductive endocrinology

Approximately 12–15% of sexually active couples are infertile. The etiology of infertility is likely multifactorial. Previous work has estimated that 50% of infertility is attributed to the female, 30% to the male, and 20% to both the male and female. Recent advances, mainly in the assisted reproductive technologies, have allowed some couples with severe male factor infertility to establish a pregnancy. The most significant advance is in vitro fertilization (IVF) with intracytoplasmic sperm injection (ICSI) [1, 2]. Previously, these couples had at best only a remote chance of establishing a pregnancy due to severely reduced sperm concentration, compromised sperm function, or uncorrectable obstruction. As a result of the high success of IVF/ICSI, and as a by-product of the fact that fertility visits are often initiated by the female partner, the practice of modern assisted reproductive technology (ART) can often times proceed without a complete evaluation of the male partner. Instead, almost any couple even those with a severe male factor, after being evaluated by only a reproductive endocrinologist, can theoretically become pregnant using IVF/ICSI [3].

A. Rhoton-Vlasak, MD (✉)
Department of Obstetrics and Gynecology, Shands Hospital, 1600 S. W. Archer Road, Room M-302, Gainesville, FL 32610, USA
e-mail: rhotona@ufl.edu

K.C. Drury, PhD
In Vitro Fertilization and Andrology Laboratory, Department of Obstetrics and Gynecology, University of Florida College of Medicine, 1600 SW Archer Road, Gainesville, FL 32610, USA
e-mail: drury k@ufl.edu

S.J. Parekattil and A. Agarwal (eds.), *Male Infertility: Contemporary Clinical Approaches, Andrology, ART & Antioxidants*, DOI 10.1007/978-1-4614-3335-4_20, © Springer Science+Business Media, LLC 2012

Male Infertility Workup and the Role of Urologists

Despite the trend toward unilateral care of the female partner in the reproductive health community, it is important to consider the couple as a whole unit. Lack of male investigation or limited urologic services can lead to potentially reversible, life-threatening, and/or genetic conditions being missed. Management of men with obstructive or nonobstructive etiologies that request or require sperm retrieval techniques also need to be coordinated [4]. A study by Kolettis and Sabanegh evaluated potential factors that could be missed by bypassing a urological evaluation in infertile male patients. The study took place at two academic infertility practices. This study showed that significant medical pathology was discovered in 33/536 (6%) of patients. A total of 27 patients had genetic abnormalities, including cystic fibrosis mutations in 24 and karyotypic abnormalities in 3. Of the remaining six patients, one had testicular cancer, one prostate cancer, three were found to have diabetes mellitus, and one with hypothyroidism [5]. Any couple proceeding directly to assisted reproductive technologies without even a basic urologic evaluation may not have their underlying medical pathology detected or discussed with them.

Despite studies such as the one just mentioned, there remains some debate in the literature about the importance of a urological evaluation. At first glance, this does not seem a problematic notion at all. If infertile couples can be afforded the opportunity to have a child in a time-efficient, cost-efficient, and safe manner, one might conclude that this event represents an ideal outcome. However, when one asks the question of whether the male patient even needs to be evaluated given the availability of IVF/ICSI, several simple, conceptual problems do present themselves. Indeed, a large volume of data exists that argues that failure to evaluate the male is less safe for male patients, their partners, and their offspring and can be less cost-efficient. There is existing medical evidence in support of the idea that an evaluation of the male partner is mandated by both patient safety and cost-effectiveness [3]. Review of this information will highlight the importance of infertility and ART centers having male infertility specialists in their practice area to optimize male factor patient's workup and treatment. The ideal infertility practice would involve a multidisciplinary clinic with both reproductive endocrinology and infertility (REI) and urology specialists. Other important clinical infertility team members would include embryologists, psychologists, and nurse specialists.

A multidisciplinary approach to male infertility patients including an urologist results in safer outcomes for couples. ART safety is a topic that has been extensively studied since the initial use of IVF in 1978 [6]. Safety issues that must be detected by a male infertility evaluation include detection of endocrine, genetic, or malignancies underlying an infertility diagnosis [5]. A thorough male infertility evaluation may also uncover genetic abnormalities which could have implications for the health of a future offspring. The identification of genetic conditions, such as Klinefelter's syndrome, microdeletions of the Y chromosome, and cystic fibrosis/congenital absence of the vas deferens, in patients with male infertility, had profound implications for offspring as well [7–9]. Complete workup of a man with obstructive azoospermia could lead to the finding of his being a carrier of a mutation for the cystic fibrosis membrane regulator (CFTR) gene. This has important implications for testing his partner to see whether she is a carrier or not. Obviously, if both were carriers of the cystic fibrosis gene, there is a higher chance they could have an offspring with this condition. Therefore, especially in the case of cystic fibrosis, it is imperative that the partner of an affected patient be screened to adequately counsel the couple on the risks of potentially passing this condition on to their anticipated offspring.

In much the same way, the identification of any genetic condition during a routine evaluation of male infertility is doubly important for the reason that couples can be counseled regarding the passage of these conditions to their offspring. Male offspring of patients with AZF region Y-chromosome microdeletions can be expected to have the same deletion and are, therefore, likely to be subfertile or infertile themselves [3]. A review of these causes of male factor infertility highlights the importance of the complete urologic evaluation of the male partner to result in a safer outcome for the male patient, his partner, and their potential offspring.

Another reason for emphasizing the importance of a urologic evaluation in male infertility is that there is proven cost-efficiency of evaluating and treating underlying male infertility, especially in cases where an anatomic or hormonal abnormality is found and correctible, thereby even allowing attempts at natural conception. Even in those patients who are unable to conceive after treatment for reversible causes of male infertility, there remains the opportunity to improve male semen parameters to the point that less costly treatment options such as intrauterine insemination (IUI) become feasible [3]. A cost analysis of effectiveness by Garso considered four studies where IUI and IVF/ICSI were compared; a significant improvement in cost-efficiency was observed in the IUI cohort across all studies, even when comparable populations were considered. The only exceptions to this conclusion were cases of severe male factor infertility or tubal obstruction. Certainly, in cases with treatable causes of male infertility and no single female factor that precludes IUI, it would seem reasonable to offer treatment to the male partner [10].

It is clear that it is cost-effective to support treatment of reversible causes of male infertility and that safety is improved where there is a male urologic evaluation. The existing evidence supports the rigorous evaluation of the male partner. Both the American Society of Reproductive Medicine (ASRM) and the American College of Obstetricians and Gynecologists (ACOG) guidelines for best practice on infertility are in agreement on this point [11, 12]. Despite the significant advances in ART technology, there remains a very important role of the urologist in the evaluation of male infertility in order to maximize the safety for both the male and female partners and their offspring. Furthermore, it is cost-effective, especially in the setting of the identification of a treatable cause of male infertility.

Multidisciplinary Clinical Care

A multidisciplinary clinic system with reproductive endocrinologists and urologists working side by side is the optimal way to offer the highest level of reproductive care to infertile couples. A multidisciplinary clinic such as this has been highly effective and well received by patients at the University of Florida. It has been an excellent collaborative clinic offering all aspects of urologic and infertility services in one location. A study by Nangia assesses the spatial distribution of assisted reproductive technology (ART) centers and male infertility specialists by location, driving distance from ART centers, and potential male population in need of these resources. It was a cross-sectional study which found that a disparity of urology male infertility specialists exist in the United States, with large areas of the country being underserved, and the overserved based on the location of the ART centers. The study identified 197 male infertility specialists and 390 ART centers (Figs. 20.1 and 20.2). On a state level, the highest male population in the reproductive years was seen in California, Texas, and Florida. The highest male populations per male specialists were found in Oregon, Tennessee, and Oklahoma [4]. The Northeast Region had the highest number of male specialists within ART service areas. The distribution of male population per male specialist was independent of a state's mandated insurance status for ART services. This study highlights the importance of proximity of male specialists to ART centers for the overall care of the couple with male factor infertility. This is related to the common understanding and interaction between urology and ART specialists regarding joint management or procedures for the couple. Lack of this joint care can potentially lead to significant biases, incomplete evaluation, and poorly informed discussion about treatment choices [13].

Offering the highest level male infertility care could involve use of an interdisciplinary clinic with a urologist and reproductive endocrinologist. Decision trees could be used to enhance interdisciplinary team work within the reproductive medicine clinic. A common scenario would involve an infertility patient coming to clinic with her partner with a known male factor infertility. If a reproductive endocrinologist and urologist hold clinic on the same day, the female history, exam, and testing could be discussed and performed while at the same time, allowing the couple to see the urologist to begin a focused male infertility evaluation. A collaborative interaction between these two specialists could allow coordination of any genetic or hormonal testing. Ultimately, collaboration between these two specialists could lead to a decision regarding whether or not the best treatment involves management of an underlying medical condition causing male infertility, a correctible underlying cause, which could be treated with medications allowing attempts at pregnancy with IUI, or whether ART and ICSI are needed. At our institution, we have such an interdisciplinary clinic, which has allowed the coordination of many complicated male infertility cases even in terms of scheduling fresh sperm aspiration procedures on the same day of oocyte retrievals and ICSI. On the other hand, having the easy availability of a urologist in our clinic has also allowed patients referred for severe oligospermia to be evaluated and then treated for underlying hormonal conditions with resolution of their poor sperm counts, therefore, allowing them the chance at a natural conception and bypassing the use of ART.

Expert Commentary

The treatment of the male, within the context of a couple, may not always allow for the months of treatment necessary for long-term hormonal interventions or, for that matter, vas reversals following years of sterility. These situations are best assessed by the urologist in collaboration with the treating infertility specialist. The array of available treatment options is also a matter of the overall abilities of the ART laboratory and its ability to bring to bear the necessary technology appropriate for the specific patient treatment requirements.

With the availability of various ART/ICSI procedures now used in clinical settings, the treatment of combined male and female infertility issues, or just those associated with the male partner only, becomes a choice of expediency. The primary road to expediency is of course due to the age of the female.

From the viewpoint of what technologies are now available for the treatment of the male's contribution to the couples' infertility issues, we should look closely at the ability to firstly assess the conditions that directly impact the patient's sperm fertilizing potential and secondly, the steps required to overcome this disability.

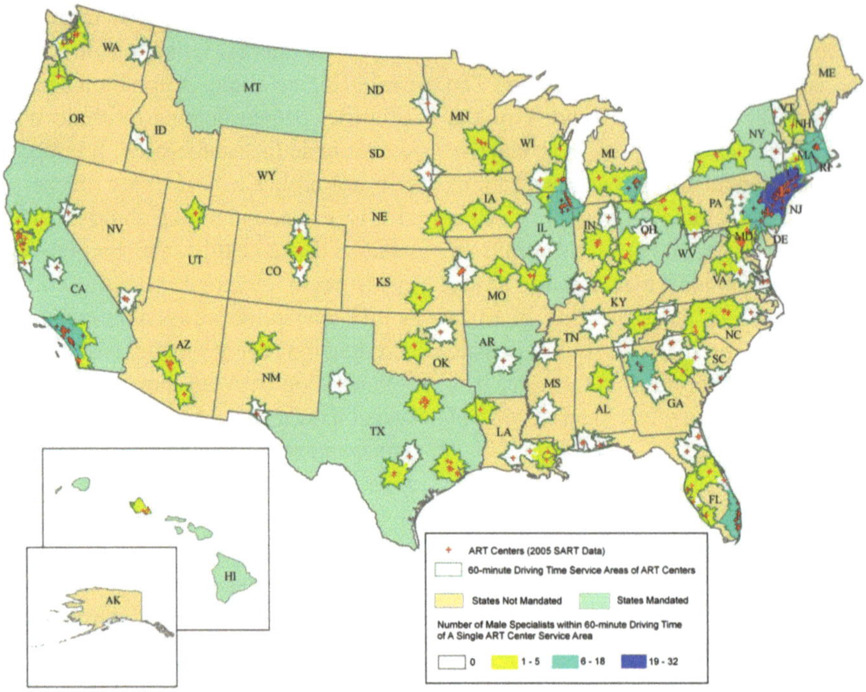

Fig. 20.1 Number of male specialists within 60-min driving distance of a single ART center (from [4], with permission from Elsevier)

The semen analysis procedure is the standard by with each male patient is initially assessed in order to categorize his level of contribution to the couples sub- or infertile condition [14]. Aspermia, azoospermia, and/or severe asthenospermia all place this patient in line for some type of assisted reproductive treatment or intervention. It may also be useful to address the use of donor sperm under these circumstances as well. However, all of the above-mentioned conditions are potentially treatable. In addition to these most serious conditions, semen volume, coloration, viscosity, and pH are often included in an initial assessment of a semen sample [15]. Further assessment of various vital sperm characteristics is also essential. These will include sperm concentration, sperm motility, sperm morphology, antisperm antibodies, and in the case of severe asthenozoospermia, vital staining tests [16]. The question however is, do these assessments in aggregate, or any one of them individually, actually allow the physician to determine whether or not these sperm can carry out normal fertilization?

There is also controversy as to the meaningfulness and significance of attached cutoff values. The newly produced WHO laboratory manual for the examination and processing of human semen, 5th edition 2010 World Health Organization, has generated debate even before publication [17–21]. However, there exists a very predictive diagnostic test that the couple could undergo in order to answer this question, and that is in vitro fertilization. Since this diagnostic test would be considered far too expensive and invasive, it tends not to be used for that purpose, although IVF procedures sometimes turn out to answer that question. Because of the potential outcome called "failed fertilization," many ART programs prefer to utilize ICSI micromanipulation as their first-round defense to treat male infertility.

If there was a validated, reasonably priced sperm diagnostic test able to determine the fertilization capacity of the male patient's sperm sample, how might it change the clinical decision making and the use of invasive ART practices like ICSI? One might also ask "what are present day ICSI usage criteria?" One very broad axiom in use today is "once ICSI always ICSI," and this covers most any treatment proposal if a program always starts a protocol using ICSI. What then are the prospects of being able to determine the ability of sperm function in order to avoid the use of micromanipulation or prepare for its expedient use if the procedure is necessary? Following a decision to employ micromanipulation, is it also necessary to choose a particular sperm within the available sample which will ensure proper embryonic development?

The following procedures or tests are offered in order to assist in the diagnosis and treatment planning of the male infertility patient. These tests may be available through collaboration with reproductive endocrinologists, urologists, and in vitro laboratory embryologists.

Oxidative Burden and the Use of Antioxidants

Since sperm function relies heavily on sperm membrane integrity and intramembrane lipid fluidity, the potential oxidative impact of oxygen radicals and peroxides, generated within the reproductive tract, can irreversibly damage the

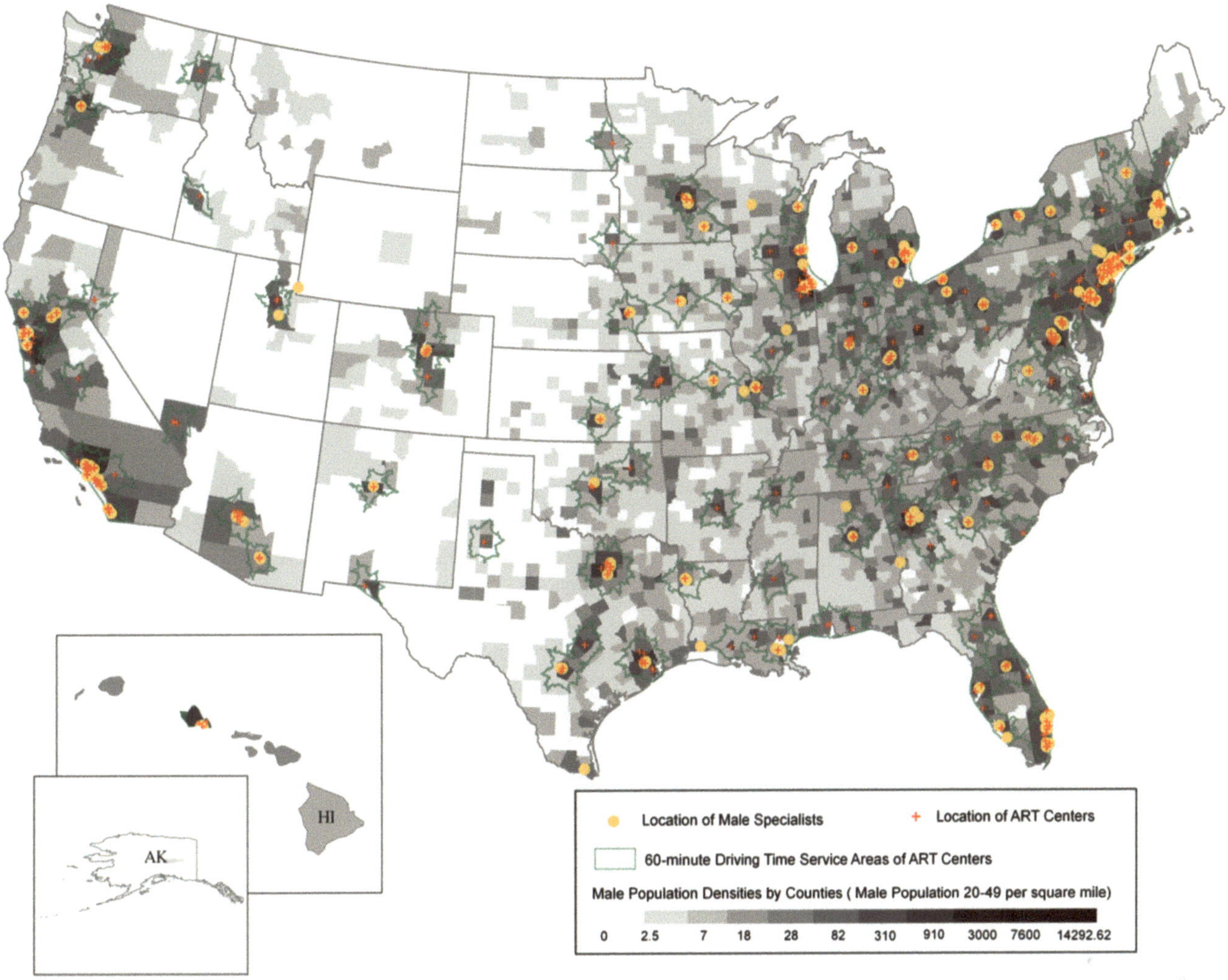

Fig. 20.2 Distribution of male population in the reproductive years, urology male infertility specialists, and ART centers with service areas within the USA (from [4], with permission from Elsevier)

sperm's ability to undergo membrane alterations and respond to ionic signaling required during spermatozoogenesis [22–24], the process of membrane fusion at the time of fertilization. It is possible to measure antioxidant levels and capacity of seminal fluid in order to determine whether sperm is adequately protected during the collection process [23, 25–27]. There have also been studies reporting favorable outcomes using oral antioxidants as enhancers of sperm function [28–30]. Low levels of vitamin C are associated with an increase in oxidatively modified DNA base, 8-hydroxydeoxyguanosine in sperm DNA [31]. However, the type and mix of antioxidants used in this prophylactic manner has not been adequately resolved [23, 32].

Use of Antioxidant Additives During Fertilization and Embryo Culture

As well as employing antioxidants during sperm production and processing, selected compounds have also been utilized to enhance culture conditions during fertilization and embryonic culture [33, 34]. In vitro culture systems lack many of the protective elements found in follicular fluids and the fallopian tube environment. The higher levels of oxygen generally found within in vitro culture systems is of concern due to its ability to overwhelm what little antioxidative capacity resides in commercial media with unlimited sources of deleterious oxygen radicals. Therefore, the first strategy that should be put into place to reduce these reactive elements is to lower the oxygen content within culture incubators to levels thought to be present in the female reproductive tract (5–7%).

The application of various compounds that may be able to destroy or convert dangerous levels of reactive radical species when added to culture systems have been explored [35]. To date, none seem to be convincibly more efficient or useful than added taurines and cysteines. Vitamins such as vitamin C and vitamin E have been used as culture antioxidants in a number of human and animal systems [33]. However, these additives have not shown consistent beneficial results in order to warrant their inclusion. Because current culture systems are static in nature, it may be that a cascade of oxidative reducing elements (vit E→lipoic acid→vit C) may be required within a dynamic culture environment such as those proposed using microfluidics where oxidized compounds can then be removed [36].

Sperm Processing for Use During IUI and ART

The major concern within the ART laboratory is the processing of sperm samples that will result in the capture of high-quality/functioning sperm which may then be used for patient treatment consisting of IUI, IVF insemination, or ICSI. Iatrogenic damage can arise through many of the processing methods used during sperm preparation.

Centrifugation

Horizontal swinging bucket centrifugation is the primary first-round processing step used to remove seminal fluids from patient or donor semen samples. Other techniques may be considered, such as raw semen sperm swim-up that does not require initial centrifugation. However, this sperm gathering method is not efficient enough for IUI procedure work-ups, and if swim-up does produce good sperm numbers and motility, then ICSI is probably not required or warranted. If raw semen is highly viscose, there will likely not be sufficient sperm for even IVF insemination. However, centrifugation has a major drawback which stems from the fact that damaged sperm tend to release harmful reactive molecules that can damage sperm membranes and disrupt fertilizing capacity [37]. Sperm gradient separation has been employed to address this problem and seems to be very efficient at separating sperm from seminal fluids and non-sperm cells while preventing the packing of sperm at the bottom of the centrifuge tube [38]. If need be, a swim-up procedure can then provide highly motile sperm for IVF insemination purposes.

Post-gradient sperm centrifugation is not benign, and close attention must be paid to centrifuge specific parameters such as radius-dependent *g*-forces. Not only does total *g*-force impact sperm membrane integrity, but the actual overall time spent under any *g*-force may negatively affect sperm function. There is a trade-off between total sperm recovery and functionality of sperm recovered. Average *g*-force (the force experience at approximately ½ the radius of the extended swinging bucket) should average around $200 \times g$, whereas the max *g*-force (that experienced at the end of the extended swinging bucket) should not exceed $600 \times g$. The corresponding speed of the rotor in terms of RPM (revolutions per minute) can then be used as an everyday guide for sperm processing. Overall time spent under centrifugal force should probably not exceed 10–15 min on any occasion [39].

In the case of testicular sperm extraction (TESE) or microsurgical epididymal sperm aspiration (MESA) (two urological sperm retrieval procedures), there may be reasons for short periods of centrifugation, but volumes of medium should be kept as small as possible so that gravity can effectively take the place of centrifugation and remove pieces of tissue from specimens, allowing samples to then be placed directly under an inverted microscope for direct observation of sperm and micromanipulation.

Cryopreservation

Semen and sperm cryopreservation has a long history in both human and domestic animal reproduction [40–43], and for the most part, glycerol has been used as the primary ingredient responsible for retention of viability post-thaw. Additional components have since been added along with glycerol in order to maintain viability and thawed motility through incorporation of processed egg yolk in an attempt to stabilize membranes through a low-cost source of phospho- and glycolipids, such as phosphatidylcholine (PtdCho), phosphatidylethanolamine (PtdEtn), phosphatidylinositol (PtdIns), and phosphatidylserine (PtdSer) [44]. Since most donor semen specimens tend to comprise many millions of sperm, a return rate of 50% of the initial live sperm content is considered very acceptable to most users.

On the other hand, if the sperm sample has been obtained through invasive urological procedures for ICSI and is to be used in expensive assisted reproductive treatments, then each sperm takes on a significant value to both the patient and the laboratory staff. Cryopreservation under these conditions requires a system that not only maintains high levels of viability but also allows for the efficient retrieval of the few thawed sperm that may have been obtained [45].

A number of semi- and/or micro-sperm freezing systems have been proposed ranging from the use of harvested empty zonae, Cryoloop, and Cryotop devices for the storage of one or two sperm [46–48] to a Cell Sleeper unit [49] housed in a standard cryo-vial, allowing storage of anywhere from a few to a few hundred collected sperm. These sperm cryo-storage systems can now provide the necessary tools in order to give infertile couples maximal use of the limited number of gametes available.

Antioxidant Supplementation

As in the case of treating a patient with supplements of vitamins and other antioxidants in order to reduce the effects of oxidative stress on systems responsible for sperm production and processing, the use of vitamins, metal chelators, proteins, and disulfide compounds in the medium used to process sperm samples for assisted reproductive treatments has shown to be useful for sperm viability and function. Sperm membrane stability is preserved by reducing or eliminating peroxidation and nuclear DNA is protected from damage by highly reactive radical species [33, 34]. These protective compounds will help to enhance in vitro embryo development.

Further Diagnostic Tests for the Selection of Sperm During ART/ICSI

Strict Morphology

The standard semen analysis incorporates sperm morphology within the category of necessary aspects linked to sperm function and as such should be taken into account along with concentration and motility parameters during the assessment of male fertility potential. This general correlation is based on the fact that structure and function are so highly linked due to the specialization of this cell. However, the criteria assigned to normal sperm morphology by WHO (50% normal) was not shown to be of high relevance for treatment composed of timed intercourse, IUI, or even IVF.

A more *strict* morphological criterion was introduced by the Tygerberg Group [50] and made popular in collaboration with the Jones Institute [51] where the impact of sperm morphology was observed during in vitro fertilization by declining fertilization rates, when the overall normal rate was less than 14%.

Due to the predictiveness of this procedure, strict morphology has become the standard for sperm morphologic determination.

DNA Fragmentation

Cellular DNA fragmentation has been associated with general cellular and tissue demise, and depending on the particular nature and profile of this fragmentation, it can be determined whether or not cellular endonucleases have been the active agents [52]. Single-strand and double-strand DNA breakpoints can be the conditions leading to total fragmentation events, and these are particularly important when considering the transmittance of an intact genome within germ cells during reproduction [52].

It has been observed that various in vitro conditions can impact sperm DNA integrity and can contribute to abnormal embryo development [53–55]. This may be a leading contributor of poor embryo development as well as spontaneous miscarriage. As such, the determination of the general integrity of sperm genomic DNA may lead to a diagnostic function during sperm testing. Sperm DNA fragmentation testing, using various technologies, has been employed for a number of years, and agreement on how those results should be factored into diagnosis or treatment options is a source of dispute [56–58]. However, the knowledge gained by being able to accurately determine the type and extent of the genomic damage being contributed by sperm during the fertilization process will play a key role in the assessment of sperm processing procedures, selection methods, and prognosis of outcome.

Hyaluronic Acid Binding

The egg–cumulus complex, obtained during aspiration of hormone-treated expanded follicles, during IVF, is without doubt the key element involved in ART. The entire process of deriving a viable embryo is based on the health of a very few mature oocytes obtained from artificially induced follicles. The initial impression of whether the egg–cumulus complex will contribute a healthy fertilizable egg comes from the organization and extent of expansion of the cumulus [59]. Expansion of the cumulus is a sign that the egg has promoted granulosa cells to manufacture extracellular matrix material in order to isolate the egg from the external environment until fertilization takes place. This extracellular matrix material is hyaluronan or hyaluronic acid [60]. If the sperm is to even get close to the egg, then it must be able to traverse this matrix. It is not surprising then that sperm were found to contain hyaluronidase which effectively dissolves the cumulus matrix and allows sperm to pass through to the egg. It is a slightly more complex arrangement for the sperm than this, but the interaction between sperm and hyaluronan has now provided a unique selection tool in order to determine which sperm are most likely to provide a good outcome during the ICSI procedure.

Sperm-hyaluronic acid binding assay is now available commercially, and experimental data indicates that selective binding of sperm increases characteristics such as minimal DNA fragmentation, normal shape, low frequency of chromosomal aneuploidies, as well as high DNA chain integrity [61, 62].

High-Magnification Sperm Selection ICSI (IMSI)

An interesting question, which is as old as ICSI itself, is how does the embryologist select any given sperm for injection? There are many answers to this question besides "the ones available." If a sperm selection method for morphology was used to produce the ICSI sample, then any sperm may be acceptable. However, many times, it is not possible to be overly selective and sustain the resulting sperm losses. As is many times the case, gross sperm morphological characteristics are used as the key selection criteria.

ICSI is not a procedure that is normally undertaken using high-magnification lenses other than a 20–40× objective. A total magnification of 200–400× is still not sufficiently high resolution to assess fine morphological features. Strict morphology assessment is normally performed at a magnification of 1,000×. If sperm could be selected at a much higher resolution than that used for ICSI, it should be possible to select the actual sperm based on a much higher standard of morphology.

With modification to the clinical inverted microscope used for ICSI, utilizing a 100× oil objective and 1.5× optical enhancer, the digital magnification enhancement can produce a high-resolution picture (6,000–10,000×) of sperm in an ICSI dish. Utilizing this enhanced visualization of sperm morphology, especially taking into account head vacuolization and midpiece abnormalities, has led to successful outcomes for severe male factor patients [63, 64]. Live birth success has been correlated to types and grades of sperm abnormalities viewed under high magnification [65–67]. Unfortunately, this procedure tends to be time-consuming and technically challenging.

Tandem Use of MICI and Sperm-Hyaluronic Acid Binding

Following the brief description of the potential roles of both hyaluronic acid sperm binding (PICI) and high-magnification selection of sperm (IMSI) for ICSI, is there any evidence for the tandem use of these selective tools to effectively increase the successful outcome of treatment for male factor infertility, especially in cases of severe male factor diagnosis based on nonobstructed azoospermia?

To date, there have been very few investigations comparing sperm selection and outcomes between conventional ICSI and procedures combining the use of MICI and PICI. The only current reference [68] found in the literature gives encouraging results in a small number of patient cycles. Of course, even the use of these highly selective technologies cannot create normal sperm if they do not exist.

Adult Autologous Stem Cell Culture Produced Sperm

The use of embryonic and adult stems in the field of regenerative medicine is creating hope that 1 day these cell types will provide the means to derive specific differentiated cells which may be used to cure diseases or renew missing or damaged body cells.

One of the most interesting findings has been that embryonic stem cells cultured under specific conditions are capable of giving rise to adult germ cells. This means that both eggs and sperm are able to be produced from these pluripotential cell lines in culture [69]. There is some question whether or not these germ cells possess all attributes of gonadally produced haploid reproductive cells [70]; however, there seems to be little doubt that other highly differentiated cells can be produced in this manner.

Particularly interesting to those engaged in reproductive medicine is the recent finding that not only embryonic stem cells can produce germ cells but that stem cells taken from individuals (in this case human fetal bone marrow stem cells) are able to give rise to male germ cell-like cell lineage [71]. In essence, this capability (if proven applicable) will be able to produce haploid reproductive cells available for the treatment of males who may not have been able to produce spermatogonia or experienced spermatogenic arrest.

Expert Commentary

It is astonishing the impact that assisted reproductive technologies have brought to the practice of reproductive medicine for both the male and female. It is beyond doubt that these highly technical laboratory procedures have provided healthy children to couples who even 20 years ago would have had no prospects of conceiving a natural offspring related to her egg and his sperm. As reviewed in this introductory chapter, future technological advances promise to extend the clinical reach of the infertility specialist and urologist to address patient specific and refractory conditions which still prevent patients from achieving successful outcomes within their first treatment cycle.

With this acknowledgment of the key role played by the advancing technologies derived from research and delivered in the ART clinical laboratory, it remains clear that initial patient assessment and diagnosis must still set the stage for treatment workup and outline the long-term algorithmic pathway. It is now appreciated more and more that there is an impact of general chronic poor nutrition on reproductive health, and attendant conditions that release systemic inflammatory by-products can often take its toll on reproductive function. High technology is not required to address these problems in initial phases of treatment. Even if a standard semen analysis cannot guarantee assessment of the total fertilizing capacity of the male's semen, if performed correctly, it is able to highlight various obvious physiological problems which need to be addressed.

Five-Year View

Looking into the future, there is now a very real prospect of utilizing adult stem cells in culture to produce authentic usable gametes. When (not if) that day arrives, every individual will have the opportunity to provide and store eggs or sperm for their future use and thereby ensure reproductive success.

Key Issues

- Uses and indications for ART and ICSI
- Importance of a complete male evaluation
- Role of the urologist in the multidisciplinary REI and urology infertility clinic
- New diagnostic laboratory testing for male infertility

Acknowledgment I would like to acknowledge Rena Fitzgerald and Cindy Puckett for their excellent editorial and transcription services.

References

1. Palermo G, Joris H, Devroey P, et al. Pregnancies after intracytoplasmic injection of single spermatozoon into an oocyte. Lancet. 1992;340:17.
2. Sherins RJ, Thorsell LP, Dorfmann A, et al. Intracytoplasmic sperm injection facilitates fertilization even in the most severe forms of male infertility: pregnancy outcome correlates with maternal age and number of eggs available. Fertil Steril. 1995;64:369.
3. Alukal JP, Lipshultz LI. Why treat the male in the era of assisted reproduction? Semin Reprod Med. 2009;27:109–14.
4. Nangia AK, Likosky DS, Wang D. Distribution of male infertility specialists in relation to the male population and assisted reproductive technology centers in the United States. Fertil Steril. 2010;94(2): 599–609.
5. Kolettis PN, Sabanegh ES. Significant medical pathology discovered during a male infertility evaluation. J Urol. 2001;166:178–80.
6. Steptoe PC, Edwards RG. Birth after the reimplantation of a human embryo. Lancet. 1978;2:366.
7. Ferguson-Smith MA, Lennox B, Mack WS, et al. Klinefelter's syndrome; frequency and testicular morphology in relation to nuclear sex. Lancet. 1957;273:167–9.
8. Tiepolo L, Zuffardi O. Localization of factors controlling spermatogenesis in the nonfluorescent portion of the human Y chromosome long arm. Hum Genet. 1976;34:119–24.
9. Anguiano A, Oates RD, Amos JA. Congenital bilateral absence of the vas deferens. A primarily genital form of cystic fibrosis. JAMA. 1992;267:1794–7.
10. Garceau L, Henderson J, Davis LJ, et al. Economic implications of assisted reproductive techniques: a systematic review. Hum Reprod. 2002;17:3090–109.
11. Practice Committee of the American Society for Reproductive Medicine. Genetic considerations related to intraductive medicine. Genetic consideration related to intracytoplasmic sperm injection (ICSI). Fertil Steril. 2006;86(5, Suppl):S103–5.
12. ACOG Committee on Practice Bulletins-Gynecology. ACOG Practice Bulletins. Clinical management guidelines for obstetrician-gynecologists number 34, February 2002. Management of infertility caused by ovulatory dysfunction. American College of Obstetricians and Gynecologists. Obstet Gynecol. 2002;99:347–58.
13. Martinez GM, Chandra A, Mosher GM, Jones J, et al. Fertility, family planning and reproductive health of US women: data from the 2002 National Survey of Family Growth. Vital Health Stat. 2005;23:1–60.
14. Keel BA, Webster BW. The semen analysis. In: Keel BA, Webster BW, editors. CRC handbook of the laboratory diagnosis and treatment of infertility. Boston: CRC Press; 1990. p. 27–69.
15. Mortimere D. Practical laboratory andrology. Oxford: Oxford University Press; 1994.
16. Keel BA. Within and between subject variation in semen parameters in infertile men and normal semen donors. Fertil Steril. 2006;85:128–34.
17. World Health Organization. WHO laboratory manual for the examination and processing of human semen. 5th ed. Geneva: WHO; 2010.
18. Handelsman DJ, Cooper TG. Foreword to semen analysis in 21st century medicine special issue in Asian Journal of Andrology. Asian J Androl. 2010;12:59–63.
19. Ford WCL. Comments on the release of the 5th edition of the WHO Laboratory manual for the Examination and Processing of Human Semen. Asian J Androl. 2010;12:59–63.
20. Boyd JC. Defining laboratory reference values and decision limits: populations, intervals, and interpretations. Asian J Androl. 2010;12: 83–90.
21. Brazil C. Practical semen analysis: from A to Z. Asian J Androl. 2010;12:14–20.
22. Sharma RK, Agarwal A. Role of reactive oxygen species in male infertility. Urology. 1996;48(6):835–50.
23. Lanzafame FM, La Vignera S, Vicari E, et al. Oxidative stress and medical antioxidant treatment in male infertility. Reprod Biomed Online. 2009;19(5):638–59.
24. James PS, Wolfe CA, Mackie A, et al. Lipid dynamics in the plasma membrane of fresh and cryopreserved human spermatozoa. Hum Reprod. 1999;14(7):1827–32.
25. Kefer JC, Agarwal A, Sabanegh E. Role of antioxidants in the treatment of male infertility. Int J Urol. 2009;16:449–57.
26. Mahfouz R, Sharma R, Thiyagarajan A, et al. Semen characteristics and sperm DNA fragmentation in infertile men with low and high levels of seminal reactive oxygen species. Fertil Steril. 2010;94:2141–6.
27. Showell MG, Brown J, Yazdani L, et al. Oral anti-oxidant use for male partners of couples undergoing fertility treatment. ESHRE annual meeting. Rome, Italy; 2010. Poster #P-006.
28. Deepinder F, Cocuzza M, Agarwal A. Should seminal oxidative stress measurement be offered routinely to men presenting for infertility evaluation? Endocr Pract. 2008;14(4):484–91.
29. Fraga CG, Motchnik PA, Wyrobek AJ, et al. Smoking and low antioxidant levels increase oxidative damage to sperm DNA. Mutat Res. 1996;351(13):199–203.
30. Fraga CG, Motchnik PA, Shigenaga MK, et al. Ascorbic acid protects against endogenous oxidative DNA damage in human sperm. Proc Natl Acad Sci USA. 1991;54(6):1302S–9.
31. Jacob RA, Kelley DS, Pianalto FS, et al. Immunocompetence and oxidant defense during ascorbate depletion of healthy men. Am J Clin Nutr. 1991;54(6):1302S–9.
32. Zini A, SanGabriel M, Baazeem A. Antioxidants and sperm DNA damage: a clinical perspective. J Assist Reprod Genet. 2009;26(8): 427–32.
33. Guérin P, Moutassim SEI, Ménézo Y. Oxidative stress and protection against reactive oxygen species in the pre-implantation embryo and its surroundings. Hum Reprod Update. 2001;7:175–89.
34. Bedaiwy M, Agarwal A, Said TM, et al. Role of total antioxidant capacity in the differential growth of human embryos in vitro. Fertil Steril. 2006;86(2):304–9.
35. Hosseini SM, Forouzanfar M, Hajian M, et al. Antioxidant supplementation of culture medium during embryo development and/or after vitrification-warming; which is the most important? J Assist Reprod Genet. 2009;26(6):355–64.
36. Smith GD, Takayama S. Gamete and embryo isolation and culture with microfluidics. Theriogenology. 2007;68(1):S190–5.
37. Aitken RJ, Clarkson JS. Significance of reactive oxygen species and antioxidants in defining the efficacy of sperm preparation techniques. J Androl. 1988;9:367–76.
38. Henkel RR, Schill WB. Sperm preparation for ART. Reprod Biol Endocrinol. 2003;1:108.

39. Sakkas D, Alavarez JG. Sperm DNA fragmentation: mechanisms of origin, impact on reproductive outcome, and analysis. Fertil Steril. 2010;93(4):1027–36.
40. Brotherton J. Cryopreservation of human semen. Arch Androl. 1990;25(2):181–95.
41. Curry MR. Cryopreservation of mammalian semen. Methods Mol Biol. 2007;368:303–11.
42. Barbas JP, Mascarenhas RD. Cryopreservation of domestic animal sperm cells. Cell Tissue Bank. 2009;10(1):49–62.
43. Curry MR. Cryopreservation of semen from domestic livestock. Rev Reprod. 2000;5(1):46–52.
44. Alvarez JG, Storey BT. Differential incorporation of fatty acids into and peroxidative loss of fatty acids from phospholipids of human spermatozoa. Mol Reprod Dev. 1995;42(3):334–46.
45. Hammitt DG, Ferrigini RG, Sattler CA, et al. Development of a new and efficient laboratory method for processing testicular sperm. J Assist Reprod Genet. 2002;19(7):335–42.
46. Cohen J, Garrisi GJ, Congedo-Ferrara TA, et al. Cryopreservation of single human spermatozoa. Hum Reprod. 1997;12:994–1001.
47. Desai NN, Blackmon H, Goldfarb J. Single sperm cryopreservation on cryoloops: an alternative to hamster zona for freezing individual spermatozoa. Reprod Biomed Online. 2004;9(1):47–53.
48. Endou Y, Fujii Y, Motoyama H. Cryopreservation of single spermatozoa using Cryo Top. Abstracts of the 26th annual meeting of ESHRE. Rome, Italy; 2010. P-004.
49. Fujii Y, Endou Y, Motoyama H. A simple method for cryopreservation of small numbers of human spermatozoa. Abstracts of the 26th annual meeting of ESHRE. Rome, Italy; 2010. P-011.
50. Kruger TF, Menkveld R, Stander FSH, et al. Sperm morphologic features as a prognostic factor in in-vitro fertilization. Fertil Steril. 1988;46(6):1118–23.
51. Kruger TF, Acosta AA, Simmons KF, et al. Predictive value of abnormal sperm morphology in in-vitro fertilization. Fertil Steril. 1988;49(1):112–7.
52. Walker NI, Harmon BV, Gobé GC, et al. Patterns of cell death. Methods Achiev Exp Pathol. 1988;13:18–54.
53. Sakkas D, Alvarez JG. Sperm DNA fragmentation: mechanisms of origin, impact on reproductive outcome, and analysis. Fertil Steril. 2010;93(4):1027–36.
54. Castro AP, Carchenilla MSC, Salvanes R, et al. Dynamics of sperm DNA fragmentation in fresh and processed samples used for ICSI cycles. Abstracts of the 26th annual meeting of ESHRE. Rome, Italy; 2010. P-014.
55. Aitken RJ, Koppers AJ. Apoptosis and DNA damage in human spermatozoa. Asian J Androl. 2011;13:36–42.
56. Aitken RJ, Delulis GN, Finnie JM, et al. Analysis of the relationships between oxidative stress, DNA damage and sperm vitality in a patient population: development of diagnostic criteria. Hum Reprod. 2010;25(10):2415–6.
57. The Practice Committee of the American Society for Reproductive Medicine. Recently reviewed Practice Committee report: the clinical utility of sperm DNA integrity testing. Fertil Steril. 2008;90(5): S178–80.
58. Barratt CLR, DeJonge CJ. Clinical relevance of sperm DNA assessment: an update. Fertil Steril. 2010;94:1958–9.
59. Gilchrist RB, Lane M, Thompson JG. Oocyte-secreted factors: regulators of cumulus cell function and oocyte quality. Hum Reprod Update. 2008;14(2):159–77.
60. Sabeur K, Cherr GN, Ashley I, et al. Hyaluronic acid enhances induction of the acrosome reaction of human sperm through interaction with the PH-20 protein. Zygote. 1998;6:103–11.
61. Huzar G. Fertility testing and ICSI sperm selection by hyaluronic acid binding: clinical and genetic aspects. Reprod BioMed Online. 2007;14(5):650–63.
62. Yagci A, Murk W, Stronk J, et al. Spermatoozoa bound to solid state hyaluronic acid show chromatin structure with high DNA chain integrity: an acridine orange fluorescence study. J Androl. 2010;31:566–72.
63. Hazout A, Dumont-Hassan M, Junca AM, et al. High-magnification ICSI overcomes paternal effect resistant to conventional ICSI. Reprod Biomed Online. 2006;12(1):19–25.
64. Vanderzwalmen P, Hiemer A, Rubner P, et al. Blastocyst development after sperm selection at high magnification is associated with size and number of nuclear vacuoles. Reprod Biomed Online. 2008;17(5):617–27.
65. Bartoov B, Berkovitz A, Eltes F, et al. Pregnancy rates are higher with introcytoplasic morphologically selected sperm injection than with conventional intracytoplasmic injection. Fertil Steril. 2003;80:1413–9.
66. Setti AS, Ferreira RC, Braga DP. Intracytoplasmic sperm injection outcome versus intracytoplasmic morphologically selected sperm injection outcome: a meta-analysis. Reprod Biomed Online. 2010;21(4):450–5.
67. Batista J, Oliveira A, Massaro FC, et al. Correlation between semen analysis by motile sperm organelle morphology examination and sperm DNA damage. Presented at the 65 Annual Meeting of the American Society for Reproductive Medicine. Fertil Steril. 2010;94(5):1937–40.
68. Parmegiani L, Cognigni GE, Bernardi S, et al. "Physiologic ICSI": hyaluronic acid (HA) favors selection of spermatozoa without DNA fragmentation and with normal nucleus, resulting in improvement of embryo quality. Fertil Steril. 2010;93(2):598–604.
69. Nagy ZP, Kerkis I, Chang C. Development of artificial gametes. Reprod Biomed Online. 2008;16(4):539–44.
70. Nolte J, Michelmann HW, Wolf M, et al. PSCDGs of mouse multipotent adult germline stem cells can enter and progress through meiosis to form haploid male germ cells in vitro. Differentiation. 2010;80:184–94.
71. Hua J, Pan S, Yang C, et al. Derivation of male germ cell-like lineage from human fetal bone marrow stem cells. Reprod Biomed Online. 2009;19(1):99–105.

Further Reading

MacLeod J. Human male infertility. Obstet Gynecol Surv. 1971;26: 325–51.

Tielemans E, Burdorf A, teVelde E, et al. Sources of bias in studies among infertility clients. Am J Epidemiol. 2002;156:86–92.

Part II

Spermatozoa Metabolism

Fuel/Energy Sources of Spermatozoa

21

M.M. Misro and T. Ramya

Abstract

While preparing for successful fertilization, spermatozoa spend a considerable amount of time maturing in the male and female reproductive tracts, which provide them the competency for fertilization. One of the most important changes attained by spermatozoa during epididymal transit is the development of forward progressive motility, which is primarily dependent on energy. The need for energy by spermatozoa comes to the center stage only when spermatozoa are either ejaculated or suspended in an artificial medium, which ultimately provide them the opportunity and the environment to move and become motile. Any transporting system can operate only when provided with an appropriate support of fuel machinery. Spermatozoa utilize energy on motility, which is primarily in the form of intracellular ATP generated by oxidation of substrates, fructose, glucose, sorbitol, lactate, or pyruvate. Spermatozoa motility is generated by the beating of the extremely long flagellum which is more than 90% of the total length of a mammalian sperm cell. Fuel is inevitable for this efficient flagellar movement leading to motility, one of the critical functional capabilities of all spermatozoa. Spermatozoa, either devoid/altered motion characteristics or depleted fuel resources or both, lose the ability to move forward and cannot fertilize the egg. It is thus important to know the energy status of spermatozoa in order to understand their mobility, survival, and the changes they undergo during their entire life cycle. Though spermatozoa of all species require energy for fuelling their movement, different species have adapted different mechanisms for obtaining this energy. The purpose of this chapter is to review our current understanding on substrates for energy production in spermatozoa, energy storage sites, and utilization with respect to the spermatozoon-specialized structure, the uptake of specific substrates, and their metabolic breakdown and the manipulation of energy resources in vitro sustaining spermatozoa motility.

Keywords

Glycolysis • Oxidative phosphorylation • ATP consumption by spermatozoa • Spermatozoa energy production • Spermatozoa metabolism • Hyperactivated sperm motility • Sperm–zona binding • Acrosome reaction • Capacitation • Glucose transporters

M.M. Misro, PhD (✉) • T. Ramya, MSc
Department of Reproductive Biomedicine, National Institute of Health and Family Welfare, Baba Gangnath Marg, Munirka, New Delhi, 0110067, India
e-mail: mm_misro@yahoo.com

S.J. Parekattil and A. Agarwal (eds.), *Male Infertility: Contemporary Clinical Approaches, Andrology, ART & Antioxidants*,
DOI 10.1007/978-1-4614-3335-4_21,

Spermatozoa are highly differentiated cells that primarily act as delivery vehicles for safe transport of paternal genetic material to the maternal counterpart. Unlike all other mammalian cells, these are specialized cells with structural features reflecting unique functional activities. In addition, spermatozoa are highly motile and consume considerable amount of energy. However, from the time of their production in the testis and till the time they reach the epididymis, spermatozoa remain mostly immobile and not dependent on any energy sources, either internal or external. While preparing for successful fertilization, spermatozoa spend a considerable amount of time maturing in the male and female reproductive tracts, which provide them the competency for fertilization. One of the most important changes attained by spermatozoa during epididymal transit is the development of forward progressive motility, which is primarily dependent on energy. However, in spite of the acquisition of this ability to move forward, epididymal spermatozoa have little scope for movement because of the highly viscous nature of the epididymal fluid in which they remain suspended until ejaculation. The need for energy by spermatozoa comes to the center stage only when spermatozoa are either ejaculated or suspended in an artificial medium, which ultimately provide them the opportunity and the environment to move and become motile.

Any transporting system can operate only when provided with an appropriate support of fuel machinery. Spermatozoa utilize energy on motility, which is primarily in the form of intracellular ATP generated by oxidation of substrates, fructose, glucose, sorbitol, lactate, or pyruvate. Spermatozoa motility is generated by the beating of the extremely long flagellum, which is more than 90% of the total length of a mammalian sperm cell. Fuel is a necessity for this efficient flagellar movement leading to motility, one of the critical functional capabilities of all spermatozoa. Spermatozoa, either devoid/altered motion characteristics or depleted fuel resources or both, lose the ability to move forward and cannot fertilize the egg. It is thus important to know the energy status of spermatozoa in order to understand their mobility, survival, and the changes they undergo during their entire life cycle. Though spermatozoa of all species require energy for fuelling their movement, different species have adapted different mechanisms for obtaining this energy. The purpose of this chapter is to review our current understanding on substrates for energy production in spermatozoa, energy storage sites, and utilization with respect to the spermatozoon-specialized structure, the uptake of specific substrates and their metabolic breakdown, and the manipulation of energy resources in vitro sustaining spermatozoa motility.

Why Do Spermatozoa Need Energy?

Energy is necessary for many purposes within spermatozoa, such as the maintenance of the intracellular milieu by the active extrusion of ions and modifications to the structure and function of the cell during maturation, capacitation, acrosome reaction, and fertilization, but the process of motility accounts for most of the ATP that is consumed by mature spermatozoa [1–3]. This energy can be derived either through glycolysis or oxidative phosphorylation occurring within the sperm cell. It is now established that glycolysis is essential to support sperm motility, hyperactivity, and protein tyrosine phosphorylation, while energy from oxidative phosphorylation may not be all that necessary for hyperactivated sperm motility, sperm–zona binding, and acrosome reaction [4]. However, as oxidative phosphorylation generates more energy than glycolysis, the preference to the later by mammalian spermatozoa has been very much debated and not completely understood [5].

Spermatozoa Structure Versus Sites of Energy Production

Spermatozoa energy production and metabolism are very much linked to its typical structure. With the advent of electron microscopy, the flagellar structure of the sperm axoneme, the position of the nucleus in the head, the acrosome, and the variations in its structure from different species are fully revealed. The structure of the sperm flagellum, the component that consumes most of the energy, is highly conserved and is composed of a number of cytoskeletal elements whose proper assembly is critical for spermatozoa motility. It consists of four major parts: the connecting piece, the mid-piece, the principal piece, and the end piece [6]. Extending posteriorly from the remnant of the centriole, the axoneme is extended throughout the length of the flagellum. The axoneme is a cytoskeletal structure composed of nine microtubule doublets surrounding a central pair of microtubule doublet. Inner and outer dynein arms project from each of the outer nine doublets, and these arms are responsible for generating the beating force of the flagellum. The mid-piece starts from the connecting piece and is characterized by the presence of nine outer dense fibers (ODFs) that lie outside each of the nine outer axonemal microtubule doublets (Fig. 21.1).

The sheath of mitochondria is localized in the helical structure wrapped around the axoneme that encloses ODFs in the mid-piece. The mid-piece terminates about one-fourth of the way down the sperm flagellum at the annulus, the beginning of the principal piece, which is devoid of mitochondria (Fig. 21.1). The sheath containing the mitochondria is the source of adenosine triphosphate (ATP) for the spermatozoa through aerobic respiration. Although their function is similar to somatic mitochondria but unlike the former, sperm mitochondria are associated with several proteins or protein isoforms like lactate dehyrogenase and hexokinase [7, 8].

As already indicated, two metabolic processes, glycolysis and oxidative phosphorylation, are the main sources of

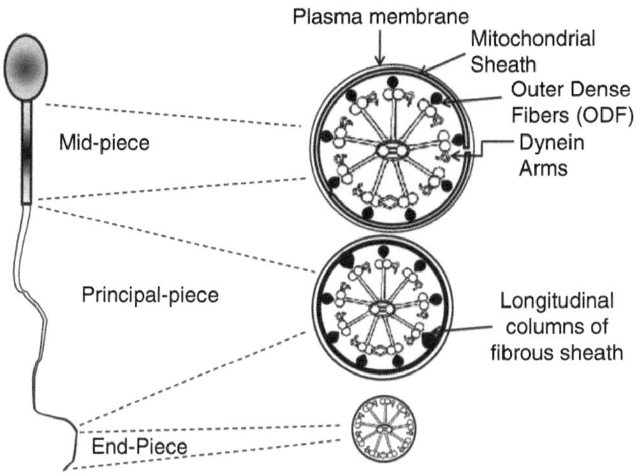

Fig. 21.1 Schematic representation of a mammalian sperm and the ultrastructure of the flagellum

energy production in spermatozoa. Because mitochondria are found only in the mid-piece, oxidative phosphorylation is restricted to this region of the cell. Spermatozoa are very minute, measuring about 25 μm, but their flagellum is extremely long. Sperm axonemal dyneins in the principal piece of the flagellum have a high requirement of ATP as an energy source for flagellar motility which is consequent to the coordinated movement of the central axoneme and surrounding flagellar structure [9]. ATP is hydrolyzed by dynein ATPases, which function as force generating molecular motors along the axoneme. As already mentioned, sperm mitochondria are restricted to the mid-piece of the flagellum, the ATP generated by these mitochondria would need to travel along the axoneme in the more distal segments of the flagellum in order to meet the needs of axonemal dyneins. It has long been suspected that this distance is too far, and the ATP originating in the mid-piece may not be able to diffuse adequately to meet these needs of a distal segment in a timely fashion [10]. On the basis of these observations, it is suggested that either other regions of the flagellum must be able to produce ATP to supply to the more distal parts of the tail, or there must be some alternative mechanism or source to meet these demands. The glycolytic pathway forms the alternative source of energy production in spermatozoa, and it is well recognized as a fuel source independent to the mitochondrial oxidative phosphorylation.

Several glycolytic enzymes have been identified in the fibrous sheath/principal piece of the spermatozoa in a large number of mammalian species. These include hexokinase, phosphoglucokinase isomerase, phosphofructokinase, lactate dehydrogenase, and glyceraldehydes 3-phosphate dehydrogenase [11–14]. All these glycolytic enzymes remain attached to the cytoskeleton even after membrane removal, which suggests that they are components of fibrous sheath or the outer dense fibers [10]. This strongly supports the contention that the problem of ATP diffusion in the entire length of the flagellum is taken care of by mammalian spermatozoa through compartmentalization of ATP production all along the fibrous sheath and through glycolysis. Since ATP production through glycolysis is also required for hyperactivated spermatozoa motility [15, 16] and the inhibition of oxidative phosphorylation does not completely block fertilization [17] indicates the significance of this pathway for sperm function [9]. There appears to be some species differences in spermatozoa metabolism (addressed in the later part of this section), but previous studies have reported that glycolysis is very much required for mouse [16–18], rat [19], and human [15, 20] spermatozoa to achieve hyperactive motility and penetrate the zona pellucida during fertilization.

Pathways of Energy Production

Oxidative Phosphorylation

Mitochondria occupy a substantial portion of the cytoplasmic volume of eukaryotic cells, and they have been essential for the evolution of complex animals. Without mitochondria, animal cells would be largely dependent on anaerobic glycolysis for all of their energy requirement. When glucose is converted to pyruvate by glycolysis, only a very small fraction of the total free energy potentially available from glucose is released. However, when pyruvate is imported into mitochondria and oxidized by molecular oxygen to CO_2 and H_2O and the metabolism of sugar is completed, the energy released is 30 molecules of ATP per molecule of glucose oxidized. This is in contrast to two molecules of ATP produced by glycolysis alone [5]. Thus, the metabolism of sugars in the mitochondria can produce 15 times more ATP than glycolysis [21]. Strong experimental evidence for oxidative phosphorylation emerged from the finding that spermatozoa remained motile for long periods despite inhibition of glycerol-3-phosphate dehydrogenase (GAPDs) by α-chlorohydrin or 6-chloro-6-deoxyglucose [22]. However, as discussed earlier, spermatozoa beating flagellum in the principal piece requires more energy to be delivered at that place which cannot be obtained from the mitochondria segregated at a spatial distance in the proximal mid-piece [23]. Thus, in spite of the fact that oxidative phosphorylation is a more efficient process, its usefulness as an alternate source for energy requirement in all species of spermatozoa continues to be still debated.

As shown in Fig. 21.2, the pathway to oxygen is common at coenzyme Q for the oxidation of either succinate or the pyridine nucleotide-linked substrates. As a mobile component of respiratory chain, coenzyme Q or ubiquinone collects the reducing equivalent from various sources and passes them on to the cytochromes leading to generation of ATP.

Glycolysis

As mitochondria are the clearly established site for respiration and oxidative phosphorylation in the cell, the mid-piece of the sperm containing these organelles is assigned this specific role. The head and principal piece lack any respiratory enzymes but possess glycolytic enzymes and thus are assigned the site of glycolytic activity. Hence, glycolysis forms an important pathway to generate ATP in spermatozoa.

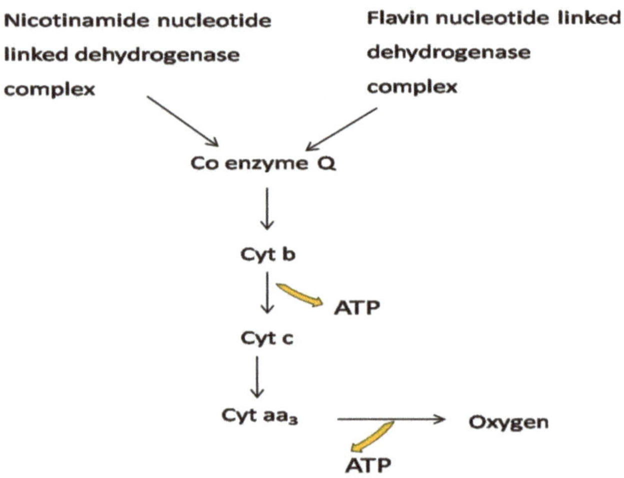

Fig. 21.2 The nicotinamide- and flavin-linked dehydrogenase complexes collect electrons from catabolic pathways and funnel them to coenzyme Q. This Co Q forms a common electron carrier which further passes electrons to oxygen

This is supported by the fact that multiple glycolytic enzymes are found tightly bound to the fibrous sheath of spermatozoa [24]. Further evidence of glycolysis supporting spermatozoa motility came from the knock-out mice which demonstrated spermatozoa immobility and lacked the expression of the gene for the sperm-specific isoform of the glycolytic enzyme, glyceraldehyde-3-phosphate dehydrogenase (GAPDH) [23].

The pathway of ATP production through glycolysis is shown in Fig. 21.3. The most important control step of hexoses metabolism through glycolysis following its uptake is sugar isomerization and phosphorylation which generates one of the most important intermediate metabolite, the glucose-6-phosphate. This phosphorylation step of the monosaccharide is modulated by hexokinase family of proteins [25] having high affinity for glucose. This would induce maximum rhythm of glucose utilization even when glucose concentration is as low as 1–5 mM as seen with spermatozoa of dog and boar [26–28]. Thus, mammalian spermatozoa have developed a very rapid and intense metabolic response to utilize glucose even at very low concentrations, thus showing one of the most efficient metabolic responses among mammalian cells [29]. The hexokinase activity has been reported to be of much lower affinity to other substrates like fructose, sorbitol, and mannose as seen in boar sperm [28]; however, this response may be species specific. Such differential sensitivity is further supported by the presence of specific substrate transporters (described later under specific sections) in spermatozoa, together making a very efficient

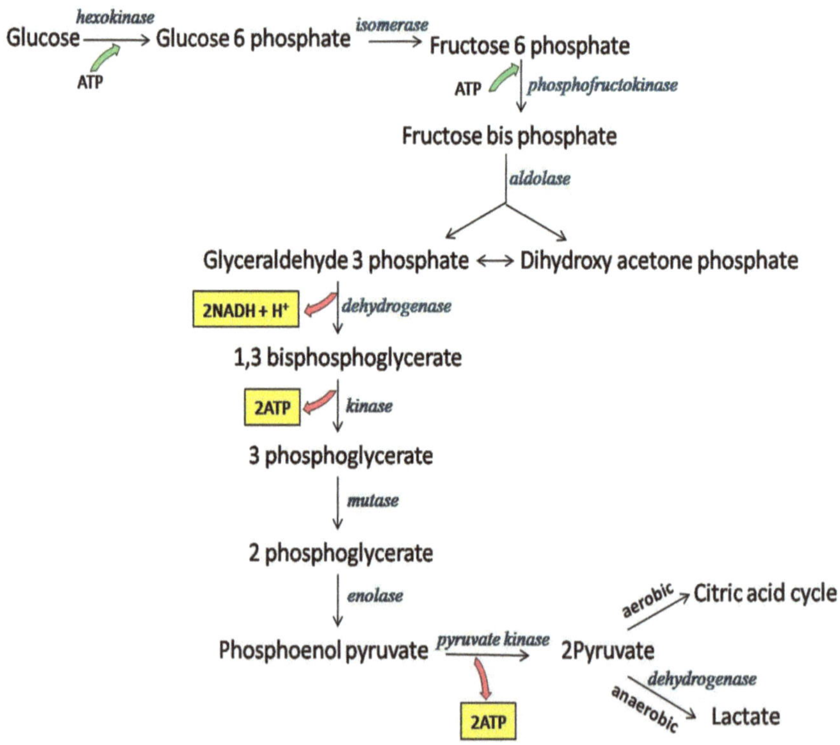

Fig. 21.3 Glycolytic pathway. Glucose molecule is lysed in a series of enzyme-catalyzed reactions to yield pyruvate molecules. During the sequential reactions of glycolysis, some of the free energy released is conserved as ATP and NADH. The fate of pyruvate varies depending upon conditions. Under aerobic condition, it enters citric acid cycle, whereas under anaerobic condition, it forms lactate

mechanism of hexose utilization. Having energy production and translocation mechanisms confined to the sperm tail suggests that spermatozoa mostly rely upon glycolysis for generation of ATP to be used for motility [30]. This is further substantiated by the fact that percent of motile human spermatozoa as well as their smothered path velocity is statistically greater when incubated in media containing glucose [20].

Substrates for Spermatozoa Energy Production in Different Species

It has been shown that species-specific differences in the metabolic capacities of mitochondria and activity of glycolytic enzymes across the fibrous sheath do exist. This, in fact, induces variations in the ability of sperms from different species to metabolize different substrates [31]. It is also hypothesized that this variation might have evolved as a result of species-specific differences in the substrate composition of oviductal fluids in the female reproductive tract through which spermatozoa traverse and thrive before fertilization.

As indicated above, two metabolic processes, namely, glycolysis and oxidative phosphorylation, stand out as prominent features of ejaculated spermatozoa; both are specifically linked with their activity [32]. However, it is apparent that there is a degree to which these metabolic processes are used by spermatozoa of different species. In human seminal plasma, unlike other body fluids, fructose, instead of glucose, is generally the principal reducing sugar and thus serves as the main glycolyzable substrate for spermatozoa in semen [33]. All tissues in the male reproductive tract, except the testis, have the ability to produce sorbitol and fructose [34]. However, depending on species, seminal plasma fructose concentrations vary from high levels in bull and ram to lower amounts in other species, such as dog and stallion where it is almost absent [35].

Following the uptake of the monosaccharide and its phosphorylation either to glucose or fructose-6-phosphate, it follows the glycolytic pathway culminating in the production of pyruvate. Pyruvate can be sent either to the formation of extracellular lactate or to its subsequent oxidation through the mitochondrial Krebs cycle. This step is controlled by the enzyme lactate dehydrogenase.

The equilibrium between sugar metabolism through simple glycolysis or through glycolysis plus the Krebs cycle depends on a great number of factors, such as intracellular levels of ATP, required oxygen availability, pH, several other signaling factors [36]. These factors enable to maintain a very fine regulation and production of intracellular levels of ATP which vary in the spermatozoa of different animal species, depending also on the substrates available for the purpose.

Sea Urchin Spermatozoa

It is known that sea urchin spermatozoa in undiluted semen are immotile. They start their flagellar movement immediately after being spawned into seawater when their respiration is activated. Sea urchin spermatozoa are not capable of using an exogenous substrate for energy metabolism, since hardly any nutrients are present in seawater. The energy for swimming is produced by mitochondrial respiration, and ATP is utilized almost exclusively by the dynein ATPase of the flagellar axoneme [37].

It was earlier assumed that the breakdown of intracellular carbohydrate provided the energy for spermatozoa movement. This assumption was mainly based on the measurements of respiratory quotient by Barron and Goldinger [38], Hayashi [39], and Spikes [40]. However, the effectiveness of glycolytic enzymes present in sea urchin is limited due to lack of utilizable substrate in the external medium and negligible quantity of carbohydrate present inside the spermatozoa. Sea urchin spermatozoa contained 4.14 mg of phospholipid per 10^{10} spermatozoa, while seminal plasma contained very small quantity of phospholipid (20 mg/ml). Later, it was proposed [41] that the oxidative breakdown of phospholipid located in the mid-piece was the principal source of the energy required for spermatozoa motility (Fig. 21.4). The transportation of high-energy phosphate from the mid-piece to the tail is associated with a creatine phosphate shuttle [42, 43]. Previous studies have shown that the endogenous phospholipid content decreases following initiation of flagellar movement in spermatozoa of several sea urchin species, such as *Echinus esculentus* [41] and *Hemicentrotus pulcherrimus* [44, 45]. *H. pulcherrimus* spermatozoa contain various phospholipids and cholesterol, and their endogenous triglyceride and glycogen contents are quite low [46, 47]. Recent studies have shown that, among the phospholipids present in *H. pulcherrimus* spermatozoa, the only one to decrease significantly during incubation is phosphatidylcholine (PC) [46].

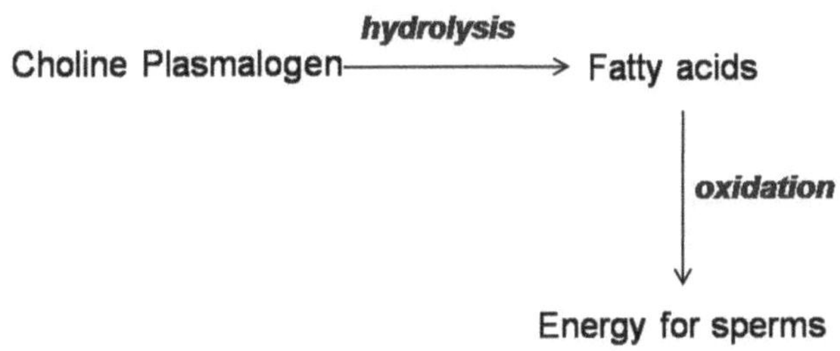

Fig. 21.4 Plasmalogen is a phosphatidyl compound. They have moieties of fatty acid, choline, serine, etc. Upon hydrolysis, they release free fatty acids. This fatty acid residue is further oxidized to yield energy for spermatozoa

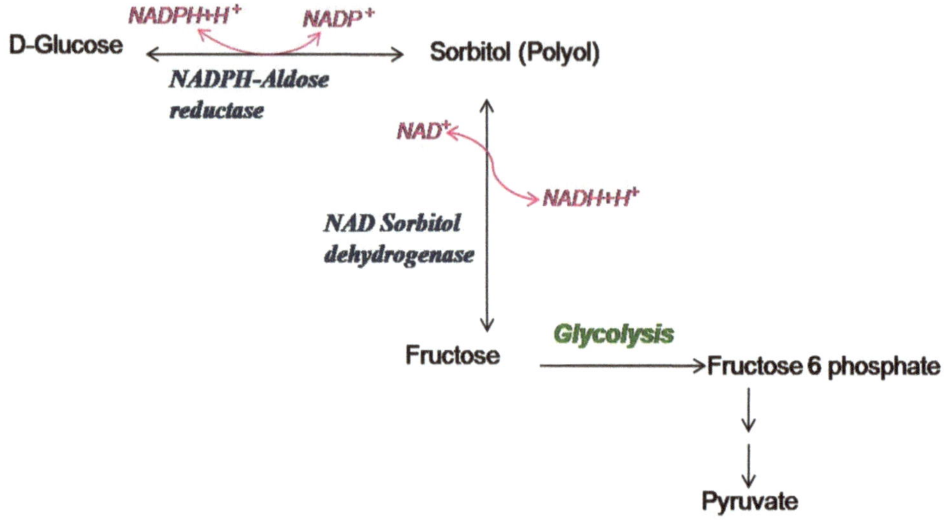

Fig. 21.5 The NAD-sorbitol dehydrogenase enzyme reversibly converts the sorbitol into fructose which is further catabolized to pyruvate. Sorbitol can also be reversibly converted to glucose by NADPH-aldose reductase

The PC available for use in energy metabolism is located in the mid-pieces of *H. pulcherrimus* spermatozoa [2]. In ultrastructural observation, the sperm mid-pieces contain the lipid bodies in a region between the mitochondrial outer- and inner-membranes—intramembrane space [48]. Following incubation of the spermatozoa in seawater, the lipid bodies become small, coincident with a decrease in the level of PC [48]. Similar findings have been obtained in other species of sea urchins of the order Echinoida [49].Thus, PC stored in the lipid bodies is considered primarily to be the endogenous substrate for energy metabolism in these sea urchin spermatozoa.

Bovine Spermatozoa

Fructose is the normal glycolyzable carbohydrate present in bovine semen, and it is broken down in part to lactic acid, which forms the main source of energy for bovine spermatozoa [33]. In addition, sorbitol has also been demonstrated as an alternate energy source for these spermatozoa, but it needs to be converted to fructose by sorbitol dehydrogenase (SORD) present in the spermatozoa plasma membrane. Sorbitol, unlike glucose and fructose, is a linear molecule, which makes it difficult to penetrate eukaryotic cell plasma membranes [50]. With SORD near the plasma membrane, it converts sorbitol into fructose [51]. Inside sperm, fructose can be converted to fructose 1-phosphate by hexokinase to enter the glycolytic pathway. The ability of sorbitol to support sperm motility can be blocked by α-chlohydrin, a blocker of glycolytic pathway, which indicates that sorbitol produces energy through the glycolytic pathway in bovine spermatozoa [52].

Ram Spermatozoa

Seminal plasma in rams is very rich in sorbitol. Like bovine spermatozoa, it is metabolized under aerobic conditions by NAD-dependent sorbitol dehydrogenase, an enzyme present in the ram's seminal vesicles and spermatozoa [35]. Due to the presence of this enzyme, the levels of fructose and sorbitol in fresh ram semen are optimally maintained. O'Shea et al. [53] studied the metabolism of sorbitol and fructose by ram spermatozoa under aerobic and anaerobic conditions using radiolabeled substrates. It was revealed that though fructose was used preferentially, the levels of declining fructose correlated well with the accumulated polyol (sorbitol) in the semen. When sorbitol alone was provided, it was utilized only under aerobic conditions (Fig. 21.5). The rate of utilization of polyol under such conditions was found dependent upon its concentration and availability to the cell through passive diffusion.

Thus, under anaerobic conditions, ram spermatozoa utilize fructose and maintain high motility, which is accompanied by the production of lactate [32]. Phospholipids present in ram spermatozoa consist mainly of choline-based plasmalogen. Ram spermatozoa, in the absence of seminal fructose, may also utilize these fatty acids as the source of energy which derive from breakdown of plasmalogen, as shown in Fig. 21.4 [54].

Boar Spermatozoa

Boar spermatozoa are unable to sustain progressive motility under anaerobic conditions, in the presence of glycolyzable substrates and exhibit only vibratory or oscillatory motions of the flagellum. In the absence of exogenous substrates, the spermatozoa remain completely immotile. In such conditions, boar spermatozoa metabolized fructose and glucose to lactate but did not produce ATP to the extent of that produced under aerobic conditions. It is also reported that the ketogenic amino acids like leucine, tryptophan, phenylalanine, and tyrosine were not oxidatively metabolized in boar spermatozoa [55]. Thus, boar spermatozoa are extremely selective in the substrates that they oxidize; these comprise glucose, fructose, glycerol, glycerol-3-phosphate, lactate and, to a limited extent, acetate, suggesting that the glycolytic pathway is the major route for the production of energy [56, 57].

The effects of chemical inhibitors on the metabolic activity of washed boar spermatozoa were examined by Jones and Porter [58]. They observed that these cells could maintain high concentrations of ATP for prolonged periods in vitro when incubated under aerobic conditions in the absence of exogenous substrates. Further investigations of the above phenomena revealed that there may be two endogenous substrates in boar spermatozoa that are degraded ultimately to produce the triose phosphates which allow the cells to produce lactate or pyruvate for mitochondrial synthesis of ATP [59]. One substrate generates minor amounts of glycerol-3-phosphate, whereas the other substrate degrades to glycerol and may be diglycerides or triglycerides or both. Studies conducted by Jones and Gillan [55] have shown that glycerol-3-phosphate was metabolized rapidly in the presence or absence of the glycolytic inhibitor, 3-chloro-1-hydroxypropanone (CHOP). In the absence of CHOP, glycerol-3-phosphate was converted to CO_2, lactate, glucose-6-phosphate, and fructose-6-phosphate, and ATP was produced. In the presence of CHOP, glycerol 3 phosphate did not produce CO_2, lactate, or ATP, but formed fructose-1,6-bisphosphate and dihydroxyacetone phosphate. With dihydroxyacetone phosphate as substrate, fructose-1,6-bisphosphate, lactate, glucose-6-phosphate, fructose-6-phosphate, and ATP were produced. Accumulation of glucose-6-phosphate and fructose appeared to depend on the production of ATP; if ATP is not produced, dihydroxyacetone phosphate and fructose-1,6-bisphosphate get accumulated. The conversion of glycerol-3-phosphate to glycolytic intermediates appears to be a mechanism for the conversion of substrates for the ultimate production of lactate.

Under aerobic conditions, high-energy charge potential (ECP) can also be maintained by boar spermatozoa with glucose and fructose [56] and with glycerol and glycerol-3-phosphate as substrates. Under anaerobic conditions, however, the ECP is not maintained; lactate accumulates from fructose and glucose, but not from glycerol and glycerol-3-phosphate, owing to the inability of the cells to regenerate oxidized cofactors. This observation together with the fact that lactate is oxidized at a far greater rate than that of pyruvate [60] suggests that boar spermatozoa are reliant on the glycolytic pathway primarily to produce lactate, rather than ATP, and that it is lactate that is the main metabolic fuel for the mitochondrial synthesis of ATP.

Dog Spermatozoa

The canine seminal plasma does not contain significant amounts of glucose, fructose, sorbitol, or mannose [61]. However, different and substantial effects of glucose and fructose on hexose metabolism in dog spermatozoa have been reported [26]. When dog spermatozoa are incubated with 10 mmol/l of glucose or fructose, increased intracellular content of glucose-6-phosphate and fructose-6-phosphate is observed although the effect of fructose is seen greater. These effects when correlated with increase in ATP, ribose phosphate, and glycogen contents and in the rates of formation of lactate and CO_2, in all cases except for ATP and glycogen, the effect of fructose is noted greater than glucose. It is thus concluded that the total hexokinase activity in dog spermatozoa is more sensitive to fructose than glucose. The differences observed in glucose and fructose metabolism of fresh dog spermatozoa are reported to be due to differences in the phosphorylation rates of these sugars. There are also reports about the difference in motility patterns of dog spermatozoa incubated with either fructose or glucose. Fructose, at concentrations from 1 to 10 mM, induced a more linear and less oscillatory motility pattern than glucose [61]. As most of spermatozoa energy consumption is devoted to maintenance of motility [62], the effect of fructose on motility is probably related to an increase in the rate of ATP consumption in these cells.

Dog spermatozoa, in spite of their low ATP content, maintain a significant degree of motility regardless of the suspension medium [61]. In dogs, glycogen was depleted in spermatozoa when incubated in substrate-free medium but accumulated in media containing millimolar concentrations of glucose or fructose with the site of deposition in the cell varying according to the hexose provided [63, 64].This indicates that these cells have a very efficient system to maintain their function. The pyruvate carboxylase inhibitor phenylacetic acid prevented glycogen synthesis and inhibited motility without a significant effect on viability. Immunocytochemical evidence indicated presence of the key gluconeogenic enzymes fructose-1,6 bisphosphatase and aldolase B [65]. Therefore, there is strong evidence that dog spermatozoa are capable of gluconeogenesis, and this may be important to maintain motility and to allow them to capacitate in glucose-free media. Such elaborate mechanisms of modulation metabolism, such as functional glycogen metabolism, [63] are perhaps responsible to optimize the capacity of dog spermatozoa to perform their functions in an unfavorable environment by maintaining constant intracellular ATP content over a long period of time [26].

There are also several transporters involved in transporting various sugars into dog spermatozoa. The main hexose transporters of dog spermatozoa are Glut 3 and Glut 5; while the former is a very high-affinity glucose transporter, the later has strong affinity for fructose [66]. The presence of two hexose transporters clearly indicates the fact that there are separate uptake pathways for glucose and fructose in dog spermatozoa. Location of Glut 5 in the peri-acrosomal head region indicates the association between energy consumption in the sperm head and the cascade of phosphorylation and dephosphorylation occurring during acrosome reaction which results net consumption of ATP.

Hamster Spermatozoa

Hamster spermatozoa are extremely sensitive to dilution in culture media which rapidly depress or even abolish their motility [67]. Glucose by itself is not an adequate energy source for the performance of Hamster spermatozoa in vitro. However, lactate and pyruvate in combination with glucose sustained motility for a longer time leading to acrosome reaction (AR) among large percentage of spermatozoa [68]. The role of glucose, lactate, and pyruvate on capacitation and acrosome reaction when examined in vitro, it was found out that in the absence of glucose or lactate, spermatozoa underwent very few AR and motility activation (whiplash-like motility characteristic of capacitated hamster sperm) compared to those preincubated in the presence of glucose, lactate, and pyruvate. Glucose and pyruvate supported more AR than glucose alone, but less than glucose, lactate, and pyruvate [69]. It is explained that exogenous glucose, presumably metabolized to pyruvate and lactate, cannot support AR alone, but requires lactate. A sperm-specific isozyme of lactate dehydrogenase (LDH-X) allows sperm mitochondria to metabolize lactate without prior oxidation to pyruvate in the cytosol and may function in the shuttle of reducing equivalents across the mitochondrial membrane. LDH-X converts cytosolic pyruvate to lactate with the concomitant conversion of NADH to NAD. Within the mitochondrion, lactate is converted by LDH-X back to pyruvate and NAD to NADH. The net result of this reaction is that the reducing power of NADH, to which the mitochondrial membrane is impermeable, is moved across the membrane and made available to the electron transport chain and oxidative phosphorylation. Pyruvate may also be made available to the Krebs cycle for further energy production. However, Dravland and Meizel [69] have shown that the generation of NADH by the Krebs tricarboxylic acid cycle is insufficient for the maintaining optimal ATP levels for capacitation and AR. Thus, the ATP levels may need to be supplemented by direct intramitochondrial oxidation of lactate to pyruvate. High levels of ATP are essential for motility of hamster spermatozoa. Pyruvate, either added exogenously or derived from glucose might be helping NADH to shuttle into the mitochondria via the LDH-X shuttle but may not be able to support AR in an optimal manner by high concentrations of pyruvate leading to inhibition of LDH-X [69].

Rabbit Spermatozoa

Mature epididymal spermatozoa from the rabbit need only dilution in a simple aerated saline buffer to show full motility. Glucose and butyrate are described as the main exogenous metabolic fuels of rabbit spermatozoa. Presence of highly active adenylate kinase and creatine kinase in rabbit spermatozoa has been reported which is characteristic in cells with high rates of ATP synthesis and utilization [70]. Rabbit spermatozoa are shown to be able to form ATP either via glycolysis or via oxidative phosphorylation. Both these metabolic pathways are active in these viable cells where creatine kinase and adenylate kinase systems are present. A dynamic balance between ATP synthesis and ATP-hydrolyzing enzymes is suggested, supported by the fact that rabbit spermatozoa in their seminal plasma preserved their motility for hours. The decrease in ATP content is mainly due to its hydrolysis by dynein ATPases coupled with movements of rabbit spermatozoa [70].

The ATP requirements of the rabbit sperm flagellum for motility have been measured from the activity of the flagellar ATPase with access to ATP. Though there are considerable variations in these activities depending on the sperm preparations, higher activities are reported in spermatozoa from more mature rabbits [31]. Pyruvate kinase remains bound to the sperm structure in all these preparations, and the activities are close to that of ATPase in any given preparation. The similarity between the activity of these two enzymes and the fact that these activities were always an order of magnitude greater than those of mitochondrial O_2 uptake provided further evidence that glycolysis alone can maintain motility in rabbit spermatozoa [71].

Mouse Spermatozoa

Mouse spermatozoa maintained vigorous motility with high beat frequency in physiological solution containing glucose. Highly motile state of spermatozoa in vivo for extended durations is essential during activation and hyperactivation [72]. It is reported that glycolysis instead of mitochondrial respiration is the main energy source supporting mouse spermatozoa motility. This is reinforced from the findings of the gapds knock-out mice [23]. These male mice are healthy but sterile as spermatozoa were nonmotile. The intracellular

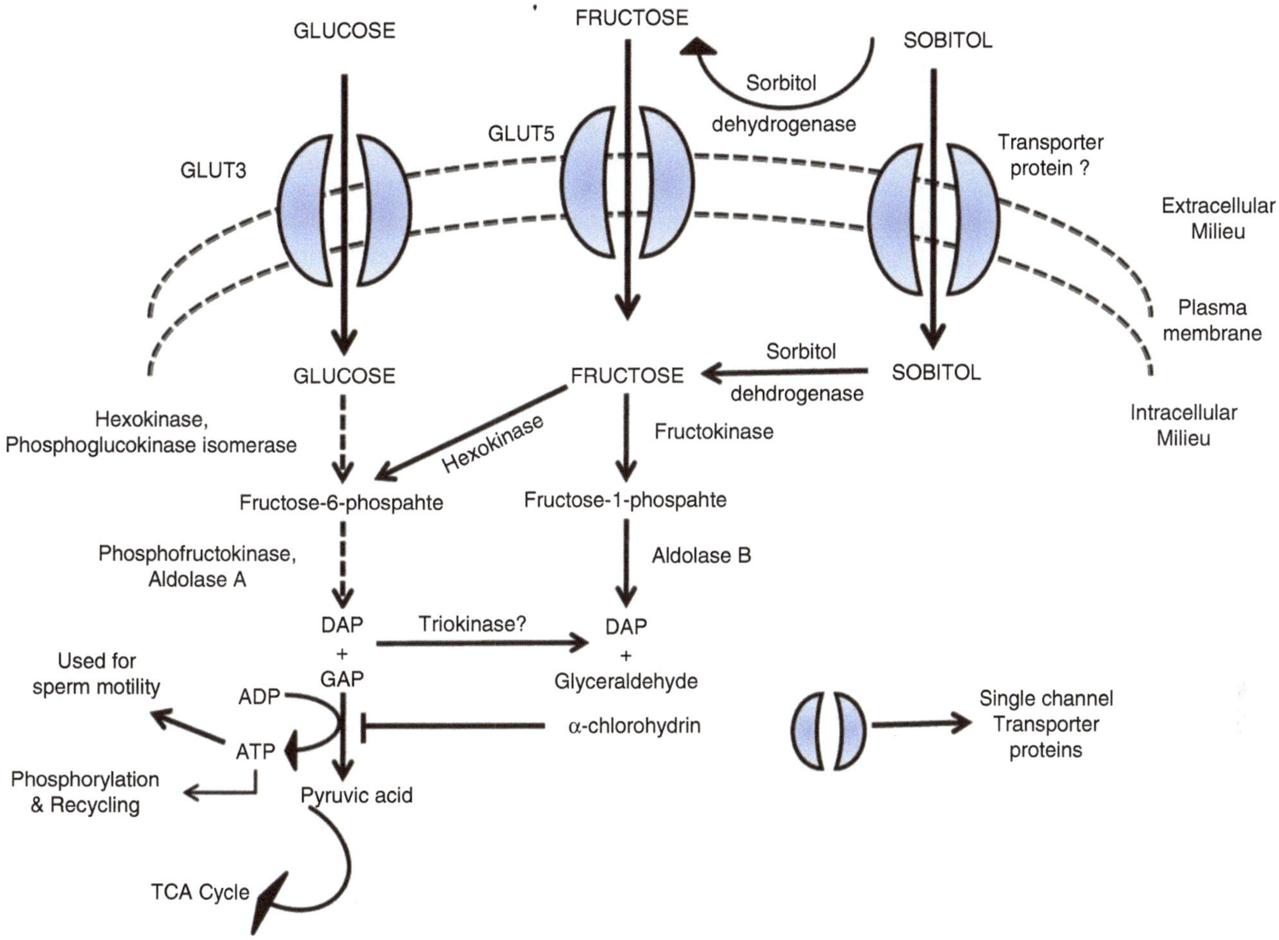

Fig. 21.6 Sorbitol can act as an alternative energy source for mouse spermatozoa. *DAP* dihydroxyacetone phosphate, *GAP* glyceraldehyde-3-phosphate

ATP content was only 10% that of the wild-type spermatozoa, but the O_2 uptake with pyruvate plus lactate in the absence of added glucose by the mid-piece mitochondria was the same as that of the wild type. The mouse spermatozoon principal piece is 87-μm long [73], so that diffusion of ATP from mid-piece to the end piece is difficult to achieve. But the glycolytic enzyme complex spread along the sperm tail can support a flux sufficient to provide for all the ATP needed for motility. The efficacy of glucose in maintaining mouse spermatozoa motility was also documented by Mukai and Okuno [74]. But these investigators also found that the spermatozoa remain motile with pyruvate or lactate in the absence of glucose and spermatozoa motility ceased in the presence of inhibitors of mitochondrial oxidative phosphorylation. These findings implied that mitochondrially generated ATP can diffuse far enough into the flagellum to power normal spermatozoa motility. The absence of GAPDH in the head implies that supply of ATP to the head cannot come from glycolysis and so must rely on ATP generated by the mid-piece mitochondria [75] with subsequent diffusion into the head [71].

Cao et al. [52] demonstrated that mouse spermatozoa are able to utilize sorbitol as an energy source to maintain baseline motility. Spermatozoa incubated for 1 h in control PBS or 5 mM sorbitol remained quite motile. Sorbitol increased the levels of protein tyrosine phosphorylation and thus can be considered as the alternate carbohydrate energy source other than glucose. However, it needs to be converted to fructose to enter the metabolic pathway (Fig. 21.6). Inside sperm, fructose can be converted to fructose 1-phosphate by hexokinase to enter the glycolytic pathway. In addition, most enzymes in fructose metabolism have also been reported to be present in the sperm flagellum, indicating that fructose metabolic pathways may be operative in mouse spermatozoa.

Human Spermatozoa

Normal motility in ejaculated human spermatozoa is a basic requirement for male fertility. Because ATP supports the chemical–mechanical coupling catalyzed by dyneins (specific

ATPases of the flagellum), it is widely accepted that both glycolysis and oxidative phosphorylation are required for optimal spermatozoa function in humans [76]. The importance of glycolyzable substrate for supporting the ability of human spermatozoa to capacitate and penetrate oocytes in vitro was investigated by Rogers and Perreault [77]. Washed spermatozoa were incubated with or without various sugars in Biggers Whitten Whittingham (BWW) culture medium containing pyruvate and lactate. After an 18-h preincubation, glucose (1 mg/ml) induced higher penetration of spermatozoa into oocytes than either mannose or fructose at the same concentration. The above study concluded that stimulation of glycolysis may promote capacitation of human spermatozoa in vitro and that optimization of penetrating ability of spermatozoa is dependent upon both the type and concentration of glycolyzable sugar present. Lactate production is found considerably low in men with sterility but presenting normal semen examination, as well as in asthenozoospermic men [78]. On the other hand, Williams and Ford [20] have shown that the percentage of motile human spermatozoa as well as their smothered path velocity is statistically greater when incubated in media containing glucose. Other fuel sources for human spermatozoa include sorbitol and fructose which are found abundantly in seminal plasma. Even human spermatozoa have the enzymes necessary to produce both sorbitol and fructose, as confirmed by Western blot analysis [79]. The enzyme aldose reductase reduces glucose to sorbitol followed by its oxidation to fructose by sorbitol dehydrogenase (see Fig. 21.5). It is also reported that human spermatozoa are able to utilize fructose for their metabolic needs while in seminal fluid and shift to glucose while in the female reproductive tract [80].

Recent studies, further confirmed that glycolysis is an important pathway to generate ATP in human spermatozoa. Analysis of the flagellum has shown that several glycolytic enzymes are localized to the fibrous sheath along with an ADP/ATP carrier protein [30].Since energy production and translocation mechanisms are localized to the spermatozoa tail, it brings home the point that spermatozoa do rely upon glycolysis for the generating ATP used for motility [30]. Comparison of glycolysis and oxidative phosphorylation as energy source for human spermatozoa has been carried out by Nascimento et al. [81] by using combination of fluorescence imaging, laser tweezers, and real-time automated tracking and trapping. Their results suggest that human spermatozoa motility is independent of mitochondrial respiration as measured by mitochondrial membrane potential. It also reveals that, with respect to spermatozoa motility, oxidative phosphorylation does not contribute as much energy to spermatozoa compared to that supplied by glycolysis. However, the study concludes that human spermatozoa may rely on alternate pathway(s) for the energy needed to sustain motility. Mitochondrial fatty acid oxidation supported by presence of L-carnitine may also act as an alternate energy source in human spermatozoa. Carnitine (b-hydroxy-c-trimethylaminobutyric acid) is a small water-soluble quaternary amine that, in conjunction with carnitine acyltransferases, plays an essential role in the transport of long-chain fatty acids, as acylcarnitine esters enter into mitochondria for β-oxidation and ATP generation [82]. Carnitine supplementation has also been seen to improve spermatozoa motility in some cases of idiopathic asthenozoospermia [83].

Transporters Involved in the Uptake of Various Substrates

Eukaryotic cells have a complete array of proteins, which are specialized in the uptake and internalization of monosaccharides through the cell membrane. A specific combination of several of these proteins can, thus, efficiently modulate monosaccharide specificity linked to the uptake of these molecules. As a general basis, the most important of these proteins are included in two main protein families. The first and most important is the GLUT family of proteins. This family comprises a great number of separate proteins, GLUT-1 to GLUT-5, which show great differences in characteristics such as uptake velocity and sugar affinity for sugars other than substrates, like vitamin C [66]. The second family is the SGLT family of proteins. This comprises several proteins that act as membrane transporters of an array of substances such as glucose and mannose. The regulation of these proteins is mainly controlled by cell plasma membrane potential. Two of the most important proteins in the SGLT family are SGLT-1 and SGLT-2.

Glucose Transporters

There are several members in the family of sugar transport facilitators. In human spermatozoa, several studies have reported the presence of various species of GLUT proteins such as GLUT-1, GLUT-2, GLUT-3, and GLUT-5. GLUT-3 protein has a very high affinity to glucose. On the other hand, GLUT-5 shows much more affinity to fructose and even mannose than other hexoses such as glucose [66]. It is preferentially located at the mid-piece of both dog and boar spermatozoa. It is now established that there are species-related differences in the expression of the observed GLUTs in the spermatozoa. Thus, whereas GLUT-1 and GLUT-2 are present in human spermatozoa, bull, and rat [84], they are absent in dog sperm, which in contrast shows the presence of SGLT family of proteins. These are related to uptake of sugars not directed to energy obtainment, but to the synthesis of glycosidic residues linked to sperm plasma membrane glycoproteins [26].

A glucose transporter 8 (GLUT-8) has been recently identified. The GLUT-8 protein was associated with spermatozoa within the seminiferous and the epididymal tubules. The GLUT-8 immunoreactivity was also detected within the head of mouse and human spermatozoa in the acrosomal region and appeared to be located at the plasma membrane, as well as within the cells. This specific expression and localization of GLUT-8 suggest that this transport facilitator plays a major role in the fuel supply of mature spermatozoa [85].

Fructose Transporters

Fructose is transported by an isoform of the glucose transporter, namely, GLUT-5. RNA and protein blotting studies have shown the presence of high levels of GLUT-5 mRNA and protein in human testis and spermatozoa. The immunocytochemical studies localize GLUT-5 to the plasma membrane of mature spermatids and spermatozoa [86]. In boar spermatozoa, GLUT-5 is located in the peri-acrosomal area.

Hexose Transporters

Dog sperm extracts show several other hexose transporters. Western blot analysis performed by Rigau et al. [26] revealed a protein that reacted against anti-GLUT-3 antibody in spermatozoa from men and dogs, thereby indicating the presence of a 45-kDa Glut-3 protein in dogs. In addition, western blot analysis performed against Glut-5 antibody also showed a positive signal in dog and human spermatozoa.

Phosphocreatine Shuttle

In sea urchins, a phosphocreatine shuttle exists which is employed in the energy transport [87]. As phosphocreatine diffuses along the flagellum, the flagellar creatine kinase regenerates ATP for use as a substrate by dynein ATPase. Mammalian spermatozoa contain only low concentrations of creatinine phosphokinases or other phosphagens; the shuttle may have very little influence in energy transport [88, 89]. This is further supported by the fact that mice in which gene for mitochondrial isotype of creatine kinase has been knocked out were fertile with spermatozoa displaying similar motility characteristics as the wild type [90].

Sperm Flagellar Energy Carrier Protein

Sperm flagellar energy carrier protein (SFEC) is a unique ADP/ATP carrier protein located on the fibrous sheath of the flagellum and is believed to be essential for human spermatozoa motility [30].

Other Shuttles

Several enzymatic shuttles like creatine kinase, adenylate kinase, and phosphoglycerate kinase shuttles have been proposed to transfer ATP away from the mitochondrion and to return ADP back [5]. Creatinine phosphokinase activity has been detected in human spermatozoa which is associated with retention of cytoplasm during spermiation. This activity is reported higher in infertile men with abnormal spermatozoa head morphology [91]. Further evidence came from the observation that demembranated human spermatozoa can be reactivated by creatine phosphate plus ADP, while creatine phosphokinase inhibitor, dinitrofluorobenzene, impaired progressive motility when lactate was only provided in the medium [76]. However, this needs to be authenticated further.

Expert Commentary

The purpose of this article is to discuss the role played by different substrates for the spermatozoa energy needs and the pathways through which these substrates are metabolized. The metabolic energy production in spermatozoa may be fuelled by glycolysis exclusively or mitochondrial phosphorylation exclusively or a combination of both pathways. Evidence supports that glycolysis is able to deliver energy to the spermatozoa distal flagellum. But the fact that spermatozoa are able to maintain motility for long periods when glycolysis is blocked by glyceraldehyde-3-phosphate dehydrogenase (GAPDH) inhibitor α-chlohydrin so long as intermediates of glycolysis do not accumulate supports the contention that this primarily may not be the pathway of ATP generation in spermatozoa. The presence of shuttles like adenylate kinase and others which have been shown to exchange ATP and ADP sufficiently between the flagellum and the mid-piece mitochondria demonstrates that motility can be sustained through oxidative phosphorylation too. The localized energy supply versus the supply through shuttle mechanisms needs further to be authenticated in more subtle experiments and to answer the question as to why *gapds* knockouts show immotile spermatozoa. On the other hand, based on its unique features and restricted expression during spermatogenesis, as well as the infertility, lack of progressive spermatozoa motility, and low-sperm ATP levels of *gapds* knock-out males, GAPDH may emerge as an excellent and highly specific target for male contraceptive strategies. Last, but not the least, is the undisputed fact that in all species, the pathway that fuels motility to speed the fertilizing spermatozoa to the oocyte depends on the substrates available in the oviduct under normal physiological conditions.

Five-Year View

A better understanding of the fuel/energy sources and their subsequent metabolism in spermatozoa may lead to the following improvements.

Better Handling of Gametes In Vitro

There are differences in the ability of spermatozoa from various species to carry out glycolysis and oxidative phosphorylation. For example, glucose inhibits capacitation in bull spermatozoa but very much essential for the same process in mouse spermatozoa [9]. The role of glucose in human spermatozoa metabolism during fertilization remains controversial. Such substrate specificity has profound importance when handling gametes in vitro and needs to be investigated further.

Storage of Sperms

Knowledge regarding the mechanisms utilized by spermatozoa to help manage its intracellular energy level is essential to optimize the conditions used to store spermatozoa for longer periods. Recently, similar attempts are made to preserve chicken spermatozoa based on bioenergetic theory [92]. Short-term storage can also be realized by manipulating mitochondrial function of spermatozoa.

Improving Sperm Function

Sugars can act on spermatozoa not only as passive metabolic substrates but also as direct function activators through mechanisms like specific changes in the tyrosine phosphorylation status of distinct proteins [29]. Mammalian spermatozoa utilize non-glucidic substrates like citrate and lactate to obtain energy. Regulation of this energy is important to maintain overall sperm function and improvements of which may lead to higher fertility rate as reported in chicken spermatozoa achieved mainly through lipid manipulation [93].

Developing Interventions for Male Contraceptive Technology

Several investigators have focused on inhibition of spermatozoa motility as a promising target for male contraceptive development. A drug targeting motility or hyperactivation would not necessarily have to be able to cross the "blood–testes" barrier as long as it is partitioned into seminal fluid and was ejaculated with the sperm. In addition, a drug working on blocking spermatozoa motility might have a very rapid onset of action, possibly allowing for administration immediately prior to intercourse or along with it [94]. Future works on impairing the sperm flagellar energy carrier (SFEC) proteins of the flagellum and inhibiting soluble adenylate cyclase (sAC) may be considered as novel targets for male contraception. Some researchers also suggest that male contraception could be achieved by exploiting the crucial dependence of spermatozoa on glycolysis for energy [23]. One such target is the selective blockage of a sperm-specific glycolytic enzyme, namely, glyceraldehydes-3-phosphate dehydrogenases (GAPDH) which may lead to spermatozoa immotility.

Key Issues

- Sperm need energy to sustain motility and perform sperm function. Mid-piece and flagellum are the sites of energy production.
- Glycolysis and oxidative phosphorylation are the major pathways involved in fuelling spermatozoa of different species. There are marked species differences in the manner by which spermatozoa generate ATP.
- Flagellum is associated with glycolysis, whereas the midpiece harboring mitochondria is actively associated with oxidative phosphorylation.
- Spermatozoa are capable of both aerobic and anaerobic glycolysis, and several glycolytic enzymes are localized along the fibrous sheath of the flagellum.
- Unlike other body fluids, fructose and not glucose is the principal reducing sugar in seminal plasma. However, its concentration varies from high levels to low or nil in various species.
- Other substrates used by spermatozoa for obtaining energy include sorbitol, glycerol, glycerol-3-phosphate, dihydroxy acetone phosphate, certain ketogenic amino acids like leucine, tyrosine, tryptophan, phenylalanine, etc. Some phospholipids like plasmalogen were also being consumed by spermatozoa to derive energy.
- Though the *gapds* knockouts demonstrate lack of spermatozoa motility, experiments with GAPDH inhibitor α-chlorohydrin indicate that spermatozoa can maintain motility for long periods when glycolysis is blocked.
- Association of GLUT-8 immunoreactivity with the acrosomal region of mouse and human spermatozoa as detected by immunofluorescence suggests that it could be interesting target for inhibition of sperm function.
- Further research utilizing the knowledge on energy sources of spermatozoa will enable manipulation of sperm

storage conditions, improvement of sperm function, better handling of gametes in vitro, and also development of a method of intervention for non hormonal male contraception.

Acknowledgments The authors are grateful to Ph.D. students Ankur, Himani, Rekha, Archana, and Shilpa for their valuable cooperation and support while preparing the manuscript.

References

1. Bohnensack R, Halangk W. Control of respiration and motility in ejaculated bull spermatozoa. Biochim Biophys Acta. 1986;850: 72–9.
2. Mita M, Harumi T, Suzuki N, Ueta N. Localization and characterization of phosphatidylcholine in sea urchin spermatozoa. J Biochem. 1991;109:238–42.
3. Hofmann R, Lehmer A, Gurster E, Harturg R. Adenosine triphosphate and adenosine diphosphate in human semen: correlation with sperm count and motility. Urol Int. 1992;48:391–4.
4. Hung PH, Miller MG, Meyers SA, VandeVoort CA. Sperm mitochondrial integrity is not required for hyperactivated motility, zona binding, or acrosome reaction in rhesus macaque. Biol Reprod. 2008;79:367–75.
5. Ford WCL. Glycolysis and sperm motility: does a spoonful of sugar help the flagellum go around? Hum Reprod Update. 2006;12: 269–74.
6. Fawcett DW. The mammalian spermatozoa. Dev Biol. 1975;44: 394–436.
7. Burgos C, Maldnado C, Gerez de Burgos NM, Aoki A, Blanco A. Intracellular localization of the testicular and sperm specific lactate dehydrogenase isozyme C4 in mice. Biol Reprod. 1995;53:84–92.
8. Travis AJ, Foster JA, Rosenbaum NA, Visconti PE, Gerton GL, Kopf GS, Moss SB. Targeting of a germ cell-specific type I hexokinase lacking a porin binding domain to the mitochondria as well as to the head and fibrous sheath of murine spermatozoa. Mol Biol Cell. 1998;9:263–76.
9. Turner RM. Tales from the tail: what do we really know about sperm motility? J Androl. 2003;24:790–803.
10. Storey BT, Kayne FJ. Energy metabolism of spermatozoa. V. The Embden-Myerhof pathway of glycolysis: activities of pathway enzymes in hypotonically treated rabbit epididymal spermatozoa. Fertil Steril. 1975;26:1257–65.
11. Mp B, Geelan A, Leitch V, Goldberg E. Cloning, sequencing and characterization of LDH-C4 from a fox testis cDNA library. Mol Reprod Dev. 1996;44:452–9.
12. Westhoff D, Kamp G. Glyceraldehyde 3-phosphate dehydrogenase is bound to the fibrous sheath of mammalian spermatozoa. J Cell Sci. 1997;110:1821–9.
13. Mori C, Nakamura N, Welch JE, Gotoh H, Goulding EH, Fujioka M, Eddy EM. Mouse spermatogenic cell-specific type 1 hexokinase (mHk1-s) transcripts are expressed by alternative splicing from the mHk1 gene and the HK1-S protein is localized mainly in the sperm tail. Mol Reprod Dev. 1998;49:374–85.
14. Bunch DO, Welch JE, Magyar PL, Eddy EM, O'Brien DA. Glyceradehyde 3-phosphate dehydrogenase-S protein distribution during mouse spermatogenesis. Biol Reprod. 1998;58:834–41.
15. Hoshi K, Tsukikawa S, Sato A. Importance of Ca2+, K+ and glucose in the medium for sperm penetration through the human zona pellucida. Tohoku J Exp Med. 1991;165:99–104.
16. Urner F, Sakkas D. Glucose participates in sperm-oocyte fusion in the mouse. Biol Reprod. 1996;55:917–22.
17. Fraser LR, Quinn PJ. A glycolytic product is obligatory for initiation of sperm acrosome reaction and whiplash motility required for fertilization in the mouse. J Reprod Fertil. 1981;61:25–65.
18. Cooper TG, Brooks DE. Entry of glycerol into the rat epididymis and its utilization by epididymal spermatozoa. J Reprod Fertil. 1981;61:163–9.
19. Bone W, Jones NG, Kamp G, Yeung CH, Cooper TG. Effect of ornidazole on fertility of male rats: inhibition of a glycolysis-related motility pattern and zona binding required for fertilization in vitro. J Reprod Fertil. 2000;118:127–35.
20. Williams AC, Ford WC. The role of glucose in supporting motility and capacitation in human spermatozoa. J Androl. 2001;22: 680–95.
21. Alberts B, Johnson A, Lewis J, Raff M, Roberts K, Walter P. Energy conversion: mitochondria and chloroplasts. In: Gibbs S, editor. The biology of the cell. 4th ed. New York: Garland Science; 2002. p. 769.
22. Ford WC, Harrison A. Futile substrate cycles in the glycolytic pathway of boar and rat spermatozoa and the effect of alpha-chlorohydrin. J Reprod Fertil. 1987;79(1):21–32.
23. Miki K, Qu W, Goulding EH, Willis WD, Bunch DO, Strader LF, Perreault SD, Eddy EM, O'Brien DA. Glyceraldehyde 3 phosphate dehydrogenase-S, a sperm specific glycolytic enzyme, is required for sperm motility and male fertility. Proc Natl Acad Sci USA. 2004;101(47):16501–6.
24. Krisfalusi m, Miki K, Magyar PL, O'Brien DA. Multiple glycolytic enzymes are tightly bound to the fibrous sheath of the mouse spermatozoa. Biol Reprod. 2006;75:270–8.
25. Cardenas ML, Cornish-Bowden A, Ureta T. Evolution and regulatory role of the hexokinases. Biochim Biophys Acta. 1998;1401: 242–64.
26. Rigau T, Rivera M, Palomo MJ, Fernandez-Novell JM, Mogas T, Ballester J, Pena A, Otaegui PJ, Guinovart JJ, Rodriguez-Gil JE. Differential effects of glucose and fructose on hexose metabolism in dog spermatozoa. Reproduction. 2002;123:579–91.
27. Marin S, Chiang K, Bassilian S, Lee W-NP, Boros LG, Fernandez-Novell JM, Centelles JJ, Medrano A, Rodrıguez-Gil JE, Cascante M. Metabolic strategy of boar spermatozoa revealed by a metabolomic characterization. FEBS Lett. 2003;554:342–6.
28. Medrano A, Pena A, Rigau T, Rodrıguez-Gil JE. Variations in the proportion of glycolytic/non-glycolytic energy substrates modulate sperm membrane integrity and function in diluted boar samples stored at 15–17°C. Reprod Domest Anim. 2005;40:448–53.
29. Rodriguez-Gil JE. Mammalian sperm energy resources management and survival during conservation in refrigeration. Reprod Domest Anim. 2006;41:11–20.
30. Kim YH, Haidl G, Schaefr M, Egner U, Mandal A, Herr JC. Compartmentalization of a unique ADP/ATP carrier protein SFEC (Sperm Flagellar Energy Carrier, AAC4) with glycolytic enzymes in the fibrous sheath of the human sperm flagellar principal piece. Dev Biol. 2007;302:463–76.
31. Storey BT, Kayne FJ. Properties of pyruvate kinase and flagellar ATPase in rabbit spermatozoa: relation to metabolic strategy of the sperm cell. J Exp Zool. 1980;211:361–7.
32. Mann T, Lutwak-Mann C. Male reproductive function and semen. Berlin: Springer; 1981. p. 196.
33. Mann T. Studies on the metabolism of semen: 3. Fructose as a normal constituent of seminal plasma. Site of formation and function of fructose in semen. Biochem J. 1946;40:481–91.
34. Kobayashi T, Kaneko T, Iuchi Y, Matsuki S, Takahashi M, Sasagawa I, Nakada T, Fujii J. Localization and physiological implication of aldose reductase and sorbitol dehydrogenase in reproductive tracts and spermatozoa of male rats. J Androl. 2002;23:674–83.
35. King TE, Mann T. Sorbitol metabolism in spermatozoa. Proc R Soc Lond B Biol Sci. 1959;151:226–43.
36. Stryer L. Biochemistry. 4th ed. New York, NY: Freeman; 1995.

37. Gibbons BH, Gibbons IR. Flagellar movement and adenosine triphosphatase activity in sea urchin sperm extracted with Triton X-100. J Biol Chem. 1972;54:75–97.
38. Barron ESG, Goldinger JM. Effect of iodoacetate and malonate on respiration of sea-urchin sperm. Proc Soc Exp Biol N Y. 1941;48: 570.
39. Hayashi T. Dilution medium and survival of the spermatozoa of *Arbacia punctulata*. II. Effect of the medium on respiration. Biol Bull Woods Hole. 1946;90:177.
40. Spikes JD. Metabolism of sea-urchin sperm. Am Nat. 1949;83: 285–301.
41. Rothschild L, Cleland KW. The physiology of sea urchin spermatozoa. The nature and location of the endogenous substrate. J Exp Biol. 1952;29:66–71.
42. Tombes RM, Shapiro BM. Metabolite channeling: a phosphocreatine shuttle to mediate high energy phosphate transport between sperm mitochondrion and tail. Cell. 1985;41:325–34.
43. Tombes RM, Brokaw CJ, Shapiro BM. Creatine kinase dependent energy transport in sea urchin spermatozoa: flagellar wave attention and theoretical analysis of P diffusion. Biophys J. 1987;4(1):66–71.
44. Mohri H. Endogenous substrates of respiration in sea urchin spermatozoa. J Fac Sci Univ Tokyo IV. 1957;8:51–63.
45. Mita M, Yasumasu I. Metabolism of lipid and carbohydrate in sea urchin spermatozoa. Gamete Res. 1983;7:133–44.
46. Mita M, Ueta N. Energy metabolism of sea urchin spermatozoa, with phosphatidylcholine as the preferred substrate. Biochim Biophys Acta. 1988;959:361–9.
47. Mita M, Ueta N. Phosphatidylcholine metabolism for energy production in sea urchin spermatozoa. Biochim Biophys Acta. 1990;1047:175–9.
48. Mita M, Nakamura M. Ultrastructural study of an endogenous energy substrate in spermatozoa of the sea urchin *Hemicentrotus pulcherrimus*. Biol Bull. 1992;182:298–304.
49. Mita M, Nakamura M. Phosphatidylcholine is an endogenous substrate for energy metabolism in spermatozoa of sea urchins of the order Echinoidea. Zool Sci. 1993;10:73–83.
50. Burg MB. Molecular basis of osmotic regulation. Am J Physiol. 1995;268:F983–96.
51. Hoshi A, Takahashi M, Fujii J, Myint T, Kaneto H, Suzuki K, Yamasaki Y, Kamada T, Taniguchi N. Glycation and inactivation of sorbitol dehydrogenase in normal and diabetic rats. Biochem J. 1996;318:119–23.
52. Cao W, Aghajanian HK, Haig-Ladewig LA, Gerton GL. Sorbitol can fuel mouse sperm motility and protein tyrosine phosphorylation via sorbitol dehydrogenase. Biol Reprod. 2009;80:124–33.
53. O'Shea T, Wales RG. Metabolism of sorbitol and fructose by ram spermatozoa. J Reprod Fertil. 1965;10:353–8.
54. Hartree EF, Mann T. Phospholipids in ram semen: metabolism of plasmalogen and fatty acids. Biochem J. 1961;80:464–75.
55. Jones AR, Gillan L. Metabolism of glycerol 3-phosphate by mature boar spermatozoa. J Reprod Fertil. 1996;106:321–7.
56. Stevenson D, Jones AR. Inhibition of fructolysis in boar spermatozoa by the male antifertility agent (5)-α chlorohydrins. Aust J Biol Sci. 1982;35:595–605.
57. Jones AR, Chantrill LA, Cokinakis A. Metabolism of glycerol by mature boar spermatozoa. J Reprod Fertil. 1992;94:129–34.
58. Jones AR, Porter LM. Inhibition of glycolysis in boar spermatozoa by α-chlorohydrin phosphate appears to be mediated by phosphatase activity. Reprod Fertil Dev. 1995;7(5):1089–94.
59. Jones AR, Milmlow D. Endogenous energy production by mature boar spermatozoa. J Reprod Fertil. 1997;111.285–90.
60. Jones AR, Chantrill LA. Oxidative metabolic activity of boar spermatozoa: a system for assessing anti-glycolytic activity of potential inhibitors in vitro. Reprod Fertil Dev. 1989;1:357–67.
61. Rigau T, Farre M, Ballester J, Mogas T, Pena A, Rodriguez-Gil JE. Effects of glucose and fructose on motility patterns of dog spermatozoa from fresh ejaculates. Theriogenology. 2001;56:801–15.
62. Roldan ERS. Signal transduction during mammalian sperm acrosomal exocytosis. In: Lauria A, Gandolfi F, Enne G, Gianaroli L, editors. Gametes: development and function. Serno Symposia: Rome; 1998. p. 219–28.
63. Bellester J, Fernandez-Novell JM, Rutlant J, García-Rocha M, Jesús Palomo M, Mogas T, Peña A, Rigau T, Guinovart JJ, Rodríguez-Gil JE. Evidence for a functional glycogen metabolism in mature mammalian spermatozoa. Mol Reprod Dev. 2000;56:207–19.
64. Palomo MJ, FernAndez-Novell JM, Pena A, Guinovart JJ, Rigau T, Rodriguez-Gil JE. Glucose- and fructose-induced dog-sperm glycogen synthesis shows specific changes in the location of the sperm glycogen deposition. Mol Reprod Dev. 2003;64:349–59.
65. Albarracin JL, Fernandez-Novell JM, Ballester J, Rauch MC, Quintero-Moreno A, Pena A, Mogas T, Rigau T, Yanez A, Guinovart JJ, et al. Gluconeogenesis-linked glycogen metabolism is important in the achievement of 'In vitro' capacitation of dog sperm in a medium without glucose. Biol Reprod. 2004;71:1437–45.
66. Mueckler M. Facilitative glucose transporters. Eur J Biochem. 1990;219:713–25.
67. Bavister BD. The effects of variations in culture conditions on the motility of hamster spermatozoa. J Reprod Fertil. 1974;38: 41–440.
68. Bavister BD, Yanagimachi R. The effects of sperm extracts and energy sources on the motility and acrosome reaction of hamster spermatozoa in vitro. Biol Reprod. 1977;16:228–37.
69. Dravland E, Meizel S. Stimulation of hamster sperm capacitation and acrosome reaction in vitro by glucose and lactate and inhibition by the glycolytic inhibitor α-chlorohydrin. Gamete Res. 1981;4:515–23.
70. Minelli MM, Castellini C, Latfaioli P, Mezzasoma I, Ronquisti G. Rabbit spermatozoa: a model system for studying ATP homeostasis and motility. J Androl. 1999;20:259–66.
71. Storey BT. Mammalian sperm metabolism: oxygen and sugar, friend and foe. Int J Dev Biol. 2008;52:427–37.
72. Austin CR. Sperm maturation in the male and female genital tracts. In: Metz CB, Monroy A, editors. Biology of fertilization, vol. 2. New York: Academic; 1985. p. 121–55.
73. Cummins JM, Woodall PF. On mammalian sperm dimensions. J Reprod Fertil. 1985;75:153–75.
74. Mukai C, Okuno M. Glycolysis plays a major role for adenosine triphosphate supplementation in mouse sperm flagellar movement. Biol Reprod. 2004;71:540–7.
75. Carey JE, Olds-Clarke P, Storey BT. Oxidative metabolism of spermatozoa from inbred and random bred mice. J Exp Zool. 1981;216:285–92.
76. Yeung CH, Majumder GC, Rolf C, Behre HM, Cooper TG. The role of phosphocreatine kinase in the motility of human spermatozoa supported by different metabolic substrates. Mol Hum Reprod. 1996;2:591–6.
77. Rogers BJ, Perreault SD. Importance of glycolysable substrates for in vitro capacitation of human spermatozoa. Biol Reprod. 1990;43:1064–9.
78. Lomage J, Guerin JF, Czyba JC. Glycolytic activity of human spermatozoa in normospermic men and in men with abnormal spermograms. Arch Androl. 1986;16:81–8.
79. Frenette G, Thabet M, Sullivan R. Polyol pathway in human epididymis and semen. J Androl. 2006;27:233–9.
80. Peterson RN, Freund M. Metabolism of human spermatozoa. In: Hafez ESE, editor. Human semen and fertility regulation in men. St Louis: Mosby; 1975. p. 176–86.

81. Nascimento JM, Shi LJ, Tam J, Chandsawangbhuwana C, Durrant B, Botvinick EL, Berns MW. Comparison of glycolysis and oxidative phosphorylation as energy source for mammalian sperm motility, using the combination of fluorescence imaging, laser tweezers and real time automated tracking and trapping. J Cell Physiol. 2008;217:745–51.
82. Xuan W, Lamhonwah AM, Librach C, Jarvi K, Tein I. Characterization of organic cation/carnitine transporter family in human sperm. Biochem Biophys Res Commun. 2003;306: 121–8.
83. Garolla A, Maiorino M, Roverato A, Roveri A, Ursini F, Foresta C. Oral carnitine supplementation increases sperm motility in asthenozoospermic men with normal sperm phospholipid hydroperoxide glutathione peroxidase levels. Fertil Steril. 2005;83:355–61.
84. Angulo C, Rauch MC, Droppelmann A, Reyes AM, Slebe JC, Delgado-Lopez F, Guaiquil VH, Vera JC, Concha II. Hexose transporter expression and function in mammalian spermatozoa: cellular localization and transport of hexoses and vitamin C. J Cell Biochem. 1998;71:189–203.
85. Schurmann A, Axer H, Scheepers A, Doege H, Joost H-G. The glucose transport facilitator GLUT8 is predominantly associated with the acrosomal region of mature spermatozoa. Cell Tissue Res. 2002;307(2):237–42.
86. Burant CF, Takeda J, Laroche EB, Bell GI, Davidson NO. Fructose transporter in human spermatozoa and small intestine is GLUT5. J Biol Chem. 1992;267:14523–6.
87. Wothe DD, Charbonneau H, Shapiro BM. The phosphocreatine shuttle of sea urchin sperm: flagellar creatine kinase resulted from a gene triplication. Proc Natl Acad Sci. 1990;87:5203–7.
88. Smith MB, Babcock DF, Lardy HA. A P-31 NMR study of the epididymis and epididymal sperm of the bull and hamster. Biol Reprod. 1985;33:1029–40.
89. Robitaille PML, Robitaille PA, Martin PA, Brown GG. P-31 nuclear magnetic resonance studies of spermatozoa from the boar, ram, goat and bull. Comp Biochem Physiol B. 1987;87:285–96.
90. Steeghs K, Oerlemans F, Weeringa B. Mice deficient of ubiquitous mitochondrial creatine-kinase are viable and fertile. Biochim Biophys Acta. 1995;1230:130–8.
91. Huszar G, Vigue L. Incomplete development of human spermatozoa is associated with increased creatine phosphokinase concentration and abnormal head morphology. Mol Reprod Dev. 1993;34: 292–8.
92. Froman DP, Feltmann AJ. New approach to sperm preservation based upon bioenergetic theory. J Anim Sci. 2010;88:1314–20.
93. Cerolini S, Pizzi F, Gliozzi T, Maldjian A, Zaniboni L, Parodi L. Lipid manipulation of chicken semen by dietary means and its relation to fertility: a review. Worlds Poult Sci J. 2003;59: 65–75.
94. Amory JK. Inhibiting sperm motility for male contraception: will the sperm tail be its "Achilles Heel"? Mol Interv. 2007;7: 68–70.

Oxidative Stress

22

Fanuel Lampiao, C.J. Opperman, Ashok Agarwal, and Stefan S. du Plessis

Abstract

The idea that free radicals are produced by spermatozoa was first proposed by Macleod in 1943. While evaluating the influence of oxygen tension on sperm cell motility, he observed that the addition of the catalase enzyme to sperm incubation medium significantly reduced their motility loss. Macleod concluded in this pioneering study that hydrogen peroxide (H_2O_2) must be produced by spermatozoa during oxygen metabolism, thereby setting the trend for future research along these lines. It was not until 1979 that Jones and colleagues resolved the underlying mechanism behind free radicals and their ability to reduce sperm motility. They reported a decrease in the flexibility of sperm membranes due to reactive oxygen species (ROS)-induced peroxidation. Nearly 70 years after the discovery of Macleod, interest is turning toward free radicals as the origin of male infertility with 1 in every 20 male individuals being infertile and accounting for half of all cases of couple infertility in the general population. It is no surprise that ROS-mediated sperm cell damage is accountable in up to 30–80% of these cases as the foremost contributing pathological factor of male infertility. A new diagnostic era is on the horizon, with over 30 advanced direct and indirect screening assays available to assess oxidative stress. Promising as it might seem, routine testing is still shortcoming due to cost contributing factors, complexity dilemmas, and the lack of a standardized routine. Nonetheless, management of oxidative stress-related male infertility is focused on the identification of underlying pathology prior to antioxidant treatment.

Keywords

Antioxidant treatment • Free radicals • Oxidative stress • Reactive oxygen species • Sperm cell damage • Male infertility • Apoptosis

The idea that free radicals are produced by spermatozoa was first proposed by Macleod in 1943 [1]. While evaluating the influence of oxygen tension on sperm cell motility, he observed that the addition of the catalase enzyme to sperm incubation medium significantly reduced their motility loss. Macleod concluded in this pioneering study that hydrogen peroxide (H_2O_2) must be produced by spermatozoa during oxygen metabolism, thereby setting the trend for future research along these lines. However, it was not until 1979 that Jones and colleagues resolved the underlying mechanism behind free radicals and their ability to reduce sperm

F. Lampiao, PhD (✉) • C.J. Opperman, BSc • S.S. du Plessis, PhD, MBA
Department of Medical Physiology, Faculty of Health Sciences, Stellenbosch University, PO Box 19063, Tygerberg, Western Cape, 7505, South Africa
e-mail: flampiao@medcol.mw; ssdp@sun.ac.za

A. Agarwal, PhD
Center for Reproductive Medicine, Glickman Urological and Kidney Institute, Cleveland Clinic Foundation, 9500 Euclid Avenue, Desk A19, Cleveland, OH 44195, USA
e-mail: agarwaa@ccf.org

S.J. Parekattil and A. Agarwal (eds.), *Male Infertility: Contemporary Clinical Approaches, Andrology, ART & Antioxidants*, DOI 10.1007/978-1-4614-3335-4_22,

motility. They reported a decrease in the flexibility of sperm membranes due to reactive oxygen species (ROS)-induced peroxidation [1].

Nearly 70 years after the discovery of Macleod, interest is turning toward free radicals as the origin of male infertility with 1 in every 20 male individuals being infertile and accounting for half of all cases of couple infertility in the general population [2]. It is no surprise that ROS-mediated sperm cell damage is accountable in up to 30–80% of these cases as the foremost contributing pathological factor of male infertility [3]. A free radical is defined as a molecule that has one or more unpaired electrons with the capability to oxidize biomolecules [4]. Leukocytes (extrinsic ROS) and spermatozoa (intrinsic ROS) are the foundation to the production of free radicals within semen [5], not excluding external origins of oxidative tension. Free radicals can attack the deoxyribose backbone of spermatozoa directly or damage its pyrimidine and purine basis [3]. Alternatively, caspase enzymatic degradation of DNA is subsequent to the initiation of ROS-mediated apoptosis [6]. Furthermore, the large amount of membrane polyunsaturated fatty acids (PUFAs) within spermatozoa makes them vulnerable to free radicals. On the other hand, the value of physiological ROS levels in the functioning of spermatozoa during the stimulation of capacitation, hyperactivation, and the acrosome reaction cannot be overlooked [7]. Therefore, the body has developed several defense mechanisms which are focused on oxidant scavenging to protect itself from the detrimental effect of ROS damage. These include enzymes such as superoxide dismutase (SOD) [8, 9] and glutathione peroxidase (GPX) [10, 11] as well as nonenzymatic molecules, for instance, ascorbic acid (vitamin C) and alpha-tocopherol (vitamin E) within the semen [12, 13]. For this reason, oxidative stress is only achieved when the balance is shifted toward the production of ROS and when its overwhelming effect on antioxidant defense mechanisms becomes prominent.

A new diagnostic era is on the horizon, with over 30 advanced direct and indirect screening assays available to assess oxidative stress. Promising as it might seem, routine testing is still shortcoming due to cost contributing factors, complexity dilemmas, and the lack of a standardized routine [14]. Nonetheless, management of oxidative stress-related male infertility is focused on the identification of underlying pathology prior to antioxidant treatment. If areas such as lifestyle modifications or lessening the burden of environmental exposures would still be attractive to pharmaceutical companies from a business perspective in the long run when compared to conventional medical supplementation regimens or possible surgery remains debatable. Finally, this overview chapter will provide the reader with the biochemistry, etiology and management of oxidative stress in male infertility. Moreover, it is hoped that the importance of free radicals in spermatogenesis is realized and further exploring into this field stimulated.

Oxidative Stress: Etiology and the Oxygen Paradox

Aerobic metabolism is associated with the production of free radical molecules. These prooxidant molecules are formed throughout the controlled enzymatic reduction of oxygen within the mitochondria during oxidative phosphorylation [11]. A free radical is defined as a molecule that has one or more unpaired electrons in its molecular or atomic orbital. Therefore, this highly reactive group of chemical molecules has the potential to significantly contribute in the oxidation of biomolecules such as amino acids in proteins and lipids in membranes, while seeking to alleviate their unpaired electron state [4, 15]. The superoxide anion radical ($O_2^{\bullet -}$), the primary form of ROS, is formed following the addition of an electron to dioxygen (O_2). However, this molecule can then be indirectly as well as directly (metal catalyzed, enzymatic) converted to secondary ROS (Fig. 22.1b), such as H_2O_2, the peroxyl radical ($ROO^{\bullet}$), and the hydroxyl radical ($^{\bullet}OH$). Interestingly, not all ROS are considered to be free radicals despite the fact that the terms are used in a common interchangeable manner [16]. For example, H_2O_2 does not contain unpaired electrons and therefore by definition is not a free radical (Table 22.1). Still, it is an addition to the ROS family since it has been derived from the metabolism of oxygen. Additionally, there is a subclass of ROS molecules derived from nitrogen. They are known as the reactive nitrogen species (RNS) and embrace a free radical nature. Some of the most common RNS are listed in Table 22.2 [17].

Leukocytes (extrinsic ROS production) and spermatozoa (intrinsic ROS production) form the two cornerstones in the production of free radicals within semen [5]. Neutrophils form the bulk of leukocytes in semen and are predominantly responsible for destroying pathogens by means of ROS production (Fig. 22.1a). A definite correlation between oxidative stress and the presence of elevated leukocyte numbers has been established [18]. It is important to remember throughout this chapter that oxidative stress only occurs when the production of free radicals overwhelms antioxidant defenses, which can ultimately lead to sperm cell damage. For this reason, spermatozoa and seminal plasma contain a selection of protective antioxidant mechanisms (Table 22.3). Nonetheless, no distinct relationship has been described between male infertility and leukocytes thus far [19]. The importance of leukocyte

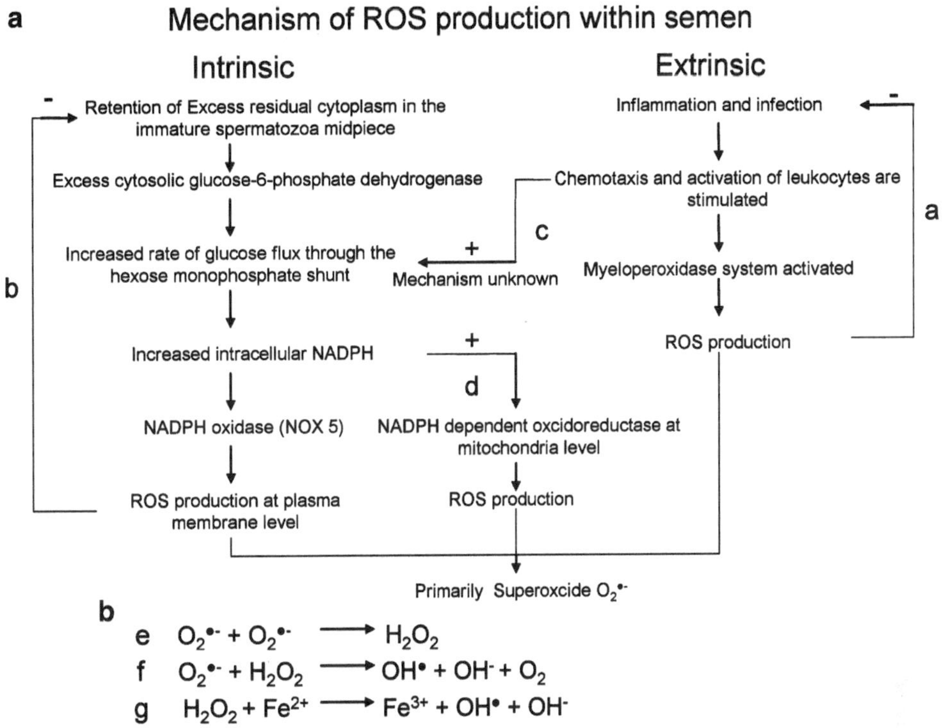

Fig. 22.1 Free radicals and their production. (**a**) Free radicals (superoxide) can be produced within the sperm cell or via leukocytes in the semen, not excluding external systemic and local genital tract sources (a, b—negative feedback; c, d—positive feedback). (**b**) Superoxide, the primary form of reactive oxygen species, is converted directly (metal catalyzed, enzymatic) and indirectly to secondary (e–g) reactive oxygen species

Table 22.1 Reactive oxygen species consist of radical and none radical oxygen derivatives

Radicals		Nonradicals	
Lipid peroxyl	$LOO^{\bullet}$	Lipid peroxide	LOOH
Thyl	$RS^{\bullet}$	Ozone	O_3
Peroxyl	$RO_2^{\bullet}$	Singlet oxygen	$^{-1}O_2$
Nitric oxide	$NO^{\bullet}$	Hydrogen peroxide	H_2O_2
Superoxide	$O_2^{\bullet-}$	Hypochloric acid	HOCl
Hydroxyl	$OH^{\bullet}$	Peroxynitrite	$ONOO^{-}$

Table 22.2 Summary of various reactive nitrogen species, also considered as a subclass of reactive oxygen species

Peroxynitrite	$OONO^{-}$	Nitrous acid	HNO_2
Peroxynitrous acid	OONOH	Dinitrogen trioxide	N_2O_3
Nitryl chloride	NO_2Cl	Nitrogen dioxide	$NO_2^{\bullet}$
Nitrous oxide	$NO^{\bullet}$	Nitroxyl cation	NO^{+}
Nitroxyl anion	NO^{-}		

activation in ROS production cannot be overlooked. Several studies have reported elevated levels of proinflammatory cytokines within seminal plasma prior to the production of large ROS amounts. These molecules include tumor necrosis factor-alpha, interleukin-8 [20], and interleukin-6 [21]. Sperm isolation and quantification techniques have not only revealed ROS production by

Table 22.3 Sperm and seminal plasma are well equipped with a collection of both enzymatic and nonenzymatic antioxidant defenses

Antioxidant	Mechanism	References
Enzyme		
Superoxide dismutase (SOD)	Inactivates superoxide radicals by converting it into oxygen	[8, 9]
Catalase	Inactivates hydrogen peroxide by converting it into water	[52, 53]
Glutathione peroxidase (GPX 1–5)	Responsible for the reduction of hydrogen peroxide through the utilizing of glutathione as an electron donor	[10, 11]
Nonenzyme		
Ascorbic acid (vitamin C) Alpha-tocopherol (vitamin E) Urate Prostasomes Albumin Carnitine Carotenoids Flavonoids	All of these antioxidants within the semen act according to the chemical principle of directly neutralizing free radicals. Additional mechanisms of protection are provided by albumin and prostasomes. Albumin can be oxidized itself, thereby intercepting free radicals. Alternatively, prostasomes (extracellular organelles secreted by the prostate) have been reported to fuse leukocytes and influence their ROS production	[12, 13]

sperm cells but also the motive behind excessive ROS output in teratozoospermia (less than the reference value for morphology) during impaired spermatogenesis. In morphologically normal spermatozoa, cytoplasm deposits in the midpiece are extruded to allow for cell elongation and condensation to occur during spermatogenesis. Furthermore, these cytoplasm deposits contain large amounts of the glucose-6-phosphate dehydrogenase enzyme which is responsible for the rate of glucose flux in the cell and the intracellular production of beta-nicotinamide adenine dinucleotide phosphate (NADPH). As a result, ROS is generated from NADPH via intermembrane-located NADPH oxidase [11]. Teratozoospermia is characterized by large amounts of cytoplasm, and therefore, higher levels of ROS is expected when these specimens are compared to normozoospermic spermatozoa. In addition, a calcium-dependent NADPH oxidase called NOX 5 has been reported in sperm. It appears to be quite distinct from the NADPH oxidase enzyme located in leukocytes. Still, expression of the NOX 5 in infertile men is unknown [3]. An interesting observation is that leukocytes are the predominant producer of ROS during spermatozoa capacitation and per cell basis. However, intrinsic production of ROS within close proximity to the sperm cell's DNA formulates a more important variable in terms of oxidative stress infertility [4].

Table 22.4 shows the origin of well-established external causes for potential oxidative stress in the male reproductive tract. A number of articles have addressed the impact of lifestyle modifications on seminal ROS levels as well as the correlation with sperm DNA damage and infertility. Alcohol has been reported to stimulate the production of ROS. Moreover, studies have shown that excessive alcohol consumers tend to be malnourished and have a diet with insufficient antioxidants [22]. A recent study reported that a person who smoked had a 48% increase in testicular leukocyte levels, accompanied by a 107% increase in seminal ROS concentration [23]. Furthermore, smokers tend to have decreased concentrations of natural antioxidants such as vitamins C and E in their seminal plasma [24], implying an additional risk for infertility. Studies confirm a link between high-antioxidant intake and an increase in motile sperm percentages as well as decreased sperm DNA damage [25]. Although, it is more likely that individual antioxidant levels within the semen reflects the biological effects before diet supplementary intake more accurately since different food preparation techniques and sources can vastly influence antioxidant levels. Unsurprisingly, fertile men with low levels of oxidative stress will not require the same amount of protection from antioxidants to maintain their DNA integrity. Therefore, insufficient antioxidants in men with a low oxidative stress risk may not lead to sperm DNA damage per se.

Table 22.4 Established causes of sperm oxidative stress potentially leading to male factor infertility

Origin of oxidative stress	References
Lifestyle	
Smoking	[24]
Insufficient diet	[25]
Psychological stress	[30]
Obesity	[27]
Alcohol	[22]
Age-related	[29]
Environmental	
Pollution	[35]
Heavy metals	[36]
Heat	[26, 28]
Plasticizers	[31]
Pesticides	[33]
Herbicides	[34]
Infection	
Systemic infection	[41–44]
Genitourinary tract	[39]
Autoimmune	
Chronic prostatitis	[45–47]
Vasectomy	[48]
Torsion	[3, 59]
Testicular	[56, 57]
Idiopathic	[49, 50]
Iatrogenic	
Medications	[54]
Centrifugation, cryopreservation	[3, 51]
Chronic disease	
Hemoglobinopathies	[63]
Diabetes	[64, 65]
Chronic kidney disease	[60–62]
Hyperhomocysteinemia	[66]

It is no surprise that extreme exercise activities lead to the production of large amounts of ROS following increased aerobic muscle metabolism [26]. Similarly, obesity provokes oxidative stress as cytokines are released from adipose tissue during proinflammatory reactions which leads to an increase in leukocytes [27]. What is more is that heating of the testicles due to the accumulation of groin region adipose tissue has shown to result in reduced sperm quality [28]. Several studies have reported a direct link between age and sperm DNA damage in both fertile as well as infertile men. In addition, systemic oxidative stress seems to increase with age [29], making oxidative stress a most likely candidate for the underlying sperm pathology in older infertile men. Finally, psychological stress has been concurrent with low sperm quality due to an increase in ROS production and reduced antioxidant defenses [30].

Numerous environmental pollutants have been linked with oxidative stress. Increased sperm DNA damage and impaired spermatogenesis have been reported in cases of phthalates absorption. Phthalates are chemicals used in personal care products and plastic food packaging [31]. Sulfur dioxide, a frequently used preservative [32], pesticides such as methoxychlor [33], and the herbicide dioxin TCDD [34] have resulted in the production of ROS during laboratory animal trials. While airborne pollutants such as diesel have the potential to increase infertility [35], exposure to heavy metals (lead, cadmium) is conclusively associated with testicular oxidative stress [36].

Male infertility might not be treated by addressing bacterial infections as a first-line therapy, but recent studies have provoked interest into this alternative treatment approach. It has been estimated that up to 50% of men will experience prostatitis [37]. The bacteria responsible for this infection can be sexually transmitted or acquired from a local urinary tract infection [38]. The recurrent entering of these pathogens into the genital tract results in an acute inflammatory response, with an influx of leukocytes and a subsequent increase in ROS production. High levels of oxidative sperm pathology have been confirmed in individuals with an elevated tendency for genitourinary tract infections, such as paraplegics [39]. Oxidative damage to sperm cells can also be linked to viral infections which include the herpes simplex virus. The DNA of this virus is found to be present in the semen of between 14% and 50% of infertile men. In addition, its presence is associated with an increase in IgM antibodies and the rate of leukocytospermia production [40]. Despite the fact that no study has directly linked chronic systemic infections with sperm oxidative stress, it remains unlikely that the male reproductive tract will be spared from these infectious diseases, which include human immunodeficiency virus [41], hepatitis B and C [42], tuberculosis [43], and malaria [44].

Chronic prostatitis of nonbacterial infectious nature affects 10% of all men [37] and has been associated with elevated levels of oxidative tension in semen [45]. An autoimmune response to prostate and seminal antigens is reported to cause a proinflammatory reaction with an increase in leukocytes [46]. Although the exact nature of this reaction is unknown, it is believed to be the result of a polymorphism linked to Th-2 cytokine interleukin-10. The production of T lymphocytes reactive against the above antigens might be due to the lack of Th-2 cytokines, shifting the balance toward the Th-1 cytokines in the immune system. Chemotaxis and activation of leukocytes are stimulated as a result of cytokines (such as tumor necrosis factor-alpha, interleukin-1-beta and interferon-gamma) liberated in response to the T lymphocytes [47]. A clinical example of this is when the blood–testis barrier is disrupted following vasectomy. Consequently, immune defense against spermatozoa is lost, leading to significant oxidative stress and infertility after vasectomy reversal [48].

Compelling results have shown that morphologically abnormal sperm cells tend to yield higher levels of ROS production and portray a lower capacity of antioxidant protection [49]. In addition, teratozoospermia is commonly observed in one of every three infertile men. For that reason, oxidative stress is frequently exhibited in the male population with idiopathic infertility. This principal association of elevated seminal ROS levels and decreased antioxidant values has also been made amongst idiopathic infertile men presenting with normozoospermia, the reason for this however is still ambiguous [50].

Sperm oxidative stress can also be augmented during assisted reproductive techniques. At some point during intrauterine insemination and similar assisted reproductive technologies, semen is separated from the seminal plasma by means of techniques that involve centrifugation. This exacerbates ROS production within the sperm cells, while removing the sperm from their antioxidant environment [3]. Cryopreservation, a commonly used technique, also increases the likelihood of an oxidative assault on sperm cells [51]. Resent chemotherapeutic agent studies on animal models have revealed a link with oxidative stress. The administration of cyclophosphamide has indicated a decrease in testicular catalase [52, 53] and an increase in testicular malondialdehyde levels, implying the presence of oxidative stress [54]. Also, commonly used household medications, such as aspirin and paracetamol, tend to elevate cytochrome P450 activity, thereby enhancing the production of free radicals [55].

It has become widely accepted that oxidative stress forms the foundation to the pathology when linking male infertility to testicular varicocele [56]. This phenomenon of increased ROS production in the presence of a varicocele is strongly correlated with the reduction in sperm DNA integrity [57]. Cryptorchidism also contributes to male infertility. Hypospermatogenesis is the primary pathology in this condition due to insufficient maturation of gonocytes to spermatogonia type A. Interestingly, men treated with orchidopexy still show a significant increase in male infertility due to DNA fragmentation and ROS production when compared to control groups [58]. Ischemia reperfusion injury in the contralateral and torted testis during torsion of the spermatic cord has been recognized in male infertility. It is no surprise that oxidative stress is generally accepted to be the underlying cause to the pathology, even when the torsion is unilateral. Spontaneous or surgical restoration of blood flow during a long-standing ischemia period leads to an influx of leukocytes in their activated form to both testes. A subsequent increase in free radicals follows with necrosis of germinal cells. Ultimately, oxidative stress results in infertility or subfertility [59].

Oxidative stress and chronic inflammation are both highly prevalent in cases of end-stage renal and chronic kidney disease [60]. In addition, patients with no evident sign of immune rejection of their graft and stable renal function following renal transplantation also reveal high levels of oxidative stress [61]. Interestingly, a constant state of oxidative stress and chronic inflammation persists after the reversal of uremia following hemodialysis [62]. Beta-thalassemia major, a hemoglobinopathy, is associated with high levels of systemic oxidative stress. Furthermore, this oxidative stress also involves sperm with iron overload forming the most likely cause due to multiple blood transfusions [63]. Diabetic men also show a tendency to produce sperm that has higher DNA fragmentation when compared to their counterparts [64]. Many of the complications of diabetes mellitus are due to an underlying oxidative pathology; therefore, it is probable that oxidative stress is responsible for sperm DNA fragmentation in these individuals as supported by recent studies [65, 66].

Physiological Importance of Reactive Oxygen Species in Normal Spermatozoa Function

Even though oxidative stress is associated with detrimental sperm function, at physiological levels, ROS play a vital role in the regulation of normal sperm function [67]. Physiological levels of ROS have been reported to act as crucial intracellular signaling molecules during the process sperm capacitation. Capacitation is the course of action whereby spermatozoa gain the ability to respond to signals presented by the oocyte–cumulus complex and initiate the cascade of cellular interactions that necessitate fertilization. Although numerous hypotheses have been developed, the precise nature of capacitation is still obscure. Changes associated with sperm capacitation include an increase in respiration and subsequent changes in the motility pattern, called hyperactivation, which is characterized by pronounced flagellar movements and a marked lateral excursion of sperm head in a nonlinear trajectory [68], removal of cholesterol from the plasma membrane, destabilization of the sperm membrane, an increase in intracellular pH and calcium levels, activation of second messenger systems and removal of zinc [69].

The most important change in sperm after capacitation is its ability to undergo the acrosome reaction in response to the *zona pellucida* protein 3 (ZP3), progesterone, and calcium ionophore [70]. Capacitation is also associated with changes in sperm plasma membrane fluidity, intracellular changes in ionic concentration, and sperm cell metabolism [71]. ROS such as O_2^- and H_2O_2 have been reported to play a role in the initiation of the processes capacitation and acrosome reaction [71].

Superoxide anion radical plays an important role during maturation of spermatozoa [72] and in the control of sperm function through the redox regulation of tyrosine phosphorylation [73]. Superoxide has been shown to promote the capacitation of human spermatozoa [74], and there is a superoxide surge in the capacitated spermatozoa during the process [75]. It has been reported that (1) exogenously generated superoxide through the xanthine/xanthine oxidase system induced hyperactivation and capacitation, (2) capacitating sperm produced elevated concentrations of superoxide over prolonged periods of time, and (3) removal of this ROS by SOD prevented hyperactivation and capacitation [76]. H_2O_2 was also shown to promote capacitation of human spermatozoa [77]. The mechanisms and targets of action of H_2O_2 are still unknown.

Nitric oxide (NO) is a free radical synthesized in vivo during the conversion of L-arginine to L-citrulline by the enzyme nitric oxide synthase (NOS). Recent reports suggested the expression of NOS in mouse and human spermatozoa [78]. NO appears to be involved in sperm hyperactivation [79] and *zona pellucida* binding [80]. However, the role of endogenous NO in human sperm capacitation still remains to be elucidated.

Pathological Effects of Oxidative Stress on Spermatozoa and Their Functional Capacity

The lipid composition of plasma membranes of mammalian spermatozoa is markedly different from those of mammalian somatic cells. They have very high levels of phospholipids, sterols, saturated fatty acids, and PUFAs. Therefore, spermatozoa are particularly susceptible to damage induced by excessive ROS release and subsequent development of oxidative stress [81]. Lipids are major substances responsible for the fluidity of membrane lipid bilayers and changes in composition of plasma membranes of spermatozoa from their epididymal maturation to their capacitation in the female reproductive tract. They are also involved as intermediates in cell fusion [82]. Lipid peroxidation of sperm plasma membranes is considered to be the key mechanism of ROS-induced sperm damage leading to infertility. As illustrated in Fig. 22.2, oxidative stress affects spermatozoa function by inducing apoptosis, membrane lipid peroxidation, and DNA fragmentation.

DNA Damage

Two factors protect the sperm DNA from an oxidative insult: (1) the characteristic tight packaging of the DNA and (2) the antioxidants present in the seminal plasma [83]. Studies in which spermatozoa was exposed to artificially produced ROS resulted in a significant increase in DNA damage in the form of modification of all bases, production of base-free sites, deletions, frameshifts, DNA cross-links, and chromosomal arrangements [84]. Oxidative stress has also been

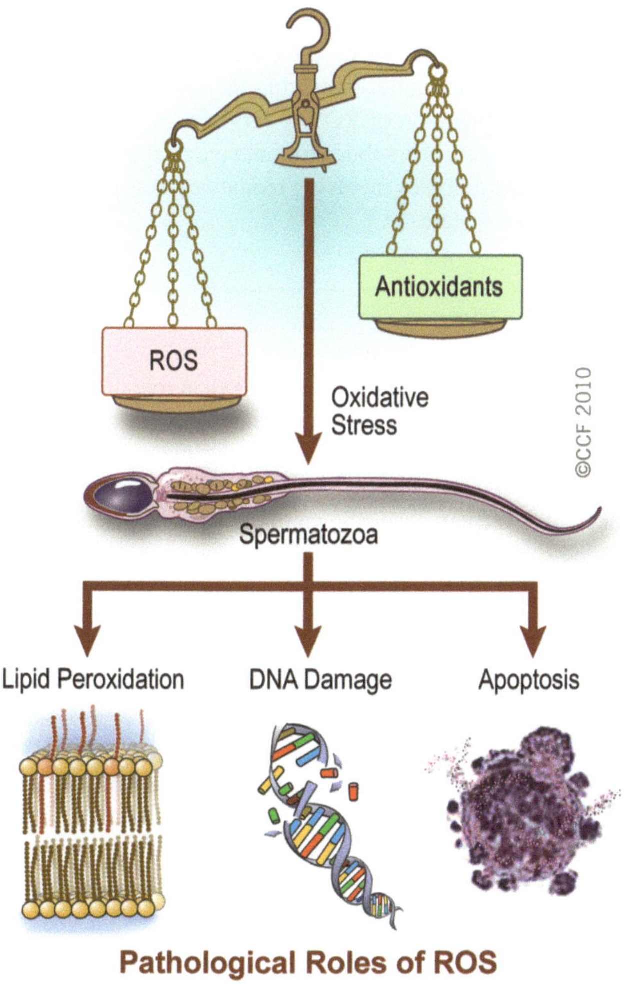

Fig. 22.2 An imbalance between reactive oxygen species (ROS) and antioxidants can lead to the development of oxidative stress which can have detrimental effects on human spermatozoa and sperm function

correlated with high frequencies of single and double DNA strand breaks [83].

Apoptosis

Apoptosis is a process of programmed cell death. It is a physiological phenomenon characterized by cellular morphological and biochemical alterations that cause a cell to commit suicide [85]. It is genetically determined and takes place to help discard cells that have an altered function or no function at all [86]. In the male reproductive system, apoptosis may be responsible for controlling the overproduction of male gametes [87]. Apoptosis appears to be strictly regulated by extrinsic and intrinsic factors and can be triggered by a wide variety of stimuli. Examples of extrinsic stimuli that are potentially important in testicular apoptosis are irradiation, chemotherapy, and toxin exposure [88]. Apoptosis-inducing genes such as p53, Bax, and Fas and apoptosis-suppressing genes such as Bcl-2 and c-kit play a prominent role in the genetic control of apoptosis [89]. Spontaneous germ cell apoptosis has been shown in spermatogonia, spermatocytes, and spermatids in the testis of normal men and in patients with nonobstructive azoospermia [90]. Ejaculated spermatozoa have also been shown to demonstrate changes consistent with apoptosis [88]. It has been shown that the levels of apoptosis in mature spermatozoa were significantly correlated with levels of seminal ROS [90]. It was also found that caspase 3 and caspase 9 levels, from infertile patients, were significantly higher in ejaculated spermatozoa when compared to normal healthy sperm donors. In addition, levels of seminal ROS were positively correlated with levels of caspase 3 and caspase 9. This caspase gene family encodes a set of proteases responsible for carrying out programmed cell death [90].

Lipid Peroxidation

Peroxidation of PUFAs in sperm cell membranes is an autocatalytic, self-propagating reaction [91] which can give rise to cell dysfunction associated with loss of membrane function and integrity. It is divided into two steps: initiation and propagation [92]. Initiation is the removal of the hydrogen atom from an unsaturated fatty acid. The second step, propagation, is the formation of a lipid alkyl radical followed by its rapid reaction with oxygen to form a lipid peroxyl radical. The peroxyl radical is capable of removing a hydrogen atom from an unsaturated fatty acid resulting in the formation of a lipid radical and lipid hydroperoxide [91]. Since the alkyl and peroxyl radicals are regenerated, the cycle of propagation could continue indefinitely or end when one of the substrates is consumed or terminated in the radical–radical reaction.

Preventive Measures, Antioxidants, and the Clinical Relevance Thereof in Male Infertility

There are a lot of preventive measures that people can take to minimize the detrimental effects of oxidative stress on the reproductive function. Lifestyle behaviors such as smoking, poor diet, alcohol abuse, obesity, or psychological stress have been implicated in increasing OS. Minimizing lifestyle triggers of oxidative stress mentioned above is therefore one way of preventing or minimizing the harmful effects of oxidative stress on the reproductive function.

It is well established that exposure to heat, pollution, and toxins (heavy metals and organic solvents) leads to an increase in OS. Avoiding activities which may heat the scrotum such as long hot water baths and saunas could help to minimize OS. Use of proper protective equipment at work places that reduces men's exposure to chemicals and vapors linked with oxidative stress is also one way of minimizing OS.

Antioxidants are the main defense mechanisms against oxidative stress induced by free radicals. Antioxidants can be preventive or scavenging. Preventive antioxidants such as metal chelators and metal-binding proteins block formation of new free radicals, whereas scavenging antioxidants remove free radicals that are already formed. Metal chelators such as transferrin, lactoferrin, and ceruloplasmin which are found in sperm plasma membrane protect spermatozoa from lipid peroxidation caused by transition metals such as iron [93]. In vitro supplementation of metal chelators such as DL-penicillamine, 2,3-dimercaptopropan-1 sulfonate and meso-2,3-dimercapto-succinimic acid showed enhancement of sperm quality during assisted reproductive technique [94].

Dietary antioxidants form an essential part of the human antioxidant defense system. These antioxidants are obtained from fruits and vegetables as well as daily dietary supplements. Oxidative stress could also be limited by using chain-breaking antioxidants such as vitamin E and vitamin C taken as drug supplements. Vitamin C has been reported to be a very potent chain-breaking antioxidant and is present in the extracellular fluid. It is capable of neutralizing hydroxyl, superoxide, and H_2O_2 radicals [95].

Vitamin E is also a chain-breaking antioxidant present within the cell membrane. It neutralizes H_2O_2 and protects the plasma membrane from lipid peroxidation. Studies have shown that vitamin E treatment directly reduced seminal ROS levels [96]. Carotenoids such as beta-carotene and lycopene have also been reported to be very potent antioxidants [97]. Beta-carotene protects the plasma membrane against lipid peroxidation. Lycopene is found in abundance in tomatoes and has been shown to be twice as potent as beta-carotene and ten times more potent than vitamin E in scavenging singlet oxygen and inhibiting lipid peroxidation in serum plasma [98].

Expert Commentary

Oxygen toxicity is an inherent challenge to cells which live under aerobic conditions including the spermatozoa. OS can be defined as the imbalance between prooxidative and antioxidative molecules in a biological system which arises as a consequence of excessive production of free radicals and impaired antioxidant defense mechanisms. The increase in oxidative damage to sperm membranes, proteins, and DNA is associated with defective sperm function. A variety of defensive mechanisms encompassing antioxidant enzymes are involved in biological systems. A balance between the benefits and risks from free radicals and antioxidants appears to be necessary for the survival and normal functioning of spermatozoa. The use of antioxidant supplements has been greatly advocated. However, the positive effects of antioxidants are still debatable. Resolving the various factors that lead to the generation of ROS is very crucial because such data will help design methods for both the treatment and prevention of pathologies involving oxidative stress in the male reproductive system.

Five-Year View

An expanding body of evidence supports a role of oxidative stress as a significant cause of male infertility. Treatment that would minimize oxidative stress in the male reproductive system is therefore highly recommended. Antioxidant supplements have now been shown in randomized placebo controlled studies to protect sperm from oxidative-related DNA damage. It may, therefore, be the right time to consider using antioxidants in all infertile men exhibiting OS. Most importantly, these antioxidants should be offered in combination with changes in lifestyle such as avoiding toxins (cigarette smoke, pollutants, heavy metals) and excessive heat.

Key Issues

- Oxidative stress occurs when antioxidant levels are overwhelmed by free radicals.
- Oxidative stress is one of the major causes of male infertility.
- Examples of free radicals that contribute to oxidative stress in human reproduction are ROS (e.g., superoxide, H_2O_2, hydroxyl radicals, etc.) and RNS (e.g., nitric oxide, peroxynitrite, etc.).
- ROS are beneficial to spermatozoa function at low physiological concentrations.
- High ROS levels cause sperm DNA damage, apoptosis, and lipid peroxidation.
- Antioxidants protect spermatozoa from OS-induced insult.

References

1. Jones R, Mann T, Sherins R. Peroxidative breakdown of phospholipids in human spermatozoa, spermicidal properties of fatty acid peroxides, and protective action of seminal plasma. Fertil Steril. 1979;31(5):531–7.
2. McLachlan RI, de Kretser DM. Male infertility: the case for continued research. Med J Aust. 2001;174(3):116–7.
3. Tremellen K. Oxidative stress and male infertility—a clinical perspective. Hum Reprod Update. 2008;14(3):243–58.
4. Agarwal A, Prabakaran S, Allamaneni S. What an andrologist/urologist should know about free radicals and why. Urology. 2006;67(1):2–8.

5. Garrido N, Meseguer M, Simon C, Pellicer A, Remohi J. Pro-oxidative and anti-oxidative imbalance in human semen and its relation with male fertility. Asian J Androl. 2004;6(1):59–65.
6. Wang X, Sharma RK, Sikka SC, Thomas Jr AJ, Falcone T, Agarwal A. Oxidative stress is associated with increased apoptosis leading to spermatozoa DNA damage in patients with male factor infertility. Fertil Steril. 2003;80(3):531–5.
7. de Lamirande E, Gagnon C. Human sperm hyperactivation and capacitation as parts of an oxidative process. Free Radic Biol Med. 1993;14(2):157–66.
8. Zini A, de Lamirande E, Gagnon C. Reactive oxygen species in semen of infertile patients: levels of superoxide dismutase- and catalase-like activities in seminal plasma and spermatozoa. Int J Androl. 1993;16(3):183–8.
9. Kobayashi T, Miyazaki T, Natori M, Nozawa S. Protective role of superoxide dismutase in human sperm motility: superoxide dismutase activity and lipid peroxide in human seminal plasma and spermatozoa. Hum Reprod. 1991;6(7):987–91.
10. Vernet P, Aitken RJ, Drevet JR. Antioxidant strategies in the epididymis. Mol Cell Endocrinol. 2004;216(1–2):31–9.
11. Williams AC, Ford WC. Functional significance of the pentose phosphate pathway and glutathione reductase in the antioxidant defenses of human sperm. Biol Reprod. 2004;71(4):1309–16.
12. Twigg J, Fulton N, Gomez E, Irvine DS, Aitken RJ. Analysis of the impact of intracellular reactive oxygen species generation on the structural and functional integrity of human spermatozoa: lipid peroxidation, DNA fragmentation and effectiveness of antioxidants. Hum Reprod. 1998;13(6):1429–36.
13. Saez F, Motta C, Boucher D, Grizard G. Antioxidant capacity of prostasomes in human semen. Mol Hum Reprod. 1998;4(7):667–72.
14. Ochsendorf FR. Infections in the male genital tract and reactive oxygen species. Hum Reprod Update. 1999;5(5):399–420.
15. Agarwal A, Gupta S, Sharma RK. Role of oxidative stress in female reproduction. Reprod Biol Endocrinol. 2005;3:28.
16. Cheesman MJ, Baer BR, Zheng YM, Gillam EM, Rettie AE. Rabbit CYP4B1 engineered for high-level expression in *Escherichia coli*: ligand stabilization and processing of the N-terminus and heme prosthetic group. Arch Biochem Biophys. 2003;416(1):17–24.
17. Agarwal A, Prabakaran SA. Mechanism, measurement, and prevention of oxidative stress in male reproductive physiology. Indian J Exp Biol. 2005;43(11):963–74.
18. Sharma RK, Pasqualotto AE, Nelson DR, Thomas Jr AJ, Agarwal A. Relationship between seminal white blood cell counts and oxidative stress in men treated at an infertility clinic. J Androl. 2001;22(4):575–83.
19. Rodin DM, Larone D, Goldstein M. Relationship between semen cultures, leukospermia, and semen analysis in men undergoing fertility evaluation. Fertil Steril. 2003;79 Suppl 3:1555–8.
20. Martinez P, Proverbio F, Camejo MI. Sperm lipid peroxidation and pro-inflammatory cytokines. Asian J Androl. 2007;9(1):102–7.
21. Nandipati KC, Pasqualotto FF, Thomas Jr AJ, Agarwal A. Relationship of interleukin-6 with semen characteristics and oxidative stress in vasectomy reversal patients. Andrologia. 2005;37(4):131–4.
22. Koch OR, Pani G, Borrello S, et al. Oxidative stress and antioxidant defenses in ethanol-induced cell injury. Mol Aspects Med. 2004;25(1–2):191–8.
23. Saleh RA, Agarwal A, Sharma RK, Nelson DR, Thomas Jr AJ. Effect of cigarette smoking on levels of seminal oxidative stress in infertile men: a prospective study. Fertil Steril. 2002;78(3):491–9.
24. Mostafa T, Tawadrous G, Roaia MM, Amer MK, Kader RA, Aziz A. Effect of smoking on seminal plasma ascorbic acid in infertile and fertile males. Andrologia. 2006;38(6):221–4.
25. Song GJ, Norkus EP, Lewis V. Relationship between seminal ascorbic acid and sperm DNA integrity in infertile men. Int J Androl. 2006;29(6):569–75.
26. Peake JM, Suzuki K, Coombes JS. The influence of antioxidant supplementation on markers of inflammation and the relationship to oxidative stress after exercise. J Nutr Biochem. 2007;18(6):357–71.
27. Singer G, Granger DN. Inflammatory responses underlying the microvascular dysfunction associated with obesity and insulin resistance. Microcirculation. 2007;14(4–5):375–87.
28. Perez-Crespo M, Pintado B, Gutierrez-Adan A. Scrotal heat stress effects on sperm viability, sperm DNA integrity, and the offspring sex ratio in mice. Mol Reprod Dev. 2008;75(1):40–7.
29. Junqueira VB, Barros SB, Chan SS, et al. Aging and oxidative stress. Mol Aspects Med. 2004;25(1–2):5–16.
30. Eskiocak S, Gozen AS, Taskiran A, Kilic AS, Eskiocak M, Gulen S. Effect of psychological stress on the L-arginine-nitric oxide pathway and semen quality. Braz J Med Biol Res. 2006;39(5):581–8.
31. Hauser R, Meeker JD, Singh NP, et al. DNA damage in human sperm is related to urinary levels of phthalate monoester and oxidative metabolites. Hum Reprod. 2007;22(3):688–95.
32. Meng Z, Bai W. Oxidation damage of sulfur dioxide on testicles of mice. Environ Res. 2004;96(3):298–304.
33. Latchoumycandane C, Mathur PP. Induction of oxidative stress in the rat testis after short-term exposure to the organochlorine pesticide methoxychlor. Arch Toxicol. 2002;76(12):692–8.
34. Latchoumycandane C, Chitra KC, Mathur PP. 2, 3, 7, 8-tetrachlorodibenzo-p-dioxin (TCDD) induces oxidative stress in the epididymis and epididymal sperm of adult rats. Arch Toxicol. 2003;77(5):280–4.
35. Alaghmand M, Blough NV. Source-dependent variation in hydroxyl radical production by airborne particulate matter. Environ Sci Technol. 2007;41(7):2364–70.
36. Acharya UR, Acharya S, Mishra M. Lead acetate induced cytotoxicity in male germinal cells of Swiss mice. Ind Health. 2003;41(3):291–4.
37. Schaeffer AJ. Epidemiology and demographics of prostatitis. Andrologia. 2003;35(5):252–7.
38. Fraczek M, Sanocka D, Kamieniczna M, Kurpisz M. Proinflammatory cytokines as an intermediate factor enhancing lipid sperm membrane peroxidation in in vitro conditions. J Androl. 2008;29(1):85–92.
39. Brackett NL, Ibrahim E, Grotas JA, Aballa TC, Lynne CM. Higher sperm DNA damage in semen from men with spinal cord injuries compared with controls. J Androl. 2008;29(1):93–9. discussion 100-101.
40. Krause W, Bohring C, Gueth A, Horster S, Krisp A, Skrzypek J. Cellular and biochemical markers in semen indicating male accessory gland inflammation. Andrologia. 2003;35(5):279–82.
41. Umapathy E, Simbini T, Chipata T, Mbizvo M. Sperm characteristics and accessory sex gland functions in HIV-infected men. Arch Androl. 2001;46(2):153–8.
42. Vicari E, Arcoria D, Di Mauro C, Noto R, Noto Z, La Vignera S. Sperm output in patients with primary infertility and hepatitis B or C virus; negative influence of HBV infection during concomitant varicocele. Minerva Med. 2006;97(1):65–77.
43. Srinivasan S, Jenita X, Kalaiselvi P, Muthu V, Chandrasekar D, Varalakshmi P. Salubrious effect of vitamin E supplementation on renal stone forming risk factors in urogenital tuberculosis patients. Ren Fail. 2004;26(2):135–40.
44. Guha M, Kumar S, Choubey V, Maity P, Bandyopadhyay U. Apoptosis in liver during malaria: role of oxidative stress and implication of mitochondrial pathway. FASEB J. 2006;20(8):1224–6.
45. Potts JM, Pasqualotto FF. Seminal oxidative stress in patients with chronic prostatitis. Andrologia. 2003;35(5):304–8.
46. Motrich RD, Maccioni M, Riera CM, Rivero VE. Autoimmune prostatitis: state of the art. Scand J Immunol. 2007;66(2–3):217–27.

47. Motrich RD, Maccioni M, Molina R, et al. Presence of INF gamma-secreting lymphocytes specific to prostate antigens in a group of chronic prostatitis patients. Clin Immunol. 2005;116(2):149–57.
48. Filippini A, Riccioli A, Padula F, et al. Control and impairment of immune privilege in the testis and in semen. Hum Reprod Update. 2001;7(5):444–9.
49. Said TM, Agarwal A, Sharma RK, Thomas Jr AJ, Sikka SC. Impact of sperm morphology on DNA damage caused by oxidative stress induced by beta-nicotinamide adenine dinucleotide phosphate. Fertil Steril. 2005;83(1):95–103.
50. Kumanov P, Nandipati K, Tomova A, Agarwal A. Inhibin B is a better marker of spermatogenesis than other hormones in the evaluation of male factor infertility. Fertil Steril. 2006;86(2):332–8.
51. Watson PF. The causes of reduced fertility with cryopreserved semen. Anim Reprod Sci. 2000;60–61:481–92.
52. Zini A, Garrels K, Phang D. Antioxidant activity in the semen of fertile and infertile men. Urology. 2000;55(6):922–6.
53. Sanocka D, Miesel R, Jedrzejczak P, Chelmonska-Soyta AC, Kurpisz M. Effect of reactive oxygen species and the activity of antioxidant systems on human semen; association with male infertility. Int J Androl. 1997;20(5):255–64.
54. Das UB, Mallick M, Debnath JM, Ghosh D. Protective effect of ascorbic acid on cyclophosphamide-induced testicular gametogenic and androgenic disorders in male rats. Asian J Androl. 2002;4(3):201–7.
55. Agarwal A, Said TM. Oxidative stress, DNA damage and apoptosis in male infertility: a clinical approach. BJU Int. 2005;95(4):503–7.
56. Ishikawa T, Kondo Y, Yamaguchi K, Sakamoto Y, Fujisawa M. Effect of varicocelectomy on patients with unobstructive azoospermia and severe oligospermia. BJU Int. 2008;101(2):216–8.
57. Chen SS, Huang WJ, Chang LS, Wei YH. 8-hydroxy-2′-deoxyguanosine in leukocyte DNA of spermatic vein as a biomarker of oxidative stress in patients with varicocele. J Urol. 2004;172(4 Pt 1):1418–21.
58. Smith R, Kaune H, Parodi D, et al. Extent of sperm DNA damage in spermatozoa from men examined for infertility. Relationship with oxidative stress. Rev Med Chil. 2007;135(3):279–86.
59. Filho DW, Torres MA, Bordin AL, Crezcynski-Pasa TB, Boveris A. Spermatic cord torsion, reactive oxygen and nitrogen species and ischemia-reperfusion injury. Mol Aspects Med. 2004;25(1–2):199–210.
60. Oberg BP, McMenamin E, Lucas FL, et al. Increased prevalence of oxidant stress and inflammation in patients with moderate to severe chronic kidney disease. Kidney Int. 2004;65(3):1009–16.
61. Moreno JM, Ruiz MC, Ruiz N, et al. Modulation factors of oxidative status in stable renal transplantation. Transplant Proc. 2005;37(3):1428–30.
62. Danielski M, Ikizler TA, McMonagle E, et al. Linkage of hypoalbuminemia, inflammation, and oxidative stress in patients receiving maintenance hemodialysis therapy. Am J Kidney Dis. 2003;42(2):286–94.
63. Carpino A, Tarantino P, Rago V, De Sanctis V, Siciliano L. Antioxidant capacity in seminal plasma of transfusion-dependent beta-thalassemic patients. Exp Clin Endocrinol Diabetes. 2004;112(3):131–4.
64. Agbaje IM, Rogers DA, McVicar CM, et al. Insulin dependent diabetes mellitus: implications for male reproductive function. Hum Reprod. 2007;22(7):1871–7.
65. Shrilatha B, Muralidhara. Early oxidative stress in testis and epididymal sperm in streptozotocin-induced diabetic mice: its progression and genotoxic consequences. Reprod Toxicol. 2007; 23(4):578–87.
66. Sonmez M, Yuce A, Turk G. The protective effects of melatonin and vitamin E on antioxidant enzyme activities and epididymal sperm characteristics of homocysteine treated male rats. Reprod Toxicol. 2007;23(2):226–31.
67. Aitken RJ, Harkiss D, Knox W, Peterson M, Irvine DS. A novel signal transduction cascade in capacitating human spermatozoa characterized by a redox-regulated cAMP-mediated induction of tyrosine phosphorylation. J Cell Sci. 1998;111:645–56.
68. Ehrenwald E, Parks JE, Foote RH. Bovine oviductal fluid components and their potential role in sperm cholesterol efflux. Mol Reprod Dev. 1990;25:195–204.
69. Andrews JC, Bavister BD. Capacitation of hamster spermatozoa with the divalent cation chelators D-penicillamine, L-histidine and L-cysteine in a protein free culture medium. Gamete Res. 1989;23:159–70.
70. Russel JH, Hale AH, Inbar D, Fisen HN. Loss of reactivity of BAL B/c myeloma tumor with allogeneic and syngeneic cytotoxic T lymphocytes. Eur J Immunol. 1978;8:640–5.
71. Yamaguchi R. Mammalian fertilization. In: Knobil E, Neill J, editors. Physiology of reproduction. New York: Raven; 1994. p. 189–317.
72. Kumar GP, Laloraya MM. Superoxide radical and superoxide dismutase activity changes in maturing mammalian spermatozoa. Andrologia. 1991;23:171–5.
73. Aitken RJ, Buckingham DW, Brindles J, Gomez E, Baker HW, Irvine DS. Analysis of sperm movement in relation to the oxidative stress created by leukocytes in washed sperm preparations and seminal plasma. Hum Reprod. 1995;10:2061–71.
74. Zhang H, Zheng RL. Promotion of human sperm capacitation by superoxide anion. Free Radic Res. 1996;24:261–7.
75. Purohit SB, Laloraya M, Kumar GP. Acrosome reaction inducers impose alterations in repulsive strain and hydration barrier in human sperm membrane. Biochem Mol Bio Int. 1998;45:227–35.
76. de Lamirande E, Gagnon C. Impact of reactive oxygen species on spermatozoa: a balancing act between beneficial and detrimental effects. Hum Reprod. 1995;10:15–21.
77. Griveau JF, Renard P, Lannou D. An *in vitro* promoting role for hydrogen peroxide in human sperm capacitation. Int J Androl. 1994;17:300–7.
78. Lewis SEM, Boyle PM, Mc Kinney KA. Total antioxidant capacity of seminal plasma is different in fertile and infertile men. Fertil Steril. 1996;64:863–9.
79. Herrero MB, Cebral E, Boquet M, Viggiano JM, Vitullo A, Ginenno MA. Effect of nitric oxide on mouse sperm hyperactivation. Acta Physiol Pharmacol Ther Latinoam. 1994;44:65–9.
80. Sengoku K, Tamate K, Yoshida T, Takaoka Y, Miyamoto T, Ishikawa M. Effects of low concentrations of nitric oxide on zona pellucida binding ability of human spermatozoa. Fertil Steril. 1998;69:522–7.
81. Alvarez JG, Storey BT. Differential incorporation of fatty acids into and peroxidative loss of fatty acids from phospholipids of human spermatozoa. Mol Reprod Dev. 1995;42:334–46.
82. Yeagle PL. Lipids and lipid-intermediate structures in the fusion of biological membranes. Curr Top Membr. 1994;4:197–214.
83. Twigg J, Irvine DS, Aitken RJ. Oxidative damage to DNA in human spermatozoa does not prelude pronucleus formation at intracytoplasmic sperm injection. Hum Reprod. 1998;103:1864–71.
84. Duru NK, Morshedi M, Oehninger S. Effects of hydrogen peroxide on DNA and plasma membrane integrity of human spermatozoa. Fertil Steril. 2000;74:1200–7.
85. Vaux DL, Flavell RA. Apoptosis genes and autoimmunity. Curr Opin Immunol. 2000;12:719–24.
86. Vaux DL, Korsmeyer SJ. Cell death in development. Cell. 1999;96:245–54.
87. Sakkas D, Mariethoz E, Manicardi G, Bizzaro D, Bianchi P, Bianchi U. Origin of DNA damage in ejaculated human spermatozoa. Rev Reprod. 1999;4:31–7.
88. Lee J, Richburg JH, Younkin SC, Boekelbeide K. The Fas system is a key regulator of germ cell apoptosis in the testis. Endocrinology. 1997;138:2081–8.
89. Sinha HAP, Swerdloff RS. Hormonal and genetic control of germ cell apoptosis in the testis. Rev Reprod. 1999;4:38–47.

90. Agarwal A, Saleh R, Bedaiwy MA. Role of reactive oxygen species in the pathophysiology of human reproduction. Fertil Steril. 2003;79:829–43.
91. Halliwell B. How to characterize a biological antioxidant. Free Radic Res Commun. 1990;9:1–32.
92. Aitken RJ, Fisher H. Reactive oxygen species generation and human spermatozoa: the balance of benefit and risk. Bioassays. 1994;16:259–67.
93. Sanocka D, Kurpisz M. Reactive oxygen species and sperm cells. Reprod Biol Endocrinol. 2004;2:12.
94. Henkel RR, Schill B. Sperm preparation for ART. Reprod Biol Endocrinol. 2003;1:108.
95. Agarwal A, Nallella KP, Allamaneni SS, Said TM. Role of antioxidants in treatment of male infertility: an overview of the literature. Reprod Biomed Online. 2004;8:616–27.
96. Comhaire FH, Christophe AB, Zalata AA, Dhooge WS, Mahmoud AM, Depuydt CE. The effects of combined conventional treatment, oral antioxidants and essential fatty acids on sperm biology in subfertile men. Prostaglandins Leukot Essent Fatty Acids. 2000;63:159–65.
97. Gupta NP, Kumar R. Lycopene therapy in idiopathic male infertility—a preliminary report. Int Urol Nephrol. 2002;34:369–72.
98. Di Mascio P, Kaiser S, Sies H. Lycopene as the most efficient biological carotenoid singlet oxygen quencher. Arch Biochem Biophys. 1989;274:532–8.

Loss of Intracellular Antioxidant Enzyme Activity During Sperm Cryopreservation: Effects on Sperm Function After Thawing

23

Juan G. Alvarez

Abstract

Despite advances in cryopreservation methodology, one of the main detrimental effects of cryopreservation on human spermatozoa is a marked reduction in motility. The primary cause of cellular damage during cryopreservation is the formation of intracellular ice. The concentration of solutes remaining in the unfrozen fraction increases, thereby both depressing the freezing point and increasing the osmotic pressure of the remaining solution. Hence, biological systems freeze progressively over a wide temperature range, during which the solute becomes gradually more concentrated as the temperature falls. This leads to irreversible rupturing of plasma and nuclear membranes and disturbance of cellular organelles. The nucleus has generally been considered to be a stable constituent of the cell. However, recent studies have suggested that this is not the case and that inappropriate chromatin condensation can occur with freezing. Cryoprotectants such as glycerol or propanediol can be added to cells to reduce freezing damage by lowering the salt concentrations and increasing the unfrozen water fraction, thereby reducing osmotic stress. Further cellular damage may be caused during the thawing process as the ice melts or recrystallizes. Slow thawing is most likely to induce injury, as it allows time for consolidation of microscopic ice crystals into larger forms which are known to be damaging. The production and dissolution of ice is associated with the actual rate of freezing and thawing. Slow freezing and gradual dehydration may accommodate cell survival, whereas rapid freezing and thawing is more likely to result in cell death.

Keywords

Sperm cryopreservation • Sperm motility • Lipid peroxidation • Sperm function after thawing • Oxygen radical production • Antioxidant enzymes • Superoxide dismutase • Glutathione peroxidase • DNA fragmentation • DNA oxidation

Despite many advances in cryopreservation methodology [1, 2], one of the main reported detrimental effects of cryopreservation on human spermatozoa is a marked reduction in sperm motility [3–5]. The primary cause of cellular damage during cryopreservation is the formation of intracellular ice [6, 7]. Whenever cells, or culture media, are cooled below their freezing point, water is removed from the solution in the form of ice. The concentration of solutes remaining in the unfrozen fraction increases, thereby both depressing the freezing point [8] and increasing the osmotic pressure of the remaining solution. Hence, biological systems freeze progressively over a wide temperature range, during which the solute becomes gradually more concentrated as the temperature falls [8]. This leads to irreversible rupturing of plasma and nuclear membranes and disturbance of cellular organelles. The nucleus has generally been considered to be

J.G. Alvarez, MD, PhD (✉)
Centro de Infertilidad Masculina Androgen, C/Fernando Macias, 8, 1C, La Coruña 15005, Spain

Harvard Medical School, Boston
e-mail: jalvarez@androgen.es

S.J. Parekattil and A. Agarwal (eds.), *Male Infertility: Contemporary Clinical Approaches, Andrology, ART & Antioxidants*,
DOI 10.1007/978-1-4614-3335-4_23,

a stable constituent of the cell. However, recent studies have suggested that this is not the case and that inappropriate chromatin condensation can occur [9, 10] with freezing. Cryoprotectants such as glycerol or propanediol can be added to cells to reduce freezing damage by lowering the salt concentrations and increasing the unfrozen water fraction, thereby reducing osmotic stress. They can also insert into phospholipid membranes to reduce the likelihood of fracture [7]. Further cellular damage may be caused during the thawing process as the ice melts or recrystallizes. Slow thawing is most likely to induce injury, as it allows time for consolidation of microscopic ice crystals into larger forms which are known to be damaging [11]. The production and dissolution of ice is associated with the actual rate of freezing and thawing. Slow freezing and gradual dehydration may accommodate cell survival, whereas rapid freezing and thawing is more likely to result in cell death [6].

Increase in Oxygen Radical Production After Cryopreservation

It has been previously reported that the production of oxygen radicals, and of superoxide anion in particular, increases during thawing of cryopreserved ejaculated sperm [12]. According to these authors, motility loss during thawing of cryopreserved sperm appears to be related, at least in part, to oxidation of reduced glutathione [12]. Recently, several groups have reported the effect of cryopreservation on sperm DNA integrity [13–17]. It has been shown that during cryopreservation, and more specifically during thawing, sperm are exposed to high levels of oxygen radicals [12, 18] due, at least in part, to a slow recovery of antioxidant enzyme activity after freezing [19]. This may help explain why membrane damage occurs during thawing [18].

Ejaculated sperm are composed of discrete subsets of spermatozoa, which can be isolated by density-gradient centrifugation. It has been previously shown that these subsets differ in their degree of maturation [20]. Cells isolated in the lowest density layer include immature germ cells, leukocytes, and immature or defective sperm with proximal cytoplasmic retention and abnormal head morphology. Spermatozoa isolated in this fraction had the highest content of docosahexaenoic acid (DHA) and sterols [20, 21] and produced the highest levels of reactive oxygen species (ROS) [21, 22]. ROS production was highest in immature sperm from males displaying abnormal semen parameters [22]. The fact that immature and defective sperm produce the highest levels of ROS and have the highest content of DHA, the primary target of ROS-induced lipoperoxidative damage explains, at least in part, the high levels of malondialdehyde observed in spermatozoa from teratozoospermic samples [23].

However, peroxidative damage in sperm not only depends on ROS production and DHA levels but also on sperm antioxidant defenses. Mammalian sperm have been reported to contain antioxidant enzyme defenses against ROS-induced damage [24–26]. These defenses include superoxide dismutase (SOD), glutathione peroxidase/reductase, and low-molecular substances [27, 28]. SOD, in particular, and glutathione peroxidase/reductase activity have been reported in sperm of several mammalian species including human [24–26, 29–31]. SOD may be the most important antioxidant factor in semen [32], and its stoichiometric relationship to DHA content, the main substrate of lipid peroxidation in sperm, will ultimately determine the susceptibility of a given sperm sample to oxidative damage. High levels of this enzyme have been associated with impaired sperm function [33]. In addition, it has been previously reported that cryopreservation increases the rate of lipid peroxidation and that this might be related, at least in part, to the loss of antioxidant enzyme activity [34]. Therefore, the increase in oxygen radical production after cryopreservation could be due to two main mechanisms: (a) a decrease in antioxidant activity, mainly in antioxidant enzyme activity or (b) an increase in the actual generation of oxygen radicals. Since oxygen radical export to the extracellular medium by mature spermatozoa is very low and relatively constant between samples from different males [20, 22] and intracellular oxygen radical production, defined as V_{int} [25, 35], is metabolized to a large extent by enzymes such as glutathione peroxidase and SOD, we can safely conclude that oxygen radical production by mature spermatozoa at constant temperature and oxygen concentration mainly depends on antioxidant enzyme activity.

Loss of Intracellular Enzyme Antioxidant Activity

It has been recently reported that motility recovery after cryopreservation can be increased by increasing thawing temperature. Sperm thawing at 40°C resulted in a 23.1% increase in sperm motility recovery compared to 37°C [19]. This finding was observed after cryopreservation of either liquefied raw semen samples or their corresponding 90% gradient fractions, known to contain spermatozoa of the highest functional quality [36]. Therefore, based on the above, it has been proposed that the increase in oxygen radical production observed after thawing would be due to damage of sperm SOD leading to partial loss of its enzyme activity. If that were the case, we should observe an increase in membrane and DNA damage after thawing. One potential explanation for the observed increase in motility recovery following thawing at 40°C would be a faster rate of recovery of sperm enzyme antioxidant activity. During thawing, there are two competing processes that determine the degree of cell damage: (a) the magnitude of oxygen radical production and (b) the rate of recovery of enzyme antioxidant activity. The higher the temperature, the faster the recovery of enzyme antioxidant activity and, therefore, at 40°C sperm will be able to neutralize more efficiently than at 37°C the increase in oxygen radical production reported during thawing [12]. However, since sperm viability, acrosomal status, DNA integrity, and sperm survival were similar at 37°C vs. 40°C, the type of cell damage prevented by a faster rate of recovery of antioxidant enzyme activity might be related to ATP production/utilization in the axoneme. That is, rather than affecting the recovery of enzyme antioxidant activity at the level of the membrane and/or nuclear compartments, it would affect antioxidant enzyme activity localized in the highly compact axonemal compartment, virtually devoid of cytoplasm. Although there were no statistically significant differences in ATP content in sperm from the 90% gradient pellet following thawing at 37°C vs. 40°C, a higher ATP content was found in samples with a motility recovery higher than 50%. Since ATP steady-state levels represent the balance between biosynthesis and consumption, it cannot be ruled out that thawing at 40°C might have an effect in increasing the ATP utilization in the axoneme by preventing more efficiently oxygen radical-induced damage to key components of the axoneme. This damage could be reversible in the first stage and become irreversible thereafter [25]. However, since long-term motility, as measured by the sperm survival test, acrosomal integrity, and DNA integrity, was not significantly different following thawing at 40°C vs. 37°C, this suggests that the beneficial effect of increasing thawing temperature only affects antioxidant enzyme activity from the axonemal compartment and not from the membrane or nuclear compartments. Therefore, the results of the study of Calamera et al., related to the effects of thawing of cryopreserved human sperm at 40°C compared to other temperatures, suggest that cryopreservation results in the loss of antioxidant enzyme activity. This damage could be reversed, at least in part, in the axonemal compartment by increasing thawing temperature. The fact that motility loss, as measured by the sperm survival test (an indicator of damage to the membrane compartment), and DNA integrity, as measured by the TUNEL test (an indicator of damage to the nuclear compartment), were not significantly different after thawing at 40°C vs. 37°C, suggests that perhaps the recovery of enzyme antioxidant activity in the axonemal compartment after thawing is slower compared to the membrane and nuclear compartments and that damage to antioxidant enzymes in the axoneme may actually occur during thawing and is mediated by oxygen radicals [19]. Previous studies have already reported the occurrence of SOD inactivation by oxygen radicals [24, 25].

Enzyme Antioxidant Activity, Rate of Lipid Peroxidation, and Sperm Damage

There is general agreement that motility loss during aerobic incubation at 37°C is significantly higher in fresh compared to cryopreserved spermatozoa. Since antioxidant enzymes, namely SOD, play a central role in protecting mammalian sperm against lipid peroxidation and motility loss during aerobic incubation, we can safely conclude that cryopreservation results in the loss of antioxidant enzyme activity. Loss of sperm motility due to lipid peroxidation during incubation becomes apparent after several hours of incubation and is mainly determined by the activation energy of lipid peroxidation reactions which requires of several hours to result in phospholipid breakdown, loss of membrane permeability, and loss of ATP leading to motility loss [25]. Assuming that mature sperm have an equivalent content of DHA and produce similar levels of oxygen radicals, the constitutive content of antioxidant enzyme activity in sperm is going to determine the rate of lipid peroxidation and the motile life span of sperm. Thus, sperm samples with a higher content of antioxidant enzyme activity would have a higher motile life span than samples with a lower content of antioxidant enzyme activity. At the same time, since motility loss of a given sperm sample does not take place in all sperm at the same time but it occurs in a stepwise fashion, it can also be concluded that, under similar conditions of DHA content and oxygen radical production, the motile life span of each individual sperm cell would be mainly determined by its antioxidant enzyme activity. At the same time, since membrane damage induced by oxygen radicals depends, for the most part, on the activation energy of lipid peroxidation which requires several hours of incubation at 37°C to be observed via motility loss, it can be postulated that oxygen radical-induced DNA damage occurring after cryopreservation should be observed earlier since the activation energy of DNA damage is much lower than that of lipid peroxidation. And this is precisely what it is observed in vitro after sperm cryopreservation. Massive DNA damage in cryopreserved mature sperm from the 80% gradient pellet can be observed as early as 2.5 h of incubation at 37°C after thawing [37], while 80% motility loss is not observed until about 24 h of aerobic incubation [19, 37]. As shown in Fig. 23.1, DNA damage after 2.5 h incubation at 37°C of thawed mature sperm from the 80% pellet isolated from samples obtained from fertile sperm donors reaches DFI values of 70–80%. Interestingly, in the sample from donor 3, the DFI value observed after 2.5 h of incubation was 20%, a value significantly lower than that observed in the other three samples. The same applies to the cryopreserved raw semen sample aliquot of sample 4. This certainly may be related to higher levels of antioxidant enzyme activity. Concerning the mechanism

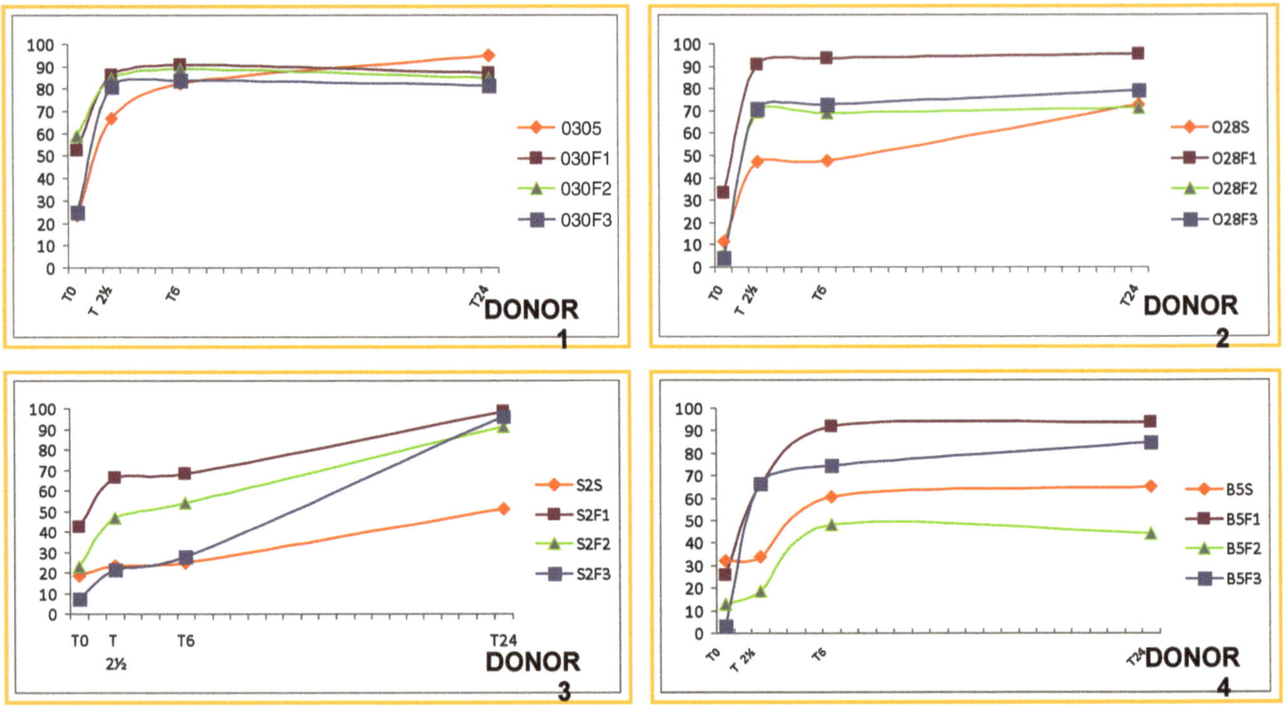

Fig. 23.1 Dynamics of sperm DNA fragmentation following incubation of subsets of cryopreserved human spermatozoa. A total of four semen samples were obtained from four fertile sperm donors and processed for density-gradient centrifugation using a 40/80% SpermFilter® gradient (Cryos International, Denmark). Following centrifugation, sperm from the seminal plasma/40% (F1) and 40/80% (F2) interfaces and from the 80% pellet (F3), were aspirated, transferred to test tubes, and washed with SpermWash® medium (Cryos International, Denmark). Aliquots of the resulting fractions and of the raw semen sample (S) were mixed 1:1, v/v with CryoProtec medium (Nidacon, Sweden), placed under liquid nitrogen vapors for 15 min, and finally immersed in liquid nitrogen. After cryopreservation, aliquots of the samples were thawed at 37°C for 5 min and incubated during 2.5, 6, and 24 h at 37°C. The percentage of sperm with DNA damage (DFI) was assessed using the Halosperm test (Halotech-DNA, Madrid) and the type of DNA strand damage using the two-dimensional COMET test

of DNA damage as shown in Fig. 23.2, all the DNA damage observed during incubation of thawed spermatozoa is oxidative in nature. All spermatozoa expressing DNA fragmentation, as measured by the SCD-Halosperm test (left panels), are positive for nucleotide damage of the 8-oxoguanine type (right panels) and indicator of oxidative damage. The same applies to the samples from the other donors. If we ask the question: why the remaining 20–30% spermatozoa still maintain intact their DNA after 2.5 h of incubation? Are these sperm better equipped with antioxidant enzyme activity? This is a question that is certainly amenable to experimental testing by looking at DNA integrity and SOD and/or GPx activity in the same sperm cell. Sperm cells with intact DNA should have a higher content and activity of SOD and/or GPx than sperm cells with damaged DNA.

Another important conclusion that can be drawn from the results of the study of the effects of cryopreservation of subsets of human sperm on DNA integrity is that related to the occurrence of intracellular vs. extracellular oxygen radical-mediated DNA damage during thawing. As shown in Fig. 23.1, the DFI value in semen from donor 2 is significantly lower than that observed in sperm from fraction 3. Since antioxidant enzymes in the extracellular medium, i.e., seminal plasma, do not have access to the intracellular compartment, it can be concluded that the DNA damage observed in semen depends mainly on the levels of oxygen radicals exported to the extracellular medium by immature sperm, while the DNA damage observed in mature sperm from fraction 3 depends mainly on the intracellular levels of oxygen radicals that normally would be metabolized by antioxidant enzymes. If constitutive intracellular SOD and/or GPx content is lower, the steady-state levels of oxygen radicals would be higher and also DNA damage. Antioxidant enzyme activity in seminal plasma would suppress oxygen radical-mediated DNA damage going from the extracellular to the intracellular compartment, while intracellular antioxidant enzyme activity in washed or mature sperm (devoid of seminal plasma) would suppress DNA damage mediated by intracellular oxygen radical levels. Therefore, the protective effect of semen is most likely determined by the presence of antioxidant enzyme activity in seminal plasma, e.g., SOD and GPx. The contribution of low-molecular-weight antioxidants in seminal plasma to this protective effect can be largely excluded, since these antioxidants are membrane permeable and would also have access to the intracellular compartment [27] thus preventing intracellular oxidative damage. Therefore, antioxidant enzyme activity in seminal plasma

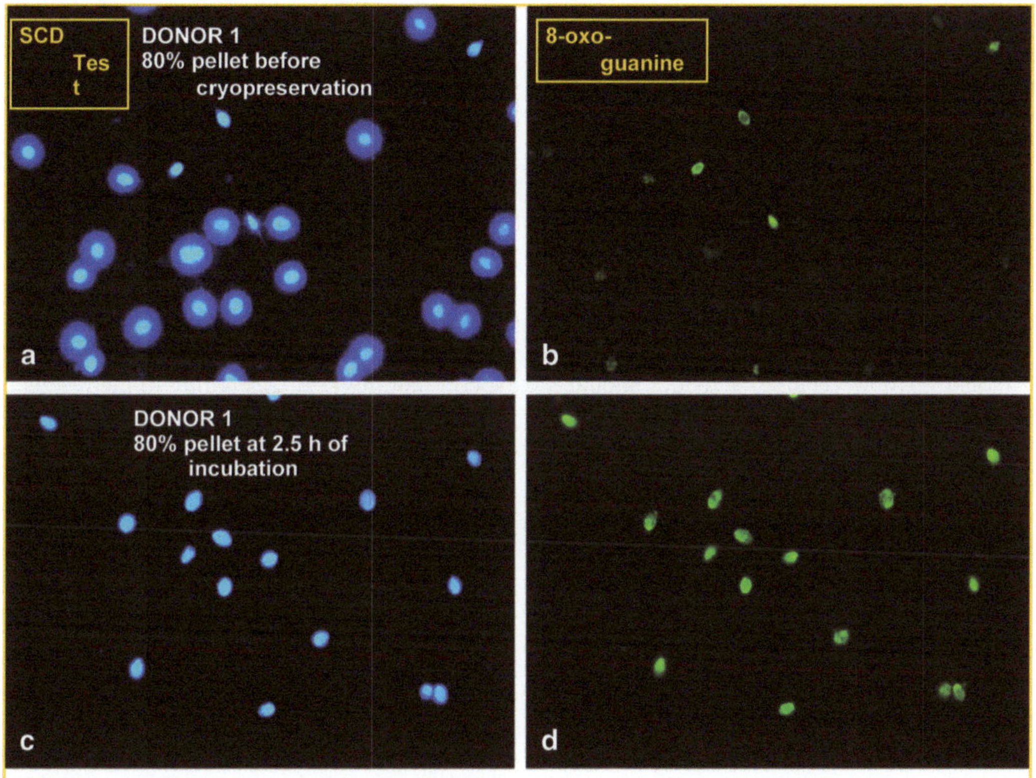

Fig. 23.2 Correlation between DNA fragmentation and DNA oxidation during incubation of cryopreserved sperm after thawing. Aliquots of testicular and ejaculated sperm obtained from the same patient were analyzed for sperm DNA fragmentation and 8-oxoguanine content using the Halosperm test (Halotech-DNA, Madrid) and the OxyDNA assay kit (Calbiochem, Barcelona, Spain), respectively

from samples 2 and 3 should be higher than that in sample 1. Since the dynamics of motility loss during incubation of cryopreserved sperm after thawing should mimic the dynamics of sperm DNA damage, although with a lower rate, this model would also predict that motility recovery following cryopreservation of raw semen may not always correlate with motility recovery after cryopreservation of mature sperm from the 80% gradient pellet obtained from the same semen sample, e.g., IUI-ready samples. That is, some samples may have higher motility recovery and lower DNA damage after semen than after cryopreservation of mature sperm from the gradient pellet and vice versa.

Therefore, sperm motility recovery, acrosomal status, and DNA integrity after cryopreservation are going to be mainly determined by the sperm content and activity of antioxidant enzymes. Sperm with a higher overall level of antioxidant enzyme activity would have a higher probability of maintaining the plasma and acrosomal membranes and the DNA intact and have a higher fertilizing ability. With regard to the sperm fertilizing ability, it should be pointed out that since sperm capacitation at 37°C may take up to 2–3 h and that sperm penetration in conventional IVF may take up to 1 h, based on these preliminary results, oocyte insemination by conventional IVF should not be recommended when using cryopreserved sperm since incubation of sperm at 37°C during 3–4 h may result in extensive DNA damage. ICSI should be used in these cases.

Sperm Enzyme Antioxidant Activity and Donor Selection

Based on the above, it can be concluded that the constitutive antioxidant enzyme content of spermatozoa should determine the motility recovery and the acrosomal and DNA integrity of cryopreserved human spermatozoa after thawing. Previous studies have shown that SOD [24] and GPx [26] play a central role in protecting mammalian spermatozoa against motility loss caused by spontaneous lipid peroxidation. SOD or GPx activity in ejaculated spermatozoa, mainly in mature sperm, should be a useful marker of sperm resistance to cryopreservation, for the selection of donor sperm samples that better withstand cryopreservation and, ultimately, for the selection of sperm to be used in in vitro fertilization. Should SOD and GPx from the paternal genome play a relevant role in embryo viability, it would be of paramount importance in preventing oxidative stress in the embryo. It is well known the beneficial effect of the use of low oxygen tension during embryo culture in preventing embryo fragmentation and improving embryo viability during long culture and IVF outcome [38–40]. It has been reported that low SOD activity in the embryo is associated with neural tube defects under high steady-state levels of oxidative stress [41]. Therefore, SOD and GPx activity may not only be important in donor selection for donor sperm cryopreservation but also for sperm selection for ICSI. Selection and microinjection of spermatozoa with higher SOD activity may result in higher embryo quality. However, SOD and GPx (and possibly catalase) activities should be stoichiometrically balanced to have a net antioxidant effect. That is, if SOD activity is exceedingly high compared to GPx activity, the high levels of H_2O_2 produced by the dismutation of the superoxide anion by SOD may not be hydrolyzed by GPx resulting in oxidative damage [26]. This explains why oxidative stress in immature spermatozoa from the lower-density layers of the gradient is higher than that in mature sperm from the 90% gradient pellet, because SOD levels (and also DHA content) are much higher in mature sperm [22]. Whether SOD and/or GPx content in spermatozoa is correlated with SOD and/or GPx content in somatic cells remains to be investigated.

Envoy

Sperm function and viability after cryopreservation appears to be determined to a great extent by the preservation of the activity of their antioxidant enzyme defenses. SOD and glutathione peroxidase are the main antioxidant enzymes present in mammalian sperm and play a central role in protecting sperm against oxidative damage. It has been shown that the motile life span and fertilizing ability of sperm in in vivo and in vitro is mainly determined by the rate of lipid peroxidation. This rate is governed by oxygen concentration in the extracellular medium, membrane-bound DHA content, and intracellular antioxidant enzyme activity. Therefore, under physiological conditions, the rate-limiting factor that determines the life span and function of sperm is the antioxidant enzyme activity. This activity not only may vary between samples from different males or even within a given male but also between individual sperm cells in a given sample. The in vitro and in vivo motile life span of mature sperm within a sample is not the same. Motility loss takes place in a stepwise fashion. That is, some sperm cells lose their motility before others and vice versa, and some sperm cells maintain their motility longer than others. Since, under these conditions, oxygen concentration and sperm DHA content are comparable [20, 21], the rate-limiting factor would be intracellular antioxidant enzyme activity. Therefore, antioxidant enzyme activity in a given sample, ΦSOD_T, could be considered the sum of the individual antioxidant enzyme activity of each sperm cell: $\Phi SOD_T = \Sigma[(\Phi SOD_1) + (\Phi SOD_2) + (\Phi SOD_3) + \ldots + (\Phi SOD_n)]$. There is evidence that during cryopreservation antioxidant enzymes may be damaged. This may be related to structural damage during the freezing process due to damage during thawing or both. Physical damage during freezing is considered irreversible and, therefore, not recoverable, while damage during thawing may be, at least in part, reversible. These are the conclusions that can be drawn from the effect of thawing temperature on sperm motility [19]. According to the results of this study, motility recovery following thawing at 40°C is significantly higher than at 37°C. Given the high levels of oxygen radicals produced intracellularly by sperm cells, defined as V_{int} [25, 30], which are normally metabolized by antioxidant enzymes, should oxygen radical production be recovered at a faster rate than antioxidant enzyme activity, during thawing the sperm cell would be exposed to supraphysiological levels of oxygen radicals leading to sperm damage. It is postulated that thawing at 40°C accelerates antioxidant enzyme activity recovery and, therefore, reduces oxidative injury to the sperm cell. However, the higher recovery of sperm motility observed after thawing at 40°C is not related to a reduction in membrane damage. Sperm viability at 40°C was identical to that observed at 37°C [19]. This is why in the study of Calamera et al., it was concluded that the reversible effect of motility loss observed was related to the production/utilization of ATP in the axoneme [19]. Should antioxidant enzyme activity in the axonemal compartment be recovered at a slower rate than that associated with the membrane compartment, the resulting damage should be preferentially observed in those axonemal components responsible for the production/utilization of ATP, a key factor that determines sperm motility. Once these components are damaged, motility loss can no longer be recovered. The main axonemal components that may be vulnerable to oxidative damage during thawing would be glycolytic enzymes, ATPase, and the microtubules. On the other hand, damage of antioxidant enzymes during thawing will have a delayed effect on membrane damage since the activation energy of lipid peroxidation reactions and phospholipid breakdown requires several hours for membrane damage to occur leading to ATP and motility loss. This explains, at least in part, why long-term motility following incubation is much higher in fresh than in cryopreserved sperm. Finally, since antioxidant enzyme activity may vary between samples, selection of sperm samples with higher antioxidant enzyme activity may result in better motility recovery after cryopreservation. Furthermore, microinjection of spermatozoa with higher antioxidant enzyme activity may result in higher embryo quality.

Key Issues

- The production of oxygen radicals, and of superoxide anion in particular, increases during thawing of cryopreserved ejaculated sperm.
- It has been recently reported that motility recovery after cryopreservation can be increased by increasing thawing temperature.
- There is general agreement that motility loss during aerobic incubation at 37°C is significantly higher in fresh compared to cryopreserved spermatozoa.
- The protective effect of semen is most likely determined by the presence of antioxidant enzyme activity in seminal plasma.
- Sperm motility recovery, acrosomal status, and DNA integrity after cryopreservation are going to be mainly determined by the sperm content and activity of antioxidant enzymes.
- Sperm function and viability after cryopreservation appears to be determined to a great extent by the preservation of the activity of their antioxidant enzyme defenses.

References

1. Centola GM, Raubertas RF, Mattox JH. Cryopreservation of human semen: comparison of cryopreservation, sources of variability, and prediction of post-thaw survival. J Androl. 1992;13:283–8.
2. Agarwal A, Tolentine MV, Sidhu Jr RS, et al. Effect of cryopreservation on semen quality in patients with testicular cancer. Urology. 1995;46:382–6.
3. Critzer JK, Huse-Benda AR, Aaker DV, et al. Cryopreservation of human spermatozoa. III. The effects of cryoprotectants on motility. Fertil Steril. 1988;50:314–20.
4. Englert Y, Delvigne A, Vekemans M, et al. Is fresh or frozen semen to be used in in vitro fertilization with donor sperm. Fertil Steril. 1989;5:661–4.
5. Yoshida H, Hoshiai H, Fukaya T, et al. Fertilizability of fresh and frozen human spermatozoa. Assist Reprod Technol Androl. 1990;1:164–72.
6. Muldrew K, McGann LE. Mechanisms of intracellular ice formation. Bio J. 1990;57:525–33.
7. Watson PF. Recent developments and concepts in cryopreservation of spermatozoa and the assessment of their post-thaw function. Reprod Fertil Dev. 1995;7:871–91.
8. Brotherton J. Cryopreservation of human semen. Arch Androl. 1990;25:181–95.
9. Royere D, Hamamah S, Nicolle JC, et al. Freezing and thawing alter chromatin stability of ejaculated human spermatozoa: fluorescence acridine orange staining and fuelgen DNA cytophotometric studies. Gamete Res. 1988;21:51–7.
10. Royere D, Hamamah S, Nicolle JC, Lansac J. Chromatin alterations induced by freeze-thawing influence the fertilizing ability of human sperm. Int J Androl. 1991;14:328–32.
11. Mazur P, Rall WF, Rigopoulos N. Relative contribution of the fraction of unfrozen water and of salt concentration to the survival of slowly frozen human erythrocytes. Biophys J. 1981;36:653–75.
12. Chatterjee S, Gagnon C. Production of reactive oxygen species by spermatozoa undergoing cooling, freezing, and thawing. Mol Reprod Dev. 2001;59:451–8.
13. Kalthur G, Adiga SK, Upadhya D, Rao S, Kumar P. Effect of cryopreservation on DNA integrity in patients with teratozoospermia. Fertil Steril. 2008;89:1723–7.
14. Zribi N, Chakroun NF, El Euch H, Gargouri J, Bahloul A, Keskes LA. Effects of cryopreservation on human sperm deoxyribonucleic acid integrity. Fertil Steril. 2010;93:159–66.
15. Gandini L, Lombardo F, Lenzi A, Spanò M, Dondero F. Cryopreservation and sperm DNA integrity. Cell Tissue Bank. 2006;7:91–8.
16. Gosalvez J, Cortes-Gutierez E, Lopez-Fernandez C, Fernandez JL, Caballero P, Nunez R. Sperm deoxyribonucleic acid fragmentation dynamics in fertile donors. Fertil Steril. 2009 Jul;92(1):170–3.
17. Toro E, Fernández S, Colomar A, Casanovas A, Alvarez JG, López-Teijón M, Velilla E. Processing of semen can result in increased sperm DNA fragmentation. Fertil Steril. 2009;92:2109–12.
18. Aitken RJ. Free radicals, lipid peroxidation and sperm function. Reprod Fertil Dev. 1995;7:659–68.
19. Calamera JC, Buffone MG, Doncel GF, Brugo-Olmedo S, de Vicentis S, Calamera MM, Storey BT, Alvarez JG. Effect of thawing temperature on the motility recovery of cryopreserved human spermatozoa. Fertil Steril. 2010;93:789–94.
20. Ollero M, Powers RD, Alvarez JG. Variation of docosahexaenoic acid content in subsets of human spermatozoa at different stages of maturation: implications for sperm lipoperoxidative damage. Mol Reprod Dev. 2000;55:326–34.
21. Ollero M, Gil-Guzman E, Lopez MC, Sharma RK, Agarwal A, Larson K, Evenson D, Thomas Jr AJ, Alvarez JG. Characterization of subsets of human spermatozoa at different stages of maturation: implications in the diagnosis and treatment of male infertility. Hum Reprod. 2001;16:1912–21.
22. Gil-Guzman E, Ollero M, Lopez MC, Sharma RK, Alvarez JG, Thomas AJ, Agarwal A. Differential production of reactive oxygen species by subsets of human spermatozoa at different stage of maturation. Hum Reprod. 2001;16:1022–30.
23. Jones R, Mann T, Sherings RJ. Adverse effects of peroxidized lipid on human spermatozoa. Proc R Soc Lond B Biol Sci. 1978; 201:413–7.
24. Alvarez JG, Storey BT. Role of superoxide dismutase in protecting rabbit spermatozoa from O2 toxicity due to lipid peroxidation. Biol Reprod. 1983;28:1129–36.
25. Alvarez JG, Storey BT. Assessment of cell damage caused by spontaneous lipid peroxidation in rabbit spermatozoa. Biol Reprod. 1984;30:323–32.
26. Alvarez JG, Storey BT. Role of glutathione peroxidise in protecting mammalian spermatozoa from loss of motility caused by spontaneous lipid peroxidation. Gamete Res. 1989;23:77–90.
27. Alvarez JG, Storey BT. Taurine, hypotaurine, epinephrine and albumin inhibit lipid peroxidation in rabbit spermatozoa and protect against loss of motility. Biol Reprod. 1983;29:548–55.
28. Kovalski NN, de Lamirande E, Gagnon C. Reactive oxygen species generated by human neutrophils inhibit sperm motility: protective effect of seminal plasma and scavengers. Fertil Steril. 1992;58: 809–15.
29. Li TK. The glutathione and thiol content of mammalian spermatozoa and seminal plasma. Biol Reprod. 1975;12:641–6.
30. Menella MRF, Jones R. Properties of spermatozoal superoxide dismutase and lack of involvement of superoxides in metals ion-catalyzed lipid peroxidation reactions in semen. Biochem J. 1980;191:289–97.
31. Alvarez JG, Touchstone JC, Blasco L, Storey BT. Spontaneous lipid peroxidation and production of hydrogen peroxide and superoxide in human spermatozoa. Superoxide dismutase as major

enzyme protectant against oxygen toxicity. J Androl. 1987;8: 338–48.
32. Gavella M, Lipovac V, Vucic M, Rocic B. Relationship of sperm superoxide dismutase-like activity with other sperm-specific enzymes' and experimentally induced lipid peroxidation in infertile men. Andrologia. 1996;28:223–9.
33. Aitken RJ, Buckingham DW, Carreras A, Irvine DS. Superoxide dismutase in human sperm suspensions: relationship with cellular composition, oxidative stress, and sperm function. Free Radic Biol Med. 1996;21:495–504.
34. Alvarez JG, Storey BT. Evidence for increased lipid peroxidative damage and loss of superoxide dismutase activity as a mode of sublethal cryodamage to human sperm during cryopreservation. J Androl. 1992;13:232–41.
35. Holland MK, Alvarez JG, Storey BT. Production of superoxide and activity of superoxide dismutase in rabbit epididymal spermatozoa. Biol Reprod. 1982;27:1109–18.
36. Buffone MG, Doncel GF, Marín Briggiler CI, Vazquez-Levin MH, Calamera JC. Human sperm subpopulations: relationship between functional quality and protein tyrosine phosphorylation. Hum Reprod. 2004;19:139–46.
37. Alvarez et al. Dynamics of DNA fragmentation after thawing of subsets of cryopreserved human spermatozoa. Annual Meeting of the Spanish Fertility Society. Valencia, Spain; 2010.
38. Bahçeci M, Ciray HN, Karagenc L, Uluğ U, Bener F. Effect of oxygen concentration during the incubation of embryos of women undergoing ICSI and embryo transfer: a prospective randomized study. Reprod Biomed Online. 2005;11:438–43.
39. Meintjes M, Chantilis SJ, Douglas JD, et al. A controlled randomized trial evaluating the effect of lowered incubator oxygen tension on live births in a predominantly blastocyst transfer program. Hum Reprod. 2009;24:300–7.
40. Kovačič B, Sajko MC, Vlaisavljević V. A prospective, randomized trial on the effect of atmospheric versus reduced oxygen concentration on the outcome of intracytoplasmic sperm injection cycles. Fertil Steril. 2010;94:511–9.
41. Weksler-Zangen S, Yaffe P, Ornoy A. Reduced SOD activity and increase neural tube defects in embryos of the sensitive but not of the resistant Cohen diabetic rats cultured under aerobic conditions. Birth Defects Res A Clin Mol Teratol. 2003;67:429–37.

Further Reading

Aitken RJ. Evaluation of human sperm function. Br Med Bull. 1990;46:654–74.
Alvarez JG, Storey BT. Differential incorporation of fatty acids into and peroxidative loss of fatty acids from phospholipids of human spermatozoa. Mol Reprod Dev. 1995;42:334–46.
Calamera JC, Brugo S, Vilar O. Relation between motility and adenosinetriphosphate (ATP) in human spermatozoa. Andrologia. 1982;14:239–41.
Calamera JC, Doncel GF, Olmedo SB, Kolm P, Acosta AA. "Modified sperm stress test: a simple assay that predicts sperm-related abnormal in-vitro fertilization. Hum Reprod. 1998;13:2484–8.
Gravance CG, Vishwanath R, Pitt C, Garner DL, Casey PJ. Effects of cryopreservation on bull sperm head morphometry. J Androl. 1998;19:704–9.
Jones R, Mann T, Sherins R. Peroxidative breakdown of phospholipids in human spermatozoa, spermicidal properties of fatty acid peroxides, and protective action of seminal plasma. Fertil Steril. 1979;31:531–7.

Part III

Common Male Infertility Disorders

Varicocele

24

Sandro C. Esteves

Abstract

This chapter discusses the current concepts and controversies regarding the epidemiology, pathophysiology, diagnosis, treatment, and significance of clinical and subclinical varicoceles in male infertility. It also reviews the management of varicoceles in azoospermic patients and the novel indications of varicocelectomy in the era of assisted reproductive technology (ART). A critical commentary, based on the author's 15-year experience treating infertile men with varicoceles, and a review of important publications from the last 5 years are included. Varicocele is still one of the most controversial issues in the field of male infertility, especially regarding why, when, and to whom treatment should be applied. Varicocele repair is considered the treatment of choice for varicocele-associated infertility, but its effectiveness has been discussed for several years. Although the ultimate end point for the treatment of male factor infertility is a live birth, efforts to maximize the couple's fertility potential by improving testicular function should be the main purpose of varicocele treatment. Approximately 8% of men in reproductive age seek for medical assistance for fertility-related problems. Of these, 1–10% carry conditions that compromise the reproductive potential, and varicocele accounts for 35% of the cases. In a group of 2,875 infertile couples attending our tertiary center for male reproduction, a varicocele was identified in 21.9% of the male partners.

Keywords

Varicocele • Male infertility • Azoospermia • Varicocele embolization • Varicocelectomy • Oligozoospermia • Pampiniform plexus • Oxidative stress • Testicular underperfusion

Approximately 8% of men in reproductive age seek for medical assistance for fertility-related problems. Of these, 1–10% carry conditions that compromise the reproductive potential, and varicocele accounts for 35% of the cases [1]. In a group of 2,875 infertile couples attending our tertiary center for male reproduction, a varicocele was identified in 21.9% of the male partners.

S.C. Esteves, MD, PhD (✉)
ANDROFERT, Andrology & Human Reproduction Clinic, Center for Male Reproduction, Av. Dr. Heitor Penteado, 1464, Campinas, SP 13075-460, Brazil
e-mail: s.esteves@androfert.com.br

The first reports of the existence of varicose veins surrounding the testis are dated to the first century AC; however, the association of varicocele and infertility was only suspected at the end of the nineteenth century when surgical occlusion of dilated veins was shown to improve semen quality [2]. Tulloch, in 1952, was the first to report that bilateral varicocele repair in a male with azoospermia resulted in an increase in sperm concentration and in a spontaneous pregnancy [3]. In 1965, MacLeod first reported that most semen specimens obtained from infertile men with varicocele had decreased sperm count, decreased motility, and increased abnormal forms [4].

S.J. Parekattil and A. Agarwal (eds.), *Male Infertility: Contemporary Clinical Approaches, Andrology, ART & Antioxidants*,
DOI 10.1007/978-1-4614-3335-4_24, © Springer Science+Business Media, LLC 2012

Varicocele is a condition involving the dilation of the veins of the pampiniform plexus that drain the testicles. Normally, the backward blood flow is prevented by small one-way valves. Valve abnormalities or vein compression by adjacent structures can cause vein dilation. The pathophysiology of varicocele and its impact on the male reproductive potential have been debated for the last 50 years. Varicocele is still one of the most controversial issues in the field of male infertility, especially regarding why, when, and to whom treatment should be applied. Varicocele repair is considered the treatment of choice for varicocele-associated infertility, but its effectiveness has been discussed for several years. Although the ultimate end point for the treatment of male factor infertility is a live birth, efforts to maximize the couple's fertility potential by improving testicular function should be the main purpose of varicocele treatment. This chapter discusses the current concepts and controversies regarding the epidemiology, pathophysiology, diagnosis, treatment, and significance of clinical and subclinical varicoceles in male infertility. It also reviews the management of varicoceles in azoospermic patients and the novel indications of varicocelectomy in the era of assisted reproductive technology (ART). A critical commentary, based on the author's 15-year experience treating infertile men with varicoceles, and a review of important publications from the last 5 years are included. Finally, a list of key issues is provided to summarize the current knowledge of varicocele-associated infertility.

Varicocele

Epidemiology

Varicoceles are identified in approximately 7% and 10–25% of prepuberal and postpuberal males, respectively [5, 6]. In older men, varicoceles can be identified in up to 43% of the individuals [7]. Prevalence of varicocele increases over time, and it is estimated that a 10% rise in incidence occurs for each decade of life. It is identified in approximately 35% of men with primary and 80% of men with secondary infertility [1, 8]. The higher frequency of varicoceles in both the elderly and in men with secondary infertility suggests that it is a progressive disease. Although the frequency of unilateral left-sided varicocele has historically been reported to be approximately 85–90%, recent data indicate that bilateral palpable varicocele may be found in more than 50% of the affected subjects [5].

An inverse relationship between the occurrence of varicocele and body mass index has been reported [9]. Also, intense physical activity (2–4 h daily, ×4–5 per week) over several years seems to worsen semen quality of men with varicoceles and abnormal semen parameters [10]. Moreover, increased incidence of varicose veins has been reported in first-degree relatives of men with varicoceles, suggesting this condition may be inherited [8].

Pathophysiology

The etiology of varicocele formation is likely to be multifactorial. The right internal spermatic vein inserts directly into the inferior vena cava at an acute angle, while the left one inserts into the left renal vein at a right angle. It is also suggested that a partial obstruction of the left spermatic vein due to the compression of the left renal vein between the aorta and the upper mesenteric artery exists (the "nutcracker" phenomenon). An increase in the hydrostatic pressure of the left spermatic vein may be transferred to the venous plexus of the spermatic cord, causing its dilation [11]. Moreover, primary insufficiency of the internal spermatic and subsequent malfunction of the external spermatic and cremasteric veins valves may lead to regression of blood [12, 13]. A fivefold increase in the hydrostatic pressure of the spermatic veins has been observed in men with varicocele as compared to controls [14]. Microscopic evaluation of the spermatic vein fragments revealed alterations in the longitudinal muscle layers and a decrease in the number of nerve elements in the vessel wall [15]. These findings suggest a defective contractile mechanism of blood transport through the pampiniform plexus that may lead to a reversal of the pressure gradient and cause a hypoxic status.

Several theories aim to explain the impact of varicoceles on testicular function, but none of them can fully elucidate the variable effect of varicocele on human spermatogenesis and male fertility [11]. Proposed mechanisms include hypoxia and stasis, testicular venous hypertension, elevated testicular temperature, and increase in spermatic vein catecholamine, leading to testicular underperfusion and increased oxidative stress [16]. It is believed that reflux of warm blood from the abdominal cavity to the varicose veins increases the scrotum temperature, but the mechanism by which temperature influences spermatogenesis is not clearly understood. Germ cell apoptosis and subsequent oligozoospermia, a common phenomenon in men with varicocele, can be attributed to increased scrotal temperature, increased intratesticular cadmium concentration, and reduced levels of androgens [17, 18]. Increased concentration of regressed toxic metabolites inside the testicles (e.g., catecholamines from the kidney and adrenal glands) can cause chronic vasoconstriction of the intratesticular arterioles, contributing, along with impaired venous return caused by valve insufficiency, to persistent testicular underperfusion and subsequent dysfunction of the spermatic epithelium [19]. Biopsies of varicocele-affected testicles showed a decrease in E-cadherin and alpha-catenin in the Sertoli-Sertoli junction and a subsequent disruption of the blood–testis barrier that can contribute to the pathology and impairment in sperm production [20]. However, histopathologic findings typical of varicocele have not been observed [21].

Excessive oxidative stress (OS) is often seen in infertile men with varicocele [22]. High production of reactive oxygen species (ROS) in the reproductive tract impairs both the fluidity of the sperm plasma membrane and the integrity of deoxyribonucleic acid (DNA) in the sperm nucleus. Abnormal high levels of sperm DNA damage are associated with a decrease in several fertility markers including fertilization rate, embryo cleavage rate, implantation rate, pregnancy rate, and live birth rate [22]. It has been recently proposed that fertility reduction in men with varicocele results from decreased pH in the spermatozoa cytosol and seminal plasma [23]. According to this hypothesis, testicular underperfusion diminishes cell oxygen and glucose supply to the metabolically active tissues. Under conditions of low glucose supply, the flux through the pentose-phosphate pathway is markedly decreased, as well as the provision of reductants to the antioxidant system. Indeed, the drastic fall in the reduced nicotinamide adenine dinucleotide phosphate/oxidized nicotinamide adenine dinucleotide phosphate (NADPH/NADP+) ratio leads to an impairment in the tissue antioxidant capacity because glutathione regeneration is retarded. Under such pathologic circumstances, ROS production surpasses the antioxidant capacity and causes increased oxidative stress [24]. Spermatozoa are susceptible to damage by oxidative stress through ROS, especially lipid peroxidation [22], owing to the low amount of cytoplasm and abundance of polyunsaturated fatty acids in the sperm plasma membrane. Lipid peroxidation damages membrane function in sperm head and midpiece, thus altering sperm morphology and impairing motility, but it also leads to a decrease in intracellular pH, partly because of malondialdehyde-mediated reactions. Malondialdehyde, produced by the peroxidation of polyunsaturated fatty acids, reacts with spermine, a polyamine essential for sperm activity, forming Schiff bases. It results in further decrease of pH as well as in a direct impairment of spermine-dependent cellular functions. The optimum pH for ROS scavenging by the enzymatic antioxidant systems ranges between neutral and slightly alkaline, but their activity is markedly depressed in low pH. It has been observed that antioxidant enzyme activity is significantly impaired in infertile men with varicocele, and it may further diminish sperm motility [22]. This novel hypothesis adds to other proposed mechanisms for defective sperm function in men with varicocele, such as the peroxidation of the unsaturated fatty acids in the sperm plasma membrane and the impaired acrosome reaction and DNA integrity induced by ROS [24]. However, it has also been speculated that individual differences may exist; therefore, the mechanism described above may not be deterministic. If, for example, the glucose supply is less restricted or the accessory glands are particularly efficient, the accumulation of ROS and acidification of the seminal plasma could be minimized. This hypothesis may help us understand the variable effect of varicocele on male fertility.

Varicocele and Infertility

The concept that varicocele causes infertility is based on three main aspects: (a) the increased incidence of this condition among infertile men, (b) the association of varicocele with reduced semen parameters and testicular size, and (c) the improvement of semen parameters and pregnancy rates after surgical repair of clinical varicoceles.

In a large observational study involving 9,034 men, it was observed that 25.6% of men with abnormal semen analysis had varicocele. It was also noted that total sperm count and testosterone levels were lower in men with varicoceles as compared to those without varicoceles. Also, testicular size was significantly reduced at the varicocele side as compared to the contralateral one in the cases of unilateral varicoceles [25]. Surgical repair of varicocele was shown to restore testicular temperature in both animals and humans [19].

The hypothesis that varicocele can cause testicular damage was further confirmed on pubertal boys in which the reduction in the size of the ipsilateral testis was restored by surgical repair of varicocele [26]. Despite the proven association between varicocele and infertility, it is still unclear the reasons why about 2/3 of men with varicocele retain their fertility [2, 27] and why fertility potential is not always improved after surgical varicocele repair [28, 29].

Diagnosis

Currently, physical examination with the patient standing in a warm room is the preferred diagnostic method. Varicoceles diagnosed by this method are termed "clinical" and may be graded according to the size. It is important to ask the patient to perform a Valsalva maneuver during examination. Large varicoceles (grade III) are varicose veins seen through the scrotal skin (Fig. 24.1). Moderate (grade II) and small-sized varicoceles (grade I) are dilated veins palpable without and with the aid of the Valsalva maneuver, respectively [30]. Physical examination is limited by a sensibility and specificity of about 70% when compared to other diagnostic modalities [31, 32]. Interobserver and intraobserver variability has been observed when diagnosing varicocele. Physical examination may be inconclusive or equivocal in cases of low-grade varicocele and in men with a history of previous scrotal surgery, concomitant hydroceles, or obesity. Imaging studies may be recommended when evaluating infertile men for varicocele

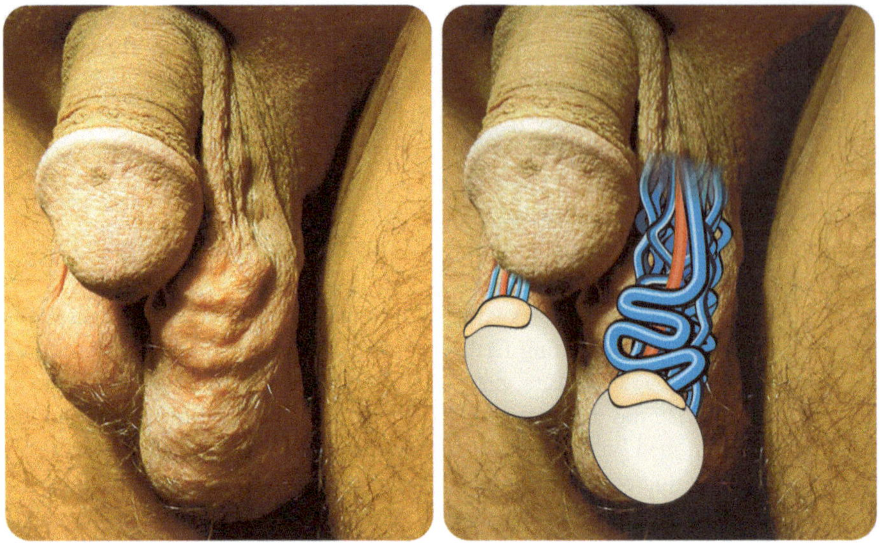

Fig. 24.1 Photograph of a large left varicocele (grade III) seen through the scrotal skin (*left*). Illustration of varicose veins on the left spermatic cord as compared to normal-sized veins on the right side (*right*)

when physical examination is inconclusive. When a varicocele is not palpable but a retrograde blood flow is detected by other diagnostic methods such as venography, Doppler examination, ultrasonography, scintigraphy, and thermography, the varicocele is termed subclinical [32, 33].

The gold standard method to diagnose blood reflux into the veins of the pampiniform plexus is the percutaneous venography of the spermatic veins; however, it is not routinely used because of its invasiveness [32, 33]. Among the noninvasive diagnostic modalities, color Doppler ultrasound (CDU) has been shown to be the best diagnostic tool. The commonly accepted CDU criterion for varicocele (maximum vein diameter of 3 mm or greater) has a sensitivity of about 50% and specificity of 90% compared to physical examination [34]. However, a scoring system, incorporating the venous diameter, the presence of a venous plexus, and the change of flow on Valsalva maneuver, yields a sensitivity and specificity greater than 85% when compared to physical examination [34] or venography [31]. A pencil-probe Doppler (9 MHz) stethoscope is an inexpensive tool that may aid in the diagnosis of the varicocele. The patient is examined in the upright position, and a venous "rush" representing blood reflux is heard with or without the Valsalva maneuver. Although simple and easily performed in the office, Hirsht et al. demonstrated that more than 50% of men without clinical varicoceles exhibited a Valsalva-maneuver Doppler-positive reflux [35]. Despite that, Doppler examination has been advocated as a useful tool to examine the contralateral spermatic cord to determine if a subclinical varicocele exists when a clinical varicocele is found on the other side [36]. Unfortunately, none of these adjunctive diagnostic methods can differentiate between clinical and subclinical varicoceles. The significance of a positive test result using any of these adjuvant techniques in infertile men remains uncertain.

Treatment

Treatment of varicocele in infertile men aims to restore or improve testicular function. Current recommendations suggest that treatment should be offered for couples with documented infertility whose male partner has a clinically palpable varicocele and abnormal semen analysis. Additionally, an adult male presenting with palpable varicocele and abnormal semen analyses who is not currently attempting to achieve conception but has a desire for future fertility is also a candidate for varicocele repair [37]. The ideal treatment must combine low complication rates with the highest seminal improvement to either increase the chances of spontaneous conception or to optimize assisted conception outcomes.

The role of medical therapy in varicocele-related infertility is poorly understood, and well-designed studies are rare. It has been reported that the use of L-carnitine in combination with the nonsteroidal anti-inflammatory agent cinnoxicam did not improve sperm parameters in infertile men with clinical varicoceles [38]. Also, the use of clomiphene citrate in men with subclinical varicoceles showed no benefit [39]. On the other hand, kallikrein therapy for 3 months (600 units orally per day) improved sperm motility and morphology in a small group of infertile men with left-sided varicocele and asthenozoospermia [40]. Early use of menotropin in association with varicocelectomy yielded to a better improvement in sperm parameters as compared to varicocelectomy alone

Table 24.1 Treatment options for varicocele repair in infertile men. Vein ligation sites and postoperative recurrence, hydrocele formation, and spontaneous pregnancy rates among different techniques

Technique	Internal spermatic vein ligation	External spermatic vein ligation	Recurrence rate	Hydrocele formation rate	Spontaneous pregnancy rate
Retroperitoneal high ligation (Palomo) [44, 45, 53]	Yes	No	7–35%	6–10%	25–55%
Laparoscopic [44, 49, 50, 53]	Yes	No	2–7%	0–9%	14–42%
Embolization [44, 45, 53]	Yes	No	2–24%	NR	20–40%
Macroscopic inguinal (Ivanissevich) [44]	Yes	Yes	0–37%	7%	34–39%
Microscopic inguinal or subinguinal [44, 50, 53, 54]	Yes	Yes	0–0.3%	0–1.6%	33–56%

NR not reported. Values are expressed as range

[41]. Recently, preliminary data from very small series have been reported on the use of vitamins and antioxidants as a medical therapy for infertile men with varicocele [42, 43]. Daily oral administration of pentoxifylline, zinc, and folic acid for 3 months was shown to improve sperm morphology in men with varicocele-associated infertility [42]. Also, a combination of vitamins and minerals was shown to significantly improve sperm count in men with persistent oligozoospermia after varicocele embolization, although spontaneous conception rates were not increased in a 1-year follow-up period [43].

Currently, varicoceles are treated either by surgery (open with or without magnification and laparoscopy) or percutaneous embolization of the internal spermatic vein. Although techniques vary, the main concept is the occlusion of the dilated veins of the pampiniform plexus. The high retroperitoneal (Palomo), radiologic, and laparoscopic approaches are performed for internal spermatic vein ligation, while the inguinal (Ivanissevich) and subinguinal approaches also allow the ligation of the external spermatic and cremasteric veins that may contribute to the varicocele (Table 24.1). Percutaneous embolization is successfully accomplished in approximately 90% of the attempts. It is associated with faster recovery and minimal pain as compared to the standard surgical approaches, but with higher recurrence rates (Table 24.1). Embolization requires interventional radiologic expertise and has potentially serious complications such as vascular perforation, coil migration, and thrombosis of pampiniform plexus [44–47]. Nonetheless, percutaneous embolization may have a role in the treatment of persistent or recurrent varicoceles previously treated by surgery [48]. Laparoscopic varicocelectomy provides higher magnification with low incidence of hydrocele formation. However, external spermatic veins, the second cause of varicocele recurrence, cannot be ligated, leading to a recurrence rate of approximately 5% [44]. Laparoscopic approach requires extensive training, and the cost of instrumentation is high. It is more invasive than an open microsurgical approach, requiring general anesthesia and placement of a urethral catheter [49, 50]. Complications include intestinal and vascular injuries that occur in approximately 8% of the cases [44].

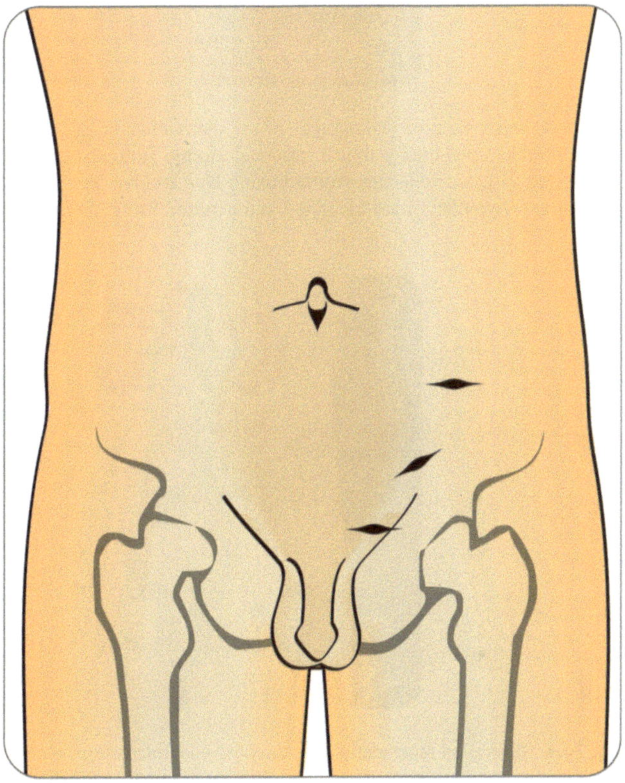

Fig. 24.2 Incision sites used for subinguinal, inguinal, and retroperitoneal open surgical varicocele repair. In the subinguinal approach, a transverse incision is made just below the level of the external inguinal ring. An oblique incision is made along the axis between the anterior superior iliac spine and the pubic tubercle for the inguinal approach. In the retroperitoneal approach, a transverse incision is made medial to the anterior superior iliac spine

Open surgical varicocele repair is often performed using a retroperitoneal, inguinal, or subinguinal approach (Fig. 24.2). High ligation of the internal spermatic vein can be easily performed via the retroperitoneal approach, but it is associated with high recurrence and hydrocele formation rates (see Table 24.1). Inguinal and subinguinal approaches offer the advantage of also allowing the ligation of the external spermatic veins. Internal and external spermatic veins can be identified via inguinal/subinguinal approaches macroscopically, but the use of magnification facilitates identification and preservation of internal spermatic artery and lymphatics,

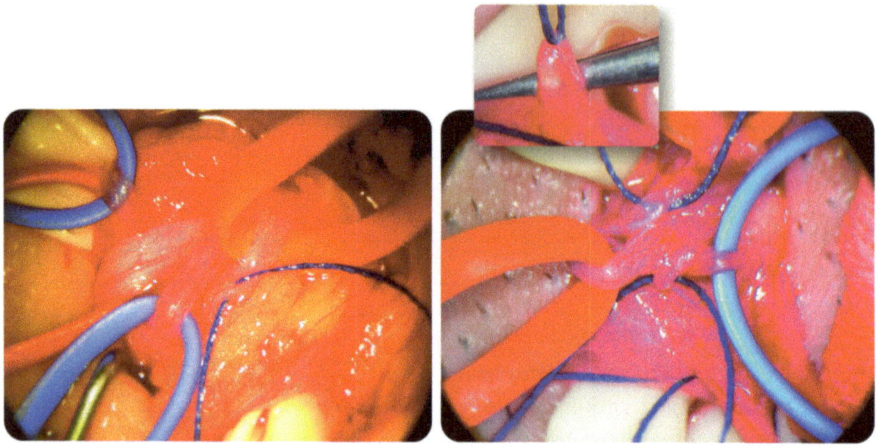

Fig. 24.3 Intraoperative photographs of the spermatic cord structures during a subinguinal microsurgical varicocelectomy. Testicular arteries and dilated varicose veins are exposed using *blue* and *red* Vessiloops, respectively. Lymphatics are identified and exposed using *blue cotton sutures*. On the left, double testicular arteries are identified. On the right, a single testicular artery is identified, and the lymphatic channels are easily seen (also highlighted on *top right image*)

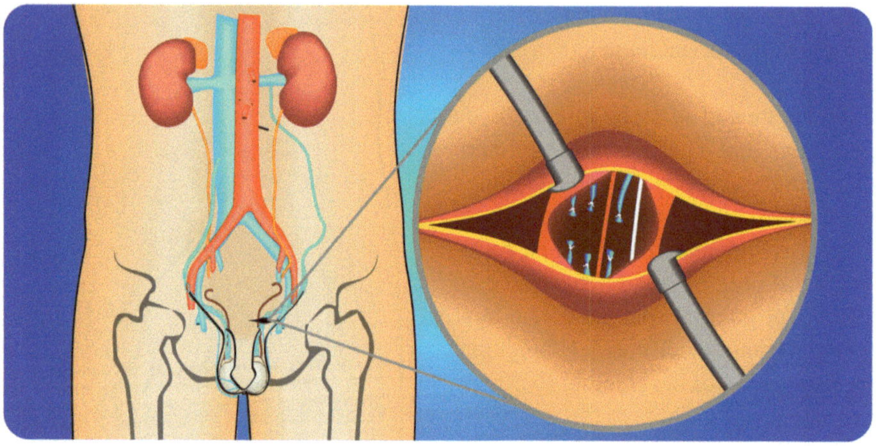

Fig. 24.4 Illustration representing a left subinguinal microsurgical varicocelectomy. A 2-cm transversal skin incision is made immediately below the external inguinal ring (*left*). The muscle layers and the inguinal canal are not violated. Dilated varicose veins are identified, transected, and ligated with nonabsorbable sutures. Testicular artery, lymphatic channels, and vas deferens are identified and preserved (*right*)

which may prevent testicular atrophy and hydrocele formation, respectively [51] (Fig. 24.3).

Microsurgical varicocelectomy can be performed via an inguinal or subinguinal approach with similar results, and reported recurrence and hydrocele formation are below 2% (Table 24.1). The main advantage of the subinguinal over the inguinal approach is that the former obviates the need to open the aponeurosis of the external oblique, which usually results in more postoperative pain and a longer time before the patient can return to work (Fig. 24.4). It is believed that subinguinal microsurgical varicocelectomy requires more microsurgery skills because it is associated with a greater number of arteries and internal spermatic veins with smaller diameter as compared to the inguinal approach [44]. However, histomorphological studies were unable to find differences in number and wall thickness of spermatic cord veins and arteries between the subinguinal and inguinal levels [52].

A recent systematic review including 4,473 individuals was performed to define the best treatment modality of palpable varicocele in infertile men [44]. The authors concluded that open microsurgical inguinal or subinguinal varicocelectomy techniques resulted in higher spontaneous pregnancy rates and fewer recurrences and postoperative complications than laparoscopic, radiologic embolization, and macroscopic inguinal or retroperitoneal varicocelectomy techniques.

Overall, varicocelectomy studies report significant improvements in one or more semen parameters in approximately 65% of men [53]. The mean time for semen improvement and spontaneous pregnancy after surgery is approximately 5 and 7 months, respectively [54]. However, it is still unknown why fertility potential is not always improved after varicocelectomy. Studies evaluating predictors for successful varicocele repair would aid in the identification of the best candidates for treatment, but to date, few reports exist and results are conflicting [44, 53, 55–60]. From the existing data, it seems that infertile men either with higher preoperative semen parameters or undergoing varicocele repair for large varicoceles are more likely to show postoperative semen parameters improvement [53, 55]. It was also shown that men who achieved a postoperative total motile sperm count greater than 20 million were more likely to initiate a pregnancy either spontaneously or via intrauterine insemination [56]. On the other hand, reduced preoperative testicular volume, elevated serum FSH levels, diminished testosterone concentrations, subclinical varicocele, as well as the presence of Y chromosome microdeletions seem to be negative predictors for fertility improvement after surgery [16, 57–62]. Interestingly, a recent report suggested that advanced paternal age does not adversely influence reproductive outcomes of men with varicocele-associated infertility. However, the authors' results might be biased by the fact that the group of men with 40 years and older had a significantly higher proportion of men with secondary infertility compared to the one with younger subjects [63]. In the presence of bilateral palpable varicocele, it is recommended to perform surgery on both sides at the same operative time [64].

Subclinical Varicocele

Subclinical varicocele refers to the presence of retrograde blood flow that cannot be detected by physical examination of the spermatic cord during Valsalva maneuver, and requires adjunctive tests for diagnosis, such as Doppler examination, color Doppler ultrasound (CDUS), scrotal thermography, isotope imaging, or venography [31–35].

The role of subclinical varicocele as a cause of male infertility remains debatable. Currently, existing evidence does not support the recommendation for treating infertile men with subclinical varicocele [28, 39, 62, 65]. The management of infertile men with a unilateral clinical varicocele and a subclinical one at the contralateral side, on the other hand, may pose a different dilemma. Zheng et al. compared the efficacy of bilateral and left unilateral varicocelectomy in a group of 104 infertile men with left clinical and right subclinical varicoceles, and found that bilateral varicocelectomy had no benefit over the left clinical varicocelectomy [66]. In their study, however, a retroperitoneal approach was used for vein ligation, which was shown to be associated with high recurrence rate [44]. Elbendary et al., in a recent prospective trial, studied a group of 145 infertile men with clinical left and subclinical right varicoceles [67]. Patients were randomized to undergo either unilateral inguinal repair of clinical varicocele or bilateral repair of both clinical and subclinical ones. Although a significant improvement in sperm parameters was observed in both groups, the magnitude of change in sperm count and motility and the spontaneous pregnancy rates were significantly higher in the group of men who had bilateral varicocele repair. Their findings are in agreement with earlier studies suggesting that bilateral varicocelectomy is more effective than unilateral for such patients [68, 69]. It is also postulated that altered blood flow after unilateral clinical varicocelectomy may unmask an underlying contralateral venous anomaly that may result in a clinically manifested varicocele [36, 68].

Varicocele and Azoospermia

Nonobstructive azoospermia (NOA) comprises a spectrum of testicular histopathology resulting from various causes that include gonadotoxins, medications, genetic and congenital abnormalities, trauma, endocrine disorders, and idiopathic. Men with NOA have historically been the infertile men most difficult to treat, but since the advent of in vitro fertilization/intracytoplasmic sperm injection (IVF/ICSI) and surgical methods for testicular sperm extraction (TESE), several pregnancies have been achieved with the use of testicular sperm. However, only about 50% of men will have sperm present at the time of TESE [70].

Varicoceles are found in approximately 5% of men with azoospermia, but it is still debatable whether varicoceles can cause or contribute to azoospermia [71]. There has been a renewed interest in varicocele repair in azoospermic men resulting from the introduction of ICSI. Success rates varied and no predictors of success have been definitively identified because of the small numbers in the case series [71–78]. A recent meta-analysis examined the impact of varicocele repair to recover spermatogenesis in NOA men [79]. A total of 233 infertile men with clinical varicocele and NOA were analyzed in a mean postoperative follow-up of 13 months. Motile sperm was found on postoperative ejaculate in 39% of men. Pregnancies were achieved in approximately 26% of men with sperm in the ejaculate, 60% unassisted, and 40% with the assistance of IVF. Postoperative mean sperm density and motility were 1.6 million and 20%, respectively. Levels of serum follicle-stimulating hormone (FSH) and testosterone, testis size, patient age, varicocele grade, and surgical technique did not appear to affect outcomes, but the limited number of patients precluded conclusions. Histopathology was the only predictor of success. Postoperatively appearance of sperm in the ejaculates was significantly higher in patients with biopsy-proven hypospermatogenesis (HS) or maturation arrest (MA) than Sertoli cell only (odds ratio 9.4; 95% confidence interval 3.2–27.3). Combined success was 48% with HS or MA compared to 11% with SCO (Table 24.2). Unfortunately, randomized

Table 24.2 Results of varicocele repair in infertile men with clinical varicocele and nonobstructive azoospermia

Publication	Esteves and Glina [71]	Weedin et al. meta-analysis [79]
Number of patients	17	233
Mean age	34.2	30.1
Treatment technique	Subinguinal; microsurgical	Microsurgical inguinal and subinguinal; embolization
Mean FSH (mIU/mL)	14.6	Range 12.3–35.0
Bilateral repair; *N* (%)	11 (64.7)	151 (64.8)
Patients with motile sperm in postoperative ejaculates; *N* (%)	6/17 (35.3)	91/233 (39.0)
Mean postoperative sperm density; $\times 10^6$/mL	0.8 (range 0.1–1.8)	1.6±1.2
Mean time to appearance of sperm in the ejaculate; months	5 (range 3–9)	NR
Outcome according to histopathology; *N* (%)[a]		
Hypospermatogenesis	5/6 (83.3)	30/55 (54.5)
Maturation arrest	3/5 (75.0)	24/57 (42.0)
Sertoli cell only	0/6 (0.0)	5/44 (11.4)
Relapse to azoospermia; *N* (%)	0 (0.0)	11 (4.6)
Spontaneous pregnancy; *N* (%)	1 (5.8)	14 (6.0)
Mean follow-up; months	18.9	13.3

NR not reported
[a]Patients with motile sperm in postoperative ejaculates

control trials are lacking, and studies included in the meta-analysis by Weedin et al. were case series lacking a control group that did not undergo varicocele repair. Although an argument can be made that a control group would remain azoospermic, it is not rare to observe that NOA men occasionally ejaculate small quantities of motile sperm despite any intervention. Therefore, one cannot exclude that the appearance of very small number of motile sperm in the ejaculates after varicocele repair may be merely coincidental. Moreover, reports including men with germinal cell aplasia who ejaculated motile sperm after varicocele repair are intriguing [78]. It is unlikely that men lacking any sperm precursor within the testicle may benefit from treatment, but the only way to investigate this relationship is to repeat the testicular biopsy after the surgical repair of varicocele. In one series [71], patients with SCO were re-biopsied 6 months after varicocele repair and testicular histopathology results remained unchanged (Table 24.2). The persistence of SCO after surgery denotes that varicocele coexisted with primary testicular failure, which of course was not affected by the surgery. However, testicular biopsy has many limitations and may not reflect the most advanced site of spermatogenesis due to the heterogeneity of sperm production within the testicle; therefore, it is still possible to retrieve sperm from men whose testicles exhibit SCO [80, 81].

Even with the improvement in spermatogenesis in up to half of the NOA patients with a favorable testicular histopathology after varicocele repair, intracytoplasmic sperm injection (ICSI) will be necessary for most couples to initiate a pregnancy [71, 79]. However, the use of motile ejaculated sperm is preferred for ICSI since their fertilizing ability is higher than that of sperm retrieved from the testis [82]. Nonetheless, continuing azoospermia after varicocele repair is still a potential problem, and sperm extraction before ICSI will be inevitable for many individuals. Results of testicular sperm extraction (TESE) for men who remain azoospermic after varicocelectomy are scarce and conflicting [83, 84]. Schlegel et al. reported sperm retrieval rates of 60% per attempt using testicular microdissection (micro-TESE) in men with NOA and varicocele, regardless of whether previous varicocelectomy had been done [83]. It is questionable, however, if the inclusion of patients with subclinical varicocele biased their results since the benefit of treating subclinical varicocele is debatable [62, 65]. On the other hand, Inci et al., also using micro-TESE, reported a 2.6-fold increase in the chances of retrieving testicular sperm for ICSI after repair of clinical varicoceles [84]. Unfortunately, testicular histopathology results were not available in their study. Therefore, it cannot be excluded that higher retrieval rates were obtained after varicocelectomy because this group was biased by patients with favorable histopathology patterns for successful sperm retrieval, such as the ones exhibiting hypospermatogenesis or maturation arrest [81, 85].

Expert Commentary

The purpose of this chapter was to discuss the current concepts and controversies of varicocele as the leading cause of male infertility. Despite of extensive investigation, approximately 4,000 peer-reviewed papers have been published on this topic, and varicocele remains one of the most debatable issues in the field of reproductive medicine. An important argument to be included in the discussion is the multifactorial origin of infertility. Several factors may interact synergistically in the same individual, and the presence of a significant condition affecting the female partner adds to the complexity of the problem. For example, many men with varicocele-associated infertility have lifestyle choices that include smoking, obesity, poor nutrition, use of gonadotoxic medication, and exposure to environmental toxins. These conditions are often associated with increased systemic or seminal oxidative stress and may have a negative synergistic effect in men with varicocele [86]. Treatment of varicocele alone in the presence of inadequate lifestyle choices is likely to solve only part of the problem. Lifestyle modifications may have an important beneficial impact on both systemic and reproductive health [87]; therefore, when considering therapeutic measures to treat varicocele-associated infertility, counseling toward lifestyle modifications should be strongly encouraged. This strategy, along with the cause-specific treatment, is more likely to lead to a marked improvement in the male reproductive health as compared to varicocele repair alone.

The treatment of varicocele in infertile men should aim to achieve the highest improvement in the male fertility status, with lower rates of complications such as recurrence or persistence, hydrocele formation, and testicular atrophy. Increase in the spontaneous pregnancy rates after the treatment of varicocele is difficult to ascertain due to a variety of factors that include the lack of a uniform posttreatment follow-up interval and the female factor parameters, such as age and reproductive health. Therefore, the ultimate treatment goal should be the improvement in the chances of conception, either unassisted or assisted. The ideal surgical technique should aim for ligation of all internal and external spermatic and cremasteric veins, with preservation of spermatic arteries and lymphatics (see Fig. 24.3). This can only be achieved by the inguinal or subinguinal microsurgical approaches. In our practice, when a clinically palpable varicocele is identified in one side, the contralateral cord is examined using a pencil-probe Doppler (9 MHz) stethoscope to determine if a subclinical varicocele exists. If so, it is treated at the same time as the coexistent clinical varicocele. This is based on the observation that altered blood flow after varicocelectomy may unmask an underlying venous anomaly and result in clinical varicocele formation [36, 68]. Although loupe magnification may be used to facilitate the ligation of the dilated varicose veins, it is insufficient for identification of both testicular arteries and lymphatics. Using this method, we found that instillation of papaverine was needed in most cases to aid in the identification of arterial pulsations. Also, recurrence seems to be higher when loupe magnification is used in association with the inguinal or subinguinal approach to repair varicoceles (Table 24.3). An intraoperative pencil Doppler examination (9 MHz) can also be used to aid in the identification of the artery pulsations. In our hands, the subinguinal microsurgical varicocelectomy using the operating microscope is the method of choice to treat varicocele-associated infertility (Table 24.3 and Fig. 24.4). The subinguinal approach provides excellent results, and the surgical intervention can be performed in an outpatient basis using intravenous anesthesia in association with spermatic cord blockade with lidocaine [71]. Although empirical, the early use of antioxidants and vitamins is a common practice after varicocelectomy in our institution [daily use of antioxidants (vitamin C 500 mg and vitamin E 400 UI)

Table 24.3 Results of microsurgical subinguinal varicocelectomies in a group of 485 men with varicocele and infertility

Type of magnification	Loupe	Operating microscope
Number of procedures	101	384
Male age in years; mean (range)	32.4 (24.0–63.0)	34.5 (24.0–52.0)
Varicocele side; *N* (%)		
Unilateral	51 (50.4)	184 (47.9)
Bilateral	50 (49.6)	200 (52.1)
Varicocele grade[a]; *N* (%)		
Grade I	14 (13.9)	73 (19.0)
Grade II	48 (47.5)	199 (51.8)
Grade III	39 (38.6)	112 (29.2)
Endocrine profile; mean ± SD		
Serum FSH (mUI/mL)	5.7 ± 8.8	6.1 ± 7.8
Serum testosterone (ng/dL)	523.8 ± 547.1	575.3 ± 677.2
Mean operative time; min (range)		
Unilateral	78.6 (50–90)	89.2 (60–105)
Bilateral	101.1 (80–150)	112.9 (90–150)
Number of veins ligated; mean (range)		
Left side	4.8 (2–7)	6.1 (2–9)
Right side	4.2 (2–6)	5.1 (2–7)
Vein diameter in millimeters; mean (range)		
Left side	3.2 (1–6)	3.1 (1–6)
Right side	2.8 (1–4)	2.5 (1–5)
Testicular artery identified; %	84.1[b]	97.6
Improvement in seminal parameters[c]; %	60.4	68.5
Recurrence rate; *N* (%)	3 (2.9)	4 (1.0)
Hydrocele formation rate; *N* (%)	1 (1.0)	0 (0.0)
Other complications; *N* (%)	2 (1.9)[d]	1 (0.2)[e]
Clinical pregnancy rate; *N*	85[f]	270[f]
Spontaneous; *N* (%)	20/69 (28.9)	58/172 (33.7)
Assisted reproduction; *N* (%)	4/16 (25.0)[g]	56/98 (57.1)[h]

[a]The largest varicocele grade is reported in cases of bilateral varicocele
[b]Instillation of papaverine for identification of artery pulsation was necessary in 85% of the cases
[c]≥15% improvement from baseline preoperative values in at least one of the semen parameters (sperm count, progressive motility, strict morphology), in a minimum of three postoperative semen analyses
[d]Scrotal hematoma (1 case); testicular atrophy (1 case)
[e]Scrotal hematoma
[f]Reported number of patients assessed for pregnancy
[g]Intrauterine insemination (IUI; $n = 12$); in vitro fertilization/intracytoplasmic sperm injection (IVF/ICSI; $n = 04$)
[h]IUI ($n = 18$); IVF/ICSI ($n = 80$)

and commercial multivitamin preparation containing minerals, β-carotene (3,500 IU), vitamin C (60 mg), vitamin E (30 UI), zinc phosphate (11 mg), selenium (55 mcg), and vitamin B complex (thiamine, riboflavin, niacin, pantothenic acid, pyridoxine, biotin, folic acid, and cyanocobalamin)].

The urologist who opts to treat varicocele using microsurgery should obtain appropriate training. It is also important to have adequate microsurgical instruments and a binocular operating microscope with foot-control zoom magnification. Microsurgical varicocelectomy, either using inguinal or subinguinal approaches, requires more skill as compared to other surgical modalities because a higher number of internal spermatic vein channels and smaller-diameter artery are seen at the level of the inguinal canal. However, the routine use of microsurgery during varicocele repair may help the urologist to master his/her microsurgical skills, which will be of great benefit when performing more demanding and less frequent microsurgical procedures, such as vasovasostomies and vasoepididymostomies.

Five-Year Review

Infertile men with clinical varicocele have a significantly lower sperm production and quality as compared to normal controls and fertile men with varicocele [88]. Such differences are observed not only in the standard sperm parameters, such as sperm count, motility, and morphology but also in the novel functional markers for oxidative stress and DNA integrity. Seminal plasma total antioxidant capacity (TAC) and sperm mitochondrial activity are decreased [22, 89, 90], while the frequency of sperm exhibiting abnormal DNA integrity and chromatin immaturity is increased [90, 91] in ejaculates of infertile men with varicocele. Although elevation of scrotal temperature is seen in both fertile and infertile men with varicocele, increase of oxidative stress (OS) is only observed in the latter, thus indicating that a disturbance of the OS scavenging system is likely to play a major role in the pathophysiology of varicocele-associated infertility [22, 92]. Using novel noninvasive contrast imaging, it has been shown that intratesticular microcirculation perfusion, which is altered in men with clinical varicocele, affects spermatogenesis [93]. Molecular biology studies show that high cadmium content and hypoxic conditions induce overexpression of metallothionein, a metal-binding protein that protects against cell apoptosis, in internal spermatic veins of infertile men with clinical varicocele [94]. Recently, Ichioka et al. determined the distribution of antioxidant enzymes genes genotype in infertile men with varicocele. Their preliminary data suggest that genetic polymorphisms in the glutathione S-transferase T1 gene may affect individual response to varicocelectomy [95].

When clinical palpable varicocele coexists with impaired semen quality, surgical repair has been shown to be the best treatment option. Varicocele repair may partially or totally restore spermatogenesis and fertility, and it offers a better cost-effectiveness as compared to ART [96]. Recent meta-analyses demonstrated a beneficial effect of varicocelectomy on the fertility status of infertile men with clinical varicocele [88, 96–98]. Agarwal et al. examined the effect of varicocelectomy on semen parameters and demonstrated that sperm concentration increased by 9.7 million/mL (95% confidence interval [CI] 7.34–12.08, $p<0.001$), sperm motility increased by 9.9% (95% CI 4.90–14.95, $p<0.001$), and WHO sperm morphology increased by 3.1% (95% CI 0.72–5.60, $p=0.01$) after varicocelectomy [88]. Ficarra et al. reviewed randomized clinical trials for varicocele repair and found a significant increase in the pregnancy rates for patients who underwent varicocele treatment (36.4%) compared to ones having no treatment (20%) [97]. Similarly, Marmar et al. reported significantly higher pregnancy rate (33%) after varicocelectomy compared to the group of patients having no surgery (15.5%) [98]. In their study, the chances of obtaining a spontaneous conception were 2.8 times higher in the varicocelectomy group as compared to the group of patients who received either no treatment or medication. Other studies demonstrated that markers of sperm function were also significantly improved after varicocele repair [99–104]. It has been reported that seminal oxidative stress may be attenuated by varicocelectomy in infertile men with varicocele, but this beneficial effect is not always associated with an improve in the conventional sperm parameters [99, 100]. Sperm DNA integrity is also increased 6 months after repair in infertile men with clinically palpable varicocele [101, 102]. Recently, Smit et al. reported a decrease in sperm DNA fragmentation after varicocelectomy and an association between DNA fragmentation index and the ability to conceive either spontaneously or via assisted reproduction [103].

Even though spontaneous pregnancy remains the litmus test for evaluating varicocele treatment success, many patients with varicocele-related infertility will require ART due to the severity of sperm abnormalities and/or the presence of a significant problem affecting the female partner. The indication of varicocele repair prior to IVF/ICSI is unusual, but in certain circumstances, varicocele treatment should be considered. Men with nonobstructive azoospermia with favorable testicular histopathology may restore sperm to the ejaculate after repair of clinical varicoceles [79]. Sperm restoration, although minimal, yields the possibility of IVF/ICSI without the need of sperm retrieval techniques (SRT). It has been shown that for patients who are still azoospermic after varicocelectomy, SRT success rates using testicular microdissection sperm extraction, and as a result the couple's chance for pregnancy, may be increased [84]. Varicocelectomy has also a potential to obviate the need for ART or to down stage the level of ART needed to bypass male factor infertility [104]. Recently, it has been shown that treatment of clinical varicoceles may also improve the outcomes of assisted reproduction in couples with varicocele-related infertility [105]. Esteves et al. studied 242 infertile men with treated and untreated clinical varicoceles who underwent intracytoplasmic sperm injection (ICSI), and found significantly higher live birth rates after ICSI in the group of men who underwent microsurgical varicocele repair before ART (46.2%) as compared to the ones undergoing ICSI in the presence of a clinical varicocele (31.4%). In their study, the chances of achieving a live birth (odds ratio = 1.87; 95% confidence interval 1.08–3.25; $p = 0.03$) by ICSI were increased, while the chances of miscarriage occurrence after obtaining a pregnancy by ICSI were reduced (odds ratio = 0.433; 95% confidence interval 0.22–0.84; $p = 0.01$) if the varicocele had been treated before assisted conception.

Key Issues

- Approximately 8% of men in reproductive age seek for medical assistance for fertility-related problems. Of these, 1–10% carries conditions that compromise the reproductive potential, and varicocele accounts for 35% of the cases.
- Epidemiology data show that approximately 35% of men with primary and 80% of men with secondary infertility have varicocele. The higher frequency of varicoceles in both the elderly and in men with secondary infertility suggests that it is a progressive disease. Bilateral varicoceles are more common than previously reported.
- Impaired drainage or pooling of blood around the testicles leading to increased scrotal temperature, hypoxia, increased testicular pressure, reflux of renal and adrenal metabolites, excessive oxidative stress, and decreased pH in the spermatozoa cytosol and seminal plasma are the leading theories to explain the detrimental effects of varicocele on spermatogenesis.
- The concept that varicocele causes infertility is based on three main aspects: (a) the increased incidence of this condition among infertile men, (b) the association of varicocele with reduced semen parameters and testicular size, and (c) the improvement of semen parameters and pregnancy rates after surgical repair of clinical varicoceles.
- Varicoceles diagnosed by physical examination, the preferred diagnostic method, are termed "clinical" and may be graded according to the size. When a varicocele is not palpable but a retrograde blood flow is detected by other diagnostic methods, the varicocele is termed subclinical. The significance of a positive test result using adjuvant diagnostic techniques in infertile men remains uncertain.
- Varicocele treatment is indicated for men with clinically palpable varicocele and abnormal semen parameters. Open microsurgical inguinal or subinguinal techniques are considered the best treatment modalities because they result in higher spontaneous pregnancy rates and fewer recurrences and postoperative complications than laparoscopic, radiologic embolization, and macroscopic inguinal or retroperitoneal varicocelectomy techniques. There are no absolute predictive factors for successful varicocele repair, and existing evidence does not support the recommendation for treating infertile men with subclinical varicocele.
- Recovery of spermatogenesis can be achieved after repair of clinical varicocele in infertile men with nonobstructive azoospermia. Testicular histopathology is predictive of success, and men with maturation arrest and hypospermatogenesis are more likely to ejaculate motile spermatozoa after surgery.
- Functional markers for oxidative stress and DNA integrity are impaired in infertile men with clinical varicocele, but may be significantly improved after varicocele repair.
- Surgical repair of varicocele increases the chance for either spontaneous or assisted conception in infertile couples whose male partner has a clinical varicocele. Also, the chance of retrieving testicular sperm for ICSI may be optimized in nonobstructive-azoospermic men with treated clinical varicocele.
- Lifestyle modifications may benefit reproductive health. When considering therapeutic measures to treat varicocele-associated infertility, counseling toward lifestyle modifications should be strongly encouraged.

Acknowledgment The author is indebted to Mrs. Fabiola Bento for her editorial assistance and text revision.

References

1. Vital and Health Statistics, series 23, no.26, CDC. http://www.cdc.gov. Accessed 10 Dec 2009.
2. Saypol DC. Varicocele. J Androl. 1981;2:61–71.
3. Tulloch WS. A consideration of sterility factors in the light of subsequent pregnancies. Edinburgh Med J. 1952;59:29–34.
4. MacLeod J. Seminal cytology in the presence of varicocele. Fertil Steril. 1965;16:735–57.
5. Akbay E, Cayan S, Doruk E, et al. The prevalence of varicocele and varicocele-related testicular atrophy in Turkish children and adolescents. BJU Int. 2000;86:490–3.
6. Callam MJ. Epidemiology of varicose veins. BJU Int. 1994;81:167–73.
7. Canales BK, Zapzalka DM, Ercole CJ, et al. Prevalence and effect of varicoceles in an elderly population. Urology. 2005;66:627–31.
8. Raman JD, Walmsley K, Goldstein M. Inheritance of varicoceles. Urology. 2005;65:1186–9.
9. Handel LN, Shetty R, Sigman M. The relationship between varicoceles and obesity. J Urol. 2006;176:2138–40.
10. Luigi L, Gentile V, Pigozzi F, et al. Physical activity as a possible aggravating factor for athletes with varicocele: impact on the semen profile. Hum Reprod. 2001;16:1180–4.
11. Nistal M, Gonzalez-Peramato P, Serrano A, et al. Physiopathology of the infertile testicle. Etiopathogenesis of varicocele. Arch Esp Urol. 2004;57:883–904.
12. Goldstein M, Eid JF. Elevation of intratesticular and scrotal skin surface temperature in men with varicocele. J Urol. 1989;142:743–5.
13. Chehval MJ, Purcell MH. Varicocelectomy: incidence of external vein involvement in the clinical varicocele. Urology. 1992;39:573–5.

14. Shafik A, Bedeir GAM. Venous tension patterns in cord veins in normal and varicocele individuals. J Urol. 1980;123:383–5.
15. Tilki D, Kilic E, Tauber R, et al. The complex structure of the smooth muscle layer of spermatic veins and its potential role in the development of varicocele testis. Eur Urol. 2007;51:1402–9.
16. Marmar JL. The pathophysiology of varicoceles in the light of current molecular and genetic information. Hum Reprod Update. 2001;7:461–72.
17. Simsek F, Turkeri L, Cevik I, et al. Role of apoptosis in testicular damage caused by varicocele. Arch Esp Urol. 1998;9:947–50.
18. Benoff S, Gilbert BR. Varicocele and male infertility: Part I. Preface. Hum Reprod Update. 2001;7:47–54.
19. Naugton DK, Nangia AK, Agarwal A. Varicocele and male infertility: Part II. Pathophysiology of varicoceles in male infertility. Hum Reprod Update. 2001;7:473–81.
20. Koksal IT, Ishak Y, Usta M, et al. Varicocele-induced testicular dysfunction may be associated with disruption of blood-testis barrier. Arch Androl. 2007;53:43–8.
21. Wang YX, Lei C, Dong SG, et al. Study of bilateral histology and meiotic analysis in men undergoing varicocele ligation. Fertil Steril. 1991;55:152–5.
22. Agarwal A, Prabakaran S, Allamaneni SS. Relationship between oxidative stress, varicocele and infertility: a meta-analysis. Reprod Biomed Online. 2006;12:630–3.
23. Kamyar Ghabili K, Shoja MM, Agutter PS, et al. Hypothesis: intracellular acidification contributes to infertility in varicocele. Fertil Steril. 2009;92:399–401.
24. Naughton CK, Nangia AK, Agarwal A. Pathophysiology of varicoceles in male infertility. Hum Reprod Update. 2001;7:473–81.
25. World Health Organization. The influence of varicocele on parameters of fertility in a large group of men presenting to infertility clinics. Fertil Steril. 1992;57:1289–93.
26. Jarow JP. Effects of varicocele on male fertility. Hum Reprod Update. 2001;7:59–64.
27. Evers JL, Collins JA. Assessment of efficacy of varicocele repair for male subfertility: a systematic review. Lancet. 2003;361: 1849–52.
28. Nieschlag E, Hertle L, Fischedick A, et al. Update on treatment of varicocele: counseling as effective as occlusion of the vena spermatica. Hum Reprod. 1998;13:2147–50.
29. Redmon JB, Carey P, Pryor JL. Varicocele-the most common cause of male factor infertility? Hum Reprod Update. 2002;8:53–8.
30. Dubin R, Amelar RD. Varicocele size and results of varicocelectomy in selected subfertile men with varicocele. Fertil Steril. 1970; 21:606–9.
31. Trum JW, Gubler FM, Laan R, van der Veen F. The value of palpation, varicoscreen contact thermography and colour Doppler ultrasound in the diagnosis of varicocele. Hum Reprod. 1996;11:1232–5.
32. Gat Y, Bachar GN, Zukerman Z, et al. Physical examination may miss the diagnosis of bilateral varicocele: a comparative study of 4 diagnostic modalities. J Urol. 2004;172:1414–7.
33. Geatti O, Gasparini D, Shapiro B. A comparison of scintigraphy, thermography, ultrasound and phlebography in grading of clinical varicocele. J Nucl Med. 1991;32:2092–7.
34. Chiou RK, Anderson JC, Wobig RK, et al. Color Doppler ultrasound criteria to diagnose varicoceles: correlation of a new scoring system with physical examination. Urology. 1997;50:953–6.
35. Hirsh AV, Cameron KM, Tyler JP, et al. The Doppler assessment of varicoceles and internal spermatic vein reflux in infertile men. Br J Urol. 1980;52:50–6.
36. Nagler HM, Luntz RK, Martinis FG. Varicocele. In: Lipshultz LI, Howards SS, editors. Infertility in the Male. 3rd ed. Mosby; 1997. pp. 336–359.
37. Practice Committee of the American Society for Reproductive Medicine. Report on varicocele and infertility. Fertil Steril. 2006 Nov;86(5 Suppl 1):S93–5. Review. PubMed PMID: 17055852.
38. Cavallini G, Ferraretti AP, Gianaroli L, et al. Cinnoxicam and L-carnitine/acetyl-L-carnitine treatment for idiopathic and varicocele-associated oligoasthenospermia. J Androl. 2004;25:761–70.
39. Unal D, Yeni E, Verit A, Karatas OF. Clomiphene citrate versus varicocelectomy in treatment of subclinical varicocele: a prospective randomized study. Int J Urol. 2001;8:227–30.
40. Micic S, Tilic C, Dotlic R. Kallikrein therapy of infertile men with varicocele and impaired sperm motility. Andrologia. 1990;22:179–83.
41. De Rose AF, Gallo F, Giglio M, et al. Early use of menotropin in the treatment of varicocele. Arch Ital Urol Androl. 2003;75:53–7.
42. Oliva A, Dotta A, Multigner L. Pentoxifylline and antioxidants improve sperm quality in male patients with varicocele. Fertil Steril. 2009;91(Suppl):1536–9.
43. Paradiso Galatioto G, Gravina GL, Angelozzi G, et al. May antioxidant therapy improve sperm parameters of men with persistent oligospermia after retrograde embolization for varicocele? World J Urol. 2008;26:97–102.
44. Cayan S, Shavakhabov S, Kadioglu A. Treatment of palpable varicocele review in infertile men: a meta-analysis to define the best technique. J Androl. 2009;30:33–40.
45. Shlansky-Goldberg RD, Van Arsdalen KN, Rutter CM, et al. Percutaneous varicocele embolization versus surgical ligation for the treatment of infertility: changes in seminal parameters and pregnancy outcomes. J Vasc Interv Radiol. 1997;8:759–67.
46. Bahren W, Lenz M, Porst H, et al. Side effects, complications and contraindications for percutaneous sclerotherapy of the internal spermatic vein in the treatment of idiopathic varicocele. Rofo. 1983;138:172–9.
47. Matthews RD, Roberts J, Walker WA, et al. Migration of intravascular balloon after percutaneous embolotherapy of varicocele. Urology. 1992;39:373–5.
48. Sharlip ID, Jarow JP, Belker AM, et al. Best practice policies for male infertility. Fertil Steril. 2002;77:873–82.
49. Sautter T, Sulser T, Suter S, et al. Treatment of varicocele: a prospective randomized comparison of laparoscopy versus antegrade sclerotherapy. Eur Urol. 2002;41:398–400.
50. Al-Kandari AM, Shabaan H, Ibrahim HM, et al. Comparison of outcomes of different varicocelectomy techniques: open inguinal, laparoscopic, and subinguinal microscopic varicocelectomy: a randomized clinical trial. Urology. 2007;69:417–20.
51. Hopps CV, Lemer ML, Schlegel PN, et al. Intraoperative varicocele anatomy: a microscopic study of the inguinal versus subinguinal approach. J Urol. 2003;170:2366–70.
52. Tuccar E, Yaman O, Erdemli E, et al. Histomorphological differences of spermatic cords regarding subinguinal versus inguinal levels: a cadaveric study. Urol Int. 2009;82:444–7.
53. Schlesinger MH, Wilets IF, Nagler HM. Treatment outcome after varicocelectomy. A critical analysis. Urol Clin North Am. 1994; 21:517–29.
54. Colpi GM, Carmignani L, Nerva F, et al. Surgical treatment of varicocele by a subinguinal approach combined with antegrade intraoperative sclerotherapy of venous vessels. BJU Int. 2006;97:142–5.
55. Steckel J, Dicker AP, Goldstein M. Relationship between varicocele size and response to varicocelectomy. J Urol. 1993;149: 769–71.
56. Matkov TG, Zenni M, Sandlow J, et al. Preoperative semen analysis as a predictor of seminal improvement following varicocelectomy. Fertil Steril. 2001;75:63–8.
57. Marks JL, McMahon R, Lipshultz LI. Predictive parameters of successful varicocele repair. J Urol. 1986;136:609–12.
58. Yoshida K, Kitahara S, Chiba K, et al. Predictive indicators of successful varicocele repair in men with infertility. Int J Fertil. 2000;45:279–84.
59. Cayan S, Lee D, Black LD, Reijo Pera RA, Turek PJ. Response to varicocelectomy in oligospermic men with and without defined genetic infertility. Urology. 2001;57:530–5.

60. Pryor JL, Kent-First M, Muallem A, et al. Microdeletions in the Y chromosome of infertile men. N Engl J Med. 1997;336:53453–9.
61. Kondo Y, Ishikawa T, Yamaguchi K, et al. Predictors of improved seminal characteristics by varicocele repair. Andrologia. 2009;41: 20–3.
62. Yamamoto M, Hibi H, Hirata Y, et al. Effect of varicocelectomy on sperm parameters and pregnancy rate in patients with subclinical varicocele: a randomized prospective controlled study. J Urol. 1996;155:1636–8.
63. Zini A, Boman J, Jarvi K, et al. Varicocelectomy for infertile couples with advanced paternal age. Urology. 2008;72:109–13.
64. Libman J, Jarvi K, Lo K, Zini A. Beneficial effect of microsurgical varicocelectomy is superior for men with bilateral versus unilateral repair. J Urol. 2006;176:2602–5.
65. Kantartzi PD, Goulis ChD, Goulis GD, et al. Male infertility and varicocele: myths and reality. Hippokratia. 2007;11:99–104.
66. Zheng YQ, Gao X, Li ZJ, Yu YL, Zhang ZG, Li W. Efficacy of bilateral and left varicocelectomy in infertile men with left clinical and right subclinical varicoceles: a comparative study. Urology. 2009;73:1236–40.
67. Elbendary MA, Elbadry AM. Right subclinical varicocele: how to manage in infertile patients with clinical left varicocele? Fertil Steril. 2009;92:2050–3.
68. Dhabuwala CB, Hamid S, Moghisi KS. Clinical versus subclinical varicocele: improvement in fertility after varicocelectomy. Fertil Steril. 1992;57:854–7.
69. Amelar RD, Dubin L. Right varicocelectomy in selected infertile patients who have failed to improve after previous left varicocelectomy. Fertil Steril. 1987;47:833–7.
70. Carpi A, Sabanegh E, Mechanick J. Controversies in the management of nonobstructive azoospermia. Fertil Steril. 2009;91:963–70.
71. Esteves SC, Glina S. Recovery of spermatogenesis after microsurgical subinguinal varicocele repair in azoospermic men based on testicular histology. Int Braz J Urol. 2005;31:541–8.
72. Matthews GJ, Matthews ED, Goldstein M. Induction of spermatogenesis and achievement of pregnancy after microsurgical varicocelectomy in men with azoospermia or severe oligoasthenospermia. Fertil Steril. 1998;70:71–5.
73. Kim ED, Leibman BB, Grinblat DM, et al. Varicocele repair improves semen parameters in azoospermic men with spermatogenic failure. J Urol. 1999;162:737–40.
74. Kadioglu A, Tefekli A, Cayan S, et al. Microsurgical inguinal varicocele repair in azoospermic men. Urology. 2001;57:328–33.
75. Cayan M, Altug U. Induction of spermatogenesis by inguinal varicocele repair in azoospermic men. Arch Androl. 2004;50:145–50.
76. Poulakis V, Ferakis N, de Vries R, et al. Induction of spermatogenesis in men with azoospermia or severe oligoteratoasthenospermia after antegrade internal spermatic vein sclerotherapy for the treatment of varicocele. Asian J Androl. 2006;8:613–9.
77. Ishikawa T, Kondo Y, Yamaguchi K, et al. Effect of varicocelectomy on patients with unobstructive azoospermia and severe oligospermia. Br J Urol Int. 2007;101:216–8.
78. Lee JS, Park HJ, Seo JT. What is the indication for varicocelectomy in men with nonobstructive azoospermia? Urology. 2007;69:352–5.
79. Weedin JW, Khera M, Lipshultz LI. Varicocele Repair in Patients with Nonobstructive Azoospermia – A Meta-Analysis. J Urol. 2010 Jun;183(6):2309–15.
80. Esteves SC. Editorial comment. J Urol. 2010 Jun;183(63):2315.
81. Esteves SC, Verza Jr S, Gomes AP. Successful retrieval of testicular spermatozoa by microdissection (micro-TESE) in nonobstructive azoospermia is related to testicular histology. Fertil Steril. 2006;86(Suppl):354.
82. Verza Jr S, Esteves SC. Sperm defect severity rather than sperm source is associated with lower fertilization rates after intracytoplasmic sperm injection. Int Braz J Urol. 2008;34:49–56.
83. Schlegel PN, Kaufmann J. Role of varicocelectomy in men with nonobstructive azoospermia. Fertil Steril. 2004;81:1585–8.
84. Inci K, Hascicek M, Kara O, et al. Sperm retrieval and intracytoplasmic sperm injection in men with nonobstructive azoospermia, and treated and untreated varicocele. J Urol. 2009;182:1500–5.
85. Esteves SC. Editorial comment. J Urol. 2009;182:1504–5.
86. Collodel G, Capitani S, Iacoponi F, et al. Retrospective assessment of potential negative synergistic effects of varicocele and tobacco use on ultrastructural sperm morphology. Urology. 2009;74:794–9.
87. Zampieri N, Zamboni C, Ottolenghi A, et al. The role of lifestyle changing to improve the semen quality in patients with varicocele. Minerva Urol Nefrol. 2008;60:199–204.
88. Agarwal A, Deepinder F, Cocuzza M, et al. Efficacy of varicocelectomy in improving semen parameters: new meta-analytical approach. Urology. 2007;70:532–8.
89. Giulini S, Sblendorio V, Xella S, et al. Seminal plasma total antioxidant capacity and semen parameters in patients with varicocele. Reprod Biomed Online. 2009;18:617–21.
90. Blumer CG, Fariello RM, Restelli AE, et al. Sperm nuclear DNA fragmentation and mitochondrial activity in men with varicocele. Fertil Steril. 2008;90:1716–22.
91. Talebi AR, Moein MR, Tabibnejad N, et al. Effect of varicocele on chromatin condensation and DNA integrity of ejaculated spermatozoa using cytochemical tests. Andrologia. 2008;40:245–51.
92. Shiraishi K, Takihara H, Naito K. Testicular volume, scrotal temperature, and oxidative stress in fertile men with left varicocele. Fertil Steril. 2009;91 Suppl 4:1388–91.
93. Caretta N, Palego P, Schipilliti M, et al. Testicular contrast harmonic imaging to evaluate intratesticular perfusion alterations in patients with varicocele. J Urol. 2010;183:263–9.
94. Jeng SY, Wu SM, Lee JD. Cadmium accumulation and metallothionein overexpression in internal spermatic vein of patients with varicocele. Urology. 2009;73:1231–5.
95. Ichioka K, Nagahama K, Okubo K, et al. Genetic polymorphisms in glutathione S-transferase T1 affect the surgical outcome of varicocelectomies in infertile patients. Asian J Androl. 2009;11:333–41.
96. Meng MV, Greene KL, Turek PJ. Surgery or assisted reproduction? A decision analysis of treatment costs in male infertility. J Urol. 2005;174:1926–31.
97. Ficarra V, Cerruto MA, Liguori G, et al. Treatment of varicocele in subfertile men: the Cochrane Review–a contrary opinion. Eur Urol. 2006;49:258–63.
98. Marmar JL, Agarwal A, Prabaskan S, et al. Reassessing the value of varicocelectomy as a treatment for male subfertility with a new meta-analysis. Fertil Steril. 2007;88:639–48.
99. Cervellione RM, Cervato G, Zampieri N, et al. Effect of varicocelectomy on the plasma oxidative stress parameters. J Pediatr Surg. 2006;41:403–6.
100. Chen SS, Huang WJ, Chang LS, Wei YH. Attenuation of oxidative stress after varicocelectomy in subfertile patients with varicocele. J Urol. 2008;179:639–42.
101. Zini A, Blumenfeld A, Libman J, et al. Beneficial effect of microsurgical varicocelectomy on human sperm DNA integrity. Hum Reprod. 2005;20:1018–21.
102. Moskovtsev SI, Lecker I, Mullen JB, et al. Cause-specific treatment in patients with high sperm DNA damage resulted in significant DNA improvement. Syst Biol Reprod Med. 2009;55:109–15.
103. Smit M, Romijn JC, Wildhagen MF, et al. Decreased sperm DNA fragmentation after surgical varicocelectomy is associated with increased pregnancy rate. J Urol. 2010;183:270–4.
104. Cayan S, Erdemir F, Ozbey I, et al. Can varicocelectomy significantly change the way couple use assisted reproductive technologies? J Urol. 2002;167:1749–52.
105. Esteves SC, Oliveira FV, Bertolla RP. Clinical outcomes of intracytoplasmic sperm injection in infertile men with treated and untreated clinical varicocele. J Urol. 2010 Oct;184(4):1442–6.

Infection in Infertility

25

Ralf Henkel

Abstract

Despite global overpopulation, human infertility is a growing concern since it is declining in both developed and developing countries. Reasons for this remarkable decline are manifold and include socio-economic changes, changes in lifestyle with higher prevalence of obesity or environmental pollution. In general, data on the prevalence of infertility, i.e. the inability of a sexually active, non-contracepting couple to achieve pregnancy within 1 year's time, vary considerably between 3% and 25%, of which 15% seek for medical assistance. Infertility is a couple problem as both men and women contribute more or less equally, with prevalence reported for male infertility between 30 and 50%. Approximately 7% of all men are confronted with fertility problems during their reproductive lifetime, thus making male infertility a problem, which has an even higher prevalence than diabetes mellitus with an overall estimate of 2.8% in the year 2000 and 4.4% in 2030, and which is considered a common disease. Potentially correctable causes of male infertility are genital tract infections, which play a major role in male infertility. Infections and inflammations are not only seriously affecting spermatogenesis and sperm transit during ejaculation as can be seen in clinical findings in cases of oligozoospermia (decreased number of sperm), asthenozoospermia (decreased sperm motility) or azoospermia (absence of sperm in the ejaculate) but are also the cause of dysfunctional male accessory glands and significantly impaired sperm functions.

Keywords

Male infertility • Male urogenital tract infection • Oligozoospermia • Asthenozoospermia • Azoospermia • *Chlamydia trachomatis* • *Ureaplasma urealyticum* • *Neisseria gonorrhoeae* • *Mycoplasma hominis* • *Mycoplasma genitalium* • *Escherichia coli* • Orchitis • Epididymitis • Prostatitis

Despite global overpopulation, human infertility is a growing concern since it is declining in both developed and developing countries [1, 2]. Reasons for this remarkable decline are manifold and include socio-economic changes [2], changes in lifestyle with higher prevalence of obesity [3, 4] or environmental pollution [5]. In general, data on the prevalence of infertility, i.e. the inability of a sexually active, non-contracepting couple to achieve pregnancy within 1 year's time [6], vary considerably between 3% and 25%, of which 15% seek for medical assistance [7, 8]. Infertility is a couple problem as both men and women contribute more or less equally, with prevalence reported for male infertility between 30 and 50% [9]. Approximately 7% of all men are confronted with fertility problems during their reproductive lifetime, thus making male infertility a problem, which has an even higher prevalence than diabetes mellitus with an overall

R. Henkel, BEd, PhD (✉)
Department of Medical Biosciences, University of the Western Cape, Private Bag X17, Bellville, 7535, South Africa
e-mail: rhenkel@uwc.ac.za

S.J. Parekattil and A. Agarwal (eds.), *Male Infertility: Contemporary Clinical Approaches, Andrology, ART & Antioxidants*,
DOI 10.1007/978-1-4614-3335-4_25,

estimate of 2.8% in the year 2000 and 4.4% in 2030, and which is considered a common disease [10, 11].

Potentially correctable causes of male infertility are genital tract infections [12], which play a major role in male infertility. Infections and inflammations are not only seriously affecting spermatogenesis and sperm transit during ejaculation as can be seen in clinical findings in cases of oligozoospermia (decreased number of sperm), asthenozoospermia (decreased sperm motility) or azoospermia (absence of sperm in the ejaculate) [13, 14], but are also the cause of dysfunctional male accessory glands [12] and significantly impaired sperm functions [15, 16]. These changes can be triggered in various ways, namely direct action of the pathogens on spermatozoa and sperm functions [17] or indirectly by inducing inflammatory processes in the seminal tract by activating leukocytes [18]. In non-selected cases, the prevalence of male genital tract infection-related infertility varies between 10 and 20% and amounts to up to 35% in a large study comprising more than 4,000 patients consulting for infertility [19]. It also appears that bacterial infections have a more detrimental effect in fertility-compromised patients than in fertile men [20], indicating that the impact of such bacterial genital tract infections may have to be differentiated.

Pathogens Causing Male Genital Tract Infections

Male urogenital tract infections can be classified according to the kind of microorganism causing the infection and the location, namely the testis (orchitis), epididymis (epididymitis), prostate (prostatitis) or urethra (urethritis). The most prevalent pathogens are *Chlamydia trachomatis*, *Ureaplasma urealyticum*, *Neisseria gonorrhoeae*, *Mycoplasma hominis*, *Mycoplasma genitalium* or *Escherichia coli*. While the first pathogens are sexually transmitted, *E. coli* is regarded as the most common cause of non-sexually transmitted urogenital tract infection, particularly of epididymo-orchitis or prostatitis where it is the cause of 65–80% of the cases [21]. Furthermore, viral infections like mumps virus, human papillomavirus (HPV), herpes simplex virus (HSV) and particularly human immunodeficiency virus (HIV) have also been associated with increased seminal leukocyte concentrations [22]. The latter virus can infect the testes and male sex accessory glands [23].

C. trachomatis

C. trachomatis is a gram-negative bacterium that is worldwide one of the most frequently sexually transmitted bacterial pathogens, accounting for an estimated 92 million new urogenital infections per year [24]. This number may be underestimated because of the high asymptomatic nature of the pathogen, with approximately 70–80% of women and up to 50% of men infected having no symptoms [25]. Reportedly, in symptomatic men, the prevalence of infections varies from 4 to 10% [26]. Another report found a prevalence of up to 35.9% in men and 38% in women [27]. Since a high number of chlamydial infections remain undiagnosed, the pathogen can even be transferred to the newborn during delivery accounting for 25–50% of conjunctivitis and 10–20% of pneumonia in newborns, thus posing a health risk on the offspring as well as enormous costs for diagnosis and treatment on countries' health systems. For the United States, these costs estimate to $2.2 million for each 500 cases [25].

In men, *C. trachomatis* has been associated with prostatitis, epididymitis and urethritis. The bacterium has been detected in the testis including Leydig cells [28], the prostate [29] and even epididymis and seminal vesicles [30]. Apart from the lesions triggered by the infection and the implications and acute inflammation can cause in the male genital tract, reports on the influence of chlamydial infections on male fertility are inconsistent. While some authors [31, 32] found no significant association, most others have shown a direct negative influence of the chlamydial infection on male fertility [29, 33, 34]. *In vitro* studies by Hosseinzadeh et al. [35–37] even indicate that the pathogen directly causes changes in sperm proteins and premature cell death induced by lipopolysaccharides secreted by Chlamydiae. These discrepancies in findings may be due to the fact that in numerous cases, chlamydial infections are accompanied by other microbial infections making a clear association of infertility with the chlamydial infection difficult [38]. Furthermore, the various detection methods for the pathogen in asymptomatic cases appear challenging, and especially the sensitivity and specificity of serological markers were regarded as problematic [25]. For that reason, recent studies suggest PCR amplification of bacterial rDNA in semen as this approach seems to be promising especially to identify asymptomatic patients [39, 40].

Mycoplasms

Mycoplasms and ureaplasms belong to the family of *Mycoplasmataceae*, which are widely distributed in vertebrate species. They are the smallest bacteria replicating in culture and are lacking a cell wall [41, 42]. Five species colonize the male urogenital tract and may contaminate the semen during ejaculation, namely *U. urealyticum*, *U. parvum*, *M. hominis*, *M. genitalium* and *M. fermentans*, of which only the first four are pathogenic.

U. urealyticum

Among other disorders, *U. urealyticum* is reported to cause non-gonococcal urethritis, pelvic inflammatory disease or infertility [12, 43]. In the past, the association between ureaplasma infections and male infertility was discussed controversially [44, 45]. In a recent large study including 346 selected subjects, Wang et al. [46] showed that ureaplasma infections cause higher seminal viscosity, decreased sperm concentrations, and lower pH. Although this study could not find any further effect on other sperm parameters, Potts et al. [45] found significantly increased seminal reactive oxygen species (ROS) levels, and Reichart et al. [47] higher levels of sperm DNA damage. These discrepancies might be due to the fact that *U. urealyticum* shows a distinct energy metabolism-dependent effect on sperm activity [48]. According to this theory, sperm motility is impaired by the bacterium at low pH since it competes with sperm mitochondrial energy production. On the other hand, at higher pH, sperm motility will even be enhanced as ureaplasma stimulates glycolysis.

The prevalence of male genital ureaplasma infections among infertile men varies considerably from 10 to 40% [49], most probably as a result of different diagnostic methods used in various studies and different population groups examined. Like in patients with chlamydial infections, detection of *U. urealyticum* and mycoplasms is particularly problematic in asymptomatic patients because these patients may shed fewer colony-forming units (CFU; organisms) to be detected in a standard culture assay. Therefore, PCR should be the diagnostic method of choice [50].

M. hominis, M. genitalium

Although the direct impact of these two bacteria on male fertility was debated in earlier years, recent reports associate *M. hominis* and *M. genitalium* with genitourinary infections [51, 52]. Frequencies of infection are with 10.8% for *M. hominis* and 5% for *M. genitalium*, respectively, reportedly lower than for other pathogens [53]. Both species can attach to and penetrate human sperm plasma membrane [54, 55], which might have a significant long-term impact on male fertility as well as on the onset of pregnancy and the health of the offspring. While the first may contribute to the distribution of the bacterium to the female to cause cervicitis and endometritis [56] or an alteration of the plasma membrane affecting acrosome reaction [57], the latter might particularly be caused by the sperm DNA damage triggered by the infection [33].

N. gonorrhoeae

These gram-negative, immotile cocci are growing in pairs (diplococci) and cause the most common infectious disease in men leading to urethritis, prostatitis and epididymitis which in turn may impair male fertility. These bacteria have pili on their surface, which facilitate attachment to other cells [58]. In sperm, an asialoglycoprotein receptor has been identified that recognizes and binds lipopolysaccharides in gonococcal membranes [59]. Since chlamydial lipopolysaccharides can cause death of human sperm by inducing apoptosis [60], it is conceivable that Neisseria lipopolysaccharides might also induce such reactions. Yet, this has not been shown thus far.

Even though its incidence has been declining in Western countries during the past decades, 150–400 new infections per 100,000 are still recorded in Europe per year [58]. Presumably due to socio-economic and behavioural factors, these numbers are much higher in Third World and developing countries, with highest numbers in Sub-Saharan Africa and South and Southeast Asia [61].

E. coli

E. coli is a gram-negative bacterium that belongs to the family of *Enterobacteriaceae* and is responsible for most urogenital tract and male accessory gland infections with up to 80% of the bacterial prostatitis [12, 62]. In contrast to other uropathogens like *Enterococcus* or *Staphylococcus saprophyticus*, *E. coli* has significant direct negative effects on sperm motility [63]. An in vitro study by Köhn et al. [57] revealed that this microbe can even negatively affect sperm functions like acrosome reaction. This might be due to morphological alterations, particularly on the acrosome and flagellum, seen after exposure of human sperm to the pathogen [64]. Due to the direct interaction of bacterial pili with the sperm plasma membrane [65], *E. coli* is interfering with sperm motility [66]. A recent study [67] demonstrated two mechanisms by which *E. coli* affects spermatozoa, the described direct interaction and by action of soluble factors that induce apoptosis and a breakdown in the mitochondrial membrane potential. Potential candidate substances causing these cellular reactions might be α-hemolysin and Shiga-like toxin as these have already been associated with sperm motility loss [68] and apoptosis in Hep-2 cells [69], respectively.

Viruses

A number of viruses are also able to infect all parts of the male genital system, namely the testes (e.g. mumps virus, HIV-1), epididymis (e.g. Coxsackie virus), seminal vesicle (e.g. cytomegalovirus [CMV]), prostate (e.g. HPV, HSV, HIV-1) and the semen (e.g. HSV, HPV, HIV) [70]. In a recent study including 241 asymptomatic patients attending an infertility clinic for semen analysis, viral DNA was detected in 45 patients; CMV (8.7%), HPV (4.5%), HHV-6 (3.7%),

HSV (3.6%), Epstein–Barr virus (0.4%) and hepatitis B virus (0.0%) [71].

While for some of the viral infections like HSV or HIV-1 are associated with poor semen and sperm quality [22, 72], this could not be confirmed for CMV and human herpesvirus type 6 (HHV6) [71]. Thus, the latter viruses appear not to cause male infertility. However, in patients with viral infections affecting male fertility, leukocytospermia (>1×10^6 leukocytes/ml ejaculate) is strongly associated with the infection as well as with elevated levels of inflammatory markers like PMN-elastase or interleukins [71].

A matter of concern is the high number of globally more than 65 million people infected with HIV, out of which a growing number of HIV-positive people are seeking for assisted reproduction to have children without infecting the partner or the offspring. Since a few years, some IVF centres started treating infected couples and received acceptable pregnancy rates; the worst in couples where both partners were infected [73]. In these patients, not only the risk of vertical transmission from a seropositive mother to her unborn child has to be dealt with but also the fact that semen is a vector of viral propagation and sperm can bind and incorporate the virus via a CD4-independent receptor and/or the HIV co-receptor CCR5 [74, 75]. Despite it is generally accepted that motile sperm are not productively infected by HIV, sperm can carry viral particles deriving from the testis or epididymis [76, 77]. In view of the fact that seminal leukocytes shed different viral strains than those in the blood [78], the question rises whether infected leukocytes and free virions contaminating the semen are of different origin, and an infected testis might represent a special reservoir for the virus as this area is resistant to antiviral drugs due to the blood–testis barrier [77]. Therefore, special care must be taken when separating sperm for assisted reproduction, particularly for ICSI.

Male Genital Tract Infections

Orchitis

As per definition by the European Association of Urology, the term *orchitis* describes an inflammatory lesion of the testicle, which is associated with a leukocytic exudate inside and outside the seminiferous tubules, resulting in tubular damage [13]. The condition can be a reason for spermatogenic arrest and testicular atrophy leading to low seminal sperm concentration and poor sperm quality [13, 79]. An acute infectious orchitis is characterized by an abrupt onset of severe pain with visible swelling of the affected testicle as well as of the inguinal lymph nodes and can be accompanied by fever. Further symptoms are similar to those of testicular torsion and can include hematuria and blood in the ejaculate. Subacute and chronic inflammatory conditions, however, often remain asymptomatic [80]. Yet, an orchitis may lead to intratesticular obstruction, which is the case in about 15% of obstructive azoospermia [13].

In contrast, non-infectious inflammations of the testis may occur in patients with testicular seminoma where predominantly CD8-positive T-lymphocytes infiltrate the tumour tissue, and macrophages are found in the fibrovascular septae and at the periphery [14]. Such infiltration of activated T-lymphocytes into testicular tissue is also indicative of a significant disturbance of the local immunoregulation [81]. Thus, due to an impaired blood–testis barrier, these inflammatory cells overcome the testicular immunosuppressive mechanisms and the formation of anti-sperm autoantibodies under such conditions would be conceivable, particularly in cases of prolonged inflammatory processes as is in chronic orchitis [82]. On the other hand, except for a few cases with positive titres for autoantibodies in patients with a mumps history, only little evidence is available for this relationship [83, 84].

As reported in a large study, the prevalence of an isolated orchitis is with 0.42% among testicular pathologies relatively low [85]. However, due to retrograde ascending infectious lesions, a "non-specific" orchitis triggered by *Pneumococcus* sp., *Salmonella* sp., *Klebsiella* sp. or *Haemophilus influenza* is in most cases associated with an epididymitis as epididymo-orchitis. The close vicinity of the different compartments as well as the ascending nature of the infections makes a distinction between inflammations, i.e. isolated epididymitis vs. epididymo-orchitis, very difficult in the clinical routine [82].

Sexually transmitted bacteria like *C. trachomatis* and *N. gonorrhoeae* are the cause of the acute infection in men younger than 35 years, while *E. coli* is the predominant trigger in older men [14]. On the other hand, an orchitis can also occur after hematogenous dissemination of pathogens like the Coxsackie B or the mumps virus [13] as a complication of a systemic viral infection. For instance, the mumps virus may affect the tests (mumps orchitis) in 20–30% [86] and lead to infertility in 13% of the cases with unilateral and in 30–87% in patients with bilateral orchitis [87, 88]. However, clinicians should realize that an epididymo-orchitis can also develop secondary to the mumps infection, even in the absence of a parotitis [89]. While bacterial infections as described above cause a 'non-specific' orchitis, *Mycobacterium tuberculosis*, *M. leprae*, *Treponema pallidum* or *Brucella* sp. may cause a 'specific', predominantly granulomatous orchitis [13].

Epididymitis

Epididymitis, an inflammation of the epididymis, is a painful, feverish, almost unilateral condition with tender swelling

Table 25.1 NIH classification and definition of the categories of prostatitis

NIH classification		Definition
Category I	Acute bacterial prostatitis	Acute infection of the prostate gland
Category II	Chronic bacterial prostatitis	Recurrent infection of the prostate
Category III	Chronic abacterial prostatitis/CPPS	No demonstrable infection
Category IIIA	Inflammatory CPPS	White blood cells in semen/EPS/voided bladder urine-3 (VB-3 or postprostatic massage)
Category IIIB	Non-inflammatory CPPS	No white blood cells in semen/EPS/VB-3
Category IV	Asymptomatic inflammatory prostatitis	No subjective symptoms detected either by prostate biopsy or the presence of white blood cells in EPS/semen during evaluation for other disorders

of epididymis and scrotum. Based on the duration of the symptoms, epididymitis may be sub-classified as acute, subacute and chronic. In the latter, symptoms are present for more than 6 weeks. In cases of acute infectious epididymitis, an involvement of the testicle due to ascending canalicular bacterial infections can represent a complication which may occur in up to 60% of affected patients as epididymo-orchitis [90]. Although up to 35% of patients consulting for fertility problems present with male genital tract infections [19, 91], data on the prevalence of epididymitis/epididymo-orchitis vary considerably from 0.29% of all consultations [92] to 20% of all urologic admissions in an US Army setup [93].

As for orchitis, *C. trachomatis* and *N. gonorrhoeae* are the most common cause of epididymitis in sexually active men younger than 35 years. In contrast, Gram-negative *Enterobacteriaceae*, of which *E. coli* is the predominant pathogen, are aetiologically responsible for the disease in older men [12, 90], which are also particularly at risk of having urethral strictures, bladder neck obstruction or benign prostate hyperplasia (BPH) resulting in increased voiding pressure to empty the bladder resulting in a reflux of contaminated urine into the excurrent genital ducts and subsequent infection [94].

Potential risk factors for epididymitis include sexual activity, strenuous activities like lifting heavy goods, bicycle and motorbike riding or extended periods of sitting at job or travelling [95]. Even traumatic events such as accidents or scrotal traumas and iatrogenic injury to the epididymis during surgeries can be a cause of epididymitis.

Major problems for male fertility may arise particularly in patients with epididymitis as this disease appears to have a greater influence on semen quality and male fertility than an infection/inflammation of the prostate or seminal vesicle [96]. In addition, in quite a number of patients, the diagnosis of chronic epididymitis is extremely difficult as these patients do not feel discomfort and their health is not compromised [82]. Due to a *silent* nature of the infection/inflammation, epididymitis will only be diagnosed once these patients appear in an andrological clinic consulting for infertility. Eventually, inflammatory lesions of the epididymis can result in dysfunction of the organ and ultimately in obstructive azoospermia, which is the most common cause for this condition [13].

In patients with an acute epididymitis, a semen analysis is not recommended [13]. In cases of chronic epididymitis, semen parameters may be dramatically affected with lower sperm count, motility and seminal α-glucosidase. In contrary, many patients can present with leukocytospermia, i.e. leukocyte count of more than 1×10^6/ml [97], elevated seminal levels of polymorphonuclear granulocyte elastase and atypically stained sperm flagella [82].

Prostatitis

Despite the numbers of urological consultations for prostatitis outnumbered those for benign prostate hyperplasia or prostate cancer [98], prostatitis has been called "the third most important disease of the prostate gland" [99]. Epidemiological studies revealed an estimated prevalence of 4–11% for prostatitis, which then represents the most common urological diagnosis in men younger than 50 years [100]. A large study with more than 600 men included suggests that only about 5–10% of these cases are of bacterial origin [101]. Yet, more recently, Bjerklund Johansen et al. [102] and Nickel et al. [103] showed that only about 50% of all patients with chronic prostatitis (bacterial and abacterial) respond positively to antibiotic treatment.

According to the classification system suggested by Drach et al. [104], prostatitis was divided into four clinical syndromes, namely acute, chronic bacterial and chronic abacterial prostatitis, and prostatodynia. Considering that this system was never validated and left clinicians confused about diagnostic and therapeutic strategies, particularly since many cases without infection have pathogenic processes outside the prostate, the National Institutes of Health (NIH) introduced a new system classifying prostatitis [105, 106] (Table 25.1). Even though many clinicians diagnose "prostatitis," it rather represents diverse clinical symptoms ranging from acute bacterial infection to chronic pelvic pain and should actually be referred to as "prostatitis syndrome" as

Table 25.2 Uropathogens causing chronic prostatitis

Established pathogens	Potential pathogens
Escherichia coli	*Staphylococcus saprophyticus*
Klebsiella pneumonia	*Staphylococcus aureus*
Proteus mirabilis	*Staphylococcus epidermidis*
Pseudomonas aeruginosa	*Mycoplasma genitalium*
Enterococcus faecalis	*Ureaplasma urealyticum*
Chlamydia trachomatis	

Data from Nickel JC, Moon T. Chronic bacterial prostatitis: an evolving clinical enigma. Urology. 2005;66:2–8

patients present with a variety of urogenital, perineal and perianal complaints [99, 107].

Considering the various problems around understanding prostatitis and its diagnosis [108], particularly chronic bacterial prostatitis, it is obvious that the diagnosis of an acute infection does not pose a problem for urologists. The symptoms for the acute bacterial prostatitis are quite clear; the patients are presenting with an urosepsis, fever, obstructive voiding symptoms and local pelvic pain [109]. In about 80% of acute bacterial prostatitis, *E. coli* can be identified as pathogen, while *Pseudomonas aeruginosa*, *Klebsiella* or Enterococci are the cause in the remaining patients [110]. Nevertheless, there are major concerns differentiating the category II from category III prostatitis. Many of these patients have a history of recurrent urinary tract infections and are asymptomatic in non-infectious intervals. If the bacterial culture is positive for an established uropathogen like *E. coli* or *Klebsiella* sp., the diagnosis is unproblematic [105] (Table 25.2). However, the classic classification scheme might fail in cases where enterococci or anaerobes are identified in prostate specimens. Even more so in patients who had initially positive cultures but are negative at the time of recurrent symptoms. Apparently, in these cases, the detection is false negative because the bacterial colonization of the prostate can be veiled as bacteria can form microcolonies or aggregates, which are surrounded by a thick protective layer [105, 111].

Urethritis

Urethritis is the infectious or non-infectious inflammation of the urethra. While non-infectious causes include injuries through traumas, masturbation, manipulation by the patient or medical treatments, acute infectious urethritis may be caused by known sexually transmitted uropathogens like *C. trachomatis*, *Mycoplasms* or *N. gonorrhoea* with incidences of 15–26%, 10–21% and 0.4–18%, respectively. In addition, among the not sexually transmitted pathogens, Enterobacteriaceae and staphylococci are causing the disease with frequencies between 20 and 31% [112]. Chronic urethritis is a rare condition, which is why the prevalence is not known [113].

The clinical symptoms of an acute urethritis vary considerably. While some patients present with distinct urethral discharge and dysuria, others are symptom-free or only show some pus prior to the first voiding of urine in the morning, which may be occur in 5–10% of the patients [12]. As can be seen for the symptoms, clinical findings also vary from inflammatory stickiness, redness and swelling at the glans penis or the urethral orifice to the absence of any clinical sign. Normally, the infection remains localized to the urethra. However, ascension of gonococci may occur in about 1% of the infected patients causing epididymitis [112].

The impact of urethritis on male fertility is debatable, particularly since the inflammatory discharge present in the anterior urethra makes an ejaculate analysis impossible as the pus contaminates the ejaculate [114] and both direct effect of bacteria [16, 37, 67, 115] and leukocytes [116, 117] demonstrated detrimental effects on sperm functions. On the other hand, obstruction due to urethral stricture or as a result of lesions in the area of the seminal colliculus may result in ejaculatory disturbances [13].

Male Accessory Gland Infection

According to definition, the male accessory glands comprise the prostate, seminal vesicles and the Cowper's glands (bulbourethral glands). However, in many publications, the term *male accessory gland infection* (MAGI) describes the clinical symptoms of the inflammation as a result of canalicularily ascending infections of the male accessory glands including the deferent duct and the epididymis via the urethra as "prostate-seminal vesiculitis," "epididymo-prostato-vesiculitis," or "male adnexitis" as long as urethritis or a urinary tract infection has been excluded [118]. Considering that in MAGI these organs are commonly inflamed, clear distinctions between prostatitis, epididymitis and glandulitis vesiculitis cannot be made [58]. General symptoms of MAGI are leukocytospermia (more than 10^6 peroxidase-positive leukocytes/ml), elevated seminal levels of polymorphonuclear granulocyte elastase (≥230 ng/ml), C3c complement (≥0.01 mg/ml), ROS and cytokines [118–120].

As a result of the leukocyte infiltration due to the infection as well as the elevated concentrations of pro-inflammatory cytokines like interleukin-6 (IL-6), IL-8 or tumour necrosis factor-α (TNF-α) into the genital system, sperm functions may be compromised by directly affecting sperm function and intensifying the level of oxidative stress, respectively [117, 121, 122]. As a result of the inflammatory processes in the male accessory glands, their secretory function may be impaired, consequently resulting in decreased seminal concentrations of citric acid, fructose, α-glucosidase, phosphatase and zinc [58, 123, 124].

In addition, there are concerns that patients presenting with MAGI are at a higher risk of developing sperm autoantibodies due to the inflammatory processes compromising the immune barrier [125, 126]. Furthermore, like in orchitis or epididymitis, stenosis or obstruction of the excurrent ducts may occur.

Consequences of Infections on Sperm Fertilizing Capacity

Apart from the specific effects on male fertility described above, male genital tract infections and inflammations cause general reactions that also negatively affect sperm fertilizing capacity by compromising specific sperm functions. In the light of spermatozoa being the most polarized cells in the body, the male germ cell has to maintain its extreme polarization, for which one of the most important prerequisites is a highly fluid plasma membrane. Therefore, sperm cells contain an extraordinary high amount of polyunsaturated fatty acids (PUFA), particularly docosahexaenoic acid, which has 6 double bonds in its molecule [127]. Since most sperm functions are dependent on membrane functionality, this high content of PUFA is also essential for normal sperm function and respective disturbances result in a loss of sperm function.

As a result of an urogenital tract infection, activated leukocytes infiltrate the infected organs releasing high amounts of ROS and cytokines such as IL-6, IL-8 or TNF-α as inflammatory mediators [128, 129]. Both ROS and cytokines have been shown to be associated not only with the impairment of sperm functions like motility but also to DNA damage and infertility by induction and stimulation of membrane lipid peroxidation through oxidative stress [120, 130–133]. By way of this mechanism, male genital tract infections/inflammations do not only damage sperm DNA and reduce sperm count and seminal volume but also impair sperm functions like motility, acrosome reaction or acrosin activity [57, 117, 134–138].

Treatment of Infections

Generally, the first choice of treatment of male genital tract infections has to be antibiotic in order to eradicate pathogenic microorganisms, normalize inflammatory parameters, prevent transmission to the female partner and decrease the risk of potential complications [12]. Considering that many urogenital tract infections are sexually transmitted, however, simultaneous treatment of the partner has to be considered, particularly in *C. trachomatis* infections. While standardized recommendations only exist for the treatment of acute bacterial epididymitis, epididymo-orchitis and specific granulomatous orchitis [13, 139], guidelines for the treatment of chronic infections and inflammations of the male genital tract have not been drawn up yet and are rather empirical and only few uncontrolled studies are available [13, 90]. For mumps orchitis, the systemic treatment with α-2β-interferon may be considered to prevent testicular atrophy and azoospermia [140].

The very same is true for acute bacterial and chronic bacterial prostatitis. While an antibiotic treatment for these conditions is mandatory, the benefit for patients with inflammatory chronic pelvic pain syndrome is questionable [102]. Still, treatment of the prostatitis syndrome with antibiotics poses the major problem of the penetrability of the agents into the prostate and its secretions, and only a few modern antibiotics like fluoroquinolones have the chemical properties to enter these compartments well [141, 142].

In order to alleviate the inflammatory lesions, therapy with both corticosteroids and non-steroidal antiphlogistic substances has shown considerable positive effects on semen quality in terms of sperm and leukocyte count and sperm motility [143–145]. Furthermore, antioxidative therapies with vitamins and/or antioxidant supplementations to reduce the oxidative stress caused by leukocytes and defective spermatozoa are currently highly debated [146, 147]. Although several studies with various antioxidants alone or in combination have shown a significant reduction in seminal ROS levels [148–150] and improvement in sperm count and motility [151–153], other studies found the opposite [154, 155]. Therefore, notwithstanding the indubitable positive effects of an antioxidative supplementation for general health purposes, no definite recommendation can be made at this point in time with regard to the treatment of male genital tract infections. Most probably, it is not only the administration of singular antioxidative substances that causes the beneficial effects but the combination of different antioxidants at very specific concentrations.

Expert Commentary

The purpose of this chapter was to discuss the contribution of male genital tract infections/inflammations to male infertility. Considering that many of these male genital tract infections are sexually transmitted, the knowledge of its impact on the female partner as well as treatment of the couple is mandatory. Moreover, as there is still a lack of knowledge about the impact of such infections on sperm functions, this chapter is dealing with impaired sperm functions as a result of the infection/inflammation. Since many patients suffer from asymptomatic, the so-called 'silent' infections, it is essential for the clinician to identify these conditions and also to urge the patients to continue with the treatment long enough. Appropriate treatment is particularly a problem in prostatitis as only few drugs penetrate the prostate and its secretions sufficiently. Therefore, this chapter was also to give an up-to-date overview on the impact of different male genital tract infections/inflammations on sperm functions and various treatment options.

Five-Year View

Although our knowledge on male genital tract infections and its impact on male fertility increased during recent years, there is still a lack of knowledge particularly regarding the impact of the infection/inflammation on sperm functions, and the following topics will have to be addressed in future studies:

- The impact of the infection/inflammation on male germ cell's DNA.
- Currently, the seminal leukocyte concentration is under debate and several authors argue that the WHO's cut-off value of leukocytospermia is too high. Therefore, clarity has to be provided for clinicians in order to properly diagnose male genital tract infections.
- Since the diagnosis, particularly of asymptomatic patients, constitutes a problem in the clinic, new test system including PCR has to be introduced into the diagnostic set-ups.
- Special care has to be taken when handling with or performing assisted reproduction with sperm deriving from HIV-positive subjects, and further investigations have to be carried out in order to clarify the possibility of transmission of the virus.
- Development of new antibiotics that penetrate more easily into the male genital tract system.

Key Issues

- Male infertility is a major issue of concern as it affects more people during their lifetime as a common disease, diabetes mellitus.
- Among these infertile men, the prevalence of male genital tract infection is reportedly between 20 and 40%.
- Many pathogens, including bacteria and viruses, are sexually transmitted and require treatment of the couple.
- Among the sexually transmitted bacteria, infections with *C. trachomatis* and mycoplasms are most prevalent, while *E. coli* causes the most prevalent non-sexually transmitted male genital tract infection.
- Among the viral infections, mumps and HIV infections are the most important as mumps can cause permanent infertility due to mumps orchitis and HIV can be carried by spermatozoa from the testis or epididymis.
- Special care must be taken in assisted reproduction with HIV-positive men since an infected testis might represent a special reservoir for the virus as this area is resistant to antiviral drugs due to the blood-testis barrier.
- Male genital tract infections can be acute or asymptomatic, which represents a particular problem as patients and doctors might not recognize the disease that impairs male fertility.
- Male genital tract infections can ascent the genital tract.
- Apart from the specific complications such as obstructions causing azoospermia, male genital tract infections/inflammations may directly impair sperm fertilizing capacity by compromising specific sperm functions through ROS and interleukins.
- Generally, infections have to be treated with appropriate antibiotics. However, antiphlogistic treatment has also been proven to have positive effects on sperm count and motility.
- Despite certain positive effects, treatment with antioxidants like vitamins to alleviate the effects of ROS is still highly debated.

References

1. Lutz W, O'Neill BC, Scherbov S. Demographics. Europe's population at a turning point. Science. 2003;299:1991–2.
2. Skakkebaek NE, Jorgensen N, Main KM, et al. Is human fecundity declining? Int J Androl. 2006;29:2–11.
3. Pasquali R, Pelusi C, Genghini S, et al. Obesity and reproductive disorders in women. Hum Reprod Update. 2003;9:359–72.
4. Sallmen M, Sandler DP, Hoppin JA, et al. Reduced fertility among overweight and obese men. Epidemiology. 2006;17:520–3.
5. Foster WG, Neal MS, Han MS, et al. Environmental contaminants and human infertility: hypothesis or cause for concern? J Toxicol Environ Health B Crit Rev. 2008;11:162–76.
6. World Health Organization. Manual for the standardized investigation and diagnosis of the infertile couple. Cambridge, UK: Cambridge University; 2000.
7. Dohle GR, Colpi GM, Hargreave TB, et al. EAU guidelines on male infertility. Eur Urol. 2005;48:703–11.
8. Henkel R. ROS and sperm DNA integrity—implications of male accessory gland infections. In: Giwercman A, Tournaye H, Björndahl L, Weidner W, editors. Clinical andrology, informa healthcare. London: Parthenon Publishing, Marcel Dekker, Taylor & Francis Medical; 2010. p. 324–8.
9. Hull MG, Glazener CM, Kelly NJ, et al. Population study of causes, treatment, and outcome of infertility. Br Med J (Clin Res Ed). 1985;291:1693–7.
10. Nieschlag E, Behre HM. Andrology. Male reproductive health and dysfunction. 2nd ed. Berlin, Heidelberg, New York: Springer; 2000.
11. Wild S, Roglic G, Green A, et al. Global prevalence of diabetes: estimates for the year 2000 and projections for 2030. Diabetes Care. 2004;27:1047–53.
12. Weidner W, Krause W, Ludwig M. Relevance of male accessory gland infection for subsequent fertility with special focus on prostatitis. Hum Reprod Update. 1999;5:421–32.
13. Weidner W, Colpi GM, Hargreave TB, et al. EAU guidelines on male infertility. Eur Urol. 2002;42:313–22.
14. Schuppe HC, Meinhardt A, Allam JP, et al. Chronic orchitis: a neglected cause of male infertility? Andrologia. 2008;40:84–91.

15. Henkel R, Schill W-B. Sperm separation in patients with urogenital infections. Andrologia. 1998;30 Suppl 1:91–7.
16. Sanocka-Maciejewska D, Ciupinska M, Kurpisz M. Bacterial infection and semen quality. J Reprod Immunol. 2005;67:51–6.
17. Monga M, Roberts JA. Spermagglutination by bacteria: receptor-specific interactions. J Androl. 1994;15:151–6.
18. Eggert-Kruse W, Kiefer I, Beck C, et al. Role for tumor necrosis factor alpha (TNF-alpha) and interleukin 1-beta (IL-1beta) determination in seminal plasma during infertility investigation. Fertil Steril. 2007;87:810–23.
19. Henkel R, Maaß G, Jung A, et al. Age-related changes in seminal polymorphonuclear elastase in men with asymptomatic inflammation of the genital tract. Asian J Androl. 2007;9:299–304.
20. Moretti E, Capitani S, Figura N, et al. The presence of bacteria species in semen and sperm quality. J Assist Reprod Genet. 2009;26:47–56.
21. Pellati D, Mylonakis I, Bertoloni G, et al. Genital tract infections and infertility. Eur J Obstet Gynecol Reprod Biol. 2008;140:3–11.
22. Umapathy E, Simbini T, Chipata T, et al. Sperm characteristics and accessory sex gland functions in HIV-infected men. Arch Androl. 2001;46:153–8.
23. Le Tortorec A, Le Grand R, Denis H, et al. Infection of semen-producing organs by SIV during the acute and chronic stages of the disease. PLoS One. 2008;3:e1792.
24. World Health Organization. Global prevalence and incidence of selected sexually transmitted diseases: overview and estimates. World Health Organization: Geneva, Switzerland; 2001.
25. Gonzales GF, Munoz G, Sanchez R, et al. Update on the impact of *Chlamydia trachomatis* infection on male fertility. Andrologia. 2004;36:1–23.
26. Rietmeijer CAM, Judson FN, van Hensbroek MB, et al. Unsuspected *Chlamydia trachomatis* infection in heterosexual men attending a sexually transmitted disease clinic: evaluation of risk factors and screening methods. Sex Transm Dis. 1991;18:28–35.
27. Gdoura R, Keskes-Ammar L, Bouzid F, et al. *Chlamydia trachomatis* and male infertility in Tunisia. Eur J Contracept Reprod Health Care. 2001;6:102–7.
28. Villegas H, Pinon M, Shor V, et al. Electron microscopy of *Chlamydia trachomatis* infection of the male genital tract. Arch Androl. 1991;27:117–26.
29. Kadar A, Bucsek M, Kardos M, et al. Detection of *Chlamydia trachomatis* in chronic prostatitis by in situ hybridization (preliminary methodological report). Orv Hetil. 1995;136:659–62.
30. Bornman MS, Ramuthaga TN, Mahomed MF, et al. Chlamydia infection in asymptomatic infertile men attending an Andrology clinic. Arch Androl. 1998;41:203–8.
31. Ruijis GJ, Kauer FM, Jager S, et al. Epidemiological aspects of chlamydial infections and tubal abnormalities in infertile couples. Eur J Obstet Gynecol Reprod Biol. 1990 Jul-Aug;36(1–2): 107–16.
32. Habermann B, Krause W. Altered sperm function or sperm antibodies are not associated with chlamydial antibodies in infertile men with leukocytospermia. J Eur Acad Dermatol Venerol. 1999;12:25–9.
33. Gallegos G, Ramos B, Santiso R, et al. Sperm DNA fragmentation in infertile men with genitourinary infection by *Chlamydia trachomatis* and *Mycoplasma*. Fertil Steril. 2008;90:328–34.
34. Mazzoli S, Cai T, Addonisio P, et al. *Chlamydia trachomatis* infection is related to poor semen quality in young prostatitis patients. Eur Urol. 2010;57(4):708–14. Epub 27 May 2009.
35. Hosseinzadeh S, Brewis IA, Pacey AA, et al. Coincubation of human spermatozoa with *Chlamydia trachomatis* in vitro causes increased tyrosine phosphorylation of sperm proteins. Infect Immun. 2000;68:4872–6.
36. Hosseinzadeh S, Brewis IA, Eley A, et al. Co-incubation of human spermatozoa with *Chlamydia trachomatis* serovar E causes premature sperm death. Hum Reprod. 2001;16:293–9.
37. Hosseinzadeh S, Pacey AA, Eley A. Chlamydia trachomatis-induced death of human spermatozoa is caused primarily by lipopolysaccharide. J Med Microbiol. 2003;52:193–200.
38. Creighton S, Tenant-Flowers M, Taylor CB, et al. Co-infection with gonorrhoea and chlamydia: how much is there and what does it mean? Int J STD AIDS. 2003;14:109–13.
39. Witkin SS, Jeremias J, Grifo JA, et al. Detection of *Chlamydia trachomatis* in semen by the polymerase chain reaction in male members of infertile couples. Am J Obstet Gynecol. 1993;168: 1457–62.
40. Kiessling AA, Desmarais BM, Yin HZ, et al. Detection and identification of bacterial DNA in semen. Fertil Steril. 2008;90: 1744–56.
41. Styler M, Shapiro SS. Mollicutes (mycoplasma) in infertility. Fertil Steril. 1985;44:1–12.
42. Kilic D, Basar MM, Kaygusuz S, et al. Prevalence and treatment of Chlamydia trachomatis, Ureaplasma urealyticum, and Mycoplasma hominis in patients with non-gonococcal urethritis. Jpn J Infect Dis. 2004 Feb;57(1):17–20.
43. Schiefer HG. Microbiology of male urethroadnexitis: diagnostic procedures and criteria for aetiologic classification. Andrologia. 1998;30 Suppl 1:7–13.
44. Kjaergaard N, Kristensen B, Hansen ES, et al. Microbiology of semen specimens from males attending a fertility clinic. APMIS. 1997;105:566–70.
45. Potts JM, Sharma R, Pasqualotto F, et al. Association of *Ureaplasma urealyticum* with abnormal reactive oxygen species levels and absence of leukocytospermia. J Urol. 2000;163: 1775–8.
46. Wang Y, Liang CL, Wu JQ, et al. Do *Ureaplasma urealyticum* infections in the genital tract affect semen quality? Asian J Androl. 2006;8:562–8.
47. Reichart M, Kahane I, Bartoov B. *In vivo* and *in vitro* impairment of human and ram sperm nuclear chromatin integrity by sexually transmitted *Ureaplasma urealyticum* infection. Biol Reprod. 2000;63:1041–8.
48. Reichart M, Levi H, Kahane I, et al. Dual energy metabolism-dependent effect of Ureaplasma urealyticum infection on sperm activity. J Androl. 2001;22:404–12.
49. Cottell E, Harrison RF, McCaffrey M, et al. Are seminal fluid microorganisms of significance or merely contaminants? Fertil Steril. 2000;74:465–70.
50. Witkin SS. Immunological aspects of genital chlamydia infections. Best Pract Res Clin Obstet Gynaecol. 2002;16:865–74.
51. Deguchi T, Maeda S. *Mycoplasma genitalium*: another important pathogen of non-gonococcal urethritis. J Urol. 2002;167:210–7.
52. Andrade-Rocha FT. *Ureaplasma urealyticum* and *Mycoplasma hominis* in men attending for routine semen analysis. Prevalence, incidence by age and clinical settings, influence on sperm characteristics, relationship with the leukocyte count and clinical value. Urol Int. 2003;71:377–81.
53. Gdoura R, Kchaou W, Chaari C, et al. Ureaplasma urealyticum, Ureaplasma parvum, Mycoplasma hominis and Mycoplasma genitalium infections and semen quality of infertile men. BMC Infect Dis. 2007;7:129.
54. Taylor-Robinson D. *Mycoplasma genitalium*—an update. Int J STD AIDS. 2002;13:145–51.
55. Diaz-Garcia FJ, Herrera-Mendoza AP, Giono-Cerezo S, et al. Mycoplasma hominis attaches to and locates intracellularly in human spermatozoa. Hum Reprod. 2006;21:1591–8.
56. Cohen CR, Manhart LE, Bukusi EA, et al. Association between *Mycoplasma genitalium* and acute endometritis. Lancet. 2002; 359:765–6.

57. Köhn FM, Erdmann I, Oeda T, et al. Influence of urogenital infections on sperm functions. Andrologia. 1998;30 Suppl 1:73–80.
58. Krause W. Male accessory gland infection. Andrologia. 2008;40: 113–6.
59. Harvey HA, Porat N, Campbell CA, et al. Gonococcal lipooligosaccharide is a ligand for the asialoglycoprotein receptor on human sperm. Mol Microbiol. 2000;36:1059–70.
60. Eley A, Hosseinzadeh S, Hakimi H, et al. Apoptosis of ejaculated human sperm is induced by co-incubation with *Chlamydia trachomatis* lipopolysaccharide. Hum Reprod. 2005;20:2601–7.
61. AVERT.org. STD Statistics Worldwide. http://www.avert.org/stdstatisticsworldwide.htm. Accessed 8 Apr 2010.
62. Weidner W, Jantos C, Schiefer HG, et al. Semen parameters in men with and without proven chronic prostatitis. Arch Androl. 1991;26:173–83.
63. Huwe P, Diemer T, Ludwig M, et al. Influence of different uropathogenic microorganisms on human sperm motility parameters in an in vitro experiment. Andrologia. 1998;30 Suppl 1:55–9.
64. Diemer T, Huwe P, Michelmann HW, et al. *Escherichia coli*-induced alterations of human spermatozoa. An electron microscopy analysis. Int J Androl. 2000;23:178–86.
65. Sanchez R, Villagran E, Concha M, et al. Ultrastructural analysis of the attachment sites of *Escherichia coli* to the human spermatozoon after *in vitro* migration through estrogenic cervical mucus. Int J Fert. 1989;34:363–7.
66. Diemer T, Huwe P, Ludwig M, et al. Influence of autogenous leucocytes and *Escherichia coli* on sperm motility parameters in vitro. Andrologia. 2003;35:100–5.
67. Schulz M, Sanchez R, Soto L, et al. Effect of *Escherichia coli* and its soluble factors on mitochondrial membrane potential, phosphatidylserine translocation, viability, and motility of human spermatozoa. Fertil Steril. 2010 Jul;94(2):619–23.
68. Diemer T, Ludwig M, Huwe P, et al. Influence of urogenital infection on sperm function. Curr Opin Urol. 2000;10:39–44.
69. Ching JC, Jones NL, Ceponis PJ, et al. *Escherichia coli* shiga-like toxins induce apoptosis and cleavage of poly(ADP-ribose) polymerase via in vitro activation of caspases. Infect Immun. 2002; 70:4669–77.
70. Dejucq N, Jegou B. Viruses in the mammalian male genital tract and their effects on the reproductive system. Microbiol Mol Biol Rev. 2001;65:208–31.
71. Bezold G, Politch JA, Kiviat NB, et al. Prevalence of sexually transmissible pathogens in semen from asymptomatic male infertility patients with and without leukocytospermia. Fertil Steril. 2007;87:1087–97.
72. Kapranos N, Petrakou E, Anastasiadou C, et al. Detection of herpes simplex virus, cytomegalovirus, and Epstein-Barr virus in the semen of men attending an infertility clinic. Fertil Steril. 2003;79 Suppl 3:1566–70.
73. Manigart Y, Rozenberg S, Barlow P, et al. ART outcome in HIV-infected patients. Hum Reprod. 2006;21:2935–40.
74. Bandivdekar AH, Velhal SM, Raghavan VP. Identification of CD4-independent HIV receptors on spermatozoa. Am J Reprod Immunol. 2003;50:322–7.
75. Muciaccia B, Padula F, Gandini L, et al. HIV-1 chemokine co-receptor CCR5 is expressed on the surface of human spermatozoa. AIDS. 2005;19:1424–6.
76. Muciaccia B, Corallini S, Vicini E, et al. HIV-1 viral DNA is present in ejaculated abnormal spermatozoa of seropositive subjects. Hum Reprod. 2007;22:2868–78.
77. Le Tortorec A, Dejucq-Rainsford N. HIV infection of the male genital tract - consequences for sexual transmission and reproduction. Int J Androl. 2010;33:e98–e108.
78. Pillai SK, Good B, Pond SK, et al. Semen-specific genetic characteristics of human immunodeficiency virus type 1 env. J Virol. 2005;79:1734–42.
79. Diemer T, Desjardins C. Disorders of spermatogenesis. In: Knobil E, Neill JD, editors. Encyclopedia of reproduction, vol. 4. San Diego: Academic; 1999. p. 546–56.
80. Schuppe HC, Meinhardt A. Immunology of the testis and the excurrent ducts. In: Schill W-B, Comhaire F, Hargreave TB, editors. Andrology for the clinician. Heidelberg: Springer; 2006. p. 292–300.
81. Schuppe HC, Meinhardt A. Immune privilege and inflammation of the testis. Chem Immunol Allergy. 2005;88:1–14.
82. Haidl G, Allam JP, Schuppe HC. Chronic epididymitis: impact on semen parameters and therapeutic options. Andrologia. 2008;40: 92–6.
83. Mazumdar S, Levine AS. Antisperm antibodies: etiology, pathogenesis, diagnosis, and treatment. Fertil Steril. 1998;70:799–810.
84. Kalaydjiev S, Dimitrova D, Nenova M, et al. Serum sperm antibodies are not elevated after mumps orchitis. Fertil Steril. 2002;77:76–82.
85. Nistal M, Paniagua R, editors. Testicular and epididymal pathology. Stuttgart, New York: Thieme; 1984.
86. Bartak V. sperm count, morphology and motility after unilateral mumps orchitis. J Reprod Fertil. 1973;32:491–3.
87. Casella R, Leibundgut B, Lehman K, et al. Mumps orchitis: Report of a mini-epidemic. J Urol. 1997;158:2158–61.
88. Behrman RE, Kliegman RM, Jenson HB, editors. Nelson textbook of pediatrics. 17th ed. Philadelphia: Saunders; 2004.
89. Philip J, Selvan D, Desmond AD. Mumps orchitis in the non-immune postpubertal male: a resurgent threat to male fertility? BJU Int. 2006;97:138–41.
90. Ludwig M. Diagnosis and therapy of acute prostatitis, epididymitis and orchitis. Andrologia. 2008;40:76–80.
91. Kopa Z, Wenzel J, Papp GK, et al. Role of granulocyte elastase and interleukin-6 in the diagnosis of male genital tract inflammation. Andrologia. 2005;37:188–94.
92. Collins MM, Stafford RS, O'Leary MP, et al. How common is prostatitis? A national survey of physician visits. J Urol. 1998;159:1224–8.
93. Vordermark JS. Acute epididymitis: experience with 123 cases. Mil Med. 1985;150:27–30.
94. Chan PT, Schlegel PN. Inflammatory conditions of the male excurrent ductal system. Part II. J Androl. 2002;23:461–9.
95. Trojian TH, Lishnak TS, Heiman D. Epididymitis and orchitis: an overview. Am Fam Physician. 2009;79:583–7.
96. Comhaire F, Mahmoud A. Infection/inflammation of the accessory sex glands. In: Schill W-B, Comhaire F, Hargreave TB, editors. Andrology for the clinician. Heidelberg: Springer; 2006. p. 72–4.
97. World Health Organization. WHO Laboratory Manual for the Examination of Human Semen and Semen-Cervical Mucus Interaction. 4th ed. Cambridge, UK: Cambridge University; 1999.
98. National Kidney and Urologic Disease Advisory Board. Long-Range Plan Window on the 21st Century. United States Department of Health and Human Services, National Institutes of Health publication. (1990) No. 90-583: 20.
99. Roberts RO, Lieber MM, Bostwick DG, et al. A review of clinical and pathological prostatitis syndromes. Urology. 1997;49:809–21.
100. Nickel JC, Downey J, Hunter D, et al. Prevalence of prostatitis-like symptoms in a population based study using the National Institutes of Health chronic prostatitis symptom index. J Urol. 2001;165:842–5.
101. Brunner H, Weidner W, Schiefer H-G. Studies on the role of *Ureaplasma urealyticum* and *Mycoplasma hominis* in prostatitis. J Infect Dis. 1983;147:807–13.
102. Bjerklund Johansen TE, Grüneberg RN, Guibert J, et al. The role of antibiotics in the treatment of chronic prostatitis: a consensus statement. Eur Urol. 1998;34:457–66.

103. Nickel JC, Downey J, Johnston B, et al. Predictors of patient response to antibiotic therapy for the chronic prostatitis/chronic pelvic pain syndrome: a prospective multicenter clinical trial. J Urol. 2001;165:1539–44.
104. Drach GW, Fair WR, Meares EM, et al. Classification of benign diseases associated with prostatic pain: prostatitis or prostatodynia? J Urol. 1978;120:266.
105. Nickel JC. Prostatitis: myths and realities. Urology. 1998;51: 362–6.
106. Krieger JN, Nyberg LJ, Nickel JC. NIH consensus definition and classification of prostatitis. JAMA. 1999;282:236–7.
107. Roberts RO, Lieber MM, Rhodes T, et al. Prevalence of a physician-assigned diagnosis of prostatitis: the Olmsted County Study of Urinary Symptoms and Health Status Among Men. Urology. 1998;51:578–84.
108. Nickel JC. Prostatitis: the last frontier. World J Surg. 2000;24: 1197–9.
109. Nickel JC. Classification and diagnosis of prostatitis: a gold standard? Andrologia. 2003;35:160–7.
110. Lopez-Plaza I, Bostwick DG. Prostatitis. In: Bostwick DG, editor. Pathology of the prostate. New York: Churchill Livingstone; 1990. p. 15–30.
111. Nickel JC, Costerton JW, McLean RJC, et al. Bacterial biofilms: influence on the pathogenesis, diagnosis and treatment of urinary tract infections. J Antimicrob Chemother. 1994;33(Suppl A):31–41.
112. Ochsendorf F. Urethritis, Sexually Transmitted Diseases (STD), Acquired Immunodeficiency Syndrome (AIDS). In: Schill W-B, Comhaire F, Hargreave TB, editors. Andrology for the clinician. Heidelberg: Springer; 2006. p. 327–38.
113. Krieger JN, Hooton TM, Brust PJ, et al. Evaluation of chronic urethritis. Defining the role for endoscopic procedures. Arch Intern Med. 1988;148:703–7.
114. Chambers RM. The mechanism of infection in the urethra, prostate and epididymis. In: Keith LG, Berger GS, Edelmann DA, editors. Infections in reproductive health. Common infections. Lancaster: MTP; 1985. p. 283–96.
115. Urata K, Narahara H, Tanaka Y, et al. Effect of endotoxin-induced reactive oxygen species on sperm motility. Fertil Steril. 2001;76:163–6.
116. Sanocka D, Fraczek M, Jedrzejczak P, et al. Male genital tract infection: an influence of leukocytes and bacteria on semen. J Reprod Immunol. 2004;62:111–24.
117. Henkel R, Kierspel E, Stalf T, et al. Effect of reactive oxygen species produced by spermatozoa and leukocytes on sperm functions in non-leukocytospermic patients. Fertil Steril. 2005;83:635–42.
118. Schiefer HG, von Graevenitz A. Clinical microbiology. In: Schill W-B, Comhaire F, Hargreave TB, editors. Andrology for the clinician. Heidelberg: Springer; 2006. p. 401–7.
119. Depuydt CE, Bosmans E, Zalata A, et al. The relation between reactive oxygen species and cytokines in andrological patients with or without male accessory gland infection. J Androl. 1996;17:699–707.
120. Kocak I, Yenisey C, Dündar M, et al. Relationship between seminal plasma interleukin-6 and tumor necrosis factor alpha levels with semen parameters in fertile and infertile men. Urol Res. 2002;30:263–7.
121. Aitken RJ, Gordon E, Harkiss D, et al. Relative impact of oxidative stress on the functional competence and genomic integrity of human spermatozoa. Biol Reprod. 1998;59:1037–46.
122. Fraczek M, Sanocka D, Kamieniczna M, et al. Proinflammatory cytokines as an intermediate factor enhancing lipid sperm membrane peroxidation in in vitro conditions. J Androl. 2008;29:85–92.
123. Cooper TG, Weidner W, Nieschlag E. The influence of inflammation of the human male genital tract on secretion of the seminal markers alpha-glucosidase, glycerophosphocholine, carnitine, fructose and citric acid. Int J Androl. 1990;13:329–36.
124. Wolff H, Bezold G, Zebhauser M, et al. Impact of clinically silent inflammation on male genital tract organs as reflected by biochemical markers in semen. J Androl. 1991;12:331–4.
125. Munoz MG, Jeremias J, Witkin SS. The 60 kDa heat shock protein in human semen: relationship with antibodies to spermatozoa and *Chlamydia trachomatis*. Hum Reprod. 1986;11:2600–1.
126. Bohring C, Krause E, Habermann B, et al. Isolation and identification of spermatozoa membrane antigens, recognized by antisperm antibodies and their possible role in immunological infertility disease. Mol Hum Reprod. 2001;7:113–8.
127. Zalata AA, Christophe AB, Depuydt CE, et al. The fatty acid composition of phospholipids of spermatozoa from infertile patients. Mol Hum Reprod. 1998;4:111–8.
128. Plante M, de Lamirande E, Gagnon C. Reactive oxygen species released by activated neutrophils, but not by deficient spermatozoa, are sufficient to affect normal sperm motility. Fertil Steril. 1994;62:387–93.
129. Comhaire FH, Mahmoud AM, Depuydt CE, et al. Mechanisms and effects of male genital tract infection on sperm quality and fertilizing potential: the andrologist's viewpoint. Hum Reprod Update. 1999;5:393–8.
130. Aitken RJ, Clarkson JS, Fishel S. Generation of reactive oxygen species, lipid peroxidation, and human sperm function. Biol Reprod. 1989;41:183–97.
131. Henkel R, Hajimohammad M, Stalf T, et al. Influence of deoxyribonucleic acid damage on fertilization and pregnancy. Fertil Steril. 2004;81:965–72.
132. Motrich RD, Maccioni M, Molina R, et al. Reduced semen quality in chronic prostatitis patients that have cellular autoimmune response to prostate antigens. Hum Reprod. 2005;20:2567–72.
133. Martinez R, Proverbio F, Camejo MI. Sperm lipid peroxidation and pro-inflammatory cytokines. Asian J Androl. 2007;9: 102–7.
134. Mortimer D. Sperm preparation techniques and iatrogenic failures of in-vitro fertilization. Hum Reprod. 1991;6:173–6.
135. Alvarez JG, Sharma RK, Ollero M, et al. Increased DNA damage in sperm from leukocytospermic semen samples as determined by the sperm chromatin structure assay. Fertil Steril. 2002;78: 319–29.
136. Henkel R, Maaß G, Hajimohammad M, et al. Urogenital inflammation: changes of leucocytes and ROS. Andrologia. 2003;35:309–13.
137. Henkel R, Ludwig M, Schuppe HC, et al. Chronic pelvic pain syndrome/chronic prostatitis affect the acrosome reaction in human spermatozoa. World J Urol. 2006;24:39–44.
138. Zalata AA, Ahmed AH, Allamaneni SSR, et al. Relationship between acrosin activity of human spermatozoa and oxidative stress. Asian J Androl. 2004;6:313–8.
139. Centres for Disease Control and Prevention. Sexually transmitted diseases treatment guidelines. Morb Mortal Wkly Rep. 2006;55: 35–62.
140. Yeniyol CO, Sorguc S, Minareci S, et al. Role of interferon-alpha-2B in prevention of testicular atrophy with unilateral mumps orchitis. Urology. 2000;55:931–3.
141. Naber KG, Madsen PO. Antibiotics: basic concepts. In: Nickel JC, editor. Textbook of prostatitis. Cambridge: Isis Medical Media; 1999. p. 83–94.
142. Naber KG, Weidner W. Chronic prostatitis-an infectious disease? J Antimicrob Chemother. 2000;46:157–61.
143. Hendry WF, Stedronska J, Hughes L, et al. Steroid treatment of male subfertility caused by antisperm antibodies. Lancet. 1979;2: 498–501.
144. Montag M, van der Ven H, Haidl G. Recovery of ejaculated spermatozoa for intracytoplasmic sperm injection after anti-inflammatory treatment of an azoospermic patient with genital tract infection: a case report. Andrologia. 1999 May;31(3):179–81.

145. Haidl G. Management strategies for male factor infertility. Drugs. 2002;62:1741–53.
146. Tremellen K. Oxidative stress and male infertility—a clinical perspective. Hum Reprod Update. 2008;14:243–58.
147. Lanzafame FM, La Vignera S, Vicari E, et al. Oxidative stress and medical antioxidant treatment in male infertility. Reprod Biomed Online. 2009;19:638–59.
148. Comhaire FH, Christophe AB, Zalata AA, et al. The effects of combined conventional treatment, oral antioxidants and essential fatty acids on sperm biology in subfertile men. Prostaglandins Leukot Essent Fatty Acids. 2000;63:159–65.
149. Comhaire FH, El Garem Y, Mahmoud A, et al. Combined conventional/antioxidant "Astaxanthin" treatment for male infertility: a double blind, randomized trial. Asian J Androl. 2005;7:257–62.
150. Vicari E, Calogero AE. Effects of treatment with carnitines in infertile patients with prostato-vesiculo-epididymitis. Hum Reprod. 2001;16:2338–42.
151. Lenzi A, Sgro P, Salacone P, et al. A placebo-controlled double-blind randomized trial of the use of combined l-carnitine and l-acetyl-carnitine treatment in men with asthenozoospermia. Fertil Steril. 2004;81:1578–84.
152. Balercia G, Regoli F, Armeni T, et al. Placebo-controlled double-blind randomized trial on the use of l-carnitine, l-acetylcarnitine, or combined l-carnitine and l-acetylcarnitine in men with idiopathic asthenozoospermia. Fertil Steril. 2005;84:662–71.
153. Akmal M, Qadri JQ, Al-Waili NS, et al. Improvement in human semen quality after oral supplementation of vitamin C. J Med Food. 2006;9:440–2.
154. Greco E, Iacobelli M, Rienzi L, et al. Reduction of the incidence of sperm DNA fragmentation by oral antioxidant treatment. J Androl. 2005;26:349–53.
155. Menezo YJ, Hazout A, Panteix G, et al. Antioxidants to reduce sperm DNA fragmentation: an unexpected adverse effect. Reprod Biomed Online. 2007;14:418–21.

Ejaculatory Dysfunction and Vasodynamics

26

Aleksey P. Ignatov and Paul J. Turek

Abstract

Ejaculation begins approximately 12 months after the onset of puberty in the male. While its importance to reproductive fitness is clear, current knowledge of the physiology of ejaculation is limited. This chapter reviews the events of ejaculation, its anatomic and neuroanatomic underpinnings, the range of ejaculatory disorders, and clinical methods for evaluating and treating ejaculatory disorders. Our current understanding of reproductive tract physiology and function as well as ejaculatory duct obstruction is also reviewed.

Keywords

Male infertility • Ejaculatory duct obstruction • Spinal cord injury • Vasodynamics • Ejaculatory dysfunction • Ejaculatory disorders • Müllerian duct cyst

Ejaculation begins approximately 12 months after the onset of puberty in the male. While its importance to reproductive fitness is clear, current knowledge of the physiology of ejaculation is limited. This chapter reviews the events of ejaculation, its anatomic and neuroanatomic underpinnings, the range of ejaculatory disorders, and clinical methods for evaluating and treating ejaculatory disorders. Our current understanding of reproductive tract physiology and function as well as ejaculatory duct obstruction (EDO) is also reviewed.

Physiology of Ejaculation

The Events

Ejaculation is two distinct processes: emission and ejaculation [1]. Although not technically considered a separate event, pre-ejaculation, occurring during foreplay, involves closure of the bladder neck that prevents retrograde ejaculation and contractions of the prostate that lubricate the urethra. Importantly, ejaculation is also distinct from orgasm, which is a purely cerebral cortical event. Most often, these two processes are coincident.

Emission combines the transport of both seminal fluid and sperm through peristalsis from the cauda epididymis, vas deferens, seminal vesicles, and prostate into the prostatic urethra. During seminal emission, the ampullary vasa deferentia contents are transported into the prostatic urethra and mixed with prostatic fluid. The expulsion of seminal vesicle contents into the prostatic urethra completes the emission phase. Subsequently, ejaculation is the forceful expulsion of the seminal mixture from the urethra. The ejaculate is expelled from the urethra in a series of spurts, 0.8 s apart, caused by the rhythmic contractions of the ischiocavernosus, bulbospongiosus, and other associated periurethral muscles [2]. The entire process is governed by the autonomic and somatic nervous systems and is considered a spinal reflex.

Neural Control

Control of the ejaculatory reflex is mediated by the sympathetic and somatic nervous systems [3]. Control of emission involves mainly the sympathetic nervous system while ejaculation is governed largely by the somatic nervous system.

A.P. Ignatov, BS (✉) • P. J. Turek, MD, FACS, FRSM
The Turek Clinic, 55 Francisco Street,
Suite 300, San Francisco, CA 94133, USA
e-mail: apignatov@yahoo.com; drpaulturek@gmail.com

S.J. Parekattil and A. Agarwal (eds.), *Male Infertility: Contemporary Clinical Approaches, Andrology, ART & Antioxidants*,
DOI 10.1007/978-1-4614-3335-4_26,

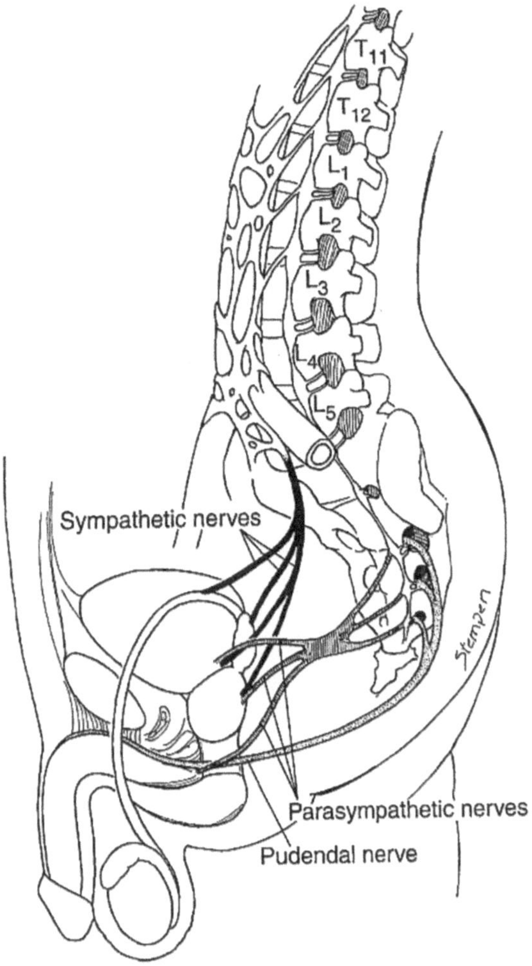

Fig. 26.1 Innervation of the male reproductive tract showing sympathetic and parasympathetic nerves and somatic innervation (from [1], with permission)

Efferent sympathetic nerves emerge from the thoracolumbar spine at T10–L2 and then merge to form the lumbar sympathetic ganglia that encircle the aorta. These nerves subsequently combine in the midline below the aortic bifurcation to form the superior hypogastric plexus (Fig. 26.1). Ultimately, these adrenergic nerves terminate as postganglionic fibers that innervate the bladder neck, prostate, vasa deferentia, and seminal vesicles [4]. The sympathetic outflow generated by these nerves is responsible for closure of the bladder neck and seminal emission.

The muscular expulsion of the ejaculate is mediated by somatic motor efferent nerves derived from the perineal branch of the pudendal nerve (S2–S4). Additional control is provided by relaxation of the external urethral sphincter and the urogenital diaphragm. Interruption at any point in this reflex arc may result in disordered ejaculation.

Definitions

Aspermia: Disordered ejaculation characterized by an inability to produce semen, despite the occurrence of climax.

Azoospermia: The absence of sperm in the ejaculate.

Anejaculation: The failure of ejaculation, including an absence of seminal emission and ejaculation. Climax is usually absent as well.

Premature ejaculation (*ejaculatio praecox*): Ejaculation which always or nearly always occurs prior to or within about 1 min of vaginal penetration, and inability to delay ejaculation on all or nearly all vaginal penetrations, and negative personal consequences, such as distress, bother, frustration, and/or the avoidance of sexual intimacy.

Delayed ejaculation (anorgasmia): A form of sexual dysfunction characterized by the inability to achieve climax, or an extreme delay in achieving climax and ejaculation.

Retrograde ejaculation: Ejaculation of semen in reverse direction into the bladder during climax due to failure of bladder neck closure.

Congenital anorgasmia: Failure of ejaculation as primary, lifelong event.

Ejaculatory anhedonia: Ejaculation associated with a lack of pleasure.

Evaluation

History

The cornerstone of evaluating ejaculatory dysfunction is a detailed patient history. Critical information to obtain from the history is whether or not normal ejaculation was ever present. This suggests a primary (congenital) or secondary (acquired) cause of sexual dysfunction. In addition, a thorough review of the medical and surgical history can be informative regarding possible neurologic etiologies for disordered ejaculation. Finally, a careful review of current medications such as alpha-blockers or antidepressants can suggest the presence of drug-induced ejaculatory dysfunction [1].

Physical Examination

A complete physical examination should include an assessment of body habitus and secondary sex characteristics, a screening neurological examination, and a thorough genital examination. Testis and epididymis size and consistency, penile length and morphology, and genital birth defects such as hypospadias, epispadias, or surgical scars suggestive of their correction should be noted. Palpation of scrotum for masses and a check for the presence of the vas deferens should also be performed. A rectal examination noting rectal tone and any masses is also important.

Laboratory Evaluation

An attempt should be made to procure a semen analysis. In cases where no ejaculate is obtained, a post-ejaculate urine sample should be retrieved and assessed for the presence of sperm, suggesting retrograde ejaculation. Blood testosterone, prolactin, and serum follicle-stimulating hormone (FSH) levels should also be assessed because low ejaculate volumes may be caused by hypoandrogenism. Further diagnostic evaluation may include imaging with transrectal ultrasonography (TRUS) to define anatomical or structural abnormalities in the prostate, seminal vesicles, or ejaculatory duct complex. If indicated, formal ejaculatory duct chromotubation, seminal vesiculography, and ejaculatory duct manometry can be performed to detect subtle ejaculatory duct abnormalities [5].

Genetic Testing

Patients with disordered ejaculation and a history of infertility or with suspected congenital abnormalities such as congenital absence of the vas deferens (CAVD) or ejaculatory duct obstruction (EDO) should be counseled on appropriate genetic testing for cystic fibrosis transmembrane regulatory (CFTR) gene mutations [6].

Management of Ejaculatory Disorders

Anatomic

Bladder Neck Incompetence

A patent bladder neck is most commonly the consequence of an incompetent internal urethral (bladder neck) sphincter. The ensuing "dry ejaculate" is due to retrograde ejaculation. It can be caused by α-blocker medication for prostate enlargement or hypertension, diabetic neuropathy, neurologic disorders such as spina bifida or multiple sclerosis, or other congenital anatomic abnormalities. It is also a common postsurgical complication of transurethral prostatectomy (TURP) [7]. Interestingly, from the patient perspective, retrograde ejaculation after TURP is often confused with anorgasmia or erectile dysfunction [8].

If retrograde ejaculation is drug induced, the offending medication should be discontinued. With neurological causes such as that associated with diabetes, alpha agonist therapy can help "close" the bladder neck and encourage antegrade ejaculation [1, 9]. Reversal after TURP is difficult. However, if fertility is sought in men after TURP, sperm in the post-masturbatory urine can be used with intrauterine insemination (IUI) or in vitro fertilization (IVF) for paternity.

Müllerian Duct Cyst

Persistence of remnants of the Müllerian ducts may exist as midline cysts associated with the prostatic utricle and ejaculatory ducts in men. If significant in size, such cysts may be occlusive and produce a low-volume ejaculate due to ejaculatory duct compression. This diagnosis is confirmed by TRUS and further investigation of ejaculatory anatomy and function with chromotubation or manometry [5, 10]. Stones, calcification, ejaculatory duct agenesis, and seminal vasculopathy resulting in acontractile, dysfunctional seminal vesicles may present similarly [11].

In patients with confirmed obstruction, transurethral unroofing of cysts, drainage of stones, or recanalizing the ejaculatory ducts effectively treats the problem [12]. In men with functional but not obstructive disorders of the reproductive tract, surgical procedures are not indicated and have no clinical value [5].

Congenital Bilateral Absence of the Vas Deferens/Cystic Fibrosis

Among men with cystic fibrosis, 99% also have Wolffian duct abnormalities that typically cause low ejaculate volume. There may be atresia or agenesis of the vas deferens, seminal vesicles, or ejaculatory ducts with this diagnosis. Low-volume ejaculation is also associated with a form fruste of cystic fibrosis, termed CAVD [11]. In this condition, there may be absence of the vas deferens but without other, systemic manifestations of cystic fibrosis.

The ejaculatory disorder associated with cystic fibrosis and CAVD is currently irreversible. Parenthood, however, can be achieved with surgical sperm retrieval procedures and assisted reproduction.

Ejaculatory Duct Obstruction

The combination of low-volume ejaculate, painful ejaculation, hematospermia, and perineal or testicular pain is highly suggestive of EDO. The diagnosis is supported by the finding of a normal physical examination and a semen analysis that shows a volume <2.0 mL, with a seminal pH <7.2, and no

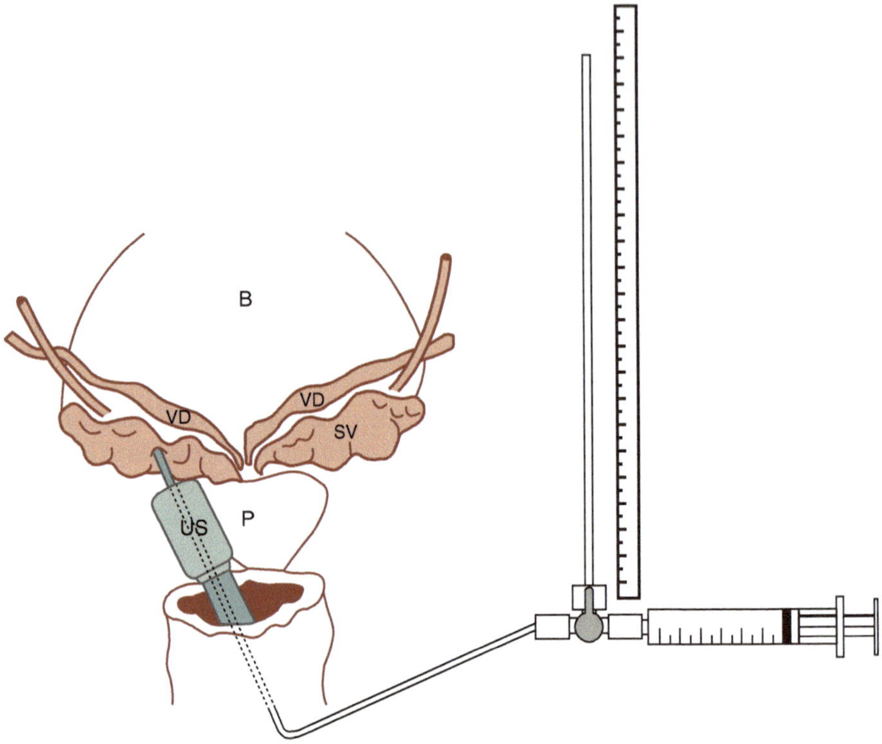

Fig. 26.2 The ejaculatory duct manometry device. Schematic representation of the intravenous tubing manometer used to measure ejaculatory duct pressure in EDO patients. After intubation of the seminal vesicle with a spinal needle attached to a three-way stopcock, the seminal vesicle is injected with saline/indigo carmine (chromotubation). The pressure at which fluid begins to traverse the ejaculatory duct orifice into the prostatic urethra cystoscopically is the "opening pressure." The pressure within the seminal vesicle is monitored by the height of the column of fluid within the IV tubing (from [5], with permission)

sperm or fructose present. Partial EDO, a variant, is harder to diagnose, but typically presents with low normal ejaculate volume and disproportionately low sperm motility. Confirmatory diagnostic tests include TRUS that shows dilated seminal vesicles (>1.5 cm) or dilated ejaculatory ducts (>2.3 mm) in association with a cyst, calcification, or stones along the ducts [13, 14]. Recently, it has become clear that *static* imaging such as TRUS cannot reliably differentiate true physical obstruction from functional disorders of the reproductive tract. TRUS, although sensitive, is not specific for the diagnosis of EDO [10]. As such, adjunctive procedures such as seminal vesicle aspiration [15], seminal vesiculography, and chromotubation can further delineate the diagnosis. Such "functional" testing has been suggested before definitive surgery on the ejaculatory duct complex [10]. To this end, a prospective study of these three adjunctive techniques in EDO patients revealed that patency with chromotubation was the most accurate way to diagnose complete or incomplete EDO [10].

With these considerations in mind and based on the concept of bladder urodynamics to assess bladder outlet obstruction, we recently described the technique of ejaculatory duct manometry to confirm the diagnosis of EDO (Fig. 26.2) [5]. This technique stemmed from the idea that the varying flow resistance patterns encountered with antegrade chromotubation in EDO patients could be more precisely quantified. We hypothesized that measuring ED "opening pressures," defined as the pressure above which fluid enters the prostatic urethra, could distinguish among the various forms of EDO.

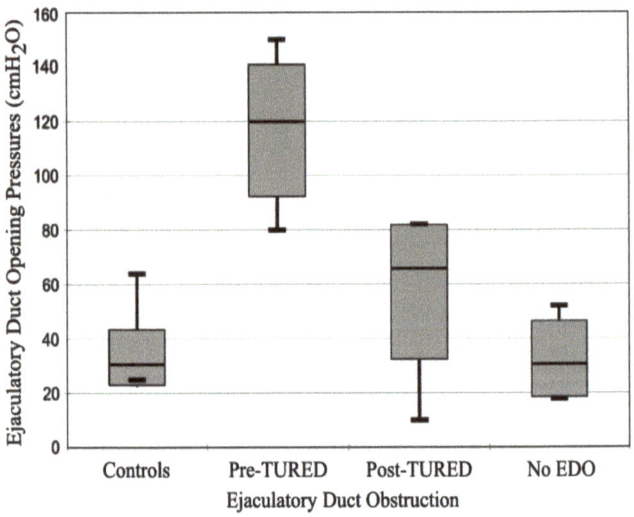

Fig. 26.3 Ejaculatory duct opening pressures in fertile men and EDO patients. *Gray boxes* represent the first to third interquartile range for measured values. *Black horizontal lines* represent the median, and whiskers represent the absolute range of values. Ejaculatory duct pressures in EDO patients were significantly higher than controls ($p<0.001$) or post-TURED ($p<0.001$) patients. Pressures after TURED were not significantly different from controls or the "No EDO" group. The "No EDO" group represents ejaculatory ducts evaluated and deemed clinically not to harbor EDO (from [5], with permission)

Indeed, in a prospective, comparative study of fertile men (vasectomy reversals) and men with confirmed EDO, ejaculatory duct opening pressures were significantly higher in untreated EDO patients (mean 116 cmH_2O) compared to fertile men (mean 33 cmH_2O) (Fig. 26.3). In addition, post-TURED duct opening pressures fell to values similar to controls. The study concluded that (1) fertile patients have consistent and low ED opening pressures with a normal pressure defined as <45 cm H_2O, (2) infertile men with EDO have significantly higher ED pressures, (3) opening pressures after EDO treatment can be lowered to that of controls, and (4) patients with suspected EDO may have other kinds of underlying pathology that will not respond to ED resection, including urethral strictures. From this analysis, ED manometry currently has the most potential to differentiate complete from partial and physical from functional forms of EDO.

Men with EDO can be treated with a transurethral resection or incision of the ejaculatory ducts, which is very effective at increasing semen volume and restoring sperm flow [13]. In cases of absence of reproductive tract organs, no remedy is currently available.

Neuropathic

Spinal Cord Injury

Most patients with spinal cord injury are young men who have sustained traumatic interruption of nerve pathways that modulate ejaculation. Spinal cord lesions at or below the level of T10–L2 level commonly lead to complete loss of ejaculation with preserved erections, whereas injuries above the T10 spinal level generally retain the ejaculatory reflex arc as the peripheral efferent nerves from T10–L2 and S2–S4 are intact. The integrity of this reflex arc can be confirmed by demonstrating an intact bulbocavernosus reflex and the ability to perform hip flexion, both of which predict successful ejaculation when sensory afferent input is increased to suprathreshold levels [16].

Fertility and paternity are commonly achieved in affected patients with the use of assisted reproductive technology in association with penile vibratory stimulation (PVS) [17] or rectal probe electroejaculation (EEJ) [18–20]. Patients with spinal cord lesions above T4 also are prone to autonomic dysreflexia from penile stimulation. Symptoms of autonomic dysreflexia include hypertension, bradycardia, sweating, chills, and headache. In some cases, autonomic dysreflexia can lead to dangerously high blood pressures and can lead to stroke, seizure, or death. Pretreatment with an oral calcium channel blocker for prophylactic management of these symptoms is advised. For patients with lower spinal cord lesions (below T10) in whom penile vibratory therapy fails, EEJ is an excellent alternative to achieve ejaculation [21]. For patients who fail electroejaculation, surgical sperm retrieval provides an excellent alternative for paternity [22].

Diabetes Mellitus

Long-standing diabetes is associated with genitourinary autonomic neuropathy. About 87% of type I diabetics have evidence of bladder dysfunction, and erectile dysfunction is observed in 35–75% of affected males [23]. The associated ejaculatory dysfunction may manifest as either retrograde ejaculation or total anejaculation depending on the degree of sympathetic autonomic neuropathy. This dysfunction results from incomplete closure of the bladder neck during ejaculation (retrograde ejaculation) or complete neurogenic "paralysis" of the reproductive tract smooth musculature (anejaculation). In cases of complete anejaculation, the post-masturbatory urine contains no sperm [24].

To aid in conception, sympathomimetic drugs may be taken to stimulate bladder neck closure in retrograde ejaculation and produce an antegrade ejaculate [9]. Because most sympathomimetics will exhibit tachyphylaxis with prolonged use, their administration should be limited to the window of timed intercourse around ovulation. Unlike with retrograde ejaculation, diabetic patients with anejaculation are more difficult to treat. If there is no conversion to retrograde or antegrade ejaculation with sympathomimetics, EEJ can be effective to induce a useful ejaculate [21]. Of note, patients with long-standing diabetes may have calcified vasa deferentia and seminal vesicles that are unable to contract and propel sperm. In such cases, surgical sperm retrieval is necessary to treat associate fertility issues [25].

Postsurgical

In general, retroperitoneal or pelvic procedures that disrupt the sympathetic nerves that course along the aorta, especially near the aortic bifurcation (hypogastric plexus), may result in ejaculatory dysfunction. The spectrum of the functional defects correlates to the degree and severity of nerve damage. Retrograde ejaculation, failure of ejaculation, or failure of ejaculation along with failure of seminal emission are all possible, depending on the degree of injury [1].

Approximately two-thirds of men will have retrograde ejaculation following TURP, and one quarter to one-third of patients will have a similar issue after bladder neck incision due to incomplete bladder neck closure. Major abdominopelvic surgery, such as colorectal resection for malignancy, ileoanal anastomosis for inflammatory bowel disease, repair of abdominal aortic aneurysm, aortoiliac bypass grafting, and retroperitoneal lymphadenectomy for testicular cancer, may lead to some damage of the lumbar sympathetic ganglia and/or the superior hypogastric plexus and result in retrograde ejaculation or anejaculation [26, 27].

The operative procedure that historically has resulted in the highest frequency of ejaculatory disorders is retroperitoneal lymph node dissection performed largely as curative treatment for metastatic testicular cancer. In its original form, the operation consisted of a bilateral suprahilar extended dissection of retroperitoneal nodes and almost uniformly resulted in ejaculatory dysfunction. Advances in surgical technique combined with newer limited surgical dissection templates have decreased the incidence of ejaculatory dysfunction [28–30].

Radical prostatectomy is another major procedure that results in functional anejaculation because the prostate and seminal vesicles are excised. Other operative procedures that can cause ejaculatory dysfunction include abdominoperineal operations for rectal cancer and spine surgeries performed anteriorly (transabdominally), which are associated with rates of ejaculatory disorders of approximately 14% [31].

Pediatric congenital anomalies of the pelvis are associated with anejaculation and retrograde ejaculation later in life. Ejaculatory disorders in these patients can be caused by the anatomic nature of the pelvic anomaly (cloacal exstrophy, imperforate anus) or the associated surgical procedure needed for its correction (exstrophy/epispadias repair, bladder neck reconstruction) [1].

The reversal of ejaculatory disorders from surgical sympathetic nerve disruption is difficult. In general, treatment with α-adrenergic stimulants can be attempted [9]. In a few instances, therapy can convert a failure of emission into simply retrograde ejaculation, or convert retrograde into antegrade ejaculation. Success with drug therapy depends on the integrity and number of residual sympathetic nerve fibers that innervate the seminal vesicles, vasa deferentia, and bladder neck areas.

If medical therapy is unsuccessful, EEJ is required to induce ejaculation and to harvest sperm for use in assisted reproductive technology. When performed under a light general anesthesia in the sensate patient, this procedure uses rhythmically applied, graded increases in voltage (0–25 V) to a rectal probe to cause contraction of the seminal vesicles and vasa ampullae directly and to induce the ejaculatory reflex. Ejaculation is obtained in virtually all men with surgically induced ejaculatory disorders through this technique [21].

Neurologic Disorders

The entire spectrum of ejaculatory disorders, ranging from premature ejaculation to anejaculation, is associated with demyelinating and inflammatory neurologic diseases including multiple sclerosis and transverse myelitis [32]. Patients with spinal dysraphism disorders (e.g., myelodysplasia, myelomeningocele, spina bifida) also harbor many of the same ejaculatory disorders. Defects above the T10–T11 cord level commonly are associated with anejaculation, whereas defects below this level allow emission without ejaculation. Patients with sacral lesions are generally spared any ejaculatory dysfunction [33].

Neurostimulatory methods may be used to induce ejaculation in men with neurogenic anejaculation. The most commonly used methods to induce ejaculation in men with neurologic disorders are PVS and EEJ [3]. PVS involves placing a vibrator on the dorsum or frenulum of the glans penis [17]. Mechanical stimulation produced by the vibrator recruits the ejaculatory reflex to induce ejaculation [34]. This method is more effective in men with an intact ejaculatory reflex, that is, men with a level of injury above the T10 cord level. Individuals who do not respond to PVS are often candidates for EEJ [35, 36]. EEJ is performed with the patient in the lateral decubitus position. A probe is placed in the rectum, and electrodes on the probe are oriented toward the prostate and seminal vesicles. Current delivered through the probe stimulates nerves that lead to emission of semen.

Pharmacologic

Antidepressants

Many common medications can cause ejaculatory dysfunction. Antidepressants, including the tricyclic antidepressants, monoamine oxidase inhibitors (MAOIs), and the newer selective serotonin reuptake inhibitors (SSRIs), are associated with sexual dysfunction and disordered ejaculation [37]. Sexual dysfunction resulting from these medications may include hypoactive sexual desire, erectile dysfunction, and delayed ejaculation. It is thought that these side effects are due to elevated central nervous system levels of serotonin (SSRIs) or catecholamines (tricyclics) [38]. In most patients, discontinuation of antidepressant therapy will restore normal sexual function.

Alpha-Adrenergic Antagonists

Both the transport of seminal fluid within the reproductive tract and bladder neck closure are controlled by α-adrenergic nerves. Therefore, α-adrenergic antagonists, given for hypertension or prostatic hypertrophy, may inhibit both seminal emission and bladder neck closure [39]. In either case, the result presents as a low-volume or dry ejaculate. In general, the sensation of orgasm is normal or nearly normal. Treatment should be directed at removal of these drugs.

Functional

Premature or Early Ejaculation

Premature ejaculation (PE) is defined as ejaculation which always or nearly always occurs prior to or within 1 min of vaginal penetration, the inability to delay ejaculation on all or nearly all vaginal penetrations and is associated with negative personal consequences, such as distress, bother, frustration, and/or the avoidance of sexual intimacy [40]. The incidence of this problem is high, affecting 35% of men between the ages of 18 and 59, and it is the most common form of male sexual dysfunction [41]. Given how commonly PE is reported, this raises the question of whether it is truly an organic, treatable disease, or merely a consequence of normal sexual function associated with abnormal expectations. Implicated etiologic factors include sexual anxiety and penile skin hypersensitivity, although there is little consensus regarding exact causes [42].

The goal of therapy for PE is to increase patient control over the ejaculation process by decreasing penile sensitivity and by adjusting the behavioral response. Treatments include oral medication, local anesthetic therapy, and sexual therapy. The most effective treatments are behavioral as drug therapy demands high compliance rates that may not be achievable with younger men who do not otherwise take medications.

Drugs that delay ejaculation are logical choices for PE treatment, and clinical trials have shown that SSRIs can effectively prolong the time to ejaculation [43–48]. However, the increase in latency time to ejaculation varies widely depending on the selected medication. Paroxetine, fluoxetine, sertraline, and clomipramine are the best tolerated medications for PE and have been observed to increase latency times to anywhere from 2 to 10 min. Treatment regimens that employ tricyclic antidepressants are generally limited by side effects such as drowsiness and insomnia.

Topical anesthetic agents such as 2% lidocaine jelly and topical 2.5% lidocaine/2.5% prilocaine cream (EMLA) [49] have been shown to decrease penile sensitivity and prolong ejaculatory latency [50]. When applied to the penile skin with a condom for 30 min before intercourse, EMLA cream has been observed to increase the time to ejaculation in 80% of men. Newer topical medications are now under investigation and include a quick-acting, local anesthetic spray and SS-cream, a herbal preparation that has been demonstrated to increase ejaculatory latency [51].

Given the frequency of PE among younger, sexually active men, durable success is an important treatment goal. Therefore, medical therapy should always be combined with behavioral modification therapy [52]. The goal of sexual therapy is to provide the patient with greater control over, and satisfaction from, sexual stimulation. Typically, the sensation of ejaculatory inevitability must be explained to patients so that they can understand, observe, and ultimately control the sensations experienced and enhance sexual pleasure. Typically, patients and their partners undergo a 6- to 20-week course of therapy in which they learn systematic relaxation techniques and acquire skills to perform prolonged self- or partner-performed sexual stimulation without the demand for erection or ejaculation. Subsequently, patients are instructed in methods of passive coitus without thrusting and, eventually, coitus with pelvic thrusting. Partner participation and cooperation with such therapy is critical for long-term success.

Seminal Megavesicles

Enlarged seminal vesicles without evidence of physical obstruction, also known as seminal megavesicles, have been reported in association with polycystic kidney disease and after surgical failure of transurethral resection of ejaculatory ducts. Dilation of the seminal vesicles may mimic obstruction of the ejaculatory ducts, and these conditions are not easily distinguishable on TRUS imaging. In a study by Hendry and coworkers, six azoospermic men with adult polycystic kidney disease had enlarged seminal vesicles. When these men were studied with seminal vesiculography, there was no obstruction found [53]. In addition, all attempts at transurethral resection of the ejaculatory ducts failed. We hypothesized that this abnormality may partly explain why 25–30% of men with presumed EDO fail to improve after TURED [54]. To demonstrate this concept, we constructed an in vivo rat model and assessed the urodynamic properties of active and resting compliance within the seminal vesicle. We found that seminal vesicles, as hollow organs lined with smooth muscle, act urodynamically like urinary bladders, thus lending credence to the concept that seminal vesicle myopathies, like bladder myopathies, can exist and result in "functional" obstruction of the reproductive tract [55]. Similar to neurogenic bladders, to date, no effective treatments have been shown to correct this type of organ dysfunction.

Retrograde Ejaculation

The true incidence of retrograde ejaculation is difficult to estimate, but approximately 14–18% of patients with ejaculatory disorders harbor this diagnosis [56]. Among 1,400 infertile couples, 0.7% of men presented with retrograde ejaculation [57]. This diagnosis is relatively straightforward and requires a history of low-volume or dry ejaculate, with a postmasturbatory voided urine sample demonstrating sperm. Systemic diseases such as diabetes mellitus, medications including alpha-blockers, neurogenic causes such as spinal cord injury, multiple sclerosis or spina bifida, and surgical treatments such as retroperitoneal lymph node dissection and transurethral prostate resection can all cause retrograde ejaculation.

Several treatment options are available for retrograde ejaculation. If the condition is drug induced, the offending medication should be discontinued if possible. In many patients without scar tissue at the bladder neck, oral therapy can be attempted with alpha-adrenergic agonists [9]. Approximately one-third of men will respond to this therapy. Sympathomimetic agents, such as imipramine, phenylpropanolamine, or pseudoephedrine, have been used with schedules ranging from interval dosing to as needed dosing immediately prior to coitus [58]. Generally, oral medical therapy is limited by side effects, including dizziness, weakness, nausea, sweating, or palpitations. If oral therapy fails, sperm-harvesting from the bladder urine obtained on post-ejaculate urination or by catheterization can be used with IUI or IVF to achieve family building goals.

Anejaculation

Congenital anorgasmia, also known as primary or psychogenic anejaculation, is a rare, well-described cause of conscious anejaculation. Despite the lack of willful orgasm, nocturnal emissions may occur [59]. The incidence of this condition is 0.14% in the general population and 0.39% among male patients seeking infertility care [60]. The cause is thought to be overly strict childhood upbringing. A classic setting includes parenting with intense performance demands and minimal physical affection. Secondary anejaculation is present in anejaculatory patients with a previous history of normal ejaculation and is generally due to neurologic disease or trauma (e.g., spinal cord injury).

Treatments that seek to reverse anejaculation are difficult. Often, affected individuals lack sensual awareness of their bodies. In addition, they often may seek a partner from a similar background and, as such, may accommodate to an asexual or minimally sexual lifestyle. Treatment is usually sought when the couple desires a pregnancy. Psychotherapy is often effective and is initiated with instruction in sex education, followed by cognitive behavioral treatment that includes systematic relaxation and sensate focus exercises [60]. At first, partners are taught to tolerate and become comfortable with touch. Later, when touch induces pleasure, sexual stimulation is encouraged as the shaping of the sexual response and orgasm occurs. Fertility issues can be managed relatively efficiently in almost all cases of anejaculation with PVS, EEJ, or/and microsurgical sperm aspiration [22, 61].

Expert Commentary

The study of disordered ejaculation remains complex since with many of these conditions an intricate web of connectivity exists between the psychosocial and neurophysiological aspects of human behavior that complicates scientific evaluation. For example, the causal etiologies ascribed to primary and secondary premature ejaculation are vastly different, yet the resultant behaviors are very similar. Likewise, the difference between normal ejaculatory latency and delayed ejaculation may simply reflect differences in learned behaviors. This complex interaction of mind and body is no more obvious than in the association between ejaculatory climax, a cerebral event, and the purely spinal reflex that defines ejaculation.

Indeed, the one physiological constant in the study of disordered ejaculation is that ejaculation is a simple spinal cord reflex. This fact makes animal models very relevant for study of the human condition, and it is precisely through model organisms that we have learned as much as we have about the neurobiology of sex. This is aptly demonstrated by the animal papers that showed that seminal vesicle physiology mimics that of the bladder as both are hollow, smooth muscular organs. This comparison led to the recent concept that seminal vesicles could be inherently dysfunctional and to the development of the "vasodynamic" study of the human reproductive tract to more accurately delineate cases of EDO.

Five-Year View

With the increasing use of animal models and surging knowledge of molecular biology and cell signaling, we look forward to improving the precision of defining many human ejaculatory disorders in the future. In addition, the continued worldwide standardization of definitions and terms used to describe ejaculatory disorders, as typified by consensus statements developed for erectile dysfunction and premature ejaculation, will further enhance our ability to work together and dissect the neurobiological from the psychosocial aspects of these conditions. This, in turn, should improve our ability to evaluate and accurately diagnose sexual disorders in the male and increase our ability to analyze response to treatment.

Although anatomical-based disorders such as EDO or spinal cord injury are fairly well understood at this point, there is ample room for an increased understanding of the effect of pharmacologic agents on ejaculation in the future. This should occur as the field of pharmacogenetics gains further traction in this field. The true value of gene and stem cell therapy has also yet to be fully realized in this discipline of medicine, especially in postsurgical, iatrogenic, or traumatic neurologically based disorders of ejaculation, but offers tremendous potential for functional restoration and cure over the next decade.

Key Issues

- A thorough history and physical examination is fundamental to defining the various behavioral and physiological bases for human ejaculatory disorders.
- There is a growing trend to develop international, consensus-based, standardized definitions of ejaculatory disorders that should help to more precisely define and study these conditions in the future.
- The anatomical and physiological basis for EDO and spinal cord injury are increasingly clear from research to date, but disorders of ejaculatory latency are currently the subject of intensive research to delineate their fundamental underlying pathophysiology.
- The psychosocial aspects of sexual and ejaculatory dysfunction are an integral part of human sexual behavior and, as such, will continue to require consideration and recognition if our understanding of these conditions is to improve in the future.

Acknowledgments The authors would like to acknowledge the contributions of Melody Lowman MA of San Francisco, California, to this chapter. As a psychotherapist with 45 years of experience, her knowledge and experience in reproductive psychology has been invaluable to us.

References

1. Master VA, Turek PJ. Ejaculatory physiology and dysfunction. Urol Clin North Am. 2001;28:363–75.
2. Yang CC, Bradley WE. Somatic innervation of the human bulbocavernosus muscle. Clin Neurophysiol. 1999;110:412.
3. Brackett NL, Ohl DA, Sønksen J, Lynne CM. Abnormalities of ejaculation. In: Lipshultz LI, Howards SS, Niederberger CS, editors. Infertility in the male. 4th ed. New York: Cambridge University Press; 2009. p. 454–73.
4. Benson GS. Erection, emission and ejaculation: physiologic mechanisms. In: Lipshultz LI, Howards SS, editors. Infertility in the male. 3rd ed. St. Louis: Mosby; 1997. p. 155.
5. Eisenberg M, Walsh TJ, Garcia M, et al. Ejaculatory duct manometry in normal men and in patients with ejaculatory duct obstruction. J Urol. 2008;180:255–60.
6. Jarvi K, Zielenski J, Wilschanski M, et al. Cystic fibrosis transmembrane conductance regulator and obstructive azoospermia. Lancet. 1995;345:1578.
7. Thorpe AC, Cleary R, Coles J, et al. Written consent about sexual function in men undergoing transurethral prostatectomy. Br J Urol. 1994;74:479.
8. Dunsmuir WD, Emberton M, Neal DE. On behalf of the steering group of the National Prostatectomy Audit. There is significant sexual dissatisfaction following TURP. Br J Urol. 1996;77:161A.
9. Lue TF, Giuliano F, Montorsi F, et al. Summary of the recommendations on sexual dysfunctions in men. J Sex Med. 2004;1:6–23.
10. Purohit RS, Wu DS, Shinohara K, Turek PJ. A prospective comparison of 3 diagnostic methods to evaluate ejaculatory duct obstruction. J Urol. 2004;171:232–5, discussion 235–36.
11. Hendry WF. Disorders of ejaculation: congenital, acquired and functional. Br J Urol. 1998;82:331.
12. Hendry WF, Pryor JP. Mullerian duct (prostatic utricle) cyst: diagnosis and treatment in subfertile males. Br J Urol. 1992;69:79.
13. Turek PJ. Seminal vesicle and ejaculatory duct surgery. In: Graham S, editor. Glenn's urologic surgery. 6th ed. Philadelphia: Lippincott, Williams & Wilkins; 2006. p. 439–45.
14. Nguyen HT, Etzell J, Turek PJ. Normal human ejaculatory duct anatomy: a study of cadaveric and surgical specimens [see comments]. J Urol. 1996;155:1639.
15. Jarow JP. Seminal vesicle aspiration in the management of patients with ejaculatory duct obstruction. J Urol. 1994;152:899.
16. Bird VG, Brackett NL, Lynne CM, et al. Reflexes and somatic responses as predictors of ejaculation by penile vibratory stimulation in men with spinal cord injury. Spinal Cord. 2001;39:514–9.
17. Brackett NL. Semen retrieval by penile vibratory stimulation in men with spinal cord injury. Hum Reprod Update. 1999;5:216–22.
18. Momose H, Hirao Y, Yamamoto M, et al. Electroejaculation in patients with spinal cord injury: first report of a large-scale experience from Japan. Int J Urol. 1995;2:326.
19. Brinsden PR, Avery SM, Marcus S, et al. Transrectal electroejaculation combined with in-vitro fertilization: effective treatment of anejaculatory infertility due to spinal cord injury. Hum Reprod. 1997;12:2687.
20. Schatte EC, Orejuela FJ, Lipshultz LI, et al. Treatment of infertility due to anejaculation in the male with electroejaculation and intracytoplasmic sperm injection. J Urol. 2000;163:1717.
21. Ohl DA. Electroejaculation. Urol Clin North Am. 1993;20:181.
22. Vanderschueren D, Spiessens C, Kiekens C, et al. Successful treatment of idiopathic anejaculation with electroejaculation after microsurgical vas aspiration. Hum Reprod. 1998;13:370.
23. Vinik AI, Maser RE, Mitchell BD, Freeman R. Diabetic autonomic neuropathy. Diabetes Care. 2003;26:1553–79.
24. Dunsmuir WD, Holmes SA. The aetiology and management of erectile, ejaculatory, and fertility problems in men with diabetes mellitus. Diabet Med. 1996;13:700.
25. Sexton WJ, Jarow JP. Effect of diabetes mellitus upon male reproductive function. Urology. 1997;49:508.
26. Hojo K, Vernava III AM, Sugihara K, et al. Preservation of urine voiding and sexual function after rectal cancer surgery. Dis Colon Rectum. 1991;34:532.
27. Nesbakken A, Nygaard K, Bull-Njaa T, et al. Bladder and sexual dysfunction after mesorectal excision for rectal cancer. Br J Surg. 2000;87:206.
28. Turek PJ, Lowther DN, Carroll PR. Fertility issues and their management in men with testis cancer. Urol Clin North Am. 1998;25:517.
29. Holman E, Kovacs G, Flasko T, et al. Hand-assisted laparoscopic retroperitoneal lymph node dissection for nonseminomatous testicular cancer. J Laparoendosc Adv Surg Tech A. 2007;17:16–20.
30. Neyer M, Peschel R, Akkad T, et al. Long-term results of laparoscopic retroperitoneal lymph-node dissection for clinical stage I nonseminomatous germ-cell testicular cancer. J Endourol. 2007;21:180–3.
31. Styblo K, Bossers GT, Slot GH. Osteotomy for kyphosis in ankylosing spondylitis. Acta Orthop Scand. 1985;56:294.
32. Berger Y, Blaivas JG, Oliver L. Urinary dysfunction in transverse myelitis. J Urol. 1990;144:103.
33. Decter RM, Furness III PD, Nguyen TA, et al. Reproductive understanding, sexual functioning and testosterone levels in men with spina bifida. J Urol. 1997;157:1466.
34. Sønksen J, Ohl DA. Penile vibratory stimulation and electroejaculation in the treatment of ejaculatory dysfunction. Int J Androl. 2002;25:324–32.
35. Kafetsoulis A, Brackett NL, Ibrahim E, et al. Current trends in the treatment of infertility in men with spinal cord injury. Fertil Steril. 2006;86:781–9.
36. Sønksen J. Assisted ejaculation and semen characteristics in spinal cord injured males. Scand J Urol Nephrol Suppl. 2003;213:1–31.
37. Kennedy SH, Dickens SE, Eisfeld BS, et al. Sexual dysfunction before antidepressant therapy in major depression. J Affect Disord. 1999;56:201.
38. Seidman S. Ejaculatory dysfunction and depression: pharmacological and psychobiological interactions. Int J Impot Res. 2006;18 Suppl 1:S33–8.
39. Hellstrom WJ, Sikka SC. Effects of acute treatment with tamsulosin versus alfuzosin on ejaculatory function in normal volunteers. J Urol. 2006;176:1529–33.
40. Sharlip ID, Hellstrom WJG, Broderick GA. The ISSM definition of premature ejaculation: a contemporary, evidence-based definition. J Urol. 2008;179(Suppl):340.
41. Read S, King M, Watson J. Sexual dysfunction in primary medical care: prevalence, characteristics and detection by the general practitioner. J Public Health Med. 1997;19:387.
42. Rowland DL, Haensel SM, Blom JH, et al. Penile sensitivity in men with premature ejaculation and erectile dysfunction. J Sex Marital Ther. 1993;19:189.
43. Biri H, Isen K, Sinik Z, et al. Sertraline in the treatment of premature ejaculation: a double-blind placebo controlled study. Int Urol Nephrol. 1998;30:611.
44. Haensel SM, Klem TM, Hop WC, et al. Fluoxetine and premature ejaculation: a double-blind, crossover, placebo-controlled study. J Clin Psychopharmacol. 1998;18:72.

45. Kara H, Aydin S, Yucel M, et al. The efficacy of fluoxetine in the treatment of premature ejaculation: a double-blind placebo controlled study [see comments]. J Urol. 1996;156:1631.
46. Kim SC, Seo KK. Efficacy and safety of fluoxetine, sertraline and clomipramine in patients with premature ejaculation: a double-blind, placebo controlled study. J Urol. 1998;159:425.
47. McMahon CG, Touma K. Treatment of premature ejaculation with paroxetine hydrochloride as needed: 2 single-blind placebo controlled crossover studies. J Urol. 1999;161:1826.
48. Waldinger MD, Hengeveld MW, Zwinderman AH. Ejaculation-retarding properties of paroxetine in patients with primary premature ejaculation: a double-blind, randomized, dose-response study. Br J Urol. 1997;79:592.
49. Berkovitch M, Keresteci AG, Koren G. Efficacy of prilocaine-lidocaine cream in the treatment of premature ejaculation [see comments]. J Urol. 1995;154:1360.
50. Andersen KH. A new method of analgesia for relief of circumcision pain. Anaesthesia. 1989;44:118.
51. Choi HK, Jung GW, Moon KH, et al. Clinical study of SS-cream in patients with lifelong premature ejaculation. Urology. 2000;55:257.
52. Annon J. The behavioral treatment of sexual problems, vol. 1. Honolulu: Enabling Systems; 1974.
53. Hendry WF, Rickards D, Pryor JP, et al. Seminal megavesicles with adult polycystic kidney disease. Hum Reprod. 1998;13:1567.
54. Turek PJ, Magana JO, Lipshultz LI. Semen parameters before and after transurethral surgery for ejaculatory duct obstruction. J Urol. 1996;155:1291.
55. Turek PJ, Aslam K, Younes AK, et al. Observations on seminal vesicle dynamics in an in vivo rat model. J Urol. 1998;159:1731.
56. Sandler B. Idiopathic retrograde ejaculation. Fertil Steril. 1979;32:474.
57. Van der Linden PJQ, Nan PM, te Velde ER, et al. Retrograde ejaculation: successful treatment with artificial insemination. Obstet Gynecol. 1992;79:126.
58. Kim SW, Paick JS. Short-term analysis of the effects of as needed use of sertraline at 5 PM for the treatment of premature ejaculation. Urology. 1999;54:544.
59. Hovav Y, Dan-Goor M, Yaffe H, et al. Nocturnal sperm emission in men with psychogenic anejaculation. Fertil Steril. 1999;72:364.
60. Geboes K, Steeno O, De Moor P. Primary anejaculation: diagnosis and therapy. Fertil Steril. 1975;26:1018.
61. Denil J, Kupker W, Al-Hasani S, et al. Successful combination of transrectal electroejaculation and intracytoplasmic sperm injection in the treatment of anejaculation. Hum Reprod. 1996;11:1247.

Environmental Factors

27

Claudio Telöken, Samuel Juncal, and Túlio M. Graziottin

Abstract

Approximately 50% of the human infertility issues involve male factors. A number of different components may result in reduction of sperm count or motility and affect sperm morphology. The etiology of male infertility is difficult to understand, due to its etiological heterogeneity, complexity, and incomplete knowledge of the underlying causes. In addition, World Health Organization (WHO) recently suggested changes in semen analysis lower limit parameters. In this way, it is not easy, based on the sperm count, to compare male infertility of earlier with the current ones. Most likely, some subfertile and/or infertile individuals under these new numbers would not be screened as such 5 years ago. Moreover, conventional reference values for seminal parameters have little diagnostic value because of their marked biological individuality variations, although seminal parameters may be useful for assessing differences in an individual's serial results, in particular of progressive motility, morphology, and vitality. Not long, we have seen significant refinement regarding male full evaluation combined with new sophisticated diagnostic techniques. Even then, most of the infertile men are described as idiopathic. Therefore, reproductive toxicity has been a topic of increasing interest and concern, as human exposure to a considerable number of potential toxicants is unavoidable due to contamination of air, water, ground, food, beverages, and household. Several lifestyle-related factors such as obesity, smoking, sedentary exposure to traffic exhaust fumes, dioxins, combustion products, cell phone electromagnetic radiations, chronic noise stress, anabolic steroids, illicit drugs, heat, notebooks use, dietary habits, oxidative stress, etc., appear to exhibit some involvement in human reproduction. Apart from this, public concern about adverse effects of environmental chemicals, pesticides, food additives, and persistent pollutants on spermatogenesis in adult men is sometimes not supported by the available data for humans. About 80,000 new chemical compounds have been introduced to human civilization in the last 100 years, and only 145 have been rigorously assessed for their reproductive health effects. The main scope of this chapter is to review some of the most frequent environmental or occupational pollutants in sex hormone levels, birth rates, and human reproduction in view of the fact that male infertility may be a surrogate marker of serious additional underlying medical problems.

C. Telöken, MD, PhD (✉) • T.M. Graziottin, MD, PhD
Andrology Section of Fertilitat – Human Reproduction Center, Department of Urology, Santa Casa Hospital and Federal University of Health Sciences, Porto Alegre, RS, Brazil
e-mail: urologyteloken@yahoo.com

S. Juncal, MS, MD
Andrology Section of Fertilitat – Human Reproduction Center, Federal University of Health Sciences, Porto Alegre, RS, Brazil

S.J. Parekattil and A. Agarwal (eds.), *Male Infertility: Contemporary Clinical Approaches, Andrology, ART & Antioxidants*, DOI 10.1007/978-1-4614-3335-4_27, © Springer Science+Business Media, LLC 2012

Keywords
Environmental factors in male infertility • Chromium • Lead • Mercury • Dioxins • Ethylene oxide • Bisphenol • Tobacco • Polychlorinated biphenyls

Approximately 50% of the human infertility issues involve male factors. A number of different components may result in reduction of sperm count or motility and affect sperm morphology. The etiology of male infertility is difficult to understand, due to its etiological heterogeneity, complexity, and incomplete knowledge of the underlying causes. In addition, World Health Organization (WHO) recently suggested changes in semen analysis lower limit parameters. In this way, it is not easy, based on the sperm count, to compare male infertility of earlier with the current ones. Most likely, some subfertile and/or infertile individuals under these new numbers would not be screened as such 5 years ago [1]. Moreover, conventional reference values for seminal parameters have little diagnostic value because of their marked biological individuality variations, although seminal parameters may be useful for assessing differences in an individual's serial results, in particular of progressive motility, morphology, and vitality [2–4].

Not long, we have seen significant refinement regarding male full evaluation combined with new sophisticated diagnostic techniques. Even then, most of the infertile men are described as idiopathic. Therefore, reproductive toxicity has been a topic of increasing interest and concern, as human exposure to a considerable number of potential toxicants is unavoidable due to contamination of air, water, ground, food, beverages, and household [5]. Several lifestyle-related factors such as obesity, smoking, sedentary exposure to traffic exhaust fumes, dioxins, combustion products, cell phone electromagnetic radiations, chronic noise stress, anabolic steroids, illicit drugs, heat, notebook use, dietary habits, oxidative stress, etc., appear to exhibit some involvement in human reproduction. Apart from this, public concern about adverse effects of environmental chemicals, pesticides, food additives, and persistent pollutants on spermatogenesis in adult men is sometimes not supported by the available data for humans [6–10].

About 80,000 new chemical compounds have been introduced to human civilization in the last 100 years, and only 145 have been rigorously assessed for their reproductive health effects [11]. It has been suggested that sperm counts have decreased and population fecundity has been declining during the second half of the last century in Western societies, in human as well as in animals, possibly due to environmental pollutants. However, studies are diverse, complex, and complicated to interpret. Whether environmental contaminants are really involved can neither be confirmed nor rejected. Under these circumstances, it should be a matter of great concern triggering studies into its causes and possibilities for prevention [12–16].

Take into account that in the field of reproductive environmental health, there remain many unanswered questions that suggest the need to include studies that target populations with high exposure to chemicals and to identify susceptibility factors and critical exposure windows (life stages) that may increase a man's risk of infertility. We need methods to better study mixtures of chemicals and develop methods to assess clinical reproductive outcomes of human exposure to the ever-growing list of chemicals [17]. Study designs need to incorporate markers of susceptibility into these chemicals and metals. Susceptibility may be increased by the timing or life stage exposure occurrences, such as neonatal or peripubertal periods. In addition, genetic susceptibility may result from differences incurred because of genetic polymorphisms in enzymes metabolizing or activating these compounds. With the current revolution in genomics and molecular biology, these techniques must be incorporated into the next generation of epidemiologic studies on male reproductive health [18], in such a way to possibly decrease the idiopathic male infertility rate.

The main scope of this chapter is to review some of the most frequent environmental or occupational pollutants in sex hormone levels, birth rates, and human reproduction in view of the fact that male infertility may be a surrogate marker of serious additional underlying medical problems.

Environmental Factors

The numerous external causes of infertility include exposure to work-related substances, contact with toxic substances like pesticides, or exposure to extremely hot areas such as blast furnace operations. Many substances such as metals, volatile organic compounds, physical factors (heat, vibration, and radiation), and pesticides including pyrethroids, organophosphates, phenoxyacetic acids, carbamates, and organochlorines have been investigated since studies reported an association between human exposure and semen quality, DNA damage, hormonal change, numerical chromosome aberrations, and possible human reproduction impairment. Studies varied widely in methods, exposures, and outcomes. Although suggestive of semen parameters, the epidemiologic evidence accumulated thus far remains equivocal as to the spermatotoxic and aneugenic potential of pesticides given the small number of published studies. This question warrants more investigation and suggestions for future studies [18, 19].

The hormone-disrupting properties of some factors have raised interest in how contemporary pollutants, which primarily take the form of low-level environmental or occupational exposures, impact human reproduction. Many studies in humans and animals support the role of environmental contaminants as potential endocrine disruptor. Endocrine-disrupting chemicals are among the most complex environmental health threat known today. By mimicking natural hormones such as estrogen and testosterone, these chemicals can interact with the body's endocrine system and exert toxic effects.

The rise in male infertility and the decline in human sperm counts could be linked with chemicals in the environment known as antiandrogens. These antiandrogens, for instance, have been linked with the feminization of fish in British rivers and could be affecting the development of male reproductive organs in humans. A link has been established between antiandrogens released into rivers from sewage outflows. There are several chemicals in widely used pharmaceuticals and pesticides that are known to have antiandrogenic activity, coming from domestic sources. One possibility is that drugs excreted from the body may end up in rivers. Antiandrogens may also be seeping into rivers as runoff from agricultural land. Sex change was first detected in fish more than 20 years ago. During the same period, human sperm counts have been falling in several countries over a period of 30 years or more. This has been matched by a corresponding rise in other male reproductive problems, such as the congenital condition testicular dysgenesis, which can affect fertility [20].

Other studies demonstrate the interaction of environmental toxicants with steroid receptors in the testis and thereby interference with proliferation and differentiation of spermatogenic cells. Acceleration of sperm transition through epididymis has been reported with several environmental contaminants such as methoxychlor, vinclozolin, etc. These and other observations support the endocrine-mediated toxic effects of environmental contaminants on reproductive and development abnormalities.

Apart from endocrine disruption, another mechanism that has emerged in the last two decades is the discovery of reactive oxygen species (ROS) and the role of associated oxidative stress in the etiology of defective sperm function and male infertility. Oxidative effects on spermatozoa may result in lower sperm motility and increased damage to sperm DNA [21]. Many contaminants have been reported to disturb the pro-oxidant/antioxidant balance, leading to excessive generation of ROS (oxidative stress). In higher concentrations, ROS may cause sperm cell damage through lipid peroxidation of the plasma membrane, single- and double-stranded DNA breaks [22], as well as the induction of germ cell apoptosis [23]. A greater proportion of infertile men demonstrate elevated levels of seminal ROS as compared with populations of fertile men. Nevertheless, other works have obtained contradictory results, indicating that these changes have not taken place homogeneously in the world. Geographical differences in semen quality also support the fact that semen quality may have declined only in some areas [24–28].

Volatile organic compounds, certain halogenated compounds, several heavy metals, xenoestrogens, organochlorine compounds, and phthalate esters may compromise reproductive male function, in vivo and in vitro. [15, 29, 30]. More studies are required for experimental confirmation of this link, which could give answer to several adverse effects of the contaminants whose mechanisms are yet to be established. Maternal beef consumption, and possibly xenobiotics (anabolic steroids) in beef, may alter a male fetus' testicular development in utero and adversely affect his reproductive capacity. In sons of "high beef consumer mothers," sperm concentration was 24.3% lower than that in the men whose mothers ate less beef [27, 31]. Ultrastructural analyses revealed damage to sperm head membranes in relation to the metal used: acrosome breakage with formation of various sized microvesicles, a large round hole and numerous folds in the acrosome membrane. Metal compounds may reduce sperm kinetic characteristics and probably fertilizing capacity by triggering specific morphological damages to the head and/or by inhibiting motility. It is worth mentioning that only specific chemical forms of a metal can pass through the blood–testis barrier and are more likely to target organs of the reproductive track and concentrate in the semen. In the testis, Leydig cells along with germ cells have been identified as the main targets of metal cytotoxicity, leading to a reduced steroidogenesis and, thereby, disruption of spermatogenesis. Inter-Sertoli tight junctions may also be affected by metals with consequent exfoliation of immature cells into the lumen of seminiferous tubules and reduction of viable sperm count in the epididymis. The underlying mechanisms of these effects are uncertain. Oxidative stress, inflammation, induced apoptosis, and ionic and molecular mimicry could be the basis of metal toxic activity. Some metals seem to experience a direct effect on sperm cells, reducing their motility and/or affecting their morphology. These adverse effects have been reported either by epidemiological studies of occupationally exposed individuals, highlighting a positive correlation, in most of the cases, with high concentrations of metals in semen [31].

Vanadium

Vanadium (V) is a rare element found combined in certain minerals and used mainly to produce certain alloys. Most of the vanadium produced is used as ferrovanadium or as a steel additive. Vanadium mixed with aluminum in titanium alloys is used in jet engines, high-speed airframes, and nuclear reactors. Vanadium oxide (V_2O_5) is used as a catalyst in manufacturing sulfuric acid and maleic anhydride and in making ceramics. Glass coated with vanadium dioxide (VO_2) can block infrared radiation at some specific temperature. Vanadium compounds are not regarded as a serious hazard; however, workers exposed to vanadium peroxide dust were found to suffer with severe eye, nose, and throat irritation. The uptake of vanadium by humans mainly takes place through foodstuffs, such as buckwheat, soya beans, olive oil, sunflower oil, apples, and eggs. Vanadium can have a number of effects on human health, when the uptake is too high.

Vanadium can be found in the environment in algae, plants, invertebrates, fishes, and many other species. In mussels and crabs, vanadium strongly bioaccumulates, which can lead to greater concentrations than those found in seawater. Vanadium causes the inhibition of certain enzymes with animals, which has several neurological effects. Laboratory tests with test animals have shown that vanadium can cause harm to the reproductive system of male animals and that it accumulates in the female placenta. In male mice, it can result in DNA damage, decrease in fertility rate, implantations, live fetuses, and fetal weight. It is concluded that vanadium pentoxide is a reprotoxic and genotoxic agent in mice [32]. Vanadium pentoxide inhalation in the human workplace enters the body through lungs, disrupts mitochondria function and the permeability of the epithelium, and promotes access of inflammatory mediators to the underlying neuronal tissue causing injury and neuronal death. Exposure results in necrosis of spermatogonium, spermatocytes, and Sertoli cells contributing to male infertility [33].

Manganese

Manganese (Mn) is a metal found as a free element in nature or often in combination with iron. It is important in the industry particularly in stainless steels. The fertility of male workers exposed to manganese dust was assessed by using a questionnaire. The manganese-exposed workers exhibited a statistically significant deficit in the number of children during their period of exposure to the metal [34]. High manganese level was associated with increased risk of low sperm motility and low sperm concentration. Ambient exposure to manganese levels is associated with a reduction in sperm motility and concentration [35].

Boron

Boron (B), the fifth element in the periodic table, is a metalloid, which occurs abundantly in the evaporite ores borax and ulexite and has widespread commercial uses. Boron is used as a dopant in the semiconductor industry, while boron compounds play specialized roles as structural and refractory materials and reagents for the synthesis of organic compounds, including pharmaceuticals.

Male reproductive toxicity has been demonstrated in experimental animals exposed to, and there is considerable interest in possible human reproductive effects. Although there appears to be considerable human exposure to boron compounds, epidemiological studies have not been sufficient for an evaluation of reproductive risk. Boron treatment of rats, mice, and dogs has been associated with testicular toxicity, characterized by inhibited spermiation at lower dose levels and a reduction in epididymal sperm count at higher dose levels. Earlier studies in human workers and populations have not identified adverse effects of boron exposure on fertility, but outcome measures in these studies were relatively insensitive, based mainly on family size and did not include an evaluation of semen end points. Reproductive outcomes in the wives of boron workers were not significantly different from those in the wives of background control men after adjustment for potential confounders. There were no statistically significant differences in semen characteristics between exposure groups, except that sperm Y:X ratio was reduced in boron workers. While boron has been shown to adversely affect male reproduction in laboratory animals, there is no clear evidence of male reproductive effects attributable to boron in studies of highly exposed workers.

In the USA, boron mine workers were not adversely affected. Overall, boron workers fathered 52.7% female offspring, compared to the US national average of 48.8%. However, the data did not indicate effects attributable to boron. Workers in the two lowest exposure categories had the highest percentage female offspring, while workers in the highest exposure category had virtually the same percentage (49.2%) as the national average. Thus, the fertility rate was not adversely affected when workers were most exposed to boron. The ratio of boys to girls was not reduced. Lack of strict epidemiological study design and the use of fertility rate to measure fecundity detracted from the utility of these papers for an evaluation of human reproductive toxicity [36] and since then, two studies that examined fertility and secondary sex ratios in response to long-term exposure to boron among the population in two geographical regions in Turkey have found no negative effects on fertility [37]. An influence of boron intake on certain human key enzymes, however, cannot be excluded [38]. There are no significant differences in sex ratio, with more females than males in the boron-rich region than in the boron-poor [38]. The only significant differences were that average pregnancies and live births among production workers exceeded those of office workers. The reported infertility rates are very low compared to usual general population infertility rates of 15%, and the reliability is questioned [39].

Cadmium

Cadmium (Cd) is one of the metallic components in the Earth's crust and oceans, and present everywhere in our environment. With the exception of its use in nickel–cadmium batteries and cadmium telluride solar panels, the use of cadmium is generally decreasing in its other applications, due to competing technologies and toxicity in certain forms and concentration. Cadmium is extremely toxic even in low concentrations and bioaccumulates in organisms and ecosystems. In the 1950s and 1960s, industrial exposure to cadmium was high, but as the toxic effects became apparent, industrial limits have been reduced in most industrialized nations. Buildup of cadmium levels in the water, air, and soil has been occurring particularly in industrial areas. Some sources of phosphate in fertilizers contain cadmium. Environmental exposure to cadmium has been particularly problematic in Japan where many people have consumed rice that was grown in cadmium-contaminated irrigation water, known under the name *itai-itai* disease [40]. Food and cigarettes are also a significant source of cadmium exposure. The general population and people living near hazardous waste sites may be exposed to cadmium in contaminated food, dust, or water from unregulated releases or accidental releases. Acute exposure to cadmium fumes may cause flu-like symptoms; kidney damage, and create hypophosphatemia causing muscle weakness. As a consequence of its unpredictability in children, in the future, it can have fertility impact. Cadmium was detected in the paint used on promotional drinking glasses for the movie *Shrek Forever After*, sold by McDonald's Restaurants, requiring the glasses to be recalled.

Chromium

Chromium (Cr) is a hard metal and was regarded with great interest because of its high corrosion resistance and hardness. A major development was the discovery that steel could be made highly resistant to corrosion and discoloration by adding chromium to form stainless steel. It is toxic in larger amounts. Cr(VI) treatment disrupted spermatogenesis, leading to accumulation of prematurely released spermatocytes, spermatids, and uni- and multinucleate giant cells in the lumen of seminiferous tubules in animals [41]. Two recent reports also correlated chronic occupational exposure to Cr(VI) to abnormal semen quality in men [42, 43], though the amount and type of Cr(VI) used were questioned [44]. The accumulation of uni- and multinucleate germ cells in the epididymal lumen of the monkeys treated with Cr(VI) causing ductal obstruction probably attributed to the disruption of spermatogenesis and testicular histoarchitecture [45]. Testes, seminal vesicle, and preputial gland weights were significantly reduced in chromium chloride- and potassium dichromate-exposed males [46].

Lead

Lead (Pb) is counted as one of the heavy metals. Lead is widely used in the production of batteries, metal products (solder and pipes), ammunition, and devices to shield X-rays, leading to its exposure to the people working in these industries. Use of lead in gasoline, paints and ceramic products, caulking, and pipe solder has been dramatically reduced in recent years because of health concerns. Ingestion of contaminated food and drinking water is the most common source of lead exposure in humans. Exposure can also occur via inadvertent ingestion of contaminated soil/dust or lead-based paints. Lead has long been known to be toxic to male fertility associated with the impairment of spermatogenesis and reduced concentrations of androgens. Male reproductive toxicity studies in humans have addressed effects on sex hormone levels, birth rates, time taken to conceive in couples not using contraception, and semen characteristics. Although findings across studies and end points are not entirely consistent, the main body of evidence points to current blood lead concentrations of about 40–50 μg/dl as a most likely no adverse effect threshold. This applies to semen characteristics such as sperm count, motility, abnormal sperm forms, as well as to fertility rate and time taken to conceive, whereas primary effects on the hormonal regulation of the male reproductive system at these exposure levels are questionable [47]. This view on male reproductive toxicity of lead is challenged by the findings on decreased fecundity among male lead workers [48, 49]. They observed an astonishing clear exposure response relation between current blood lead level and time taken to conceive among male battery workers in Taiwan. The fecundability ratio, in exposed compared to nonexposed, declined steadily from 0.9 in men with blood lead levels below 20 μg/dl to 0.4 among men with a blood level above 40 μg/dl. The fact that a large European study failed to show effects of lead on time to pregnancy in any of the three independent study populations is not reassuring if the consistency of findings across countries reflects repetition of errors inherent in the study design. The divergent findings in Europe and Taiwan could of course also be due to differences in susceptibility to the toxic effects of lead. During recent years, it has been shown that lead may interfere with the reorganization and tight packaging of sperm DNA during spermatogenesis by competition with zinc on protamine-binding sites. This results in reduced stability of the chromatin, and abnormal chromatin structure is strongly related to reduced fertility in humans. And there is indeed limited evidence that chromatin structure abnormalities are related to lead exposures in the lower range of blood lead values in men with high concentrations of lead within spermatozoa. Other mechanisms might be of significance as well. Thus, it was recently found that lead at environmental levels strongly interferes with the sperm acrosome reaction, which is essential for fertilization and negatively affects outcomes of artificial insemination [50]. There is definitely a need to keep open this line of research [48]. However, unlike many other metals such as zinc, chromium, manganese, copper, and iron, lead has no known essential effects for living organisms, and current exposure levels are still high compared to preindustrial populations.

Mercury

Mercury (Hg) is a heavy metal, proven to be toxic, has a very long half-life in the body, and is found in air, water, and soil. Mercury is in many products: metallic mercury is used in thermometers, dental fillings, and batteries. Mercury salts may be used in skin creams and ointments. It is also used in many industries. Mercury in the air settles into water and can pass through the food chain and build up in fish, shellfish, and animals that eat fish. The nervous system is sensitive to all forms of mercury. Exposure to high levels can damage the brain and kidneys. Pregnant women can pass the mercury in their bodies to their babies. Research on mercury increased after the accident in Japan in 1968, when more was learned about exposure to this metal in rats and humans. Mercury can concentrate in the kidneys, cerebellum, testes, and epididymis, leading to neurological disorders, kidney failure, and infertility, particularly on susceptible individuals and on susceptible groups such as fetuses and young children [51]. Young syndrome can be associated with obstructive lesion of the upper epididymis [52]. Some authors reviewed that existing scientific evidence does not demonstrate that mercury from dental amalgam poses a public health hazard, although there exists some controversy on this issue [51, 53, 54].

Copper

Copper (Cu) levels in serum and seminal plasma in the subfertile male group were significantly higher than those in the fertile male group [55]. Copper can act on FSH receptors, interfering in spermatogenesis. In animals, the main endocrine alterations are in testosterone, LH, and FSH secretion. Significant correlations between copper concentrations in semen and sperm concentration ($P<0.001$), percentage progressive motility ($P<0.005$), and normal morphology ($P<0.005$) were observed. However, semen copper concentrations of infertile men and fertile men did not differ significantly [56].

Dioxins

Dioxins are the result of various industrial processes and are considered the most toxic anthropogenic agents. Exposure to dioxins decreases spermatogenesis and testicular weight, reduces fertility, and can affect libido, causing changes in the sexual behavior of male fish, birds, and mammals, and also reptiles when in uterus. Sperm counts have dropped, and alterations in the male reproductive tract have increased since the 1950s.

2,3,7,8-Tetrachlorodibenzo-*p*-dioxin (TCDD) is a polychlorinated dibenzodioxin, the most toxic dioxin. The half-life in rodents is usually 2–4 weeks, but in humans has been estimated to be 7–11 years although with wide individual variation. It became known as a contaminant in Agent Orange, a herbicide used in the Vietnam War. It is a colorless solid with no distinguishable odor formed as an unintentional by-product of incomplete combustion of fossil fuels and wood, and during incineration processes of municipal and industrial wastes. Human studies have shown an association between TCDD and soft-tissue sarcomas, lymphomas, and stomach carcinomas. TCDD may be formed during the chlorine bleaching process used by pulp and paper mills and as a by-product from the manufacture of certain chlorinated organic chemicals, such as chlorinated phenols. Very low levels of TCDD are found throughout the environment, but most of the exposure of the general population is from food, mainly meat, dairy products, and fish. Chloracne is also the major effect seen from long-term exposure to TCDD in humans. Animal studies have reported hair loss, loss of body weight, and a weakened immune system from oral exposure to TCDD. The results of available reproductive and developmental studies in humans are inconclusive. Reproductive effects, including altered levels of sex hormones, reduced production of sperm, and increased rates of miscarriages, have been seen in animals exposed to TCDD.

Ethylene Oxide

Ethylene oxide (EtO) is produced in large volumes and is both flammable and highly reactive. Primarily used as an intermediate in the production of several industrial chemicals, the most notable of which is ethylene glycol. It is also used as a fumigant in certain agricultural products and as a sterilant for medical equipment and supplies and possesses several physical and health hazards that merit special attention. Acute exposures to EtO gas may result in respiratory irritation and lung injury, headache, nausea, vomiting, diarrhea, shortness of breath, and cyanosis. Chronic exposure has been associated with the occurrence of cancer, sperm damaging, reduces fertility, mutagenic changes, neurotoxicity, and sensitization [57].

Arsenic

Arsenic (As), a chemical element, is found in the uncombined condition in various localities but more generally in combination with other metals and sulfur. Arsenic is a constituent of the minerals arsenical iron, arsenical pyrites or mispickel, tin-white cobalt or smaltite, arsenical nickel, realgar, orpiment, pharmacolite, and cobalt bloom. The ordinary commercial arsenic is either the naturally occurring form, which is, however, more or less contaminated with other metals, or is the product obtained by heating arsenical pyrites, out of contact with air, in earthenware retorts which are fitted with a roll of sheet iron at the mouth and an earthenware receiver. A very high exposure to inorganic arsenic can cause infertility and miscarriages with women, and it can cause skin disturbances, declined resistance to infections, heart disruptions, and brain damage. Finally, inorganic arsenic can damage DNA, peripheral neuritis, sensory disturbances, tingling, numbness, formication, and occasionally cutaneous anesthesia. Later, the affected muscles become tender, and then atrophy, while the knee jerk or an other reflex is lost. Arsenic and most of its soluble compounds are very poisonous. Despite their toxic effect, inorganic arsenic bonds occur on earth naturally in small amounts. Humans may be exposed through food, water, and air. Exposure may also occur through skin contact with soil or water that contains arsenic. Levels of arsenic in food are fairly low, but levels in fish and seafood may be high because fish absorb arsenic from the water they live in. Luckily, this is mainly the fairly harmless organic form, but fish that contain significant amounts of inorganic arsenic may be a danger to human health [57]. Human data are limited to a few studies of populations exposed to arsenic from drinking water or from working at or living near smelters. Associations with spontaneous abortion and stillbirth have been reported in more than one of these studies. Interpretation of most of these studies is complicated because study populations were exposed to multiple chemicals.

Methoxychlor

Methoxychlor is a synthetic organochlorine used as an insecticide to protect crops, ornamentals, livestock, and pets against fleas, mosquitoes, and other insects and has been used to some degree as a replacement for dichlorodiphenyltrichloroethane (DDT) as it is metabolized faster and does not lead to bioaccumulation. The amount of methoxychlor in the environment changes seasonally due to its use in farming and foresting. It does not dissolve readily in water, so it is mixed with a petroleum-based fluid and sprayed, or used as a dust. Sprayed methoxychlor settles on the ground or in aquatic ecosystems, where it can be found in sediments. Its degradation may take many months. Methoxychlor is ingested and absorbed by living organisms, but is readily released and does not accumulate in the food chain. Some metabolites may have unwanted side effects. The use of methoxychlor as a pesticide was banned in the USA in 2003 and in Europe in 2002.

Human exposure to methoxychlor occurs via air, soil, and water [58], primarily in people who work with the substance or who are exposed to air, soil, or water that has been contaminated. Some of the agent's metabolites have an estrogenic effect as shown in adult and developing animals before and after birth [58]. One metabolite is 2,2-bis(*p*-hydroxyphenyl)-1,1,1-trichloroethane (HPTE) which is considered to have reproductive toxicity in the animal model by reducing testosterone biosynthesis. Such effects adversely affect both the male and female reproductive systems. It is expected that this "could occur in humans" but has not been proven.

Little information is available regarding effects on human pregnancy and children, but it is assumed from animal studies that methoxychlor crosses the placenta, and it has been detected in human milk. Exposure to children may be different from that in adults because they tend to play on the ground; further, their reproductive system may be more sensitive to the effects of methoxychlor as an endocrine disruptor. Food contamination may occur at low levels. Some of the breakdown products of methoxychlor cause effects similar to those produced by estrogen. Exposure in animals to methoxychlor adversely affects prostate and fertility in adults and in developing animals exposed prenatally or shortly after birth. Likewise, it is expected that reproductive effects seen in animals could occur in humans exposed to methoxychlor but yet has not been clearly reported [59, 60].

Glycol Ether

"Glycol ethers" is a name for a large group of chemicals. Most glycol ether compounds are clear, colorless liquids. Some have mild, pleasant odors, or no smell at all; others (mainly the acetates) have strong odors. The common belief that glycol ethers never evaporate fast enough to create harmful levels in the air is false. Some evaporate quickly and can easily reach hazardous levels in the air; others evaporate very slowly and therefore are less hazardous by inhalation. The glycol ethers are widely used industrial solvents. Each of them may be used alone or as an ingredient in products such as coatings (paints, varnishes, dyes, stains, inks, and semiconductor chip coatings), cleaners (for degreasing, dry cleaning, film cleaning, and circuit board manufacture), jet fuel deicing additives, brake fluids, and perfumes and cosmetics. Certain glycol ethers have been found to cause birth defects and damage to the testicles in laboratory animals [61]. Glico exposure is related to low motile sperm count in men attending fertility clinics. This suggests that it continues to be a hazard to male fertility [62]. Controversially, another study suggests that most glycol ethers currently used do not have an impact on human semen characteristics. Those that were more prevalent from the 1960s until recently may have long-lasting negative effects on human semen quality [63].

Polychlorinated Biphenyls

Polychlorinated biphenyls (PCBs) are a class of synthetic, persistent, lipophilic, halogenated aromatic compounds that were widely used in industrial and consumer products for decades before their production was banned in the USA in the late 1970s. PCBs were used in cutting oils, lubricants, and as electrical insulators. As a result of their extensive use and persistence, PCBs remain ubiquitous environmental contaminants. They are distributed worldwide and have been measured in air, water, aquatic and marine sediments, fish, and wildlife. Furthermore, they are biologically concentrated and stored in human adipose tissue. The general population is exposed primarily through the ingestion of contaminated foods (fish, meat, and dairy products), as PCBs bioaccumulate up the food chain. As a result of their persistence and ubiquity, measurable levels of serum PCBs are found in the majority of the US general population. Epidemiologic data support an inverse association of PCBs with reduced semen quality, specifically reduced sperm motility. The associations found were generally consistent across studies despite a range of PCB levels [64].

Bisphenol

Bisphenol (BPA) is an organic compound with two phenol functional groups and is a dysfunctional building block of several important plastic additives. BPA is an industrial chemical used to make a hard, clear plastic known as polycarbonate, which has been used in many consumer products, including reusable water bottles and baby bottles and is also found in epoxy resins, which act as a protective lining on the inside of metal-based food and beverage cans. These uses of BPA are subject to premarket approval by FDA as indirect food additives or food contact substances. The original approvals were issued under FDA's food additive regulations and date from the 1960s. Suspected of being hazardous to humans since the 1930s, concerns about the use of bisphenol A in consumer products were regularly reported in the news media in 2008 after several governments issued reports questioning its safety, thus prompting some retailers to remove products containing it from their shelves. A 2010 report by the United States Food and Drug Administration (FDA) raised further concerns regarding exposure of fetuses, infants, and young children [65–67].

Tobacco

The data on the effect of cigarette smoking on semen parameters and fertility is conflicting as well as uncertain [68]. However, increasing evidence suggests detrimental effects on human reproduction. DNA-binding carcinogens from cigarette smoke have been found in spermatozoa of smokers as well as in the embryos formed during IVF cycles using the sperm from these men. The heavy metal cadmium and the toxic alkaloid nicotine are present in increased amounts in the semen of smokers [69–71]. Studies have correlated smoking with adverse effects on parameters such as seminal volume, sperm count, motility, morphology, and increased numbers of white blood cells (WBCs) in the semen [72]. Other studies, however, have found no influence of cigarette smoking on semen parameters [73]. Smoking has been found to have an adverse effect on female fecundity, whereas the effects on male fecundity are conflicting [73, 74]. There is a delay in conception when the male smoked and evidence of a delay in conception when the nonsmoking female was passively exposed to cigarette smoke. In addition, a decreased success of IVF was reported when the male was a smoker [75]. We believe the evidence is strong enough that smoking should be considered a risk factor for infertility, and patients should be strongly encouraged to discontinue this habit [68]. However, smoking may serve as a cofactor for patients with other causes of male infertility. Effects of tobacco on fertility include higher levels of seminal oxidative stress than infertile nonsmokers, increasing DNA damage, higher sperm aneuploidy, testicular atrophy, and poorer sperm quality such as abnormal semen volume, concentration, motility, morphology, and elevated prolactin and estrogen. That is why some IVF programs restrict smoking for 3 months prior to semen collection. It is well established that smoking has a detrimental effect on male reproductive system. Cigarette smoking leads to an increase in ROS level. The high level of leukocytospermia in smokers suggests that oxidative stress is due to ROS generation by activated leukocytes. Various compounds of cigarette smoke (polycyclic aromatic hydrocarbons) and smoking metabolites may act as chemotactic stimuli and thereby induce an inflammatory response, recruitment of leukocyte, and subsequent generation of ROS. Motility is one of the first sperm parameters affected, and asthenozoospermia may be an early indicator of reduced semen quality in light smokers. There is a significantly high teratozoospermia in heavy smokers compared to nonsmokers. Studies have shown that maternal smoking affects reproductive parameters of their offspring (male) during adolescence [74, 76]. Remains unclear possible impact on nonsmokers industrial cigarette workers.

Radiation

Exposure to radiation, even small amounts of ionizing, particularly that used for medical therapy, may destroy sperm-forming cells. Spermatogonia are particularly radiosensitive and may further impair testicular function [77]. Spermatogenesis may take up to 4–5 years to return after radiation therapy or chemotherapy [78].

Radiation induces chromosomal changes, which lead to congenital abnormalities. Radiation exposure of pregnant mothers and the developing male fetus can cause adverse reproductive effects. A study among more than 100 prisoners, who volunteered themselves for testicle X-irradiation, showed that a dose of 0.11 Gy could cause substantial suppression of sperm counts and a dose of 3–5 Gy could lead to permanent sterility.

Phthalates

Phthalates are chemical substances widespread in general population exposure. High molecular weight phthalates, for example, di-[2-ethylhexylphthalate (DEHP)], are primarily used as plasticizers in the manufacture of flexible vinyl plastic, which in turn is used in consumer products, flooring and wall coverings, food contact applications, manufacturing of automobiles plastics, beverage containers, coating of metal cans, etc., and medical devices. Manufacturers use low molecular weight phthalates (e.g., diethyl phthalate [DEP] and dibutyl phthalate [DBP]) in personal care products (perfumes, lotions, and cosmetics), as solvents and plasticizers for cellulose acetate, and in making lacquers, varnishes, and coatings, including those used to provide timed releases in some pharmaceuticals [55, 57, 58]. Parenteral exposure from medical devices containing phthalates is an important source of high exposure to phthalates, primarily DEHP [79]. As a result of the ubiquitous use of phthalates in personal forms of exposure are food, inhalation, dermal contact, medical materials and by occupational contact. Phthalates can alter reproductive development regardless of binding to androgen or estrogen receptors. Some phthalate esters inhibit steroidogenesis in Leydig cells. Through exposure by food, they can cause testicular atrophy and reduced fertility. Data have demonstrated that perinatal exposure to a variety of phthalate esters alters the development of the male reproductive tract in an antiandrogenic way, causing underdevelopment and agenesis of the epididymis at relatively low doses [80]. Compared with the laboratory animal data on the reproductive toxicity of phthalates, the human data is limited. In spite of that, the concentration of the sum of several phthalate diesters in seminal plasma was inversely correlated with sperm morphology in infertile males, an inverse relationship with sperm concentration [81]. DBP concentrations have been found in the cellular fractions of ejaculates [80, 82]. In contrast, according to a singular study, there were no relationships of MBP or MBzP with any of the semen parameters. However, this particular study should not be compared with others due to many epidemiologic differences, such as median age, type of recruitment, etc. Phthalates can alter reproductive development regardless of binding to androgen or estrogen receptors.

Xenoestrogens

Xenoestrogens are industrially made compounds that have estrogenic effects and differ chemically from naturally occurring estrogenic substances produced by living organisms. In the environment, xenoestrogens can be divided into natural compounds (plants or fungi) and synthetically derived agents including pesticides and industrial by-products. They have been a ubiquitous part of the environment even before the existence of the human race [83]. Xenoestrogens have also been identified as endocrine disruptors that might not only cause the "testicular dysgenesis syndrome" (TDS) but also disturb meiosis in developmental germinal cells. The male reproductive system is most vulnerable to estrogenic agents during the critical period of cell differentiation and organ development in fetal and neonatal life [84]. In this period, the testes are structurally organized, establishing Sertoli cell and spermatogonia numbers to support spermatogenesis that will be initiated at puberty. The maintenance of tightly regulated estrogen levels is essential for its completion. Analysis of maternal and fetal biological fluids has shown that xeno- and phytoestrogens cross the placenta barrier into fetal circulation and that they can bioaccumulate in fetal organs. An analysis on amniotic fluid samples undergoing routine amniocentesis shows that overall, one in three amniotic fluid samples tested positive for at least one environmental contaminant. It has also been described that these compounds cross a blood–tissue barrier similar to that of the testis, suggesting that intratubular germ cells might be exposed [21]. Soy isoflavone has estrogenic activity and is widely consumed. The association of soy foods and soy isoflavones suggested that higher intake of these foods was associated with lower sperm concentration. A further recent change in diet is the increased use of soy-based infant formula milk (SFM) for the increasing numbers of babies that are lactose intolerant. Infants on SFM have a phytoestrogen (a kind of xenoestrogen) intake per kg body weight 6–11 times higher than that seen in adults consuming a high phytoestrogen diet. Plasma levels of phytoestrogen are 1,000-fold greater than those of endogenous estradiol raising concerns about the effects of prolonged neonatal exposure to such high concentrations of estrogens and its potential disruption of male reproductive tract. Consequently, it is being speculated that male reproductive anomalies (hypospadias and cryptorchidism) and the global fall in sperm counts have both a causal link in the marked increase of phytoestrogens in our diet, causing a disruption of the male reproductive system. Endogenous hormones have a vital role in fetal life and ensure future fertility. Exposure to the wrong hormones (female fetus exposed to male hormones) or inadequate amounts of these could affect the reproductive system, and genitalia may not develop correctly, resulting in fertility problems in adulthood. Children are extremely sensitive to estradiol and may respond with increased growth and/or breast development even at serum levels below the current detection limits and that those changes in hormone levels during fetal and prepubertal development may have severe effects in adult life [85]. A cautionary approach should be taken in order to avoid unnecessary exposure of fetuses and children to exogenous sex steroids and endocrine disruptors, even at very low levels. That caution includes food intake, as possible adverse effects on human health may be expected by consumption of meat from hormone-treated animals [86]. Maternal beef consumption, and possibly xenobiotics (anabolic steroids) in beef, may alter testicular development in uterus and adversely affect reproductive capacity. Sperm concentration was inversely related to mother's weekly beef intake. In sons of "high consumers" (>7 beef meals/week), sperm concentration was 24.3% lower than that of men whose mothers ate less beef [87]. General population is exposed to many potential endocrine disruptors concurrently. Studies have shown that the action of estrogenic compounds is additive, but little is known about the possible synergistic or additive effects of these compounds in humans [88]. There is a human association between maternal exposure to pesticides and cryptorchidism among male children. Pesticide levels in breast milk were significantly higher in boys with cryptorchidism [89].

Vinclozolin

Vinclozolin [3-(3-5-dichlorophenyl)-5-methyl-oxazolidine-2,4-dione] is a common dicarboximide fungicide used to control various diseases on raspberries, chicory grown, lettuce, kiwi, canola, snap beans, dry bulb onions, ornamentals, and turf. Vinclozolin is formulated as a dry flowable and extruded granule which may be applied with aerial, chemigation, or ground equipment (broadcast, band, or soil drench); as a dip treatment on ornamental bulbs and corms, cut flowers, rose budwood, or nursery stock; and with thermal fogger in greenhouses. It has been registered since 1981 in the USA for use as fungicide [90]. Vinclozolin is a known environmental endocrine disrupter. Endocrine disruptors are hormonally active environmental compounds that have been shown to influence both male and female reproductive development and function [91]. Endocrine disruptors affect normal reproductive physiological development and functions by acting as antiandrogenic compounds. Vinclozolin binds with high affinity to the androgen receptor and blocks the action of gonadal hormones on male reproductive organs [92]. An alternative mechanism of the action of vinclozolin involves transgenerational effects on the male reproductive tract, which has defined as the transfer of heritable material from parents to offspring continuing through multiple subsequent generations. To transfer an induced phenotype from parents to offspring, the germ line genetic material must be altered via an epigenetic (DNA methylation) or stable genetic alteration (mutation, change in the DNA sequence) mechanism [93]. Vinclozolin exposure results in a transgenerational effect on spermatogenic capacity and testis function. The embryonic exposure to vinclozolin resulted in a reduced spermatogenetic capacity in the male offspring, and this phenotype was transmitted to the subsequent generations. This is a stage where spermatogenesis has advanced germ cell maturation and early spermatocyte development [94, 95].

Heat

There is human and animal experimental evidence that heat exposure may be detrimental to spermatogenesis [96]. The relationship between fertility and occupations in which heat exposure occurs reported a detrimental effect on sperm morphology and time to conception [97]. Occupations that have been reported to be associated with heat exposure and infertility include bakers, drivers (industrial machinery, taxis, and trucks), ceramic oven operators, welders, and workers in submarines [98]. In addition, a dysfunction of testicular thermoregulation has been suggested to occur in paraplegic men in wheelchairs [99].

Expert Commentary

Environmental factors that affect male fertility are complex and not well known. Our modern society produces several new chemical components each day, and the impact of these products in male fertility is not enough studied. This plethora of new components can affect the environmental delicate equilibrium, and the results for humans, flora, and fauna are a public issue. The twenty-first century is still facing many controversial issues, although there is a growing body of literature relating to the deleterious effect of substances on semen quality, fertility, offspring anomalies, etc. There is no question that it is a matter of interest for any country in spite of its economic, social, and cultural background. However, there is a tremendous heterogeneous situation among countries: some carrying out outstanding control as well as studies, some working as slaves and do not have "the right" to learn what they are working with, no personal exposure protection, and the family living just behind the industry plant, without any water, soil, and dust control. It is well known that the protection against exposure may prevent diseases with high morbidity and mortality, such as tumors, endocrine disruption, immune and endocrine system, behavior, human reproduction, etc.

In spite of all the growing progress, we need to learn more about food additives, toxicants, contaminants, outdoor and indoor air pollutants, pesticides, xenoestrogens, and hazardous substances in the workplace and particularly the possible synergism among many substances.

We are already aware that sperm DNA fragmentation, ROS (oxidative stress), apoptosis, and some Y microdeletions are happening, and the scientific world is running far behind.

Five-Year View

For the next 5 years, strong education development program worldwide for lay population is crucial since some pollutant hazards can be prevented. The scientific community should be committed to share the data, information, not just among the sophisticated academic world but to make it happen across the target general population.

Many current study designs present bias and need to be reviewed in such a way that we can compare data and come across with definite conclusions. Certainly, they need to incorporate markers of susceptibility for several chemicals, metals, etc. Thus, it will be possible temporally be sure when and what hazards chemicals are being deleterious to human being as well as determined in what life stage (fetus, neonatal, peripubertal, etc.) will be risky.

Regarding preventive attitudes, it depends upon the country policy. Most of the situations are just government investment and decisions because pollutants are spread all over (air, water, soil, food, etc.).

Key Issues

- Male infertility and pollutants.
- Environmental hazards can cause infertility as well as many serious diseases.
- The decreasing male fertility can be a consequence of external toxic ubiquitous substances.

References

1. World Health Organization, Department of Reproductive Health and Research. WHO laboratory manual for the examination and processing of human semen. 5th ed. Geneva: World Health Organization; 2010. p. 1–271. ISBN 978 92 4 154778 9 (NLM classification: QY 190).
2. Lampiao F. Variation of semen parameters in healthy medical students due to exam stress. Malawi Med J. 2009;21(4):166–7.
3. Castilla JA, Álvarez C, Aguilar J, et al. Influence of analytical and biological variation on the clinical interpretationol seminal parameters. Hum Reprod. 2006;21(4):847–51.
4. Oshio S, Ashizawa Y, Yotsukura M, et al. Individual variation in semen parameters of healthy young volunteers. Arch Androl. 2004;50:417–25.
5. Public Health Statement about Methoxychlor. Agency for toxic substances and disease registry (ATSDR). 2002. http://www.atsdr.cdc.gov/toxprofiles/tp47-c1-b.pdf. Accessed 22 Aug 2010.
6. Agarwal A, Desai NR, Ruffoli N, et al. Lifestyle and testicular dysfunction: a brief update. Biomed Pharmacother. 2008;62:550–3.
7. Ausmees K, Zarkovski M, Timberg G, et al. Associations between semen quality, body mass index and metabolic syndrome parameters in young male. Eur Urol Suppl. 2010;9:I–XX.
8. Phillips KP, Tanphaichitr N. Mechanisms of obesity-induced male infertility. Expert Rev Endocrinol Metabol. 2010;5(2):229–51.
9. Sharpe RM. Environmental/lifestyle effects on spermatogenesis. Philos Trans R Soc Lond B Biol Sci. 2010;365(1546):1697–712.
10. Sheynkin Y, Jung M, Yoo P, et al. Increase in scrotal temperature in laptop computer users. Hum Reprod. 2004. http://humrep.oxfordjournals.org/cgi/content/abstract/deh616v1. Accessed 1 June 2011.
11. Turek PJ. Does the male infertility clinical evaluation adequately assess toxicologic exposures? Fertil Steril. 2008;89:69.
12. Velde E, Burdorf A, Nieschlag E, et al. Is human fecundity declining in Western countries? Hum Reprod. 2010;25(6):1348–53.
13. Sinclair S. Male infertility nutritional and environmental considerations. Altern Med Rev. 2000;5(1):28–38.
14. Rhind SM, Evans NP, Bellingham M. Effects of environmental pollutants on the reproduction and welfare of ruminants. Animal. 2010;4(7):1227–39.
15. Toppari J, Larsen JC, Christiansen P, et al. Male reproductive health and environmental xenoestrogens. Environ Health Perspect. 1996;104(4):741–803.
16. Carlsen E, Giwercman A, Keiding N, et al. Evidence for decreasing quality of semen during past 50 years. BMJ. 1992;305:609–13.
17. Hauser R, Sokol R. Science linking environmental contaminant exposures with fertility and reproductive health impacts in the adult male. Fertil Steril. 2008;89:59–65.
18. Saradha B, Mathur PP. Effect of environmental contaminants on male reproduction. Environ Toxicol Pharmacol. 2006;21:34–41.
19. Perry MJ. Effects of environmental and occupational pesticide exposure on human sperm: a systematic review. Hum Reprod Update. 2008;14(3):233–42.
20. Jobling S, Burn RW, Thorpe R, et al. Statistical modelling suggests that anti-androgens in wastewater treatment works effluents are contributing causes of widespread sexual disruption in fish living in English rivers. Environ Health Perspect. 2009. doi: 10.1289/ehp.0800197. http://dx.doi.org/. Accessed 7 Jan 2009.
21. Sharpe RM, Skakkebaek NE. Are oestrogens involved in falling sperm counts and disorders of the male reproductive tract? Lancet. 1993;341:1392–5.
22. Aitken RJ, Baker MA. Reactive oxygen species generation by human spermatozoa: a continuing. Int J Androl. 2002;25(4):191–4.

23. Agarwal A, Said TM. Role of sperm chromatin abnormalities and DNA damage in male infertility. Hum Reprod Update. 2003;9: 331–45.
24. Rao AVSK, Shaha C. Role of glutathione S-transferases in oxidative stress–induced male germ cell apoptosis. Free Radic Biol Med. 2000;29(10):1015–27.
25. Iwasaki A, Gagnon C. Formation of reactive oxygen species in spermatozoa of infertile patients. Fertil Steril. 1992;57(2):409–16.
26. Chen SS, Chang LS, Wei YH. Oxidative damage to proteins and decrease of antioxidant capacity in patients with varicocele. Free Radic Biol Med. 2001;30:1328–34.
27. Agarwal A, Saleh RA, Bedaiwy MA. Role of recreative oxygen species in the pathophysiology of human reproduction. Fertil Steril. 2003;79:829–43.
28. Mendiola J, Torres-Cantero AM, Moreno-Grau JM, et al. Food intake and its relationship with semen quality: a case-control study. Fertil Steril. 2009;91:812–8.
29. Agarwal A. New insights in molecular mechanisms of male infertility. Urology News. 2003;22(8):22.
30. Akingbemi BT, Ge RS, Klinefelter GR, et al. A metabolite of methoxychlor, 2,2-bis(p-hydroxyphenyl)-1,1,1-trichloroethane, reduces testosterone biosynthesis in rat Leydig cells through suppression of steady-state messenger ribonucleic acid levels of the cholesterol side-chain cleavage enzyme. Biol Reprod. 2000;62:571–8.
31. Cummings AW. Methoxychlor as a model for environmental estrogens. Crit Rev Toxicol. 1997;27:367–79.
32. Castellini C, Mourvakia E, Sartini B, et al. In vitro toxic effects of metal compounds on kinetic traits and ultrastructure of rabbit spermatozoa. Reprod Toxicol. 2009;27:46–54.
33. Altamirano-Lozano M, Alvarez-Barrera L, Basurto-Alcántara F, et al. Reprotoxic and genotoxic studies of vanadium pentoxide in male mice. Teratog Carcinog Mutagen. 1996;16:7–17.
34. Cooper RG. Vanadium pentoxide inhalation. Indian J Occup Environ Med. 2007;11:97–102.
35. Lauwerys R, Roels H, Genet P, et al. Fertility of male workers exposed to mercury vapor or to manganese dust: a questionnaire study. Am J Ind Med. 1985;7(2):171–6.
36. Wirth JJ, Rossano MG, Daly DC, et al. Ambient manganese exposure is negatively associated with human sperm motility and concentration. Epidemiology. 2007;18:270–3.
37. ayli BS. Assessment of fertility and infertility in boron-exposed turkish subpopulations. Evaluation of fertility among sibs and in "borate families". Biol Trace Elem Res. 2001;81(3):255–67.
38. ayli BS. Low frequency of infertility among workers in a borate processing facility. Biol Trace Elem Res. 2003;93:19–29.
39. Tüccar EAH, Yavuz Y, et al. Comparison of infertility rates in communities from boron-rich and boron-poor territories. Biol Trace Elem Res. 1998;66:401–7.
40. Scialli AR, Bondeb JP, Bruske-Hohlfeldc I, et al. An overview of male reproductive studies of boron with an emphasis on studies of highly exposed Chinese workers. Reprod Toxicol. 2010;29: 10–24.
41. Jarup L. Health effects of cadmium exposure—a review of the literature and a risk estimate. Scand J Work Environ Health. 1998;24:11–51.
42. Aruldhas MM, Subramanian S, Sekar P, et al. Chronic chromium exposure-induced changes in testicular histoarchitecture are associated with oxidative stress: study in a non-human primate (Macaca radiata Geoffroy). Hum Reprod. 2005;20(10):2801–13.
43. Danadevi K, Roya R, Reddy PP, et al. Semen quality of Indian welders occupationally exposed to nickel and chromium. Reprod Toxicol. 2003;17(4):451–6.
44. Li H, Chen Q, Li S, et al. Effect of Cr(VI) exposure on sperm quality: human and animal studies. Ann Occup Hyg. 2001;45:505–11.
45. Duffus JH. Effect of Cr(VI) exposure on sperm quality. Ann Occup Hyg. 2002;46(2):269–70.
46. Aruldhas MM, Subramanian S, Sekar P, et al. Microcanalization in the epididymis to overcome ductal obstruction caused by chronic exposure to chromium—a study in the mature bonnet monkey (Macaca radiata Geoffroy). Reproduction. 2004;128:127–37.
47. Bataineh H, Al-Hamood MH, Elbetieha A. Effect of long-term ingestion of chromium compounds on aggression, sex behavior and fertility in adult male rat. Drug Chem Toxicol. 1997;20(3):133–49.
48. Sheiner EK, Sheiner E, Hammel RD, et al. Effect of occupational exposures on male fertility. Ind Health. 2003;41:55–62.
49. Shiau CY, Wang JD, Chen PC. Decreased fecundity among male lead workers. Occup Environ Med. 2004;61:915–23.
50. Benoff S, Centola GM, Millan C, et al. Increased seminal plasma lead levels adversely affect the fertility potential of sperm in IVF. Hum Reprod. 2003;18(2):374–83.
51. Bonde JP, Apostoli P. Any need to revisit the male reproductive toxicity of lead? Occup Environ Med. 2005;62:2–3.
52. Podzimek S, Prochazkova J, Bultasova L, et al. Sensitization to inorganic mercury could be a risk factor for infertility. Neuroendocrinology. 2005;4(26):277–82.
53. Hendry WF, A'Hern RP, Cole PJ. Was Young's syndrome caused by exposure to mercury in childhood? BMJ. 1993;307(6919): 1579–82.
54. Mutter J, Naumann J, Sadaghiani C, et al. Amalgam studies: disregarding basic principles of mercury toxicity. Int J Hyg Environ Health. 2004;207(4):391–7.
55. Choy CM, Lam CW, Cheung LT, et al. Infertility, blood mercury concentrations and dietary seafood consumption. BJOG. 2002;109(10):1121–5.
56. Aydemir B, Kiziler AR, Onaran I, et al. Impact of Cu and Fe concentrations on oxidative damage in male infertility. Biol Trace Elem Res. 2006;112(3):193–203.
57. Jockenhövel F, Bals-Pratsch M, Bertram HP, et al. Seminal lead and copper in fertile and infertile men. Andrologia. 1990;22(6): 503–11.
58. Water treatment solutions—Lenntech. http://www.lenntech.com/periodic/elements/as.htm. Accessed 22 Aug 2010.
59. Public Health Statement about Methoxychlor. Atlanta, GA: ATSDR. http://www.atsdr.cdc.gov/toxprofiles/tp47-c1-b.pdf. Accessed 1 June 2011.
60. EU Pesticides Database. European Union—DG SANCO. http://ec.europa.eu/sanco_pesticides/public/index.cfm?event=activesubstance.selection&a=1. Accessed 2 Oct 2009.
61. Consumer Factsheet on: METHOXYCHLOR. United States Environmental Protection Agency. http://www.epa.gov/safewater/contaminants/dw_contamfs/methoxyc.html. Accessed 26 Nov 2006.
62. Hazard Evaluation System and Information Service. http://www.dhs.ca.gov/ohb/HESIS/glycols.htm. Accessed 5 Jun 2008.
63. Cherry N, Moore H, McNamee R, et al. Occupation and male infertility: glycol ethers and other exposures. Occup Environ Med. 2008;65:708–14.
64. Multigner L, Brik EB, Arnaud I, et al. Glycol ethers and semen quality: a cross-sectional study among male workers in the Paris municipality. Occup Environ Med. 2007;64:467–73.
65. Hauser R. The environment and male fertility: recent research on emerging chemicals and semen quality. Semin Reprod Med. 2006;24(3):156–67.
66. Update on bisphenol A for use in food contact applications. U.S. Food and Drug Administration—FDA. 2010. http://www.fda.gov/NewsEvents/PublicHealthFocus/ucm197739.htm. Accessed 1 June 2011.
67. European Food Safety Authority. Toxicokinetics of bisphenol A, scientific opinion of the panel on food additives, flavourings, processing aids and materials in contact with food. EFSA J. 2008. http://www.efsa.europa.eu/cs/BlobServer/Scientific_Opinion/afc_ej759_bpa_%20toxicokinetics_op_en.pdf?ssbinary=true. Accessed 9 Jul 2008.

68. Vom Saal VS, Akingbemi BT, Belcher SM, et al. Bisphenol A expert panel consensus statement: integration of mechanisms, effects in animals and potential to impact human health at current levels of exposure. Reprod Toxicol. 2007;24:131–8.
69. Sigman M, Jarow JP. Male infertility. In: Wein AJ, Kavoussi LR, Novick AC, Partin AW, Peters CA, editors. Campbell-Walsh urology, vol. 9. Philadelphia: Elsevier Inc; 2007. p. 609–50.
70. Sepaniak S, Forges T, Gerard H, et al. The influence of cigarette smoking on human sperm quality and DNA fragmentation. Toxicology. 2006;223:54–60.
71. Zenzes MT. Smoking and reproduction: gene damage to human gametes and embryos. Hum Reprod Update. 2000;6:122–31.
72. Zenzes MT, Bielecki R, Reed TE. Detection of benzo(a)pyrene diol epoxide-DNA adducts in sperm of men exposed to cigarette smoke. Fertil Steril. 1999;72:330–5.
73. Vine MF, Tse CK, Hu P, et al. Cigarette smoking and semen quality. Fertil Steril. 1996;65:835–42.
74. Goverde HJM, Dekker HS, Janssen HJ, et al. Semen quality and frequency of smoking and alcohol consumption—an explorative study. Int J Fertil. 1995;40(3):135–8.
75. Bolumar F, Olsen J, Boldsen J. Smoking reduces fecundity: a European multicenter study on infertility and subfecundity. The European Study Group on Infertility and Subfecundity. Am J Epidemiol. 1996;143:578–87.
76. Zitzmann M, Rolf C, Nordhoff V, et al. Male smokers have a decreased success rate for in vitro fertilization and intracytoplasmic sperm injection. Fertil Steril. 2003;79(3):1550–4.
77. Bonde JP, Kold JT, Brixen LS, et al. Year of birth and sperm count in 10 Danish occupational studies. Scand J Work Environ Health. 1998;24:407–13.
78. Turek PJ. Evaluation and treatment of male factor infertility. In: Shoskes DA, Morey AF, editors. The American Urological Association Educational review manual in urology. 2nd ed. New York: Castle Connolly Graduate Medical Publishing; 2009. p. 825–52.
79. Costabile RA. The effects of cancer and cancer therapy on male reproductive function. J Urol. 1993;149:1327–30.
80. Hauser R, Meeker JD, Duty S, et al. Altered semen quality in relation to urinary concentrations of phthalate monoester and oxidative metabolites. Epidemiology. 2006;17:682–91.
81. Queiroz EKR, Waissmann W. Occupational exposure and effects on the male reproductive system. Cad Saúde Pública. 2006;22(3): 485–93.
82. Murature DA, Tang SY, Steinhardt G, Dougherty RC. Phthalate esters and semen quality parameters. Biomed Environ Mass Spectrom. 1987;14:473–7.
83. Singleton DW, Sohaib AK. Xenosestrogen exposure and mechanisms of endocrine disruption. Front Biosci. 2003;8:110–8.
84. Skakkebæk NE, Rajpert-De Meyts E, Main KM. Testicular dysgenesis syndrome: an increasingly common developmental disorder with environmental aspects. Hum Reprod. 2001;16:972–8.
85. Aksglaede L, Juul A, Leffers H, et al. The sensitivity of the child to sex steroids: possible impact of exogenous estrogens. Hum Reprod. 2006;12:341–9.
86. Andersson AM, Skakkebæk NE. Exposure to exogenous estrogens in food: possible impact on human development and health. Eur J Endocrinol. 1999;140:477–85.
87. Swan SH, Liu F, Overstreet JW, et al. Semen quality of fertile US males in relation to their mothers beef consumption during pregnancy. Hum Reprod. 2007;22:1497–502.
88. Toppari J, Skakkebæk NE. Sexual differentiation and environmental endocrine disrupters. Baillieres Clin Endocrinol Metab. 1998;12:143–56.
89. Damgaard IN, Skakkebæk NE, Toppari J, et al. Persistent pesticides in human breast milk and cryptorchidism. Environ Health Perspect. 2006;114:1133–8.
90. Kelce WR, Monosson E, Gamcsik MP, Laws SC, et al. Environmental hormone disruptors: evidence that vinclozolin developmental toxicity is mediated by antiandrogenic metabolites. Toxicol Appl Pharmacol. 1994;126:276–85.
91. Bayley M, Junge M, Baatrup E. Exposure of juvenile guppies to three antiandrogens causes demasculinization and a reduced sperm count in adult males. Aquat Toxicol. 2002;56(4):227–39.
92. Blake LS, et al. Reproductive toxicity of vinclozolin in the fathead minnow: confirming an anti-androgenic mode of action. Environ Toxicol Chem. 2008;27(2):478–88.
93. Anway MD, Cupp AS, Uzumcu M, et al. Epigenetic transgenerational actions of endocrine disruptors and male fertility. Science. 2005;308:1466–9.
94. Uzumcu M, Suzuki H, Skinner MK. Effect of the anti-androgenic endocrine disruptor vinclozolin on embryonic testis cord formation and postnatal testis development and function. Reprod Toxicol. 2004;18(6):765–74.
95. Anway MD, Memon AM, Uzumcu M, et al. Transgenerational effect of the endocrine disruptor vinclozolin on male spermatogenesis. J Androl. 2006;27(6):868–79.
96. MacLeod J, Hotchkiss RS. The effect of hyperpyrexia upon spermatozoa counts in men. Endocrinology. 1941;28:780–4.
97. Thonneau P, Bujan L, Multigner L, et al. Occupational heat exposure and male fertility: a review. Hum Reprod. 1998;13(8):2122–5.
98. de la Calle JF Velez, Rachou E, le Martelot MT, et al. Male infertility risk factors in a French military population. Hum Reprod. 2001;16(3):481–6.
99. Brindley GS. Deep scrotal temperature and the effect on it of clothing, air temperature, activity, posture and paraplegia. Br J Urol. 1982;54(1):49–55.

28 Exogenous Medication or Substance-Induced Factors

Kubilay Inci and Levent Mert Gunay

Abstract

Half of all the primary causes of infertility are attributed to males. These causes vary by individual, and the exact etiology is difficult to determine in many infertile men. Spermatogenesis is a long and complex process controlled by gonadotropins [FSH (follicle-stimulating hormone) and LH (luteinizing hormone)], which stimulates Sertoli and Leydig cells. Some men seem to be genetically subfertile, while epigenetic causes play an important role in others. Men are exposed to environmental pollutants, toxic substances, and drugs that compromise fertility. Medications and other substances that provoke infertility may affect sperm development or the pituitary–hypothalamic axis. Unfortunately, most physicians do not inform patients about the potential for infertility associated with the drugs they prescribe. This chapter aims to highlight the adverse effect profile on fertility of frequently used drugs.

Keywords

Male infertility • Effect of medications • Factors affecting fertility • Exogenous substances • Environmental pollutants • Toxic substances • Drugs compromising fertility

Half of all the primary causes of infertility are attributed to males. These causes vary by individual, and the exact etiology is difficult to determine in many infertile men. Spermatogenesis is a long and complex process controlled by gonadotropins [FSH (follicle-stimulating hormone) and LH (luteinizing hormone)], which stimulates Sertoli and Leydig cells. Some men seem to be genetically subfertile, while epigenetic causes play an important role in others. Men are exposed to environmental pollutants, toxic substances, and drugs that compromise fertility. Medications and other substances that provoke infertility may affect sperm development or the pituitary–hypothalamic axis. Unfortunately, most physicians do not inform patients about the potential for infertility associated with the drugs they prescribe. This chapter aims to highlight the adverse effect profile on fertility of frequently used drugs.

Exogenous Medications with the Potential to Compromise Male Fertility

Gastrointestinal Medications

Medications for Acid-Related Disorders

Wong et al. drew attention to these commonly used medications by reporting that their use increased the prevalence of oligozoospermia, as compared to normal men (OR 6.2) [1].

Histamine receptor blockers are the mainstay of peptic disorders treatment. Their direct gonadotoxic and centrally acting properties on reproductive hormones have been reported. Cimetidine, ranitidine, and famotidine cause Ca^{2+} influx into sperm cells in vitro and cause sperm death [2]. Additionally, cimetidine decreases testosterone levels in vivo [3, 4].

K. Inci, MD (✉) • L.M. Gunay, MD
Department of Urology, Hacettepe University School of Medicine, Ankara 06100, Turkey
e-mail: kuinci@hacettepe.edu.tr; mertgunay@yahoo.com

S.J. Parekattil and A. Agarwal (eds.), *Male Infertility: Contemporary Clinical Approaches, Andrology, ART & Antioxidants*,
DOI 10.1007/978-1-4614-3335-4_28, © Springer Science+Business Media, LLC 2012

Medications for Functional Gastrointestinal Disorders

Among the antiemetics and gastroprokinetic agents, metoclopramide was investigated for its adverse effects on sperm function. It is a benzamide derivative dopamine antagonist. Prolactin levels increase with metoclopramide use (10 mg q.i.d.), whereas testosterone levels decrease [5]. In a relatively old and small cohort study, its hyperprolactinemic effect was found to be responsible for reductions in seminal volume and total sperm count [6]. In contrast, a more recent study reported that after administration of 10 mg of metoclopramide three times daily for 12 weeks, serum prolactin levels increased fivefold, semen volume and abnormal sperm forms decreased, while spermatozoa velocity increased. No influence was noted on the number of spermatozoa per milliliter, total number of spermatozoa, percentage of motile spermatozoa, or index of motility. The authors concluded that hyperprolactinemia seemed to improve spermatozoal velocity and morphology, although a direct effect of metoclopramide on these parameters could not be excluded [7].

Antidiarrheal and Intestinal Anti-inflammatory/ Anti-infective Agents

Sulfasalazine, a drug primarily used for its anti-inflammatory effect and inhibition of chemotaxis and lysosomal protease, is widely used in gastroenterology and rheumatology and is associated with oligospermia, asthenospermia, and teratozoospermia when given for a course of at least 2 months, regardless of dosage level [8]. The sulfasalazine metabolite sulfapyridine is responsible for impaired spermatogenesis. As initial contact between the drug and sperm occurs via secretions from the testicular adnexa, its effect is thought to occur during epididymal transport of the sperm pass or after ejaculation in ejaculate [9]. Sulfasalazine also inhibits acrosomal reactions in sperm. Impaired spermatogenesis reverses as soon as 3 weeks after discontinuation of therapy, and children can be conceived without congenital anomalies [8].

Cardiovascular Medications

Beta-Blocking Agents

In addition to the known effects of beta-blockers on erectile function, their negative effects on sperm quality have also been studied. Propranolol has been investigated in vitro, and its inhibitory effect on sperm motility was delineated [10, 11]. cAMP was shown to increase the duration of motility in vitro; however, addition of cAMP to the medium in which propranolol is found did not increase sperm motility [10].

Calcium Channel Blockers

Regulation of protein function via Ca^{++} is involved in a range of male germ cell activities, including hyperactivation, chemotaxis, and acrosome reaction [12]. In vitro studies report that nifedipine and verapamil inhibit sperm motility and acrosome reaction [13, 14]. These features of calcium channel blockers gave rise to the idea of developing male contraceptives; however, both the wide spectrum of involved Ca^{++} channel types and the obscurity of the identity of the molecular mechanisms involved limit their use for contraceptive purposes [15].

Agents that Act on the Renin-Angiotensin System

The renin-angiotensin system (RAS) is best known for its role in fluid–electrolyte and blood pressure homeostasis. There are numerous mammalian tissues that implement their own RAS for local needs, including those of the male reproductive system. The angiotensins in the male reproductive system are frequently of paracrine/autocrine origin [16]. Angiotensin receptors are present in the acrosomal region of spermatozoa [17]. ACE-lacking knockout male mice have been shown to be infertile [18]. Inhibition of ACE activity by captopril reduces the binding of spermatozoa to zona-free hamster oocytes, but not their binding to the zona pellucida; however, only in vitro studies have reported alterations in sperm function [19].

Nervous System Medications

Analgesics

Human semen is rich in prostaglandins; however, the consequences of a reduction in seminal prostaglandins by non-steroidal anti-inflammatory drug (NSAID) usage have not received the level of attention that was expected. NSAIDs, such as acetylsalicylic acid, lysine salicylate, naproxen, and flurbiprofen, have been shown to lower the prostaglandin content of semen [20]. In experimental studies on male rabbits and rats, administration of several NSAIDs (acetylsalicylic acid, indomethacin, naproxen, and phenylbutazone) caused a reduction in seminal prostaglandins, but their affects did not appear to critically impair male fertility [20, 21].

There is limited knowledge concerning the detrimental effects of aspirin on spermatogenesis. In a placebo-controlled, single-blinded study, Conte et al. reported that aspirin treatment inhibited androgen response to chorionic gonadotropin stimulation in normal humans and speculated that this effect was mediated via an effective arachidonate cyclooxygenase block [22].

Antiepileptics

Most antiepileptics used for schizophrenia were reported to decrease serum testosterone levels and increase sex hormone-binding globulin (SHBG) levels. Carbamazepine is a hepatic enzyme inducer, and patients treated with it have lower serum testosterone levels than those treated with valproate (a hepatic enzyme inhibitor) [23].

Mikkonen et al. studied two groups, each including 70 pubertal boys. Group 1 consisted of boys with epilepsy that were taking various antiepileptics (carbamazepine: $n=28$; valproate: $n=25$; oxcarbazepine: $n=12$; lamotrigine: $n=5$), and group 2 consisted of healthy boys. They evaluated serum androgen levels and testicular structure during pubertal maturation and observed that serum testosterone levels were normal in the epilepsy patients; however, androstenedione levels in the patients treated with valproate remained high throughout puberty. In boys treated with carbamazepine, serum SHBG levels increased and serum dehydroepiandrosterone sulfate concentrations decreased. There was no significant difference in testicular volume between the two groups. Mikkonen et al. concluded that changes in serum sex hormone levels in boys and young men with epilepsy are associated with carbamazepine and valproate, but not with oxcarbazepine or lamotrigine [24].

Isojärvi et al. studied the effects of both epilepsy and antiepileptics on sperm parameters and reproductive hormones. They concluded that carbamazepine or partial epilepsy may affect sperm concentrations, morphology, and motility, that valproate or generalized epilepsy may affect sperm morphology and motility, and that oxcarbazepine may affect sperm morphology; however, the exact mechanism by which sperm is affected is not known. It can be hypothesized that the direct effects on spermatogenesis and the effects mediated by drug-induced endocrine changes are possible causes; yet the observed sperm abnormalities in this study were not correlated with the observed endocrine effects of the drugs. Moreover, temporal lobe epilepsy itself may be associated with altered serum gonadotropin and testosterone levels, which may affect sperm quality and be the cause of the sperm abnormalities seen in men with partial epilepsy [25]. Isojärvi et al. also reported reduced testicular volume in valproate-treated men with generalized epilepsy, but not in men with partial epilepsy treated with carbamazepine or oxcarbazepine. In non-epileptic animal experiments, abnormal spermatogenesis and testicular atrophy were linked to valproate [25].

Psycholeptics and Psychoanaleptics

Psychotropic agents can affect male reproductive functions by acting on central neurotransmitters. For example, increased serotonin levels inhibit dopamine and augment increases in serum prolactin, which may suppress gonadotropin release. In a double-blind trial that included 20 male patients with major depression, neither trimipramine nor imipramine caused significant changes in plasma testosterone concentrations [26]. Sperm motility problems were reported to be associated with clomipramine, imipramine, desmethylimipramine, and nortriptyline [27, 28]. Tanrikut and Schlegel reported two cases treated with antidepressant medication that were referred for evaluation of male infertility [29]. The patients' initial semen analyses revealed oligospermia, impaired motility, and abnormal morphology while they were taking serotonin reuptake inhibitors. Semen analyses performed 1–2 months after discontinuation of the antidepressants showed marked improvements in sperm concentration and motility. While the mechanism by which antidepressants may affect spermatogenesis or sperm motility is not clear, possible mechanisms include effects on sperm pH or viscosity, on nitric oxide concentrations (an inhibitor of sperm motility), or on GABA (a physiologic regulator of sperm motility) [30]. There is little evidence that lithium salts affect male reproductive functions. In animal models, subchronic exposure to lithium salts had toxic effects on sperm, germinal, and somatic cells in the testis and reduced fertility in male rats [31, 32].

Other Nervous System Medications

Opiates. Only a limited number of controlled trials concerning the effects of opiates on fertility have been published; however, several studies reported that they negatively affect semen parameters. Ragni et al. studied 80 heroin addicts and dual heroin–methadone users and reported that there were semen abnormalities in all of the heroin addicts, whereas only 10 of the 22 (45%) methadone-only users in the study had low semen quality. The most common abnormalities were asthenospermia (100%), teratozoospermia (24%), and oligozoospermia (17%). All the heroin addicts and dual heroin–methadone users had normal plasma levels of FSH, LH, and testosterone. The authors concluded that decreased forward motility under normal hormonal control might be an early indication of heroin toxicity to the male reproductive tract [33]. The timing of the normalization of these effects has not been sufficiently studied.

Tetrahydrocannabinol. Tetrahydrocannabinol (THC) will be discussed in Sect. 2.4

Dermatological Medications

Retinoids have anti-inflammatory and immunomodulatory effects, stimulate cell differentiation, and inhibit cutaneous keratinization and sebum production. All-trans retinoic acid has been reported to increase apoptosis, with a subsequent decline in germinal epithelial cells; however, at therapeutic doses, no association was observed between systemic retinoid therapy and significantly low sperm analysis parameters [34, 35]. Isotretinoin, which is used for treating acne, can impair spermatogenesis [36]. Animal models have shown that retinoid treatment is not associated with a risk to reproductive safety [37]. Nonetheless, maternal usage of this agent is highly teratogenic though, to the best of our knowledge, no evidence of spontaneous abortion or teratogenicity induced by paternal isotretinoin therapy has been reported.

Chloroquine is a protease inhibitor and lysosome stabilizer. It has been reported that chloroquine can be detected in human seminal plasma in patients undergoing chloroquine therapy, and it has the potential to impair sperm function; however, there is no evidence that it has a negative effect on acrosomal reaction [38]. In vitro studies have shown that chloroquine decreased sperm motility and in animal models, negatively affected male fertility after a 2-week course of treatment at a minimum daily dose of 5 mg kg^{-1} of body weight [39, 40].

Endocrine Therapy

Estrogen-Like Compounds

Estrogens can cause impaired sperm function and are associated with the risk of testicular tumors in adult males.

Gonadotropin-Releasing Hormone

Gonadotropin-releasing hormone (GnRH) is widely used for the treatment of hypogonadism. Exogenous pulsatile use can boost testosterone surges and the maturation of spermatogenesis; however, continuous use of long-acting agonists and antagonists inhibits gonadotropin secretion, resulting in a decrease in testosterone production and spermatogenic activity [41–43]. GnRH was successfully used to lower testosterone secretion in prostate cancer patients [44]. Testosterone levels returned to normal within 10–12 months after cessation of GnRH treatment, but LH levels remained high [44]. Long-term use of agonists and antagonists also decreases spermatogenesis.

Anabolic Steroids

Anabolic androgenic steroids are commonly used by competitive body builders; interestingly, their use is becoming more popular among nonathletes for cosmetic purposes [45]. Impaired spermatogenesis is related to the androgenic effects of the drugs. Sperm counts improve after cessation of steroid intake [46].

Sex Hormones and Modulators of the Genital System

Progestogens

As the progestogens (medroxyprogesterone, etonogestrel, levonorgestrel, and desogestrel, etc.) are capable of decreasing testicular function by causing oligospermia and hypogonadism, they have been studied for use in male contraception [47]. Their suppressive effects on Leydig cell function are balanced with testosterone combinations.

Androgens

Androgens also have the ability to inhibit spermatogenesis via a negative feedback loop. Nevertheless, the same problems concerning progestogens mentioned above limit their use as contraceptives. Additionally, the rebound action may be a choice for spermatogenesis induction.

Quinagolide is a dopamine agonist used to inhibit prolactin synthesis. As a male contraceptive, it is combined with testosterone implants. Quinagolide + testosterone treatment can reversibly suppress spermatogenesis [48].

Estrogens

Based on experimental models, inappropriate exposure to estrogenic compounds could lead to free radical-mediated damage that significantly disrupts the structure of spermatozoan DNA [49].

Other Progestins

Many trials have been conducted with many different drugs (gestrinone, gossypol, 7-α-methyl-nortestosterone, nonoxynol, and α-chlorohydrin) for use in male contraception; however, none were adequately efficient and/or safe.

Antiandrogens

Cyproterone acetate negatively affects libido and erectile function via its antiandrogenic activity; however, its negative effects on spermatogenesis appear to be related to the suppression of gonadotropin secretion [50].

Antineoplastic Agents

In 80% of young cancer survivors, fertility and sexual function are considered principal quality of life measures [51]. Antineoplastics used for systemic malignancies and autoimmune disorders usually cause spermatogenic dysfunction. Their effects are mostly dose related, and some alternative combination therapies can be used in order to avoid these highly toxic doses; however, polychemotherapy can be expected to severely impair spermatogenesis, even in prepubescent patients. Radiotherapy combined with chemotherapy potentiates the damage to gametes.

As the management of childhood malignancies has improved during the last decades, the number of patients that survive to adulthood following chemotherapy has increased. As such, preservation of fertility in these boys became a quality of life issue, both for physicians and parents. Polychemotherapy in older adolescent male patients was shown to cause oligospermia in up to 75% of cases 10 years after treatment; however, it did not affect puberty, and 30% of patients subsequently fathered children without any congenital abnormalities [52, 53]. Shielding the gonads may help limit radiation exposure; however, the impact of chemotherapeutics on the gonads is more difficult to control.

Some commonly used chemotherapeutic agents have been categorized according to their associated risk to the recovery of spermatogenesis as follows: good, moderate, and poor [54]. Agents categorized as good include adriamycin, methotrexate, prednisone, estrogens, androgens, cisplatin, thioguanine, doxorubicin, and 6-mercaptopurine. Agents associated with moderate risk include vincristine, PEB (cisplatin, etoposide, and bleomycin), and ABVD (doxorubicin, bleomycin, vinblastine, and dacarbazine). Poor-risk agents include cyclophosphamide, chlorambucil, mechlorethamine, procarbazine, and MOPP (nitrogen mustard, vincristine, procarbazine, and prednisone).

Among the cytostatic chemotherapy agents, the most severe gonadal damage is caused by alkylating agents; foremost among these is cyclophosphamide. Cyclophosphamide has been used for decades primarily by oncologists and rheumatologists for its antineoplastic and immunosuppressive characteristics. Sperm counts begin to drop 2–3 weeks after beginning chemotherapy and reach their lowest levels 2–3 months after the initiation of treatment [55]. Testicular damage was reported to be dose dependent in children and adults in both oncologic and rheumatologic use. Azoospermia was reported in pediatric sarcoma patients after cumulative doses as low as 6.0 $g\,m^{-2}$ [56]. In about 50% of patients treated with cyclophosphamide, spermatogenesis recovered after an average of 31 months [55]. In addition to oligospermia and azoospermia, deterioration of Leydig cell function resulting in reduced serum testosterone levels was also reported [57].

Green et al., in the Childhood Cancer Survivor Study, reported that sperm parameters were negatively affected in patients treated with procarbazine or cyclophosphamide and in those with a cumulative alkylating agent score ≥2 (described by Tucker et al. [58]).

Chlorambucil is another alkylating agent used in lymphoma and leukemia chemotherapy protocols and for nephrotic syndrome. In a study that included 16 boys with minimal-change nephrotic syndrome, doses that were safe for the gonads were investigated; 0.2 mg kg^{-1} day^{-1} of chlorambucil for 6 weeks was reported to be safe [59]. Sperm parameter recovery took about 20 years in two boys that received long-term treatment for nephrotic syndrome with high-dose chlorambucil (cumulative doses: 3,000 and 6,500 mg) [60]. Nonetheless, the prediction of the level of recovery after various doses is not possible.

Dacarbazine was evaluated in young adults with childhood malignancies and caused azoospermia in all seven patients [61]. Hydroxyurea has been linked to significantly reduced cell counts during spermatogenesis. Vincristine has been shown to impair spermatogenesis by inhibiting the incorporation of thymidine, uridine, and leucine in spermatogenic cells but without causing permanent infertility [62].

Methotrexate is a folic acid antagonist known to cause chromosomal anomalies and point mutations that result in mutagenesis; however, oligospermia caused by methotrexate is reversible [63, 64]. Methotrexate toxicity has mostly been studied in psoriasis patients, and some case studies reported that it can damage spermatogenesis [63]. Yet in a study that included 26 psoriasis patients whose semen parameters, testicular histology, and spermatogenic function were evaluated using radioactive phosphorus, methotrexate's negative effects on fertility were not observed [65].

Cisplatin is a chemotherapeutic widely used in combination therapies and is an important drug in germ cell cancer treatment. It can cause long-lasting deterioration in sperm parameters and testicular atrophy. Experiments on rodents showed that Leydig cells, Sertoli cells, and germ cells were all affected by cisplatin-induced toxicity, suggesting that the drug targets multiple cell types and molecular pathways while producing testicular injury [66]. Active spermatogenesis enabling fatherhood has been observed in over 50% of patients 1–3 years after cisplatin therapy [67]. One rat study reported that testosterone substitution with cisplatin therapy can reduce the damage caused to sperm cells by the drug [68].

Paclitaxil is a natural product used in the treatment of several cancers. When given in doses of 300 mg m^{-2}, paclitaxil will induce infertility [57]. Additionally, docetaxel, a microtubule stabilizer, is known to induce testicular atrophy and Leydig cell hyperplasia in mice, and it induced infertility and azoospermia in humans when administered in total doses of around 100 mg m^{-2} [57].

Different combination therapies and their azoospermic potential were reviewed by Puscheck et al. [69]. The rate of azoospermia in response to combinations used for Hodgkin's disease (MOPP, MVPP, ChlVPP, and MOPP/ABVD) was 85–100% and 0–67% for non-Hodgkin's lymphoma (CHOPP-Bl and COPP). Combination chemotherapy protocols for sarcomas (CyVADIC and CyADIC) can also cause azoospermia in 30–90% of patients. Moreover, 14–28% of patients treated for testicular cancers using a PVB protocol remained azoospermic. Tempest et al. studied sperm aneuploidy in Hodgkin's lymphoma and testicular cancer survivors for chromosomes 13, 21, X, and Y [70]. Both cancer groups had chromosomal abnormalities in the selected chromosomes. Aneuploidy seemed to recover in 18 months. In a similar patient group, sperm DNA parameters were adversely affected by combination chemotherapy protocols lasting approximately 2 years [71]. These studies emphasize the importance of genetic counseling in young male cancer survivors.

Today, the only accepted standard option for preserving male fertility before chemotherapy is cryopreservation of spermatozoa. Usually, the quality of sperm obtained at the time cancer is diagnosed is poor. Due to improvements in assisted reproduction techniques, this problem is somewhat solved by intracytoplasmic sperm injection, provided that some spermatozoa can be harvested from the patient; however, currently there are no fertility preservation options for prepubertal males scheduled for chemotherapy.

Immunosuppressive Agents

In addition to the previously discussed sulfur compounds, the effects of immunosuppressives (primarily transplantation drugs) on male fertility have been studied. A few human studies reported a decrease in serum testosterone levels and an increase in gonadotropin levels in renal and heart transplant recipients [72, 73]. The fathered pregnancy rates that were reported highlight the negative effects on the fertility of patients treated with sirolimus [74].

Animal-based research on cyclosporin A, which selectively inhibits T-helper cells, shows that it damaged the germinal epithelium of the testes and decreased testosterone production, spermatogenesis, and fertility, even at low doses [75]. Cyclosporin inhibits testosterone synthesis by inhibiting delta 5–3 beta-hydroxysteroid dehydrogenase/delta 5–4 isomerase (3 beta-HSD) activity, without damaging Leydig cells [76].

High-dose tacrolimus decreased sperm counts and motility rates in rats, though they were reversible; however, it had no significant effect on fertility parameters. To date, there are no studies on the influence of mycophenolate mofetil or fumarate on spermatogenesis and male fertility [77].

Antimicrobial Agents

Antibacterials

Antibiotics from all the major classes have significant adverse effects on sperm production and sperm function [78]. In a study by Hargreaves et al., tetracycline, chloroquine, erythromycin, and co-trimoxazole irreversibly impaired sperm movement characteristics and significantly reduced sperm viability in vitro; amoxicillin was observed to be the least toxic of the drugs tested [40]. A recent study conducted with male rats reported that tetracycline was toxic to the gonads and that antioxidants (vitamin C and *N*-acetylcysteine) had chemoprotective effects [79]. The administration of tetracycline reduced the epididymal sperm motility, the percentage of live spermatozoa, and the sperm count and increased abnormal sperm morphology; it also induced adverse changes to gonadal histology. Vitamin C and *N*-acetylcysteine significantly lessen the extent of the toxic effect of tetracycline on sperm parameters, whereas the antioxidants had no effect on the adverse histopathologic changes induced by tetracycline. The researchers posited that the inability of the antioxidants to fully protect the testes against tetracycline-induced toxicity suggests that antibiotics could mediate testicular damage via mechanisms apart from oxidative stress. Another animal study reported that oxidative stress was induced by gentamicin [80]. Gentamicin increased free radical formation and lipid peroxidation, decreased antioxidant reserves, and caused structural and cytotoxic damage in rat testes. Adverse effects of aminoglycoside on sperm motility, count, and morphology were observed.

Antimycotics

Ketoconazole decreases testosterone levels via its central effects; this suppression is not prominent in other antimycotics, such as itraconazole, fluconazole, and terbinafin [81–83].

Antiparasitic Products

Niridazole was observed to induce defective spermatogenesis in the form of focal spermatocyte arrest and/or germinal cell hypoplasia in 20 male bilharzial patients [84]. The effect was transient, with active recovery occurring 3 months after therapy.

Treatment with 1–40 mg of lindane kg^{-1} of body weight disrupts testicular morphology, decreases spermatogenesis, inhibits testicular steroidogenesis, and reduces plasma androgen concentrations, and may adversely affect reproductive performance in males [85].

Antigout Agents

Colchicine is used for the treatment of gouty arthritis, familial Mediterranean fever (FMF), Peyronie's disease, and Behcet's disease. The adverse effect profile of colchicine on spermatogenesis and sperm shows a spectrum of pathology, from oligospermia and azoospermia to normospermia with disturbances in sperm motility [86]. With regard to the different pathophysiological mechanisms of these diseases, it is difficult to comment on the direct adverse effect on sperm production and function. Additionally, the literature still lacks evidence that allopurinol affects fertility.

Exogenous Substances

Alcohol

Chronic ethanol consumption may result in changes in gonadal structure and decreased testosterone levels, which may be a cause of infertility. It has toxic effects on Leydig cells [87]. Muthusami et al. studied a group of 66 alcoholic, nonsmoking, non-drug-abusing men that consumed a minimum of 180 ml of alcohol day^{-1} (brandy and whisky, both 40–50% alcohol content) at a minimum of 5 days $week^{-1}$ for 1 year. Thirty nonsmoking nonalcoholics were also selected as controls. In the alcoholic group, gonadotropins (FSH and LH) and estradiol levels were significantly higher, and testosterone levels were significantly lower. In addition, semen parameters (volume, sperm count, motility, and number of morphologically normal sperm) were significantly lower [88].

On the other hand, the Center for the Evaluation of Risks to Human Reproduction (CERHR) organized an expert panel on the impact of methanol on human reproduction. The panel reported that that there was not enough evidence to conclude that methanol was a male reproductive toxicant under dietary conditions or in response to occupational exposure that results in blood methanol concentrations <10 mg l^{-1}. Moreover, they highlighted the lack of sufficient data to rule out the possibility that high or acutely toxic doses of methanol might affect male reproduction [89].

Cigarette Smoking

The effect of cigarette smoking on spermatogenesis remains unclear. A meta-analysis of 21 studies on the effect of cigarette smoking on semen quality reported that smokers' sperm density is on average 13–17% (95% confidence interval = 8.0, 21.5) lower than that of nonsmokers, although 14 of the studies did not document an effect [90]. Nonetheless, smoking may be a cofactor in patients with other causes of male infertility. It should be kept in mind that the adverse effects of smoking may not be due to nicotine alone. Cigarette smoking negatively affects sperm production, motility, and morphology and is associated with an increased risk of DNA damage; however, in most cases, the exact molecular mechanisms are not well understood. In contrast, some studies report that there aren't any significant differences between sperm parameters in smokers, ex-smokers, and nonsmokers, and therefore the Practice Committee of the American Society for Reproductive Medicine, in collaboration with the Society for Reproductive Endocrinology and Infertility, reported that currently available data do not conclusively demonstrate that smoking decreases male fertility rates [91–95].

Cocaine

Only a few studies on the effects of cocaine on human spermatogenesis have been conducted. Bracken et al. reported that a history of cocaine use within 2 years of an initial semen analysis was twice as common among men with oligospermia [96]. George et al. studied the effects of long-term cocaine exposure on spermatogenesis and fertility in peripubertal male rats [97]. They observed that rats administered cocaine had smaller seminiferous tubule diameters. Furthermore, when the rats were bred after 100 days of cocaine administration, the pregnancy rate was much lower than that in the controls (33% vs. 86%) and germinal epithelium was thinner.

Marijuana

Cannabinoids are the main active components of marijuana. They have negative effects on hypothalamic-hypophyseal reproductive hormone secretion (decrease plasma LH levels) and testicular functions. Cannabinoids depress sperm motility and acrosome reaction. These inhibitory effects are mediated by the direct action of cannabinoids on sperm via the activation of the cannabinoid receptor subtype CNR1, which was shown to be expressed in mature sperm [98].

Expert Commentary

The purpose of this chapter was to evaluate the role of exogenous medication and substance-induced factors in male infertility. Many drugs and medications can impair male fertility via two basic mechanisms: direct gonadotoxic effect and alteration of the hypothalamic–pituitary–gonadal (HPG) axis. The direct gonadotoxic effect primarily occurs at the molecular level, and the mechanism of action is generally unclear. Clinicians that treat patients of reproductive age and urologists that treat infertile men should be familiar with the more commonly encountered drugs and medications that may affect fertility.

While available evidence suggests the importance of these undesired effects, the evidence base supporting them is limited, as studies conducted with healthy participants are not available due to ethical concerns. In addition, animal models are not always applicable to humans, in vitro models do not always mimic in vivo studies, and retrospective studies do not constitute a sufficient basis for definitive conclusions. Cessation of most medications may reverse the process and restore fertility in some otherwise infertile men; however, it can be permanent in some cases following treatment and cessation of medication is not always possible. As such, the most effective way to preserve fertility is to switch to a class of medication that has no effect on fertility before planning treatment for an existent disease or sperm cryopreservation in cases of expected permanent infertility and in patients with suboptimal semen quality. After cryopreservation, stored spermatozoa can be used for in vitro fertilization, and the potential for fatherhood can be preserved.

Five-Year View

The search for improved treatment modalities is never ending. New drugs, combinations of existing treatments, and new methods of treatment with proven safety and efficacy—possibly better than standard treatments—are discovered almost daily. Gaining a better understanding of the safety profile of a drug and the ability to discriminate more precisely between drugs within the same class could result in a decrease in the occurrence of medication-induced infertility. Safety is dependent upon a continuum of knowledge, including an understanding of a drug's mechanism of action, as well as data obtained from clinical trials and in vitro studies.

As the treatment of cancers has become highly successful, more and more attention is focused upon quality of life in cancer survivors. Fertility preservation is an important issue for cancer survivors. In recent years, improvements in cancer treatment have resulted in both prolongation of life and a reduction in the risk of infertility, and new treatment options are making fertility a possibility for many cancer survivors. Although cryopreservation of spermatozoa is accepted as a standard option for preserving male fertility before chemotherapy, it can be highly demanding to sperm cells and reduces the number of viable sperm capable of fertilization. With advances in assisted reproduction techniques, especially intracytoplasmic sperm injection, the problems of low sperm count and poor motility can be circumvented; however, novel cryopreservation approaches are needed to improve recovery and post-thaw parameters, especially in cases of low sperm count.

Well-designed clinical trials with appropriate sample sizes are needed to assess the feasibility and efficiency of various low sperm count freezing methodologies. Thus, there remains room for improvement in the methodology of sperm cryopreservation. Fertility is also a very important issue for childhood cancer survivors. The only fertility preservation option for prepubertal boys lies within the spermatogonial stem cells; many researchers have therefore focused on these particular cells. Autologous spermatogonial stem cell transplantation is a technique with great potential for clinical application.

Key Issues

- Numerous drugs and medications have negative effects on male fertility, acting via different mechanisms; however, the molecular-level mechanisms are not always well understood. Impaired fertility varies by individual and is dependent on the medications used, dose used, and duration of therapy.
- Mechanisms involved in impaired fertility include direct effects on germ cells or their supporting cells or effects on the HPG axis.
- There are some limitations in the current literature. Controlled studies on the association between these drugs and male fertility are not available.
- It is essential for urologists treating infertile men to be familiar with the more commonly used drugs and medications that may potentially affect fertility.
- Clinicians must be well informed about the adverse effects on spermatogenesis that these medications have. Patients should be informed before administration of potentially gonadotoxic drugs.
- Most adverse effects of drugs and medications can be reversed by simple cessation or switching to another medication with less or no effect on spermatogenesis. Nevertheless, in some instances particular medications cannot be discontinued; in such cases, sperm cryopreservation before treatment is crucial.

References

1. Wong WY, Zielhuis GA, Thomas CM, Merkus HM, Steegers-Theunissen RP. New evidence of the influence of exogenous and endogenous factors on sperm count in man. Eur J Obstet Gynecol Reprod Biol. 2003;110(1):49–54.
2. Gupta A, Gupta S, Tiwary AK. Spermicidal efficacy of H2-receptor antagonists and potentiation with 2′, 4′-dichlorobenzamil hydrochloride: role of intrasperm Ca^{2+}. Contraception. 2003;68(1):61–4.
3. Van Thiel DH, Gavaler JS, Smith Jr WI, Paul G. Hypothalamic-pituitary-gonadal dysfunction in men using cimetidine. N Engl J Med. 1979;300(18):1012–5.
4. Knigge U, Dejgaard A, Wollesen F, Ingerslev O, Bennett P, Christiansen PM. The acute and long term effect of the H2-receptor antagonists cimetidine and ranitidine on the pituitary-gonadal axis in men. Clin Endocrinol (Oxf). 1983;18(3):307–13.
5. Nakagawa K, Obara T, Matsubara M, Kubo M. Relationship of changes in serum concentrations of prolactin and testosterone during dopaminergic modulation in males. Clin Endocrinol (Oxf). 1982;17(4):345–52.
6. Falaschi P, Frajese G, Sciarra F, Rocco A, Conti C. Influence of hyperprolactinaemia due to metoclopramide on gonadal function in men. Clin Endocrinol (Oxf). 1978;8(5):427–33.
7. Panidis D, Rousso D, Skiadopoulos S, Panidou E, Mamopoulos M. Evaluation of semen parameters in man with hyperprolactinemia induced by metoclopramide. Arch Androl. 1997;39(3):237–42.
8. O'Morain C, Smethurst P, Dore CJ, Levi AJ. Reversible male infertility due to sulphasalazine: studies in man and rat. Gut. 1984;25(10):1078–84.
9. Niederberger C. The adverse effect of sulfasalazine on spermatogenesis and male reproductive potential. J Androl. 2002;23(2):180.
10. De Turner E, Aparicio NJ, Turner D, Schwarzstein L. Effect of two phosphodiesterase inhibitors, cyclic adenosine 3′:5′-monophosphate, and a beta-blocking agent on human sperm motility. Fertil Steril. 1978;29(3):328–31.
11. White DR, Clarkson JS, Ratnasooriya WD, Aitken RJ. Complementary effects of propranolol and nonoxynol-9 upon human sperm motility. Contraception. 1995;52(4):241–7.
12. Publicover S, Harper CV, Barratt C. [Ca2+]i signalling in sperm—making the most of what you've got. Nat Cell Biol. 2007;9(3):235–42.
13. Kanwar U, Anand RJ, Sanyal SN. The effect of nifedipine, a calcium channel blocker, on human spermatozoal functions. Contraception. 1993;48(5):453–70.
14. Kirkman-Brown JC, Barratt CL, Publicover SJ. Nifedipine reveals the existence of two discrete components of the progesterone-induced [Ca2+]i transient in human spermatozoa. Dev Biol. 2003;259(1):71–82.
15. Goodwin LO, Leeds NB, Hurley I, Mandel FS, Pergolizzi RG, Benoff S. Isolation and characterization of the primary structure of

testis-specific L-type calcium channel: implications for contraception. Mol Hum Reprod. 1997;3(3):255–68.
16. Leung PS, Sernia C. The renin-angiotensin system and male reproduction: new functions for old hormones. J Mol Endocrinol. 2003;30(3):263–70.
17. Kohn FM, Dammshauser I, Neukamm C, et al. Ultrastructural localization of angiotensin-converting enzyme in ejaculated human spermatozoa. Hum Reprod. 1998;13(3):604–10.
18. Krege JH, John SW, Langenbach LL, et al. Male-female differences in fertility and blood pressure in ACE-deficient mice. Nature. 1995;375(6527):146–8.
19. Kohn FM, Muller C, Drescher D, et al. Effect of angiotensin converting enzyme (ACE) and angiotensins on human sperm functions. Andrologia. 1998;30(4–5):207–15.
20. Loscher W, Luttgenau H, Schlegel W, Kruger S. Pharmacokinetics of non-steroidal anti-inflammatory drugs in male rabbits after acute and chronic administration and effect of chronic treatment on seminal prostaglandins, sperm quality and fertility. J Reprod Fertil. 1988;82(1):353–64.
21. Loscher W, Blazaki D. Effect of non-steroidal anti-inflammatory drugs on fertility of male rats. J Reprod Fertil. 1986;76(1):65–73.
22. Conte D, Romanelli F, Fillo S, et al. Aspirin inhibits androgen response to chorionic gonadotropin in humans. Am J Physiol. 1999;277(6 Pt 1):E1032–7.
23. Bauer J, Blumenthal S, Reuber M, Stoffel-Wagner B. Epilepsy syndrome, focus location, and treatment choice affect testicular function in men with epilepsy. Neurology. 2004;62(2):243–6.
24. Mikkonen K, Tapanainen P, Pakarinen AJ, Paivansalo M, Isojarvi JI, Vainionpaa LK. Serum androgen levels and testicular structure during pubertal maturation in male subjects with epilepsy. Epilepsia. 2004;45(7):769–76.
25. Isojarvi JI, Lofgren E, Juntunen KS, et al. Effect of epilepsy and antiepileptic drugs on male reproductive health. Neurology. 2004;62(2):247–53.
26. Sonntag A, Rothe B, Guldner J, Yassouridis A, Holsboer F, Steiger A. Trimipramine and imipramine exert different effects on the sleep EEG and on nocturnal hormone secretion during treatment of major depression. Depression. 1996;4(1):1–13.
27. Maier U, Koinig G. Andrological findings in young patients under long-term antidepressive therapy with clomipramine. Psychopharmacology (Berl). 1994;116(3):357–9.
28. Levin RM, Amsterdam JD, Winokur A, Wein AJ. Effects of psychotropic drugs on human sperm motility. Fertil Steril. 1981;36(4): 503–6.
29. Tanrikut C, Schlegel PN. Antidepressant-associated changes in semen parameters. Urology. 2007;69(1):185.e5–7.
30. Hendrick V, Gitlin M, Altshuler L, Korenman S. Antidepressant medications, mood and male fertility. Psychoneuroendocrinology. 2000;25(1):37–51.
31. Thakur SC, Thakur SS, Chaube SK, Singh SP. Subchronic supplementation of lithium carbonate induces reproductive system toxicity in male rat. Reprod Toxicol. 2003;17(6):683–90.
32. Zarnescu O, Zamfirescu G. Effects of lithium carbonate on rat seminiferous tubules: an ultrastructural study. Int J Androl. 2006;29(6):576–82.
33. Ragni G, De Lauretis L, Bestetti O, Sghedoni D, Gambaro V. Gonadal function in male heroin and methadone addicts. Int J Androl. 1988;11(2):93–100.
34. Comitato R, Esposito T, Cerbo G, Angelini F, Varriale B, Cardone A. Impairment of spermatogenesis and enhancement of testicular germ cell apoptosis induced by exogenous all-trans-retinoic acid in adult lizard Podarcis sicula. J Exp Zool A Comp Exp Biol. 2006;305(3):288–98.
35. Parsch EM, Ruzicka T, Przybilla B, Schill WB. Andrological investigations in men treated with acitretin (Ro 10-1670). Andrologia. 1990;22(5):479–82.
36. Coleman R, MacDonald D. Effects of isotretinoin on male reproductive system. Lancet. 1994;344(8916):198.
37. Geiger JM, Walker M. Is there a reproductive safety risk in male patients treated with acitretin (neotigason/soriatane)? Dermatology. 2002;205(2):105–7.
38. Adeeko AO, Dada OA. Chloroquine excretion in semen following antimalarial-drug administration. Andrologia. 1994;26(3):165–6.
39. Grunewald S, Paasch U, Glander HJ. Systemic dermatological treatment with relevance for male fertility. J Dtsch Dermatol Ges. 2007;5(1):15–21 (A review of systemic therapies in current dermatologic practice that raises questions of temporary or permanent damage to sperm cells.).
40. Hargreaves CA, Rogers S, Hills F, Rahman F, Howell RJ, Homa ST. Effects of co-trimoxazole, erythromycin, amoxycillin, tetracycline and chloroquine on sperm function in vitro. Hum Reprod. 1998; 13(7):1878–86.
41. Christiansen P, Skakkebaek NE. Pulsatile gonadotropin-releasing hormone treatment of men with idiopathic hypogonadotropic hypogonadism. Horm Res. 2002;57(1–2):32–6.
42. Lunglmayr G, Girsch E, Meixner EM, Viehberger G, Bieglmayer C. Effects of long term GnRH analogue treatment on hormone levels and spermatogenesis in patients with carcinoma of the prostate. Urol Res. 1988;16(4):315–9.
43. Bagatell CJ, Matsumoto AM, Christensen RB, Rivier JE, Bremner WJ. Comparison of a gonadotropin releasing-hormone antagonist plus testosterone (T) versus T alone as potential male contraceptive regimens. J Clin Endocrinol Metab. 1993;77(2):427–32.
44. Shahidi M, Norman AR, Gadd J, Huddart RA, Horwich A, Dearnaley DP. Recovery of serum testosterone, LH and FSH levels following neoadjuvant hormone cytoreduction and radical radiotherapy in localized prostate cancer. Clin Oncol (R Coll Radiol). 2001;13(4):291–5.
45. Parkinson AB, Evans NA. Anabolic androgenic steroids: a survey of 500 users. Med Sci Sports Exerc. 2006;38(4):644–51.
46. Knuth UA, Maniera H, Nieschlag E. Anabolic steroids and semen parameters in bodybuilders. Fertil Steril. 1989;52(6):1041–7.
47. Amory JK. Progress and prospects in male hormonal contraception. Curr Opin Endocrinol Diabetes Obes. 2008;15(3):255–60.
48. Hair WM, Wu FC, Lincoln GA. An investigation of the effectiveness of testosterone implants in combination with the prolactin inhibitor quinagolide in the suppression of spermatogenesis in men. Hum Reprod. 2003;18(4):749–55.
49. Anderson D, Schmid TE, Baumgartner A, Cemeli-Carratala E, Brinkworth MH, Wood JM. Oestrogenic compounds and oxidative stress (in human sperm and lymphocytes in the Comet assay). Mutat Res. 2003;544(2–3):173–8.
50. Meriggiola MC, Bremner WJ, Costantino A, et al. Twenty-one day administration of dienogest reversibly suppresses gonadotropins and testosterone in normal men. J Clin Endocrinol Metab. 2002; 87(5):2107–13.
51. Schover LR. Psychosocial aspects of infertility and decisions about reproduction in young cancer survivors: a review. Med Pediatr Oncol. 1999;33(1):53–9.
52. Longhi A, Macchiagodena M, Vitali G, Bacci G. Fertility in male patients treated with neoadjuvant chemotherapy for osteosarcoma. J Pediatr Hematol Oncol. 2003;25(4):292–6.
53. Nijman JM, Schraffordt Koops H, Kremer J, Sleijfer DT. Gonadal function after surgery and chemotherapy in men with stage II and III nonseminomatous testicular tumors. J Clin Oncol. 1987; 5(4):651–6.
54. Nudell DM, Monoski MM, Lipshultz LI. Common medications and drugs: how they affect male fertility. Urol Clin North Am. 2002;29(4):965–73 (Review examining commonly encountered medications and drugs that affect male fertility.).
55. Buchanan JD, Fairley KF, Barrie JU. Return of spermatogenesis after stopping cyclophosphamide therapy. Lancet. 1975;2(7926):156–7.

56. Green DM, Kawashima T, Stovall M, et al. Fertility of male survivors of childhood cancer: a report from the Childhood Cancer Survivor Study. J Clin Oncol. 2010;28(2):332–9.
57. Jequier AM. Male infertility: a guide for the clinician. Malden: Blackwell Science; 2000.
58. Tucker MA, D'Angio GJ, Boice Jr JD, et al. Bone sarcomas linked to radiotherapy and chemotherapy in children. N Engl J Med. 1987;317(10):588–93.
59. Callis L, Nieto J, Vila A, Rende J. Chlorambucil treatment in minimal lesion nephrotic syndrome: a reappraisal of its gonadal toxicity. J Pediatr. 1980;97(4):653–6.
60. Marmor D, Grob-Menendez F, Duyck F, Delafontaine D. Very late return of spermatogenesis after chlorambucil therapy: case reports. Fertil Steril. 1992;58(4):845–6.
61. Thomson AB, Anderson RA, Irvine DS, Kelnar CJ, Sharpe RM, Wallace WH. Investigation of suppression of the hypothalamic-pituitary-gonadal axis to restore spermatogenesis in azoospermic men treated for childhood cancer. Hum Reprod. 2002; 17(7):1715–23.
62. Meistrich ML, Wilson G, Mathur K, et al. Rapid recovery of spermatogenesis after mitoxantrone, vincristine, vinblastine, and prednisone chemotherapy for Hodgkin's disease. J Clin Oncol. 1997;15(12):3488–95.
63. Morris LF, Harrod MJ, Menter MA, Silverman AK. Methotrexate and reproduction in men: case report and recommendations. J Am Acad Dermatol. 1993;29(5 Pt 2):913–6.
64. Sussman A, Leonard JM. Psoriasis, methotrexate, and oligospermia. Arch Dermatol. 1980;116(2):215–7.
65. El-Beheiry A, El-Mansy E, Kamel N, Salama N. Methotrexate and fertility in men. Arch Androl. 1979;3(2):177–9.
66. Boekelheide K. Mechanisms of toxic damage to spermatogenesis. J Natl Cancer Inst Monogr. 2005;34:6–8.
67. Fossa SD, Ous S, Abyholm T, Norman N, Loeb M. Post-treatment fertility in patients with testicular cancer. II. Influence of cis-platin-based combination chemotherapy and of retroperitoneal surgery on hormone and sperm cell production. Br J Urol. 1985;57(2):210–4.
68. Vawda AI. Effect of testosterone on cisplatin-induced testicular damage. Arch Androl. 1994;32(1):53–7.
69. Puscheck E, Philip PA, Jeyendran RS. Male fertility preservation and cancer treatment. Cancer Treat Rev. 2004;30(2):173–80 (Review of available methods to maintain male fertility in patients undergoing chemotherapy or radiation therapy.).
70. Tempest HG, Ko E, Chan P, Robaire B, Rademaker A, Martin RH. Sperm aneuploidy frequencies analysed before and after chemotherapy in testicular cancer and Hodgkin's lymphoma patients. Hum Reprod. 2008;23(2):251–8.
71. O'Flaherty C, Hales BF, Chan P, Robaire B. Impact of chemotherapeutics and advanced testicular cancer or Hodgkin lymphoma on sperm deoxyribonucleic acid integrity. Fertil Steril. 2010; 94(4):1374–9.
72. Fritsche L, Budde K, Dragun D, Einecke G, Diekmann F, Neumayer HH. Testosterone concentrations and sirolimus in male renal transplant patients. Am J Transplant. 2004;4(1):130–1.
73. Kaczmarek I, Groetzner J, Adamidis I, et al. Sirolimus impairs gonadal function in heart transplant recipients. Am J Transplant. 2004;4(7):1084–8.
74. Zuber J, Anglicheau D, Elie C, et al. Sirolimus may reduce fertility in male renal transplant recipients. Am J Transplant. 2008; 8(7):1471–9.
75. Seethalakshmi L, Flores C, Carboni AA, Bala R, Diamond DA, Menon M. Cyclosporine: its effects on testicular function and fertility in the prepubertal rat. J Androl. 1990;11(1):17–24.
76. Iwasaki M, Fuse H, Katayama T. Histological and endocrinological investigations of cyclosporine effects on the rat testis. Andrologia. 1995;27(3):185–9.
77. Hisatomi A, Fujihira S, Fujimoto Y, Fujii T, Mine Y, Ohara K. Effect of Prograf (FK506) on spermatogenesis in rats. Toxicology. 1996;109(2–3):75–83.
78. Schlegel PN, Chang TS, Marshall FF. Antibiotics: potential hazards to male fertility. Fertil Steril. 1991;55(2):235–42.
79. Farombi EO, Ugwuezunmba MC, Ezenwadu TT, Oyeyemi MO, Ekor M. Tetracycline-induced reproductive toxicity in male rats: effects of vitamin C and N-acetylcysteine. Exp Toxicol Pathol. 2008;60(1):77–85.
80. Narayana K. An aminoglycoside antibiotic gentamycin induces oxidative stress, reduces antioxidant reserve and impairs spermatogenesis in rats. J Toxicol Sci. 2008;33(1):85–96.
81. Queiroz-Telles F, Purim KS, Boguszewski CL, Afonso FC, Graf H. Adrenal response to corticotrophin and testosterone during long-term therapy with itraconazole in patients with chromoblastomycosis. J Antimicrob Chemother. 1997;40(6):899–902.
82. Touchette MA, Chandrasekar PH, Milad MA, Edwards DJ. Contrasting effects of fluconazole and ketoconazole on phenytoin and testosterone disposition in man. Br J Clin Pharmacol. 1992; 34(1):75–8.
83. Nashan D, Knuth UA, Weidinger G, Nieschlag E. The antimycotic drug terbinafine in contrast to ketoconazole lacks acute effects on the pituitary-testicular function of healthy men: a placebo-controlled double-blind trial. Acta Endocrinol (Copenh). 1989;120(5):677–81.
84. El-Beheiry AH, Kamel MN, Gad A. Niridazole and fertility in bilharzial men. Arch Androl. 1982;8(4):297–300.
85. Pages N, Sauviat MP, Bouvet S, Goudey-Perriere F. Reproductive toxicity of lindane. J Soc Biol. 2002;196(4):325–38.
86. Haimov-Kochman R, Ben-Chetrit E. The effect of colchicine treatment on sperm production and function: a review. Hum Reprod. 1998;13(2):360–2.
87. Van Thiel DH, Gavaler JS, Cobb CF, Santucci L, Graham TO. Ethanol, a Leydig cell toxin: evidence obtained in vivo and in vitro. Pharmacol Biochem Behav. 1983;18 Suppl 1:317–23.
88. Muthusami KR, Chinnaswamy P. Effect of chronic alcoholism on male fertility hormones and semen quality. Fertil Steril. 2005; 84(4):919–24.
89. Shelby M, Portier C, Goldman L, et al. NTP-CERHR expert panel report on the reproductive and developmental toxicity of methanol. Reprod Toxicol. 2004;18(3):303–90.
90. Vine MF, Margolin BH, Morrison HI, Hulka BS. Cigarette smoking and sperm density: a meta-analysis. Fertil Steril. 1994;61(1):35–43.
91. Zenzes MT, Bielecki R, Reed TE. Detection of benzo(a)pyrene diol epoxide-DNA adducts in sperm of men exposed to cigarette smoke. Fertil Steril. 1999;72(2):330–5.
92. Kunzle R, Mueller MD, Hanggi W, Birkhauser MH, Drescher H, Bersinger NA. Semen quality of male smokers and nonsmokers in infertile couples. Fertil Steril. 2003;79(2):287–91.
93. Arabi M, Moshtaghi H. Influence of cigarette smoking on spermatozoa via seminal plasma. Andrologia. 2005;37(4):119–24.
94. Rodriguez-Rigau LJ, Smith KD, Steinberger E. Cigarette smoking and semen quality. Fertil Steril. 1982;38(1):115–6.
95. Practice Committee of American Society for Reproductive Medicine in collaboration with Society for Reproductive Endocrinology and Infertility. Optimizing natural fertility. Fertil Steril. 2008;90 Suppl 5:S1–6.
96. Bracken MB, Eskenazi B, Sachse K, McSharry JE, Hellenbrand K, Leo-Summers L. Association of cocaine use with sperm concentration, motility, and morphology. Fertil Steril. 1990;53(2):315–22.
97. George VK, Li H, Teloken C, Grignon DJ, Lawrence WD, Dhabuwala CB. Effects of long-term cocaine exposure on spermatogenesis and fertility in peripubertal male rats. J Urol. 1996; 155(1):327–31.
98. Rossato M, Pagano C, Vettor R. The cannabinoid system and male reproductive functions. J Neuroendocrinol. 2008;20 Suppl 1:90–3.

Sperm DNA Damage and Antioxidant Use: Roles in Male Fertility

29

Ashok Agarwal and Aspinder Singh

Abstract

The physician and andrologist must turn to semen characteristics to decipher what are the causative agents behind male infertility. WHO standard protocol requires analysis of sperm concentration, motility, and morphology. However, recent evidence implies that such parameters are not enough to fully depict the basis for male infertility and suggests that the effects of sperm DNA damage on infertility be considered. With the increasing ease of artificial reproduction, factors effecting embryo viability and offspring health related to the quality of sperm used and associated integrity of DNA must be strictly analyzed. Male gametes with damaged DNA can transmit genetic defects, lead to pregnancy loss, infant mortality, birth defects, and genetic diseases in the offspring. Furthermore, studies report that more than 80% of the structural de novo chromosome aberrations are of paternal origin. This chapter will begin by reviewing the process of spermatozoal chromatin organization and DNA packaging. Next, DNA damage and the factors causing this damage shall be considered. Subsequently, recognition and detection of this DNA damage will be reviewed followed finally by the effects of this damage and clinical applications encompassing treatment options involving antioxidants.

Keywords

Male infertility • Antioxidants • Sperm DNA damage • Spermatozoal chromatin organization • DNAse-insensitive toroids • Reactive oxygen species • Florescent in situ hybridization

Introduction

Infertility is formally characterized as a state in which a couple of reproductive age desiring a child is unable to conceive following 12 months of unprotected intercourse. Infertility represents one of the most common diseases and affects between 17% and 25% of couples [1]. Of these, male factor infertility is responsible for approximately 50% of the infertility cases. With female parameters held constant, the physician and andrologist must turn to semen characteristics to decipher what are the causative agents behind male infertility. WHO standard protocol requires analysis of sperm concentration, motility, and morphology [2]. However, recent evidence implies that such parameters are not enough to fully depict the basis for male infertility and suggests that the effects of sperm DNA damage on infertility be considered [3].

With the increasing ease of artificial reproduction, factors affecting embryo viability and offspring health related to the quality of sperm used and associated integrity of DNA must be strictly analyzed. Male gametes with damaged DNA can transmit genetic defects, lead to pregnancy loss, infant mortality, birth defects, and genetic diseases in the offspring [4, 5]. Furthermore, studies report that more than 80% of the structural de novo chromosome aberrations are of paternal origin [6].

A. Agarwal, PhD (✉) • A. Singh, BS
Center for Reproductive Medicine, Glickman Urological and Kidney Institute, Cleveland Clinic Foundation, 9500 Euclid Avenue, Desk A19, Cleveland, OH 44195, USA
e-mail: agarwaa@ccf.org

S.J. Parekattil and A. Agarwal (eds.), *Male Infertility: Contemporary Clinical Approaches, Andrology, ART & Antioxidants*, DOI 10.1007/978-1-4614-3335-4_29, © Springer Science+Business Media, LLC 2012

This chapter begins by reviewing the process of spermatozoal chromatin organization and DNA packaging. Next, DNA damage and the factors causing this damage shall be considered. Subsequently, recognition and detection of this DNA damage will be reviewed followed finally by the effects of this damage and clinical applications encompassing treatment options involving antioxidants.

How Is Spermatozoal Chromatin Organized and the DNA Packaged?

Unlike somatic cells, the packaging of sperm DNA utilizes a special procedure and also very unique proteins [7]. It is packed and coiled such that proper condensation and augmentation can be controlled in a time appropriate manner at specific stages of embryo development. Mammalian sperm chromatic can be divided into three major structural domains: the majority of sperm DNA is coiled into toroids by protamines, a smaller percent remains bound to histones, and the remaining DNA is attached to the sperm nuclear matrix at matrix attachment regions (MARs) at intervals of roughly 50 kb throughout the genome (Fig. 29.1) [8].

Protamine-bound DNA is the most condensed and therefore most protected form of sperm DNA. The tight condensation of the DNA allows for it to exist in toroids in a semicrystalline state, making it resistant to nuclease digestion. This type of DNA must first be uncoiled such that a reading frame can be exposed for protein synthesis. Protamines have large tracts of positively charged arginine residues that neutralize the negative phosphodiester backbone of the DNA. The presence of these arginine residues allows for the repulsion between the DNA residues and backbone to be reduced such that DNA can be packaged tighter and wound into highly compact toroids. This neutralization mimics that of divalent cations that can also cause DNA to form similar toroids with smaller amounts of DNA [8]. Furthermore, mammalian protamines contain several cysteines, which confer an increased stability on sperm chromatin by intermolecular disulfide cross-links, making sperm chromatin rich in protamines resistant to much greater mechanical disruption than somatic cells, supporting protamines' role in DNA protection [9]. Once compacted, protamine toroids are stacked side to side like a package of lifesavers (Fig. 29.1). Variations of this theme are two adjacent lines of toroids, which are aligned together with alternating toroids on the same chromatin being packaged on one line [10].

Protamine toroids are unique to mature sperm cells, and their major function is entirely for fertilization and not for embryonic development. Protamine binding causes gene expression silencing during spermiogenesis [11]. Postfertilization, the protamines are completely replaced in the first 4 h by histone proteins by the oocyte such that the paternal chromatin has increased accessibility. Ogura et al. found that when round spermatids lacking protamine condensation are used for injection into mouse oocytes, normal fertile mice developed. This therefore indicates that this level of sperm chromatin structure is sufficient for proper fertilization [12, 13].

The second largest manner of sperm DNA organization is sperm chromatin bound to histones. Depending on the mammalian species, between 2% and 15% of the sperm chromatin is bound to histones rather than protamines [14]. These histones are nonrandomly distributed throughout the sperm genome and are associated with specific genes. They are primarily found associated with gene promoter sites. Furthermore, entire functional gene families that are important for spermiogenesis and early fertilization development are preferentially associated with histones in human spermatozoa [14–16]. This correlates with the properties of DNA bound to histones. This type of association allows for a more easily accessible reading frame; however, this accessibility also makes the DNA more prone to nuclease activity. Sotolongo et al. proposed that histone-bound DNA made up the linker regions between each protamine toroid in the chromatin fiber because these were the regions that were the most nuclease sensitive [9].

Unlike protamines, which are replaced by histones provided by the oocyte postfertilization, the regions where histones are already present are not substituted. The transmission of sperm histones, and the associated chromatin structures, suggests that it is possible for the newly fertilized oocyte to inherit histone-based chromatin structural organization from the sperm [13, 17]. This also implies that any histone-based chromatin damage and malformation is also transmitted without modification.

The third and final form of sperm DNA organization is that bound to the nuclear matrix. The chromatin is organized into loop domains which are attached every 20–120 kb in length to a proteinaceous structure termed the nuclear matrix. This organizes the chromatin into functional loops of DNA that help regulate DNA replication and gene transcription [8, 18]. These MARs, no larger than 1,000 bp, occur between each protamine toroid and are termed the toroid linkers. These regions are nuclease sensitive and point to the presence of histones, which is consistent with the wide yet regular distribution of histones throughout the sperm genome [8, 16]. The two main roles of the sperm nuclear matrix are to properly associate DNA to the nuclear matrix such that paternal pronuclear DNA can be replicated in the one-cell embryo and to function as a checkpoint for sperm DNA integrity after fertilization [8].

A study conducted by Shaman et al. utilized sperm halos to determine the role of the loop domains in fertilization and development. Sperm halos can be constructed by treating

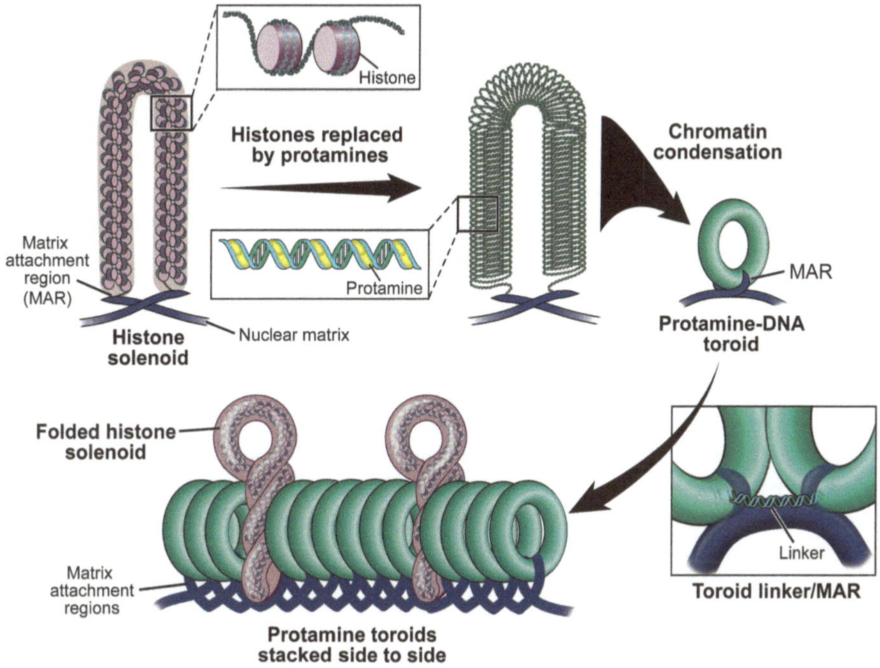

Fig. 29.1 Organization of sperm DNA. Majority of the sperm DNA is coiled into DNAse-insensitive toroids which are stacked side to side to maximize compaction. This toroid structure is held stable due to the presence of protamines which neutralize the repulsion between the phosphodiester backbone. Sperm maturation involves histone replacement via these protamines, allowing for proper transmission of paternal genetic material (adapted from [8], with permission)

the chromatin with high salt and reducing agents, which remove the protamine condensation and histone-bound nucleosomes, leaving only the sperm nuclear matrix with its associated loop domains. The study found that these halos were able to sufficiently form pronuclear and allow DNA replication, yet do not support developmental progression to the blast cyst stage, suggesting the loop domain DNA organization is required for replication but is not sufficient for development [18].

Of the three types of packaging, histones and MARs are inherent to the embryo and needed for proper development. Therefore, methods to manipulate spermatozoa for ART should be developed that maintain the integrity of the sperm chromatin structure, as well as taking into consideration the integrity of the paternal DNA [8].

One final consideration that must be taken when studying sperm DNA is that of the organization of the sperm chromosome. These chromosomes are relatively homogenous in structure, with the sperm chromatin folded into hairpin-like structures with the controlees positioned near the center of the sperm cell and the telomeres of each chromosome paired and arrayed around the periphery of the sperm nucleus [19, 20]. These individual chromosomes are positioned into nonoverlapping regions within the nucleolus entitled "chromosome territories." Existence of a preferred chromosome territory positioning in the human spermatozoa implies that deviation from the regular localization may be possibly deleterious for proper fertilization and development [21, 22]. Determination of such sperm chromosome territories is done using 2D and 3D florescent in situ hybridization (FISH) in interphone cells.

What Is DNA Damage and What Can Cause It?

Strand Breaks

DNA damage comes in a wide array of forms. Besides cardinal damage such as point deletions or frame shifts, the main categories of DNA damage are stand breaks, base modifications, and telomere shortening via oxidation. These damages can modify the genome and lead to translational errors, ultimately affecting the viability of the embryo and health of progeny. Single-strand breaks are found at regular intervals throughout the sperm genome. These breaks may be needed to reduce the torsion stress caused by chromatin packaging. Unfortunately, the single-strand breaks often lead to double-strand breaks via the activities of topoisomerase II variants [23]. These double-strand breaks can be difficult to repair by the limited innate DNA polymerase activity in the spermatozoa.

Base Modifications

In addition to stand breaks, base modifications are the second most common form of DNA damage. These modifications come in the forms of base supplementation such as fappy G, thymidine glycols addition, or adducts and also via chemical modifications such as oxidation of guanosine sites and abasic sites [24]. Furthermore, improper residue linkage resulting in mismatched pairs, formation of intrastrand bridges, and pyridine dimerization can lead to erroneous transmission of genetic data [24, 25].

Telomere Shortening

Finally, reactive oxygen species insult can yield mitochondrial dysfunction which leads to telomere attrition [26]. Liu et al. found that when mitochondrial function was disrupted in one-cell zygotes, cascade events led to a significant increase in measurements of reactive oxygen species which negatively correlated with embryo survival in a time-dependent manner [26].

Mitochondrial dysfunction was accompanied by telomere shortening, telomere loss, and chromosome fusion, leading to development arrest [26]. This halted developmental progress, thereby leading to apoptosis. However, they discovered that the antioxidant *N*-acetylcysteine scavenges ROS, forms cysteine by deacetylation, and increases glutathione (a free radical scavenger) production [26]. This antioxidant therefore prevented ROS accumulation, telomere shortening, and cell death.

Agents of DNA Damage

As important as it is to understand the forms of DNA damage, it is equally as crucial to comprehend what can cause this damage. There are internal factors, such as spermatozoal maturation processes and oxidative stress, and there are external causes, including lifestyle choices and environmental factors, which can all contribute to sperm DNA damage.

One major causative agent of damage is the inherent nature of spermatozoal maturation, which leaves the sperm enable to protect themselves from stress factors. This maturation eliminates the spermatozoal cytoplasm, without which it is difficult to combat oxidative stress. Furthermore, the human sperm plasma membrane contains an abundance of unsaturated fatty acids, which not only provide membrane fluidity necessary for motility and fusion, but also leaves the sperm predisposed to free radical attach and peroxidation of the plasma membrane lipids [27]. In addition to cytoplasmic reduction, maturation encompasses DNA remodeling. Irregular protamination can result in increased torsional stress, leading to strand breaks [28].

Another main instigator of DNA damage is oxidative stress. The presence of highly reactive oxygen species has been reported in the semen of between 25% and 40% of infertile men, resulting in cascade events which lead to lipid peroxidation and breakdown of macromolecules including nucleotides [29]. Furthermore, leukocytospermia yields an increased presence of pro-inflammatory mediators, such as cytokines, which lead to alterations in the regulation of spermiogenesis and subsequent DNA aberration [30].

Formation of 8-hydroxy-2-deoxyguanosine (8-OHdG) is a biological marker for oxidative stress DNA damage. 8-OHdG is an adduct which represents a modified DNA structure potentially leading to a DNA break [31]. The likelihood of pregnancy occurring with a single menstrual cycle is inversely associated with the levels of 8-OHdG [32]. Additional information concerning the roles of oxidative stress is provided in greater detail in proceeding chapters.

Many external factors are associated with DNA damage, including smoking, and certain medication and treatments. Cigarettes and their associated smoke contain many chemicals which can increase oxidative stress. Saleh et al. found that the percent DNA fragmentation index is significantly higher in infertile men who smoke [33]. The metabolites found in cigarette smoke, including vinyl chloride and benzopyrene, result in DNA adducts and can induce the increased activity of chemical inflammation mediators such as interleukin-6 and interleukin-8 which lead to the subsequent recruitment of leukocytes [24, 34, 35]. Also, there is an association between paternal smoking and an increased risk of childhood cancer in the offspring [34].

Many cancer treatments utilize chemicals which are detrimental to sperm DNA, primarily alkylating agents. Chemotherapy drugs such as fludarabine, cyclophosphamide, and busulfan can cause testicular damage, manifested as reduced volume, oligozoospermia, elevated follicle-stimulating hormone and luteinizing hormone, and decreased testosterone concentrations [36]. The effects of these drugs may persist for several months after cessation of their use [37].

How Can Sperm DNA Damage Be Detected and Prevented?

There are plethora tests available to measure DNA damage in sperm samples. These include TUNEL, Comet, in situ nick translation assay, sperm chromatin structure assay, 3D FISH, and many more. Each can measure and quantify the DNA damage in terms of strand breaks, denaturation of strands, or presence of adducts. Unfortunately, many of these tests lack useful thresholds or a clinically accepted standardization. Also, in order to determine the damage levels, the spermatozoa must first be prepared, which often involves strenuous conditions such as high heat which can lead to additional damage.

There are, however, many test preparation methods with reduced levels of DNA damage. The swim-up method is associated with an increased high post-in vitro fertilization (IVF) rate and better DNA integrity compared to density gradient centrifugation [38, 39]. Furthermore, glass wool preparation significantly decreases the percent DNA fragmentation index when compared with raw semen samples [40].

One must also keep in mind that minute DNA damage can be repaired with appropriate mechanisms available inside the oocyte; however, extensive damage is not reparable. Between the period of sperm entry into the cytoplasm and the beginning of the next S phase, DNA damage can be repaired by the oocyte via pre- and post-replication mechanisms [24]. Both nonhomologous end joining (NHEJ) and homologous recombination are used by the oocyte to rescue the genetic integrity of the paternal genome after fertilization [3]. However, maternal age must be considered as it is found that increased age reflects decreased stores of mRNA for many repair mechanisms, as well as reduced efficient of DNA repair [41].

Clinical Applications

DNA Damage in Relation to Artificial Reproduction Techniques

Today's advances in artificial reproduction techniques (ART) make it easier to transmit defective spermatozoa and still induce pregnancy [25]. Intracytoplasmic sperm injection (ICSI) and IVF allow couples to bypass natural selection processes inherent to the fusion procedure which would normally select out abnormal spermatozoa. These abnormal spermatozoa thereby may hinder implantation, embryo development, and progeny viability.

It has been shown that during the first mitotic cycle, 77.5% of ROSI-generated embryos exhibit abnormal chromosome segregation. This abnormal segregation originates from double-strand breaks of the male-derived genomic DNA [12]. ICSI and ROSI procedures resulted in no embryonic development when chromosome segregation was abnormal during this first mitotic division [42]. Therefore, residual DNA breaks in spermatozoa, those which fall outside the realm of oocyte repair, may lead to impaired zygote development via abnormal chromosome segregation and genetic impairment. This supports the notion that the use of genetically abnormal spermatozoa could lead to unsuccessful reproductive outcomes [43].

Furthermore, Duran et al. found that in intrauterine insemination (IUI), there is a failure of pregnancy in women inseminated with a semen sample with >12% sperm with fragmented DNA and miscarriages in those with moderate degrees of DNA damage (10–12%). One hundred and nineteen couples comprised a total of 154 IUI cycles, both natural and stimulated. From these it was found that an increased rate of miscarriages and pregnancy failure correlated with the level of DNA damage brought upon by defective chromatin organization, ineffective apoptosis, and oxidative stress [44].

Bungum et al. also conducted a similar study, looking at 387 IUI cycles. Their results confirmed those of Duran, with a pregnancy success rate of 19% when the DFI value was <30%, compared to a success rate of only 1.5% when the DFI was >30%. These results indicate that the DFI value can be used a reliable and accurate independent predictor of fertility [45].

Moreover, many studies have found that sperm DNA damage is associated with a significantly higher rate of pregnancy loss after IVF or ICSI (Table 29.1) [46]. Specifically, Zini et al. derived, from a meta-analysis of 2,549 cycles of IVF or ICSI involving 640 pregnancies and 122 failures, that sperm genetic damage was significantly associated with pregnancy loss ($p<0.0001$). They also concluded that in populations with abnormal sperm DNA damage, the miscarriage rate increases to 37% compared to an average rate of 18%. In comparison, this rate fell to only 10% when the testing results are negative [46]. This difference may be valuable for patients when making decisions regarding ART procedures.

Also, Virro et al. concluded in their study of 249 couples undergoing IVF/ICSI that men with a DFI <33% has a significantly greater chance of initiating a pregnancy, lower miscarriage rates, and an increase of ongoing pregnancies at week 12 (47% vs. 28%) than those compared with a DFI <33% [47]. These results indicate the importance of DNA integrity in not only pregnancy initiation but also in long-term viability of the embryo.

Table 29.1 Study parameters on sperm DNA damage and pregnancy loss after IVF and ICSI

Study	*n*	ART	Assay	PL%	Abn test %	Sens	Spec	PPV	NPV
Check, '05	104	ICSI	SCSA	47	24	0.31	0.83	0.63	0.58
Zini, '05	60	ICSI	SCSA	16	19	0.40	0.85	0.33	0.88
Borini, '06	82	IVF	TUNEL	6	11	0.91	0.94	0.50	0.99
Borini, '06	50	ICSI	TUNEL	25	25	0.97	0.99	0.97	0.99
Benchaib, '07	84	IVF	TUNEL	15	15	0.50	0.91	0.50	0.91
Benchaib, '07	218	ICSI	TUNEL	12	15	0.38	0.88	0.30	0.91
Bungum, '07	388	IVF	SCSA	24	14	0.11	0.85	0.19	0.76
Bungum, '07	223	ICSI	SCSA	19	40	0.50	0.63	0.24	0.84
Frydman, '08	117	IVF	TUNEL	19	32	0.64	0.75	0.37	0.90
Lin, '08	137	IVF	SCSA	10	17	0.29	0.84	0.17	0.92
Lin, '08	86	ICSI	SCSA	18	23	0.50	0.83	0.40	0.88

From [46], with permission
ART assisted reproductive technology, *Abn test* proportion of abnormal sperm DNA test among documented pregnancies, *PL* pregnancy loss, *Sens* sensitivity, *Spec* specificity, *PPV* positive predictive value, *NPV* negative predictive value

Table 29.2 Effect of dietary antioxidant supplements on sperm DNA integrity

Study	Patients/test	Treatment(s)	*n*	Results
Infertile men with high sperm DNA fragmentation levels or oxidative stress				
Greco '05	1 failed ICSI TUNEL > 15%	Vits C 1 g, E 1 g	38	Rx (2 months): ↓ DD in 76%, 48% ICSI pregnancy No control group
Greco '05	Infertility TUNEL > 15%	Vits C 1 g, E 1 g	32 32	Rx (2 months): ↓ DD (22% → 9%) Placebo group: no effect on DD (22% → 22%)
Menezo '07	2 failed ICSI DFI > 15% Decond > 15%	Vits C, E (400 mg), zinc, Se, β-carotene	57	Rx (90 days):↓ sperm%DFI (32 → 26%: by 19%) but ↑ sperm%HDS (17.5 → 25.5%: by 23%) No control group
Tremellen '07	Male Infert TUNEL > 25%	Menevit (lycopene, vits C, E, zinc, Se, folate, garlic)	36 16	Rx (3 months): 39% ICSI pregnancy rate, but no ↑ in embryo quality, no post-Rx DD Placebo group: 16% ICSI pregnancy rate
Gil Villa '09	Pregn. loss ↑ LPO or DFI	Vits C, E, zinc, β-carotene	9	Rx (3 months): 6 (of 9) couples got pregnancy No control group
Unselected infertile men				
Piomboni '08	Asthenosp. AO stain	Vits C, E, β-glucan, papaya, lactoferrin	36 15	Rx (90 days): ↑ motility and morph but not DD Control group: no effect
Kodama '97	Male infert 8-OHdG	Vits C, E (200 mg), glutathione (400 mg)	14 7	Rx (2 months): ↓ in 8-OHdG (1.5 → 1.1/10^5 dG) Control group: no change in 8-OHdG levels

From [52], with permission
8-OHdG 8-hydroxy-2-deoxyguanosine, *AO* acridine orange, *DD* DNA damage, *Decond* decondensation, *DFI* DNA fragmentation index, *LPO* lipid peroxidation, *OS* oxidative stress, *Rx* treatment, *ROS* reactive oxygen species, *Se* selenium, *TUNEL* terminal nucleotidyl transferase dUTP nick end labeling, *vit* vitamin

Lastly, the Y chromosome is particularly susceptible to damage because the haploid genome is unable to retrieve genetic information, leading to gene deletion. In turn, Y chromosome deletions and microdeletions can lead to infertility in the offspring [48].

Unfortunately, when selecting spermatozoa for ART, often, morphology is the main determinant between "normal" and "abnormal" cells. This selection cannot however illustrate the DNA properties, and one cannot avoid selecting spermatozoa undergoing chromatin remodeling, or those with substantial genetic damage.

Antioxidant Use in DNA Repair

An antioxidant regimen to reduce damage is an integral part of applications to repair sperm DNA. tocopherols (TP), vitamin E, and tocotrienol are the main chain breaking antioxidants in biological membranes. TP prevents the dissemination of lipid peroxidation by directly interacting with lipid-free radicals derived from oxidation of polyunsaturated fatty acids. These radicals include the alkoxy radical (LO•), lipid peroxyl radical (LOO•), and alkyl radicals (L•) [49].

TP interacts with these radical in the lipid/water interphase and results in the formation of lipid hydroperoxides (LO–OH) and also the tocopheroxyl radical (TO•), which can be reduced by ascorbic acid, or coenzyme Q. The reaction is as follows: TP–OH + L–OO• → L–OOH + TO• [49].

Fraga et al. demonstrated that oral vitamin C intake leads to an increase in semen vitamin C levels. They found decreased dietary ascorbic acid intake (from 250 to 5 mg/day) in a group of 24 men ages 20–50 correlated with decreased seminal ascorbic acid levels by half and increased (91%) levels of 8-OHdG in sperm DNA. The increased intake was associated with a decrease in DNA oxidation and therefore improved sperm DNA integrity [50].

Also, Greco et al. concluded that an antioxidant regimen of vitamin C and E was associated with a significant decrease in the percentage of DNA-fragmented spermatozoa. Oral intake of vitamin C 500 mg and vitamin E 500 mg (twice a day for 2 months) resulted in a significant decrease in sperm DNA fragmentation in a group of 64 men with unexplained infertility (pretreatment 22.1% ± 7.7 vs. posttreatment 9.1% ± 7.2) [51]. The incidence of DNA fragmentation was lower in all men after the antioxidant treatment, showing a generalized benefit of these compounds in preventing damage to sperm DNA.

In addition to vitamin C and vitamin E, there are many other antioxidants which are effective in reducing the free radical load and preventing subsequent damage. Glutathione, selenium, alpha-tocopherol, folic acid, and zinc have all been proven to be effective (Table 29.2) [52] and are generally prescribed to help reduce sperm DNA damage and assist men in their reproductive goals. Many of these oral antioxidants come in compact, multicompound pill form; popular brand names include Fertibiol®, Procrelia®, Condesyl®, and Nurelia® [53].

Recently, Zini et al. demonstrated a reduction in sperm DNA damage when lycopene was added to the semen sample. In this study, sperm samples from 12 fertile men were pretreated with 5 μmol/L lycopene prior to 2-h incubation with hydrogen peroxide. The authors demonstrated a significant decrease in% DFI (treatment 8.0% ± 7.9 vs. control 29.8% ± 39.4; $p < 0.05$), suggesting that pretreatment with lycopene could protect sperm against DNA damage in vitro [54].

The authors recently conducted a survey of three leading male infertility urologists concerning their use of antioxidants in male infertility treatment. These urologists agree that patients should begin an antioxidant regimen upon initial evaluation, along with any other treatments necessary to treat their overlying conditions, such as antibiotics or a varicocelectomy. They recommend their patients to follow an antioxidant regimen for an average of 3–6 months. This typically includes 500 mg of vitamin C, 400 IU of vitamin E, 50 mg of zinc, 100 μg of selenium, and 5 mg of folate. Their patients reported only minor side effects with the antioxidant use, such as mild stomachaches, and were usually very responsive to the suggestion. Overall, patients showed good responsiveness to these antioxidants, and these compounds have become a critical part of the armamentarium used to treat DNA damage-based male infertility.

Future Considerations and Conclusions

Artificial reproduction is a growing field, with daily advancements and technological breakthroughs. However, the implications of DNA damage on male fertility and pregnancy outcomes are substantially overlooked. ART procedures allow spermatozoa to bypass natural selection processes and initiate pregnancy; therefore, one must be cautious concerning the quality of the spermatozoa utilized. Multiple studies have shown a strong correlation between DNA damage and pregnancy failure, implying the importance of this parameter on fertility.

Clinical antioxidant use in the repair of DNA damage has increased over the past years. Growing data shows antioxidants protect the DNA from free radicals, thereby increasing the rate of successful pregnancies. Scavenging of these radicals reduces strand breaks, decreases chemical modifications, and halts telomere attrition, thereby increasing the viability of the embryo. Therefore, antioxidant use is strongly recommended in couples facing male factor infertility.

Future studies are needed to further validate the benefits of dietary antioxidants for male fertility. Although the results in the literature suggest that sperm DNA integrity testing may potentially have the greatest clinical utility for predicting pregnancy loss after IVF and ICSI, properly controlled studies are urgently required to confirm these results [55]. Larger studies with longer treatments could help clearly define the correlation between antioxidants and DNA repair. Also, designations of the repair mechanisms are of utmost necessity. Additional in vitro supplementation studies will better define a standard protocol for antioxidant use and will allow for the clinicians, andrologist, and the patients to work together to achieve an increased rate of successful pregnancy. Also, additional studies involving antioxidant benefits on normal, fertile men would serve as guidelines for comparison with the infertile cases.

Expert Commentary

The aim of this chapter is to elucidate the necessity of DNA integrity tests prior to ART techniques and to highlight the benefits of antioxidant use in damage prevention and repair. Current advances in artificial reproduction are making it easier to utilize damaged spermatozoa, yet techniques for detection of this damage are not routinely used. This chapter brings forth clinical evidence which supports the association between sperm DNA damage, male fertility, and pregnancy success, as well as the benefits derived from a prescribed antioxidant regimen. This data will provide information to both clinicians and patients to make choices which will maximize the patients' fertility and pregnancy success. The techniques and methods necessary to detect and improve the DNA damage are already integral parts of semen analysis, and only a standardized protocol for testing and antioxidant use needs to be evolved. Artificial reproduction is a very distinct and specialized field, one that requires further research and standards of practice which may lead to better patient care and improved pregnancy success.

Key Issues

- With the increasing ease of artificial reproduction, factors effecting embryo viability and health of offspring related to the quality of sperm used and associated integrity of DNA must be strictly analyzed.
- Sperm chromatin packaging and DNA organization requires special proteins and unique structural organization to provide accurate folding and unfolding in a time-specific manner.
- Sperm histones and MARs are inherent to the embryo and are needed for proper development.
- DNA damage comes in three main forms: strand breaks, base modifications, and telomere shortening.
- This damage can be attributed to various internal and external factors including spermatozoal maturation, oxidative stress, smoking, and medical treatments.

- Various test procedures are used to elucidate the levels of DNA damage. These include TUNEL, Comet, in situ nick translation, and 3D FISH.
- There have been many studies relating the effects of DNA damage on IVF, ICSI, and IUI. These studies show a strong correlation between increased DNA damage and decreased pregnancy success.
- Antioxidants are clinically prescribed to reduce the levels of DNA damage and repair the genetic integrity. Antioxidants function to reduce the oxidative load and are correlated with an increased pregnancy success rate in ART as shown.

References

1. Dunson DB, Baird DD, Colombo B. Increased infertility with age in men and women. Obstet Gynecol. 2004;103:51–6.
2. World Health Organization, DoRHaR. WHO laboratory manual for the examination and processing of human semen. 5th ed. Geneva: World Health Organization; 2010.
3. Marchetti F, Essers J, Kanaar R, Wyrobek AJ. Disruption of maternal DNA repair increases sperm-derived chromosomal aberrations. Proc Natl Acad Sci USA. 2007;104:17725–9.
4. Aitken RJ, De Iuliis GN, McLachlan RI. Biological and clinical significance of DNA damage in the male germ line. Int J Androl. 2009;32:46–56.
5. Fernandez-Gonzalez R, Moreira PN, Perez-Crespo M, et al. Long-term effects of mouse intracytoplasmic sperm injection with DNA-fragmented sperm on health and behavior of adult offspring. Biol Reprod. 2008;78:761–72.
6. Thomas NS, Durkie M, Van Zyl B, et al. Parental and chromosomal origin of unbalanced de novo structural chromosome abnormalities in man. Hum Genet. 2006;119:444–50.
7. Fuentes-Mascorro G, Serrano H, Rosado A. Sperm chromatin. Arch Androl. 2000;45:215–25.
8. Ward WS. Function of sperm chromatin structural elements in fertilization and development. Mol Hum Reprod. 2010;16:30–6.
9. Sotolongo B, Lino E, Ward WS. Ability of hamster spermatozoa to digest their own DNA. Biol Reprod. 2003;69:2029–35.
10. Mudrak O, Chandra R, Jones E, Godfrey E, Zalensky A. Reorganisation of human sperm nuclear architecture during formation of pronuclei in a model system. Reprod Fertil Dev. 2009;21:665–71.
11. Martins RP, Ostermeier GC, Krawetz SA. Nuclear matrix interactions at the human protamine domain: a working model of potentiation. J Biol Chem. 2004;279:51862–8.
12. Ogura A, Matsuda J, Yanagimachi R. Birth of normal young after electrofusion of mouse oocytes with round spermatids. Proc Natl Acad Sci USA. 1994;91:7460–2.
13. Ajduk A, Yamauchi Y, Ward MA. Sperm chromatin remodeling after intracytoplasmic sperm injection differs from that of in vitro fertilization. Biol Reprod. 2006;75:442–51.
14. Hammoud SS, Nix DA, Zhang H, Purwar J, Carrell DT, Cairns BR. Distinctive chromatin in human sperm packages genes for embryo development. Nature. 2009;460:473–8.
15. Ostermeier GC, Goodrich RJ, Diamond MP, Dix DJ, Krawetz SA. Toward using stable spermatozoal RNAs for prognostic assessment of male factor fertility. Fertil Steril. 2005;83:1687–94.
16. Arpanahi A, Brinkworth M, Iles D, et al. Endonuclease-sensitive regions of human spermatozoal chromatin are highly enriched in promoter and CTCF binding sequences. Genome Res. 2009;19:1338–49.
17. van der Heijden GW, Ramos L, Baart EB, et al. Sperm-derived histones contribute to zygotic chromatin in humans. BMC Dev Biol. 2008;8:34.
18. Shaman JA, Yamauchi Y, Ward WS. The sperm nuclear matrix is required for paternal DNA replication. J Cell Biochem. 2007;102:680–8.
19. Zalenskaya IA, Bradbury EM, Zalensky AO. Chromatin structure of telomere domain in human sperm. Biochem Biophys Res Commun. 2000;279:213–8.
20. Churikov D, Siino J, Svetlova M, et al. Novel human testis-specific histone H2B encoded by the interrupted gene on the X chromosome. Genomics. 2004;84:745–56.
21. Mudrak O, Tomilin N, Zalensky A. Chromosome architecture in the decondensing human sperm nucleus. J Cell Sci. 2005;118:4541–50.
22. Zalensky A, Zalenskaya I. Organization of chromosomes in spermatozoa: an additional layer of epigenetic information? Biochem Soc Trans. 2007;35:609–11.
23. Spano M, Seli E, Bizzaro D, Manicardi GC, Sakkas D. The significance of sperm nuclear DNA strand breaks on reproductive outcome. Curr Opin Obstet Gynecol. 2005;17:255–60.
24. Menezo Y, Dale B, Cohen M. DNA damage and repair in human oocytes and embryos: a review. Zygote. 2010;18(4):357–65.
25. Erenpreiss J, Spano M, Erenpreisa J, Bungum M, Giwercman A. Sperm chromatin structure and male fertility: biological and clinical aspects. Asian J Androl. 2006;8:11–29.
26. Liu L, Trimarchi JR, Smith PJ, Keefe DL. Mitochondrial dysfunction leads to telomere attrition and genomic instability. Aging Cell. 2002;1:40–6.
27. Twigg J, Fulton N, Gomez E, Irvine DS, Aitken RJ. Analysis of the impact of intracellular reactive oxygen species generation on the structural and functional integrity of human spermatozoa: lipid peroxidation, DNA fragmentation and effectiveness of antioxidants. Hum Reprod. 1998;13:1429–36.
28. Agarwal A, Said TM. Role of sperm chromatin abnormalities and DNA damage in male infertility. Hum Reprod Update. 2003;9:331–45.
29. Padron OF, Sharma RK, Thomas Jr AJ, Agarwal A. Effects of cancer on spermatozoa quality after cryopreservation: a 12-year experience. Fertil Steril. 1997;67:326–31.
30. Kasturi SS, Charles Osterberg E, Tannir J, Brannigan RE. The effect of genital tract infection and inflammation on male infertility. In: Lipshiltz LI, Howards SS, Neiderberger CS, editors. Infertility in the male. 4th ed. Cambridge: Cambridge University Press; 2009. p. 295–330.
31. Aitken RJ, De Iuliis GN. On the possible origins of DNA damage in human spermatozoa. Mol Hum Reprod. 2010;16:3–13.
32. Loft S, Kold-Jensen T, Hjollund NH, et al. Oxidative DNA damage in human sperm influences time to pregnancy. Hum Reprod. 2003;18:1265–72.
33. Saleh RA, Agarwal A, Sharma RK, Nelson DR, Thomas Jr AJ. Effect of cigarette smoking on levels of seminal oxidative stress in infertile men: a prospective study. Fertil Steril. 2002;78:491–9.
34. Chang JS. Parental smoking and childhood leukemia. Methods Mol Biol. 2009;472:103–37.
35. Sepaniak S, Forges T, Gerard H, Foliguet B, Bene MC, Monnier-Barbarino P. The influence of cigarette smoking on human sperm quality and DNA fragmentation. Toxicology. 2006;223:54–60.
36. Chatterjee R, Haines GA, Perera DM, Goldstone A, Morris ID. Testicular and sperm DNA damage after treatment with fludarabine for chronic lymphocytic leukaemia. Hum Reprod. 2000;15:762–6.
37. Campbell DI, Bunn JE, Weaver LT, Harding M, Coward WA, Thomas JE. Human milk vacuolating cytotoxin A immunoglobulin

A antibodies modify *Helicobacter pylori* infection in Gambian children. Clin Infect Dis. 2006;43:1040–2.
38. Younglai EV, Holt D, Brown P, Jurisicova A, Casper RF. Sperm swim-up techniques and DNA fragmentation. Hum Reprod. 2001;16:1950–3.
39. Zini A, Finelli A, Phang D, Jarvi K. Influence of semen processing technique on human sperm DNA integrity. Urology. 2000;56:1081–4.
40. Larson KL, Brannian JD, Timm BK, Jost LK, Evenson DP. Density gradient centrifugation and glass wool filtration of semen remove spermatozoa with damaged chromatin structure. Hum Reprod. 1999;14:2015–9.
41. Hamatani T, Falco G, Carter MG, et al. Age-associated alteration of gene expression patterns in mouse oocytes. Hum Mol Genet. 2004;13:2263–78.
42. Yamagata K, Suetsugu R, Wakayama T. Assessment of chromosomal integrity using a novel live-cell imaging technique in mouse embryos produced by intracytoplasmic sperm injection. Hum Reprod. 2009;24:2490–9.
43. Kimura Y, Yanagimachi R. Mouse oocytes injected with testicular spermatozoa or round spermatids can develop into normal offspring. Development. 1995;121:2397–405.
44. Duran EH, Morshedi M, Taylor S, Oehninger S. Sperm DNA quality predicts intrauterine insemination outcome: a prospective cohort study. Hum Reprod. 2002;17:3122–8.
45. Bungum M, Humaidan P, Axmon A, et al. Sperm DNA integrity assessment in prediction of assisted reproduction technology outcome. Hum Reprod. 2007;22:174–9.
46. Zini A, Boman JM, Belzile E, Ciampi A. Sperm DNA damage is associated with an increased risk of pregnancy loss after IVF and ICSI: systematic review and meta-analysis. Hum Reprod. 2008;23:2663–8.
47. Virro MR, Larson-Cook KL, Evenson DP. Sperm chromatin structure assay (SCSA) parameters are related to fertilization, blastocyst development, and ongoing pregnancy in in vitro fertilization and intracytoplasmic sperm injection cycles. Fertil Steril. 2004;81:1289–95.
48. Aitken RJ, Krausz C. Oxidative stress, DNA damage and the Y chromosome. Reproduction. 2001;122:497–506.
49. Fagerstedt K, Blokhina O. Oxidative stress and antioxidant defences in plants. In: Singh KK, editor. Oxidative stress, disease and cancer. London: Imperial College Press; 2006. p. 151–99.
50. Fraga CG, Motchnik PA, Shigenaga MK, Helbock HJ, Jacob RA, Ames BN. Ascorbic acid protects against endogenous oxidative DNA damage in human sperm. Proc Natl Acad Sci USA. 1991;88:11003–6.
51. Greco E, Iacobelli M, Rienzi L, Ubaldi F, Ferrero S, Tesarik J. Reduction of the incidence of sperm DNA fragmentation by oral antioxidant treatment. J Androl. 2005;26:349–53.
52. Zini A, San Gabriel M, Baazeem A. Antioxidants and sperm DNA damage: a clinical perspective. J Assist Reprod Genet. 2009;26:427–32.
53. Boxmeer JC, Smit M, Utomo E, et al. Low folate in seminal plasma is associated with increased sperm DNA damage. Fertil Steril. 2009;92:548–56.
54. Zini A, San Gabriel M, Libman J. Lycopene supplementation in vitro can protect human sperm deoxyribonucleic acid from oxidative damage. Fertil Steril. 2010;94:1033–6.
55. Barratt CL, De Jonge CJ. Clinical relevance of sperm DNA assessment: an update. Fertil Steril. 2010;94(6):1958–9.

Common Male Infertility Disorders: Aging

30

Fabio Pasqualotto, Edson Borges Jr.,
and Eleonora Pasqualotto

Abstract

As life expectancy increases and our lives become busiest every day, many couples are waiting longer to establish their families. The recent trend toward delayed parenthood raises concerns because of the adverse effects of aging on fertility. The effects of paternal age on a couple's fertility are real and may be greater than has previously been thought. After adjustments for other factors, it has been demonstrated that the probability of a fertile couple will take >12 months to conceive nearly doubles from 8% when the man is <25 years to 15% when he is >35 years; thus, paternal age is a further factor to be taken into account when deciding the prognosis for infertile couples. Also, the increased male age is associated with a significant decline in fertility (fivefold longer time to pregnancy at age of 45 years), which is independent of the woman's age, coital frequency, and lifestyle effect, as well as the effect of other subfertility risk factors. Furthermore, fathering at older ages may have significant effects on the viability and genetic health of human pregnancies and offspring, primarily as a result of structural chromosomal aberrations in sperm. The evidence for sex chromosomal aneuploidy suggests that there may be about a twofold increase in risk at the age of 50. In fact, the risk for a father over 40 years old to have a child with an autosomal dominant mutation equals the risk of Down syndrome for a child whose mother is 35–40 years old. Although individual Leydig cell volume does change with age, the total Leydig cell volume and the absolute number of Leydig cells decrease significantly with age. Also, anatomical studies of Sertoli cell populations in humans show that the young adult male testis is endowed with 500 million Sertoli cells and appears to decline to approximately 300 million cells in the older adult, suggesting that there is also an age-related decline in Sertoli cell numbers in the human testis. Furthermore, estimations of individual

F. Pasqualotto, MD, PhD (✉)
Department of Anatomy and Urology, Institute of Biotechnology,
University of Caxias do Sul, Caxias do Sul, RS, Brazil

CONCEPTION—Center for Human Reproduction,
Caxias do Sul, RS, Brazil
e-mail: fabio@conception-rs.com.br

E. Borges Jr., MD
Department of Fertility, Center for Assisted Fertilization,
Av. Brigadeiro Luis Antomiom 4545, Sao Paulo, 01401-002, Brazil

E. Pasqualotto, MD, PhD
CONCEPTION—Center for Human Reproduction,
Caxias do Sul, RS, Brazil

Department of Clinical Medicine, Center of Health Science,
University of Caxias do Sul, R. Pinheiro Machado, 2569, sl 23/24,
Caxias do Sul, RS, Brazil

S.J. Parekattil and A. Agarwal (eds.), *Male Infertility: Contemporary Clinical Approaches, Andrology, ART & Antioxidants*,
DOI 10.1007/978-1-4614-3335-4_30, © Springer Science+Business Media, LLC 2012

germ cell populations in younger and older men suggest that the age-related decrease in sperm production results from an absolute decrease in primary spermatocytes or a decrease in spermatogonial proliferation rather than a cellular degeneration in the spermatid to mature sperm stages of spermatogenesis. Mechanisms of cell senescence have been well studied in the last decade, and the free radical theory has brought considerable attention. Despite the established effects of free radicals on biomarkers of aging, their correlation with male infertility is not completely elucidated.

Keywords
Cell senescence • Male infertility • Paternal age and infertility • Antioxidants and aging • Mitochondrial reactive oxygen species • Structural chromosomal anomalies • Lipofuscin • Amyloid accumulation

Approximately 15% of couples of reproductive age experience infertility, and approximately 1/3 to half of infertility cases may be attributed to male factors [1]. It is well known that maternal age is a significant contributor to human infertility [2], due primarily to the precipitous loss of functional oocytes in women by their late 30s [3]. Human spermatogenesis, on the other hand, continues well into advanced ages, allowing men to reproduce during senescence. Although very little is known about the topic, paternal age may also contribute to human infertility.

It is well known that practically no children are born to mothers aged >50 years and it is common to all older fathers that they have younger partners. The discrepancy in the reproductive arena between males and females is astonishing, and reduced fertility and higher reproductive risks associated with advancing maternal age raise the question whether advanced paternal age is also associated with compromised fertility and increasing risks. In addition, it is well documented a progressive decrease of fertility due to both quantitative and qualitative loss of oocytes, eventually ending in menopause, women experience an age-dependent increase of miscarriages, obstetric morbidities, and chromosomal anomalies of the fetus [4]. This question should be discussed to younger age groups since increasing numbers of couples postpone parenthood into their fourth or fifth decade of life.

In contrast to the female, male reproductive functions do not cease abruptly, but androgen production and spermatogenesis continue lifelong. However, to evaluate a possible decline in the semen quality is a little bit difficult. Some men are reluctant to provide semen samples unless actively concerned about their fertility. For instance, population-based studies typically recruit at least 20% of young men willing to provide semen samples [5] constituting an inevitable participation bias in such studies [6, 7]. In addition, most of the published studies about sperm output in older men are largely restricted to patients attending infertility clinics, where few are older than 50 years [8]. An uncertain, but probably high proportion of such men have unrecognized defects in sperm production and/or function. Furthermore, access to such specialized medical services may be strongly influenced by nonbiological factors, and the results from infertility clinics may not be reliably extrapolated to the general male population.

Anyway, the effects of paternal age on a couple's fertility are real and may be greater than has previously been thought. Ford et al. stated that, after adjustments for other factors, the probability that a fertile couple will take >12 months to conceive nearly doubles from 8% when the man is <25 years to 15% when he is >35 years; thus, paternal age is a further factor to be taken into account when deciding the prognosis for infertile couples [9].

To explain the age-dependent changes observed in semen quality, two issues should be considered [8–11]. First, cellular or physiological changes due to aging have been described in testicles, seminal vesicles, prostate, and epididymis. Age-related narrowing and sclerosis of the testicular tubular lumen, decreases in spermatogenic activity, increased degeneration of germ cells, and decreased numbers and function of Leydig cells have been found in autopsies of men who died from accidental causes [12]. Smooth muscle atrophy and a decrease in protein and water content, which occur in the prostate with aging, may contribute to decreased semen volume and sperm motility. Also, the epididymis, a hormonally sensitive tissue, may undergo age-related changes. The hormonal or epididymal senescence may lead to decreased motility in older men. Secondly, increasing age implies more frequent exposure to exogenous damage or disease [8]. In addition to age per se, factors such as urogenital infections, vascular diseases, or an accumulation of toxic substances (cigarettes) may be responsible for worsening semen parameters. Indeed, a retrospective cross-sectional study in 3,698 infertile men showed an infection rate of the accessory glands in 6.1% in patients aged <25 years but in 13.6% of patients >40 years, and total sperm counts were significantly lower in patients with an infection of the accessory glands [13]. In addition, an age-dependent increase of polychlorinated biphenyls (PBC) in men has been described, and in men with normal semen parameters, the PBC concentration is inversely correlated with sperm count and progressive motility [14]. The concentration of cadmium also increases with age in the human testis, epididymis, and prostate, although lead and selenium remain constant over the whole age range in reproductive organs [15, 16].

Handelsman and Staraj demonstrated that, after exclusion of men with different diseases associated with diminishing testicular size, the specific effects of age on testicular volume appear only in the eighth decade of life [17]. In healthy men of this age group, the testis volume is 31% lower than in 18–40-year-old men [18]. However, recently a study showed a decline in testicular volume over time, especially after the age of 45 years old [19].

Morphological characteristics of aging testes vary from Sertoli cells accumulating cytoplasmic lipid droplets and are reduced in number [20], as are the Leydig cells [21], which may also be multinucleated [22]. Tubule involution is associated with an enlargement of the tunica propria, leading to progressive sclerosis parallel to a reduction of the seminiferous epithelium with complete tubular sclerosis as an endpoint [23]. Testicular sclerosis is associated with defective vascularization of the testicular parenchyma and with systemic arteriosclerosis of affected men [24]. Arteriographic patterns of the epididymis and the testes support these findings and are correlated with the degree of systemic arteriosclerosis [24]. In addition, age-dependent alterations of the prostate are well known [25] and are detectable histologically in 50% of 50-year-old men, but in 90% of men aged >90 years [26].

Common Male Infertility Disorders: Aging

Semen Analysis

Considering the age-dependent changes in reproductive organs of men, variations in semen parameters over time are not surprising; however, only few studies are controlled for abstinence time and other possible factors that may influence semen quality such as hypertension or smoking habits. Most studies are retrospective and rarely include males with more than 60 or 70 years old. Pasqualotto et al. recently described a decrease in semen volume across the groups evaluated in the study [19]. In fact, reports in the literature have shown a decrease in semen volume with aging [8, 27, 28]. The higher number of days' abstinence in men over 50 years old could explain these results. In the studies where the analyses were adjusted for days' abstinence, a decrease in semen volume of 3–22% was observed [9].

Regarding sperm motility, many studies adjusted for time of abstinence found a significant decrease in sperm motility associated with age and a yearly decrease ranging between 0.17% [29] and 0.7% [30]. However, these studies were performed in sperm donors [29–32] as well as in infertile patients [33, 34]. Pasqualotto et al. are in the same page than others showing that sperm motility tends to decrease as time goes by. Those studies that have been adjusted for duration of abstinence have reported statistically significant effects, such as negative linear relationships and decreases in motility ranging from 0.17 to 0.6% for each year of age [8, 29, 35, 36].

A computer-assisted semen analysis (CASA) has been developed as a specific tool to make the assessment of semen quality more objective and detailed [37]. Several specific motility parameters describing the movements of spermatozoa in a more detailed manner can be obtained with CASA. In addition, the classification into motile and immotile spermatozoa can be based on well-defined velocity thresholds. However, no correlations are detected between specific motion parameters as evaluated with CASA and the aging effect in the study by Pasqualotto et al. [19].

When focusing on sperm concentration, abstinence-adjusted studies do not provide a uniform picture. Even though some studies have reported a decrease in sperm concentration with increased age, several other studies have reported an increase in sperm concentration with age or found little or no association between age and sperm concentration [9, 11, 35, 38]. In fact, there are two different populations that we have to consider before evaluating the results: fertile vs. infertile men. A significant age-dependent decrease [30, 32] as well as constant values over the age range [31] or even a nonsignificant age-dependent increase with age [29] has been detected in healthy men. Regarding the infertile population, sperm concentration increases [33, 34] or remains unaltered [13], as indicated in abstinence-adjusted studies.

One of the good indicators of the germinal epithelium status is the sperm morphology. Degenerative changes in the germinal epithelium because of aging may affect spermatogenesis and thus sperm morphology. Pasqualotto et al., based on a linear regression analysis, stated that normal sperm morphology tends to decrease by 0.039% each year [19]. Auger et al., in a linear regression model, have shown that the normal sperm morphology decreases 0.9% yearly [32]. Thus, as compared to the average 30-year-old man, an average 50-year-old had an 18% decrease in normally shaped sperm [33]. Ng et al. showed that older men had more abnormal sperm morphology with decreasing numbers of normal forms and reduced vitality, as well as increased numbers of cytoplasmic droplets and sperm tail abnormalities (30% vs. 17%) compared to younger men [28]. The aberrant sperm morphology in older men was most evident in defects of tail morphology, possibly reflecting the complex cellular structural assembly process of the axoneme. Such increasing proportion of defects may reflect degenerative changes with aging in the germinal epithelium and/or in the intrinsic program directing spermiogenesis. In fact, the decrease per year varies from 0.2% [36] to 0.9% [32].

All reported changes of histological and seminal parameters develop gradually without a sudden age threshold. The alterations in semen parameters fall within normal ranges. Nevertheless, the age-dependent alterations of testicular histology and semen parameters are accompanied by a significant increase in FSH [19, 39] and a slight but significant decrease in inhibin B [18, 40], which are also found in men with apparently normal semen parameters.

Fertility of Aging Men

Without any type of doubt, male fertility is basically maintained until very late in life, and it has been documented scientifically up to more than 90 years old [41]. Besides female age, further confounders, such as reduced coital frequency, an increasing incidence of erectile dysfunction, and smoking habits, have to be considered in studies analyzing male fertility. All studies focused on a nonclinical population found a significant negative relationship between male age and couples' fertility.

A retrospective study of a large sample of European couples analyzed the risk of difficulties (due to adverse pregnancy outcomes, such as ectopic pregnancy, miscarriage, or stillbirth, or due to delayed conception) and the risk of delay in pregnancy onset [42]. Age-related changes were also found in a prospective study that estimated day-specific probabilities for pregnancy relative to ovulation [43]. Frequency of sexual intercourse was monitored by sexual diaries, and ovulation was based on basal body temperature measurements. According to this study, fertility for men aged >35 years is significantly reduced, and the age effect of men aged 35–40 years is about the same as when intercourse frequency drops from twice per week to once per week [44]. In studies dealing with subfertile couples, a significant decrease in pregnancy rates [33] or increase in TTP [45] was observed with female but not with male age, possibly indicating that male age-dependent alterations are masked by the infertility as such.

With methods of assisted reproduction, prerequisites for natural conception such as motility or fertilizing capacity are circumvented. In fact, the more invasive the treatment, the less important male age appears. Therefore, the success rates of ICSI [46] or IVF [47–49] are not associated with male age. On the other hand, the success rate of intrauterine insemination (IUI), a method requiring much higher quality and capability of sperm, is without question related to male age [50, 51].

Reactive Oxygen Species Generation from Mitochondria

Mitochondria play an important role in cellular energy generation, apoptosis regulation, and calcium homeostasis [52]. Coupled to the tricarboxylic acid cycle, the electron transport chain (ETC), and adenosine triphosphate (ATP) synthase in the mitochondria generate ATP, being the source of most cellular energy [52]. The mitochondrial respiratory chain is also the major intracellular source of reactive oxygen species (ROS) and free radicals under normal physiological and pathological conditions. It has been thought that loss of mitochondrial function and increased mitochondrial ROS production are important causal factors in aging.

In fact, it has been suggested that loss of mitochondrial function and increased mitochondrial ROS production are important causal factors in aging. Superoxide is continually produced as a by-product of normal cellular respiration [53]. As electrons are passed from complexes I to IV in the mitochondrial ETC, continuous leakage of electrons occurs, forming superoxide (1% of total rate of electron transport) [54]. This superoxide is converted to hydrogen peroxide by manganese superoxide dismutase in the mitochondrial matrix under physiological conditions. Superoxide oxidizes iron–sulfur-containing enzymes such as aconitase that produces hydrogen peroxide and ferrous iron [55].

A wide spectrum of alterations in mitochondria and mitochondrial DNA (mtDNA) with aging has been observed in animals and humans. These include (1) decline in mitochondrial respiratory function; (2) increase in mitochondrial production of reactive oxygen species (ROS) and the extent of oxidative damage to DNA, proteins, and lipids; (3) accumulation of point mutations and large-scale deletions of mtDNA; and (4) enhanced apoptosis. Recent studies have provided abundant evidence to substantiate the importance of mitochondrial production of ROS in aging. On the other hand, somatic mtDNA mutations can cause premature aging without increasing ROS production.

The formation of hydrogen peroxide from superoxide and its transformation to hydroxyl radical is apparent, especially when the mitochondrial ETC is abnormal or compromised [53]. Among five mitochondrial ETC complexes, complex I (nicotinamide adenine dinucleotide [NADH]–ubiquinone oxidoreductase) and complex III (ubiquinol–cytochrome c oxidoreductase) are responsible for most of the superoxide production [53]. Numerous studies have used mitochondrial ETC inhibitors such as rotenone and antimycin to disrupt the flow of electrons through complexes I and III, respectively [53, 56, 57].

Mitochondrial superoxide production generally is greater in antimycin-inhibited complex III than in rotenone-inhibited complex I [53, 57]. However, superoxide production by complex III is minimal in the absence of antimycin. Therefore, it can be suggested that in vivo complex I is responsible for much of mitochondrial ROS production through reversed and forward electron transport [57].

Mitochondria are more susceptible to oxidative damage because of the active production of ROS, as mtDNA is not protected by histones [58]. Therefore, mitochondrial ROS production damages the mitochondria themselves [58]. A dysfunctional mitochondrial respiratory chain would lead to more ROS production. Oxidative damage is more prevalent in mtDNA and protein in vivo than in other cell components [59].

Reactive Oxygen Species and Human Spermatozoa

Small amounts of ROS are continuously produced in the spermatozoa and play a crucial role in capacitation, hyperactivation, motility, acrosome reaction, and ultimately, normal fertilization [60]. Located at the midpiece of the spermatozoa, mitochondria function to produce ATP necessary for the physiological functions of the spermatozoa. Electrons leaking out of the mitochondrial respiratory chain contribute to superoxide production in the spermatozoa, such as in somatic cells [56]. Koppers et al. [56] recently were the first to demonstrate that mitochondria in human spermatozoa can produce ROS. They also suggested that rotenone-inhibited complex I, not antimycin, induces peroxidative damage in the midpiece of the spermatozoa. This continuous ROS generation induces oxidative damage to somatic cells as well as the spermatozoa, although spermatozoa are particularly susceptible to ROS damage because of the high content of polyunsaturated fatty acid in the cell membrane. ROS also reduces sperm motility by decreasing axonemal protein phosphorylation, as well as by lipid peroxidation [60].

Effects of Reactive Oxygen Species on Telomere and Telomerase

Telomeres are noncoding, repetitive DNA sequences (TTAGGG) at the ends of eukaryotic chromosomes that function to stabilize and protect chromosome ends. Progressive reduction of telomere length to a critically short size is related to the cessation of cell division and the onset of senescence. The enzyme telomerase adds specific TTAGGG sequence at the chromosome ends and maintains the length of the telomere. Telomerase is found in the male germ line cell, activated lymphocytes, and certain stem cell populations [61, 62]. It has been demonstrated telomerase activity in human testes [63]. Telomerase activity is highly critical in the germ line cells because of its task of preserving the full length chromosome of the cells and to maintain spermatogenesis. Yashima et al. observed variations in the location of telomerase expression in testis between the end of the prepubertal period and adulthood [64]. Before puberty, telomerase is highly expressed in immature Sertoli cells, whereas in advanced age groups, the telomerase expression is highest in the germ cells of the seminiferous tubules [64].

ROS also can affect telomere function indirectly by their interaction with telomerase. Intracellular ROS leads to a loss of activity of telomerase reverse transcriptase (TERT), whereas antioxidants such as *N*-acetylcysteine delay the onset of cell senescence [65]. Zhu et al. [66] suggested that decreased activity of catalytic subunits of telomerase (mTERT) increases neuronal cell susceptibility to oxidative stress. In embryonic mouse neurons, the decrease in the level of mTERT is directly related to the increase in apoptosis induced by oxidative stress. Consequently, the catalytic subunits of telomerase may play a role in protection from oxidative stress.

In fact, ROS-induced telomere shortening may be due to direct injury to guanine repeat telomere DNA by ROS. Using a culture of human WI-38 fibroblasts, it was demonstrated that telomere shortening is significantly increased under mild oxidative stress compared with that observed under normal conditions [67]. Most importantly, the addition of an antioxidant suppresses the rate of telomere shortening in somatic cells. Furumoto et al. [68] showed that the telomere shortening rate slowed after enrichment by ascorbic acid, a very potent antioxidant. Overexpression of the extracellular superoxide dismutase gene in human fibroblasts decreased the peroxide content, decreasing the rate of telomere shortening [67]. The rate of telomere shortening in sheep and humans is directly related to the cellular oxidative stress levels [69].

Deleterious conditions including dysfunctional telomeres, genomic instability, and apoptosis may result from telomere shortening in both somatic and germ line cells. A shortened telomere may limit the potential for replication and could result in genomic instability. This instability may play a role in about 70% of conceptions lost before birth, and also in 50% of spontaneous abortions with detectable chromosomal abnormalities [70]. It is important to notice here that to keep the telomere length may be essential for healthy human spermatozoa.

Antioxidants and Aging

Antioxidants play a crucial role in minimizing oxidative damage in the spermatozoa [71]. Studies have shown that along with an increase in ROS with aging, the antioxidant defense activity decreases with aging [58, 72]. Using Brown Norway rats, the antioxidant enzymatic activities in epididymal spermatozoa from young (4 months old) and old (21 months old) rats were compared [73]. Their study showed that antioxidants such as glutathione peroxidase (Gpx1, Gpx4) and superoxide dismutase had decreased activity in aging spermatozoa. However, ROS production and lipid peroxidation were significantly increased in their study. Examining the activities of copper–zinc superoxide dismutase, manganese superoxide dismutase, and glutathione peroxidase in Leydig cells isolated from 4- and 20-month-old rats, noting related to the age were detected regarding all antioxidant enzymatic activities [74]. The authors concluded that age-related reductions in testosterone levels of Leydig cells may be associated with the impairment of the antioxidant defense system of these cells.

Lipofuscin, also known as *age pigment*, is a marker of cell senescence [75]. It is an intralysosomal polymeric material that cannot be degraded or exocytosed, and continuously accumulates during the lifespan of a cell. It has been suggested that lipofuscin formation is ruled by ROS, mainly hydrogen peroxide, which is a by-product of normal oxygen metabolism that is produced continuously by mitochondria, peroxisomes, and cytosolic oxidases [75]. It is largely eliminated by catalase and glutathione peroxidase. However, it partially diffuses through lysosomal membranes. The Fenton reaction takes place in the presence of iron; the reducing substance (such as cysteine) is converted into a hydroxyl radical that induces peroxidation of various degraded macromolecules in the lysosome. This results in the formation of a cross-linked, nondegradable material called lipofuscin. Thus, ROS are involved in the accumulation of lipofuscin inside lysosomes. An age-related increase in lipofuscin formation is also seen in male reproductive tissue. Significantly large amounts of lipofuscin were found in the Leydig cells of elderly men compared to younger men [76]. Lipofuscin content matches the apoptosis indices of Leydig and epididymal cells. ROS are produced by Leydig cells during steroid genesis (testosterone synthesis), as well as by spermatozoal mitochondria [77]. Therefore, the ROS production might be responsible for Leydig cell aging, considering that inhibition of steroid genesis has been shown to prevent Leydig cell aging [76].

Amyloid (an insoluble protein aggregate) deposition affords additional evidence that ROS play a significant role in the loss of testicular function with aging. An age-dependent increase in oxidative stress and oxidative damage to cellular proteins contribute to amyloid formation [77]. Amyloid formation is associated with age-related increases in the fraction of abnormally folded proteins and decreases in the functions of proteases that degrade newly synthesized proteins [77]. The protein aggregates promote formation of protein aggregates by serving as nucleation sites. Amyloid fibrils with protease resistant structures are not reversible when formed, and promote further formation of protein aggregates. Age-related accumulation of intracellular amyloid fibrils in Sertoli cells of atrophic testes has also been described [78].

Aging and Sperm DNA Damage: Genomic Defects in Offspring

The association between aging and sperm DNA damage has been extensively studied during the last decade. Singh et al. showed higher levels of double-stranded DNA breaks in older men (by comet assay) [79]. Wyrobek et al. found an inverse relationship between DNA fragmentation index and male age by sperm chromatin structure assay [80]. Increasing oxidative stress levels associated with aging might be responsible for this increase in DNA damage with age [81, 82]. Oxidative stress-mediated DNA damage may be an etiology for repeated assisted reproductive technology failures in older men [83]. Increasing male age may have an influence on DNA fragmentation in the form of single-strand breaks [84]. This may not have any effect on fertilization because the oocyte can repair single-strand breaks. However, if the oocyte repair mechanisms are dysfunctional, this may result in poor, if not failed, blastocyst formation. Thus, oxidative stress-induced DNA damage can lead to various genomic defects [85, 86]. Plas et al. [85] showed a positive correlation between structural chromosomal abnormalities and paternal age, with a fourfold increase in abnormalities in the 45 years of age group compared with men aged 20–24 years. They also suggested that children of men in advanced age groups have 20% higher risk of carrying autosomal dominant diseases potentially due to increasing germ cell mitoses and meioses [85].

It is interesting to note that a direct relationship occurs between paternal aging and offspring development [87]. Considerable evidence shows a connection between aging and offspring learning and cognition. One study conducted by Reichenberg et al. [88] examining Israeli births over a consecutive 6-year period concluded that offspring of men 40 years or older were 5.75 times more likely to develop autism spectrum disorder than those of men younger than 30 years. The possible etiologies for that consisted of genetic imprinting alterations and de novo mutations. An older study showed an increased risk of trisomy 21 in offspring of older men, with a profound age effect in men 41 years and older [89]. Because the spontaneous mutation rate closely relates to paternal age, it can be suggested that the progeny of advanced male age groups experience a higher frequency of imprinting disorders such as Beckwith–Wiedemann syndrome, which involves insulin-like growth factor 2, a paternally expressed gene. This syndrome is characterized by abdominal wall defects, macroglossia, and susceptibility to embryonic tumor.

Structural Chromosomal Anomalies

Structural chromosomal anomalies result from chromosomal breakage and the following abnormal rearrangement within the same or within different chromosomes. In 84% of cases, de novo structural aberrations are of paternal origin [90], and they are found in 2% of spontaneous abortions and in 0.6% of live births [91]. Cytogenetic studies on structural chromosomal anomalies in sperm are rare but consistently describe an increase of mutations with age [92].

FISH was used for the structural analysis of individual chromosomes: duplications and deletions for the centromeric and subtelomeric regions of chromosome 9 increase significantly with age [93]. In spite of these age-dependent structural alterations in sperm, no increase of de novo structural chromosomal anomalies has been detected in newborns from older fathers [94].

Autosomal Dominant Diseases

Achondroplasia, the most common form of dwarfism, is the first genetic disorder that was hypothesized to have a paternal age component [95]. Apert's syndrome and achondroplasia have been amenable to direct sperm DNA mutation analysis [96, 97], and both are characterized by an age-dependent increase of mutations in sperm, but there are some peculiarities. For sporadic cases of Crouzon's or Pfeiffer's syndrome, 11 different mutations of the FGFR 2 gene are responsible, indicating that, unlike Apert's syndrome or achondroplasia, these are genetically heterogeneous conditions [98]. These mutations also arise in the male germ line, and advanced paternal age was noted for fathers of those patients.

The relationship between mutation frequency and paternal age is heterogeneous among autosomal dominantly inherited diseases [99]. In contrast to the above-mentioned diseases, osteogenesis imperfecta, neurofibromatosis, or bilateral retinoblastoma shows a weak paternal age effect [100]. Many of the mutations of the neurofibromatosis gene are intragenic deletions. These deletions are not age-dependent because they occur by mechanisms other than the base substitutions and are maternally derived in 16 of 21 cases [101].

Due to this heterogeneity of the paternal age effect in autosomal dominant diseases, the risk estimates proposed by Friedman for paternal age and autosomal dominant mutations may be overestimated [102]. Friedman calculated a risk for autosomal dominant diseases of 0.3–0.5% among offspring of fathers aged >40 years. This risk is comparable with the risk of Down's syndrome for 35–40-year-old women. However, the calculation was based on the assumption that the paternal age effect found in achondroplasia is typical of all autosomal dominant diseases.

There are conflicting data for Alzheimer's disease. Few studies conclude that paternal age is a risk factor [103]. However, the inconsistent results may be due to small sample sizes of the studies or due to the genetic heterogeneity of the disease.

Regarding schizophrenia, there are more conclusive data. In fact, the studies identified an increased risk of schizophrenia with paternal age [104]. Patients without a family history of schizophrenia had significantly older fathers than familial patients, so that de novo mutations were considered responsible [105]. Preeclampsia, which is considered to be a risk factor for schizophrenia, is also associated with paternal age [106].

One very important point we should never forget is that advanced paternal age increases the risk of other cancers in offspring. According to the Swedish Family-Cancer Database, there is an effect of paternal age on the incidence of sporadic breast and sporadic nervous system cancer in offspring [107]. Interestingly, an association between paternal age and the son's risk of prostate cancer was found [108]. The association of paternal age with early-onset prostate cancer (<65 years) was greater than that with late onset.

Expert Commentary

Most studies suggest that reduced fertility begins to become evident in the late 30s in men. Increased male age is associated with an increased risk of miscarriages, and both the risk of infertility and the risk of miscarriage strongly depend on female age. Couples should be aware of these age-dependent alterations in fertility and predisposition to genetic risks. Although at the moment increased paternal age is not an indication for prenatal diagnosis, there may be further developments in the future.

Although abundant experimental data have been gathered in the past decade to support the concept that mitochondrial dysfunction, ROS overproduction, and accumulation of mtDNA mutations in tissue cells are important contributors to human aging, the detailed mechanisms by which these biochemical events cause human aging remain to be established. The functional genomics and proteomics approaches to study aging on a genome-wide basis will provide novel information so we may gain a deeper understanding of the age-related alterations in the structure and function of mitochondria in the aging process. This is critical for the elucidation of the molecular mechanism of aging and for better management of aging and age-related diseases in humans.

Five-Year View

ROS are continually produced in the mitochondria of spermatozoa and play an important role in age-related male reproductive pathophysiology. The increased ROS level in semen observed with aging is associated with a possible decrease in antioxidant enzyme activity. This imbalance between prooxidants and antioxidants induces oxidative damage, resulting in abnormalities in telomeres and telomerase in male germ line cells. This sequence of events may explain the decrease in sperm concentration seen with aging. Oxidative stress in aging male reproductive system may inhibit sperm axonemal phosphorylation and increase lipid peroxidation, which can decrease sperm motility. This oxidative stress can also lead to lipofuscin and amyloid accumulation in the male reproductive tract, potentially the cause of decreased Leydig cell function and a subsequent decrease in blood testosterone levels. A higher rate of lipofuscin accumulation, in turn, may increase the amount of dysfunctional mitochondria in spermatozoa, thus increasing ROS formation.

Along with its negative effect on the fertilizing potential of spermatozoa, ROS also leads to offspring malformation (if fertilization is successful). Oxidative stress-induced mtDNA damage and nuclear DNA damage in aging men may put them at a higher risk for transmitting multiple genetic and chromosomal defects. Thus, this review suggests that ROS might play a central role in decreased male fertility with aging. This hypothesis provides guidance for future study and experiments, focusing on specific biomarkers of aging in men (telomere function, lipofuscin, amyloid) and their comparison with semen parameters and male fertility.

Key Issues

As life expectancy increases and our lives become busiest every day, many couples are waiting longer to establish their families. The recent trend toward delayed parenthood raises concerns because of the adverse effects of aging on fertility. Aging is a biological process that is characterized by the gradual loss of physiological function and increases in the susceptibility to disease of an individual. During the aging process, a wide spectrum of alterations in mitochondria and mitochondrial DNA (mtDNA) has been observed in somatic tissues of humans and animals. This is associated with the decline in mitochondrial respiratory function; excess production of the reactive oxygen species (ROS); increase in the oxidative damage to mtDNA, lipids, and proteins in mitochondria; accumulation of point mutations and large-scale deletions of mtDNA; and altered expression of genes involved in intermediary metabolism.

It has been demonstrated that the ROS may cause oxidative damage and mutations of mtDNA and alterations of the expression of several clusters of genes in aging tissues and senescent cells. The respiratory function decline and increase in the production of the ROS in mitochondria, accumulation of mtDNA mutation and oxidative damage, and altered expression of a few clusters of genes that culminated in the metabolic shift from mitochondrial respiration to glycolysis for major supply of ATP were key contributory factors in the aging process in the human and animals.

Although based on a small number of cases, the data presented for testicular morphology, semen parameters and fertility in aging males are conclusive and reflect a gradual deterioration with age within a broad individual spectrum. Most studies suggest that reduced fertility begins to become evident in the late 30s in men. Increased male age is associated with an increased risk of miscarriages, and both the risk of infertility and the risk of miscarriage strongly depend on female age. Advancing paternal age is associated with an increased risk for trisomy 21 and with diseases of complex etiology such as schizophrenia. Couples should be aware of these age-dependent alterations in fertility and predisposition to genetic risks. Although at the moment increased paternal age is not an indication for prenatal diagnosis, there may be further developments in the future.

References

1. Templeton A. Infertility-epidemiology, aetiology and effective management. Health Bull (Edinb). 1995;53(5):294–8.
2. Joffe M, Li Z. Male and female factors in fertility. Am J Epidemiol. 1995;141(11):1107–8.
3. Lansac J. Delayed parenting. Is delayed childbearing a good thing? Hum Reprod. 1995;10(5):1033–5.
4. Lubna P, Santoro N. Age-related decline in fertility. Endocrinol Metab Clin North Am. 2003;32:669–88.
5. Jensen TK, Jorgensen N, Punab M, Haugen TB, Suominen J, Zilaitiene B, Horte A, Andersen AG, Carlsen E, Magnus O, et al. Association of in utero exposure to maternal smoking with reduced semen quality and testis size in adulthood: a cross-sectional study of 1 770 young men from the general population in five European countries. Am J Epidemiol. 2004;159:49–58.
6. Handelsman DJ. Sperm output of healthy men in Australia: magnitude of bias due to self-selected volunteers. Hum Reprod. 1997;12:2701–5.
7. Cohn BA, Overstreet JW, Fogel RJ, Brazil CK, Baird DD, Cirillo PM. Epidemiologic studies of human semen quality: considerations for study design. Am J Epidemiol. 2002;155:664–71.
8. Kidd SA, Eskenazi B, Wyrobek AJ. Effects of male age on semen quality and fertility: a review of the literature. Fertil Steril. 2001; 75:237–48.
9. Ford WCL, North K, Taylor H, Farrow A, Hull MGR, Golding J, The ALSPAC Study Team. Increasing paternal age is associated with delayed conception in a large population of fertile couples: evidence for declining fecundity in older men. Hum Reprod. 2000;15:1703–8.
10. Eskenazi B, Wyrobek AJ, Sloter E, Kidd SA, Moore L, Young S, Moore D. The association of age and semen quality in healthy men. Hum Reprod. 2003;18:447–54.

11. Hassan MAM, Killick SR. Effect of male age on fertility: evidence for the decline in male fertility with increasing age. Fertil Steril. 2003;79:1520–7.
12. Neaves WB, Johnson L, Porter JC, Parker CR, Petty S. Leydig cell numbers, daily sperm production and serum gonadotropin levels in aging men. J Clin Endocrinol Metab. 1984;59:756–63.
13. Rolf C, Kenkel S, Nieschlag E. Age-related disease pattern in infertile men: increasing incidence of infections in older patients. Andrologia. 2002;34:209–17.
14. Dallinga JW, Moonen EJ, Dumoulin JC, Evers JL, Geraedts JP, Kleinjans JC. Decreased human semen quality and organochlorine compounds in blood. Hum Reprod. 2002;17:1973–99.
15. Oldereid NB, Thomassen Y, Attramadal A, Olaisen B, Purvis K. Concentrations of lead, cadmium and zinc in the tissues of reproductive organs of men. J Reprod Fertil. 1993;99:421–55.
16. Oldereid NB, Thomassen Y, Purvis K. Selenium in human male reproductive organs. Hum Reprod. 1998;13:2172–6.
17. Handelsman DJ, Staraj S. Testicular size: the effects of aging, malnutrition, and illness. J Androl. 1985;6:144–51.
18. Mahmoud AM, Goemaere S, El-Garem Y, Van Pottelbergh I, Comhaire FH, Kaufman JM. Testicular volume in relation to hormonal indices of gonadal function in community-dwelling elderly men. J Clin Endocrinol Metab. 2003;88:179–84.
19. Pasqualotto FF, Sobreiro BP, Hallak J, Pasqualotto EB, Lucon AM. Sperm concentration and normal sperm morphology decrease and follicle-stimulating hormone level increases with age. BJU Int. 2005;96:1087–91.
20. Harbitz TB. Morphometric studies of the Sertoli cells in elderly men with special reference to the histology of the prostate. Acta Pathol Microbiol Scand Sect A Pathol. 1973;81:703–17.
21. Johnson L. Spermatogenesis and aging in the human. J Androl. 1986;7:331–54.
22. Paniagua R, Amat P, Nistal M, Martin A. Ultrastructure of Leydig cells in human ageing testes. J Anat. 1986;146:173–83.
23. Paniagua R, Nistal M, Amat P, Rodriguez MC, Martin A. Seminiferous tubule involution in elderly men. Biol Reprod. 1987;36:939–47.
24. Regadera J, Nistal M, Paniagua R. Testis, epididymis, and spermatic cord in elderly men. Correlation of angiographic and histologic studies with systemic arteriosclerosis. Arch Pathol Lab Med. 1985;109:663–7.
25. Hermann M, Untergasser G, Rumpold H, Berger P. Aging of the male reproductive system. Exp Gerontol. 2000;35:1267–79.
26. Coffey DS, Berry SJ, Ewing LL. An overview of current concepts in the study of benign prostate hyperplasia. In: Rodgers CH, Coffey DS, Cunha G, Grayhack JT, Hinman F, Horton R, editors. Benign prostatic hyperplasia. Vol II. Washington, DC: US Department of Health and Human Services, NIH publication No. 87-2881; 1987. p. 1–13.
27. Kühnert B, Nieschlag E. Reproductive functions of the ageing male. Hum Reprod Update. 2004;10:327–39.
28. Ng KK, Donat R, Chan L, Lalak A, Di Pierro I, Handelsman DJ. Sperm output of older men. Hum Reprod. 2004;8:1811–5.
29. Fisch H, Goluboff ET, Olson JH, Feldshuh J, Broder SJ, Barad DH. Semen analysis in 1,283 men from the United States over a 25-year period: no decline in quality. Fertil Steril. 1996;65:1009–14.
30. Eskenazi B, Wyrobek AJ, Kidd SA, Lowe X, Moore II D, Weisiger K, Aylstock M. Sperm aneuploidy in fathers of children with paternally and maternally inherited Klinefelter syndrome. Hum Reprod. 2002;17:576–83.
31. Schwartz D, Mayaux MJ, Spira A, Moscato ML, Jouannet P, Czyglik F, David G. Semen characteristics as a function of age in 833 fertile men. Fertil Steril. 1983;39:530–5.
32. Auger J, Kunstmann JM, Czyglik F, Jouannet P. Decline in semen quality among fertile men in Paris during the past 20 years. N Engl J Med. 1995;332:281–5.
33. Rolf C, Behre HM, Nieschlag E. Reproductive parameters of older couples compared to younger men of infertile couples. Int J Androl. 1996;19:135–42.
34. Andolz P, Bielsa MA, Vila J. Evolution of semen quality in North-eastern Spain: a study in 22 759 infertile men over a 36 year period. Hum Reprod. 1999;14:731–5.
35. Berling S, Wolner-Hanssen P. No evidence of deteriorating semen quality among men in infertile relationships during the last decade: a study of males from southern Sweden. Hum Reprod. 1997; 12:1002–5.
36. Carlsen E, Giwercman A, Keiding N, Skakkebaek NE. Evidence for decreasing quality of semen during past 50 years. BMJ. 1992;305:609–13.
37. Agarwal A, Ozturk E, Loughlin KR. Comparison of semen analysis between the two Hamilton-Thorn semen analysers. Andrologia. 1992;24:327–9.
38. Hommonai ZT, Fainman N, David MP, Paz GF. Semen quality and sex hormone pattern of 29 middle aged men. Andrologia. 1982;14:164–70.
39. Nieschlag E, Lammers U, Freischem CW, Langer K, Wickings EJ. Reproductive functions in young fathers and grandfathers. J Clin Endocrinol Metab. 1982;55:676–81.
40. Baccarelli A, Morpurgo PS, Corsi A, Vaghi I, Fanelli M, Cremonesi G, Vaninetti S, Beck-Peccoz P, Spada A. Activin A serum levels and aging of the pituitary–gonadal axis: a cross-sectional study in middle-aged and elderly healthy subjects. Exp Gerontol. 2001;36:1403–12.
41. Seymour FI, Duffy C, Koerner A. A case of authenticated fertility in a man, aged 94. J Am Med Assoc. 1935;105:1423–4.
42. de la Rochebrochard E, Thonneau P. Paternal age and maternal age are risk factors for miscarriage; results of a multicentre European study. Hum Reprod. 2002;17:1649–56.
43. Dunson DB, Colombo B, Baird D. Changes with age in the level and duration of fertility in the menstrual cycle. Hum Reprod. 2002;17:1399–403.
44. Dunson DB, Baird DD, Colombo B. Increased infertility with age in men and women. Obstet Gynecol. 2004;103:51–6.
45. Olson J. Subfecundity according to the age of the mother and the father. Dan Med Bull. 1990;37:281–2.
46. Spandorfer SD, Avrech OM, Colombero LT, Palermo GD, Rosenwaks Z. Effect of parental age on fertilization and pregnancy characteristics in couples treated by intracytoplasmic sperm injection. Hum Reprod. 1998;13:334–8.
47. Piette C, de Mouzon J, Bachelot A, Spira A. In-vitro fertilization: influence of women's age on pregnancy rates. Hum Reprod. 1990;5:56–9.
48. Gallardo E, Sim6n C, Levy M. Effect of age on sperm fertility potential: oocyte donation as a model. Fertil Steril. 1996; 66:260–4.
49. Paulson RJ, Milligan RC, Sokol RZ. The lack of influence of age on male fertility. Am J Obstet Gynecol. 2001;184:818–22.
50. Mathieu C, Ecochard R, Bied V, Lornage J, Czyba JC. Cumulative conception rate following intrauterine artificial insemination with husband's spermatozoa: influence of husband's age. Hum Reprod. 1995;10:1090–7.
51. Brzechffa PR, Buyalos RP. Female and male partner age and menotrophin requirements influence pregnancy rates with human menopausal gonadotrophin therapy in combination with intrauterine insemination. Hum Reprod. 1997;12:29–33.
52. McBride HM, Neuspiel M, Wasiak S. Mitochondria: more than just a powerhouse. Curr Biol. 2006;16:R551–60.
53. Raha S, Robinson BH. Mitochondria, oxygen free radicals, disease and ageing. Trends Biochem Sci. 2000;25:502–8.
54. Green K, Brand MD, Murphy MP. Prevention of mitochondrial oxidative damage as a therapeutic strategy in diabetes. Diabetes. 2004;53 suppl 1:S110–8.

55. Vasquez-Vivar J, Kalyanaraman B, Kennedy MC. Mitochondrial aconitase is a source of hydroxyl radical. An electron spin resonance investigation. J Biol Chem. 2000;275:14064–9.
56. Koppers AJ, De Iuliis GN, Finnie JM, et al. Significance of mitochondrial reactive oxygen species in the generation of oxidative stress in spermatozoa. J Clin Endocrinol Metab. 2008;93:3199–207.
57. St-Pierre J, Buckingham JA, Roebuck SJ, et al. Topology of superoxide production from different sites in the mitochondrial electron transport chain. J Biol Chem. 2002;277:44784–90.
58. Wei YH, Ma YS, Lee HC, et al. Mitochondrial theory of aging matures—roles of mtDNA mutation and oxidative stress in human aging. Zhonghua Yi Xue Za Zhi (Taipei). 2001;64:259–70.
59. James AM, Murphy MP. How mitochondrial damage affects cell function. J Biomed Sci. 2002;9:475–87.
60. Agarwal A, Makker K, Sharma R. Clinical relevance of oxidative stress in male factor infertility: an update. Am J Reprod Immunol. 2008;59:2–11.
61. Zalenskaya IA, Zalensky AO. Telomeres in mammalian male germline cells. Int Rev Cytol. 2002;218:37–67.
62. Liu Y, Bohr VA, Lansdorp P. Telomere, telomerase and aging. Mech Ageing Dev. 2008;129:1–2.
63. Fujisawa M, Tanaka H, Tatsumi N, et al. Telomerase activity in the testis of infertile patients with selected causes. Hum Reprod. 1998;13:1476–9.
64. Yashima K, Maitra A, Rogers BB, et al. Expression of the RNA component of telomerase during human development and differentiation. Cell Growth Differ. 1998;9:805–13.
65. Haendeler J, Hoffmann J, Diehl JF, et al. Antioxidants inhibit nuclear export of telomerase reverse transcriptase and delay replicative senescence of endothelial cells. Circ Res. 2004;94:768–75.
66. Zhu H, Fu W, Mattson MP. The catalytic subunit of telomerase protects neurons against amyloid beta-peptide-induced apoptosis. J Neurochem. 2000;75:117–24.
67. Serra V, von Zglinicki T, Lorenz M, et al. Extracellular superoxide dismutase is a major antioxidant in human fibroblasts and slows telomere shortening. J Biol Chem. 2003;278:6824–30.
68. Furumoto K, Inoue E, Nagao N, et al. Age-dependent telomere shortening is slowed down by enrichment of intracellular vitamin C via suppression of oxidative stress. Life Sci. 1998;63:935–48.
69. Richter T, von Zglinicki T. A continuous correlation between oxidative stress and telomere shortening in fibroblasts. Exp Gerontol. 2007;42:1039–42.
70. Baird DM, Britt-Compton B, Rowson J, et al. Telomere instability in the male germline. Hum Mol Genet. 2006;15:45–51.
71. Cocuzza M, Sikka SC, Athayde KS, et al. Clinical relevance of oxidative stress and sperm chromatin damage in male infertility: an evidence based analysis. Int Braz J Urol. 2007;33:603–21.
72. Cao L, Leers-Sucheta S, Azhar S. Aging alters the functional expression of enzymatic and non-enzymatic antioxidant defense systems in testicular rat Leydig's cells. J Steroid Biochem Mol Biol. 2004;88:61–7.
73. Weir CP, Robaire B. Spermatozoa have decreased antioxidant enzymatic capacity and increased reactive oxygen species production during aging in the Brown Norway rat. J Androl. 2007;28:229–40.
74. Luo L, Chen H, Trush MA, et al. Aging and the brown Norway rat Leydig cell antioxidant defense system. J Androl. 2006; 27:240–7.
75. Terman A, Gustafsson B, Brunk UT. The lysosomal-mitochondrial axis theory of postmitotic aging and cell death. Chem Biol Interact. 2006;163:29–37.
76. Zirkin BR, Chen H. Regulation of Leydig cell steroidogenic function during aging. Biol Reprod. 2000;63:977–81.
77. Squier TC. Oxidative stress and protein aggregation during biological aging. Exp Gerontol. 2001;36:1539–50.
78. Bohl J, Steinmetz H, Storkel S. Age-related accumulation of congophilic fibrillar inclusions in endocrine cells. Virchows Arch A Pathol Anat Histopathol. 1991;419:51–8.
79. Singh NP, Muller CH, Berger RE. Effects of age on DNA doublestrand breaks and apoptosis in human sperm. Fertil Steril. 2003;80:1420–30.
80. Wyrobek AJ, Eskenazi B, Young S, et al. Advancing age has differential effects on DNA damage, chromatin integrity, gene mutations, and aneuploidies in sperm. Proc Natl Acad Sci USA. 2006;103:9601–6.
81. Desai N, Sabanegu Jr E, Kim T, Agarwal A. Free radical theory of aging: implications in male infertility. Urology. 2010;75:14–9.
82. Angelopoulou R, Lavranos G, Manolakou P. ROS in the aging male: model diseases with ROS-related pathophysiology. Reprod Toxicol. 2009;28:167–71.
83. Shamsi MB, Kumar R, Dada R. Evaluation of nuclear DNA damage in human spermatozoa in men opting for assisted reproduction. Indian J Med Res. 2008;127:115–23.
84. Schmid TE, Eskenazi B, Baumgartner A, et al. The effects of male age on sperm DNA damage in healthy non-smokers. Hum Reprod. 2007;22:180–7.
85. Plas E, Berger P, Hermann M, et al. Effects of aging on male fertility? Exp Gerontol. 2000;35:543–51.
86. Bosco L, Ruvolo G, Morici G, et al. Apoptosis in human unfertilized oocytes after intracytoplasmic sperm injection. Fertil Steril. 2005;84:1417–23.
87. Lawrence LT, Moley KH. Epigenetics and assisted reproductive technologies: human imprinting syndromes. Semin Reprod Med. 2008;26:143–52.
88. Reichenberg A, Gross R, Weiser M, et al. Advancing paternal age and autism. Arch Gen Psychiatry. 2006;63:1026–32.
89. Stene J, Stene E, Stengel-Rutkowski S, et al. Paternal age and Down's syndrome: data from prenatal diagnoses (DFG). Hum Genet. 1981;59:119–24.
90. Olsen SD, Magenis RE. Preferential paternal origin of de novo structural chromosome rearrangements. In: Daniel A, editor. The cytogenetics of mammalian autosomal rearrangements. New York: Alan R. Liss; 1988. p. 583–99.
91. Jacobs PA. The chromosome complement of human gametes. Oxf Rev Reprod Biol. 1992;14:47–72.
92. Sartorelli EM, Mazzucatto LF, de Pina-Neto JM. Effect of paternal age on human sperm chromosomes. Fertil Steril. 2001;76: 1119–23.
93. Bosch M, Rajmil O, Martinez-Pasarell O, Egozcue J, Templado C. Linear increase of diploidy in human sperm with age: a four-colour FISH study. Eur J Hum Genet. 2001;9:533–8.
94. Hook EB, Regal RR. A search for a paternal-age effect upon cases of 47, þ 21 in which the extra chromosome is of paternal origin. Am J Hum Genet. 1984;36:413–21.
95. Penrose LS. Parental age and mutation. Lancet. 1955;269:312–3.
96. Glaser RL, Broman KW, Schulman RL, Eskenazi B, Wyrobek AJ, Jabs EW. The paternal-age effect in Apert syndrome is due, in part, to the increased frequency of mutations in sperm. Am J Hum Genet. 2003;73:939–47.
97. Goriely A, McVean GA, Rojmyr M, Ingemarsson B, Wilkie AO. Evidence for selective advantage of pathogenic FGFR2 mutations in the male germ line. Science. 2003;301:643–6.
98. Glaser RL, Jiang W, Boyadjiev SA, Tran AK, Zachary AA, Van Maldergem L, Johnson D, Walsh S, Oldridge M, Wall SA, et al. Paternal origin of FGFR2 mutations in sporadic cases of Crouzon syndrome and Pfeiffer syndrome. Am J Hum Genet. 2000;66: 768–77.
99. Risch N, Reich EW, Wishnick MM, McCarthy JG. Spontaneous mutation and parental age in humans. Am J Hum Genet. 1987;41: 218–48.

100. Sivakumaran TA, Ghose S, Kumar HAS, Kucheria K. Parental age in Indian patients with sporadic hereditary retinoblastoma. Ophthalmic Epidemiol. 2000;7:285–91.
101. Lazaro C, Gaona A, Ainsworth P, Tenconi R, Vidaud D, Kruyer H, Ars E, Volpini V, Estivill X. Sex differences in the mutation rate and mutational mechanism in the NF gene in neurofibromatosis type 1 patients. Hum Genet. 1996;98:696–9.
102. Friedman JM. Genetic disease in the offspring of older fathers. Obstet Gynecol. 1981;57:745–9.
103. Bertram L, Busch R, Spiegl M, Lautenschlager NT, Müller U, Kurz A. Paternal age is a risk factor for Alzheimer disease in the absence of a major gene. Neurogenetics. 1998;1:277–80.
104. Dalman C, Allebeck P. Paternal age and schizophrenia: further support for an association. Am J Psychiatry. 2002;159:1591–2.
105. Malaspina D, Corcoran C, Fahim C, Berman A, Harkavy-Friedman J, Yale S, Goetz D, Goetz R, Harlap S, Gorman J. Paternal age and sporadic schizophrenia: evidence for de novo mutations. Am J Med Genet. 2002;114:299–303.
106. Harlap S, Paltiel O, Deutsch L, Knaanie A, Masalha S, Tiram E, Caplan LS, Malaspina D, Friedlander Y. Paternal age and preeclampsia. Epidemiology. 2002;13:660–7.
107. Hemminki K, Kyyronen P. Parental age and risk of sporadic and familial cancer in offspring: implications for germ cell mutagenesis. Epidemiology. 1999;10:747–51.
108. Zhang Y, Kreger BE, Dorgan JF, Cupples LA, Myers RH, Splansky GL, Schatzkin A, Ellison RC. Parental age at child's birth and son's risk of prostate cancer. The Framingham Study. Am J Epidemiol. 1999;150:1208–12.

Apoptosis and Male Infertility

31

S. Vaithinathan, Shereen Cynthia D'Cruz, and P.P. Mathur

Abstract

The pathogenesis of male infertility can be reflected by defective spermatogenesis due to pituitary disorders, testicular cancer, germ cell aplasia, varicocele, and environmental factors or due to defective sperm transport resulting from congenital abnormalities, immunological or neurological factors. Recent findings show that male infertility could be increased incidence of genetic disorders and apoptosis. Of these, apoptosis has been identified as a major factor contributing to male infertility and has been studied extensively in recent years. Apoptosis, also known as *programmed cell death* (PCD), is required for normal spermatogenesis in mammals and is believed to ensure cellular homeostasis, and an adequate number of germ cells are eliminated via the process of apoptosis in order to maintain a precise germ cell population in compliance with the supportive capacity of the Sertoli cells. This chapter briefs both physiological and pathological events that can trigger apoptosis and their effects on the male reproductive system.

Keywords

Apoptosis • Male infertility • Programmed cell death • Sertoli cells • Tumor necrosis factor • Cytoplasmic pathway • Extrinsic pathway • Steroidogenesis • Environmental contaminants

Infertility is a global health issue, affecting approximately 8–10% of couples worldwide. In India, the infertility rate is between 7 and 10% [1]. In the United States, approximately 15% of the couples are infertile, and although infertility is considered to be preliminarily a women's problem, men are responsible in about 20% of infertile cases [2]. In some African countries, one-third of the couples are infertile [3, 4]. Increasing evidence from clinical and epidemiological studies suggests that male infertility could be due to increasing incidence of male reproductive problems. The pathogenesis of male infertility can be reflected by defective spermatogenesis due to pituitary disorders, testicular cancer, germ cell aplasia, varicocele, and environmental factors or due to defective sperm transport resulting from congenital abnormalities, immunological or neurological factors. Recent findings show that male infertility could be increased incidence of genetic disorders and apoptosis [5]. Of these, apoptosis has been identified as a major factor contributing to male fertility and has been studied extensively in recent years.

Apoptosis, also known as programmed cell death (PCD), is required for normal spermatogenesis in mammals and is believed to ensure cellular homeostasis, and an adequate amount of germ cells are eliminated via the process of apoptosis in order to maintain a precise germ cell population in compliance with the supportive capacity of the Sertoli cells. This chapter briefs both physiological and pathological events that can trigger apoptosis and their effects on the male reproductive system.

S. Vaithinathan, PhD • S.C. D'Cruz, MPhil
Department of Biochemistry and Molecular Biology, Pondicherry University, Kalapet, Pondicherry 605014, India
e-mail: msvnathan@gmail.com; shereencynthia@gmail.com

P.P. Mathur, PhD (✉)
Department of Biochemistry & Molecular Biology and Center for Bioinformatics, School of Life Sciences, Pondicherry University, Kalapet, Pondicherry, 605014, India
e-mail: ppmathur@gmail.com

S.J. Parekattil and A. Agarwal (eds.), *Male Infertility: Contemporary Clinical Approaches, Andrology, ART & Antioxidants*,
DOI 10.1007/978-1-4614-3335-4_31,

Physiological Role of Apoptosis in Male Reproduction

Testes perform one of the complex events called spermatogenesis, which is necessary for life. Spermatogenesis, a highly dynamic and synchronized process of germ cell maturation from diploid spermatogonia to mature haploid spermatozoa, takes place in the seminiferous epithelium of the testis. This highly intricate cellular development is fostered by the somatic cell type, the Sertoli cells, which envelope the germ cells [6]. During testicular development, the Sertoli cell number increases gradually and thereafter their proliferative capacity declines to produce a stable population of nondividing Sertoli cells [7]. On the other hand, the germ cells continuously proliferate and differentiate to become mature spermatozoa. Overproliferation of germ cells is tempered by selective apoptosis of their progeny in order to maintain a precise germ cell population in compliance with the supportive capacity of the Sertoli cells [7]. Apoptosis also occurs as a defense mechanism such as in immune reactions or when cells are damaged by disease or environmental agents.

Necrosis and apoptosis are the two major mechanisms of cell death. Necrosis occurs in cells that are damaged by external injury, whereas apoptosis occurs in cells that are induced to commit programmed death from internal or external stimuli. Apoptosis consists of highly intricate, sophisticated, and energy-dependent cascade mechanisms and occurs through two main pathways (Fig. 31.1). The first, referred to as the extrinsic or cytoplasmic pathway, is triggered through the Fas death receptor, a member of the tumor necrosis factor (TNF) receptor superfamily [8]. The second pathway is the intrinsic or mitochondrial pathway that when stimulated leads to the release of cytochrome c from the mitochondria and activation of the death signal [9]. Both pathways converge to a final common pathway involving the activation of a cascade of proteases called caspases that cleave regulatory and structural molecules, culminating in the death of the cell. The pathways are linked; thus, the distinction between the two pathways is simplistic. Overexpression of Bcl-2 in the intrinsic pathway may lead to the inhibition of extrinsic mediated apoptosis [10]; conversely, TNFα may increase the expression of NF-κB and stimulates antiapoptotic members of the Bcl-2 family proteins.

Extrinsic Pathway

This pathway comprises several protein members including the death receptors, the membrane-bound Fas ligand, the Fas complexes, the Fas-associated death domain, and caspases 8 and 10, which ultimately activate the rest of the downstream caspases leading to apoptosis. Activation of the extrinsic pathway is initiated with the ligation of cell surface receptors called death receptors (DRs). Fas is a member of the tumor necrosis factor receptor superfamily and is also called Apo-1. Fas signaling play an important role in apoptosis.

The Fas ligand (FasL)–Fas system is mainly recognized for its death-related functions. When a death stimulus triggers the pathway, the membrane-bound FasL interacts with the inactive Fas complexes and forms the death-inducing signaling complex. The Fas death-inducing signaling complex contains

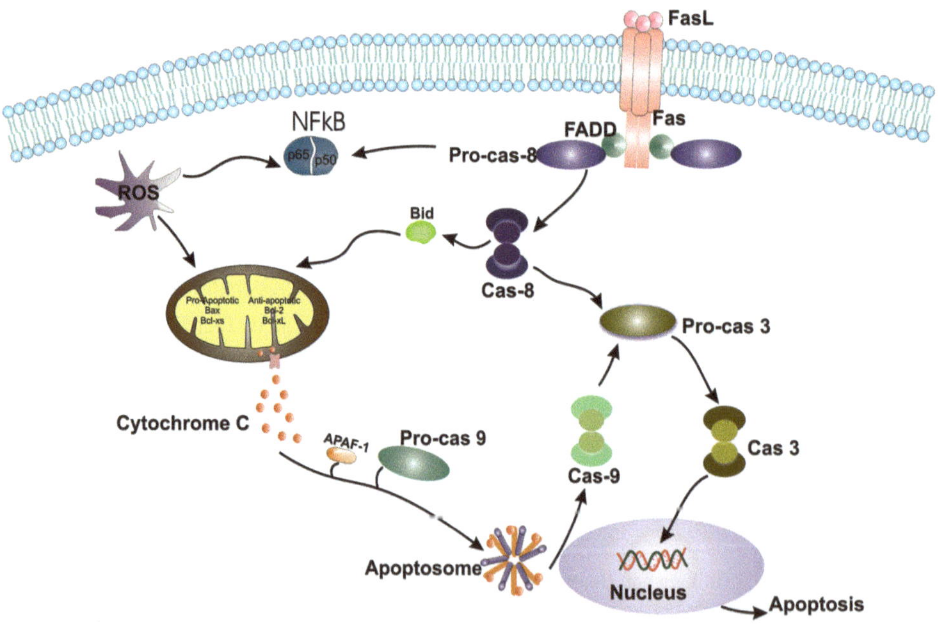

Fig. 31.1 Pathways of apoptosis: Components involved in the mitochondrial and cell death-mediated pathways are shown in the figure

the adaptor protein Fas-associated death domain protein and caspases 8 and 10 and leads to activation of caspase 8, which, in turn, can activate the rest of the downstream caspases. Caspase 8 interacts with the intrinsic apoptotic pathway by cleaving Bid (a proapoptotic member of the Bcl-2 family), leading to the subsequent release of cytochrome c [11].

Intrinsic Pathway

One of the most important regulators of this pathway is the Bcl-2 family of proteins. The Bcl-2 family includes proapoptotic members such as Bax, Bak, Bad, Bcl-Xs, Bid, Bik, Bim, and Hrk and antiapoptotic members. Antiapoptotic Bcl-2 members act as repressors of apoptosis by blocking the release of cytochrome c, whereas proapoptotic members act as promoters. Following a death signal, proapoptotic proteins undergo posttranslational modifications that include dephosphorylation and cleavage resulting in their activation and translocation to the mitochondria leading to apoptosis [9]. In response to apoptotic stimuli, the outer mitochondrial membrane becomes permeable, leading to the release of cytochrome c. Once cytochrome c is released into the cytosol, it interacts with Apaf-1, leading to the activation of caspase 9 proenzymes. Active caspase 9 then activates caspase 3, which subsequently activates the rest of the caspase cascade and leads to apoptosis [10].

Fas/FasL

Fas is a type I transmembrane receptor protein that belongs to the TNF/nerve growth factor family [12, 13], and Fas ligand (FasL) has been identified as a tumor necrosis factor-related type II transmembrane protein [14]. Ligation of FasL to Fas induces apoptosis of the Fas-bearing cells by the activation of various caspases. Caspase 8 is involved in the upstream of the apoptosis process [15, 16]. Activation of caspase 8 is followed by activation of caspase 3, which is known as executioner protease in this process [17]. In the rodent testis, apoptosis, induced by FasL, has been suggested to be one of the mechanisms that limit the number of germ cells during normal spermatogenesis or after testicular injuries. Recent studies have also shown that, in the human testis, apoptosis is a conspicuous event during spermatogenesis; Fas–FasL interaction is reportedly involved in the regulation of this event [18, 19].

Caspases Family

Caspases (cysteinyl aspartate-specific proteinases) are aspartic acid-directed cysteine proteases. These proteases are synthesized as precursors that have insignificant catalytic activity. The precursor caspase is converted to the active enzyme by proteolytic processing either by another protease or by autocatalysis and triggered by the binding of cofactors or removal of inhibitors. Caspases share similarities in amino acid sequence, structure, and substrate specificity. They are all expressed as proenzymes (30–50 kDa). It contains three domains and they are: the amino terminal domain, a large subunit (~20 kDa), and a small subunit (~10 kDa). During the caspase activation, proteolytic processing occurs between domains, followed by association of the large and small subunits to form a heterodimer [20]. Although the majority of caspases are situated within the cytoplasm, some of the members can be found in the Golgi apparatus (caspase 12) or in association with the mitochondria (caspases 2, 3, and 9) [21, 22]. Caspase 3 is the most important effector caspase. Its activation is important in PCD signaling [23].

Cytochrome c

Different proapoptotic proteins, such as cytochrome c and Smac/Diablo that are normally present in the intermembrane space of mitochondria, are released during the early stages of apoptosis. In the cytosol, cytochrome c participates in the formation of the apoptosome complex together with its adaptor molecule, Apaf-1, resulting in the recruitment, processing, and activation of procaspase 9 in the presence of adenosine triphosphate [24]. Subsequently, caspase 9 cleaves and activates procaspases 3 and 7; these effector caspases are responsible for the cleavage of various proteins leading to the biochemical and morphological features characteristic of apoptosis [25]. The release of cytochrome c is, therefore, considered a key initiative step in the apoptotic process.

Nuclear Factor-κB

The classical NF-κB transcriptional factors are composed of homodimers or heterodimers of Rel protein, of which p65/p50 heterodimer is the predominant complex, in testicular germ cells [26]. In unstimulated cells, NF-κB dimers are sequestered in the cytoplasm by inhibitory kappa B (IκB) protein. Upon exposure to various extracellular signals that leads to phosphorylation and degradation of IκB, free NF-κB dimers rapidly translocate to the nucleus, wherein they activate transcription of target genes [27].

Spermatogenesis

Spermatogenesis is a dynamic and well-regulated process that involves multiplication, maturation, and differentiation of germ cells resulting in the formation of mature spermatozoa. The process is subdivided into spermatogoniogenesis known as mitotic multiplication of spermatogonia, maturation of spermatocytes, spermiogenesis, and spermiation. Germ cells undergo mitosis to produce primary spermatocytes. The primary spermatocytes enter meiosis to form secondary spermatocytes and proceed through meiosis to produce haploid spermatids. These in turn undergo a complex process of morphological and functional differentiation resulting in the production of mature spermatozoa, which is known as spermiogenesis. In mammalian species, to maintain proper germ cell numbers, apoptosis takes place in the testis [16]. Loss of spermatogenic cells is incurred mostly during maturity of spermatogonia and to a lesser extent during maturation of spermatocytes and spermatid in adult rat testis. The sign of spermatogenesis is started when germ cells differentiate into spermatogonia. While some spermatogonia become self-renewing spermatogonial stem cells, most differentiate into spermatocytes and at ~10 days after birth in mice and at puberty in man, initiate meiosis and is accompanied by extensive germ cell apoptosis [16].

Steroidogenesis

Leydig cells are the principal cells involved in the process of steroidogenesis. Leydig cells secrete androgens, particularly testosterone, which is extremely essential for the initiation and maintenance of spermatogenesis [28]. Any factor affecting the Leydig cell viability, in turn, can interrupt the endocrine regulation of spermatogenesis and consequently affect the reproductive performance. Aroclor 1254, a commercial mixture of polychlorinated biphenyls, brought about a state of oxidative stress in cultured Leydig cells characterized by decline in the levels of enzymatic and nonenzymatic antioxidants accompanied by an elevation in the levels of lipid peroxidation and ROS. In addition, the activities of steroidogenic enzymes were inhibited at the level of gene expression causing diminished testosterone production [29]. Exposure of rats to single dose of cadmium (0.20 mg/100 g body weight) inhibited the activities of testicular 3β and 17β-hydroxysteroid dehydrogenase along with reduced expression of StAR protein resulting in lowered serum testosterone levels. The observed effects have been attributed to the excess generation of ROS in the testis resulting from depletion of antioxidant enzymes like SOD and glutathione peroxidase. Exposure of primary cultured Leydig cells to cadmium at the concentrations of 10 mM, caused increased oxidative DNA damage resulting in decreased viability of cells and testosterone secretion [30]. The high levels of corticosterone associated with stress are known to induce apoptosis in Leydig cells. The activation of Fas system, cleavage of procaspase 3, loss of mitochondrial membrane potential, and increased ROS generation are reported to be the possible mechanisms involved in corticosterone-induced Leydig cell death [31]. The decline in testosterone production following toxicant exposure may be in part due to apoptosis of Leydig cells caused by induction of stress by corticosterone.

Effect of Environmental Contaminants

Apoptosis of spermatogenic cells is essential for the maintenance of testicular homeostasis, although increased cell death can result in defective spermatogenesis leading to infertility [32]. In the testis, apoptotic death is a common programmed event that reduces 75% of germ cells [33]. However, excessive or inadequate apoptosis of testicular cells results in abnormal spermatogenesis or testicular tumors [34]. Various testicular toxicants have been reported to induce massive germ cell apoptosis indicating that the seminiferous epithelium responds to most of the adverse stimuli by eliminating germ cells through programmed cell death [35]. Recent studies showed that DDT and its metabolite to induce apoptosis either in vitro or in vivo experiments [36, 37]. Song et al. demonstrated that exposure to *p,p′*-DDE, a metabolite of DDT, at over 30 μM dose level showed induction of apoptotic cell death in cultured rat Sertoli cells by inducing mitochondria-mediated apoptotic changes including elevation in reactive oxygen species (ROS) generation, decrease in mitochondrial membrane potential, and release of cytochrome c into the cytosol which could be blocked by *N*-acetyl-L-cysteine, an antioxidant with an elevated ratios of Bax/Bcl-w and Bak/Bcl-w, and cleavages of procaspases 3 and 9 were induced by *p,p′*-DDE [38]. Metabolite of DDT (*p,p′*-DDE) at a dose of 30 μM for 24 h exposure could induce apoptosis of Sertoli cells through a FasL-dependent pathway including nuclear translocation of NF-κB, increase of the FasL mRNA, and protein expression, which could be blocked by an antioxidant agent *N*-acetyl-L-cysteine. In addition, caspases 3 and 8 were activated by *p,p′*-DDE treatment in these cells [39]. Ichimura et al. demonstrated the expression and localization of FasL, Fas, and caspase 3 proteins in mouse testis 12 h after the exposure to 4–0.004 mg/g of di(2-ethylhexyl) phthalate (DEHP) and correlated the expression of these proteins with TUNEL labeling of the DNA-fragmented nucleus [40]. Immunocytochemical examination of the DEHP-exposed (4 mg/g) mouse revealed a distribution of FasL in Sertoli cell and Fas in nearby spermatocyte and Fas and caspase 3 in the same spermatocyte. Exposure to mono(2-ethylhexyl) phthalate (MEHP), a Sertoli cell-specific toxicant, induced massive germ cell apoptosis associated with increased expression of both the Fas and FasL genes in rat testis.

Mono(2-ethylhexyl) phthalate (MEHP), a well-known Sertoli cell toxicant, could decrease the levels of procaspase 8 and increase the levels of procaspase 8 cleavage products in mice testis [41]. β-Benzene hexachloride (β-BHC), a major metabolite of benzene hexachloride induced apoptosis by activation of JNKs, translocation of NF-κB, expression of FasL, and further activation of caspase cascade [42].

Spontaneous germ cell apoptosis has been observed in several species of mammalian testis. Vaithinathan et al. demonstrated a significant increase in the levels of cytosolic cytochrome c and procaspase 9 as early as 6 h following exposure to a single dose of methoxychlor at 50 mg/kg body weight. Time-dependent elevations in the levels of Fas, FasL, and pro- and cleaved caspase 3 demonstrate induction of testicular apoptosis in adult rats following a single dose of methoxychlor [43]. Recent findings reveal that methoxychlor exposure to pregnant female rats from embryonic day 8–15 at the dosage of 100 and 200 mg/kg/day showed an increase in spermatogenic cell apoptosis and decreased sperm number and motility in adult animals of F1 and F2 generation [44, 45]. In another study, oral administration of bisphenol A 480 and 960 mg/kg/day induces apoptosis of Leydig and germ cells in the mouse testis through the Fas-signaling pathway [46].

Lindane, a well-known endocrine disruptor could induce apoptosis of testicular cells by stimulating the mitochondrion-dependent pathway by elevating the levels of cytochrome c with a parallel increase in procaspase 9. A time-dependent elevation of Fas, FasL, and caspase 3 in peritubular germ cells illustrates induction of testicular apoptosis in adult rats following exposure to a single dose of lindane [47].

Oxidative Stress

Control of apoptosis may involve various pathways such as the mitochondria-mediated, Bcl-2 family, Fas/FasL system that can be engaged by oxidative stress. Reactive oxygen species (ROS) is considered a potential signal for apoptosis. Elevated levels of ROS can cause oxidation of the mitochondrial pores thereby disrupting the mitochondrial membrane potential and releasing cytochrome c thereby activating the mitochondrial mediated pathway of apoptosis. In addition, ROS have been shown to induce the expression of Fas receptor and ligand stimulating the Fas/FasL-mediated apoptotic signal transduction pathway. Several environmental disruptors are known to inappropriately activate apoptosis in testicular locale by increasing the levels of ROS [47, 48].

During the transit from undifferentiated germ cells to mature spermatozoa, the sperms are vulnerable to multitudinous threats which are counteracted by the powerful antioxidant defense system of the testis [49]. Many toxicants have been shown to damage this protective shield, thus increasing the susceptibility of this organ to oxidative stress [50, 51]. Experimental studies have demonstrated that exposure to hexachlorocyclohexane (i.p., 20 mg/kg/day) during the critical stages of testicular development induces elevation in the levels of lipid peroxidation and hydrogen peroxide (H_2O_2) along with reduction in the levels of superoxide dismutase (SOD), catalase, and ascorbic acid [52]. Doreswamy et al. have demonstrated the induction of oxidative stress, DNA damage, and apoptosis in testis following exposure to multiple doses of nickel chloride [53]. Our earlier studies on various toxicants in rodent models have exemplified the role of oxidative stress in mediating its effects on testis. Oral exposure to lindane (5 mg/kg body weight/day) for 30 days resulted in elevated levels of hydrogen peroxide and lipid peroxidation with concomitant decline in the activities of antioxidant and steroidogenic enzymes in testis [54]. Similar impairment of antioxidant system and mitochondria-dependent apoptosis of rat testis have been observed with lindane following single dose of lindane [47, 55]. Methoxychlor, at dose levels of 50 mg/kg body weight, caused significant diminution in testicular antioxidant enzymes along with Fas–FasL and mitochondria-mediated apoptosis in a time-dependent manner [43]. Compilation of these studies indicates the generation of free radicals and associated oxidative stress as the pathological mechanism underpinning the adverse effects of testicular toxicants. Innumerable studies have disclosed the involvement of oxidative stress in carrying out the malicious role of apoptosis in testis.

Most of the toxicants have been reported to perturb the testicular locale either directly or indirectly targeting pivotal constituents of testis—the germ cells, Sertoli cells, and Leydig cells. A study on the exposure of testis to a single dose (2 g/kg body weight) of di(2-ethylhexyl) phthalate revealed an augmented generation of ROS with simultaneous decrement in the concentrations of glutathione and ascorbic acid leading to selective apoptosis of spermatocytes [56]. Further exploration revealed the accrual of mono(2-ethylhexyl) phthalate, a toxic metabolite of DEHP, in testis causing mitochondrial respiratory damage and release of cytochrome c inciting apoptosis [56]. Exposure of spermatogenic cells to synthetic organic chemical, methyl *tert*-butyl ether (MTBE), enervated cell viability and induced generation of ROS and enhanced lipid peroxidation [57]. Similar damaging effects of oxidative stress followed by apoptosis in maturing germ cells have been observed with multifarious array of toxicants including metals.

The uninterrupted close association of germ cells with Sertoli cells is yet another obligatory factor in spermatogenesis. Apart from its fostering role, Sertoli cells play a highly remarkable phagocytic role in eliminating spermatogenic cells undergoing apoptosis in response to chemical insult [58]. Consequently, any agent that confronts Sertoli cells may have a profound effect on spermatogenesis. In vitro exposure to β-BHC can enhance ROS and oxidative stress and then induce activation of JNKs and NF-κB, expression of FasL in rat Sertoli cells. Upon ligation of FasL to Fas, an FasL-mediated apoptotic death is stimulated in a target cell leading to the activation of caspase 8. Finally, apoptosis of Sertoli cells is mediated by executioner caspase 3, thereby disturbing the spermatogenic process [42]. Sertoli cells on exposure to an environmental contaminant, nonylphenol (10–40 μM), caused accumulation of ROS within 2 h of exposure which subsequently resulted in the loss of mitochondrial membrane potential and enhanced lipid peroxidation at 12 h posttreatment [59]. In vitro studies on the effects of 4-*tert*-octylphenol, a degradation product of alkylphenol polyethoxylate, at a concentration of 30–60 μM for 6–24 h decreased the viability of Sertoli cells and increased apoptosis via caspase 3 pathway in a concentration- and time-dependent manner [60]. Diverse studies have cumulated over time which accentuates the feasible role of oxidative stress in Sertoli cell apoptosis in response to toxicants.

Mechanisms Involved in Inducing Apoptosis

The possible mechanisms involved in the action of various factors in mediating testicular apoptosis are summed up. Most of the studies involving toxicants illustrate the undoubted role of ROS in executing its detrimental effects [29, 54, 59, 61]. These elevated levels of ROS can cause oxidation of the mitochondrial pores thereby disrupting the mitochondrial membrane potential and releasing cytochrome c [56, 62]. Once free of the mitochondrial membrane, cytochrome c rapidly assembles a multi-protein complex involving Apaf-1 and procaspase 9 leading to the activation of the caspase 9, which subsequently triggers the effector caspase 3, 6, and/or 7 [24, 25]. These caspases, in turn, activate endonucleases and proteases resulting in DNA fragmentation and degradation of nuclear and cytoskeletal proteins [63, 64]. Apart from ROS, Bax, a member of proapoptotic Bcl-2 family, can directly influence the release of cytochrome c from mitochondria [65]. It is interesting to note that the Bcl-2 protein family is itself regulated by ROS [66]; however, whether this regulation has any role in toxicant-mediated apoptosis is not known.

ROS have been shown to induce the expression of Fas receptor and ligand stimulating the Fas/FasL-mediated apoptotic signal transduction pathway [67]. Interaction of Fas with FasL leads to a cascade of events which begins with the proteolytic cleavage of procaspase 8 to its active form, which consequently activates downstream effector caspase 3, 6, or 7 [68–70]. These caspases execute the cells by degrading the constituent proteins [71]. Elimination of apoptotic action of Fas by antioxidants further emphasizes the role of ROS in Fas-mediated death process [72, 73]. Therefore, deprivation of antioxidants and/or generation of free radicals by toxicants is capable of reducing the Fas pathway. In addition, they impair steroidogenesis and may deprive germ cell of the essential growth factor, testosterone, and increase their susceptibility to ROS attack [74].

Key Issues

The purpose of this chapter was to evaluate the role of various factors influencing apoptosis in male infertility. Mounting evidence suggests that apoptosis occurs as the predominant cell death mechanism in testis in response to several diseases and toxic injuries. Research implies that reactive oxygen species and other secondary free radicals such as nitric oxide and hydroperoxides could be inducers or mediators of apoptosis in testis through downregulation of antioxidant defense system or increased expression of apoptosis-related proteins. However, the exact mechanism of action of apoptosis in inducing male infertility remains a mystery. Further studies are warranted to evaluate the adverse effects of apoptosis on testis.

Acknowledgments P. P. Mathur acknowledges the receipt of financial support from the Department of Science and Technology, Government of India, under the projects (1) SP/SO/B-65/99, (2) DST-FIST-2009, and (3) Indian Council of Medical Research, New Delhi. Shereen Cynthia D'Cruz acknowledges the Indian Council of Medical Research, New Delhi, India, for Senior Research Fellowship. The authors also thank the staff of Bioinformatics Center, Pondicherry University, Pondicherry, for providing various facilities.

References

1. Nene UA, Coyaji K, Apte H. Infertility: a label of choice in the case of sexually dysfunctional couples. Patient Educ Couns. 2005;59(3):234–8.
2. Sharlip ID, Jarow JP, Belker AM, et al. Best practice policies for male infertility. Fertil Steril. 2002;77(5):873–82.
3. Larsen U. Sterility in sub-Saharan Africa. Popul Stud. 1994;48:459–74.
4. Larsen U. Infertility in Central Africa. Trop Med Int Health. 2003;8:354–67.
5. Iammarrone E, Balet R, Lower AM, Gillott C, Grudzinskas JG. Male infertility. Best Pract Res Clin Obstet Gynaecol. 2003;17(2): 211–29.
6. Saez JM, Avallet O, Lejeune H, Chatelain PG. Cell-cell communication in the testis. Horm Res. 1991;36(3–4):104–15.
7. McLachlan RI, Wreford NG, Meachem SJ, De Kretser DM, Robertson DM. Effects of testosterone on spermatogenic cell populations in the adult rat. Biol Reprod. 1994;51(5):945–55.
8. Zapata JM, Pawlowski K, Haas E, et al. A diverse family of proteins containing tumor necrosis factor receptor-associated factor domains. J Biol Chem. 2001;276(26):24242–52.

9. Scorrano L, Korsmeyer SJ. Mechanisms of cytochrome c release by proapoptotic BCL-2 family members. Biochem Biophys Res Commun. 2003;304(3):437–44.
10. Reed JC. Bcl-2 family proteins: regulators of apoptosis and chemoresistance in hematologic malignancies. Semin Hematol. 1997;34(4 Suppl 5):9–19.
11. Wajant H. The Fas signaling pathway: more than a paradigm. Science. 2002;296(5573):1635–6.
12. Watanabe-Fukunaga R, Brannan CI, Copeland NG, Jenkins NA, Nagata S. Lymphoproliferation disorder in mice explained by defects in Fas antigen that mediates apoptosis. Nature. 1992;356(6367):314–7.
13. Nagata S, Golstein P. The Fas death factor. Science. 1995;267(5203): 1449–56.
14. Suda T, Takahashi T, Golstein P, Nagata S. Molecular cloning and expression of the Fas ligand, a novel member of the tumor necrosis factor family. Cell. 1993;75(6):1169–78.
15. Tanaka M, Itai T, Adachi M, Nagata S. Downregulation of Fas ligand by shedding. Nat Med. 1998;4(1):31–6.
16. Sinha Hikim AP, Lue Y, Diaz-Romero M, et al. Deciphering the pathways of germ cell apoptosis in the testis. J Steroid Biochem Mol Biol. 2003;85(2-5):175–82.
17. Philchenkov AA. Caspases as regulators of apoptosis and other cell functions. Biochemistry (Mosc). 2003;68(4):365–76.
18. Pentikainen V, Erkkila K, Dunkel L. Fas regulates germ cell apoptosis in the human testis in vitro. Am J Physiol. 1999;276(2 Pt 1): E310–6.
19. Francavilla S, D'Abrizio P, Rucci N, et al. Fas and Fas ligand expression in fetal and adult human testis with normal or deranged spermatogenesis. J Clin Endocrinol Metab. 2000;85(8):2692–700.
20. Thornberry NA, Lazebnik Y. Caspases: enemies within. Science. 1998;281(5381):1312–6.
21. Nicholson DW. Caspase structure, proteolytic substrates, and function during apoptotic cell death. Cell Death Differ. 1999;6(11):1028–42.
22. Cohen GM. Caspases: the executioners of apoptosis. Biochem J. 1997;326(Pt 1):1–16.
23. Earnshaw WC, Martins LM, Kaufmann SH. Mammalian caspases: structure, activation, substrates, and functions during apoptosis. Annu Rev Biochem. 1999;68:383–424.
24. Hengartner MO. The biochemistry of apoptosis. Nature. 2000;407(6805):770–6.
25. Budihardjo I, Oliver H, Lutter M, Luo X, Wang X. Biochemical pathways of caspase activation during apoptosis. Annu Rev Cell Dev Biol. 1999;15:269–90.
26. Delfino F, Walker WH. Stage-specific nuclear expression of NF-kappaB in mammalian testis. Mol Endocrinol. 1998;12(11): 1696–707.
27. Karin M. How NF-kappaB is activated: the role of the IkappaB kinase (IKK) complex. Oncogene. 1999;18(49):6867–74.
28. Huleihel M, Lunenfeld E. Regulation of spermatogenesis by paracrine/autocrine testicular factors. Asian J Androl. 2004;6(3): 259–68.
29. Murugesan P, Balaganesh M, Balasubramanian K, Arunakaran J. Effects of polychlorinated biphenyl (Aroclor 1254) on steroidogenesis and antioxidant system in cultured adult rat Leydig cells. J Endocrinol. 2007;192(2):325.
30. Yang JM, Arnush M, Chen QY, et al. Cadmium-induced damage to primary cultures of rat Leydig cells. Reprod Toxicol. 2003; 17(5):553–60.
31. Gao HB, Tong MH, Hu YQ, et al. Mechanisms of glucocorticoid-induced Leydig cell apoptosis. Mol Cell Endocrinol. 2003; 199(1–2):153–63.
32. Print CG, Loveland KL. Germ cell suicide: new insights into apoptosis during spermatogenesis. Bioessays. 2000;22(5):423.
33. Allan DJ, Harmon BV, Roberts SA. Spermatogonial apoptosis has three morphologically recognizable phases and shows no circadian rhythm during normal spermatogenesis in the rat. Cell Prolif. 1992;25(3):241–50.
34. Lin WW, Lamb DJ, Wheeler TM, Lipshultz LI, Kim ED. In situ end-labeling of human testicular tissue demonstrates increased apoptosis in conditions of abnormal spermatogenesis. Fertil Steril. 1997;68(6):1065–9.
35. Blanchard TL, Johnson L. Increased germ cell degeneration and reduced germ cell:Sertoli cell ratio in stallions with low sperm production. Theriogenology. 1997;47(3):665–77.
36. Tebourbi O, Rhouma KB, Sakly M. DDT induces apoptosis in rat thymocytes. Bull Environ Contam Toxicol. 1998;61(2):216–23.
37. Perez-Maldonado IN, Diaz-Barriga F, de la Fuente H, et al. DDT induces apoptosis in human mononuclear cells in vitro and is associated with increased apoptosis in exposed children. Environ Res. 2004;94(1):38–46.
38. Song Y, Liang X, Hu Y, et al. p, p′-DDE induces mitochondria-mediated apoptosis of cultured rat Sertoli cells. Toxicology. 2008;253(1–3):53–61.
39. Shi Y, Song Y, Wang Y, et al. p, p′-DDE induces apoptosis of rat Sertoli cells via a FasL-dependent pathway. J Biomed Biotechnol. 2009;2009:181282.
40. Ichimura T, Kawamura M, Mitani A. Co-localized expression of FasL, Fas, Caspase-3 and apoptotic DNA fragmentation in mouse testis after oral exposure to di(2-ethylhexyl)phthalate. Toxicology. 2003;194(1–2):35–42.
41. Giammona CJ, Sawhney P, Chandrasekaran Y, Richburg JH. Death receptor response in rodent testis after mono-(2-ethylhexyl) phthalate exposure. Toxicol Appl Pharmacol. 2002;185(2):119–27.
42. Shi Y, Song Y, Wang Y, et al. Beta-benzene hexachloride induces apoptosis of rat sertoli cells through generation of reactive oxygen species and activation of JNKs and FasL. Environ Toxicol. 2011;26(2):124–35. doi:10.1002/tox.20536.
43. Vaithinathan S, Saradha B, Mathur PP. Methoxychlor induces apoptosis via mitochondria- and FasL-mediated pathways in adult rat testis. Chem Biol Interact. 2010;185(2):110–8.
44. Anway MD, Cupp AS, Uzumcu M, Skinner MK. Epigenetic transgenerational actions of endocrine disruptors and male fertility. Science. 2005;308(5727):1466–9.
45. Cupp AS, Uzumcu M, Suzuki H, et al. Effect of transient embryonic in vivo exposure to the endocrine disruptor methoxychlor on embryonic and postnatal testis development. J Androl. 2003;24(5): 736–45.
46. Li YJ, Song TB, Cai YY, et al. Bisphenol A exposure induces apoptosis and upregulation of Fas/FasL and caspase-3 expression in the testes of mice. Toxicol Sci. 2009;108(2):427–36.
47. Saradha B, Vaithinathan S, Mathur PP. Lindane induces testicular apoptosis in adult Wistar rats through the involvement of Fas-FasL and mitochondria-dependent pathways. Toxicology. 2009;255(3): 131–9.
48. Lee J, Richburg JH, Shipp EB, Meistrich ML, Boekelheide K. The Fas system, a regulator of testicular germ cell apoptosis, is differentially up-regulated in Sertoli cell versus germ cell injury of the testis. Endocrinology. 1999;140(2):852–8.
49. Peltola V, Huhtaniemi I, Ahotupa M. Antioxidant enzyme activity in the maturing rat testis. J Androl. 1992;13(5):450–5.
50. Abdollahi M, Ranjbar A, Shadnia S, Nikfar S, Rezaie A. Pesticides and oxidative stress: a review. Med Sci Monit. 2004;10(6):RA141–7.
51. Saradha B, Mathur PP. Effect of environmental contaminants on male reproduction. Environ Toxicol Pharmacol. 2006;21(1):34.
52. Samanta L, Roy A, Chainy GB. Changes in rat testicular antioxidant defence profile as a function of age and its impairment by hexachlorocyclohexane during critical stages of maturation. Andrologia. 1999;31(2):83–90.
53. Doreswamy K, Shrilatha B, Rajeshkumar T, Muralidhara. Nickel-induced oxidative stress in testis of mice: evidence of DNA damage and genotoxic effects. J Androl. 2004;25(6):996–1003.

54. Sujatha R, Chitra KC, Latchoumycandane C, Mathur PP. Effect of lindane on testicular antioxidant system and steroidogenic enzymes in adult rats. Asian J Androl. 2001;3(2):135–8.
55. Saradha B, Vaithinathan S, Mathur PP. Lindane alters the levels of HSP70 and clusterin in adult rat testis. Toxicology. 2008;243(1–2): 116–23.
56. Kasahara E, Sato EF, Miyoshi M, et al. Role of oxidative stress in germ cell apoptosis induced by di(2-ethylhexyl)phthalate. Biochem J. 2002;365(Pt 3):849–56.
57. Li D, Yin D, Han X. Methyl tert-butyl ether (MTBE)-induced cytotoxicity and oxidative stress in isolated rat spermatogenic cells. J Appl Toxicol. 2007;27(1):10–7.
58. Tay TW, Andriana BB, Ishii M, et al. Phagocytosis plays an important role in clearing dead cells caused by mono(2-ethylhexyl) phthalate administration. Tissue Cell. 2007;39(4):241–6.
59. Gong Y, Han XD. Nonylphenol-induced oxidative stress and cytotoxicity in testicular Sertoli cells. Reprod Toxicol. 2006;22(4): 623–30.
60. Qian J, Bian Q, Cui L, et al. Octylphenol induces apoptosis in cultured rat Sertoli cells. Toxicol Lett. 2006;166(2):178–86.
61. Sen Gupta R, Sen Gupta E, Dhakal BK, Thakur AR, Ahnn J. Vitamin C and vitamin E protect the rat testes from cadmium-induced reactive oxygen species. Mol Cells. 2004;17(1):132–9.
62. Zamzami N, Marchetti P, Castedo M, et al. Sequential reduction of mitochondrial transmembrane potential and generation of reactive oxygen species in early programmed cell death. J Exp Med. 1995;182(2):367–77.
63. Bortner CD, Oldenburg NB, Cidlowski JA. The role of DNA fragmentation in apoptosis. Trends Cell Biol. 1995;5(1):21–6.
64. Kohler C, Hakansson A, Svanborg C, Orrenius S, Zhivotovsky B. Protease activation in apoptosis induced by MAL. Exp Cell Res. 1999;249(2):260–8.
65. Green DR, Reed JC. Mitochondria and apoptosis. Science. 1998;281(5381):1309–12.
66. Tripathi P, Hildeman D. Sensitization of T cells to apoptosis—a role for ROS? Apoptosis. 2004;9(5):515–23.
67. Krammer PH. CD95(APO-1/Fas)-mediated apoptosis: live and let die. Adv Immunol. 1999;71:163–210.
68. Scaffidi C, Fulda S, Srinivasan A, et al. Two CD95 (APO-1/Fas) signaling pathways. EMBO J. 1998;17(6):1675–87.
69. Sanchez-Gomez MV, Alberdi E, Ibarretxe G, Torre I, Matute C. Caspase-dependent and caspase-independent oligodendrocyte death mediated by AMPA and kainate receptors. J Neurosci. 2003;23(29):9519–28.
70. Henkler F, Behrle E, Dennehy KM, et al. The extracellular domains of FasL and Fas are sufficient for the formation of supramolecular FasL-Fas clusters of high stability. J Cell Biol. 2005;168(7):1087–98.
71. Riedl SJ, Shi Y. Molecular mechanisms of caspase regulation during apoptosis. Nat Rev Mol Cell Biol. 2004;5(11):897–907.
72. Um HD, Orenstein JM, Wahl SM. Fas mediates apoptosis in human monocytes by a reactive oxygen intermediate dependent pathway. J Immunol. 1996;156(9):3469.
73. Chiba T, Takahashi S, Sato N, Ishii S, Kikuchi K. Fas-mediated apoptosis is modulated by intracellular glutathione in human T cells. Eur J Immunol. 1996;26(5):1164.
74. Chainy GB, Samantaray S, Samanta L. Testosterone-induced changes in testicular antioxidant system. Andrologia. 1997;29(6): 343–9.

Impact of Spinal Cord Injury

32

Viacheslav Iremashvili, Nancy L. Brackett, and Charles M. Lynne

Abstract

Although medical advances have greatly improved the prognosis for people who sustain spinal cord injury, it remains a major social and health-care problem. There are estimated 10,000–12,000 spinal cord injuries every year in the USA alone. More than a quarter of a million Americans are currently living with spinal cord injury, with many millions more worldwide. The cost of managing the care of patients with spinal cord injury is approximately $4 billion per year. The majority of spinal cord injury victims are young adults. Of them, more than 80% are men. As a result, young males constitute the largest part of this patient population. Reproductive function is essential for men with spinal cord injury, but unfortunately, less than 10% of them can father children without medical assistance. Infertility in male patients with spinal cord injury results from a combination of erectile dysfunction, ejaculatory dysfunction, and poor semen quality. As a result of advancements in assisted ejaculation techniques including electroejaculation and high-amplitude penile vibratory stimulation, semen can be safely obtained from nearly all men with spinal cord injury without resorting to surgical procedures; however, semen quality is poor in the majority of cases.

Keywords

Male infertility • Spinal cord injury • Semen abnormalities • Scrotal hyperthermia • Elevated reactive oxygen activity • Assisted ejaculation techniques • Penile vibratory stimulation

Although medical advances have greatly improved the prognosis for people who sustain spinal cord injury, it remains a major social and health-care problem. There are estimated 10,000–12,000 spinal cord injuries every year in the USA alone. More than a quarter of a million Americans are currently living with spinal cord injury [1], with many millions more worldwide. The cost of managing the care of patients with spinal cord injury is approximately $4 billion per year. Car accidents are the most common cause of spinal cord injury followed by violent encounters, sporting and work-related accidents, and falls [1].

The majority of spinal cord injury victims are young adults. Of them, more than 80% are men. As a result, young males constitute the largest part of this patient population. Reproductive function is essential for men with spinal cord injury, but unfortunately, less than 10% of them can father children without medical assistance [2]. Infertility in male patients with spinal cord injury results from a combination of erectile dysfunction, ejaculatory dysfunction, and poor

V. Iremashvili, MD, PhD.
Department of Urology, University of Miami Miller School of Medicine, P.O. Box 016960 (M-14), Miami, Florida 33101, USA
e-mail: iremashvili@hotmail.com

N.L. Brackett, PhD, HCLD (✉)
The Miami Project to Cure Paralysis, University of Miami Miller School of Medicine, Lois Pope Life Center,
Room 2-17, 1095 NW 14th Terrace, Miami, Florida 33136, USA
e-mail: nbrackett@miami.edu

C.M. Lynne, MD
Department of Urology, University of Miami Miller School of Medicine, Miami, FL, USA
e-mail: clynne@miami.edu

S.J. Parekattil and A. Agarwal (eds.), *Male Infertility: Contemporary Clinical Approaches, Andrology, ART & Antioxidants*,
DOI 10.1007/978-1-4614-3335-4_32, © Springer Science+Business Media, LLC 2012

semen quality [3]. As a result of advancements in assisted ejaculation techniques including electroejaculation and high-amplitude penile vibratory stimulation, semen can be safely obtained from nearly all men with spinal cord injury without resorting to surgical procedures [4]; however, semen quality is poor in the majority of cases [5].

Semen Abnormalities in Men with Spinal Cord Injury

The origin and/or cause of low sperm quality in men with spinal cord injury has not been clearly defined. Several possible etiologies have been postulated, including hormonal dysfunction [6], elevated scrotal temperature [7], methods of bladder management [8], and alterations in sperm transport and storage due to reproductive tract stasis [9, 10], but none of these causes has been conclusively proven.

Role of Hormonal Alterations

Alterations in the hypothalamic-pituitary-gonadal axis may result in the disruption of spermatogenesis. The endocrine status of men with spinal cord injury has been examined in several studies that provided contradictory results (Table 32.1). Some studies in humans [6, 11–14] and animals [15, 16] have reported different hormone abnormalities associated with spinal cord injury, but no consistent correlation with the semen quality was shown. It was also not clear if these abnormalities were primary or secondary. We studied this problem in a group of 66 men with spinal cord injury and found no association between semen quality and serum levels of luteinizing hormone, follicle-stimulating hormone, testosterone, or prolactin [17]. The only exception was in a subgroup of subjects who had elevated levels of follicle-stimulating hormone. In each case, the patient was azoospermic, even patients with only small elevations of follicle-stimulating hormone. Hormonal alterations are unlikely to be a major contributor to poor semen quality in men with spinal cord injury [18–22].

Role of Scrotal Temperature

Elevated scrotal/testicular temperature was one of the first hypotheses to explain the origin of semen abnormalities in men with spinal cord injury. It is common knowledge that spermatogenesis is temperature-sensitive and proceeds optimally at 35°C. Higher scrotal temperatures could have detrimental effects on sperm production [23]. It was assumed that men with spinal cord injury could have scrotal hyperthermia as a result of generalized scrotal thermoregulatory dysfunction or because of sitting in a wheelchair for prolonged periods [24]. Some studies showed that men with spinal cord injury sitting in wheelchairs had higher scrotal temperatures compared to able-bodied men, sitting in armchairs [7, 25]. Brindley reported an inverse correlation between scrotal temperatures and motile sperm counts in men with spinal cord injury [25]. However, we did not find any difference between control subjects and spinal cord-injured subjects in oral temperature, scrotal temperature, or the difference between these two parameters [26]. Furthermore, men with spinal cord injury who were ambulatory (i.e., not in wheelchairs) still had impaired semen quality [26], indicating that some aspect of spinal cord injury, other than the simple act of sitting in a wheelchair, contributes to abnormal semen quality in these men. Supporting this idea is the fact that no study has found improvement in semen quality by cooling the scrotum of men with spinal cord injury.

Studies of scrotal temperature in noninjured men have suggested that short-term versus long-term exposure to elevated temperature causes reversible versus irreversible changes in the seminiferous tubules [27, 28]. In men with

Table 32.1 Principal results of the studies of endocrine status of men with spinal cord injuries

	Luteinizing hormone	Follicle-stimulating hormone	Testosterone	Prolactin
No difference from control	Huang et al. [19]	Naftchi et al. [21]	Huang et al. [19]	Huang et al. [19]
	Tsitouras et al. [20]	Tsitouras et al. [20]	Naftchi et al. [21]	Brackett et al. [17]
		Huang et al. [22]	Brackett et al. [17]	Naderi et al. [14]
Lower than control	Naftchi et al. [21]	Brackett et al. [17]	Tsitouras et al. [20]	
	Brackett et al. [17]	Safarinejad et al. [23]	Safarinejad et al. [23]	
	Safarinejad et al. [23]	Naderi et al. [14]	Naderi et al. [14]	
	Naderi et al. [14]	Kostovski et al. [12]	Kostovski et al. [12]	
	Kostovski et al. [12]			
Higher than control		Huang et al. [19]		
Outside reference range		Huang et al.—high, 18.7% [19]	Tsitouras et al. low, 45% [20]	Huang et al.—high, 25% [19]

Endocrine profiles of spinal cord injury men were not shown to follow any specific pattern and/or to be significantly related to impairments in semen quality spinal cord injury, however, both cross-sectional [29] and longitudinal [30] studies have shown that semen parameters were not significantly related to the duration of the postinjury period, suggesting a stable (and null) pattern for the measures across time. In light of these facts, it appears that no strong evidence exists to support the role of elevated scrotal temperature as a leading etiologic factor of the semen abnormalities in men with spinal cord injury.

Role of Bladder Management

No bladder management regime has been associated with normal semen quality in men with spinal cord injury. However, some studies have shown that the use of intermittent catheterization is associated with better sperm motility than the use of indwelling urethral catheters, suprapubic catheters, or spontaneous voiding [31, 32]. Although semen quality is improved with intermittent catheterization versus the other methods mentioned, it does not become normalized. Bladder management, then, does not seem to be a significant cause of impaired semen quality in men with spinal cord injury.

Role of Ejaculation Frequency

The majority of men with spinal cord injury cannot ejaculate without medical assistance. It has been hypothesized that long periods between ejaculations may result in reproductive tract stasis which can negatively affect sperm. However, most studies investigating the effect of repeated ejaculation on semen quality in men with spinal cord injury found no improvement in semen parameters [9, 33–37]. Only one group reported a moderate increase in sperm motility and sperm morphology after 3 months of weekly ejaculations with penile vibratory stimulation [38]. These findings indicate that frequency of ejaculation is not the sole factor causing abnormal semen quality in men with spinal cord injury.

Interesting data were presented by Ohl et al., suggesting that spinal cord injury could result in fundamental changes in sperm transport and storage [10]. In eight patients with spinal cord injury, bilateral seminal vesicle aspiration was performed immediately before electroejaculation or penile vibratory stimulation. The seminal vesicle aspirates contained large numbers of poor quality sperm. It should be noted that normal men do not have large numbers of sperm in seminal vesicles. Duration of abstinence did not correlate with the number of seminal vesicle sperm. Furthermore, the semen parameters in samples obtained immediately after seminal vesicle aspiration were significantly better compared to historical ejaculated parameters [10]. The authors concluded that altered transport with stagnation of sperm in the seminal vesicles could be a primary source of semen with poor quality in men with spinal cord injury. It was also shown that factors within the seminal plasma contribute to poor semen abnormalities in men with spinal cord injury [39]. This issue will be discussed in more detail later in this chapter.

Studies of Oxidative Stress in Men with Spinal Cord Injury

In addition to the aforementioned putative causes that have been investigated, there is increasing evidence that oxidative stress is an important mechanism contributing to sperm damage in this group of patients. The generation of reactive oxygen species and their relation to semen quality in men with spinal cord injury has been investigated in several studies.

Electric Current and Reactive Oxygen Species

The first study to demonstrate elevated reactive oxygen activity in the semen of men with spinal cord injury was published by Rajasekaran et al. [40]. At the time of this study, electroejaculation was the most common method of obtaining semen from men with spinal cord injury. Electroejaculation is performed by passing electricity into the pelvic region via a probe inserted into the rectum. The authors hypothesized that this electric current induced sperm damage in men with spinal cord injury. The study consisted of two experiments. During the first experiment, sperm from healthy men was incubated with a normal and an electrolyzed medium. Reactive oxygen species levels and sperm motility were measured in each group of samples. The second experiment included measurements of reactive oxygen species generation in semen from men with spinal cord injury and normal controls. In subjects with spinal cord injury, semen samples were obtained by electroejaculation.

The results of this study showed that when sperm from control subjects were incubated with electrolyzed physiologic medium, a significant and time-dependent decrease occurred in sperm motility and sperm viability, and these decreases were associated with an increased generation of reactive oxygen species. The second experiment demonstrated a significant increase in reactive oxygen species generation in semen from subjects with spinal cord injury compared to semen from control subjects. In the latter group, ejaculates were collected by masturbation. These findings led authors to conclude that, in patients with spinal cord injury, the electric current was a likely cause of poor quality of semen obtained by electroejaculation, and this effect was mediated by increased the production of reactive oxygen species [40].

Reactive Oxygen Species in Whole Semen Versus Washed Sperm

The purpose of a study performed by de Lamirande et al. was to determine whether whole semen samples versus washed spermatozoa obtained from men with spinal cord injury produced excessive amounts of reactive oxygen species [41]. This study included three groups of men: healthy volunteers ($n=20$), infertile able-bodied men ($n=166$), and subjects with spinal cord injury ($n=21$). In the latter group, semen was obtained by masturbation after butylbromide and physostigmine injections in 19 patients and by electroejaculation in the remaining two men. Formation of reactive oxygen species was measured in neat semen and in Percoll-washed spermatozoa of all subjects.

The presence of reactive oxygen species in whole semen was detected in 97% of subjects with spinal cord injury compared to 40% and 15% in infertile able-bodied men and volunteers, respectively. Compared to a threshold value of 10 mV/s/10^9, reactive oxygen species production was elevated in 81% of patients with spinal cord injury, 25% of infertile able-bodied men, and 10% of healthy controls. In healthy controls and in infertile able-bodied subjects, the levels of reactive oxygen species measured in semen were, respectively, 40 and 14 times lower than those detected in semen from subjects with spinal cord injury. No correlation was found between reactive oxygen species production and level or duration of injury.

After centrifugation on Percoll gradients, sperm from men with spinal cord injury continued to generate large amounts of reactive oxygen species. High reactive oxygen species production by Percoll-washed spermatozoa was found in 75% of men with spinal cord injury, 20% of infertile men, and 5% of healthy controls. The mean reactive oxygen species levels in washed spermatozoa from men with spinal cord injury were sixfold higher than that of infertile patients, and 140-fold higher than that of normal volunteers. There was a significant inverse relationship between levels of reactive oxygen species and percentage of motile sperm in Percoll-washed specimens from patients with spinal cord injury.

Results of this study showed that semen samples and Percoll-washed sperm samples from subjects with spinal cord injury produced reactive oxygen species at a higher frequency and at higher levels than equivalent samples from normal men or infertile men. In men with spinal cord injury, levels of reactive oxygen species correlated negatively with sperm motility [41]. These data suggest that the role of reactive oxygen species as a mechanism of sperm damage leading to infertility could be more important in men with spinal cord injury compared to the general population or to the infertile population.

Reactive Oxygen Species and Sperm Characteristics

The generation of reactive oxygen species and its relation to semen characteristics in men with spinal cord injury was investigated by our group [42]. This study included 24 men with spinal cord injury and 19 able-bodied controls. In the spinal cord-injured patients, semen was obtained by penile vibratory stimulation ($n=15$), electroejaculation ($n=8$), and masturbation ($n=1$). Measurements of reactive oxygen species formation were performed before and after stimulation with *N*-formyl-methionyl-leucyl-phenylalanine and 12-myristate 13-acetate phorbol ester. These two substances trigger the generation of reactive oxygen species by leukocytes and spermatozoa, respectively.

The study showed that mean levels of reactive oxygen species in unstimulated and stimulated specimens were significantly higher in spinal cord-injured men compared to controls. The actual values reflecting the reactive oxygen species activity in the spinal cord-injured group were from 250 to 2,000 times higher than that of the control group. The incidence of samples positive for reactive oxygen species specimens in unstimulated controls and men with spinal cord injury was 47.3% versus 100%, respectively. It was also found that the levels of reactive oxygen species in semen from men with spinal cord injury correlated negatively with sperm motility and positively with white blood cell concentrations. Interestingly, the levels of reactive oxygen species did not differ between antegrade and retrograde samples or between different methods of ejaculation (penile vibratory stimulation and electroejaculation). Therefore, the high levels of reactive oxygen species in semen specimens obtained by electroejaculation and vibratory stimulation may not be due exclusively to the effects of electrical current, as was suggested by Rajasekaran et al. [40].

As can be seen from the above studies, reactive oxygen species production is elevated in semen from patients with spinal cord injury, and increased oxidative stress may be an important mechanism of impaired sperm quality in this group of men. The next section of this chapter will discuss the potential sources of reactive oxygen species in semen from patients with spinal cord injury, the consequences of this oxidative stress elevation, and the possible applications of this information in the treatment of infertility in these patients.

Sources of Reactive Oxygen Radicals in Semen from Men with Spinal Cord Injury

Human ejaculate consists of several types of cells including mature and immature spermatozoa, germ cells from different stages of the spermatogenic process, epithelial cells, and

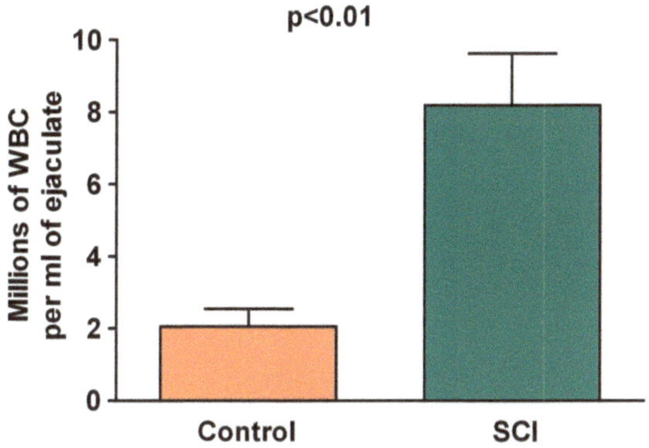

Fig. 32.1 Studies have shown that semen of men with spinal cord injury (SCI) contains higher concentrations of white blood cells (WBC) compared with semen of able-bodied, healthy control subjects

leukocytes. Of these different cell types, leukocytes and spermatozoa have been shown to be the two principal sources of production of free radicals [43].

Effect of Leukocytes

Ejaculates from men with spinal cord injury are known to have increased leukocyte counts (Fig. 32.1) [44, 45]. The main sources of leukocytes in human ejaculate are the prostate gland, seminal vesicles, and epididymis [46]. Our studies showed no evidence of chronic or acute prostate gland inflammation in leukocytospermic patients with spinal cord injury [47], and we did not find any white blood cells in the vasal aspirates from these subjects [48]. These data indicate that the seminal vesicles are the most likely origin of leukocytospermia in men with spinal cord injury.

Leukocytes can produce large amounts of reactive oxygen species. A positive correlation has been reported between seminal leukocyte counts and reactive oxygen species production [49, 50]. Of different leukocyte subtypes, peroxidase-positive cells, namely, neutrophils and macrophages, are predominant sources of reactive oxygen species production [51]. In ejaculates obtained by electroejaculation, these two leukocyte subpopulations, as identified by immunohistochemical staining, were the predominant contributors to leukocytospermia in men with spinal cord injury [52]. Lymphocytes were also found to be significant contributors to leukocytospermia in men with spinal cord injury. Immunophenotypic analysis by flow cytometry showed that the greater fraction were T cells, many of which coexpressed the human leukocyte antigen HLA-DR and CD25, suggesting they were in an activated state. No significant B-cell population was evident. [44].

The activation state of leukocytes plays a crucial role in determining reactive oxygen species output because activated white blood cells can produce up to 100-fold increases in reactive oxygen species compared with nonactivated cells [53]. This effect is mediated by an increase in reduced nicotinamide adenine dinucleotide phosphate production via the hexose monophosphate shunt [54]. The myeloperoxidase system of neutrophils and macrophages is also activated, resulting in a respiratory burst and production of large amounts of superoxide and other reactive oxygen species.

Effect of Cytokines

Elevated concentrations of the proinflammatory cytokines interleukin 1 beta, interleukin 6, and tumor necrosis factor α have been detected in the semen of men with spinal cord injury [55], reflecting activation of T-lymphocytes [44] (Fig. 32.2). Inactivation of these cytokines, by adding monoclonal antibodies or receptor blockers to semen from men with spinal cord injury, improves sperm motility [56, 57]. Interleukins are important mediators of free radical generation in many tissues, and the role of cytokines as mediators of oxidative stress is well known. Supporting this notion is the observation of a positive correlation between seminal reactive oxygen species production and seminal plasma concentrations of cytokines such as interleukin 6 [58, 59], interleukin 1, and tumor necrosis factor α [60, 61] in infertile able-bodied men. Interleukin 1 and tumor necrosis factor α were also shown to stimulate reactive oxygen species production in fertile donor semen [62]. Thus, activated seminal leukocytes have the potential to cause oxidative stress elevation in men with spinal cord injury.

Effect of Immature Sperm

Even after complete separation of sperm from leukocytes by density gradient and magnetic beads coated with leukocyte-specific CD45 antibodies, reactive oxygen species production can still be recorded, indicating the ability of spermatozoa to generate reactive oxygen species [63]. This ability inversely correlates with the sperm maturation state. During the process of spermiogenesis, surplus cytoplasm is normally extruded, and the sperm cell takes on a condensed, elongated form. If this process is defective, residual cytoplasm forms a cytoplasmic droplet in the sperm mid-region. These spermatozoa are immature and functionally and morphologically abnormal. The residual cytoplasm contains high concentration of cytosolic enzyme glucose-6-phosphate dehydrogenase, which controls the rate of glucose flux and intracellular production of b-nicotinamide adenine dinucleotide phosphate through the hexose monophosphate shunt [64].

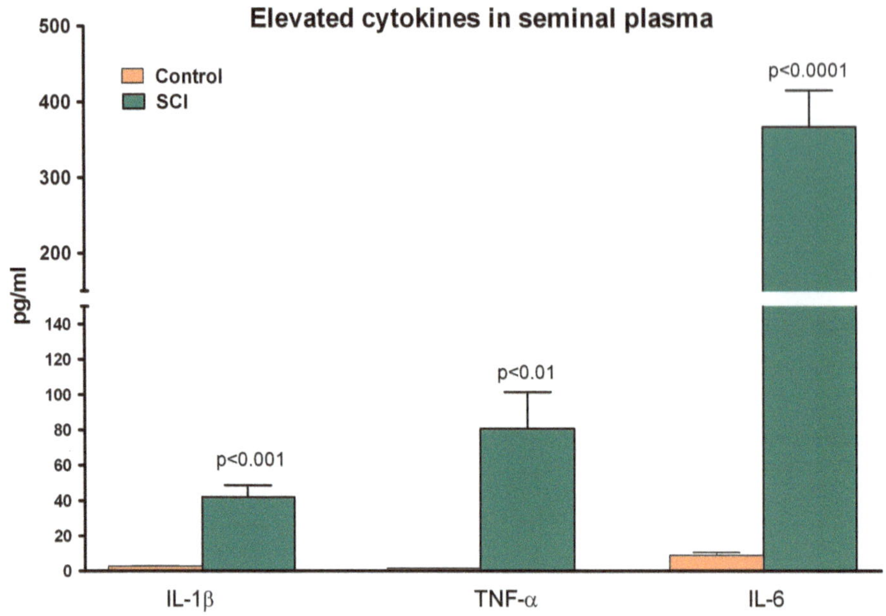

Fig. 32.2 Cytokines can be detrimental to sperm cells. Concentrations of the proinflammatory cytokines, interleukin 1β (1IL-1β), tumor necrosis factor α (TNF-α), and interleukin 6 (IL-6) were significantly elevated in semen of men with spinal cord injury (SCI) compared with semen of control subjects. pg/ml = picograms per milliliter

Nicotinamide adenine dinucleotide phosphate contributes to the reactive oxygen species production by nicotinamide adenine dinucleotide phosphate oxidase located within the sperm membrane and nicotinamide adenine dinucleotide phosphate-dependent oxidoreductase in the mitochondria [65, 66]. As a consequence, immature sperm with retained cytoplasm produce increased amounts of reactive oxygen species as compared to mature, morphologically normal sperm.

The maturation stage of sperm development, along with cytoplasmic extrusion, involves other changes, including the sperm plasma membrane remodeling. This remodeling step facilitates the formation of the sites for zona pellucida and hyaluronic acid binding [67]. Immature sperm with cytoplasmic retention are characterized by low densities of zona pellucida binding sites and also of hyaluronic acid receptors [68]. Sperm hyaluronic acid binding capacity can be tested by evaluating the percentage binding of sperm to hyaluronic acid-coated slides. After application of semen to the slide, it can be seen that mature sperm with a high density of hyaluronic acid receptors exhibits permanent binding, while the immature sperm does not bind and swims freely. We have recently investigated hyaluronic acid binding in sperm from men with spinal cord injury and compared it to that of healthy noninjured control subjects [69]. This study included 13 spinal cord-injured subjects and 13 control subjects. Hyaluronic acid binding was significantly lower in spinal cord-injured subjects compared to the control group (55.7 ± 3.8 versus 82.0 ± 2.8, $p < 0.001$) (Fig. 32.3). A hyaluronic acid binding score of 65% is considered a threshold value, as evidenced by clinical outcomes [70]. Only 3 of the 13 spinal cord-injured subjects had a hyaluronic acid binding score higher than 65%, while in the control group, 12 of 13 had a hyaluronic acid binding score higher than 65%.

These results seem to be consistent with an earlier finding showing increased incidences of sperm with cytoplasmic droplets compared to ejaculates of healthy control subjects [42]. It should be noted that the effects of reactive oxygen species produced by immature sperm (intrinsic) and leukocytes (extrinsic) may differ. For example, it has been shown that while both intrinsic and extrinsic reactive oxygen species negatively affect the integrity of deoxyribonucleic acid (DNA) (explained below), the correlation is much stronger for the intrinsic reactive oxygen species production [71]. These data suggest that reactive oxygen species produced by immature sperm cells may have a greater capacity to damage fertility potential.

Consequences of Oxidative Stress in Semen of Men with Spinal Cord Injury

While leukocytospermia and increased numbers of immature sperm are potential sources of reactive oxygen species in semen of men with spinal cord injury, several other abnormalities, which are characteristic of this group of patients, could be the result of sperm oxidative stress.

Seminal reactive oxygen species, when present in excess, damage different molecules, including lipids, proteins, nucleic acids, and sugars [72]. These toxic effects can result in decreased sperm motility, decreased sperm viability, DNA

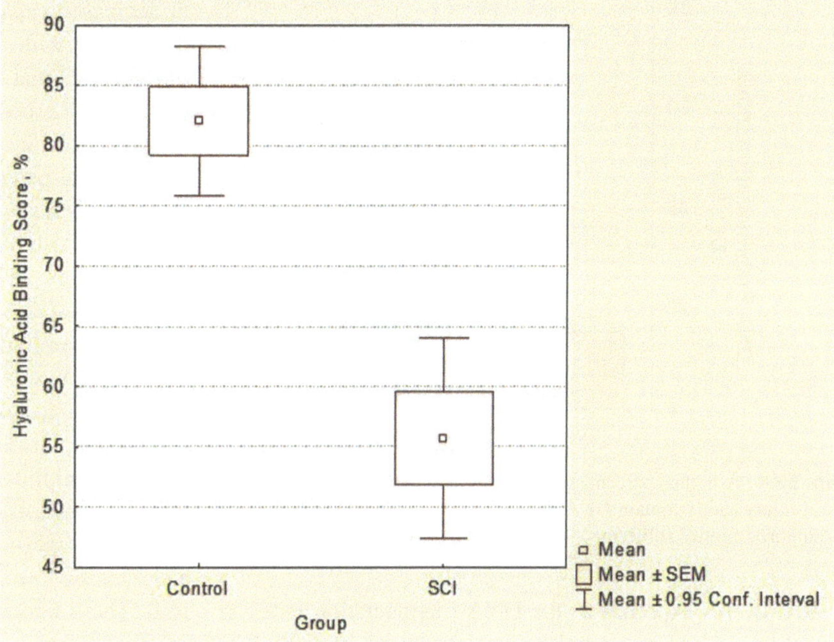

Fig. 32.3 The hyaluronic acid binding score was significantly lower in sperm from men with spinal cord injury (SCI) compared with that of healthy, noninjured control subjects. This deficit may be related to the presence of increased reactive oxygen species and immature sperm in the semen. *SEM* standard error of the mean, *Conf. Interval* confidence interval

damage, impaired sperm function, and hyperviscosity of the semen [73]. All of these defects are highly prevalent in semen of men with spinal cord injury [24].

Spermatozoa are vulnerable to the damage induced by excessive reactive oxygen species because their plasma membranes contain large quantities of polyunsaturated fatty acids. Peroxidation of these molecules could decrease membrane flexibility and therefore tail motion. This mechanism of how reactive oxygen species exert their effects on sperm motility was introduced in 1979 [74]. The levels of lipid peroxidation in spermatozoa have been shown to be directly correlated with the loss of motility [75]. Later, several other hypotheses explaining the link between increased reactive oxygen species levels and impaired sperm motility were proposed, including inhibition of enzymes important for intracellular energy production by H_2O_2 and axonemal protein phosphorylation [76, 77]. High levels of seminal reactive oxygen species can also disrupt the inner and outer mitochondrial membranes in spermatozoa [78, 79]. They cause the release of the cytochrome-C protein and activate caspases and apoptosis, resulting in necrospermia [80].

Decreased sperm motility and decreased sperm viability are characteristic features of semen from men with spinal cord injury [24]. In contrast to able-bodied men, most immotile spermatozoa in the semen from men with spinal cord injury are dead. It was shown that the dead-to-live immotile sperm ratio in spinal cord-injured subjects was more than double that in able-bodied subjects (7:3 versus 3:7) [81]. Apoptosis may play an important role in these changes. Experimental data shows that spinal cord injury in rats is associated with decreased sperm mitochondrial transmembrane potential and decreased sperm viability, suggesting excessive apoptosis [16, 82].

Increased levels of reactive oxygen species can also negatively affect the integrity of DNA in the sperm nucleus. Several forms of sperm DNA damage could be caused by reactive oxygen species, including chromatin cross-linking, chromosome deletion, DNA single- and double-stranded breaks, and base oxidation [83, 84]. With the use of flow cytometry, it has been shown that sperm from men with spinal cord injury have a high degree of abnormal chromatin condensation and reduced binding [85].

DNA fragmentation in sperm from men with spinal cord injury was also investigated by our group [86]. The study consisted of three experiments. In experiment 1, we compared the DNA fragmentation index in sperm from men with spinal cord injury to that of able-bodied controls. This experiment showed that the mean DNA fragmentation index was fourfold higher in the spinal cord injury group compared with the control group, and there was no overlap in the DNA fragmentation index between these two groups (Fig. 32.4). As was discussed earlier, chronic anejaculation is considered to be one of the possible explanations of semen abnormalities seen in men with spinal cord injury. To examine this possibility, we performed experiment 2 in which we compared the sperm DNA fragmentation index in two semen specimens obtained 3 days apart from the same spinal cord-injured subjects. No significant difference was found in the sperm DNA fragmentation between the two specimens. The purpose of experiment 3 was to determine if necrospermia, leukocytospermia, or semen processing in men with spinal cord injury contribute to their sperm DNA fragmentation

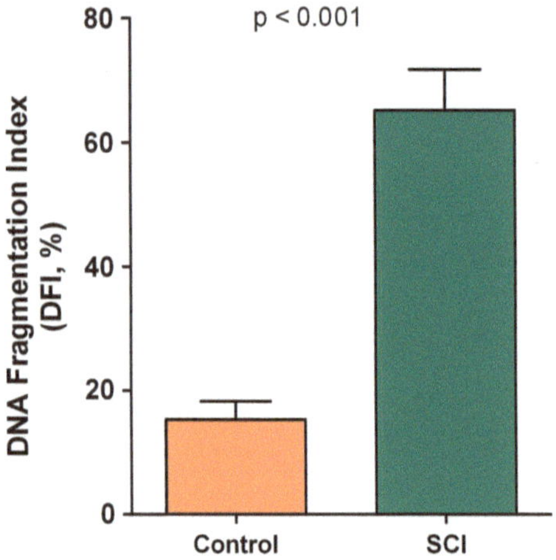

Fig. 32.4 A higher percentage of sperm cells from men with spinal cord injury (SCI) contain DNA damage compared with sperm cells of noninjured control subjects

index. In this experiment, the DNA fragmentation index in unprocessed semen samples was compared with that of semen samples processed on a gradient (i.e., free of dead sperm and leukocytes). The results of experiment 3 found no significant difference between mean DNA fragmentation index in aliquots of neat versus processed semen in spinal cord-injured subjects. Although removal of leukocytes did not result in a change in the DNA fragmentation index, it is possible that their negative effects on sperm were exerted prior to sperm processing. Thus, it appears that men with spinal cord injury have significantly greater sperm DNA damage, which may be related to high levels of oxidative stress in semen.

Complex relationships exist between apoptosis and DNA damage. Induction of apoptosis by reactive oxygen species results in a high frequency of single- and double-stranded DNA strand breaks in a process referred to karyorrhexis. Severe DNA damage can initiate the apoptosis pathway. Agarwal and Said have suggested that in the context of male infertility, there may be an interaction between seminal reactive oxygen species, sperm DNA damage, and apoptosis, and this interaction may constitute a unified pathogenic molecular mechanism [87].

Semen from men with spinal cord injury is typically highly viscous. Hyperviscous seminal plasma has been reported to be associated with elevated levels of malondialdehyde, an unsaturated carbonyl product of oxidative stress indicating excessive lipid peroxidation [88]. Hyperviscosity has also been shown to be linked to reduced seminal plasma antioxidant capacity [89]. The mechanism of change in seminal viscosity due to oxidative stress could be related to altered interactions between oxidized proteins in the seminal plasma [90].

Increased levels of reactive oxygen species can impair sperm function. This phenomenon could be attributed to changes in membrane fluidity and acrosome integrity, resulting in decreased capacity for sperm-oocyte fusion [91]. Sperm from men with spinal cord injury may have functional impairments that could diminish the capacity of their sperm to bind to the oocyte. A study by Denil et al. [92] showed decreased bovine cervical mucus penetration and decreased hamster egg penetration in sperm from spinal cord-injured subjects versus controls. Sperm from men with spinal cord injury was also reported to have a high degree of acrosomal abnormalities [85].

Acrosin (EC 3.4.21.10) is a sperm acrosomal proteinase with trypsin-like substrate specificity, located in the acrosomal matrix as an enzymatically inactive zymogen. Evidence suggests that its active form, acrosin, is necessary for normal fertilization in humans. If acrosin is reduced, absent, or inhibited, sperm binding to, and penetration of, the zona pellucida is severely impaired [93]. We have recently shown that sperm from men with spinal cord injury is characterized by lower acrosin activity compared to healthy men [69]. These findings indicate that sperm from men with spinal cord injury could have functional defects in sperm-oocyte fusion, resulting from oxidative damage.

Oxidative stress results from an imbalance between the production of reactive oxygen species and their efficient removal by available antioxidant systems. Seminal plasma contains different reactive oxygen species scavengers which take part in the protection of spermatozoa from oxidative stress [94]. Exhaustion of seminal plasma antioxidant capacity facilitates sperm damage by free radicals.

It is known that seminal plasma is a major contributor to semen abnormalities in men with spinal cord injury (Fig. 32.5). For example, seminal plasma of spinal cord-injured men rapidly inhibits motility of sperm from normal men. Similarly, seminal plasma from normal men improves the motility of sperm from spinal cord-injured men [39]. Further evidence that an abnormal seminal plasma environment impairs sperm of men with spinal cord injury comes from a study measuring sperm motility and sperm viability in ejaculates versus vas deferens aspirates of the same group of spinal cord-injured subjects [48]. Because sperm from the vas deferens has not yet been subjected to the effects of the seminal plasma, a direct comparison of these sperm sources provided information about the effects of seminal plasma on sperm function in men with spinal cord injury. The results of this study showed that in patients with spinal cord injury, but not in able-bodied controls, sperm motility and sperm viability were significantly lower in ejaculated specimens (Fig. 32.6). These data provided evidence that, in men with spinal cord injury, seminal plasma is toxic to the sperm.

Interestingly, sperm obtained from men with spinal cord injury not only lose motility more rapidly than sperm from normal men but this deterioration is also exacerbated when semen is stored at body temperature compared to room temperature. In normal men, this correlation was not found [95].

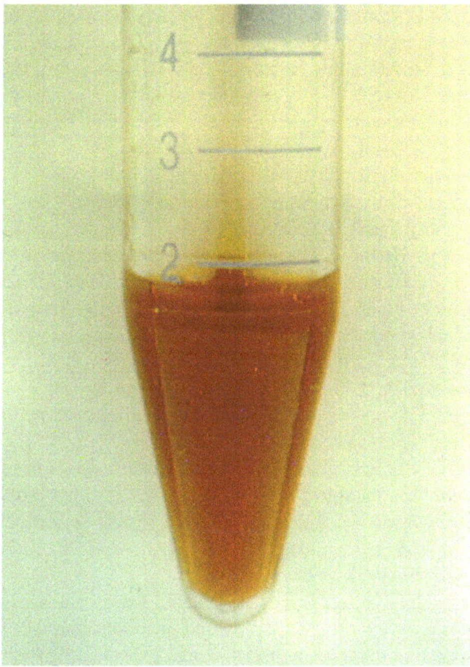

Fig. 32.5 Twenty-seven percent of men with spinal cord injury have *brown-colored semen*. The cause of the *brown color* is unknown and may be related the presence of abnormal constituents in the seminal plasma. Evidence suggests that an abnormal seminal plasma environment contributes to sperm impairments in men with spinal cord injury

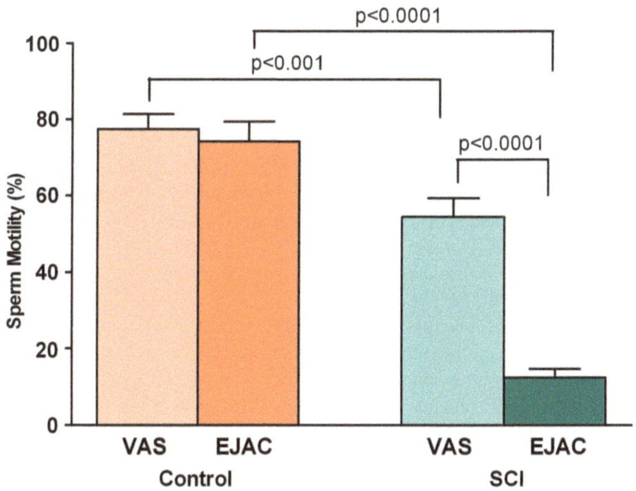

Fig. 32.6 In men with spinal cord injury (SCI), sperm motility was significantly higher when obtained from the vas deferens (VAS) than from the ejaculate (EJAC). In contrast, in control subjects, there was little difference in sperm motility between the two sites. This study provided definitive evidence that seminal plasma was a major contributor to low sperm motility in men with spinal cord injury

The possible explanation of this discrepancy is higher reactive oxygen species production by activated leukocytes at body temperature in the semen from men with spinal cord injury. Based on the information presented above, it is reasonable to suggest that elevated levels of reactive oxygen species and/or decreased antioxidant capacity of seminal plasma could be, at least in part, responsible for its detrimental effects on sperm motility and viability. Table 32.2 summarizes characteristics of the semen in men with spinal cord injury, suggesting the presence of increased oxidative stress.

Table 32.2 Characteristics of semen from men with spinal cord injury suggesting increased oxidative stress

Leukocytospermia
Teratozoospermia, increased numbers of immature sperm
Low sperm motility
Necrospermia
Increased DNA fragmentation
Semen hyperviscosity

Treating Oxidative Stress to Improve Semen Quality in Men with Spinal Cord Injury

Understanding the role of increased levels of reactive oxygen species in the pathogenesis of semen quality deterioration after spinal cord injury indicates that correction of oxidative stress could be used in the treatment of these abnormalities. Some studies have shown that treatment with antioxidants, including vitamin E, was associated with improvement of semen parameters in infertile men [96]. The effects of vitamin E on semen quality were studied by Wang et al. in an experimental rat model of spinal cord injury [97]. In this study, rats were given oral vitamin E in two different schedules starting immediately after the injury (maintenance) or 8–10 weeks postinjury (restoration) for 8 weeks. Various sperm parameters were studied including motility, viability, mitochondrial potential, and head decondensation. Weight of accessory glands was also measured.

Feeding with vitamin E was associated with improvement in sperm motility only during the chronic phase of injury, while viability and mitochondrial potential were partially preserved as a result of vitamin E treatment in both restoration and maintenance groups. Sperm head decondensation was significantly less pronounced in rats receiving vitamin E feeding compared to a sham control group. Vitamin E feeding during the chronic phase of injury also resulted in a significant increase in the weight of prostate and seminal vesicles. The results of this study indicate that the antioxidant, vitamin E, attenuated some of the damaging effects of spinal cord injury on sperm motility, viability, and morphology. The partial restoration of prostate and seminal vesicle weight during the chronic phase implies that reactive oxygen species could also be involved in the damaging effects of the spinal cord injury on these glands [97].

Taken together, these data provide evidence that oxidative stress may account for decreased semen quality after spinal cord injury. It is hoped that further laboratory and clinical studies will help to define the role of vitamin E and other antioxidants in the management of male infertility resulting from spinal cord injury.

Expert Commentary

A number of reasons for poor sperm quality in men with spinal cord injury have been postulated; however, we believe that no single factor has been convincingly shown to be the root cause, and it is likely that the etiology of this problem is multifactorial. Among these factors, elevated oxidative stress is clearly an important pathogenic mechanism leading to sperm damage and subsequent infertility resulting from spinal cord injury. Several studies have established a relationship between poor semen quality and the generation of reactive oxygen species in men with spinal cord injury. Semen from these men is characterized by higher concentrations of leukocytes and immature spermatozoa, both of which are important sources of reactive oxygen species production. Furthermore, other seminal abnormalities found in men with spinal cord injury may result from increased oxidative stress.

Five-Year View and Key Issues

While much evidence indicates a role for oxidative stress in the development of infertility in men with spinal cord injury, many unanswered questions still remain. Future research should take several directions:

- Measurements of the changes in antioxidant capacity of seminal plasma in men with spinal cord injury.
- Analysis of the correlation between antioxidant capacity of the seminal plasma and its toxic effects on sperm.
- Studies of the effects of antioxidants as an in vitro and possibly in vivo treatment of semen abnormalities in men with spinal cord injury.
- Investigation of the role of apoptosis in the development of semen abnormalities in men with spinal cord injury and the correlation between apoptosis and seminal plasma antioxidant capacity/seminal reactive oxygen species production.

Results of these studies will enhance our understanding of the underlying pathophysiology of infertility in men with spinal cord injury. These data may also help introduce new methods of treatment of semen abnormalities in this group of patients.

References

1. Spinal cord injury—facts and figures at a glance. 2009. http://www.spinalcord.uab.edu/. Accessed 10 Sept 2009.
2. Elliott S. Sexual dysfunction and infertility in men with spinal cord disorders. In: Lin V, editor. Spinal cord medicine: principles and practice. New York: Demos Medical Publishing; 2003. p. 349–65.
3. Brackett NL, Lynne CM, Ibrahim E, Ohl DA, Sonksen J. Treatment of infertility in men with spinal cord injury. Nat Rev Urol. 2010;7:162–72.
4. Brackett NL, Ibrahim E, Iremashvili V, Aballa TC, Lynne CM. Treatment of ejaculatory dysfunction in men with spinal cord injury: a single-center experience of more than 18 years. J Urol. 2010;183:2304–8.
5. Kafetsoulis A, Brackett NL, Ibrahim E, Attia GR, Lynne CM. Current trends in the treatment of infertility in men with spinal cord injury. Fertil Steril. 2006;86:781–9.
6. Wang YH, Huang TS, Lien IN. Hormone changes in men with spinal cord injuries. Am J Phys Med Rehabil. 1992;71:328–32.
7. Wang YH, Huang TS, Lin MC, Yeh CS, Lien IN. Scrotal temperature in spinal cord injury. Am J Phys Med Rehabil. 1993;72:6–9.
8. Ohl DA, Denil J, Fitzgerald-Shelton K, et al. Fertility of spinal cord injured males: effect of genitourinary infection and bladder management on results of electroejaculation. J Am Paraplegia Soc. 1992;15:53–9.
9. Beretta G, Chelo E, Zanollo A. Reproductive aspects in spinal cord injured males. Paraplegia. 1989;27:113–8.
10. Ohl DA, Menge AC, Jarow JP. Seminal vesicle aspiration in spinal cord injured men: insight into poor sperm quality. J Urol. 1999;162:2048–51.
11. Kostovski E, Iversen PO, Birkeland K, Torjesen PA, Hjeltnes N. Decreased levels of testosterone and gonadotrophins in men with long-standing tetraplegia. Spinal Cord. 2008;46:559–64.
12. Bors E, Engle ET, Rosenquist RC, Holliger VH. Fertility in paraplegic males; a preliminary report of endocrine studies. J Clin Endocrinol Metab. 1950;10:381–98.
13. Shim HB, Kim YD, Jung TY, Lee JK, Ku JH. Prostate-specific antigen and prostate volume in Korean men with spinal cord injury: a case-control study. Spinal Cord. 2008;46:11–5.
14. Naderi AR, Safarinejad MR. Endocrine profiles and semen quality in spinal cord injured men. Clin Endocrinol (Oxf). 2003;58:177–84.
15. Huang HF, Li MT, Giglio W, Anesetti R, Ottenweller JE, Pogach LM. The detrimental effects of spinal cord injury on spermatogenesis in the rat is partially reversed by testosterone, but enhanced by follicle-stimulating hormone. Endocrinology. 1999;140:1349–55.
16. Huang HF, Li MT, Wang S, Barton B, Anesetti R, Jetko JA. Effects of exogenous testosterone on testicular function during the chronic phase of spinal cord injury: dose effects on spermatogenesis and Sertoli cell and sperm function. J Spinal Cord Med. 2004;27:55–62.
17. Brackett NL, Lynne CM, Weizman MS, Bloch WE, Abae M. Endocrine profiles and semen quality of spinal cord injured men. J Urol. 1994;151:114–9.
18. Huang HF, Linsenmeyer TA, Li MT, et al. Acute effects of spinal cord injury on the pituitary-testicular hormone axis and Sertoli cell functions: a time course study. J Androl. 1995;16:148–57.
19. Tsitouras PD, Zhong YG, Spungen AM, Bauman WA. Serum testosterone and growth hormone/insulin-like growth factor-I in adults with spinal cord injury. Horm Metab Res. 1995;27:287–92.
20. Naftchi NE, Viau AT, Sell GH, Lowman EW. Pituitary-testicular axis dysfunction in spinal cord injury. Arch Phys Med Rehabil. 1980;61:402–5.
21. Huang HF, Linsenmeyer TA, Anesetti R, Giglio W, Ottenweller JE, Pogach L. Suppression and recovery of spermatogenesis following spinal cord injury in the rat. J Androl. 1998;19:72–80.
22. Safarinejad MR. Level of injury and hormone profiles in spinal cord-injured men. Urology. 2001;58:671–6.
23. Zorgniotti A, Reiss H, Toth A, Sealfon A. Effect of clothing on scrotal temperature in normal men and patients with poor semen. Urology. 1982;19:176–8.
24. Brackett NL, Nash MS, Lynne CM. Male fertility following spinal cord injury: facts and fiction. Phys Ther. 1996;76:1221–31.
25. Brindley GS. Deep scrotal temperature and the effect on it of clothing, air temperature, activity, posture and paraplegia. Br J Urol. 1982;54:49–55.

26. Brackett NL, Lynne CM, Weizman MS, Bloch WE, Padron OF. Scrotal and oral temperatures are not related to semen quality of serum gonadotropin levels in spinal cord-injured men. J Androl. 1994;15:614–9.
27. Dada R, Gupta NP, Kucheria K. Spermatogenic arrest in men with testicular hyperthermia. Teratog Carcinog Mutagen. 2003;(Suppl 1): 235–43.
28. Morgentaler A, Stahl BC, Yin Y. Testis and temperature: an historical, clinical, and research perspective. J Androl. 1999;20:189–95.
29. Brackett NL, Ferrell SM, Aballa TC, Amador MJ, Lynne CM. Semen quality in spinal cord injured men: does it progressively decline postinjury? Arch Phys Med Rehabil. 1998;79:625–8.
30. Iremashvili V, Brackett NL, Ibrahim E, et al. Semen quality remains stable during the chronic phase of spinal cord injury: a longitudinal study. J Urol. 2010;184(5):2073–7.
31. Rutkowski SB, Middleton JW, Truman G, Hagen DL, Ryan JP. The influence of bladder management on fertility in spinal cord injured males. Paraplegia. 1995;33:263–6.
32. Ohl DA, Bennett CJ, McCabe M, Menge AC, McGuire EJ. Predictors of success in electroejaculation of spinal cord injured men. J Urol. 1989;142:1483–6.
33. Momen MN, Fahmy I, Amer M, Arafa M, Zohdy W, Naser TA. Semen parameters in men with spinal cord injury: changes and aetiology. Asian J Androl. 2007;9:684–9.
34. Das S, Dodd S, Soni BM, Sharma SD, Gazvani R, Lewis-Jones DI. Does repeated electro-ejaculation improve sperm quality in spinal cord injured men? Spinal Cord. 2006;44:753–6.
35. Siosteen A, Forssman L, Steen Y, Sullivan L, Wickstrom I. Quality of semen after repeated ejaculation treatment in spinal cord injury men. Paraplegia. 1990;28:96–104.
36. Heruti RJ, Katz H, Menashe Y, et al. Treatment of male infertility due to spinal cord injury using rectal probe electroejaculation: the Israeli experience. Spinal Cord. 2001;39:168–75.
37. Sonksen J, Ohl DA, Giwercman A, Biering-Sorensen F, Skakkebaek NE, Kristensen JK. Effect of repeated ejaculation on semen quality in spinal cord injured men. J Urol. 1999;161:1163–5.
38. Hamid R, Patki P, Bywater H, Shah PJ, Craggs MD. Effects of repeated ejaculations on semen characteristics following spinal cord injury. Spinal Cord. 2006;44:369–73.
39. Brackett NL, Davi RC, Padron OF, Lynne CM. Seminal plasma of spinal cord injured men inhibits sperm motility of normal men. J Urol. 1996;155:1632–5.
40. Rajasekaran M, Hellstrom WJ, Sparks RL, Sikka SC. Sperm-damaging effects of electric current: possible role of free radicals. Reprod Toxicol. 1994;8:427–32.
41. de Lamirande E, Leduc BE, Iwasaki A, Hassouna M, Gagnon C. Increased reactive oxygen species formation in semen of patients with spinal cord injury. Fertil Steril. 1995;63:637–42.
42. Padron OF, Brackett NL, Sharma RK, Lynne CM, Thomas Jr AJ, Agarwal A. Seminal reactive oxygen species and sperm motility and morphology in men with spinal cord injury. Fertil Steril. 1997;67:1115–20.
43. Aitken RJ, Buckingham DW, Brindle J, Gomez E, Baker HW, Irvine DS. Analysis of sperm movement in relation to the oxidative stress created by leukocytes in washed sperm preparations and seminal plasma. Hum Reprod. 1995;10:2061–71.
44. Basu S, Lynne CM, Ruiz P, Aballa TC, Ferrell SM, Brackett NL. Cytofluorographic identification of activated T-cell subpopulations in the semen of men with spinal cord injuries. J Androl. 2002;23:551–6.
45. Aird IA, Vince GS, Bates MD, Johnson PM, Lewis-Jones ID. Leukocytes in semen from men with spinal cord injuries. Fertil Steril. 1999;72:97–103.
46. Pentyala S, Lee J, Annam S, et al. Current perspectives on pyospermia: a review. Asian J Androl. 2007;9:593–600.
47. Randall JM, Evans DH, Bird VG, Aballa TC, Lynne CM, Brackett NL. Leukocytospermia in spinal cord injured patients is not related to histological inflammatory changes in the prostate. J Urol. 2003;170:897–900.
48. Brackett NL, Lynne CM, Aballa TC, Ferrell SM. Sperm motility from the vas deferens of spinal cord injured men is higher than from the ejaculate. J Urol. 2000;164:712–5.
49. Whittington K, Harrison SC, Williams KM, et al. Reactive oxygen species (ROS) production and the outcome of diagnostic tests of sperm function. Int J Androl. 1999;22:236–42.
50. Sharma RK, Pasqualotto AE, Nelson DR, Thomas Jr AJ, Agarwal A. Relationship between seminal white blood cell counts and oxidative stress in men treated at an infertility clinic. J Androl. 2001;22:575–83.
51. Agarwal A, Makker K, Sharma R. Clinical relevance of oxidative stress in male factor infertility: an update. Am J Reprod Immunol. 2008;59:2–11.
52. Trabulsi EJ, Shupp-Byrne D, Sedor J, Hirsch IH. Leukocyte subtypes in electroejaculates of spinal cord injured men. Arch Phys Med Rehabil. 2002;83:31–4.
53. Plante M, de Lamirande E, Gagnon C. Reactive oxygen species released by activated neutrophils, but not by deficient spermatozoa, are sufficient to affect normal sperm motility. Fertil Steril. 1994;62:387–93.
54. Said TM, Agarwal A, Sharma RK, Thomas Jr AJ, Sikka SC. Impact of sperm morphology on DNA damage caused by oxidative stress induced by beta-nicotinamide adenine dinucleotide phosphate. Fertil Steril. 2005;83:95–103.
55. Basu S, Aballa TC, Ferrell SM, Lynne CM, Brackett NL. Inflammatory cytokine concentrations are elevated in seminal plasma of men with spinal cord injuries. J Androl. 2004;25:250–4.
56. Brackett NL, Cohen DR, Ibrahim E, Aballa TC, Lynne CM. Neutralization of cytokine activity at the receptor level improves sperm motility in men with spinal cord injuries. J Androl. 2007;28:717–21.
57. Cohen DR, Basu S, Randall JM, Aballa TC, Lynne CM, Brackett NL. Sperm motility in men with spinal cord injuries is enhanced by inactivating cytokines in the seminal plasma. J Androl. 2004;25:922–5.
58. Nandipati KC, Pasqualotto FF, Thomas Jr AJ, Agarwal A. Relationship of interleukin-6 with semen characteristics and oxidative stress in vasectomy reversal patients. Andrologia. 2005;37:131–4.
59. Camejo MI, Segnini A, Proverbio F. Interleukin-6 (IL-6) in seminal plasma of infertile men, and lipid peroxidation of their sperm. Arch Androl. 2001;47:97–101.
60. Martinez P, Proverbio F, Camejo MI. Sperm lipid peroxidation and pro-inflammatory cytokines. Asian J Androl. 2007;9:102–7.
61. Sanocka D, Jedrzejczak P, Szumala-Kaekol A, Fraczek M, Kurpisz M. Male genital tract inflammation: the role of selected interleukins in regulation of pro-oxidant and antioxidant enzymatic substances in seminal plasma. J Androl. 2003;24:448–55.
62. Buch JP, Kolon TF, Maulik N, Kreutzer DL, Das DK. Cytokines stimulate lipid membrane peroxidation of human sperm. Fertil Steril. 1994;62:186–8.
63. Aitken RJ, Buckingham DW, West K, Brindle J. On the use of paramagnetic beads and ferrofluids to assess and eliminate the leukocytic contribution to oxygen radical generation by human sperm suspensions. Am J Reprod Immunol. 1996;35:541–51.
64. Fisher HM, Aitken RJ. Comparative analysis of the ability of precursor germ cells and epididymal spermatozoa to generate reactive oxygen metabolites. J Exp Zool. 1997;277:390–400.
65. Aitken RJ, Buckingham DW, West KM. Reactive oxygen species and human spermatozoa: analysis of the cellular mechanisms involved in luminol- and lucigenin-dependent chemiluminescence. J Cell Physiol. 1992;151:466–77.
66. Gavella M, Lipovac V. NADH-dependent oxidoreductase (diaphorase) activity and isozyme pattern of sperm in infertile men. Arch Androl. 1992;28:135–41.

67. Huszar G, Sbracia M, Vigue L, Miller DJ, Shur BD. Sperm plasma membrane remodeling during spermiogenetic maturation in men: relationship among plasma membrane beta 1,4-galactosyltransferase, cytoplasmic creatine phosphokinase, and creatine phosphokinase isoform ratios. Biol Reprod. 1997;56:1020–4.
68. Huszar G, Ozenci CC, Cayli S, Zavaczki Z, Hansch E, Vigue L. Hyaluronic acid binding by human sperm indicates cellular maturity, viability, and unreacted acrosomal status. Fertil Steril. 2003;79 Suppl 3:1616–24.
69. Iremashvili V, Brackett NL, Ibrahim E, Aballa TC, Bruck D, Lynne CM. Hyaluronic acid binding and acrosin activity are decreased in sperm from men with spinal cord injury. Fertil Steril. 2010; 94(5):1925–7.
70. Huszar G, Ozkavukcu S, Jakab A, Celik-Ozenci C, Sati GL, Cayli S. Hyaluronic acid binding ability of human sperm reflects cellular maturity and fertilizing potential: selection of sperm for intracytoplasmic sperm injection. Curr Opin Obstet Gynecol. 2006; 18:260–7.
71. Henkel R, Kierspel E, Stalf T, et al. Effect of reactive oxygen species produced by spermatozoa and leukocytes on sperm functions in non-leukocytospermic patients. Fertil Steril. 2005;83:635–42.
72. Pryor WA, Houk KN, Foote CS, et al. Free radical biology and medicine: it's a gas, man! Am J Physiol Regul Integr Comp Physiol. 2006;291:R491–511.
73. Tremellen K. Oxidative stress and male infertility–a clinical perspective. Hum Reprod Update. 2008;14:243–58.
74. Jones R, Mann T, Sherins R. Peroxidative breakdown of phospholipids in human spermatozoa, spermicidal properties of fatty acid peroxides, and protective action of seminal plasma. Fertil Steril. 1979;31:531–7.
75. Gomez E, Irvine DS, Aitken RJ. Evaluation of a spectrophotometric assay for the measurement of malondialdehyde and 4-hydroxyalkenals in human spermatozoa: relationships with semen quality and sperm function. Int J Androl. 1998;21:81–94.
76. Aitken RJ, Fisher HM, Fulton N, et al. Reactive oxygen species generation by human spermatozoa is induced by exogenous NADPH and inhibited by the flavoprotein inhibitors diphenylene iodonium and quinacrine. Mol Reprod Dev. 1997;47:468–82.
77. de Lamirande E, Gagnon C. Reactive oxygen species and human spermatozoa. I. Effects on the motility of intact spermatozoa and on sperm axonemes. J Androl. 1992;13:368–78.
78. de Lamirande E, Gagnon C. Reactive oxygen species and human spermatozoa. II. Depletion of adenosine triphosphate plays an important role in the inhibition of sperm motility. J Androl. 1992;13:379–86.
79. de Lamirande E, Jiang H, Zini A, Kodama H, Gagnon C. Reactive oxygen species and sperm physiology. Rev Reprod. 1997;2:48–54.
80. Wang X, Sharma RK, Sikka SC, Thomas Jr AJ, Falcone T, Agarwal A. Oxidative stress is associated with increased apoptosis leading to spermatozoa DNA damage in patients with male factor infertility. Fertil Steril. 2003;80:531–5.
81. Brackett NL, Bloch WE, Lynne CM. Predictors of necrospermia in men with spinal cord injury. J Urol. 1998;159:844–7.
82. Nunez R, Murphy TF, Huang HF, Barton BE. Use of SYBR14, 7-amino-actinomycin D, and JC-1 in assessing sperm damage from rats with spinal cord injury. Cytometry A. 2004;61:56–61.
83. Aitken RJ, Krausz C. Oxidative stress, DNA damage and the Y chromosome. Reproduction. 2001;122:497–506.
84. Kemal Duru N, Morshedi M, Oehninger S. Effects of hydrogen peroxide on DNA and plasma membrane integrity of human spermatozoa. Fertil Steril. 2000;74:1200–7.
85. Engh E, Clausen OP, Purvis K, Stien R. Sperm quality assessed by flow cytometry and accessory sex gland function in spinal cord injured men after repeated vibration- induced ejaculation. Paraplegia. 1993;31:3–12.
86. Brackett NL, Ibrahim E, Grotas JA, Aballa TC, Lynne CM. Higher sperm DNA damage in semen from men with spinal cord injuries compared with controls. J Androl. 2008;29:93–9. discussion 100–101.
87. Agarwal A, Said TM. Oxidative stress, DNA damage and apoptosis in male infertility: a clinical approach. BJU Int. 2005;95:503–7.
88. Aydemir B, Onaran I, Kiziler AR, Alici B, Akyolcu MC. The influence of oxidative damage on viscosity of seminal fluid in infertile men. J Androl. 2008;29:41–6.
89. Siciliano L, Tarantino P, Longobardi F, Rago V, De Stefano C, Carpino A. Impaired seminal antioxidant capacity in human semen with hyperviscosity or oligoasthenozoospermia. J Androl. 2001; 22:798–803.
90. Traverso N, Menini S, Maineri EP, et al. Malondialdehyde, a lipoperoxidation-derived aldehyde, can bring about secondary oxidative damage to proteins. J Gerontol A Biol Sci Med Sci. 2004;59:B890–5.
91. Agarwal A, Saleh RA. Role of oxidants in male infertility: rationale, significance, and treatment. Urol Clin North Am. 2002;29: 817–27.
92. Denil J, Ohl DA, Menge AC, Keller LM, McCabe M. Functional characteristics of sperm obtained by electroejaculation. J Urol. 1992;147:69–72.
93. Liu DY, Baker HW. Inhibition of acrosin activity with a trypsin inhibitor blocks human sperm penetration of the zona pellucida. Biol Reprod. 1993;48:340–8.
94. Balercia G, Armeni T, Mantero F, Principato G, Regoli F. Total oxyradical scavenging capacity toward different reactive oxygen species in seminal plasma and sperm cells. Clin Chem Lab Med. 2003;41:13–9.
95. Brackett NL, Santa-Cruz C, Lynne CM. Sperm from spinal cord injured men lose motility faster than sperm from normal men: the effect is exacerbated at body compared to room temperature. J Urol. 1997;157:2150–3.
96. Kessopoulou E, Powers HJ, Sharma KK, et al. A double-blind randomized placebo cross-over controlled trial using the antioxidant vitamin E to treat reactive oxygen species associated male infertility. Fertil Steril. 1995;64:825–31.
97. Wang S, Wang G, Barton BE, Murphy TF, Huang HF. Beneficial effects of vitamin E in sperm functions in the rat after spinal cord injury. J Androl. 2007;28:334–41.

Obesity and Male Fertility

33

Stephanie Cabler, Ashok Agarwal, and Stefan S. du Plessis

Abstract

The obesity pandemic has grown to concerning proportions in recent years, not only in the Western world but in developing countries as well. The corresponding decrease in male fertility and fecundity may be explained partially by obesity, and obesity should be considered an etiology of male subfertility. Studies show that obesity contributes to infertility by reducing semen quality, changing sperm proteomes, contributing to erectile dysfunction, and inducing other physical problems related to obesity. Mechanisms for explaining the effect of obesity on male infertility include abnormal reproductive hormone levels, an increased release of adipose-derived hormones and adipokines associated with obesity, and other physical problems including sleep apnea and increased scrotal temperatures. Recently, genetic factors and markers for an obesity-related infertility have been discovered and may explain the difference between fertile obese and infertile obese men. Treatments are available for not only infertility related to obesity but also for the other comorbidities arising from obesity. Natural weight loss and bariatric surgery are options for obese patients and have shown promising results in restoring fertility and normal hormonal profiles. Therapeutic interventions including aromatase inhibitors, exogenous testosterone replacement therapy and maintenance, and regulation of adipose-derived hormones, particularly leptin, may also be able to restore fertility in obese males. The increasing prevalence of obesity calls for greater clinical awareness of its effects on fertility, better understanding of underlying mechanisms, and exploration into avenues of treatment.

Keywords

Male fertility • Male obesity • Abnormal semen parameters • Body mass index • Hypothalamic–pituitary–gonadal axis • Aromatase polymorphism • ALMS1 mutation • Hypogonadism

In the past 5–10 years, obesity has become a worldwide epidemic that has brought attention to learning more about the various causes, effects, and treatments. A combination of an increasingly acceptable sedentary lifestyle and unhealthy diet in the Western world has resulted in an increasing number of overweight and obese children and adults. According to the WHO, approximately 1.6 billion adults were classified as being overweight and 400 million adults were obese in 2005 [1]. It is predicted that globally, in the next 5 years, more than 700 million adults will suffer from obesity [1]. Once considered a problem only in high-income countries, overweight and

S. Cabler, BSc (✉) • A. Agarwal, PhD
Center for Reproductive Medicine, Glickman Urological and Kidney Institute, Cleveland Clinic Foundation, 9500 Euclid Avenue, Desk A19, Cleveland, OH 44195, USA
e-mail: agarwaa@ccf.org

S.S. du Plessis, PhD, MBA
Department of Medical Physiology, Faculty of Health Sciences, Stellenbosch University, PO Box 19063, Tygerberg, Western Cape 7505, South Africa
e-mail: ssdp@sun.ac.za

S.J. Parekattil and A. Agarwal (eds.), *Male Infertility: Contemporary Clinical Approaches, Andrology, ART & Antioxidants*, DOI 10.1007/978-1-4614-3335-4_33, © Springer Science+Business Media, LLC 2012

obesity are now dramatically on the rise in all countries. Evidence of this is the five unit increase in body mass index (BMI) for the period 1997–2006 in the 95th percentile BMI level among children aged 6–9 years in China. These children in the 95th percentile have a BMI of 24.8, which is surprisingly higher than that of the USA (22.2), Australia (20.1), and the UK (20.1) [2].

Parallel to the global increase in obesity is the reported world decrease in male fertility and fecundity [3]. Interestingly, men with increased BMI were significantly more likely to be infertile than normal-weight men, according to research conducted at the National Institute of Environmental Health Sciences (NIEHS) [4]. According to Carlson et al., the quality of semen has substantially declined, which has subsequently lead to decreased male fertility [5]. This could likely contribute to overall reduced male reproductive potential. Some studies estimate that male sperm counts continue to decrease at a rate of approximately 1.5% per year in the USA and similar findings have been found in other Western countries as well [3]. In addition, there is also a significant increase in the incidence of obesity in patients with male factor infertility, and couples with obese male partners are more likely to experience subfecundity, a correlation that seems necessary to address [6]. Due to the fact that this decline has occurred in close parallel with increasing rates of obesity, it is necessary to focus on the possibility of obesity as an etiology of male infertility and reduced fecundity. The "obesity pandemic" seen in many countries is a serious threat to public health, and a reduced capacity to reproduce is a potential but less well-known health hazard that can often be attributed to obesity.

It is therefore necessary to explore the links between obesity and male infertility, as well as to explain how it disrupts the male reproductive system at a mechanistical level. Treatment and prevention of obesity and associated fertility disorders will also be discussed in a clinical context.

What Is Obesity?

Obesity is a medical condition in which excess body fat, or white adipose tissue, accumulates in the body to the extent that the excess fat adversely affects health, often reducing life expectancy. The fundamental cause of obesity and overweight is an energy imbalance, where the energy consumed exceeds the energy expended. Global increases in overweight and obesity are attributable to a number of factors, including a shift in diet toward increased intake of energy-dense foods that are high in fat and sugars, and a trend toward decreased physical activity, resulting from increasingly sedentary nature of work, changing modes of transportation, and increasing urbanization.

Currently, overweight and obesity are defined more broadly as abnormal or excessive fat accumulation that may impair health. However, there are other specific requirements that qualify an individual as obese. The most accurate measures are to weigh a person underwater or to use an X-ray test called dual energy X-ray absorptiometry. These methods are not practical for the average individual and are conducted only in research centers with special equipment. There are simpler methods to estimate body fat such as BMI, skin fold measurements, waist-to-hip ratio (WHR), waist circumference, and also methods such as bioelectrical impedance analysis, risk factors and comorbidities [7]. The two tools that are most commonly used to identify obese patients are BMI, a waist-to-height ratio, and waist circumference. An individual is normally defined as being overweight if their BMI is between 25 and 30 kg/m^2 and obese if it exceeds 30 kg/m^2.[8] A problem with this method is that individuals with a high BMI may be mesomorphic and have a high amount of muscle mass. Therefore, BMI may not be the most accurate marker for total body fat percentage and is an even less suitable tool to assess body fat distribution.

Waist circumference is a slightly less common method used to predict obesity in an individual, but may be more accurate in predicting obesity-related health issues. For females, a waist circumference of 88 cm or greater is considered unhealthy. For men, a waist circumference of 102 cm or greater is considered unhealthy. If waist circumference is used as the criterion, then according to a study conducted in 2006, the prevalence of being overweight among Australian adults, and probably other Caucasian populations, may be significantly greater than indicated by surveys relying on self-reported height and weight. The development of valid self-reported measures of waist circumference for use in population surveys may allow more accurate monitoring of overweight and obesity and should be considered instead of BMI [9]. A WHR can also be used to predict unhealthy consequences as a result of increased body fat (normal WHR: males=<0.9; females <0.85), especially related to the risk of coronary heart disease as it relates to obesity [10]. WHR may be the most useful measure of obesity and the best simple anthropometric index in predicting a wide range of risk factors and related health conditions [11].

How Does Obesity Affect Male Fertility?

The relationship between male infertility and obesity has more concrete evidence than solely studies showing reduced fecundity among couples, one of whom is an obese male. Although spermatogenesis and fertility are not impaired in a majority of obese men, a disproportionate number of men seeking infertility treatment are obese. There have been a number of studies analyzing the relationship between semen quality and obesity, with a common finding that there is an inverse correlation between BMI and quality of semen parameters.

Abnormal Semen Parameters

Altered semen parameters attributed to obesity include decreased sperm concentration, abnormal morphology, compromised chromatin integrity, and abnormal motility. Although there is a convincing amount of evidence to demonstrate the adverse affect of excess body fat on spermatogenesis, not all studies have come to the same conclusions. Individual studies are conflicting in evidence, but a recent meta-analysis by MacDonald et al. [12] combined 31 studies containing data and information relating to obesity and male infertility. Reproductive hormones studied included testosterone, free testosterone, estradiol, FSH, LH, inhibin B, and sex hormone-binding globulin (SHBG). This meta-analysis was conducted to investigate sperm concentration and total sperm count but found no evidence for a relationship between BMI and sperm concentration or total sperm count. Examination of further studies may present more insight in the relationship between obesity and semen quality.

Decreased Sperm Count and Concentration

Obese men are three times more likely than healthy men of normal weight to have a sperm count of fewer than 20 million/ml, an indicator of oligospermia [13]. In one of the largest studies on male fertility and obesity, done on Danish men, Jensen et al. measured BMI in relation to semen quality and reproductive hormones and found significant relationships between sperm concentration and BMI. A lower sperm concentration was observed in not only obese and overweight males, but also in males who were significantly underweight. This could serve as an indication that there may be an ideal range of BMI for normal spermatogenesis. Subjects whose BMI was within the normal range showed a higher sperm concentration, as well as a higher total sperm count, and a lower percentage of abnormal spermatozoa [14]. According to a study by Chavarro et al., men with a BMI greater than 25 kg/m^2 had a lower total sperm count than men of normal weight, and the measured volume of ejaculate decreased steadily with an increasing BMI [15]. These findings have been corroborated by other studies as well [16, 17]. Although several reports exist indicating a considerable negative effect of BMI on sperm count and concentration, some discrepancies have been noted. In these studies, a correlation between sperm count and concentration in obese men compared to controls was demonstrated but was not deemed significant [18–20].

Sperm Motility

Some consensus on the effects of obesity on sperm motility has been established, but there is no overall agreement. Hofny et al. [16] found that the BMI correlated negatively with sperm motility, and Hammoud et al. [21] concluded that the incidence of low progressively motile sperm count increased with increasing BMI. Fejes et al. [22] found that the waist to hip circumference of men correlated negatively with the total motile sperm count as well as the rapid progressively motile sperm count. Another study by Martini et al. found that there was a negative association between BMI and motility and rapid motility [23]. Despite this evidence, not all studies have come to the same conclusions, and a majority of studies measuring semen parameters neglect to include motility in their measurements [14, 19].

DNA Fragmentation

Kort et al. found that an increase in the DNA fragmentation index (DFI) accompanied an increase in BMI, demonstrating that obesity might compromise the integrity of sperm chromatin, their only genetic material [24].

DFI is the percent of sperm in a semen sample that have increased levels of single or double strand breaks in their nuclear DNA. A young and healthy man has about 3–5% of sperm with fragmented DNA while a level of 25–30% DFI places a man attempting natural conception at a statistical risk for infertility [25]. An increase in the BMI above 25 kg/m^2 causes an increase in sperm DFI and a decrease in the number of normal chromatin-intact sperm cells per ejaculate, relative to the degree of obesity [24]. Men with type 2 diabetes also present with a significantly higher number of severe structural defects in sperm compared with sperm from controls ($p<0.05$) [47]. Typically, males presenting with a high DFI will have reduced fertility, and their partners will display an increased incidence of miscarriage as a consequence [26].

A new breakthrough in gel electrophoresis has been used to identify obesity-associated changes of the sperm proteome. Semen samples from obese males differed from those of normal-weight men with 12 spots seen after running their "difference gel electrophoresis" (DIGE) of fluorescently labeled human sperm proteins. Tryptic digestion of the 12 spot proteins and mass spectrometric analysis of the corresponding peptides identified nine sperm proteins associated with obesity. This can now be considered a noninvasive experimental tool in the diagnosis of male infertility and monitoring device for fertility-restoring therapy in obese males [27]. This finding clearly demonstrates the differences or changes in the protein composition of spermatozoa in obese men.

Sperm Morphology

Measuring differences in the morphology of sperm between obese and normal-weight men can be difficult owing to differences in what is classified as "normal" morphology and high individual variability within individual patient samples. However, most studies have shown no correlation between obesity and abnormal sperm morphology [14, 15, 19, 21]. In the large retrospective study of Danish military recruits, no association between obesity and poor sperm motility or morphology was reported [14]. Identifying the specific hormones, proteins, and mechanisms involved in regulating sperm morphology might help to explain how and why obesity affects normal spermatogenesis.

What Are the Proposed Mechanisms?

Although environmental and lifestyle factors might help to explain the growing numbers of obese adults and children, there is less evidence explaining how obesity causes male infertility. The mechanisms responsible for effects on male infertility are mostly ambiguous and undefined. Several mechanisms have been proposed, all of which are described below (Fig. 33.1). Most of these mechanisms contribute to the dysregulation of the hypothalamic–pituitary–gonadal (HPG) axis, one of the most important functions of which is to regulate aspects of reproduction.

Genetic Link

Despite there being a known effect of obesity on infertility, many obese males are fertile and have normal reproductive function and fecundity. However, because obesity can result from an unfavorable genotype and because obesity can cause infertility, a genetic link between these two factors might explain this discrepancy. Patients with Klinefelter, Prader–Willi, or Laurence–Moon–Bardet–Biedel syndromes all display, to varying degrees, both obesity and infertility. In addition, men who are both infertile and obese show significantly lower testosterone levels than obese fertile men [28]. Although the specific genes involved and mechanism(s) explaining these syndromes are quite well understood, it is possible that other, less severe genetic mutations exist. These small mutations might explain the discrepancies between obese fertile and infertile men and shed light upon a possible genetic link between obesity and infertility.

Also, mutations in the human ALMS1 gene are responsible for Alström syndrome, a disorder in which key metabolic and endocrinological features include childhood-onset obesity, metabolic syndrome, and diabetes, as well as infertility [29]. Scientists still need to accurately establish which particular genes are involved in these different syndromes, but the linkage of obesity and infertility in extreme genetic cases points to some degree of genetic linkage. Hammoud et al. [30] recently discovered that an aromatase polymorphism modulates the relationship between weight and estradiol levels in obese men. This could explain why only certain obese men experience this rise in estradiol and subsequent fertility problems, while others experience no fertility issues. It seems possible that there are other less severe genetic markers than chromosome 15 abnormalities and ALMS1 mutations that may clarify the discrepancies between obese fertile and infertile men and give explanation to a possible genetic link between obesity and infertility.

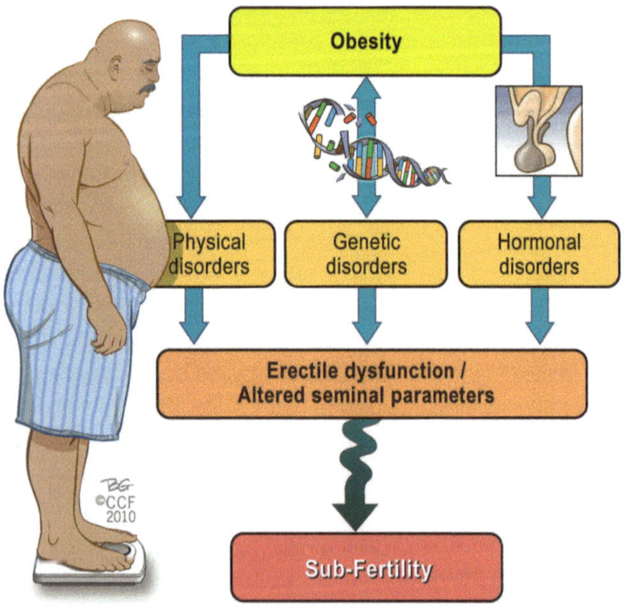

Fig. 33.1 The effects of obesity on male fertility and mechanisms involved

Hormonal Mechanisms

Abdominal or visceral fat is more likely to lead to changes in hormone levels and to cause inflammation than fat stored in other parts of the body. This is predominantly due to the fact that white adipose tissue, found in high levels in obese men, exhibits elevated aromatase activity and secretes adipose-derived hormones, as well as adipokines.

Reproductive Hormones

The reproductive hormonal profiles of most obese men deviate from what is considered the norm. Obese men tend to present with elevated estrogen and low testosterone and FSH levels. Androgen deficiency or hypogonadism found in males who are obese or have metabolic syndrome can account for problems with erectile dysfunction and spermatogenesis. However, many other hormones associated with obesity may alter the male reproductive potential. In morbidly obese individuals, reduced spermatogenesis associated with severe hypotestosteronemia may contribute to infertility [31, 32]. This estrogen excess is explained by overactivity of the aromatase cytochrome P450 enzyme, which is expressed at high levels in white adipose tissue and is responsible for a key step in the biosynthesis of estrogens. High levels of estrogens in obese males result from the increased conversion of androgens into estrogens, owing to the high bioavailability of these aromatase enzymes [33]. Visceral obesity can serve as a major endocrine disrupter and can also influence the endocrine interactions by reducing the levels of luteinizing hormone (LH) and testosterone, resulting in hypogonadotropic hypogonadism, a condition which contributes to male infertility. Although abnormal levels of reproductive hormones could be the source of fertility problems in obese males, Qin et al. [34] established that the associations between BMI and semen quality were found statistically significant even after an adjustment for reproductive hormones, thereby demonstrating that reproductive hormones cannot fully explain the association between BMI and semen quality [35]. Perhaps, altered hormone levels themselves do not explain poor semen quality but instead a deregulation of the normal HPG axis. Regardless, studies indicate that the association between BMI and semen quality is clearly more complex than what can be accounted for simply by reproductive hormones. Instead, it may be a result of other factors such as one's lifestyle and increased adipokine release.

White Adipose Tissue as an Endocrine Organ

White adipose tissue is a major secretory and endocrine organ that secretes approximately 30 biologically active peptides and proteins that can be grouped as either adipose-derived hormones (e.g., leptin, adiponectin, resistin) or adipokines (immunomodulating agents). Adipose-derived hormones play a central role in body homeostasis including the regulation of food intake and energy balance, insulin action, lipid and glucose metabolism, angiogenesis and vascular remodeling, coagulation, and the regulation of blood pressure [35]. Due to an excess of white adipose tissue in obese men, levels of adipose-derived hormones are often elevated, and their action is thought to modify many obesity-related diseases, including reproductive functioning.

Leptin

One such adipose-derived hormone is leptin, which is best known as a regulator of food intake and energy expenditure via hypothalamic-mediated effects [36, 37]. Although normal levels of leptin are required for overall reproductive health, excess leptin may be an important contributor to the development of reduced androgens in male obesity [38]. In addition to a higher prevalence of infertility, obese individuals are reported to have higher circulating levels of leptin than nonobese individuals [39, 40]. An increasing body of data suggests that leptin is also involved in glucose metabolism as well as in normal sexual maturation and reproduction [16]. Leptin receptors are not only present in testicular tissue but also on the plasma membrane of sperm suggesting that leptin may directly affect sperm via the endocrine system, independent of changes in the HPG axis [38, 41]. Diet-induced obesity in mice caused a significant reduction in male fertility and resulted in a fivefold increase in leptin levels compared to control mice. Sperm from these obese males exhibited decreased motility and reduced hyperactivated progression compared to the lean mice.

Oxidative Stress and Reactive Oxygen Species

As mentioned previously, adipocytes secrete various adipokines (e.g., tumor necrosis factor α (TNF-α), interleukin 6 (IL-6), plasminogen activator inhibitor-1 (PAI-1), and tissue factor) [37]. A number of these adipokines have been connected to infertility and testicular cancer. Bialas et al. [42] found that changes in the activity of intratesticular cytokines may promote various distinct pathologies such as testicular cancer or infertility. Also, increased release of adipokines from excess white adipose tissue, resulting in inflammation, can have a toxic effect on spermatozoa through the release of excess reactive oxygen species (ROS) and reactive nitrogen species (RNS) [43]. ROS and RNS are free radicals, highly reactive and unstable molecules that arise as a consequence of oxidative stress or inflammation that can induce significant cellular damage throughout the body. Two adipocyte-released adipokines, TNF-α and IL-6, significantly reduced human sperm progressive motility in a dose- and time-dependent manner by promoting the elevation of nitric oxide production to pathological levels [44]. Numerous authors have noted that obesity and several of its causative agents, namely insulin resistance and dyslipidemia, are associated with increased oxidative stress [45, 46]. This association is most likely the result of the elevated metabolic rates that are required to maintain normal biological processes and increased levels of stress in the local testicular environment, both of which naturally produce ROS. The local influences of biologically active substances (cytokines) released by activated leukocytes in the course of the inflammatory response to obesity may damage sperm and inhibit spermatogenesis. Agarwal et al. [47] found that abnormal patterns of increased ROS were associated with male factor infertility and are responsible for abnormal sperm concentration, motility, and morphology found in obese males.

Inhibin B

In a study by Winters et al. [48], the levels of inhibin B, another hormone involved in the HPG axis, declined with increasing obesity in young adult men, and values were 26% lower in men who were obese compared to normal-weight men. As it was shown that inhibin B is positively correlated with the number of Sertoli cells in normal adult rhesus monkeys, the reduced levels of inhibin B in the Winters study may indicate that obese men have fewer Sertoli cells than men of normal weight [49]. Since each Sertoli cell is thought to support a finite number of germ cells, fewer Sertoli cells as a result of obesity may result in a lower sperm count [48].

Resistin Secretion and Insulin Resistance

Resistin is another adipose-tissue-specific factor, which is reported to induce insulin resistance. Almost 80% of men with type 2 diabetes are also obese, and an increase in resistin secretion owing to a higher number of adipocytes links obesity to type 2 diabetes [37, 50]. As a consequence of insulin resistance in patients with type 2 diabetes, increased renin secretion was associated with increased oxidative stress [45, 46]. This association is most likely the result of the higher-than-usual metabolic rates required to maintain normal biological processes and an increased level of stress in the local testicular environment. Hyperinsulinemia, which often occurs in obese men, has an inhibitory effect on normal spermatogenesis and can be linked to decreased male fertility. In a group of diabetic men, semen parameters (concentration, motility, and morphology) did not differ from the control group, but the amount of nuclear and mitochondrial DNA damage in the sperm was significantly higher [51]. This sperm DNA damage can impair male fertility and reproductive health. In addition to inducing sperm DNA damage, high insulin levels also have been shown to influence the levels of sex hormone binding globulin (SHBG), a glycoprotein that binds to sex hormones, specifically testosterone and estradiol, thereby inhibiting their biologic activity as a carrier.

High circulating insulin levels inhibit SHBG synthesis in the liver, whereas weight loss has been shown to increase SHBG levels [52]. In obese males, the decrease in SHBG means that less estrogen will be bound, resulting in more biologically active, free estrogen. In addition to the conversion of testosterone to estrogen in obese patients, the decreased ability of SHBG to sustain homeostatic levels of free testosterone also contributes to abnormal testosterone levels [14]. This failure to maintain homeostatic levels might magnify the negative feedback effect of elevated total estrogen levels. Even when the presence of SHBG is accounted for, an independent relationship between insulin resistance and testosterone production can still be demonstrated [32]. Therefore, the levels of SHBG might be important only as a marker of altered hormone profiles in obese infertile men.

Environmental Toxins

Most environmental toxins are fat soluble and therefore accumulate in fatty tissue. Their accumulation not only around the scrotum and testes, but also elsewhere in the body may disrupt the normal hormone profile because they are proven endocrine disruptors in male fertility [21]. Since morbidly obese males present with excess scrotal fat, environmental toxins accumulating in white adipose tissue surrounding the scrotum may also have a direct localized effect on spermatogenesis in the testes. Lipophilic contaminants such as organochlorines, organic compounds containing at least one covalently bonded chlorine atom whose uses are controversial because of the often toxic effects of these compounds on the environment, are associated with decreased sperm production and thus decreased male reproductive potential, even if fat is not localized in the scrotal area [34]. Other toxic species that may induce abnormal spermatogenesis are ROS discussed in the previous section. Despite reports that certain toxins can negatively affect fertility, Magnusdottir et al. [6] found that poor semen quality was found to be associated with sedentary work and obesity, but not with increased plasma levels of persistent organochlorines.

Dysregulation of HPG Axis

Excess body weight can impair the feedback regulation of the HPG axis, and all of the factors above might contribute to or be a result of this dysregulation, contributing to apparent semen quality abnormalities. Sex steroids and glucocorticoids control the interaction between the hypothalamic–pituitary–adrenal (HPA) and the HPG axes, and any amount of disturbance might, in turn, affect spermatogenesis and male reproductive function. Men of normal weight with low levels of testosterone regularly present with elevated levels of LH and FSH, in contrast with obese men, who usually present with low LH and FSH levels [53]. Inhibin B, a growth-like factor, is produced by Sertoli cells in the testis and normally acts to inhibit both FSH production and stimulation of testosterone production by Leydig cells in the testis. Surprisingly, the expected compensatory increase in FSH levels in response to low levels of inhibin B is not observed in obese men. A low level of inhibin B might result from the suppressive effects of elevated estrogen levels. A study by Globerman et al. [54] also found that there was no increase in FSH levels in obese men whose inhibin B levels remained low after weight loss. Obese, infertile men exhibit endocrine changes that are not observed in men with either obesity or infertility alone. This defective response to hormonal changes might be explained by partial or complete dysregulation of the HPG axis.

Physical Mechanisms

Many obese men face physical problems that could be related to their decreased fecundity and fertility, including erectile dysfunction, scrotal lipomatosis leading to increased scrotal temperatures, and sleep apnea that can cause disruptions in the nightly testosterone rise.

Sleep Apnea

Sleep apnea is a disorder affecting 4% of middle-aged men. The disorder is characterized by repetitive collapse of the pharyngeal airway during sleep resulting in hypoxia and hypercapnia. About two-thirds of middle-aged men with obstructive sleep apnea suffer from obesity, particularly central obesity [55]. Sleep apnea is characterized by a fragmented sleep course owing to repeated episodes of upper airway obstructions and hypoxia and is often diagnosed in obese and diabetic males. Patients with sleep apnea have a disrupted nightly rise in testosterone levels and, therefore, lower mean levels of testosterone and LH compared with controls. In a study analyzing sleep apnea in obese, control, and lean patients, Luboshitzky et al. [56] concluded that the condition is associated with decreased pituitary–gonadal function and that the accompanying decline in testosterone concentrations is the result of obesity and, to a lesser degree, sleep fragmentation and hypoxia. This disruption has been associated with abnormal spermatogenesis and male reproductive potential.

Erectile Dysfunction

Whereas the effects of sleep apnea on reproduction are confounding owing to obesity itself being a cause of infertility, erectile dysfunction is significantly associated with obesity. Patients who are overweight or obese make up of 76% of men who report erectile dysfunction and a decrease in libido [19]. Many studies have found an association between an increased incidence of erectile dysfunction and an increase in BMI; hormonal dysfunction is central to the connection between obesity and erectile dysfunction [57]. Erectile dysfunction is highly prevalent in men with both type 2 diabetes and obesity and might act as a forerunner to cardiovascular disease in this high-risk population. Conversely, improved diabetes control and weight loss have been found to improve erectile function [58].

Elevated Scrotal Temperature

An elevated BMI can impair or arrest spermatogenesis by causing an increase in scrotal temperature. Increased fat distribution in the upper thighs, suprapubic area, and scrotum in conjunction with the sedentary lifestyle often associated with obesity can result in increased testicular temperature [21, 28]. Studies of cyclists, truck drivers, and individuals that almost constantly experience elevated scrotal temperatures, like those with undescended testes demonstrate a negative influence of genital heat stress on spermatogenesis. Many studies have focused on genital heat stress as a potential cause of impaired semen quality in cases of sedentary occupations, the occurrence of frequent fever, and varicocele [59]. Hjollund et al. [60] concluded that even a moderate physiological elevation in scrotal skin temperature is associated with substantially reduced sperm concentrations. Additionally Magnusdottir et al. [6] found that the duration of sedentary posture correlated positively with increased scrotal temperatures, leading to a decrease in sperm density.

Are There Solutions?

From the preceding literature, it is evident that obesity is an influencing factor in male infertility. Many experts believe that overweight and obesity, as well as their related chronic diseases, are largely preventable and measures can be taken to reverse both the unhealthy consequences associated with obesity and the negative impacts on male fertility. Therefore, the treatment approach will predominantly focus on the management of obesity (Fig. 33.2). The rapid rise in the incidence of obesity has prompted researchers to not only look at natural treatment methods but also for new treatments in order to manage the pandemic and its subsequent comorbidities.

Lifestyle Changes

Lifestyle changes that can lead to weight loss can include diet modifications (eating smaller meals, cutting down on certain types of food) as well as making a conscious effort to exercise more in order to achieve a normal energy balance. Several studies showed that natural weight loss through diet and/or exercise resulted in an increase in androgen, inhibin B, and SHBG levels and decreased serum concentrations of insulin and leptin, thereby improving semen parameters in obese men [15, 28, 38, 61, 62]. In addition, reducing adipose tissue mass through weight loss in association with exercise or a low-energy and low-fat diet decreases levels of TNF-α, IL-6, and other inflammatory cytokines associated with infertility [63, 64].

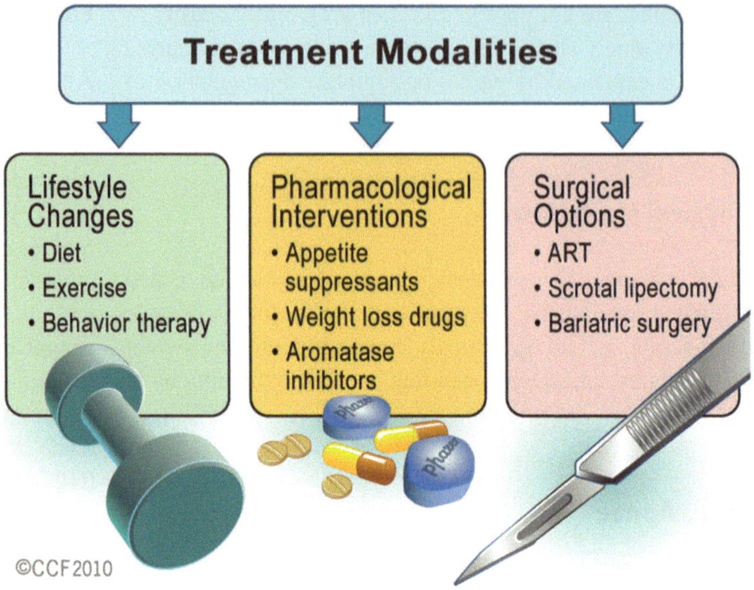

Fig. 33.2 Approaches to treatment of obesity-related male infertility

Gradual weight loss is best achieved through a sensible eating plan that can be maintained over long periods of time. The likelihood of maintaining weight loss is increased when the diet is combined with regular exercise, cognitive behavior therapy, and connecting with a supportive group environment [65]. Therefore, adoption of these principles in a primary health-care setting can aid in the treatment of infertility related to obesity.

Pharmacological Interventions

Medication can be used to either treat obesity by addressing weight loss or dealing with the obesity-related effects on the male reproductive system. Currently, only two anti-obesity medications are approved by the Food and Drug Administration for long-term use [66]. One is orlistat (Xenical), which reduces intestinal fat absorption by inhibiting pancreatic lipase; the other is sibutramine (Meridia), which acts in the brain to inhibit deactivation of the neurotransmitters norepinephrine, serotonin, and dopamine, therefore decreasing appetite. However, weight loss with these drugs is modest.

Aromatase inhibitors are an option for obese males facing infertility problems, especially if they have elevated estrogen and lowered testosterone levels. Aromatase inhibitors prevent the excess aromatase enzymes from converting testosterone to estrogen. They interfere with the aromatase p450 enzyme that is highly expressed in white adipose tissue. Currently available aromatase inhibitors include anastrozole, testolactone, and letrozole. Numerous case studies have found this to be an effective treatment in not only restoring normal hormone levels, but also fertility. Raman and Schlegel tested the effects of anastrozole on nonobstructive azoospermic patients who presented with normal or decreased levels of testosterone and elevated levels of estradiol. Anastrozole treatment normalized the testosterone-to-estrogen ratio and total testosterone levels and improved semen parameters [67]. Zumoff et al. [68] found that by inhibiting estrogen biosynthesis (through administration of the aromatase inhibitor testolactone), there was an alleviation of possible infertility as a result of hypogonadotropic hypogonadism in obese male subjects. In another case study, a patient diagnosed with infertility secondary to morbid obesity was treated with the aromatase inhibitor, anastrozole. This led to normalization of the patient's testosterone, LH, and FSH hormone levels, as well as suppression of the serum estradiol levels and the normalization of spermatogenesis and fertility [33]. In a study including normal, overweight, and obese men, treatment with anastrozole led to an increase in testosterone-to-estradiol ratio that occurred in association with increased semen parameters. Anastrozole and testolactone have similar effects on hormonal profiles and semen analysis, but anastrozole appears to be at least as effective as testolactone for treating men with abnormal testosterone-to-estradiol ratio [67].

New directions in pharmacological treatment might include testosterone replacement therapy and maintenance and regulation of adipose-derived hormones, particularly leptin, which is produced by fat cells in the body and known to affect appetite and the body's energy balance, and also reproductive function. Testosterone replacement therapy has been shown to suppress the levels of circulating leptin, although no information regarding the effect of the treatment on semen parameters was reported [16, 38]. Quennell et al. [69] discovered that leptin indirectly regulates gonadotropin-releasing hormone neuronal function, affecting forebrain neurons that induce infertility. By decreasing elevated leptin levels in obese patients, it might be possible to reverse some of the potential suppressive effects of excess leptin on the HPG axis and restore normal spermatogenesis and sperm function. Ghrelin, a hormone which is secreted by cells in the lining of the stomach, also affects appetite and the body's energy balance. Further studies of these hormones may lead to the development of new medications to control appetite and provide an option for treating obesity-related health problems and infertility.

Surgical Options

In vitro fertilization may be an option for obese patients facing problems such as erectile dysfunction or other purely physical fertility problems. Although morbid obesity is associated with unfavorable IVF/ICSI cycle outcome as evidenced by lower pregnancy rates in females, there is no evidence for a contributing male factor when assisted reproductive methods are used [70]. It is recommended that morbidly obese patients undergo appropriate counseling before the initiation of this expensive and invasive therapy. Fortunately, studies show that obesity in men may not adversely affect the results of their partners who are undergoing in vitro fertilization or embryo transfer [71].

Scrotal lipectomy is a treatment option available for infertility in obese men whose excess fat accumulation may be contributing to their infertility, either through increased scrotal temperature or excess toxin accumulation. One-fifth of patients who were previously considered infertile and underwent scrotal lipectomy to remove excess fat were able to achieve a successful pregnancy [28]. For individuals who are severely obese, dietary changes and behavior modification may be accompanied by surgery to reduce or bypass portions of the stomach or small intestine. The risks of obesity surgery have declined in recent years, but it is still only performed on patients for whom other strategies have failed and whose obesity seriously threatens their health.

Bariatric surgery ("weight loss surgery") is the use of surgical intervention in the treatment of obesity by reducing or bypassing portions of the stomach or small intestine. As it is a rather extreme intervention, it is only recommended for severely obese people (BMI>40) who have failed to lose sufficient weight following dietary modification and pharmacological treatment [72]. Gastric bypass and banding surgeries are very effective in the treatment of morbid obesity and its comorbid conditions. One study reported that a significant decrease in estrogen, increase in testosterone, and normalization of other hormone and adipokine levels were experienced by patients who underwent vertical banded gastroplasty [73]. However, others speculate that the drastic weight loss that accompanies this procedure might induce secondary infertility, even though natural weight loss has shown promising results in terms of restoring fertility [74]. Bariatric surgery should therefore not be recommended as a treatment for obesity-linked infertility until extensive, long-term studies have been performed to determine the definite effects on male fertility.

Conclusion

Obesity is a modern-day pandemic with serious comorbidities, both physical and psychological. Studies clearly show that obese men have an increased chance of subfertility and subfecundity due to various mechanisms (physical, genetic, hormonal, adipokine, cytokine) that ultimately lead to ED and abnormal semen parameters. The central factor behind these mechanisms is the abnormal regulation of the HPG axis. An abnormal hormonal profile, and more specifically increased adipose-derived hormones and adipokine levels, may explain the association between BMI, altered semen parameters, and infertility more accurately as it is clearly more complex than can be accounted for simply by abnormal levels of reproductive hormones.

New studies point to many causes for abnormal semen parameters, including genetic markers, excess adipose-derived hormone and adipokine release, as well as oxidative stress. The consistent decrease in inhibin B levels and increase in leptin levels, and specific proteomic sperm changes observed in obese infertile males, all may have negative impacts on spermatogenesis. Increased cytokines, such as IL-6, are connected with oxidative stress and impaired reproduction and could also contribute to abnormal semen parameters observed in obese men facing fertility issues. These markers point toward a true suppression of normal spermatogenesis and sperm quality despite some inconsistency in the results of studies performed to measure the effects of obesity on semen parameters.

In treating obesity-linked sperm disorders and male infertility, few controlled studies have been performed, and effective therapeutic treatments, advice for lifestyle changes, and surgical options should be explored further. In addition, determining the most accurate measure for qualifying patients as obese could more accurately clarify the cause of their infertility and other health issues that might accompany their state of obesity. Neither the reversibility of obesity-associated male infertility in response to weight loss nor effective therapeutic treatments or interventions have been extensively studied. The increasing prevalence of obesity worldwide in conjunction with a perceived declining male sperm count in modern man calls for more research and attention to obesity as an etiology of male infertility. Clinicians should consider obesity when a male patient with idiopathic infertility is confronted.

Expert Commentary

The purpose of this chapter was to discuss obesity as a newly discovered etiology of male infertility. The growing incidence of obesity and apparent decrease in male fertility makes this topic especially relevant when dealing with the infertile male. Although data on abnormal semen parameters and evidence of obesity as it affects male fertility is abundant, the specific mechanisms are not definite. Discovery and proof of significant mechanisms that contribute to this issue are important when treating infertility as it relates to obesity. The recognition of adipose tissue as an endocrine organ and discovery of adipokines have contributed greatly in explaining the mechanisms behind this problem, but it appears many other factors are involved. Further understanding of adipose tissue as an endocrine organ and obesity as a constant inflammatory state will allow in-depth studies of specific adipokines and cytokines involved.

In addition, controlled studies demonstrating all effects of aromatase inhibitors, weight loss, bariatric surgery, and other new therapeutic measures are needed to definitely recommend specific treatments. Lifestyle changes leading to weight loss appear to be of the greatest benefit, not only for male fertility issues but overall health. The pressing issue of obesity in not only the USA, but the whole world, has lasting effects that now appear to include male infertility. It is important, especially as the obesity epidemic grows, for clinicians to consider obesity as explanation to a male with idiopathic infertility or subfertility.

Key Issues

- Obesity is a medical condition in which excess body fat, or white adipose tissue, accumulates in the body to the extent that the excess fat often adversely affects health, often reducing life expectancy.
- The obesity epidemic is growing to concerning proportions, affecting an estimated 700 million people in the next 5 years.
- Altered semen parameters ascribed to obesity include decreased sperm concentration, abnormal morphology, compromised chromatin integrity, and abnormal sperm motility.
- The discovery of adipose tissue as an active endocrine organ secreting adipokines and adipose-derived hormones is an important mechanistic link explaining infertility in many obese men.
- Mechanisms that may explain abnormal semen parameters in obese men include genetic, hormonal, and physical mechanisms that may all contribute to the deregulation of the HPG axis that controls normal spermatogenesis.
- Physical problems that may contribute to decreased fertility include ED; scrotal lipomatosis, which leads to increased scrotal temperatures; and sleep apnea that can cause disruptions in the essential nightly testosterone rise.
- There are solutions that often lead to restoration of fertility that include lifestyle changes, pharmacological interventions, and surgical procedures.
- The rapid rise in the incidence of obesity has prompted researchers to not only look at natural or lifestyle changes as treatment methods but also for new therapies in order to manage the pandemic and its subsequent comorbidities.
- Clinicians should consider obesity when a male patient with idiopathic infertility is confronted.

References

1. World Health Organization. http://www.who.int/nmh/publications/fact_sheet_diet_en.pdf (2009). Accessed 15 July 2009.
2. Popkin BM. Recent dynamics suggest selected countries catching up to US obesity. Am J Clin Nutr. 2010;91(1):284S–8.
3. Swan SH, Elkin EP, Fenster L. The question of declining sperm density revisited: an analysis of 101 studies published 1934–1996. Environ Health Perspect. 2000;108(10):961–6.
4. Sallmen M, Sandler DP, Hoppin JA, Blair A, Baird DD. Reduced fertility among overweight and obese men. Epidemiology. 2006; 17(5):520–3.
5. Carlsen E, Giwercman A, Keiding N, Skakkebaek NE. Evidence for decreasing quality of semen during past 50 years. BMJ. 1992; 305(6854):609–13.
6. Magnusdottir EV, Thorsteinsson T, Thorsteinsdottir S, Heimisdottir M, Olafsdottir K. Persistent organochlorines, sedentary occupation, obesity and human male subfertility. Hum Reprod. 2005;20(1): 208–15.
7. Kyle UG, Bosaeus I, De Lorenzo AD, et al. Bioelectrical impedance analysis—part I: review of principles and methods. Clin Nutr. 2004;23(5):1226–43.
8. World Health Organization. Diet, nutrition and the prevention of chronic diseases. World Health Organ Tech Rep Ser. 2003;916:i–viii. 1–149, backcover.
9. Booth ML, Hunter C, Gore CJ, Bauman A, Owen N. The relationship between body mass index and waist circumference: implications for estimates of the population prevalence of overweight. Int J Obes Relat Metab Disord. 2000;24(8):1058–61.
10. Huxley R, Mendis S, Zheleznyakov E, Reddy S, Chan J. Body mass index, waist circumference and waist: hip ratio as predictors of cardiovascular risk–a review of the literature. Eur J Clin Nutr. 2010;64(1):16–22.
11. Akpinar E, Bashan I, Bozdemir N, Saatci E. Which is the best anthropometric technique to identify obesity: body mass index, waist circumference or waist-hip ratio? Coll Antropol. 2007; 31(2):387–93.
12. Macdonald AA, Herbison GP, Showell M, Farquhar CM. The impact of body mass index on semen parameters and reproductive hormones in human males: a systematic review with meta-analysis. Hum Reprod Update. 2010;16(3):293–311.
13. Hammoud AO, Gibson M, Peterson CM, Hamilton BD, Carrell DT. Obesity and male reproductive potential. J Androl. 2006;27(5): 619–26.
14. Jensen TK, Andersson AM, Jorgensen N, et al. Body mass index in relation to semen quality and reproductive hormones among 1,558 Danish men. Fertil Steril. 2004;82(4):863–70.

15. Chavarro JE, Toth TL, Wright DL, Meeker JD, Hauser R. Body mass index in relation to semen quality, sperm DNA integrity, and serum reproductive hormone levels among men attending an infertility clinic. Fertil Steril. 2010;93(7):2222–31.
16. Hofny ER, Ali ME, Abdel-Hafez HZ, et al. Semen parameters and hormonal profile in obese fertile and infertile males. Fertil Steril. 2010;94(2):581–4.
17. Fejes I, Koloszar S, Szollosi J, Zavaczki Z, Pal A. Is semen quality affected by male body fat distribution? Andrologia. 2005;37(5): 155–9.
18. Aggerholm AS, Thulstrup AM, Toft G, Ramlau-Hansen CH, Bonde JP. Is overweight a risk factor for reduced semen quality and altered serum sex hormone profile? Fertil Steril. 2008;90(3):619–26.
19. Pauli EM, Legro RS, Demers LM, Kunselman AR, Dodson WC, Lee PA. Diminished paternity and gonadal function with increasing obesity in men. Fertil Steril. 2008;90(2):346–51.
20. Nicopoulou SC, Alexiou M, Michalakis K, et al. Body mass index vis-a-vis total sperm count in attendees of a single andrology clinic. Fertil Steril. 2009;92(3):1016–7.
21. Hammoud AO, Wilde N, Gibson M, Parks A, Carrell DT, Meikle AW. Male obesity and alteration in sperm parameters. Fertil Steril. 2008;90(6):2222–5.
22. Fejes I, Koloszar S, Zavaczki Z, Daru J, Szollosi J, Pal A. Effect of body weight on testosterone/estradiol ratio in oligozoospermic patients. Arch Androl. 2006;52(2):97–102.
23. Martini AC, Tissera A, Estofan D, et al. Overweight and seminal quality: a study of 794 patients. Fertil Steril. 2010;94(5):1739–43.
24. Kort HI, Massey JB, Elsner CW, et al. Impact of body mass index values on sperm quantity and quality. J Androl. 2006;27(3):450–2.
25. Evenson D, Wixon R. Meta-analysis of sperm DNA fragmentation using the sperm chromatin structure assay. Reprod Biomed Online. 2006;12(4):466–72.
26. Gopalkrishnan K, Padwal V, Meherji PK, Gokral JS, Shah R, Juneja HS. Poor quality of sperm as it affects repeated early pregnancy loss. Arch Androl. 2000;45(2):111–7.
27. Kriegel TM, Heidenreich F, Kettner K, et al. Identification of diabetes- and obesity-associated proteomic changes in human spermatozoa by difference gel electrophoresis. Reprod Biomed Online. 2009;19(5):660–70.
28. Kasturi SS, Tannir J, Brannigan RE. The metabolic syndrome and male infertility. J Androl. 2008;29(3):251–9.
29. Arsov T, Silva DG, O'Bryan MK, et al. Fat aussie–a new Alstrom syndrome mouse showing a critical role for ALMS1 in obesity, diabetes, and spermatogenesis. Mol Endocrinol. 2006;20(7): 1610–22.
30. Hammoud A, Carrell DT, Meikle AW, et al. An aromatase polymorphism modulates the relationship between weight and estradiol levels in obese men. Fertil Steril. 2010;94(5):1734–8.
31. Strain GW, Zumoff B, Kream J, et al. Mild hypogonadotropic hypogonadism in obese men. Metabolism. 1982;31(9):871–5.
32. Tsai EC, Matsumoto AM, Fujimoto WY, Boyko EJ. Association of bioavailable, free, and total testosterone with insulin resistance: influence of sex hormone-binding globulin and body fat. Diabetes Care. 2004;27(4):861–8.
33. Roth MY, Amory JK, Page ST. Treatment of male infertility secondary to morbid obesity. Nat Clin Pract Endocrinol Metab. 2008;4(7):415–9.
34. Qin DD, Yuan W, Zhou WJ, Cui YQ, Wu JQ, Gao ES. Do reproductive hormones explain the association between body mass index and semen quality? Asian J Androl. 2007;9(6):827–34.
35. Athyros VG, Tziomalos K, Karagiannis A, Anagnostis P, Mikhailidis DP. Should adipokines be considered in the choice of the treatment of obesity-related health problems? Curr Drug Targets. 2010; 11(1):122–35.
36. Bhat GK, Sea TL, Olatinwo MO, et al. Influence of a leptin deficiency on testicular morphology, germ cell apoptosis, and expression levels of apoptosis-related genes in the mouse. J Androl. 2006;27(2):302–10.
37. Trayhurn P, Beattie JH. Physiological role of adipose tissue: white adipose tissue as an endocrine and secretory organ. Proc Nutr Soc. 2001;60(3):329–39.
38. Isidori AM, Caprio M, Strollo F, et al. Leptin and androgens in male obesity: evidence for leptin contribution to reduced androgen levels. J Clin Endocrinol Metab. 1999;84(10):3673–80.
39. Wang P, Mariman E, Renes J, Keijer J. The secretory function of adipocytes in the physiology of white adipose tissue. J Cell Physiol. 2008;216(1):3–13.
40. Wozniak SE, Gee LL, Wachtel MS, Frezza EE. Adipose tissue: the new endocrine organ? A review article. Dig Dis Sci. 2009; 54(9):1847–56.
41. Jope T, Lammert A, Kratzsch J, Paasch U, Glander HJ. Leptin and leptin receptor in human seminal plasma and in human spermatozoa. Int J Androl. 2003;26(6):335–41.
42. Bialas M, Fiszer D, Rozwadowska N, Kosicki W, Jedrzejczak P, Kurpisz M. The role of IL-6, IL-10, TNF-alpha and its receptors TNFR1 and TNFR2 in the local regulatory system of normal and impaired human spermatogenesis. Am J Reprod Immunol. 2009;62(1):51–9.
43. Fraczek M, Kurpisz M. Inflammatory mediators exert toxic effects of oxidative stress on human spermatozoa. J Androl. 2007;28(2): 325–33.
44. Lampiao F, du Plessis SS. TNF-alpha and IL-6 affect human sperm function by elevating nitric oxide production. Reprod Biomed Online. 2008;17(5):628–31.
45. Dandona P, Aljada A, Chaudhuri A, Mohanty P, Garg R. Metabolic syndrome: a comprehensive perspective based on interactions between obesity, diabetes, and inflammation. Circulation. 2005; 111(11):1448–54.
46. Davi G, Falco A. Oxidant stress, inflammation and atherogenesis. Lupus. 2005;14(9):760–4.
47. Agarwal A, Sharma RK, Nallella KP, Thomas Jr AJ, Alvarez JG, Sikka SC. Reactive oxygen species as an independent marker of male factor infertility. Fertil Steril. 2006;86(4):878–85.
48. Winters SJ, Wang C, Abdelrahaman E, Hadeed V, Dyky MA, Brufsky A. Inhibin-B levels in healthy young adult men and prepubertal boys: is obesity the cause for the contemporary decline in sperm count because of fewer Sertoli cells? J Androl. 2006; 27(4):560–4.
49. Ramaswamy S, Marshall GR, McNeilly AS, Plant TM. Evidence that in a physiological setting Sertoli cell number is the major determinant of circulating concentrations of inhibin B in the adult male rhesus monkey (Macaca mulatta). J Androl. 1999;20(3):430–4.
50. Bener A, Al-Ansari AA, Zirie M, Al-Hamaq AO. Is male fertility associated with type 2 diabetes mellitus? Int Urol Nephrol. 2009;41(4):777–84.
51. Agbaje IM, Rogers DA, McVicar CM, et al. Insulin dependant diabetes mellitus: implications for male reproductive function. Hum Reprod. 2007;22(7):1871–7.
52. Lima N, Cavaliere H, Knobel M, Halpern A, Medeiros-Neto G. Decreased androgen levels in massively obese men may be associated with impaired function of the gonadostat. Int J Obes Relat Metab Disord. 2000;24(11):1433–7.
53. Jarow JP, Kirkland J, Koritnik DR, Cefalu WT. Effect of obesity and fertility status on sex steroid levels in men. Urology. 1993;42(2):171–4.
54. Globerman H, Shen-Orr Z, Karnieli E, Aloni Y, Charuzi I. Inhibin B in men with severe obesity and after weight reduction following gastroplasty. Endocr Res. 2005;31(1):17–26.
55. Vgontzas AN, Papanicolaou DA, Bixler EO, et al. Sleep apnea and daytime sleepiness and fatigue: relation to visceral obesity, insulin resistance, and hypercytokinemia. J Clin Endocrinol Metab. 2000;85(3):1151–8.

56. Luboshitzky R, Lavie L, Shen-Orr Z, Herer P. Altered luteinizing hormone and testosterone secretion in middle-aged obese men with obstructive sleep apnea. Obes Res. 2005;13(4):780–6.
57. Cheng JY, Ng EM. Body mass index, physical activity and erectile dysfunction: an U-shaped relationship from population-based study. Int J Obes (Lond). 2007;31(10):1571–8.
58. Tamler R. Diabetes, obesity, and erectile dysfunction. Gend Med. 2009;6 Suppl 1:4–16.
59. Jung A, Schuppe HC. Influence of genital heat stress on semen quality in humans. Andrologia. 2007;39(6):203–15.
60. Hjollund NH, Bonde JP, Jensen TK, Olsen J. Diurnal scrotal skin temperature and semen quality. The Danish First Pregnancy Planner Study Team. Int J Androl. 2000;23(5):309–18.
61. Kaukua J, Pekkarinen T, Sane T, Mustajoki P. Sex hormones and sexual function in obese men losing weight. Obes Res. 2003; 11(6):689–94.
62. Niskanen L, Laaksonen DE, Punnonen K, Mustajoki P, Kaukua J, Rissanen A. Changes in sex hormone-binding globulin and testosterone during weight loss and weight maintenance in abdominally obese men with the metabolic syndrome. Diabetes Obes Metab. 2004;6(3):208–15.
63. Sharman MJ, Volek JS. Weight loss leads to reductions in inflammatory biomarkers after a very-low-carbohydrate diet and a low-fat diet in overweight men. Clin Sci (Lond). 2004;107(4):365–9.
64. Ziccardi P, Nappo F, Giugliano G, et al. Reduction of inflammatory cytokine concentrations and improvement of endothelial functions in obese women after weight loss over one year. Circulation. 2002;105(7):804–9.
65. Lau DC, Douketis JD, Morrison KM, et al. 2006 Canadian clinical practice guidelines on the management and prevention of obesity in adults and children [summary]. CMAJ. 2007;176(8): S1–13.
66. National Institute of Diabetes and Digestive and Kidney Diseases. Prescription medications for the treatment of obesity. http://win.niddk.nih.gov/publications/prescription.htm#fdameds. Accessed 20 July 2009.
67. Raman JD, Schlegel PN. Aromatase inhibitors for male infertility. J Urol. 2002;167(2 Pt 1):624–9.
68. Zumoff B, Miller LK, Strain GW. Reversal of the hypogonadotropic hypogonadism of obese men by administration of the aromatase inhibitor testolactone. Metabolism. 2003;52(9):1126–8.
69. Quennell JH, Mulligan AC, Tups A, et al. Leptin indirectly regulates gonadotropin-releasing hormone neuronal function. Endocrinology. 2009;150(6):2805–12.
70. Awartani KA, Nahas S, Al Hassan SH, Al Deery MA, Coskun S. Infertility treatment outcome in sub groups of obese population. Reprod Biol Endocrinol. 2009;7:52.
71. Dechaud H, Anahory T, Reyftmann L, Loup V, Hamamah S, Hedon B. Obesity does not adversely affect results in patients who are undergoing in vitro fertilization and embryo transfer. Eur J Obstet Gynecol Reprod Biol. 2006;127(1):88–93.
72. Encinosa WE, Bernard DM, Chen CC, Steiner CA. Healthcare utilization and outcomes after bariatric surgery. Med Care. 2006; 44(8):706–12.
73. Bastounis EA, Karayiannakis AJ, Syrigos K, Zbar A, Makri GG, Alexiou D. Sex hormone changes in morbidly obese patients after vertical banded gastroplasty. Eur Surg Res. 1998;30(1):43–7.
74. di Frega AS, Dale B, Di Matteo L, Wilding M. Secondary male factor infertility after Roux-en-Y gastric bypass for morbid obesity: case report. Hum Reprod. 2005;20(4):997–8.

Part IV

Nutrition and Antioxidants Role for Male Reproductive Health

Nutritional Pathways to Protect Male Reproductive Health

34

Tung-Chin Hsieh and Paul Shin

Abstract

Nutrition is an essential component of one's overall health. Many common disease processes can be alleviated or prevented by a healthy diet. First reports of antioxidant deficiency and decreased male fertility can be traced back to over 50 years ago. With the understanding of oxidative damage to spermatogenesis, most of the nutritional research has focused on the role of antioxidants in improving male fertility. However, there are not any randomized controlled trials studying whole food diet in infertile male patients. Patients are often counseled based on data extrapolated from antioxidant supplement studies. This chapter is intended to give an overview of contemporary research on nutrient and male reproductive health with guidance to natural food source that contain high levels of antioxidants.

Keywords

Nutrition • Male infertility • Oxidative damage to spermatogenesis • Antioxidants and male fertility • Nutritional supplements • L-Carnitine • Zinc • Selenium • Vitamin C

Nutrition is an essential component of one's overall health. Many common disease processes can be alleviated or prevented by a healthy diet. First reports of antioxidant deficiency and decreased male fertility can be traced back to over 50 years ago [1]. With the understanding of oxidative damage to spermatogenesis, most of the nutritional research has focused on the role of antioxidants in improving male fertility. However, there are not any randomized controlled trials studying whole food diet in infertile male patients. Patients are often counseled based on data extrapolated from antioxidant supplement studies. This chapter is intended to give an overview of contemporary research on nutrient and male reproductive health with guidance to natural food source that contain high levels of antioxidants.

T.-C. Hsieh, MD (✉)
Department of Urology, Baylor College of Medicine,
One Baylor Plaza, Houston, TX 77030, USA

P. Shin, MD
Urologic Surgeons of Washington,
2021 K St. NW, Suite 408, Washington, DC 20006, USA
e-mail: pshin@dcurology.net

Nutrients and Male Reproductive Health

Arginine

Arginine is a semi-essential amino acid because it can be synthesize by the human body from glutamine, glutamate, and praline. It plays an important role in cell division, wound healing, immune function, hormone production, and ammonia metabolism. Arginine has significant effects on endothelial function since it is a precursor for nitric oxide synthesis. Not surprisingly, it is involved in the pathophysiology of many vascular disorders including vasogenic erectile dysfunction [2].

Arginine is required for normal spermatogenesis. Researchers found that adult men on an arginine-deficient diet had decreased sperm counts and increased percentage of nonmotile sperm [1]. Oral administration of arginine (500 mg/day) to infertile men for 6–8 weeks has shown improvement in sperm counts, motility, and conception rates [3–6]. However, similar benefits were not observed in patients with baseline sperm counts less than 10 million/mL [7].

S.J. Parekattil and A. Agarwal (eds.), *Male Infertility: Contemporary Clinical Approaches, Andrology, ART & Antioxidants*,
DOI 10.1007/978-1-4614-3335-4_34, © Springer Science+Business Media, LLC 2012

De novo biosynthesis, however, does not produce sufficient arginine; dietary intake remains the primary determinant of plasma arginine levels. It is considered an essential nutrient for human children but not adults by US Department of Agriculture (USDA) [8]. Currently, there is not a consensus on the daily recommended intake of arginine; studied doses range from 1 to 15 g/day. Although no significant adverse effects have been observed in studied doses, patients with impaired renal or hepatic dysfunction might be unable to metabolize arginine properly. The effect of arginine on the airway is also unclear, therefore precaution should be taken in asthmatic patients. Animal sources of arginine include dairy products, turkey, pork, and beef. Vegetable sources include seeds, soybeans, and nuts.

Zinc

Zinc is an essential micromineral. There are 2–4 g of zinc throughout the human body with the highest concentrations in the prostate and parts of the eye [9]. It is a metalloprotein cofactor for DNA binding. Copper/zinc superoxide dismutase is involved in the repair of damaged DNA. It has an important role in testes development and sperm physiologic functions. Zinc deficiency is associated with hypogonadism, testicular/seminiferous tubular atrophy, and inadequate development of secondary sexual characteristics [10].

Semen analysis of fertile and infertile men showed a positive correlation between low zinc levels and poor sperm quality [11]. Treating asthenozoospermia men with zinc (200 mg twice per day) for 3 months showed an improvement in sperm parameters: an increase in the seminal antioxidant capacity and a reduction of oxidative status. Researchers postulated that poor zinc nutrition can impair antioxidant defenses, be a risk factor in oxidant release, and compromise the mechanism of DNA repair, making the sperm cells highly susceptible to oxidative damage [11, 12]. Currently, there is limited data available in humans to establish a dietary dosage to achieve adequate concentration in seminal plasma.

The daily recommended dietary allowance of zinc is 8 mg/day for women and 11 mg/day for men [8]. Excess zinc absorption (>15 mg/day) can interfere with copper and iron absorption, cholesterol metabolism, and cause anosmia. Animal sources of zinc include red meat, oysters, and liver. Vegetable sources include seeds, nuts, and whole grains.

Selenium

Similar to zinc, selenium is also an essential micronutrient. It functions as a cofactor for reduction of antioxidant enzymes, such as glutathione peroxidase. Although rare in healthy, well-nourished adult human, selenium deficiency is associated with reduced or impaired reproduction throughout the animal kingdom [13]. A sperm-specific selenoprotein has been identified and suspected to play a key role in selenium deficiency-induced subfertility [14].

Population studies on the effect of selenium in subfertile men have yielded conflicting data [15, 16]. In a randomized, double-blinded study, treating subfertile men with selenium (100 μg) once a day showed no influence on sperm count but an improvement on sperm motility with 56% response rate when compared with placebo [17].

The recommended daily allowance for selenium is 70 μg/day for men. Selenosis can occur when intake reaches the level greater than 400 μg resulting in cirrhosis, pulmonary edema, and death. In Europe, there is a documented decline in the mean intake from 60 μg/day in the 1970s to 30 μg/day in the 1990s due to a change in the source of cereals for bread making [18]. Animal sources of selenium include meat, fish, and eggs. Vegetable sources include Brazil nuts, wheat/cereals, and soy products.

Vitamin C

Ascorbic acid is an essential nutrient for humans and other animal species. It has been associated with fertility for many years, since it is a key compound in gonadal physiology. However, the precise mechanism of action has not been elucidated. Most consider the effect of vitamin C on fertility is related to its three principal functions: promotion of collagen synthesis, role in hormone production, and protection or prevention against oxidation.

Early reports on the effects of vitamin C on male fertility were based on animal studies. Ascorbate deficiency was associated with poor breeding performance and degeneration of the testicular germinal epithelium [19, 20]. The gonadal growth-enhancing effects of gonadotropins were enhanced by simultaneous treatment with ascorbic acid [21]. In human studies, low ascorbate level has been associated with low sperm counts, increased number of abnormal sperm, reduced motility, and agglutination [22]. Dietary treatment with vitamin C has yielded mixed data on improvement of sperm parameters [23]. No current randomized controlled trials show an improvement in the semen parameters or pregnancy rates of healthy infertile men who take oral supplementation of vitamin C.

The recommended daily allowance of vitamin C is 90 mg/day for adult male and 75 mg/day for adult female [8]. Dietary intake should not exceed 2,300 mg/day since intoxication can lead to gastrointestinal disturbances, iron poisoning, and hemolytic anemia in patients with glucose-6-phosphate dehydrogenase deficiency. A well-balanced diet without supplementation is generally accepted to be sufficient to meet the daily requirement for vitamin C, except those who are pregnant or smoke tobacco. The highest natural sources are fruit and vegetables: black currant, red pepper, and guava.

Vitamin E

Vitamin E is a lipid-soluble antioxidant. It protects cell membranes from oxidation by reacting with free radicals generated during lipid peroxidation. Various forms of vitamin E had been identified, and the exact role and importance of the isoforms is still unclear. The motility of the spermatozoa depends on the integrity of the mitochondrial sheath which is composed of phospholipids and can be damaged by lipid peroxidation [24]. Therefore, vitamin E has been hypothesized to be an important factor in maintaining overall health of sperm.

In asthenozoospermic male, an increased concentration of the peroxidation by-product (malondialdehyde, MDA) was observed in the semen plasma. Treating these patients with vitamin E (100 mg/day) in a randomized, double-blinded fashion showed improvement in sperm motility and decreased MDA concentration, resulting in 11/52 successful pregnancies [25]. Although more well-designed studies were available, randomized controlled trials showed conflicting results regarding improvement in semen parameters [26, 27]. Since vitamin C has been shown to work synergistically with vitamin E, many studies examining the effect of combination therapy in infertile men were conducted and failed to show any improvement in semen parameters [28, 29].

The recommended daily intake of vitamin E is 15 mg (30 international unit, IU) per day for adults [8]. At dose of greater than 1,000 mg (1,500 IU) per day, there is an increased risk of hemorrhage and death. The best sources of vitamin E are nuts, seeds, and vegetable oil along with green leafy vegetable and fortified cereals.

L-Carnitine

Carnitine is a semi-essential nutrient that can be biosynthesized from lysine and methionine by the liver and kidneys. Two stereoisoforms exist with L-carnitine as the bioactive form. It is involved in the metabolism of long-chain fatty acids and serves as an antioxidant by removing acetyl-CoA that is responsible for mitochondrial lipid peroxidation [30]. The highest concentration of carnitine occurs in the epididymis with epididymal concentrations 2,000-fold higher than in plasma [31].

A multicenter, uncontrolled trial showed that oral administration of L-carnitine (3 g/day) for 4 months in asthenozoospermia patients resulted in an improvement in sperm motility, linearity index, rapid linear progression, and mean velocity [32]. In a randomized, double-blinded, placebo-controlled trial, L-carnitine therapy (2 g/day) for 4 months in infertile men showed improvements in sperm concentration and motility [33]. Despite the observed success in improvement of sperm motility with carnitine supplementation, other randomized controlled studies have not been able to replicate similar results [34].

Seventy-five percent of carnitine that is present in human is derived from diet [35]. Currently, there is not a recommended daily allowance of carnitine intake or any detrimental reports of carnitine overdose. Oral intake greater than 1 g/day did not show any advantage since absorption studies indicate saturation at this level. The highest concentration of carnitine is found in red meat and dairy products. Vegetable sources include nuts, seeds, and asparagus.

Factors Contributing to Subfertility

Obesity

Multiple population-based studies suggest an increased risk of subfertility among obese couples [36]. In women, there are extensive researches performed on the effects of extremes of body composition on fertility by altered menstrual function [37]. Epidemiologic studies have observed a higher incidence of male factor infertility in obese male [38, 39]. Obese men often exhibit an altered reproductive hormonal profile: decreased androgen, sex hormone-binding globulin (SHBG), and inhibin B levels along with elevated estrogen levels [40, 41]. Other contributing factors of male obesity to increase risk for infertility are altered lifestyle and increased risks for sexual dysfunction.

Currently, there is limited data on the reversibility of obesity-associated male infertility with weight loss. A small randomized controlled trial showed increases in SHBG and testosterone (free and total) after weight loss from 10 weeks of a very low-energy diet and behavior modification program [42]. Other studies on the effect of weight loss, both surgical and diet/lifestyle modification programs showed mixed improvement in hormonal profile and sperm parameters [36].

Alcohol

Alcohol abuse has been shown to cause impaired testosterone production and testicular atrophy resulting in impotence, infertility, and reduced secondary sexual characteristics [43]. It has been shown to have a deleterious effect on all levels of male reproductive system: altered hypothalamic–pituitary–gonadal axis, Leydig and Sertoli cell dysfunction, and altered spermatogenesis leading to spermatogenic arrest and Sertoli cell-only syndrome in advanced cases [44, 45].

In an uncontrolled study, semen analysis of alcohol users showed altered sperm count, morphology, and motility. The association is most significant when alcohol consumption is greater than 40 g/day [43]. There are reports of potential beneficial effects of alcohol consumption on fertility since red wine has been shown to exert protection against oxidative damage [46]. Currently, the dose-dependent effects of alcohol on male factor infertility are not well understood.

Five-Year View and Key Issues

Diet modification and nutritional supplements are popular areas of concern for the infertile patient. Altering one's nutritional habits is usually easily done, for little to no cost, and empowers the patient to feel as though they are actively tackling their condition. From our survey of the literature, it should be apparent that no consensus exists regarding the best pathway or outcome to maximize fertility. Basic science research does not always translate into clinical success especially in the arena of infertility. Avoidance of toxicants, proper nutrition, and stress reduction are general guidelines that all infertile men should follow.

References

1. Holt L, Albanese A. Observations on amino acid deficiencies in man. Trans Assoc Am Physicians. 1944;58:143–56.
2. Appleton J. Arginine: clinical potential of a semi-essential amino acid. Altern Med Rev. 2002;7(6):512–22.
3. Tanimura J. Studies on arginine in human semen. Part II. The effects of medication with L-arginine-HCl on male infertility. Bull Osaka Med School. 1967;13:84–9.
4. De Aloysio D, Mantuano R, Mauloni M, et al. The clinical use of arginine aspartate in male infertility. Acta Eur Fertil. 1982;13: 133–67.
5. Scibona M, Meschini P, Capparelli S, et al. L-Arginine and male infertility. Minerva Urol Nefrol. 1994;46:251–3.
6. Schacter A, Goldman JA, Zukerman Z, et al. Treatment of oligospermia with amino acid arginine. J Urol. 1973;110:311–3.
7. Pryor J, Blandy J, Evans P, et al. Controlled clinical trail of arginine for infertile men with oligozoospermia. Br J Urol. 1978;50:47–50.
8. National Academy of Sciences. Institute of Medicine. Food and Nutrition Board. Dietary guidance: DRI tables. US Department of Agriculture, National Agricultural Library and National Academy of Sciences, Institute of Medicine, Food and Nutrition Board. 2009.
9. Wapnir R. Protein nutrition and mineral absorption. Boca Raton: CRC; 1990.
10. Bedwal R, Bahuguna A. Zinc, copper and selenium in reproduction. Experientia. 1994;50:626–40.
11. Colagar A, Marzony E, Chaichi M. Zinc levels in seminal plasma are associated with sperm quality in fertile and infertile men. Nutr Res. 2009;29:82–8.
12. Omu A, Al-Azemi M, Kehinde E, et al. Indications of the mechanism involved in improved sperm parameters by zinc therapy. Med Princ Pract. 2008;17:108–16.
13. Underwood EJ. Trace elements in human and animal nutrition. New York: Academic; 1977. p. 321.
14. Behne D, Weiss-Nowak C, Kalcklosch M, et al. Studies on new mammalian seleno proteins. Proceedings of the eighth international symposium on trace elements in man and animals. 1993. pp. 516–23.
15. Abou-Shakra R, Ward N, Everard D. The role of trace elements in male infertility. Fertil Steril. 1989;52:502–10.
16. Xu B, Chia SE, Tsakok M, et al. Trace elements in blood and seminal plasma and the relationship to sperm quality. Reprod Toxicol. 1993;7:613–8.
17. Scott R, Macpherson A, Yates R, et al. The effect of oral selenium supplementation on human sperm motility. Br J Urol. 1998;82: 76–80.
18. Macpherson A, Barclay M, Dixon J, et al. Decline in dietary selenium intake in Scotland and effects on plasma concentrations. In: Anke M, Meissner D, Mills CF, editors. Trace elements in man and animals—TEMA 8. Gersdorf Germany: Verlag Media Tourisik; 1993. p. 269–70.
19. Counsell J, Hornig D. Vitamin C ascorbic acid. Essex: Applied Science; 1981.
20. Phillips P, Lardy H, Jeizer E, et al. Sperm stimulation in the bull through the subcutaneous administration of ascorbic acid. J Dairy Sci. 1040;23:873–8.
21. Dr Cio A, Schteingart M. The influence of ascorbic acid on the activity of gonadotrophic hormones. Endocrinology. 1942;30: 263–4.
22. Wilson I. Sperm agglutinations in human serum and blood. Proc Soc Exp Biol Med. 1954;85:652–5.
23. Luck M, Jeyaseelan I, Scholes R. Ascorbic acid and fertility. Biol Reprod. 1995;52:262–6.
24. Palamanda J, Kehrer J. Involvement of vitamin E and protein thiols in the inhibition of microsomal lipid peroxidation by glutathione. Lipids. 1993;28:427–31.
25. Suleiman S, Ali M, Zaki Z, et al. Lipid peroxidation and human sperm motility: protective role of vitamin E. J Androl. 1996;17(5):530–7.
26. Geva E, Bartoov B, Zabuludovsky N, et al. The effect of antioxidant treatment on juman spermatozoa and fertilization rate in and in vitro fertilization program. Fertil Steril. 1996;66:430–4.
27. Kessopoulou E, Powers H, Sharma K, et al. A double-blind randomized placebo cross-over controlled trail using the antioxidant vitamin E to treat reactive oxygen species associated male infertility. Fertil Steril. 1996;64:825–31.
28. Greco E, Iacobelli M, Rienzi I, et al. Reduction of the incidence of sperm DNA fragmentation by oral antioxidant treatment. J Androl. 2005;26:349–53.
29. Rolf C, Cooper T, Yeung C, et al. Antioxidant treatment of patients with asthenozoospermia or moderate oligoasthenozoospermia with high-dose vitamin C and vitamin E: a randomized, placebo-controlled, double-blind study. Hum Repod. 1999;14:1028–33.
30. Agarwal A, Said T. Carnitines and male infertility. Reprod Biomed Online. 2004;8:376–84.
31. Hinton B, Snoswell A, Setchell B. The concentration of carnitine in the luminal fluid of the testis and epididymis of the rate and some other mammals. J Reprod Fertil. 1979;56:105–11.
32. Costa M, Canale D, Filicori M, et al. L-Carnitine in idiopathic asthenozoospermia: a multicenter study Italian study group on carnitine and male infertility. Andrologia. 1994;26:155–9.
33. Lenzi A, Lombardo F, Sgro P, et al. Use of carnitine therapy in selected cases of male factor infertility: a double-blind crossover trail. Fertil Steril. 2003;79(2):292–300.
34. Patel S, Sigman M. Antioxidant therapy in male infertility. Urol Clin North Am. 2008;35:319–30.
35. Peluso G, Nicolai R, Reda E, et al. Cancer and anticancer therapy-induced modifications on metabolism mediated by carnitine system. J Cell Physiol. 2000;182:339–50.
36. Hammoud A, Gibson M, Peterson M, et al. Impact of male obesity on infertility: a critical review of the current literature. Fertil Steril. 2008;90(4):897–904.

37. Pasquali R, Pelusi C, Genghini S, et al. Obesity and reproductive disorders in women. Hum Reprod Update. 2003;9:359–72.
38. Alavanja M, Sandler D, McMaster S, et al. The agricultural health study. Environ Health Perspect. 1996;104:362–9.
39. Magnusdottir E, Thorsteinsson T, Thorsteinsdottir S, et al. Persistent organochlorines, sedentary occupation, obesity and human male subfertility. Hum Reprod. 2005;20:208–15.
40. Stellato R, Feldman H, Hamdy O, et al. Testosterone, sex hormone-binding globulin and the development of type 2 diabetes in middle-age men: prospective results from the massachussetts male aging study. Diabetes Care. 2000;23:490–4.
41. Globerman H, Shen-Orr Z, Karnieli E, et al. Inhibin B in men with severe obesity and after weight reduction following gastroplasty. Endocr Res. 2005;31:17–26.
42. Kaukua J, Pekkarinen T, Sane T, et al. Sex hormones and sexual function in obese men losing weight. Obes Res. 2003;11:689–94.
43. Gaur D, Talekar M, Pathak V. Alcohol intake and cigarette smoking: impact of 2 major lifestyle factors on male infertility. Indian J Pathol Microbiol. 2010;53(1):35–40.
44. Emanuele M, Emanuele N. Alcohol's effects on male reproduction. Alcohol Health Res World. 1998;22:195–218.
45. Zhu Q, Van Thiel D, Gavaler J. Effect of ethanol on rate sertoli cell function: studies in vitro and in vivo. Alcohol Clin Exp Res. 1997;21:1409–17.
46. Perez D, Strobel P, Foncea R, et al. Wine, diet, antioxidant defenses and oxidative damage. Ann N Y Acad Sci. 2002;957:136–45.

Natural Antioxidants

35

Giancarlo Balercia, Antonio Mancini, and Gian Paolo Littarru

Abstract

An excess of reactive oxygen species (ROS) and other oxidant radicals has been associated with male infertility. The total oxyradical scavenging capacity (TOSC) is a recently developed assay measuring the overall capability of biological fluids or cellular antioxidants to neutralize the toxicity of various oxyradicals. The TOSC assay can discriminate between different forms of ROS, allowing to identify the role of specific antioxidants, or their pathway of formation in the onset of toxicological or pathological processes. The previous application of TOSC assay in andrology led us to show a reduced antioxidant efficiency in seminal fluid of infertile men with a significant correlation between the scavenging capacity towards hydroxyl radicals and parameters of sperm cell motility. Despite the fact that oxidative stress is well recognized as a cause of male infertility, the use of antioxidants as a treatment is still debated, and it is considered as a "supplementation" therapy, rather than an etiological or physiopathological therapy, since no clear correlation has been investigated between a real deficiency of a specific antioxidant and the effect of oral supplementation. Various models have been introduced to explore the protective role of different antioxidants in vitro, and some differences can be discovered regarding the protective effects exerted by specific enzymatic or non-enzymatic molecules. We focus our attention on two main natural antioxidants, the efficacy of which has been supported by clinical trials: coenzyme Q_{10} and carnitine.

Keywords

Male infertility • Reactive oxygen species • Coenzyme Q_{10} • L-Carnitine • Antioxidant radicals • Total oxyradical scavenging capacity • Oxidative stress

An excess of reactive oxygen species (ROS) and other oxidant radicals has been associated with male infertility [1–8]. The total oxyradical scavenging capacity (TOSC) is a recently developed assay measuring the overall capability of biological fluids or cellular antioxidants to neutralize the toxicity of various oxyradicals [9, 10]. The TOSC assay can discriminate between different forms of ROS, allowing to identify the role of specific antioxidants, or their pathway of formation in the onset of toxicological or pathological processes. The previous application of TOSC assay in andrology led us to show a reduced antioxidant efficiency in seminal fluid of infertile men with a significant correlation between the scavenging capacity towards hydroxyl radicals and parameters of sperm cell motility [11].

G. Balercia, MD (✉)
Andrology Unit, Endocrinology, Department of Internal Medicine and Applied Biotechnologies, Umberto I Hospital, School of Medicine, Marche Polytechnic University, Ancona 60100, Italy

Department of Clinical and Molecular Sciences, Marche Polytechnic University, Umberto I Hospital, Ancona 60100, Italy
e-mail: g.balercia@univpm.it

A. Mancini, MD
Department of Internal Medicine, Division of Endocrinology, The Catholic University of Sacred Heart, Rome, Italy
e-mail: amancini@rm.unicatt.it

G.P. Littarru, MD
Dipartimento di Scienze Cliniche Specialistiche ed Odontostomatologiche, Universita Politecnica delle Marche, Vai Ranieri 65, Ancona 60131, Italy
e-mail: g.littarru@univpm.it

S.J. Parekattil and A. Agarwal (eds.), *Male Infertility: Contemporary Clinical Approaches, Andrology, ART & Antioxidants*,
DOI 10.1007/978-1-4614-3335-4_35, © Springer Science+Business Media, LLC 2012

Despite the fact that oxidative stress is well recognized as a cause of male infertility, the use of antioxidants as a treatment is still debated, and it is considered as a "supplementation" therapy, rather than an etiological or physiopathological therapy, since no clear correlation has been investigated between a real deficiency of a specific antioxidant and the effect of oral supplementation. Various models have been introduced to explore the protective role of different antioxidants in vitro, and some differences can be discovered regarding the protective effects exerted by specific enzymatic or non-enzymatic molecules [12]. We focus our attention on two main natural antioxidants, the efficacy of which has been supported by clinical trials: coenzyme Q_{10} and carnitine.

Coenzyme Q_{10}

Among natural antioxidant, a special role is covered by coenzyme Q_{10} (CoQ_{10}), also called ubiquinone for its wide distribution in different plants, animals and tissues. Coenzyme Q_{10} is a crucial component of the mitochondrial oxidative phosphorylation process because of its role in redox link between flavoproteins and cytochromes in the inner mitochondrial membrane; it has also many other functions, first of all the antioxidant activity, and new roles in different cellular functions were recently highlighted: this molecule can participate in redox reactions also in not mitochondrial cellular reactions, as in lysosomes, in Golgi apparatus and in plasma membranes [13], also contributing to membrane fluidity. Moreover, coenzyme Q_{10} can participate in many aspects of the redox control of the cellular signalling origin and transmission; in fact, the auto-oxidation of semiquinone, formed in various membranes during electron transport, can be a primary source for H_2O_2 generation, which activates some transcription factors, such as NF-kB, to induce gene expression [14]. It is also possible that ROS generation could suppress other genes. There are also some indications regarding the involvement of CoQ_{10} in cellular proliferation at least two aspects of the stimulation of cellular proliferation concerning CoQ_{10} [15, 16]. Cellular growth stimulation could be based on the activation of an oxidase in the plasma membrane, which should need CoQ_{10} to transfer electrons through the membrane, in which is also present [17]. On the other hand, the exposition of cells to serum can induce the apoptosis, while coenzyme Q_{10} allows a normal cellular division. Further studies are necessary to better understand these issues.

Clinically, the significance of antioxidant action of ubiquinone has been clarified by many studies concerning lipoproteins both in vitro and in vivo. In fact, LDLs are molecules particularly susceptible to the oxidative damage with cytotoxic products generation, associated with vascular contractile response alterations and atherosclerosis [18]. It has been demonstrated that reduced CoQ_{10} present in LDL is oxidized before vitamin E and the appearance of fatty acid hydroperoxides occurs only after the oxidation of ubiquinol [19]. Moreover, treatment per os with CoQ_{10} in normal subjects induces an increase of ubiquinol in plasma and lipoproteins and an augmented resistance to LDL peroxidation [20]. Based on the above-mentioned biochemical features, different trials have highlighted the potential therapeutic usefulness of CoQ_{10} in the treatment of various diseases (cardiovascular, neurologic, muscular, immunologic, diabetic endotheliopathy).

There is a relationship between low concentrations of CoQ_{10} and coronary pathologies, even if this correlation is not so strong to be considered a casual relation [21]. The ubiquinol/ubiquinone ratio is considered an oxidative stress marker in coronary pathologies, and the LDL/CoQ_{10} ratio is an index of coronary risk [22].

Both the bioenergetic and the antioxidant roles of CoQ_{10} suggested a possible involvement in male fertility: it is known that a large amount of mitochondria are present in spermatozoa, in which motility requires a high energy expenditure [23]; moreover, as shown in the previous paragraph, the protection of membrane from oxidative stress could play a role in preserving sperm integrity; finally, the biosynthetic machinery for CoQ_{10} is present at remarkably high levels in rat testis [24].

Original studies on CoQ_{10} administration in unselected population of infertile patients showed an amelioration of the results in membrane integrity tests (swelling test) [25] and an improvement in seminal parameters in men with sperm pathology [26]; however, these studies did not report the endogenous CoQ_{10} levels in such patients.

The first analytical data on CoQ_{10} levels in seminal fluid were produced by our group [27], in a sample including 60 subjects (21 patients with normozoospermia, 15 patients with azoospermia or oligozoospermia, 2 patients with germ-free genital tract inflammation and 22 subjects with varicocele, 7 of whom presented oligo-azoospermia). We showed that CoQ_{10} was assayable in total seminal fluid and in seminal plasma; its levels showed a good correlation with sperm count ($R=0.504$, $p<0.0005$) and motility ($R=0.261$, $p<0.05$), except in the population of varicocele patients, in whom the correlation with sperm count was maintained ($R=0.666$, $p<0.0005$) but that with sperm motility was completely lacking ($R=0.008$, N.S.). Moreover, in the varicocele patients, a significantly higher proportion of total CoQ_{10} was present in seminal plasma when compared with normal subjects or other infertile patients without varicocele (the ratio plasma/seminal fluid Q_{10} was $69\pm7.1\%$ vs 41. $2\pm5.6\%$, $p<0.01$, respectively).

These data were also confirmed in larger series of patients [12, 28, 29].

Since CoQ_{10} in seminal plasma did not correlate with LDH levels, we concluded that the amounts of CoQ_{10} in plasma were not due to spermatozoa damage and to a consequent release of ubiquinone from the cells. We hypothesized that seminal plasma CoQ_{10} levels reflect an interchange between cellular and extracellular compartments, with a pathophysiological meaning similar to serum ubiquinone values [30, 31]; a relative deficiency or utilization of CoQ in sperm cell was therefore supposed to be present in varicocele condition [32]. We also hypothesized that the significantly higher percent of CoQ_{10} in plasma found in VAR patients could reflect an altered compartment distribution: the intracellular, bioenergetic use of CoQ_{10} could be defective in these patients, and its shift towards the plasma compartment could be considered of relevance regarding a possible antioxidant role in that environment.

Finally, we studied VAR patients, after surgical repair; only a partial reversion was observed, since the ratio plasma-to-total CoQ_{10} decreased, but the correlation between total CoQ_{10} and motility was not restored; on the contrary, the peculiar correlation between cellular CoQ_{10} and motility was no more detectable in post-operative VAR patients [33]. Vitamin E has also been demonstrated to be positively affected by surgical VAR repair [34].

In order to explore physiological hormone control of seminal CoQ_{10}, since FSH seemed to be involved in regulation of total antioxidant capacity of seminal plasma [35], another trial was conducted in 13 oligoasthenozoospermic subjects, studied before and after 3 months of rh-FSH (225 UI/week) [36]. Following FSH treatment, CoQ_{10} showed an increase, although not significant, in seminal plasma levels (0.035 ± 0.010 μg/ml vs 0.028 ± 0.005). A possible role of systemic thyroid hormones has been also hypothesized, since seminal TAC inversely correlated with free T3 in infertile patients [37].

All these referred studies consider total CoQ_{10} levels, irrespective of its redox status. The first report on the assay of reduced and oxidized forms of ubiquinone was performed in our laboratory [38]. We showed a significant correlation between the reduced form (ubiquinol) and sperm count in seminal plasma, an inverse correlation between ubiquinol and hydroperoxide levels both in seminal plasma and seminal fluid, a strong correlation—using multiple regression analysis—between sperm count, motility and ubiquinol-10 content in seminal fluid, and, finally, an inverse correlation between ubiquinol/ubiquinone ratio and the percentage of abnormal forms. These results indicate an important role of ubiquinol-10 in inhibiting hydroperoxide formation. We also found a lower ubiquinol/ubiquinone ratio in sperm cells from idiopathic asthenozoospermic (IDA) patients and in seminal plasma from IDA and varicocele-associated asthenozoospermic (VARA) patients compared to controls [39]. The important conclusion was that the QH_2/Q_{OX} ratio may be an index of oxidative stress and its reduction a risk factor for semen quality. Sperm cells characterized by low motility and abnormal morphology, equipped with low CoQ_{10} content, could be less capable in counteracting oxidative stress, which could lead to a reduced QH_2/Q_{OX} ratio.

Concerning the therapeutic role of CoQ_{10}, it should be mentioned that CoQ_{10} was first introduced as an ethical drug for heart failure patients but its use has grown since its recognition as a food supplement aimed at improving cellular bioenergetics, counteracting oxidative stress and slowing down some age-related pathologies. Numerous clinical studies have shown its efficacy as an adjunctive therapy in cardiovascular and neurodegenerative diseases and in mitochondrial myopathies [40]. The above-mentioned studies constitute a rationale that eventually led us to treat infertile subjects with exogenous CoQ_{10}.

Lewin and Lavon [41] originally reported the effect of CoQ_{10} on sperm motility in vitro: a significant increase in motility was observed in sperm obtained from asthenozoospermic men, incubated with exogenous CoQ_{10}, while no significant variation was reported in the motility of sperm cells from normal subjects. The same study also reports the effect of exogenous CoQ_{10} in vivo, in a group of patients with low fertilization rates, after in vitro fertilization with intracytoplasmatic sperm injection for male factor infertility: no significant changes were reported in most sperm parameters, but a significant improvement was noticed in fertilization rates after a treatment with 60 mg/day for a mean of 103 days.

CoQ_{10} is one of the compounds contributing to the total antioxidant buffer capacity of semen, and its decrease could lead to an impairment of the system in counteracting oxidative stress [11]; exogenous administration of CoQ_{10} could increase its content in semen and improve sperm cell function. In different clinical models, such as hypogonadism and hypoadrenalism, a significant correlation has been discovered between CoQ_{10} and total antioxidant capacity, measured as latency phase in the formation of radicals using the system H_2O_2-metmyoglobin/ABTS system [42, 43].

To investigate a potential therapeutic role, we administered CoQ_{10} to a group of 22 idiopathic asthenozoospermic infertile patients [44], classified according to the WHO 1999 criteria [45] as having <50% forward motile forms at two distinct sperm analyses and normal sperm morphology >30%.

Patients were given CoQ_{10} (Pharma Nord, Denmark), 200 mg/day divided into two doses, for 6 months. Semen analysis, including computer-assisted sperm analysis and motility (C.A.S.A.), CoQ_{10} and phosphatidylcholine assays, was performed at baseline and after 6 months of therapy. A semen analysis was further performed after 6 months from interruption of therapy (washout). CoQ_{10} levels were assayed in sperm cells and seminal plasma using a Beckman Gold HPLC System HPLC (Beckman Instruments, San Ramon, CA, USA) equipped with an electrochemical detector (EC, ESA 5100, Bedford, MA, USA) [39]. PC was essentially determined according to Frei et al. [46].

An increase of CoQ_{10} was found in seminal plasma after treatment, the mean value rising significantly from 42.0±5.1 at baseline to 127.1±1.9 ng/ml after 6 months of exogenous CoQ_{10} administration ($p<0.005$). A significant increase of CoQ_{10} content was also detected in sperm cells (from 3.1±0.4 to 6.5±0.3 ng/10^6 cells; $p<0.05$). Similarly, PC levels increased significantly both in seminal plasma and sperm cells after treatment (from 1.49±0.50 to 5.84±1.15 μM, $p<0.05$; and from 6.83±0.98 to 9.67±1.23 nmoles/10^6 cells, $p<0.05$, respectively).

Regarding semen, a significant difference was found in forward (class a+b) motility of sperm cells after 6 months of CoQ_{10} dietary implementation (from 9.13±2.50 to 16.34±3.43%, $p<0.05$). The improvement of motility was also confirmed by means of computer-assisted determination of kinetic parameters. A significant increase of VCL (from 26.31±1.50 to 46.43±2.28 μm/s, $p<0.05$) and VSL (from 15.20±1.30 to 20.40±2.17 μm/s, $p<0.05$) was found after treatment. No significant differences were found in sperm cell concentration and morphology.

Although a direct correlation was not found (data not shown), a positive dependence (using the Cramer's index of association) was evident among the relative variations, baseline and after treatment, of seminal plasma or intracellular CoQ_{10} content and of C.A.S.A. (VCL and VSL) kinetic parameters (Cramer's $V=0.4637$; 0.3818; 0.3467; 0.5148, respectively) [44].

A significant reduction in sperm forward motility was reported after 6 months of washout (from 16.34±3.43 to 9.50±2.28%, $p<0.001$), while no significant differences were found in sperm cell concentration and morphology.

In order to find out whether different responses were age related, the relative variations (before and after treatment) of CoQ_{10} and PC content in seminal plasma and sperm cells, as well as forward motility, were analysed, but no correlation was found (data not shown).

Wives of 3 out of 22 patients (13.6%) achieved spontaneous pregnancy within 3 months from the discontinuation of therapy (2.4% pregnancy rate per cycle).

This study indicates a significant improvement of kinetic features of sperm cells after 6 months of administration of CoQ_{10}, both on the basis of manual and computer-assisted evaluation. Moreover, these results constitute the first demonstration that exogenous administration of CoQ_{10} increases its levels in seminal plasma and in spermatozoa.

The increment was important, especially in seminal plasma where post-treatment levels were three times higher than basal ones. Similar increases of CoQ_{10} concentration (two–three times higher than baseline value) are commonly found in blood plasma after chronic administration of the quinone [47]. As CoQ_{10} is a highly lipophilic molecule, we could reasonably hypothesize its diffusion through the phospholipid bilayer of cellular membranes, but we presently do not know whether transport from blood plasma to testicular and accessory male genital glands is passive or involves an active mechanism.

Statistical analysis did not reveal any significant functional relationship among the therapy-induced variations of CoQ_{10} and kinetic parameters of spermatozoa, probably due to the low number of samples. Nevertheless, the good degree of association among these variables, according to Cramer's *V* index of association, supports the hypothesis of a pathogenetic role of CoQ_{10} in asthenozoospermia, according to previously reported data [39]. Improvement of the spontaneous pregnancy rate also suggests that this therapeutic approach is beneficial.

These results were confirmed by a double-blind, placebo-controlled clinical trial, also from our group [48]. The selected patients underwent a double-blind therapy with CoQ_{10} (Q-absorb soft gels, Jarrow Formulas LA, USA), containing 100 mg of CoQ_{10}, lecithin and medium-chain glycerides Placebo had the same composition but the soft gels did not contain any CoQ_{10}. All patients were given a total of two soft capsules in two separate daily administrations, with meals. The CoQ_{10} dose was similar to that used in our previous open trial on male infertility.

The study design was 1-month run-in, 6 months of therapy (30 patients) or placebo (30 patients), and further 3 months' follow-up (controls at months T−1, T0, T+3, T+6, T+9).

CoQ_{10} levels increased in seminal plasma after treatment, the mean value rising significantly from 61.29±20.24 at baseline to 99.39±31.51 ng/ml after 6 months of exogenous CoQ_{10} administration ($p<0.0001$). A significant increase of CoQ_{10} content was also detected in sperm cells (from 2.44±0.97 to 4.57±2.46 ng/10^6 cells, $p<0.0001$). Similarly, QH_2 levels increased significantly both in seminal plasma and sperm cells after treatment (from 31.54±10.05 to 51.93±16.44 ng/ml, $p<0.0001$; and from 0.95±0.46 to 1.84±1.03 ng/10^6 cells, $p<0.0001$, respectively). No statistically significant modifications were found in the placebo group.

A significant improvement of sperm cell total motility (from 33.14±7.12 to 39.41±6.80%, $p<0.0001$) and forward motility (from 10.43±3.52 to 15.11±7.34%, $p=0.0003$) was observed in the treated group after 6 months (T+6) of CoQ_{10} administration. The improvement of sperm cell kinetic parameters was also confirmed after computer-assisted analysis, with an increase both in VCL (from 27.99±5.32 to 33.18±4.22 μm/s, $p<0.0001$) and VSL (from 10.76±2.63 to 13.13±2.86 μm/s $p<0.0001$) after treatment. No statistically significant modifications in kinetic parameters were found in placebo group.

A significant inverse correlation between baseline (T0) and T+6 relative variations of seminal plasma or intracellular

CoQ_{10} or QH_2 content and kinetic parameters was also found in treated group. In fact, patients with lower baseline value of motility and levels of CoQ_{10} had a statistically significant higher probability to be responders to the treatment.

After washout (T+9), sperm cell kinetic features (total and forward motility, VSL) resulted significantly reduced in treatment groups when compared with month T+6.

Nine spontaneous pregnancies were achieved during the observation period. After opening the randomization list, it was found that six of the patients who had impregnated their female partner had undergone CoQ_{10} therapy (three of them after 4 months, one after 5 months and one after 6 months of treatment). Three out of the nine pregnancies occurred in partners of patients undergoing placebo treatment, respectively, one after 2 months of therapy and the other two after 3 months of washout. Recently, a positive effect of CoQ_{10} treatment on sperm motility was also confirmed in a study by Safarinejad et al. [49].

Carnitine

L-Carnitine (LC) plays a central role in cellular energetic metabolism, being the shuttle of the activated long-chain fatty acids (acyl-CoA) into the mitochondria, where the beta-oxidation takes place [50–52]. An important role in sperm cell metabolism is strongly suggested by the high levels found in epididymal fluid due to an active secretory mechanism [53], and there are evidences that the initiation of the sperm motility is related to an increase of L-carnitine in the epididymal lumen and L-acetyl-carnitine (LAC) in sperm cells [54–56].

We have performed a 6-month double-blind randomized placebo-controlled trial using LC or LAC or combined LC and LAC treatment in infertile males affected by idiopathic asthenozoospermia [57]. The evaluation of the effectiveness of these treatments in improving semen kinetic parameters and the variation of TOSC in semen after treatment were the end points of the study. Sixty patients (mean age, 30 years; range, 24–38 years) affected by idiopathic asthenozoospermia have been enrolled in the study. The patients were selected at the Andrology Unit of Endocrinology, Umberto I Hospital, Polytechnic University of Marche, Ancona (Italy). All subjects underwent medical screening, including history and clinical examination, and presented a clinical history of primary infertility >2 years. Testicular volume was evaluated in each patient using Prader's orchidometer. To accomplish a complete diagnosis, the following investigations were also performed: semen analysis; Mar-test (SperMar test, CGA, Florence, Italy) for anti-spermatozoa antibodies (Ab); sperm culture and urethral specimens collection for *Chlamydia* and *Mycoplasma ureoliticum* detection; FSH, LH, testosterone (T), estradiol (E2) and prolactin (PRL) assays, using commercial radioimmunoassay kits; testicular, prostatic and seminal vesicle ultrasonography and echo-colour Doppler of venous spermatic plexus, for anatomical abnormalities and varicocele detection. No female-related factor was apparently involved in sterility, since all partners (mean age, 26 years; range, 21–32 years) were ovulating regularly, as formally proven by biphasic basal body temperature and luteal phase progesterone levels; no anatomical abnormalities were detected after ultrasound ovary and uterus evaluation; no abnormal Fallopian tube anatomy was detected after hysterosalpingography.

The selected patients were submitted to a double-blind therapy of LC (10 ml phials containing 3 g/day orally of Carnitene—Sigma Tau, Italy, n. 15 patients), LAC (tablets containing 3 g/day orally of Zibren—Sigma Tau, n. 15 patients), a combination of LC (10 ml phials containing 2 g/day orally of Carnitene) and LAC (tablets containing 1 g/day orally of Zibren) (n. 15 patients) or a seemingly identical placebo (each 10 ml placebo phial contains malic acid, sodium benzoate, sodium saccharinate dihydrate, anhydrous sodium citrate, pineapple flavouring, demineralised water; each placebo tablet contains a core with 1-hydro lactose, magnesium stearate, polyvinylpyrrolidone, cornstarch and a coating with cellulose acetophtalate, dimethicone, ethyl phthalate, Sigma Tau, Pomezia, Rome, Italy). All patients assumed a total of one phial and two tablets three times a day. The study design was 1-month run-in, 6 months of therapy (45 patients) or placebo (15 patients), and further 3 months' follow-up (controls at months T−1, T0, T+3, T+6, T+9). Monthly evaluation of two semen samples before the beginning of treatment (T−1, T0) was carried out to test semen parameter stability in each patient, as recommended by the WHO [45]. At various time points, the following analyses were carried out: semen analysis at months T−1, T0, T+3, T+6, T+9, including computer-assisted sperm analysis (C.A.S.A.) at months T0, T+3, T+6, T+9 to evaluate modifications in semen parameters and TOSC of the seminal fluid towards different ROS at months T0 and T+6 to evaluate any variations during therapy [9, 10].

Table 35.1 reports mean and standard deviation of sperm variables at each time; the percentage variations with respect to baseline are reported in Table 35.2. The percentage variations of total and forward sperm motility in all groups at each time are shown in Figs. 35.1 and 35.2. The univariate analysis of variance performed on variables (percentage variations with respect to T−1 or T0) for the homogeneity at baseline (T0 or T+3) showed that there were no significant differences between groups regarding motility (total and forward), sperm concentration, atypical sperm cells, semen volume and VCL. On the contrary, the percentage variation of VSL between T0 and T+3 was significantly higher in the

Table 35.1 Descriptive statistics of sperm variables at each time: mean ± standard deviation

Treatment	Month T−1	Month T0	Month T+3	Month T+6	Month T+9
Sperm total motility					
Placebo	43.73 ± 10.06	43.93 ± 10.26	44.60 ± 7.68	43.40 ± 9.85	42.73 ± 10.02
LC	54.33 ± 8.59	51.67 ± 11.08	59.93 ± 8.04	64.53 ± 8.41	54.27 ± 8.96
LAC	45.07 ± 12.01	43.87 ± 11.36	56.47 ± 11.56	60.43 ± 10.46	50.57 ± 5.71
LC and LAC	46.73 ± 10.10	44.53 ± 11.84	55.13 ± 10.15	61.07 ± 9.07	49.00 ± 7.80
Sperm forward motility					
Placebo	24.33 ± 7.93	24.13 ± 7.74	22.33 ± 7.76	24.00 ± 8.50	23.20 ± 8.96
LC	33.47 ± 6.55	31.20 ± 7.43	38.93 ± 7.09	43.80 ± 7.12	34.00 ± 7.02
LAC	27.00 ± 10.87	25.53 ± 10.43	34.93 ± 9.24	37.50 ± 9.20	30.21 ± 7.84
LC and LAC	25.47 ± 8.90	24.60 ± 9.40	33.87 ± 8.37	38.13 ± 8.23	28.47 ± 8.27
Sperm concentration					
Placebo	35.27 ± 21.98	29.53 ± 10.07	31.40 ± 12.85	33.73 ± 14.36	30.13 ± 9.30
LC	35.47 ± 9.21	39.00 ± 10.39	41.00 ± 17.34	45.53 ± 21.42	39.40 ± 13.93
LAC	27.07 ± 6.47	30.40 ± 10.80	39.33 ± 18.05	39.57 ± 19.99	31.21 ± 8.60
LC and LAC	29.93 ± 10.57	29.40 ± 9.39	36.93 ± 19.71	37.40 ± 16.42	33.27 ± 13.62
Atypical sperm cells					
Placebo	66.40 ± 6.50	68.20 ± 5.86	67.40 ± 6.42	67.27 ± 6.71	67.53 ± 7.42
LC	63.13 ± 5.04	62.87 ± 4.69	58.47 ± 6.20	54.87 ± 7.27	58.07 ± 11.82
LAC	65.93 ± 8.19	67.13 ± 7.06	61.73 ± 6.82	58.93 ± 5.62	60.93 ± 10.12
LC and LAC	65.40 ± 6.22	67.13 ± 6.01	61.73 ± 5.86	59.60 ± 5.82	61.53 ± 8.84
Semen volume					
Placebo	2.97 ± 1.36	3.01 ± 0.83	3.08 ± 0.85	2.75 ± 0.68	2.82 ± 0.45
LC	2.96 ± 0.74	3.12 ± 1.04	3.10 ± 0.68	3.18 ± 0.93	3.03 ± 0.83
LAC	2.89 ± 0.85	2.59 ± 0.63	2.71 ± 0.62	3.03 ± 0.66	2.76 ± 0.51
LC and LAC	3.05 ± 0.94	2.87 ± 0.88	2.75 ± 0.80	2.69 ± 0.78	2.50 ± 0.41
Curvilinear velocity					
Placebo		41.67 ± 14.14	39.67 ± 14.07	42.87 ± 6.83	46.33 ± 11.96
LC		41.73 ± 13.47	47.73 ± 13.43	57.13 ± 13.95	41.67 ± 6.28
LAC		39.73 ± 12.57	45.87 ± 11.33	51.79 ± 6.17	44.79 ± 8.19
LC and LAC		43.00 ± 12.02	44.93 ± 15.72	51.40 ± 13.71	42.53 ± 7.78
Straight progressive velocity					
Placebo		24.47 ± 15.19	21.80 ± 12.23	15.87 ± 2.47	17.67 ± 2.58
LC		20.00 ± 7.87	21.13 ± 6.92	21.47 ± 3.52	16.80 ± 2.11
LAC		23.27 ± 16.28	18.60 ± 5.78	20.36 ± 3.41	16.36 ± 2.41
LC and LAC		18.60 ± 6.93	25.67 ± 12.75	22.53 ± 10.26	16.73 ± 2.89

subgroup C (LC–LAC combined) than in the placebo group (44.86 ± 76.24 vs −4.59 ± 37.05; $F=3.077$; $p=0.035$)[1]; in other words, patients treated with the combination of the two molecules improved significantly during the first 3-month period of the administration. A significant improvement in total sperm motility was found in patients to whom LAC was administered, either alone or combined with LC (from −3.3 ± 17.4 at T0 to 37.7 ± 27.8 at T+6; $F=11.19$; $p=0.001$)[2] (Fig. 35.3). The analysis of forward sperm cell motility showed the same results (from −2.9 ± 26 at T0 to 63 ± 66.8 at T+6; $F=12.68$; $p=0.001$)[2] (Fig. 35.4). An improvement of forward motility was found when combined LC–LAC was compared with LC or LAC therapy alone, although the variations of kinetics sperm parameters were not significantly (see Fig. 35.2). No significant modifications were found in placebo group. In all carnitine therapy groups, a significant dependence of the total and forward motility variations on the baseline values was found, and patients with lower baseline values of motility had a significantly higher probability to be responders to the treatment (Table 35.3). After washout (T+9), sperm cell kinetic features (total and forward motility, VSL) resulted significantly reduced in treatment groups when compared with month T+6. In the group to whom LAC was administered (alone or combined), the sperm concentration varied significantly during the treatment period (from 6.95 ± 22.06 at T0 to 42.88 ± 50.80 at T+6; $F=3.611$; $p=0.015$)[2]. A significant reduction of atypical sperm cells was also evident between T0 and T+6; in particular, the improvement was significantly different in the LC. No significantly different variations in semen volume and in VCL were detected in the studied patients. Table 35.4 reports mean and standard

Table 35.2 Percentage variations with respect to baseline for sperm variables at each time: mean ± standard deviation

Treatment	Month T0	Month T+3	Month T+6	Month T+9
Sperm total motility[a]				
Placebo	1.30 ± 13.52	4.15 ± 16.26	0.52 ± 17.43	−1.22 ± 17.46
LC	−5.32 ± 9.91	11.01 ± 8.46	19.90 ± 12.74	0.31 ± 9.65
LAC	−1.31 ± 23.07	30.83 ± 33.99	41.25 ± 29.98	19.78 ± 28.51
LC and LAC	−5.16 ± 10.18	19.59 ± 16.63	34.46 ± 26.20	6.75 ± 14.77
Sperm forward motility[a]				
Placebo	−0.30 ± 9.81	−8.31 ± 13.73	−0.96 ± 18.99	−4.38 ± 21.71
LC	−5.92 ± 16.21	18.01 ± 15.73	33.08 ± 17.29	2.67 ± 14.40
LAC	−4.09 ± 27.85	41.85 ± 42.43	56.84 ± 54.23	22.32 ± 29.03
LC and LAC	−1.86 ± 25.16	43.58 ± 46.86	68.70 ± 78.18	17.73 ± 28.61
Sperm concentration[a]				
Placebo	−4.31 ± 27.75	−3.17 ± 20.06	6.98 ± 36.97	−1.51 ± 32.69
LC	11.14 ± 19.58	17.18 ± 45.99	24.50 ± 35.55	12.06 ± 30.16
LAC	6.95 ± 22.06	46.55 ± 55.80	42.88 ± 50.80	15.90 ± 32.10
LC and LAC	1.26 ± 19.66	24.34 ± 41.83	29.78 ± 41.53	15.18 ± 42.83
LC and LAC	1.26 ± 19.66	24.34 ± 41.83	29.78 ± 41.53	15.18 ± 42.83
Atypical sperm cells[a]				
Placebo	2.92 ± 4.68	1.70 ± 5.97	1.49 ± 6.50	1.86 ± 7.46
LC	−0.28 ± 4.86	−7.42 ± 6.23	−13.24 ± 7.66	−7.51 ± 19.46
LAC	2.20 ± 5.48	−6.00 ± 6.90	−10.24 ± 5.56	−6.32 ± 18.07
LC and LAC	2.90 ± 6.66	−5.27 ± 8.13	−8.35 ± 10.41	−5.36 ± 14.29
Semen volume[a]				
Placebo	11.76 ± 31.35	19.75 ± 62.98	4.85 ± 40.95	12.03 ± 51.56
LC	3.63 ± 12.23	6.58 ± 16.51	11.26 ± 34.40	4.91 ± 25.76
LAC	−5.20 ± 22.97	0.00 ± 27.80	11.63 ± 39.62	2.09 ± 32.97
LC and LAC	−4.94 ± 17.07	−8.27 ± 16.62	−5.38 ± 38.63	−13.74 ± 19.09
Curvilinear velocity[b]				
Placebo		2.20 ± 54.71	20.16 ± 67.30	31.24 ± 79.69
LC		26.00 ± 67.68	64.99 ± 104.59	20.06 ± 74.02
LAC		30.95 ± 66.09	53.31 ± 71.53	31.23 ± 62.97
LC and LAC		8.27 ± 33.87	29.05 ± 51.98	8.44 ± 44.17
Straight progressive velocity[b]				
Placebo		−4.59 ± 37.05	−22.08 ± 27.78	−15.04 ± 25.40
LC		11.08 ± 22.75	17.07 ± 32.81	−6.45 ± 29.37
LAC		2.05 ± 43.11	15.57 ± 46.31	−6.78 ± 38.88
LC and LAC		44.86 ± 76.24	33.26 ± 70.62	−3.15 ± 26.25

[a]Percentage variations with respect to month T−1
[b]Percentage variations with respect to month T0

deviation of TOSC values at each time and the percentage variations with respect to baseline.

TOSC assay of the seminal fluid towards different ROS showed a significant improvement for both hydroxyl and peroxyl radicals in the treatment groups (Table 35.4), while no significant modifications were found in placebo group. The increase in TOSC values, between T0 and T+6, was positively correlated with the improvement of kinetic features, i.e. with total motility, forward motility for both the radicals, and with VCL or VSL respectively for hydroxyl radicals and peroxyl radicals (Tables 35.5 and 35.6). Moreover, forward motility variation was found depending on the baseline values of TOSC (hydroxyl radicals) (see Table 35.3) [11].

Several controlled and uncontrolled studies support at present a potential positive effect of therapy with LC and its acyl derivatives in selected forms of oligo-astheno-teratozoospermia [58–61]. In particular, a very recent controlled study reports the efficacy of LC and LAC combined treatment in improving sperm motility, especially in patients with lower baseline levels [62]. The main rationale is based on the central role of carnitine in energetic metabolism and its accumulation in epididymal fluid and spermatozoa, both as free and acetylated LC [53]. Although some evidences suggest a secondary role of carnitine as antioxidant [63], its effective role and the mechanism of action still remain an interesting open question.

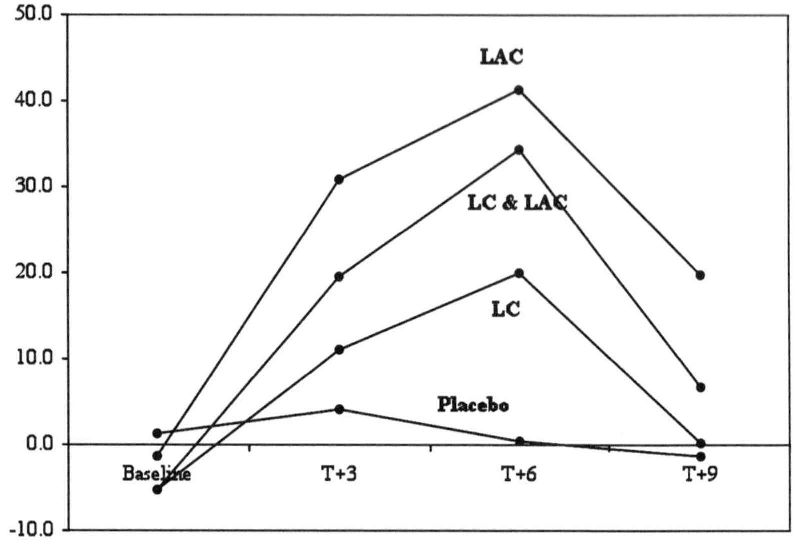

Fig. 35.1 Total sperm motility at each time in the four treatment groups: percentage variations with respect to T−1

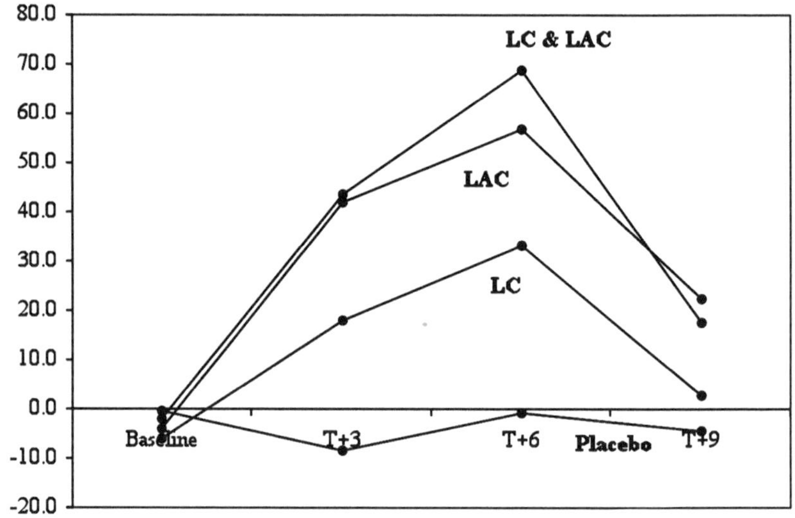

Fig. 35.2 Forward sperm motility at each time in the four treatment groups: percentage variations with respect to T−1

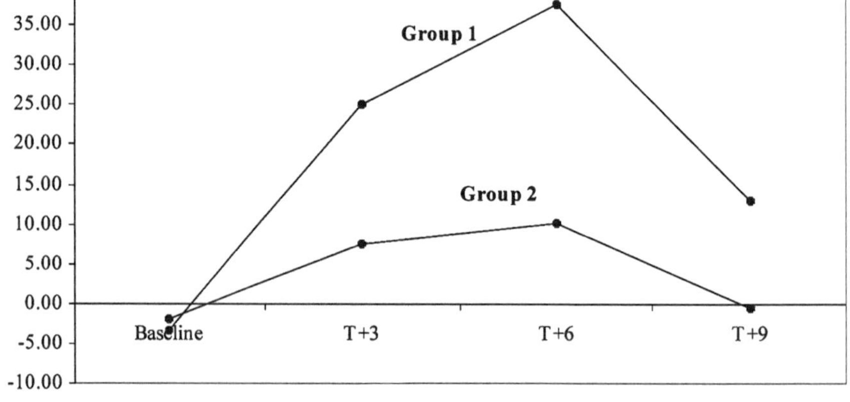

Fig. 35.3 Total sperm motility at each time: percentage variations with respect to T−1 (Time: $p<0.001$; LACTX: $p=0.001$). Group 1: patients treated with LAC, alone or combined. Group 2: patients treated with LC or placebo

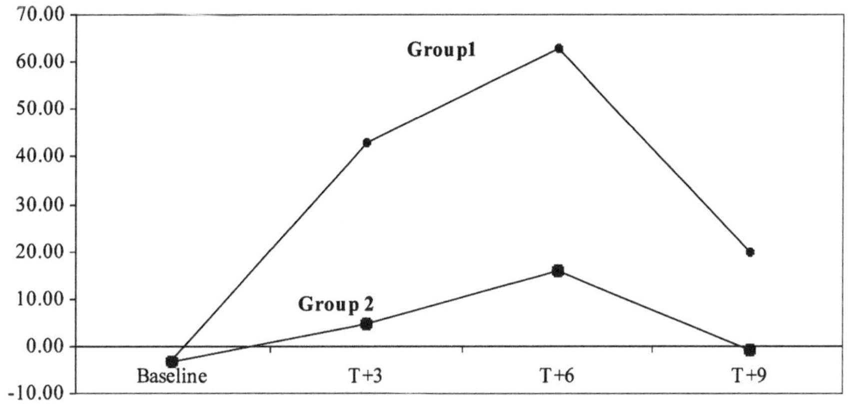

Fig. 35.4 Forward sperm motility at each time: percentage variations with respect to T−1 (Time: $p<0.001$; LACTX: $p=0.001$). Group 1: patients treated with LAC, alone or combined. Group 2: patients treated with LC or placebo

Table 35.3 Analysis of responders: logistic models

	Estimate	Pr(>|z|)	Exp(Estimate)
Dependent variable: variation of total motility during treatment period			
Sperm total motility (T0)	−0.434	0.016	0.648
Dependent variable: variation of forward motility during treatment period			
Sperm forward motility (T0)	−0.235	0.039	0.790
TOSC hydroxyl (T0)	0.0002	0.030	1.0002

Table 35.4 Descriptive statistics of TOSC variables at each time; absolute and percentage variations: mean ± standard deviation

Treatment	Month T0	Month T+6	Absolute variations	Percentage variations
TOSC (hydroxyl)				
PLAC	31,276.67±5,467.22	28,931±5,351.48	−2,345.53±3,516.51	−7.12±10.43
LC	26,301±6,127.76	30,636.20±5,646.47	4,334.67±3,454.27	18.52±15.49
LAC	27,566±6,139.02	31,645.79±4,680.91	4,382.64±3,883.61	20.67±27.19
LC*LAC	27,207±6,061.33	31,712.67±5,933.07	4,505.07±2,730.59	18.38±12.72
TOSC (peroxyl)				
PLAC	24,879.20±4,865.76	24,358.20±5,466.47	−521.00±1,548.79	−2.43±6.79
LC	26,035.60±4,098.60	31,003.00±5,841.52	4,967.40±3,900.30	19.48±15.86
LAC	22,775.73±5,675.95	24,924.07±6,344.44	2,474.00±3,487.15	12.32±21.72
LC*LAC	25,899.00±5,147.45	29,889.13±5,345.76	3,990.13±2,533.32	16.53±12.64

Table 35.5 Correlations between the increase of TOSC values (hydroxyl radicals) and the variation of sperm variables

		Total motility
TOSC	Pearson correlation	0.391
	Sig. (two-tailed)	0.002
		Forward motility
TOSC	Pearson correlation	0.455
	Sig. (two-tailed)	<0.001
		Curvilinear velocity (VCL)
TOSC	Pearson correlation	0.357
	Sig. (two-tailed)	0.006

Table 35.6 Correlations between the increase of TOSC values (peroxyl radicals) and the variation of sperm variables

		Total motility
TOSC	Pearson correlation	0.410
	Sig. (two-tailed)	0.001
		Forward motility
TOSC	Pearson correlation	0.439
	Sig. (two-tailed)	0.001
		Straight progressive velocity (VSL)
TOSC	Pearson correlation	0.316
	Sig. (two-tailed)	0.015

Key Issues

Endogenous CoQ_{10} is significantly related to sperm count and motility, as one could expect considering its important cellular compartmentalization; furthermore, it appears to be one of the most important antioxidants in seminal plasma. Its presence in this compartment does not depend on sperm lysis, as it does not correlate with LDH [27]; moreover, its distribution between intra- and extracellular compartments seems to be an active process, which is profoundly disturbed in VAR patients [28]. CoQ_{10} levels in seminal plasma do correlate with sperm motility. It can be hypothesized that, in certain circumstances, the increased oxidative stress in sperm cells can somehow over-consume CoQ_{10} to the detriment of its bioenergetic role.

Improved sperm motility upon exogenous CoQ_{10} administration could be explained on the basis of the well-known involvement of CoQ_{10} in mitochondrial bioenergetics and of its widely recognized antioxidant properties. Regarding the first point, it is well known that mitochondrial concentration of CoQ_{10} in mammals is close to its K_M, as far as NADH oxidation is concerned, and therefore is not kinetically saturated [64]. In these conditions, one might reasonably hypothesize that a small increase in mitochondrial CoQ_{10} leads to a relevant rise in respiratory velocity. The resulting improvement of oxidative phosphorylation might well affect sperm cells. Since low PC levels in semen were found to be related to a reduction of the phospholipid pool and to low antioxidant capacity [65], the increased PC content in semen after treatment might reasonably involve the restoration of scavenger equilibrium. Another possible reason for this finding is that increased levels of CoQ_{10} also need an appropriate, highly concentration of a lipid carrier.

Thus, the administration of CoQ_{10} may play a positive role in the treatment of asthenozoospermia, probably related not only to its function in mitochondrial respiratory chain but also to its antioxidant properties. The increased concentration of CoQ_{10} in seminal plasma and sperm cells, the improvement of semen kinetic features after treatment, and the evidence of a direct correlation between CoQ_{10} concentrations and sperm motility strongly support a cause/effect relationship.

As far as carnitine is concerned, taken together, the data available in literature seem to suggest that long-term carnitine therapy is effective in improving sperm function and fertilization capacity. Following the end points of our study, we can conclude that (1) LC and LAC administration is effective in increasing sperm motility in patients affected by idiopathic asthenozoospermia, especially in the ones with lower baseline values and lower baseline TOSC; combined LC–LAC therapy improves sperm cell forward motility (not significantly) and VSL (significantly) when compared with LC or LAC therapy alone and (2) the administration of both LAC and LC improves the total scavenging capacity of the seminal fluid in the same population.

A deeper insight into these molecular mechanisms could lead to a greater knowledge of the so-called unexplained infertility.

References

1. Alvarez JG, Storey B. Spontaneous lipid peroxidation in rabbit epididymal spermatozoa: its effect on sperm motility. Biol Reprod. 1982;27:1102–8.
2. Aitken RJ, Clarkson JS. Cellular basis of defective sperm function and its association with the genesis of reactive oxygen species by human spermatozoa. J Reprod Fertil. 1987;81:459–69.
3. Aitken RJ, Clarkson JS, Fishel S. Generation of reactive oxygen species, lipid peroxidation and human sperm function. Biol Reprod. 1989;40:183–97.
4. Rao B, Soufir JC, Martin M, David G. Lipid peroxidation in human spermatozoa as related to midpiece abnormalities and motility. Gamete Res. 1989;24:127–34.
5. Suleiman SA, Ali ME, Zaki ZM, El-Malik EM, Nasr MA. Lipid peroxidation and human sperm motility: protective role of vitamin E. J Androl. 1996;17:530–7.
6. Aitken RJ, Krausz C. Oxidative stress, DNA damage and the Y chromosome. Reproduction. 2001;122:497–506.
7. Aitken RJ, Baker MA. Reactive oxygen species generation by human spermatozoa: a continuing enigma. Int J Androl. 2002;25:191–4.
8. Balercia G, Moretti S, Vignini A, Magagnini M, Mantero F, Boscaro M, et al. Role of nitric oxide concentration on human sperm motility. J Androl. 2004;25:245–9.
9. Winston GW, Regoli F, Dugas Jr AJ, Fong JH, Blanchard KA. A rapid chromatographic assay for determining oxyradical scavenging capacity of antioxidants and biological fluids. Free Radic Biol Med. 1998;24:480–93.
10. Regoli F, Winston GW. Quantification of total antioxidant scavenging capacity (TOSC) of antioxidants for peroxynitrite, peroxyl radicals and hydroxyl radicals. Toxicol Appl Pharmacol. 1999;156:96–105.
11. Balercia G, Mantero F, Armeni T, Principato G, Regoli F. Oxyradical scavenging capacity toward different reactive species in seminal plasma and sperm cells. A possible influence on kinetic parameters. Clin Chem Lab Med. 2003;41:13–9.
12. Mancini A, Meucci E, Bianchi A, Milardi D, De Marinis L, Littarru GP. Antioxidant systems in human seminal plasma: physiopathological meaning and new perspectives. In: Panglossi HV, editor. New perspective in antioxidant research. New York: Nova Science; 2006. p. 131–47.
13. Crane FL. Biochemical functions of coenzyme Q10. J Am Coll Nutr. 2001;20(6):591–8.
14. Kaltschmidt B, Sparna T, Kaltschmidt C. Activation of NF-kB by reactive oxygen intermediates in the nervous system. Antioxid Redox Signal. 1999;1:129–44.
15. Crane FL, Navas P. The diversity of coenzyme Q function. Mol Aspects Med. 1997;18:S1–6.
16. Sun IL, Sun EE, Crane FL. Comparison of growth stimulation of HeLa cells, HL-60 cells and mouse fibroblasts by coenzyme Q. Protoplasma. 1995;184:214–9.
17. Sun IL, Sun EE, Crane FL, Morrè DJ, Lindgren A, Low H. A requirement for coenzyme Q in plasma membrane electron transport. Proc Natl Acad Sci USA. 1990;89:11126–30.
18. Thomas SR, Witting PK, Stocker R. A role for reduced coenzyme Q in atherosclerosis? Biofactors. 1999;9:207–24.
19. Stocker R, Bowry VW, Frei B. Ubiquinol-10 protects human low density lipoprotein more efficiently against lipid peroxidation than does alpha-tocopherol. Proc Natl Acad Sci USA. 1991;88(5):1646–50.
20. Mohr D, Bowry VW, Stocker R. Dietary supplementation with coenzyme Q10 results in increased levels of ubiquinol-10 within circulating lipoproteins and increased resistance of human low-density lipoprotein to the initiation of lipid peroxidation. Biochim Biophys Acta. 1992;1126(3):247–54.
21. Yalcin A, Kilinc E, Sagcan A, Kultursay H. Coenzyme Q10 concentrations in coronary artery disease. Clin Biochem. 2004;37:706–9.
22. Hughes K, Lee BL, Feng X, Lee J, Ong CN. Coenzyme Q10 and differences in coronary heart disease risk in Asian Indians and Chinese. Free Radic Biol Med. 2002;32(2):132–8.
23. Fawcett DW. The mammalian spermatozoon. Dev Biol. 1975;44: 394–436.
24. Kalen A, Appelkvist EL, Chojnacki T, Dallner G. Nonaprenyl-4-hydroxybenzoate transferase, an enzyme involved in ubiquinone biosynthesis in endoplasmic reticulum-Golgi system of rat liver. J Biol Chem. 1990;265:1158–64.

25. Mazzilli F, Cerasaro M, Bisanti A, Rossi T, Dondero F. Seminal parameters and the swelling test in patients with sperm before and after treatment with ubiquinone (CoQ10). 2nd International Symposium on Reproductive Medicine, Fiuggi, vol. 71. Rome: Acta Medica, Edizioni e Congressi; 1988.
26. Mazzilli F, Bisanti A, Rossi T, DeSantis L, Dondero F. Seminal and biological parameters in dysspermic patients with sperm hypomotility before and after treatment with ubiquinone (CoQ10). J Endocrinol Invest. 1990;13S(1):88.
27. Mancini A, De Marinis L, Oradei A, Hallgass ME, Conte G, Pozza D, Littarru GP. Coenzyme Q10 concentrations in normal and pathological human seminal fluid. J Androl. 1994;15:591–4.
28. Mancini A, Milardi D, Conte G, Bianchi A, Balercia G, De Marinis L, Littarru GP. Coenzyme Q10: another biochemical alteration linked to infertility in varicocele patients? Metabolism. 2003;52:402–6.
29. Angelitti AG, Colacicco L, Callà C, Arizzi M, Lippa S. Coenzyme Q: potentially useful index of bioenergetic and oxidative status of spermatozoa. Clin Chem. 1995;41:217–9.
30. Mancini A, Conte G, De Marinis L, Hallgass ME, Pozza D, Oradei A, Littarru GP. Coenzyme Q10 levels in human seminal fluid: diagnostic and clinical implications. Mol Aspects Med. 1994;15:s249–55.
31. Littarru GP, Lippa S, Oradei A, Fiorini RM, Mazzanti L. Metabolic and diagnostic implications of human blood CoQ10 levels. In: Folkers K, Littarru GP, Yamagami T, editors. Biomedical and clinical aspects of coenzyme Q, vol. 6. Amsterdam: Elsevier; 1991. p. 167–78.
32. Mancini A, Conte G, Milardi D, De Marinis L, Littarru GP. Relationship between sperm cell ubiquinone and seminal parameters in subjects with and without varicocele. Andrologia. 1998; 30:1–4.
33. Mancini A, Milardi D, Conte G, Festa R, De Marinis L, Littarru GP. Seminal antioxidants in humans: preoperative and postoperative evaluation of coenzyme Q10 in varicocele patients. Horm Metab Res. 2005;37:428–32.
34. Mostafa T, Anis TH, El-Nashar A, Imam H, Othman IA. Varicocelectomy reduces reactive oxygen species levels and increases antioxidant attività of seminal plasma from infertile men with varicocele. Int J Androl. 2001;24:261–5.
35. Meucci E, Milardi D, Mordente A, Martorana GE, Giacchi E, De Marinis L, Mancini A. Total antioxidant capacity in patients with varicocele. Fertil Steril. 2003;79:1577–83.
36. Mancini A, Milardi D, Festa R, Balercia G, De Marinis L, Pontecorvi A, Principi F, Littarru GP. Seminal CoQ10 and male infertility: effects of medical or surgical treatment on endogenous seminal plasma concentrations. Abstracts of the 4th International Coenzyme Q10 association. Los Angeles, 14–17 Apr 2005. p. 64–65.
37. Mancini A, Festa R, Silvestrini A, Nicolotti N, Di Donna V, La Torre G, Pontecorvi A, Meucci E. Hormonal regulation of total antioxidant capacity in seminal plasma. J Androl. 2009;30(5): 534–40. e-pub 19 Feb 2009.
38. Alleva R, Scaramucci A, Mantero F, Bompadre S, Leoni L, Littarru GP. The protective role of ubiquinol-10 against formation of lipid hydroperoxides in human seminal fluid. Mol Aspects Med. 1997;18:s221–8.
39. Balercia G, Arnaldi G, Fazioli F, Serresi M, Alleva R, Mancini A, Mosca F, Lamonica GR, Mantero F, Littarru GP. Coenzyme Q10 levels in idiopathic and varicocele-associated asthenozoospermia. Andrologia. 2002;34:107–11.
40. Littarru GP, Tiano L. Clinical aspects of coenzyme Q_{10}: an update. Curr Opin Clin Nutr Metab Care. 2005;8:641–6.
41. Lewin A, Lavon H. The effect of coenzyme Q_{10} on sperm motility and function. Mol Aspects Med. 1997;18:s213–9.
42. Mancini A, Leone E, Festa R, Grande G, Silvestrini A, De Marinis L, Pontecorvi A, Maira G, Littarru GP, Meucci E. Effects of testosterone on antioxidant systems in male secondary hypogonadism. J Androl. 2008;29(6):622–9.
43. Mancini A, Leone E, Silvestrini A, Festa R, Di Donna V, De Marinis L, Pontecorvi A, Littarry GP, Meucci E. Evaluation of antioxidant systems in pituitary-adrenal axis diseases. Pituitary. Heidelberg: Springer; 2010;13:138–45.
44. Balercia G, Mosca F, Mantero F, Boscaro M, Mancini A, Ricciardo-Lamonica G, Littarru GP. Coenzyme Q(10) supplementation in infertile men with idiopathic asthenozoospermia: an open, uncontrolled pilot study. Fertil Steril. 2004;81:93–8.
45. World Health Organization (WHO). Laboratory manual for the examination of human semen and semen-cervical mucus interaction. 4th ed. Cambridge: Cambridge University; 1999.
46. Frei B, Kim MC, Ames BN. Ubiquinol-10 is an effective lipid-soluble antioxidant at physiological concentrations. Proc Natl Acad Sci USA. 1990;87(12):4879–83.
47. Langsjoen P, Langsjoen A, Willis R, Folkers K. Treatment of hypertrophic cardiomyopathy with coenzyme Q_{10}. Mol Aspects Med. 1997;8:145–151s.
48. Balercia G, Buldreghini E, Vignini A, Tiano L, Paggi F, Amoroso S, Ricciardo-Lamonica G, Boscaro M, Lenzi A, Littarru GP. Coenzyme Q_{10} treatment in infertile men with idiopathic asthenozoospermia: a placebo-controlled, double-blind randomized trial. Fertil Steril. 2009;91:1785–92.
49. Safarinejad MR. Efficacy of coenzyme Q_{10} on semen parameters, sperm function and reproductive hormones in infertile men. J Urol. 2009;182:237–48.
50. Bremer J. Carnitine-metabolism and functions. Physiol Rev. 1983; 63:1420–80.
51. Jeulin C, Dacheux JL, Soufir JC. Uptake and release of free L-carnitine by boar epididymal spermatozoa in vitro and subsequent acetylation rate. J Reprod Fertil. 1994;100:263–71.
52. DiLisa F, Barbato R, Manebo R, Siliprandi N. Carnitine and carnitine esters in mitochondrial metabolism and function. In: DeJong JW, Ferrari R, editors. The carnitine system. A new therapeutical approach to cardiovascular diseases. Dordrecht: Kluwer Academic; 1995. p. 21–38.
53. Enomoto A, Wempe MF, Tsuchida H, Shin HJ, Cha SH, Anzai N, et al. Molecular identification of a novel carnitine transporter specific to human testis. J Biol Chem. 2002;39:36262–71.
54. Bohmer T, Johansen L. Carnitine-binding related suppressed oxygen uptake by spermatozoa. Arch Androl. 1978;1:321–4.
55. Jeulin C, Lewin LM. Role of free L-carnitine and acetyl-L-carnitine in post-gonadal maturation of mammalian spermatozoa. Hum Reprod Update. 1996;2:87–102.
56. Radigue C, Es-Slami S, Soufir JC. Relationship of carnitine transport across the epididymis to blood carnitine and androgens in rats. Arch Androl. 1996;37:27–31.
57. Balercia G, Regoli F, Armeni T, Koverech A, Mantero F, Boscaro M. Placebo-controlled double-blind randomized trial on the use of L-carnitine, L-acetylcarnitine, or combined L-carnitine and L-acetylcarnitine in men with idiopathic asthenozoospermia. Fertil Steril. 2005;84:662–71.
58. Costa M, Canale D, Filicori M, D'Iddio S, Lenzi A. L-Carnitine in idiopathic asthenozoospermia: a multicenter study. Andrologia. 1994;26:155–9.
59. Vicari E, Calogero AE. Effects of treatment with carnitines in infertile patients with prostato-vesiculo-epididymitis. Hum Reprod. 2001;16:2338–42.
60. Vicari E, La Vignera S, Calogero AE. Antioxidant treatment with carnitine is effective in infertile patients with prostato-vesiculo-epididymitis and elevated seminal leukocyte concentration after treatment with nonsteroidal anti-inflammatory compounds. Fertil Steril. 2002;78:1203–8.
61. Lenzi A, Lombardo F, Sgro P, Salacone P, Caponecchia L, Dondero F, et al. Use of carnitine therapy in selected cases of male factor infertility: a double blind cross-over trial. Fertil Steril. 2003;79:292–300.

62. Lenzi A, Sgrò P, Salacone P, Paoli D, Gilio B, Lombardo F, et al. Placebo controlled double blind randomised trial on the use of L-carnitine and L-acetyl-carnitine combined treatment in asthenozoospermia. Fertil Steril. 2004;81:1578–84.
63. Kobayashi A, Fujisawa S. Effect of L-carnitine on mitochondrial acyl-carnitine, acyl-coenzyme A and high energy phosphate in ischemic dog heart. J Mol Cell Cardiol. 1994;26:499–508.
64. Fato R, Cavazzoni M, Castelluccio C, Parenti Castelli G, Lenaz G. Steady-state kinetics of ubiquinol-cytochrome *c* reductase in bovine heart submitochondrial particles: diffusional effects. Biochem J. 1993;290:225–36.
65. Kelso KA, Redpath A, Noble RC, Speake BK. Lipid and antioxidant changes in spermatozoa and seminal plasma throughout the reproductive period of bulls. J Reprod Fertil. 1997;109:1–6.

Synthetic Antioxidants

36

Edmund Y. Ko, John C. Kefer, Ashok Agarwal, and Edmund Sabanegh

Abstract

Male factor infertility is responsible for approximately 25% of all infertility issues. Suboptimal semen quality is a common yet poorly understood cause of male infertility. Environmental factors such as oxidative stress induced by reactive oxygen species (ROS) have recently been implicated as contributing to poor semen quality. Causes of increased ROS in men with suboptimal sperm quality are actively being investigated, and multiple options for therapy are under investigation. ROS have been shown to induce cellular injury by a variety of mechanisms, and antioxidants, which scavenge and inactivate ROS, have been shown to protect cells against this damage. There are many forms of antioxidants. Antioxidants are found endogenously in cell signaling pathways, as well as exogenously in fruits and vegetables (natural antioxidants), or as individual isolates in vitamins or dietary supplements (synthetic antioxidants) Here, we define synthetic antioxidants and discuss the cellular mechanisms by which synthetic antioxidants may improve oxidative stress levels and semen quality in men with male factor infertility.

Keywords

Male factor infertility • Reactive oxygen species • Oxidative stress • Natural antioxidants • Synthetic antioxidants • Leukocytospermia • Vitamin toxicity

Infertility is a major clinical concern, affecting 15% of all reproductive-aged couples, and male factors, including decreased semen quality, are responsible for 25% of these cases [1, 2]. Many men with male factor infertility have suboptimal semen quality, the etiology of which is poorly understood. Many environmental, genetic, and physiological factors, including oxidative stress induced by reactive oxygen species (ROS), have been implicated [3–5]. Oxidative stress induces significant damage to sperm, including decreased motility [6–8], increased DNA damage [9–11], lipid peroxidation [12–14], and decreased oocyte–sperm fusion [15]. Interestingly, while excessive levels of ROS can negatively impact sperm quality, lower levels of ROS have been shown to be required for sperm capacitation, hyperactivation, sperm–oocyte fusion, and other critical cellular processes [16, 17].

E.Y. Ko, MD (✉) • E. Sabanegh, MD
Section of Male Fertility, Department of Urology, Center for Reproductive Medicine, Glickman Urological and Kidney Institute, Cleveland Clinic, 9500 Euclid Avenue/Q10-1, Cleveland, OH, 44195, USA
e-mail: edko05@gmail.com; sabanee@ccf.org

J.C. Kefer, MD, PhD
Alpine Urology, 1155 Alpine Ave, Suite 180, Boulder, CO, 80304, USA
e-mail: johnckefer@gmail.com

A. Agarwal, PhD
Center for Reproductive Medicine, Glickman Urological and Kidney Institute, Cleveland Clinic Foundation, 9500 Euclid Avenue, Desk A19, Cleveland, OH 44195, USA
e-mail: agarwaa@ccf.org

Reactive Oxygen Species and Antioxidants

The majority of aerobic metabolism utilizes oxidative phosphorylation within mitochondria. During the enzymatic reduction of oxygen by cells to produce energy, free radicals

S.J. Parekattil and A. Agarwal (eds.), *Male Infertility: Contemporary Clinical Approaches, Andrology, ART & Antioxidants*, DOI 10.1007/978-1-4614-3335-4_36, © Springer Science+Business Media, LLC 2012

are a constant by-product [18]. Free radicals are defined as oxygen molecules containing one or more unpaired electrons. Free radicals induce cellular damage when they pass this unpaired electron onto nearby structures, resulting in the oxidation of cell membrane lipids, amino acids in proteins, or within nucleic acids [19]. Oxygen is especially susceptible to free radical formation, as it normally has two unpaired electrons. For example, the addition of one electron to molecular oxygen (O_2) forms a superoxide anion radical ($O_2^{\bullet -}$), the primary form of ROS. Superoxide is then directly or indirectly converted to secondary ROS, including hydroxyl radical ($^{\bullet}OH$), peroxyl radical ($ROO^{\bullet}$) or hydrogen peroxide (H_2O_2) [20].

ROS are also formed during the normal enzymatic reactions of inter- and intracellular signaling [21]. ROS generation by leukocytes as a cytotoxic mechanism of host defense, hypoxic states, and a wide array of drugs with oxidizing effects all generate oxidative stress. Significant cellular damage by oxidative stress is prevented through enzymatic and nonenzymatic antioxidant pathways which scavenge excess ROS. This oxidant–antioxidant system allows a critically important balance to be achieved, allowing beneficial oxidant generation for proper cell function, while preventing damaging oxidative stress.

Under normal conditions, antioxidants maintain an overall low level of oxidative stress in the semen, allowing for normal cell signaling processes and normal spermatic function while avoiding oxidant-induced cell damage. In contrast, the pathological effects of oxidative stress arise under conditions where levels of unscavenged ROS increase, thus perturbing the delicate oxidant/antioxidant balance, significantly impacting both sperm quality and function [15, 19, 21]. ROS-induced sperm damage may be a significant contributing factor in 30–80% of all cases of male infertility [19, 21].

Free radicals are scavenged most commonly by one tripeptide, glutathione, and three enzymes important for cellular metabolism. Glutathione, which contains a sulfhydryl group that directly scavenges free radicals, is the most important intracellular defense against ROS. Once oxidized, glutathione is then regenerated/reduced by glutathione reductase and NADPH to complete the cycle [15]. Of the three antioxidant enzymes, superoxide dismutase (SOD) is a metal-containing enzyme that catalyzes two superoxides into oxygen and hydrogen peroxide, which is less toxic than superoxide [18]. Catalase, an enzyme found in peroxisomes, then degrades hydrogen peroxide to water and oxygen, thereby completing the reaction started by SOD. Glutathione peroxidase, as well as other enzymes such as glutathione transferase, ceruloplasmin, or heme oxygenase, also acts to degrade hydrogen peroxide.

Vitamins E and C also play critical roles as nonenzymatic antioxidants [15]. Vitamin E protects cell membranes from oxidative damage by scavenging free radicals within the cellular membrane. Vitamin C is a water-soluble antioxidant that reduces a variety of free radicals and also recycles oxidized vitamin E.

Oxidative Damage to Sperm Cells

Unfortunately, spermatozoa are particularly sensitive to oxidative stress. The majority of the antioxidant enzymatic buffering capacity (i.e., SOD, glutathione peroxidase and catalase, vitamins E and C) is contained in the seminal fluid [15]. Conversely, levels of these antioxidants in the sperm cytoplasm are minimal due to the extremely low volume of spermatic cytoplasm. The polyunsaturated fatty acids found in sperm cell membranes are exquisitely sensitive to peroxidation, making sperm more susceptible to lipid membrane damage than other nongerm cells [12, 13]. Maintaining a healthy antioxidant level is further exacerbated by the ROS production by spermatozoa during its lifespan.

Two principal sources of free radicals are found in the semen: leukocytes and spermatozoa. Most semen specimens contain variable numbers of leukocytes, with neutrophils noted as the predominant type [22–24]. Neutrophils function by generating and releasing high concentrations of ROS to form cytotoxic reactions against nearby cells and pathogens. Over the last 10 years, many studies have investigated the correlation between leukocytospermia and oxidative stress injury to sperm [20, 22, 25–33]. Nonetheless, the relationship between leukocytes in the semen and male infertility remains incompletely defined.

Leukocytospermia has long been associated with decreased sperm concentration, motility, and morphology, as well as decreased hyperactivation and defective fertilization. In a thorough review of the literature, Wolff et al. delineated several studies describing the association between the presence of white blood cells (WBC) in semen and overall sperm quality [34]. The majority of these epidemiologic, clinical, and experimental studies all described a significant negative association between the number of WBCs and overall sperm function. Moskovtsev et al. recently analyzed the relationship between leukocytospermia and sperm DNA damage in 1,230 unselected nonazoospermic infertility patients [35]. While the authors found no significant relationship between leukocytospermia and DNA integrity, a significant negative effect was again noted between the presence of leukocytospermia and corresponding sperm concentration, motility, and morphology. Nonetheless, this relationship is not definitive, as several other studies have found no evidence of a correlation between leukocytospermia and abnormal sperm parameters [25–27]. In a review of these studies, Aitken et al. emphasized that seminal leukocytes may not necessarily affect the fertilizing potential of spermatozoa, as not all men with leukocytospermia demonstrate abnormal sperm parameters [25]. The authors also postulate that the role of leukocytes on male fertility might instead be related to the etiology of the leukocytospermia or the degree of inflammation in the seminal fluid, as well as the leukocyte subtypes present. This relationship therefore remains controversial. While the significance of leukocytospermia on the fertility potential of the individual patient remains difficult to quantify, it can nonetheless be considered a marker of urological or systemic inflammation and possible sperm dysfunction.

Synthetic Antioxidants

Due to the cost and difficulty of chemically extracting and isolating vitamins and other dietary antioxidants from their food source, a broad number of antioxidants are chemically synthesized and packaged in a pill form as an isolate compound. Several studies have suggested that these synthetic antioxidants offer suboptimal antioxidant properties due to their chemical composition and the fact that they are isolated from other synergistic compounds present in normal food sources [36, 37]. Several large clinical studies have been performed investigating the efficacy of these antioxidant isolates, used alone or in combination with one other antioxidant supplement. Bjelakovic et al. postulated that antioxidant supplements have not been proven to have a distinct benefit and may be overall harmful [38, 39]. They investigated the role of synthetic vitamin A, vitamin E, and beta-carotene supplements on overall morbidity and mortality in patients with gastrointestinal cancers. The authors concluded that antioxidant supplements, with the potential exception of selenium, were without significant effects on gastrointestinal cancers and, most concerning, were noted to increase all-cause mortality.

This same group went further and performed a meta-analysis of 68 randomized trials with 232,606 participants to assess the effect of antioxidant supplements on all-cause mortality in randomized primary and secondary prevention trials [40]. All included trials used beta-carotene, vitamin A, vitamin C (ascorbic acid), vitamin E, and selenium, either alone or in combination, and compared these groups to placebo or no intervention. In 47 trials with 180,938 participants, the antioxidant supplements were overall shown to significantly increase mortality (RR, 1.05; 95% CI, 1.02–1.08). After exclusion of the trials involving selenium, patients taking synthetic beta-carotene, vitamin A, and vitamin E, either singly or combined, had significantly increased overall mortality, whereas patients taking synthetic Vitamin C and selenium had no significant effect on mortality.

These authors and others have postulated that natural antioxidants obtained from food sources contribute multiple antioxidants and other co-factors that have been shown to act in synergy with each other, potentially increasing the overall antioxidant capabilities of these agents.

It has been suggested that antioxidant supplements may show interdependency and may have effects only if given in combination [41]. A single antioxidant in high doses will donate a single electron to scavenge free radicals. However, if these "spent" or oxidized antioxidants are present in higher concentrations than the enzymes and cofactors that "recycle" or reduce the antioxidants through reduction–oxidation pathways, these high-dose synthetic antioxidants may actually increase the oxidative stress environment within the relevant physiological system. These theories offer a possible explanation for the broad number of antioxidant trials noting no improvement in a disease process or showing a detrimental effect [42].

Based on numerous clinical studies reviewed in detail by subsequent authors, our center has derived an empiric regimen of synthetic vitamin C, vitamin E, L-carnitine, and coenzyme-Q10 for men demonstrating elevated oxidative stress levels in their semen profile. Here, we review the mechanism of action of these vitamins and further delineate the potential toxicities and side effects of these vitamins and dietary supplements. Overall, it should be noted that most of these dietary supplements can have a detrimental effect on patient health, and our center counsels all patients regarding the potential for toxicity with misuse of these supplements.

Mechanisms of Action

Vitamin E

Synthetic vitamin E is a general term used to describe a group of tocopherols, of which α-tocopherol has the highest biological activity. It is interesting to note that vitamin E has been recognized as an essential nutrient for reproduction since its discovery in 1922 [43]. The cellular and antioxidant functions of vitamin E have yet to be fully described, and it is unlikely they are solely limited to antioxidant functions [44]. Of importance, synthetic vitamin E is mainly comprised of α-tocopherol. The other three tocopherols (β, γ, δ) found in vitamin E-containing foods having antioxidant properties may be present at much lower concentrations in synthesized nutritional supplements [45]. Vitamin E is a chain-breaking antioxidant that prevents oxidant-induced lipid peroxidation, thereby preventing damage to cellular membranes and related structures [46]. It should be noted, however, that like all other oxidative–reductive compounds, vitamin E can act as an oxidant as well. Prooxidative functions of α-tocopherol have been noted in healthy volunteers [47].

A potential role for γ-tocopherol has also been postulated. Unlike α-tocopherol, γ-tocopherol is a powerful nucleophile that traps electrophilic mutagens in lipophilic compartments [48]. Thus, protection of cellular lipids, DNA, and proteins from oxidant damage may occur from several isomers of vitamin E.

In regard to maintaining proper male fertility, optimal function of vitamin E has been linked to optimal levels of selenium. It has been shown that vitamin E scavenges free radicals generated from lipid peroxidation and that the by-products of these scavenged radicals, hydrogen peroxide molecules, are in turn reduced by glutathione peroxidase, a selenium-dependent enzyme [44].

Synthetic vitamin E may therefore indeed play an injurious role in cellular function. Since the optimal role of vitamin E has been linked to the presence of optimal concentrations of other micronutrients, and vitamin E has been shown to have prooxidant properties, supplementing with high levels of synthetic vitamin E may in fact lead to a prooxidant state due to a buildup of hydrogen peroxides or oxidized tocopherols. These data should be considered when prescribing vitamin E and when discussing dietary supplements with patients.

Vitamin A

Vitamin A is a fat-soluble vitamin that is required for proper vision. Vitamin A also functions as a hormone-like growth factor and has been postulated to have antioxidant properties; however, the mechanism by which vitamin A may act as an antioxidant is currently unknown [49]. In clinical trials, vitamin A was shown to improve semen parameters, most significantly by improving the oxidant stress levels of the semen [49–51]. One caveat in these trials is that vitamin A was given in conjunction with other antioxidants, which could have induced the changes in parameters.

Vitamin C

Vitamin C, or ascorbic acid, is considered a vitamin because humans cannot synthesize it enzymatically and must instead obtain it through dietary intake. Vitamin C is a monosaccharide catalyst of oxidation–reduction (redox) reactions in human physiology. Vitamin C is also required for the conversion of procollagen to collagen through the oxidation of proline residues to hydroxyproline, where vitamin C deficiency leads to scurvy [52].

Vitamin C has a unique function as an antioxidant. When vitamin C loses one electron (oxidation), it remains very stable, allowing it to gain an electron from a more aggressive free radical while also not damaging critical cellular structures. When vitamin C is oxidized, it is recycled back to an antioxidant form via the NADPH pathway, as well as by the glutathione pathways [52]. It is therefore not surprising then that synthetic vitamin C in very high doses has been shown to have a detrimental clinical effect. Regular intake of high doses of synthetic vitamin C as a single dietary supplement may overwhelm the recycling pathways, leading to excessive levels of oxidized vitamin C, which would then act as an oxidant on the cellular system.

Safety, Dosing, and Side Effects of Toxicity

Adverse events due to vitamin toxicity are almost exclusively seen due to overconsumption of synthetic vitamin supplements [53]. Adverse events associated with vitamin use can be of similar magnitude as conventional pharmaceuticals, and significant toxicity can occur with high intake of most vitamins [53]. For some vitamins, such as the B vitamins, vitamin C, and vitamin K, adverse reactions are minor and usually reversible. Other vitamins, such as vitamin A or vitamin E, have been shown to cause serious, irreversible adverse events. We will review these significant adverse events and safety profiles of potentially dangerous antioxidant vitamins.

Vitamin E

Vitamin E is a lipid-soluble vitamin, capable of reaching toxic tissue concentrations through storage in the liver and fatty tissues of the body. In healthy adults, 200–800 mg/day may cause gastrointestinal distress, and 800–1,200 mg/day may induce antiplatelet effects and bleeding disturbances, thrombophlebitis, elevated creatinine, and gonadal dysfunction [54, 55]. In several large, multicenter, randomized trials investigating the potential benefits of vitamin E on prevention of cardiac events; many of these studies described a significant increased risk of congestive heart failure. In the multicenter GISSI-Prevenzione trial studying 11,000 patients with previous myocardial infarction, vitamin E showed no benefit for all study endpoints but a 20% increased risk of developing congestive heart failure [56, 57]. In over 9,500 patients studied under the multicenter Heart Outcomes Prevention Evaluation randomized trial, treatment with vitamin E did not prevent cancer or cardiovascular events but did increase the risk of heart failure [58, 59]. High-dose vitamin E can also synergistically interact with vitamin K to exacerbate bleeding diatheses, especially in patients already taking anticoagulation or antiplatelet therapy [60, 61]. Among 30,000 male smokers, a higher incidence of hemorrhagic stroke was noted in men taking vitamin E [62].

High-dose vitamin E has also been shown to increase the overall cancer and mortality risk. The largest study, a meta-analysis of 19 clinical trials investigating over 135,000 patients, demonstrated that 400 IU/day or higher may increase all-cause mortality [63]. In the Women's Health Study, multivariate analysis demonstrated that vitamin E serum levels were associated with increased risk of both invasive and noninvasive breast cancers [64]. In men, supplementation with vitamin E has been demonstrated to increase risk of prostate cancer, especially in men taking vitamin E in conjunction with other supplements [65].

Vitamin A

Vitamin A is mainly stored in the liver. Vitamin A toxicity ranges from elevated liver function tests to cirrhosis, hepatic fibrosis, and death due to liver failure. Vitamin A has been associated with nausea, blurred vision, anorexia, and mental status changes, as well as electrolyte disorders [66]. In over 2,000 men aged 49–51 years old, risk of fracture was seven times higher in men with elevated serum retinol compared to men with the lowest retinol levels [67]. In a large, randomized, multicenter, double-blind, placebo-controlled trial of more than 29,000 male smokers receiving beta-carotene in Finland, as well as in a study of 18,000 men in the USA, the incidence of lung cancer in these men was 18% higher in men taking beta-carotene supplementation compared to the placebo group [62]. In women, two large studies demonstrated conflicting results in female smokers taking beta-carotene supplementation. In one study, the Women's Health Study, no benefit or harm was noted in the incidence of lung cancer and cardiovascular events, whereas a similar study design in another trial also showed no benefit with a significant increased incidence of mortality from both lung cancer and cardiovascular disease [68, 69].

Vitamin C

Vitamin C is generally well tolerated, but in large doses, such as the several gram doses frequently found in supplements, acute adverse effects can include nausea, vomiting, esophagitis, heartburn, fatigue, insomnia, and diarrhea [70]. Long-term vitamin C intake can induce crystallization of urate, oxalate, cysteine, and other drugs in the urinary tract [71]. In a prospective cohort study of over 45,000 men without history of nephrolithiasis, vitamin C intake may increase risk of stones, and multivariate analysis in men consuming 1,000 mg/day versus those consuming less than the recommended daily allowance showed a significant risk of stone formation [72]. However, the comparison of men taking below the recommended amount versus men taking doses measured in grams could be questioned.

In a study of over 1,900 postmenopausal diabetic women, vitamin C was associated with a dose-related increased risk of coronary artery disease, stroke, and overall cardiovascular mortality [73]. More significantly, in the Los Angeles Atherosclerosis Study, investigating the impact of vitamin C supplementation on more than 500 men and women without symptomatic cardiovascular disease, carotid inner wall thickening was noted in males taking 500 mg of vitamin C per day [74].

While these studies demonstrate a significant increased risk of morbidity and mortality in vitamin C supplementation, Lee et al. directly demonstrated that vitamin C can break down lipids in cell membranes into compounds that act as genotoxins, leading to increased levels of DNA damage [73]. This study is one of the few to directly demonstrate the potential mechanisms by which vitamin C may induce cell injury and decreased cellular function.

Management of Oxidative Stress in Male Infertility

Vitamin/Antioxidant Supplementation

The antioxidants α-tocopherol (vitamin E), ascorbic acid (vitamin C), and the retinoids (vitamin A) are all potent scavengers of ROS. Many studies have investigated the role of these and other antioxidants on improvements in sperm parameters. However, the majority of these studies are uncontrolled, focus on healthy men without infertility, or have indirect endpoints of success. Several other studies are noted due to the quality of their study design and demonstrate compelling evidence regarding efficacy of antioxidants toward improving semen parameters.

Silver et al. surveyed 97 healthy nonsmoking men aged 20–80 years old regarding antioxidant intake using a dietary questionnaire and subsequently examined semen samples [75]. Those with high daily intake of antioxidants were noted to have improved semen quality compared to men with low or moderate intake, thereby demonstrating some correlation between increased dietary antioxidant intake and improved semen parameters.

Keskes-Ammar et al. examined the therapeutic efficacy of increased antioxidant intake on semen parameters [76]. They randomized 54 men to either vitamin E and selenium or vitamin B for 3 months, with examination of semen samples quantifying the lipid peroxidation marker, malondialdehyde (MDA), as well as measurement of serum vitamin E levels. Although only 20 patients completed the study protocol, results indicated that vitamin E and selenium supplementation produced a significant decrease in MDA concentrations with improved sperm motility, whereas vitamin B showed no impact.

Suleiman et al. randomized their cohort of asthenozoospermic men with normal female partners to vitamin E or placebo for 6 months, noting decreased MDA levels and increased motility, as well as increased pregnancy rates in the vitamin E arm [77].

Conversely, Rolf et al. randomized 31 men with asthenospermia to either 2 months of high-dose oral treatment with vitamins C and E or placebo and investigated semen parameters [78]. The authors found no changes in semen parameters during treatment, and no pregnancies were initiated during this period.

Most recently, the best designed trial by Greco et al. examined the impact of increased antioxidant intake in a randomized, prospective manner [79]. A group of 64 infertile men with >15% DNA-fragmented spermatozoa were randomized into two groups to receive either 1 g of vitamin C and E daily or placebo for 2 months. While no differences in basic sperm parameters were noted, the antioxidant cohort demonstrated a significantly reduced percentage of DNA-fragmented spermatozoa. The authors further went on to demonstrate that supplementation with Vitamins E and C significantly increased rates of clinical pregnancy and implantation following ICSI [80].

Although these data from different centers are potentially conflicting, direct comparison of these results is difficult given the varying nature of the dose, duration of treatment, and study end points in each of these trials. Nonetheless, these studies provide compelling evidence toward the efficacy of vitamin antioxidants on improving overall sperm quality and possibly improved pregnancy following ICSI.

Five-Year View and Key Issues

Our understanding of the role of ROS on male fertility continues to increase supported by an expanding literature refining this relationship. The etiology of suboptimal semen quality due to oxidative stress is becoming elucidated. The origin of ROS generation and the etiologies of increased ROS in men with suboptimal sperm quality are increasingly clear, offering multiple pathways for potential therapy.

Nonetheless, more well-designed, randomized controlled trials will be required to assess the potential of these antioxidant regimens. Without further studies to test the treatments of best efficacy, it is difficult to derive cohesive clinical guidelines of therapy from these studies. Nonetheless, the initial data demonstrating efficacy in improving sperm quality and conception rates are indeed encouraging.

References

1. Rowe PJ, Comhaire FH, Hargreave TB, Mahmoud AMA. WHO manual for the standardized investigation, diagnosis and management of the infertile male. Cambridge: Cambridge University Press; 2004.
2. Sharlip ID, Jarow JP, Belker AM, Lipshultz LI, Sigman M, Thomas AJ, et al. Best practice policies for male infertility. Fertil Steril. 2002;77:873–82.
3. Vine MF. Smoking and male reproduction: a review. Int J Androl. 1996;19:323–37.
4. Auger J, Eustache F, Andersen AG, Irvine DS, Jorgensen N, Skakkebaek NE, et al. Sperm morphological defects related to environment, lifestyle and medical history of 1001 male partners of pregnant women from four European cities. Hum Reprod. 2001;16:2710–7.
5. Kenkel S, Rolf C, Nieschlag E. Occupational risks for male fertility: an analysis of patients attending a tertiary referral centre. Int J Androl. 2001;24:318–26.
6. Eskenazi B, Wyrobek AJ, Sloter E, Kidd SA, Moore L, Young S, et al. The association of age and semen quality in healthy men. Hum Reprod. 2003;18:447–54.
7. Lenzi A, Lombardo F, Gandini L, Alfano P, Dondero F. Computer assisted sperm motility analysis at the moment of induced pregnancy during gonadotropin treatment for hypogonadotropic hypogonadism. J Endocrinol Invest. 1993;6:683–6.
8. Agarwal A, Ikemoto I, Loughlin KR. Relationship of sperm parameters with levels of reactive oxygen species in semen specimens. J Urol. 1994;152:107–10.
9. Armstrong JS, Rajasekaran M, Chamulitrat W, Gatti P, Hellstrom WJ, Sikka SC. Characterization of reactive oxygen species induced effects on human spermatozoa movement and energy metabolism. Free Radic Biol Med. 1999;26:869–80.
10. Kodama H, Yamaguchi R, Fukuda J, Kasai H, Tanaka T. Increased oxidative deoxyribonucleic acid damage in the spermatozoa of infertile male patients. Fertil Steril. 1997;68:519–24.
11. Barroso G, Morshedi M, Oehninger S. Analysis of DNA fragmentation, plasma membrane translocation of phosphatidylserine and oxidative stress in human spermatozoa. Hum Reprod. 2000; 15:1338–44.
12. Aitken RJ, Harkiss D, Buckingham DW. Analysis of lipid peroxidation mechanisms in human spermatozoa. Mol Reprod Dev. 1993;35:302–15.
13. Kemal Duru N, Morshedi M, Oehninger S. Effects of hydrogen peroxide on DNA and plasma membrane integrity of human spermatozoa. Fertil Steril. 2000;74:1200–7.
14. Smith R, Vantman D, Ponce J, Escobar J, Lissi E. Total antioxidant capacity of human seminal plasma. Hum Reprod. 1996;11(8):1655–60.
15. Agarwal A, Prabhakaran SA, Sikka SC. Clinical relevance of oxidative stress in patients with male factor infertility: evidence-based analysis. AUA Update Series. 2007;26:1–12.
16. Aitken RJ, Ryan AL, Baker MA, McLaughlin EA. Redox activity associated with the maturation and capacitation of mammalian spermatozoa. Free Radic Biol Med. 2004;36:994–1010.
17. de Lamirande E, Tsai C, Harakat A, Gagnon C. Involvement of reactive oxygen species in human sperm acrosome reaction induced by A23187, lysophosphatidylcholine, and biological fluid ultrafiltrates. J Androl. 1998;19:585–94.
18. Tremellen K. Oxidative stress and male infertility—a clinical perspective. Hum Reprod Update. 2008;14:243–58.
19. Ochsendorf FR. Infections in the male genital tract and reactive oxygen species. Hum Reprod Update. 1999;5:399–420.
20. Warren JS, Johnson KJ, Ward PA. Oxygen radicals in cell injury and cell death. Pathol Immunopathol Res. 1987;6:301–15.
21. Agarwal A, Prabakaran S, Allamaneni S. What an andrologist/urologist should know about free radicals and why. Urology. 2006;67:2–8.
22. Saleh RA, Agarwal A. Oxidative stress and male infertility: from research bench to clinical practice. J Androl. 2002;23:737–52.
23. Aitken RJ, Fisher H. Reactive oxygen species and human spermatozoa: the balance of benefit and risk. Bioessays. 1994;16:259–67.
24. Aitken RJ. Free radicals, lipid peroxidation, sperm function. Reprod Fertil Dev. 1995;7:659–68.
25. Aitken RJ, Baker HW. Seminal leukocytes: passengers, terrorists or good Samaritans? Hum Reprod. 1995;10:1736–9.
26. Tomlinson JM, Barratt CL, Cooke ID. Prospective study of leukocytes and leukocyte subpopulations in semen suggests they are not the cause of a male infertility. Fertil Steril. 1993;60:1069–75.
27. Christiansen E, Tollefsrud A, Purvis K. Sperm quality in men with chronic abacterial prostatovesiculitis verified by rectal ultrasonography. Urology. 1991;38:545–9.
28. Aitken RJ, Fisher HM, Fulton N, Gomez E, Knox W, Lewis B, et al. Reactive oxygen species generation by human spermatozoa is induced by exogenous NADPH and inhibited by the flavoprotein inhibitors diphenylene iodonium and quinacrine. Mol Reprod Dev. 1997;47:468–82.
29. Hendin B, Kolletis PN, Sharma RK, Thomas AJ, Agarwal A. Varicocele is associated with elevated spermatozoal reactive oxygen species production and diminished seminal plasma antioxidant capacity. J Urol. 1999;161:1831–4.
30. Shalika S, Dugan K, Smith RD, Padilla SL. The effect of positive semen bacterial and ureaplasma cuture on in-vitro fertilization success. Hum Reprod. 1996;11:2789–92.
31. Sharma RK, Pasqualetto AE, Nelson DR, Thomas AJ, Agarwal A. Relationship between seminal white blood cell counts and oxidative stress in men treated at an infertility clinic. J Androl. 2001;22:575–83.
32. Kao SH, Chao HT, Chen HW, Hwang TI, Liao TL, Wei YH. Increase of oxidative stress in human sperm with lower motility. Fertil Steril. 2008;89(5):1183–90.

33. Saleh RA, Agarwal A, Kandirali E, Sharma RK, Thomas AJ, Nada E, et al. Leukocytospermia is associated with increased reactive oxygen species production by human spermatozoa. Fertil Steril. 2002;78:1215–24.
34. Wolff H. The biologic significance of white blood cells in semen. Fertil Steril. 1995;63:1143–57.
35. Moskovtsev SI, Willis J, White J, Mullen BM. Leukocytospermia: relationship to sperm deoxyribonucleic acid integrity in patients evaluated for male factor infertility. Fertil Steril. 2007;88(3):737–40.
36. Herbert V. The value of antioxidant supplements vs their natural counterparts. J Am Diet Assoc. 1997;97:375–6.
37. Stanner SA, Hughes J, Kelly CN, Buttriss J. A review of the epidemiological evidence or the "antioxidant hypothesis". Public Health Nutr. 2004;7:407–22.
38. Bjelakovic G, Nikolova D, Simonetti RG, Gluud C. Antioxidant supplements for preventing gastrointestinal cancers. Cochrane Database Syst Rev. 2004;(4):CD004183.
39. Bjelakovic G, Nikolova D, Simonetti RG, Gluud C. Antioxidant supplements for preventing gastrointestinal cancers. Cochrane Database Syst Rev. 2008;(3):CD004183.
40. Bjelakovic G, Nikolova D, Simonetti RG, Gluud C. Antioxidant supplements for prevention of gastrointestinal cancers: a systematic review and meta-analysis. Lancet. 2004;364(9441):1219–28.
41. Hercberg S, Galan P, Preziosi P, Alfarez MJ, Vazquez C. The potential role of antioxidant vitamins in preventing cardiovascular diseases and cancers. Nutrition. 1998;14:513–20.
42. Bjelakovic G, Nikolova D, Gluud LL, Simonetti RG, Gluud C. Mortality in randomized trials of antioxidant supplements for primary and secondary prevention: systematic review and meta-analysis. JAMA. 2007;197:842–57.
43. Evans HM, Bishop KS. On the existence of a hitherto unrecognized dietary factor essential for reproduction. Science. 1922;56:650–1.
44. Brigelius-Flohe R, Traber MG. Vitamin E: function and metabolism. FASEB J. 1999;13:1145–55.
45. Esterbauer H, Dieber-Rotheneder M, Striegl G, Waeg G. Role of vitamin E in preventing the oxidation of low-density lipoprotein. Am J Clin Nutr. 1991;53(1 Suppl):314S–21.
46. Kamal-Eldin A, Appelqvist LA. The chemistry and antioxidant properties of tocopherols and tocotrienols. Lipids. 1996; 31:671–701.
47. Bowry VW, Ingold KU, Stocker R. Vitamin E in human low-density lipoprotein. When and how this antioxidant becomes a pro-oxidant. Biochem J. 1992;288:341–4.
48. Cooney RV, Franke AA, Harwood PJ, Hatch-Pigott V, Custer LJ, Mordan LJ. Gamma-tocopherol detoxification of nitrogen dioxide: superiority to alpha-tocopherol. Proc Natl Acad Sci USA. 1993;90:1771–5.
49. Comhaire FH, Christophe AB, Zalata AA, Dhooge WS, Mahmoud AM, Depuydt CE. The effects of combined conventional treatment, oral antioxidants and essential fatty acids on sperm biology in subfertile men. Prostaglandins Leukot Essent Fatty Acids. 2000;63:159–65.
50. Scott R, MacPherson A, Yates RW, Hussain B, Dixon J. The effect of oral selenium supplementation on human sperm motility. Br J Urol. 1998;82:76–80.
51. Galatioto GP, Gravina GL, Angelozzi G, Sacchetti A, Innominato PF, Pace G, et al. May antioxidant therapy improve sperm parameters of men with persistent oligospermia after retrograde embolization for varicocele? World J Urol. 2008;26:97–102.
52. Linster CL, Van Schaftingen E, Vitamin C. Biosynthesis, recycling and degradation in mammals. FEBS J. 2007;274:1–22.
53. Mulholland CA, Benford DJ. What is known about the safety of multivitamin-multimineral supplements for the generally healthy population? Theoretical basis for harm. Am J Clin Nutr. 2007;85:318S–22.
54. Bendich A, Machlin LJ. Safety of oral intake of vitamin E. Am J Clin Nutr. 1988;48:612–9.
55. Weber P, Bendich A, Machlin LJ. Vitamin E and human health: rationale for determining recommended intake levels. Nutrition. 1997;13:450–60.
56. Dietary supplementation with n-3 polyunsaturated fatty acids and vitamin E after myocardial infarction: results of the GISSI-Prevenzione trial. Gruppo Italiano per lo Studia della Sopravvivenza nell'Infarto miocardico. Lancet. 1999;354:447–55. http://www.ncbi.nlm.nih.gov/pubmed/10465168.
57. Marchioli R, Levantesi G, Macchia A, Marfisi RM, Nicolosi GL, Tavazzi L, et al. Vitamin E increases the risk of developing heart failure after myocardial infarction: results from the GISSI-Prevenzione trial. J Cardiovasc Med (Hagerstown). 2006;7: 347–50.
58. Yusuf S, Dagenais G, Pogue J, Bosch J, Sleight P. Vitamin E supplementation and cardiovascular events in high-risk patients. The Heart Outcomes Prevention Evaluation Study Investigators. N Engl J Med. 2000;342:154–60.
59. Lonn E, Bosch J, Yusuf S, Sheridan P, Pogue J, Arnold JM, et al. Effects of long-term vitamin E supplementation on cardiovascular events and cancer: a randomized controlled trial. JAMA. 2005; 293:1338–47.
60. Greenblatt DJ, von Moltke LL. Interaction of warfarin with drugs, natural substances, and foods. J Clin Pharmacol. 2005;45:127–32.
61. Mousa SA. Antithrombotic effects of naturally derived products on coagulation and platelet function. Methods Mol Biol. 2010;663:229–40.
62. The Alpha-Tocopherol, Beta Carotene Cancer Prevention Study Group. The effect of vitamin E and beta carotene on the incidence of lung cancer and other cancers in male smokers. N Engl J Med. 1994;330:1029–35.
63. Miller 3rd ER, Pastor-Barriuso R, Dalal D, Riemersma RA, Appel LJ, Guallar E. Meta-analysis: high-dosage vitamin E supplementation may increase all-cause mortality. Ann Intern Med. 2005;142:37–46.
64. Kabat GC, Kim M, Adams-Campbell LL, Caan BJ, Chlebowski RT, Neuhouser ML, et al. Longitudinal study of serum carotenoid, retinol, and tocopherol concentrations in relation to breast cancer risk among postmenopausal women. Am J Clin Nutr. 2009; 90:162–9.
65. Lawson KA, Wright ME, Subar A, Mouw T, Hollenbeck A, Schatzkin A, et al. Multivitamin use and risk of prostate cancer in the National Institutes of Health-AARP Diet and Health Study. J Natl Cancer Inst. 2007;99:754–64.
66. Food and Nutrition Board, Institute of Medicine. Dietary reference intakes for vitamin A, vitamin K, arsenic, boron, chromium, copper, iodine, iron, manganese, molybdenum, nickel, silicon, vanadium, and zinc. Washington, DC: National Academy Press; 2002.
67. Michaelsson K, Lithell H, Vessby B, Melhus H. Serum retinol levels and the risk of fracture. N Engl J Med. 2003;348:287–94.
68. Lee IM, Cook NR, Manson JE, Buring JE, Hennekens CH. Beta-carotene supplementation and incidence of cancer and cardiovascular disease: the Women's Health Study. J Natl Cancer Inst. 1999; 91:2102–6.
69. Goodman GE, Thornquist MD, Balmes J, Cullen MR, Meyskens Jr FL, Omenn GS, et al. The Beta-Carotene and Retinol Efficacy Trial: incidence of lung cancer and cardiovascular disease mortality during 6-year follow-up after stopping beta-carotene and retinol supplements. J Natl Cancer Inst. 2004;96:1743–50.
70. Food and Nutrition Board, Institute of Medicine. Dietary reference intakes for vitamin C, vitamin E, selenium, and carotenoids. Washington, DC: National Academy Press; 2000.
71. Levine M, Rumsey SC, Daruwala R, Park JB, Wang Y. Criteria and recommendations for vitamin C intake. JAMA. 1999;281: 1415–23.

72. Taylor EN, Stampfer MJ, Curhan GC. Dietary factors and the risk of incident kidney stones in men: new insights after 14 years of follow-up. J Am Soc Nephrol. 2004;15:3225–32.
73. Lee DH, Folsom AR, Harnack L, Halliwell B, Jacobs Jr DR. Does supplemental vitamin C increase cardiovascular disease risk in women with diabetes? Am J Clin Nutr. 2004;80:1194–200.
74. Dwyer JH, Paul-Labrador MJ, Fan J, Shircore AM, Merz CN, Dwyer KM. Progression of carotid intima-media thickness and plasma antioxidants: the Los Angeles Atherosclerosis Study. Arterioscler Thromb Vasc Biol. 2004;24:313–9.
75. Silver EW, Eskenazi B, Evenson DP, Block G, Young S, Wyrobek AJ. Effect of antioxidant intake on sperm chromatin stability in healthy nonsmoking men. J Androl. 2005;26:550–6.
76. Keskes-Ammar L, Feki-Chakroun N, Rebai T, Sahnoun Z, Ghozzi H, Hammami S, et al. Sperm oxidative stress and the effect of an oral vitamin E and selenium supplement on semen quality in infertile men. Arch Androl. 2003;49:83–94.
77. Suleiman SA, Ali ME, Zaki ZM, el-malik EM, Nasr MA. Lipid peroxidation and human sperm motility: protective role of vitamin E. J Androl. 1996;17:530–7.
78. Rolf C, Cooper TG, Yeung CH, Nieschlag E. Antioxidant treatment of patients with asthenozoospermia or moderate oligoastheozoospermia with high dose vitamin C and E: a randomized, placebo-controlled, double blinded study. Hum Reprod. 1999;14:1028–33.
79. Greco E, Iacobelli M, Rienzi L, Ubaldi F, Ferrero S, Tesarik J. Reduction of the incidence of sperm DNA fragmentation by oral antioxidant treatment. J Androl. 2005;26:349–53.
80. Greco E, Romano S, Iacobelli M, Ferrero S, Baroni E, Minasi MG, et al. ICSI in cases of sperm DNA damage: beneficial effect of oral antioxidant treatment. Hum Reprod. 2005;20:2590–4.

Antioxidant Therapy for the Enhancement of Male Reproductive Health: A Critical Review of the Literature

37

Kelton Tremellen

Abstract

Oxidative stress is a well-established cause of male infertility. Reactive oxygen species (ROS) are present at higher concentrations in infertile men's semen and have been conclusively shown to impair sperm function through oxidative damage to the sperm membrane and paternal DNA. It, therefore, would appear logical that amelioration of oxidative stress through the use of antioxidant therapy could improve infertile men's sperm function, potentially leading to better reproductive performance. In this chapter, we critically analyse the published studies that have used in vivo antioxidant therapy to treat male infertility. Overall, there is reasonably good evidence that some antioxidant therapies can significantly reduce oxidative damage to sperm DNA and improve sperm membrane function (motility and sperm fertilization capacity). However, what is more contentious is whether these improvements in sperm quality actually translate into an increase in pregnancy rates or an improvement in pregnancy outcomes for infertile couples. While a few small studies have shown antioxidant therapy to assist both in vivo and in vitro conception, larger more definitive studies will be required before antioxidant therapy becomes established medical practice for infertile males.

Keywords

In vivo antioxidant therapy • Male infertility • Impaired sperm function • Oxidative damage to sperm • Reactive oxygen species • Oxidative stress • Antioxidant supplementation • Antioxidant trials

Infertility is a condition that affects one in six couples, with impaired sperm quality playing a role in at least half of all cases of infertility. Of even more concern is the evidence suggesting that sperm quality has actually been decreasing over the last 50 years [1, 2], leading to more and more couples requiring expensive fertility treatments such as in vitro fertilization (IVF) and intracytoplasmic sperm injection (ICSI). This trend has prompted researchers to focus more on identifying the underlying causes of male infertility, allowing treatments to be tailored to pathology, rather than a reliance on generic "mechanical" solutions such as ICSI.

While the identifiable causes of male infertility are many and varied (reviewed in Chap. 1), oxidative stress has been identified as a very significant cause. MacLeod [3] was first to link oxidative stress with impaired sperm function when, in 1943, he published his observations that sperm cultured under conditions of high oxygen tension lost their motility, yet in vitro supplementation with the antioxidant catalase could preserve sperm motility. Since this pioneering work, there has been an exponential increase in our knowledge of how oxidative stress may impair male reproductive function and how various treatments may help combat this damage.

K. Tremellen, MBBS (Hons.), PhD, FRANZCOG, CREI (✉)
School of Pharmacy and Medical Sciences, University of South Australia, Repromed, 180 Fullarton Road, Dulwich, SA 5065, Australia
e-mail: ktremellen@repromed.com.au

S.J. Parekattil and A. Agarwal (eds.), *Male Infertility: Contemporary Clinical Approaches, Andrology, ART & Antioxidants*,
DOI 10.1007/978-1-4614-3335-4_37,

The focus of this chapter is to critically review the evidence for use of antioxidant supplements to improve male reproductive function.

Rationale for Antioxidant Therapy in Male Infertility

Before examining the clinical studies exploring the use of antioxidant supplements in male infertility, one should ask if there is a reasonable scientific basis behind such treatment. To assess the biological plausibility of using antioxidant therapy in the setting of male infertility, three key questions need to be considered.

- Is oxidative stress more common in the infertile population?
- Do in vitro studies suggest a mechanism whereby oxidative stress could impair male reproductive function?
- Do in vitro studies suggest that antioxidant supplementation may restore normal male reproductive function in the presence of oxidative stress?

As is evident from discussions in the preceding chapters, there does appear to be an abundance of evidence supporting the biological rationale behind the use of antioxidant supplements to treat male factor infertility. Infertile men's semen does contain higher levels of reactive oxygen species (ROS) and lower levels of protective antioxidants than fertile men, thereby placing these men's sperm at increased risk of oxidative damage. It has been estimated that between 30 and 80% of infertile men have some evidence of oxidative stress damage to their sperm, even when routine semen analysis results (concentration, motility and morphology) are within the normal WHO prescribed range [4]. Secondly, in vitro studies have confirmed that the direct application of ROS to sperm or the stimulation of sperm's own production of ROS can reduce sperm motility, membrane integrity and DNA quality, all linked with reduced male reproductive capacity. Finally, direct application of antioxidants in vitro can block the harmful effects of ROS on sperm motility and DNA integrity, confirming a causal association between oxidative stress and impaired male reproductive function. In summary, there appears to be a sound scientific rationale to the use of antioxidants to treat male infertility.

Antioxidant Therapy for the Treatment of Male Infertility

A MEDLINE search for the key words antioxidant, sperm and male infertility, combined with a manual search of references contained in several key review papers [4–6] located a total of 31 primary studies exploring the in vivo use of antioxidant supplements for the enhancement of male reproductive performance. Trials with poorly defined treatments such as the use of "general multi-vitamins" or studies that employed botanical preparations with poorly defined antioxidant action were excluded from this review.

With the existence of such a large body of evidence one would expect that definitive conclusions on the value of antioxidant supplements in the treatment of male infertility would be able to be made. Unfortunately this is not the case for several reasons. Firstly, there is a huge variation in the different types and dosages of antioxidants used in the published studies. Secondly, most of the studies are small and therefore underpowered, making meaningful analysis of differences in pregnancy outcomes with antioxidant therapy very difficult. The following discussion group together clinical studies which use a similar antioxidant treatment protocol, facilitating an assessment of which antioxidants may prove to be useful for the treatment of male infertility.

Vitamin E

Vitamin E is an essential fat soluble vitamin, with α-tocopherol being the most common form of vitamin E available in food. Vitamin E is the major chain breaking antioxidant that directly neutralizes superoxide anions, hydrogen peroxide and the hydroxyl radical. As sperm membranes contain abundant phospholipids which are prone to oxidative damage, it is believed that vitamin E plays a critical role in protecting cellular structures from damage caused by free radicals and reactive products of lipid peroxidation. Secondly, vitamin E exhibits some anti-inflammatory activity and therefore may reduce leukocyte initiated sperm oxidative stress.

The recommended dietary allowance (RDA) for vitamin E is suggested to be 15 mg (equivalent to 22.4 IU) of α-tocopherol per day for adult men, with the tolerable upper intake being suggested as 1,000 mg (1,500 IU) by the US National Institute of Health [7]. However, a meta-analysis of 19 clinical trials using long-term vitamin E supplementation in patients with chronic disease has reported that at dosages of 400 IU or greater per day, vitamin E may actually increase overall mortality compared to placebo [8]. Furthermore, it is known to inhibit platelet aggregation and has been linked with an increased risk of haemorrhagic stroke. Therefore its use in infertile men on anticoagulants or at risk of serious haemorrhagic illness is probably contraindicated.

Two studies have analysed the ability of vitamin E to decrease sperm membrane oxidative damage by measurement of sperm malondialdehyde (MDA) levels before and after vitamin E supplementation. Geva et al. [9] reported that 200 mg a day of vitamin E was able to significantly reduce MDA levels within 1 month of therapy, while Suleiman et al. [10] found that the use of 300 mg of vitamin E per day for

6 months also produced a significant drop in MDA levels. Consistent with a reduction in sperm membrane oxidative damage, two studies have also reported an improvement in in vitro sperm fertilization capacity assessed by routine insemination IVF or the use of sperm–zona binding assays [9, 11]. No study to date has analysed the ability of vitamin E monotherapy to improve sperm DNA quality.

Two small non-controlled studies of vitamin E supplementation have reported no effect on sperm count, motility or morphology [12, 13]. In addition, a well-conducted placebo-controlled trial of 3 months therapy with 600 mg of vitamin E per day reported no significant effect on sperm concentration, motility or morphology [11]. Conversely, another randomized controlled trial (RCT) using 6 months of vitamin E (300 mg/day) or placebo reported a statistically significant improvement in sperm motility, but no change in concentration or morphology [10]. However, in this last trial the "drop out" rate for patients in the placebo arm was significantly greater than that seen in the active treatment arm (20 vs. 3 patients, respectively, from a starting number of 55 patients in each study arm). This selective "drop out" from placebo raises the possibility that patients or their treating physicians became unblinded to treatment allocation during the trial, biasing the final results.

No study to date using vitamin E monotherapy for the treatment of male infertility has been adequately powered to analyse pregnancy outcomes. While some studies have reported pregnancies [10, 11], the small number of pregnancies makes clear conclusions impossible. In the Suleiman study 17% of patients allocated to vitamin E therapy achieved a live birth, compared to none in the placebo [10]. However, as previously outlined, the large "drop out" rate in the placebo arm suggests the potential for significant bias, thereby making it impossible to make firm conclusions on the value of Vitamin E to assist pregnancy in the setting of male infertility.

Vitamin C

Vitamin C (ascorbic acid) is an important water soluble antioxidant that competitively protects lipoproteins from peroxyl radical attack, while also enhancing the antioxidant activity of vitamin E by assisting in its recycling. Seminal plasma vitamin C levels are tenfold higher than serum [14], suggesting a very important protective role for vitamin C in the male reproductive tract. The RDA for vitamin C in the adult male is 75 mg, with the tolerable upper intake limit being suggested as 2,000 mg/day [7]. However, the use of high dosages of vitamin C (≥1 gm/day) may be harmful since at these high concentrations vitamin C can act as a pro-oxidant and may predispose to kidney stone formation [7, 15].

A small placebo-controlled randomized study of 30 infertile men allocated an equal number of participants to either placebo, 200 mg or 1,000 mg of vitamin C per day for a total of 4 weeks [16]. Both dosages of vitamin C were able to significantly increase seminal plasma vitamin C levels, with the magnitude of the increase being more significant in the 1,000 mg treated group. Sperm motility, morphology and viability all significantly improved within 1 week of vitamin C therapy. While the average sperm concentration doubled on vitamin C, this did not reach statistical significance. More critical analysis of the baseline characteristics in this study suggests that randomization may not have been successful in creating study groups that were equal. For example, at entry the percentage of abnormal sperm morphology was 45, 64 and 62% for the placebo, 200 and 1,000 mg vitamin C groups, respectively. At the end of the study the placebo group abnormal morphology was 41% and the vitamin C groups had decreased to 35 and 36%, a statistically significant decrease. Critical analysis of these results suggests that the two vitamin C group's morphology results were significantly inferior to the placebo at study entry and that following 4 weeks of vitamin C treatment these poor morphology results simply returned to levels equivalent to that seen in the placebo. This raises the possibility of selection bias or at least a "regression to the mean" spontaneous improvement in sperm morphology. Pregnancy outcomes were not reported in this trial, although, in the introduction, the authors comment that a prior unpublished pilot study had achieved a 100% pregnancy rate with vitamin C therapy ($n=20$ patients).

A larger study that randomly allocated 75 smokers to either a placebo, 200 or 1,000 mg vitamin C per day reported significant improvements in sperm morphology in the 1,000 mg subgroup, but no significant changes in the 200 mg treated group [17]. As these men were not infertile and were not trying for pregnancy, the implications of this study for the infertile population are uncertain.

Combined Vitamin C and Vitamin E Therapy

Two excellent placebo-controlled randomized studies have examined the ability of a combination of vitamins C and E to alter sperm quality. Rolf et al. [18] reported a small RCT in which infertile patients were allocated to either placebo ($n=16$) or 2 months treatment with 800 mg vitamin E and 1,000 mg vitamin C ($n=15$). The inclusion criteria for this study were impaired motility, not the presence of confirmed oxidative stress. No significant difference in sperm concentration, motility or morphology was observed during therapy and no direct assessment of oxidative damage was made. Furthermore, no pregnancies were seen in either study group during the treatment period. A similar RCT using 2 months therapy with both 1,000 mg of vitamin C and E daily or placebo also found no significant changes in sperm count, motility or morphology [19]. However, this group did observe a very significant drop in sperm DNA damage. Unfortunately pregnancy outcomes were not reported in this study, but were reported for a non-controlled study using the same treatment protocol by the same clinical group [20]. In this later report, patients who had failed to have a successful pregnancy after at least one cycle of IVF and who had documented elevated levels of sperm DNA damage were given 2 months of vitamins C/E combination therapy before a further cycle of IVF treatment. A total of 76.3% of participants experienced a normalization of their sperm DNA damage and this "improved" subgroup achieved an implantation rate of 19.6%. As the study did not include a concurrent placebo control, firm conclusions on pregnancy effect are not possible.

Several small non-placebo-controlled trials have also examined the effect of vitamins C and E combinations on sperm quality. Kodama et al. [21] was able to show a significant drop in sperm DNA oxidative damage (8-OHdG) and MDA with 2 months of therapy (200 mg vitamin C, 200 mg vitamin E and 400 mg glutathione). They also reported a small but significant increase in sperm concentration, but no effect of antioxidant supplementation on sperm motility or morphology. Menezo [22] observed a significant drop in sperm DNA fragmentation with 2 months of antioxidant therapy (400 mg vitamins C and E per day, plus low dosages of vitamin A, zinc and selenium) but no change in sperm concentration, motility or morphology. Interestingly, these investigators also noted a significant increase in sperm DNA decondensation. They believed that this was due to the high redox potential of vitamin C interfering with the reduction of cystine to two cysteine moieties, thereby opening protamine disulphide bridges. Decondensation of the sperm DNA may make the DNA more susceptible to ROS attack and may interfere with proper embryo development. Menezo, therefore, cautions against the use of antioxidant preparations containing high dosages of vitamin C in infertile men with sperm decondenstation levels exceeding 20% at baseline. Finally, a small placebo-controlled study of 45 infertile men allocated to placebo, vitamin C 5 mg/vitamin E 10 mg/200 mg zinc or zinc alone observed a non-significant trend in improvement in sperm motility and a decrease in MDA in each treatment group [23]. The magnitude of improvement in MDA and motility was similar in the zinc alone group compared to those treated with zinc and vitamins C/E. This observation suggests that the low dosages of vitamins C and E used in this study are likely to be subtherapeutic, with any improvement in sperm quality more likely to reflect the action of zinc.

Overall, the high-quality placebo-controlled studies suggest that vitamins C and E do not produce significant improvements in sperm concentration, motility or morphology. The significant drop in sperm DNA damage seen in two trials [19, 22], together with observations of a drop in 8-OHdG and MDA in a non-placebo-controlled study [21] suggests that vitamins C and E can still have positive reproductive effects even if they do not alter routine sperm parameters. The ability of vitamins C and E to improve pregnancy rates is still debatable until future adequately powered studies are conducted in this area.

Coenzyme Q_{10}

Coenzyme Q_{10} is primarily concentrated in the mitochondria of the sperm mid-piece and plays an important antioxidant and energy production role in sperm. Coenzyme Q_{10} transports electrons from complexes I and II to complex III in the mitochondrial respiratory chain, leading to ATP synthesis in the mitochondrial membrane. In its reduced form (ubiquinol), coenzyme Q_{10} acts as a strong antioxidant preventing lipid peroxidation in biological membranes.

A small non-controlled study involving 38 men with male factor infertility and previous poor fertilization during IVF–ICSI therapy reported on the use of 60 mg of coenzyme Q_{10} per day for a period of 3 months [24]. This study found that coenzyme Q_{10} produced no significant changes in sperm concentration, motility or morphology, yet IVF–ICSI fertilization rates did improved significantly. This study did not measure sperm lipid peroxidation or DNA damage and offered no explanation how coenzyme Q_{10} supplementation may boost fertilization without altering routine sperm parameters.

A large and very well-conducted placebo RCT recently reported on the use of 300 mg coenzyme Q_{10} or placebo per day for a period of 6 months in 212 men with male factor infertility [25]. At this dose of supplementation a significant increase in seminal plasma coenzyme Q_{10} concentration was observed. Furthermore, significant improvements in sperm count and motility were also observed, together with an increase in serum inhibin B levels and a corresponding fall in FSH concentration. This would suggest that coenzyme Q_{10} therapy is capable of enhancing Sertoli cell function, not just sperm function. This study did not report on oxidative end-points but did report a significant increase in the calcium ionophore induced acrosome reaction, suggesting some improvement in sperm membrane function. Unfortunately no significant difference in pregnancy rates was observed between the placebo and coenzyme Q_{10} supplement groups over a 12-month period of observation. This is not surprising when one recognizes that the magnitude of the statistically significant changes in sperm parameters observed were very small and unlikely to be of clinical significance. For example, total sperm motility after 6 months of coenzyme Q_{10} therapy was 27.6%, compared to 23.1% in the placebo group. While for the coenzyme Q_{10} group this was a statistically significant increase in sperm motility from baseline (22.2%), the final motility result was not significantly different from the placebo and highly unlikely to be of any clinical significance. As only men, who had partners with no evidence for female factor infertility, were enrolled in the study, the lack of differences in pregnancy outcomes in such a large study suggests that coenzyme Q_{10} monotherapy is not of major benefit in assisting in vivo conception.

Selenium

Selenium is an essential trace element required for normal male reproductive function. The antioxidant glutathione peroxidase 4 (GPX-4) is present within sperm and requires the presence of selenium to function. Not only does GPX-4 play an antioxidant role, it is also involved in augmenting sperm chromatin stability by acting as a protein thiol peroxidase. The adult male RDA for selenium is 55 μg/day, with the upper tolerable limit being 400 μg/day [7]. An individual's dietary intake of selenium depends on the selenium content in the local soil where food is grown. Men living in countries such as China where the soil is commonly selenium deficient are more likely to benefit from selenium supplementation. Conversely, excess supplementation of selenium may lead to toxicity and have detrimental effects on sperm quality [26].

Iwanier et al. [27] gave 200 μg of selenium per day to a group of men (33 infertile and 9 fertile) for a period of 2 months and measured sperm quality before and after treatment. The investigators observed a significant increase in seminal plasma selenium concentration and GPX activity during the trial, but no significant improvement in sperm concentration, motility or morphology. In a small placebo RCT ($n=18$ placebo, 46 active treatment) the supplementation of infertile men exhibiting low sperm motility with 100 μg of selenium (+/– very low dosages of vitamins A, C and E) produced no significant change in sperm concentration but a small significant improvement in motility (20.6–28.2%) [28]. The clinical significance of this improvement is questionable, as no significant difference in pregnancy rates was observed (no pregnancies in the placebo vs. 11% selenium group).

A very large placebo-controlled study randomized 468 infertile men to either placebo, 200 μg/day of selenium, with or without *N*-acetyl-cysteine, for a period of 6 months [29]. Sperm quality and male reproductive hormones were then assessed during supplementation and for a further 6 months. This study observed statistically significant increases in sperm concentration, motility and morphology in all treatment arms, together with an increase in serum inhibin B and testosterone. However, the magnitude of these improvements was again very small and unlikely to be of any clinical significance. Unfortunately pregnancy outcomes were not reported for this study, making it impossible to draw any firm conclusions on the benefits of selenium supplementation to boost pregnancy rates.

Glutathione

Glutathione is an antioxidant released in large amounts by the epididymis that in turn can neutralize the damaging effects of superoxide anions, thereby preventing lipid peroxidation. Two trials by a single group of investigators have examined the effect of glutathione supplementation (600 μg intramuscular alternate days for 2 months) on two separate groups of infertile men. The first trial involved 20 infertile men with likely oxidative stress (past genitourinary tract infection with residual inflammation, varicocele) in a placebo crossover trial design [30]. This study observed no significant changes in sperm concentration, but significant improvements in sperm motility and morphology. These improvements were observed within 1 month of supplementation, suggesting an epididymal rather than a testicular mode of action. A second smaller non-controlled study using identical inclusion criteria examined changes in sperm lipid peroxidation with glutathione treatment [31]. This study observed improvement in all routine sperm parameters and a significant decrease in sperm MDA concentration, confirming an antioxidant effect. Neither study reported pregnancy outcome, making conclusions about the fertility promoting effect of glutathione treatment impossible. However, the requirement for intramuscular administration of glutathione therapy is certainly likely to limit its clinical application.

L-Carnitine

Carnitine is produced in the liver and then passes via the circulation to the epididymis, where it is taken up by the epididymal epithelium and actively transported into the luminal fluid bathing sperm. In the epididymis carnitine is taken up by sperm, where it involves in energy metabolism by transporting fatty acids from the cytosolic compartment to the mitochondrial matrix.

Costa et al. [32] were the first to examine the effects of L-carnitine supplementation in the setting of male infertility. Their study group of 100 infertile men with unexplained impaired motility were given L-carnitine (3 g/day) for a period of 4 months, while measuring changes in sperm function. They reported small but statistically significant improvements in sperm concentration and motility, but no changes in sperm morphology. Lenzi et al. [33] used an active medication/washout/placebo study design to determine if 2 months of L-carnitine therapy (2 g/day) could alter sperm quality. Analysis of the raw outcome data indicated no significant difference in sperm quality after L-carnitine therapy. However, when the researchers excluded several "outliers" from the analysis a borderline statistically significant increase in sperm concentration and motility was reported. The subjective removal of "outliers" to create statistical significance, plus the failure of L-carnitine therapy to improve either epididymal function (alpha-glucosidase) or reduce levels of sperm lipid peroxidation casts significant doubt on whether L-carnitine therapy has any beneficial effect on male reproductive performance.

Vicari et al. [34] studied the ability of L-carnitine in combination with non-steroidal inflammatory medication (NSAID) to alter sperm function in a group of 98 infertile men with confirmed oxidative stress. The use of 2 months pre-treatment with NSAIDs, followed by 2 months of L-carnitine (2 g/day) produced a significant reduction in seminal ROS production and an improvement in sperm motility and viability. A total of 23% of patients on NSAID/L-carnitine therapies achieved pregnancy, but the absence of a control group makes firm conclusions on these therapies effect on pregnancy rates impossible.

N-Acetyl Cysteine

N-acetyl cysteine (NAC) is believed to act as a precursor to glutathione, increasing the tissue concentration of this potent antioxidant. Recently, several good quality placebo-controlled studies have examined the ability of NAC to alter sperm quality in infertile men with presumed oxidative pathology. Galatioto et al. [35] conducted a RCT in which 42 men with oligospermia were allocated to receive either 600 mg NAC a day plus a vitamin–mineral supplement for 3 months or no therapy at all. This small study reported a significant increase in sperm concentration but no change in sperm motility or morphology. A larger placebo-controlled study using 600 mg/day of NAC for a period of 3 months reported no change in sperm concentration or morphology, but a small improvement in motility [36]. Finally, one arm of a multi-therapy RCT compared sperm quality between men with idiopathic male factor infertility on 600 mg NAC per day with placebo [29]. This study reported very minor, although statistically significant, improvements in sperm concentration and morphology but no changes in sperm motility.

The conflicting sperm quality outcomes for these three trials using an identical dose of NAC, and the failure to report pregnancy outcomes makes it impossible to conclude that NAC therapy has any clinically meaningful effect on male reproductive performance.

Miscellaneous Antioxidant Monotherapies

Astaxanthin is a carotenoid extract from the algae *Haematococcus pluvialis* with reported potent antioxidant qualities. A small placebo-controlled RCT reported on the effect of 3 months therapy with this antioxidant in men with idiopathic male factor infertility [37]. Astaxanthin produced no change in sperm concentration or morphology but did produce a significant reduction in seminal ROS levels and improvement in sperm motility. Furthermore, the researchers observed a significant increase in natural or intrauterine insemination-assisted conceptions in the antioxidant-treated group, suggesting that the small improvement in sperm motility was of clinical significance.

Lycopene, an antioxidant found in high concentrations in fruits such as tomatoes and watermelon is a powerful natural antioxidant. A non-controlled trial of 30 men with male factor infertility reported a significant improvement in sperm quality with 3 months of lycopene therapy at a dose of 4 mg/day [38]. However, upon further analysis of this study, it appears that the researchers only analysed sperm outcomes for the 14–20 men who showed an improvement in either sperm concentration, motility or morphology. Such an analysis is obviously flawed since excluding half the study participants who did not respond to treatment is clearly going to result in a significant difference being concluded. Therefore, this study provides no scientific support for the use of lycopene in male factor infertility.

Combination Therapies

The combination of vitamins C and E would appear to be the most commonly studied combinational antioxidant therapies for male factor infertility. However, other unique combinations have been trialled in the hope that using several different antioxidants with different modes of action may be more beneficial than antioxidant monotherapy.

The combination of vitamin E (400 mg/day) and selenium (225 μg/day) has been trialled in a placebo-controlled study of 54 men with male factor infertility [39]. Antioxidant therapy produced a small increase in sperm mobility and a drop in sperm MDA levels, confirming an antioxidant effect. No changes in sperm concentration or morphology were observed, and pregnancy outcomes were not reported. A significant weakness in this study was that out of a total of 54 initial participants, only 20 completed the study. This raises the possibility of bias and makes firm conclusions difficult.

Piomboni et al. [40] performed a controlled study comparing 3 months therapy with an antioxidant combination (β-glucan 20 mg, papaya 50 mg, lactoferrin 97 mg, vitamin C 30 mg, vitamin E 5 mg) or no therapy. They observed a significant improvement in sperm motility, viability and morphology, but no change in sperm DNA quality. Pregnancy outcomes were not reported in this study.

A small uncontrolled study of 33 men reported on the use of a combination of 600 mg NAC, 30 mg β-carotene, 180 mg vitamin E and a mixture of essential fatty acids for a period of 6 months as treatment for male factor infertility [41]. This combination produced no change in sperm concentration, motility or morphology, but a drop in seminal ROS levels and sperm DNA oxidative damage (8-OHdG) was observed, together with an increase in the ionophore induce acrosome reaction. A total off 22.2% of couples who completed the 6-month therapy did successfully conceive, but the absence of a control arm makes it impossible to determine if this is a clinical improvement above non-treatment levels.

A small case series reported on the success of using a combinational antioxidant (β-carotene 5,000 IU, vitamin C 60 mg, vitamin E 30 IU, zinc 15 mg) for the treatment of early embryo loss related to sperm oxidative damage [42]. Out of 17 men screened, 9 men were confirmed to have oxidative stress-related sperm pathology which could be amenable to antioxidant therapy. In six of these nine cases the partners subsequently fell pregnant. When antioxidants had been taken by the male before conception all pregnancies were viable ($n=4$), whereas all the pregnancies conceived by men who refused antioxidant therapy miscarried. Such a small case series precludes definitive conclusions, yet does suggest that oxidative pathology may be a significant cause of early pregnancy wastage.

One of the most widely studied combinational antioxidants in the field of male infertility is Menevit®. This preparation consists of a combination of several natural antioxidants (vitamin C 100 mg, vitamin E 400 IU, lycopene 6 mg, selenium 26 μg, garlic oil 333 μg) and other ingredients involved in sperm DNA synthesis and packaging (zinc 25 mg, folate 500 mg). Three-month therapy with the Menevit® antioxidant has been reported to produce no significant change in sperm concentration, motility or morphology, but did produce a significant reduction in seminal ROS levels and sperm DNA fragmentation [43]. Interestingly, while a dose of 400 mg of vitamin C has been shown to produce sperm chromatin decondensation by interfering with protamine disulphide bonds [22], the Menevit® antioxidant containing one-quarter the dose of vitamin C has been reported to significantly increase sperm DNA protamination [43]. The Menevit® antioxidant has also been shown to improve pregnancy outcomes when compared with placebo in a RCT of 60 patients undergoing IVF–ICSI treatment [44]. Finally, recent preliminary studies have linked male infertility and sperm oxidative stress with impaired sperm DNA methylation, a possible risk factor for epigenetic disease in the next generation [45]. The treatment of infertile men with 3 months of Menevit® resulted in an improvement in the levels of sperm global DNA methylation [45]. This pilot study will require replication, and large epidemiological studies will need to confirm the link between sperm DNA methylation defects and childhood illness before definitive conclusions can be made regarding the utility of antioxidant supplements to prevent epigenetic disease in the next generation.

Overview of Common Methodological Weaknesses in Antioxidant Trials

Before making a final conclusion on the value of antioxidant therapy to enhance male reproductive function, a brief overview of the common methodological flaws in the previously described studies is warranted. Table 37.1 summarizes the methodological strengths and weaknesses of the placebo-controlled antioxidant trials and should allow the reader to make their own conclusions on the merits of antioxidant supplement therapy for the treatment of male infertility.

Lack of Appropriate Inclusion Criteria

It is obvious that only infertile men experiencing oxidative stress are likely to benefit from antioxidant therapy. Despite this, the vast majority of studies do not actually screen potential study participants for oxidative stress before enrolling only those with confirmed oxidative pathology. If we assume that only half of all infertile men have significant oxidative stress, the indiscriminate enrolment of all infertile men will result in half of the study participants having no significant oxidative pathology and no real chance of improvement in sperm function on antioxidant therapy. This will of course weaken the study power and increase the chance of missing a true biological effect of antioxidant supplementation.

Lack of an Appropriate Control

The most common study design in the field of antioxidant therapy is where patients act as their own historical control, with comparison of sperm quality or pregnancy outcomes before and after antioxidant therapy. While these studies are easy to conduct, they can be fraught with significant bias that can invalidate the study outcomes. One common bias is the so-called regression to the mean phenomenon. When study participants are selected for enrolment in a study by the presence of a single extreme sperm result (e.g., rapidly progressive motility $<25\%$), there is a natural tendency for the results to normalize on retesting with no active treatment. Sperm parameters such as concentration and motility are prone to large day-to-day intra-individual fluctuations and, therefore, are especially susceptible to this regression to the mean bias.

Table 37.1 Placebo controlled studies examining the effect of antioxidant therapy on male reproductive heath

Study reference	Therapy used per day	Duration of therapy (months)	Oxidative stress as an inclusion criteria	Positive changes in semen quality	Positive changes in sperm OS endpoints	Positive changes in reproductive outcomes
[10]	Vitamin E 300 mg	6	No	↑ motility	↓ MDA	Pregnancy 17% active group vs. 0% placebo
[11]	Vitamin E 300 mg	3	No	Nil	Nil	Improved sperm zona binding
[16]	Vitamin C (200 or 1,000 mg)	1	No	↑ motility, morph and viability	Not tested	Not reported
[18]	Vit E 800 mg, Vit C 1,000 mg	2	No	Nil	Not tested	None
[19]	Vitamin E and C (1,000 mg each)	2	No	Nil	↓ sperm DNA damage	Not reported
[23]	Vit E 10 mg, Vit C 5 mg, zinc 200 mg	3	No	Nil	Trend for ↓ MDA	Not reported
[25]	Coenzyme Q_{10} 300 mg	6	No	↑ conc and motility	Not tested	No difference in pregnancy rates
[28]	Sn 100 μg, Vit A 1 mg, Vit C 10 mg, Vit E 15 mg	3	No	↑ motility	Not tested	No difference in pregnancy rates (11% vs. 0% placebo)
[29]	Sn 200 μg, NAC 600 mg	6	No	↑ conc, motility and morph.	Not tested	Not reported
[30]	Glutathione 600 mg	2	No	↑ motility and morph	Not tested	Not reported
[33]	L-carnitine 2 g	2	No	Nil (raw data analysis)	No change in MDA	No difference
[35]	NAC 600 mg	3	No	↑ sperm conc	Not tested	Not reported
[36]	NAC 600 mg	3	No	↑ motility	Not tested	Not reported
[37]	Astaxanthin 16 mg	3	No	↑ motility	↓ semen ROS	↑ natural + IUI conceptions
[39]	Vitamin E 400 mg, Sn 225 μg	3	No	↑ motility	↓ MDA	Not reported
[40]	Vit C 30 mg, Vit E 5 mg, β-glucan 20 mg, papaya 50 mg, lactoferrin 97 mg	3	No	↑ motility and morph	No change in DNA quality	Not reported
[44]	Menevit® (Vit C, Vit E, Sn, lycopene, folate, zinc and garlic oil)	3	No	Not reported	Not tested	↑ IVF–ICSI conceptions on active antioxidant (38.5% vs. 16% placebo)

OS oxidative stress, *MDA* malondialdehyde, *ROS* reactive oxygen species, *Sn* selenium, *NAC N*-acetyl cysteine, *IUI* intrauterine insemination

Any study that reports an improvement in sperm motility or concentration when deficiencies in either were used as inclusion criteria for entry into the study has the potential to be biased by regression to the mean. More stable sperm parameters such as morphology and sperm DNA integrity are less prone to this type of bias due to their minimal intra-individual day-to-day variation and may be studied adequately using patients own historical controls.

Insufficient Sample Size

Unfortunately, the majority of studies examining the effect of antioxidant supplements are relatively small, most containing less than 100 subjects in total. Such small underpowered studies predispose to type II statistical errors where the null hypothesis is not rejected (i.e., no statistical significant difference), despite a true biological effect of treatment being present. For relatively low-probability outcomes such as pregnancy in an infertile cohort, very large sample sizes are required to reach statistical significance. Only large multicentre studies or a meta-analysis of many similar studies can negate this size weakness and allow for true biological effects of antioxidant therapy to become apparent.

Inappropriate Study Endpoints

Due to their ease of measurement, the majority of male infertility antioxidant studies use sperm parameters such as count, motility and morphology as their primary study endpoint. While this approach has some scientific merit, there are major drawbacks from this approach. Firstly, infertile men are not primarily interested in improving their sperm quality. They and their partner wish to have a child and, therefore, the only clinically relevant primary endpoint should be live birth, not sperm quality. Secondly, the presence of a statistically significant improvement in sperm count, motility and morphology may not have any clinical significance in regard to fertility potential. The link between routine sperm parameters and fertility potential is relatively weak until sperm quality reaches low levels. For example, the ability of an antioxidant therapy to improve average sperm concentration from 40 to 45 million/ml is not likely to be clinically relevant, even if statistical significance has been reached.

Expert Commentary

The current evidence clearly identifies oxidative stress as a cause of impaired sperm function and a significant underlying pathology in many cases of male factor infertility. While many studies have been conducted examining the ability of various antioxidants to improve male reproductive function, it is still unclear if male pre-conception antioxidant therapy can actually improve a couples chances of becoming parents. Critical analysis of the various higher quality trials suggest that a combination of vitamins C and E can produce a reduction in sperm membrane and DNA oxidation, an increase in sperm IVF capacity but no clinically meaningful improvements in sperm concentration, motility and morphology. The ability of vitamins C and E to boost natural conception is questionable, but may assist in vitro conception. The use of a supplement exceeding 400 IU of vitamin E or 1,000 mg of vitamin C per day is probably not advisable, due to potential general health concerns and the ability of high dose vitamin C to produce sperm DNA decondensation. The addition of other antioxidants such as selenium, lycopene and astaxanthin to a combination of vitamins C and E antioxidant therapy has some possible additional benefit as these ingredients have also been shown to produce meaningful decreases in sperm oxidative damage and improvement in natural and assisted conception. However, evidence supporting the use of other antioxidants such as coenzyme Q_{10}, glutathione, L-carnitine and NAC as effective therapies for male infertility is presently weak.

Five-Year View

Before pre-conception antioxidant therapy becomes routine medical care for the infertile couple, three main issues need to be addressed. Firstly, the present diagnostic tests for oxidative stress are cumbersome and expensive and therefore not available in the majority of infertility clinics. Without the availability of a "quick and easy" test for oxidative stress, many doctors are unwilling to offer empirical antioxidant therapy. Future studies aimed at developing such assays for sperm oxidative stress are urgently required. Secondly, there is a pressing need for large multi-centre trials using a single antioxidant combination therapy to confirm that pre-conception supplementation can boost live birth rates. Without such a definitive trial, antioxidant therapy will be relegated to the "promising but never proven" basket of medical treatments, never receiving widespread medical support. Finally, research suggests that oxidative damage to sperm DNA may result in miscarriage and possibly even affect the health of the next generation [46, 47]. If such inter-generational effects of sperm oxidative damage are confirmed, pre-conception antioxidant supplementation for the male will become standard medical practice, just as pre-conception folate supplementation for the prevention of neural tube defects is standard care for women. If these three goals can be realized, the next 5 years promise to be a very exciting time for the discipline of andrology.

Key Issues

- Oxidative stress is a significant cause of impaired sperm function resulting in infertility, miscarriage and possibly even long-term health consequences for the next generation.
- The present body of evidence surrounding the treatment of male factor infertility with antioxidants is difficult to critically interpret because of less than ideal study design (not screening for oxidative stress at enrolment, sperm quality as a primary endpoint instead of pregnancy and a lack of concurrent placebo controls). Furthermore, the use of a large number of different types and dosages of antioxidant and the lack of adequately powered studies to analyse pregnancy outcomes precludes definitive conclusions being made.
- Antioxidants such as vitamin E, vitamin C, selenium, lycopene and astaxanthin all appear to have the ability to improve sperm health by reducing seminal ROS levels, decreasing sperm membrane peroxidation and oxidative DNA damage. Furthermore, there is some evidence that these antioxidants may also result in improved natural and in vitro conception.
- Firm conclusions relating to antioxidant therapies ability to improve sperm concentration, motility and morphology is presently impossible due to the abundance of contradictory results and inadequately controlled studies. However, an analysis of the better quality placebo-controlled studies suggests that antioxidants may produce a small improvement in sperm motility but are unlikely to improve sperm concentration or morphology.
- Future studies must use confirmed oxidative stress-related damage to sperm as their inclusion criteria and should have pregnancy outcome as their primary endpoint. These studies must be adequately powered to detect meaningful improvements in pregnancy rates.

References

1. Carlsen E, Giwercman A, Keiding N, Skakkebaek NE. Evidence for decreasing quality of semen during past 50 years. BMJ. 1992;305(6854):609–13.
2. Auger J, Kunstmann JM, Czyglik F, Jouannet P. Decline in semen quality among fertile men in Paris during the past 20 years. N Engl J Med. 1995;332(5):281–5.
3. MacLeod J. The role of oxygen in the metabolism and motility of human spermatozoa. Am J Physiol. 1943;138:512–8.
4. Tremellen K. Oxidative stress and male infertility—a clinical perspective. Hum Reprod Update. 2008;14(3):243–58.
5. Agarwal A, Nallella KP, Allamaneni SS, Said TM. Role of antioxidants in treatment of male infertility: an overview of the literature. Reprod Biomed Online. 2004;8(6):616–27.
6. Lanzafame FM, La Vignera S, Vicari E, Calogero AE. Oxidative stress and medical antioxidant treatment in male infertility. Reprod Biomed Online. 2009;19(5):638–59.
7. Panel of Dietary Antioxidants and Related Compounds, Food and Nutrition Board, Institute of Medicine. Dietary reference intakes for vitamin C, vitamin E, selenium and crotenoids. Washington, DC: National Academy Press; 2000.
8. Miller 3rd ER, Pastor-Barriuso R, Dalal D, Riemersma RA, Appel LJ, Guallar E. Meta-analysis: high-dosage vitamin E supplementation may increase all-cause mortality. Ann Intern Med. 2005;142(1):37–46.
9. Geva E, Bartoov B, Zabludovsky N, Lessing JB, Lerner-Geva L, Amit A. The effect of antioxidant treatment on human spermatozoa and fertilization rate in anin vitro fertilization program. Fertil Steril. 1996;66(3):430–4.
10. Suleiman SA, Ali ME, Zaki ZM, el-Malik EM, Nasr MA. Lipid peroxidation and human sperm motility: protective role of vitamin E. J Androl. 1996;17(5):530–7.
11. Kessopoulou E, Powers HJ, Sharma KK, Pearson MJ, Russell JM, Cooke ID, Barratt CL. A double-blind randomized placebo cross-over controlled trial using the antioxidant vitamin E to treat reactive oxygen species associated male infertility. Fertil Steril. 1995;64(4):825–31.
12. Moilanen J, Hovatta O. Excretion of alpha-tocopherol into human seminal plasma after oral administration. Andrologia. 1995;27(3):133–6.
13. Giovenco P, Amodei M, Barbieri C, Fasani R, Carosi M, Dondero F. Effects of kallikrein on the male reproductive system and its use in the treatment of idiopathic oligozoospermia with impaired motility. Andrologia. 1987;19:238–41.
14. Jacob RA, Pianalto FS, Agee RE. Cellular ascorbate depletion in healthy men. J Nutr. 1992;122(5):1111–8.
15. Wayner DD, Burton GW, Ingold KU. The antioxidant efficiency of vitamin C is concentration-dependent. Biochim Biophys Acta. 1986;884(1):119–23.
16. Dawson EB, Harris WA, Rankin WE, Charpentier LA, McGanity WJ. Effect of ascorbic acid on male fertility. Ann N Y Acad Sci. 1987;498:312–23.
17. Dawson EB, Harris WA, Teter MC, Powell LC. Effect of ascorbic acid supplementation on the sperm quality of smokers. Fertil Steril. 1992;58(5):1034–9.
18. Rolf C, Cooper TG, Yeung CH, Nieschlag E. Antioxidant treatment of patients with asthenozoospermia or moderate oligoasthenozoospermia with high-dose vitamin C and vitamin E: a randomized, placebo-controlled, double-blind study. Hum Reprod. 1999;14(4):1028–33.
19. Greco E, Iacobelli M, Rienzi L, Ubaldi F, Ferrero S, Tesarik J. Reduction of the incidence of sperm DNA fragmentation by oral antioxidant treatment. J Androl. 2005;26(3):349–53.
20. Greco E, Romano S, Iacobelli M, Ferrero S, Baroni E, Minasi MG, Ubaldi F, Rienzi L, Tesarik J. ICSI in cases of sperm DNA damage: beneficial effect of oral antioxidant treatment. Hum Reprod. 2005;20(9):2590–4.
21. Kodama H, Yamaguchi R, Fukuda J, Kasai H, Tanaka T. Increased oxidative deoxyribonucleic acid damage in the spermatozoa of infertile male patients. Fertil Steril. 1997;68(3):519–24.
22. Ménézo YJ, Hazout A, Panteix G, Robert F, Rollet J, Cohen-Bacrie P, Chapuis F, Clément P, Benkhalifa M. Antioxidants to reduce sperm DNA fragmentation: an unexpected adverse effect. Reprod Biomed Online. 2007;14(4):418–21.
23. Omu AE, Al-Azemi MK, Kehinde EO, Anim JT, Oriowo MA, Mathew TC. Indications of the mechanisms involved in improved sperm parameters by zinc therapy. Med Princ Pract. 2008;17(2):108–16.
24. Lewin A, Lavon H. The effect of coenzyme Q_{10} on sperm motility and function. Mol Aspects Med. 1997;18(Suppl):S213–9.
25. Safarinejad MR. Efficacy of coenzyme Q_{10} on semen parameters, sperm function and reproductive hormones in infertile men. J Urol. 2009;182(1):237–48.

26. Bleau G, Lemarbre J, Faucher G, Roberts KD, Chapdelaine A. Semen selenium and human fertility. Fertil Steril. 1984;42(6):890–4.
27. Iwanier K, Zachara BA. Selenium supplementation enhances the element concentration in blood and seminal fluid but does not change the spermatozoal quality characteristics in subfertile men. J Androl. 1995;16(5):441–7.
28. Scott R, MacPherson A, Yates RW, Hussain B, Dixon J. The effect of oral selenium supplementation on human sperm motility. Br J Urol. 1998;82(1):76–80.
29. Safarinejad MR, Safarinejad S. Efficacy of selenium and/or *N*-acetyl-cysteine for improving semen parameters in infertile men: a double-blind, placebo controlled, randomized study. J Urol. 2009;181(2):741–51.
30. Lenzi A, Culasso F, Gandini L, Lombardo F, Dondero F. Placebo-controlled, double-blind, cross-over trial of glutathione therapy in male infertility. Hum Reprod. 1993;8(10):1657–62.
31. Lenzi A, Picardo M, Gandini L, Lombardo F, Terminali O, Passi S, Dondero F. Glutathione treatment of dyspermia: effect on the lipoperoxidation process. Hum Reprod. 1994;9(11):2044–50.
32. Costa M, Canale D, Filicori M, D'lddio S, Lenzi A. L-carnitine in idiopathic asthenozoospermia: a multicenter study. Italian Study Group on Carnitine and Male Infertility. Andrologia. 1994;26(3):155–9.
33. Lenzi A, Lombardo F, Sgrò P, Salacone P, Caponecchia L, Dondero F, Gandini L. Use of carnitine therapy in selected cases of male factor infertility: a double-blind crossover trial. Fertil Steril. 2003; 79(2):292–300.
34. Vicari E, La Vignera S, Calogero AE. Antioxidant treatment with carnitines is effective in infertile patients with prostatovesiculoepididymitis and elevated seminal leukocyte concentrations after treatment with nonsteroidal anti-inflammatory compounds. Fertil Steril. 2002;78(6):1203–8.
35. Paradiso Galatioto G, Gravina GL, Angelozzi G, Sacchetti A, Innominato PF, Pace G, Ranieri G, Vicentini C. May antioxidant therapy improve sperm parameters of men with persistent oligospermia after retrograde embolization for varicocele? World J Urol. 2008;26(1):97–102.
36. Ciftci H, Verit A, Savas M, Yeni E, Erel O. Effects of *N*-acetylcysteine on semen parameters and oxidative/antioxidant status. Urology. 2009;74(1):73–6.
37. Comhaire FH, El Garem Y, Mahmoud A, Eertmans F, Schoonjans F. Combined conventional/antioxidant "Astaxanthin" treatment for male infertility: a double blind, randomized trial. Asian J Androl. 2005;7(3):257–62.
38. Gupta NP, Kumar R. Lycopene therapy in idiopathic male infertility—a preliminary report. Int Urol Nephrol. 2002;34(3):369–72.
39. Keskes-Ammar L, Feki-Chakroun N, Rebai T, Sahnoun Z, Ghozzi H, Hammami S, Zghal K, Fki H, Damak J, Bahloul A. Sperm oxidative stress and the effect of an oral vitamin E and selenium supplement on semen quality in infertile men. Arch Androl. 2003;49(2):83–94.
40. Piomboni P, Gambera L, Serafini F, Campanella G, Morgante G, De Leo V. Sperm quality improvement after natural anti-oxidant treatment of asthenoteratospermic men with leukocytospermia. Asian J Androl. 2008;10(2):201–6.
41. Comhaire FH, Christophe AB, Zalata AA, Dhooge WS, Mahmoud AM, Depuydt CE. The effects of combined conventional treatment, oral antioxidants and essential fatty acids on sperm biology in subfertile men. Prostaglandins Leukot Essent Fatty Acids. 2000; 63(3):159–65.
42. Gil-Villa AM, Cardona-Maya W, Agarwal A, Sharma R, Cadavid A. Role of male factor in early recurrent embryo loss: do antioxidants have any effect? Fertil Steril. 2009;92(2):565–71.
43. Tunc O, Thompson J, Tremellen K. Improvement in sperm DNA quality using an oral antioxidant therapy. Reprod Biomed Online. 2009;18(6):761–8.
44. Tremellen K, Miari G, Froiland D, Thompson J. A randomised control trial examining the effect of an antioxidant (Menevit) on pregnancy outcome during IVF-ICSI treatment. Aust N Z J Obstet Gynaecol. 2007;47(3):216–21.
45. Tunc O, Tremellen K. Oxidative DNA damage impairs global sperm DNA methylation in infertile men. J Assist Reprod Genet. 2009;26(9–10):537–44.
46. Zini A, Boman JM, Belzile E, Ciampi A. Sperm DNA damage is associated with an increased risk of pregnancy loss after IVF and ICSI: systematic review and meta-analysis. Hum Reprod. 2008; 23(12):2663–8.
47. Aitken RJ, Koopman P, Lewis SE. Seeds of concern. Nature. 2004;432(7013):48–52.

In Vitro Studies of Antioxidants for Male Reproductive Health

38

Armand Zini and Maria San Gabriel

Abstract

The detrimental effects of oxidants on spermatozoa were suggested close to 70 years ago with the demonstration that oxygen is sperm toxic. Later studies confirmed the susceptibility of spermatozoa to oxidative stress and the fact that human spermatozoa and semen leukocytes can generate reactive oxygen species (ROS). These observations have led to studies on the role of antioxidants in protecting spermatozoa from oxidative stress in vitro. The purpose of this chapter is to discuss the rationale for antioxidant therapy in male infertility and evaluate the data on the efficacy of in vitro antioxidant preparations on sperm function. A review of the literature demonstrates a beneficial effect of in vitro antioxidants in protecting spermatozoa from exogenous oxidants and cryopreservation (and subsequent thawing). However, the protective effect of in vitro antioxidants on sperm preparations subjected to endogenous ROS and gentle sperm processing has not been established.

Keywords

Sperm DNA fragmentation • Oxidative stress • Vitamins • Sperm washing • Male infertility • Reactive oxygen species • Semen antioxidants

The detrimental effects of oxidants on spermatozoa were suggested close to 70 years ago with the demonstration that oxygen is sperm toxic. Later studies confirmed the susceptibility of spermatozoa to oxidative stress and the fact that human spermatozoa and semen leukocytes can generate reactive oxygen species (ROS). These observations have led to studies on the role of antioxidants in protecting spermatozoa from oxidative stress in vitro.

The purpose of this chapter is to discuss the rationale for antioxidant therapy in male infertility and evaluate the data on the efficacy of in vitro antioxidant preparations on sperm function. A review of the literature demonstrates a beneficial effect of in vitro antioxidants in protecting spermatozoa from exogenous oxidants and cryopreservation (and subsequent thawing). However, the protective effect of in vitro antioxidants on sperm preparations subjected to endogenous ROS and gentle sperm processing has not been established.

Reactive Oxygen Species and Male Infertility

The relationship between seminal ROS and male infertility is the basis for proposing treatment with antioxidants in these men [1, 2]. High levels of ROS have been detected in the semen of 25% of infertile men but not in the semen of fertile men [3, 4]. Semen ROS levels are inversely related to the probability of achieving a spontaneous pregnancy [5]. Moreover, the levels of sperm DNA oxidation (a marker of oxidative stress) are higher in infertile men compared with fertile men [6, 7]. Semen ROS are generated by spermatozoa (especially, defective or immature) and semen leukocytes [8–12].

A. Zini, MD (✉)
Department of Surgery, St. Mary's Hospital,
3830 Lacombe Avenue, Room B270, Montreal, QC, Canada, H3T 1M5
e-mail: ziniarmand@yahoo.com

M.S. Gabriel, PhD
Division of Urology, Department of Surgery, Royal Victoria Hospital,
McGill University, Montreal, QC, Canada, H3A 1A1

S.J. Parekattil and A. Agarwal (eds.), *Male Infertility: Contemporary Clinical Approaches, Andrology, ART & Antioxidants*,
DOI 10.1007/978-1-4614-3335-4_38, © Springer Science+Business Media, LLC 2012

In contrast to the pathologic effects of excess ROS production, small amounts of ROS may be necessary for the initiation of critical sperm functions, including capacitation and the acrosome reaction [13–15]. Therefore, there is a finely tuned balance between ROS scavenging and low, physiologic levels of ROS that are necessary for normal sperm function and maturation.

The susceptibility of human spermatozoa to oxidative stress stems primarily from the abundance of unsaturated fatty acids localized within the sperm plasma membrane. These fatty acids provide fluidity necessary for membrane fusion events, such as the acrosome reaction and sperm–egg interaction, and for sperm motility. However, the unsaturated nature of these fatty acids predisposes them to oxidative stress and lipid peroxidation. Once the lipid peroxidation cascade has been initiated, sperm dysfunction (e.g., loss of motility) ensues as a result of accumulation of lipid peroxides on the sperm membrane, depletion of ATP, and oxidative damage to the DNA [16–19]. It has been shown that ROS can cause damage to the sperm DNA, directly or indirectly via production and subsequent translocation of lipid peroxides [19–22].

Semen Antioxidants and Sperm Function

Seminal fluid is an important source of antioxidants (both enzymatic and nonenzymatic) that can protect spermatozoa from oxidative injury [4, 23, 24]. This feature of seminal plasma is of critical importance in view of the inherent susceptibility of spermatozoa to oxidative stress and also because spermatozoa themselves have little cytoplasmic fluid and minimal antioxidant capacity [4]. There are several endogenous antioxidant enzymes in the male reproductive tract and in seminal fluid: superoxide dismutase (SOD), catalase, and glutathione peroxidase (GPX) [4, 23, 25–28]. Moreover, there are several small, nonenzymatic antioxidants (e.g., vitamins C and E, hypotaurine, taurine, L-carnitine, and lycopene) in semen and, in fact, this nonenzymatic fraction represents most of the total seminal antioxidant activity [4, 29].

A number of investigators have proposed that oxidative sperm dysfunction may be secondary to reduced semen antioxidant capacity. However, clinical studies have reported conflicting results in this respect. Several studies have found that seminal antioxidant activity is reduced in infertile men with high levels of seminal ROS (relative to those with normal levels of ROS), whereas others have not shown this [4, 30–32]. Studies have also reported that a deficiency in semen antioxidants is related to sperm dysfunction (including DNA damage), whereas other studies have not observed this relationship [17, 33–37].

There is no evidence to suggest that male infertility is caused by systemic antioxidant or vitamin deficiency. Silver et al. evaluated a group of fertile men and did not identify any relationships between dietary antioxidant intake (vitamins C, E, or β-carotene) and sperm DNA damage [38]. Nonetheless, it is very likely that a subgroup of infertile men may have specific antioxidant deficiency, particularly, vitamin C deficiency [35, 39, 40]. Moreover, infertile men with various lifestyles (smoking, excessive alcohol intake, and dieting) may also be at high risk for antioxidant or vitamin deficiency [41, 42].

In Vitro Antioxidants in Male Infertility

Several studies have examined the role of in vitro antioxidant supplementation in protecting spermatozoa from oxidative injury and resulting sperm dysfunction (i.e., loss of motility and viability). This is clinically relevant as sperm washing is routinely performed prior to ARTs (e.g., intrauterine insemination and in vitro fertilization) and the process may result in the generation of ROS with ensuing sperm dysfunction [43]. During semen processing, spermatozoa are particularly vulnerable to oxidative stress because seminal plasma (rich in antioxidants) has been removed in the process [44, 45]. For assisted reproductive techniques that require the use of spermatozoa with progressive motility (e.g., IUI and IVF), minimizing sperm dysfunction during semen processing is critical for fertilization and subsequent pregnancy.

Recently, a number of studies have examined the role of in vitro antioxidant supplementation in protecting the sperm DNA from oxidative damage because of the concern that unrepaired oxidative sperm DNA damage may be transmitted to the offspring when used in the context of ARTs (e.g., intrauterine insemination and in vitro fertilization) [46]. However, it is important to note that subpopulations of spermatozoa will exhibit variable susceptibility to oxidative stress: the DNA of normal spermatozoa is reportedly less susceptible to gentle processing techniques than the DNA of abnormal or immature spermatozoa [10, 47]. It is likely that the susceptibility of the sperm DNA to oxidative injury is related to the degree of sperm chromatin compaction (i.e., level of protamination) [48, 49]. Experimental (animal) studies suggest that the spermatozoa of infertile men may be more susceptible to oxidative injury in vitro but benefit more so from antioxidants than the spermatozoa of fertile men [50].

Role of In Vitro Antioxidants in Protecting Spermatozoa from Exogenous Reactive Oxygen Species

This is of clinical relevance as many of the semen samples contain leukocytes and these cells have the potential to generate exogenous ROS [51]. Antioxidants such as vitamin E, catalase, and glutathione have been shown to protect sperm motility from the effects of exogenous ROS (Table 38.1)

Table 38.1 Role of in vitro antioxidants in protecting spermatozoa from the loss of motility and DNA damage due to exogenous ROS

Study	Exogenous ROS	Antioxidant supplement and results
Sperm motility		
de Lamirande (1992)	X+XO	Catalase protects spz from X+XO-induced loss of motility
		SOD, DTT, or GSH less effective in protecting spz motility from ROS
Griveau (1995)	X+XO	Catalase protects spz from X+XO-induced loss of motility
		SOD or mannitol ineffective in protecting spz motility from ROS
Sperm DNA		
Lopes (1998)	X+XO	GSH+hypotaurine protect spz from X+XO-induced DD
		Catalase protects spz from X+XO-induced DD
		n-Acetylcysteine protects spz from X+XO-induced DD
Potts (2000)	H_2O_2+Fe+ADP	S. plasma (>60%v/v) lowers oxidative spz damage (↓DD, LPO)
Sierens (2002)	H_2O_2	Isoflavones, vit C and E protect spz from H_2O_2-induced DD (Isoflavones: genistein, equol). Dose effect noted
Russo (2006)	(1) H_2O_2	Propolis lowers oxidative spz damage (↓LPO, DD, LDH) (Propolis—a natural resinous hive product)
	(2) Benzopyrene	
	(3) H_2O_2+Fe+ADP	

ADP adenosine diphosphate, *COMET* single-cell gel electrophoresis, *DD* DNA damage, *DFI* DNA fragmentation index, *Fe* iron, *GSH* glutathione, *LDH* lactate dehydrogenase, *LPO* lipid peroxidation, *S. plasma* seminal plasma, *Spz* sperm, *TUNEL* terminal deoxynucleotidyl transferase dUTP nick end labeling, *X* xanthine, *XO* xanthine oxidase

Table 38.2 Role of in vitro antioxidant supplements in protecting sperm DNA from stimulated endogenous ROS generation

Study	Assay	ROS stimulant	Antioxidant supplement and results
Twigg (1998)	IS NTL	NADPH	Vit E, SOD, catalase, hypotaurine, albumin all ineffective in protecting spz DNA from endogenous ROS
Anderson (2003)	COMET	Estrogens	Catalase protects spz from estrogen-induced oxidative DD SOD and vit C less effective (Estrogens: equol, daidzein, genistein, DES, E2)
Cemeli (2004)	COMET	Estrogens (1 h 37 C)	Flavonoid (Kaempferol) protects sperm from estrogen-induced oxidative DD
Dobrzynska (2004)	COMET	DES, T3, T4, NA (1 h 37 C)	Flavonoids and catalase protect spz from stimulant-induced oxidative DD (Flavonoids: Kaempferol, Quercetin)

COMET alkaline single-cell gel electrophoresis, *DD* DNA damage, *ISNTL* in situ nick translation assay, *LPO* lipid peroxidation, *NA* noradrenaline, *ROS* reactive oxygen species, *SOD* superoxide dismutase, *Spz* sperm, *T3* triiodothyronine, *T4* thyroxine, *vit* vitamin

[19, 52]. In contrast, superoxide dismutase is less effective in preventing the loss of motility due to exogenous oxidants [19, 52]. Altogether, these data suggest that H_2O_2 is the most sperm-toxic exogenous ROS.

Antioxidants have also been shown to protect the sperm DNA from the effects of exogenous ROS (Table 38.1) [44, 53–55]. This is of clinical relevance as sperm DNA damage may impact on reproductive outcomes after ARTs [56]. Indeed, sperm DNA damage has been associated with reduced pregnancy rates with IUI, and, to a lesser extent with conventional IVF.

Role of In Vitro Antioxidants in Protecting Spermatozoa from Endogenous Reactive Oxygen Species

Spermatozoa can be stimulated to generate ROS using a variety of agents (e.g., NADPH and estrogens) and this ROS production can impair sperm function [57]. In contrast to the beneficial effect of antioxidants in protecting spermatozoa from exogenous ROS, antioxidants appear to be of limited value in protecting spermatozoa from endogenous ROS production [58]. Twigg et al. demonstrated that SOD, catalase, or both are ineffective, whereas albumin is effective in protecting spermatozoa from loss of motility due to endogenous ROS generation [58]. These findings stress the importance of using gentle (brief, little centrifugation) semen processing protocols so as to minimize the production and adverse impact of low levels of endogenous ROS.

Similarly, antioxidants appear to be of limited value in protecting the DNA of normal spermatozoa (with normal chromatin compaction) from endogenous ROS production (e.g., NADPH induced or centrifugation induced) (Table 38.2) [58–61]. In samples with poor morphology and poor sperm chromatin compaction, antioxidants may protect the sperm DNA from endogenous ROS production, as these samples are more vulnerable to oxidative stress [10, 47].

Table 38.3 The effect of in vitro antioxidants on sperm motility during semen processing

Study	Parameter	Semen processing	Antioxidant supplement and results
Griveau (1994)	Motility	1. CF at 400 *g* × 2	DTT, Catalase, SOD or GSH improve motility
		2. Swim-up	
		3. 24-h Incubation	
Zheng (1997)	Motility	2- and 3-h Incubation (fertile and infertile)	Ferulic acid improves sperm motility and reduces LPO
	LPO		Ferulic acid increases sperm cAMP and cGMP
Oeda (1997)	Motility	2-h Incubation	NAC lowers semen ROS levels
	ROS		NAC improves sperm motility
Verma (1999)	Motility	6-h Incubation	Vitamin E lowers sperm LPO and protects spermatozoa from loss of motility
	LPO		
Donnelly (2000)	Motility	Percoll DGC + 4 h incubation	GSH or hypotaurine do not protect spermatozoa from loss of motility
Calamera (2001)	Motility	2–47-h Incubation	Catalase did not protect spermatozoa from loss of motility
	ROS		
Chi (2008)	Motility	Centrifugation (1,000 rpm × 2) + 1 h incubation	EDTA or catalase lower CF-induced sperm ROS
	ROS		EDTA (but not catalase) protects spermatozoa from CF-induced loss sperm motility

CF centrifugation, *COMET* alkaline single-cell gel electrophoresis, *DD* DNA damage, *DGC* density-gradient centrifugation, *DTT* dithiotreitol, *GSH* glutathione, *LPO* lipid peroxidation, *NAC* *N*-acetyl-L-cysteine, *ROS* reactive oxygen species, *SOD* superoxide dismutase

Table 38.4 Role of in vitro antioxidant supplements in protecting sperm DNA from semen processing

Study	Assay	Semen processing	Antioxidant supplement and results
Hughes (1998)	COMET	Percoll DGC	Vits C, E, or urate lower sperm DD after DGC
			Vits C + E or AC increase sperm DD after DGC
Donnelly (1999)	COMET	Percoll DGC	Vit C or E does not lower baseline sperm ROS and DD
			Vit C or E protect sperm from H_2O_2-induced ROS and DD
			Vits C + E induce sperm DD and increase H_2O_2-induced DD
Donnelly (2000)	COMET	Percoll DGC ± H_2O_2	GSH, hypotaurine or both do not alter baseline sperm DD
			GSH, hypotaurine or both do not alter sperm motility at 4 h
			GSH and/or hypotaurine lower H_2O_2-induced sperm DD
Chi (2008)	COMET	Centrifugation (1,000 rpm × 2) + 1 h incubation	EDTA or catalase lower centrifugation-induced sperm ROS
			EDTA or catalase lower centrifugation-induced sperm DD
			EDTA or datalase have no protective effect on LPO

AC Acetyl cysteine, *COMET* alkaline single-cell gel electrophoresis, *DD* DNA damage, *DGC* density-gradient centrifugation, *GSH* glutathione, *LPO* lipid peroxidation, *ROS* reactive oxygen species, *vit* vitamin

Role of In Vitro Antioxidants in Protecting Spermatozoa from Semen Processing

Several studies have reported on the effects of antioxidants in preventing the decline in sperm motility after semen processing and incubation (Table 38.3). These studies have clinical relevance because it is important to maximize sperm motility prior to assisted reproductive techniques such as IUI and standard IVF. The available studies report conflicting results regarding the effects of antioxidants in preventing the loss of sperm motility during sperm processing such as centrifugation and incubation. Some studies have shown that antioxidants (e.g., vitamin E, glutathione, *n*-acetylcysteine, catalase, and ferulic acid) are effective in reducing ROS levels and in preventing the decline in sperm motility during sperm processing [62–65]. In contrast, other studies have reported that antioxidants (e.g., glutathione and catalase) are ineffective in protecting spermatozoa from the loss of motility during sperm processing [66–68]. It is important to note that sperm samples from infertile men may be more susceptible to oxidative injury (from semen processing) and be afforded greater protection by antioxidants than samples from fertile men [50].

Antioxidants appear to be of limited value in protecting sperm DNA from gentle semen processing (e.g., incubation or density-gradient centrifugation) (Table 38.4) [67–70]. In some cases, antioxidants supplementation in vitro (e.g., combination of vitamins C and E) may cause sperm DNA damage [68, 70].

Role of In Vitro Antioxidants in Protecting Spermatozoa from Cryopreservation and Thawing

Several studies have evaluated the role of antioxidants in protecting spermatozoa from the loss of motility that occurs following cryopreservation and thawing. Most studies have reported on the use of pentoxifylline (an antioxidant and phosphodiesterase inhibitor). Some studies have shown that

Table 38.5 The role of in vitro antioxidants in protecting human sperm DNA from injury caused by cryopreservation and thawing

Study	Assay	Antioxidant	Effect of antioxidant on cryopreservation and thawing
Taylor '09	TUNEL	Vitamin E	No effect on sperm DNA integrity
			Improved post-thaw motility
Li '09	COMET	Catalase or ascorbic acid	Improved sperm DNA integrity
			Reduced ROS production
Branco '09	COMET	Resveratrol or ascorbic acid	Improved sperm DNA integrity
Martinez-Soto '09	TUNEL	Genistein	Improved sperm DNA integrity
			Reduced ROS production, improved post-thaw motility
Thompson '09	8-OHdG TUNEL	Genistein	Improved sperm DNA integrity (reduced oxidative damage)

8-OHdG 8-hydroxy-2-deoxyguanosine, *COMET* alkaline single-cell gel electrophoresis, *ROS* reactive oxygen species, *TUNEL* terminal deoxynucleotidyl transferase dUTP nick end labeling

pentoxifylline improves post-thaw sperm motility and/or sperm function [71–74], whereas others have demonstrated that this antioxidant does not have a beneficial effect [75]. Other antioxidants (vitamins E and C and rebamipide) have been used to enhance post-thaw motility, however the results have been modest [76, 77].

Several studies have also evaluated the role of antioxidants in protecting sperm DNA from injury following cryopreservation and thawing. Most studies have shown that antioxidants (vitamin C, catalase, resveratrol, and genistein) can protect the sperm DNA from oxidative injury during cryopreservation and subsequent thawing [78–81] (Table 38.5). In contrast, Taylor et al. reported that the antioxidant vitamin E does not protect sperm DNA during cryopreservation [82].

Taken together, the data suggest that antioxidants are generally effective in protecting spermatozoa from the effects of cryopreservation and thawing. However, the technique of cryopreservation and type of cryoprotectant are also important in improving post-thaw sperm function [83].

Summary

Oxidative stress plays an important role in the pathophysiology of male infertility. The study of in vitro antioxidants is highly relevant in the era of assisted reproduction because of the susceptibility of human spermatozoa to oxidative injury and the vulnerability of these cells during semen processing. Most studies have demonstrated a beneficial effect of in vitro antioxidant supplements in protecting spermatozoa from exogenous oxidants and cryopreservation (with subsequent thawing). In contrast, the effect of these antioxidants in protecting normal spermatozoa from endogenous ROS and gentle sperm processing has not been established conclusively. Additional studies are needed to determine the optimal antioxidant preparation to protect spermatozoa from oxidative stress in vitro.

Expert Commentary

The biological basis for the use of in vitro antioxidants in male infertility is sound and is based on the body of literature showing that sperm dysfunction is strongly related to oxidative stress. Furthermore, the inherent susceptibility of human spermatozoa to oxidative stress is particularly relevant during semen processing as a result of the removal of seminal plasma, a natural antioxidant. Clinical studies of in vitro antioxidants support the use of antioxidants in protecting spermatozoa (particularly abnormal spermatozoa) from exogenous ROS and cryopreservation. However, the optimal antioxidant and concentration has not been established.

Five-Year View

To see a real advance in the field of in vitro antioxidants for male infertility, we need to undertake additional, comparative studies to assess the differential effect of various antioxidants. We also need to undertake studies to better define the differential treatment response between normal (fertile) and subnormal (infertile) semen samples and identify the optimal protocol (type and concentration of antioxidant).

References

1. Agarwal A, Saleh RA. Role of oxidants in male infertility: rationale, significance, and treatment. Urol Clin North Am. 2002;29:817–27.
2. Agarwal A, Sharma RK, Nallella KP, Thomas Jr AJ, Alvarez JG, Sikka SC. Reactive oxygen species as an independent marker of male factor infertility. Fertil Steril. 2006;86:878–85.
3. Iwasaki A, Gagnon C. Formation of reactive oxygen species in spermatozoa of infertile patients. Fertil Steril. 1992;57:409–16.
4. Zini A, de Lamirande E, Gagnon C. Reactive oxygen species in semen of infertile patients: levels of superoxide dismutase-and catalase-like activities in seminal plasma and spermatozoa. Int J Androl. 1993;16:183–8.
5. Aitken RJ, Irvine DS, Wu FC. Prospective analysis of sperm-oocyte fusion and reactive oxygen species generation as criteria for the diagnosis of infertility. Am J Obstet Gynecol. 1991;164:542–51.
6. Kodama H, Yamaguchi R, Fukuda J, Kasai H, Tanaka T. Increased oxidative deoxyribonucleic acid damage in the spermatozoa of infertile male patients. Fertil Steril. 1997;68:519–24.

7. Shen HM, Chia SE, Ong CN. Evaluation of oxidative DNA damage in human sperm and its association with male infertility. J Androl. 1999;20:718–23.
8. Barroso G, Morshedi M, Oehninger S. Analysis of DNA fragmentation, plasma membrane translocation of phosphatidylserine and oxidative stress in human spermatozoa. Hum Reprod. 2000;15:1338–44.
9. Gomez E, Buckingham DW, Brindle J, Lanzafame F, Irvine DS, Aitken RJ. Development of an image analysis system to monitor the retention of residual cytoplasm by human spermatozoa: correlation with biochemical markers of the cytoplasmic space, oxidative stress, and sperm function. J Androl. 1996;17:276–87.
10. Muratori M, Piomboni P, Baldi E, Filimberti E, Pecchioli P, Moretti E, et al. Functional and ultrastructural features of DNA-fragmented human sperm. J Androl. 2000;21:903–12.
11. de Lamirande E, Jiang H, Zini A, Kodama H, Gagnon C. Reactive oxygen species and sperm physiology. Rev Reprod. 1997;2:48–54.
12. Aitken RJ, West K, Buckingham D. Leukocytic infiltration into the human ejaculate and its association with semen quality, oxidative stress, and sperm function. J Androl. 1994;15:343–52.
13. Aitken RJ, Paterson M, Fisher H, Buckingham DW, van Duin M. Redox regulation of tyrosine phosphorylation in human spermatozoa and its role in the control of human sperm function. J Cell Sci. 1995;108(Pt 5):2017–25.
14. de Lamirande E, Gagnon C. Impact of reactive oxygen species on spermatozoa: a balancing act between beneficial and detrimental effects. Hum Reprod. 1995;10 Suppl 1:15–21.
15. Griveau JF, Le Lannou D. Reactive oxygen species and human spermatozoa: physiology and pathology. Int J Androl. 1997;20:61–9.
16. Aitken RJ, Gordon E, Harkiss D, Twigg JP, Milne P, Jennings Z, et al. Relative impact of oxidative stress on the functional competence and genomic integrity of human spermatozoa. Biol Reprod. 1998;59:1037–46.
17. Alvarez JG, Touchstone JC, Blasco L, Storey BT. Spontaneous lipid peroxidation and production of hydrogen peroxide and superoxide in human spermatozoa. Superoxide dismutase as major enzyme protectant against oxygen toxicity. J Androl. 1987;8:338–48.
18. de Lamirande E, Gagnon C. Reactive oxygen species and human spermatozoa. II. Depletion of adenosine triphosphate plays an important role in the inhibition of sperm motility. J Androl. 1992;13:379–86.
19. de Lamirande E, Gagnon C. Reactive oxygen species and human spermatozoa. I. Effects on the motility of intact spermatozoa and on sperm axonemes. J Androl. 1992;13:368–78.
20. Yang MH, Schaich KM. Factors affecting DNA damage caused by lipid hydroperoxides and aldehydes. Free Radic Biol Med. 1996;20:225–36.
21. Aitken RJ, Clarkson JS. Cellular basis of defective sperm function and its association with the genesis of reactive oxygen species by human spermatozoa. J Reprod Fertil. 1987;81:459–69.
22. Lewis SE, Aitken RJ. DNA damage to spermatozoa has impacts on fertilization and pregnancy. Cell Tissue Res. 2005;322:33–41.
23. Jeulin C, Soufir JC, Weber P, Laval-Martin D, Calvayrac R. Catalase activity in human spermatozoa and seminal plasma. Gamete Res. 1989;24:185–96.
24. Gagnon C, Iwasaki A, De Lamirande E, Kovalski N. Reactive oxygen species and human spermatozoa. Ann N Y Acad Sci. 1991;637:436–44.
25. Jow WW, Schlegel PN, Cichon Z, Phillips D, Goldstein M, Bardin CW. Identification and localization of copper-zinc superoxide dismutase gene expression in rat testicular development. J Androl. 1993;14:439–47.
26. Zini A, Schlegel PN. Expression of glutathione peroxidases in the adult male rat reproductive tract. Fertil Steril. 1997;68:689–95.
27. Zini A, Schlegel PN. Catalase mRNA expression in the male rat reproductive tract. J Androl. 1996;17:473–80.
28. Zini A, Schlegel PN. Identification and characterization of antioxidant enzyme mRNAs in the rat epididymis. Int J Androl. 1997;20:86–91.
29. Holmes RP, Goodman HO, Shihabi ZK, Jarow JP. The taurine and hypotaurine content of human semen. J Androl. 1992;13:289–92.
30. Lewis SE, Boyle PM, McKinney KA, Young IS, Thompson W. Total antioxidant capacity of seminal plasma is different in fertile and infertile men. Fertil Steril. 1995;64:868–70.
31. Sanocka D, Miesel R, Jedrzejczak P, Kurpisz MK. Oxidative stress and male infertility. J Androl. 1996;17:449–54.
32. Smith R, Vantman D, Ponce J, Escobar J, Lissi E. Total antioxidant capacity of human seminal plasma. Hum Reprod. 1996;11:1655–60.
33. Aitken RJ, Buckingham DW, Carreras A, Irvine DS. Superoxide dismutase in human sperm suspensions: relationship with cellular composition, oxidative stress, and sperm function. Free Radic Biol Med. 1996;21:495–504.
34. Appasamy M, Muttukrishna S, Pizzey AR, Ozturk O, Groome NP, Serhal P, et al. Relationship between male reproductive hormones, sperm DNA damage and markers of oxidative stress in infertility. Reprod Biomed Online. 2007;14:159–65.
35. Fraga CG, Motchnik PA, Shigenaga MK, Helbock HJ, Jacob RA, Ames BN. Ascorbic acid protects against endogenous oxidative DNA damage in human sperm. Proc Natl Acad Sci USA. 1991;88:11003–6.
36. Song GJ, Norkus EP, Lewis V. Relationship between seminal ascorbic acid and sperm DNA integrity in infertile men. Int J Androl. 2006;29:569–75.
37. Verit FF, Verit A, Kocyigit A, Ciftci H, Celik H, Koksal M. No increase in sperm DNA damage and seminal oxidative stress in patients with idiopathic infertility. Arch Gynecol Obstet. 2006;274:339–44.
38. Silver EW, Eskenazi B, Evenson DP, Block G, Young S, Wyrobek AJ. Effect of antioxidant intake on sperm chromatin stability in healthy nonsmoking men. J Androl. 2005;26:550–6.
39. Fraga CG, Motchnik PA, Wyrobek AJ, Rempel DM, Ames BN. Smoking and low antioxidant levels increase oxidative damage to sperm DNA. Mutat Res. 1996;351:199–203.
40. Hampl JS, Taylor CA, Johnston CS. Vitamin C deficiency and depletion in the United States: the Third National Health and Nutrition Examination Survey, 1988 to 1994. Am J Public Health. 2004;94:870–5.
41. Jacob RA. Assessment of human vitamin C status. J Nutr. 1990;120 Suppl 11:1480–5.
42. Ryle PR, Thomson AD. Nutrition and vitamins in alcoholism. Contemp Issues Clin Biochem. 1984;1:188–224.
43. Aitken RJ, Clarkson JS. Significance of reactive oxygen species and antioxidants in defining the efficacy of sperm preparation techniques. J Androl. 1988;9:367–76.
44. Potts RJ, Notarianni LJ, Jefferies TM. Seminal plasma reduces exogenous oxidative damage to human sperm, determined by the measurement of DNA strand breaks and lipid peroxidation. Mutat Res. 2000;447:249–56.
45. Twigg J, Irvine DS, Houston P, Fulton N, Michael L, Aitken RJ. Iatrogenic DNA damage induced in human spermatozoa during sperm preparation: protective significance of seminal plasma. Mol Hum Reprod. 1998;4:439–45.
46. Aitken RJ. Founders' Lecture. Human spermatozoa: fruits of creation, seeds of doubt. Reprod Fertil Dev. 2004;16:655–64.
47. Said TM, Agarwal A, Sharma RK, Thomas Jr AJ, Sikka SC. Impact of sperm morphology on DNA damage caused by oxidative stress induced by beta-nicotinamide adenine dinucleotide phosphate. Fertil Steril. 2005;83:95–103.
48. Cho C, Willis WD, Goulding EH, Jung-Ha H, Choi YC, Hecht NB, et al. Haploinsufficiency of protamine-1 or -2 causes infertility in mice. Nat Genet. 2001;28:82–6.

49. De Iuliis GN, Thomson LK, Mitchell LA, Finnie JM, Koppers AJ, Hedges A, et al. DNA damage in human spermatozoa is highly correlated with the efficiency of chromatin remodeling and the formation of 8-hydroxy-2′-deoxyguanosine, a marker of oxidative stress. Biol Reprod. 2009;81:517–24.
50. Libman J, Gabriel MS, Sairam MR, Zini A. Catalase can protect spermatozoa of FSH receptor knock-out mice against oxidant-induced DNA damage in vitro. Int J Androl. 2010;33:818–22.
51. Aitken RJ, Buckingham DW, Brindle J, Gomez E, Baker HW, Irvine DS. Analysis of sperm movement in relation to the oxidative stress created by leukocytes in washed sperm preparations and seminal plasma. Hum Reprod. 1995;10:2061–71.
52. Aitken RJ, Buckingham D, Harkiss D. Use of a xanthine oxidase free radical generating system to investigate the cytotoxic effects of reactive oxygen species on human spermatozoa. J Reprod Fertil. 1993;97:441–50.
53. Lopes S, Jurisicova A, Sun JG, Casper RF. Reactive oxygen species: potential cause for DNA fragmentation in human spermatozoa. Hum Reprod. 1998;13:896–900.
54. Russo A, Troncoso N, Sanchez F, Garbarino JA, Vanella A. Propolis protects human spermatozoa from DNA damage caused by benzo[a] pyrene and exogenous reactive oxygen species. Life Sci. 2006;78:1401–6.
55. Sierens J, Hartley JA, Campbell MJ, Leathem AJ, Woodside JV. In vitro isoflavone supplementation reduces hydrogen peroxide-induced DNA damage in sperm. Teratog Carcinog Mutagen. 2002; 22:227–34.
56. Zini A, Sigman M. Are tests of sperm DNA damage clinically useful? Pros and cons. J Androl. 2009;30:219–29.
57. Aitken RJ, Fisher HM, Fulton N, Gomez E, Knox W, Lewis B, et al. Reactive oxygen species generation by human spermatozoa is induced by exogenous NADPH and inhibited by the flavoprotein inhibitors diphenylene iodonium and quinacrine. Mol Reprod Dev. 1997;47:468–82.
58. Twigg J, Fulton N, Gomez E, Irvine DS, Aitken RJ. Analysis of the impact of intracellular reactive oxygen species generation on the structural and functional integrity of human spermatozoa: lipid peroxidation, DNA fragmentation and effectiveness of antioxidants. Hum Reprod. 1998;13:1429–36.
59. Anderson D, Schmid TE, Baumgartner A, Cemeli-Carratala E, Brinkworth MH, Wood JM. Oestrogenic compounds and oxidative stress (in human sperm and lymphocytes in the Comet assay). Mutat Res. 2003;544:173–8.
60. Cemeli E, Schmid TE, Anderson D. Modulation by flavonoids of DNA damage induced by estrogen-like compounds. Environ Mol Mutagen. 2004;44:420–6.
61. Dobrzynska MM, Baumgartner A, Anderson D. Antioxidants modulate thyroid hormone- and noradrenaline-induced DNA damage in human sperm. Mutagenesis. 2004;19:325–30.
62. Griveau JF, Le Lannou D. Effects of antioxidants on human sperm preparation techniques. Int J Androl. 1994;17:225–31.
63. Oeda T, Henkel R, Ohmori H, Schill WB. Scavenging effect of N-acetyl-L-cysteine against reactive oxygen species in human semen: a possible therapeutic modality for male factor infertility? Andrologia. 1997;29:125–31.
64. Verma A, Kanwar KC. Effect of vitamin E on human sperm motility and lipid peroxidation in vitro. Asian J Androl. 1999;1:151–4.
65. Zheng RL, Zhang H. Effects of ferulic acid on fertile and asthenozoospermic infertile human sperm motility, viability, lipid peroxidation, and cyclic nucleotides. Free Radic Biol Med. 1997;22:581–6.
66. Calamera JC, Fernandez PJ, Buffone MG, Acosta AA, Doncel GF. Effects of long-term in vitro incubation of human spermatozoa: functional parameters and catalase effect. Andrologia. 2001;33: 79–86.
67. Chi HJ, Kim JH, Ryu CS, Lee JY, Park JS, Chung DY, et al. Protective effect of antioxidant supplementation in sperm-preparation medium against oxidative stress in human spermatozoa. Hum Reprod. 2008;23:1023–8.
68. Donnelly ET, McClure N, Lewis SE. Glutathione and hypotaurine in vitro: effects on human sperm motility, DNA integrity and production of reactive oxygen species. Mutagenesis. 2000;15:61–8.
69. Donnelly ET, McClure N, Lewis SE. The effect of ascorbate and alpha-tocopherol supplementation in vitro on DNA integrity and hydrogen peroxide-induced DNA damage in human spermatozoa. Mutagenesis. 1999;14:505–12.
70. Hughes CM, Lewis SE, McKelvey-Martin VJ, Thompson W. The effects of antioxidant supplementation during Percoll preparation on human sperm DNA integrity. Hum Reprod. 1998;13:1240–7.
71. Esteves SC, Sharma RK, Thomas Jr AJ, Agarwal A. Cryopreservation of human spermatozoa with pentoxifylline improves the post-thaw agonist-induced acrosome reaction rate. Hum Reprod. 1998; 13:3384–9.
72. Brennan AP, Holden CA. Pentoxifylline-supplemented cryoprotectant improves human sperm motility after cryopreservation. Hum Reprod. 1995;10:2308–12.
73. Bell M, Wang R, Hellstrom WJ, Sikka SC. Effect of cryoprotective additives and cryopreservation protocol on sperm membrane lipid peroxidation and recovery of motile human sperm. J Androl. 1993;14:472–8.
74. Wang R, Sikka SC, Veeraragavan K, Bell M, Hellstrom WJ. Platelet activating factor and pentoxifylline as human sperm cryoprotectants. Fertil Steril. 1993;60:711–5.
75. Check DJ, Kiefer D, Katsoff D, Check JH. Effect of pentoxifylline added to freezing media on subsequent post-thaw hypoosmotic swelling test and other semen parameters. Arch Androl. 1995; 35:161–3.
76. Park NC, Park HJ, Lee KM, Shin DG. Free radical scavenger effect of rebamipide in sperm processing and cryopreservation. Asian J Androl. 2003;5:195–201.
77. Askari HA, Check JH, Peymer N, Bollendorf A. Effect of natural antioxidants tocopherol and ascorbic acids in maintenance of sperm activity during freeze-thaw process. Arch Androl. 1994;33:11–5.
78. Branco CS, Garcez ME, Pasqualotto FF, Erdtman B, Salvador M. Resveratrol and ascorbic acid prevent DNA damage induced by cryopreservation in human semen. Cryobiology. 2010;60:235–7.
79. Li Z, Lin Q, Liu R, Xiao W, Liu W. Protective effects of ascorbate and catalase on human spermatozoa during cryopreservation. J Androl. 2010;31:437–44.
80. Martinez-Soto JC, de Dioshourcade J, Gutierrez-Adan A, Landeras JL, Gadea J. Effect of genistein supplementation of thawing medium on characteristics of frozen human spermatozoa. Asian J Androl. 2010;12:431–41.
81. Thomson LK, Fleming SD, Aitken RJ, De Iuliis GN, Zieschang JA, Clark AM. Cryopreservation-induced human sperm DNA damage is predominantly mediated by oxidative stress rather than apoptosis. Hum Reprod. 2009;24:2061–70.
82. Taylor K, Roberts P, Sanders K, Burton P. Effect of antioxidant supplementation of cryopreservation medium on post-thaw integrity of human spermatozoa. Reprod Biomed Online. 2009;18: 184–9.
83. Nallella KP, Sharma RK, Allamaneni SS, Aziz N, Agarwal A. Cryopreservation of human spermatozoa: comparison of two cryopreservation methods and three cryoprotectants. Fertil Steril. 2004; 82:913–8.

Part V

Assisted Reproductive Technologies in Male Fertility

Micro-testicular Sperm Extraction

39

Doron Sol Stember and Peter Schlegel

Abstract

The condition of nonobstructive azoospermia (NOA) defines men with testicular failure who have severely deficient sperm production with no sperm in the ejaculate. NOA is the underlying diagnosis in approximately 10% of men seeking fertility evaluation. On testicular biopsy, these patients demonstrate hypospermatogenesis, maturation arrest, or Sertoli cell-only pattern (germinal cell aplasia). NOA may be related to genetic causes, as in Klinefelter syndrome (KS) and XX-male syndrome, or may be acquired, as in testicular failure secondary to cryptorchidism or systemic chemotherapy administration. As recently as two decade ago, techniques were not available to assist patients with NOA conceive offspring and their options were limited to donor spermatazoa or child adoption.

A remarkable series of four distinct advancements have dramatically advanced the field in recent years, and today it is possible even for men with Sertoli cell-only pattern to initiate pregnancy with medical assistance. In this chapter, we focus on considerations and techniques related to microscopic testicular sperm extraction (microdissection TESE). Microdissection TESE, a technique developed at New York-Presbyterian Hospital–Weill Cornell Medical Center, yields superior sperm retrieval rates and requires a minimal amount of testicular tissue compared with other methods of sperm extraction.

Keywords

Nonobstructive azoospermia • Micro-testicular sperm extraction • Microdissection TESE • Male infertility • Klinefelter syndrome • Cryptorchidism • Hypospermatogenesis

The condition of nonobstructive azoospermia (NOA) defines men with testicular failure who have severely deficient sperm production with no sperm in the ejaculate. NOA is the underlying diagnosis in approximately 10% of men seeking fertility evaluation. On testicular biopsy, these patients demonstrate hypospermatogenesis, maturation arrest, or Sertoli cell-only pattern (germinal cell aplasia). NOA may be related to genetic causes, as in Klinefelter Syndrome (KS) and XX-male syndrome, or may be acquired, as in testicular failure secondary to cryptorchidism or systemic chemotherapy administration. As recently as two decade ago, techniques were not available to assist patients with NOA conceive offspring and their options were limited to donor spermatazoa or child adoption.

A remarkable series of four distinct advancements have dramatically advanced the field in recent years, and today it is possible even for men with Sertoli cell-only pattern to initiate pregnancy with medical assistance. The first factor was the clinical recognition that successful fertilization is not dependent on epididymal transit. This concept allows for the direct retrieval of seminiferous tubules harboring sperm from the testes. Multiple techniques have been developed for this purpose and will be discussed herein.

D.S. Stember, MD (✉)
Division of Urology, Department of Surgery,
Memorial Sloan-Kettering Cancer Center, 1275 York Avenue,
Box 435, New York, 10065, NY, USA
e-mail: stemberd@mskcc.org

P. Schlegel, MD
Department of Urology, Weill Cornell Medical College,
New York-Presbyterian/ Weill Cornell Hospital,
525 East 6th Street, Starr 900, New York, 10065, NY, USA
e-mail: pnschleg@med.cornell.edu

S.J. Parekattil and A. Agarwal (eds.), *Male Infertility: Contemporary Clinical Approaches, Andrology, ART & Antioxidants*,
DOI 10.1007/978-1-4614-3335-4_39, © Springer Science+Business Media, LLC 2012

The second critical factor was the development of intracytoplasmic sperm injection (ICSI), a technique that involves in vitro injection of a single sperm into an oocyte. ICSI removes many natural barriers to fertilization, theoretically requires only a single sperm, and can be successful even with nonmotile sperm. Surgical sperm retrieval by testicular extraction followed by ICSI was first described as a treatment for obstructive azoospermia in 1993 [1].

The third factor was the recognition, and histologic demonstration, that heterogeneity of seminiferous tubules exists in testis biopsy specimens. Patients with Sertoli cell-only as a predominant testicular histologic pattern, for example, often have microscopic foci of spermatogenesis. The indications for testicular sperm extraction (TESE) were thus expanded as it became apparent that viable spermatozoa could be obtained from men with sperm production so poor that no sperm or no viable sperm in the ejaculate, as with NOA [2, 3].

The fourth major factor has been the introduction of a sperm extraction technique that takes advantage of the opportunity to identify even microscopic sites of recognition of sperm production. Use of an operating microscope during TESE enables selective extraction of tubules that may harbor active spermatogenesis, even in a testicle otherwise overwhelmingly composed of nonproductive tubules. Under microscopic vision, normal-appearing tubules that are more likely to contain sperm can be selectively removed, while abnormal and sclerotic tubules are avoided. Microscopic vessels can also be avoided or coagulated, thereby significantly reducing the risk of post-procedure hematoma or devascularization of remaining testicular tissue.

In this chapter, we focus on considerations and techniques related to microscopic testicular sperm extraction (microdissection TESE). Microdissection TESE, a technique developed at New York-Presbyterian Hospital–Weill Cornell Medical Center, yields superior sperm retrieval rates and requires a minimal amount of testicular tissue compared with other methods of sperm extraction.

Microdissection Testicular Sperm Extraction

Nonobstructive Azoospermia

The formal diagnosis of NOA requires a histological diagnosis, but a preoperative biopsy is not required for microdissection TESE. A clinical diagnosis of NOA may be established with reasonable certainty on the combined basis of history, azoospermia on semen analysis, small testes, flat/empty epididymides, and elevated serum FSH levels. Definitive confirmation of the diagnosis can then be made by histologic analysis of tissue extracted at the time of the sperm retrieval procedure.

Multiple strategies to retrieve testicular tissue in patients with NOA have been developed. Successfully retrieved sperm may be cryopreserved for future use or, if timed to coincide with oocyte retrieval, can be immediately used for ICSI. Each technique has associated advantages and disadvantages. A variety of these sperm retrieval techniques can also be utilized by patients with obstructive azoospermia as an alternative to surgical reconstruction.

Conventional TESE

This procedure involves standard single or multiple testicular biopsies are performed in an open fashion under local or general anesthesia. A scrotal incision is made and the tunica albuginea is opened. Samples of testicular parenchyma are excised and subsequently evaluated for the presence of sperm by an embryologist. Conventional TESE may take less time than microdissection TESE and also has the advantage of not requiring microsurgical training.

However, in contrast to microdissection TESE, blind open tunical incision has a greater risk of interrupting the vascular supply and devascularizing testicular tissue. Loss of functional tissue is particularly problematic in NOA patients who tend to have limited volume and function of testicular tissue in the first place. The testicular blood vessels course under the tunica albuginea before penetrating the testicular parenchyma can be visualized and avoided by using an operating microscope.

Open or conventional TESE is also a relatively inefficient process, since a large proportion of tubules removed in patients with NOA are sclerotic. Multiple blind biopsies are obtained since there is no way to intraoperatively identify normal-appearing tubules. Open TESE does not allow for thorough dissection of all areas of the testis, because to do so would essentially remove all testicular tissue. Even though relatively large amounts of tissue may be removed in open TESE, the possibility of missing foci of spermatogenesis deep within the testes is high compared with microdissection TESE.

Fine Needle Aspiration/Testicular Mapping

Testicular fine needle aspiration (FNAB) is a cytologic technique that does not directly evaluate seminiferous tubule architecture, in contrast to standard biopsies, conventional or microdissection TESE. With a diagnosis of NOA made by other means, however, FNAB is generally considered the least involved procedure for acquiring sperm. It is usually performed under local anesthesia in the office setting, requires little time, is well tolerated by patients, and does not require advanced training on the practitioner's part. It is performed by inserting a 23-19-gage butterfly needle directly into the testis and aspirating contents into plastic tubing that is connected to the needle. Multiple needle punctures are usually directed into various parts of the testis.

A report of patients who underwent FNAB with 2–3 tissue samples from each testis demonstrated spermatozoa recovered in 47% of patients with NOA [4]. An earlier report of FNAB in which up to 15 samples were taken from each testis (with the patients under general anesthesia) had a 60% sperm retrieval rate in men with testicular failure, although most patients had too few sperm retrieved to inject all partner's eggs [5]. Other studies, however, have found that FNAB yields a much lower rate of sperm retrieval than TESE [6, 7]. The scarcity of high-quality literature regarding FNAB, the lack of careful follow-up of patients for potential complications, as well as a wide disparity of specific techniques used, makes critical evaluation of the technique difficult.

One major disadvantage of FNAB is the potential risk of disruption of testicular vascular supply. Given the nature of the procedure, subtunical or intraparenchymal testicular bleeding cannot by directly stopped and must be allowed to progress and resolve with the treatment of ice packs. Another negative consequence of the blind technique is the high chance of failure simply due to sampling error. To address this issue in a systematic manner, Turek et al. introduced the technique of testis mapping. In mapping, the skin over the testis is pulled taut and secured in place by wrapping a sponge or Penrose drain behind the testis. A sterile pen is used to mark four to nine evenly distributed sites, depending on testis size, on the skin surface. A wheal of local anesthesia, and subsequently the aspiration needle, is then directed at each marked site [8].

Microdissection TESE

Microdissection TESE is most sophisticated method for retrieving spermatozoa in men with NOA (Fig. 39.1). With the benefit of optical magnification, an avascular area of the tunica albuginea is incised. Small areas of bleeding are selectively coagulated with a microbipolar forceps, thereby minimizing damage to adjacent testicular tissue. Microscopic foci of healthy-appearing tubules are identified among the much great numbers of Sertoli cell-only or sclerotic tubules typically seen in NOA patients. Normal tubules are usually larger and more opaque than those that do not produce sperm and can be carefully dissected out along their entire length. They are then extracted from the testis without disrupting blood supply or removing an extraneous volume of tissue.

Microdissection TESE requires advanced microsurgical training. An embryologist must be available in the operating room to evaluate processed testis parenchymal samples as they are passed off the operating table. By evaluating testicular samples in real time, the surgeon can stop the procedure as soon as the presence of sperm is confirmed, thereby minimizing operative time and removal of testicular tissue. Patience is essential since thorough microscopic dissection of tubules can require up to 2 h per testis if sperm is not identified during the case.

In an early prospective study from our institution, we reported a sperm retrieval rate of 45% (10/22) for patients undergoing conventional TESE compared with 63% (17/27) of patients undergoing microdissection TESE ($p<0.05$). We further showed that the average of 64,000 spermatozoa yielded by conventional TESE biopsy samples were obtained in samples averaging 720 mg each, compared with an average of 160,000 spermatozoa in the microdissected samples that averaged only 9.4 mg ($p<0.05$ for all comparisons) [6].

In another prospective study, Amer et al. performed conventional TESE on one testis and microdissection TESE on the other in 100 patients with NOA. The authors reported a

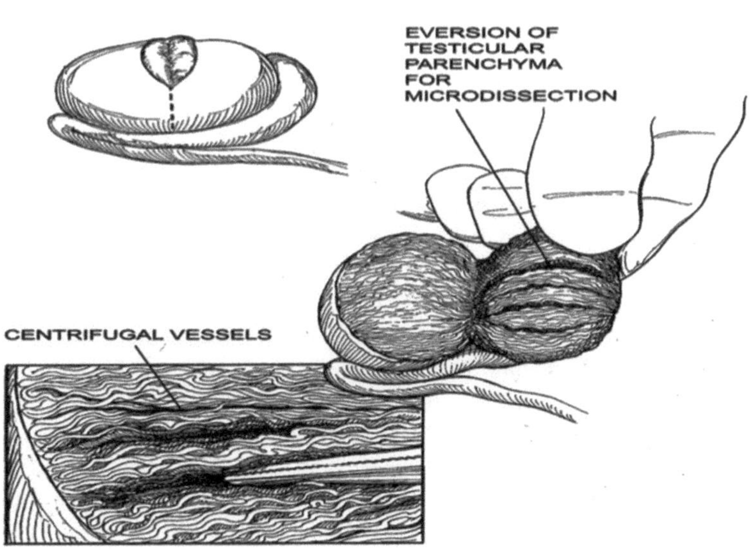

Fig. 39.1 An area of the tunica albuginea is incised and microdissected (Copyright, Brady Urology Foundation 2005)

significantly higher sperm retrieval of 47% for the microdissection TESE side than the 30% obtained on conventional TESE side ($p<0.05$). Follow-up ultrasound testing demonstrated significantly fewer acute and chronic complications on the testes that underwent microdissection TESE, presumably due to improved vascular control as described previously [7].

Genetic Screening for TESE Candidates

Along with the revolutionary fertility potential associated with the development of advanced reproductive techniques, some novel concerns have been raised. In previous eras, the state of azoospermia ensured that some abnormalities (including infertility traits themselves) would not typically be vertically transmitted. With the advent of ICSI in conjunction with sperm retrieval techniques, it has now become possible for these genetic abnormalities to be passed on to offspring.

In an analysis of patients with NOA who were candidates for the TESE procedure at Weill-Cornell Medical College, 17% (33/190) were found to have abnormalities on Y chromosome microdeletion analysis and/or karyotypic testing [9]. Given the high rate of genetic abnormalities and the potential for genetic transmission, we believe that genetic screening should be conducted on all TESE candidates prior to attempts at assisted reproduction. Karyotype evaluation and Y chromosome microdeletion analysis are indicated for men with sperm concentrations lower than 10 × 106/cc.

Men with abnormalities detected on screening should be referred to a genetics counselor for a thorough discussion of the potential implications prior to undergoing attempts at sperm retrieval. Depending on the likelihood of passing infertility traits or other genetic defects to offspring, some couples may elect to use donor spermatozoa. Genetic screening of TESE candidates also facilitates specific genetic testing and subsequent selection of embryos prior to implantation.

Cytogenetic Analysis

Common abnormalities seen in men with NOA includes Klinefelter's syndrome (47, XXY), autosomal translocations, and other sex chromosome anomalies (i.e., 46, XX). Of 190 sequential TESE candidates who underwent genetic evaluation at Weill-Cornell, 33 (17%) were found to have genetic abnormalities. Of the 183 patients who underwent molecular analysis for Y chromosome partial deletion, defects were detected in 17 (9%). Of the 101 patients who underwent karyotype testing, 21 (21%) had cytogenetic abnormalities, including 13 with KS. Five men with sex chromosome anomalies detected on karyotyping also had Y chromosome deletions. Of the 33 men in whom genetic anomalies were discovered, 31 underwent counseling with their partners. Knowledge of the specific genetic defect affected the choice of clinical treatment in 7 of the 33 couples (21%), including donor insemination, adoption, and delaying treatment. The remainder of couples elected to proceed with TESE/ICSI and was counseled to use the preimplantation diagnosis to allow selection of genetically unaffected embryos for transfer [9].

Y Microdeletion Testing

Submicroscopic deletions in the Y chromosome are also prevalent in patients with severely impaired spermatogenesis. Partial Y chromosome deletion analysis is recommended for all men who are TESE candidates, along with karyotype testing, prior to assisted reproductive techniques. Microdeletion testing involves polymerase chain reaction (PCR) of Y-chromosome sequence-tagged sites.

Girardi et al. evaluated 160 men referred for evaluation of male factor infertility who consented to genetic testing. Azoospermic men had a 7% prevalence of Y chromosome microdeletions. Microdeletions were found in 10% of men who were virtually azoospermic ($0–1 \times 10^6$) and 8% of severely oligozoospermic ($1–5 \times 10^6$) men. No microdeletions were detected in men who were moderately oligozoospermic ($5–20 \times 10^6$) or normozoospermic ($>20 \times 10^6$) [10].

Most Y chromosome deletions relating to fertility partially or completely involve the AZFa, AZFb, or AZFc regions. Although sperm can often be found with isolated AZFc deletions, complete AZFa and AZFb deletions are highly associated with failure of sperm retrieval techniques. The likelihood of failing to find sperm in these patients is high enough that we do not recommend that they undergo sperm retrieval procedures. If they do elect to undergo a sperm retrieval procedure despite the dismal odds of sperm retrieval, then the couple should plan to use donor sperm as a backup.

Microdissection TESE Technique

Surgical Approach

After general anesthesia is induced and a first-generation cephalosporin is given the scrotum is shaved. The decision of which testicle to address first is based on several factors. The intent is to maximize the likelihood of adequate sperm retrieval on the first testicle incised and avoid opening the contralateral testis. If the patient has had previous biopsy or TESE that shows better sperm production on a given side then that side is addressed first. Size of the testes is also considered, since larger testis size is usually attributed to fuller seminiferous tubules and thereby better sperm production. Patients who have had previous testis procedures have preoperative ultrasounds to determine scarring. Significantly more scar tissue in one side gives cause for starting with the contralateral side. If all else is equal, the right testis is incised first. It has been our clinical experience that the likelihood of finding sites of spermatogenesis tends to be better on the right than the left side. In some cases, this may be related to left-sided varicoceles, but we have found it to be true even in the absence of a clinical varicocele.

The testis is manipulated to the midline and an incision is made along the median raphe with a 15-blade. The scrotal layers are divided and the testis is delivered. Many patients undergoing microdissection TESE have had one or more prior testicular procedures. In these patients, the tunica vaginalis is often scarred down and adherent to the tunica albuginea. In some cases, a curved scissors can be used to separate the layers. If there are extensive adhesions, then it is advisable to bring in the microscope with 6–8× power magnification. The flat edge of the bovie is used to gently cauterize the lateral aspect of the vaginalis, which is gently spread away from the testis by the assistant. Care is taken to avoid thermal injury to the epididymis. It is necessary to completely free the testis from any adhesions so that it is easily manipulated in the latter parts of the operation.

Under microscopic vision, a 15° ultrasharp ophthalmic knife is used to widely incise the testis horizontally. Subtunical blood vessels can be visualized under the microscope and should be avoided. The testis should be divided as evenly as possible since an eccentric cut makes it harder to subsequently separate the tissue. The knife is lightly applied so that the albuginea is completely incised but the tubules are not disrupted. Using both hands, with the thumbs applying pressure outwards, the testis is bivalved. Bipolar electrocautery is used sparingly to coagulate bleeding at the cut edges of the tunica layer. It is crucial to avoid separating the tubules from the inner surface of the tunica albuginea, since this can result in bleeding that is difficult to control. With extremely small testes that are difficult to manipulate (as is characteristic of azoospermic patients) clamps are placed on either edge of the tunica albuginea inclusive of a portion of the underlying tubules to help prevent such separation.

The testis, which is held in the nondominant hand with the arm in a comfortable position, can be safely twisted up to 180° on its cord as needed during the procedure. The entire superficial surface of exposed seminiferous tubules is scanned to identify large and/or opaque tubules. Magnification is then increased to 20× for tubule dissection. Applying pressure with fingers helps expose tubules of interest while excluding others. The thumb and index finger is used to apply gentle pressure to expose an area of interest between them, while the middle finger of the same hand applies gentle counterpressure in the opposite direction (see diagram). Tension should be applied parallel to the wedges of seminiferous tubules (that is, perpendicular to the incision).

Maintaining proper visualization and exposure of the tubules is a dynamic process that requires constant readjustment so that the entire field, with the tubule of interest in the center, remains in focus. Since the depth of focus is very shallow, and perfect focus is imperative for visualization of the tubules, only one pair of gloves should be used to ensure sufficient tactile feedback. When a promising tubule is found among sclerotic tubules, it is individually teased and dissected out for its entire length before being passed off the field. This is best accomplished by repetitive lateral sweeping motions, whereby a portion of the tubule is mobilized to the side before another area is picked up and then also brought over the side.

In addition to the tubules that are dissected for fertility purposes, a random sample of testis should be separately removed with Iris scissors for histological diagnosis. This sample should be removed without individual dissection of tubules or any other distortion to preserve the histological architecture.

An important adjunct to the treatment of microdissection TESE is the presence of a skilled embryologist in the operating room during the procedure. The role of the embryologist is to review tissue under microscopic vision and immediately give an initial report on the presence, number, and quality of sperm directly to the surgeon. The yield of the tissue is greatly increased by surgeon processing just prior to evaluation by the embryologist. Once a sufficient sperm concentration has been identified (taking into account morphology and motility), there is no need for further dissection. The procedure is terminated with the expectation that more sperm will be found by the laboratory after formal processing. If adequate sperm numbers are not seen by the embryologist, then the surgeon proceeds until the entire testis has been thoroughly sampled. At that point, the contralateral testis will be opened and the procedure will be repeated on that side.

Once extraction is finished on a given testis, the edges of the tunica albuginea are brought together and secured with multiple mosquito clamps under 4× magnification. Special care is taken to avoid manipulation of testicular tissue, since movement of the testicular tissue off of the tunica albuginea can result in substantial bleeding under the surface of the testis resulting in substantial scarring. The incision is closed with nonabsorbable 5-0 polypropylene suture, which is chosen to minimize inflammatory response as well as for immediate identification of the original incision site if repeat microdissection TESE is required. The dartos layer and skin are then closed in the usual fashion. Bacitracin ointment and fluff dressing is applied within a scrotal support.

Sperm Processing

Excised biopsy samples are placed in human tubular fluid with 5% Plasmanate and the testicular region from which it was dissected is noted. Since spermatazoa are normally within seminiferous tubules, mechanical dispersion with fine scissors is required to release sperm that can be identified on wet-prep in the operating room. A few drops of medium (up to 400 μL) should be used to keep the tissue from drying out during the mincing process, but adding too much saline will dilute the sample. Once the tissue has been processed into a suspension, the entire sample drawn up into a syringe and passed several times through a 24-gage angiocatheter to assure adequate dispersion of the tubules. This aggressive dispersion approach can increase sperm yield up to 300-fold [11].

To avoid accidentally losing any volume, the tubule suspension is injected into a 1.5-cc microfuge Eppendorf tube instead of a Petri dish. Since each drop of processed tissue placed on the wet-prep slide contains about 5 μL, the number of sperm in the Eppendorf can be estimated based on number of sperm counted under the microscope. For example, if three spermatozoa are seen on microscopy from a sample containing a total volume of 500 μL, multiplying by 100 gives an estimated number of 300 individual sperm in the Eppendorf.

Each specimen containing the best-appearing seminiferous tubules should be sequentially processed and examined in the operating room until either sperm is found or all areas of the testis have been thoroughly examined. Detecting sperm under the microscope allows the procedure to stop and unnecessary removal of testicular tissue can be avoided. Since there may be a sampling error, the procedure continues until the surgeon feels that adequate sperm have been confirmed to be present to inject all eggs. It is fruitful to continue working in the same area of the testis where sperm were positively identified. However, even a single spermatozoon seen on wet-prep can represent sufficient spermatozoa to end the procedure and proceed with ICSI.

Microdissection TESE Results

Sperm Retrieval Rates

Rates of successful microdissection TESE–ICSI rates have been encouraging in our institutional experience of more than 1,000 attempted treatment cycles for couples in whom the man had NOA. Sperm were retrieved by microdissection TESE in 58% of attempts. For cycles in which sperm was retrieved, clinical pregnancies (as defined by fetal heartbeat on ultrasound) were established at a 45% rate.

The risk of postoperative hematoma following microdissection TESE is approximately 2–3%. One of the advantages of the microdissection TESE technique is the ability of removing minimal tissue with an eye toward future attempts at retrieval, if necessary. Indeed, repeat microdissection TESE following prior successful retrieval has been shown to yields a 96% sperm retrieval rate. If sperm are not found on prior microdissection TESE, the success rate of repeat microdissection TESE drops to 33% [12].

Histopathologic Findings

Three testicular histologic architectural patterns can be found in men with NOA: hypospermatogenesis, maturation arrest, and Sertoli cell-only [11]. Hypospermatogenesis represents abnormally decreased sperm production. While germ cells are present within seminiferous tubules, only limited numbers of spermatids are found. In maturation arrest, both Sertoli cells and immature germ cells are present in seminiferous tubules. In Sertoli cell-only pattern, no germ cells are present and, the name describes, only Sertoli cells are seen.

As previously noted, men with NOA often have heterogeneous patterns of function within their testicular tissue, with microscopic foci of germ cells present among more abundant sclerotic or Sertoli cell-only tubules. The results of microdissection TESE are related to the most advanced pattern of spermatogenesis, rather than the predominant pattern. In 2007, we evaluated the microdissection results for men who have had prior unsuccessful biopsies in men with NOA. There was no statistically significant difference between sperm retrieval rates in men with a diagnosis of either maturation or hypospermatogenesis regardless of the number of previous biopsies. In men with at least one prior biopsy showing Sertoli cell-only as the most advanced pattern, however, sperm retrieval rates were lower than among men with no previous biopsies [13].

Endocrine Evaluation and Treatment

Men with testicular failure typically have abnormally elevated follicle-stimulating hormone (FSH) levels, as a result of decreased germ cells within the testis as a result of poor sperm production, along with normal or near-normal testosterone levels. Randomized controlled trials have not yet demonstrated a fertility benefit from manipulating hormone levels in these patients. Many men with severely impaired spermatogenesis, however, demonstrate a low testosterone (ng/dL) to estradiol (pg/mL) ratio (T/E2 ratio). High intratesticular testosterone is necessary for spermatogenesis, and estradiol may impair spermatogenesis by decreasing pituitary luteinizing hormone (LH) and FSH [14].

Pavlovich et al. compared T/E2 ratios of fertile men, used for reference, with those with of patients with severe male factor infertility [15]. While normal fertile men had an average T/E2 ratio of 16 ± 3, men with NOA had a ratio of 7. Men with KS had an even lower T/E2 ratio of 4. Patients were treated with the aromatase inhibitor testolactone, resulting in significant increases in testosterone and decreases in estradiol. In 12 oligozoospermic patients, the hormonal changes were associated with significant improvements in sperm concentration and motility compared with initial semen analyses. Of men who were initially azoospermic, however, no sperm was found on semen analysis after testolactone treatment.

At our institution, we now routinely obtain testosterone and estradiol levels in men with NOA or severe oligozoospermia. Men with KS who have low testosterone and low T/E2 ratios are treated with 50–100 mg of testolactone per day. For non-KS patients, T/E2 ratio and estradiol level responses were found to be even better with the use of 1 mg of oral anastrazole per day [16].

Structural Changes to Testis

The testicular artery enters the posterior aspect of the testis beneath the epididymis and continues inferiorly to the lower pole of the testis. From the lower pole, multiple branches course superiorly along the anterior testis in the subtunical space, periodically giving off branches to supply the parenchyma. FNAB and conventional TESE provide no visualization of these microscopic vessels and therefore engender an increased risk of vascular disruption. Testicular bleeding can lead to hematoma, devascularization of parenchymal tissue, and intratesticular scarring. Since the tunica albuginea is a non-flexible enclosure surrounding the testicular parenchyma, even small amounts of bleeding can quickly lead to significant increases in intratesticular pressure with possible subsequent atrophy [17].

In contrast with conventional open TESE or FNAB, microdissection TESE allows avoidance of subtunical and parenchymal blood vessels and provides for microscopic hemostatic control of any bleeding sites that may arise. Several comparative studies have confirmed that the likelihood of acute and chronic complications, including structural changes to the testes, is minimized with microdissection TESE compared with conventional TESE. FNAB has not been directly compared with TESE. Okada et al. retrospectively reviewed 147 TESE consecutive cases for patients with azoospermia [18]. All patients underwent testicular ultrasound 1 month after TESE. Diffuse heterogeneic patterns or hypoechoic areas on ultrasound were considered indicative of hematoma. Among the 47 patients who underwent conventional TESE, 24 (51%) showed evidence of hematoma. Of the 100 patients who underwent microdissection TESE, only 12 (12%) had ultrasound findings consistent with hematoma. On follow-up ultrasound at 6 months, hematoma was identified in 3/40 (7.5%) of conventional TESE patients and 2/80 (2.5%) of microdissection patients.

The relative safety of microdissection TESE was also demonstrated in a retrospective review from our institution. Patients who underwent either conventional (83 attempts) or microdissection TESE (460 attempts) were followed with serial Color Doppler scrotal ultrasounds at 3 and 6 months following surgery. Diffuse heterogeneic patterns or hypoechoic areas were considered acute changes consistent with hematoma or inflammation and calcifications were considered chronic changes. Acute and chronic changes were both significantly lower for the patients who underwent microdissection TESE than conventional TESE. Although no evidence of permanent testicular devascularization was found in any patient after 6 months, the findings suggest that microdissection TESE is relatively safer than conventional TESE, in addition to providing an improved sperm retrieval rate as discussed elsewhere [19].

Predictors of Microdissection TESE Success

Microdissection TESE is typically carried out in conjunction with ICSI cycles. TESE–ICSI cycles are financially and emotionally burdensome for infertile couples. It is, therefore, important to predict, to the extent possible, the likelihood of successful sperm retrieval before committing to treatment. Several factors, including prior biopsy, FSH levels, and genetic screening have been studied to determine their associated probabilities of spermatozoa retrieval.

Effect of Prior Biopsy or Conventional TESE Procedure

Successful sperm retrieval is often possible in men with NOA in the context of multiple prior negative biopsies. Ostad et al. reported that half of these patients required multiple (range 2–14) biopsies to retrieve sperm. The effect of prior negative biopsies or conventional TESE procedures on the rate of sperm retrieval with microdissection TESE in men with NOA was evaluated at Weill-Cornell. Successful sperm retrieval in patients who underwent no prior biopsies (56%) was higher than for those who had underwent 1–2 biopsies per testis (51%) or 3–4 biopsies per testis (23%) ($p=0.04$) [20]. In addition to causing scarring and parenchymal fibrosis, diagnostic testicular biopsy can cause parenchymal devascularization with subsequent deleterious effects for spermatogenesis. These changes may explain the lower rates of sperm retrieval associated with more frequent prior biopsies.

Testicular Histology on Diagnostic Biopsy

Diagnosis of NOA can only be definitively made with histopathologic evaluation. Testis biopsy can also be useful for ruling out intratubular germ cell neoplasia (carcinoma in situ). Although the diagnosis is unlikely, it is more common (approximately 3%) in patients who present with NOA and biopsy is required by some centers prior to surgical sperm retrieval procedures. In addition to the role of diagnostic biopsy in identifying rare cases of intratubular germ cell neoplasia (carcinoma in situ) and confirming the diagnosis of nonobstructive azoospermia, diagnostic biopsy helps to predict the chance that a TESE procedure will obtain sperm.

Since diagnostic biopsy does not sample all areas of the testis, small foci of more advanced patterns may be missed. However, the likelihood of successful sperm retrieval in patients with NOA who undergo microdissection TESE can be estimated by the most advanced (as opposed to most predominant) histopathological pattern if a previous testicular biopsy has been performed. In men with at least on area of hypospermatogenesis on prior biopsy, 81% had sperm retrieved with microdissection TESE. In patients with maturation arrest at the most advanced pattern on biopsy, 44% had successful sperm retrieval [19].

If Sertoli cell-only is found without any areas of more advanced spermatogenesis, however, the sperm retrieval rate drops to approximately 30%. This has important counseling implications. Some couples may decide that they would be unwilling to undergo the process of a simultaneous IVF cycle, if there is only a 30% chance of sperm retrieval on microdissection TESE. In these cases, then the male partner is encouraged to undergo a diagnostic biopsy. If a uniform pattern of Sertoli-cell only is demonstrated the couple can avoid ovarian hyperstimulation as well as the financial and emotional costs of an IVF cycle.

Microdissection TESE in Setting of Elevated FSH Levels

Elaboration of inhibin by Sertoli cells is decreased in the setting of testicular failure. With less negative feedback mediated by inhibin, production of FSH from the anterior pituitary increases. Elevated serum FSH levels are, therefore, generally associated with impaired spermatogenesis. In a study of predictors of sperm retrieval and probability of fertilization using open testis biopsy methods, receiver operating characteristics showed FSH levels of ≥20 IU/L as cutoff for treatment success [21].

FSH levels are less relevant for predicting success of microdissection TESE. Even though serum FSH indirectly reflects the global histology of the testes, microdissection TESE is much more sensitive than open techniques for finding isolated foci of sperm production. This hypothesis was tested in a retrospective study of nearly 800 men with NOA who underwent microdissection TESE. Patients were divided into four groups by serum FSH levels (<15, 15–30, 31–45, and >45 IU/mL). Contrary to previous studies and conventional wisdom, sperm retrieval rates were higher for men with elevated FSH than those in the normal range. Sperm retrieval rates were maintained as FSH levels increased, furthermore, even for several patients with FSH >90 IU/mL [22]. These findings underscore the concept that FSH levels are not predictive of sperm retrieval success with microdissection. Azoospermic patients with normal FSH may represent a distinct infertility population. Indeed, it has been reported that many patients with diffuse maturation arrest have normal FSH levels (and testicular volume) [23].

AZF Deletions

PCR-based analysis of Y chromosome sequence-tagged sites is prognostically important and is routinely performed for microdissection TESE candidates. The Y chromosome microdeletions that relate to infertility are seen in part of all of the AZFa, AZFb, or AZFc regions of the DAZ (deleted in azoospermia) gene and are found in 6–18% of men presenting with NOA or oligozoospermia [24–28].

In men with isolated AZFc deletions, sperm production found in the testis at similar rates as for other patients with NOA. At Weill-Cornell, in a retrospective analysis of 1,591 men with sperm concentrations less than 5 million sperm/mL, a total of 149 microdeletions (9.4%) were found. Of the 718 patients who underwent microdissection TESE, sperm retrieval failed in all men with AZFa, AZFb, AZFb+c, and complete Yq deletions. Presence of an AZFc microdeletion, in contrast, was associated with a 71.4% sperm retrieval rate. Of the 15 patients with AZFc deletions with successful sperm retrieval, ten of them achieved clinical pregnancy. AZFc deletions were reported, for the first time, to be favorably associated with sperm retrieval in comparison to 385 patients with idiopathic azoospermia, for whom microdissection TESE yielded a 48.8% retrieval rate [28].

Given the exceptionally poor prognosis for patients with complete AZFa or AZFb deletions, primary utilization of donor sperm is advised rather than microdissection TESE. Men with AZFc deletions have a favorable likelihood of successful sperm retrieval on microdissection TESE. Genetics counseling for the couple is mandatory, since potential male offspring will carry the same infertility trait as the father and eventually face similar reproductive issues themselves.

Microdissection TESE in Patient Subpopulation

There are several patient populations that may particularly benefit from microdissection TESE, including KS, post-chemo, and NOA associated with cryptorchidism.

Klinefelter's Syndrome

Klinefelter syndrome (KS), or hypergonadotropic hypogonadism, is, at 11%, the most frequently diagnosed karyotype abnormality in infertile men [29]. The genetic abnormality results from a meiotic nondisjunction event that most commonly results in classic 47,XXY genotype, but 3% of men with KS are mosaic 46,XX/47,XXY.

KS patients have traditionally been considered a difficult group to retrieve sperm from because they typically have small testes, high FSH levels, and a predominant expression of tubular sclerosis on testicular histopathology. Several studies, however, have reported successful sperm retrieval rates of 40–48% in KS patients [30, 31]. More recently, even higher retrieval rates have been demonstrated.

In 2005, we retrospectively reviewed a series of 42 patients with KS who underwent 54 attempted microdissection TESE procedures. Patients had mean FSH levels of 33.2 IU/L and a successful sperm retrieval rate of 72% per TESE attempt [32].

There are several possible reasons for the higher sperm retrieval rates in these data. Patients at Weill-Cornell are screened for hormonal profiles prior to undergoing sperm retrieval procedures. We have previously shown that while normal fertile men have an average T/E2 ratio of 16 ± 3, patients with KS have an average T/E2 ratio of 4 [15]. We treat patients with low T/E2 ratios with aromatase inhibition and have found dramatic improvements in sperm concentration and motility in patients with severe oligozoospermia. Men with KS respond best to testolactone 50–100 mg per day. Testolactone treatment of these KS patients boosted testosterone levels from 190 to 332 ng/dL during the time period evaluated, with improved T/E2 ratios [16].

Other possible explanations for our better sperm retrieval rates in KS patients include use of the more effective surgical technique of microdissection TESE versus standard TESE procedures, as well as the benefits inherent with a substantial single-surgeon (PNS) experience. It has been our clinical observation that in KS patients, for example, there may often exist short segments of normal-appearing tubule, even while the vast majority of the remainder of the tubule (and adjacent tubules) is sclerotic. In this situation, when such tiny amounts of partial tubule is recovered, we do not process the sample as we usually do in the operating room, to avoid any risk of losing or otherwise disrupting the miniscule sample.

Postchemotherapy Azoospermia

As screening and treatment for various types of cancer continue to improve, a growing population of cancer patients has the opportunity to focus on quality of life issues. Cancer patients are at an elevated risk for infertility even before treatment. Although the importance of the concept of fertility preservation has been increasingly recognized, fewer than half of oncologists in the USA routinely refer patients to a fertility specialist prior to treatment that will further threaten their patients' fertility potential [33]. Systemic chemotherapeutic agents are associated with dose-dependent toxicity to germinal epithelium and posttreatment azoospermia. Alkylating agents are particularly destructive and typically render patients severely oligozoospermic or azoospermic. Vinca alkaloids and high-dose antimetabolites are also known to impair spermatogenesis.

A subset of patients with persistent postchemotherapy azoospermia has been treated at Weill-Cornell. Of 20 attempts of TESE–ICSI performed for 17 patients, 45% (9/20) resulted in successful sperm retrieval. The mean time period between chemotherapy and microdissection TESE was 16.3 years (range, 6–34 years). Clinical pregnancy was established for a third (3/9) of patients with sperm retrieved with two live deliveries. Sertoli cell-only pattern was demonstrated in 76% of patients, with hypospermatogenesis as the most advanced pattern seen in the remainder of patients. Sperm was retrieved in 23% of men with Sertoli cell-only pattern. There was no association between TESE–ICSI outcomes and the underlying conditions treated with chemotherapy or with the specific chemotherapeutic agents used [34].

The use of microdissection TESE–ICSI for patients with persistent postchemotherapy azoospermia has enabled conception and delivery of healthy children. Although certain chemotherapeutic agents are known to be particularly toxic to germinal epithelium, specific regimens have not been shown to clearly predict success of sperm retrieval with microdissection TESE. Although patients should ideally be encouraged to bank ejaculated sperm prior to treatment when possible, patients with long-standing postchemotherapy azoospermia may now be successfully treated with advanced reproductive techniques.

NOA Associated with Cryptorchidism

Failure of testicular descent is associated with impaired sperm counts, sperm quality, and fertility rates. Cryptorchid testes are also associated with an increased risk of eventual testicular germ cell cancer. There is no evidence that orchiopexy decreases the risk of cancer, although it facilitates testicular examination. In contrast, subfertility secondary to cryptorchidism appears to be a duration-dependent process. Increased age at orchiopexy, as well as bilateral cryptorchidism, has been associated with worsening fertility parameters [35].

Several mechanisms for increased risk of infertility with cryptorchidism have been proposed. The first step in postnatal spermatogenic development, maturation of gonocytes to type A spermatagonia, is believed to be deficient in infants with cryptorchidism. Gonocytes that fail to mature degenerate and consequently yield a decreased total number of germ cells. Androgen production, which is impaired within a few months of birth, may be either a primary or secondary defect. Poor steroid secretion may be a causative factor in deficient germ cell maturation. Finally, cryptorchid testes are subject to an increased temperature of several degrees compared with the normal scrotal temperatures. Temperature-induced degeneration of germ cells in cryptorchid testes is also believed to contribute to decreased fertility potential [36].

In 2003, we reported on our experience at Weill-Cornell with sperm retrieval in men with NOA associated with cryptorchidism. A total of 38 men (mean age 36.7 ± 6.5 years) underwent a total of 8 conventional and 39 microdissection TESE procedures. Successful retrieval of sperm was achieved in 63% (5/8) of open TESE procedures and 77% (30/39) of microdissection TESE procedures for a combined rate of 74%.

Spermatozoa were retrieved in all patients (9/9) with a history of unilateral cryptorchidism and 68% (26/38) patients with a history of bilateral cryptorchidism. For cases of successful sperm retrieval, couples achieved clinical pregnancy for 46% (16/35) cycles. Serum FSH was not correlated with successful sperm retrieval, but larger testicular volume ($p<0.05$) and younger patient age at orchiopexy ($p<0.001$) were independent predictors for spermatozoa recovery [37].

Expert Commentary

The story of fertility potential for men with NOA represents one of the greatest medical achievements of the past 20 years. In the recent past, men with NOA were considered sterile and child adoption was the only option for those seeking parenthood. Today many patients with the same diagnosis have successfully fathered children with the assistance of advanced reproductive techniques.

The advent of ICSI has created the potential for oocyte fertilization even with a single, nonmotile sperm. It has been clinically recognized that sperm retrieved directly from the testis can be used for ICSI. It has also been recognized that many men with NOA, including even those with Sertoli cell-only pattern as the most advanced pattern on diagnostic biopsy, may harbor microscopic testicular foci of spermatogenesis. Finally, microdissection TESE has yielded a superior sperm retrieval rate compared with open TESE by allowing visualization of individual normal-appearing seminiferous tubules. The use of an operating microscope that characterizes microdissection TESE, a procedure we developed at Weill-Cornell, also serves to minimize bleeding, devascularization, and subsequent scarring, thereby increasing likelihood of successful sperm retrieval should a repeat procedure become necessary.

Five-Year View

The use of an operating microscope represents a significant improvement in the mechanical sophistication of sperm retrieval compared with conventional TESE or FNAB. Improving fertility rates in the near future will likely relate to the development of treatment options in gene therapy, stem cell therapies, and possibly in vitro spermatogenesis. Screening TESE candidates for Y chromosome microdeletions associated with infertility has facilitated appropriate counseling and improved prognostic accuracy, but approximately 50% of men with NOA have an idiopathic underlying etiology. At Weill-Cornell, we are currently addressing this issue by seeking to identify novel underexpressed genes and mutations in testicular tissue samples taken from men with idiopathic NOA.

Key Issues

Microdissection TESE is an advanced sperm retrieval procedure that requires microscopic training. Patience in the operating room is necessary as the procedure almost invariably takes more time to perform than conventional TESE or FNAB. An embryologist is required in the operating room for real-time wet-prep analysis of testicular specimens so that the procedure may be terminated as soon as spermatozoa are identified. Increased operating times, use of an operating room, and utilization of an operating room embryologist increase financial costs for the procedure.

However, microdissection TESE has distinct and important advantages over conventional open TESE or FNAB. These include a superior sperm retrieval rate with much less tissue removed. Microscopic vision allows for avoidance of subtunical vessels and improved hemostasis with a resultant decrease in acute and chronic complications. Microdissection TESE is an extremely sensitive technique for locating isolated foci of spermatogenesis in testes that may otherwise be overwhelmingly composed of sclerotic tubules. The most advanced pattern of spermatogenesis, rather than the predominant one, is associated with likelihood of sperm retrieval on microdissection TESE.

Microdissection TESE has been successfully used in subpopulations with NOA, including postchemotherapy patients and those with KS. Correctable factors that may impact spermatogenesis, such as presence of a varicocele or low T/E2 ratio, should be treated prior to attempting microdissection TESE. There are no absolute predictors of successful sperm retrieval on microdissection TESE, but genetic screening of operative candidates and results from prior biopsies provide useful prognostic information.

References

1. Schoysman R, Vanderzwalmen P, Nijs M, Segal L, Segal-Bertin G, Geerts L, van Roosendaal E, Schoysman D. Pregnancy after fertilisation with human testicular spermatozoa. Lancet. 1993; 342(8881):1237.
2. Silber SJ, Nagy Z, Liu J, et al. The use of epididymal and testicular spermatozoa for intracytoplasmic sperm injection: the genetic implications for male infertility. Hum Reprod. 1995;10:2031–43.
3. Devroey P, Liu J, Nagy Z. Pregnancies after testicular sperm extraction and intracytoplasmic sperm injection in non-obstructive azoospermia. Hum Reprod. 1995;10:1457–60.
4. Levine LA, Dimitri RJ, Fakouri B. Testicular and epididymal percutaneous sperm aspiration in men with either obstructive or non-obstructive azoospermia. Urology. 2003;62:328–32.
5. Lewin A, Reubinoff B, Porat-Katz A, Weiss D, Eisenberg V, Arbel R, et al. Testicular fine needle aspiration: the alternative method for sperm retrieval in nonobstructive azoospermia. Hum Reprod. 1999;14:1785–90.

6. Schlegel PN. Testicular sperm extraction: microdissection improves sperm yield with minimal tissue excision. Hum Reprod. 1999;14:131–5.
7. Amer M, Ateyah A, Hany R, Zohdy W. Prospective comparative study between microsurgical and conventional testicular sperm extraction in nonobstructive azoospermia: follow-up by serial ultrasound examinations. Hum Reprod. 2000;15:653–6.
8. Turek PJ, Cha I, Ljung BM. Systematic fine-needle aspiration of the testis: correlation to biopsy and results of organ "mapping" for mature sperm in azoospermic men. Urology. 1997;49:734–48.
9. Rucker GB, Mielnik A, King P, Goldstein M, Schlegel PN. Preoperative screening for genetic abnormalities in men with nonobstructive azoospermia prior to testicular sperm extraction. J Urol. 1998;160:2068–71.
10. Girardi SK, Mielnik A, Schlegel PN. Submicroscopic deletions in the Y chromosome of infertile men. Hum Reprod. 1997;12:1635.
11. Ostad M, Liotta D, Ye Z, Schlegel PN. Testicular sperm extraction (TESE) for non-obstructive azoospermia: results of a multi-biopsy approach with optimized tissue dispersion. Urology. 1998; 52:692–7.
12. Haimov-Kochman R, Lossos F, Nefesh I, et al. The value of repeat testicular sperm retrieval in azoospermic men. Fertil Steril. 2009;91(4 Suppl):1401–3.
13. Ramasamy R, Schlegel PN. Microdissection testicular sperm extraction: effect of prior biopsy on success of sperm retrieval. J Urol. 2007;177(4):1447–9.
14. Veldhuis JD, Sowers JR, Rogol AD, et al. Pathophysiology of male hypogonadism associated with endogenous hyperestrogenism. Evidence for dual defects in the gonadal axis. N Engl J Med. 1985;312:1371.
15. Pavlovich CP, King P, Goldstein M, Schlegel PN. Evidence for a treatable endocrinopathy in infertile men. J Urol. 2001;165:837–41.
16. Raman JD, Schlegel PN. Aromatase inhibitors for male infertility. J Urol. 2002;167(2 Pt 1):624–9.
17. Silber SJ. Microsurgical TESE the distribution of spermatogenesis in non-obstructive azoospermia. Hum Reprod. 2000;15:2278.
18. Okada H, Dobashi M, Yamazaki T, Hara I, Fujisawa M, Arakawa S, Kamidono S. Conventional versus microdissection testicular sperm extraction for nonobstructive azoospermia. J Urol. 2002;168(3):1063–7.
19. Ramasamy R, Yagan N, Schlegel PN. Structural and functional changes to the testis after conventional versus microdissection testicular sperm extraction. J Urol. 2007;177(4):1447–9.
20. Ostad M, Liotta D, Ye Z, et al. Testicular sperm extraction with optimized tissue dispersion. Urology. 1998;52:692–7.
21. Zitzmann M, Nordhoff V, von Schönfeld V, Nordsiek-Mengede A, Kliesch S, Schüring AN, Luetjens CM, Kamischke A, Cooper T, Simoni M, Nieschlag E. Elevated follicle-stimulating hormone levels and the chances for azoospermic men to become fathers after retrieval of elongated spermatids form cryopreserved testicular tissue. Fertil Steril. 2006;86(2):339–47.
22. Ramasamy R, Lin K, Gosden LV, Rosenwaks Z, Palermo GD, Schlegel PN. High serum FSH levels in men with nonobstructive azoospermia does not affect success of microdissection testicular sperm extraction. Fertil Steril. 2009;92(2):590–3.
23. Tsujimura A, Matsumiya K, Miyagawa Y, Takao T, Fujita K, Koga M, et al. Prediction of successful outcome of microdissection testicular sperm extraction in men with idiopathic nonobstructive azoospermia. J Urol. 2004;172:1944–7.
24. Vogt PH, Edelmann A, Kirsch S, Henegariu O, Hirschmann P, Kiesewetter F, Kohn FM, Schill WB, Farah S, Ramos C, et al. Human Y chromosome azoospermia factors (AZF) mapped to different subregions in Yq11. Hum Mol Genet. 1996;5:933–43.
25. Reijo R, Alagappan R, Patrizio P, Page DC. Severe oligospermia resulting from deletions of asoospermia factor gene on Y chromosome. Lancet. 1996;347:1290–3.
26. Brandell RA, Mielnik A, Liotta D, Ye Z, Veeck LL, Palermo GD, Schlegel PN. AZFb deletions predict the absence of spermatozoa with testicular sperm extraction: preliminary report of a prognostic genetic test. Hum Reprod. 1998;13:2812–5.
27. Martinez MC, Bernabe MJ, Gomez E, Ballesteros A, Landeras J, Glover G, Gil-Salom M, Remohi J, Pellicer A. Screening for AZF deletion in a large series of severely impaired spermatogenesis patients. J Androl. 2000;21:651–5.
28. Stahl PJ, Masson P, Mielnik A, Marean MB, Schlegel PN, Paduch DA. A decade of experience emphasizes that testing for Y microdeletions is essential in American men with azoospermia and severe oligozoospermia. Fertil Steril. 2010;94:1753–6.
29. Foresta C, Galeazzi C, Bettella A, Marin P, Rossato M, Grolla A, Ferlin A. Analysis of meiosis in intratesticular germ cells from subjects affected by classic Klinefelter's syndrome. J Clin Endocrinol Metab. 1999;84:3807–10.
30. Friedler S, Raziel A, Strassburger D, Schachter M, Bern O, Ron-El R. Outcome of ICSI using fresh and cryopreserved thawed testicular spermatozoa in patients with non-mosaic Klinefelter's syndrome. Hum Reprod. 2001;16:2616–20.
31. Levron J, Aviram-Goldring A, Madgar I, Raviv G, Barkai G, Dor J. Sperm chromosome analysis and outcome of IVF in patients with non-mosaic Klinefelter's syndrome. Fertil Steril. 2000;74:925–9.
32. Schiff JD, Palermo GE, Veeck LL, Goldstein M, Rosenwaks Z, Schlegel PN. Success of testicular sperm extraction and intracytoplasmic sperm injection in men with Klinefelter syndrome. J Clin Endocrinol Metab. 2005;90(11):6263–7.
33. Quinn GP. Physician referral for fertility preservation in oncology patients: a national study of practice behaviors. J Clin Oncol. 2009;27(35):5952.
34. Chan PT, Palermo GD, Veeck LL, Rosenwaks Z, Schlegel PN. Testicular sperm extraction combined with intracytoplasmic sperm injection in the treatment of men with persistent azoospermia postchemotherapy. Cancer. 2001;92(6):1632–7.
35. Trsinar B, Muravec UR. Fertility potential after unilateral and bilateral orchidopexy for cryptorchidism. World J Urol. 2009;27(4):513–9.
36. Hutson JM, Hasthorpe S, Heyns CF. Anatomical and functional aspects of testicular descent and cryptorchidism. Endocr Rev. 1997;18(2):259–80.
37. Raman JD, Schlegel PN. Testicular sperm extraction with intracytoplasmic sperm injection is successful for the treatment of nonobstructive azoospermia associated with cryptorchidism. J Urol. 2003;170:1287–90.

Further Reading

Goldstein M, Tanrikut C. Microsurgical management of male infertility. Nat Clin Pract Urol. 2006;3(7):381–91.

Turunc T, Gul U, Haydardedeoglu B, Bal N, Kuzgunbay B, Peskircioglu L, Ozkardes H. Conventional testicular sperm extraction combined with the microdissection technique in nonobstructive azoospermic patients: a prospective comparative study. Fertil Steril. 2010;94:2157–60.

Schlegel PN. Nonobstructive azoospermia: a revolutionary surgical approach and results. Semin Reprod Med. 2009;27:165–70.

Su LM, Palermo GD, Goldstein M, Veeck LL, Rosenwaks Z, Schlegel PN. Testicular sperm extraction with intracytoplasmic sperm injection for nonobstructive azoospermia: testicular histology can predict chance of sperm retrieval. J Urol. 1999;161(1):112–6.

Sperm Processing and Selection

40

Sonja Grunewald and Uwe Paasch

Abstract

The application of assisted reproductive techniques (ART) has provided help to many men seeking to father a child, although the current success rates of these procedures remain suboptimal. Since raw semen cannot be used in most ART, a workup of the ejaculate is needed to extract those sperm that are capable to fertilize the egg. However, the seminal fluid has a high antioxidant capacity, and sperm processing and separation could profoundly increase oxidative stress. High levels of reactive oxygen species (ROS, e.g., superoxide, hydroxyl, hydrogen peroxide, nitric oxide, peroxynitrite) endanger sperm motility, viability, and function by interacting with membrane lipids, proteins, and nuclear and mitochondrial DNA. Normally, there exists a balance between free radical generating and scavenging systems. It is well known that high levels of ROS are generated by immature and abnormal spermatozoa, contaminating leukocytes and sperm processing, for example, excessive centrifugation and cryopreservation/thawing. Naturally, high antioxidant levels in seminal plasma are the major scavenging mechanism. This chapter gives an overview on sperm selection techniques and their impact on oxidative stress to the sperm.

Keywords

Sperm processing • Sperm selection • Assisted reproductive techniques • Reactive oxygen species • Molecular glass wool • Hyaluronic acid-mediated sperm selection • Oxidative stress

The application of assisted reproductive techniques (ART) has provided help to many men seeking to father a child, although the current success rates of these procedures remain suboptimal [1]. Since raw semen cannot be used in most ART, a workup of the ejaculate is needed to extract those sperm that are capable to fertilize the egg. However, the seminal fluid has a high antioxidant capacity, and sperm processing and separation could profoundly increase oxidative stress [2].

High levels of reactive oxygen species (ROS, e.g., superoxide, hydroxyl, hydrogen peroxide, nitric oxide, peroxynitrite) endanger sperm motility, viability, and function by interacting with membrane lipids, proteins, and nuclear and mitochondrial DNA. Normally, there exists a balance between free radical generating and scavenging systems. It is well known that high levels of ROS are generated by immature and abnormal spermatozoa, contaminating leukocytes and sperm processing, for example, excessive centrifugation and cryopreservation/thawing [3–8]. Naturally, high antioxidant levels in seminal plasma are the major scavenging

S. Grunewald, MD (✉)
Department of Dermatology, Venerology and Allergology, European Training Center of Andrology, University of Leipzig, Philipp-Rosenthal-Strasse 23-25, Leipzig 04103, Germany
e-mail: sonja.grunewald@medizin.uni-leipzig.de

U. Paasch, MD, PhD
Division of Dermatopathology, Division of Aesthetics and Laserdermatology, Department of Dermatology, Venerology and Allergology, European Training Center of Andrology, University of Leipzig, Philipp-Rosenthal-Strasse 23-25, Leipzig 04103, Germany
e-mail: uwe.paasch@medizin.um-leipzig.de

S.J. Parekattil and A. Agarwal (eds.), *Male Infertility: Contemporary Clinical Approaches, Andrology, ART & Antioxidants*,
DOI 10.1007/978-1-4614-3335-4_40, © Springer Science+Business Media, LLC 2012

mechanism [9]. This chapter gives an overview on sperm selection techniques and their impact on oxidative stress to the sperm.

Impact of Sperm Processing and Separation on Oxidative Stress Levels

The Protective Effects of Seminal Plasma

During maturation, the extrusion of the sperm cytoplasm leads to a loss of cytoplasmic enzymes and as a consequence to diminished endogenous repair mechanisms and enzymatic defenses to oxidative stress [8, 10]. This loss of defense mechanisms is compensated by seminal plasma that contains an array of antioxidants acting as free radical scavengers to protect spermatozoa against oxidative stress [7, 11–13].

Enzymatic antioxidants detected in human seminal plasma are superoxide dismutase (SOD), catalase [14], and glutathione peroxidase [15]. In addition, a variety of nonenzymatic antioxidants are present, among them, alpha-tocopherol, ascorbate, glutathione, pyruvate, taurine, hypotaurine, and urate [16].

The removal of seminal fluid during sperm processing and separation procedures leads to a significant reduction of ROS scavenging mechanisms. ROS generation is a major source of sperm DNA damage leading to significantly impaired male fertility [17, 18]. Consequently, many studies investigated the effect of antioxidant supplementation to sperm media during ART procedures [11].

When added in vitro during IVF preparation or as sperm wash media, ascorbic acid (600 mM), alpha-tocopherol (30 and 60 mM), and urate (400 mM) have each been reported to provide significant protection from subsequent sperm DNA damage [19, 20]. Also isoflavones (e.g., genistein and equol) show antioxidant activity and may prevent sperm damage. Compared with ascorbic acid and alpha-tocopherol, genistein was shown to be more potent antioxidant when added to culture media [21].

However, human serum albumin which is routinely added to ART media acts as a powerful antioxidant that prevents oxidative stress-induced damage [22]. This might explain other studies showing no effect of additional antioxidant supplementation of sperm media [10].

Conventional Sperm Selection Methods

Over the last decades, a variety of standard procedures have been developed with certain modifications (conventional selection strategies). These sperm selection techniques can be classified by their basis on centrifugation, filtration, or sperm migration. Among the centrifugation techniques, density gradient centrifugation has been proposed as the gold standard for sperm preparation. The latest developments in sperm selection are focused on the sperm surface combined with or without a standard preparation protocol (see advanced selection strategies).

Simple Sperm Wash

The first sperm separation methods developed embraced one- or two-step washing procedures with subsequent resuspension of the male germ cells [23]. The one-step washing technique is considered a good alternative for processing certain compromised samples [24]. However, simple centrifugation is known to cause definite harm to the cells. In addition, it does not reduce the number of leukocytes or immature sperm cells (producing ROS) while removing the protecting seminal plasma. Therefore, simple sperm wash should not be applied for routine sperm processing.

Density Gradient Centrifugation

Density gradient centrifugation is currently the gold standard technique for sperm preparation. The semen sample is placed on top of a density gradient and centrifuged for 15–30 min. During this procedure, highly motile spermatozoa move actively in the direction of the sedimentation gradient and therefore can reach lower areas quicker than poorly motile or immotile cells. Finally, highly purified motile sperm cells are enriched in the soft pellet at the bottom. This method allows for the enrichment of mature and motile sperm (multiplicative factor of 1.2–2.1) as well as morphologically normal sperm (multiplicative factor of 1.2–1.8) [25]. In principle, continuous or discontinuous gradients are in use for density gradient centrifugation [26, 27]. The former comprises a gradually increasing density toward the bottom of the vial, whereas the latter is characterized by clear boundaries between the different densities of the gradient. Comparison of the discontinuous gradient with other techniques for separation of motile sperm indicated that the discontinuous gradient has advantages in terms of recovery, enhancement of motility, and increased ability to penetrate zona-free hamster ova [28]. In principle, density gradient centrifugation has been demonstrated to be beneficial for all methods of assisted reproduction [29].

Various separation media for density gradient centrifugation have been introduced over time, such as Percoll®—polyvinylpyrrolidone (PVP)-coated silica particles (GE Healthcare Life Sciences, Uppsala, Sweden) [30], Ficoll® (Chalfont St. Giles, UK) [31, 32], Nycodenz (Gentaur, Brussels, Belgium) [33], and dextran-visotrast [34].

Although the discontinuous Percoll® gradient centrifugation significantly reduced the number of seminal bacteria and oxidative stress levels [35], the risk of contamination with endotoxins, possible membrane alterations, and inflammatory responses that could be induced by the insemination of sperm populations contaminated with Percoll® led to its withdrawn from the market for clinical use in assisted reproduction in October 1996 [36]. Replacement products such as IxaPrep® (Medicult, Copenhagen, Denmark), PureSperm® (Sepal Reproductive Devices, Boston, MA), SilSelect® (FertiPro NV, Beernem, Belgium), and Isolate® (Irvine Scientific, Santa Ana, CA) have been introduced [37, 38]. All these less toxic media contain silane-coated silica particles. The quality of sperm prepared with these media seems to be equivalent compared with Percoll® with regard to recovery rate, motility, viability, normal sperm morphology, and velocity [39–41], although there are conflicting reports for motility [42].

Nitric oxide production has been found to be lower using IxaPrep® as a replacement for Percoll®. High levels of nitric oxide adversely affect sperm motility, zona binding, and embryonic development, suggesting that using these newer replacement media offers an overall advantage in ART [43].

Several studies have demonstrated that samples processed by density gradient centrifugation displayed fewer secondary indicators of oxidative stress, for example, membrane changes, disruption of transmembrane mitochondrial membrane potential, in line with activated apoptosis signal transduction and DNA fragmentations [3, 44–46], emphasizing the—although nonspecific—superior selection quality of the procedure.

Glass Wool Filtration

The principle of glass wool filtration for sperm preparation was introduced more than 30 years ago [47, 48]. This technique offers the advantage of processing the whole ejaculate, followed by an additional centrifugation step to remove the seminal plasma. Highly motile spermatozoa are separated from immotile sperm cells by means of densely packed glass wool fibers using gravitational forces and the self-propelled motion of the cells. The efficacy of each filtration run is directly dependent on the properties of the glass wool used [49]. The chemical nature of the glass (i.e., borate, silicate, or quartz), its surface structure, charge, and thickness of the glass wool fibers directly influence filtration efficacy [50].

Potential risks of the technique are damages to the membrane and acrosome or the transmission of glass particles into the filtrate [51]. Glass wool filtration separates human spermatozoa according to motility and size of the sperm head. Sperm head size is closely correlated with the chromatin condensation quality and DNA fragmentation as measured by the sperm chromatin structure assay (SCSA) [45, 52]. In addition, glass wool filtration eliminates up to 90% of leukocytes [49], leading to a significant reduction of ROS in the sample, thereby also contributing to the prevention of sperm DNA damage [53]. A very recent study proved that sperm filtrated through glass wool columns contains significantly lower levels of intracellular H_2O_2 compared to sperm prepared with Percoll® density gradient centrifugation [54]. These findings are underlined by molecular analyses of semen samples filtrated through glass wool, which show significantly reduced activation of apoptosis signaling and DNA fragmentation [45, 54, 55]. Particularly, the glass wool (Code # 112, Manville Fiber Glass Corp., Denver, CO) available as SpermFertil® columns (TransMIT GmbH, Giessen, Germany) has been tested extensively [55] for those subcellular separation effects.

From a clinical perspective, glass wool filtration yields functionally intact spermatozoa of superior quality [56], without affecting fertilization rate and embryo quality in an ICSI program [57]. However, these positive effects are partly related to initial sperm concentration [58].

Swim-Up Procedure

The self-propelled movement of spermatozoa is an essential prerequisite for all migration methods and guarantees a very clean workup. The swim-up procedure uses the active motion of spermatozoa. Intact moving cells swim out of a pellet derived by a simple washing step into an overlaid media for 30–60 min. Highly motile, morphologically intact spermatozoa are enriched in the absence of other cells, proteins, and debris within the supernatant [59]. Antigravitational centrifugation has been proposed to shorten preparation time [60].

The benefit of this method may be limited due to the fact that close cell-to-cell contacts of sperm with each other, debris, and other substances may lead to extensive ROS production and consecutive DNA damage [9], although this could not be verified [61].

To further improve the quality of sperm, a combination of density gradient centrifugation and swim-up is widely used [62] and should overcome the discussed cell-to-cell contacts causing oxidative stress. On a subcellular level, post density gradient centrifugation swim-up was proven to eliminate sperm with activated apoptosis signaling in a routine IVF setting [63].

Other Conventional Sperm Selection Techniques

Over the years, several other sperm selection techniques were introduced, among them, glass bead filtration [64], Sephadex columns (SpermPrep®, ZDL, Lexington, KY) [65], as well as migration-sedimentation [66] and transmembrane migration techniques [67]. None of them is used routinely for sperm preparation in ART procedures due to several limitations and disadvantages compared to the previously described sperm preparation procedures. No data on oxidative stress levels are available for those less known techniques with the exception of one study on the transmembrane migration technique. This sperm preparation technique uses filtration through a membrane filter with cylindrical pores at right angles to the plane of the membrane. Due to a low ratio of the total cross-sectional area of the pores to the overall membrane area, the yield is extremely low. Nevertheless, L4 membranes selective for spermatozoa with normal membrane integrity have been introduced with a simultaneous increase in motility and significant depletion of leukocytes implicating reduced oxidative stress levels [68].

Advanced Sperm Selection Methods

Advanced protocols that allow sperm to be selected according to their ultrastructural morphology [69] or surface charges by electrophoresis [70] overcome the limitations of the classical separation procedures. New insights into the molecular biology of spermatozoa have prompted the development of molecular selection strategies, including hyaluronic acid-mediated sperm selection [71], annexin V magnetic-activated cell sorting (MACS), and annexin V molecular glass wool filtration [55, 72, 73].

Ultrastructural Sperm Selection

The ultrastructural morphology of the sperm head components has been correlated with sperm fertilizing capacity in vitro [74]. The examination is performed in real time using an inverted light microscope equipped with high-power Nomarski optics enhanced by digital imaging to achieve a magnification up to 6,300×. The motile sperm organelle morphology examination (MSOME) method was significantly and positively associated with both fertilization rate and pregnancy outcome [69]. ICSI performed with selected spermatozoa with strictly defined, morphologically normal nuclei significantly improves the incidence of pregnancy in couples with previous ICSI failures [75], particularly when the use of sperm with vacuoles is avoided. Recently, a conductive correlation between an increase in DNA fragmentation and the presence of spermatozoa with large nuclear vacuoles (LNV) was shown. A correlation between a higher fraction of denatured DNA and LNV also was found. These results support the routine selection of spermatozoa by MSOME [76]. On the other hand, the procedure is very time consuming, and although the sperm are placed after density gradient centrifugation in a sperm medium containing human serum albumin, they might be exposed to oxidative stress. Currently, there are no studies available on ROS levels during the MSOME procedure.

Electrophoretic Sperm Isolation

A novel approach to sperm isolation based on electrophoretic sperm separation by size and charge has been described recently [70]. Electrophoresis-based microflow technology for the separation of spermatozoa by size and charge consists of two outer chambers separated from two inner chambers by polyacrylamide restriction membranes with a pore size of 15 kDa. The pore size allows directional movement of competent spermatozoa in the applied electric field and the size exclusion of contaminating cell populations. Large numbers of spermatozoa that are viable and morphologically normal can be selected. The resulting cell population shows a low incidence of DNA damage and contaminating cells and compared favorably with density gradient centrifugation in purity of the sperm population and lack of ROS generation, as well as the viability and morphological integrity of the isolated cells. However, although the method was successfully applied to reduce the percentage of DNA-fragmented sperm before ICSI [77], electrophoresis of spermatozoa is detrimental to their motility [70].

Hyaluronic Acid-Mediated Sperm Selection

Hyaluronic acid-mediated sperm selection is a novel technique that is comparable to sperm-zona pellucida binding. The presence of hyaluronic acid receptor on the plasma membrane of mature acrosome-intact sperm, coupled with hyaluronic acid-coated glass or plastic surfaces, facilitates selection of single mature sperm [71]. The frequencies of sperm with chromosomal disomy are reduced approximately four- to fivefold in hyaluronic acid-selected sperm compared with semen sperm compensating the increase in such abnormalities in intracytoplasmic sperm injection offspring. Hyaluronic acid binding also excludes immature sperm with cytoplasmic extrusion, persistent histones, and DNA chain breaks [78] implicating reduced oxidative stress to the selected sperm. However, no studies are available measuring directly oxidative stress levels before, during, and after the hyaluronic acid binding assay.

Annexin V MACS Separation

The effects of oxidative stress on human sperm comprise impairment of sperm motility, viability, and subcellular function by interacting with membrane lipids, proteins, and nuclear and mitochondrial DNA [79]. In addition, recent studies indicate not only direct effects of ROS on sperm DNA integrity [18, 80, 81] but also activation of parts of the apoptosis signaling cascade as known from somatic cells [82, 83].

Externalization of phosphatidylserine from inner to outer leaflet of the plasma membrane is a main apoptosis event detectable at the sperm surface [84, 85]. Annexin V is a phospholipid-binding protein that has high affinity for PS and lacks the ability to pass through an intact sperm membrane. Therefore, annexin V binding to spermatozoa may be used to label sperm that have compromised membrane integrity and that are less capable to fertilize eggs [85].

Annexin V-conjugated super-paramagnetic microbeads can effectively separate nonapoptotic spermatozoa from those with deteriorated plasma membranes based on the externalization of phosphatidylserine using magnetic-activated cell sorting (MACS). Annexin V MACS separation of sperm yields two fractions: EPS negative (no *e*xternalized *p*hosphatidyl*s*erine, intact membranes, nonapoptotic) and EPS positive (externalized *p*hosphatidyl*s*erine, apoptotic) which is retained in the magnetic field [72, 73, 86].

Sperm preparation that combines density gradient centrifugation with annexin V MACS enhances the advantages of both methods. While density gradient centrifugation removes immature sperm cells, debris, and leukocytes, the annexin V MACS depletes already damaged sperm cells with membrane changes, activated apoptosis signaling, and (in part) DNA fragmentations [87]. Although it was not measured directly, these results implicate a tremendous reduction of oxidative stress.

This is of clinical relevance as it leads to a selection of sperm showing not only improved motility, viability, and morphology but also superior oocyte penetration and fertilization rates [88, 89]. First clinical application revealed also improved pregnancy rates compared to sperm preparation by density gradient centrifugation only [90].

Annexin V Molecular Glass Wool Separation

Although integrating annexin V MACS into standard semen preparation protocols offers a potentially major advantage, the possibility of an accidental transmission of super-paramagnetic microbeads into eggs cannot be fully excluded. A sperm preparation system using the binding properties of annexin V by avoiding free flotation of super-paramagnetic microbeads in a liquid-phase buffer might help to avoid side effects. The annexin V glass filtration technique is a promising step toward the development of a molecular-based preparation system with an enhanced capability for selecting vital spermatozoa with superior fertilizing capacity. We recently demonstrated the feasibility of combining the classical glass wool filtration method with phosphatidylserine-binding properties of annexin V and a solid-phase molecular filtration system. We investigated the apoptosis markers of sperm following simple wash, glass wool filtration, molecular glass wool filtration, and MACS. We demonstrated that the application of annexin V glass wool may improve the outcome of ART. Our data support the highly efficient filtration capacity of a solid-phase annexin V-coated glass wool filter [55].

Expert Commentary

Immature sperm and leukocytes are main sources of ROS in human semen. In general, sperm separation techniques depleting these cells are able to reduce oxidative stress that may harm the sperm during assisted reproduction techniques. Conventional sperm preparation techniques fulfilling those criteria are density gradient centrifugation and glass wool filtration. The swim-up procedure as well as advanced sperm separation methods, like ultrastructural and electrophoretic sperm selection as well as methods based on binding to hyaluronic acid and annexin V, should be used after density gradient centrifugation or glass wool filtration to prevent longer contacts of mature sperm with immature sperm and leukocytes.

In contrast, the removal of the seminal fluid during sperm processing and separation procedures leads to a significant reduction of ROS scavenging mechanisms. However, human serum albumin in ART media acts as a powerful antioxidant that prevents oxidative stress-induced damage. Thus, any substitution of IVF media with other potential antioxidants may not have additional benefits to improve the success of ART procedures.

Five-Year View

Further development of specific molecular-based sperm selection methods requires careful analysis of oxidative stress levels to avoid side effects. All new sperm separation procedures must include depletion of immature sperm and leukocytes. Possibly, the combination with standard sperm preparation techniques like density gradient centrifugation or glass wool filtration will be still of advantage. The supplementation of sperm preparation and incubation media with human serum albumin or other antioxidants should not be forgotten during the development of such "high-end" techniques.

Key Issues

- Sperm separation from seminal fluid removes also the natural protective antioxidants contained in the seminal fluid. To prevent excessive oxidative stress to the sperm, antioxidants like human serum albumin must be added to sperm preparation and incubation media for assisted reproduction.
- Standard sperm selection techniques like density gradient centrifugation and glass wool filtration, but not swim-up, are able to reduce oxidative stress by depletion of immature sperm and leukocytes.
- Advanced sperm separation techniques focus rather on the depletion of already damaged sperm; until now, a combination with gradient centrifugation or glass wool filtration is advisable.

Acknowledgment The authors are grateful to Prof. emeritus Hans-Juergen Glander for his support and encouragement.

References

1. Society for Assisted Reproductive Technology and American Society for Reproductive Medicine. Assisted reproductive technology in the United States: 2000 results generated from the American Society for Reproductive Medicine/Society for Assisted Reproductive Technology Registry. Fertil Steril. 2004;81:1207–20.

2. Sikka SC. Relative impact of oxidative stress on male reproductive function. Curr Med Chem. 2001;8:851–62.
3. Ollero M, Gil-Guzman E, Lopez MC, Sharma RK, Agarwal A, Larson K, Evenson D, Thomas Jr AJ, Alvarez JG. Characterization of subsets of human spermatozoa at different stages of maturation: implications in the diagnosis and treatment of male infertility. Hum Reprod. 2001;16:1912–21.
4. Mazzilli F, Rossi T, Sabatini L, Pulcinelli FM, Rapone S, Dondero F, Gazzaniga PP. Human sperm cryopreservation and reactive oxygen species (ROS) production. Acta Eur Fertil. 1995;26:145–8.
5. Ludwig M, Kummel C, Schroeder-Printzen I, Ringert RH, Weidner W. Evaluation of seminal plasma parameters in patients with chronic prostatitis or leukocytospermia. Andrologia. 1998;30 Suppl 1:41–7.
6. Iwasaki A, Gagnon C. Formation of reactive oxygen species in spermatozoa of infertile patients. Fertil Steril. 1992;57:409–16.
7. Aitken RJ, Buckingham DW, Brindle J, Gomez E, Baker HW, Irvine DS. Analysis of sperm movement in relation to the oxidative stress created by leukocytes in washed sperm preparations and seminal plasma. Hum Reprod. 1995;10:2061–71.
8. Aitken RJ, West K, Buckingham D. Leukocytic infiltration into the human ejaculate and its association with semen quality, oxidative stress, and sperm function. J Androl. 1994;15:343–52.
9. Twigg J, Irvine DS, Houston P, Fulton N, Michael L, Aitken RJ. Iatrogenic DNA damage induced in human spermatozoa during sperm preparation: protective significance of seminal plasma. Mol Hum Reprod. 1998;4:439–45.
10. Donnelly ET, McClure N, Lewis SE. Antioxidant supplementation in vitro does not improve human sperm motility. Fertil Steril. 1999;72:484–95.
11. Agarwal A, Saleh RA. Role of oxidants in male infertility: rationale, significance, and treatment. Urol Clin North Am. 2002; 29:817–27.
12. Lewis SE, Boyle PM, McKinney KA, Young IS, Thompson W. Total antioxidant capacity of seminal plasma is different in fertile and infertile men. Fertil Steril. 1995;64:868–70.
13. Potts RJ, Notarianni LJ, Jefferies TM. Seminal plasma reduces exogenous oxidative damage to human sperm, determined by the measurement of DNA strand breaks and lipid peroxidation. Mutat Res. 2000;447:249–56.
14. Zini A, de Lamirande E, Gagnon C. Reactive oxygen species in semen of infertile patients: levels of superoxide dismutase- and catalase-like activities in seminal plasma and spermatozoa. Int J Androl. 1993;16:183–8.
15. Alvarez JG, Storey BT. Role of glutathione peroxidase in protecting mammalian spermatozoa from loss of motility caused by spontaneous lipid peroxidation. Gamete Res. 1989;23:77–90.
16. Saleh RA, Agarwal A. Oxidative stress and male infertility: from research bench to clinical practice. J Androl. 2002;23:737–52.
17. Twigg J, Fulton N, Gomez E, Irvine DS, Aitken RJ. Analysis of the impact of intracellular reactive oxygen species generation on the structural and functional integrity of human spermatozoa: lipid peroxidation, DNA fragmentation and effectiveness of antioxidants. Hum Reprod. 1998;13:1429–36.
18. Lopes S, Jurisicova A, Sun JG, Casper RF. Reactive oxygen species: potential cause for DNA fragmentation in human spermatozoa. Hum Reprod. 1998;13:896–900.
19. Hughes CM, Lewis SE, McKelvey-Martin VJ, Thompson W. The effects of antioxidant supplementation during Percoll preparation on human sperm DNA integrity. Hum Reprod. 1998;13:1240–7.
20. Geva E, Lessing JB, Lerner GL, Amit A. Free radicals, antioxidants and human spermatozoa: clinical implications. Hum Reprod. 1998;13:1422–4.
21. Sierens J, Hartley JA, Campbell MJ, Leathem AJ, Woodside JV. In vitro isoflavone supplementation reduces hydrogen peroxide-induced DNA damage in sperm. Teratog Carcinog Mutagen. 2002;22:227–34.
22. Armstrong JS, Rajasekaran M, Hellstrom WJ, Sikka SC. Antioxidant potential of human serum albumin: role in the recovery of high quality human spermatozoa for assisted reproductive technology. J Androl. 1998;19:412–9.
23. Edwards RG, Bavister BD, Steptoe PC. Early stages of fertilization in vitro of human oocytes matured in vitro. Nature. 1969;221: 632–5.
24. Srisombut C, Morshedi M, Lin MH, Nassar A, Oehninger S. Comparison of various methods of processing human cryopreserved-thawed semen samples. Hum Reprod. 1998;13:2151–7.
25. Glander HJ. Effektivität der Methoden für die Spermienkollektion, -separation und -konzentrierung. Hautnah Derm. 1993;9:52–60.
26. Bolton VN, Braude PR. Preparation of human spermatozoa for in vitro fertilization by isopycnic centrifugation on self-generating density gradients. Arch Androl. 1984;13:167–76.
27. Pousette A, Akerlof E, Rosenborg L, Fredricsson B. Increase in progressive motility and improved morphology of human spermatozoa following their migration through Percoll gradients. Int J Androl. 1986;9:1–13.
28. Berger T, Marrs RP, Moyer DL. Comparison of techniques for selection of motile spermatozoa. Fertil Steril. 1985;43:268–73.
29. Mortimer D, Mortimer ST. Methods of sperm preparation for assisted reproduction. Ann Acad Med Singapore. 1992;21:517–24.
30. Gorus FK, Pipeleers DG. A rapid method for the fractionation of human spermatozoa according to their progressive motility. Fertil Steril. 1981;35:662–5.
31. Bongso A, Ng SC, Mok H, Lim MN, Teo HL, Wong PC, Ratnam S. Improved sperm concentration, motility, and fertilization rates following Ficoll treatment of sperm in a human in vitro fertilization program. Fertil Steril. 1989;51:850–4.
32. Harrison RA. A highly efficient method for washing mammalian spermatozoa. J Reprod Fertil. 1976;48:347–53.
33. Gellert-Mortimer ST, Clarke GN, Baker HW, Hyne RV, Johnston WI. Evaluation of Nycodenz and Percoll density gradients for the selection of motile human spermatozoa. Fertil Steril. 1988;49: 335–41.
34. Glander HJ, Schaller J, Ladusch M. [A simple method for separating fresh and cryopreserved human sperm using dextran-visotrast density gradient centrifugation] Eine einfache Methode zur Separierung von frischen und kryokonservierten Humanspermien mittels einer Dextran-Visotrast- Dichtegradientenzentrifugation. Zentralbl Gynakol. 1990;112:91–7.
35. Aitken RJ, Clarkson JS. Significance of reactive oxygen species and antioxidants in defining the efficacy of sperm preparation techniques. J Androl. 1988;9:367–76.
36. Pharmacia Biotech I. Important notice: Percoll® NOT to be used in Assisted Reproduction Technologies in Humans. 1996.
37. Makkar G, Ng HY, Yeung SB, Ho PC. Comparison of two colloidal silica-based sperm separation media with a non-silica-based medium. Fertil Steril. 1999;72:796–802.
38. Sills ES, Wittkowski KM, Tucker MJ, Perloe M, Kaplan CR, Palermo GD. Comparison of centrifugation- and noncentrifugation-based techniques for recovery of motile human sperm in assisted reproduction. Arch Androl. 2002;48:141–5.
39. Chen MJ, Bongso A. Comparative evaluation of two density gradient preparations for sperm separation for medically assisted conception. Hum Reprod. 1999;14:759–64.
40. Claassens OE, Menkveld R, Harrison KL. Evaluation of three substitutes for Percoll in sperm isolation by density gradient centrifugation. Hum Reprod. 1998;13:3139–43
41. Soderlund B, Lundin K. The use of silane-coated silica particles for density gradient centrifugation in in-vitro fertilization. Hum Reprod. 2000;15:857–60.
42. McCann CT, Chantler E. Properties of sperm separated using Percoll and IxaPrep density gradients. A comparison made using CASA, longevity, morphology and the acrosome reaction. Int J Androl. 2000;23:205–9.

43. Wu TP, Huang BM, Tsai HC, Lui MC, Liu MY. Effects of nitric oxide on human spermatozoa activity, fertilization and mouse embryonic development. Arch Androl. 2004;50:173–9.
44. Barroso G, Taylor S, Morshedi M, Manzur F, Gavino F, Oehninger S. Mitochondrial membrane potential integrity and plasma membrane translocation of phosphatidylserine as early apoptotic markers: a comparison of two different sperm subpopulations. Fertil Steril. 2006;85:149–54.
45. Larson KL, Brannian JD, Timm BK, Jost LK, Evenson DP. Density gradient centrifugation and glass wool filtration of semen remove spermatozoa with damaged chromatin structure. Hum Reprod. 1999;14:2015–9.
46. Said TM, Grunewald S, Paasch U, Glander HJ, Baumann T, Kriegel C, Li L, Agarwal A. Advantage of combining magnetic cell separation with sperm preparation techniques. Reprod Biomed Online. 2005;10:740–6.
47. Paulson JD, Polakoski KL. A glass wool column procedure for removing extraneous material from the human ejaculate. Fertil Steril. 1977;28:178–81.
48. Paulson JD, Polakoski KL, Leto S. Further characterization of glass wool column filtration of human semen. Fertil Steril. 1979;32:125–6.
49. Sanchez R, Concha M, Ichikawa T, Henkel R, Schill WB. Glass wool filtration reduces reactive oxygen species by elimination of leukocytes in oligozoospermic patients with leukocytospermia. J Assist Reprod Genet. 1996;13:489–94.
50. Ford WC, McLaughlin EA, Prior SM, Rees JM, Wardle PG, Hull MG. The yield, motility and performance in the hamster egg test of human spermatozoa prepared from cryopreserved semen by four different methods. Hum Reprod. 1992;7:654–9.
51. Sherman JK, Paulson JD, Liu KC. Effect of glass wool filtration on ultrastructure of human spermatozoa. Fertil Steril. 1981;36:643–7.
52. Henkel RR, Franken DR, Lombard CJ, Schill WB. Selective capacity of glass-wool filtration for the separation of human spermatozoa with condensed chromatin: a possible therapeutic modality for male-factor cases? J Assist Reprod Genet. 1994;11:395–400.
53. Henkel R, Schill WB. Sperm separation in patients with urogenital infections. Andrologia. 1998;30 Suppl 1:91–7.
54. Kim SH, Yu DH, Kim YJ. Apoptosis-like change, ROS, and DNA status in cryopreserved canine sperm recovered by glass wool filtration and Percoll gradient centrifugation techniques. Anim Reprod Sci. 2010;119:106–14.
55. Grunewald S, Miska W, Miska G, Rasch M, Reinhardt M, Glander HJ, Paasch U. Molecular glass wool filtration as a new tool for sperm preparation. Hum Reprod. 2007;22:1405–12.
56. Johnson DE, Confino E, Jeyendran RS. Glass wool column filtration versus mini-Percoll gradient for processing poor quality semen samples. Fertil Steril. 1996;66:459–62.
57. Van den Bergh M, Revelard P, Bertrand E, Biramane J, Vanin AS, Englert Y. Glass wool column filtration, an advantageous way of preparing semen samples for intracytoplasmic sperm injection: an auto-controlled randomized study. Hum Reprod. 1997;12:509–13.
58. Rhemrev J, Jeyendran RS, Vermeiden JP, Zaneveld LJ. Human sperm selection by glass wool filtration and two-layer, discontinuous Percoll gradient centrifugation. Fertil Steril. 1989;51:685–90.
59. Dominguez LA, Burgos MH, Fornes MW. Morphometrical comparison of human spermatozoa obtained from semen and swim-up methodology. Andrologia. 1999;31:23–6.
60. Babbo CJ, Hecht BR, Jeyendran RS. Increased recovery of swim-up spermatozoa by application of "antigravitational" centrifugation. Fertil Steril. 1999;72:556–8.
61. Younglai EV, Holt D, Brown P, Jurisicova A, Casper RF. Sperm swim-up techniques and DNA fragmentation. Hum Reprod. 2001;16:1950–3.
62. Scott Jr RT, Oehninger SC, Menkveld R, Veeck LL, Acosta AA. Critical assessment of sperm morphology before and after double wash swim-up preparation for in vitro fertilization. Arch Androl. 1989;23:125–9.
63. Grunewald S, Reinhardt M, Blumenauer V, Hmeidan AF, Glander HJ, Paasch U. Effects of post-density gradient swim-up on apoptosis signalling in human spermatozoa. Andrologia. 2010;42: 127–31.
64. Daya S, Gwatkin RB. Improvement in semen quality using glass bead column. Arch Androl. 1987;18:241–4.
65. Drobnis EZ, Zhong CQ, Overstreet JW. Separation of cryopreserved human semen using Sephadex columns, washing, or Percoll gradients. J Androl. 1991;12:201–8.
66. Yavetz H, Hauser R, Homonnai ZT, Paz GF, Lessing JB, Amit A, Yogev I. Separation of sperm cells by sedimentation technique is not suitable for in vitro fertilization purposes. Andrologia. 1996;28:3–6.
67. Chijioke PC, Crocker PR, Gilliam M, Owens MD, Pearson RM. Importance of filter structure for the trans-membrane migration studies of sperm motility. Hum Reprod. 1988;3:241–4.
68. Agarwal A, Manglona A, Loughlin KR. Improvement in semen quality and sperm fertilizing ability after filtration through the L4 membrane: comparison of results with swim up technique. J Urol. 1992;147:1539–41.
69. Bartoov B, Berkovitz A, Eltes F, Kogosowski A, Menezo Y, Barak Y. Real-time fine morphology of motile human sperm cells is associated with IVF-ICSI outcome. J Androl. 2002;23:1–8.
70. Ainsworth C, Nixon B, Aitken RJ. Development of a novel electrophoretic system for the isolation of human spermatozoa. Hum Reprod. 2005;20:2261–70.
71. Huszar G, Ozkavukcu S, Jakab A, Celik-Ozenci C, Sati GL, Cayli S. Hyaluronic acid binding ability of human sperm reflects cellular maturity and fertilizing potential: selection of sperm for intracytoplasmic sperm injection. Curr Opin Obstet Gynecol. 2006;18: 260–7.
72. Grunewald S, Paasch U, Glander HJ. Enrichment of non-apoptotic human spermatozoa after cryopreservation by immunomagnetic cell sorting. Cell Tissue Bank. 2001;2:127–33.
73. Paasch U, Grunewald S, Fitzl G, Glander HJ. Deterioration of plasma membrane is associated with activation of caspases in human spermatozoa. J Androl. 2003;24:246–52.
74. Mashiach R, Fisch B, Eltes F, Tadir Y, Ovadia J, Bartoov B. The relationship between sperm ultrastructural features and fertilizing capacity in vitro. Fertil Steril. 1992;57:1052–7.
75. Bartoov B, Berkovitz A, Eltes F, Kogosovsky A, Yagoda A, Lederman H, Artzi S, Gross M, Barak Y. Pregnancy rates are higher with intracytoplasmic morphologically selected sperm injection than with conventional intracytoplasmic injection. Fertil Steril. 2003;80:1413–9.
76. Franco Jr JG, Baruffi RL, Mauri AL, Petersen CG, Oliveira JB, Vagini LD. Significance of large nuclear vacuoles in human spermatozoa: implications for ICSI. Reprod Biomed Online. 2008; 17:42–5.
77. Ainsworth C, Nixon B, Jansen RP, Aitken RJ. First recorded pregnancy and normal birth after ICSI using electrophoretically isolated spermatozoa. Hum Reprod. 2007;22:197–200.
78. Huszar G, Ozenci CC, Cayli S, Zavaczki Z, Hansch E, Vigue L. Hyaluronic acid binding by human sperm indicates cellular maturity, viability, and unreacted acrosomal status. Fertil Steril. 2003;79 Suppl 3:1616–24.
79. Sikka SC, Rajasekaran M, Hellstrom WJ. Role of oxidative stress and antioxidants in male infertility. J Androl. 1995;16:464–8.
80. Barroso G, Morshedi M, Oehninger S. Analysis of DNA fragmentation, plasma membrane translocation of phosphatidylserine and oxidative stress in human spermatozoa. Hum Reprod. 2000;15:1338–44.
81. Moustafa MH, Sharma RK, Thornton J, Mascha E, Abdel-Hafez MA, Thomas Jr AJ, Agarwal A. Relationship between ROS production, apoptosis and DNA denaturation in spermatozoa from patients examined for infertility. Hum Reprod. 2004;19: 129–38.

82. Wang X, Sharma RK, Sikka SC, Thomas Jr AJ, Falcone T, Agarwal A. Oxidative stress is associated with increased apoptosis leading to spermatozoa DNA damage in patients with male factor infertility. Fertil Steril. 2003;80:531–5.
83. Aitken RJ, De Iuliis GN. On the possible origins of DNA damage in human spermatozoa. Mol Hum Reprod. 2010;16:3–13.
84. Vermes I, Haanen C, Steffens-Nakken H, Reutelingsperger CP. A novel assay for apoptosis: flow cytometric detection of phosphatidylserine expression of early apoptotic cells using fluorescein labelled Annexin V. J Immunol Methods. 1995;184:39–51.
85. Glander HJ, Schaller J. Binding of annexin V to plasma membranes of human spermatozoa: a rapid assay for detection of membrane changes after cryostorage. Mol Hum Reprod. 1999;5:109–15.
86. Paasch U, Grunewald S, Agarwal A, Glandera HJ. Activation pattern of caspases in human spermatozoa. Fertil Steril. 2004;81 Suppl 1:802–9.
87. Said TM, Paasch U, Grunewald S, Baumann T, Li L, Glander HJ, Agarwal A. Advantage of combining magnetic cell separation with sperm preparation techniques. Reprod Biomed Online. 2005;10:740–6.
88. Said TM, Agarwal A, Grunewald S, Rasch M, Baumann T, Kriegel C, Li L, Glander HJ, Thomas Jr AJ, Paasch U. Selection of nonapoptotic spermatozoa as a new tool for enhancing assisted reproduction outcomes: an in vitro model. Biol Reprod. 2006;74:530–7.
89. Grunewald S, Reinhardt M, Blumenauer V, Said TM, Agarwal A, Abu HF, Glander HJ, Paasch U. Increased sperm chromatin decondensation in selected nonapoptotic spermatozoa of patients with male infertility. Fertil Steril. 2009;92:572–7.
90. Dirican EK, Ozgun OD, Akarsu S, Akin KO, Ercan O, Ugurlu M, Camsari C, Kanyilmaz O, Kaya A, Unsal A. Clinical outcome of magnetic activated cell sorting of non-apoptotic spermatozoa before density gradient centrifugation for assisted reproduction. J Assist Reprod Genet. 2008;25:375–81.

41 Antioxidants in Sperm Cryopreservation

Tamer Said and Ashok Agarwal

Abstract

Cryopreservation and thawing exposes spermatozoa to various stresses that could lead eventually to loss of fertilizing potential. Despite various advances in cryopreservation methodology, the recovery rate of functional post-thaw spermatozoa remains to be improved. The use of cryoprotectants such as glycerol, ethylene glycol, dimethyl sulfoxide (DMSO), and 1,2-propanediol (PROH) marked one of the most significant advancements in cryopreservation. Cryoprotectants are low molecular weight, highly permeable chemicals that serve to protect spermatozoa from freeze damage induced by ice crystallization. Cryoprotectants act by decreasing the freezing point of a substance, reducing the amount of salts and solutes present in the liquid phase of the sample, and decreasing ice formation within the spermatozoa. Oxidative stress (OS), resulting from an imbalance between reactive oxygen species (ROS) and antioxidants, is detrimental to human spermatozoa resulting in significant loss of function. Increased ROS production and decreased antioxidant levels is known to occur during sperm cryopreservation and thawing. Therefore, OS does play a role in injury sustained by spermatozoa during cryopreservation. Subsequently antioxidants which counteract the effects of ROS could be of use in preventing OS-induced cryoinjury.

Keywords

Antioxidants • Sperm cryopreservation • Male infertility • Reactive oxygen species • Oxidative stress • Sperm motility • Lipid peroxidation • DNA fragmentation

Cryopreservation of human spermatozoa and achieving successful fertilization via assisted reproductive techniques (ART) has been well established [1]. The indications for sperm cryopreservation are many. The technique could be of help in many scenarios encountered in infertility management. It provides an option for storing spermatozoa while maintaining their functional capabilities. Subsequent uses of the preserved semen include intrauterine insemination, in vitro fertilization (IVF), and intracytoplasmic sperm injection (ICSI). Sperm cryopreservation is the only standardized and most feasible method for fertility preservation in men. Cancer patients often resort to cryopreservation to preserve their fertility prior to treatments such as radiation and chemotherapy. Additionally, men undergoing vasectomy procedures will use this method to maintain fertility [2].

Cryopreservation and thawing exposes spermatozoa to various stresses that could eventually lead to the loss of fertilizing potential. Therefore, several improvements have been made to the process of cryopreservation [3]. Despite various advances in cryopreservation methodology, the recovery rate of functional post-thaw spermatozoa remains to be improved [4]. The use of cryoprotectants such as glycerol, ethylene

T. Said, MD, PhD (✉)
Andrology Laboratory and Reproductive Tissue Bank, The Toronto Institute for Reproductive Medicine, Toronto, ON, Canada
e-mail: tsaid@rerpromed.ca

A. Agarwal, PhD
Center for Reproductive Medicine, Glickman Urological and Kidney Institute, Cleveland Clinic Foundation, 9500 Euclid Avenue, Desk A19, Cleveland, OH 44195, USA
e-mail: agarwaa@ccf.org

S.J. Parekattil and A. Agarwal (eds.), *Male Infertility: Contemporary Clinical Approaches, Andrology, ART & Antioxidants*, DOI 10.1007/978-1-4614-3335-4_41, © Springer Science+Business Media, LLC 2012

glycol, dimethyl sulfoxide (DMSO), and 1,2-propanediol (PROH) marked one of the most significant advancements in cryopreservation. Cryoprotectants are low molecular weight, highly permeable chemicals that serve to protect spermatozoa from freeze damage induced by ice crystallization. Cryoprotectants act by decreasing the freezing point of a substance, reducing the amount of salts and solutes present in the liquid phase of the sample, and decreasing ice formation within the spermatozoa [5].

Oxidative stress (OS), resulting from an imbalance between reactive oxygen species (ROS) and antioxidants, is detrimental to human spermatozoa resulting in significant loss of function. Increased ROS production and decreased antioxidant levels is known to occur during sperm cryopreservation and thawing [6, 7]. Therefore, OS does play a role in injury sustained by spermatozoa during cryopreservation. Subsequently, antioxidants that counteract the effects of ROS could be of use in preventing OS-induced cryoinjury.

Oxidative Stress and Male Infertility

Oxidative stress has become the focus of interest as a potential cause of male infertility [8–12]. Under physiological conditions, spermatozoa produce small amounts of ROS, which are needed for capacitation, acrosome reaction, and fertilization [13]. However, excessive amounts of ROS produced by leukocytes and immature spermatozoa can cause damage to the normal spermatozoa by inducing lipid peroxidation and DNA damage [14–16]. The primary product of the spermatozoon's free radical generating system appears to be the superoxide anion, which secondarily dismutase to hydrogen peroxide (H_2O_2) through the catalytic action of superoxide dismutase (SOD) [17]. The combination of superoxide anion and H_2O_2 is potentially harmful, and in the presence of transition metals, it can precipitate the generation of hydroxyl radicals [18]. Sperm damage induced by OS includes membrane and DNA damage leading to necrozoospermia, asthenozoospermia, and DNA fragmentation [19].

Normally, an equilibrium exists between ROS production and antioxidant scavenging activities in the male reproductive tract. However, the production of excessive amounts of ROS in semen may overwhelm the antioxidant defense mechanisms of spermatozoa and seminal plasma, leading to OS [20, 21]. The OS status of an individual can be identified by measuring the ROS levels and antioxidants [22]. ROS levels are usually measured by chemiluminescence method [23], while total antioxidant capacity is measured by enhanced chemiluminescence assay or colorimetric assay [22, 24].

Spermatozoa are naturally surrounded by the seminal plasma, which is well endowed with an array of antioxidants that act as free radical scavengers to protect spermatozoa against OS [25]. This defense mechanism compensates for the loss of sperm cytoplasmic enzymes occurring when the cytoplasm is extruded during spermiation, which in turn diminishes endogenous repair mechanisms and enzymatic defenses. Seminal plasma contains a number of enzymatic antioxidants such as SOD, catalase, and glutathione peroxidase. In addition, it contains a variety of nonenzymatic antioxidants such as vitamin C (ascorbic acid), vitamin E (α-tocopherol), pyruvate, glutathione, and carnitine [26].

Effects of Antioxidants on Spermatozoa In Vitro

Reduction of Reactive Oxygen Species Levels

In vitro supplementation with antioxidants is responsible for decreasing ROS in sperm suspensions. Significant reduction of H_2O_2 was achieved by adding different concentrations of vitamin C (300 and 600 μM) and vitamin E (40 and 60 μM) to the sperm preparation medium [27, 28]. Significant reduction in the release of superoxide anion by 29–72% was also achieved following the addition of 10 mM pentoxifylline [29, 30]. Other studies also showed that pentoxifylline is capable of reducing spermatozoal generation of ROS and subsequent lipid peroxidation in asthenozoospermic men [31, 32]. The addition of *N*-acetyl-L-cysteine (NAC) (1 mg/mL) also effectively reduced ROS levels. It may of importance to note that samples with initially high ROS showed the greatest tendency for reduction of ROS in response to antioxidant supplementation in vitro [33].

Effects on Sperm Motility

Antioxidants counteract lipid peroxidation which has a negative effect on sperm motility [34]. In vitro exposure of spermatozoa to several antioxidants proved to be of benefit in terms of promoting sperm motility. Vitamin C (ascorbate) is one of the main antioxidants in seminal plasma; it is a chain breaking antioxidant that protects the lipoproteins from peroxyl radicals. Data show that vitamin C can preserve sperm motility, yet in a dose-dependant manner. Motility was highest after 6-h of incubation in 800 μM vitamin C. However, motility was decreased with concentrations exceeding 1,000 μM [35]. Vitamin E is another major chain breaking antioxidant that when added in vitro can efficiently protect sperm motility and morphology by suppressing the lipid peroxidation [36]. The effects of supplementation with vitamins C and E on sperm motility are also dose dependant. Higher concentrations of vitamins C and E are not protective against H_2O_2-induced peroxidative damage of motility, instead they increase the damage in both normozoospermic and asthenozoospermic patients [37].

Glutathione (GSH) appears also to have a protective effect on sperm motility. In samples characterized with leukocytospermia, the addition of GSH during sperm Percoll preparation and during 24-h incubation resulted in higher recovery of motile spermatozoa [38]. Similarly, a significant improvement was observed in sperm motility after 2-h of incubation with 1.0 mg/mL NAC [33]. Albumin is another antioxidant/extender that has been extensively used in sperm preparation. Its antioxidant properties are due to its ability to react against peroxyl radicals and prevent the propagation of peroxidative damage in sperm [39, 40]. When albumin was used in sperm preparation media, it resulted in significant improvement in motility and viability compared to Percoll [41]. Other antioxidants that proved to be of benefit to the sperm motility include Coenzyme Q10 (50 μM) [42], hypotaurine, and catalase [43].

Enzymatic antioxidants such as SOD and catalase protect spermatozoa from superoxide anion and H_2O_2. Sperm suspension treated with SOD (400 U/mL) had significantly reduced motility loss and malondialdehyde concentration [44]. Similarly, the addition of catalase (0.008 mg/mL) to sperm suspension offered protection against H_2O_2-induced toxicity [45]. In support, the role of catalase and SOD against sperm-intracellular (mitochondrial and plasma membrane) and -extracellular (leukocytes) ROS and their beneficial effects on sperm motility has been consistently reported [46, 47]. Different concentrations of vitamin E (800 μM, 10 mM) can protect sperm against lipid peroxidation [35, 44]. Similarly, pentoxifylline at the dosage of 3.6 and 7.2 mM was proven to limit lipid peroxidation in asthenozoospermic men [31].

Protection of Sperm DNA Integrity

Sperm preparation protocols that are routinely applied during ART and cryopreservation, involve repeated high-speed centrifugation and the isolation of spermatozoa from the protective antioxidant environment provided by seminal plasma. This has been shown to result in sperm DNA damage via pathways that are mediated by increased ROS generation [48]. Swim-up media supplemented with antioxidants [NAC (0.01 mM), catalase (500 U/mL), reduced GSH (10 mM), and hypotaurine (10 mM)] lead to significant reduction of DNA damage induced by ROS generation [49]. Similarly, albumin in doses from 0.3% to 10% protected sperm DNA integrity by neutralizing peroxides produced during lipid peroxidation [40].

It is critical to consider the approach for adding antioxidants as protective agents for the sperm DNA integrity whether single or in combination. Protection of the sperm DNA against H_2O_2-induced damage was demonstrated by vitamin E and vitamin C individually in both normozoospermic and asthenozoospermic samples [27]. However, when a combination of vitamin C and vitamin E was used for spermatozoa incubation, DNA damage was increased. This may be due to vitamin E and vitamin C acting as pro-oxidants [50].

Oxidative Stress During Cryopreservation

Oxidative stress occurring during sperm cryopreservation may be due to increased ROS production and/or decreased antioxidant scavenging activities. The process of ROS generation during cryopreservation and thawing of spermatozoa has been well documented. Data shows that freezing and thawing of spermatozoa causes an increase in the generation of superoxide radicals. A sudden burst of nitric oxide radicals was also observed during thawing [7]. In samples with initially detectable ROS levels, these levels are further significantly increased after cryopreservation/thawing. On the other hand, when samples with no detectable ROS are subjected cryopreservation/thawing, ROS become detected [51].

The decrease in recovery of motile viable sperm following cryopreservation/thawing may be due to damage caused by OS leading to lipid peroxidation of the sperm membrane. The increase in lipid peroxidation sometimes appears to be more significant than the increase in ROS levels when comparing fresh to cryopreserved/thawed sperm [52]. This was demonstrated by the detection of ROS-induced membrane lipid damage in frozen spermatozoa [53]. It has been also reported that the extent of lipid peroxidation is negatively correlated with post-thaw sperm motility [54]. Thus, cryopreservation does enhance lipid peroxidation in human sperm, and this enhancement may be mediated at least in part by the loss of SOD activity occurring during the process [53].

The cryopreservation process has been shown to diminish the antioxidant activity of the spermatozoa making them more susceptible to the ROS-induced damage. In bovine spermatozoa, levels of antioxidants were diminished during freeze/thaw cycles. Cryopreservation significantly reduced sperm GSH levels by 78% and SOD activity by 50% [6]. In humans, one consistent effect of cryopreservation is loss of the enzymatic activity of the peroxidation defense enzyme, SOD [55].

Recent body of evidence suggests that sperm DNA fragmentation occurs as a result of increased OS during sperm cryopreservation [56]. The alteration in the mitochondrial membrane fluidity that occurs during cryopreservation will lead to rise in mitochondrial membrane potential and the release of ROS. Subsequently, released ROS causes DNA damage in sperm. It has been reported that ROS production by both human sperm and seminal leukocytes increases on cooling to 4°C [57]. Thus, cryopreserved semen samples containing leukocytes may be more prone to DNA fragmentation.

The increase in sperm DNA damage during cryopreservation remains to be fully elucidated. While some studies documented it [4, 58], others found that it does not occur [59, 60]. The oxidative DNA biomarker 8-oxoguanine was recently used to assess oxidative DNA damage. Results showed that oxidative sperm DNA damage increased significantly after cryopreservation/thawing [61].

DNA damage resulting from OS is not limited to only ejaculated sperm. Aerobic incubation of testicular sperm results in a significant increase in DNA fragmentation. DNA fragmentation was noted to be higher in cryopreserved sperm than in fresh testicular sperm, and it was maximal after 4-h of incubation [62]. Therefore, care must be taken to avoid incubating cryopreserved and fresh testicular sperm for prolonged periods of time before ICSI is performed.

ROS-induced DNA damage that occurs during sperm cryopreservation may be further increased by other technical procedures that are concurrently conducted. Current laboratory cryopreservation protocols include freezing of raw semen and freezing of washed sperm without seminal plasma [63]. The removal of the antioxidant-rich seminal plasma by sperm preparation prior to cryopreservation may lead to the deterioration in sperm motility post-thaw, cryosurvival rates, and DNA integrity [58, 64]. However, it is important to note that the decrease in quality in pre-washed frozen-thawed sperm may not result in an actual decrease in cycle fecundity [65]. A potential source of ROS in the ART media during semen preparation is the activation of ROS production by immature spermatozoa by either centrifugation or leukocytes contamination. Recent studies suggest that DNA fragmentation in sperm is induced, for the most part, during sperm transport through the seminiferous tubules and the epididymis [15]. This could be mediated by ROS produced by immature sperm. A similar mechanism occurs in the pellet of centrifuged semen where sperm would be also highly packed.

Antioxidants as In Vitro Supplements During Sperm Cryopreservation

Antioxidants act as the main defense against OS induced by free radicals. Thus, the concept of their integration during the cryopreservation/thawing procedure to protect sperm against OS has been extensively evaluated. The addition of vitamin E and vitamin C along to cryoprotectants during cryopreservation resulted in limited if no preservation to the sperm motility when compared with cryoprotectant alone [66]. In contrast, a recent study tested if the addition of an antioxidant to cryopreservation medium could improve the post thaw integrity of cryopreserved human spermatozoa. It was found that vitamin E was significantly associated with post-thaw motility but neither sperm viability nor DNA fragmentation was affected [67]. The protective effects reported for vitamin E on post-thaw motility may be due to its ability to counteract lipid peroxidation [68]. The benefits of vitamin E supplementation in the cryoprotectant appear to be most marked in semen samples compromised with higher levels of ROS. These include samples from older men and samples with high prevalence of abnormal forms [67, 69].

It is important to note that in other studies, the preserving effects of antioxidants on sperm DNA integrity following cryopreservation and thawing were reported. Supplementation with the antioxidants ascorbate and catalase with the cryoprotectant pre-freeze resulted in a decrease in ROS levels and sperm DNA damage during cryopreservation/thawing [70]. Similarly, the addition of GSH to the thawing medium resulted in: (1) a higher number of non-capacitated viable spermatozoa, (2) a reduction in ROS generation, (3) lower chromatin condensation, (4) lower DNA fragmentation, (5) higher oocyte penetration rate in vitro, and (6) higher in vitro embryo production compared with control group [71].

The positive impact for adding the antioxidant rebamipide was also validated in an in vitro study. Rebamipide effectively scavenged ROS during sperm processing and cryopreservation. This was manifested by lesser decrease in sperm motility following cryopreservation in sperm exposed to 100 and 300 μmol/L of rebamipide. The levels of ROS and lipid peroxidation in semen were also significantly decreased in proportion to the concentrations of rebamipide both after incubation and after cryopreservation [72].

In support of using antioxidant combinations, a study was performed to evaluate the in vitro supplementation of cryoprotectant media with SOD and catalase using samples from 25 male partners of infertile couples. No significant variation in the recovery of progressive motility after freezing and thawing was seen in the aliquots with added SOD or catalase alone, compared with the control group. On the other hand, a significant improvement in sperm parameter recovery was seen in the aliquot with both SOD and catalase supplementation. This may be due to their combined and simultaneous action on superoxide anion and H_2O_2. Therefore, the combination of SOD and catalase prevented lipid peroxidation of the sperm plasma membrane by ROS and contributed to the recovery of high-quality spermatozoa after freezing–thawing procedures [73].

Expert Commentary

The purpose of this chapter was to discuss the role played by antioxidants in human sperm cryopreservation. Normally, a balance exists between levels of ROS and antioxidant scavenging systems. Disturbance of such balance results in OS, which is known to cause peroxidative damage of the sperm plasma membrane and loss of its motility, viability, and DNA integrity. Cryopreservation and thawing is a widely used technique in conjunction with ART that has been reported to cause OS when applied to human spermatozoa. Moreover, cryopreserved spermatozoa tend to lose their limited antioxidant defenses. Subsequent sperm preparation for ART involves several steps of centrifugation and incubation, which further induces OS and affects ART outcome.

One of the rational strategies to counteract OS is to increase the scavenging capacity of the medium that surrounds spermatozoa. This approach has been documented to improve the recovery of spermatozoa with higher quality and fertilization potential following cryopreservation and thawing. Thus, the use of antioxidants in cryopreservation and post-thaw media should be advised.

Five-Year View

Strong body of evidence currently supports the use of antioxidants in cryoprotectant and post-thaw media to prevent OS during cryopreservation and thawing. However, there is no consensus as to the type, combination and concentrations of antioxidants to be added. Therefore, caution should be exercised as excessive amounts of antioxidants may act paradoxically to damage spermatozoa. Well-controlled studies are still needed to identify the ideal antioxidant combination/concentration to supplement the cryoprotectant and post-thaw media.

Key Issues

- Sperm damage is likely to occur during cryopreservation and thawing due to OS. The occurrence of OS during these processes is due to increased ROS production and decrease in scavenging antioxidants.
- Sperm damage occurring during cryopreservation and thawing is manifested by decrease in motility, viability, and DNA integrity. Such changes will ultimately affect the sperm fertilization potential.
- The use of antioxidants is proven to counteract ROS preventing their damaging effect on the sperm.
- The supplementation of cryoprotectant and post-thaw media with antioxidants prevents to a certain extent the damage inflicted on spermatozoa during cryopreservation and thawing.
- Further studies are still needed to identify the optimum antioxidant supplementation protocol during sperm cryopreservation and thawing.

References

1. Bunge RG, Sherman JK. Fertilizing capacity of frozen human spermatozoa. Nature. 1953;172:767–8.
2. Anger JT, Gilbert BR, Goldstein M. Cryopreservation of sperm: indications, methods and results. J Urol. 2003;170:1079–84.
3. Nawroth F, Rahimi G, Isachenko E, et al. Cryopreservation in assisted reproductive technology: new trends. Semin Reprod Med. 2005;23:325–35.
4. Donnelly ET, Steele EK, McClure N, Lewis SE. Assessment of DNA integrity and morphology of ejaculated spermatozoa from fertile and infertile men before and after cryopreservation. Hum Reprod. 2001;16:1191–9.
5. Royere D, Barthelemy C, Hamamah S, Lansac J. Cryopreservation of spermatozoa: a 1996 review. Hum Reprod Update. 1996;2:553–9.
6. Bilodeau JF, Chatterjee S, Sirard MA, Gagnon C. Levels of antioxidant defenses are decreased in bovine spermatozoa after a cycle of freezing and thawing. Mol Reprod Dev. 2000;55:282–8.
7. Chatterjee S, Gagnon C. Production of reactive oxygen species by spermatozoa undergoing cooling, freezing, and thawing. Mol Reprod Dev. 2001;59:451–8.
8. Sharma RK, Agarwal A. Role of reactive oxygen species in male infertility. Urology. 1996;48:835–50.
9. Pasqualotto FF, Sharma RK, Nelson DR, Thomas AJ, Agarwal A. Relationship between oxidative stress, semen characteristics, and clinical diagnosis in men undergoing infertility investigation. Fertil Steril. 2000;73:459–64.
10. Aitken RJ, Baker MA, Sawyer D. Oxidative stress in the male germ line and its role in the aetiology of male infertility and genetic disease. Reprod Biomed Online. 2003;7:65–70.
11. Wang X, Sharma RK, Gupta A, et al. Alterations in mitochondria membrane potential and oxidative stress in infertile men: a prospective observational study. Fertil Steril. 2003;80 Suppl 2:844–50.
12. Wang X, Sharma RK, Sikka SC, Thomas Jr AJ, Falcone T, Agarwal A. Oxidative stress is associated with increased apoptosis leading to spermatozoa DNA damage in patients with male factor infertility. Fertil Steril. 2003;80:531–5.
13. Griveau JF, Le Lannou D. Reactive oxygen species and human spermatozoa: physiology and pathology. Int J Androl. 1997;20:61–9.
14. Aitken RJ, Gordon E, Harkiss D, et al. Relative impact of oxidative stress on the functional competence and genomic integrity of human spermatozoa. Biol Reprod. 1998;59:1037–46.
15. Ollero M, Gil-Guzman E, Lopez MC, et al. Characterization of subsets of human spermatozoa at different stages of maturation: implications in the diagnosis and treatment of male infertility. Hum Reprod. 2001;16:1912–21.
16. Moustafa MH, Sharma RK, Thornton J, et al. Relationship between ROS production, apoptosis and DNA denaturation in spermatozoa from patients examined for infertility. Hum Reprod. 2004;19:129–38.
17. Alvarez J, Touchstone J, Blasco L, Storey B. Spontaneous lipid peroxidation and production of hydrogen peroxide and superoxide in human spermatozoa. Superoxide dismutase as major enzyme protectant against oxygen toxicity. J Androl. 1987;8:338–48.
18. Aitken RJ, Fisher HM, Fulton N, et al. Reactive oxygen species generation by human spermatozoa is induced by exogenous NADPH and inhibited by the flavoprotein inhibitors diphenylene iodonium and quinacrine. Mol Reprod Dev. 1997;47:468–82.
19. Agarwal A, Said TM, Bedaiwy MA, Banerjee J, Alvarez JG. Oxidative stress in an assisted reproductive techniques setting. Fertil Steril. 2006;86:503–12.
20. Sikka SC, Rajasekaran M, Hellstrom WJ. Role of oxidative stress and antioxidants in male infertility. J Androl. 1995;16:464–8.
21. Sikka SC. Role of oxidative stress and antioxidants in andrology and assisted reproductive technology. J Androl. 2004;25:5–18.
22. Sharma RK, Pasqualotto FF, Nelson DR, Thomas Jr AJ, Agarwal A. The reactive oxygen species-total antioxidant capacity score is a new measure of oxidative stress to predict male infertility. Hum Reprod. 1999;14:2801–7.
23. Kobayashi H, Gil-Guzman E, Mahran AM, et al. Quality control of reactive oxygen species measurement by luminol-dependent chemiluminescence assay. J Androl. 2001;22:568–74.

24. Said TM, Kattal N, Sharma RK, et al. Enhanced chemiluminescence assay vs colorimetric assay for measurement of the total antioxidant capacity of human seminal plasma. J Androl. 2003;24: 676–80.
25. Smith R, Vantman D, Ponce J, Escobar J, Lissi E. Total antioxidant capacity of human seminal plasma. Hum Reprod. 1996;11:1655–60.
26. Saleh RA, Agarwal A. Oxidative stress and male infertility: from research bench to clinical practice. J Androl. 2002;23:737–52.
27. Donnelly ET, McClure N, Lewis SE. The effect of ascorbate and alpha-tocopherol supplementation in vitro on DNA integrity and hydrogen peroxide-induced DNA damage in human spermatozoa. Mutagenesis. 1999;14:505–12.
28. Donnelly ET, McClure N, Lewis SE. Glutathione and hypotaurine in vitro: effects on human sperm motility, DNA integrity and production of reactive oxygen species. Mutagenesis. 2000;15:61–8.
29. Gavella M, Lipovac V, Marotti T. Effect of pentoxifylline on superoxide anion production by human sperm. Int J Androl. 1991;14: 320–7.
30. Gavella M, Lipovac V. Pentoxifylline-mediated reduction of superoxide anion production by human spermatozoa. Andrologia. 1992;24:37–9.
31. McKinney KA, Lewis SE, Thompson W. The effects of pentoxifylline on the generation of reactive oxygen species and lipid peroxidation in human spermatozoa. Andrologia. 1996;28:15–20.
32. Okada H, Tatsumi N, Kanzaki M, Fujisawa M, Arakawa S, Kamidono S. Formation of reactive oxygen species by spermatozoa from asthenospermic patients: response to treatment with pentoxifylline. J Urol. 1997;157:2140–6.
33. Oeda T, Henkel R, Ohmori H, Schill WB. Scavenging effect of N-acetyl-L-cysteine against reactive oxygen species in human semen: a possible therapeutic modality for male factor infertility? Andrologia. 1997;29:125–31.
34. Jones R, Mann T, Sherins R. Peroxidative breakdown of phospholipids in human spermatozoa, spermicidal properties of fatty acid peroxides, and protective action of seminal plasma. Fertil Steril. 1979;31:531–7.
35. Verma A, Kanwar KC. Human sperm motility and lipid peroxidation in different ascorbic acid concentrations: an in vitro analysis. Andrologia. 1998;30:325–9.
36. Aitken RJ, Clarkson JS. Significance of reactive oxygen species and antioxidants in defining the efficacy of sperm preparation techniques. J Androl. 1988;9:367–76.
37. Donnelly ET, McClure N, Lewis SE. Antioxidant supplementation in vitro does not improve human sperm motility. Fertil Steril. 1999;72:484–95.
38. Parinaud J, Le Lannou D, Vieitez G, Griveau JF, Milhet P, Richoilley G. Enhancement of motility by treating spermatozoa with an antioxidant solution (Sperm-Fit) following ejaculation. Hum Reprod. 1997;12:2434–6.
39. Hong CY, Lee MF, Lai LJ, Wang CP. Effect of lipid peroxidation on beating frequency of human sperm tail. Andrologia. 1994;26:61–5.
40. Twigg J, Fulton N, Gomez E, Irvine DS, Aitken RJ. Analysis of the impact of intracellular reactive oxygen species generation on the structural and functional integrity of human spermatozoa: lipid peroxidation, DNA fragmentation and effectiveness of antioxidants. Hum Reprod. 1998;13:1429–36.
41. Armstrong JS, Rajasekaran M, Hellstrom WJ, Sikka SC. Antioxidant potential of human serum albumin: role in the recovery of high quality human spermatozoa for assisted reproductive technology. J Androl. 1998;19:412–9.
42. Lewin A, Lavon H. The effect of coenzyme Q10 on sperm motility and function. Mol Aspects Med. 1997;18(Suppl):S213–9.
43. Baker HW, Brindle J, Irvine DS, Aitken RJ. Protective effect of antioxidants on the impairment of sperm motility by activated polymorphonuclear leukocytes. Fertil Steril. 1996;65:411–9.
44. Kobayashi T, Miyazaki T, Natori M, Nozawa S. Protective role of superoxide dismutase in human sperm motility: superoxide dismutase activity and lipid peroxide in human seminal plasma and spermatozoa. Hum Reprod. 1991;6:987–91.
45. Gagnon C, Iwasaki A, De Lamirande E, Kovalski N. Reactive oxygen species and human spermatozoa. Ann N Y Acad Sci. 1991; 637:436–44.
46. Kovalski NN, de Lamirande E, Gagnon C. Reactive oxygen species generated by human neutrophils inhibit sperm motility: protective effect of seminal plasma and scavengers. Fertil Steril. 1992;58: 809–16.
47. Griveau JF, Le Lannou D. Effects of antioxidants on human sperm preparation techniques. Int J Androl. 1994;17:225–31.
48. Zalata A, Hafez T, Comhaire F. Evaluation of the role of reactive oxygen species in male infertility. Hum Reprod. 1995;10: 1444–51.
49. Lopes S, Jurisicova A, Sun JG, Casper RF. Reactive oxygen species: potential cause for DNA fragmentation in human spermatozoa. Hum Reprod. 1998;13:896–900.
50. Hughes CM, Lewis SE, McKelvey-Martin VJ, Thompson W. The effects of antioxidant supplementation during Percoll preparation on human sperm DNA integrity. Hum Reprod. 1998;13:1240–7.
51. Mazzilli F, Rossi T, Sabatini L, et al. Human sperm cryopreservation and reactive oxygen species (ROS) production. Acta Eur Fertil. 1995;26:145–8.
52. Kadirvel G, Kumar S, Kumaresan A. Lipid peroxidation, mitochondrial membrane potential and DNA integrity of spermatozoa in relation to intracellular reactive oxygen species in liquid and frozen-thawed buffalo semen. Anim Reprod Sci. 2009;114: 125–34.
53. Alvarez JG, Storey BT. Evidence for increased lipid peroxidative damage and loss of superoxide dismutase activity as a mode of sublethal cryodamage to human sperm during cryopreservation. J Androl. 1992;13:232–41.
54. Bell M, Wang R, Hellstrom WJ, Sikka SC. Effect of cryoprotective additives and cryopreservation protocol on sperm membrane lipid peroxidation and recovery of motile human sperm. J Androl. 1993;14:472–8.
55. Lasso JL, Noiles EE, Alvarez JG, Storey BT. Mechanism of superoxide dismutase loss from human sperm cells during cryopreservation. J Androl. 1994;15:255–65.
56. Thomson LK, Fleming SD, Aitken RJ, De Iuliis GN, Zieschang JA, Clark AM. Cryopreservation-induced human sperm DNA damage is predominantly mediated by oxidative stress rather than apoptosis. Hum Reprod. 2009;24:2061–70.
57. Wang AW, Zhang H, Ikemoto I, Anderson DJ, Loughlin KR. Reactive oxygen species generation by seminal cells during cryopreservation. Urology. 1997;49:921–5.
58. Donnelly ET, McClure N, Lewis SE. Cryopreservation of human semen and prepared sperm: effects on motility parameters and DNA integrity. Fertil Steril. 2001;76:892–900.
59. Duru NK, Morshedi MS, Schuffner A, Oehninger S. Cryopreservation-thawing of fractionated human spermatozoa is associated with membrane phosphatidylserine externalization and not DNA fragmentation. J Androl. 2001;22:646–51.
60. Isachenko V, Isachenko E, Katkov II, et al. Cryoprotectant-free cryopreservation of human spermatozoa by vitrification and freezing in vapor: effect on motility, DNA integrity, and fertilization ability. Biol Reprod. 2004;71:1167–73.
61. Zribi N, Chakroun NF, El Euch H, Gargouri J, Bahloul A, Keskes LA. Effects of cryopreservation on human sperm deoxyribonucleic acid integrity. Fertil Steril. 2010;93:159–66.
62. Dalzell LH, McVicar CM, McClure N, Lutton D, Lewis SE. Effects of short and long incubations on DNA fragmentation of testicular sperm. Fertil Steril. 2004;82:1443–5.

63. Saritha KR, Bongso A. Comparative evaluation of fresh and washed human sperm cryopreserved in vapor and liquid phases of liquid nitrogen. J Androl. 2001;22:857–62.
64. Grizard G, Chevalier V, Griveau JF, Le Lannou D, Boucher D. Influence of seminal plasma on cryopreservation of human spermatozoa in a biological material-free medium: study of normal and low-quality semen. Int J Androl. 1999;22:190–6.
65. Wolf DP, Patton PE, Burry KA, Kaplan PF. Intrauterine insemination-ready versus conventional semen cryopreservation for donor insemination: a comparison of retrospective results and a prospective, randomized trial. Fertil Steril. 2001;76:181–5.
66. Askari HA, Check JH, Peymer N, Bollendorf A. Effect of natural antioxidants tocopherol and ascorbic acids in maintenance of sperm activity during freeze-thaw process. Arch Androl. 1994;33:11–5.
67. Taylor K, Roberts P, Sanders K, Burton P. Effect of antioxidant supplementation of cryopreservation medium on post-thaw integrity of human spermatozoa. Reprod Biomed Online. 2009;18:184–9.
68. Verma A, Kanwar KC. Effect of vitamin E on human sperm motility and lipid peroxidation in vitro. Asian J Androl. 1999;1:151–4.
69. Weir CP, Robaire B. Spermatozoa have decreased antioxidant enzymatic capacity and increased reactive oxygen species production during aging in the Brown Norway rat. J Androl. 2007;28:229–40.
70. Li ZL, Lin QL, Liu RJ, Xie WY, Xiao WF. Reducing oxidative DNA damage by adding antioxidants in human semen samples undergoing cryopreservation procedure. Zhonghua Yi Xue Za Zhi. 2007;87:3174–7.
71. Gadea J, Gumbao D, Canovas S, Garcia-Vazquez FA, Grullon LA, Gardon JC. Supplementation of the dilution medium after thawing with reduced glutathione improves function and the in vitro fertilizing ability of frozen-thawed bull spermatozoa. Int J Androl. 2008;31:40–9.
72. Park NC, Park HJ, Lee KM, Shin DG. Free radical scavenger effect of rebamipide in sperm processing and cryopreservation. Asian J Androl. 2003;5:195–201.
73. Rossi T, Mazzilli F, Delfino M, Dondero F. Improved human sperm recovery using superoxide dismutase and catalase supplementation in semen cryopreservation procedure. Cell Tissue Bank. 2001;2:9–13.

Antioxidants in ICSI

42

Nicolas Garrido, Sandra García-Herrero, Laura Romany, José Remohí, Antonio Pellicer, and Marcos Meseguer

Abstract

Male factor infertility can be caused by reasons, either related or not with total sperm production. Among causes of male infertility in cases of normal sperm count and motility, oxidative stress is one of the most relevant processes influencing fertility in vivo or in assisted reproduction treatments' results. This chapter provides the most updated information regarding the oxidative stress situation in sperm and the relevance of antioxidants use in intracytoplasmic sperm injection results.

Keywords

Sperm • Oxidative stress • ICSI • DNA oxidation • DNA fragmentation • Antioxidant • Male factor infertility • Intracytoplasmic sperm injection • Free radicals

Male factor infertility can be caused by reasons, either related or not with total sperm production. Among causes of male infertility in cases of normal sperm count and motility, oxidative stress is one of the most relevant processes influencing fertility in vivo or in assisted reproduction treatments' results. This chapter provides the most updated information regarding the oxidative stress situation in sperm and the relevance of antioxidants use in intracytoplasmic sperm injection (ICSI) results.

Male Factor Infertility Relevance and Its Measurement

Male factor infertility is responsible of almost 50% of all cases of infertility affecting approximately 5% of the general population [1]. Male fertility is defined as the ability of a man to impregnate a healthy, fertile woman of reproductive age, although tagging a male as fertile or infertile is extremely difficult since it can vary in short periods of time and with different partners. These assertions also may be considered from the point of view of a single ejaculate. In this sense, it is interesting to define a sperm sample as able to accomplish a pregnancy or not, instead of defining a male as fertile or infertile [2].

The only accepted tool to estimate the male potential to become a father is the basic sperm analysis as stated by the

N. Garrido, PhD (✉)
Andrology Laboratory and Sperm Bank, Instituto Universitario IVI Valencia, Plaza de la Policía Local 3, Valencia 46015, Spain
e-mail:nicolas@garrido@ivi.es

S. García-Herrero, PhD
IVIOMICS, Valencia, Spain

L. Romany, MD
Clinical Embryology Laboratory, Instituto Universitario IVI Valencia, Plaza de la Policía Local 3, Valencia 46015, Spain

J. Remohí, MD
Department of Gynecology and Obstetrics, School of Medicine, Universidad de Valencia, Assisted Reproduction Unit, Instituto Universitario IVI Valencia,Universidad de Valencia, Plaza de la Policía Local 3, Valencia 46015, Spain

A. Pellicer, MD
Department of Gynecology and Obstetrics, School of Medicine, Universidad de Valencia, Assisted Reproduction Unit, Instituto Universitario IVI Valencia, Universidad de Valencia, Plaza de la Policía Local 3, Valencia 46015, Spain

M. Meseguer, PhD
Clinical Embryology Laboratory, Instituto Universitario IVI Valencia, Plaza de la Policía Local 3, Valencia 46015, Spain

S.J. Parekattil and A. Agarwal (eds.), *Male Infertility: Contemporary Clinical Approaches, Andrology, ART & Antioxidants*,
DOI 10.1007/978-1-4614-3335-4_42, © Springer Science+Business Media, LLC 2012

World Health Organization [3] based on the volume, sperm concentration and percentage of spermatozoa with progressive motility and normal shape. Subsequently, subfertile men can also be defined as those unable to achieve a natural conception and able to conceive with the help of assisted reproduction techniques.

The inability of a man to conceive can be either related with sperm count or related with molecular defects within sperm cells, thus leading to a reproductive failure at different levels. Those related with sperm count, that are frequently associated with genetic defects in the most severe cases, have been successfully treated to a certain extent by the implementation of IVF and ICSI techniques. These assisted reproduction technologies permit the avoidance of natural barriers in the selection of sperm that makes conception in the cases presenting low sperm count impossible and that otherwise would be necessary to select the most adequate sperm to fertilize an egg. Then, the collateral and involuntary harm caused by assisted reproduction could be related to the lack of natural selection, thus being able to create embryos with the wrong sperm cell, thus leading to embryo blockage during development or implantation failure.

Subsequently, the most intriguing male infertility causes are those independent of the total number of sperm cells produced by the testis. These defects are not detected by the routine sperm analysis and are responsible of a huge amount of infertility problems [4].

Apart from the basic sperm analysis, few tests are available to conclude the real possibilities of a male to become a father. From the existing evidences, several factors in sperm seem to be related to male fertility, but none of them seem enough to cause themselves inability to procreate, leading us to the hypothesis that male idiopathic infertility is caused by the combination of different factors. Among them, those related to the reactive oxygen species overload have been demonstrated to be crucial in the last years.

Oxidative Stress and Male Factor Infertility

OS is the disequilibrium between oxidative and antioxidative molecules in a biological system where the oxidants overcome the defensive systems [5] where reactive oxygen species (ROS) are primarily produced by O_2 metabolism in cells living under aerobic conditions, being able to produce interferences or destruction on cell biological functions and properties.

Cells in a biological environment (as ejaculates are) contribute to the maintenance of the oxidative homeostasis by complex systems, and a small and controlled level of ROS is necessary for normal cell function. This control is produced by antioxidants. These molecules are grouped depending on their nature into enzymatic and non-enzymatic or endogenous and exogenous.

Endogenous Sources of Free Radicals

Human semen is a complex mixture of cells comprised within the seminal plasma, including both mature and immature spermatozoa, round cells from different initial steps of the spermatogenesis process, leukocytes and other occasional cell types as epithelial cells. Among them, immature spermatozoa and leukocytes are the main ROS sources. Abnormal spermatozoa, with aberrant cytoplasmic droplet retention, are very important ROS producers since an excess of cytoplasmic enzymes involved in the glucose metabolism (such as glucose-6-phosphate dehydrogenase, NADPH oxidase system and NADH-dependent oxidoreductase) can be found at two different sites: plasma membrane and mitochondria due to the above mentioned cytoplasmic retention. Serial steps of different ROS production and combination results in the formation of distinct ROS such as superoxide anion, hydrogen peroxide and hydroxyl radicals initiate molecular damage by their interaction with cellular macromolecules [5].

Moreover, it has been confirmed that when seminal ejaculate is fractioned by a density gradient, the layer with immature sperm, with the highest percentage of immature cells, is the most ROS-producing group. This ROS production by immature spermatozoa is also directly correlated with the extent of DNA damage to mature sperm, and the higher ROS production, the lower percentage of mature spermatozoa [6].

The second ROS-producer cell type is the immune white cell lineage of the ejaculate. Their presence on the raw semen exerts a part of their defensive activity when activated by the direct production of ROS and the indirect stimulation of ROS production of neighbouring immune cells via soluble factors

Furthermore, two considerations must be done: first, leukocyte concentrations considered normal by WHO criteria can produce damaging ROS levels [7, 8], and second, leukocytes can stimulate sperm ROS production [9]. Overwhelming the protective capacity of the system by the ROS attack provokes an irreparable damage on DNA molecules, acrosome reaction is also deregulated, and finally, sperm/oocyte fusion is disturbed or impeded [10]. Some studies revealed the importance of these phenomena on male fertility since almost 40% of infertile males displayed abnormally increased ROS levels, and human spermatozoa are extremely sensitive to ROS-induced damage due to their particular plasma membrane composition [11, 12].

Exogenous Sources of Free Radicals

Lab Protocols

Centrifugation of a semen sample to select sperm for intrauterine insemination (IUI) or in vitro fertilization (IVF) can aggravate sperm oxidative stress, although it is easily reduced by decreasing the centrifugation time [13, 14], use of swim-up or glass-wool filtration, and limiting the time in which sperm are cultured in media with seminal plasma. Furthermore, culturing sperm under low oxygen tension (5% O_2/95% CO_2 vs. 20% atmospheric O_2 content) has been shown to also improve sperm quality by reducing seminal leukocyte ROS production [15–17]. Avoiding use of cryopreserved sperm for fertilization is also ideal since ROS are produced during freezing and thawing of the sperm, thereby decreasing sperm quality [17]. Sperm preparation media may also be supplemented with a variety of antioxidants to guard against oxidative stress. The addition of catalase/SOD [18], vitamin C [19], vitamin E [20], ferulic acid [21], EDTA [22], glutathione/hypotaurine [23], albumin [24] and *N*-acetyl-cysteine [25] to sperm preparation media has all been shown to protect sperm from oxidative attack. Recently, the use of genistein as antioxidant has been demonstrated to improve sperm quality after freezing [26].

Lifestyle

Smoking results in a 48% increase in seminal leukocyte concentrations and a 107% in seminal ROS [27]. Smokers have decreased levels of seminal plasma antioxidants such as vitamin E [28] and vitamin C [29], placing their sperm at additional risk of oxidative damage, confirmed by the finding of a significant increase in levels of 8-OHdG within smoker's seminal plasma [28] and deficiencies in sperm capacitation [30, 31].

Dietary deficiencies have been linked with sperm oxidative damage by several research groups. The Age and Genetic Effects in Sperm (AGES) study examined the self-reported dietary intake of various antioxidants and nutrients (vitamins C and E, b-carotene, folate and zinc) [32], finding a significant correlation between vitamin C intake and sperm concentration and also in vitamin E intake and total progressively motile sperm and reinforcing previous reports [33]. However, the AGES study did not found a link between low intake of antioxidants and sperm DNA damage [34] against the results of others [35].

Excessive alcohol consumption increases systemic oxidative stress as ethanol stimulates the production of ROS [36, 37]. Within the testicle, this implies a significant reduction in plasma testosterone, increase in serum lipid peroxidation by-products and a drop in antioxidants [38].

Extremes of exercise activity, at both ends of the spectrum, have been linked with oxidative stress; this link is explained by the increased muscle aerobic metabolism creating a large amount of ROS [39]. On the other side, obesity produces oxidative stress as adipose tissue releases proinflammatory cytokines that increase leukocyte production of ROS [40] and heating of the testicle [41] also linked with oxidative stress and reduced sperm quality.

Environment

Phthalates are chemicals used as plastics softener contained in food packaging and personal care products that have been linked with impaired spermatogenesis and increased sperm DNA damage [42, 43]. In rat model, it increases the generation of ROS within the testis and decreases antioxidant levels, impairing spermatogenesis [44].

Several environmental pollutants have been linked with testicular oxidative stress in rodent models, such as lindane [45], methoxychlor and the herbicide dioxin-TCDD [46, 47]. The commonly used preservative sulphur dioxide has also been shown to produce testicular oxidative stress in laboratory animals [48]. Air pollutants such as diesel particulate matter act as potent stimuli for leukocyte ROS generation [49].

Also, heavy metal exposure has reported with a similar effect: cadmium and lead are linked with an increase in testicular oxidative stress [50, 51] and a resultant increase in sperm DNA oxidation [52].

The Control of Free Radicals: The Antioxidants

Seminal plasma is especially relevant to protect spermatozoa against ROS; because of the diminished cytoplasm of sperm decreasing the capacity to retain adequate loads of protective molecules inside the cell, they will depend on the extracellular environment [53].

The SOD family presents three different classes, depending on the catalytic metal at the active site. This enzyme works catalyzing the dismutation of superoxide into hydrogen peroxide and oxygen, with a direct relationship between sperm SOD activity in spermatozoa, but not plasma, and sperm motility [54]. Adding exogenous SOD significantly decreased the loss of motility with time, and the increase of the malonyldialdehyde concentration was initially related with the percentage of immotile cells. These data suggest a significant role for SOD in sperm motility. It seems that lipid peroxidation of human spermatozoa may cause loss of motility and that SOD may inhibit this lipid peroxidation, although some authors are unable to correlate SOD activity with seminal parameters or reproductive success in vivo or in vitro [55, 56].

Catalase has also been found in human spermatozoa and seminal plasma [57] of normal and infertile males and works preventing oxygen-derived free radicals induced damage inactivating H_2O_2 and yielding water and O_2.

Other ROS scavenging enzymes such as glutathione (GPx) have also been measured in seminal plasma and correlated with the fertility status of the male [58]. GPx family of enzymes is composed by several forms, able to detoxify organic or hydrogen peroxides, converting them to stable alcohols or water, thus protecting cells from oxidative damage. The GPx enzymes containing Se (as a selenocysteine on the molecule) need the presence of reduced glutathione (GSH) that restores the oxidized Se.

Coupled to this reaction, glutathione reductase (GR) catalyzes the step from oxidized to reduced glutathione to restore the GSH stock available for GPX. We evaluated the GPX system finding a significant relationship of GPX1 and GPX4 with male infertility [59, 60].

Non-enzymatic protection was described by Alvarez and Storey in classical experiments, where demonstrated in vitro the effect of protection against lipid peroxidation of rabbit spermatozoa of some molecules as taurine, hypotaurine, epinephrine, pyruvate, lactate and bovine serum albumin [61]. Taurine and hypotaurine molecules have been lately found to be present in seminal plasma of infertile males at a different rate and have also been associated with concrete sperm defects [62].

Other non-enzymatic molecules are probably acting in seminal plasma in the scavenging of free radicals, and lots of information regarding their scavenging activity are available about other biological systems, but they have not been yet studied in semen. Some examples are ferritin, alpha lipoic acid, L-ergothioneine, ebselen, etc.

The Main Consequences of OS in Sperm Affecting ICSI Results

DNA Oxidation

We determined the relevance of sperm DNA oxidation as one of the main consequences of OS on embryo quality and reproductive outcome by prospectively studying pairs of oocyte donation cycles, i.e. the same oocyte donors, donating to two recipients, where the only difference between the two treatments was the use of a different sperm sample [63].

Regarding embryo morphology, an important association of embryo asymmetry and increased DNA oxidation was observed. In addition, in the later in vitro phase, those embryos reaching the blastocyst stage were associated with lower sperm DNA oxidation levels (Fig. 42.1).

On the other hand, we observed a minor decrease of DNA oxidation in patients who did not achieve pregnancy, although this difference was not found to be significant. This could be due to embryo selection before transfer, so poor prognosis embryos are no longer chosen. In the absence of such selection, it may be possible that pregnancy could be associated with sperm DNA oxidation.

In addition, if we analyse the differences in the IVF outcome parameters of the couples who shared the oocyte cohort (same donor) with the differences in the OXI DNA values, we observed increased and further relationships with embryo quality. Oxidative damage in the DNA is clearly increased in samples with lower sperm motility. An association between early and late embryo quality and sperm DNA oxidation supports the relevance of the hydroxylation of 8-oxoguanine like a biomarker of sperm quality reflecting the free radical damage in human sperm.

We also determined the relevance of sperm DNA oxidation caused by free radicals in sperm retrieved from testicular biopsies, analysing their corresponding assisted reproduction treatments and using ovum donation to standardize female's characteristics. We studied its effect on embryo quality and reproductive success [64].

Testicular cells present a wide variety of types and ploidies (haploids: sperm and round spermatids; diploids: spermatogonia, secondary spermatocytes, Sertoli and Leydig cells, immune cells, myoid cells, fibroblasts, etc.; and tetraploids: primary spermatocytes). To determine cell ploidy, a dual staining with propidium iodide was conducted. This staining permitted the evaluation of cell ploidy and 8-OH staining within each category. Also, the appropriate gates to distinguish between spermatic and non-spermatic haploid cells were created by using shape properties analysed by the flow cytometer.

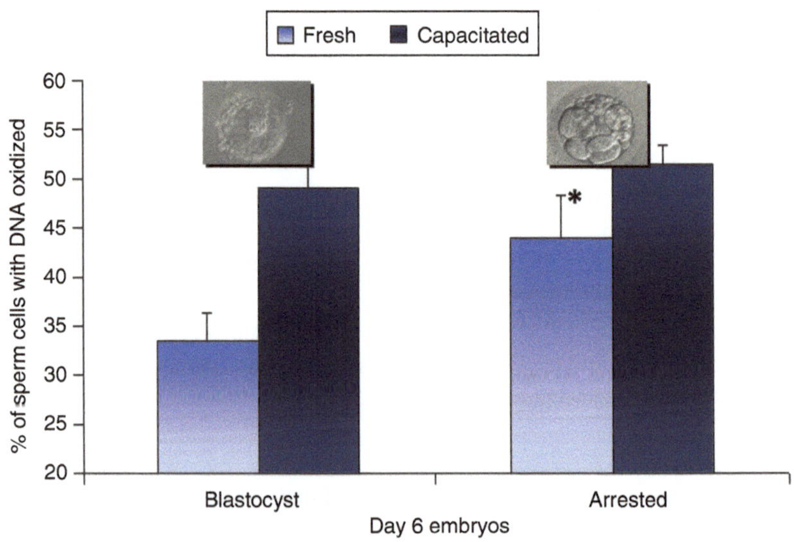

Fig. 42.1 Embryo quality on day 6 depending on the sperm DNA oxidation status

When the percentage of 8-OhdG+cells is considered within each ploidy group among testicular cells, there is an increase in non-obstructive azoospermia (NOA) in comparison with obstructive azoospermia (OA) when considering only haploid sperm cells. About 4% more haploid cells are presenting oxidized DNA, while there is almost a 1.5% increase in the cells with presence of DNA oxidation in NOA vs. OA considering only diploid cells, and only a 0.16% difference in 8-OH staining is found in tetraploid cells between groups, being again more oxidized those sperm from males with NOA.

We found a very low clinical relevance of the status of sperm DNA oxidation on several embryo quality parameters and fertilization rate, early (days 2–3) and late (days 5–6) development, and pregnancy achievement (Fig. 42.2). The studies where DNA oxidation was studied, the employment of ovum donation models to standardize clinical parameters of the female, as well as employing regression analysis controlling possible confounding factors add strength to these results.

DNA Fragmentation

One of the OS consequences is a direct damage to sperm DNA integrity. Male infertility is associated with poor sperm DNA integrity, and it has been suggested that these DNA abnormalities may affect procreation in couples having natural intercourse and in those treated by IUI, IVF and ICSI. The increase in the use of ART has enlarged the emphasis on the sperm chromatin quality.

We prospectively evaluated the predictive value of the SCD test on pregnancy outcome by IUI [65] in couples with non-severe male factor, and to correlate DNA fragmentation with sperm parameters, and found that chromatin dispersion in sperm, as measured by the SCD test, is not correlated with pregnancy outcome in IUI. However, a significant correlation was found between sperm motility and SCD test values.

In a subsequent work, we also determined the prognostic value of sperm DNA fragmentation levels, as measured by the SCD test, in predicting IVF and ICSI outcome [66]. The degree of DNA fragmentation was inversely correlated with fertilization rate, synchrony of the nucleolar precursor bodies' pattern in pronuclei, embryo ability to achieve blastocyst stage and embryo morphological quality. Because SCD test values were correlated with embryo quality and blastocyst rate, the lack of correlation between sperm DNA fragmentation and pregnancy outcome in IVF might be due to embryo selection before transfer.

Essentially, two interesting details are derived from the results. First, the determination of sperm DNA fragmentation could be of special interest for patients with low gestation rates and low embryonic quality. The DNA damage could then be a possible causal factor. Second, the observation of nucleolar asynchrony in a zygote from a sperm sample with a high DNA fragmentation level could orientate to the possibility of a future low-quality blastocyst. Therefore, this could be an interesting criterion to orientate for embryo selection when the transfer occurs on day 3. In conclusion, the degree of DNA fragmentation, established using the SCD test, appears to be related to the ability of sperm to fertilize as well as the ability of the embryo to achieve the blastocyst stage until day 6 (Fig. 42.3). The correlation of the sperm DNA fragmentation level with embryo health appears quite promising in predicting the quality of the embryo cohort after IVF/ICSI procedures [66].

Recently, we prospectively analysed sperm DNA fragmentation in testicular sperm samples from azoospermic

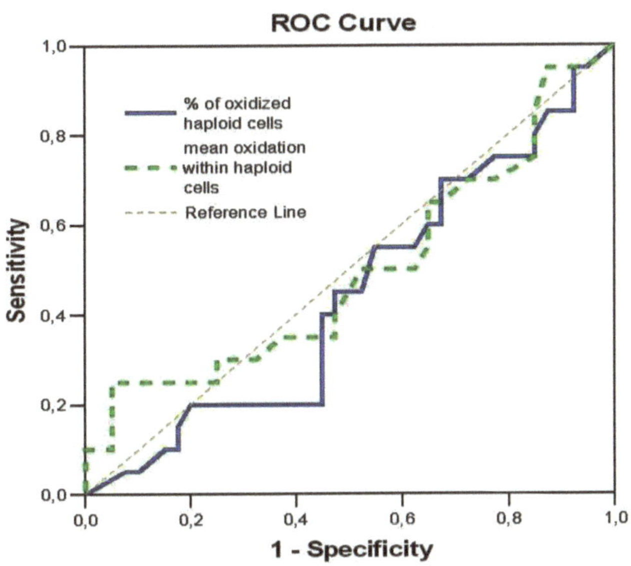

Fig. 42.2 ROC curve analysis of the predictive value of sperm DNA oxidation on pregnancy forecast

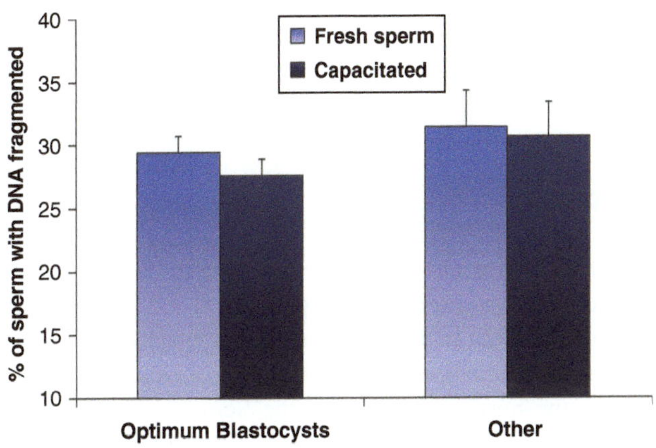

Fig. 42.3 Blastocyst quality depending on the % of sperm with fragmented DNA

patients, either by spermatogenic failure or by duct obstruction [67]. Levels of fragmentation in testes presented an average of 37.70% of positive spermatozoa. The number of sperm cells recovered per field and motile sperm recovered per field after testicular biopsy was not related with fragmentation, but these levels were associated with defective spermatogenesis (non-obstructive azoospermia), being significantly higher (46.92%) than in patients with normal testicular sperm production (obstructive azoospermia) (35.96%). A moderate relationship between embryo morphological parameters and testicular DNA fragmentation was detected. Nevertheless and like we observed in previous studies, no significant relation was detected with pregnancy outcome. This result suggests that spermatogenesis failure may result in a more severe affectation of sperm DNA integrity. The degree of DNA fragmentation using SCD test is related to the sperm production and embryo quality parameters, but finally, this is not reflected in pregnancy chances, probably due to the embryo selection bias. Our observations reveal that average values are comparable to levels of ejaculated sperm from infertile and cancer patients and higher than sperm donors [68].

Further studies are also needed to determine those mechanisms involved in the increased DNA fragmentation observed in those patients with defective spermatogenesis process. Current research projects are undergoing the study of the direct effect of the oxidative stress on DNA oxidation process produced in testicular sperm cells. This could be the primary source of DNA fragmentation and maybe the administration of antioxidants to the patients could overcome this defect. Nevertheless, reports concerning the clinical usefulness of antioxidants in the treatment of male infertility are controversial. Even more, our results are against post-testicular damage on DNA because we have observed high levels of fragmentation in testicular sperm and even higher in those coming from a non-obstructive azoospermia. In consequence, we are not completely agreed with those that postulate the utility of testicular sperm extraction in patients with high levels of DNA fragmentation combined with severe oligozoospermia. In this situation, spermatogenesis is clearly affected, and in consequence, those sperm cells coming from testes would have also high fragmentation.

Apoptosis

Apoptosis is a noninflammatory response to tissue damage characterized by a series of morphological and biochemical changes [69–72]. In the context of male reproductive tissue, it helps in elimination of abnormal spermatozoa, thus maintaining the nursing capacity of the Sertoli cells [71, 72]. High levels of ROS disrupt the inner and outer mitochondrial membranes, inducing the release of the cytochrome-C protein and activating the caspases and apoptosis. Apoptosis in sperm also may be initiated by ROS-independent pathways involving the cell surface protein Fas. Fas is a type I membrane protein that belongs to the tumour necrosis factor-nerve growth factor receptor family and mediates apoptosis [73]. When Fas ligand or agonistic anti-Fas antibody binds to Fas, apoptosis occurs. On the other hand, bcl-2, the inhibitor gene of apoptosis, protects the cell, most likely by mechanisms that reduce ROS production.

Although the Fas protein often leads to apoptosis, some of the Fas-labelled cells may escape apoptosis through abortive apoptosis. This results in a failure to clear all of the spermatozoa destined for elimination and thus leads to a large population of abnormal spermatozoa in the semen. This failure to clear Fas-positive spermatozoa may be due to a dysfunction at one or more levels. First, the production of spermatozoa may not be enough to trigger apoptosis in men with hypospermatogenesis. In this case, Fas-positive spermatogonia may escape the signal to undergo apoptosis. Second, Fas-positive spermatozoa also may exist because of problems in activating Fas-mediated apoptosis. In this scenario, apoptosis is aborted and fails to clear spermatozoa that are earmarked for elimination by apoptosis [69]. In men with abnormal sperm parameters (oligozoospermia and azoospermia), the percentage of Fas-positive spermatozoa can be as high as 50%. Samples with low sperm concentrations are more likely to have a high proportion of Fas-positive spermatozoa [52].

Mitochondrial exposure to ROS results in the release of apoptosis-inducing factor (AIF), which directly interacts with the DNA and leads to DNA fragmentation. In another study by our group, a positive correlation was demonstrated between increased sperm damage by ROS and higher levels of cytochrome C and caspases 9 and 3, which indicate positive apoptosis in patients with male factor infertility [74]. Activation of caspases 8, 9, 1 and 3 in human ejaculated spermatozoa has been studied to examine the main pathways of apoptosis [75, 76]. Potential functional impact of this phenomenon and possible activation mechanisms were examined by subjecting cells to freezing and thawing, and testing the dependence of caspase activity on membrane integrity [77].

In an earlier study carried out by our group, annexin V staining assay was used to study the externalization of phosphatidylserine, a marker of early apoptosis. It was shown that mature spermatozoa from infertility patients had significantly higher levels of apoptosis compared with the mature spermatozoa from a control group of normal sperm donors [78].

The Effects of Antioxidant Supplementation in ICSI Results

To date, several studies have been published examining the effect of antioxidant treatments on sperm parameters and pregnancy outcome by means of sexual intercourse, but the use of different types and doses of antioxidants and lack of placebos or randomized and controlled studies, together with very small sample sizes, made unable to draw valid conclusions about the benefits of an antioxidant therapy.

Several studies have reported that levels of ROS within semen can be reduced by increasing the protective capacity of seminal plasma with oral antioxidants, such as Astaxanthin [79], carnitine [80] or a combination of antioxidants such as acetylcysteine, b-carotene, vitamin E and essential fatty acids [81]. A randomized control study comparing vitamin E [82, 83] and vitamin E+selenium [84] decreases sperm malonyldialdehyde levels, while the use of vitamin C and vitamin E reported a very significant reduction in sperm DNA damage in other reports [85, 86].

The ability of these changes to improve pregnancy chances is dubious. Suleiman et al. [82] reported that treatment with vitamin E resulted in a significant fall in ROS damage to sperm and an improvement in spontaneous pregnancy rates during the next 6 months (21% pregnancy rate in the vitamin E group vs. 0% in the placebo group), but these results have not been confirmed later.

The only work with an adequate design as randomized and controlled trial available comparing the ability of an antioxidant treatment to improve IVF/ICSI results was published by Tremellen, comparing Menevit with placebo and demonstrating a significant increase in clinical pregnancy rate if the antioxidant was taken for 3 months prior to IVF/ICSI treatment [87].

The main components of Menevit were Vitamins C and E, selenium and lycopene, garlic to reduce seminal leukocyte ROS production and also zinc, selenium and folate to increase protamine and DNA packaging.

Otherwise, there is very little information available about the benefits on having an antioxidant therapy in the male prior to IVF/ICSI treatments, with a design that limited the possibility to draw definitive conclusions [86].

Couples undergoing assisted reproduction treatments due to a prolonged infertility may be benefited from the use of antioxidants in order to improve OS situation in sperm. This hypothesis can be formulated from data supporting a deleterious effect of the ROS in sperm and ICSI treatments, although very few direct evidences are available so far and well-designed randomized controlled trials to demonstrate their effectiveness are needed.

Expert Commentary

Oxidative stress is undoubtedly one of the most prevalent problems contributing to male infertility when infertile males do not present non-severe spermatogenesis impairment. Several molecules and processes are involved, thus forming a complex system, where it is really difficult to measure the OS status of the man and even design the therapies to solve it.

Moreover, these men are investigated regarding infertility causes at the time of attempting assisted reproduction, after they consumed more than 1 year of unprotected intercourse seeking parenthood. This fact makes also relevant to characterize the link between assisted reproduction treatments and results depending on oxidative stress in semen.

Then, it is fundamental to consider the effects of OS in sperm cells as a real difficulty to succeed and consider the possibility of the supplementation of the culture media to treat the biological samples or administering the male antioxidative agents to avoid this damage and improve their chances. Nevertheless, RCTs are needed in order to confirm which antioxidants can effectively improve live birth rates, in both natural conception and after assisted reproduction treatments.

Five-Year View

Actually several scientific efforts are being conducted aiming to establish the way to accurately evaluate the OS situation in a sperm sample and its relevance on the modulation of sperm physiology together with clinical trials aiming to test the effectiveness of several molecules with antioxidant abilities to increase the fertility potential of men susceptible to be affected by OS.

In a recent future, there will be available laboratory tests to determine OS and identify infertile males affected, together with the most convenient antioxidant treatments in each case, which directly will improve conception by either natural or assisted reproduction.

Key Issues

- Oxidative stress is the biochemical situation where the presence of reactive oxygen species overcomes the antioxidant capacity, thus leading to abnormally high amounts of these species causing cellular damage.
- Male infertility has been related to oxidative stress situations in many research works, by means of different origins, processes and systems.
- There are several features of sperm caused by oxidative stress able to be microscopically identified in vivo and others that remain 'occult'.
- When attempting assisted reproduction by ICSI, these 'occult' effects of oxidative stress can be impairing the success chances, given that sperm cells are only selected by their morphology.
- The potential benefits of the antioxidant molecules in these cases are not fully accepted, although theoretically, they can have a positive effect.

- These molecules to scavenge free radicals can be employed either in the male or in the culture media in contact with reproductive cells.
- Further research is needed to develop the adequate diagnostic tools for oxidative stress determination and the design of efficient therapies.

Acknowledgments The authors want to acknowledge all the staff of IVI Valencia, in both the IVF and Andrology laboratory for their support in the research projects during the recent years, especially for the lab technicians.

References

1. McLachlan RI, de Kretser DM. Male infertility: the case for continued research. Med J Aust. 2001;174(3):116–7.
2. Braundmeier AG, Miller DJ. The search is on: finding accurate molecular markers of male fertility. J Dairy Sci. 2001;84(9):1915–25.
3. World Health Organization. WHO laboratory manual for the examination of human semen and sperm-cervical mucus interaction. 4th ed. Cambridge: Cambridge University Press; 1999.
4. Garrido N, Remohi J, Martinez-Conejero JA, Garcia-Herrero S, Pellicer A, Meseguer M. Contribution of sperm molecular features to embryo quality and assisted reproduction success. Reprod Biomed Online. 2008;17(6):855–65.
5. Garrido N, Meseguer M, Simon C, Pellicer A, Remohi J. Pro-oxidative and anti-oxidative imbalance in human semen and its relation with male fertility. Asian J Androl. 2004;6(1):59–65.
6. Saleh RA, Agarwal A. Oxidative stress and male infertility: from research bench to clinical practice. J Androl. 2002;23(6):737–52.
7. Sharma RK, Pasqualotto AE, Nelson DR, Thomas Jr AJ, Agarwal A. Relationship between seminal white blood cell counts and oxidative stress in men treated at an infertility clinic. J Androl. 2001;22(4):575–83.
8. Sharma RK, Pasqualotto FF, Nelson DR, Thomas Jr AJ, Agarwal A. The reactive oxygen species-total antioxidant capacity score is a new measure of oxidative stress to predict male infertility. Hum Reprod. 1999;14(11):2801–7.
9. Saleh RA, Agarwal A, Kandirali E, et al. Leukocytospermia is associated with increased reactive oxygen species production by human spermatozoa. Fertil Steril. 2002;78(6):1215–24.
10. Agarwal A, Gupta S, Sikka S. The role of free radicals and antioxidants in reproduction. Curr Opin Obstet Gynecol. 2006;18(3):325–32.
11. Padron OF, Brackett NL, Sharma RK, Lynne CM, Thomas Jr AJ, Agarwal A. Seminal reactive oxygen species and sperm motility and morphology in men with spinal cord injury. Fertil Steril. 1997;67(6):1115–20.
12. Sikka SC. Role of oxidative stress and antioxidants in andrology and assisted reproductive technology. J Androl. 2004;25(1):5–18.
13. Shekarriz M, DeWire DM, Thomas Jr AJ, Agarwal A. A method of human semen centrifugation to minimize the iatrogenic sperm injuries caused by reactive oxygen species. Eur Urol. 1995;28(1):31–5.
14. Shekarriz M, Thomas Jr AJ, Agarwal A. Incidence and level of seminal reactive oxygen species in normal men. Urology. 1995;45(1):103–7.
15. Griveau JF, Le Lannou D. Influence of oxygen tension on reactive oxygen species production and human sperm function. Int J Androl. 1997;20(4):195–200.
16. Whittington K, Ford WC. The effect of incubation periods under 95% oxygen on the stimulated acrosome reaction and motility of human spermatozoa. Mol Hum Reprod. 1998;4(11):1053–7.
17. Watson PF. The causes of reduced fertility with cryopreserved semen. Anim Reprod Sci. 2000;60–61:481–92.
18. Rossi T, Mazzilli F, Delfino M, Dondero F. Improved human sperm recovery using superoxide dismutase and catalase supplementation in semen cryopreservation procedure. Cell Tissue Bank. 2001;2(1):9–13.
19. Donnelly ET, McClure N, Lewis SE. The effect of ascorbate and alpha-tocopherol supplementation in vitro on DNA integrity and hydrogen peroxide-induced DNA damage in human spermatozoa. Mutagenesis. 1999;14(5):505–12.
20. Yenilmez E, Yildirmis S, Yulug E, et al. Ham's F-10 medium and Ham's F-10 medium plus vitamin E have protective effect against oxidative stress in human semen. Urology. 2006;67(2):384–7.
21. Zheng RL, Zhang H. Effects of ferulic acid on fertile and asthenozoospermic infertile human sperm motility, viability, lipid peroxidation, and cyclic nucleotides. Free Radic Biol Med. 1997;22(4):581–6.
22. Gomez E, Aitken J. Impact of in vitro fertilization culture media on peroxidative damage to human spermatozoa. Fertil Steril. 1996;65(4):880–2.
23. Donnelly ET, McClure N, Lewis SE. Glutathione and hypotaurine in vitro: effects on human sperm motility, DNA integrity and production of reactive oxygen species. Mutagenesis. 2000;15(1):61–8.
24. Twigg J, Fulton N, Gomez E, Irvine DS, Aitken RJ. Analysis of the impact of intracellular reactive oxygen species generation on the structural and functional integrity of human spermatozoa: lipid peroxidation, DNA fragmentation and effectiveness of antioxidants. Hum Reprod. 1998;13(6):1429–36.
25. Oeda T, Henkel R, Ohmori H, Schill WB. Scavenging effect of N-acetyl-L-cysteine against reactive oxygen species in human semen: a possible therapeutic modality for male factor infertility? Andrologia. 1997;29(3):125–31.
26. Martinez-Soto JC, de DiosHourcade J, Gutierrez-Adan A, Landeras JL, Gadea J. Effect of genistein supplementation of thawing medium on characteristics of frozen human spermatozoa. Asian J Androl. 2010;12(3):431–41.
27. Saleh RA, Agarwal A, Sharma RK, Nelson DR, Thomas Jr AJ. Effect of cigarette smoking on levels of seminal oxidative stress in infertile men: a prospective study. Fertil Steril. 2002;78(3):491–9.
28. Fraga CG, Motchnik PA, Wyrobek AJ, Rempel DM, Ames BN. Smoking and low antioxidant levels increase oxidative damage to sperm DNA. Mutat Res. 1996;351(2):199–203.
29. Mostafa T, Tawadrous G, Roaia MM, Amer MK, Kader RA, Aziz A. Effect of smoking on seminal plasma ascorbic acid in infertile and fertile males. Andrologia. 2006;38(6):221–4.
30. Viloria T, Meseguer M, Martinez-Conejero JA. et al. Fertil Steril: Cigarette smoking affects specific sperm oxidative defense but does not cause oxidative DNA damage in infertile men; 2009.
31. Viloria T, Garrido N, Fernandez JL, Remohi J, Pellicer A, Meseguer M. Sperm selection by swim-up in terms of deoxyribonucleic acid fragmentation as measured by the sperm chromatin dispersion test is altered in heavy smokers. Fertil Steril. 2007;88(2):523–5.
32. Eskenazi B, Kidd SA, Marks AR, Sloter E, Block G, Wyrobek AJ. Antioxidant intake is associated with semen quality in healthy men. Hum Reprod. 2005;20(4):1006–12.
33. Therond P, Auger J, Legrand A, Jouannet P. Alpha-tocopherol in human spermatozoa and seminal plasma: relationships with motility, antioxidant enzymes and leukocytes. Mol Hum Reprod. 1996;2(10):739–44.
34. Silver EW, Eskenazi B, Evenson DP, Block G, Young S, Wyrobek AJ. Effect of antioxidant intake on sperm chromatin stability in healthy nonsmoking men. J Androl. 2005;26(4):550–6.

35. Song GJ, Norkus EP, Lewis V. Relationship between seminal ascorbic acid and sperm DNA integrity in infertile men. Int J Androl. 2006;29(6):569–75.
36. Wu D, Cederbaum AI. Alcohol, oxidative stress, and free radical damage. Alcohol Res Health. 2003;27(4):277–84.
37. Koch OR, Pani G, Borrello S, et al. Oxidative stress and antioxidant defenses in ethanol-induced cell injury. Mol Aspects Med. 2004;25(1–2):191–8.
38. Maneesh M, Dutta S, Chakrabarti A, Vasudevan DM. Alcohol abuse-duration dependent decrease in plasma testosterone and antioxidants in males. Indian J Physiol Pharmacol. 2006;50(3):291–6.
39. Peake JM, Suzuki K, Coombes JS. The influence of antioxidant supplementation on markers of inflammation and the relationship to oxidative stress after exercise. J Nutr Biochem. 2007; 18(6):357–71.
40. Singer G, Granger DN. Inflammatory responses underlying the microvascular dysfunction associated with obesity and insulin resistance. Microcirculation. 2007;14(4–5):375–87.
41. Perez-Crespo M, Pintado B, Gutierrez-Adan A. Scrotal heat stress effects on sperm viability, sperm DNA integrity, and the offspring sex ratio in mice. Mol Reprod Dev. 2008;75(1):40–7.
42. Agarwal DK, Maronpot RR, Lamb IV JC, Kluwe WM. Adverse effects of butyl benzyl phthalate on the reproductive and hematopoietic systems of male rats. Toxicology. 1985;35(3):189–206.
43. Srivastava SP, Srivastava S, Saxena DK, Chandra SV, Seth PK. Testicular effects of di-n-butyl phthalate (DBP): biochemical and histopathological alterations. Arch Toxicol. 1990;64(2):148–52.
44. Lee E, Ahn MY, Kim HJ, et al. Effect of di(*n*-butyl) phthalate on testicular oxidative damage and antioxidant enzymes in hyperthyroid rats. Environ Toxicol. 2007;22(3):245–55.
45. Chitra KC, Sujatha R, Latchoumycandane C, Mathur PP. Effect of lindane on antioxidant enzymes in epididymis and epididymal sperm of adult rats. Asian J Androl. 2001;3(3):205–8.
46. Latchoumycandane C, Chitra KC, Mathur PP. 2,3,7,8-Tetrachlorodibenzo- *p*-dioxin (TCDD) induces oxidative stress in the epididymis and epididymal sperm of adult rats. Arch Toxicol. 2003;77(5):280–4.
47. Latchoumycandane C, Mathur PP. Induction of oxidative stress in the rat testis after short-term exposure to the organochlorine pesticide methoxychlor. Arch Toxicol. 2002;76(12):692–8.
48. Meng Z, Bai W. Oxidation damage of sulfur dioxide on testicles of mice. Environ Res. 2004;96(3):298–304.
49. Gonzalez-Flecha B. Oxidant mechanisms in response to ambient air particles. Mol Aspects Med. 2004;25(1–2):169–82.
50. Hsu PC, Guo YL. Antioxidant nutrients and lead toxicity. Toxicology. 2002;180(1):33–44.
51. Acharya UR, Acharya S, Mishra M. Lead acetate induced cytotoxicity in male germinal cells of Swiss mice. Ind Health. 2003;41(3):291–4.
52. Xu DX, Shen HM, Zhu QX, et al. The associations among semen quality, oxidative DNA damage in human spermatozoa and concentrations of cadmium, lead and selenium in seminal plasma. Mutat Res. 2003;534(1–2):155–63.
53. Garrido N, Meseguer M, Simon C, Pellicer A, Remohi J. Pro-oxidative and anti-oxidative imbalance in human semen and its relation with male fertility. Asian J Androl. 2004;6(1):59–65.
54. Kobayashi T, Miyazaki T, Natori M, Nozawa S. Protective role of superoxide dismutase in human sperm motility: superoxide dismutase activity and lipid peroxide in human seminal plasma and spermatozoa. Hum Reprod. 1991;6(7):987–91.
55. Hsieh YY, Sun YL, Chang CC, Lee YS, Tsai HD, Lin CS. Superoxide dismutase activities of spermatozoa and seminal plasma are not correlated with male infertility. J Clin Lab Anal. 2002;16(3):127–31.
56. Miesel R, Jedrzejczak P, Sanocka D, Kurpisz MK. Severe antioxidase deficiency in human semen samples with pathological spermiogram parameters. Andrologia. 1997;29(2):77–83.
57. Jeulin C, Soufir JC, Weber P, Laval-Martin D, Calvayrac R. Catalase activity in human spermatozoa and seminal plasma. Gamete Res. 1989;24(2):185–96.
58. Meseguer M, Martinez-Conejero JA, Muriel L, Pellicer A, Remohi J, Garrido N. The human sperm glutathione system: a key role in male fertility and successful cryopreservation. Drug Metab Lett. 2007;1(2):121–6.
59. Meseguer M, Garrido N, Simon C, Pellicer A, Remohi J. Concentration of glutathione and expression of glutathione peroxidases 1 and 4 in fresh sperm provide a forecast of the outcome of cryopreservation of human spermatozoa. J Androl. 2004;25(5):773–80.
60. Garrido N, Meseguer M, Alvarez J, Simon C, Pellicer A, Remohi J. Relationship among standard semen parameters, glutathione peroxidase/glutathione reductase activity, and mRNA expression and reduced glutathione content in ejaculated spermatozoa from fertile and infertile men. Fertil Steril. 2004;82 Suppl 3:1059–66.
61. Alvarez JG, Storey BT. Taurine, hypotaurine, epinephrine and albumin inhibit lipid peroxidation in rabbit spermatozoa and protect against loss of motility. Biol Reprod. 1983;29(3):548–55.
62. Holmes RP, Goodman HO, Shihabi ZK, Jarow JP. The taurine and hypotaurine content of human semen. J Androl. 1992;13(3):289–92.
63. Meseguer M, Martinez-Conejero JA, O'Connor JE, Pellicer A, Remohi J, Garrido N. The significance of sperm DNA oxidation in embryo development and reproductive outcome in an oocyte donation program: a new model to study a male infertility prognostic factor. Fertil Steril. 2008;89(5):1191–9.
64. Aguilar C, Meseguer M, Garcia-Herrero S, Gil-Salom M, O'Connor JE, Garrido N. Relevance of testicular sperm DNA oxidation for the outcome of ovum donation cycles. Fertil Steril. 2010;94:979–88.
65. Muriel L, Goyanes V, Segrelles E, Gosalvez J, Alvarez JG, Fernandez JL. Increased aneuploidy rate in sperm with fragmented DNA as determined by the sperm chromatin dispersion (SCD) test and fish analysis. J Androl. 2007;28:38–49.
66. Muriel L, Garrido N, Fernandez JL, et al. Value of the sperm deoxyribonucleic acid fragmentation level, as measured by the sperm chromatin dispersion test, in the outcome of in vitro fertilization and intracytoplasmic sperm injection. Fertil Steril. 2006;85(2):371–83.
67. Meseguer M, Santiso R, Garrido N, Gil-Salom M, Remohi J, Fernandez JL. Sperm DNA fragmentation levels in testicular sperm samples from azoospermic males as assessed by the sperm chromatin dispersion (SCD) test. Fertil Steril. 2009;92(5):1638–45.
68. Meseguer M, Santiso R, Garrido N, Fernandez JL. The effect of cancer on sperm DNA fragmentation as measured by the sperm chromatin dispersion test. Fertil Steril. 2008;90(1):225–7.
69. Sakkas D, Mariethoz E, Manicardi G, Bizzaro D, Bianchi PG, Bianchi U. Origin of DNA damage in ejaculated human spermatozoa. Rev Reprod. 1999;4(1):31–7.
70. Sakkas D, Mariethoz E, St John JC. Abnormal sperm parameters in humans are indicative of an abortive apoptotic mechanism linked to the Fas-mediated pathway. Exp Cell Res. 1999;251(2):350–5.
71. Sinha Hikim AP, Swerdloff RS. Hormonal and genetic control of germ cell apoptosis in the testis. Rev Reprod. 1999;4(1):38–47.
72. Shen HM, Dai J, Chia SE, Lim A, Ong CN. Detection of apoptotic alterations in sperm in subfertile patients and their correlations with sperm quality. Hum Reprod. 2002;17(5):1266–73.
73. Krammer PH, Dhein J, Walczak H, et al. The role of APO-1-mediated apoptosis in the immune system. Immunol Rev. 1994;142:175–91.
74. Cande C, Cecconi F, Dessen P, Kroemer G. Apoptosis-inducing factor (AIF): key to the conserved caspase-independent pathways of cell death? J Cell Sci. 2002;115(Pt 24):4727–34.
75. Paasch U, Sharma RK, Gupta AK, et al. Cryopreservation and thawing is associated with varying extent of activation of apoptotic machinery in subsets of ejaculated human spermatozoa. Biol Reprod. 2004;71(6):1828–37.

76. Paasch U, Agarwal A, Gupta AK, et al. Apoptosis signal transduction and the maturity status of human spermatozoa. Ann N Y Acad Sci. 2003;1010:486–8.
77. Paasch U, Grunewald S, Agarwal A, Glandera HJ. Activation pattern of caspases in human spermatozoa. Fertil Steril. 2004;81 Suppl 1:802–9.
78. Agarwal A, Said TM. Role of sperm chromatin abnormalities and DNA damage in male infertility. Hum Reprod Update. 2003;9(4):331–45.
79. Comhaire FH, El Garem Y, Mahmoud A, Eertmans F, Schoonjans F. Combined conventional/antioxidant "Astaxanthin" treatment for male infertility: a double blind, randomized trial. Asian J Androl. 2005;7(3):257–62.
80. Vicari E, La Vignera S, Calogero AE. Antioxidant treatment with carnitines is effective in infertile patients with prostatovesiculoepididymitis and elevated seminal leukocyte concentrations after treatment with nonsteroidal anti-inflammatory compounds. Fertil Steril. 2002;78(6):1203–8.
81. Comhaire FH, Christophe AB, Zalata AA, Dhooge WS, Mahmoud AM, Depuydt CE. The effects of combined conventional treatment, oral antioxidants and essential fatty acids on sperm biology in subfertile men. Prostaglandins Leukot Essent Fatty Acids. 2000;63(3): 159–65.
82. Suleiman SA, Ali ME, Zaki ZM, el-Malik EM, Nasr MA. Lipid peroxidation and human sperm motility: protective role of vitamin E. J Androl. 1996;17(5):530–7.
83. Kessopoulou E, Powers HJ, Sharma KK, et al. A double-blind randomized placebo cross-over controlled trial using the antioxidant vitamin E to treat reactive oxygen species associated male infertility. Fertil Steril. 1995;64(4):825–31.
84. Keskes-Ammar L, Feki-Chakroun N, Rebai T, et al. Sperm oxidative stress and the effect of an oral vitamin E and selenium supplement on semen quality in infertile men. Arch Androl. 2003;49(2): 83–94.
85. Greco E, Iacobelli M, Rienzi L, Ubaldi F, Ferrero S, Tesarik J. Reduction of the incidence of sperm DNA fragmentation by oral antioxidant treatment. J Androl. 2005;26(3):349–53.
86. Greco E, Romano S, Iacobelli M, et al. ICSI in cases of sperm DNA damage: beneficial effect of oral antioxidant treatment. Hum Reprod. 2005;20(9):2590–4.
87. Tremellen K, Miari G, Froiland D, Thompson J. A randomised control trial examining the effect of an antioxidant (Menevit) on pregnancy outcome during IVF-ICSI treatment. Aust N Z J Obstet Gynaecol. 2007;47(3):216–21.

Antioxidants in IMSI

43

Monica Antinori

Abstract

Sperm morphology as assessed according to strict criteria demonstrated to have an influence on rates of fertilization, pregnancy, implantation, embryo quality and blastocyst formation.

ICSI visual assessment of sperm morphology, limited by its low magnification (200x-400x) and concomitant low resolution, overlooks minor morphologic defects potentially related to sperm functional impairment.

With the development of a new method for a real-time ,detailed morphological evaluation of motile spermatozoa under high magnification (6600x), called MSOME, the sperm's nucleus turned out to be the most important parameter influencing ICSI outcome particularly in the form of large nuclear vacuoles that were proposed to reflect damages in the nuclear DNA content and organization.

Keywords

Male infertility • Sperm morphology • DNA chain fragmentation • Intracytoplasmic morphologically selected sperm injection • IMSI • Motile sperm organellar morphology examination • Reactive oxygen species • Oxidative stress

Conventional semen parameters are used routinely by clinicians to obtain a reliable overview of the male reproductive potential. However, in most cases, little information on the fertility status is provided unless semen parameters are evidently abnormal [1, 2]. Among them the best fertilization predictor is morphology as assessed according to strict criteria [3]. Since its introduction [4], several advantages to the outcomes of conventional in vitro fertilization [3, 5], intrauterine insemination [6], and in vivo reproduction have been shown [7]. Therefore, it is clear how a spermatozoon's morphological normality reflects its function, in relation to its ability to reach, recognize, bind to, penetrate, and deliver its genome to the oocyte. Differently, once the sperm is mechanically injected into the oocyte by an intracytoplasmic sperm injection (ICSI) method, and crossed both barriers of the zona pellucida and oolemma, its abnormal morphology does not seem to interfere with its fertilizing capacity. In this respect, most authors do not notice any correlation between ICSI outcomes and the strict morphology of the sperm used for microinjection [8–11]. In patients with a poor prognosis (4% normal sperm morphology) [12], and even in extreme, specific cases of total teratozoospermia [13], globozoospermia [14, 15], and megalozoospermia [16] fertilization can be achieved with ICSI.

Nonetheless major morphological anomalies lead to decreased rates of fertilization, pregnancy, implantation [16, 17], embryo quality [18–20], and blastocyst formation [17, 21]. Hence, during an ICSI procedure, the embryologist selects a motile, normal-looking spermatozoon and discards the most distorted forms. This visual assessment of sperm morphology, limited by its low magnification (200–400×) and concomitant low resolution [17], overlooks minor morphologic defects potentially related to sperm functional impairment. At the same time, since men with oligoasthenozooteratozoospermia were shown to have significantly elevated levels of sperm numerical chromosomal aberrations,

M. Antinori, MD (✉)
Department of Infertility, RAPRUI Day Surgery, Via Timavo 2, 00193 Rome, Italy
e-mail: monica.antinori@raprui.com

S.J. Parekattil and A. Agarwal (eds.), *Male Infertility: Contemporary Clinical Approaches, Andrology, ART & Antioxidants*, DOI 10.1007/978-1-4614-3335-4_43, © Springer Science+Business Media, LLC 2012

and DNA chain fragmentation [22–28], and the incidence of chromosome aneuploidy in spermatozoa relates with the severity of sperm defects [29], there is great concern regarding the increased risk of chromosomal abnormalities in infants conceived with ICSI [30–32].

Moreover, although ICSI has provided treatment for new groups of couples with male infertility who were previously untreatable by IVF, at present the resulting pregnancy rates are only between 30 and 45% [33–35], while the European average "take-home baby" rates remain similar (36) to those of a decade ago [37, 38]. The rising demand for the improvement of success rates compels us to reassess male fertility potential through the development of new tests with clinical relevance for each ART procedure.

Conventional light microscopy cannot identify the entire variety of sperm morphological defects, especially in the head structure [39–41]. In 1999, electron microscopy [scanning electron microscopy (SEM) and transmission electron microscopy (TEM)] allowed Bartoov [42] to correctly identify the ultramorphological state of seven sperm subcellular organelles (acrosome, postacrosomal lamina, nucleus, neck, axoneme, mitochondrial sheath, and outer dense fibers) that were found to be highly predictive factors for male fertility potential. Due to the fact that the applied technology was expensive and often not available in conventional laboratories, its application was limited only to those cases in which the male infertility factor could not be clearly identified by routine tests or following repeated ART failures. As for conventional sperm morphology, this new evaluation turned out to be useful only in the context of a reproductive prognosis assessment, considering that the single sperm used for fertilization might not reflect the peculiarity of the analyzed sample. Moreover, being that the evaluation was based on the examination of fixed and stained sperm cells, it did not provide the patient with any information about the single sperm used for ICSI. To solve all these problems at the same time, a few years later the same group developed a new method for a real-time, detailed morphological evaluation of motile spermatozoa. The sperm analysis called motile sperm organellar morphology examination (MSOME) was performed using an invertoscope equipped with interferential contrast Normaski optics that combines maximal optical magnification(100×), magnification selector (×1.5) and a video-coupled magnification to reach a final video magnification of 6,600×. This analysis, developed to allow the visualization of subtle sperm morphological malformations which might remain unnoticed to the embryologist during the routine sperm selection performed prior to microinjection, has been introduced to improve the success of ICSI. Of the six sperm subcellular organelles examined (Fig. 43.1) (acrosome, postacrosomal lamina, neck, mitochondria, tail, and nucleus), the sperm's nucleus turned out to be the most important parameter influencing ICSI outcome [43] particularly in the form of large nuclear vacuoles (LNV) that were proposed to reflect damages in the nuclear DNA content and organization [44].

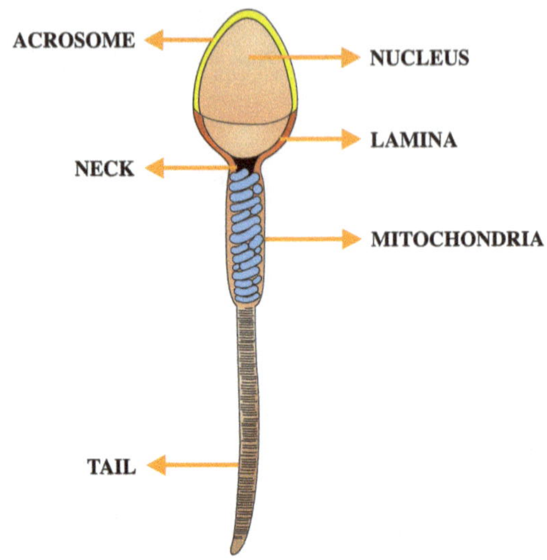

Fig. 43.1 The six subcellular organelles analyzed by MSOME

Sperm DNA integrity is a prerequisite to normal fertilization and transmission of paternal genetic information [45]. Higher levels of nuclear DNA strand breaks are detected in the ejaculated spermatozoa of men with abnormal semen parameters [46, 47] due to a great sensitivity to DNA damage caused by oxidative stress and increased production of reactive oxygen species (ROS) [48].

ROS are active oxidizing agents that render the chromatin vulnerable to external oxidative attacks [49]. An increase in oxidative damage to the sperm membrane, intracellular proteins and DNA, is associated with alterations in signal transduction mechanisms that can affect fertility [50].

Several studies have examined the relationship of sperm nuclear DNA fragmentation with cleavage, embryo quality, and pregnancy rates for both natural [51, 52] and ART conceptions [53–60]. Overall, these studies have suggested that a high content of fragmented DNA may have a negative influence on embryo development and pregnancy rates [61]. Sperm DNA damage further increases the incidence of pregnancy loss after IVF and ICSI [51, 62–64].

For the above reasons, DNA fragmentation can be considered closely related to male-derived repeated ICSI failures, the only marker of such failures that can be objectively and reproducibly diagnosed nowadays [60, 65].

According to the current literature, the incidence of DNA fragmentation in ejaculated spermatozoa can be reduced by oral antioxidant treatment [66–71] and by antioxidant media supplementation [72–74]. The hypothesis that vacuolization of the sperm nucleus may reflect some underlying DNA defects [75] which could impair male fertility potential, such as DNA strand breaks, is a promising perspective considering that, at present, we are not able to offer our patients a selection technique, whereby spermatozoa used for fertilization are preventively tested for DNA integrity. If confirmed, this speculation would enable the verification of potential morphological effects of the antioxidants exogenous administration, validating the further implementation of this therapeutic approach.

Presently, only a few studies have been carried out to investigate the real meaning of nuclear vacuolization and its relationship to DNA status. The opportunity to carry on additional research justifies the application of this new technique which requires specially trained personnel and can be considered extremely time-consuming and expensive, to be routinely inserted in the ART laboratory.

Intracytoplasmic Morphologically Selected Sperm Injection Procedure

According to the current literature, a sterile glass-bottomed dish suitable for MSOME evaluation can be prepared as follows (Fig. 43.2).

- On the left side, 4 μl *observation droplets* made up of sperm culture medium containing between 0 and 10% polyvinyl pyrrolidone (PVP) solution. Small bays extruding from the rim of the droplets are created to capture the heads of motile spermatozoa .The temperature of the sperm sample and the PVP concentration are coordinated with the intensity of the sperm motility.
- In the middle, 4 μl *selection droplets* of sperm culture, where selected sperm cells are located after MSOME evaluation. Three distinct drops are made to host spermatozoa with different morphological features.
- On the left side, 4 μl droplets of sperm culture medium that will host the oocytes to be injected in the following ICSI procedure (*injection droplets*), one for each oocyte available for microinjection.
- All microdroplets are placed under sterile liquid paraffin.

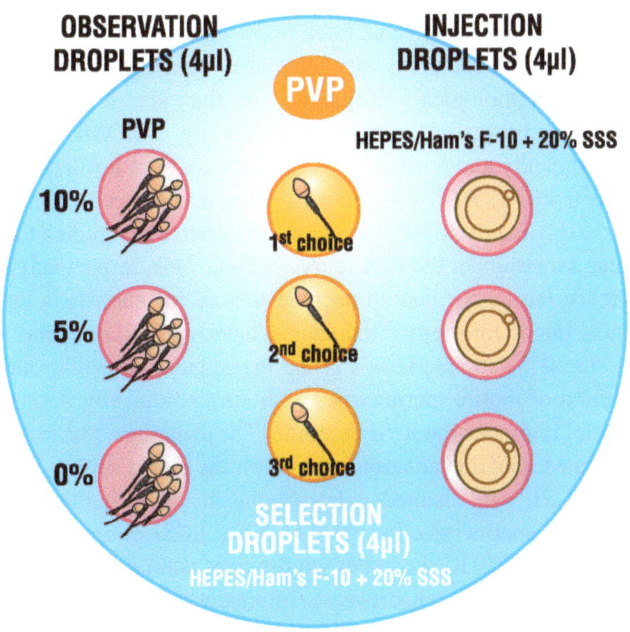

Fig. 43.2 The IMSI Petri dish

Motile Sperm Organellar Morphology Examination Criteria and Evaluation Procedure

Based on data collected by electron microscopy [42] that supplied both external and internal information, the MSOME criteria for normally shaped nuclei were defined as size (average length and width to be 4.75 ± 0.28 and 3.28 ± 0.20 μm, respectively), smoothness, symmetry, oval configuration (an extrusion or invagination of the nuclear mass was defined as a regional nuclear shape malformation), and homogeneity of the nuclear chromatin mass containing no more than one vacuole, which occupies less than 4% of the nuclear area (0.78 ± 0.18 μm). Spermatozoa with abnormal head size are excluded by superimposing a transparent celluloid form on the motile examined gametes (Fig. 43.3), representing the correct sperm size, which is calculated by the ratio of expected normal sperm size to the actual size visualized on the monitor screen. Spermatozoa with severe malformations, such as a pin, amorphous, tapered, round, or multinucleated head, which can be identified clearly even by low magnification (200–400×), are not assessed by MSOME. Spermatozoa with a doubtful determination are excluded from selection. To perform a correct sperm evaluation, the embryologist follows each motile single sperm cell with apparent suitability by moving the microscopic stage in the *x*, *y*, and *z* directions until also the smallest details are visualized.

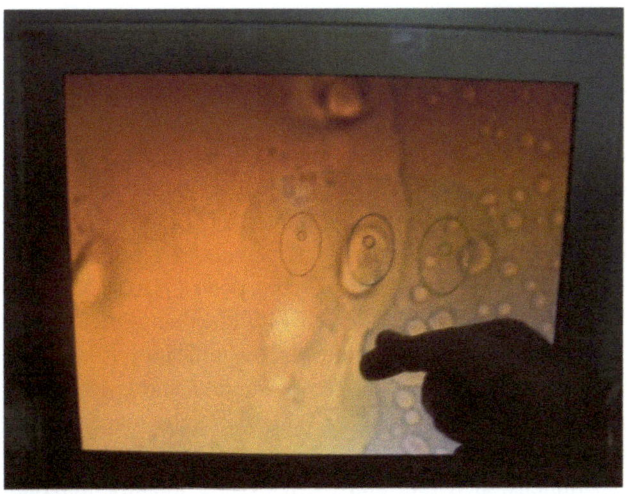

Fig. 43.3 The "chablon" superimposed on the motile sperm to verify its correct size

Actually some morphological defects, such as large vacuoles, can be revealed only during sperm movement, and therefore motility can be advantageous to the morphological observation. Furthermore, it is relevant to emphasize how a single sperm evaluation is reliable only when it is carried out on a motile sperm cell; on the other hand, static sperm images only allow evaluation of the visible part, leaving some morphological alterations undiscovered.

The sperm selection procedure does not require any computer application since the automated sperm morphology analysis systems available on the market allow only morphology evaluation of immotile sperms on stained slides. On the other hand, two embryologists working together at the same time on the analysis of the same sample, are recommended to minimize the subjective nature of sperm evaluation. Finding normal-looking spermatozoa is variable according to the quality of the semen sample.

Intracytoplasmic Morphologically Selected Sperm Injection Step by Step

Freshly ejaculated semen is subjected to routine morphological selection of motile spermatozoa on the basis of a two-layer density gradient system: 1 mL of postejaculated liquefied semen is placed onto the gradient and centrifuged at $375 \times g$ for 15 min at 25°C. The sperm cell pellet is suspended by adding 3 mL of sperm culture medium and then recentrifuged for 10 min. The supernatant is removed and replaced by sperm culture medium to bring the final concentration of motile sperm cells to about 4×10^6 spermatozoa per milliliter. In severe oligozoospermic cases with sperm density below 1×10^6 spermatozoa per ejaculate, liquefied semen is placed onto 1 mL of the low density layer only, centrifuged as previously mentioned, and the final sperm cell pellet is suspended in 0.1–0.2 mL of sperm culture medium.

The sperm cell suspension obtained after semen preparation is used for real-time high-magnification MSOME [76] that is performed on the observation droplets by means of an inverted microscope (e.g., Olympus IX81, Tokyo, Japan) equipped with Nomarski differential interference contrast optics, an Uplan Apo X 100 oil/1.50 objective lens previously covered by a droplet of immersion oil, and a 0.55 NA condenser lens. The images are captured by a DXC-990P video camera having ½-in., 3-chip power HAD CCD and visualized on a monitor screen with diagonal dimension of 355.6 mm. Calculation of the total magnification is based on four parameters: (1) objective magnification 100×, (2) magnification selector 1.5×, (3) video coupler magnification 0.99 (UPMTV X 0.3, PE X 3.3), and (4) (*a*) CCD chip diagonal dimension 8 mm and (*b*) television monitor diagonal dimension for a calculated video magnification (*b*/*a*) of 44.45. Thus, total magnification = microscope magnification (150×) × video coupler magnification (0.99×) × video magnification (44.45×) = 6,600×.

Only motile spermatozoa with morphologically normal nuclei are retrieved from the observation droplets and aspirated into a sterilized glass pipette with a 9-μm inner diameter tip. Sperm cells are then placed into the selection droplet and finally used to be injected into the oocytes for the classical ICSI procedure [77]. This procedure is performed using a motorized micropanipulator system (e.g., TransferMan NK2, Eppendorf Germany).

Intracytoplasmic Morphologically Selected Sperm Injection in Assisted Reproductive Technology

Assuming that spermatozoa with severely impaired morphology show reduced fertilization, pregnancy, and implantation rates [16, 17], having identified a variety of morphological anomalies that conventional light microscopy cannot detect at 200–400×, Bartoov [43] developed a new technique for real-time motile sperm morphological evaluation (MSOME) in 2002. The study aimed to determine whether subtle morphological anomalies affected ICSI outcome and identified those that are relevant. Based on the analysis of a total of 10,000 spermatozoa (100 sperm samples and 100 spermatozoa each), it was demonstrated that in routine IVF–ICSI cycles patients who exhibited less than 20% spermatozoa with a morphologically normal nucleus, defined by MSOME, did not achieve any pregnancy. With respect to the ICSI fertilization rate, the morphological normalcy of the entire sperm cell, according to MSOME criteria, showed a positive and significant correlation ($r=0.52$, $P \le 0.01$) and a very high predictive value (area under the ROC curve, 88%), whereas no association with pregnancy outcome was found. The morphological normalcy of the sperm nucleus (shape + chromatin content), defined by MSOME, was significantly and positively correlated with both fertilization rate ($r=0.42$, $P \le 0.01$) and pregnancy occurrence ($r=0.38$, $P \le 0.01$). Even its predictive value turned out to be significantly high (areas under the ROC curve, 72 and 74%, respectively). Hence the author could conclude that sperm nucleus is the most important sperm parameter influencing ICSI outcome.

In 2003 [76], Bartoov's group investigated how microinjection of motile spermatozoa with morphologically normal nuclei improves the ICSI outcome. The technique, that combines MSOME evaluation applied on the single sperm used for ICSI, was named intracytoplasmic morphologically selected sperm injection (IMSI). Fifty IMSI couples were compared to 50 ICSI couples wit the same number of previous ICSI failures (matched cases). Implantation and pregnancy rates after IMSI were significantly higher, and the abortion rate was significantly lower, compared to the current ICSI trial ($F=18.0$, $P\leq 0.01$; $\chi^2=4.4$, $P\leq 0.01$; and $\chi^2=4.4$, $P\leq 0.05$). In addition, the IMSI attempt produced a significantly higher value of top embryo percentage, compared to the current ICSI treatment ($F=6.5$, $P\leq 0.01$). Even 12 unmatched IMSI cases with over eight previous failed routine ICSI attempts (9.1 ± 1.2 ICSI cycles in average) achieved a 50% pregnancy rate after one IMSI trial. The obtained results demonstrated that IMSI improves significantly the success rate in couples with previous ICSI failures.

To exclude that the increased pregnancy outcome was linked to the sperm preparation technique adapted for IMSI and not to the nuclear morphology of the selected spermatozoa special, in 2005 Berkovitz [44] published a comparative study between IVF and IMSI cycles involving 38 transfers of embryos derived from "second best" morphologically evaluated sperm cells (negative group) versus 38 derived from morphologically normal nuclei (positive group). Comparison between the groups revealed that fertilization rate, percentage of top embryos, and implantation rates were significantly higher in the positive group than in the negative group.

Out of six pregnancies achieved in the negative group, four turned into a first trimester missed abortion. Interestingly, three missed abortion cases occurred when microinjection was conducted with spermatozoa exhibiting LNV, whereas in the other abortion case the sperm cells exhibited combined malformations: large vacuoles associated with narrow formed head shape.

Thus, implantation and pregnancy achieved by ICSI seem associated with morphological nuclear normalcy of the sperm. Nonetheless spermatozoa with a morphologically abnormal nucleus show low fertility potential, even if some with certain nuclear abnormalities may still be able to produce pregnancy following ICSI.

This result confirmed previous reports [78, 79] that had already shown a clear negative association between the existence of sperm nuclear vacuoles and natural male fertility potential, and led the author to assume that within the category of specific morphological malformations the existence of large vacuoles in the sperm nuclei indicates more damage to the nuclear DNA content and organization than nuclear shape or size impairment.

The impact of sperm with normal nuclear shape but LNV on pregnancy outcome compared to those with strictly defined morphologically normal nuclei, including shape and content, was investigated in 2006 by Berkovitz [75]. The pregnancy rate per cycle in the experimental group was significantly lower, and the early spontaneous abortion rate per pregnancy significantly higher, than those of the control group (18 versus 50%, Pearson's Chi-square = 6.4 and 80 versus 7%, Pearson's Chi-square = 10.9, respectively, $P=0.01$).

MSOME revealed that the ejaculates of males routinely referred for ICSI exhibit on average 30–40% spermatozoa with a vacuolated nucleus. This sperm malformation, identifiable as a pregnancy risk factor on the basis of the previous findings, can easily be missed by the standard selection prior to ICSI, and have, therefore, a chance to be chosen for microinjection of at least 30%.

A subsequent enlarged study by the same group [80] confirmed all the previous findings as follows: pregnancy rate was significantly higher and the abortion rate significantly lower following IMSI compared with ICSI attempts ($\chi^2=20.1$, $P\leq 0.01$ and $\chi^2=5.1$, $P\leq 0.03$, respectively). Furthermore, IMSI provided a higher percentage of top quality embryos with a better implantation rate than ICSI.

A comparison between a "best group," in which embryos were obtained from microinjection exclusively performed using spermatozoa with intact nuclei, and a "second best group," in which only sperm cells with minimal impairment were used for microinjection, since no "best" sperm cells were available in those cases, demonstrated that fertilization rate, percentage of top embryos, implantation, pregnancy, and delivery rates per cycle were significantly higher, and the abortion rate was significantly lower in the "best" group than in the "second best" one ($F=10.5$, $P\leq 0.01$; $F=4.6$, $P\leq 0.03$; $F=23.4$, $P\leq 0.01$; $\chi^2=15.5$, $P\leq 0.05$; $\chi^2=19.6$, $P\leq 0.01$; and $\chi^2=5.5$, $P\leq 0.02$, respectively).

The authors restricted the application of this new procedure to cases with over two previous implantation failures. Actually, according to recent publications, those couples seem to have the worst reproductive prognosis with a dramatic reduction in pregnancy and implantation rates as against couples with no or one previous failed IVF attempts [81, 82]. Hence Antinori [83] designed a prospective randomized controlled protocol to assess the potential advantages of the IMSI procedure in the treatment of patients with severe oligoasthenoteratozoospermia regardless of their previous failed ICSI attempts, followed by a subgroup splitting according to the number of previous failed attempts (Subgroup *A*: no previous attempts; Subgroup *B*: 1 previous failed attempt; Subgroup *C*: ≥2 previous failed attempts). The comparisons between the two different techniques were made in terms of pregnancy, abortion, and implantation rates.

Pregnancy and implantation rates resulted statistically better in IMSI than ICSI cycles (PR: 39.2 versus 26.5%; $P=0.004$) (IR: 17.3 versus 11.3%; $P=0.007$).

However, cases with two or more failed attempts benefited most from IMSI, with a statistically significant doubling of pregnancy rate (12.9 versus 29.8%; $P=0.017$) and a remarkable 50% reduction in the abortion rate (17 versus 35%). Comparisons did not show any statistical difference in terms of abortions but the clinical trend was clearly in favor of the IMSI method in cases with two or more previous failed attempts. Based on the above results, it is likely that in those couples the male factor could be featured by semen impairment, undetected by conventional diagnostic tools, thus reducing the effectiveness of previous ICSI treatments.

Intracytoplasmic Morphologically Selected Sperm Injection and DNA Fragmentation

When strict criteria were originally introduced [84], it was not already possible to correlate poor sperm morphology with fragmented DNA [85, 86], whereas later reports found that abnormal spermatozoa negatively correlates with DNA integrity [72], especially in the case of head malformations [87–89]. The potential relationship between sperm shape and genetic integrity has become very relevant with the introduction of ICSI, which gives chances of fertilization even to those male gametes affected by severe malformations [13, 90, 91] and might overlook some subtle defects because of the low magnification and low resolution of its sperm morphology assessment. In both cases, there is a great concern about the fact that morphologically abnormal spermatozoa, which were shown to have significantly elevated levels of sperm numerical chromosomal aberrations, ROS production, and DNA chain fragmentation [46, 47] would participate to impair fertilization, embryogenesis, or fetal development and nonetheless to the birth of infants with higher prevalence of chromosomal abnormalities and birth defects compared to natural conception [92].

The introduction of a new visual method applied on conventional ICSI technique, called IMSI, that clearly identifies sperm characteristics, undetected by conventional microscopy, allows investigation into their relationship with ICSI outcome that provide reliable evidence that a morphologically normal sperm nucleus is the most important sperm parameter showing high correlation with both fertilization and pregnancy occurrence [43]. Based on preliminary unpublished data by Bartoov' s group, which reported a significant negative correlation between the size of the nuclear vacuoles and chromatin stability assessed by the sperm chromatin structure assay (SCSA) and according to Lee [93], who demonstrated that no increase in chromosome aberrations was found in spermatozoa with large or small heads, it has been proposed that the existence of large sperm vacuoles in the sperm nuclei indicates more damage to nuclear DNA content and organization than nuclear shape or size impairment [44]. Furthermore, LNV injection resulted in a normal early embryonic development (normal fertilization, development of top-quality embryos, and implantation) followed by an impaired embryo survival (low pregnancy and high abortion rates) [75].

To verify that the LNV in the sperm cell reflect some underlying chromosomal or DNA defect, sperm cells with and without large vacuoles were recommended be selected from the same ejaculate and examined by different biochemical methods from an external analytic system [75]. Therefore, a recent paper from Hazout [94] compared the outcomes of 125 couples with at least two previous ICSI failures and an undetected female infertility factor that underwent conventional and high-magnification ICSI in two sequential attempts. Following sperm injection into the oocytes without nuclear alterations a double pregnancy rate and a 50% decrease in the abortion rate were recorded as against similar cases treated by conventional ICSI. In 72 out of 125 patients involved in the study, the degree of sperm DNA fragmentation was determined by TUNEL and the outcomes of high-magnification ICSI were compared in cases with different sperm DNA fragmentation degrees. However, this test was not performed directly with the sperm samples used for ICSI. A marked rise in clinical implantation and birth rates was observed in patients with normal (<30%), moderately (30–40%), and highly (>40%) increased percentage of DNA-fragmented spermatozoa in the ejaculate.

To stress the above assumptions, Franco [95] evaluated the extent of DNA fragmentation (TUNEL assay) and the presence of denatured single-stranded or normal double-stranded DNA (acridine orange fluorescence method; AOT) in spermatozoa with LNV selected by high magnification compared with those with normal nucleus. The percentage of positive DNA fragmentation was significantly higher ($P<0.0001$) in LNV spermatozoa (29.1%) than in NN spermatozoa (15.9%). Similarly, the percentage of denatured-stranded DNA was significantly higher ($P<0.0001$) in the former (67.9%) than in the latter (33.1%).

So far a direct correlation in DNA quality has not been tested in single-selected spermatozoa. With that in mind as their major aim, Garolla [96] analyzed the chromatin structure (sperm DNA integrity by acridine orange; DNA fragmentation by TUNEL assay) and sperm aneuploidies (FISH test) in ten patients affected by severe testicular damage (severe oligozoospermia) on single immotile sperm cells morphologically selected by high-magnification microscopy (13×,161). From the sample of each patient, ten spermatozoa with normal morphology and no vacuoles (group A) and ten spermatozoa with normal morphology and at least one large head vacuole (group B) were selected. Single cells from group A showed a more physiological status of DNA integrity and DNA fragmentation than cells from group B. Furthermore, FISH analysis showed that no chromosomal alteration was present in cells from group A. Moreover, the authors reported that considering together spermatozoa with normal morphology and both presence and absence of large head vacuoles the mean results (data not shown) from all tests were significantly better with respect to those of unselected cells performed in the first part of the study (all $P<0.001$).

These results seem to suggest a strong relationship between high-magnification morphology and the DNA status of spermatozoa and the chromatinic origin of nuclear vacuoles visualized by MSOME. Based on technical limits of differential interference contrast (DIC),which does not allow intracellular evaluation [97–99], since to detect chromatin vacuoles the evaluation has to be performed at 20,000× magnification by electron microscopy, and because of their main localization, in the anterior part of the sperm head, the acrosomal origin of these vacuoles was theorized [100]. The first experiment of this study consisted of MSOME evaluation on immotile spermatozoa followed by acrosomal status assessment of the same spermatozoon using pisum sativum agglutinin (PSA)–fluorescein isothiocyanate (FITC). The complete acrosome reaction corresponded in most of the cases (70.9%) to spermatozoa of regular shape and absent or slight vacuolization, whereas those sperms with incomplete or missing acrosome reaction showed 60.7% of vacuole presence. The second experiment involved ten patients whose immotile sperms were analyzed according to MSOME before and after the acrosome had been induced by ionophore A23587. Vacuole-free spermatozoa increased from 41.2 to 63.8% ($P>0.005$) with a concomitant rise of the acrosome-reacted gamete from 17.4+7.8 to 36.1+12.7 ($P<0.001$).The last part of the study was performed on motile acrosome-reacting spermatozoa that were analyzed by MSOME. It was possible to visualize large protruding bleds that become similar to vacuoles when seen upfront as well as a sort of invagination that look like a vacuole in the following image. The author came to the conclusion that the vacuole-free spermatozoa microinjected during IMSI are mostly acrosome-reacted spermatozoa.

Expert Commentary

This chapter was designed to provide practical guidelines and clinical application data of a new method of sperm morphological selection recently introduced to increase ART success in overcoming severe male infertility. In the last two decades, male infertility diagnosis has been based on conventional sperm parameters of questionable clinical value and other tests that investigate properties of sperm function that have been perceived useless after ICSI introduction. This was valid especially regarding strict morphological criteria, since it was demonstrated that even sperm with major malformations can obtain pregnancies and healthy deliveries. Moreover, most of these tests investigate the entire sperm population rather than providing information about the single sperm that would be employed in the conventional ICSI technique.

Even if a rising number of studies have reported IMSI as having remarkable clinical advantages in terms of fertilization, embryo quality, pregnancy occurrence, and prosecution until delivery, this new method continues to be debated regarding its routine application in the ART laboratory. Although, currently, IMSI has not been standardized and requires further validation, one promising perspective is the potential correlation between MSOME sperm nuclear malformations of the male gamete and its abnormal DNA status in terms of fragmentation. The future confirmation of this hypothesis warrants the investment of great quantities of human and technical resources needed to enable a reliable identification of only those spermatozoa that possess low levels of DNA damage in their nuclei for assisted conception purposes. Consequently this research would provide explanations regarding:

- Unexplained sterility
- Repeated abortions
- Repeated implantation failures.

Five-Year View

In light of its confirmed effectiveness and prospective reliability concerning DNA fragmentation, ultramorphological selection could be applied to demonstrate if, and how, antioxidants play a role in male reproductive potential. Hence, new antioxidant therapies with different morphological targets, at different concentrations could be tested and modified according to the attained quality of nuclear morphology, as well as the corresponding implantation and pregnancy rates.

The main challenges resulting from this research are:

- The number of microinjected oocytes/patient could be reduced in relation with improved fertilization ability of selected spermatozoa.
- It would be possible to decrease the number of embryos/transfer and the consequent risk of going through multiple pregnancy, due to the greater implantation potential of those embryos.
- Higher pregnancy rates would enable to reduce the number of ART attempts to achieve pregnancy and consequently hormonal intakes and costs.

Key Issues

- IMSI allows the selection, under 6,600× magnification, of a best motile sperm to be injected into the oocyte.
- According to current literature, IMSI provides better results than ICSI in defeating severe male infertility.
- Among the different malformation identified by MSOME, those affecting nuclear structure, such as vacuoles, demonstrated to be highly correlated with a male reproductive prognosis impairment.
- Nuclear vacuoles has been proposed to be morphological signs of DNA fragmentation. At present, the few published data are limited and in some cases discordant, at the point that we are still far from the clear identification of their meaning.
- Further studies are necessary to come to an adequate standardization of this new technique to clarify what are the biological mechanisms involved and give enlarged confirmations of effectiveness.

Acknowledgment The author gratefully acknowledge the help of Mrs Stella Antinori in the preparation of the manuscript.

References

1. Dumpy BC, Neal LM, Cooke ID. The clinical value of conventional semen analysis. Fertil Steril. 1989;51:324–9.
2. Aitkin RJ, Irvine DS, Wu FC. Prosecute analysis of spermoocyte fusion and reactive oxygen species generation as catena for the diagnosis of infertility. Am J Obstet Gynecol. 1991;164:542–51.
3. Vawda AI, Gunby J, Younglai EV. Semen parameters as predictors of in-vitro fertilization: the importance of strict criteria morphology. Hum Reprod. 1996;11:1445–50.
4. Kruger TF, Menkveld R, Stander FSH, Lombard CJ, Van der Merwe JP, van Zyl JA, et al. Sperm morphologic features as a prognostic factor in vitro fertilization. Fertil Steril. 1986;46: 1118–23.
5. Grow DR, Oehninger S, Seltman HJ, Toner JP, Swanson RJ, Kruger TF, et al. Sperm morphology as diagnosed by strict criteria: probing the impact of teratozoospermia on fertilization rate and pregnancy outcome in a large in vitro fertilization population. Fertil Steril. 1994;62:559–67.
6. Burr RW, Siegberg R, Flaherty SP, Wang X-J, Matthews CD. The influence of sperm morphology and the number of motile sperm inseminated on the outcome of intrauterine insemination combined with mild ovarian stimulation. Fertil Steril. 1996;65:127–32.
7. Eggert-Kruse W, Schwartz H, Rohr G, Demirakca T, Tilgen W. Runnebaum sperm morphology assessment using strict criteria and male fertility under in-vivo conditions of conception. Hum Reprod. 1996;11:139–46.
8. Oehninger S, Veeck L, Lanzendorf S, Maloney M, Toner J, Muasher S. Intracytoplasmic sperm injection: achievement of high pregnancy rates in couples with severe male factor infertility is dependent primarily upon female not male factors. Fertil Steril. 1995;64: 977–81.
9. Kupker W, Schulze W, Diedrich K. Ultrastructure of gametes and intracytoplasmic sperm injection: the significance of sperm morphology. Hum Reprod. 1998;13 Suppl 1:99–106.
10. Host E, Ernst E, Lindenberg S, Smidt-Jensen S. Morphology of spermatozoa used in IVF and ICSI from oligozoospermic men. Reprod Biomed Online. 2001;3:212–5.
11. Celik-Ozenci C, Jakab A, Kovacs T, Catalanotti J, Demir R, Bray-Ward P, Ward D, Huszar G. Sperm selection for ICSI: shape properties do not predict the absence or presence of numerical chromosomal aberrations. Hum Reprod. 2004;19:2052–9.
12. Sukcharoen N, Sithipravej T, Promviengchai S, Chinpilas V, Boonkasemsanti W. Sperm morphology evaluated by computer (IVOS) cannot predict the fertilization rate in vitro after intracytoplasmic sperm injection. Fertil Steril. 1998;69:564–8.
13. Tasdemir I, Tasdemir M, Tavukcuog S, Kahraman S, Biberog K. Effect of abnormal sperm head morphology on the outcome of intracytoplasmic sperm injection in humans. Hum Reprod. 1997;12:1214–7.
14. Liu J, Nagy Z, Joris H, Tournaye H, Devroey P, Van Steirteghem A. Successful fertilization and establishment of pregnancies after intracytoplasmic sperm injection in patients with globozoospermia. Hum Reprod. 1995;10:626–9.
15. Battaglia DE, Koehler JK, Klein NA, Tucker MJ. Failure of oocyte activation after intracytoplasmic sperm injection using round-headed sperm. Fertil Steril. 1997;68:118–22.
16. Kahraman S, Akarsu C, Cengiz G, Dirican K, Sozen E, Can B, Guven C, Vanderzwalmen P. Fertility of ejaculated and testicular megalohead spermatozoa with intracytoplasmic sperm injection. Hum Reprod. 1999;14:726–30.
17. De Vos A, Van De Velde H, Joris H, Verheyen G, Devroey P, Van Steirteghem A. Influence of individual sperm morphology on fertilisation, embryo morphology, and pregnancy outcome of intracytoplasmic sperm injection. Fertil Steril. 2003;79:42–8.
18. Cohen J, Alikani M, Malter H, et al. Partial zona dissection or subzonal sperm insertion: microsurgical fertilization alternatives based on evaluation of sperm and embryo morphology. Fertil Steril. 1991;56:696–706.
19. Parinaud J, Mieusset R, Vieitez G, Labal B, Richoilley G. Influence of sperm parameters on embryo quality. Fertil Steril. 1993;60: 888–92.
20. Loutradi KE, Tarlatzis BC, Goulis DG, Zepiridis L, Pagou T, Chatziioannou E, et al. The effects of sperm quality on embryo development after intracytoplasmic sperm injection. J Assist Reprod Genet. 2006;23:69–74.
21. Miller JE, Smith TT. The effect of intracytoplasmic sperm injection and semen parameters on blastocyst development in vitro. Hum Reprod. 2001;16:918–24.
22. Huszar G, Vigue L. Incomplete development of human spermatozoa is associated with increased creatine phosphokinase concentration and abnormal head morphology. Mol Reprod Dev. 1993;34:292–8.
23. Aitken J, Krausz C, Buckingham D. Relationships between biochemical markers for residual sperm cytoplasm, reactive oxygen species generation, and the presence of leukocytes and precursor germ cells in human sperm suspensions. Mol Reprod Dev. 1994;39:268–79.
24. Bernardini L, Borini A, Preti S, Conte N, Flamigni C, Capitanio GL, Venturini PL. Study of aneuploidy in normal and abnormal germ cells from semen of fertile and infertile men. Hum Reprod. 1998;13:3406–13.
25. Rubes J, Lowe X, Moore II D, Perreault S, Slott V, Evenson D, Selevan SG, Wyrobek AJ. Smoking cigarettes is associated with increased sperm disomy in teenage men. Fertil Steril. 1998;70: 715–23.
26. Twigg JP, Irvine DS, Aitken RJ. Oxidative damage to DNA in human spermatozoa does not preclude pronucleus formation at intracytoplasmic sperm injection. Hum Reprod. 1998;13:1864–71.
27. Sakkas D, Mariethoz E, Manicardi G, Bizzaro D, Bianchi PG, Bianchi U. Origin of DNA damage in ejaculated human spermatozoa. Rev Reprod. 1999;4:31–7.
28. Griffin DK, Hyland P, Tempest HG, Homa ST. Safety issues in assisted reproduction technology: should men undergoing ICSI be screened for chromosome abnormalities in their sperm? Hum Reprod. 2003;18:229–35.
29. Carrell DT, Emery BR, Wilcox AL, Campbell B, Erickson L, Hatasaka HH, Jones KP, Peterson CM. Sperm chromosome aneuploidy as related to male factor infertility and some ultrastructure defects. Arch Androl. 2004;50:181–5.
30. Rubio C, Simon C, Blanco V, Vidal F, Minguez Y, Egozcue J, et al. Implications of sperm chromosome abnormalities in recurrent miscarriage. J Assist Reprod Genet. 1999;16:253–8.
31. Van Steirteghem A, Bonduelle M, Devroey P, Liebaers I. Follow up of children born after ICSI. Hum Reprod Update. 2002;8:111–6.
32. Hansen M, Kurinczuk JJ, Bower C, Webb S. The risk of major birth defects after intracytoplasmic sperm injection and in vitro fertilization. N Engl J Med. 2002;346:725–30.
33. Kuczynski W, Dhont M, Grygoruk C, Grochowski D, Wolczynski S, Szamatowicz M. The outcome of intracytoplasmic injection of fresh and cryopreserved ejaculated spermatozoa – a prospective randomized study. Hum Reprod. 2001;16:2109–13.
34. Stolwijk AM, Wetzels AM, Braat DD. Cumulative probability of achieving an ongoing pregnancy after in-vitro fertilization and intracytoplasmic sperm injection according to a woman's age, subfertility diagnosis and primary or secondary subfertility. Hum Reprod. 2000;15:203–9.

35. Olivius K, Friden B, Lundin K, Bergh C. Cumulative probability of live birth after three in vitro fertilization/intracytoplasmic sperm injection cycles. Fertil Steril. 2002;77:505–10.
36. European IVF-monitoring programme (EIM), for the European Society of Human Reproduction and Embryology (ESHRE) Assisted reproductive technology in Europe, Results generated from European registers by ESHRE. Human reproduction 2002; 21:1680–97.
37. Van Steirteghem AC, Liu J, Joris H, Nagy Z, Janssenswillen C, Tournaye H, Derde MP, Van Assche E, Devroey P. Higher success rate by intracytoplasmic sperm injection than by subzonal insemination. Report of a second series of 300 consecutive treatment cycles. Hum Reprod. 1993;8:1055–60.
38. Harari O, Bourne H, McDonald M, Richings N, Speirs AL, Johnston WIH, Baker HWG. Intracytoplasmic sperm injection—a major advance in the management of severe male subfertility. Fertil Steril. 1995;64:360–8.
39. Glezerman M, Bartoov B. Semen analysis. In: Insler V, Lunenfeld B, editors. Infertility: male and female. Edinburgh: Churchill Livingstone; 1993. p. 285–315.
40. Piomboni P, Strehler E, Capitani S, Collodel G, De Santo M, Gambera L, Moretti E, Baccetti B, Sterzic K. Submicroscopic mathematical evaluation of spermatozoa in assisted reproduction, in vitro fertilization (notulae seminologicae 7). J Assist Reprod Genet. 1996;13:635–46.
41. Zamboni L. The ultrastructural pathology of the spermatozoan as a course of infertility: the role of electron microscopy in the evaluation of sperm quality. Fertil Steril. 1987;48:711–34.
42. Bartoov B, Eltes F, Reichart M, Langzam J, Lederman H, Zabludovsky N. Quantitative ultramorphological analysis of human sperm: fifteen years of experience in the diagnosis and management of male factor infertility. Arch Androl. 1999;43(1): 13–25.
43. Bartoov B, Berkovitz A, Eltes F, Kogosowski A, Menezo Y, Barak Y. Real-time fine morphology of motile human sperm cells is associated with IVF-ICSI outcome. J Androl. 2002;23:1–8.
44. Berkovitz A, Eltes F, Yaari S, Katz N, Barr I, Fishman A, Bartoov B. The morphological normalcy of the sperm nucleus and pregnancy rate of intracytoplasmic injection with morphologically selected sperm. Hum Reprod. 2005;20:185–90.
45. Agarwal A, Said TM. Role of sperm chromatin abnormalities and DNA damage in male infertility. Hum Reprod Update. 2003;9: 331–45.
46. Irvine DS, Twigg JP, Gordon EL, Fulton N, Milne PA, Aitken RJ. DNA integrity in human spermatozoa: relationships with semen quality. J Androl. 2000;21:33–44.
47. Seli E, Moffatt O, Kayisli UA, Nijs M, Ombelet W and Sakkas D. Apoptosis in testis of normal and azoospermic males: a Fas mediated phenomenon. Annual meeting of the society for gynecologic investigation, Los Angeles, CA, 2002.
48. Aitken RJ, De Iuliis GN. Origins and consequences of DNA damage in male germ cells. Reprod Biomed Online. 2007;14:727–33.
49. Aitken RJ, De Iuliis GN, McLachlan RI. Biological and clinical significance of DNA damage in the male germ line. Int J Androl. 2008;32(1):46–56. ISSN 0105-6263.
50. Sikka A, Rajasekaran M, Hellstrom W. Role of oxidative stress and antioxidants in male infertility. J Androl. 1995;16:464–8.
51. Evenson DP, Jost LK, Marshall D, Zinaman MJ, Clegg E, Purvis K, de Angelis P, Claussen OP. Utility of the sperm chromatin structure assay as a diagnostic and prognostic tool in the human fertility clinic. Hum Reprod. 1999;14:1039–49.
52. Spano M, Bonde JP, Hjollund HI, Kolstad HA, Cordelli E, Leter G. Sperm chromatin damage impairs human fertility. The Danish First Pregnancy Planner Study Team. Fertil Steril. 2000;73:43–50.
53. Morris ID, Ilott S, Dixon L, Brison DR. The spectrum of DNA damage in human sperm assessed by single cell gel electrophoresis (Comet assay) its relationship to fertilization and embryo development. Hum Reprod. 2002;17:990–8.
54. Benchaib M, Braun V, Lornage J, Hadj S, Salle B, Lejeune H, Guerin JF. Sperm DNA fragmentation decreases the pregnancy rate in an assisted reproductive technique. Hum Reprod. 2003;18:1023–8.
55. Larson-Cook KL, Brannian JD, Hansen KA, Kasperson KM, Aamold ET, Evenson DP. Relationship between the outcomes of assisted reproductive techniques and sperm DNA fragmentation as measured by the sperm chromatin structure assay. Fertil Steril. 2003;80:895–902.
56. Loft S, Kold-Jensen T, Hjollund NH, Giwercman A, Gyllemborg J, Ernst E, Olsen J, Scheike T, Poulsen HE, Bonde JP. Oxidative DNA damage in human sperm influences time to pregnancy. Hum Reprod. 2003;18:1265–72.
57. Bungum M, Humaidan P, Spano M, Jepson K, Bungum L, Giwercman A. The predictive value of sperm chromatin structure assay (SCSA) parameters for the outcome of intrauterine insemination, IVF and ICSI. Hum Reprod. 2004;19:1401–8.
58. Gandini L, Lombardo F, Paoli D, Caruso F, Eleuteri P, Leter G, Ciriminna R, Culasso F, Dondero F, Lenzi A, et al. Full-term pregnancies achieved with ICSI despite high levels of sperm chromatin damage. Hum Reprod. 2004;19:1409–17.
59. Seli E, Gardner DK, Schoolcraft WB, Moffatt O, Sakkas D. Extent of nuclear DNA damage in ejaculated spermatozoa impacts on blastocyst development after in vitro fertilization. Fertil Steril. 2004;82:378–83.
60. Tesarik J, Greco E, Mendoza C. Late, but not early, paternal effect on human embryo development is related to sperm DNA fragmentation. Hum Reprod. 2004;19:611–5.
61. Smith R, Kaune H, Parodi D, Madariaga M, Rios R, Morales I, Castro A. Increased sperm DNA damage in patients with varicocele: relationship with seminal oxidative stress. Hum Reprod. 2006; 21(4):986–93.
62. Borini A, Tarozzi N, Bizzaro D, Bonu MA, Fava L, Flamigni C, Coticchio G. Sperm DNA fragmentation: paternal effect on early post-implantation embryo development in ART. Hum Reprod. 2006;21:2876–81.
63. Lin MH, Kuo-Kuang Lee R, Li SH, Lu CH, Sun FJ, Hwu YM. Sperm chromatin structure assay parameters are not related to fertilization rates, embryo quality, and pregnancy rates in in vitro fertilization and intracytoplasmic sperm injection, but might be related to spontaneous abortion rates. Fertil Steril. 2007;90:352–9.
64. Zini A, Boman JM, Belzile E, Ciampi A. Sperm DNA damage is associated with an increased risk of pregnancy loss after IVF and ICSI: systematic review and metaanalysis. Hum Reprod. 2008;23:2663–8.
65. Tesarik J. Paternal effects on cell division in the human preimplantation embryo. Reprod Biomed Online. 2005;10:370–5.
66. Suleiman SA, Ali ME, Zaki ZM, el-Malik EM, Nasr MA. Lipid peroxidation and human sperm motility: protective role of vitamin E. J Androl. 1996;17:530–7.
67. Kodama H, Yamaguchi R, Fukuda J, Kasai H, Tanaka T. Increased oxidative deoxyribonucleic acid damage in the spermatozoa of infertile male patients. Fertil Steril. 1997;68:519–24.
68. Geva E, Lessing JB, Lerner-Geva L, Amit A. Free radicals, antioxidants and human spermatozoa: clinical implications. Hum Reprod. 1998;13:1422–4.
69. Comhaire FH, Christophe AB, Zalata AA, Dhooge WS, Mahmoud AM, Depuydt CE. The effects of combined conventional treatment, oral antioxidants and essential fatty acids on sperm biology in subfertile men. Prostaglandins Leukot Essent Fatty Acids. 2000;63:159–65.
70. Keskes-Ammar L, Feki-Chakroun N, Rebai T, Sahnoun Z, Ghozzi H, Hammami S, Zghal K, Fki H, Damak J, Bahloul A. Sperm oxidative stress and the effect of an oral vitamin E and selenium supplement on semen quality in infertile men. Arch Androl. 2003; 49:83–94.

71. Greco E, Scarselli F, Iacobelli M, Rienzi L, Ubaldi F, Ferrero S, Franco G, Anniballo N, Mendoza C, Tesarik J. Efficient treatment of infertility due to sperm DNA damage by ICSI with testicular spermatozoa. Hum Reprod. 2005;20:226–30.
72. Lopes S, Jurisicova A, Sun JG, Casper RF. Reactive oxygen species: potential cause for DNA fragmentation in human spermatozoa. Hum Reprod. 1998;13(4):896–900.
73. Krausz C, Mills C, Rogers S, et al. Stimulation of oxidant generation by human sperm suspensions using phorbol esters and formyl peptides: relationships with motility and fertilization in vitro. Fertil Steril. 1994;62:599–605.
74. Lenzi A, Picardo M, Gandini L, et al. Glutathione treatment of dyspermia: effect on the lipoperoxidation process. Hum Reprod. 1994;9:2044–50.
75. Berkovitz A, Eltes F, Ellenbogen E, Peer S, Feldberg D, Bartoov B. Does the presence of nuclear vacuoles in human sperm selected for ICSI affect pregnancy outcome? Hum Reprod. 2006;21:1787–90.
76. Bartoov B, Berkovitz A, Eltes F, Kogosovsky A, Yagoda A, Lederman H, Artzi S, Gross M, Barak Y. Pregnancy rates are higher with intracytoplasmic morphologically selected sperm injection than with conventional intracytoplasmic injection. Fertil Steril. 2003;80:1413–9.
77. Palermo G, Joris H, Devroey P, Van Steirteghem A. Pregnancies after intracytoplasmic injection of single spermatozoon into an oocyte. Lancet. 1992;340:17.
78. Bartoov B, Eltes F, Pansky M, Langzam J, Reichart M, Soffer Y. Improved diagnosis of male fertility potential via a combination of quantitative ultramorphology and routine semen analyses. Hum Reprod. 1994;9:2069–75.
79. Mundy AJ, Ryder TA, Edmonds DK. A quantitative study of sperm head ultrastructure in subfertile males with excess sperm precursors. Fertil Steril. 1994;61:751–4.
80. Berkovitz A, Eltes F, Lederman H, Peer S, Ellenbogen A, Feldberg B, Bartoov B. How to improve IVF–ICSI outcome by sperm selection. Reprod Biomed Online. 2006;12:634–8.
81. Shapiro BS, Richter KS, Harris DC, Daneshmand ST. Dramatic declines in implantation and pregnancy rates in patients who undergo repeated cycles of in vitro fertilization with blastocyst transfer after one or more failed attempts. Fertil Steril. 2001;76:538–42.
82. Silberstein T, Trimarchi JR, Gonzalez L, Keefe D, Blazar AS. Pregnancy outcome in in vitro fertilization decreases to a plateau with repeated cycles. Fertil Steril. 2005;84:1043–5.
83. Antinori M, Licata E, Dani G, Cerusico C, Versaci C, D'Angelo D, Antinori S. Intracytoplasmic morphologically selected sperm injection :a prospective randomized trial. Reprod Biomed Online. 2008;16:835–41.
84. Kruger TF, Acosta AA, Simmons KF, Swanson JR, Matta JF, Oehninger S. Predictive value of sperm morphology in in vitro fertilization. Fertil Steril. 1988;49:112–7.
85. Martin RH, Rademaker A. The relationship between sperm chromosomal abnormalities and sperm morphology in humans. Mutat Res. 1988;207:159–64.
86. Rosenbusch B, Strehler E, Sterzik K. Cytogenetics of human spermatozoa: correlations with sperm morphology and age of fertile men. Fertil Steril. 1992;58:1071–2.
87. Sailer BL, Jost LK, Evenson DP. Bull sperm head morphometry related to abnormal chromatin structure and fertility. Cytometry. 1996;24:167–73.
88. Virro MR, Larson-Cook KL, Evenson DP. Sperm chromatin structure assay (SCSA) parameters are related to fertilization, blastocyst development, and ongoing pregnancy in in vitro fertilization and intracytoplasmic sperm injection cycles. Fertil Steril. 2004;81:1289–95.
89. Vicari E, de Palma A, Burrello N, Longo G, Grazioso C, Barone N, Zahi M, D'Agata Check JH, Graziano V, Cohen R, et al. Effect of an abnormal sperm chromatin structural assay (SCSA) on pregnancy outcome following (IVF) with ICSI in previous IVF failures. Arch Androl. 2005;51:121–4.
90. Nagy ZP, Liu J, Joris H, et al. The result of intracytoplasmic sperm injection is not related to any of the three basic sperm parameters. Hum Reprod. 1995;10:1123–9.
91. McKenzie LJ, Kovanci E, Amato P, et al. Pregnancy outcome of in vitro fertilization/intracytoplasmic sperm injection with profound teratospermia. Fertil Steril. 2004;82:847–9.
92. Bonduelle M, Aytoz A, Van Assche E, et al. Incidence of chromosomal aberrations in children born after assisted reproduction through intracytoplasmic sperm injection. Hum Reprod. 1998;13:781–2.
93. Lee JD, Kamiguchi Y, Yanagimachi R. Analysis of chromosome constitution of human spermatozoa with normal and aberrant head morphologies after injection into mouse oocytes. Hum Reprod. 1996;11:1942–6.
94. Hazout A, Dumont-Hassan M, Junca AM, Bacrie PC, Tesarik J. High-magnifi cation ICSI overcomes paternal effect resistant to conventional ICSI. Reprod Biomed Online. 2006;12:19–25.
95. Franco Jr JG, Baruffi RL, Mauri AL, et al. Significance of large nuclear vacuoles in human spermatozoa: implications for ICSI. Reprod Biomed Online. 2008;17:42–5.
96. Garolla A, Fortini D, Menegazzo M, et al. High-power microscopy for selecting spermatozoa for ICSI by physiological status. Reprod Biomed Online. 2008;17:610–6.
97. Hoffman R, Gross L. Reflected light differential-interference microscopy: principles, use and image interpretation. J Microsc. 1970;91:149–72.
98. Hoffman R, Gross L. Demodulation contrast microscope. Nature. 1975;254:586–8.
99. Padawer J. The Nomarski interference-contrast microscope. An experimental basis for image interpretation. J R Microsc Soc. 1968;88:305–49.
100. Kacem O, Sifer C, Barraud-Lange V, Ducot B, De Ziegler D, Poirot C, Wolf JP. Sperm nuclear vacuoles, assessed by motile sperm organellar morphological examination, are mostly of acrosomal origin. Reprod Biomed Online. 2010;20:132–7.

Role of Sperm–Hyaluronic Acid Binding in the Evaluation and Treatment of Subfertile Men with ROS-Affected Semen: Assessment of Sperm with Oxidative Damage and HA-Mediated ICSI Sperm Selection

44

Ciler Celik-Ozenci and Gabor Huszar

Abstract

Excessive levels of free radicals diminish the functional integrity of spermatozoa. Levels of reactive oxygen species (ROS) produced by spermatozoa were negatively correlated with the quality of sperm in the original semen. "Intrinsic," "extrinsic," and "iatrogenic" sources of ROS production were identified within semen: Intrinsic sources of ROS in semen are morphologically abnormal and arrested maturity spermatozoa and leukocytes. Poor sperm quality showing attributes of arrested sperm maturation is linked to increased ROS generation as a consequence of excess residual cytoplasm related to the arrest of cytoplasmic extrusion in terminal spermiogenesis. In the past 20 years, the Huszar lab has studied several key events of sperm maturation, including cytoplasmic extrusion and expression of the HspA2 chaperone protein. In experiments related to sperm function and fertilizing potential, Huszar et al. have established that, simultaneously with cytoplasmic extrusion during terminal spermiogenesis, there is a remodeling of the plasma membrane that facilitates the formation of the zona pellucida- and hyaluronic acid (HA)-binding sites. The studies with HA immobilized to glass slides or Petri dishes showed that sperm firmly bind to HA. However, not all sperm exhibited HA binding ability. These data supported the hypothesis that the ability of sperm–HA binding is related to sperm cellular maturity.

Keywords

Sperm–hyaluronic acid binding • Sperm function • Oxidative stress • Reactive oxygen species • Male infertility • Intracytoplasmic sperm selection • DNA chain degradation

Sperm Function and Oxidative Damage

Reactive oxygen species (ROS) are free radicals that have one or more unpaired electrons capable of oxidizing adjacent biomolecules. ROS have a significant role in many of the sperm physiological processes and in causing sperm damage [1, 2]. Low levels of free radical production by sperm may also facilitate sperm capacitation. Hydrogen peroxide stimulates the acrosome activation and hyperactivated motility, thus improving the transit of sperm through the cumulus and zona pellucida during fertilization. Low concentrations of hydrogen peroxide also cause tyrosine phosphorylation, which is associated with sperm membrane binding to the zona pellucida and sperm–oocyte interaction [3, 4].

However, excessive levels of free radicals diminish the functional integrity of spermatozoa [3–5]. Gomez et al. demonstrated that levels of ROS produced by spermatozoa were negatively correlated with the quality of sperm in the original semen [6]. "Intrinsic," "extrinsic," and "iatrogenic" sources of ROS production were identified within semen: Intrinsic sources of ROS in semen are morphologically abnormal and

C. Celik-Ozenci, DDS, PhD (✉)
Department of Histology and Embryology, School of Medicine, Akdeniz University, Campus, Antalya 07070, Turkey
e-mail: cilerozenci@akdeniz.edu.tr

G. Huszar, MD
Sperm Physiology Laboratory, Yale School of Medicine, 333 Cedar Street, New Haven 06510, CT, USA
e-mail: gabor.huszar@yale.edu

S.J. Parekattil and A. Agarwal (eds.), *Male Infertility: Contemporary Clinical Approaches, Andrology, ART & Antioxidants*,
DOI 10.1007/978-1-4614-3335-4_44,

arrested maturity spermatozoa and leukocytes. Poor sperm quality showing attributes of arrested sperm maturation is linked to increased ROS generation as a consequence of excess residual cytoplasm related to the arrest of cytoplasmic extrusion in terminal spermiogenesis. Simultaneously, in the Aitken and Huszar laboratory, correlations were found between sperm creatine phosphokinase (CK) content (a component of the residual cytoplasm) and lipid peroxidation in semen specimens. This suggested that both the increased ROS production and CK activity are related to increased cytoplasmic content of sperm [7, 8]. Further, in the Huszar and Vigue paper [8], sperm fractions with high and low (normal) cytoplasmic content and malondialdehyde (MDA) levels (representing ROS production) were co-incubated and co-centrifuged, to test the potential propagation of high ROS production from the high to the low ROS producing (normal) sperm fractions. In these experiments, the ROS production did not increase in the combined low and high MDA sperm fractions. Thus, the conclusion was drawn that the increased ROS production (and higher cytoplasmic content) was an "inborn" rather than an acquired attribute of individual spermatozoa [8–11]. Others also confirmed the results of Aitken and Huszar laboratories, further suggesting that sperm with excess cytoplasmic droplets were classified as immature and functionally defective cells and were sources of increased ROS production [12, 13].

In the past 20 years, the Huszar lab has studied several key events of sperm maturation, including cytoplasmic extrusion and expression of the HspA2 chaperone protein [9, 14–19]. In experiments related to sperm function and fertilizing potential, Huszar et al. have established that, simultaneously with cytoplasmic extrusion during terminal spermiogenesis, there is a remodeling of the plasma membrane that facilitates the formation of the zona pellucida- and hyaluronic acid (HA)-binding sites [20]. The studies with HA immobilized to glass slides or Petri dishes showed that sperm firmly bind to HA. However, not all sperm exhibited HA-binding ability [21, 22]. These data supported the hypothesis that the ability of sperm–HA binding is related to sperm cellular maturity. The validity of this concept was confirmed, by studying the properties of both HA-binding and HA-nonbinding spermatozoa, utilizing the various biochemical markers of spermatozoa maturity and function. It has became clear that only spermatozoa that fully completed the maturation process in late spermiogenesis, along with membrane remodeling, cytoplasmic extrusion, and the steps of histone–transition protein–protamine replacement, are able to bind to HA. However, high levels of ROS may damage the sperm membrane and DNA, leading to a decline in sperm fertilizing potential and paternal contribution to the embryo. Thus, the sperm–HA binding assay may be an excellent test for the proportion of sperm with arrested cellular maturation, damaged sperm membrane integrity, and thus diminished fertilizing function in semen samples. Also HA-mediated sperm binding facilitates the selection of individual, fully developed spermatozoa without ROS generation and with high DNA chain integrity [23, 24].

The role of additional factors, such as leukocytes, particularly neutrophils and macrophages in semen, was shown to be associated with excessive ROS production and diminished sperm function [25]. In addition, the mechanical stress of centrifugation during sperm preparation was also reported to enhance ROS production. There are data to indicate that sperm preparation during assisted reproductive technologies (ART) has the potential to aggravate sperm oxidative stress. [26–28]. A contributing factor may arise from the removal of seminal fluid with its protective antioxidant content [29, 30]. In addition, cryopreservation of sperm, another commonly used technique in ART, is associated with an increase in sperm oxidative stress and an increase in DNA chain degradation. Indeed, with the method of in situ DNA nick translation, the degradation of DNA chain integrity following sperm cryopreservation and thawing has been observed [31, 32].

Studies on ROS production have indicated that >30% of unselected infertile patients show high levels of seminal ROS. Reactive oxygen species formation was detected in 40% of the semen with spermatozoa from infertile patients with a related damage to the sperm membrane and DNA [26, 33, 34]. Jones et al. reported that ROS-induced peroxidation of the sperm membrane diminishes its fluidity and increases membrane rigidity, thus reducing tail motion [35]. Sperm membranes are vulnerable to this type of damage as they contain large amounts of unsaturated fatty acids. Direct ROS damage to mitochondria, decreasing energy availability, may also attenuate the rate of sperm motility. By either mechanism, oxidative stress impairs sperm motility and the chances for reaching the oocyte and initiates fertilization [36–38].

Free radicals may directly damage sperm DNA by attacking the purine and pyrimidine base integrity as well as the deoxyribose interlink of the DNA chain. Normally, sperm DNA is tightly packaged with protamine, and the complex withstand well the attacks by free radicals. However, sperm with diminished maturity often exhibit excess histones detected by aniline blue staining [39, 40], and the consequential deficient protamination leaves the sperm DNA more vulnerable to ROS attack [41]. Several investigators have now demonstrated connections between oxidative stress and sperm DNA damage by the sperm chromatin structure (SCSA) and TUNEL assays and by measurement of 8-hydroxydeoxyguanosine (8-OHdG), a product of DNA oxidation [38, 42–46].

Sperm Properties Associated with ROS-Related DNA Defects and Improved Cellular Attributes of HA-Selected Sperm

In the past two decades, major advances have been made regarding the biochemical markers of human sperm development and function. Human sperm cellular maturation, which is necessary to attain zona pellucida binding and oocyte fertilizing potential, comprises three interrelated and synchronized events: (1) Extrusion of excess cytoplasm, which is a phase of normal development, facilitates the formation of normal sperm shape; (2) sperm membrane remodeling in spermiogenesis and acquisition of ZP and HA-binding sites; and (3) sperm nuclear maturation with the milestones of histone–transition protein–protamine transition with the associated changes in DNA packing and chromosomal status.

In the initial phase of our studies, we intended to search for an objective biochemical marker that might predict sperm fertility, independently from the classical semen parameters. In these experiments directed to sperm development and fertility, elevated sperm cytoplasmic and CK content, reflecting surplus cytoplasm, as well as a low expression of the HspA2 chaperone was identified in spermatozoa of subfertile men, in association with arrested sperm cellular maturation and diminished cytoplasmic extrusion [9, 10, 14, 16, 18, 24, 47, 48]. Regarding arrested cytoplasmic extrusion, significantly higher frequencies of sperm with excess cytoplasmic CK content were found in subfertile men. The CK-immunostaining patterns indicated that the high sperm CK activity was a direct consequence of increased cytoplasmic protein concentrations. This suggested a sperm developmental defect in terminal spermiogenesis when the surplus cytoplasm is normally extruded from elongating spermatid as "residual bodies." Another sperm protein with ATP content has also been found in mature spermatozoa [16]. This protein was later characterized as the 70-kDa testis-expressed heat shock protein HspA2 chaperone [15]. Thus, the proportional concentrations of CK and HspA2 (or HspA2 ratio) reflect the proportion of mature and immature spermatozoa in semen samples In three studies with approximately 500 men, there were similarly close correlations between sperm CK activities and decreased HspA2 ratios ($r=-0.69$, -0.71, and -0.76, $p<0.001$) [9–11, 14, 15, 47].

The recognition that low sperm HspA2 affects the meiotic process as well as the cytoplasmic and nuclear downstream events of cellular maturation was a key advance [15, 19, 47, 49–51]. The spermatid phase seems to be the key dividing point between normally developing and diminished maturity spermatozoa. In elongating spermatids with a substantial upregulation of HspA2, normal sperm maturation occurs with the orderly extrusion of cytoplasm in the form of residual bodies and simultaneous plasma membrane remodeling, yielding a spermatozoon with normal head morphology and fully formed binding sites for the zona pellucida and hyaluronic acid. In contrast, in spermatids with low HspA2 expression, there are increased frequencies of chromosomal aneuploidies due to synaptonemal complex defects. These cells are characterized by cytoplasmic retention, abnormal head morphology, increased concentrations of ROS, and consequential additional DNA fragmentation. Further, there were three characteristic features: (1) Sperm with the highest level of cytoplasmic retention, demonstrated with CK immunocytochemistry, also had the very high occurrence of amorphous spermatozoa [14]; (2) In 159 couples with oligozoospermic husbands treated with intrauterine insemination, in the subject couples who did or did not achieve pregnancy, a logistic regression analysis revealed no differences in sperm concentration or motility; only the sperm cytoplasmic CK content demonstrated a significant contributing factor [8]. (3) Further, due to the incomplete spermatogenesis, the process of plasma membrane remodeling and diminished formation of the zona pellucida and hyaluronic acid binding sites failed to occur. Thus, arrested maturity spermatozoa are unable to bind to the zona pellucida or fertilize via natural or in vitro fertilization (IVF) conception, only by intracytoplasmic sperm injection (ICSI). This was demonstrated in sperm–hemizona complexes immunostained with CK, in which all hemizona-bound spermatozoa were clear-headed without cytoplasmic retention [19]. This selectivity was attributed to a sperm plasma membrane remodeling step during spermiogenesis which facilitates the formation of the sperm binding sites for the zona pellucida and hyaluronic acid [15, 20]. The probable reason why such aberrant spermatozoa do not demise prior to ejaculation is the presence of anti-apoptotic protein Bclx2 in the surviving germ cells [52].

In addition to the increased rates of lipid peroxidation and consequential DNA fragmentation in sperm with arrested development, DNA chain fragmentation may also be promoted by the decreased HspA2 chaperone activity and consequential inadequate delivery of DNA repair enzymes and arrest of the histone–transition protein–protamine replacement process [1, 8, 15, 49, 50, 53, 54]. Indeed, a close relationship was established recently between HspA2 function and transition protein transport [55]. However, sperm with normal development, that are able to bind to hyaluronic acid and to the zona pellucida, show high DNA integrity and a high proportion of sperm with normal Tygerberg morphology [23, 48, 52, 56].

Earlier studies have suggested a relationship between oligozoospermia, male infertility, and increased frequencies of chromosomal aneuploidies [57–64]. Further, as a novel idea, Kovanci et al. suggested that, because HspA2 is part of the synaptonemal complex and immature spermatozoa with cytoplasmic retention are deficient in HspA2, synaptic anomalies may occur in association with arrested sperm maturation. In a study that tested this concept, there was a close correlation between the frequencies of immature spermatozoa with cytoplasmic retention and with spermatozoa of disomic nuclei at the levels of approximately $r=0.7$ ($p<0.001$), indicating that disomies primarily originate in immature sperm [65]. There was no relationship with diploidies; thus, chromosomal disomy and diploidy are arising from different cellular mechanisms [66].

Following up on the attributes of arrested sperm cellular maturation and the potential relationships between early and late spermatogenetic failures and between nuclear and cytoplasmic events, solid support was developed by studies of double-stained individual spermatozoa [48]. Testing of the same spermatozoa with aniline blue staining and CK immunostaining (persistent histones and cytoplasmic retention), aniline blue staining and caspase-3 immunostaining (persistent histones and apoptosis), and aniline blue staining and DNA nick translation (persistent histones and DNA chain degradation) demonstrated that the presence or absence of such cellular attributes shows an impressive, approximately 70% agreement [48]. Thus, there is a solid relationship between the early and late spermatogenetic events within the same sperm whether one probes the nuclear or cytoplasmic compartment. This experiment is also important for sperm fertilizing potential from two points of view: (1) Late spermiogenetic events maybe disturbed by upstream spermiogenesis; (2) The lack of sperm membrane remodeling and diminished zona pellucida or hyaluronic acid binding (thus sperm–oocyte interactions or sperm selection) are also influenced by upstream events, as the hyaluronic acid-bound spermatozoa does not exhibit staining with probes detecting arrested sperm development, such as aniline blue staining of retained histones, cytoplasmic retention, or presence of the apoptosis marker caspase-3, and DNA chain degradation, attributes that are collectively related and reflect elevated sperm ROS levels.

Sperm Selection and Biochemical Markers

With respect to the sperm selection for ICSI, the potential relationship between abnormal sperm morphology and chromosomal aberrations has been of a long-term interest. A deficiency of earlier data is the fact that the data were based on the frequencies of abnormal or aneuploid spermatozoa in semen samples, which fails to show whether the aneuploidy and abnormal morphology occurs within the same spermatozoa [65, 66]. Recently, examination of the same individual spermatozoa for both shape and chromosomal aneuploidy has become possible because it was established that spermatozoa preserve their shape following the decondensation and denaturation steps that are a prerequisite of fluorescence in situ hybridization analysis [67]. These experiments revealed that visual shape assessment, i.e., choosing the "best-looking" spermatozoon, is an unreliable method for ICSI selection of mature haploid spermatozoa [68, 69].

The increased rates of chromosomal aberrations and other potential consequences of using immature spermatozoa for ICSI are of major concern [70–73]. Based on the presence of the HA receptors in fully developed but not in spermatozoa with arrested development and absence of plasma membrane remodeling….. selection of single mature (out) spermatozoa with high DNA integrity and low frequencies of chromosomal aneuploidies for ICSI [74]. Regarding ICSI sperm selection, the efficiency of elimination of aneuploid and diploid spermatozoa from the HA-bound population has been tested in three experiments. In the HA-bound spermatozoa vs. unselected spermatozoa, the chromosomal disomy frequencies, with the three probes studied, were reduced to 0.16 from 0.52%, diploidy to 0.09 from 0.51%, and sex chromosome disomy to 0.05 from 0.27%. No matter how high the aneuploidy frequencies in the semen sperm fractions were, the respective frequencies were within the narrow low 0.04–0.10% range per probed chromosome in HA-bound spermatozoa, comparable with the range of normozoospermic fertile men. The fivefold decline in X, Y, and XY disomies is consistent with the reported increase in chromosomal aberrations (approximately fourfold) in ICSI children conceived with visually selected spermatozoa [70, 74, 75].

Further attributes of HA-bound spermatozoa indicated that the HA-selected sperm fraction are devoid of cytoplasmic retention, persistent histones (and probably the associated protamine 1/protamine 2 ratio disorder), DNA fragmentation, and the apoptotic marker caspase-3 [24]. These properties are very important because nuclear and cytoplasmic immaturities, particularly the presence of DNA fragmentation, are known to adversely affect the paternal contribution of sperm to the zygote [1, 23, 47, 48, 52, 53, 67, 76–78].

The above-discussed concepts confirmed that sperm–HA binding provides a new approach in andrology testing and in ICSI sperm selection as follows:

- The sperm–HA-binding test provides a 15-min microscopic assay for the assessment of the proportion of spermatozoa that would bind to the zona pellucida.
- In spermatozoa selected by the HA-mediated approach, the frequencies of chromosomal disomies and diploidies are within the normal range, independently from the aneuploidy frequencies of the initial semen.
- Mature spermatozoa selected by virtue of HA binding are also viable and devoid of DNA fragmentation, persistent histones, and of apoptotic markers, such as caspase-3. Thus, HA selection is a feasible compensatory approach for men who did not respond to antioxidant treatment yet wishes to father children [24].

Detection and Assessment of Sperm and Seminal Reactive Oxygen Species

Whereas screening for oxidative stress would be important in andrology laboratories, there are three factors that hinder this development: (a) cost of the tests, (b) the complexity of testing, and (c) the lack of a generally accepted assessment/approach of oxidative stress measurements [79].

Direct assays focus upon measurement of damage on the integrity of the sperm lipid membrane or DNA, created by the free radicals. The primary approach is the assessment of sperm malondialdehyde (MDA) levels with the thiobarbituric acid assay [80–82]. Other assessment methods of sperm membrane lipid peroxidation based on isoprostane 8-iso-PGF2a or the c11-BODIPY assays are promising but are not used widely [39, 83, 84].

Oxidative stress seems to be one of the primary causes of damaged sperm DNA [5, 44, 85, 86]. Because sperm DNA can also be damaged by nonoxidative mechanisms, such as aberrant apoptosis and incomplete sperm protamination, the TUNEL or SCSA measurements for oxidative stress are likely to be inaccurate [87]. Sperm DNA oxidative damage is reflected by the levels of oxidized deoxynucleoside, 8-oxo-7,8-dihydro-20-deoxyguanosine (8-OHdG), in sperm or seminal plasma [39, 88]. Indeed, a study has reported that chances of natural conception are inversely related with sperm 8-OHdG levels [89]. As DNA damage reduces the paternal contribution of sperm to the embryo, the study reporting that 8-OHdG levels and the chances for conception are inversely related is not surprising [89].

Luminescence methods based on either luminol or lucigenin are used frequently to detect ROS production within semen. However, these methods do not take into consideration the sperm-to-sperm variability in susceptibility to ROS damage. The general introduction of these assays in clinical andrology laboratories has been slow because of the expensive equipment (luminometer) and high levels of necessary quality control, to standardize incubation times and seminal plasma contaminations [90–93].

Measurement of total antioxidant capacity within semen is another approach via the inhibition of chemiluminescence generated by a constant added source of ROS (e.g., horseradish peroxidase). The total antioxidant capacity may be quantified against a vitamin E analogue and expressed as a ROS–TAC value [94]. However, other colorimetric techniques are now being introduced, and they are easier to perform and more cost-effective [95, 96]. A further important consideration, independently from TAC assessment, is the susceptibility of the spermatozoa to ROS damage which, as discussed above, varies greatly depending on the developmental and cellular state of the individual cells. Indeed, a correlation was found between the degree of cytoplasmic retention and ROS production in sperm samples, whether detected with chemiluminescence or MDA production [8, 91].

Due to the complexity of the above-mentioned ROS measurements in human semen, a clinically useful test is needed for assessment of sperm DNA fragmentation or selection of sperm with high DNA integrity. Since sperm affected by free radicals have both DNA fragmentation and lipid peroxidation, the ability of HA binding can well discriminate the identity of mature vs. ROS-damaged immature sperm. The data powerfully indicate that diminished maturity sperm with DNA damage and arrested membrane remodeling (unable to bind to HA or the oocyte), thus handicapped in natural conception, are eliminated by HA-mediated sperm selection [23, 24].

Medical Aspects

Causes of Oxidative Stress

There are several potential causes of sperm oxidative stress, including various pathogenetic origins, such as idiopathic, iatrogenic, lifestyle, environmental, infection, chronic diseases, autoimmune, and testicular, as reviewed recently [97]. Oxidative stress is often associated with idiopathic male factor infertility. One key factor that supports this association is the finding that sperm with cytoplasmic retention and abnormal shape have an increased capacity for ROS generation [7, 8]. For instance, men with sperm concentrations in the normal range but with high frequency of sperm with arrested development, and thus diminished fertility, may show increased ROS levels and lower antioxidant capacity compared to fertile men. This may be easily detected by a lower sperm–HA-binding score (<65%). Regarding iatrogenic reasons, the mechanical stress of centrifugation and the removal of sperm from the protective seminal plasma antioxidants during sperm preparation have the potential to increase sperm oxidative stress. Moreover, lifestyle factors, including strenuous exercise, smoking, dietary deficiencies or obesity, excessive alcohol consumption, psychological stress, and aging can account for a reduction in sperm quality mediated by an increase in seminal plasma ROS generation. These changes in semen/sperm quality and function (including zona pellucida binding ability) are detectable by the 10-min-long sperm–HA-binding assay.

Further, exposure to environmental toxicants, such as various pesticides and herbicides, paint solvents, heavy metals (cadmium), and various airborne pollutants, was linked with sperm oxidative damage. Moreover, genitourinary tract infections are well-established causes of sperm oxidative stress. Also, several chronic systemic infections, such as human immunodeficiency virus (HIV) and hepatitis B and C, have been linked with leukocytosis within semen. Increased ROS production may also occur after postvasectomy vasovasostomy repairs which promote antisperm immune responses by exposing the immune system to intrinsic sperm proteins and by disruption of the blood–testis barrier. As a testicular cause of oxidative stress, varicocele is now widely believed to be a major underlying pathology linking oxidative stress with male infertility. Recently, it has been reported that even after corrective surgery, men with cryptorchid testis have markedly elevated sperm ROS production and DNA fragmentation [98]. Chronic diseases, such as diabetes or metabolic imbalance, are also increasingly recognized as possible causes.

A recent review recognizes eight laboratory signs identifiable and suggestive of the possible presence of sperm oxidative stress [97]. These include poor sperm motility, teratozoospermia, high number of leukocytes in semen, increased semen viscosity, poor sperm membrane integrity on hypoosmotic swelling test (HOST), poor fertilization in routine IVF, poor sperm motility after overnight incubation, and poor blastocyst development (diminished paternal contribution by sperm), in addition to potential female factors (such as maternal age of >40 years or poor ovarian reserve, etc.). In the light of the research by the Huszar, Aitken, and other laboratories, one can also consider increased sperm cytoplasmic retention, caspase-3 apoptotic activity, increased proportion of sperm with aniline blue staining, and a low HA binding score [7, 8, 14, 48, 52, 53]. Sperm selected by the HA-mediated methods of either by the hyaluronic acid-coated glass slide or by the IVF–ICSI "PICSI" dish have high sperm DNA integrity whether tested by the DNA nick translation method or by the acridine orange fluorescence assay [24, 99].

Management of Oxidative Stress in Male Infertility

Upon establishment that a man has oxidative stress-related infertility, the appropriate treatment strategy is the identification and elimination of the underlying cause, whether lifestyle and environmental factors that enhance oxidative stress; antibiotic treatment of infections; and optimizing laboratory procedures by reducing the mechanical stress of sperm preparation or using sperm for IVF–ICSI arising from testicular sperm extraction (TESE sperm), as well as diminished application of cryopreserved spermatozoa which are very susceptible for DNA damage [97]. Regarding the utilization of TESE sperm, it is established that sperm in contact with Sertoli cells are relatively protected from oxidative attack, whereas most ROS-mediated sperm DNA fragmentation occurs during epididymal storage [100]. Three studies have compared sperm DNA quality of the same individuals in either ejaculated sperm or surgically aspirated epididymal sperm [100–102]. There were significant improvements in sperm DNA quality in the testicle-extracted samples. This seems reassuring; however, it is presently uncertain if protection from epididymal oxidative stress is the sole reason for the observed improvements in DNA quality. Thus, the use of TESE sperm in men with poor DNA quality should only be introduced, if more conservative treatments such as lifestyle modification and antioxidant therapy have failed.

Vitamin and antioxidant supplementation has also been suggested to treat oxidative stress related to diminished male fertility. To date, numerous studies have focused upon the effects of various antioxidant treatments on sperm parameters and pregnancy outcome. With such a large body of emerging evidence, one would expect solid data on the clinical efficacy of oral antioxidants toward better sperm function and pregnancy outcome. Unfortunately, this is not the case because (a) there are variations in the types and doses of antioxidants used; (b) some of the studies had small patient populations; (c) there is a lack of proper prospective placebo-controlled studies; (d) other factors of uncertainty have been added by the variations in the length of therapy and posttherapy observation periods; (e) further, the male germ cell developmental stage during which the ROS is most damaging is yet to be clarified, along with the related question of the most beneficial duration of antioxidant therapy.

A recent trend introduced is the enhancement of the seminal plasma resistance to ROS by administration of oral antioxidant supplements. Such agents include astaxanthin [103], carnitine [104], or a combination of acetylcysteine, β-carotene, vitamin E, and essential fatty acids [105]. Although the study by Comhaire et al. [103] suggests a positive effect of astaxanthin on sperm parameters and fertility, the results need to be confirmed in a larger trial before introducing astaxanthin for the treatment of infertile men. The 2000 study by Comhaire et al. [105] is a prospective pilot study for the assessment of oral antioxidant treatment and fatty acid administration on sperm functional integrity. The study is not placebo-controlled, and thus one cannot well ascertain the possible effects and efficacy of this approach. However, the study is considered important because negative findings would have discouraged a full-scale randomized trial.

In other approaches, vitamin E (400 mg) and selenium (225 μg) or vitamin E alone were reported to reduce sperm MDA levels in placebo-controlled studies [106, 107]. For instance, an 8-week treatment with vitamins C and E resulted in a very significant reduction in sperm DNA damage [100, 101]. Therefore, antioxidant treatment of infertile men, or addition of antioxidants to sperm media, may significantly improve sperm DNA quality [42, 100, 101, 105–119]. Antioxidants may also protect sperm from ROS-mediated impaired motility [109–115]. However, the lack of improvement in sperm DNA integrity in response to oral antioxidant treatment in subgroups of patients suggests that this treatment may provide benefits in men affected by seminal oxidative stress. It is possible that oxidative damage may be involved in the etiology of early or late spermatogenetic damage in some patients and not in others. A larger prospective controlled study is necessary to confirm these data, employing various diagnostic and therapeutic protocols to evaluate cases of male infertility combined with sperm DNA damage for an optimal management.

While many studies featured antioxidant supplements to improve sperm concentration and sperm morphology, the majority of good-quality studies focus upon improvements in sperm motility. However, the associated benefits in improved pregnancy rates are less well documented [120]. Suleiman et al. found that treatment with vitamin E resulted in a significant decline in sperm ROS damage and an increase in pregnancy rates during a 6-month period (21% pregnancy rate achieved in the vitamin E group vs. 0% in the placebo group) [107, 120]. However, in another study, there was a lack of improvement in pregnancy outcome following a 60-day-long regimen with a combination of vitamins C and E [121]. A recent randomized placebo-controlled trial, evaluating the antioxidant complex Menevit, administered at least 3 months, showed an increase in clinical pregnancy rates following IVF–ICSI. [122]. An editorial by Baker and Edgar has criticized the claims of this study [123]. In a related study by Tremellen et al., the proportions of embryos with excellent, good, and poor quality were similar in the placebo and Menevit groups, further probing (out) exploring the claims of improved embryo quality [122].

Key Issues and Recommendations in Light of Current Concepts

- Oxidative stress and consequential DNA chain degradation has been recognized as one of the most relevant contributor to male factor infertility, affecting the fertilization process and/or the paternal contribution of sperm to the embryo.
- The combination of diminished antioxidant activity in the seminal plasma and in other parts of the male reproductive tract, along with arrested development of individual spermatozoa, causes oxidative stress with consequential damages of DNA integrity and sperm function.
- Since antioxidants counteract the action of reactive oxygen species, these agents have been used in the medical treatment of male infertility. In assisted reproduction, antioxidants were supplemented in culture media during sperm preparation techniques which expose spermatozoa to the mechanical stress of centrifugation.
- Nevertheless, the efficacy of antioxidant treatment has been reported to be very limited. This may relate to (1) patient selection bias; (2) vague diagnosis of male infertility and incomplete understanding of the underlying pathophysiology; (3) wide man-to-man variations in the proportion of sperm with arrested cellular development and consequential attributes (lipid peroxidation, cytoplasmic retention, low HspA2 expression, persistent histones, DNA damage, etc.); (4) lack of double-blinded, placebo-controlled clinical trials; (5) use of imprecise clinical endpoints or targets in evaluating antioxidant treatment benefits; or (6) uncertainty regarding the duration of necessary or optimal treatment length.

Benefits of sperm–hyaluronic acid binding assay and the hyaluronic acid-mediated selection of single sperm without DNA chain fragmentation:

- Regarding the diagnostic and therapeutic benefits of sperm–hyaluronic acid interaction, until the development of the sperm–hyaluronic acid binding assay, there were no objective methods for the assessment of the proportion of native sperm that were affected by ROS, including plasma membrane damage and DNA chain degradation. The sperm–hyaluronic acid-binding assay introduced in the past 5 years for semen laboratories and for reproductive physicians is a welcome addition for four reasons: (1) Sperm–HA-binding assay is a 15-min objective test that probes the integrity of the plasma membrane with respect to HA and zona pellucida receptors. Thus, in addition to measuring the HA-binding score, the test reflects the proportion of sperm with full cellular maturity that are candidates for zona pellucida binding. (2) The HA-bound sperm population was shown to lack DNA degradation by both the DNA nick translation assay and by acridine orange fluorescence [24, 99]. The HA-bound sperm also lack attributes of arrested sperm development, such as cytoplasmic retention, persistent histones detectable by aniline blue, and apoptotic processes. (3) Further, the HA-bound sperm fraction is enriched in sperm with Tygerberg normal morphology to the same extent as zona pellucida-bound spermatozoa, and (4) the genetic integrity of the HA-selected sperm supported by the findings that, in addition to the lack of DNA fragmentation, such sperm exhibit aneuploidy frequencies comparable to that of normozoospermic fertile man. [24, 56, 74] (Fig. 44.1).
- Because sperm that are damaged by oxidative stress do not bind to the zona pellucida, these male infertility patients are often treated with ICSI. However, following ICSI fertilization with visually selected spermatozoa, increased rates of de novo numerical and structural chromosomal aberrations and increased rates of spontaneous abortions and birth defects may occur [71, 73, 75]. However, HA-mediated sperm selection for ICSI may optimize fertilization rates and sperm quality with respect to the various parameters. Therefore, HA-mediated selection of mature sperm facilitates the avoidance of sperm affected by oxidative damage. Thus, HA-mediated sperm selection is a treatment opportunity for male infertility due to oxidative damage of sperm. This is also an approach that provides a means of reduction of the potential genetic consequences evidenced following ICSI with visually selected sperm. We can

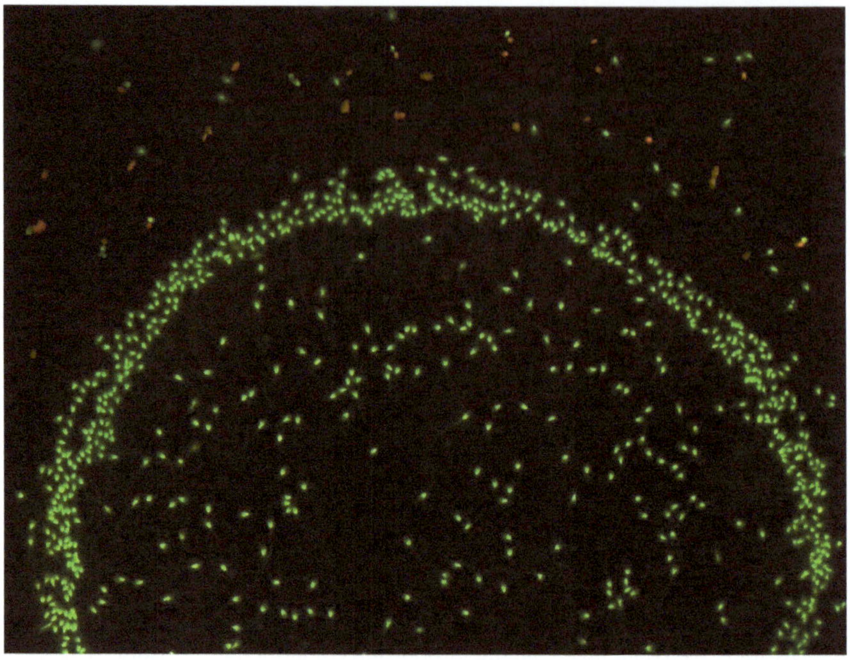

Fig. 44.1 Acridine orange stained sperm (DNA damaged sperm - red, high DNA chain integrity sperm - green). Perimeter of hyaluronic acid spot - washed semen sperm (red and green), middle of PICSI (inside the hyaluronic acid spot ring). Hyaluronic acid selected sperm - green (hundreds of them!)

conclude that HA-mediated sperm ICSI selection will increase the fertilization rates and paternal contribution of sperm to the embryo even if the man to be treated is affected by seminal ROS production. For these reasons, the sperm–HA binding assay and the PICSI dish introduced in andrology and embryology laboratories are excellent devices for both the diagnosis and fertility assessment of men with seminal ROS-related infertility, as well as in treatment of such couples by ejaculated sperm, rather than the currently recommended approach with TESE sperm of testicular origin.

References

1. Aitken RJ, Koopman P, Lewis SE. Seeds of concern. Nature. 2004;432:48–52.
2. De Laminardi E, Tsai C, Harakat A, et al. Involvement of reactive oxygen species in human sperm acrosome reaction induced by A 23187, lysophosphatidylcholine, and biological Xuid ultraWltrates. J Androl. 1998;19:585–94.
3. de Lamirande E, Gagnon C. Human sperm hyperactivation and capacitation as parts of an oxidative process. Free Radic Biol Med. 1993;14:157–66.
4. Aitken RJ, Paterson M, Fisher H, et al. Redox regulation of tyrosine phosphorylation in human spermatozoa and its role in the control of human sperm function. J Cell Sci. 1995;108:2017–25.
5. Aitken RJ, Gordon E, Harkiss D, et al. Relative impact of oxidative stress on the functional competence and genomic integrity of human spermatozoa. Biol Reprod. 1998;59:1037–46.
6. Gomez E, Irvine DS, Aitken RJ. Evaluation of a spectrophotometric assay for the measurement of malondialdehyde and 4-hydroxyalkenals in human spermatozoa: relationships with semen quality and sperm function. Int J Androl. 1998;21:81–94.
7. Aitken J, Krausz C, Buckingham D. Relationships between biochemical markers for residual sperm cytoplasm, reactive oxygen species generation, and the presence of leukocytes and precursor germ cells in human sperm suspensions. Mol Reprod Dev. 1994;39:268–79.
8. Huszar G, Vigue L. Correlation between the rate of lipid peroxidation and cellular maturity as measured by creatine kinase activity in human spermatozoa. J Androl. 1994;15:71–7.
9. Huszar G, Corrales M, Vigue L. Correlation between sperm creatine phosphokinase activity and sperm concentrations in normospermic and oligospermic men. Gamete Res. 1988;19:67–75.
10. Huszar G, Vigue L, Corrales M. Sperm creatine phosphokinase activity as a measure of sperm quality in normospermic, variablespermic, and oligospermic men. Biol Reprod. 1988;38:1061–6.
11. Huszar G, Vigue L. Spermatogenesis-related change in the synthesis of the creatine kinase B-type and M-type isoforms in human spermatozoa. Mol Reprod Dev. 1990;25:258–62.
12. Aziz N, Saleh RA, Sharma RK. et al; Novel association between sperm reactive oxygen species production, sperm morphological defects, and the sperm deformity index. Fertil Steril. 2004;81:349–54.
13. Sikka SC. Relative impact of oxidative stress on male reproductive function. Curr Med Chem (Rev). 2001;8:851–62.
14. Huszar G, Vigue L. Incomplete development of human spermatozoa is associated with increased creatine phosphokinase concentration and abnormal head morphology. Mol Reprod Dev. 1993;34:292–8.
15. Huszar G, Stone K, Dix D, et al. Putative creatine kinase M-isoform in human sperm is identified as the 70-kilodalton heat shock protein HspA2. Biol Reprod. 2000;63:925–32.
16. Lalwani S, Sayme N, Vigue L, et al. Biochemical markers of early and late spermatogenesis: relationship between the lactate dehydrogenase-X and creatine kinase-M isoform concentrations in human spermatozoa. Mol Reprod Dev. 1996;43:495–502.
17. Huszar G, Vigue L, Corrales M. Sperm creatine kinase activity in fertile and infertile oligospermic men. J Androl. 1990;11:40–6.
18. Huszar G, Vigue L, Morshedi M. Sperm creatine phosphokinase M-isoform ratios and fertilizing potential of men: a blinded study of 84 couples treated with in vitro fertilization. Fertil Steril. 1992;57:882–8.
19. Huszar G, Vigue L, Oehninger S. Creatine kinase immunocytochemistry of human sperm-hemizona complexes: selective bind-

ing of sperm with mature creatine kinase-staining pattern. Fertil Steril. 1994;61:136–42.
20. Huszar G, Sbracia M, Vigue L, et al. Sperm plasma membrane remodeling during spermiogenetic maturation in men: relationship among plasma membrane beta 1,4-galactosyltransferase, cytoplasmic creatine phosphokinase, and creatine phosphokinase isoform ratios. Biol Reprod. 1997;56:1020–4.
21. Huszar G, Willetts M, Corrales M. Hyaluronic acid (Sperm Select) improves retention of sperm motility and velocity in normospermic and oligospermic specimens. Fertil Steril. 1990;54:1127–34.
22. Sbracia M, Grasso J, Sayme N, et al. Hyaluronic acid substantially increases the retention of motility in cryopreserved/thawed human spermatozoa. Hum Reprod. 1997;12:1949–54.
23. Huszar G, Ozenci CC, Cayli S, et al. Hyaluronic acid binding by human sperm indicates cellular maturity, viability, and unreacted acrosomal status. Fertil Steril. 2003;79:1616–24.
24. Huszar G, Jakab A, Sakkas D, et al. Fertility testing and ICSI sperm selection by hyaluronic acid binding: clinical and genetic aspects. RBM Online. 2007;14:650–63.
25. Sharma RK, Pasqualotto AE, Nelson DR, et al. Relationship between seminal white blood cell counts and oxidative stress in men treated at an infertility clinic. J Androl. 2001;22:575–83.
26. Iwasaki A, Gagnon C. Formation of reactive oxygen species in spermatozoa of infertile patients. Fertil Steril. 1992;57:409–16.
27. Shekarriz M, Thomas Jr AJ, Agarwal A. Incidence and level of seminal reactive oxygen species in normal men. Urology. 1995;45:103–7.
28. Shekarriz M, DeWire DM, Thomas Jr AJ, et al. A method of human semen centrifugation to minimize the iatrogenic sperm injuries caused by reactive oxygen species. Eur Urol. 1995;28:31–5.
29. Potts JM, Sharma R, Pasqualotto F, et al. Association of ureaplasma urealyticum with abnormal reactive oxygen species levels and absence of leukocytospermia. J Urol. 2000;163:1775–8.
30. Potts RJ, Notarianni LJ, Jefferies TM. Seminal plasma reduces exogenous oxidative damage to human sperm, determined by the measurement of DNA strand breaks and lipid peroxidation. Mutat Res. 2000;447:249–56.
31. Watson PF. The causes of reduced fertility with cryopreserved semen. Anim Reprod Sci. 2000;60–61:481–92.
32. Irvine DS, Twigg JP, Gordon EL, et al. DNA integrity in human spermatozoa: relationships with semen quality. J Androl. 2000; 21:33–44.
33. Ochsendorf FR, Thiele J, Fuchs J, et al. Chemiluminescence in semen of infertile men. Andrologia. 1994;26:289–93.
34. Agarwal A, Prabakaran S, Allamaneni S. What an andrologist/urologist should know about free radicals and why. Urology. 2006;67:2–8.
35. Jones R, Mann T, Sherins R. Peroxidative breakdown of phospholipids in human spermatozoa, spermicidal properties of fatty acid peroxides, and protective action of seminal plasma. Fert Steril. 1979;31:531–7.
36. de Lamirande E, Gagnon C. Reactive oxygen species and human spermatozoa. II Depletion of adenosine triphosphate plays an important role in the inhibition of sperm motility. J Androl. 1992;13:379–86.
37. Whittington K, Harrison SC, Williams KM, et al. Reactive oxygen species (ROS) production and the outcome of diagnostic tests of sperm function. Int J Androl. 1999;22:236–42.
38. Kao SH, Chao HT, Chen HW, et al. Increase of oxidative stress in human sperm with lower motility. Fertil Steril. 2008;89:1183–90.
39. Dadoune JP, Mayaux MJ, Guihard-Moscato ML. Correlation between defects in chromatin condensation of human spermatozoa stained by aniline blue and semen characteristics. Andrologia. 1988;20:211–7.
40. Foresta C, Zorzi M, Rossato M, et al. Sperm nuclear instability and staining with aniline blue: abnormal persistence of histones in spermatozoa in infertile men. Int J Androl. 1992;15:330–7.
41. Oliva R. Protamines and male infertility. Hum Reprod Update. 2006;12:417–35.
42. Kodama H, Yamaguchi R, Fukuda J, et al. Increased oxidative deoxyribonucleic acid damage in the spermatozoa of infertile male patients. Fertil Steril. 1997;68:519–24.
43. Saleh RA, Agarwal A, Kandirali E, et al. Leukocytospermia is associated with increased reactive oxygen species production by human spermatozoa. Fertil Steril. 2002;78:1215–24.
44. Oger I, Da Cruz C, Panteix G, et al. Evaluating human sperm DNA integrity: relationship between 8-hydroxydeoxyguanosine quantification and the sperm chromatin structure assay. Zygote. 2003;11:367–71.
45. Wang X, Sharma RK, Sikka SC, et al. Oxidative stress is associated with increased apoptosis leading to spermatozoa DNA damage in patients with male factor infertility. Fertil Steril. 2003;80: 531–5.
46. Henkel R, Kierspel E, Stalf T, et al. Effect of reactive oxygen species produced by spermatozoa and leukocytes on sperm functions in nonleukocytospermic patients. Fertil Steril. 2005;83:635–42.
47. Cayli S, Jakab A, Ovari L, et al. Biochemical markers of sperm function: male fertility and sperm selection for ICSI. RBM Online. 2003;7:462–8.
48. Sati L, Ovari L, Bennett D, et al. Double probing of human spermatozoa for persistent histones, surplus cytoplasm, apoptosis and DNA fragmentation. RBM Online. 2008;16:570–9.
49. Allen JW, Dix DJ, Collins BW, et al. HSP70-2 is part of the synaptonemal complex in mouse and hamster spermatocytes. Chromosoma. 1996;104:414–21.
50. Eddy EM. Role of heat shock protein HSP70-2 in spermatogenesis. Rev Reprod. 1999;4:23–30.
51. Tsunekawa N, Matsumoto M, Tone S, et al. The Hsp70 homolog gene, Hsc70t, is expressed under translational control during mouse spermiogenesis. Mol Reprod Dev. 1999;52:383–91.
52. Cayli S, Sakkas D, Vigue L, et al. Cellular maturity and apoptosis in human sperm: creatine kinase, caspase-3 and Bcl-XL levels in mature and diminished maturity sperm. Mol Hum Reprod. 2004;10:365–72.
53. Aitken RJ, Baker MA, Sawyer D. Oxidative stress in the male germ line and its role in the aetiology of male infertility and genetic disease. RBM Online. 2003;7:65–70.
54. Alvarez JG. DNA fragmentation in human spermatozoa: significance in the diagnosis and treatment of infertility. Minerva Ginecol. 2003;55:233–9.
55. Govin J, Caron C, Escoffier E, et al. Post-meiotic shifts in HSPA2/HSP70.2 chaperone activity during mouse spermatogenesis. J Biol Chem. 2006;281:37888–92.
56. Prinosilova P, Kruger T, Sati L, et al. Selectivity of hyaluronic acid binding for spermatozoa with normal Tygerberg strict morphology. RBM Online. 2009;18:177–83.
57. Bernardini L, Borini A, Preti S, et al. Study of aneuploidy in normal and abnormal germ cells from semen of fertile and infertile men. Hum Reprod. 1998;13:3406–13.
58. Colombero LT, Hariprashad JJ, Tsai MC, et al. Incidence of sperm aneuploidy in relation to semen characteristics and assisted reproductive outcome. Fertil Steril. 1999;72:90–6.
59. Calogero AE, De Palma A, Grazioso C, et al. Aneuploidy rate in spermatozoa of selected men with abnormal semen parameters. Hum Reprod. 2001;16:1172–9.
60. Templado C, Hoang T, Greene C, et al. Aneuploid spermatozoa in infertile men: teratozoospermia. Mol Reprod Dev. 2002;61:200–4.
61. Aran B, Vidal F, Vendrell JM, et al. Outcome of intracytoplasmic sperm injection in relation to the meiotic pattern in patients with severe oligoasthenozoospermia. Fertil Steril. 2003;80:91–5.
62. Pang MG, Kim YJ, Lee SH, et al. The high incidence of meiotic errors increases with decreased sperm count in severe male factor infertilities. Hum Reprod. 2005;20:1688–94.

63. Lee JD, Kamiguchi Y, Yanagimachi R. Analysis of chromosome constitution of human spermatozoa with normal and aberrant head morphologies after injection into mouse oocytes. Hum Reprod. 1996;11:1942–6.
64. Yakin K, Kahraman S. Certain forms of morphological anomalies of spermatozoa may reflect chromosomal aneuploidies. Hum Reprod. 2001;16:1779–80.
65. Kovanci E, Kovacs T, Moretti E, et al. FISH assessment of aneuploidy frequencies in mature and immature human spermatozoa classified by the absence or presence of cytoplasmic retention. Hum Reprod. 2001;16:1209–17.
66. Egozcue S, Blanco J, Vidal F, et al. Diploid sperm and the origin of triploidy. Hum Reprod. 2002;17:5–7.
67. Celik-Ozenci C, Catalanotti J, Jakab A, et al. Human sperm maintain their shape following decondensation and denaturation for fluorescent in situ hybridization: shape analysis and objective morphometry. Biol Reprod. 2003;69:1347–55.
68. Celik-Ozenci C, Jakab A, Kovacs T, et al. Sperm selection for ICSI: shape properties do not predict the absence or presence of numerical chromosomal aberrations. Hum Reprod. 2004;19:2052–9.
69. Zavaczki Z, Celik-Ozenci C, Ovari L, et al. Dimensional assessment of X-bearing and Y-bearing haploid and disomic human sperm with the use of fluorescence in situ hybridization and objective morphometry. Fertil Steril. 2006;85:121–7.
70. Simpson JL, Lamb DJ. Genetic effects of intracytoplasmic sperm injection. Semin Reprod Med. 2001;19:239–49.
71. Van Steirteghem A, Bonduelle M, Liebaers I, et al. Children born after assisted reproductive technology. Am J Perinatol. 2002;19:59–65.
72. Barri PN. Multiple pregnancies: a plea for informed caution. Hum Reprod Update. 2005;11:1–2.
73. Bonduelle M, Wennerholm UB, Loft A, et al. A multi-centre cohort study of the physical health of 5-year-old children conceived after intracytoplasmic sperm injection, in vitro fertilization and natural conception. Hum Reprod. 2005;20:413–9.
74. Jakab A, Sakkas D, Delpiano E, et al. Intracytoplasmic sperm injection: a novel selection method for sperm with normal frequency of chromosomal aneuploidies. Fertil Steril. 2005;84:1665–73.
75. Bonduelle M, Van Assche E, Joris H, et al. Prenatal testing in ICSI pregnancies: incidence of chromosomal anomalies in 1586 karyotypes and relation to sperm parameters. Hum Reprod. 2002;17:2600–14.
76. Huszar G, Patrizio P, Vigue L, et al. Cytoplasmic extrusion and the switch from creatine kinase B to M isoform are completed by the commencement of epididymal transport in human and stallion spermatozoa. J Androl. 1998;19:11–20.
77. Seli E, Sakkas D. Spermatozoal nuclear determinants of reproductive outcome: implications for ART. Hum Reprod Update. 2005;11:337–49.
78. Borini A, Tarozzi N, Bizzaro D, et al. Sperm DNA fragmentation: paternal effect on early post-implantation embryo development in ART. Hum Reprod. 2006;21:2876–81.
79. Ochsendorf FR. Infections in the male genital tract and reactive oxygen species. Hum Reprod Update. 1999;5:399–420.
80. Li K, Shang X, Chen Y. High-performance liquid chromatographic detection of lipid peroxidation in human seminal plasma and its application to male infertility. Clin Chim Acta. 2004;346: 199–203.
81. Shang XJ, Li K, Ye ZQ, et al. Analysis of lipid peroxidative levels in seminal plasma of infertile men by high-performance liquid chromatography. Arch Androl. 2004;50:411–6.
82. Aitken RJ, Harkiss D, Buckingham D. Relationship between iron-catalysed lipid peroxidation potential and human sperm function. J Reprod Fertil. 1993;98:257–65.
83. Khosrowbeygi A, Zarghami N. Levels of oxidative stress biomarkers in seminal plasma and their relationship with seminal parameters. BMC Clin Pathol. 2007;7:6.
84. Aitken RJ, Wingate JK, De Iuliis GN, et al. Analysis of lipid peroxidation in human spermatozoa using BODIPY C11. Mol Hum Reprod. 2007;13:203–11.
85. Saleh RA, Agarwal A, Nada EA, et al. Negative effects of increased sperm DNA damage in relation to seminal oxidative stress in men with idiopathic and male factor infertility. Fertil Steril. 2003;79: 1597–605.
86. Simon L, Brunborg G, Stevenson M, Lutton D, McManus J, Lewis SE. Clinical significance of sperm DNA damage in assisted reproduction outcome. Human Reprod. 2010;25(7):1594–608.
87. Ozmen B, Koutlaki N, Youssry M, et al. DNA damage of human spermatozoa in assisted reproduction: origins, diagnosis, impacts and safety. RBM Online. 2007;14:384–95.
88. Nakamura H, Kimura T, Nakajima A, et al. Detection of oxidative stress in seminal plasma and fractionated sperm from subfertile male patients. Eur J Obstet Gynecol Reprod Biol. 2002;105: 155–60.
89. Loft S, Kold-Jensen T, Hjollund NH, et al. Oxidative DNA damage in human sperm influences time to pregnancy. Hum Reprod. 2003;18:1265–72.
90. Kobayashi H, Gil-Guzman E, Mahran AM, et al. Quality control of reactive oxygen species measurement by luminol-dependent chemiluminescence assay. J Androl. 2001;22:568–74.
91. Aitken RJ, Baker MA, O'Bryan M. Shedding light on chemiluminescence: the application of chemiluminescence in diagnostic andrology. J Androl. 2004;25:455–65.
92. Liochev SI, Fridovich I. Lucigenin (Bis-N-methylacridinium) as a mediator of superoxide anion production. Archives Biochem Biophys. 1997;337:115–20.
93. World Health Organization. Laboratory manual for the examination of human semen and sperm-cervical mucous interaction. 4th ed. New York: Cambridge University; 1999.
94. Sharma RK, Pasqualotto FF, Nelson DR, et al. The reactive oxygen species-total antioxidant capacity score is a new measure of oxidative stress to predict male infertility. Hum Reprod. 1999;14: 2801–7.
95. Said TM, Kattal N, Sharma RK, et al. Enhanced chemiluminescence assay vs colorimetric assay for measurement of the total antioxidant capacity of human seminal plasma. J Androl. 2003;24:676–80.
96. Erel O. A novel automated direct measurement method for total antioxidant capacity using a new generation, more stable ABTS radical cation. Clin Biochem. 2004;37:277–85.
97. Tremellen K. Oxidative stress and male infertility—a clinical perspective. Hum Reprod Update. 2008;14:243–58.
98. Smith GR, Kaune GH, Ch Parodi D, et al. Extent of sperm DNA damage in spermatozoa from men examined for infertility. Relationship with oxidative stress. Rev Med Chil. 2007;135: 279–86.
99. Yagci A, Murk W, Stronk J, et al. Spermatozoa bound to solid state hyaluronic acid show chromatin structure with high DNA chain integrity: an Acridine Orange Fluorescence Study. J Androl. 2010; 31:566–72.
100. Greco E, Scarselli F, Iacobelli M, et al. Efficient treatment of infertility due to sperm DNA damage by ICSI with testicular spermatozoa. Hum Reprod. 2005;20:226–30.
101. Greco E, Iacobelli M, Rienzi L, et al. Reduction of the incidence of sperm DNA fragmentation by oral antioxidant treatment. J Androl. 2005;26:349–53.
102. O'Connell M, McClure N, Lewis SE. Mitochondrial DNA deletions and nuclear DNA fragmentation in testicular and epididymal human sperm. Hum Reprod. 2002;17:1565–70.
103. Comhaire FH, El El Garem Y, Mahmoud A, et al. Combined conventional/antioxidant 'Astaxanthin' treatment for male infertility: a double blind, randomized trial. Asian J Androl. 2005;7: 257–62.
104. Vicari E, Calogero AE. Effects of treatment with carnitines in infertile patients with prostato-vesiculo-epididymitis. Hum Reprod. 2001;16:2338–42.

105. Comhaire FH, Christophe AB, Zalata AA, et al. The effects of combined conventional treatment, oral antioxidants and essential fatty acids on sperm biology in subfertile men. Prostaglandins Leukot Essent Fatty Acids. 2000;63:159–65.
106. Keskes-Ammar L, Feki-Chakroun N, Rebai T, et al. Sperm oxidative stress and the effect of an oral vitamin E and selenium supplement on semen quality in infertile men. Arch Androl. 2003;49:83–94.
107. Suleiman SA, Ali ME, Zaki ZM, et al. Lipid peroxidation and human sperm motility: protective role of vitamin E. J Androl. 1996;17:530–7.
108. Menezo YJ, Hazout A, Panteix G, et al. Antioxidants to reduce sperm DNA fragmentation: an unexpected adverse effect. RBM Online. 2007;14:418–21.
109. MacLeod J. The role of oxygen in the metabolism and motility of human spermatozoa. Am J Physiol. 1943;138:512–8.
110. Kobayashi T, Miyazaki T, Natori M, et al. Protective role of superoxide dismutase in human sperm motility: superoxide dismutase activity and lipid peroxide in human seminal plasma and spermatozoa. Hum Reprod. 1991;6:987–91.
111. Oeda T, Henkel R, Ohmori H, et al. Scavenging effect of N-acetyl-L-cysteine against reactive oxygen species in human semen: a possible therapeutic modality for male factor infertility? Andrologia. 1997;29:125–31.
112. Zheng RL, Zhang H. Effects of ferulic acid on fertile and asthenozoospermic infertile human sperm motility, viability, lipid peroxidation, and cyclic nucleotides. Free Radic Biol Med. 1997;22:581–6.
113. Donnelly ET, McClure N, Lewis SE. Glutathione and hypotaurine in vitro: effects on human sperm motility, DNA integrity and production of reactive oxygen species. Mutagenesis. 2000;15:61–8.
114. Rossi T, Mazzilli F, Delfino M, et al. Improved human sperm recovery using superoxide dismutase and catalase supplementation in semen cryopreservation procedure. Cell Tissue Bank. 2001;2:9–13.
115. Yenilmez E, Yildirmis S, Yulug E, et al. Ham's F-10 medium and Ham's F-10 medium plus vitamin E have protective effect against oxidative stress in human semen. Urology. 2006;67:384–7.
116. Lenzi A, Culasso F, Gandini L, et al. Placebo-controlled, double-blind, cross-over trial of glutathione therapy in male infertility. Hum Reprod. 1993;8:1657–62.
117. Lenzi A, Sgro P, Salacone P, et al. A placebo-controlled double-blind randomized trial of the use of combined l-carnitine and l-acetyl-carnitine treatment in men with asthenozoospermia. Fertil Steril. 2004;81:1578–84.
118. Scott R, MacPherson A, Yates RW, et al. The effect of oral selenium supplementation on human sperm motility. Br J Urol. 1998;82:76–80.
119. Balercia G, Regoli F, Armeni T, et al. Placebo-controlled double-blind randomized trial on the use of L-carnitine, L-acetylcarnitine, or combined L-carnitine and L-acetylcarnitine in men with idiopathic asthenozoospermia. Fertil Steril. 2005;84:662–71.
120. Agarwal A, Nallella KP, Allamaneni SS, et al. Role of antioxidants in treatment of male infertility: an overview of the literature. RBM Online. 2004;8:616–27.
121. Rolf C, Cooper TG, Yeung CH, et al. Antioxidant treatment of patients with asthenozoospermia or moderate oligoasthenozoospermia with high-dose vitamin C and vitamin E: a randomized, placebo-controlled, double-blind study. Hum Reprod. 1999;14:1028–33.
122. Tremellen K, Miari G, Froiland D, et al. A randomised control trial examining the effect of an antioxidant (Menevit) on pregnancy outcome during IVF-ICSI treatment. Aust N Z J Obstet Gynaecol. 2007;47:216–21.
123. Baker HW, Edgar D. Trials of antioxidants for male infertility. Aust N Z J Obstet Gynaecol. 2008;48:125–6.

Further Reading

Gomez E, Buckingham DW, Brindle J, et al. Development of an image analysis system to monitor the retention of residual cytoplasm by human spermatozoa: correlation with biochemical markers of the cytoplasmic space, oxidative stress, and sperm function. J Androl. 1996;17:276–87.

Pasqualotto FF, Sharma RK, Kobayashi H, et al. Oxidative stress in normospermic men undergoing infertility evaluation. J Androl. 2001;22:316–22.

Garrido N, Meseguer M, Alvarez J, et al. Relationship among standard semen parameters, glutathione peroxidase/glutathione reductase activity, and mRNA expression and reduced glutathione content in ejaculated spermatozoa from fertile and infertile men. Fertil Steril. 2004;82:1059–66.

Said TM, Agarwal A, Sharma RK, et al. Human sperm superoxide anion generation and correlation with semen quality in patients with male infertility. Fertil Steril. 2004;82:871–7.

Said TM, Agarwal A, Sharma RK, et al. Impact of sperm morphology on DNA damage caused by oxidative stress induced by beta-nicotinamide adenine dinucleotide phosphate. Fertil Steril. 2005;83: 95–103.

Agarwal A, Sharma RK, Nallella KP, et al. Reactive oxygen species as an independent marker of male factor infertility. Fertil Steril. 2006;86:878–85.

Sakkas D, Alvarez JG. Sperm DNA fragmentation: mechanism of origin, impact onreproductive outcome, and analysis. Fertil Steril. 2010; 93(4): 1027–1036.

Extended Embryo Culture Supplementation

45

Alex C. Varghese, Eulalia Pozo-Guisado, Ignacio S. Alvarez, and Francisco Javier Martin-Romero

Abstract

This chapter reviews the recent knowledge regarding the role of reactive oxygen and nitrogen species on reproductive physiology. We also focus on the strategies that researches have considered to neutralize the oxidative stress that develops in parallel to the in vitro culture and cryopreservation. Basic studies using mammalian oocytes have been designed to investigate the benefits of the use of antioxidants during in vitro culture, and this chapter summarizes the most important findings regarding the supplementation of culture media to minimize the harmful effects of oxidative stress. However, an intense line of investigation is required to transfer the recent knowledge to the particular conditions of the human embryo, to ensure high rates of successful IVF cycles.

Keywords

Reactive oxygen species • Fertilization • Oxidative stress • Oogenesis • Blastocyst • Embryo • Heavy metal chelation • EDTA • Embryo culture supplementation • Cryopreservation

Reactive oxygen and nitrogen species are generated by diverse enzyme activities, and they act as modulators of many physiological functions. However, due to the intrinsic reactivity of these species, they can trigger undesired reactions threatening optimal cell functions. As a result, there are enzymatic and nonenzymatic mechanisms that counteract these reactive species, by means of scavenging, chelation, or modification of reactive species. These mechanisms constitute the antioxidant system, and the imbalance between the generation of reactive oxygen species (ROS) and the antioxidant system is known as oxidative stress. Oxidative stress is often the origin of a large number of diseases, including some cases of male and female infertility. However, the handling and in vitro culture of gametes and embryos during assisted reproductive techniques generate a significant level of oxidative stress. This oxidative stress is associated to the culture conditions because gametes and embryos are not within their physiological environment, but in a much more oxidative milieu. This chapter reviews the recent knowledge regarding the role of reactive oxygen and nitrogen species on reproductive physiology. We also focus on the strategies that researches have considered to neutralize the oxidative stress that develops in parallel to the in vitro culture and cryopreservation. Basic studies using mammalian oocytes have been designed to investigate the benefits of the use of antioxidants during in vitro culture, and this chapter summarizes the most important findings regarding the supplementation of culture media to minimize the harmful effects of oxidative stress. However, an intense line of investigation is required to transfer the recent knowledge to the particular conditions of the human embryo to ensure high rates of successful IVF cycles.

A.C. Varghese, PhD (✉)
Montreal Reproductive Centre,
2110, Boul Decarie, Montreal, QC, Canada, H4A 3J3
e-mail: alexcv2008@gmail.com; alex@lifeinvitro.com

E. Pozo-Guisado, PhD • F.J. Martin-Romero, PhD
Department of Biochemistry and Molecular Biology,
School of Life Sciences, University of Extremadura,
Avenida De Elvas S/N, 06006, Badajoz, Spain
e-mail: epozo@unex.es; fjmartin@unex.es

I.S. Alvarez, PhD
Department of Cell Biology, School of Life Sciences, University of Extremadura, Avenida de Elvas s/n, 06006, Badajoz, Spain
e-mail: ialvarez@unex.es

S.J. Parekattil and A. Agarwal (eds.), *Male Infertility: Contemporary Clinical Approaches, Andrology, ART & Antioxidants*,
DOI 10.1007/978-1-4614-3335-4_45, © Springer Science+Business Media, LLC 2012

Role of Reactive Species During Oogenesis

Many of the physiological processes in the ovary function are influenced by ROS and reactive nitrogen species (RNS), including folliculogenesis and oocyte maturation. Regarding oogenesis, it is known that members of the *Bcl-2* gene family are expressed in ovarian granulosa cells during follicular maturation, this expression being controlled by the follicle-stimulating hormone (FSH) [1]. FSH prevents apoptosis in granulosa cells in vitro in the same extent as some antioxidants, either enzymatic or nonenzymatic, such as superoxide dismutase (SOD), ascorbic acid, or *N*-acetyl-L-cysteine. In this regard, granulosa cells from rats primed with equine chorionic gonadotropin (eCG), which promotes antral follicular growth and survival, show an upregulation of the expression of extracellular SOD (ecSOD) isoforms and Mn-SOD, although the expression of Cu/Zn-SOD, glutathione peroxidase (GPx), or catalase is not influenced by gonadotropin priming [1]. In sum, gonadotropins support granulosa cell survival in developing follicles through the activation of antioxidant defenses, strongly strengthening the hypothesis that ROS/RNS modulate the kinetics of the folliculogenesis.

Physiological Role of Superoxide and Hydrogen Peroxide

ROS/RNS play a physiological role during ovulation, as suggested by the finding that the perfusion of in vitro cultured ovaries with SOD or catalase, delays ovulation [2]. In fact, the ovary and uterus of cycling and pregnant mice generate NADPH-dependent superoxide [3], and both ovarian and uterine NADPH-dependent superoxide production are likely to be luteinizing hormone (LH) inducible [4, 5]. Therefore, it is possible that ovulation could be associated with a significant increase in ROS levels, and these reactive species are generated by different cells. Macrophages, neutrophilic granulocytes, and T lymphocytes are present in the ovary at ovulation. There is an increase in the level of macrophages and neutrophilic granulocytes in the medullar region and in the thecal layer as the ovulatory period progresses [6], and macrophages and neutrophils in the ovary could be a major source of significant amounts of ROS during ovulation [7].

The superoxide generation can eventually lead to an increase in the hydrogen peroxide (H_2O_2) concentration, due to the presence of different isoforms of SOD in developing antral follicles. The location of these isoforms has been described recently, and cumulus–oocyte complexes (COCs) exhibit a compartmentalized and varying distribution of SOD1 (or Cu, Zn-SOD), SOD2 (or Mn-SOD), and SOD3 (extracellular or ecSOD) protein expression. SOD1, SOD2, and SOD3 proteins and activities are found in high levels in the follicular fluid from small follicles, while SOD3 protein is more abundant in oocytes and cumulus cells from large follicles relative to small follicles [8]. Regardless, SOD2 is the predominant isoform expressed in the ooplasm, while SOD3 is the main SOD type in cumulus cells. SOD1 shows enhanced nuclear accumulation in oocytes [8]. It is important to highlight that oocytes or COCs accumulate the diverse SOD isoforms during folliculogenesis to neutralize potential oxidative bursts, like the suggested during ovulation [7].

Role of Nitric Oxide

Nitric oxide (NO$^\bullet$) is biologically produced form L-arginine in a reaction catalyzed by the nitric oxide synthase (NOS). There are three isoforms of the NOS: NOS1, or neuronal constitutive NOS (nNOS); NOS2, the inducible NOS (iNOS); and NOS3, endothelial constitutive NOS (eNOS). They are all expressed in the ovary [9–11] and the oviduct [12], and NOS has been immunolocalized in thecal and stromal cells of the ovary during follicular development and ovulation [10]. NO$^\bullet$ synthesis increases during follicular development [13] and a parallel increase of nitrites and nitrates in the follicular fluid has been reported, as a direct consequence of the upregulated NO$^\bullet$ synthesis [14]. Moreover, the NO$^\bullet$ synthesis kinetics correlates well with estradiol concentration, suggesting that NO$^\bullet$ is an important signaling molecule during folliculogenesis. The role of NO$^\bullet$ as endocrine regulator is supported by the significant reduction in the folliculogenesis rate and the production of mature oocytes by aminoguanidine and L-NMMA, two well-known NOS inhibitors [15]. In contrast, sodium nitroprusside (SNP), a NO$^\bullet$ donor, reverses the effects of the NOS inhibitors, further supporting the conclusion that the production of NO$^\bullet$ by NOS is required for the follicle development. However, the production of NO$^\bullet$ during folliculogenesis and ovulation shows a relevant spatiotemporal dynamics. The concentration of nitrate/nitrite (NO$^\bullet$ metabolites) in preovulatory follicles is high before hCG injection but then decreases after hCG stimulation [16]. Accordingly, NO$^\bullet$ donors prevent germinal vesicle breakdown (GVBD) and inhibitors of iNOS counteract this prevention [17, 18]. Expression of eNOS, mainly localized in the thecal layer, increases after hCG priming [9, 11, 17, 19], and iNOS expression, largely found in granulosa cells, significantly decreases after hCG injection, reducing the levels of NO$^\bullet$ in the preovulatory follicular fluid. Thus, NO$^\bullet$ levels are reflecting significant alterations of iNOS expression, but not for eNOS in preovulatory follicles [17]. The iNOS-produced NO$^\bullet$ is an inhibitor of the oocyte maturation, because it induces an increase in cGMP concentration in preovulatory follicles, leading to an increase in the cGMP concentration in the oocyte via gap junctions, where it plays a key role in the meiotic arrest of oocytes [20]. cGMP keeps the meiotic arrest of preovulatory oocytes, increasing cAMP levels by the inhibition of the phosphodiesterase activity and by the activation of cGMP-dependent protein kinase [20], demonstrating that the iNOS–NO•–cGMP pathway modulates meiotic progression of the oocyte.

The dual effect of NO• is also observed with maturing mammalian oocytes in vitro. High concentrations of SNP or *S*-nitroso-*N*-acetylpenicillamine (SNAP) in culture induce the blockade of the meiotic progression at MI, and these arrested MI oocytes remain at this stage even after eliminating the NO• donor from the culture [21, 22]. On the contrary, SNP, within the nano- to low micromolar range, increases cytoplasmic maturation [21, 23] and the percentage of blastocyst rates [21]. There is some controversy in the recent data from studies using NO• donors, since micromolar SNP has been shown to induce embryotoxicity by reducing the glucose and pyruvate uptake and by depleting the levels of essential amino acids, leading to a much lower developmental rate [24]. The major reason for the contradictory data could be that the NO• concentration produced by NO• donors, which depends on the decomposition time, is strongly influenced by the redox status in the cytosolic compartment, or by the presence of other reactive species. Thus, it is absolutely required to study further the effect of ROS/RNS after a meticulous determination of the ROS/RNS-releasing kinetics from donors in every experimental condition. These methods are available from the recent literature and they are easy to adapt to particular experimental conditions. This is the case for the quantification of peroxynitrite, a strong oxidant and nitrating agent formed after the reaction of NO• with superoxide anion. Decomposition of SIN-1, a peroxynitrite releasing agent, can be continuously monitored, with a sensitivity lower to 0.1 μM, from the kinetics of NADH fluorescence quenching [25], giving the opportunity to calculate the concentration of the donor required for the exposure of oocytes/embryos to specific peroxynitrite concentrations.

Endothelium-derived NOS is a key modulator of oocyte meiotic maturation in vitro, as it has been described that ovaries from eNOS-knockout (eNOS-KO) mice contain a smaller amount of COCs, relative to wild-type mice, and that in vitro maturation of COCs from eNOS-KO mice results in a high percentage of abnormal MI oocytes with a low rate of fully mature oocytes [26]. Inhibition of NOS activity, attained with L-NAME, during in vitro maturation leads to an increase in apoptosis during embryo development, supporting a role for NO• in the meiotic progression of the mammalian oocyte [27].

In addition, NO• is probably related with implantation, as iNOS is differentially expressed in pre-, peri-, and postimplantation blastocysts. The enhanced NO• production, supported by the upregulation of iNOS, would act as a vasodilator and an angiogenic mediator [28]. However, it is known that embryo development is downregulated under high NO• concentrations in vivo and in vitro [29]. Finally, the eNOS-KO mice show severe defects in different aspects of reproduction, such as ovulation, fertilization, and early embryo development, further demonstrating the physiological role of NO• [30].

Oxidative Stress in Fertilization and Preimplantation Development

Role of Reactive Oxygen Species at Fertilization

In some marine invertebrate species, fertilization is accompanied by a significant increase in the rate of ROS/RNS formation in eggs, producing a respiratory burst [31–36]. After insemination, a rise in $[Ca^{2+}]_i$ takes place in the zygote, similarly to other species, which induces the secretion of the cortical granules content into the perivitelline space [37]. In addition, there is a calcium-dependent activation of PKC that slightly increases the intracellular pH (pH_i) through the activation of the Na^+/H^+ antiporter. Both the PKC activation and the increase in the pH_i are required for the subsequent activation of Udx1, a plasma membrane protein with NADPH oxidase activity that produces H_2O_2 [38]. The function of this increased production of H_2O_2 is to drive the formation of dityrosyl bonds to harden the fertilization envelope and block polyspermy [39]. Thus, the oxidative burst observed in invertebrates plays a well-known physiological role during the first stages of fertilization. However, a similar oxidative burst in mammalian oocytes at fertilization has not been reported.

It has been suggested that the lipid peroxidation has an enhancing effect on the ability of human spermatozoa to bind to zona pellucida, by a mechanism that can be reverted by α-tocopherol [40], suggesting a possible physiological role for the ROS generated by human spermatozoa in mediating sperm–zona interaction. The lipid peroxidation is indicative of an oxidative burst in spermatozoa, and the generation of superoxide has been demonstrated in sperm cells [41, 42]. Mammalian oocytes have been found to produce superoxide, this production being inhibited by SOD and diphenyleneiodonium (DPI), therefore suggesting that superoxide is generated by a NADPH oxidase activity in oocytes [41]. During the fusion of gametes at fertilization the measured production of superoxide is somehow diminished, probably due to the SOD release from gametes [41], weakening any potential physiological role of the oocyte superoxide at fertilization.

Although the occurrence of a sustained increase in ROS generation at fertilization is unlikely, a physiological role of some of these reactive species in the cell signaling in embryos has been suggested. This is the case for H_2O_2 in the blastocysts. Hydrogen peroxide in the blastocele fluid triggers apoptosis in the trophectoderm, while inner cell mass cells are protected from this oxidative-triggered apoptosis by a mechanism dependent on reduced glutathione (GSH) [43].

Thus, the balance of H_2O_2/GSH has a role in early development for the maintenance of the balance between inner cell mass and trophectoderm.

NO•, which plays a role during oocyte maturation, embryonic development, and implantation, has a signaling role at fertilization. Although it has been reported that the concentration of intracellular NO• does not change, globally or locally, during sperm-induced calcium waves [44], the role of NO• remains controversial, since Goud et al. observed that SNAP prevents deterioration of oocytes in culture, extending the temporal window for optimal fertilization, while NOS inhibitors reverse this action [45]. Those authors propose that the role of NO• in oocytes at fertilization is closely related with the regulation of calcium waves. The transient increase in the NO• concentration within the oocyte leads to the activation of guanylyl cyclase, and the increased cGMP levels would trigger the phosphorylation of inositol 1,4,5-trisphosphate receptor at the endoplasmic reticulum, which in turn regulates Ca^{2+} release from the ER. Thus, NO• would be required for the optimal generation of calcium waves, and the insufficient availability of NO• may accelerate oocyte aging.

Oxidative Burst in Embryos

The role of ROS as signal molecules at early embryo development is controversial. While many works reported the inhibition of embryo development under exposure to ROS (reviewed in [46]), a considerable oxidative burst has been monitored in the mouse embryos during blastocyst hatching [47]. The blastocyst hatching, i.e., the emergence of the mammalian blastocysts from their glycoprotein envelope previous to the implantation, generates ROS for a short period of time. This observation is supported by the induction of the hatching by the treatment of blastocysts with extracellular superoxide, within the low micromolar range, an experimental approach that shows no detrimental effects on the viability of the embryo. In fact, this level of superoxide (1–1.5 μM) is close to that observed in peri-hatching blastocyst suspension, and supports the hypothesis that superoxide has a key role in cell signaling at peri-hatching stages. Conversely, superoxide scavengers (*N*-*t*-butyl-α-phenyl nitrone, SOD, and menadione) inhibit the hatching of blastocysts in vitro, and the implantation in vivo, further strengthening the potential role of superoxide [47]. However, the molecular mechanism underlying the superoxide generation at this stage of the early development remains unknown.

As stated above, NO• is related with implantation. Inducible NOS is expressed in pre-, peri-, and postimplantation blastocysts with a differential profile, and the NO• production helps in the implantation while acting as vasodilator and angiogenic mediator [28]. However, it has been demonstrated that high concentrations of NO• inhibit development in vivo and in vitro [29], showing that the excess of NO• is harmful for embryo development.

On the other hand, the superoxide-dependent oxidative burst in embryos is probably limited to this particular stage of the development. A similar increase in other ROS, such as H_2O_2, has been found in the transition from the two- to four-cell stages, but it reflects, at least in part, the consequences of the exposure of embryos to in vitro culture conditions. Thus, the rise of H_2O_2 levels in the transition to the two- and four-cell stages is much reduced for in vivo developed embryos [48]. The increase of H_2O_2 observed in vitro is also involved in the two-cell block that has been reported in embryos developed in vitro from several mouse strains.

Sensitivity of Embryos to Oxidative Stress

The preimplantation embryo development, i.e., between one-cell and blastocyst stage, is highly sensitive to oxidative stress [7, 49, 50]. The oxygen tension in the female reproductive organ varies during the preimplantation stage, although it is always significantly lower than the atmospheric concentration. While this tension is $pO_2 \sim 5$–10% in the oviduct [51], the uterus shows a lower O_2 tension [52]. The decreasing O_2 tension suggests that embryos develop in healthy conditions under low O_2 concentrations, and that the almost anoxic conditions that surround the trophectoderm at the moment of implantation are not harmful for the developing embryo. In this regard, a number of reports demonstrate that the rate of in vitro development to the blastocyst stage increases when pO_2 is lowered to 5–10% in culture [53–57], and it is assumed that the benefits of the low O_2 tension culture are due to a probable lower generation of ROS in these experimental conditions. At the onset of compaction in vivo, embryos increase ATP production [51]. This metabolic switch is characterized by the increase in glycolytic activity and the dependence on ATP generated by glycolysis during preimplantation development, i.e., the preferred use of glucose as major energy substrate, rather than pyruvate [58]. The metabolic shift at the compaction stage is associated with a shift in the embryo redox state that accommodates the embryo to a more reduced environment in vivo, with decreasing O_2 concentration in the oviduct–uterus transition.

ROS have been associated with detrimental effects during maturation and early cleavage, since they account for alterations in the segregation of chromosomes during meiosis, the blockade of embryos in the two-cell stage, and low pregnancy rates [59, 60]. ROS are also responsible for an increased rate of cell death in both oocytes and embryos [50], and the H_2O_2 generation rate in fragmented embryos developed in vitro is significantly higher compared to non-fragmented embryos and unfertilized oocytes, while apoptosis is usually observed in fragmented embryos, but it is absent in non-fragmented embryos [49, 50]. ROS are also responsible for an increase of cell death in spermatozoa, and for severe molecular and biochemical alterations in oocytes such as lipid peroxidation [61, 62] and disruption of the intracellular calcium homeostasis [63, 64]. In this regard, we have shown that extracellular H_2O_2, within the micromolar range, enhances Ca^{2+} influx through store-operated calcium channels, impairing calcium homeostasis in human oocytes [64]. This impairment could eventually lead to a decrease in fertilization rates.

Role of Glutathione in Early Development

The tripeptide glutathione (γ-L-glutamyl-L-cysteinyl-glycine) is the most abundant thiol in mammalian cells. Reduced glutathione (GSH) is an important regulator of the intracellular redox state, and helps in the protection of cells against oxidative damage. Consequently, a depletion of the intracellular pool of GSH has been observed in cells under intensive oxidative insult [65]. Glutathione metabolism of early embryos is dependent on the developmental stage [66–68]. Although glutathione consumption and synthesis increase during oocyte maturation, early development is characterized by a significant decrease of glutathione metabolism. GSH content decreases approximately tenfold from that in the unfertilized oocyte to 0.12 pmol/blastocyst, representing an estimated change in concentration from 7 to 0.7 mM [66]. The GSH content falls quite rapidly, i.e., by 20–25% at fertilization and by approximately 45% by the late two-cell and early four-cell stages [69]. In parallel, NAD(P)H levels remain constant during oocyte maturation, but decrease after fertilization, concomitantly with GSH levels. NADPH is required for the GSH recycling by the glutaredoxin–glutathione reductase system (Grx–GR) and the thioredoxin system, suggesting that the restricted NADPH availability may be responsible of the GSH production after fertilization [68]. Moreover, the lowered GSH level is consistent with a diminished de novo synthesis of GSH after fertilization, compared with the high ATP-dependent de novo synthesis during oocyte maturation [67, 70].

Antioxidant Supplementation in Culture Media and Outcomes

Reactive Oxygen Species Generated by Culture Media

As stated above, the presence of ROS in the fluids and organs involved in the reproductive processes has been well documented in vivo and in vitro. The sources for these ROS in vivo are mainly the follicular fluid and the tubal and uterine milieu where gametes and embryos undergo maturation, fertilization, and early development [46, 71, 72]. During IVF, the gametes and embryos are not in their physiological environment but in culture media. Although culture media composition differs significantly, almost all commercial media are supplemented with serum or serum synthetic replacements, albumin [73], and other molecules that may potentially generate reactive species. Among these molecules we find HEPES buffer, metal chelators, ethylenediaminetetraacetic acid (EDTA), and ferric and cupric salts. Therefore, the medium itself is a putative source of ROS [74–76], and the contribution of this external source to the overall oxidative stress in gametes and embryos should not be considered as negligible.

The most important ROS generated in this manner are superoxide anion, hydroxyl radical, alkyl hydroperoxide, and hydrogen peroxide [74, 77–79]. In buffers and culture media with ferric or cupric salts or with molecules that have been shown to have photodynamic activities, such as flavins, the exposure to visible light and the mixing with atmospheric oxygen may generate significant amounts of superoxide [46, 78].

The effects of ROS on oocytes and early embryos have been tested experimentally, and we have described above the detrimental effects of elevated levels of ROS on oocyte maturation and early development. To gain knowledge on the effect of ROS on these biological processes, most of the studies have been developed on the basis of the addition of fluxes or single bolus of ROS to the culture medium. However, it should be considered that ROS can be chemically produced by buffers and enriched culture media, with superoxide anion, hydroxyl radical, hydroxide anion, and hydrogen peroxide, being the most important ROS generated [74, 77–79]. Serum and serum replacements include oxidases that usually enhance the production of ROS in buffers [46] and commercial media used during IVF generate ROS at different rates depending on the composition, i.e., different brands generate diverse rates of ROS [74, 76]. Common culture media used in IVF are able to produce levels of ROS high enough to damage human oocytes, being easily detectable with ROS-sensitive dyes, like H_2DCF-DA, dihydroethidium, or Amplex red combined with horseradish peroxidase. As we mentioned, superoxide is one of the major ROS generated by the culture media, but many other ROS are derived from superoxide, and they can be detected in significant levels. This is the case for H_2O_2. The production of H_2O_2 in buffers and media used for culturing human oocytes can be as high as 1–15 μM/h, i.e., reaching subtoxic levels that may account for decreased rates of fertilization, or early embryo development [74]. Interestingly, the amount of H_2O_2 produced by follicular fluid, and measured following the same experimental procedure, is extremely low and not significantly different when compared to phosphate-buffered saline buffers. Since H_2O_2 is one of the reactive species involved in the initial steps of lipid peroxidation in biological membranes [61], and taking into consideration that micromolar H_2O_2 induces lipid peroxidation in other cell types [80], we have reported that lipid peroxidation of COCs cultured following a protocol that mimics human IVF protocols is significantly higher compared to COCs in follicular fluid, and that lipid peroxidation can be prevented with the supplementation of the culture medium with either α-tocopherol or catalase [74].

Exposure of human oocytes to extracellularly added H_2O_2, within the micromolar range, enhances Ca^{2+} influx through store-operated calcium channels, and this Ca^{2+} overload leads to a sustained increase of the intracellular free Ca^{2+} concentration ($[Ca^{2+}]_i$) and eventually to the deregulation of the Ca^{2+} homeostasis [64]. Because Ca^{2+} signaling is tightly regulated in the mature oocyte, and the progression from fertilized oocyte to one-cell embryo requires the controlled generation of repetitive and transient increases of the $[Ca^{2+}]_i$, known as calcium waves, H_2O_2 may impairs the cytosolic Ca^{2+} spiking at fertilization [63]. Thus, fluxes of micromolar H_2O_2, generated by culture media, could severely affect Ca^{2+} signaling in human IVF during extended cultures, as the accumulated oxidative damage originated by culture media may not be neutralized by oocytes. In fact, this is the assumed explanation for the deficient generation of calcium waves in aged oocytes. The culture-induced aging of oocytes impairs Ca^{2+} oscillations patterns [81], similarly to what is found in H_2O_2-treated oocytes, leading in both cases to a poor embryo development [82]. In parallel, oocytes exposed to H_2O_2 concentrations within the micromolar range show a deficient readjusting of the intracellular ATP concentration at fertilization [83]. As a consequence of the extracellular oxidative stress generated by in vitro culture protocols with long incubation times in IVF media (at least 6–8 h), a partial but significant depletion in oocyte GSH content has been found, which is not observed when the incubation time is shorter than 2 h, suggesting that a long exposure to the oxidative stress associated with in vitro culture is required to induce this partial depletion in the intracellular pool of GSH [74]. Since GSH reacts with reactive species such as superoxide anion [84] and singlet oxygen [85], a decreased concentration of GSH promotes nonselective disulfide bond formation within proteins [86], having tremendous effects on cell physiology and endangering cell viability, enhancing the effects of a sustained oxidative stress, risking oocyte fertilization and even oocyte viability.

Consequently, we can conclude that exposure of oocytes and embryos to in vitro culture conditions should be considered as a low intensity, but long-lasting oxidative insult, and it seems reasonable to suggest that the proper management of the period time of incubation of oocytes in the culture medium could improve oocyte quality, thereby increasing the success rate of the IVF cycle. For this reason, the future design of culture media and protocols for IVF should consider the role of ROS generation by these media. Although some antioxidants show a rapid decay in cell-free buffers at physiological pH (e.g., ascorbate), the addition of combined lipophilic and hydrosoluble antioxidants arises as a reasonable method for decreasing the impact of the oxidative stress triggered by the culture media.

Addition of Antioxidants to the Culture Medium: Reduced Glutathione

Reduced glutathione represents the oocyte's main nonenzymatic defense against oxidative stress. The data regarding the way in which the in vitro culture conditions affect GSH levels in embryos are not consistent, although it is accepted that in vitro developed embryos show lower GSH levels when compared with in vivo developed embryos [66]. Contradictory data are found when the effect of the addition of extracellular GSH to the culture medium shows either no effect [69] or the improvement in the development through the two-cell block to the morula or blastocyst stages [66, 87, 88]. However, diverse treatments designed to modify the GSH concentration within oocytes or embryos reveal that this intracellular ROS scavenger is a key regulator of both the oocyte maturation and the embryo early development. Thus, the inhibition of GSH synthesis reduces the development of bovine embryos through the eight-cell block to the blastocyst stage [89], and the treatment with *t*-butyl hydroperoxide (*t*BH) to induce the depletion in GSH levels significantly blocks development in the two-cell stage [66]. Conversely, the treatment of mammalian oocytes with the flavonoid anthocyanin during IVM that leads to the increase in GSH levels and significant lowering of intracellular ROS generation is accompanied by higher rates of blastocyst formation after parthenogenetic activation or somatic cell nuclear transfer (SCNT) [90].

The addition of GSH to the culture medium (~1 mM) has also been tested during in vitro maturation of mammalian oocytes, increasing significantly the rate of fertilization [91] and the proportion of oocytes undergoing cleavage and morula/blastocyst development [92]. In all cases, GSH was used in the micromolar to low millimolar range, and the success in the observed parameters strongly suggest that oocytes and embryos undergo oxidative stress during in vitro culture, which can be prevented by supplementation with ROS scavengers. The addition of micromolar cysteine increases the content of GSH in oocytes during IVM, but has no effect on sperm penetration, pronuclear formation, or blastocyst formation [93].

The addition of cysteine and cystine, for de novo synthesis of GSH, triggers the increase of the oocyte GSH content, and represents a classic strategy to improve the maturation rate in porcine oocytes [94, 95]. However, a better strategy to increase intracellular GSH levels, without stimulating the γ-glutamyl cycle, is the esterification of GSH to increase the loading of GSH into cells in culture. In this sense, it has been reported recently that the use of extracellularly added GSH ethyl ester does not affect fertilization, or day 3 cleavage rate, but increases blastocyst total cell number [96]. Thus, GSH ethyl ester represents an effective approach to elevate oocyte GSH in vitro and to improve blastocyst cell number.

α-Tocopherol and Ascorbate

The active form of vitamin E, α-tocopherol, is one of the major antioxidants that protects mammalian cells against lipid peroxidation. The recycling of α-tocopherol in cell membranes is achieved by a direct mechanism using intracellular ascorbate [97]. Thus, the cooperation of both antioxidants prevents lipid peroxidative damage due to oxidative stress in culture. Moreover, the concentration of α-tocopherol in COC membranes significantly decreases during IVM, but remains constant in the presence of ascorbic acid in the culture media [98]. Once again, there is some controversy with the recent data regarding the effect of the addition of α-tocopherol to the IVM medium on the subsequent percentage of blastocysts obtained after IVF. Some studies found that the addition of α-tocopherol has no positive effect [98] or even a detrimental effect on fertilization rates in bovine oocytes [99].

Vitamin C or ascorbic acid has long been recognized to be one of the most relevant antioxidants in mammals. Because ascorbic acid has a pK_a value of 4.1 [100], ascorbate is the ionic form of ascorbic acid predominant at physiological pH. The combined addition of both α-tocopherol and ascorbate decreases the rate of development, thus suggesting that high levels of α-tocopherol may impair the acquisition of oocyte developmental competence [98]. On the contrary, using a similar mammalian model of study, it has been reported that 100–400 μM α-tocopherol in the oocyte maturation media increased the rates of cleavage, morula, and blastocyst [101]. Moreover, 100 μM α-tocopherol is found to be beneficial for improving embryo quality by decreasing the number of apoptotic cells in the blastocyst and improving the tolerance of embryos to freezing–thawing cycles [102, 103]. Interestingly, there is some consensus in the use of this concentration of α-tocopherol (100 μM) for the embryo culture [101, 102, 104], as it yields higher rates of blastocyst development, decreasing the number of apoptotic nuclei. Also, the addition of ascorbate along with α-tocopherol does not seem to be beneficial to embryo development [98, 104], although the addition of ascorbate alone (100 μM) reduces the number of apoptotic cells significantly [104]. However, all these studies were carried out in animal models and the blastocyst formation rate was strongly dependent on the concentrations of both the antioxidants. The particular conditions for human embryo culture require further investigation to set the optimal concentrations of both antioxidants with a significant effect on blastocyst formation, although the mentioned reports have narrowed this range to 50–200 μM for either α-tocopherol or ascorbate.

Superoxide Dismutase and Catalase

Superoxide and H_2O_2 are generated by cell-free culture medium, under atmospheric O_2 tension [74]. It has been reported that the addition of ecSOD to the culture medium decreases the intracellular oxidative stress in other cell types [105, 106], and constitutes a potential strategy to be used with oocytes and embryos in culture for assisted reproductive techniques. The dismutation of superoxide however leads to the increased levels of H_2O_2, and the combined use of both SOD and catalase should be considered to completely remove these ROS from the culture.

The probable beneficial effects of the addition of SOD to the embryo culture are supported by the finding that human follicular cells express a number of antioxidant enzymes, including SOD, and this expression negatively correlates with age [107]. In addition, fertilization and cleavage rates correlate with levels of SOD and catalase in the human follicular fluid [108]. The effect of the addition of SOD to the embryo culture has been studied in a mammalian animal model, but the improving of the developmental rate is significant only in high glucose media [109], i.e., SOD protects cells from ROS generation in high glucose conditions. Similarly, when bovine oocytes are matured in vitro in the presence of 10–1,000 U/ml SOD, the developmental competence of the oocytes after IVF is not improved [110]. This lack of SOD effectiveness is also found when SOD is added to a defined culture medium [synthetic oviductal fluid (SOF)+BSA] and the development to the morula and blastocyst stages is monitored [92], or when SOD is added to the synthetic medium KSOM [111]. There are some reports that show the attenuation of the two-cell block for mouse embryos when the cultured medium is supplemented with SOD, expanding the blastocyst formation in a basic synthetic medium but with no improvement of fertilization rate [112, 113].

Heavy Metal Chelation

As we stated above, heavy metals are mediators of ROS generation, mainly through Haber–Weiss reactions. This is the major reason for the requirement of heavy metals chelation in culture media for oocytes and embryos, and several studies have addressed this important question. Jinno et al. [114]

reported more than two decades ago the efficiency of the addition of EDTA (10 μM) in the composition of culture media for mammalian oocytes, enhancing the development of embryos derived from oocytes maturated in vitro, both to two-cell embryos and to blastocysts. Mouse oocytes and embryos have let us know that the requirement of EDTA is higher in embryos compared to oocytes, and that the increase of EDTA increases blastocysts frequency and day-4 hatching in a low-glucose Earle's balanced salt solution [115]. The optimal concentration of EDTA in the modified Ham's F-10 medium was studied in the 10–100 μM range, strongly supporting the development of in vivo conceived mouse zygotes to the blastocyst stage and increasing the implantation rate, especially when 100 μM EDTA was used [116]. This concentration of EDTA, either alone or in combination with hemoglobin, decreases the accumulation of ROS in porcine embryos, reducing the incidence of apoptosis, demonstrating that EDTA improves mammalian embryo development [117]. Similar conclusions were achieved with human embryos when the HTF medium was modified for the inclusion of EDTA and glutamine, further supporting a key role for heavy metal chelation during handling and culture of embryos in assisted reproductive techniques [118].

Other Antioxidants

The beneficial properties of other compounds, in most cases with radical scavenging properties, have been studied thoroughly in mammalian oocytes, although the information using human oocytes in IVF cycles is much more restricted. Taurine (1 mM) and melatonin (10–50 μM) improve embryo development in vitro [119]. Epigallocatechin-3-gallate, a polyphenolic compound, increases fertilization rates in porcine oocytes when this catechin is added to the IVF medium, although the enhancing effect is found in a narrow range of micromolar concentrations [120], and the efficacy of this supplement should be tested in other species.

Selenium is an inorganic element present in some enzymes with antioxidants activities (reviewed in [121]). The absence of selenium in the diet leads to several diseases, shortens life span, and has a significant effect in fertilization [121, 122]. As a consequence selenium supplementation, usually as sodium selenite, is required in synthetic culture media, although the level of selenium in culture medium needs to be addressed carefully. High concentrations (10 μM or higher) of selenium are toxic because selenium replaces sulfur in many enzyme active sites, modifying the chemistry of the catalyzed reactions. Within the nanomolar range, selenium in the diet of laboratory animals normalizes fertilization [122]. As a supplement of embryo culture media selenite increases levels of GPx, which is a selenoprotein, and reduces levels of caspase 3, improving development rates and preventing apoptosis in embryos [123].

Oxygen Tension

An interesting point regarding oxidative stress and in vitro handling of gametes and embryos during assisted reproductive techniques is the O_2 tension during in vitro culture. In mammals, early embryo development takes place in vivo in a low-oxygen environment (<10%) that may serve to protect the embryo from free radical damage. Culture conditions with defined media are usually carried out under humidified atmosphere of 5% CO_2: 95% air. Under these conditions the partial pressure of O_2 is ~150 mmHg, which is equivalent to ~20% O_2 [124]. Exposure of early embryos to atmospheric oxygen concentrations (20%) may explain reduced viability and the increased rate of fragmentation in embryos [53]. Although no significant differences have been found between 5 and 20% O_2 in the rate of human fertilization and embryo development at day 2 or 3, higher rates of blastocyst formation has been reported with lower O_2 tension at day 5–6. Therefore, it is believed that low O_2 concentrations may exert a beneficial effect only during the later stages of preimplantation development [54]. This conclusion is further supported by extending the culture of mammalian oocytes in 5% O_2 for 7 days in vitro, because the development to the blastocyst stage is significantly higher and the H_2O_2 generation by these embryos is much lower when compared to that measured with 20% O_2 culture [55]. In mammalian models of study, similar results have been reported culturing rabbit embryos in low O_2 (5–10%) [57], bovine embryos [125], or mouse embryos [126] in 5% O_2, suggesting a role for redox regulation during preimplantation development. However, the implantation rate and embryo viability after intrauterine transplantation to pseudopregnant females do not show significant differences for embryos cultured in low or atmospheric oxygen [126].

Antioxidants During Cryopreservation of Embryos

The ability to cryopreserve embryos without critical loss of viability has a profound effect on the success of assisted conception techniques. However, the survival of cryopreserved in vitro produced embryos, as measured either by post-warming survival in culture or by established pregnancies after embryo transfer, has lagged behind that of in vivo-derived embryos. Since oocyte/embryo plasma membranes contain significant amounts of polyunsaturated fatty acids (PUFA), they are particularly vulnerable to oxidative attack. It has been reported that lipid peroxidation of membrane phospholipids is adversely involved in embryo development. Some reports have indicated that cryopreservation can damage the antioxidant enzymes that protect against lipid peroxidation and that freeze–thaw stress can be modified by incubating the embryo in the presence of inhibitors of membrane lipid peroxidation [127]. These results suggest the possibility that the process of cryopreservation induces the production of ROS or alters the antioxidant enzyme potential of oocytes, leading to lipid peroxidation of the plasma membrane and resulting in reduced survival potential by a perturbation of membrane structure and permeability. Loss of cell membrane function via lipid peroxidation might interfere with transport systems such as pH regulatory systems on the cell membrane, and disruption of organelle membranes could affect transport systems such as mitochondrial transport systems essential for oxidative phosphorylation—the major energy-generating pathway of the early embryo [128]. Cells contain antioxidants such as GSH and SOD to protect against the production of oxygen radicals. However, it has been shown in sperm that the levels of these antioxidants are reduced by >50% following cryopreservation [129].

In addition, it is possible that loss of survival and fertilization competence following oocyte cryopreservation is also mediated by the mechanism of cytotoxic action of peroxynitrite and NO•. Supplementation of 50 IU/ml of SOD or the concomitant addition of SOD and hemoglobin to the freezing and thawing media has been shown to improve survival and fertilization of mouse oocytes [130]. It has been suggested that peroxynitrite, formed by the interaction of superoxide and NO•, may exert a cytotoxic effect on mouse metaphase II oocytes during cryopreservation. In addition, inclusion of LBP (*Lycium barbarum* polysaccharide) in the vitrification solution has shown to reduce the production of ROS, thereby preventing plasma membranes from lipid peroxidation and stabilizing membrane structure and permeability of porcine oocytes [131].

The work by Lane et al. demonstrates that including 0.1 mM ascorbate when cryopreserving mouse cleavage-stage and blastocyst-stage embryos is beneficial to subsequent embryo development and maintenance of normal cell function [128]. Ascorbate assists embryo development by stimulating development of the ICM following cryopreservation. Ascorbate is a very potent hydrophilic antioxidant that is able to scavenge H_2O_2, superoxide anion, hydroxyl free radical, and singlet oxygen [132]. Additionally, ascorbate is found in follicular fluid [133], indicating that it may have a physiological role as an antioxidant in oocyte and embryo development. The beneficial effects of ascorbate were most evident in mouse embryos that were slow frozen compared with those that were vitrified, which substantiates the fact that the damage from oxygen radicals is greater after slow freezing. Ascorbate also reduced H_2O_2 generation significantly among vitrified embryos but failed to do so in the slow-frozen ones. These results indicate that the slow-freezing procedure augments the production of H_2O_2 considerably more than the ultrarapid vitrification procedure. Peng et al. evaluated the effect of α-tocopherol on blastocyst development and subsequent cryosurvival of the somatic cell nuclear transferred ovine embryos [102]. α-Tocopherol (100 μg/ml) was added to the culture medium for the SCNT embryos, and the blastocysts from the α-tocopherol and untreated groups were then freeze–thawed, and their cryosurvival was assessed by in vitro culture for 48 h. The addition of α-tocopherol to the culture medium significantly decreased the apoptotic cell number (3.4% vs. 5.5%) and significantly increased the cryosurvival of SCNT blastocysts (66.8% vs. 50.7%). Future studies on antioxidative enzymes and NO• scavengers may lead to a better understanding of the biochemical processes that occur during oocyte/embryo cryopreservation. Estimating the total antioxidant capacity (TAC) and measuring ROS levels in the follicular fluid of retrieved oocytes will help select patients whose gametes require antioxidant supplementation during cryopreservation and subsequent in vitro cultures. The implication that ROS-mediated damage to oocyte developmental competence and embryo viability in infertile patients have raised concerns and need further research to identify the pathology behind this condition [134].

The expression of various biomarkers of oxidative stress such as SOD (both Cu/Zn-SOD and Mn-SOD), GPx, glutamyl synthetase, and lipid peroxides has been demonstrated in normal cycling human ovaries [134]. A delicate balance exists between ROS and antioxidant enzymes in ovarian tissue. Since hypoxic conditions following retransplantation of ovarian tissues induce ROS generation and the freeze–thaw cycle depletes the tissue's antioxidant capacity, antioxidant treatment might arguably be warranted. Studies of bovine ovarian tissue transplantation showed that ascorbate reduced apoptosis [135].

Another molecule with potent antioxidant activity is resveratrol. Studies are required for its effect on embryo freezing. However, when embryos produced by in vitro fertilization were incubated with 0.5 μM resveratrol, the treatment led to higher frequencies of blastocyst formation (8.6% vs. 13.3%) and elevated total cell numbers (37.1 ± 2.4 vs. 43.2 ± 1.7) by the end of the 7-day culture period ($p < 0.05$). The results indicate that 0.5 μM resveratrol during culture has a positive effect on early embryonic development of porcine embryos [136]. When embryos produced by in vitro fertilization were incubated with 0.5 μM resveratrol, the treatment led to higher frequencies of blastocyst formation (8.6% vs. 13.3%) and elevated total cell number (37.1 ± 2.4 vs. 43.2 ± 1.7) by the end of the 7-day culture period ($p < 0.05$). The results indicate that 0.5 μM resveratrol during culture has a positive effect on early embryonic development of porcine embryos. More studies are required to find the optimal candidate antioxidants and the required concentration that could nullify the deleterious effect of oxidative stress associated with embryo cryopreservation.

Five-Year View and Key Issues

From what we have stated regarding oxidative stress it should not be inferred that the complete elimination of ROS is a requirement for the success in IVF cycles. ROS/RNS play a physiological role in many aspects of female reproduction, including folliculogenesis, ovulation, as well as fertilization, early embryo development, and implantation. Thus, every culture condition intended to diminish the occurrence of ROS/RNS should be carefully studied and clinically evaluated. We have reviewed in this chapter the recent research focused on the use of antioxidants as supplements of culture media, and we have shown that most of the studies have used mammalian models (mouse, pig, and bovine oocytes). There is though a long way to transfer this knowledge to human-assisted reproductive techniques, i.e., oocyte in vitro maturation, IVF and embryo culture, and cryoconservation. Enriched culture media are well-known producers of ROS/RNS upon exposure of the media to daylight and atmospheric oxygen [74, 79]. This finding can be explained also by the presence in the media of riboflavin or amine pyridine nucleotides, or even glucose plus metals, which can lead to the photochemical production of superoxide and singlet oxygen [78, 137, 138]. The generation of superoxide enhances the formation of H_2O_2, especially when HEPES buffer or other amines are present in the medium. Then, H_2O_2 may react with superoxide (Haber–Weiss reaction) to produce hydroxyl radicals and peroxyl radicals [78], amplifying the production of radicals and reactive species by chain reactions. In this scenario, a plausible strategy to neutralize the oxidative stress induced by the culture medium could be the drastic reduction of the concentration in those metabolites not required by oocytes/embryos. In this regard, the requirement of glucose increases during early development, and the combination of glucose and metals act as an enhancer of ROS production (see above). For this reason, EDTA needs to be increased in cultured media for embryos, compared to what is found for oocytes.

In addition to EDTA, GSH has been successfully used as culture media supplement. However, GSH ethyl ester arises as an alternative for GSH loading in oocytes. In fact, GSH ethyl ester has been used to increase GSH levels in oocytes during IVM, an experimental approach that increases blastocyst cell number [96, 139]. Although addition of α-tocopherol, within the micromolar range improved IVM, fertilization, and blastocysts rates, the effects of the combination of ascorbate and α-tocopherol are still obscure. We have reviewed here the recent data and it has been proved that the presence of both antioxidants does not sustain embryo viability. One explanation could be that ascorbate is oxidized rapidly at physiological pH to dehydroascorbate (DHA) and ascorbate free radical (AFR), and cells need to recycle the oxidized forms of ascorbate, to keep high intracellular levels [140]. Both AFR and DHA are recycled by coupling to other cellular redox systems, being NAD(P)H/NAD(P)$^+$ and GSH/GSSG, particularly relevant to this end (reviewed in [141]). Recycling of ascorbate at the AFR oxidation stage is advantageous for living cells, because DHA is unstable at physiological pH (half-life of approximately 6 min) undergoing irreversible ring opening to form 2,3-diketo-1-gulonic acid. Intracellular ascorbate recycling from AFR and DHA in cells is achieved at the expense of NAD(P)H oxidation, and is mostly carried out by specific NADH- and NADPH oxidases [141]. However, this recycling requires the consumption of ATP, and the excessive drop in ATP levels could block oocyte maturation or early development. Besides, in the absence of DHA recycling to ascorbate, DHA can rapidly react with many other dithiol systems, including dithiols in unfolded or partially folded proteins [142], with likely adverse consequences.

Although this chapter does not review the recent bibliography regarding clinical trials to counteract infertility, it should be highlighted that one of the most promising therapies to this end is the oral administration of dietary antioxidants, like the combination of α-tocopherol and selenium [143], further supporting a role of antioxidants in the overall treatment of infertility.

There is still an open question regarding the effect of short-term mild oxidative stress on fertilization and development rates. In this sense, data from recent studies suggest that a short-term osmotic stress during maturation of mammalian oocytes improves embryo development [144, 145], and short-term exposure of oocytes to micromolar H_2O_2 results in a significant blastocyst yield, without altering fertilization rate or the GSH content [146]. However, it is required to address this phenomenon in human embryos.

In conclusion, current protocols for handling and in vitro culture of gametes and embryos generate a significant oxidative stress which may risk the outcome of the IVF cycle. The shortening of incubation times in pro-oxidant culture media and the appropriate addition of antioxidants that have been reported to work well in the improvement of embryo development, such as GSH (as its derivative GSH ethyl ester), α-tocopherol, and EDTA as heavy metal chelator, will lead to the significant reduction of oxidative stress, without affecting the viability of gametes and embryos. The concentration of these antioxidants and the stage of addition need to be addressed carefully for human embryos, but the recent literature showed narrowed ranges of concentrations for all these antioxidants when mammalian embryos were used.

Acknowledgments Funded by Grant PDT08A021 of the Junta de Extremadura—Fondo Europeo de Desarrollo Regional (FEDER), and Grant BFU2008-00104 of the Spanish Ministerio de Ciencia e Innovación.

References

1. Tilly JL, Tilly KI. Inhibitors of oxidative stress mimic the ability of follicle-stimulating hormone to suppress apoptosis in cultured rat ovarian follicles. Endocrinology. 1995;136(1):242–52.
2. Miyazaki T, et al. Effect of inhibition of oxygen free radical on ovulation and progesterone production by the in-vitro perfused rabbit ovary. J Reprod Fertil. 1991;91(1):207–12.
3. Jain S, et al. NADPH dependent superoxide generation in the ovary and uterus of mice during estrous cycle and early pregnancy. Life Sci. 2000;66(12):1139–46.
4. Carlson JC, et al. Stimulation of progesterone secretion in dispersed cells of rat corpora lutea by antioxidants. Steroids. 1995; 60(3):272–6.
5. Sawada M, Carlson JC. Intracellular regulation of progesterone secretion by the superoxide radical in the rat corpus luteum. Endocrinology. 1996;137(5):1580–4.
6. Brannstrom M, Mayrhofer G, Robertson SA. Localization of leukocyte subsets in the rat ovary during the periovulatory period. Biol Reprod. 1993;48(2):277–86.
7. Fujii J, Iuchi Y, Okada F. Fundamental roles of reactive oxygen species and protective mechanisms in the female reproductive system. Reprod Biol Endocrinol. 2005;3:43.
8. Combelles CM, et al. Profiling of superoxide dismutase isoenzymes in compartments of the developing bovine antral follicles. Reproduction. 2010;139(5):871–81.
9. Jablonka-Shariff A, Olson LM. Hormonal regulation of nitric oxide synthases and their cell-specific expression during follicular development in the rat ovary. Endocrinology. 1997;138(1):460–8.
10. Zackrisson U, et al. Cell-specific localization of nitric oxide synthases (NOS) in the rat ovary during follicular development, ovulation and luteal formation. Hum Reprod. 1996;11(12):2667–73.
11. Van Voorhis BJ, et al. Expression and localization of inducible and endothelial nitric oxide synthase in the rat ovary. Effects of gonadotropin stimulation in vivo. J Clin Invest. 1995;96(6):2719–26.
12. Lapointe J, et al. Hormonal and spatial regulation of nitric oxide synthases (NOS) (neuronal NOS, inducible NOS, and endothelial NOS) in the oviducts. Endocrinology. 2006;147(12):5600–10.
13. Rosselli M, et al. Circulating nitrite/nitrate levels increase with follicular development: indirect evidence for estradiol mediated NO release. Biochem Biophys Res Commun. 1994;202(3):1543–52.
14. Anteby EY, et al. Human follicular nitric oxide pathway: relationship to follicular size, oestradiol concentrations and ovarian blood flow. Hum Reprod. 1996;11(9):1947–51.
15. Shukovski L, Tsafriri A. The involvement of nitric oxide in the ovulatory process in the rat. Endocrinology. 1994;135(5):2287–90.
16. Nakamura Y, et al. Nitric oxide inhibits oocyte meiotic maturation. Biol Reprod. 2002;67(5):1588–92.
17. Yamagata Y, et al. Alterations in nitrate/nitrite and nitric oxide synthase in preovulatory follicles in gonadotropin-primed immature rat. Endocr J. 2002;49(2):219–26.
18. Dave S, Farrance DP, Whitehead SA. Evidence that nitric oxide inhibits steroidogenesis in cultured rat granulosa cells. Clin Sci (Lond). 1997;92(3):277–84.
19. Nakamura Y, et al. Changes in nitric oxide synthase activity in the ovary of gonadotropin treated rats: the role of nitric oxide during ovulation. Endocr J. 1999;46(4):529–38.
20. Tornell J, Billig H, Hillensjo T. Regulation of oocyte maturation by changes in ovarian levels of cyclic nucleotides. Hum Reprod. 1991;6(3):411–22.
21. Viana KS, et al. Effect of sodium nitroprusside, a nitric oxide donor, on the in vitro maturation of bovine oocytes. Anim Reprod Sci. 2007;102(3–4):217–27.
22. Schwarz KR, et al. Influence of nitric oxide during maturation on bovine oocyte meiosis and embryo development in vitro. Reprod Fertil Dev. 2008;20(4):529–36.
23. Bilodeau-Goeseels S. Effects of manipulating the nitric oxide/cyclic GMP pathway on bovine oocyte meiotic resumption in vitro. Theriogenology. 2007;68(5):693–701.
24. Orsi NM. Embryotoxicity of the nitric oxide donor sodium nitroprusside in preimplantation bovine embryos in vitro. Anim Reprod Sci. 2006;91(3–4):225–36.
25. Martin-Romero FJ, et al. Fluorescence measurements of steady state peroxynitrite production upon SIN-1 decomposition: NADH versus dihydrodichlorofluorescein and dihydrorhodamine 123. J Fluoresc. 2004;14(1):17–23.
26. Jablonka-Shariff A, Olson LM. Nitric oxide is essential for optimal meiotic maturation of murine cumulus-oocyte complexes in vitro. Mol Reprod Dev. 2000;55(4):412–21.
27. Schwarz KR, et al. Consequences of nitric oxide synthase inhibition during bovine oocyte maturation on meiosis and embryo development. Reprod Domest Anim. 2010;45(1):75–80.
28. Saxena D, et al. Increased appearance of inducible nitric oxide synthase in the uterus and embryo at implantation. Nitric Oxide. 2000;4(4):384–91.
29. Barroso RP, et al. Nitric oxide inhibits development of embryos and implantation in mice. Mol Hum Reprod. 1998;4(5):503–7.
30. Pallares P, et al. Disruption of the endothelial nitric oxide synthase gene affects ovulation, fertilization and early embryo survival in a knockout mouse model. Reproduction. 2008;136(5):573–9.
31. Turner E, Somers CE, Shapiro BM. The relationship between a novel NAD(P)H oxidase activity of ovoperoxidase and the CN--resistant respiratory burst that follows fertilization of sea urchin eggs. J Biol Chem. 1985;260(24):13163–71.
32. Heinecke JW, Shapiro BM. Respiratory burst oxidase of fertilization. Proc Natl Acad Sci USA. 1989;86(4):1259–63.
33. Schomer B, Epel D. Redox changes during fertilization and maturation of marine invertebrate eggs. Dev Biol. 1998;203(1):1–11.
34. Heinecke JW, et al. A specific requirement for protein kinase C in activation of the respiratory burst oxidase of fertilization. J Biol Chem. 1990;265(14):7717–20.
35. Heinecke JW, Shapiro BM. The respiratory burst oxidase of fertilization. A physiological target for regulation by protein kinase C. J Biol Chem. 1992;267(12):7959–62.
36. Schomer Miller B, Epel D. The roles of changes in NADPH and pH during fertilization and artificial activation of the sea urchin egg. Dev Biol. 1999;216(1):394–405.
37. Steinhardt RA, Epel D. Activation of sea-urchin eggs by a calcium ionophore. Proc Natl Acad Sci USA. 1974;71(5):1915–9.
38. Wong JL, Creton R, Wessel GM. The oxidative burst at fertilization is dependent upon activation of the dual oxidase Udx1. Dev Cell. 2004;7(6):801–14.
39. Foerder CA, Shapiro BM. Release of ovoperoxidase from sea urchin eggs hardens the fertilization membrane with tyrosine crosslinks. Proc Natl Acad Sci USA. 1977;74(10):4214–8.
40. Aitken RJ, Clarkson JS, Fishel S. Generation of reactive oxygen species, lipid peroxidation, and human sperm function. Biol Reprod. 1989;41(1):183–97.
41. Miesel R, Drzejczak PJ, Kurpisz M. Oxidative stress during the interaction of gametes. Biol Reprod. 1993;49(5):918–23.
42. Ellis JA, Mayer SJ, Jones OT. The effect of the NADPH oxidase inhibitor diphenyleneiodonium on aerobic and anaerobic microbicidal activities of human neutrophils. Biochem J. 1988;251(3): 887–91.
43. Pierce GB, Parchment RE, Lewellyn AL. Hydrogen peroxide as a mediator of programmed cell death in the blastocyst. Differentiation. 1991;46(3):181–6.

44. Hyslop LA, et al. Simultaneous measurement of intracellular nitric oxide and free calcium levels in chordate eggs demonstrates that nitric oxide has no role at fertilization. Dev Biol. 2001;234(1): 216–30.
45. Goud PT, et al. Nitric oxide extends the oocyte temporal window for optimal fertilization. Free Radic Biol Med. 2008;45(4):453–9.
46. Guerin P, El Mouatassim S, Menezo Y. Oxidative stress and protection against reactive oxygen species in the pre-implantation embryo and its surroundings. Hum Reprod Update. 2001;7(2):175–89.
47. Thomas M, et al. A programmed oxyradical burst causes hatching of mouse blastocysts. J Cell Sci. 1997;110(Pt 14):1597–602.
48. Nasr-Esfahani MH, Aitken JR, Johnson MH. Hydrogen peroxide levels in mouse oocytes and early cleavage stage embryos developed in vitro or in vivo. Development. 1990;109(2):501–7.
49. Johnson MH, Nasr-Esfahani MH. Radical solutions and cultural problems: could free oxygen radicals be responsible for the impaired development of preimplantation mammalian embryos in vitro? Bioessays. 1994;16(1):31–8.
50. Yang HW, et al. Detection of reactive oxygen species (ROS) and apoptosis in human fragmented embryos. Hum Reprod. 1998;13(4): 998–1002.
51. Leese HJ. Metabolic control during preimplantation mammalian development. Hum Reprod Update. 1995;1(1):63–72.
52. Fischer B, Bavister BD. Oxygen tension in the oviduct and uterus of rhesus monkeys, hamsters and rabbits. J Reprod Fertil. 1993;99(2):673–9.
53. Burton GJ, Hempstock J, Jauniaux E. Oxygen, early embryonic metabolism and free radical-mediated embryopathies. Reprod Biomed Online. 2003;6(1):84–96.
54. Dumoulin JC, et al. Effect of oxygen concentration on human in-vitro fertilization and embryo culture. Hum Reprod. 1999;14(2): 465–9.
55. Kitagawa Y, et al. Effects of oxygen concentration and antioxidants on the in vitro developmental ability, production of reactive oxygen species (ROS), and DNA fragmentation in porcine embryos. Theriogenology. 2004;62(7):1186–97.
56. Thompson JG, et al. Effect of oxygen concentration on in-vitro development of preimplantation sheep and cattle embryos. J Reprod Fertil. 1990;89(2):573–8.
57. Li J, Foote RH. Culture of rabbit zygotes into blastocysts in protein-free medium with one to twenty per cent oxygen. J Reprod Fertil. 1993;98(1):163–7.
58. Houghton FD, et al. Oxygen consumption and energy metabolism of the early mouse embryo. Mol Reprod Dev. 1996;44(4):476–85.
59. Hu Y, et al. Effects of low O2 and ageing on spindles and chromosomes in mouse oocytes from pre-antral follicle culture. Hum Reprod. 2001;16(4):737–48.
60. Bedaiwy MA, et al. Differential growth of human embryos in vitro: role of reactive oxygen species. Fertil Steril. 2004;82(3):593–600.
61. Agarwal A, Saleh RA, Bedaiwy MA. Role of reactive oxygen species in the pathophysiology of human reproduction. Fertil Steril. 2003;79(4):829–43.
62. Tarin JJ. Potential effects of age-associated oxidative stress on mammalian oocytes/embryos. Mol Hum Reprod. 1996;2(10): 717–24.
63. Takahashi T, et al. Impact of oxidative stress in aged mouse oocytes on calcium oscillations at fertilization. Mol Reprod Dev. 2003;66(2):143–52.
64. Martin-Romero FJ, et al. Store-operated calcium entry in human oocytes and sensitivity to oxidative stress. Biol Reprod. 2008;78(2): 307–15.
65. Shivakumar BR, Kolluri SV, Ravindranath V. Glutathione and protein thiol homeostasis in brain during reperfusion after cerebral ischemia. J Pharmacol Exp Ther. 1995;274(3):1167–73.
66. Gardiner CS, Reed DJ. Status of glutathione during oxidant-induced oxidative stress in the preimplantation mouse embryo. Biol Reprod. 1994;51(6):1307–14.
67. Gardiner CS, Reed DJ. Synthesis of glutathione in the preimplantation mouse embryo. Arch Biochem Biophys. 1995;318(1):30–6.
68. Dumollard R, et al. Regulation of redox metabolism in the mouse oocyte and embryo. Development. 2007;134(3):455–65.
69. Nasr-Esfahani MH, Johnson MH. Quantitative analysis of cellular glutathione in early preimplantation mouse embryos developing in vivo and in vitro. Hum Reprod. 1992;7(9):1281–90.
70. Luberda Z. The role of glutathione in mammalian gametes. Reprod Biol. 2005;5(1):5–17.
71. Pasqualotto EB, et al. Effect of oxidative stress in follicular fluid on the outcome of assisted reproductive procedures. Fertil Steril. 2004;81(4):973–6.
72. Bedaiwy MA, et al. Relationship between oxidative stress and embryotoxicity of hydrosalpingeal fluid. Hum Reprod. 2002;17(3): 601–4.
73. Blake D, et al. Protein supplementation of human IVF culture media. J Assist Reprod Genet. 2002;19(3):137–43.
74. Martin-Romero FJ, et al. Contribution of culture media to oxidative stress and its effect on human oocytes. Reprod Biomed Online. 2008;17(5):652–61.
75. Esfandiari N, et al. Protein supplementation and the incidence of apoptosis and oxidative stress in mouse embryos. Obstet Gynecol. 2005;105(3):653–60.
76. Miguel-Lasobras EM, et al. Oxidative stress in human oocytes during IVF handling. Fertil Steril. 2004;82(Supp 2):S56–7.
77. Cohen G. The fenton reaction. In: Greenwald RA, editor. Handbook of methods for oxygen radical research. Boca Raton: CRC Press; 1987. p. 55–64.
78. Michelson AM. Photochemical production of oxy radicals. In: Greenwald RA, editor. Handbooks of methods for oxygen radical research. Boca Raton: CRC Press; 2000. p. 71–5.
79. Grzelak A, Rychlik B, Bartosz G. Light-dependent generation of reactive oxygen species in cell culture media. Free Radic Biol Med. 2001;30(12):1418–25.
80. Gutierrez-Martin Y, et al. Alteration of cytosolic free calcium homeostasis by SIN-1: high sensitivity of L-type Ca2+ channels to extracellular oxidative/nitrosative stress in cerebellar granule cells. J Neurochem. 2005;92(4):973–89.
81. Igarashi H, et al. Aging-related changes in calcium oscillations in fertilized mouse oocytes. Mol Reprod Dev. 1997;48(3):383–90.
82. Takahashi T, et al. Poor embryo development in mouse oocytes aged in vitro is associated with impaired calcium homeostasis. Biol Reprod. 2009;80(3):493–502.
83. Igarashi H, et al. Aged mouse oocytes fail to readjust intracellular adenosine triphosphates at fertilization. Biol Reprod. 2005;72(5): 1256–61.
84. Winterbourn CC, Metodiewa D. The reaction of superoxide with reduced glutathione. Arch Biochem Biophys. 1994;314(2):284–90.
85. Devasagayam TP, et al. Activity of thiols as singlet molecular oxygen quenchers. J Photochem Photobiol B. 1991;9(1):105–16.
86. Cumming RC, et al. Protein disulfide bond formation in the cytoplasm during oxidative stress. J Biol Chem. 2004;279(21): 21749–58.
87. Legge M, Sellens MH. Free radical scavengers ameliorate the 2-cell block in mouse embryo culture. Hum Reprod. 1991;6(6): 867–71.
88. Ozawa M, et al. Addition of glutathione or thioredoxin to culture medium reduces intracellular redox status of porcine IVM/IVF embryos, resulting in improved development to the blastocyst stage. Mol Reprod Dev. 2006;73(8):998–1007.
89. Takahashi M, et al. Effect of thiol compounds on in vitro development and intracellular glutathione content of bovine embryos. Biol Reprod. 1993;49(2):228–32.
90. You J, et al. Anthocyanin stimulates in vitro development of cloned pig embryos by increasing the intracellular glutathione level and inhibiting reactive oxygen species. Theriogenology. 2010;74: 777–85.

91. Fukui Y, et al. Fertilizability and developmental capacity of individually cultured bovine oocytes. Theriogenology. 2000;53(8): 1553–65.
92. Luvoni GC, Keskintepe L, Brackett BG. Improvement in bovine embryo production in vitro by glutathione-containing culture media. Mol Reprod Dev. 1996;43(4):437–43.
93. Viet Linh N, et al. Effects of cysteine during in vitro maturation of porcine oocytes under low oxygen tension on their subsequent in vitro fertilization and development. J Reprod Dev. 2009; 55(6):594–8.
94. Grupen CG, Nagashima H, Nottle MB. Cysteamine enhances in vitro development of porcine oocytes matured and fertilized in vitro. Biol Reprod. 1995;53(1):173–8.
95. Bing YZ, et al. In vitro maturation and glutathione synthesis of porcine oocytes in the presence or absence of cysteamine under different oxygen tensions: role of cumulus cells. Reprod Fertil Dev. 2002;14(3–4):125–31.
96. Curnow EC, et al. Developmental potential of bovine oocytes following IVM in the presence of glutathione ethyl ester. Reprod Fertil Dev. 2010;22(4):597–605.
97. May JM, Qu ZC, Mendiratta S. Protection and recycling of alpha-tocopherol in human erythrocytes by intracellular ascorbic acid. Arch Biochem Biophys. 1998;349(2):281–9.
98. Dalvit G, et al. Effect of alpha-tocopherol and ascorbic acid on bovine oocyte in vitro maturation. Reprod Domest Anim. 2005; 40(2):93–7.
99. Marques A, et al. Effect of alpha-tocopherol on bovine in vitro fertilization. Reprod Domest Anim. 2010;45(1):81–5.
100. Bors W, Buettner GR. The vitamin C radical and its reactions. In: Packer L, Fuchs J, editors. Vitamin C in health and disease. New York: Marcel Dekker, Inc; 1997. p. 75–94.
101. Natarajan R, Shankar MB, Munuswamy D. Effect of alpha-tocopherol supplementation on in vitro maturation of sheep oocytes and in vitro development of preimplantation sheep embryos to the blastocyst stage. J Assist Reprod Genet. 2010;27:483–90.
102. Peng XR, Liu T, Zhang Y. Addition of alpha-tocopherol to culture medium improves the quality and cryosurvival of nuclear-transferred ovine embryos. J Reprod Dev. 2008;54(6):403–7.
103. Hossein MS, et al. Temporal effects of alpha-tocopherol and L-ascorbic acid on in vitro fertilized porcine embryo development. Anim Reprod Sci. 2007;100(1–2):107–17.
104. Jeong YW, et al. Antiapoptotic and embryotrophic effects of alpha-tocopherol and L-ascorbic acid on porcine embryos derived from in vitro fertilization and somatic cell nuclear transfer. Theriogenology. 2006;66(9):2104–12.
105. Martin-Romero FJ, Garcia-Martin E, Gutierrez-Merino C. Inhibition of oxidative stress produced by plasma membrane NADH oxidase delays low-potassium-induced apoptosis of cerebellar granule cells. J Neurochem. 2002;82(3):705–15.
106. Martin-Romero FJ, Garcia-Martin E, Gutierrez-Merino C. Involvement of free radicals in signalling of low-potassium induced apoptosis in cultured cerebellar granule cells. Int J Dev Biol. 1996;(Suppl 1):197S–8S.
107. Carbone MC, et al. Antioxidant enzymatic defences in human follicular fluid: characterization and age-dependent changes. Mol Hum Reprod. 2003;9(1):639–43.
108. Pasqualotto EB, et al. The role of enzymatic antioxidants detected in the follicular fluid and semen of infertile couples undergoing assisted reproduction. Hum Fertil (Camb). 2009;12(3): 166–71.
109. Iwata H, et al. Effects of antioxidants on the development of bovine IVM/IVF embryos in various concentrations of glucose. Theriogenology. 1998;50(3):365–75.
110. Blondin P, Coenen K, Sirard MA. The impact of reactive oxygen species on bovine sperm fertilizing ability and oocyte maturation. J Androl. 1997;18(4):454–60.
111. Liu Z, Foote RH, Yang X. Development of early bovine embryos in co-culture with KSOM and taurine, superoxide dismutase or insulin. Theriogenology. 1995;44(5):741–50.
112. Nonogaki T, et al. Protection from oxidative stress by thioredoxin and superoxide dismutase of mouse embryos fertilized in vitro. Hum Reprod. 1991;6(9):1305–10.
113. Nonogaki T, et al. Effects of superoxide dismutase on mouse in vitro fertilization and embryo culture system. J Assist Reprod Genet. 1992;9(3):274–80.
114. Jinno M, Sandow BA, Hodgen GD. Enhancement of the developmental potential of mouse oocytes matured in vitro by gonadotropins and ethylenediaminetetraacetic acid (EDTA). J In Vitro Fert Embryo Transf. 1989;6(1):36–40.
115. Hentemann M, Bertheussen K. New media for culture to blastocyst. Fertil Steril. 2009;91(3):878–83.
116. Mehta TS, Kiessling AA. The developmental potential of mouse embryos conceived in Ham's F-10 medium containing ethylenediaminetetraacetic acid. Fertil Steril. 1993;60(6):1088–93.
117. Kim JH, et al. Embryotrophic effects of ethylenediaminetetraacetic acid and hemoglobin on in vitro porcine embryos development. Theriogenology. 2006;66(2):449–55.
118. Quinn P. Enhanced results in mouse and human embryo culture using a modified human tubal fluid medium lacking glucose and phosphate. J Assist Reprod Genet. 1995;12(2):97–105.
119. Manjunatha BM, et al. Effect of taurine and melatonin in the culture medium on buffalo in vitro embryo development. Reprod Domest Anim. 2009;44(1):12–6.
120. Spinaci M, et al. Effects of epigallocatechin-3-gallate (EGCG) on in vitro maturation and fertilization of porcine oocytes. Theriogenology. 2008;69(7):877–85.
121. Hatfield DL, editor. Selenium: its molecular biology and role in human health. Boston: Kluwer; 2001.
122. Martin-Romero FJ, et al. Selenium metabolism in drosophila: selenoproteins, selenoprotein mRNA expression, fertility, and mortality. J Biol Chem. 2001;276(32):29798–804.
123. Uhm SJ, et al. Selenium improves the developmental ability and reduces the apoptosis in porcine parthenotes. Mol Reprod Dev. 2007;74(11):1386–94.
124. Sullivan M, Galea P, Latif S. What is the appropriate oxygen tension for in vitro culture? Mol Hum Reprod. 2006;12(11):653.
125. Liu Z, Foote RH. Development of bovine embryos in KSOM with added superoxide dismutase and taurine and with five and twenty percent O2. Biol Reprod. 1995;53(4):786–90.
126. Umaoka Y, et al. Developmental potentiality of embryos cultured under low oxygen tension with superoxide dismutase. J In Vitro Fert Embryo Transf. 1991;8(5):245–9.
127. Tarin JJ, Trounson AO. Effects of stimulation or inhibition of lipid peroxidation on freezing-thawing of mouse embryos. Biol Reprod. 1993;49(6):1362–8.
128. Lane M, Maybach JM, Gardner DK. Addition of ascorbate during cryopreservation stimulates subsequent embryo development. Hum Reprod. 2002;17(10):2686–93.
129. Bilodeau JF, et al. Levels of antioxidant defenses are decreased in bovine spermatozoa after a cycle of freezing and thawing. Mol Reprod Dev. 2000;55(3):282–8.
130. Dinara S, et al. Effects of supplementation with free radical scavengers on the survival and fertilization rates of mouse cryopreserved oocytes. Hum Reprod. 2001;16(9):1976–81.
131. Huang J, et al. Effect of sugars on maturation rate of vitrified-thawed immature porcine oocytes. Anim Reprod Sci. 2008; 106(1–2):25–35.
132. Meister A. On the antioxidant effects of ascorbic acid and glutathione. Biochem Pharmacol. 1992;44(10):1905–15.
133. Paszkowski T, Clarke RN. The graafian follicle is a site of L-ascorbate accumulation. J Assist Reprod Genet. 1999;16(1):41–5.
134. Agarwal A, Gupta S, Sharma RK. Role of oxidative stress in female reproduction. Reprod Biol Endocrinol. 2005;3:28.

135. Kim SS. Ovarian tissue banking for cancer patients. To do or not to do? Hum Reprod. 2003;18(9):1759–61.
136. Lee K, et al. Effect of resveratrol on the development of porcine embryos produced in vitro. J Reprod Dev. 2006;56(3):330–5.
137. Cunningham ML, et al. Superoxide anion is generated from cellular metabolites by solar radiation and its components. Free Radic Biol Med. 1985;1:381–5.
138. Wells-Knecht KJ, et al. Mechanism of autoxidative glycosylation: identification of glyoxal and arabinose as intermediates in the autoxidative modification of proteins by glucose. Biochemistry. 1995;34(11):3702–9.
139. Curnow EC, et al. Bovine in vitro oocyte maturation as a model for manipulation of the gamma-glutamyl cycle and intraoocyte glutathione. Reprod Fertil Dev. 2008;20(5):579–88.
140. May JM. Is ascorbic acid an antioxidant for the plasma membrane? FASEB J. 1999;13(9):995–1006.
141. Martin-Romero FJ, S.G. Pandalai, et al. Molecular biology of mammalian AFR reductases. In: Recent research developments in molecular biology. Kerala: Research Signpost; 2004. pp. 29–52.
142. Saaranen MJ, et al. The role of dehydroascorbate in disulfide bond formation. Antioxid Redox Signal. 2010;12(1):15–25.
143. Westphal LM, et al. A nutritional supplement for improving fertility in women: a pilot study. J Reprod Med. 2004;49(4):289–93.
144. Pribenszky C, et al. Increased stress tolerance of matured pig oocytes after high hydrostatic pressure treatment. Anim Reprod Sci. 2008;106(1–2):200–7.
145. Lin L, et al. Osmotic stress induced by sodium chloride, sucrose or trehalose improves cryotolerance and developmental competence of porcine oocytes. Reprod Fertil Dev. 2009;21(2):338–44.
146. Vandaele L, et al. Short-term exposure to hydrogen peroxide during oocyte maturation improves bovine embryo development. Reproduction. 2010;139(3):505–11.

Part VI

Special Topics

Best Practice Guidelines for the Use of Antioxidants

46

Francesco Lanzafame, Sandro La Vignera, and Aldo E. Calogero

Abstract

Oxidative stress (OS) has been recognized as one important cause of male infertility because it may damage sperm function and DNA integrity. Since antioxidants counteract the action of OS, these compounds are used in the medical treatment of male infertility. This chapter reviews the effects of the main antioxidants used in the clinical practice. We applied an evidence-based medicine criterion in the attempt to identify the molecules provided with the most effective action. Some antioxidants have been clearly shown to be effective by many well-conducted studies. These include vitamins C and E and carnitines which may be considered as a first-line treatment. Other molecules, such as glutathione and coenzyme Q10, may be proposed as a second-line treatment because their use is supported by few but well-performed studies. Lastly, the efficacy of other antioxidants, such as pycnogenol, lycopene, etc., is not yet supported by a sufficient number of studies.

Keywords

Oxidative stress • Spermatozoa • Male infertility • Antioxidant treatment • Reactive oxygen species • DNA fragmentation • Vitamin supplementation

Oxidative stress (OS), an imbalance between the production of radical oxygen species (ROS) and antioxidant scavenging activities in which the former prevails [1], causes male and female infertility. The role of OS in the pathophysiology of human sperm function has been extensively explored. Indeed, spermatozoa are extremely sensitive to ROS because of their high content of polyunsaturated fatty acids (PUFA) and their limited ability to repair deoxyribonucleic acid (DNA) damage [2, 3]. Therefore, the administration of many antioxidants has been proposed in the attempt to improve sperm quality.

Treatments varied over the years involving the use of many different compounds, such as carnitines, phosphatidylcholine, kallikrein, pentoxifylline, vitamins A, C, and E, etc. [4]. The administration of antioxidants to infertile men represents a great challenge for the andrologist. Indeed, a correct and complete diagnostic workup should not be ignored, because different andrological diseases may respond to antioxidant administration in a different manner [5]. Furthermore, it should be reminded that no standardized markers have been developed which may help to identify those patients who may benefit from a scavenging treatment. Additionally, reliable, prognostic, and not expensive tests to evaluate the effects of ROS exposure or to determine the total antioxidant capacity (TAC) are not yet available for the clinical practice.

The efficacy of antioxidant administration would positively benefit from the use of markers that can reliable measure OS before treatment and/or of markers which evaluate the damage caused by OS on the sperm membrane and DNA. Many studies have not taken into account the measurement of OS, a primary endpoint, and its possible improvement after treatment. Moreover, diseases which impair fertility have often a subclinical development with few or no symptoms

F. Lanzafame, MD (✉)
Centro Territoriale di Andrologia, Siracusa, Italy

S. La Vignera, MD • A.E. Calogero, MD
Department of Medical and Pediatrics Sciences, University of Catania, Catania,
e-mail: sandrolavignera@unict.it; acaloger@unict.it

S.J. Parekattil and A. Agarwal (eds.), *Male Infertility: Contemporary Clinical Approaches, Andrology, ART & Antioxidants*,
DOI 10.1007/978-1-4614-3335-4_46, © Springer Science+Business Media, LLC 2012

and/or signs; consequently, the patients consult the andrologist after a long time from their onset. This often causes an irreparable OS-induced damage in spermatozoa. Therefore, the first step before antioxidant treatment is prescribed is to consider the clinical history and laboratory and instrumental data to understand if a given patient is suitable for a scavenging treatment or whether a different therapeutic strategy should be undertaken. A suitable strategy would be to eradicate all causes that increase ROS production and/or reduce seminal plasma scavenging action.

Regardless of the discrepant findings reported [6, 7], antioxidant treatment seems to be helpful both in vitro and in vivo to ameliorate sperm quality. Experimental data in laboratory and farm animals support this statement [8, 9]. Glutathione (GSH) administration plays a fundamental action in improving sperm motility and consequently in enhancing fertilization in asthenozoospermic bulls with varicocele and in rabbits with dispermy due to cryptorchidism [10]. In lead-injected mice, vitamin C, given at a concentration equivalent to the human therapeutic dose (10 mg/kg body weight), significantly decreases the concentration of testicular malondialdehyde (MDA), an important marker of OS, with a concurrent improvement in sperm concentration and a significant decline in the percentage of abnormal spermatozoa. Vitamin E (100 mg/kg body weight) and vitamin C have a comparable scavenging power, but the later has a poorer effectiveness. Given together at the above indicated dosages, they induce the most important fall in the content of MDA in lead-treated mice with a concomitant increase of sperm concentration and amelioration of sperm morphology [11]. An analogous protective effect of vitamin E has been shown in mice with mercury-induced decrease of sperm count and function [12]. The presence of an elevated quantity of α-tocopheryl acetate in the rabbit diet significantly enhances the semen amount of vitamin E and its oxidative firmness subsequent to cryoconservation [13].

Thus, many studies carried out in animal models suggest a possible successful use of antioxidants in humans. Unfortunately, contrasting evidences have been published from several uncontrolled trials, often performed to sustain the efficacy of some type of treatment even if its usefulness is not yet confirmed [14]. Heavy smokers [15] and infertile patients [16] who received an antioxidant supplementation ameliorate sperm quality. In addition, antioxidants improve the fertilizing capability of healthy men with elevated seminal ROS concentration [17] and also increase the fertilization rate of fertile normozoospermic men with low fertilization rates in previous IVF cycles, by significantly decreasing the levels of MDA [18].

Despite many studies describe positive results following administration of antioxidants on sperm parameters, there is no well-defined scavenging approach during OS, and often the experimental design is not double blind and/or placebo controlled. In addition, nonhomogenous cohorts of patients have been enrolled in some studies. Finally, a large number of compounds have been used, but for many of them scanty evidence have been published. All this makes the scientific literature very heterogeneous with a consequent difficulty to get an ultimate conclusion, as also suggested by Patel and Sigman [19].

Because of the difficulty to clearly distinguish which antioxidant can play a better role to reduce OS generation and/or to prevent the OS-mediated sperm damage, we decided to use an evidence-based medicine (EBM) method to better understand the role of the various antioxidants for the treatment of infertile men (Table 46.1). To accomplish this, the National Clinical Guidelines for Type 2 Diabetes, formulated by the Royal College of General practitioners, Effective Clinical Practice Unit, ScHAAR University of Sheffield (http://www.nice.org.uk/nicemedia/live/10911/28998/28998.pdf) were used as they seem more suitable for pharmacological trials. These guidelines classify evidences and recommendations as indicated in Tables 46.2 and 46.3.

Ascorbic Acid

Ascorbic acid (vitamin C) is ten times higher in the seminal plasma than in serum [20]. It is an effective scavenger if peroxyl radicals are in an aqueous phase [21], but does not have the some powerful scavenging action within membrane lipids [22]. The amount of vitamin C in the seminal plasma decreases significantly when the concentration of ROS increases [23]. Also in leucocytospermic samples, the

Table 46.1 List of antioxidants reviewed in this chapter

Antioxidants
Ascorbic acid (vitamin C)
α-Tocopherol (vitamin E)
Ascorbic acid (vitamin C) plus α-tocopherol (vitamin E)
α-Tocopherol (vitamin E) plus selenium
Glutathione
L-Carnitine plus L-acetyl-carnitine
Coenzyme Q10
Lycopene
Pycnogenol
N-acetyl-cysteine
Vitamin A and vitamin E
Pentoxifylline
Selenium
Shao-Fu-Zhu-Yu-Tang
Astaxanthin
Lepidium meyenii
α-Linolenic acid and lignans
Vitamin C and E, lycopene, selenium, folic acid, garlic oil plus zinc
Morindae officinalis extract

Table 46.2 Evidence classification of the National Clinical Guidelines for Type 2 Diabetes (the Royal College of General practitioners, Effective Clinical Practice Unit, ScHAAR University of Sheffield) used in this chapter

Evidence classification	
Ia	Evidence from meta-analysis of randomized controlled trials
Ib	Evidence from at least one randomized controlled trial
IIa	Evidence from at least one controlled study without randomization
IIb	Evidence from at least one other type of quasi-experimental study
III	Evidence from nonexperimental descriptive studies, such as comparative studies, correlation studies, and case–control studies
IV	Evidence from expert committee reports or opinions and/or clinical experience of respected authorities

Adapted from [82]

Table 46.3 Recommendations of the National Clinical Guidelines for Type 2 Diabetes (the Royal College of General practitioners, Effective Clinical Practice Unit, ScHAAR University of Sheffield) used in this chapter

Recommendation grading	
A	Directly based on category I evidence
B	Directly based on category II evidence, or extrapolated recommendation from category I evidence
C	Directly based on category III evidence, or extrapolated recommendation from category I or II evidence
D	Directly based on category IV evidence, or extrapolated recommendation from category I, II, or III evidence

From [83], with permission

concentrations of seminal ascorbic acid are significantly depleted. When this occurs, a significantly higher sperm DNA fragmentation index (DFI) has been found compared with semen samples with normal or high levels of ascorbic acid [24]. Interestingly, vitamin C plays an antioxidant action at low concentrations, but it can start auto-oxidation processes at higher concentrations [25]. Moreover, in humans, the vitamin C plasma saturation occurs at daily dose of 1 g. Higher amounts could promote the development of kidney stones, because of the enhanced excretion of oxalate [26].

The administration of vitamin C (1 g/day) increases ascorbic acid level by 2.2-fold [27]. In addition, it has been reported that seminal plasma vitamin C concentrations positively correlates with the number of normal spermatozoa, in a controlled clinical trial [28] (IIa). In earlier studies, vitamin C (1 g/day) supplementation has been proposed to ameliorate the sperm quality in infertile men [15] (III), [29] (Ib). Sperm parameters also increases with a higher vitamin C intake, as shown by the higher sperm concentration and total progressive motile sperm count (TPMS) [30] (IIb). In a placebo-controlled study, vitamin C given to heavy smokers at a dose of 200 or 1,000 mg/day for 4 months improved sperm parameters. The group that took the dose of 1,000 mg/day had a higher increase [29] (Ib). In addition, vitamin C safeguards human spermatozoa from endogenous oxidative DNA damage [31] (IIb).

α-Tocopherol

In a single-blinded study, eight patients treated with 100 mg of α-tocopherol (vitamin E) tid for 4 months did not show any sperm parameter improvement [32] (IIa). While the administration of vitamin E, at a dose of 100 mg three times per day, produced a slight increase in seminal plasma vitamin E concentration. In a study performed on 15 subjects, the number of spermatozoa, the percentage of sperm with forward motility, the half-life of the percentage displaying forward motility, and the rate of swollen spermatozoa in hypo-osmotic medium did not show any significant enhancement during vitamin E administration. The authors explained the lack of effects on these parameters with the small increase of vitamin E achieved in the seminal plasma. They hypothesized that higher doses of vitamin E may be more effective [33] (Ib). When in infertile men, vitamin E is administered at doses ranging between 300 and 1,200 mg/day for 3 weeks, seminal plasma vitamin E levels increase weakly [34]. The sperm α-tocopherol concentration is independent from the concentration or the total amount present in the seminal plasma. On the other hand, the percentage of motile spermatozoa relates significantly with sperm α-tocopherol content [35].

Many trials have been conducted to ameliorate sperm parameters of infertile men by vitamin E administration. In a double-blind randomized, placebo-controlled, crossover trial, 30 healthy men with elevated semen ROS concentrations and healthy female partners were given vitamin E (600 mg/day) or placebo for 3 months. Vitamin E increased significantly blood serum α-tocopherol concentrations and sperm function evaluated by the zona-binding assay [17] (Ia). Some other reports utilized lower doses of vitamin E. For example, a single-blinded study took into account eight patients receiving 300 mg/day of vitamin E, divided in three daily doses of 100 mg each, for 4 months. Patients receiving α-tocopherol did not show any improvement [32] (IIa). A placebo-controlled, double-blind study showed that the elevated sperm MDA in asthenozoospermic and oligoasthenozoospermic men decreased significantly after vitamin E administration which also improved sperm motility in asthenozoospermic patients. In addition, 11 of 52 wives (21%) of the treated group got pregnant during the 6-month treatment; 9 of them had a normal term deliveries, while 2 aborted in the first trimester. No pregnancy was reported in the placebo group [36]

(IIa). Moreover, elevated MDA concentrations significantly dropped to normal levels, and the fertilization rate per cycle increased significantly following administration of 200 mg/day of vitamin E for 3 months, in a prospective study conducted in 15 fertile normozoospermic men. The elevated MDA concentrations significantly decreased to normal levels and the fertilization rate per cycle increased significantly after 1 month of treatment [18] (IIa). Furthermore, an elevated consumption of daily nutrients with scavenging potential (food and nutraceutical complements such as zinc, folate, vitamins C and E, and β-carotene) proposed to 97 healthy, nonsmoking men, showed that the vitamin E intake correlated with the highest progressive motility and TPMS [30] (IIb).

Ascorbic Acid and α-Tocopherol

Ascorbic acid (vitamin C) and α-tocopherol (vitamin E) may be administered together to decrease the peroxidative injury on spermatozoa, taking advantage of their hydrophilicity and lipophilicity, respectively. In addition, if these compounds act directly on spermatozoa to prevent ROS-induced damage, the improvement could be rapid, given that the two vitamins reach spermatozoa both within the epididymis and following ejaculation.

A double-blind, placebo-controlled, randomized trial has been performed in asthenozoospermic or moderate oligoasthenozoospermic men. Vitamins C (1 g) and E (800 mg) were prescribed simultaneously for 2 months, but no improvement of semen parameters was reported [37] (Ib). These unsatisfactory findings match with the findings of other studies [32, 33], but diverge from other published data [18, 38]. It is also possible that the duration of the treatment was too short to produce an effect, particularly if the action occurs within the testis.

Sixty-four patients with idiopathic infertility and an increased (≥15%) proportion of spermatozoa with DNA fragmentation were randomly divided into two groups: one was given vitamins C (1 g) and E (1 g) daily and the other one, placebo. After a 2 months of treatment, the proportion of DNA-fragmented spermatozoa decreased significantly in the antioxidant-treated group, while no variation was detected in the placebo group [39] (Ib). An additional trial was performed on 38 patients with a raised (≥15%) percentage of DNA-fragmented spermatozoa in the ejaculate. They were prescribed vitamins C (1 g) and E (1 g) daily for 2 months following one ICSI cycle failure. In 29 of them (76%), the scavenging therapy led to a decline in the proportion of DNA-fragmented spermatozoa and a successful ICSI attempt with a higher clinical pregnancy (48.2% vs. 6.9%) and implantation (19.6% vs. 2.2%) rates [40] (IIb).

α-Tocopherol and Selenium

Few studies have been performed using the association between vitamin E and selenium [41, 42]. A trial was conducted in nine oligoasthenoteratozoospermic patients who were prescribed vitamin E (400 mg) plus selenium (100 μg) daily for 1 month. Thereafter, selenium supplementation was increased to 200 μg/day for the next 4 months. This kind of association produced a significant improvement of sperm motility, morphology, and vitality [41] (Ib). The other study, using the same association, was performed in 28 men who were given vitamin E (400 mg) and selenium (225 μg) daily for 3 months. Other 26 patients assumed vitamin B (4.5 g/day) for the same length of time, as control. The administration of vitamin E and selenium resulted in a significant reduction in MDA concentrations and an enhancement of sperm kinetic parameters [42] (Ib).

Glutathione

GSH is one of the most commonly used drugs; thanks to its antitoxic and scavenging action in different diseases. Although it cannot cross cell membranes, its concentration increases in biological fluids following a systemic intake. GSH is able to reach the seminal plasma and to play an action at this level. Here, it safeguards spermatozoa from ROS attack; hence, GSH may play a beneficial function in several andrological diseases, particularly during male genital tract inflammation [16].

GSH (600 mg/day i.m.) has been prescribed to 11 patients with dyspermia associated with different andrological diseases in a 2-month pilot trial. Sperm kinetics improved, particularly in men with male accessory gland infections (MAGI) and in men with varicocele [43] (III), two circumstances wherein ROS or other noxious substance may play a pathogenic role. Following these encouraging findings, the same investigators conducted a placebo-controlled, double-blind, crossover study on infertile men experiencing unilateral varicocele and amicrobial MAGI. The patients were allocated to treatment with GSH, 600 mg i.m. on alternate days, or placebo ampoules. Men who received GSH showed higher sperm number, motility, kinetic parameters, and percentage of normal forms. These effects on sperm motility and morphology lasted for some time after the treatment was discontinued. The authors hypothesized that these findings may relate to a post-spermatocyte action of GSH, since the length of the treatment did not cover the full length of a complete spermatogenesis [16] (Ib). This kind of sperm modification can be partially corrected by GSH administration when cell membrane injury is not too critical [44] (IIa).

The above reported data indicate that, at least to some extent, the beneficial effect of GSH is suitable for the biochemical changes in membrane organization and its following defensive action on the lipid components of the cell membrane. The decline of lipoperoxide levels in seminal plasma let to consider that GSH minimizes the consequence of lipoperoxidation generated by vascular or inflammatory diseases.

Carnitines

Carnitines are involved in many metabolic pathways in several cellular organelles. These compounds play a primary function in sperm maturation within the male genital organs and a relevant role in the metabolism of spermatozoa by furnishing immediately accessible energy to be utilized by spermatozoa. This positively correlates with sperm motility and concentration [45]. An increase of sperm progressive motility occurs simultaneously to L-carnitine augmentation and storage in the epididymal lumen [46].

Several different kinds of studies (controlled, uncontrolled, human, and animal) have been carried out to evaluate the potential application of carnitines as scavenging molecules. In 1992, a study was conducted on the male partners of 20 couples affected by idiopathic oligoasthenozoospermia (concentration $<20\times10^6$ spermatozoa/ml, progressive motility <50%) who were given 4 g/day of L-acetyl-carnitine for 2 months. No significant effect on sperm concentration, total motility, and morphology resulted, whereas a significant improvement of progressive motility ($21.7\pm3.2\%$ vs. 38.2 ± 4.7) was appreciated [47] (IIb). Afterwards, a multicentre open study was performed on 100 men with idiopathic asthenozoospermia. L-carnitine was administered orally at the dose of 1 g three times per day for 4 months with a significant improvement of several sperm kinetic parameters [48] (IIb). Another study came to similar results, giving an oral solution of L-carnitine (1 g) three times a day for 3 months, to 47 patients with idiopathic asthenozoospermia [49] (IIb). A review article proposed carnitines treatment as an alternative method in the broader medical treatment of patients with infertility due to OS [50].

Some clinical evidence suggests that a selective group of infertile men, those with prostato-vesiculo-epididymitis (PVE), benefit from carnitines administration, since antimicrobial and/or nonsteroidal anti-inflammatory drugs, although effective to eliminate microbial infection, have a poor scavenging action [51] (Ib). Another study conducted on 98 men with PVE and leukocytospermia showed that carnitine scavenging treatment was totally successful once they were pretreated with nonsteroidal anti-inflammatory compounds [52] (Ib).

In a placebo-controlled, double-blind, crossover trial, L-carnitine was capable to enhance sperm parameters, even if it was unsuccessful to reduce LPO concentrations. These findings suggested an incomplete action of L-carnitine to counteract the ROS attack [53] (Ib). The same group proposed a double-blind, randomized, placebo-controlled trial. They gave a combined treatment with L-carnitine (2 g/day) and L-acetyl-carnitine (1 g/day) or placebo to 60 infertile males with oligoasthenoteratozoospermia. All sperm parameters improved, but the most important enhancement was found in both progressive and total sperm motility particularly in men with the highest degree of asthenozoospermia [54] (Ib). Another placebo-controlled study conducted also in patients with oligoasthenoteratozoospermia showed that the same treatment improved sperm concentration, motility, and morphology, particularly when cinnoxicam (1 suppository every 4 days) was added [55] (Ib). Furthermore, 60 patients with asthenozoospermia were enrolled in a double-blind clinical trial with L-carnitine (3 g/day), L-acetyl-carnitine (3 g/day), a combination of L-carnitine (2 g/day) plus L-acetyl-carnitine (1 g/day), or placebo, for 6 months. Total and forward motility, including kinetic parameters analyzed by computer-assisted sperm analysis, improved in men receiving either L-acetyl-carnitine alone or in association with L-carnitine. The total oxyradical scavenging capacity of the semen towards hydroxyl and peroxyl radicals also improved and correlated with the enhancement of sperm kinetics. Patients with lower motility and total oxyradical scavenging capacity of the seminal fluid had more chances of responding to the treatment [56] (Ib). In another trial, L-carnitine (2 g/day) and L-acetyl-carnitine (1 g/day) were given orally tid for 3 months to 90 men with oligoasthenozoospermia. In the treatment group, ten female partners (11.6%) achieved pregnancy, whereas only two pregnancies (3.7%) were recorded in the control group. Moreover, their percentage of forward and total motile spermatozoa increased significantly [57] (Ib). In the trial lead by De Rosa and colleagues, 66 patients with <50% motility receiving L-carnitine (1 g/day) and L-acetyl-carnitine (500 mg tid), for 6 months, had a significant increase in sperm total motility, viability, membrane integrity, and linearity of sperm movement, both after 3 and 6 months of treatment, and in the ability to penetrate the cervical mucus increased after 6 months [58] (IIb). Twenty-one patients with infertility and with sperm motility ranging from 10 to 50% were given carnitines (2 g of L-carnitine and 1 g of L-acetyl-carnitine per day) orally for 6 months, but differently from the other studies, no significant effects on sperm motility resulted [59] (Ib). In a further trial, L-carnitine (2 g/day) and L-acetyl-carnitine (1 g/day) were administered for 3 months in men with PVE and increased ROS production. Carnitines showed to be a successful treatment once seminal leukocytes were within the normal range [60] (IIb).

On the light of the many studies exploring the effects of carnitines on sperm parameters, a systematic review has been recently published. The meta-analysis compared L-carnitine and/or L-acetyl-carnitine treatment to placebo reported significant improvement in total and forward sperm motility, atypical sperm cells, and pregnancy rate. No significant difference has been found in sperm concentration [61] (Ia).

Coenzyme Q10

Coenzyme Q10 (CoQ10) is a lipid-soluble constituent of the respiratory chain. Ubiquinol is the reduced form and the active one. It behaves as a powerful scavenger in some biological components, for instance lipoproteins and membranes. The concentrations of reduced and oxidized forms of CoQ10 (ubiquinol/ubiquinone) and of hydroperoxide have been measured in the seminal plasma and seminal fluid of 32 infertile men. A positive correlation between ubiquinol concentration and sperm count has been observed, whereas a negative correlation was reported between sperm count and ubiquinol concentration or hydroperoxide levels. An important correlation between sperm concentration, motility, and seminal fluid ubiquinol-10 content has been found, whereas, in total fluid, an inverse correlation between ubiquinol/ubiquinone ratio and the severity of teratozoospermia has been reported. These findings indicate that ubiquinol-10 impedes hydroperoxide occurrence in seminal fluid and in seminal plasma [62].

CoQ10 has been given orally at the dose of 60 mg/day to 17 men with low fertilization rate after ICSI for male infertility for an average of 103 days previously the subsequent ICSI procedure. The results showed a significant enhancement of the fertilization rate [63] (IIb).

In the human seminal fluid, CoQ10 has been found at relevant concentrations and it shows a direct association with sperm concentrations and kinetics. Differently, in men with varicocele, despite a higher proportion of CoQ10 in the seminal plasma, the correlation with sperm motility was not observed [64]. Elevated CoQ10 levels have been found in spermatozoa of oligozoospermic and asthenozoospermic patients without varicocele. This correlation was not detected in men with varicocele, who additionally showed slightly lower intracellular absolute concentrations of CoQ10. Higher intracellular levels could be linked to a spermatozoa protective system. In men with varicocele, this kind of system could be inadequate, leading to an excessive susceptivity to OS [64].

Very recently a double-blind, randomized trial has been carried out in 60 infertile patients with idiopathic asthenozoospermia. Patients underwent a double-blind therapy with CoQ10 (200 mg/day) or placebo for 6 months. After treatment, CoQ10 and ubiquinol raised appreciably in seminal plasma as well as in spermatozoa. Interestingly, spermatozoa improved their motility. Men with a poorer sperm motility and lower concentrations of CoQ10 had a statistically significant more elevated chance to better respond to its administration [65] (Ib).

Lycopene

Lycopene is an element of human redox defensive system against oxidative stress. Oral lycopene administration appears to have a function in the treatment of patients with idiopathic infertility. Following the administration of 2 g of lycopene, twice a day, for 3 months, a significant increase occurs in the sperm number and motility, but the sperm concentration increase is present only in men with a sperm concentration >5 million/ml [66] (IIb).

Pycnogenol

Pycnogenol is a substance obtained from the bark of the "Pinus maritima." Pycnogenol's constituents inhibit cyclooxygenases that release inflammatory prostaglandins [67]. A study has been conducted in subfertile men who were administered pycnogenol (200 mg/day), for 3 months. The results showed a mean sperm morphology improvement by 38% of the pretreatment values and the mannose receptor binding assay score augmented by 19% [68] (IIb).

Other Compounds

N-Acetyl-Cysteine or Vitamin A Plus Vitamin E and Essential Fatty Acids

An open, prospective study, conducted in 27 infertile men who were given a combined oral antioxidants treatment with *N*-acetyl-cysteine (NAC) or vitamin A plus vitamin E and essential fatty acids, showed an increase of sperm concentration in oligozoospermic patients. Moreover, this treatment significantly reduced ROS and 8-OH-dG production, and in the mean time, it increased the percentage of acrosome-reacted spermatozoa, the quantity of PUFA in phospholipids, and sperm membrane [69] (IIb). Very recently, 120 idiopathic infertile men were randomly given NAC alone (600 mg/day orally) or placebo for 3 months. NAC increased semen volume and sperm motility as well as semen viscosity [70] (Ib).

Pentoxifylline

Spermatozoa from 15 patients with asthenozoospermia and high ROS levels were treated in vitro with pentoxifylline to evaluate the effects of this compound on ROS generation and sperm movement. Pentoxifylline was able to reduce the production of ROS by spermatozoa, and it slowed down the in vitro decline of the curvilinear velocity and the beat cross frequency for 6 h. These same 15 patients and 18 asthenozoospermic patients, whose spermatozoa did not generate ROS at steady state, were then prescribed pentoxifylline at two distinct doses (300 and 1,200 mg daily) to validate its in vivo outcome on ROS generation, sperm kinetics, and sperm fertilizing competence. Pentoxifylline administration had no effects on spermatozoon-induced ROS formation, and it did not show any effect on sperm motility and fertilizing capacity. Nevertheless, it increased motility and beat cross frequency at the dose of 1,200 mg daily [71].

Selenium

Selenium supplementation has been given alone to 33 subfertile men for 3 months, but it did not produced any improvement of sperm count, motility, and morphology [72] (IIa). Subsequently, a trial was performed on 69 asthenozoospermic patients who received either placebo, selenium alone, or selenium plus vitamins A, C, and E daily for 3 months. Treatment did not show any improvement of sperm concentration, while sperm motility increased in both selenium-treated groups. This study showed that oral selenium administration is effective especially in patients with a low selenium [73] (Ib). Recently, a clinical trial investigated the usefulness of selenium (200 μg) and/or NAC (600 mg) in 468 infertile men with idiopathic oligoasthenoteratozoospermia for 6 months. This treatment showed to be effective on all sperm parameters measured and a clear correlation between seminal plasma selenium concentrations, NAC, and semen characteristics [74] (Ib).

Shao-Fu-Zhu-Yu-Tang

Shao-Fu-Zhu-Yu-Tang has been proposed to have antiaging and sperm scavenging properties. Its administration, for 60 days to 36 patients with chronic prostatitis, revealed a significant increase in sperm motility as evaluated by computer-assisted semen analysis [75] (IIb).

Zinc, Folic Acid, Astaxanthin, and Acetyl-Carnitine

Sperm parameters ameliorated following the intake of a mixture of zinc and folic acid, or the antioxidant astaxanthin, or a so-called energy-providing combination including (acetyl)-carnitine (Proxeed®). Furthermore, a double-blind study showed that the latter two compounds increase spontaneous or intrauterine insemination-assisted conception rates [76] (Ib). Astaxanthin appears to act significantly to decrease ROS and inhibin B concentrations and to increase sperm linear velocity and pregnancy rate [76].

Lepidium meyenii

The extracts of the Peruvian plant *Lepidium meyenii* appeared to be useful to increase the sperm number and the percentage of normal forms in an uncontrolled trial [77] (IIb).

α-Linolenic Acid and Lignans

Linseed oil includes α-linolenic acid and lignans. α-linolenic acid adjusts the poor intake of omega-3 essential fatty acids that is connected with reduced sperm motility among patients with fertility problems [78] (IV).

Vitamin C and Vitamin E, Lycopene, Selenium, Folic Acid, Garlic Oil, and Zinc

Menevit®, an antioxidant preparation combining vitamins C and E, lycopene, selenium, folic acid, garlic oil, and zinc, has been prescribed to 60 couples with severe male infertility for 3 months, in a prospective randomized, double-blind, placebo-controlled trial, before undergoing an IVF cycle. Men who assumed Menevit® had a significant improvement of the pregnancy rate (38.5%) compared to the men who assumed placebo (16% pregnancy) [79] (Ib).

Morindae Officinalis Extract

Morindae officinalis extract, taken at the concentrations of 0.25 or 0.5 g/ml, showed to be more effective than vitamin C in enhancing SOD vitality of sperm suspension and in decreasing MDA concentration. It has been shown to take part in a defensive action in the ROS-mediated damage of sperm membrane. Moreover, at higher dosage (0.5 mg/ml), Morindae officinalis particularly safeguards sperm membrane function [80].

Zinc

Zinc therapy has been shown valuable in decreasing OS, sperm apoptosis, and DFI in asthenozoospermic men. Zinc associated with vitamin E or with vitamin E plus vitamin C did not result in any further significant effect [81].

Table 46.4 Summary of the evidences and grading of the recommendations of the effects of each antioxidant used alone or combination on sperm quality and function, according to the National Clinical Guidelines for Type 2 Diabetes (the Royal College of General practitioners, Effective Clinical Practice Unit, ScHAAR University of Sheffield)

Compound	Classification of evidences		Grading of recommendations
	Positive effects	Any or negative effects	
Vitamin C			
Dawson et al. [15]	III		B
Fraga et al. [31]	IIb		C
Dawson et al. [29]	Ib		A
Thiele et al. [28]	IIa		B
Eskenazi et al. [30]	IIb		B
Vitamin E			
Giovenco et al. [32]		IIa	B
Moilanen et al. [33]		Ib	A
Kessopoulou et al. [17]	Ia		A
Suleiman et al. [36]	IIa		B
Geva et al. [18]	IIa		B
Eskenazi et al. [30]	IIb		B
Vitamin C plus vitamin E			
Rolf et al. [37]		Ib	A
Greco et al. [39]	Ib		A
Greco et al. [40]	IIb		B
Vitamin E plus selenium			
Vezina et al. [41]	Ib		A
Keskes-Ammar et al. [42]	Ib		A
N-acetyl-cysteine plus vitamin E			
Comhaire et al. [69]	IIb		
Selenium plus *N*-acetyl-cysteine			
Safarinejad and Safarinejad [74]	Ib		
N-acetyl-cysteine			
Ciftci et al. [70]	Ib		
Glutathione			
Lenzi et al. [43]	III		C
Lenzi et al. [16]	Ib		A
Carnitines			
Moncada et al. [47]	IIb		B
Costa et al. [48]	IIb		B
Vitali et al. [49]	IIb		B
Vicari et al. [51]	Ib		A
Vicari and Calogero [60]	IIb		A
Vicari et al. [52]	Ib		A
Lenzi et al. [53]	Ib		A
Lenzi et al. [54]	Ib		A
Cavallini et al. [55]	Ib		A
Balercia et al. [56]	Ib		A
Li et al. [57]	Ib		B
De Rosa et al. [58]	IIb		A
Sigman et al. [59]		Ib	A
Zhou et al. [61]	Ia		B
Coenzyme Q10			
Lewin and Lavon [63]	IIb		B
Balercia et al. [65]	Ib		A
Lycopene			
Gupta and Kumar [66]	IIb		B
Pycnogenol			

(continued)

Table 46.4 (continued)

Compound	Classification of evidences		Grading of recommendations
	Positive effects	Any or negative effects	
Roseff [68]	IIb		B
Selenium			
Iwanier and Zachara [72]		IIa	B
Scott [73]	Ib		A
Shao-Fu-Zhu-Yu-Tang			
Yang et al. [75]	IIb		B
Astacarox®			
Comhaire et al. [76]	Ib		A
Proxeed®			
Comhaire et al. [76]	Ib		A
Lepidium meyenii			
Gonzales et al. [77]	IIb		B
Linseed oil			
Comhaire and Mahmoud [78]	IV		None
Menevit®			
Tremellen et al. [79]	Ib		A

Expert Commentary

Many studies have been performed using different antioxidant compounds with the aim to improve semen parameters. Unfortunately the endpoints taken into account by these studies are often different and this does not help in understanding the efficacy of a given antioxidant. Moreover, it should be kept in mind that any andrological disease, independently of the OS, may be reversible or not according to the degree of the damage that has developed at the time of the therapeutical intervention. A prolonged exposure to OS can also cause an extensive damage that, over the time, can compromise the efficiency of the male accessory glands on sperm function. This represents an additional bias for many trials that have not considered the duration of the disease. All these reasons make the scavenging therapy a great challenge for the andrologist.

Bearing this in mind, we attempted a primary distinction dividing the antioxidants into compounds which play positive effects and compounds which play negative effects, as reported in Table 46.4. Using an EBM method, we proposed that some compounds may be considered as first-line treatment, because of the extensive investigation and the higher EBM evidences. These include vitamins C and E and carnitines. The efficacy of other antioxidants is not yet supported by a sufficient number of studies. These include pycnogenol, lycopene, etc., which need additional controlled trials. Other scavenging molecules, such as CoQ10 and GSH, can be proposed as second-line treatment because of the well-done, though few, studies performed on them. Nevertheless, also for these compounds studies that can clarify the dark points previously analyzed are welcome.

Five-Year View and Key Issues

Two major issues need, in the next future, to be further implemented to allow a clear evaluation of the true effectiveness of the antioxidant treatment in the infertile men. First of all, studies should be carried out in homogeneous cohorts of patients. This requires a careful andrological screening aimed at exactly diagnosing the disease which increases the oxidative stress. The second issue relates to the development of more precise and hopefully inexpensive and non-cumbersome methods to estimate the oxidative stress in the semen samples. Finally, keeping these issues in mind, there is an absolute need of additional double-blind, placebo-controlled, randomized, crossover, multicenter clinical trials to gain more information about the effectiveness of an antioxidant (or a combination of them) over another one in men with infertility due to an increased oxidative stress.

References

1. Sikka SC. Relative impact of oxidative stress on male reproductive function. Curr Med Chem. 2001;8:851–62.
2. Griveau JF, Le Lannou D. Reactive oxygen species and human spermatozoa: physiology and pathology. Int J Androl. 1997;20:61–9.
3. Shen H, Ong C. Detection of oxidative DNA damage in human sperm and its association with sperm function and male infertility. Free Radic Biol Med. 2000;28:529–36.
4. Lanzafame F, Chapman MG, Guglielmino A, et al. Pharmacological stimulation of sperm motility. Hum Reprod. 1994;9:192–9.
5. Vicari E. Effectiveness and limits of antimicrobial treatment on seminal leukocyte concentration and related reactive oxygen species production in patients with male accessory gland infection. Hum Reprod. 2000;15:2536–44.

6. Ten J, Vendrell FJ, Cano A, et al. Dietary antioxidant supplementation did not affect declining sperm function with age in the mouse but did increase head abnormalities and reduced sperm production. Reprod Nutr Dev. 1997;37:481–92.
7. Ménézo YJ, Hazout A, Panteix G, et al. Antioxidants to reduce sperm DNA fragmentation: an unexpected adverse effect. Reprod Biomed Online. 2007;14:418–21.
8. Chew BP. Effects of supplemental β-carotene and vitamin A on reproduction in swine. J Anim Sci. 1993;71:247–52.
9. Luck MR, Jeyaseelan I, Scholes RA. Ascorbic acid and fertility. Biol Reprod. 1995;52:262–6.
10. Tripodi L, Tripodi A, Mammi C, et al. Pharmacological action and therapeutic effects of glutathione on hypokinetic spermatozoa for enzymatic-dependent pathologies and correlated genetic aspects. Clin Exp Obstet Gynecol. 2003;30:130–6.
11. Mishra M, Acharya UR. Protective action of vitamins on the spermatogenesis in lead-treated Swiss mice. J Trace Elem Med Biol. 2004;18:173–8.
12. Rao MV, Sharma PS. Protective effect of vitamin E against mercuric chloride reproductive toxicity in male mice. Reprod Toxicol. 2001;15:705–12.
13. Castellini C, Lattaioli P, Bernardini M, et al. Effect of dietary alpha-tocopheryl acetate and ascorbic acid on rabbit semen stored at 5 degrees C. Theriogenology. 2000;54:523–33.
14. Agarwal A, Said TM. Carnitines and male infertility. Reprod Biomed Online. 2004;8:376–84.
15. Dawson EB, Harris WA, Rankin WE, et al. Effect of ascorbic acid on male fertility. Ann N Y Acad Sci. 1987;498:312–23.
16. Lenzi A, Culasso F, Gandini L, et al. Placebo-controlled, double blind, cross-over trial of glutathione therapy in male infertility. Hum Reprod. 1993;8:1657–62.
17. Kessopoulou E, Powers HJ, Sharma KK, et al. A double-blind randomized placebo cross-over controlled trial using the antioxidant vitamin E to treat reactive species associated male infertility. Fertil Steril. 1995;64:825–31.
18. Geva E, Bartoov B, Zabludovsky N, et al. The effect of antioxidant treatment on human spermatozoa and fertilization rate in an in vitro fertilization program. Fertil Steril. 1996;66:430–4.
19. Patel SR, Sigman M. Antioxidant therapy in male infertility. Urol Clin North Am. 2008;35:319–30.
20. Jacob RA, Pianalto FS, Agee RE. Cellular ascorbate depletion in healthy men. J Nutr. 1992;122:1111–8.
21. Frei B, England L, Ames BN. Ascorbate is an outstanding antioxidant in human blood plasma. Proc Natl Acad Sci USA. 1989;86: 6377–81.
22. Doba T, Burton GW, Ingold KU. Antioxidant and co-antioxidant activity of vitamin C. The effect of vitamin C, either alone or in the presence of vitamin E or a water-soluble vitamin E analogue, upon the peroxidation of aqueous multilamellar phospholipid liposomes. Biochim Biophys Acta. 1985;835:298–303.
23. Lewis SE, Sterling ES, Young IS, et al. Comparison of individuals antioxidants of sperm and seminal plasma in fertile and infertile men. Fertil Steril. 1997;67:142–7.
24. Song GJ, Norkus EP, Lewis V. Relationship between seminal ascorbic acid and sperm DNA integrity in infertile men. Int J Androl. 2006;29:569–75.
25. Wayner DD, Burton GW, Ingold KU. The antioxidant efficiency of vitamin C is concentration-dependent. Biochim Biophys Acta. 1986;884:119–23.
26. Levine M, Conry-Cantilena C, Wang Y, et al. Vitamin C pharmacokinetics in healthy volunteers: evidence for a recommended dietary allowance. Proc Natl Acad Sci USA. 1996;93:3704–9.
27. Wen Y, Cooke T, Feely J. The effect of pharmacological supplementation with vitamin C on low-density lipoprotein oxidation. Br J Clin Pharmacol. 1997;44:94–7.
28. Thiele JJ, Friesleben HJ, Fuchs J, et al. Scorbic acid and urate in human seminal plasma: determination and interrelationships with chemiluminescence in washed semen. Hum Reprod. 1995;10: 110–5.
29. Dawson EB, Harris WA, Teter MC, et al. Effect of ascorbic acid supplementation on the sperm quality of smokers. Fertil Steril. 1992;58:1034–9.
30. Eskenazi B, Kidd SA, Marks AR, et al. Antioxidant intake is associated with semen quality in healthy men. Hum Reprod. 2005;20:1006–12.
31. Fraga CG, Motchnik PA, Shigenaga MK, et al. Ascorbic acid protects against endogenous oxidative DNA damage in human sperm. Proc Natl Acad Sci USA. 1991;88:11003–6.
32. Giovenco P, Amodei M, Barbieri C, et al. Effects of kallikrein on the male reproductive system and its use in the treatment of idiopathic oligozoospermia with impaired motility. Andrologia. 1987;19 Spec No:238–41.
33. Moilanen J, Hovatta O, Lindroth L. Vitamin E levels in seminal plasma can be elevated by oral administration of vitamin E in infertile men. Int J Androl. 1993;16:165–6.
34. Moilanen J, Hovatta O. Excretion of alpha-tocopherol into human seminal plasma after oral administration. Andrologia. 1995;27: 133–6.
35. Therond P, Auger J, Legrand A, et al. Alpha-tocopherol in human spermatozoa and seminal plasma: relationships with motility, antioxidant enzymes and leukocytes. Mol Hum Reprod. 1996;2:739–44.
36. Suleiman SA, Ali ME, Zaki ZM, et al. Lipid peroxidation and human sperm motility: protective role of vitamin E. J Androl. 1996;17:530–7.
37. Rolf C, Cooper TG, Yeung CH, et al. Antioxidant treatment of patients with asthenozoospermia or moderate oligoasthenozoospermia with high-dose vitamin C and vitamin E: a randomized, placebo-controlled, double-blind study. Hum Reprod. 1999;14: 1028–33.
38. De Lamirande E, Gagnon C. Reactive oxygen species and human spermatozoa. I. Effects on the motility of intact spermatozoa and on sperm axonemes. J Androl. 1992;13:368–78.
39. Greco E, Iacobelli M, Rienzi L, et al. Reduction of the incidence of sperm DNA fragmentation by oral antioxidant treatment. J Androl. 2005;26:349–53.
40. Greco E, Romano S, Iacobelli M, et al. ICSI in cases of sperm DNA damage: beneficial effect of oral antioxidant treatment. Hum Reprod. 2005;20:2590–4.
41. Vézina D, Mauffette F, Roberts KD, et al. Selenium-vitamin E supplementation in infertile men. Effects on semen parameters and micronutrient levels and distribution. Biol Trace Elem Res. 1996;53:65–83.
42. Keskes-Ammar L, Feki-Chakroun N, Rebai T, et al. Sperm oxidative stress and the effect of an oral vitamin E and selenium supplement on semen quality in infertile men. Arch Androl. 2003; 49:83–94.
43. Lenzi A, Lombardo F, Gandini L, et al. Glutathione therapy for male infertility. Arch Androl. 1992;29:65–8.
44. Lenzi A, Picardo M, Gandini L, et al. Glutathione treatment of dyspermia: effect on the lipoperoxidation process. Hum Reprod. 1994;9:2044–50.
45. Tang LF, Jiang H, Shang XJ, et al. Seminal plasma levocarnitine significantly correlated with semen quality. Zhonghua Nan Ke Xue. 2008;14:704–8.
46. Jeulin C, Soufir JC, Marson J, et al. Acetylcarnitine and spermatozoa: relationship with epididymal maturation and motility in the boar and man. Reprod Nutr Dev. 1988;28:1317–27.
47. Moncada ML, Vicari E, Cimino C, et al. Effect of acetyl carnitine treatment in oligoasthenospermic patients. Acta Eur Fertil. 1992; 23:221–4.

48. Costa M, Canale D, Filicori M, et al. L-carnitine in idiopathic asthenozoospermia: a multicenter study. Andrologia. 1994;26:155–9.
49. Vitali G, Parente R, Melotti C. Carnitine supplementation in human idiopathic asthenospermia: clinical results. Drugs Exp Clin Res. 1995;21:157–9.
50. Dokmeci D. Oxidative stress, male infertility and the role of carnitines. Folia Med (Plovdiv). 2005;47:26–30.
51. Vicari E, Rubino C, De Palma A, et al. Antioxidant therapeutic efficiency after the use of carnitine in infertile patients with bacterial or non bacterial prostato-vesiculo-epididymitis. Arch Ital Urol Androl. 2001;73:15–25.
52. Vicari E, La Vignera S, Calogero AE. Antioxidant treatment with carnitines is effective in infertile patients with prostatovesiculoepididymitis and elevated seminal leukocyte concentrations after treatment with nonsteroidal anti-inflammatory compounds. Fertil Steril. 2002;78:1203–8.
53. Lenzi A, Lombardo F, Sgrò P, et al. Use of carnitine therapy in selected cases of male factor infertility: a double-blind crossover trial. Fertil Steril. 2003;79:292–300.
54. Lenzi A, Sgrò P, Salacone P, et al. A placebo-controlled double-blind randomized trial of the use of combined L-carnitine and L-acetyl-carnitine treatment in men with asthenozoospermia. Fertil Steril. 2004;81:1578–84.
55. Cavallini G, Ferraretti AP, Gianaroli L, et al. Cinnoxicam and L-carnitine/acetyl-L-carnitine treatment for idiopathic and varicocele associated oligoasthenospermia. J Androl. 2004;25:761–70.
56. Balercia G, Regoli F, Armeni T, et al. Placebo-controlled double-blind randomized trial on the use of L-carnitine, L-acetylcarnitine, or combined L-carnitine and L-acetylcarnitine in men with idiopathic asthenozoospermia. Fertil Steril. 2005;84:662–71.
57. Li Z, Chen GW, Shang XJ, et al. A controlled randomized trial of the use of combined L-carnitine and acetyl-L-carnitine treatment in men with oligoasthenozoospermia. Zhonghua Nan Ke Xue. 2005;11:761–4.
58. De Rosa M, Boggia B, Amalfi B. Correlation between seminal carnitine and functional spermatozoal characteristics in men with semen dysfunction of various origins. Drugs R&D. 2005;6:1–9.
59. Sigman M, Glass S, Campagnone J, et al. Carnitine for the treatment of idiopathic asthenospermia: a randomized, double-blind, placebo-controlled trial. Fertil Steril. 2006;85:1409–14.
60. Vicari E, Calogero AE. Effects of treatment with carnitines in infertile patients with prostato-vesiculo-epididymitis. Hum Reprod. 2001;16:2338–42.
61. Zhou X, Liu F, Zhai S. Effect of L-carnitine and/or L-acetyl-carnitine in nutrition treatment for male infertility: a systematic review. Asia Pac J Clin Nutr. 2007;16 Suppl 1:383–90.
62. Alleva R, Scararmucci A, Mantero F, et al. Protective role of ubiquinol content against formation of lipid hydroperoxide in human seminal fluid. Mol Aspects Med. 1997;18:S221–8.
63. Lewin A, Lavon H. The effect of coenzyme Q10 on sperm motility and function. Mol Aspects Med. 1997;18:S213–9.
64. Mancini A, Conte G, Milardi D, et al. Relationship between sperm cell ubiquinone and seminal parameters in subjects with and without varicocele. Andrologia. 1998;30:1–4.
65. Balercia G, Buldreghini E, Vignini A, et al. Coenzyme Q10 treatment in infertile men with idiopathic asthenozoospermia: a placebo-controlled, double-blind randomized trial. Fertil Steril. 2009;91:1785–92.
66. Gupta NP, Kumar R. Lycopene therapy in idiopathic male infertility: a preliminary report. Int Urol Nephrol. 2002;34:369–72.
67. Baumann J, Wurm G, von Bruchhausen F. Prostaglandin synthetase inhibition by flavonoids and phenolic compounds in relation to their O_2—scavenging properties. Arch Pharm (Weinheim). 1980; 313:330–7.
68. Roseff SJ. Improvement in sperm quality and function with French maritime pine tree bark extract. J Reprod Med. 2002;47:821–4.
69. Comhaire FH, Christophe AB, Zalata AA, et al. The effects of combined conventional treatment, oral antioxidants and essential fatty acids on sperm biology in subfertile men. Prostaglandins Leukot Essent Fatty Acids. 2000;63:159–65.
70. Ciftci H, Verit A, Savas M, et al. Effects of *N*-acetylcysteine on semen parameters and oxidative/antioxidant status. Urology. 2009; 74:73–6.
71. Okada H, Tatsumi N, Kanzaki M, et al. Formation of reactive oxygen species by spermatozoa from asthenospermic patients: response to treatment with pentoxifylline. J Urol. 1997;157:2140–6.
72. Iwanier K, Zachara BA. Selenium supplementation enhances the element concentration in blood and seminal fluid but does not change the spermatozoal quality characteristics in subfertile men. J Androl. 1995;16:441–7.
73. Scott R, MacPherson A, Yates RW, et al. The effect of oral selenium supplementation on human sperm motility. Br J Urol. 1998; 82:76–80.
74. Safarinejad MR, Safarinejad S. Efficacy of selenium and/or *N*-acetyl-cysteine for improving semen parameters in infertile men: a double-blind, placebo controlled, randomized study. J Urol. 2009;181:741–51.
75. Yang CC, Chen JC, Chen GW, et al. Effects of Shao-Fu-Zhu-Yu-Tang on motility of human sperm. Am J Chin Med. 2003;31:573–9.
76. Comhaire FH, El Garem Y, Mahmoud A, et al. Combined conventional/antioxidant "Astaxanthin" treatment for male infertility: a double blind, randomized trial. Asian J Androl. 2005;7:257–62.
77. Gonzales GF, Cordova A, Gonzales C, et al. *Lepidium meyenii* (Maca) improved semen parameters in adult men. Asian J Androl. 2001;3:301–3.
78. Comhaire FH, Mahmoud A. The role of food supplements in the treatment of the infertile man. Reprod Biomed Online. 2003; 7:385–91.
79. Tremellen K, Miari G, Froiland D, et al. A randomised control trial examining the effect of an antioxidant (Menevit) on pregnancy outcome during IVF-ICSI treatment. Aust N Z J Obstet Gynaecol. 2007;47:216–21.
80. Yang X, Zhang YH, Ding CF, et al. Extract from Morindae officinalis against oxidative injury of function to human sperm membrane. Zhongguo Zhong Yao Za Zhi. 2006;31:1614–7.
81. Omu AE, Al-Azemi MK, Kehinde EO, et al. Indications of the mechanisms involved in improved sperm parameters by zinc therapy. Med Princ Pract. 2008;17:108–16.
82. Agency for Health Care Policy and Research. Acute pain management: operative or medical procedures and trauma. Rockville: Agency for Health Care Policy and Research/US Department of Health and Human Services, Public Health Service; 1992.
83. Eccles M, et al. North of England evidence based guideline development project: guideline for angiotensin converting enzyme inhibitors in primary care management of adults with symptomatic heart failure. BMJ. 1998;316:1369.

Harmful Effects of Antioxidant Therapy

47

Adam F. Stewart and Edward D. Kim

Abstract

Oxidative stress has an integral role in the pathophysiology of most human diseases. With a rapidly aging population, increased attention and study have been directed toward the use of antioxidant therapy. The appeal is that these agents are considered "natural" substances and are associated with a healthy diet. The hypothesis has been that decreasing oxidative stress may prevent disease processes such as cancer or coronary heart disease. Because much of the general population use is in relatively healthy patients, it is critically important that these supplements are free of toxicity and side effects. While initial studies of antioxidant supplementation suggested a beneficial role in disease prevention, more recent clinical trials and a meta-analysis have questioned the benefit of these therapies. Several studies have suggested that excess supplementation may in fact be harmful. Recent attention has also focused on the use of antioxidants for the treatment of male infertility. The focus of this chapter is the potentially harmful effects of antioxidant therapy.

Keywords

Oxidative stress • Male infertility • Antioxidant therapy • Harmful effects of supplementation • Risks of antioxidant supplements • Antioxidant side effects

Oxidative stress has an integral role in the pathophysiology of most human diseases. With a rapidly aging population, increased attention and study have been directed toward the use of antioxidant therapy. The appeal is that these agents are considered "natural" substances and are associated with a healthy diet. The hypothesis has been that decreasing oxidative stress may prevent disease processes such as cancer or coronary heart disease [1, 2]. Because much of the general population use is in relatively healthy patients, it is critically important that these supplements are free of toxicity and side effects.

While initial studies of antioxidant supplementation suggested a beneficial role in disease prevention, more recent clinical trials and a meta-analysis have questioned the benefit of these therapies. Several studies have suggested that excess supplementation may in fact be harmful [3–6]. Recent attention has also focused on the use of antioxidants for the treatment of male infertility. The focus of this chapter is the potentially harmful effects of antioxidant therapy.

A.F. Stewart, MD (✉) • E.D. Kim, MD
Division of Urology, Department of Surgery, University of Tennessee Medical Center, 1928 Alcoa Highway, Suite 222, Knoxville, TN 37920, USA
e-mail: afstewart@utmck.edu; ekim@utmck.edu

Risks of Dietary Antioxidants

Certain vegetables have high contents of oxalic acid, phytic acid, and tannins. These relatively strong reducing acids may have antinutrient effects by binding to dietary minerals in the gastrointestinal tract and diminishing their absorption [7, 8]. Calcium and iron deficiencies are not uncommon in developing countries where less meat is eaten, and there is high consumption of phytic acid from beans and unleavened whole grain bread [9]. In modern, industrialized nations where balanced diets are more common, the adverse effects of excessive dietary antioxidant intake are minimal. Table 47.1 lists foods containing oxalic acid, phytic acid, and tannins.

S.J. Parekattil and A. Agarwal (eds.), *Male Infertility: Contemporary Clinical Approaches, Andrology, ART & Antioxidants*,
DOI 10.1007/978-1-4614-3335-4_47,

Table 47.1 Dietary antioxidants

Foods	Reducing acid present
Cocoa bean and chocolate, spinach, turnip, and rhubarb	Oxalic acid
Whole grains, maize, legumes	Phytic acid
Tea, beans, cabbage	Tannins

Oxalic Acid

Oxalic acid impairs calcium absorption by forming an insoluble salt of calcium oxalate. Cases of calcium deficiency have been associated with a high content of oxalates in foods [10]. A high level of oxalate intake constitutes a health risk for infants and metabolically disposed adults. Spinach is among the vegetables richest in oxalate. Sweet potatoes and peanut greens are also high in oxalic acid [11].

Phytic Acid

Phytic acid is a strong inhibitor of iron absorption in both infants and adults [12]. Iron and zinc deficiencies are widespread in infants and young children in developing countries where vegetable protein sources are often mixed with cereals. This iron deficiency in infants can lead to reduced psychomotor and mental development. Complementary foods increase the protein content and improve the protein quality of cereal-based foods. Cereals and common legumes, such as soybean, mung bean, black bean, lentils, and chick peas, are high in phytic acid. Decreasing phytic acid by 90% (approximately 100 mg/100 g dried product) would be expected to increase iron absorption about twofold. Complete enzymatic degradation of phytic acid with cooking methods such as blanching has been recommended for at-risk populations [7, 11].

Tannins

Tannins, which include condensed tannins (proanthocyanidins) and derived tannins, belong to the flavonoid family [13]. Tannins are found in a wide variety of foods, that is, apples, berries, chocolate, red wines, and nuts. Derived tannins are formed during food handling and processing and are found primarily in black and oolong teas, red wine, and coffee. Flavonoids and tannins are quite sensitive to oxidative enzymes and cooking conditions.

Condensed tannins inhibit herbivore digestion by binding to consumed plant proteins and making them more difficult for animals to digest and by interfering with protein absorption and digestive enzymes. Tannins have traditionally been considered antinutritional, but it is now known that their beneficial or antinutritional properties depend upon their chemical structure and dosage. If ingested in excessive quantities, tannins inhibit the absorption of minerals such as iron, which may, if prolonged, lead to anemia [14]. In sensitive individuals, a large intake of tannins may cause bowel irritation, kidney irritation, liver damage, irritation of the stomach, and gastrointestinal pain.

Others

Nonpolar antioxidants such as eugenol, a major component of oil of cloves, have toxicity limits that can be exceeded with the misuse of undiluted essential oils. Toxicity associated with high doses of water-soluble antioxidants such as ascorbic acid is less of a concern as these compounds can be excreted rapidly in urine.

Risk of Antioxidant Supplements

It is well established that certain amounts of antioxidants, vitamins, and minerals are required in the diet. However, the benefit, dosing requirements, and risk profile of most antioxidant supplements are largely unknown. When used for disease prevention, the doses are severalfold greater than the recommended daily allowance (RDA). The hypothesis that antioxidant supplements can prevent diseases has been proven false by researchers. In spite of this information, many companies manufacture and sell dietary supplements with antioxidants in a variety of different formulations. Common ones include the "ACES" (Vitamins A, C, E, and selenium), resveratrol (found in grape seeds and knotweed roots), and herbs like green tea and jiaogulan.

The potential for harmful effects of antioxidant therapy has been suggested, such as in the β-Carotene and Retinol Efficacy Trial (CARET) which was a randomized, double-blinded, placebo-controlled chemoprevention trial in 18,314 men and women at high risk of developing lung cancer [15]. The study was initiated due to the observation of other studies that found people who have high serum β-carotene concentrations had lower rates of lung cancer [15]. The hypothesis of the CARET study was that these antioxidants would decrease the risk of lung cancer in an already high-risk population. Subjects were treated for up to 6 years. This study demonstrated that smokers who ingested a combination of 30 mg β-carotene and 25,000 IU retinyl palmitate (vitamin A) taken daily had 28% more lung cancer and 17% more deaths than placebo subjects. The CARET intervention was stopped 21 months early because of clear evidence of no benefit and substantial evidence of possible harm.

Other studies have found similar findings of adverse events. Table 47.2 lists observed side effects of supplemental

Table 47.2 Observed side effects with supplemental antioxidants

Antioxidant metabolite	Recommended daily allowance (RDA)	Reported side effects
Glutathione	250 mg/day or 600 mg IM QOD for male infertility	Acute: gastrointestinal disturbances
Carotenes	15–30 mg/day	Acute: skin color changes
		Chronic: possible increased risk of death and certain cancers
α-Tocopherol (vitamin E)	22.4 IU/day	Acute: headache, fatigue, muscle weakness, creatinuria
		Chronic: impaired bone mineralization, increased bleeding, cardiovascular disease; increased overall mortality
Ascorbic acid (vitamin C)	75–90 mg/day	Acute: diarrhea
		Chronic: hyperoxaluria, urinary stone formation, iron overload
Ubiquinol (coenzyme Q)	60–90 mg/day	Acute: gastrointestinal disturbances, heartburn, abdominal discomfort
		Chronic: hemorrhagic toxicity
Selenium	55 mcg/day	Acute: fatigue, gastrointestinal disturbances, skin rashes, irritability
		Chronic: concern for diabetes, loss of hair and nails, neuropathy
Melatonin	10 mg/day (bedtime)	Acute: diarrhea, rash, dizziness, headache, heartburn, nausea
		Chronic: sleep disturbance
Zinc	8–11 mg/day	Acute: gastrointestinal disturbance, anosmia (intranasal)
		Chronic: concern for increased risk prostate cancer, copper deficiency, suppression of immune system, anemia

antioxidants. The α-Tocopherol (Vitamin E) β-Carotene Cancer Prevention Study Group (ATBC) reported on a randomized, double-blind, placebo-controlled primary prevention trial [16]. The objective was to determine whether daily supplementation with vitamin E, β-carotene, or both would reduce the incidence of lung cancer and other cancers. A total of 29,133 male smokers 50–69 years of age from southwestern Finland were randomly assigned to one of four regimens: α-tocopherol (50 mg/day) alone, β-carotene (20 mg/day) alone, both α-tocopherol and β-carotene, or placebo. These patients were followed for 5–8 years. There was no reduction in the incidence of lung cancer among male smokers after 5–8 years of dietary supplementation with vitamin E. Those men given β-carotene had an 18% increase in the incidence of lung cancer compared to placebo. There was also an increased number of deaths due to ischemic heart disease and lung cancer in the β-carotene group compared to placebo. The vitamin E group had an increased incidence of death due to hemorrhagic stroke and an increased incidence of other cancers compared to placebo. While these data suggest that there may be harmful effects of these supplements, the authors state that further studies would need to be performed in order to validate these results [16].

Observation of these adverse effects was not limited to smokers. Bjelakovic's meta-analysis from 2007 included 68 randomized trials with 232,606 participants. This publication showed that treatment with β-carotene, vitamin A, and vitamin E may increase all-cause mortality and the potential roles of vitamin C and selenium on mortality may need further study [3].

These results were later confirmed by the same authors with an additional publication using the Cochrane Colla-boration methodology [3]. In this systematic review, several key findings were noted: (1) β-carotene, vitamin A, and vitamin E given singly or combined with other antioxidant supplements appeared to significantly increase mortality, (2) there was no evidence that vitamin C increases longevity, (3) selenium tended to reduce mortality, and (4) trials with inadequate bias control overestimated intervention effects [17–20]. It should be noted that only all-cause, not the cause of the increased mortality, was assessed. It is likely that increased cancer and cardiovascular mortality are the main reasons for the increased all-cause mortality [21, 22].

Several other publications have disagreed with the Bjelakovic meta-analysis [17, 21, 23, 24] and reported no effect on all-cause mortality. The Supplementation en Vitamines et Mineraux Antioxydants (SU.VI.MAX) study by Hercberg et al. was a randomized, double-blind, placebo-controlled primary prevention trial. A total of 13,017 participants took a single daily capsule of a combination of 120 mg of ascorbic acid, 30 mg of vitamin E, 6 mg of β-carotene, 100 mcg of selenium, and 20 mg of zinc or a placebo. After a mean of 7.5 years, there were no major differences found between the groups in total cancer incidence, ischemic cardiovascular disease incidence, or all-cause mortality [23].

Miller et al. performed a meta-analysis on the dose–response relationship between vitamin E supplementation and total mortality by evaluating randomized, controlled trials. Vitamin E doses ranged from 16.5 to 2,000 IU/day, and there were 135,967 who took vitamin E alone or in combination with other vitamins and minerals. While the results showed that there very well may be an increased risk of all-cause mortality with high doses of vitamin E (greater than or equal to 400 IU/day), lower doses did not reveal this same concern [24].

Although Bjelakovic et al. found no compelling evidence that antioxidant supplements have a significant beneficial effect on primary or secondary prevention of colorectal adenoma formation, in their meta-analysis of eight randomized clinical trials comparing antioxidant supplements with placebo or no intervention, they found no statistically significant effects of supplementation with β-carotene, vitamins A, C, E, and selenium alone or in combination. Antioxidant supplements seemed to increase the development of colorectal adenoma in three low-bias risk trials (1.2, 0.99–1.4) and significantly decrease its development in five high-bias risk trials (0.59, 0.47–0.74). There was also no significant difference between the intervention groups regarding adverse events including mortality (0.82, 0.47–1.4) [17].

The mechanism of the possible negative impact of antioxidant supplements is speculative. First, it is known that oxidative stresses are a part of the pathogenesis of different chronic diseases; however, could the oxidative stress be the cause of the chronic disease or the chronic disease causing the oxidative stress [25]? Second, some essential defense mechanisms, such as phagocytosis, detoxification, and apoptosis, depend on free radicals. If impaired, a negative impact on homeostasis may ensue [26–28]. Third, unlike prescription drugs, antioxidant supplements are not put through the same thorough toxicity studies in order to be sold to consumers [29]. A better understanding of the mechanisms and actions of antioxidants toward specific disease processes is needed [30].

Finally, if antioxidants reduce the redox stress in cancer cells, then they may decrease the effectiveness of chemotherapy and radiation therapy. However, other researchers argue that the antioxidants would reduce the unintentional side effects of the cancer treatment and increase survival times [31, 32].

β-Carotene

α-Carotene, β-carotene, and β-cryptoxanthin are provitamin A carotenoids. In the human body, these carotenoids can be converted to retinol (vitamin A). The essential function of carotenoids is that of provitamin A carotenoids (α-carotene, β-carotene, and β-cryptoxanthin) to serve as a source of vitamin A. Because of its vitamin A activity, β-carotene may be used to provide all or part of the vitamin A in multivitamin supplements. The vitamin A activity of β-carotene from supplements is much higher than that of β-carotene from foods [33].

As previously mentioned, the use of β-carotene was tested for its ability to prevent lung cancer in two large trials, the ATBC trial and the CARET trial. Surprisingly, an increased incidence of lung cancer was observed in the study groups. In CARET, it was not feasible to distinguish whether β-carotene or vitamin A was to blame for the negative results. In ATBC, there was a clear distinction that β-carotene was responsible for the increased incidence of lung cancers and increased overall mortality. Of note, there was no benefit of preventing other cancers, including gastric, pancreatic, breast, bladder, colorectal, and prostate cancer as well as leukemia, mesothelioma, and lymphoma [15, 16].

A large randomized, double-blind, placebo-controlled trial of β-carotene (50 mg on alternate days) involved 22,071 United States male physicians. The results after 12 years showed practically no early or late differences in the overall incidence of malignant neoplasms, cardiovascular disease, or in overall mortality. At initial glance, it seemed there was an increased incidence of thyroid cancer (16 vs. 2) and bladder cancer (62 vs. 41) in the β-carotene versus placebo group; however, after adjustment for multiple comparisons, neither of these differences was statistically significant. Overall, the only adverse side effects reported in the β-carotene group were yellowing of the skin and upset stomach [34].

Two other trials [35, 36] studied the ability of β-carotene to prevent nonmelanoma skin cancer. Neither found a beneficial effect on subsequent skin cancer incidence or reported any adverse effects of β-carotene supplementation.

The Women's Health Study was a large study of 39,876 healthy American women over 45 years old which found no effect of β-carotene on cancer incidence, but there was a suggestion of increased stroke risk during the study duration of 4.1 years (2.1 years treatment plus another 2.0 years follow-up). While it did not show statistical significance, the number of women who suffered a stroke was 61 (0.31%) for the β-carotene group versus 43 (0.22%) for the placebo group [37].

Minor side effects associated with the use of β-carotene include yellowing of the skin, also known as hypercarotenemia, when doses of greater than 30 mg/day are used for more than several weeks. This side effect is reversible upon cessation and has been observed in patients with photosensitivity disorders using these doses. Infrequently, mild gastrointestinal distress with gas and bloating may be seen.

Carotenemia is the ingestion of excessive amounts of vitamin A precursors in food, mainly carrots. It is manifested by a yellow–orange coloring of the skin. This differs from jaundice because the sclerae are still white in carotenemia. Other than the cosmetic effect, carotenemia has no adverse consequences because the conversion of carotenes to retinol is not sufficient to cause toxicity [38].

Tocopherols and Tocotrienols

Vitamin E, also known as α-tocopherol, refers to a set of eight related tocopherols and tocotrienols which are fat-soluble vitamins with antioxidant properties. Adequate amounts of this vitamin are typically present in Western diets. Multivitamins often contain about 30 international units (IU) of vitamin E, but supplements often contain 200, 400, or 1,000 IU. While research suggests that taking vitamin E supplements may boost immune systems and prevent heart disease and some types of cancer [39], large amounts of vitamin E may increase the risk for bleeding problems and death.

Miller et al. published a meta-analysis in 2005 that included 135,967 adults who had participated in 19 placebo-controlled studies of over a 1-year duration [24]. Approximately 60% of subjects had heart disease or risk factors for heart disease. Vitamin E in amounts of 400 IU or more daily for longer than 1 year increased the risk for death compared with placebo or no treatment. Limitations of the study were that trials which tested high amounts of vitamin E often involved older adults with chronic diseases. Therefore, findings from these trials may not apply to younger adults. Also, multivitamin combinations rather than vitamin E alone were often studied. This meta-analysis also did not find the exact lowest amount of vitamin E that was associated with increased risk for death.

In the HOPE and HOPE-TOO trials, the daily administration of 400 IU of natural source vitamin E for a median of 7 years had no clear impact on fatal and nonfatal cancers, major cardiovascular events, or deaths [39]. Unexpectedly, a consistent increase in the risk of heart failure was observed. A regression analysis identified vitamin E as an independent predictor of heart failure and supportive mechanistic evidence from an echocardiographic substudy of the HOPE trial found that vitamin E decreased left ventricular ejection fraction. Based on these findings, the authors recommended that vitamin E supplements should not be used in patients with vascular disease or diabetes mellitus.

A double-blind, placebo-controlled trial by Hemila et al. evaluated 652 Dutch subjects aged greater than or equal to 60 years [40]. These authors identified a greater severity of respiratory infections among participants supplemented with 200 mg vitamin E daily than among those not given vitamin E. These findings suggest that some population groups may be harmed by vitamin E supplementation. In contrast, Hathcock et al. supported the safety of vitamin E supplementation. They concluded that "at present, the evidence is not convincing that vitamin E supplementation up to the UL (i.e., the tolerable upper intake level, or 1,000 mg/day) increases the risk of death due to cardiovascular disease or other causes" [41].

Adults should consider avoiding taking vitamin E preparations in amounts of 400 IU or more. In November 2004, the American Heart Association stated that high amounts of vitamin E can be harmful. Taking 400 IU/day, or higher, may increase the risk of death. Taking smaller amounts, such as those found in a typical multivitamin, was not harmful.

Ascorbic Acid

Ascorbic acid, also known as vitamin C, is a monosaccharide antioxidant that is found in plants and animals. It functions specifically as a substrate for the antioxidant enzyme ascorbate peroxidase. Reactive oxygen species can be neutralized by ascorbic acid because it is a reducing agent [42, 43]. In humans, vitamin C is required for the synthesis of collagen. It is also a component of blood vessels, tendons, ligaments, and bone. Vitamin C also plays an important role in the synthesis of the neurotransmitter norepinephrine. Also, vitamin C is required for the synthesis of carnitine, a small molecule that is essential for the transport of fat into cellular organelles called mitochondria where the fat is converted to energy [44].

In a 15-year study of postmenopausal women, Lee et al. found that diabetic women who reported taking at least 300 mg/day of vitamin C from supplements were at significantly higher risk of death from coronary heart disease and stroke than those who did not take vitamin C supplements. Overall, vitamin C supplement use was not associated with a significant increase in cardiovascular disease mortality in the cohort as a whole [45]. Although a number of observational studies have found that higher dietary intakes of vitamin C are associated with lower cardiovascular disease risk, randomized controlled trials have not found antioxidant supplementation that included vitamin C to reduce the risk of cardiovascular disease in diabetic or other high-risk individuals [46].

Some studies have attempted to reveal if vitamin C supplementation would benefit athletes. While there does not seem to be an increased demand for vitamin C in athletes, there is the idea that if vitamin C is taken, it can allow the athlete a longer more strenuous exercise with less muscle damage. In fact, some research has found that amounts of vitamin C as high as 1,000 mg inhibits recovery theoretically by causing a decrease in mitochondria production and hampering endurance capacity [47].

Excessive doses of vitamin C not absorbed by the gastrointestinal tract can lead to mild diarrhea and indigestion. Large doses of ascorbic acid over prolonged periods can lead to urinary oxalate stone formation, although this effect is minimal and inconsistent [48, 49].

Glutathione

Glutathione is a cysteine-containing peptide made in human cells from specific amino acids. Glutathione is an endogenous intracellular antioxidant. Glutathione has a thiol group in its molecular structure which gives it antioxidant properties that allows it to be reversibly oxidized and reduced [50, 51]. Glutathione is touted by some to be the most important cellular antioxidant due to its high concentration and its main role in keeping the cell's redox state. Thorough literature review has failed to find any reported adverse effects of taking glutathione [52].

Melatonin

Melatonin is a unique antioxidant in that it can easily cross cell membranes including the blood–brain barrier. Another reason it is unique is because it does not undergo redox cycling which allows the antioxidant to undergo repeated reduction and oxidation. This repeated reduction and oxidation functions as a prooxidant and may allow the formation of free radicals [53, 54].

The recommended dose for melatonin is 10 mg by mouth at bedtime. Melatonin has been reported to cause sleep disruption, daytime fatigue, irritability, mood changes, depression, paranoia, hyperglycemia, headaches, dizziness, abdominal cramps, chest pain, and even tachycardia or seizures at higher doses [55, 56]. Several drug interaction precautions should be noted. First, there is caution with the use of systemic steroids and melatonin due to the interference with immunosuppressive activity of the steroid. Second, there is a caution with the use of ginkgo biloba due to the increased risk of seizures. Third, there is a caution with the use of melatonin and other CYP1A2 substrates. Lastly, a caution should be given with concomitant use of any CNS depressing drugs, sedatives, or hypnotics [56].

Antioxidant Nutrients: Selenium and Zinc

Selenium and zinc, commonly referred to as antioxidant nutrients, have no antioxidant action themselves and are instead required for the activity of some antioxidant enzymes. Selenium protects against oxidative damage by means of selenium-dependent proteins called selenoproteins including glutathione peroxidase. At serum levels of 70–90 ng/mL, a maximum level of activity is reached for the selenoproteins with the possible exception of one named selenoprotein P. The dietary intake of selenium in the USA is enough so that 99% of Americans have a serum selenium level greater than 90 ng/mL [57].

The Selenium and Vitamin E Cancer Prevention Trial (SELECT) was a large randomized, placebo-controlled trial set up to evaluate the potential benefit of selenium and vitamin E for the prevention of prostate cancer. Over 35,000 men enrolled and were divided into four groups (selenium, vitamin E, selenium + vitamin E, or placebo). After a mean follow-up of 5.46 years, there was no significant effect on the prevention of prostate cancer or any other prespecified cancer end points. However, in the selenium alone group, there was a statistically insignificant increased risk of type 2 diabetes mellitus. Further prospective, randomized studies would be needed to delineate the effect of supplemental and dietary selenium on the risk for developing diabetes. Because of the lack of benefit on prostate cancer and the potential risk of therapy, this trial was terminated [58].

While long-term use of zinc supplements at the upper limit of tolerability (40 mg/day) in adults is not considered unsafe, there are some common adverse effects of excessive zinc intake. These include metallic taste, nausea, vomiting, abdominal cramping, urinary tract infection, and diarrhea. Extended intake of amounts above the tolerable upper intake level may suppress immunity, decrease high-density lipoprotein cholesterol levels, and cause hypochromic microcytic anemia and copper deficiency [59, 60]. Interestingly, Leitzmann et al. evaluated zinc intake and the risk of prostate cancer in the Health Professionals Follow-up Study [61]. Results showed that in the 46,974 adult men studied, there was a 2.3 increased relative risk of advanced prostate cancer in men using elemental zinc in amounts of 100 mg/day or more. There was not an associated risk of prostate cancer in men who consumed less than 100 mg/day. While the authors could not rule out residual confounding by supplemental calcium intake or some unmeasured correlate of zinc supplement use, the evidence that chronic zinc ingestion above 100 mg/day may play a role in prostate carcinogenesis justifies further investigation.

Zinc may alter the way the body processes some drugs and other vitamins and minerals. For example, it may inhibit the absorption of tetracyclines, penicillamine, and quinolones. On the other hand, the absorption of zinc can be impaired by iron supplements and phytates, which are found in grains and legumes. Therefore, zinc supplements should be taken at least 2 h from iron and phytate ingestion [60].

Expert Commentary

Antioxidant supplements are widely used with the belief their use may improve health and have beneficial effects on disease prevention. These supplements are used in addition to the adequate amounts obtained in the typical Western diet. Recent meta-analyses and large-scale placebo-controlled trials suggest that long-term antioxidant supplements such as β-carotene, vitamin A, and vitamin E may increase overall all-cause mortality. The significance of these adverse effects is controversial as other meta-analyses, using many of the same studies, have provided mixed results depending on the criteria used for study inclusion. Although speculative, it is likely that increased cancer risk and cardiovascular disease risks are the main reasons for the increased mortality. Long-term indiscriminate use of antioxidant supplements should be avoided as the true benefit cannot be determined without further study.

Five-Year View

Although antioxidant supplements have been extensively studied, further large-scale randomized clinical trials with sufficient safety analyses will be necessary to determine their true long-term safety profile. Short-term use appears to be without significant adverse events. While supplements are not subject to the rigorous study required for FDA labeling of pharmaceutical agents, numerous clinical trials of their efficacy in specific disease states are ongoing as indicated by a search of http://www.clinicaltrials.gov. Some insight into harmful effects can be obtained from these studies, although clinical safety is usually a secondary end point of such investigations.

Key Issues

- Antioxidant supplements are increasingly used in the general population. Very high doses of some antioxidants, both dietary and as supplements, may have harmful long-term effects.
- β-Carotene, vitamin A, and vitamin E given singly or combined with other antioxidant supplements may increase all-cause mortality. Meta-analyses on this topic have yielded mixed results.
- Although speculative, it is likely that increased cancer and cardiovascular mortality are the main reasons for the increased all-cause mortality seen with β-carotene, vitamin A, and vitamin E.

Acknowledgment We acknowledge Joy Nicely and Kathy Gribble for manuscript preparation.

References

1. Halliwell B. Antioxidant defense mechanisms: from the beginning to the end (of the beginning). Free Radic Res. 1999;31:261–72.
2. Willcox JK, Ash SL, Catignani GL. Antioxidants and prevention of chronic disease. Crit Rev Food Sci Nutr. 2004;44:275–95.
3. Bjelakovic G, Nikolova D, Gluud LL, Simonetti RG, Gluud C. Mortality in randomized trials of antioxidant supplements for primary and secondary prevention: systematic review and meta-analysis. JAMA. 2007;297(8):842–57.
4. Bjelakovic G, Nikolova D, Simonetti RG, Gluud C. Antioxidant supplements for preventing gastrointestinal cancers. Cochrane Database Syst Rev. 2004;1(4):CD004183.
5. Bjelakovic G, Nikolova D, Simonetti RG, Gluud C. Antioxidant supplements for prevention of gastrointestinal cancers: a systematic review and meta-analysis. Lancet. 2004;364:1219–28.
6. Stanner SA, Hughes J, Kelly CN, Buttriss J. A review of the epidemiological evidence for the "antioxidant hypothesis". Public Health Nutr. 2004;7:407–22.
7. Hurrell R. Influence of vegetable protein sources on trace element and mineral bioavailability. J Nutr. 2003;133(9):2973S–7S.
8. Hunt J. Bioavailability of iron, zinc, and other trace minerals from vegetarian diets. Am J Clin Nutr. 2003;78(3 Suppl):633S–9S.
9. Gibson R, Perlas L, Hotz C. Improving the bioavailability of nutrients in plant foods at the household level. Proc Nutr Soc. 2006;65(2):160–8.
10. Kelsay JL. Effect of oxalic acid on bioavailability of calcium. In: Kies C, editor. Nutritional bioavailability of calcium. Washington, DC: American Chemical Society; 1985.
11. Mosha TC, Gaga HE, Pace RD, Laswai HS, Mtebe K. Effect of blanching on the content of antinutritional factors in selected vegetables. Plant Foods Hum Nutr. 1995;47:361–7.
12. Hallberg L, Brune M, Rossander L. Iron absorption in man: ascorbic acid and dose-dependent inhibition by phytate. Am J Clin Nutr. 1989;49:140–4.
13. Beecher G. Overview of dietary flavonoids: nomenclature, occurrence and intake. J Nutr. 2003;133(10):3248S–54S.
14. Brune M, Rossander L, Hallberg L. Iron absorption and phenolic compounds: importance of different phenolic structures. Eur J Clin Nutr. 1989;43(8):547–57.
15. Omenn GS, Goodman GE, Thornquist MD, et al. Risk factors for lung cancer and for intervention effects in CARET, the Beta-Carotene and Retinol Efficacy Trial. J Natl Cancer Inst. 1996;88(21):1550–9.
16. Heinonen OP, Huttuten JK, Albanes D, et al. The effect of vitamin E and beta carotene on the incidence of lung cancer and other cancers in male smokers. The Alpha-Tocopherol, Beta Carotene Cancer Prevention Study Group. N Engl J Med. 1994;330(15):1029–35.
17. Bjelakovic G, Nagorni A, Nikolova D, Simonetti RG, Bjelakovic M, Gluud C. Meta-analysis: antioxidant supplements for primary and secondary prevention of colorectal adenoma. Aliment Pharmacol Ther. 2006;24:281–91.
18. Moher D, Pham B, Jones A, et al. Does quality of reports of randomized trials affect estimates of intervention efficacy reported in meta-analysis. Lancet. 1998;352:609–13.
19. Schulz KF, Chalmers I, Hayes RJ, Altman DG. Empirical evidence of bias: dimensions of methodological quality associated with estimates of treatment effects in controlled trials. JAMA. 1995;273:408–12.
20. Kjaergard LL, Villumsen J, Gluud C. Reported methodologic quality and discrepancies between large and small randomized trials in meta-analyses. Ann Intern Med. 2001;135:982–9.
21. Caraballoso M, Sacristan M, Serra C, Bonfill X. Drugs for preventing lung cancer in healthy people. Cochrane Database Syst Rev. 2003;2:CD002141.
22. Vivekananthan DP, Penn MS, Sapp SK, Hsu A, Topol EJ. Use of antioxidant vitamins for the prevention of cardiovascular disease: meta-analysis of randomized trials. Lancet. 2003;361:2017–23.
23. Hercberg S, Galan P, Preziosi P, et al. The SU.VI.MAX Study: a randomized, placebo-controlled trial of the health effects of antioxidant vitamins and minerals. Arch Intern Med. 2004;164(21):2335–42.
24. Miller E, Pastor-Barriuso R, Dalal D, Riemersma R, Appel L, Guallar E. Meta-analysis: high-dosage vitamin E supplementation may increase all-cause mortality. Ann Intern Med. 2005;142(1):37–46.
25. Halliwell B. Free radicals, antioxidants, and human disease: curiosity, cause, or consequence? Lancet. 2000;344:721–4.
26. Salganik RI. The benefits and hazards of antioxidants: controlling apoptosis and other protective mechanisms in cancer patients and the human population. J Am Coll Nutr. 2001;20(5 Suppl):464S–72S.
27. Simon HU, Haj-Yehia A, Levi-Schaffer F. Role of reactive oxygen species (ROS) in apoptosis induction. Apoptosis. 2000;5:415–8.
28. Kimura H, Sawada T, Oshima S, Kozawa K, Ishioka T, Kato M. Toxicity and roles of reactive oxygen species. Curr Drug Targets Inflamm Allergy. 2005;4:489–95.
29. Bast A, Haenen GR. The toxicity of antioxidants and their metabolites. Environ Toxicol Pharmacol. 2002;11:251–8.

30. Ratnam DV, Ankola DD, Bhardwaj V, Sahana DK, Kumar MN. Role of antioxidants in prophylaxis and therapy: a pharmaceutical perspective. J Control Release. 2006;113:189–207.
31. Seifried H, McDonald S, Anderson D, Greenwald P, Milner J. The antioxidant conundrum in cancer. Cancer Res. 2003;63(15): 4295–8.
32. Lawenda BD, Kelly KM, Ladas EJ, Sagar SM, Vickers A, Blumberg JB. Should supplemental antioxidant administration be avoided during chemotherapy and radiation therapy? J Natl Cancer Inst. 2008;100(11):773–83.
33. Institute of Medicine, Food and Nutrition Board. Beta-carotene, other carotenoids. Dietary reference intakes for vitamin C, vitamin E, selenium, and carotenoids. Washington, DC: National Academy; 2000. p. 325–400.
34. Hennekens CH, Buring JE, Manson JE, et al. Lack of effect of long-term supplementation with beta carotene on the incidence of malignant neoplasms and cardiovascular disease. N Engl J Med. 1996; 334:1145–9.
35. Green A, Williams G, Neale R, et al. Daily sunscreen application and betacarotene supplementation in prevention of basal-cell and squamous-cell carcinomas of the skin: a randomized controlled trial. Lancet. 1999;354:723–9.
36. Greenberg ER, Baron JA, Karagas MR, et al. Mortality associated with low plasma concentration of beta carotene and the effect of oral supplementation. JAMA. 1996;275:699–703.
37. Lee IM, Cook NR, Manson JE, Buring JE, Hennekens CH. Beta-carotene supplementation and incidence of cancer and cardiovascular disease: the Women's Health Study. J Natl Cancer Inst. 1999;91:2102–6.
38. Penniston KL, Tanumihardjo S. The acute and chronic toxic effects of vitamin A. Am J Clin Nutr. 2006;83:191–201.
39. Lonn E, Bosch J, Yusuf S, et al. Effects of long-term vitamin E supplementation on cardiovascular events and cancer: a randomized controlled trial. JAMA. 2005;293:1338–47.
40. Hemila H. Potential harm of vitamin E supplementation [letter]. Am J Clin Nutr. 2005;82(5):1141–2.
41. Hathcock JN, Azzi A, Blumberg J, et al. Vitamins E and C are safe across a broad range of intakes. Am J Clin Nutr. 2005;81:736–45.
42. Padayatty S, Katz A, Wang Y, et al. Vitamin C as an antioxidant: evaluation of its role in disease prevention. J Am Coll Nutr. 2003;22(1):18–35.
43. Linster CL, Van Schaftingen E. Vitamin C biosynthesis, recycling and degradation in mammals. FEBS J. 2007;274(1):1–22.
44. Carr AC, Frei B. Toward a new recommended dietary allowance for vitamin C based on antioxidant and health effects in humans. Am J Clin Nutr. 1999;69(6):1086–107.
45. Lee DH, Folsom AR, Harnack L, Halliwell B, Jacobs Jr DR. Does supplemental vitamin C increase cardiovascular disease risk in women with diabetes? Am J Clin Nutr. 2004;80(5): 1194–200.
46. Waters DD, Alderman EL, Hsia J, et al. Effects of hormone replacement therapy and antioxidant vitamin supplements on coronary atherosclerosis in postmenopausal women: a randomized controlled trial. JAMA. 2002;288(19):2432–40.
47. Mastaloudis A, Traber M, Carstensen K, Widrick J. Antioxidants did not prevent muscle damage in response to an ultramarathon run. Med Sci Sports Exerc. 2006;38(1):72–80.
48. Peake J. Vitamin C: effects of exercise and requirements with training. Int J Sport Nutr Exerc Metab. 2003;13(2):125–51.
49. Massey LK, Liebman M, Kynast-Gales SA. Ascorbate increases human oxaluria and kidney stone risk. J Nutr. 2005;135(7): 1673–7.
50. Meister A. Glutathione metabolism and its selective modification. J Biol Chem. 1988;263(33):17205–8.
51. Meister A. Glutathione-ascorbic acid antioxidant system in animals. J Biol Chem. 1994;269(13):9397–400.
52. Meister A, Anderson M. Glutathione. Annu Rev Biochem. 1983;52:711–60.
53. Reiter RJ, Carneiro RC, Oh CS. Melatonin in relation to cellular antioxidative defense mechanisms. Horm Metab Res. 1997;29(8): 363–72.
54. Tan DX, Manchester LC, Reiter RJ, Qi WB, Karbownik M, Calvo JR. Significance of melatonin in antioxidative defense system: reactions and products. Biol Signals Recept. 2000;9(3–4):137–59.
55. Taylor SR, Weiss JS. Review of insomnia pharmacotherapy options for the elderly: implications for managed care. Popul Health Manag. 2009;12(6):317–23.
56. Buscemi N, Vandermeer B, Hooton N, et al. Efficacy and safety of exogenous melatonin for secondary sleep disorders and sleep disorders accompanying sleep restriction: meta-analysis. BMJ. 2006;332(7538):385–93.
57. Bleys J, Navas-Acien A, Guallar E. Selenium and diabetes: more bad news for supplements. Ann Intern Med. 2007;147(4):271–2.
58. Lippman SM, Klein EA, Goodman PJ, et al. Effect of selenium and vitamin E on risk of prostate cancer and other cancers: the Selenium and Vitamin E Cancer Prevention Trial (SELECT). JAMA. 2009;301(1):39–51.
59. Fosmire GJ. Zinc toxicity. Am J Clin Nutr. 1990;51(2):225–7.
60. Saper RB, Rash R. Zinc: an essential micronutrient. Am Fam Physician. 2009;79(9):768–72.
61. Leitzmann MF, Stampfer MJ, Wu K, Colditz GA, Willet WC, Giovannucci EL. Zinc supplement use and risk of prostate cancer. J Natl Cancer Inst. 2003;95(13):1004–7.

Index

S.J. Parekattil and A. Agarwal (eds.), *Male Infertility: Contemporary Clinical Approaches, Andrology, ART & Antioxidants*,
DOI 10.1007/978-1-4614-3335-4, © Springer Science+Business Media, LLC 2012

P

R

S

Kolibri 360 GmbH
Ferdinand-Jühlke-Straße 7
99095 Erfurt, Deutschland
produktsicherheit@kolibri360.de